Springer Monographs in Mathematics

For other titles published in this series, go to
http://www.springer.com/series/3733

Igor Chueshov · Irena Lasiecka

Von Karman Evolution Equations

Well-Posedness and Long-Time Dynamics

Springer

Igor Chueshov
Kharkov National University
Department of Mechanics & Mathematics
61077 Kharkov
Ukraine
chueshov@univer.kharkov.ua

Irena Lasiecka
University of Virginia
Department of Mathematics
Charlottesville, Virginia 22904
USA
il2v@virginia.edu

ISBN 978-0-387-87711-2 e-ISBN 978-0-387-87712-9
DOI 10.1007/978-0-387-87712-9
Springer New York Dordrecht Heidelberg London

Library of Congress Control Number: 2010924290

Mathematics Subject Classification (2010): 35Q74; 37L05, 37L30, 74B20, 74K20, 74K25

Printed on acid-free paper

Springer is part of Springer Science+Business Media (www.springer.com)

Contents

Preface

In the study of mathematical models that arise in the context of concrete applications, the following two questions are of fundamental importance: (i) well-posedness of the model, including existence and uniqueness of solutions; and (ii) qualitative properties of solutions. A positive answer to the first question, being of prime interest on purely mathematical grounds, also provides an important test of the viability of the model as a description of a given physical phenomenon. An answer or insight to the second question provides a wealth of information about the model, hence about the process it describes. Of particular interest are questions related to long-time behavior of solutions. Such an evolution property cannot be verified empirically, thus any in a-priori information about the long-time asymptotics can be used in predicting an ultimate long-time response and dynamical behavior of solutions. In recent years, this set of investigations has attracted a great deal of attention. Consequent efforts have then resulted in the creation and infusion of new methods and new tools that have been responsible for carrying out a successful analysis of long-time behavior of several classes of nonlinear PDEs.

The main goal of this book is to present recently developed mathematical meth ods of interest in the study of models that arise in continuum (nonlinear) mechanics of elastic bodies and are subject to different external influences and loads. In order to focus and streamline the exposition, we use the well-known von Karman model for the dynamics of plates and shells with large deflections as a benchmark prototype for the description of nonlinear oscillations. The theory of well-posedness and long-time behavior associated with von Karman evolutions has reached a high level of maturity, if not completeness. Thus, the time seems ripe to collect relevant results and to present them in a complete and also self-contained manner. In addition, the von Karman model epitomizes many important features and mathematical difficulties that arise in the study of attractors for various nonlinear PDEs. These include: (i) predominantly hyperbolic or hyperbolic-like type of dynamics combined with nonlinearity of the dissipation; (ii) critical (or potentially supercritical, with respect to Sobolev's embeddings) level of nonlinear terms which, in turn, results in the loss of compactness; (iii) loss of gradient flow structure due to the presence of external forces that are nonconservative. In order to cope with these difficulties, we develop and present an array of new methods that are capable of handling some of the issues discussed above. The novelty of the methods proposed is twofold, at two levels: (i) an abstract level within the realm of dynamical systems and (ii) a more concrete PDE level, within the context of establishing the validity of certain inverse-type inequalities. The book contains a number of new original results that appear in print for the first time.

Although our final results on well-posedness and long-time behavior of solutions to the problem of nonlinear oscillations of plates and shells are presented in the context of von Karman evolutions, the techniques developed transcend this particular model and are applicable to a broad variety of nonlinear models in mechanics that share similar features and obstacles in their mathematical treatment.

We hope that the methods presented and the ideas developed in this book will be useful not only to the interested mathematical community working with differential equations and dynamical systems, but also to physicists and engineers interested in both the mathematical background and the asymptotic analysis of infinite-dimensional dissipative systems that arise in continuum mechanics. Much of the analysis in this book is devoted to a rigorous reduction of infinite-dimensional (PDE) systems to some finite-dimensional structures, which are described only by finitely many degrees of freedom. These finite-dimensional structures should be of interest to application-oriented scientists, who pursue the actual design and implementation of algorithmic schemes, aiming at an accurate reconstruction and simulation of real infinite-dimensional phenomena.

Key words and phrases: Von Karman equations, well-posedness, long-time behavior, global attractors, rates of stabilization, inertial manifolds

AMS 2010 subject classification: primary 35Q74; secondary 37L05,37L30,74B20, 74K20, 74K25

Acknowledgments

During the several years of writing this book, along with the original papers on which it is based, our work has been partially supported by the National Science Foundation, Division of Mathematical Sciences, whose support is greatly appreciated. In addition, the second author acknowledges with gratitude support by the Army Research Office and, more recently, by the Air Force Office of Scientific Research.

It is our great pleasure to express our thanks to our co-authors: Francesca Bucci, Matthias Eller, and Daniel Toundykov for collaborative work that has contributed to several new results presented in this manuscript.

We are also indebted to the Springer editorial staff, in particular to Achi Dosanjh and Donna Chernyk, Springer editors, for their interest and much appreciated assistance with this project.

Kharkov, Charlottesville,
September 2009

Igor Chueshov
Irena Lasiecka

Introduction

Recent advances in science and technology have brought forward a number of mathematical problems and models that are critical to modern technology, including "smart material technology." Some particular examples include: structural acoustic problems with "smart sensors/actuators," fluid–structure interactions with "smart" controls, flexible structures such as antennas, space stations deployed in space, and so on. All these models involve equations of dynamic plates or shells (or combinations thereof) which constitute the prototype for the "structure." One of the main engineering concerns is to reduce the vibrations/oscillations of the structure. This can be accomplished by a combination of passive and active damping mechanisms or by manipulating the shape of the underlying spatial domain (shape design). From the mathematical point of view, this means stability analysis and, more generally, qualitative analysis of long-time behavior of the corresponding solutions. On the other hand, some very fundamental mathematical issues related to these models are poorly understood. Indeed, until recently, very little was known about the basic issue such as well-posedness of finite energy solutions to nonlinear dynamic shell or plate models. This particular question is not only a preliminary foundational step of the analysis but a cornerstone of the entire stability/bifurcation theory. Only very recently have a number of important results in this direction been obtained that shed light on more general models and more general situations. There is a widespread interest in the engineering and applied science community in the study of well-posedness and qualitative properties of solutions to elastic shell models.

The focus of this book is on von Karman evolution equations. Von Karman evolutions provide an established model that describes nonlinear deformations of elastic plates and shells. These equations, both in static and dynamic form, have attracted a great deal of interest in the literature [158, 279, 220, 237, 84, 23, 24, 25, 85, 173]. The derivation of equations goes back to the pioneering work of von Karman [158]. The hypotheses made there allow us to consider vertical displacements with large deflections but suitably small strains and negligible influences of plane (horizontal) movements on the transversal dynamics. A rigorous derivation of von Karman equations as a limit of nonlinear elasticity where the thickness of a body goes to zero is given in [83, 85, 84].

In the case where both vertical and in-plane displacements along with in-plane accelerations are accounted for, then the so-called *Vlasov–Marguerre system* provides for an adequate description of nonlinear oscillations (see, e.g., [279, 257] and the references therein). In what follows below, we focus on physical problems where in-plane accelerations can be neglected. In such a case, the system satisfied by nonlinear stress resultants can be decoupled by introducing the Airy stress function $v(u)$. The vertical displacement of a body is then described by von Karman equations. The latter can be seen as a system coupling a plate equation (vertical displacement) with a fourth-order elliptic problem (Airy stress function) driven by a nonlinear term referred to as the von Karman bracket (sometimes it is also called the Monge–Ampere form).

The main goal of this book is to discuss and present results on well-posedness, regularity, and long-time behavior of nonlinear dynamic plate (shell) models described by von Karman evolutions. Many of the results presented here are the outgrowth of very recent studies by the authors, including a number of new original results here in print for the first time. We have made an effort to provide for a comprehensive and reasonably self-contained exposition of the general topic outlined above. This includes supplying the complete functional analytic framework along with function space theory as pertinent in the study of nonlinear plate models. We also wish to add that although von Karman evolutions are the object under consideration, the methods developed transcend this specific model and may be applied to many other equations, systems that exhibit similar hyperbolic or ultrahyperbolic behavior (e.g., Berger's plate equations, Mindlin–Timoschenko systems, Kirchhoff–Boussinesq equations etc., see, for instance, [37, 38, 71, 72, 75] and the references therein).

In order to achieve a reasonable level of generality, the theoretical tools presented in the book are fairly abstract and tuned to general classes of second-order (in time) evolution equations, which are defined on abstract spaces. The mathematical machinery needed to establish well-posedness of these dynamical systems, their regularity and long-time behavior is developed at the abstract level, where the needed hypotheses are axiomatized. This approach allows us to look at von Karman evolutions as just one of the examples of a much broader class of evolutions. The generality of the approach and techniques developed are applicable (as shown in the book) to many other dynamics sharing certain rather general properties.

In order to focus our discussion, we begin with a description of the basic model in its simplest form. Later on, we consider several generalizations that account for additional physical effects.

0.1 Von Karman evolutions

The nonlinear oscillations of plates can be described by the following equations:

$$u_{tt} - \alpha\Delta u_{tt} + \Delta^2 u - [u, v + F_0] + Lu = p(x,t), \quad x = (x_1;x_2) \in \Omega, t > 0, \quad (0.1.1)$$

where $v = v(u)$ is a solution of the problem

$$\Delta^2 v + [u,u] = 0 \text{ in } \Omega, \quad v|_{\partial\Omega} = \left.\frac{\partial v}{\partial n}\right|_{\partial\Omega} = 0. \tag{0.1.2}$$

Here Ω is a smooth bounded domain in $\mathbb{R}^2$, and Δ is the Laplace operator. We set

$$[u,v] = \partial_{x_1}^2 u \cdot \partial_{x_2}^2 v + \partial_{x_2}^2 u \cdot \partial_{x_1}^2 v - 2 \cdot \partial_{x_1 x_2}^2 u \cdot \partial_{x_1 x_2}^2 v. \tag{0.1.3}$$

The unknown function $u = u(x,t)$ measures the transverse displacement of the plate at the point x and time t; $v(x,t)$ is the Airy stress function; $u_0(x)$ and $u_1(x)$ are initial data. The coefficient $\alpha \geq 0$ corresponds to rotational forces (which may or may not be accounted for). This coefficient changes the character of the dynamics from hyperbolic (with finite speed of propagation) $\alpha > 0$ to Petrowsky-type $\alpha = 0$. Function F_0 represents in-plane forces within the plate, function $p(x,t)$ is a given transverse force and Lu stands for a first-order differential operator modeling a nonconservative force (see below).

As mentioned above, the model (0.1.1) does not account for in-plane accelerations. For this, one should consider a system consisting of a plate equation (vertical displacements) coupled with a nonlinear system of dynamic elasticity (horizontal displacements). The interested reader may be referred to [279] or [173, 184].

This book is devoted to a study of well-posedness and long-time behavior associated with problem (0.1.1)–(0.1.3) supplemented with initial data and boundary conditions for the displacement $u(x,t)$. Several natural generalizations and modifications of these equations are considered as well. In addition to the most popular clamped boundary conditions, we also consider other types of boundary conditions, such as *hinged* or *free* boundary conditions associated with stationary von Karman equations. For the reader's convenience we list them below.

- **Clamped (Dirichlet) boundary conditions:**

$$u|_{\partial\Omega} = \nabla u|_{\partial\Omega} = 0.$$

- **Hinged boundary conditions:**

$$u|_{\partial\Omega} = \Delta u|_{\partial\Omega} = 0.$$

- **Free boundary conditions:**

$$[\Delta u + (1-\mu)B_1 u]\big|_{\partial\Omega} = \left.\left[\frac{\partial}{\partial n}\Delta u + (1-\mu)B_2 u - \nu u - \alpha \frac{\partial}{\partial n} u_{tt}\right]\right|_{\partial\Omega} = 0,$$

where the boundary operators B_1 and B_2 have the following forms

$$B_1 u = 2 n_1 n_2 u_{x_1 x_2} - n_1^2 u_{x_2 x_2} - n_2^2 u_{x_1 x_1},$$

$$B_2 u = \tfrac{\partial}{\partial \tau}\left[\left(n_1^2 - n_2^2\right) u_{x_1 x_2} + n_1 n_2 \left(u_{x_2 x_2} - u_{x_1 x_1}\right)\right].$$

Here $n = (n_1; n_2)$ is the outer normal to $\partial\Omega$, $\tau = (-n_2; n_1)$ is the unit tangent vector along $\partial\Omega$, the constant $0 < \mu < 1$ has a meaning of the Poisson modulus and $\nu \geq 0$ is a parameter.

Mixed boundary conditions are considered as well. This corresponds to the case where the boundary $\partial\Omega$ is split in two parts and on each of them different types of boundary conditions are imposed.

0.2 Sources

The von Karman evolution, without source and damping terms, conserves the energy of the system. In the case $\alpha = 0$ the energy is topologically equivalent to $H^2(\Omega) \times L_2(\Omega)$ (H^2 for the position and L_2 for the velocity u_t). The presence of the sources, such as the in-plane force F_0 or external transversal force p and feedback force Lu modeled by the first-order differential operator do provide a destabilizing effect on the system. The term Lu models potentially existing nonconservative loads on the system. For instance, this term arises in the case where the plate interacts with the supersonic flow of gas. Its simplest canonical form may be given by

$$Lu \equiv \rho(x)\partial_{x_1} u,$$

where $\rho(x)$ depends on the velocity—which is assumed in the direction of x_1—of the flow in which the plate is immersed. It is worth noting that the source Lu makes the problem nonconservative, in the sense that the force Lu cannot be presented as a gradient of some potential function and thus the energy is not conserved (unlike the in-plane forces F_0 that usually are accounted for by the energy function).

The presence of the sources is a prime reason for a need to stabilize the motion. This is done by introducing suitable forms of damping.

0.3 Damping

Our prime interest is in long-time behavior of the model described by (0.1.1)–(0.1.3), which is perturbed by the forces F_0, p, and Lu affecting the plate. Therefore, we must account for some sort of dissipative/damping mechanism. Instabilities in the free model are intrinsically infinite-dimensional (infinite number of eigenvalues on the imaginary axis), therefore the damping used must be strong enough in order to suppress instabilities. By strong enough, we mean that it cannot be modeled by a compact operator (with respect to the phase space). There are several physically significant ways of introducing dissipation into the system. These include:

- Internal, possibly nonlinear, mechanical dissipation affecting the entire interior of the domain. This type of dissipation is modeled by a function $g(u_t)$, where $g(s)$ is a nonlinear function subject to suitable hypotheses.

- Still interior, mechanical dissipation but localized to a certain subset of Ω. In this case the damping term takes the form $\xi(x)g(u_t)$ with $\xi(x)$ supported in some suitable set of Ω.
- Mechanical dissipation that acts only on the boundary. In that case some function of velocity u_t is added to the boundary conditions.
- Thermal dissipation. In this case there is coupling with another equation that describes the temperature in the plate.

Our aim is to discuss various forms of dissipation by analyzing their effect on long-time behavior and attractors that characterize the ultimate dynamics. The emphasis is placed on two particular aspects of the damping: geometric and topological.

Topological aspects are related to strong nonlinearity of the damping. In fact, any deviation from linearity (or more precisely a linearly bounded case) diminishes the decay rates for the energy function corresponding to a free system (without a source). This is true with both sublinear and superlinear growth conditions imposed on $g(s)$ both at the origin and at infinity. Consequently, the effect of "damping out" the source is reflected by these growth conditions. Ironically enough, it is not true that more damping (stronger damping) provides better stability properties. For the second-order systems "overdamping" may cause an imbalance of equipartition of energy, with the effect of destroying the overall rates of the total energy (potential and kinetic). As shown in this book, special care must be given to nonlinear damping that exceeds linear bounds (be it at the origin or at infinity). Superlinear damping at the origin slows down the rate of decay of the energy in the neighborhood of zero. Sublinear damping at infinity slows down (and possibly prevents) the transfer of energy from far away into a neighborhood of zero. In fact, uniform decay rates with sublinear damping do require higher-energy initial data.

The geometric aspect is connected with a dissipation that is geometrically constrained. This happens when the dissipation is placed on the boundary, or is supported in a strict subset of the domain Ω. These problems, although physically very attractive, are more involved from the mathematical point of view. In order to suppress the energy of the entire system, geometrically constrained dissipation must be propagated through the entire domain. This, of course, requires suitable tools dealing with propagation problems. In addition, the geometric features of the spatial domain play critical role for the results obtained, which require suitable geometric hypotheses.

0.4 Goals

The nonlinear theory of elasticity and long-time behavior of the corresponding dynamical systems is a topic of current interest and intense research. New results and new techniques have been contributed to the field over the last decade or so. Although long-time behavior of parabolic-like structures, with an inherent smoothing mechanism, has been studied for a long time, much less is known about hyperbolic-like dynamics where instability is inherently an infinite-dimensional phenomenon. The lack of smoothing and the lack of natural dissipation (unlike the case of parabolic equations) makes the study of asymptotic behavior quite challenging. In fact, for hyperbolic systems, ultimate behavior that is both smooth and finite-dimensional is not a property to be expected. In order to understand the mechanism responsible for long-time dynamics and to exhibit smoothing and finite-dimensional phenomena, new techniques, both at the abstract and concrete PDE level have been developed over the years. It is the interplay between "abstract methods" and their PDE implementation that is responsible for recent successes in the field. Concrete PDE examples lead to questions that can be asked at the abstract level. An answer at this level then provides an unified way of treating entire classes of problems embracing similar features and difficulties. This is very much the case of the von Karman evolution, which is a model for several classes of hyperbolic and hyperbolic-like nonlinear problems. The study of the asymptotic behavior for the von Karman model (which is an inherently noncompact dynamics) led to the creation of a number of tools that are formulated at the abstract level. As such, they can then be applied to large classes of dynamical systems generated by second-order in-time evolution equations.

The aim of this work is first to collect these general abstract results pertaining to fundamental dynamical system properties such as generation of semiflows, compactness of attractors, their dimensionality, and so on. In the next step, the abstract results just obtained are specified to second-order evolutions, thus providing a ready set of tools and criteria to be employed. It turns out that a large number of criteria that need to be verified are reminiscent of some properties and inequalities of interest to PDE control theory. This link between dynamical systems and control theory, via inverse-type estimates, is thoroughly exploited and used throughout the entire book. In the third step we develop PDE techniques that enable the verification of the abstract criteria. The analysis thus carried out leads to new results in the area of von Karman evolutions. This refers to both well-posedness and long-time behavior. It turns out that the von Karman model under consideration is a very rich model, which is representative of, and epitomizes, a number of general properties exhibited by other nonlinear equations such as shell equations, Berger's plate models, and Kirchhoff–Boussinesq equations among others (see [75] and the references therein).

0.5 Brief outline of the book

As mentioned above, the main theme is on von Karman evolution subject to different damping mechanisms. In addition to a single von Karman evolution, we also consider various interactive coupled systems (structure–acoustic interaction; structure–gas–flow interactions), where the von Karman evolution constitutes just one component of such a system. In this latter case, the emphasis is placed on propagating smoothness properties from one component to another and on transferring stability properties.

The main results presented in this book pertain to

- Well-posedness of solutions in a finite energy space, including Hadamard well-posedness.
- Regularity of solutions corresponding to more regular initial data.
- Long-time behavior of solutions when $t \to \infty$. Here we study questions such as
  - Existence of global attractors
  - Properties of attractors such as structure, dimensionality and smoothness
  - Rate of convergence of solutions to points of equilibria
  - Inertial manifolds
  - Determining functionals

Part I of the book deals with the issues of well-posedness and regularity of solutions, and Part II presents results on long-time behavior. A chapter-by-chapter analysis of the book is given below.

Chapter 1 provides the mathematical foundations and preliminary background needed for the analysis of von Karman evolutions. These include properties of the relevant Sobolev–Hardy–Lizorkin spaces, elements of nonlinear monotone operator theory, theory of biharmonic equations, and properties of von Karman brackets defining the Airy stress functions.

Chapter 2 provides an abstract treatment of second-order nonlinear evolution equations with nonlinear damping. This is done within the framework of nonlinear semigroup theory. Because this treatment depends heavily (via the choice of the respective spaces) on the boundary conditions imposed, a detailed analysis of various boundary conditions and their impact on well-posedness and regularity theory is presented at the end of Chapter 2.

Chapter 3 deals with the well-posedness issues that pertain to von Karman evolutions accounting for rotational inertia (parameter $\alpha > 0$ in (0.1.1)). This corresponds to hyperbolic-like dynamics with finite speed of propagation. Nonlinear dissipation both in the interior and on the boundary is treated. The model is recast as a special case of an abstract second-order evolution treated in Chapter 2.

Chapter 4 repeats the same program for the von Karman evolutions without the rotational inertia term ($\alpha = 0$). In that case the dynamics has infinite speed of propagation. Because of the lack of regularizing effect on the velocity caused by rotational terms, the well-posedness for the model requires more subtle tools that involve some elements of harmonic analysis.

Chapter 5 takes up the analysis of coupled structures involving von Karman evolutions. In this chapter thermoelastic coupling with a heat equation is considered. Well-posedness and regularity theory along with analyticity of dynamics in the nonrotational case are presented.

Chapter 6 deals with two other coupled interactive structures that involve von Karman evolutions as one of the components. These are: structural acoustic interaction and plates in a potential flow of gas. In the first case the plate is coupled with an acoustic wave equation on an interface of the two media. The interface is a two-dimensional manifold and the acoustic equation is given in a bounded three-dimensional domain. In the second case, the coupling occurs between the plate and linearized equation for potential flow defined on $\mathbb{R}^3_+$. For both models, questions related to well-posedness and regularity of solutions are discussed. Both structures share a number of common features, however, technical details needed for deriving critical estimates are of a different nature.

Chapter 7 opens Part II and pertains to long-time behavior of dynamical systems. Foundation of a general abstract theory developed for autonomous dynamical systems is provided. This includes discussion of compactness, fractal dimensionality of compact sets, gradient flows, inertial manifolds, and more.

Chapter 8 specializes and further develops, in the direction of second-order systems, the material presented in Chapter 7. Various methods and concrete criteria for compactness, finite dimensionality, and smoothness of attractors are developed. Applicability or superiority of one method over another depends on the type of assumptions imposed. These phenomena are illustrated through a sequence of examples. New criteria obtained for long-time behavior are motivated by recent inverse-type estimates developed in the context of control theory of PDEs.

Chapter 9 presents long-time behavior results for von Karman evolutions equipped with nonlinear internal damping. First, models with rotational inertia are treated. For these, existence of attractors, and properties of attractors such as smoothness and fractal dimension, are presented. The results presented are derived from the abstract theory of Chapter 8 with, however, the critical support of several technical estimates capturing, and specific to, the dynamics of von Karman evolutions. In the second part of Chapter 9, models without rotational inertia are considered. In that case, the theory is more demanding because the effect of the source is no longer compact with respect to the dynamics. New developments based on derivation of certain inverse-type inequalities are responsible for optimal results.

Chapter 10 takes up the same problem within the context of boundary damping. In this case, propagation of damping from the boundary into the interior plays a major role in all results obtained. Both models with and without rotational inertia are considered. In order to deal with criticality of the source in the case of rotational free models, new approaches combining multipliers techniques with some specialized arguments in dynamical systems and related gradient flow dynamics are developed.

Chapter 11 describes long-time behavior of thermoelastic plates with von Karman nonlinearity. Here an interesting phenomenon is associated with the fact that rotational terms change the linearized dynamics from hyperbolic to parabolic. The challenge of the problem is to obtain characterization of long-time behavior parameters (i.e., fractal dimension of the attractor) that is uniform with respect to parabolicity or hyperbolicity encoded in the model. This, again, is achieved by using novel criteria in terms of inverse estimates for the difference of the two solutions.

Chapter 12 presents three examples of interactive dynamical systems involving von Karman evolutions. The first two are acoustic–structure interaction (isothermal and heat generating), whereas the third one deals with structure–gas–flow interaction. Interface coupling between the von Karman plate and acoustic wave equation describes a model of acoustic structure interactions. For the gas–flow case, the von Karman plate is coupled on the boundary with a linearized potential flow equation. For all models, long-time dynamics is studied. The role of the coupling and transfer of stability properties from one component of the structure into another is the main theme considered here.

Chapter 13 deals with an existence of inertial manifolds. As is known, the results are strongly dependent on the spectral properties of the operator describing the model. Thus, the geometry of the domain plays a dominant role. Existence of inertial manifolds for von Karman evolutions is established under various types of damping such as: viscous damping, strong (structural damping), and thermal damping.

Appendix provides the necessary background and preliminary material used throughout the book.

Part I
Well-Posedness

Chapter 1
Preliminaries

This chapter collects several preliminary results pertaining to the theory of function spaces and general theory of nonlinear operators. Properties of the biharmonic operator endowed with various boundary conditions (clamped, hinged, free, and mixed) are discussed and the structure and properties of the von Karman bracket, representing the main nonlinearity in von Karman evolutions, are analyzed. The chapter concludes with application of the results mentioned above to stationary (time-independent) von Karman equations.

1.1 Function spaces and embedding theorems

Below $\mathcal{O}$ is a domain in $\mathbb{R}^n$ whose boundary $\partial\mathcal{O}$ is an $(n-1)$-dimensional sufficiently smooth manifold. It is assumed that $\mathcal{O}$ lies locally on one side of the boundary $\partial\mathcal{O}$.

1.1.1 Sobolev spaces

For any integer $k \geq 0$ and for $1 \leq p \leq \infty$ we denote by $W_p^k(\mathcal{O})$ the Sobolev space:

$$W_p^k(\mathcal{O}) = \left\{ u \in L_p(\mathcal{O}) \; : \; \partial^\alpha u \in L_p(\mathcal{O}) \; \text{ for all } |\alpha| \leq k \right\},$$

where $\partial = \partial_x = (\partial_{x_1}, \ldots, \partial_{x_n})$ is the gradient operator, and $\alpha = (\alpha_1, \ldots, \alpha_n)$ is a multi-index of nonnegative integers, $|\alpha| = \alpha_1 + \cdots + \alpha_n$ and $\partial^\alpha = \partial_{x_1}^{\alpha_1} \cdot \ldots \cdot \partial_{x_n}^{\alpha_n}$. We denote by $\|\cdot\|_{W_p^k(\mathcal{O})}$ the norm in $W_p^k(\mathcal{O})$. We also define the Sobolev space $W_p^s(\mathcal{O})$ for positive real superscripts $s \notin \mathbb{N}$ and $1 \leq p < \infty$ by the formula

I. Chueshov, I. Lasiecka, *Von Karman Evolution Equations*,
Springer Monographs in Mathematics, DOI 10.1007/978-0-387-87712-9_1,

$$W_p^s(\mathcal{O}) = \left\{ u \in W_p^k(\mathcal{O}) \;:\; \|u\|_{W_p^s(\mathcal{O})}^p \equiv \|u\|_{W_p^k(\mathcal{O})}^p + \sum_{|\alpha|=k} I_{\sigma,p}(\partial^\alpha u) < \infty \right\},$$

where $s = k + \sigma$ with $k \in \mathbb{N}$ and $0 < \sigma < 1$ and

$$I_{\sigma,p}(u) = \int_{\mathcal{O}\times\mathcal{O}} \frac{|u(x)-u(y)|^p}{|x-y|^{n+\sigma p}}\, dxdy.$$

We denote $H^s(\mathcal{O}) \equiv W_2^s(\mathcal{O})$ and consider the space $H_0^s(\mathcal{O})$ defined as the closure in $H^s(\mathcal{O})$ of the space of infinite differentiable functions on $\mathcal{O}$ with compact support in $\mathcal{O}$ and the space $H^{-s}(\mathcal{O}) \equiv \left[H_0^s(\mathcal{O})\right]'$ of distributions on $\mathcal{O}$. In this book we often use the notation $\|\cdot\|_{s,\mathcal{O}}$ for the norm in $H^s(\mathcal{O})$ for each $s \in \mathbb{R}$ (sometimes we omit the subscript $\mathcal{O}$ if no ambiguity may arise). By $\|\cdot\|_{\mathcal{O}}$ we denote the norm in $L_2(\Omega)$.

Let $C^k(\overline{\mathcal{O}})$ be the space of k times continuously differentiable functions on $\overline{\mathcal{O}}$. For $0 < \mu < 1$ we denote by $C^{k+\mu}(\overline{\mathcal{O}})$ the subset of $C^k(\overline{\mathcal{O}})$ that consists of functions whose derivatives of order k are locally μ-Hölder continuous. Standard Sobolev embeddings [1] give:

$$C^{k+\sigma}(\overline{\mathcal{O}}) \subset W_p^{k+\beta}(\mathcal{O}) \;\text{ if }\; k \in \mathbb{N},\; 0 \le \beta < \sigma \le 1,\; 1 < p < \infty. \tag{1.1.1}$$

Some of the properties of Sobolev spaces, needed in the sequel, are collected below (see, e.g., [1, 250, 275] for the proofs).

1.1.1. Theorem. *Assume that $\partial\mathcal{O} \in C^\infty$ and $\mathcal{O}$ lies locally on one side of the boundary $\partial\mathcal{O}$. Then*

- *The following continuous embeddings are valid,*

$$W_p^s(\mathcal{O}) \subset C^\sigma(\overline{\mathcal{O}}) \;\; \textit{if}\;\; s - \frac{n}{p} > \sigma,\; 1 < p < \infty,\; s,\sigma \ge 0, \tag{1.1.2}$$

(if σ is not an integer the embedding holds also for $\sigma = s - n/p$) and

$$W_p^s(\mathcal{O}) \subset W_{p^*}^{s^*}(\mathcal{O}) \;\; \textit{if}\;\; s - \frac{n}{p} \ge s^* - \frac{n}{p^*},\; 1 < p \le p^* < \infty,\; s^* \ge 0. \tag{1.1.3}$$

- *The trace operator $u \mapsto u\big|_{\partial\mathcal{O}}$ is continuous from $W_p^s(\mathcal{O})$ into $W_p^{s-1/p}(\partial\mathcal{O})$ for every $s > 1/p$ and $1 < p < \infty$.*

In particular, from (1.1.3) we have that

$$H^s(\mathcal{O}) \subset L_p(\mathcal{O}) \;\text{ if }\; s = \frac{n}{2} - \frac{n}{p},\; p \ge 2,\; n = \dim\mathcal{O}. \tag{1.1.4}$$

In the case $n = 2$ Theorem 1.1.1 implies the following embeddings:

$$H^s(\mathcal{O}) \subset L_\infty(\mathcal{O}), \quad s > 1,\; \mathcal{O} \subset \mathbb{R}^2, \tag{1.1.5}$$

and

$$H^s(\mathscr{O}) \subset L_{2/(1-s)}(\mathscr{O}), \quad 0 \le s < 1,\ \mathscr{O} \subset \mathbb{R}^2. \tag{1.1.6}$$

1.1.2. Remark. The regularity of the boundary assumed in Theorem 1.1.1 is conservative. For our applications, we use Sobolev spaces with finite smoothness, so that the regularity of the boundary can be significantly relaxed. In fact, for our purposes $\partial\mathscr{O} \in C^4$ would be a sufficient condition for most of the results. Inasmuch as the regularity of the boundary is not a focus in this book, we do not pay much attention to this issue. ■

1.1.2 Besov spaces

Assume that $1 \le p,q < \infty$ and $s = [s]^- + [s]^+ > 0$ with $[s]^-$ integer and $0 < [s]^+ \le 1$. We define Besov space $B^s_{p,q}(\mathbb{R}^n)$ as a subset of functions u from the Sobolev space $W_p^{[s]^-}(\mathbb{R}^n)$ such that

$$\|u\|_{B^s_{p,q}(\mathbb{R}^n)} \equiv \|u\|_{W_p^{[s]^-}(\mathbb{R}^n)} + \sum_{|\alpha|=[s]^-} \left(I^{(2)}_{[s]^+,p,q}(\partial^\alpha u)\right)^{1/q} < \infty,$$

where

$$I^{(2)}_{\sigma,p,q}(u) = \int_{\mathbb{R}^n} \frac{dy}{|y|^{n+\sigma q}} \left(\int_{\mathbb{R}^n} |u(x+y)+u(x-y)-2u(x)|^p dx\right)^{q/p}.$$

If $s = k+\sigma$ with $k = 0,1,2,\dots$ and $0 < \sigma < 1$, then we can define an equivalent norm in $B^s_{p,q}(\mathbb{R}^n)$ by the formula

$$\|u\|_{B^s_{p,q}(\mathbb{R}^n)} = \|u\|_{W^k_p(\mathbb{R}^n)} + \sum_{|\alpha|=k} \left(I^{(1)}_{\sigma,p,q}(\partial^\alpha u)\right)^{1/q}, \tag{1.1.7}$$

where

$$I^{(1)}_{\sigma,p,q}(u) = \int_{\mathbb{R}^n} \frac{dy}{|y|^{n+\sigma q}} \left(\int_{\mathbb{R}^n} |u(x+y)-u(x)|^p dx\right)^{q/p}.$$

Therefore relation (1.1.7) implies that

$$B^s_{p,p}(\mathbb{R}^n) = W^s_p(\mathbb{R}^n) \quad \text{for all } s > 0,\ s \notin \mathbb{N},\ 1 \le p < \infty. \tag{1.1.8}$$

We also have the following continuous embedding

$$B^s_{p,q}(\mathbb{R}^n) \subset B^{s^*}_{p^*,q^*}(\mathbb{R}^n) \ \text{ if } s - \frac{n}{p} = s^* - \frac{n}{p^*}, \tag{1.1.9}$$

where $1 \le p \le p^* < \infty$ and $1 \le q \le q^* < \infty$. In particular, for $n = 2$, we have

$$H^s(\mathbb{R}^2) \subset B^{s^*}_{p,2}(\mathbb{R}^2) \text{ for } s = 1 + s^* - \frac{2}{p},\ s > 0,\ s \notin \mathbb{N},\ p \geq 2. \tag{1.1.10}$$

We also note that the Besov space $B^s_{p,q}(\mathscr{O})$ on a domain $\mathscr{O} \subseteq \mathbb{R}^n$ is defined as a set of restrictions on $\mathscr{O}$ of functions from $B^s_{p,q}(\mathbb{R}^n)$. The relations (1.1.8) and (1.1.9) remain true if we change $\mathbb{R}^n$ into a smooth domain $\mathscr{O} \subset \mathbb{R}^n$.

We refer to [26] and [275] for the proofs and for other facts from the theory of Sobolev and Besov spaces.

1.1.3 Lizorkin and real Hardy spaces

In what follows we introduce another type of Sobolev spaces which are referred to as Lizorkin spaces. In order to provide a self-contained definition, we need some notation.

Let $\mathscr{S}(\mathbb{R}^n)$ be a Schwartz space of all rapidly decreasing infinitely differentiable functions on $\mathbb{R}^n$ and $\mathscr{S}'(\mathbb{R}^n)$ be the space of tempered distributions on $\mathbb{R}^n$. Denote by $\Phi(\mathbb{R}^n)$ a collection of all systems $\phi = \{\phi_j(x)\}_{j=0}^\infty \subset \mathscr{S}(\mathbb{R}^n)$ of real-valued, even functions with respect to the origin, such that

- $\operatorname{supp} \phi_0 \subset \{x : |x| \leq 2\}$, $\operatorname{supp} \phi_j \subset \{x : 2^{j-1} \leq |x| \leq 2^{j+1}\}$, $j = 1, 2, 3, \ldots$.
- For every multi-index α there exists a positive number c_α such that $2^{j|\alpha|}|D^\alpha \phi_j(x)| \leq c_\alpha$ *for all* $j = 0, 1, 2, \ldots$ and $x \in \mathbb{R}^n$.
- $\sum_{j=0}^\infty \phi_j(x) = 1$ for every $x \in \mathbb{R}^n$.

1.1.3. Definition (spaces $F^s_{p,q}$). Let $0 < p < \infty$, $0 < q \leq \infty$, $s \in \mathbb{R}$ and $\{\phi_j\} \in \Phi(\mathbb{R}^n)$. Then

$$F^s_{p,q}(\mathbb{R}^n) = \left\{ f \in \mathscr{S}'(\mathbb{R}^n) \,:\, \| f \|_{F^s_{p,q}(R^n)} = \| \{2^{js}\mathscr{F}^{-1}(\phi_j \mathscr{F} f)\} \|_{L_p(l_q)} < \infty \right\},$$

where $\mathscr{F}$ denotes the Fourier transform and

$$\| \{f_j\} \|_{L_p(l_q)} = \left[\int_{\mathbb{R}^n} \left(\sum_j |f_j(x)|^q \right)^{p/q} dx \right]^{1/p}.$$

If $\mathscr{O}$ is a bounded C^∞-domain in $\mathbb{R}^n$, we suppose that $F^s_{p,q}(\mathscr{O})$ is the restriction of $F^s_{p,q}(R^n)$ on $\mathscr{O}$ quasi-normed (normed for $p \geq 1$) by

$$\| f \|_{F^s_{p,q}(\mathscr{O})} = \inf \ \| g \|_{F^s_{p,q}(\mathbb{R}^n)},$$

where the infimum is taken over all $g \in F^s_{p,q}(\mathbb{R}^n)$ such that the restriction of g to $\mathscr{O}$ coincides with f. ■

We note that spaces $F^s_{p,q}$ do not depend on $\{\phi_j\}$ (another choice of $\{\phi_j\} \in \Phi(\mathbb{R}^n)$ leads to an equivalent quasi-norm). We have the following relations between Lizorkin and Sobolev spaces (see, e.g., [276] and [250]),

$$F^s_{p,p}(\mathscr{O}) = W^s_p(\mathscr{O}) \ \left(= B^s_{p,p}(\mathscr{O})\right) \quad \text{for all } s>0,\ s \notin \mathbb{N},\ 1 \le p < \infty, \tag{1.1.11}$$

and

$$F^m_{p,2}(\mathscr{O}) = W^m_p(\mathscr{O}) \ \text{ for } m = 0,1,2,\dots \ \text{ and } \ 1<p<\infty. \tag{1.1.12}$$

We also have the following continuous embeddings (see, e.g., [276] and [250]),

$$F^s_{p,q}(\mathscr{O}) \subset F^{s^*}_{p^*,q^*}(\mathscr{O}) \ \text{ if } \ s - \frac{n}{p} \ge s^* - \frac{n}{p^*}, \quad \mathscr{O} \subseteq \mathbb{R}^n, \tag{1.1.13}$$

where $0 < p < p^* < \infty$ and $0 < q, q^* < \infty$ and also (see [262] and [250]):

$$F^s_{p,q}(\mathscr{O}) \subset L_\infty(\mathscr{O}) \ \text{ if either } \ s > \frac{n}{p} \ \text{ or } \ s = \frac{n}{p},\ 0 < p \le 1, \tag{1.1.14}$$

where $0 < q < \infty$ and $\mathscr{O} \subseteq \mathbb{R}^n$, and

$$F^s_{p,q}(\mathbb{R}^n) \subset L^1_{loc}(\mathbb{R}^n) \ \text{ if either } \ s > 0,\ q > 0 \ \text{ or } \ s = 0,\ 0 < q \le 2, \tag{1.1.15}$$

with $1 \le p < \infty$. In particular, for $n = 2$ (1.1.13) implies:

$$F^4_{1,2}(\mathbb{R}^2) \subset F^{2+\frac{2}{p}}_{p,q}(\mathbb{R}^2) \ \text{ for } \ 1 < p < \infty \ \text{ and } \ q > 0, \tag{1.1.16}$$

hence

$$F^4_{1,2}(\Omega) \subset W^{2+2/p}_p(\Omega) \ \text{ for all } \ 1 \le p \le \infty, \tag{1.1.17}$$

where Ω is a smooth bounded domain in $\mathbb{R}^2$. Indeed, for $1 < p < \infty$ (1.1.17) follows from (1.1.11), (1.1.12), and (1.1.16). Using the relation (see, e.g., [276]),

$$\partial^\alpha F^s_{p,q}(\mathscr{O}) \subset F^{s-|\alpha|}_{p,q}(\mathscr{O}) \ \text{ for every } \ s \in \mathbb{R},\ 0 < p,q < \infty, \tag{1.1.18}$$

we find from (1.1.15) that $\partial^\alpha F^4_{1,2}(\Omega) \subset F^0_{1,2}(\Omega) \subset L^1(\Omega)$ for every multi-index α such that $|\alpha| \le 4$. This implies (1.1.17) for $p = 1$. Similarly, from (1.1.14) and (1.1.18) we have $\partial^\alpha F^4_{1,2}(\Omega) \subset F^2_{1,2}(\Omega) \subset L_\infty(\Omega)$ for every multi-index α such that $|\alpha| \le 2$. This provides the result claimed in (1.1.17) for $p = \infty$.

In the sequel we use the fact that $F^0_{1,2}(\mathbb{R}^n)$ coincides with the local Hardy space $h_1(\mathbb{R}^n)$ (see, e.g., [276, Theorem 2.5.8/1]). The space h_1 can be defined in the following way.

1.1.4. Definition (Hardy spaces). Let $\varphi(x)$ be an infinitely differentiable function with a compact support such that $\varphi(0) = 1$. We define *local Hardy's spaces*

$$h_1(\mathbb{R}^n) = \left\{ f \in \mathscr{S}'(\mathbb{R}^n) \ : \| f \|_{h_1(\mathbb{R}^n)} = \| \sup_{0<t<1} |\mathscr{F}^{-1}(\varphi_t \mathscr{F} f)| \, \|_{L_1(\mathbb{R}^n)} < \infty \right\},$$

where $\varphi_t(x) = \varphi(tx)$, and as above $\mathscr{F}$ denotes the Fourier transform. Replacing $\sup_{0<t<1}$ by $\sup_{t>0}$, leads to the definition of (homogeneous) Hardy space $H_1(\mathbb{R}^n)$.

∎

We have the following relation [250, 276],

$$H_1(\mathbb{R}^n) \subset h_1(\mathbb{R}^n) = F^0_{1,2}(\mathbb{R}^n). \tag{1.1.19}$$

Other equivalent characterizations of local ($h_1(\mathbb{R}^n)$) and homogeneous ($H_1(\mathbb{R}^n)$) Hardy spaces can be given and some of them are reported below [127].

1.1.5. Definition **(h_1 and H_1 spaces).**
• *Riesz transform definition:* By [127, Theorem 2] $f \in h_1(\mathbb{R}^n)$ iff $f \in L_1(\mathbb{R}^n)$ and $r_j f \in L_1(\mathbb{R}^n)$, $j = 1,2,\dots,n$, where smooth quasi-homogeneous multiplier r_j has the form $\mathscr{F}(r_j f) = \mathrm{i}(1-\phi(\xi))(\xi_j/|\xi|)\mathscr{F}f$ with $\phi \in \mathscr{S}$ such that $\phi = 1$ in a neighborhood of the origin. Note, the kernel of r_j is a distribution of compact support plus a rapidly decaying Schwarz function.
• *Convolutional definition:* $f \in h_1(\mathbb{R}^n)$ iff $\sup_{0<t<1} |h_t * f| \in L_1(\mathbb{R}^n)$, where $h_t \equiv t^{-n}h(x/t)$, $h \in \mathscr{S}$ and $\int h \neq 0$ (see [127, Theorem 1]). Similarly, replacing $\sup_{0<t<1}$ by $\sup_{t>0}$ we arrive at the convolution definition for homogeneous Hardy space $H_1(\mathbb{R}^n)$. The norm in $H_1(\mathbb{R}^n)$ can be defined as $L_1(\mathbb{R}^n)$ of $\sup_{t>0}|h_t * f|$, similarly for $h_1(\mathbb{R}^n)$.
• *Atomic definition:* A function $a(x)$ is an atom if it is bounded, has compact support, and mean value zero (moment condition) in the case when the measure of the support of atom $\mathrm{mes}(Q) < 1$. Here Q denotes the smallest cube containing the support of a. The atomic norm $|a|_{\mathrm{atom}}$ is defined as $|a|_{\mathrm{atom}} \equiv \|a\|_{L_\infty}\mathrm{mes}(Q)$. We have (see Lemma 5 [127]) that any atom a belongs to $h_1(\mathbb{R}^n)$, $|a|_{h_1} \le C|a|_{\mathrm{atom}}$ and

$$f \in h_1(\mathbb{R}^n) \ \text{ iff } \ f = \lim_k \sum^k a_i \ \text{ in } \ h_1(\mathbb{R}^n).$$

In this case $\sum |a_i|_{\mathrm{atom}} \le C\|f\|_{h_1}$. ■

The properties of the space $h_1(\mathbb{R}^n)$, that are used later on, are reported below (see Theorem 4 and proof of Lemma 4 in [127]).

1.1.6. Lemma.

- *The Schwarz space $\mathscr{S}$ is dense in h_1.*
- *Pseudodifferential operators of order zero are bounded on h_1. In particular $\|\psi f\|_{h_1} \le C\|f\|_{h_1}$ for any $\psi \in \mathscr{S}$.*

Recall that zero-order PDO although bounded on L_p spaces, $1 < p < \infty$. *are not bounded* at the end of the scale, $p = 1$ and $p = \infty$.

We conclude this subsection with the following property of Jacobians in $\mathbb{R}^2$ that we need for regularity of the von Karman bracket (see Section 1.4 below).

1.1.7. Lemma. *Let $W = (u;v) \in \left[H^1(\mathbb{R}^2)\right]^2$. Then* $\det\{\nabla W\} \equiv u_{x_1}v_{x_2} - u_{x_2}v_{x_1}$ *belongs to the homogeneous Hardy space $H_1(\mathbb{R}^2)$. Here $H^1(\mathbb{R}^2)$ is the first-order Sobolev space.*

This lemma follows from [89, Theorem II]. However, for self-containedness the corresponding argument that we adapt from [89] is given in Appendix A.

1.1.4 Vector-valued spaces

Let X be a Banach space and $[a,b] \subset \mathbb{R}$. We denote by $C^m(a,b;X) \equiv C^m([a,b];X)$ the space of m-differentiable (in the norm topology) functions on $[a,b]$ with values in X. If $[a,b]$ is a finite interval, then $C^m(a,b;X)$ equipped with the norm

$$\|u\|_{C^m(a,b;X)} = \max\left\{ \|u^{(k)}(t)\|_X \, : \, t \in [a,b], k = 0,1,\dots,m \right\}$$

becomes a Banach space. Here $u^{(k)}(t) = \partial_t^k u(t)$ is the strong derivative of u of order k. We denote by $C^m([a,b);X)$ the space of functions $u : [a,b) \mapsto X$ such that $u \in C^m([a,b'];X)$ for any $b' \in (a,b)$. A similar meaning has the notations $C^m((a,b];X)$ and $C^m((a,b);X)$. We also use the notation $C_w(a,b;X)$ for the space of the functions on $[a,b]$ that are continuous with respect to weak topology on X.

$L_p(a,b;X)$, $1 \le p \le \infty$ are classical L_p spaces defined as sets of (classes of almost everywhere equal) strongly Bochner-measurable functions $f(t)$ with values in X such that $\|f(\cdot)\|_X \subset L_p(a,b;\mathbb{R})$. Each $L_p(a,b;X)$ is a Banach space with the norm

$$\|f\|_{L_p(a,b;X)} = \left(\int_a^b \|f(t)\|_X^p dt \right)^{1/p}, \quad 1 \le p < \infty,$$

$$\|f\|_{L_\infty(a,b;X)} = \operatorname{esssup}\{\|f(t)\|_X \, : \, t \in [a,b]\}.$$

Let $\mathscr{I} \subseteq \mathbb{R}$ be a measurable set. We denote by $L_p^{\mathrm{loc}}(\mathscr{I};X)$ the set of measurable functions $u : \mathscr{I} \mapsto X$ such that $u \in L_p(\mathscr{K};X)$ for any measurable compact set $\mathscr{K} \subset \mathscr{I}$. For more details concerning $L_p(a,b;X)$ spaces we refer to [93] and [280].

Let $X \subseteq Y$ be a couple of Banach spaces. Given $f \in L_p(a,b;X)$, $p \ge 1$, the function $g \in L_q(a,b;Y)$ is called the derivative of f in the distributional sense, if

$$\int_a^b g(t)\phi(t)dt = -\int_a^b f(t)\phi'(t)dt \quad \text{for any} \quad \phi \in C_0^\infty(a,b;\mathbb{R}).$$

This relation is equivalent to the equality

$$f(t) = f_0 + \int_a^t g(\tau)d\tau \quad \text{in } Y \text{ for almost every } t \in [a,b],$$

where $f_0 \in Y$. We use notation $g = \partial_t f = f_t = f'$.

For every $1 \le p,q \le \infty$ we define the Banach space

$$W_{p,q}^1(a,b;X,Y) = \{f \in L_p(a,b;X) \, : \, f' \in L_q(a,b;Y)\} \tag{1.1.20}$$

with the norm

$$\|f\|_{W_{p,q}^1(a,b;X,Y)} = \|f\|_{L_p(a,b;X)} + \|f'\|_{L_q(a,b;Y)}.$$

For brevity, we use the notation $W_p^1(a,b;X) = W_{p,p}^1(a,b;X,X)$. Below we also need higher-order spaces of L_p-differentiable functions

$$W_p^m(a,b;X) = \left\{ f \in L_p(a,b;X) : f^{(k)} \in L_p(a,b;X),\ k = 1,\dots,m \right\}, \quad m \geq 1. \tag{1.1.21}$$

The following compactness theorem is important in further considerations.

1.1.8. Theorem. *Let $X \subset Y \subset Z$ be a triple of Banach spaces such that X is compactly embedded in Y. Then*

- *The space $W_{p,q}^1(a,b;X,Z)$ is compactly embedded in $L_p(a,b;Y)$ for every $1 \leq p,q < \infty$.*
- *The space $W_{\infty,q}^1(a,b;X,Z)$ is compactly embedded in $C(a,b;Y)$ for every $q > 1$.*

We refer to Appendix A.2 for the proof of this theorem which is based on the argument given in [263] (see Corollary 4). We also note that particular cases of Theorem 1.1.8 (when $p < \infty$) can be found in [3] and [98] (see also [220]).

1.2 Nonlinear operators and related operator equations

This section collects several results on nonlinear operators in Banach spaces that are used in the study of stationary von Karman equations.

1.2.1 Monotone and pseudomonotone operators

There are several techniques developed in PDEs that are used for proving existence and uniqueness of solutions to nonlinear equations. In the case of von Karman models the two approaches that appear particularly suitable are the following: monotone operator theory and Galerkin methods. In what follows we provide a brief exposition of these methods. For more detailed presentation we refer to the books [18, 260, 281].

We consider operators acting from V into V', where V is a reflexive Banach space and V' its dual. This kind of setup is very useful in dealing with elliptic theory where we choose $V \subset H \subset V'$, with selected Hilbert space H. Below we denote by $(v,v^*)_{V,V'}$ the value of functional $v^* \in V'$ on the element $v \in V$. If V is a Hilbert space, we can identify V' with V and suppose that $(v,v^*)_{V,V'} = (v,v^*)_V$ is an inner product in V.

1.2.1. Definition. The operator $A : \mathscr{D}(A) \subseteq V \mapsto V'$ is said to be *monotone* iff

$$(v_1 - v_2, Av_1 - Av_2)_{V,V'} \geq 0 \quad \text{for all } v_1, v_2 \in \mathscr{D}(A).$$

We say A is *maximal monotone (m-monotone)* iff A is monotone and the relation

$$(v-u, Av-u^*)_{V,V'} \geq 0$$

for some $u \in V$ and $u^* \in V'$ and for all $v \in \mathscr{D}(A)$ implies that $u \in \mathscr{D}(A)$ and $u^* = Au$. ■

Note that if V is a Hilbert space and V' is identified with V, then monotone operators are usually called *accretive* (see also Section 2.2 in Chapter 2).

1.2.2. Example (Nemytskij operator). Let X be a measurable space with σ-finite measure μ. Assume that $g : X \times \mathbb{R} \mapsto \mathbb{R}$ satisfies the Carathéodory conditions:

- $x \mapsto g(x,\xi)$ is measurable for each $\xi \in \mathbb{R}$.
- $\xi \mapsto g(x,\xi)$ is continuous for almost all $x \in X$.

Assume additionally that $|g(x,\xi)| \leq c|\xi|^{p-1} + k(x)$ for almost all $x \in X$ and for all $\xi \in \mathbb{R}$, where $p > 1$ and $k(x) \in L_q(X,\mu)$, $p^{-1} + q^{-1} = 1$. Then the superposition operator defined by the formula

$$A(u)(x) = g(x,u(x)), \quad u(x) \in L_p(X,\mu),$$

maps $V = L_p(X,\mu)$ into $V' = L_q(X,\mu)$. If the function $\xi \mapsto g(x,\xi)$ is nondecreasing for almost all $x \in X$, then it is easy to see that A is m-monotone. ■

We introduce next several notions of boundedness and continuity [260, 18, 281].

1.2.3. Definition. Let $A : V \mapsto V'$ be an operator acting on a reflexive Banach space V.

- The operator A is said to be *locally bounded at* $v_0 \in V$ iff there exists a neighborhood $U(v_0)$ such that $A(U) = \{Av : v \in U\}$ is a bounded set of V'.
- The operator A is said to be *bounded* iff it maps every bounded subset of V into a bounded subset of V'.
- A is said to be *proper* iff the preimage of every compact subset of V' is again compact.
- A is said to be *continuous* iff $Au_n \to Au$ strongly in V' for any sequence $\{u_n\} \subset V$ such that $u_n \to u$ strongly in V.
- The operator A is said to be *compact* iff A is continuous and maps bounded subsets of V into relatively compact subsets of V'.
- A is said to be *strongly continuous* iff $Au_n \to Au$ strongly in V' for any sequence $\{u_n\} \subset V$ such that $u_n \rightharpoonup u$ weakly in V.
- A is said to be *hemicontinuous* iff for each $u,v \in V$ the real-valued function $t \to (v, A(u+tv))_{V,V'}$ is continuous.
- A is said to be *demicontinuous* on V iff it is strongly-weakly continuous from V to V'; that is, w-lim $Av_n = Av$ for all $v_n \to v$ strongly in V.

■

It turns out that for m-monotone operators defined on the entire space V the two concepts, hemicontinuity and demicontinuity are equivalent, see, e.g., [281, vol. II/B, p. 596]. Also, for this class of operators demicontinuity is equivalent to local boundedness and hemicontinuity. Moreover, monotonicity and hemicontinuity imply local boundedness.

1.2.4. Definition. We say that $A : V \mapsto V'$ is *coercive* iff

$$\frac{(v, Av)_{V,V'}}{|v|_V} \to \infty \quad \text{when } |v|_V \to \infty,$$

where $|\cdot|_V$ denotes the norm in the space V. ∎

We collect some more properties for a class of monotone operators such that $\mathscr{D}(A) \equiv V$ (see, e.g., [260, 18, 281]).

1.2.5. Proposition. *Let $A : V \mapsto V'$ where V is a reflexive Banach space.*

- *If A is monotone and hemicontinuous, then it is m-monotone.*
- *If A is linear and monotone, then it is continuous.*
- *If A is monotone, then A is locally bounded at any point of V.*
- *If A is monotone and hemicontinuous, then it is demicontinuous.*
- *Let A and B be m-monotone from $V \to V'$. Then $A+B$ is m-monotone.*
- *If A is monotone, hemicontinuous, and coercive and the space V is separable, then $R(A) = V'$; that is the equation $Au = f$ has solutions for any $f \in V'$.*
- *Let A be m-monotone and B be monotone and hemicontinuous from $V \to V'$. Then $A+B$ is m-monotone. Assuming in addition that $A+B$ is coercive, we have $R(A+B) = V'$.*

The following *weak convergence* result is fundamental within the context of *m*-monotone operators [18].

1.2.6. Proposition. *Assume that V is a reflexive Banach space. Let $A : \mathscr{D}(A) \subseteq V \mapsto V'$ be maximal monotone and $\{v_n\}$ be a sequence in $\mathscr{D}(A)$ such that $v_n \rightharpoonup v$ weakly in V and $Av_n \rightharpoonup f$ weakly in V'.*

- *If*
$$\limsup_{n\to\infty} (v_n - v, Av_n - f)_{V,V'} \le 0, \tag{1.2.1}$$
then $v \in \mathscr{D}(A)$, $f = Av$ and
$$\lim_{n\to\infty} (v_n, Av_n)_{V,V'} = (v, Av)_{V,V'}. \tag{1.2.2}$$
- *The same conclusion remains true, if instead of (1.2.1) we assume that*
$$(v_n - v_m, Av_n - Av_m)_{V,V'} \to 0 \quad \textit{when } n,m \to \infty. \tag{1.2.3}$$

Proof. It is contained in [18]. For readers' convenience we reproduce the details.
Part 1: One can see that (1.2.1) is equivalent to

$$\limsup_{n\to\infty} (v_n, Av_n)_{V,V'} \le (v, f)_{V,V'}. \tag{1.2.4}$$

Therefore, because

$$(v_n - w, Av_n - Aw)_{V,V'} \ge 0 \quad \text{for any } w \in \mathscr{D}(A), \tag{1.2.5}$$

taking lim sup yields $(v-w, f-Aw)_{V,V'} \geq 0$. Hence maximality of A implies that $v \in \mathscr{D}(A)$ and $f = Av$. Taking now lim inf in the relation (1.2.5) with $w = v$, we obtain that $\liminf_{n\to\infty}(v_n, Av_n)_{V,V'} \geq (v, f)_{V,V'}$ which, along with (1.2.4), implies (1.2.2).

Part 2: Assume (1.2.3) instead of (1.2.1). We first prove that

$$\lim_{n\to\infty} (v_n, Av_n)_{V,V'} = (v, f)_{V,V'}. \tag{1.2.6}$$

It follows from (1.2.3) that for any $\varepsilon > 0$ there exists N such that

$$0 \leq (v_n - v_m, Av_n - Av_m)_{V,V'} \leq \varepsilon \quad \text{for all } \ n, m \geq N. \tag{1.2.7}$$

Let a be a limit point of the sequence $\{(v_n, Av_n)_{V,V'}\}$. Then $a = \lim_{k\to\infty}(v_{n_k}, Av_{n_k})_{V,V'}$ for some sequence $\{n_k\}$. From (1.2.7) we have

$$0 \leq (v_{n_k}, Av_{n_k})_{V,V'} + (v_{n_l}, Av_{n_l})_{V,V'} - (v_{n_k}, Av_{n_l})_{V,V'} - (v_{n_l}, Av_{n_k})_{V,V'} \leq \varepsilon$$

for all $k, l \geq \tilde{N}$. Thus, if we let $l \to \infty$ in the last relation, we obtain that

$$0 \leq (v_{n_k}, Av_{n_k})_{V,V'} + a - (v, Av_{n_k})_{V,V'} - (v_{n_k}, f)_{V,V'} \leq \varepsilon$$

for all $k \geq \tilde{N}$. Therefore after limit transition $k \to \infty$ we obtain that

$$0 \leq 2a - 2(v, f)_{V,V'} \leq \varepsilon \quad \text{for any } \ \varepsilon > 0.$$

Thus $a = (v, f)_{V,V'}$ and hence (1.2.6) holds.

Now using (1.2.6) after limit transition in (1.2.5) we find that $(v-w, f-Aw)_{V,V'} \geq 0$ for any $w \in \mathscr{D}(A)$ and hence, by maximality of A, $v \in \mathscr{D}(A)$ and $f = Av$. Relation (1.2.2) follows from (1.2.6). □

Some extensions of the concept of monotonicity are given below. These objects are used for the study of stationary von Karman equations. Following [281] (see also [121, 220, 260]), we introduce the following definition.

1.2.7. Definition. Let $A : V \mapsto V'$ be an operator on the reflexive Banach space V.

- The operator A is said to be *pseudomonotone* iff the relations

  $$u_n \rightharpoonup u \ \text{ weakly in } V \text{ and } \limsup_{n\to\infty}(u_n - u, Au_n)_{V,V'} \leq 0$$

  imply that

  $$(u-w, Au)_{V,V'} \leq \liminf_{n\to\infty}(u_n - w, Au_n)_{V,V'} \ \text{ for all } \ w \in V.$$

- The operator $A : V \mapsto V'$ satisfies the condition $(S)_+$ iff the relations

  $$u_n \rightharpoonup u \ \text{ weakly in } \ V \ \text{ and } \ \limsup_{n\to\infty}(u_n - u, Au_n - Au)_{V,V'} \leq 0 \tag{1.2.8}$$

  imply that $u_n \to u$ strongly in V.

■

The following proposition is important in applications of the theory of pseudomonotone operators to quasilinear elliptic PDEs.

1.2.8. Proposition. *Let $A : V \longrightarrow V'$ be operator defined on reflexive Banach space V.*

- *If A is monotone and hemicontinuous, then it is pseudomonotone.*
- *If A is strongly continuous, then A is pseudomonotone.*
- *If A is demicontinuous and satisfies $(S)_+$, then A is pseudomonotone.*
- *If $A, B : V \longrightarrow V'$ are pseudomonotone, then $A + B$ is pseudomonotone.*
- *If A satisfies $(S)_+$ and $B : V \longrightarrow V'$ is strongly continuous, then $A + B$ satisfies $(S)_+$.*
- *If A is pseudomonotone and locally bounded on V, then A is demicontinuous.*

For the proof we refer to [281, Chapter 27].

We consider next the following operator equation

$$Au = b, \quad u \in V, \tag{1.2.9}$$

in a real separable Banach space V. Let $\{e_k\}$ be a basis in V. We define the Galerkin approximation for (1.2.9) with respect to the basis $\{e_k\}$ as the problem: to find $u_n \in V_n \equiv \text{ span } \{e_1, \ldots, e_n\}$ satisfying the equations

$$(e_k, Au_n)_{V,V'} = (e_k, b)_{V,V'} \quad \text{for all } \ k = 1, \ldots, n.$$

1.2.9. Theorem. *Let $A : V \longrightarrow V'$ be a pseudomonotone, bounded, and coercive operator. Then*

- *For each $b \in V'$ the equation* (1.2.9) *has a solution. If A is proper, then the set of solutions to* (1.2.9) *is a compact set in V.*
- *For each $b \in V'$ and $n \in \mathbb{N}$ the Galerkin approximation has a solution. There exists a subsequence $\{u_{n_k}\}$ that converges weakly to a solution of the original equation* (1.2.9). *If A satisfies $(S)_+$, then a subsequence $\{u_{n_k}\}$ converges strongly.*

For the proof we refer to [281, Chapter 27].

1.2.2 Proper Fredholm operators

In order to study existence and properties of stationary solutions, some elements of the theory of nonlinear Fredholm operators are used. Following [281, Chapter 29], see also [17, Chapter 6], we begin by recalling several definitions.

1.2.10. Definition. Let V and W be Banach spaces.

- The linear operator $L : V \mapsto W$ is said to be *Fredholm* iff L is continuous, the image $R(L) = \{Lu : u \in V\}$ is closed, and
$$\dim N(L) < \infty \quad \text{and} \quad \operatorname{codim} R(L) < \infty,$$
where $N(L) = \{u \in V : Lu = 0\}$. The *index* of L is defined by the formula:
$$\operatorname{ind} L = \dim N(L) - \operatorname{codim} R(L).$$

- The nonlinear operator $A : \mathscr{D}(A) \subseteq V \mapsto W$, which is defined on the open (in V) set $\mathscr{D}(A)$, is called *Fredholm* iff (i) A is Fréchet differentiable, (ii) its linearization (Fréchet derivative) $A'(u) : V \mapsto W$ is Fredholm for all $u \in \mathscr{D}(A)$, and (iii) it continuously depends on u in the uniform operator topology. The number $\operatorname{ind} A'(u)$ is called the index of A at u.

■

The following results are well known; see, for example, [281, Section 29.10].

1.2.11. Proposition. *Let $A : \mathscr{D}(A) \subseteq V \mapsto W$ be Fredholm on an open set $\mathscr{D}(A)$. Then*

- *If $B : \mathscr{D}(A) \subseteq V \mapsto W$ is a compact Fréchet differentiable operator and the derivative $B'(u)$ is continuous with respect to u, then $A + B$ is a Fredholm operator with the same index as A at each point of $\mathscr{D}(A)$.*
- *Let $\mathscr{M}$ be a connected subset of $\mathscr{D}(A)$. Then $\operatorname{ind} A'(u) = const$ for all $u \in \mathscr{M}$.*

1.2.12. Definition. Let V and W be Banach spaces and $A : \mathscr{D}(A) \subseteq V \mapsto W$ be Fredholm on the open set $\mathscr{D}(A)$. The point $u \in \mathscr{D}(A)$ is called a *singular point* of A, iff $A'(u) : V \mapsto W$ isnot surjective. Otherwise, u is called a *regular point* of A. The point $b \in W$ is called a *singular value* of A, iff there exists a singular point $u \in \mathscr{D}(A)$ such that $b = Au$. Otherwise, b is called a *regular value* of A. ■

We note that if A is Fredholm with index zero and $\mathscr{D}(A) = V$, then

$$D_{\text{sing}} = \left\{u \in V : A'(u)h = 0 \text{ has a solution } h \neq 0\right\} \tag{1.2.10}$$

is the set of all singular points of A. Moreover $D_{\text{reg}} = V \setminus D_{\text{sing}}$ is the set of all regular points, $R_{\text{sing}} = A(D_{\text{sing}})$ is the set of singular values, and $R_{\text{reg}} = W \setminus A(D_{\text{sing}})$ is the set of regular values.

The results from the theory of Fredholm operators used in the sequel are collected below (for the proof we refer to [281, Section 29.10]).

1.2.13. Theorem. *Let $A : V \mapsto W$ be a proper (see Definition 1.2.3) Fredholm operator of index zero. Then*

- *For each $b \in R_{\text{reg}} = W \setminus A(D_{\text{sing}})$ the number $n_V(b) \geq 0$ of solutions to the equation*
$$Au = b, \quad u \in V, \tag{1.2.11}$$
is finite. The function $n_V(\cdot)$ is a constant on each connected subset of R_{reg}.

- *If* b_0 *is a regular value of* A *and* $b_0 \in R(A)$*, then* $b_0 \in \mathrm{int}R(A)$*; that is, equation* (1.2.11) *has a solution for every* b *in a neighborhood of* b_0.
- *The set* R_{reg} *of regular values of* A *is open and dense in* W *and hence of second Baire category in* W*. The set* $R_{\mathrm{sing}} = A(D_{\mathrm{sing}})$ *of singular values of* A *is closed and nowhere dense in* W *and hence of first Baire category in* W.
- *The set* D_{reg} *of regular points of* A *is open* V.

This theorem is related to the infinite-dimensional version of the Sard–Smale theorem. The main fact stated in Theorem 1.2.13 and of use in the future is that the equation (1.2.11) *generically* has a finite number of solutions. We also refer to [17] for some application of this theorem to the theory of infinite-dimensional dissipative dynamical systems.

1.3 Biharmonic operator

We collect here several results related to the biharmonic operator Δ^2 and the associated boundary value problems defined on a smooth bounded domain $\Omega \subset \mathbb{R}^2$. We consider several types of self-adjoint boundary conditions associated with Δ^2. The boundary conditions considered—clamped, hinged, simply supported and free—have strong physical motivation. We refer to [100] and [216, Chapter 3] for relevant discussions of the physical meaning (from the point of view of mechanics) of the boundary conditions considered.

We begin the exposition by collecting number of "Green's type" formulas that provide a connection between the biharmonic operator and various boundary conditions. To accomplish this we first explain the meaning of boundary operators and associated boundary traces. Let $\Gamma := \partial\Omega$ be a C^1 curve. Then for every $\eta \in \Gamma$ we can define the unit outward normal vector $n = n(\eta) = (n_1(\eta); n_2(\eta))$ and the corresponding unit tangent vector $\tau = \tau(\eta) = (-n_2(\eta); n_1(\eta))$. Moreover, for every $\eta_0 \in \Gamma$ there exists a (tabular) neighborhood (collar) $\mathcal{U} \ni \eta_0$ in $\mathbb{R}^2$ such that every $x \in \mathcal{U} \cap \Omega$ can be uniquely presented in the form

$$x = \eta - \delta n(\eta) \ \text{ for some } \ \eta \in \mathcal{U} \cap \Gamma \text{ and } \delta > 0. \tag{1.3.1}$$

This it makes possible to extend the vector fields n and τ defined on Γ inside Ω by the formulas

$$n(x) = n(\eta) \ \text{ and } \ \tau(x) = \tau(\eta), \quad x \in \mathcal{U} \cap \Omega,$$

where x is given by (1.3.1). Therefore we can define normal and tangential derivatives inside $\mathcal{U} \cap \Omega$ by the formulas

$$\frac{\partial u}{\partial n}(x) = (\nabla u(x), n(x))_{\mathbb{R}^2} \ \text{ and } \ \frac{\partial u}{\partial \tau}(x) = (\nabla u(x), \tau(x))_{\mathbb{R}^2}, \quad x \in \mathcal{U} \cap \Omega,$$

where $(\cdot,\cdot)_{\mathbb{R}^2}$ is the inner product in $\mathbb{R}^2$. Moreover, by induction we can also define higher-order derivatives, such as

$$\frac{\partial^2 u}{\partial^2 n}, \ \frac{\partial^2 u}{\partial^2 \tau}, \ \text{or} \ \frac{\partial^2 u}{\partial n \partial \tau} \quad \text{in } \mathscr{U} \cap \Omega,$$

for instance. Letting $\delta \to 0$ leads to the definition of the corresponding boundary values. For more details concerning calculations involving boundary operators and values we refer to [216, Chapter 3, Section 3C].

1.3.1. Proposition. *For any u and v from $C^3(\overline{\Omega})$, the following identity*

$$\int_\Omega [u,v]\,dx = -\int_\Gamma \left[(B_1 u)\frac{\partial v}{\partial n} - (B_2 u) v \right] d\Gamma$$

holds true. Here

$$[u,v] = \partial_{x_1}^2 u \cdot \partial_{x_2}^2 v + \partial_{x_2}^2 u \cdot \partial_{x_1}^2 v - 2 \cdot \partial_{x_1 x_2}^2 u \cdot \partial_{x_1 x_2}^2 v$$

is the von Karman bracket, the boundary operators B_1 and B_2 defined by the relations

$$\begin{aligned} B_1 u &= 2 n_1 n_2 u_{x_1 x_2} - n_1^2 u_{x_2 x_2} - n_2^2 u_{x_1 x_1}, \\ B_2 u &= \tfrac{\partial}{\partial \tau}\left[\left(n_1^2 - n_2^2\right) u_{x_1 x_2} + n_1 n_2 \left(u_{x_2 x_2} - u_{x_1 x_1}\right) \right], \end{aligned} \tag{1.3.2}$$

where $n = (n_1; n_2)$ is the outer normal to $\Gamma = \partial\Omega$, and $\tau = (-n_2; n_1)$ is the unit tangent vector along $\partial\Omega$. We assume that Γ is oriented along τ.

Proof. See [216, Chapter 3, Lemma 3C.2, p. 300], for instance. □

Proposition 1.3.1 and Green's formula for the Laplace operator imply (see also [173]) the following

1.3.2. Proposition. *For any $u \in H^4(\Omega)$, $v \in H^2(\Omega)$, and $\mu \in \mathbb{R}$ we have that*

$$\begin{aligned} \int_\Omega \Delta^2 u v\,dx = a_0(u,v) + \int_\Gamma \Bigg[& \left(\frac{\partial \Delta u}{\partial n} + (1-\mu) B_2 u \right) v \\ & - (\Delta u + (1-\mu) B_1 u) \frac{\partial v}{\partial n} \Bigg] d\Gamma, \end{aligned} \tag{1.3.3}$$

where

$$\begin{aligned} a_0(u,v) &= \int_\Omega (\Delta u \Delta v - (1-\mu)[u,v])\,dx \qquad (1.3.4) \\ &= \int_\Omega (\mu \Delta u \Delta v + (1-\mu)(u_{x_1 x_1} v_{x_1 x_1} + 2 u_{x_1 x_2} v_{x_1 x_2} + u_{x_2 x_2} v_{x_2 x_2}))\,dx. \end{aligned}$$

1.3.3. Remark. We note that from the point of view of modeling, the parameter μ represents the Poisson ratio, hence its value lies in the interval $(0, \frac{1}{2})$. However, mathematical arguments do not depend on this restriction and are valid for the full range of parameters $\mu \in (0,1)$. Therefore, in what follows we use formula (1.3.3) for $0 < \mu < 1$. The advantage of doing this is that the "extreme" case $\mu = 1$ corresponds to the form $a_0(u,v) = \int_\Omega \Delta u \Delta v dx$, which is a familiar bilinear form associated with either clamped or hinged boundary conditions (see below). Similarly, (1.3.3) is just a classical Green's formula used in standard plate theory. ■

In the study of long time behavior the following "multipliers" identity is fundamental (see [173, p.80]).

1.3.4. Proposition.

$$\begin{aligned}\int_\Omega \Delta^2 u \cdot (h,\nabla)u dx = a_0(u,u) + \frac{1}{2} a_\Gamma(u,u) \\ + \int_\Gamma \left[\left(\frac{\partial \Delta u}{\partial n} + (1-\mu)B_2 u \right) (h,\nabla)u \right. \\ \left. - (\Delta u + (1-\mu)B_1 u) \frac{\partial (h,\nabla)u}{\partial n} \right] d\Gamma,\end{aligned} \tag{1.3.5}$$

for any $u \in H^4(\Omega)$, *where* $h = x - x_0$ *and*

$$a_\Gamma(u,v) = \int_\Gamma (h,n) (\Delta u \Delta v - (1-\mu)[u,v]) \, d\Gamma.$$

A geometric representation of the boundary operators B_1 and B_2 is both insightful for physics and helpful for calculations. In fact, the following representation in terms of normal and tangential directions is used in Chapter 10.

1.3.5. Proposition. *Let* $\Omega \subset \mathbb{R}^2$ *be a domain with* C^1*-boundary* Γ. *Then the boundary operators* B_1 *and* B_2 *have the form*

$$B_1 u = -\frac{\partial^2 u}{\partial \tau^2} - \operatorname{div} n(x) \cdot \frac{\partial u}{\partial n} \quad \text{for } x \in \Gamma, \tag{1.3.6}$$

and

$$B_2 u = \frac{\partial}{\partial \tau} \frac{\partial}{\partial n} \frac{\partial u}{\partial \tau} = \frac{\partial}{\partial \tau} \left[\frac{\partial}{\partial \tau} \frac{\partial u}{\partial n} - \operatorname{div} n(x) \cdot \frac{\partial u}{\partial \tau} \right] \quad \text{for } x \in \Gamma, \tag{1.3.7}$$

where we use the same notations as in Proposition 1.3.1. *We also have the representation*

$$\Delta u \Big|_\Gamma = \frac{\partial^2 u}{\partial n^2} + \frac{\partial^2 u}{\partial \tau^2} + \operatorname{div} n(x) \cdot \frac{\partial u}{\partial n}, \quad x \in \Gamma. \tag{1.3.8}$$

Proof. We refer to [216, Chapter 3]. See Proposition 3C.1 (p. 298) for (1.3.6), Propositions 3C.8 and 3C.9 (pp. 306, 307) for (1.3.7), Proposition 3C.6 (p. 305) for (1.3.8). □

1.3.1 Clamped (Dirichlet) boundary conditions

Clamped, often referred to as the Dirichlet boundary conditions are the most widely known and frequently used boundary conditions in plate theory.

We denote by $\Delta_D^2 : L^2(\Omega) \mapsto L^2(\Omega)$ biharmonic operator with the zero clamped conditions:

$$u|_{\partial\Omega} = \nabla u|_{\partial\Omega} = 0. \tag{1.3.9}$$

This is to say

$$\Delta_D^2 u \equiv \Delta^2 u, \quad u \in \mathscr{D}(\Delta_D^2) \equiv H^4(\Omega) \cap H_0^2(\Omega). \tag{1.3.10}$$

The operator Δ_D^2 is self-adjoint and strictly positive. It also possesses a discrete spectrum. We recall the following definition.

1.3.6. Definition. A positive self-adjoint operator A in a Hilbert space H is said to be an *operator with a discrete spectrum* iff there exists an orthonormal basis $\{e_k\}$ in H consisting of eigenvectors of the operator A:

$$Ae_k = \lambda_k e_k, \quad e_k \in \mathscr{D}(\mathscr{A}), \quad k = 1, 2, \ldots,$$

and the corresponding eigenvalues $\{\lambda_k\}$ have the properties $0 \le \lambda_1 \le \lambda_2 \le \cdots$, and $\lim_{k\to\infty} \lambda_k = \infty$. ■

The operator Δ_D^2 generates on $H_0^2(\Omega)$ the bilinear form

$$a(u,v) = \int_\Omega \Delta u \Delta v dx = \left(\left(\Delta_D^2\right)^{1/2} u, \left(\Delta_D^2\right)^{1/2} v \right)_{L^2(\Omega)}, \quad u, v \in H_0^2(\Omega).$$

1.3.7. Remark. We note that $(\Delta_D^2)^{1/2} \neq \Delta_D$. This is due to the boundary conditions that enter the definition of Δ_D^2. Only in the commutative case (hinged boundary conditions; see (1.3.14) below) does the equivalence $(\Delta_H^2)^{1/2} = \Delta_D$ take place.

We also note that by Proposition 1.3.1 the form $a(u,v)$ also admits the following representation

$$a(u,v) = a_0(u,v), \quad u, v \in H_0^2(\Omega),$$

where $a_0(u,v)$ is defined by (1.3.4) with the constant $0 < \mu < 1$, where μ is the Poisson modulus. The above representation is important in the context of analyzing mixed boundary conditions. ■

Let $(\Delta_D^2)^{-1}$ denote the inverse of Δ_D^2, which is defined as a bounded operator from $L_p(\Omega)$ into $W_p^4(\Omega)$ for all $1 < p < \infty$. To study properties of the Airy stress function (see Section 1.4) we use the fact that the operator Δ_D^2 is an isomorphism from $H^s(\Omega) \cap H_0^2(\Omega)$ onto $H^{s-4}(\Omega)$ for $s \ge 2$ (see, e.g., [222]) and, therefore,

$$(\Delta_D^2)^{-1} : H^s(\Omega) \to H^{s+4}(\Omega) \cap H_0^2(\Omega), \quad s \ge -2. \tag{1.3.11}$$

To establish sharp regularity of the Airy stress function we also invoke the following elliptic regularity result which can be easily derived from the more general elliptic theory presented in [120] (see also [250]).

1.3.8. Theorem. *Let $\Omega \subset \mathbb{R}^2$ be a smooth bounded domain. Assume that $f(x)$ belongs to the Lizorkin space $F^s_{p,q}(\Omega)$ (see Definition* 1.1.3*), where*

$$s > \frac{1+\max(1,p)}{p} - 4, \quad 0 < p < \infty, \quad 0 < q < \infty.$$

Then the problem

$$\Delta^2 v = f, \quad v|_{\partial\Omega} = \frac{\partial v}{\partial n}\Big|_{\partial\Omega} = 0,$$

has a unique solution v in $F^{s+4}_{p,q}(\Omega)$ and

$$\|v\|_{F^{s+4}_{p,q}(\Omega)} \le C\|f\|_{F^s_{p,q}(\Omega)}$$

for some constant C independent of f.

This theorem and relation (1.1.17) imply the following assertion.

1.3.9. Corollary. *Let $\Omega \subset \mathbb{R}^2$ be a smooth bounded domain. Assume that $f(x)$ belongs to the Lizorkin space $F^0_{1,2}(\Omega)$. Then $(\Delta^2_D)^{-1} f$ lies in the Sobolev space $W^{2+2/p}_p(\Omega)$ for every $p \in [1,\infty]$ and*

$$\|(\Delta^2_D)^{-1} f\|_{W^{2+2/p}_p(\Omega)} \le C\|f\|_{F^0_{1,2}(\Omega)}, \quad 1 \le p \le \infty. \tag{1.3.12}$$

1.3.10. Remark. The assertion in Corollary 1.3.9 remains true in the case when $\Omega = (0,a) \times (0,b)$ is a rectangle. To justify it we first note that the regularity described in Theorem 1.3.8 holds [250] for an arbitrary fourth-order strongly elliptic operator in the case of either smooth domains or half-planes. Thus, in order to obtain the result for rectangular domains, we need to justify applicability of the relation

$$\left(\Delta^2_D\right)^{-1} : F^0_{1,2}(\Omega) \to F^4_{1,2}(\Omega) \subset W^{2+2/p}_p(\Omega) \text{ is bounded,} \tag{1.3.13}$$

to this class of domains. To this end, we consider the problem

$$\Delta^2 u = f \in F^0_{1,2}(\Omega) \text{ in } \Omega \text{ and } u = \nabla u = 0 \text{ on } \partial\Omega\,.$$

We first extend the solution of the elliptic problem across the boundary of the rectangle $[0,a] \times [0,b]$. Let u_e denote an extension to $[0,a] \times [0,2b]$. We set $u_e(x_1,x_2) = u(x_1, 2b - x_2)$ for $b \le x_2 \le 2b$ and $0 \le x_1 \le a$. Then u_e satisfies clamped boundary conditions and the original equation. By extending the solution in both directions we obtain the problem (biharmonic) defined on the half-space $\Delta^2 u_e = f_e$, with $f \in F^0_{1,2}(\Omega)$. We then consider

$$\Delta^2(u_e e^{-|x|^2}) - \{\Delta^2, e^{-|x|^2}\} u_e = f_e e^{-|x|^2} \text{ in } \mathbb{R}^2_+\,,$$

$$u_e e^{-|x|^2} = \nabla(u_e e^{-|x|^2}) = 0 \text{ on } \{(0;x_2) : x_2 \in \mathbb{R}\},$$

where $\{A,B\} = AB - BA$ denotes the commutator and $|x|^2 = x_1^2 + x_2^2$. We note, that after computing commutators, the equation above is nothing else but a fourth-order strongly elliptic equation with variable smooth coefficients in lower-order terms for the solution $e^{-|x|^2}u_e$. By [250, Theorem 4.4.4] the function $f_e e^{-|x|^2}$ belongs to $F^0_{1,2}(\mathbb{R}^2_+)$. Thus the half-space elliptic estimates in [250] apply and lead to the desired elliptic estimate (1.3.13) valid also for a rectangle. ∎

1.3.11. Remark. The results similar to Theorem 1.3.8 (with other ranges of the parameter s) can be obtained for all the boundary conditions specified below. The corresponding results follow from the application of the theory developed in [120] (see also [250]). We do not specify these results here because we do not use them in the subsequent analysis. ∎

In order to provide an abstract-functional analytic representation of nonhomogeneous boundary value problems (see Section 2.5.2 below), it is convenient to introduce the so-called Green's maps that provide for suitable extensions of boundary values into the interior.

In the case of clamped (Dirichlet) boundary conditions (1.3.9) this Green's map is given by $u = G_D g$ iff u satisfies

$$\Delta^2 u = 0 \quad \text{for } x \in \Omega, \quad u\big|_{\partial\Omega} = 0, \quad \frac{\partial}{\partial n}u\Big|_{\partial\Omega} = g.$$

Classical elliptic theory [222] implies that

$$G_D : L_2(\partial\Omega) \mapsto H^{3/2}(\Omega) \cap H^1_0(\Omega) \quad \text{is continuous}$$

and therefore

$$G_D : L_2(\partial\Omega) \mapsto \mathscr{D}\left((\Delta_D^2)^{3/8-\varepsilon}\right) = H^{3/2-4\varepsilon}(\Omega) \cap H^1_0(\Omega)$$

is continuous for every $\varepsilon > 0$ small enough.[1]

1.3.2 Hinged boundary conditions

Hinged boundary conditions, owing to their symmetry, are the simplest ones to treat from the mathematical point of view. This is because the biharmonic operator becomes just a square of the regular Dirichlet–Laplace operator. Indeed, hinged boundary conditions are of the following form

$$u|_{\partial\Omega} = \Delta u|_{\partial\Omega} = 0. \tag{1.3.14}$$

[1] The identification with the fractional power of the operator Δ_D^2 is short of ε.

As above the operator $\Delta_H^2 : L_2(\Omega) \mapsto L_2(\Omega)$ defined by the relations

$$\Delta_H^2 u \equiv \Delta^2 u, \quad u \in \mathscr{D}(\Delta_H^2) \equiv \left\{ u \in H^4(\Omega) \cap H_0^1(\Omega) \, : \, \Delta u|_{\partial\Omega} = 0 \right\} \tag{1.3.15}$$

is a self-adjoint strictly positive operator with a discrete spectrum. It is clear that $\Delta_H^2 = (\Delta_D)^2$, where Δ_D is the Laplace operator with the Dirichlet boundary conditions, $\mathscr{D}(\Delta_D) = H^2(\Omega) \cap H_0^1(\Omega)$. Thus $\left(\Delta_H^2\right)^{1/2} = -\Delta_D$.

Green's operator associated with hinged boundary conditions is defined by the relation $u = G_H g$, where u satisfies the following boundary value problem

$$\Delta^2 u = 0 \;\text{ for }\; x \in \Omega, \quad u\big|_{\partial\Omega} = 0, \quad \Delta u\Big|_{\partial\Omega} = g.$$

In this case G_H can be identified with $-A_D^{-1}D$, where $A_D = -\Delta_D$ and D is a classical Dirichlet map defined as a harmonic extension of g from the boundary $\partial\Omega$ into the interior Ω. This is to say, $w = Dg$, where

$$\Delta w = 0 \;\text{ in }\; \Omega \quad\text{and}\quad w|_{\partial\Omega} = g.$$

Because D is bounded from $L_2(\partial\Omega)$ into $H^{1/2}(\Omega) \subset \mathscr{D}(A_D^{1/4-\varepsilon})$, we infer that $G_H = -A_D^{-1}D$ is a continuous mapping from $L_2(\partial\Omega)$ into $\mathscr{D}(A_D^{5/4-\varepsilon}) = \mathscr{D}\left((\Delta_H^2)^{5/8-\varepsilon/2}\right)$. Therefore

$$G_H \, : \, L_2(\partial\Omega) \mapsto \mathscr{D}\left(\left(\Delta_H^2\right)^{5/8-\varepsilon}\right) \equiv H^{5/2-4\varepsilon}(\Omega) \cap H_0^1(\Omega)$$

is continuous for every $\varepsilon > 0$ small enough. We also note that as in the clamped case we can get rid of that ε and obtain the relation

$$G_H \, : \, L_2(\partial\Omega) \mapsto H^{5/2}(\Omega) \cap H_0^1(\Omega) \;\text{ is continuous.}$$

For details we refer to [216, Chapter 3, Section 3.6] and to the references therein.

We also note that

$$\int_\Omega \Delta^2 u G_H g dx = \int_\Gamma \frac{\partial}{\partial n} u g d\Gamma \quad \text{for any } \; u \in \mathscr{D}(\Delta_H^2), \; g \in L_2(\Gamma). \tag{1.3.16}$$

Indeed, one can see that Green's formula (1.3.3) with $\mu = 1$ can be written in the form

$$\begin{aligned} &\int_\Omega \Delta^2 u w dx - \int_\Omega u \Delta^2 w dx \\ &= \int_\Gamma \left[\frac{\partial \Delta u}{\partial n} w - \Delta u \frac{\partial w}{\partial n} \right] d\Gamma - \int_\Gamma \left[u \frac{\partial \Delta w}{\partial n} - \frac{\partial u}{\partial n} \Delta w \right] d\Gamma. \end{aligned} \tag{1.3.17}$$

Substituting $u \in \mathscr{D}(\Delta_H^2)$ and $w = G_H g$ in (1.3.17) and taking into account the boundary conditions for u and w yield (1.3.16).

It is also worth to mention that G_H is a particular case of Green's map G_S considered in the next subsection ($G_H = G_S$ when $\mu = 1$).

1.3.3 Simply supported (hinged revisited) boundary conditions

Simply supported boundary conditions are perturbed (through unbounded perturbation) hinged boundary conditions. These are defined below.

$$u\Big|_{\partial\Omega} = [\Delta u + (1-\mu)B_1 u]\Big|_{\partial\Omega} = 0, \tag{1.3.18}$$

where the boundary operator B_1 has the following form,

$$B_1 u = 2n_1 n_2 u_{x_1x_2} - n_1^2 u_{x_2x_2} - n_2^2 u_{x_1x_1}$$

and $n = (n_1, n_2)$ is the outer normal to $\partial\Omega$. The parameter $0 < \mu < 1$ has a meaning of the Poisson modulus.

We define the operator $\Delta_S^2 : L_2(\Omega) \mapsto L_2(\Omega)$ by the formula $\Delta_S^2 u \equiv \Delta^2 u$ for $u \in \mathscr{D}(\Delta_S^2)$, where

$$\mathscr{D}(\Delta_S^2) = \left\{ u \in H^4(\Omega) \cap H_0^1(\Omega) \; : \; [\Delta u + (1-\mu)B_1 u]\Big|_{\partial\Omega} = 0 \right\}.$$

The operator Δ_S^2 is a self-adjoint strictly positive operator with a discrete spectrum. The operator $\left(\Delta_S^2\right)^{1/2}$ has the domain $\mathscr{D}(\left(\Delta_S^2\right)^{1/2}) = H^2(\Omega) \cap H_0^1(\Omega)$ and

$$\left(\left(\Delta_S^2\right)^{1/2} u, \left(\Delta_S^2\right)^{1/2} v\right)_{L_2(\Omega)} = a_0(u,v), \quad u, v \in H^2(\Omega) \cap H_0^1(\Omega),$$

where $a_0(u,v)$ is defined by (1.3.4).

In the case of simply supported boundary conditions, Green's map G_S is defined by the formula $u = G_S g$, where u solves the following problem

$$\begin{cases} \Delta^2 u = 0, & x \in \Omega, \\ u\Big|_{\partial\Omega} = 0, & [\Delta u + (1-\mu)B_1 u]\Big|_{\partial\Omega} = g, \end{cases}$$

is a continuous mapping from $L_2(\partial\Omega)$ into $H^{5/2}(\Omega) \cap H_0^1(\Omega)$ and, in particular,

$$G_S \; : \; L_2(\partial\Omega) \mapsto \mathscr{D}\left(\left(\Delta_S^2\right)^{5/8-\varepsilon}\right) \equiv H^{5/2-4\varepsilon}(\Omega) \cap H_0^1(\Omega).$$

is continuous for every $\varepsilon > 0$ small enough. For details we refer to [216, Chapter 3, p. 239] and to the references therein.

1.3.12. Remark. Clearly, one could define other types of Green's maps that support the nonhomogeneity in the lower-order boundary conditions. Because these are not emphasized much in the present book, we do not go into detail but we invite the reader to consult [216]. In the case when both boundary conditions are nonhomogeneous, it suffices to consider superposition of the corresponding Green's maps. We use this idea in the next section. ∎

1.3.4 Free-type boundary conditions

Free boundary conditions are defined via moments and shear forces prescribed on the boundary [173]:

$$[\Delta u+(1-\mu)B_1u]\big|_{\partial\Omega}=\left[\frac{\partial}{\partial n}\Delta u+(1-\mu)B_2u-\nu u\right]\Bigg|_{\partial\Omega}=0, \qquad (1.3.19)$$

where, as above, the boundary operators B_1 and B_2 defined by the relations

$$\begin{aligned} B_1u &= 2n_1n_2u_{x_1x_2}-n_1^2u_{x_2x_2}-n_2^2u_{x_1x_1}, \\ B_2u &= \tfrac{\partial}{\partial\tau}\left[\left(n_1^2-n_2^2\right)u_{x_1x_2}+n_1n_2\left(u_{x_2x_2}-u_{x_1x_1}\right)\right]. \end{aligned} \qquad (1.3.20)$$

Here $n=(n_1;n_2)$ is the outer normal to $\partial\Omega$, $\tau=(-n_2;n_1)$ is the unit tangent vector along $\partial\Omega$, the constant $0<\mu<1$ has a meaning of the Poisson modulus, and $\nu\ge 0$ is a parameter.

We define the operator $\Delta_F^2 : L_2(\Omega)\mapsto L_2(\Omega)$ by the formula $\Delta_F^2u\equiv\Delta^2u$ for $u\in\mathscr{D}(\Delta_F^2)$, where the domain $\mathscr{D}(\Delta_F^2)$ consists of functions u from $H^4(\Omega)$ satisfying (1.3.19). One can see that Δ_F^2 is a regular elliptic operator (in the sense of [275, Section 5.2]). Therefore for any $0<\mu<1$ and $\nu\ge 0$ the operator Δ_F^2 is a self-adjoint positive operator (it is strictly positive when $\nu>0$) with a discrete spectrum. The corresponding bilinear form $a_F(u,v)$ is defined on $\mathscr{D}\left(\left(\Delta_F^2\right)^{1/2}\right)\equiv H^2(\Omega)$ and the relation

$$a_F(u,v)=\left(\left(\Delta_F^2\right)^{1/2}u,\left(\Delta_F^2\right)^{1/2}v\right)_{L_2(\Omega)}=a_0(u,v)+\nu\int_{\partial\Omega}uv\,d\Gamma,$$

holds for any $u,v\in H^2(\Omega)$ with $a_0(u,v)$ given by (1.3.4).

Green's operators G_1 and G_2 associated with free boundary conditions are defined as follows: $u=G_1g$ and $w=G_2g$, where u and w solve the problems

$$\begin{cases} \Delta^2u=0, \quad x\in\Omega, \\ [\Delta u+(1-\mu)B_1u]\Big|_{\partial\Omega}=g, \quad \left[\frac{\partial}{\partial n}\Delta u+(1-\mu)B_2u-\nu u\right]\Bigg|_{\partial\Omega}=0, \end{cases} \qquad (1.3.21)$$

and

$$\begin{cases} \Delta^2 w = 0, \quad x \in \Omega, \\ \left[\Delta w + (1-\mu)B_1 w\right]\Big|_{\partial\Omega} = 0, \quad \left[\frac{\partial}{\partial n}\Delta w + (1-\mu)B_2 w - \nu w\right]\Big|_{\partial\Omega} = g. \end{cases} \tag{1.3.22}$$

Standard elliptic theory (see [216, Chapter 3] and the references therein) implies the following regularity result.

1.3.13. Proposition.

$$G_1 \,:\, L_2(\partial\Omega) \mapsto H^{5/2}(\Omega) \subset \mathscr{D}\left(\left(\Delta_F^2\right)^{5/8-\varepsilon}\right) \equiv H^{5/2-4\varepsilon}(\Omega)$$

and

$$G_2 \,:\, L_2(\partial\Omega) \mapsto \left\{u \in H^{7/2}(\Omega) \,:\, [\Delta u + (1-\mu)B_1 u]\Big|_{\partial\Omega} = 0\right\}$$

continuously. In particular $G_2 \,:\, L_2(\partial\Omega) \mapsto \mathscr{D}\left(\left(\Delta_F^2\right)^{7/8-\varepsilon}\right)$, *where*

$$\mathscr{D}(\left(\Delta_F^2\right)^{7/8-\varepsilon}) = \left\{u \in H^{7/2-4\varepsilon}(\Omega) \,:\, [\Delta u + (1-\mu)B_1 u]\Big|_{\partial\Omega} = 0\right\}.$$

Here $\varepsilon > 0$ *is small enough. Moreover, we also have*

$$\begin{aligned} G_1 \,&:\, H^s(\partial\Omega) \mapsto H^{s+5/2}(\Omega), \quad s \geq 0, \\ G_2 \,&:\, H^s(\partial\Omega) \mapsto H^{s+7/2}(\Omega), \quad s \geq 0. \end{aligned} \tag{1.3.23}$$

We also note that

$$\int_\Omega \Delta^2 u G_1 g dx = \int_\Gamma \frac{\partial}{\partial n} u g d\Gamma \quad \text{for any } u \in \mathscr{D}(\Delta_F^2),\ g \in L_2(\Gamma), \tag{1.3.24}$$

and

$$\int_\Omega \Delta^2 u G_2 g dx = -\int_\Gamma u g d\Gamma \quad \text{for any } u \in \mathscr{D}(\Delta_F^2),\ g \in L_2(\Gamma), \tag{1.3.25}$$

Indeed, Green's formula (1.3.3) can be written as

$$\begin{aligned} &\int_\Omega \Delta^2 u w dx - \int_\Omega u \Delta^2 w dx \\ &= \int_\Gamma \left[\left(\frac{\partial \Delta u}{\partial n} + (1-\mu)B_2 u\right) w - (\Delta u + (1-\mu)B_1 u)\frac{\partial w}{\partial n}\right] d\Gamma \\ &\quad - \int_\Gamma \left[u\left(\frac{\partial \Delta w}{\partial n} + (1-\mu)B_2 w\right) - \frac{\partial u}{\partial n}(\Delta w + (1-\mu)B_1 w)\right] d\Gamma. \end{aligned} \tag{1.3.26}$$

Therefore, substituting $u \in \mathscr{D}(\Delta_F^2)$ and $w = G_i g$ in (1.3.26) and taking into account the boundary conditions for u and w yield (1.3.24) and (1.3.25).

1.3.5 Mixed boundary conditions

In this section we consider boundary value problems that are governed by a mixture of the boundary conditions (1.3.9), (1.3.18), and (1.3.19). That is to say, we consider part (Γ_1) of the boundary to be clamped, part (Γ_2) simply supported, and the remainder (Γ_3) free.

Assume that the boundary $\partial\Omega$ is divided into three nonoverlapping parts Γ_i, each open relative to $\partial\Omega$ and such that $\partial\Omega = \cup_i \overline{\Gamma}_i$. The case is allowed where only some of these three (one or two) boundary conditions are imposed.

Let $0 < \mu < 1$ and B_1 and B_2 be given by (1.3.20). We consider the following boundary conditions

$$\left.\begin{aligned} u|_{\Gamma_1} = \nabla u|_{\Gamma_1} &= 0, && \text{(clamped)} \\ u|_{\Gamma_2} = [\Delta u + (1-\mu)B_1 u]\big|_{\Gamma_2} &= 0, && \text{(simply supported)} \\ [\Delta u + (1-\mu)B_1 u]\big|_{\Gamma_3} = \left[\tfrac{\partial}{\partial n}\Delta u + (1-\mu)B_2 u - \nu u\right]\bigg|_{\Gamma_3} &= 0, && \text{(free)} \end{aligned}\right\} \tag{1.3.27}$$

and define the bilinear form

$$a_M(u,v) = a_0(u,v) + \nu \int_{\Gamma_3} uv\, d\Gamma\,, \tag{1.3.28}$$

where $a_0(u,v)$ is given by (1.3.4), on the space

$$V = \left\{ u \in H^2(\Omega) \;:\; u|_{\Gamma_1 \cup \Gamma_2} = 0,\ \nabla u|_{\Gamma_1} = 0 \right\}.$$

The bilinear form $a_0(u,v)$ is bounded on V. Moreover, one can easily see from (1.3.4) that

$$a_M(u,u) \geq (1-\mu) \sum_{|\alpha|=2} \int_\Omega |\partial^\alpha u|^2 dx + \nu \int_{\Gamma_3} |u|^2 d\Gamma, \quad u \in V. \tag{1.3.29}$$

Therefore a_M is strictly positive, if either $\Gamma_1 \cup \Gamma_2 \neq \emptyset$ and $\nu \geq 0$ or $\Gamma_1, \Gamma_2 = \emptyset$ and $\nu > 0$.

The following proposition is used to prove the existence of stationary solutions to von Karman equations.

1.3.14. Proposition. *Let V' be the dual to V with respect to the inner product $(\cdot,\cdot)_{L_2(\Omega)}$. Then the form* (1.3.28) *generates an m-monotone continuous operator*

$\mathscr{A} : V \mapsto V'$ such that $(u, \mathscr{A} v)_{L_2(\Omega)} = a_M(u,v)$ for all $u,v \in V$. This operator satisfies the condition $(S)_+$.

In the case of clamped–hinged boundary conditions with $\Gamma_3 = \emptyset$, the same result remains valid after setting $\mu = 1$ in (1.3.27) (see Remark 1.3.3).

Proof. As before, the key property is the coercivity estimate in (1.3.29). Because $\mathscr{A}$ is a linear (positive) operator, we easily obtain that $\mathscr{A}$ is an m-monotone continuous operator from V into V'. To prove the condition $(S)_+$, we note that for the case considered, relations (1.2.8) are equivalent to

$$u_n \rightharpoonup u \quad \text{weakly in } V \quad \text{and} \quad \limsup_{n\to\infty} a_M(u_n - u, u_n - u) \le 0.$$

Therefore (1.3.29) implies that

$$\lim_{n\to\infty} \sum_{|\alpha|=2} \int_\Omega |\partial^\alpha (u_n - u)|^2 dx = 0.$$

Because $u_n \rightharpoonup u$ weakly in V, we have that $u_n \to u$ strongly in $H^1(\Omega)$. Thus u_n converges to u strongly in V. □

Using the operator $\mathscr{A} : V \mapsto V'$ we can define a self-adjoint nonnegative the operator $\Delta_M^2 : \mathscr{D}(\Delta_M^2) \subset L_2(\Omega) \mapsto L_2(\Omega)$ by the formula

$$\Delta_M^2 u = \mathscr{A} u, \qquad u \in \mathscr{D}(\Delta_M^2) := \{u \in V \,:\, \mathscr{A} u \in L_2(\Omega)\}\,.$$

Because V is compactly embedded in $L_2(\Omega)$, it can be proved that for any $0 < \mu < 1$ and $\nu \ge 0$ the operator Δ_M^2 is a self-adjoint nonnegative operator with a discrete spectrum such that $\mathscr{D}\left(\left(\Delta_M^2\right)^{1/2}\right) = V$ and

$$a_M(u,v) = \left(\left(\Delta_M^2\right)^{1/2} u, \left(\Delta_M^2\right)^{1/2} v\right)_{L_2(\Omega)}, \qquad u,v \in V.$$

Therefore by (1.3.29) Δ_M^2 is strictly positive, if either $\Gamma_1 \cup \Gamma_2 \neq \emptyset$ and $\nu \ge 0$ or $\Gamma_1, \Gamma_2 = \emptyset$ and $\nu > 0$.

1.3.15. Remark. We note that there exists a natural embedding $V \subset L_2(\Omega) \subset V'$ and we can consider operator $\mathscr{A}_\lambda = \mathscr{A} + \lambda I$ for any $\lambda \in \mathbb{R}_+$. It is clear that $\mathscr{A}_\lambda$ is a proper Fredholm operator of index zero for any $\lambda > 0$. This fact remains true for $\lambda = 0$ when Δ_M^2 is strictly positive, which is the case for clamped, hinged, and simply supported boundary conditions. In the case of free boundary conditions one needs to assume that either $\nu > 0$ or $\Gamma_1 \cup \Gamma_2 \neq \emptyset$ and $\nu \ge 0$. It is also clear that $\mathscr{A}_\lambda$ is an m-monotone continuous operator satisfying $(S)_+$ for every $\lambda > 0$ (cf. Proposition 1.3.14). ■

1.3.16. Remark. If we assume that portions $\overline{\Gamma}_i$ of the boundary corresponding to a different type of boundary condition are disjoint, then using standard elliptic estimates it is easy to show that the domain $\mathscr{D}(\Delta_M^2)$ consists of functions u from $H^4(\Omega)$

satisfying (1.3.27). In general case we only have the relation

$$\mathscr{D}(\Delta_M^2) \supset \{u \in H^4(\Omega) \; : \; u \;\; \text{satisfies (1.3.27)}\}.$$

In this case to give an exact description instead of standard elliptic estimates, one must invoke more technical arguments due to [130], which explore the interplay between the geometry of the domain and regularity of higher derivatives (above variational level). However, using abstract elliptic theory presented in [4] it is possible to prove that the boundary operators in (1.3.27) are well defined on $\mathscr{D}(\Delta_M^2)$ and

$$\mathscr{D}(\Delta_M^2) = \{u \in H^2(\Omega) \; : \; \Delta^2 u \in L_2(\Omega) \;\; \text{and} \;\; u \;\; \text{satisfies (1.3.27)}\}.$$

In the case of mixed boundary conditions we can also consider global Green's maps consisting of the superposition of the corresponding local Green's maps. If the boundaries $\overline{\Gamma}_i$ do not overlap, then the regularity of the overall Green's map is according to the local chart. However, in the case of overlap, a singularity may appear at the intersection points [130]. These singularities need to be considered on a one-by-one basis. ∎

1.4 Properties of the von Karman bracket

The von Karman bracket describes the main nonlinearity in the scalar von Karman equations. These nonlinearities—of nonlocal character—arise in models that do not account for in-plane accelerations. If the latter are accounted for, the resulting system is referred to as a *full von Karman system*, which consists of nonlinearly coupled plate and dynamic wave equations. In that case, the nonlinearity is of local character.

The goal of this section is to provide some results pertaining to the structure and regularity of the von Karman bracket, which is defined by the equality

$$[u,v] = \partial_{x_1}^2 u \cdot \partial_{x_2}^2 v + \partial_{x_2}^2 u \cdot \partial_{x_1}^2 v - 2 \cdot \partial_{x_1 x_2}^2 u \cdot \partial_{x_1 x_2}^2 v. \tag{1.4.1}$$

We provide, in particular, a sharp estimate for this bracket expressed in terms of Lizorkin spaces. This estimate is critical in order to establish sharp regularity of the Airy stress function which, in turn, is key in proving uniqueness and continuous dependence of solutions to von Karman evolutions with respect to variations of finite energy initial data.

The following notations are used.

$$\| u \|_s \equiv \| u \|_{H^s(\Omega)}, \quad \| u \| \equiv \| u \|_{L_2(\Omega)}, \;\; \text{and} \;\; (u,v) \equiv (u,v)_{L_2(\Omega)}.$$

We begin by collecting several estimates for products of functions from Sobolev spaces. There are many known "product" estimates that determine Sobolev-type regularity of the product of two (or more) functions. We refer the interested reader to [229, 250, 266] and others. However, in the context of von Karman brackets more specific product estimates are needed (which not always can be derived from more general rules). The proposition stated below lists several such estimates that are frequently used later on.

1.4.1. Lemma. *Let Ω be a smooth bounded domain in $\mathbb{R}^2$.*

1. *If $f \in H^s(\Omega)$ for some $0 < s < 1$, then*

$$\| f \cdot g \| \leq C \| f \|_s \cdot \| g \|_{1-s} \tag{1.4.2}$$

provided that $g \in H^{1-s}(\Omega)$, and

$$\| f \cdot g \|_{-1+s} \leq C \| f \|_s \cdot \| g \| \tag{1.4.3}$$

provided that $g \in L_2(\Omega)$.

2. *If $f \in H^{s+\sigma}(\Omega)$ and $g \in H^{1-\sigma}(\Omega)$ for some $0 < s < 1$ and $0 < \sigma < 1 - s$, then $f \cdot g \in H^s(\Omega)$ and*

$$\| f \cdot g \|_s \leq C \cdot \| f \|_{s+\sigma} \cdot \| g \|_{1-\sigma} . \tag{1.4.4}$$

3. *If $f \in H^s(\Omega)$ and $g \in H^1(\Omega) \cap L_\infty(\Omega)$ for some $0 \leq s < 1$, then $f \cdot g \in H^s(\Omega)$ and*

$$\| f \cdot g \|_s \leq C \cdot \| f \|_s \cdot \left(\| g \|_{L_\infty(\Omega)} + \| g \|_{H^1(\Omega)} \right) . \tag{1.4.5}$$

4. *If $f \in H^s(\Omega)$ and $g \in H^{1+\delta}(\Omega)$ for some $0 \leq s \leq 1$ and $\delta > 0$, then $f \cdot g \in H^s(\Omega)$ and*

$$\| f \cdot g \|_s \leq C \cdot \| f \|_s \cdot \| g \|_{1+\delta} . \tag{1.4.6}$$

It should be noted that strict inequalities for the parameters involved are critical. For instance, (1.4.2) does not hold with $s = 1$, as $H^1(\Omega)$ is not a multiplier on $L_2(\Omega)$ ($\dim \Omega = 2$).

Proof. It follows from the Hölder inequality that

$$\| f \cdot g \| \leq C \| f \|_{L^{2/(1-s)}} \cdot \| g \|_{L^{2/s}} , \quad 0 < s < 1.$$

Therefore the embedding (1.1.6) implies (1.4.2).

Relation (1.1.6) implies that the embedding $L^{2/(2-s)}(\Omega) \subset H^{-1+s}(\Omega)$ is continuous. Therefore using Hölder's inequality we have

$$\| f \cdot g \|_{-1+s} \leq C \| f \cdot g \|_{L^{2/(2-s)}} \leq C \| f \|_{L^{2p/(2-s)}} \cdot \| g \|_{L^{2q/(2-s)}} ,$$

where $p^{-1} + q^{-1} = 1$. Setting $q = 2 - s$ and $p = (2 - s)(1 - s)^{-1}$, we obtain (1.4.3) from the embedding result (1.1.6).

Using the extension procedure (see, e.g., [222]) one can easily show that estimates (1.4.4)–(1.4.6) follow from the same inequalities established for $\Omega = \mathbb{R}^2$ and for f and g lying in $C_0^\infty(\mathbb{R}^2)$.

Let us prove (1.4.4) for $\Omega = \mathbb{R}^2$. By the definition of $H^s(\mathbb{R}^2)$,

$$\| fg \|_{H^s(\mathbb{R}^2)}^2 = \| fg \|_{L^2(\mathbb{R}^2)}^2 + \int_{\mathbb{R}^2} \frac{dy}{|y|^{2+2s}} \int_{\mathbb{R}^2} |(fg)(x+y) - (fg)(x)|^2 \, dx,$$

the Hölder inequality yields

$$\| f \cdot g \|_s \leq \| f \cdot g \|_{L^2} + \sqrt{2} \left(\| f \|_{L^{2p_1}} \cdot \| g \|_{s,2q_1} + \| g \|_{L^{2p_2}} \cdot \| f \|_{s,2q_2} \right), \quad (1.4.7)$$

where $p_j^{-1} + q_j^{-1} = 1$ ($p_j = \infty$ is allowed) and

$$\| h \|_{s,p} = \| h \|_{L^p} + \left(\int_{\mathbb{R}^2} \frac{dy}{|y|^{2+2s}} \left(\int_{\mathbb{R}^2} |h(x+y) - h(x)|^p \, dx \right)^{2/p} \right)^{1/2}$$

is the norm in the Besov space $B^s_{p,2}(\mathbb{R}^2)$ (cf. (1.1.7)). Using the continuity of the embedding (1.1.10) we have

$$\| g \|_{s,2q_1} \leq C \| g \|_{1+s-1/q_1}, \quad \| f \|_{s,2q_2} \leq C \| f \|_{1+s-1/q_2}, \quad q_j \geq 1. \quad (1.4.8)$$

Let $p_2 = \sigma^{-1}$ and $q_1 = (s+\sigma)^{-1}$. Then inequality (1.4.4) follows from (1.4.2) and (1.1.6).

To prove (1.4.5) we set $q_1 = s^{-1}$ and $p_2 = \infty$ in (1.4.7) and (1.4.8) and use relations (1.4.2) and (1.1.6).

Inequality (1.4.6) for $0 \leq s < 1$ follows from (1.4.5) and (1.1.5). In the case $s = 1$ we additionally invoke (1.4.2). □

Next we analyze properties of the von Karman bracket. We start with the following assertion which reveals some symmetry properties of the bracket.

1.4.2. Proposition. *The mapping $\{u; v\} \mapsto [u,v]$ is a symmetric bilinear mapping from $H^2(\Omega) \times H^2(\Omega)$ into $L_1(\Omega)$. The trilinear form $([u,v],w)$ is symmetric on $H^2(\Omega)$ if either at least one of the elements u, v, or w belongs to $H_0^2(\Omega)$ or all of them belong to $H^2(\Omega) \cap H_0^1(\Omega)$. Moreover, we have the relation*

$$\int_\Omega [u,v] w dx = \int_\Omega [u,w] v dx \quad \text{for any } u \in H^2(\Omega) \quad (1.4.9)$$

provided the functions $w, v \in H^2(\Omega)$ possess the properties

$$w|_{\Gamma_1 \cup \Gamma_2} = 0, \quad \nabla w|_{\Gamma_1} = 0, \quad v|_{\Gamma_2 \cup \Gamma_3} = 0, \quad \nabla v|_{\Gamma_3} = 0, \quad (1.4.10)$$

where $\{\Gamma_1, \Gamma_2, \Gamma_3\}$ is a division of the boundary $\partial\Omega$ into three nonoverlapping parts Γ_i, each open relative to $\partial\Omega$ and such that $\partial\Omega = \cup_i \overline{\Gamma}_i$. The case when some of the parts (one or two) are empty is allowed. In particular (1.4.9) *is true under the condition $w, v \in H^2(\Omega) \cap H_0^1(\Omega)$ which corresponds to the case $\Gamma_1 = \Gamma_3 = \emptyset$.*

Proof. If one of the functions u, v, or w from $H^2(\Omega)$ belongs to $C_0^\infty(\Omega)$, then using the representation

$$[u,v] = \partial_{x_1}^2 (u \cdot \partial_{x_2}^2 v) + \partial_{x_2}^2 (u \cdot \partial_{x_1}^2 v) - 2 \cdot \partial_{x_1 x_2}^2 (u \cdot \partial_{x_1 x_2}^2 v), \quad (1.4.11)$$

it is easy to see that $([u,v],w)$ is symmetric. Therefore a simple density argument gives symmetry of this trilinear form on $H^2(\Omega)$ provided one of the elements u, v, or w belongs to $H_0^2(\Omega)$.

From the representation

$$[u,v] = \partial_{x_1}(\partial_{x_1}v\cdot\partial_{x_2}^2u - \partial_{x_2}v\cdot\partial_{x_1x_2}^2u) + \partial_{x_2}(\partial_{x_2}v\cdot\partial_{x_1}^2u - \partial_{x_1}v\cdot\partial_{x_1x_2}^2u) \quad (1.4.12)$$

we obtain

$$\int_\Omega [u,v]w\,dx = -b(u|v,w) + \int_{\partial\Omega} w\cdot\left(v_{x_1}\frac{\partial}{\partial\tau}u_{x_2} - v_{x_2}\frac{\partial}{\partial\tau}u_{x_1}\right)d\Gamma \quad (1.4.13)$$

for smooth functions u, v and w, where we use the notations

$$b(u|v,w) = \int_\Omega \left[u_{x_2x_2}v_{x_1}w_{x_1} + u_{x_1x_1}v_{x_2}w_{x_2} - u_{x_1x_2}(v_{x_1}w_{x_2} + v_{x_2}w_{x_1})\right]dx$$

and $(\partial/\partial\tau)f = -n_2 f_{x_1} + n_1 f_{x_2}$. Here, as above, $n = (n_1;n_2)$ is the outer normal to $\Gamma = \partial\Omega$ and $\tau = (-n_2;n_1)$ is the unit tangent vector along $\partial\Omega$. We assume that Γ is oriented along τ. The symmetry of the form $b(u|v,w)$ with respect to v and w and relation (1.4.13) imply that

$$\int_\Omega [u,v]w\,dx - \int_\Omega [u,w]v\,dx$$
$$= \int_{\partial\Omega}\left\{(wv_{x_1} - vw_{x_1})\frac{\partial u_{x_2}}{\partial\tau} - (wv_{x_2} - vw_{x_2})\frac{\partial u_{x_1}}{\partial\tau}\right\}d\Gamma \quad (1.4.14)$$

for any $u \in H^3(\Omega)$ and $v, w \in H^2(\Omega)$. This implies relation (1.4.9) under condition (1.4.10). The symmetry of $([u,v],w)$ on $H^2(\Omega)\cap H_0^1(\Omega)$ follows from (1.4.9). □

Our main result in this section is the following theorem.

1.4.3. Theorem. *The von Karman bracket*

$$[u,v] = \partial_{x_1}^2 u\cdot\partial_{x_2}^2 v + \partial_{x_2}^2 u\cdot\partial_{x_1}^2 v - 2\cdot\partial_{x_1x_2}^2 u\cdot\partial_{x_1x_2}^2 v$$

gives a continuous bilinear mapping from $H^2(\Omega)$ into the Lizorkin space $F_{1,2}^0(\Omega)$ and

$$\|\,[u,v]\,\|_{F_{1,2}^0(\Omega)} \le C\,\|\,u\,\|_2\|\,v\,\|_2\ . \quad (1.4.15)$$

Furthermore, we have the following estimates

$$\|\,[u,v]\,\|_{-j-\theta} \le C\,\|\,u\,\|_{2-\theta+\beta}\cdot\|\,v\,\|_{3-j-\beta}, \quad (1.4.16)$$

where $j = 0, 1$ and $0 < \beta \le \theta < 1$, and

$$\|\,[u,v]\,\|_{-j} \le C\,\|\,u\,\|_{2-\beta}\cdot\|\,v\,\|_{3-j+\beta}, \quad (1.4.17)$$

where $j = 1, 2$ and $0 \le \beta < 1$.

Proof. To prove the property $[u,v] \in F^0_{1,2}(\Omega)$ and estimate (1.4.15) we use the compensated compactness method (see [89]) along with the theory of real Hardy spaces.

STEP 1: We first extend u and v to $\mathbb{R}^2$ preserving H^2 regularity. Let u_e and v_e be extensions of u and v that are compactly supported in $\mathbb{R}^2$ and such that

$$\| u_e \|_{H^2(\mathbb{R}^2)} \leq C \| u \|_2 \quad \text{and} \quad \| v_e \|_{H^2(\mathbb{R}^2)} \leq C \| v \|_2 .$$

STEP 2: Let $H_1(\mathbb{R}^2)$ denote the Hardy space (for the definition see Section 1.1). The following regularity holds,

$$[u_e, v_e] \in H_1(\mathbb{R}^2) \ \text{ if } \ u, v \in H^2(\Omega) \tag{1.4.18}$$

To see this we write the von Karman bracket in the following form,

$$[u,v] = \det(\nabla U) + \det(\nabla V),$$

where $U \equiv (\partial_{x_1} u_e; \partial_{x_2} v_e)$ and $V \equiv (\partial_{x_1} v_e; \partial_{x_2} u_e)$. By Lemma 1.1.7 we obtain

$$\det(\nabla U) \in H_1(\mathbb{R}^2), \quad \det(\nabla V) \in H_1(\mathbb{R}^2)$$

which proves (1.4.18).

STEP 3: Using (1.1.19) we have that

$$[u_e, v_e] \in F^0_{1,2}(\mathbb{R}^2) \ \text{ when } \ u, v \in H^2(\Omega). \tag{1.4.19}$$

By the continuity (with control of the norms) of the restriction operator from $F^0_{1,2}(\mathbb{R}^2)$ into $F^0_{1,2}(\Omega)$, the closed graph theorem, and the estimate in Step 1 we obtain

$$\| [u,v] \|_{F^0_{1,2}(\Omega)} \leq \| [u_e, v_e] \|_{F^0_{1,2}(\mathbb{R}^2)}$$

$$\leq C \| u_e \|_{H^2(\mathbb{R}^2)} \| v_e \|_{H^2(\mathbb{R}^2)} \leq C \| u \|_2 \| v \|_2,$$

which proves (1.4.15).

To prove estimate (1.4.16) and (1.4.17) we use the representation in (1.4.12) for the von Karman bracket and also the relation

$$[u,v] = -\partial^2_{x_1} (u_{x_2} v_{x_2}) - \partial^2_{x_2} (u_{x_1} v_{x_1}) + \partial^2_{x_1 x_2} (u_{x_1} v_{x_2} + u_{x_2} v_{x_1}) \tag{1.4.20}$$

which follows from (1.4.12).

Let D and D^2 denote differential operators of the first and second order with constant coefficient. Using (1.4.4) with $s = 1 - \theta$ and $\sigma = \beta$ we have

$$\|D^2 (Du \cdot Dv) \|_{-1-\theta} \leq C \|Du \cdot Dv\|_{1-\theta} \leq C \|Du\|_{1-\theta+\beta} \|Dv\|_{1-\beta}.$$

Similarly,

$$\|D\left(Du \cdot D^2 v\right)\|_{-\theta} \leq C\|Du\|_{1-\theta+\beta}\|D^2 v\|_{1-\beta}.$$

Thus estimate (1.4.16) applied for $0 < \beta < \theta < 1$ follows from (1.4.20) and (1.4.12).

In the case $\beta = \theta$, using the embedding $L^{2/(1+\theta)}(\Omega) \subset H^{-\theta}(\Omega)$, which follows from (1.1.6), we obtain

$$\|D^2 u \cdot D^2 v\|_{-\theta} \leq C\|D^2 u \cdot D^2 v\|_{L^{2/(1+\theta)}}.$$

Applying the Hölder inequality with $p = 1+\theta$ and $q = (1+\theta)/\theta$ and using (1.1.6) we find that

$$\|D^2 u \cdot D^2 v\|_{-\theta} \leq C\|D^2 u\|\|D^2 v\|_{L^{2/\theta}} \leq C\|D^2 u\|\|D^2 v\|_{1-\theta}.$$

Similarly,

$$\|D\left(D^2 u \cdot Dv\right)\|_{-1-\theta} \leq C\|D^2 u\|\|Dv\|_{L^{2/\theta}} \leq C\|D^2 u\|\|Dv\|_{1-\theta}.$$

These inequalities along with (1.4.12) imply (1.4.16) for $\beta = \theta$.

Using (1.4.2) we have

$$\|D\left(Du \cdot D^2 v\right)\|_{-1} \leq C\|Du \cdot D^2 v\| \leq C\|Du\|_{1-\beta}\|D^2 v\|_{\beta}$$

and

$$\|D^2\left(Du \cdot Dv\right)\|_{-2} \leq C\|Du \cdot Dv\| \leq C\|Du\|_{1-\beta}\|Dv\|_{\beta},$$

where $0 < \beta < 1$. Combining with (1.4.12) and (1.4.20) yields (1.4.17) with $0 < \beta < 1$.

Consider the case $\beta = 0$ in (1.4.17). For $u \in H^2(\Omega)$ we denote by $\mathscr{A}_u$ the operator from $H^2(\Omega)$ into $H^{-2}(\Omega)$ defined by the formula $\mathscr{A}_u v = [u,v]$. Note that A_u is a linear operator for a fixed u. The estimate (1.4.16) with $j = 0, \beta = \theta = 1-\alpha$ and with $j = 1, \beta = \theta = \alpha$ gives

$$\mathscr{A}_u : H^{2+\alpha}(\Omega) \mapsto H^{-1+\alpha}(\Omega) \text{ and } \mathscr{A}_u : H^{2-\alpha}(\Omega) \mapsto H^{-1-\alpha}(\Omega),$$

respectively. Therefore from interpolation theory (see, e.g., [26]) we have that $\mathscr{A}_u : H^2(\Omega) \mapsto H^{-1}(\Omega)$ and, consequently, we have (1.4.17) with $j = 1$ and $\beta = 0$. Because

$$|([u,v],\phi)| = |([u,\phi],v)| \leq \| [u,\phi] \|_{-1} \cdot \| v \|_1$$

for any $\phi \in H_0^2(\Omega)$, we obtain (1.4.17) for $j = 2$, $\beta = 0$ from (1.4.17) with $j = 1$ and $\beta = 0$. □

Corollary 1.3.9 (regularity of elliptic problems in Lizorkin spaces) and relation (1.4.15) imply the following regularity of $[u,v]$.

1.4.4. Corollary. *Let $u,v \in H^2(\Omega)$ and Δ_D^2 be a biharmonic operator with the zero clamped boundary conditions defined by* (1.3.10). *Then*

$$(\Delta_D^2)^{-1}[u,v] \in W_p^{2+2/p}(\Omega) \cap H_0^2(\Omega) \quad \textit{for any } 1 \leq p \leq \infty \tag{1.4.21}$$

and

$$\| (\Delta_D^2)^{-1}[u,v] \|_{W_p^{2+2/p}(\Omega)} \le C \| u \|_2 \| v \|_2, \quad 1 \le p \le \infty. \tag{1.4.22}$$

In particular

$$\| (\Delta_D^2)^{-1}[u,v] \|_{W_\infty^2(\Omega)} \le C \| u \|_2 \| v \|_2 . \tag{1.4.23}$$

We note that estimate (1.4.23) is critically invoked in the study of uniqueness and long-time behavior of von Karman evolutions.

Now we are in a position to state and prove the main results pertaining to "sharp" regularity of the Airy function.

1.4.5. Corollary. *Let $f \in H^2(\Omega)$. Then for any $u \in H^2(\Omega)$ the problem*

$$\Delta^2 v + [u+2f, u] = 0, \quad v|_{\partial\Omega} = \frac{\partial v}{\partial n}\Big|_{\partial\Omega} = 0,$$

has a unique solution $v = v(u)$ in $H_0^2(\Omega) \cap W_p^{2+2/p}(\Omega)$, for any $1 \le p \le \infty$. This solution (often referred as the Airy stress function with respect to the displacement u) has the following properties,

$$\| v(u_1) - v(u_2) \|_{W_p^{2+2/p}(\Omega)} \le C_p \| u_1 + u_2 + 2f \|_2 \cdot \| u_1 - u_2 \|_2 \tag{1.4.24}$$

for every $1 \le p \le \infty$ and

$$\| v(u_1) - v(u_2) \|_{3-\delta} \le C_\delta \| u_1 + u_2 + 2f \|_2 \cdot \| u_1 - u_2 \|_{2-\delta} \tag{1.4.25}$$

for every $0 \le \delta \le 1$. We also have the estimate

$$\begin{aligned} &\| [u_1, v(u_1)] - [u_2, v(u_2)] \|_{-\delta} \\ &\le C_\delta (\| u_1 \|_2^2 + \| u_2 \|_2^2 + \| f \|_2^2) \cdot \| u_1 - u_2 \|_{2-\delta} \end{aligned} \tag{1.4.26}$$

for every $0 \le \delta \le 1$.

Proof. Due to (1.4.21) we only need to prove estimates (1.4.24)–(1.4.26). It is clear that $\bar{v} = v(u_1) - v(u_2)$ solves the problem

$$\Delta^2 \bar{v} + [u_1 + u_2 + 2f, u_1 - u_2] = 0, \quad \bar{v}|_{\partial\Omega} = \frac{\partial \bar{v}}{\partial n}\Big|_{\partial\Omega} = 0. \tag{1.4.27}$$

Therefore (1.4.24) follows from (1.4.22).

Relations (1.3.11) and (1.4.27) imply that

$$\|\bar{v}\|_{3-\delta} \le C \| [u_1 + u_2 + 2f, u_1 - u_2] \|_{-1-\delta}, \quad 0 \le \delta \le 1.$$

Therefore using (1.4.16) with $\beta = \theta = \delta$ and $j = 1$ we obtain (1.4.25) for $0 < \delta < 1$. To obtain (1.4.25) with $\delta = 0$ and $\delta = 1$ we apply (1.4.17) with $\beta = 0$.

To prove (1.4.26) with $\delta = 0$ we note that

$$\|[u_1, v(u_1)] - [u_2, v(u_2)]\|$$
$$\leq C\left(\|u_1 - u_2\|_2 \cdot \|v(u_1)\|_{W^2_\infty} + \|u_2\|_2 \cdot \|v(u_1) - v(u_2)\|_{W^2_\infty}\right).$$

Thus (1.4.26) for $\delta = 0$ follows from (1.4.23) and (1.4.24).

Let $0 < \delta \leq 1$. We obviously have that

$$\|[u_1, v(u_1)] - [u_2, v(u_2)]\|_{-\delta} \leq \|[u_1 - u_2, v(u_1)]\|_{-\delta} + \|[u_2, v(u_1) - v(u_2)]\|_{-\delta}.$$

Using (1.4.12) we obtain that

$$\|[u_1 - u_2, v(u_1)]\|_{-\delta} \leq C \sum_{i,j} \|D_i(u_1 - u_2) \cdot D_j^2 v(u_1)]\|_{1-\delta},$$

where D_i and D_j^2 are differential operators of the first and second order with constant coefficient. Therefore relation (1.4.5) implies that

$$\|[u_1 - u_2, v(u_1)]\|_{-\delta} \leq C\|u_1 - u_2\|_{2-\delta} \cdot \left(\|v(u_1)\|_3 + \|v(u_1)\|_{W^2_\infty}\right), \quad 0 < \delta \leq 1.$$

Applying inequalities (1.4.24) with $p = \infty$ and $u_2 \equiv 0$ and (1.4.25) with $\delta = 0$ and $u_2 \equiv 0$, we obtain that

$$\|[u_1 - u_2, v(u_1)]\|_{-\delta} \leq C\|u_1 - u_2\|_{2-\delta} \cdot \left(\|u_1\|_2^2 + \|f\|_2^2\right), \quad 0 < \delta \leq 1. \quad (1.4.28)$$

Using (1.4.16) with $j = 0$ and $\theta = \beta = \delta > 0$ (or (1.4.17) with $j = 1$ and $\beta = 0$ in the case $\delta = 1$), we find that

$$\|[u_2, v(u_1) - v(u_2)]\|_{-\delta} \leq C\|u_2\|_2 \cdot \|v(u_1) - v(u_2)\|_{3-\delta}, \quad 0 < \delta \leq 1.$$

Therefore (1.4.26) for $0 < \delta \leq 1$ follows from (1.4.25) and (1.4.28) □

1.4.6. Remark. The results concerning the von Karman bracket presented in this section remain true for the case when $\Omega = (0,a) \times (0,b)$ is rectangle. Here we only use the possibility to extend functions from Ω to $\mathbb{R}^2$ while preserving their smoothness and the regularity result from Corollary 1.3.9 (see Remark 1.3.10). More details can be found in [66]. ■

1.5 Stationary von Karman equations

In this section we consider the properties of solutions to the following stationary problem

$$\Delta^2 u - [u + f, v(u) + F_0] + L(u) = p(x), \quad x \in \Omega, \quad (1.5.1)$$

where u satisfies the mixed boundary conditions:

$$\left.\begin{aligned} u|_{\Gamma_1} = \nabla u|_{\Gamma_1} = 0, \qquad\qquad\qquad\qquad & \text{(clamped)} \\ u|_{\Gamma_2} = [\Delta u + (1-\mu)B_1 u]\big|_{\Gamma_2} = 0, \quad & \text{(simply supported)} \\ [\Delta u + (1-\mu)B_1 u]\big|_{\Gamma_3} = \left[\tfrac{\partial}{\partial n}\Delta u + (1-\mu)B_2 u - \nu u\right]\Big|_{\Gamma_3} = 0, & \text{ (free)} \end{aligned}\right\} \qquad (1.5.2)$$

and $v(u) \in H_0^2(\Omega)$ solves the problem

$$\Delta^2 v + [u + 2f, u] = 0, \quad v|_{\partial\Omega} = \frac{\partial v}{\partial n}\Big|_{\partial\Omega} = 0. \qquad (1.5.3)$$

Here Ω is a smooth bounded domain in $\mathbb{R}^2$ with the boundary $\partial\Omega$ divided into three nonoverlapping parts Γ_i, each open relative to $\partial\Omega$ and such that $\partial\Omega = \cup_i \overline{\Gamma}_i$. The case when some parts (one or two) are empty is allowed.

- $F_0 \in H^3(\Omega)$ is a given function determined by mechanical loads. For some of the results less regularity on F_0 is required.
- The function $f \in H^2(\Omega)$ determines the initial form of the shell.
- $L(\cdot)$ is a linear bounded operator from $H^{2-\sigma}(\Omega)$ into $H^{-\sigma}(\Omega)$ for all $0 \le \sigma \le \sigma_0$ with some $\sigma_0 > 0$. In typical applications L represents a "drift" and as such is a first-order differential operator with smooth coefficients.
- As above we denote by $[u, v]$ the von Karman bracket (see (1.4.1)). The boundary operators B_1 and B_2 are defined by (1.3.20), $n = (n_1; n_2)$ is the outer normal to $\partial\Omega$, the constant $0 < \mu < 1$ has a meaning of the Poisson modulus, and $\nu \ge 0$ is a given parameter. We also consider the clamped–hinged case ($\Gamma_3 = \emptyset$, $\mu = 1$).

This stationary problem, equipped with a variety of boundary conditions has been studied extensively in the literature (see, e.g., [85, 220, 237, 23, 24, 25] and the references therein). We begin with a proper definition of weak solution.

1.5.1. Definition. A function u from the space

$$V = \left\{u \in H^2(\Omega) \,:\, u|_{\Gamma_1 \cup \Gamma_2} = 0,\ \nabla u|_{\Gamma_1} = 0\right\} \qquad (1.5.4)$$

is said to be a *weak solution* to the problem (1.5.1)–(1.5.3) iff

$$a(u, w) + \int_\Omega (-[u + f, v(u) + F_0] + L(u) - p) \cdot w\,dx = 0 \qquad (1.5.5)$$

for any $w \in V$. Here

$$a(u, w) = a_0(u, w) + \nu \int_{\Gamma_3} uw\,d\Gamma, \qquad (1.5.6)$$

with $a_0(u, w)$ defined by (1.3.4) and $v(u) \in H_0^2(\Omega)$ is determined from (1.5.3). ∎

To prove the existence of a solution to (1.5.5), we use the theory of pseudomonotone operators (see Section 1.2). One may also use other methods of nonlinear analysis, such as the compactness method (see [220]), or m-monotone operator theory [18, 260].

Let V' be the dual to V with respect to the inner product $(\cdot,\cdot) \equiv (\cdot,\cdot)_{L_2(\Omega)}$. We also note that the bilinear form $a(u,w)$ determines the H^2 semi-norm on V. By Proposition 1.3.14 this form $a(u,w)$ generates an m-monotone continuous (linear) operator $\mathscr{A} : V \mapsto V'$ such that

$$(w, \mathscr{A} u)_{L^2(\Omega)} = a(w,u) \quad \text{for all } u, v \in V. \tag{1.5.7}$$

Moreover the operator $\mathscr{A}$ satisfies the condition $(S)_+$ (see Section 1.2 for the definition). This follows from the linearity of $\mathscr{A}$ and the fact that $a(u,u)$ is an equivalent H^2 seminorm defined on V. It is a trivial fact that $\mathscr{A} + I$ is Fredholm with index zero.

We also define operator $B : V \mapsto V'$ by the formula

$$(w, Bu) = \int_\Omega (-[u+f, v(u)+F_0] + L(u)) \cdot w\, dx.$$

If $f \in H^2(\Omega)$ and $F_0 \in H^3(\Omega)$, then it is clear from Theorem 1.4.3 and Corollary 1.4.5 that

$$Bu = -[u+f, v(u)+F_0] + L(u) \in H^{-\delta}(\Omega) \tag{1.5.8}$$

for every $u \in H^2(\Omega)$, where $\delta > 0$ is arbitrary. Because $V \subset H_0^\delta(\Omega)$ for $\delta \in (0, \frac{1}{2})$, we have that $H^{-\delta}(\Omega) \subset V'$ for sufficiently small $\delta > 0$. Thus $B : V \mapsto H^{-\delta}(\Omega) \subset V'$. Now equation (1.5.5) can be rewritten as the operator equation

$$\mathscr{A} u + Bu = p, \quad u \in V. \tag{1.5.9}$$

We want to apply Theorems 1.2.9 and 1.2.13. For this we need to study the properties of the operator B.

1.5.2. Lemma. *Let $f \in H^2(\Omega)$, $F_0 \in H^3(\Omega)$, and $L(\cdot)$ be a linear bounded operator from $H^{2-\sigma}(\Omega)$ into $H^{-\sigma}(\Omega)$ for all $0 \le \sigma \le \sigma_0$ with some $\sigma_0 > 0$. Then the operator $B : V \mapsto V'$ is strongly continuous and Fréchet differentiable. The derivative $B'(u)$ is a bounded linear operator that depends continuously on $u \in V$ in uniform operator topology. The operator B is also compact.*

If $\Gamma_3 = \emptyset$ (i.e., $\partial\Omega = \Gamma_1 \cup \Gamma_2$), then the same properties of B remain valid under the condition $F_0 \in H^2(\Omega)$ in both cases $0 < \mu < 1$ (clamped/simply supported plate), and $\mu = 1$ (clamped/hinged plate); see Remark 1.3.3.

Proof. Let $u_n \rightharpoonup u$ weakly in V. Then $\|u_n\|_2 \le C$ and $u_n \to u$ strongly in $H^{2-\delta}(\Omega)$ for any $\delta > 0$. It is clear from Theorem 1.4.3 that

$$\begin{aligned}\|B(u_n) - B(u)\|_{-\delta} \le \|[u_n, v(u_n)] - [u, v(u)]\|_{-\delta} \\ + C_1 \|f\|_2 \cdot \|v(u_n) - v(u)\|_{3-\delta} + C_2 \|u_n - u\|_{2-\delta'}\end{aligned}$$

for $0 < \delta' < \delta < 1$. Therefore Corollary 1.4.5 and the boundedness of $\|u_n\|_2$ imply that

$$\|B(u_n) - B(u)\|_{-\delta} \le C \|u_n - u\|_{2-\delta'} \tag{1.5.10}$$

with $0 < \delta' < \delta$ small enough. Because $H^{-\delta}(\Omega) \subset V'$ for sufficiently small $\delta > 0$, (1.5.10) implies that $B(u_n) \to B(u)$ strongly in V'. Thus the operator $B : V \mapsto V'$ is strongly continuous. This operator is also compact due to the compactness of the embedding $H^{-\delta}(\Omega) \subset V'$.

It is easy to see that B is Fréchet differentiable with the derivative $B'(u) : V \mapsto V'$ given by the formula

$$\langle B'(u); w\rangle = -[w, v(u) + F_0] - [u + f, v(u,w)] + L(w),$$

where $v(u,w) \in H_0^2(\Omega)$ solves the problem

$$\Delta^2 v(u,w) + 2[w, u+f] = 0, \quad v(u,w)|_{\partial\Omega} = \frac{\partial v(u,w)}{\partial n}\Big|_{\partial\Omega} = 0.$$

For brevity, below we also use the notation $B'(u)w \equiv \langle B'(u); w\rangle$.

From (1.4.16) with $j = 0$ and $\theta = \beta = \delta \in (0,1)$ one obtains:

$$\|B'(u_1)w - B'(u_2)w\|_{-\delta} \le C_1\|w\|_2\|v(u_1) - v(u_2)\|_{3-\delta}$$
$$+ C_2\|u_1 + f\|_2\|v(u_1,w) - v(u_2,w)\|_{3-\delta} + C_3\|u_1 - u_2\|_2\|v(u_2,w)\|_{3-\delta}.$$

The function $\bar{v} = v(u_1,w) - v(u_2,w) \in H_0^2(\Omega)$ solves the problem

$$\Delta^2\bar{v} + 2[w, u_1 - u_2] = 0, \quad \bar{v}|_{\partial\Omega} = \frac{\partial\bar{v}}{\partial n}\Big|_{\partial\Omega} = 0,$$

therefore (1.4.16) yields $\|v(u_1,w) - v(u_2,w)\|_{3-\delta} \le C \cdot \|u_1 - u_2\|_2 \cdot \|w\|_{2-\delta}$. Similarly,

$$\|v(u_2,w)\|_{3-\delta} \le C \cdot \|u_2 + f\|_2 \cdot \|w\|_{2-\delta}.$$

Thus using (1.4.25) we obtain that

$$\|B'(u_1)w - B'(u_2)w\|_{-\delta} \le C \cdot (1 + \|u_1\|_2 + \|u_2\|_2) \cdot \|u_1 - u_2\|_2 \cdot \|w\|_2.$$

In a similar way, (1.4.16) applied with $j = 0$, $\theta = \beta = \delta$ yields:

$$\|B'(u)w\|_{-\delta} \le C \cdot \left(1 + \|u\|_2^2\right) \cdot \|w\|_2.$$

Because $H^{-\delta}(\Omega) \subset V'$ for small $\delta > 0$, these relations imply that $B'(u)$ is a bounded linear operator from V to V' depending continuously on $u \in V$.

In the case $\Gamma_3 = \emptyset$ we have that $V \subset H_0^{1+\delta}(\Omega)$ for every $\delta < \frac{1}{2}$. Therefore, by using (1.4.16) applied with $j = 1, \theta = \beta = \delta$ we obtain that $H^{-1-\delta}(\Omega) \subset V'$ for $\delta < \frac{1}{2}$. This makes it possible to obtain the desired estimates for the terms containing F_0 under the condition $F_0 \in H^2(\Omega)$. □

Lemma 1.5.2 yields the following

1.5.3. Proposition. *Under the hypotheses of Lemma 1.5.2 the operator $\mathscr{A}+B$ is a pseudomonotone bounded continuous operator from V into V' satisfying the condition $(S)_+$. Furthermore, $\mathscr{A}+B$ is a Fredholm operator of index zero.*

Proof. Because by Lemma 1.5.2 B is strongly continuous, Proposition 1.2.8 part (ii) implies that B is pseudomonotone. $\mathscr{A}$ being monotone and continuous $V \to V'$ is pseudomonotone. Thus, by parts (iv) and (v) of Proposition 1.2.8 $\mathscr{A}+B$ is a pseudomonotone and it satisfies the property S_+.

By Lemma 1.5.2 $B-I : V \mapsto V'$ is also compact and continuously Frechet differentiable. By Remark 1.3.15 we have that $\mathscr{A}+I$ is a Fredholm operator of index zero. Therefore Proposition 1.2.11 implies that $\mathscr{A}+B$ is Fredholm of index zero.

□

Now we consider coercivity of the operator $\mathscr{A}+B$. We first deal with the clamped and hinged boundary conditions. The case of general mixed boundary conditions is more complex and we discuss this separately.

1.5.1 Clamped and hinged boundary conditions

We suppose now that $\Gamma_3 = \emptyset$ and $\mu = 1$ in (1.5.2), which configuration corresponds to clamped and/or hinged boundary conditions. In a study of long-time behavior an important property of the system is that the potential energy be bounded from below (i.e., it does not go to $-\infty$ on some configurations). The presence of the sources may have a destabilizing effect on potential energy driving it to $-\infty$. In line with variational principles, it is expected that nonlinear terms in the equation should balance this undesirable effect by the forcing existence of a minimal point for the energy. The lemma stated below provides a key inequality responsible for this phenomenon.

1.5.4. Lemma. *The functional*

$$\Psi(u;A,\delta) = \|u\|_2^2 + \|\Delta v_0(u)\|^2 - A \cdot \|u\|_{2-\delta}^2$$

is bounded from below on $H^2(\Omega)\cap H_0^1(\Omega)$ for any $A>0$ and $0<\delta\le 2$. Here $v_0(u)$ is the solution to (1.5.3) *with $f\equiv 0$.*

Proof. Arguing by contradiction, we assume that there exists a sequence $\{u_n\}$ in the space $H^2(\Omega)\cap H_0^1(\Omega)$ such that

$$M_n \equiv \Psi(u_n;A,\delta) \to -\infty \quad \text{as } n \to \infty.$$

We can assume that $M_n < 0$ for all n. It should be $\|u_n\|_{2-\delta} \to \infty$. Let $w_n = u_n/\|u_n\|_{2-\delta}$. Then we have $\|w_n\|_{2-\delta} = 1$ and

$$\|u_n\|_{2-\delta}^2 \cdot \left(\|w_n\|_2^2 + \frac{\|\Delta v_0(u_n)\|^2}{\|u_n\|_{2-\delta}^2} - A \right) = M_n \to -\infty.$$

Consequently

$$\|w_n\|_2^2 + \|u_n\|_{2-\delta}^2 \cdot \|\Delta v_0(w_n)\|^2 = A + \frac{M_n}{\|u_n\|_{2-\delta}^2} < A.$$

This implies that the sequence $\{w_n\}$ contains a subsequence converging weakly in $H^2(\Omega)$. Thus we can suppose that $w_n \rightharpoonup w$ weakly in $H^2(\Omega)$ for some $w \in H^2(\Omega) \cap H_0^1(\Omega)$. We also have

$$\|\Delta v_0(w_n)\|^2 \le \frac{A}{\|u_n\|_{2-\delta}^2} \to 0 \quad \text{as} \quad n \to \infty. \tag{1.5.11}$$

Because $w_n \to w$ strongly in $H^{2-\varepsilon}(\Omega)$ for any $\varepsilon > 0$, we have from (1.4.25) with $\delta = 1$ and $f \equiv 0$ that

$$\|v_0(w_n) - v_0(w)\|_2^2 \to 0 \quad \text{as} \quad n \to \infty.$$

Therefore by (1.5.11) $v_0(w) = 0$ and hence $[w,w] = 0$. In order to reach the contradiction, we wish to show that $w \equiv 0$. It is precisely this part of the argument that depends on the boundary conditions. The simplest case is when $w \in H_0^2(\Omega)$, which corresponds to clamped boundary conditions. In this case, we have by using symmetry of the von Karman bracket

$$0 = \int_\Omega [w,w]\theta_0 dx = \int_\Omega [w,\theta_0] w dx = \int_\Omega \Delta w w dx = -\int_\Omega |\nabla w|^2 dx,$$

where $\theta_0 = (x_1^2 + x_2^2)/2$. Thus $w \equiv 0$. Therefore $w_n \to 0$ strongly in $H^{2-\delta}(\Omega)$ which contradicts the property $\|w_n\|_{2-\delta} = 1$. In the remaining case we have only Dirichlet boundary conditions equal to zero. In some cases of special geometries (small curvature, e.g., rectangles) the same argument as above still provides the desired uniqueness property. However, for general domains we evoke the result in elliptic theory which is based on the maximum principle. In fact, the following inequality results from specializing Lemma 9.2 in [125] to the case of functions vanishing on the boundary.

1.5.5. Lemma.

$$\max_\Omega |u| \le \frac{\text{diam}(\Omega)}{\sqrt{\pi}} \left(\int_\Omega |[w,w]| dx \right)^{1/2}, \quad \forall\, w \in H_0^1(\Omega) \cap H^2(\Omega).$$

Thus, as before we have that $[w,w] = 0 \to w \equiv 0$ which leads to the desired contradiction. □

Lemma 1.5.4 allows us to prove coercivity of problem (1.5.9). This is given below.

1.5.6. Proposition. *Let $f \in H^2(\Omega)$, $F_0 \in H^2(\Omega)$, and $L(\cdot)$ be a linear bounded operator from $H^{2-\sigma}(\Omega)$ into $H^{-\sigma}(\Omega)$ for all $0 \le \sigma \le \sigma_0$ with some $\sigma_0 > 0$. Assume that $\Gamma_3 = \emptyset$ and $\mu = 1$ in (1.5.2). Then there exist positive constants α and β such* *that*

$$(u,\mathscr{A}u+Bu)\geq\alpha\|u\|_2^2-\beta,\quad u\in V, \tag{1.5.12}$$

where $V=\{w\in(H^2\cap H_0^1)(\Omega)\,:\,\nabla w|_{\Gamma_1}=0\}$. *Thus the operator* $\mathscr{A}+B:V\mapsto V'$ *is coercive. Furthermore* $\mathscr{A}+B$ *is proper.*

Proof. Let $v_0(u)$ be the solution to (1.5.3) with $f\equiv 0$ and $v_f(u)\in H_0^2(\Omega)$ solves the problem

$$\Delta^2 v_f+2[f,u]=0,\quad v_f|_{\partial\Omega}=\frac{\partial v_f}{\partial n}\Big|_{\partial\Omega}=0. \tag{1.5.13}$$

Because $v(u)=v_0(u)+v_f(u)$, the symmetry of the von Karman bracket implies:

$$-([u+f,v(u)],u)=-([u,u]+[u,f],v(u))=\left(\Delta^2\left(v_0(u)+\frac{1}{2}v_f(u)\right),v(u)\right).$$

Therefore

$$-([u+f,v(u)],u)\geq\frac{1}{4}\left(\|\Delta v_0(u)\|^2-\|\Delta v_f(u)\|^2\right).$$

Consequently

$$(u,Bu)\geq\frac{1}{4}\left(\|\Delta v_0(u)\|^2-\|\Delta v_f(u)\|^2\right)-([u+f,F_0]+L(u),u).$$

Using Theorem 1.4.3 and exploiting the regularity of F_0, it is easy to see that

$$|([u+f,F_0]+L(u),u)|\leq C\left(1+\|u\|_{2-\delta}^2\right)$$

for some $\delta>0$. Similarly, from (1.4.17) with $j=2$ and $\beta=0$ we have

$$\|\Delta v_f(u)\|\leq C\cdot\|[f,u]\|_{-2}^2\leq C\cdot\|f\|_2^2\cdot\|u\|_1^2.$$

Therefore there exist constants D_1 and D_2 such that

$$(u,Bu)\geq\frac{1}{4}\|\Delta v_0(u)\|^2-D_1\|u\|_{2-\delta}^2-D_2.$$

for some $0<\delta<1$. Because $(u,\mathscr{A}u)=\|\Delta u\|^2\geq\gamma_0\|u\|_2^2$ for any $u\in V$ with some $\gamma_0>0$, from Lemma 1.5.4 we obtain estimate (1.5.12) which implies coercivity of $\mathscr{A}+B$.

Next we assert that $\mathscr{A}+B$ is proper. It is sufficient to establish that for any sequence $\{u_n\}\subset V$ such that $\mathscr{A}u_n+Bu_n$ converges to some element $h\in V'$ we can find subsequence $\{u_{n_k}\}$ which converges strongly in V.

The coercivity estimate (1.5.12) implies that the sequence $\{u_n\}$ is bounded in V. Therefore there exists subsequence $\{u_{n_k}\}$ and an element $u\in V$ such that $u_{n_k}\to u$ strongly in $H^{2-\varepsilon}(\Omega)$ for any $\varepsilon>0$. Therefore an analogue of (1.5.10) in the clamped–hinged case implies that $B(u_{n_k})\to B(u)$ in $H^{-1-\delta}(\Omega)\subset V'$. Consequently $\mathscr{A}u_{n_k}\to -B(u)+h$ in V'. Because $\mathscr{A}^{-1}$ is continuous from V' into V, we find that $u_{n_k}\to u$ strongly in V. □

Propositions 1.5.3 and 1.5.6 allow us to obtain the following assertion.

1.5.7. Theorem. *Let $p \in V'$, $f \in H^2(\Omega)$, $F_0 \in H^2(\Omega)$, and $L(\cdot)$ be a linear bounded operator from $H^{2-\sigma}(\Omega)$ into $H^{-\sigma}(\Omega)$ for all $0 \le \sigma \le \sigma_0$ with some $\sigma_0 > 0$. Assume that $\Gamma_3 = \emptyset$ and $\mu = 1$. Then*

- *Problem* (1.5.1)–(1.5.3) *has a weak solution. This solution can be obtained as a strong limit of a subsequence of Galerkin approximations. The set*[2] *of all weak solutions is compact in $H^2(\Omega)$. Furthermore there exists an open dense set $\mathscr{R}$ in V' such that for any $p \in \mathscr{R}$ problem* (1.5.1)–(1.5.3) *has a finite number of weak solutions.*
- *If $p \in L^2(\Omega)$ and $F_0 \in W^2_\infty(\Omega)$, then any weak solution belongs to the space*

$$V(\Omega,\Delta^2) = \left\{ u \in V \;:\; \Delta^2 u \in L^2(\Omega),\ \Delta u\big|_{\Gamma_2} = 0 \right\}.$$

Moreover there exists an open dense set $\mathscr{R}_0$ in $L^2(\Omega)$ such that for any $p \in \mathscr{R}_0$ problem (1.5.1)–(1.5.3) *has a finite number of solutions in $V(\Omega,\Delta^2)$. If $\overline{\Gamma_1} \cap \overline{\Gamma_2} = \emptyset$, then*

$$V(\Omega,\Delta^2) = \left\{ u \in (H^4 \cap H^1_0)(\Omega) :\ \nabla u\big|_{\Gamma_1} = 0,\ \Delta u\big|_{\Gamma_2} = 0 \right\}.$$

Proof. The first part easily follows from Theorems 1.2.9 and 1.2.13 and from Propositions 1.5.3 and 1.5.6.

Let us prove the regularity of weak solutions. Let $u \in V$ be a weak solution with $p \in L^2(\Omega)$. By "sharp" regularity of Airy's stress function in Corollary 1.4.5 we infer that $-B(u) + p \in L^2(\Omega)$. Therefore by Remark 1.3.16 we have that

$$u = \mathscr{A}^{-1}(-B(u) + p) \in V(\Omega,\Delta^2).$$

We note that $\mathscr{A} + B$ maps $V(\Omega,\Delta^2)$ onto $L^2(\Omega)$. It is also easy to see $\mathscr{A} + B$ is a Fredholm proper operator of index zero as a mapping from $V(\Omega,\Delta^2)$ into $L^2(\Omega)$. Thus we can apply Theorem 1.2.13 to conclude the finiteness of the set of solutions for generic p from $L^2(\Omega)$. □

1.5.8. Remark. The assertion similar to Theorem 1.5.7 can also be proved for the problem (1.5.1) and (1.5.3) with the simply supported (instead of hinged) boundary conditions (1.3.18) on Γ_2. ■

1.5.2 General mixed boundary conditions

We consider boundary conditions prescribed by (1.5.2), which also include a "free" component on Γ_3.

In this case we have the following result.

[2] It is known [85] that the problem may have several solutions; see also Remark 1.5.11(1).

1.5.9. Theorem. *Let $p \in V'$, $f \in V$, and $L(\cdot)$ be a linear bounded operator from $H^{2-\sigma}(\Omega)$ into $H^{-\sigma}(\Omega)$ for all $0 \le \sigma \le \sigma_0$ with some $\sigma_0 > 0$ such that $(L(u), u) \ge 0$ for every $u \in V$. The space V is given by* (1.5.4). *Assume that F_0 belongs to $H^3(\Omega)$ and satisfies the boundary condition*

$$F_0\Big|_{\partial\Omega} = 0 \ \ and \ \ \frac{\partial F_0}{\partial n}\Big|_{\Gamma_3} = 0. \tag{1.5.14}$$

Assume that either $\Gamma_1 \cup \Gamma_2 \neq \emptyset$ or $\nu > 0$ in (1.5.2). *Then*

- *Problem* (1.5.1)–(1.5.3) *has a weak solution. This solution is a strong limit of a subsequence of Galerkin approximations. The set of all weak solutions is compact in V. Furthermore there exists an open dense set $\mathscr{R}$ in V' such that for any $p \in \mathscr{R}$ problem* (1.5.1)–(1.5.3) *has a finite number of weak solutions.*
- *If $p \in L^2(\Omega)$ and $F_0 \in W^2_\infty(\Omega)$, then any weak solution belongs to*

$$W(\Omega, \Delta^2) = \left\{ u \in H^2(\Omega) \ : \ \Delta^2 u \in L^2(\Omega) \ \ and \ \ u \ \ satisfies \ (1.5.2) \right\}.$$

Moreover there exists an open dense set $\mathscr{R}_0$ in $L^2(\Omega)$ such that for any $p \in \mathscr{R}_0$ problem (1.5.1)–(1.5.3) *has a finite number of solutions in the space $W(\Omega, \Delta^2)$. If $\overline{\Gamma_i} \cap \overline{\Gamma_j} = \emptyset$ for every $i \neq j$, then*

$$W(\Omega, \Delta^2) = \left\{ u \in H^4(\Omega) : \ u \ \ \text{satisfies} \ (1.5.2) \right\}.$$

Proof. We look for a solution u of the form $u = w - f$; that is instead of (1.5.9) we consider the problem

$$\mathscr{A} w + B_f w = p + \mathscr{A} f, \quad w \in V,$$

where V is given by (1.5.4) and

$$B_f w = B(w - f) = -[w, v(w - f) + F_0] + L(w - f).$$

It is clear that $v(w - f) = v_0(w) + F_{00}$, where $v_0(u)$ is the solution to (1.5.3) with $f \equiv 0$ and $F_{00} = \left(\Delta_D^2\right)^{-1}[f, f]$. It follows from (1.3.11) and (1.4.17) with $j = 1$ and $\beta = 0$ that $F_{00} \in H^3(\Omega) \cap H_0^2(\Omega)$. Thus we arrive at the problem

$$\mathscr{A} w + \tilde{B} w = q, \quad w \in V,$$

where

$$\tilde{B} w = -[w, v_0(w) + \tilde{F}_0] + L(w), \quad \tilde{F}_0 = F_{00} + F_0, \quad q = p + \mathscr{A} f + L(f) \in V'.$$

By Proposition 1.5.3 the operator $\mathscr{A} + \tilde{B}$ is a pseudomonotone bounded continuous operator from V into V' satisfying the condition $(S)_+$. Moreover $\mathscr{A} + \tilde{B}$ is a Fredholm operator of index zero. Therefore to prove the first part of the theorem we need to show that $\mathscr{A} + \tilde{B}$ is coercive and proper.

From properties of the von Karman bracket we obviously have that

$$(w,\tilde{B}w) = \|\Delta v_0(w)\|^2 - ([w,w],F_{00}) - ([w,F_0],w) + (L(w),w).$$

It follows from (1.5.14) and (1.4.9) that $([w,F_0],w) = ([w,w],F_0)$ for any $w \in V$. Therefore using (1.3.29) we obtain that

$$(w,\mathscr{A}w+\tilde{B}w) \geq \alpha \cdot \|w\|_2^2 + \|\Delta v_0(w)\|^2 - ([w,w],\tilde{F}_0), \quad w \in V, \tag{1.5.15}$$

with some positive constant α. Coercivity in (1.5.15) depends on the following bound from below.

1.5.10. Lemma. *Let $\tilde{F}_0 \in H^3(\Omega)$ and satisfy* (1.5.14). *Then*

$$\Psi(w) = \frac{\alpha}{2} \cdot \|w\|_2^2 + \|\Delta v_0(w)\|^2 - ([w,w],\tilde{F}_0)$$

is bounded from below on V, where V is given by (1.5.4).

Proof. We follow the line of the argument given in the proof of Lemma 1.5.4. Assume that there exists a sequence $\{w_n\} \subset V$ such that

$$0 > M_n \equiv \Psi(w_n) \to -\infty \quad \text{as } n \to \infty.$$

It should be $([w_n,w_n],\tilde{F}_0) \to +\infty$. Let $\tilde{w}_n = w_n \cdot |([w_n,w_n],\tilde{F}_0)|^{-1/2}$. We have the relation

$$\frac{\alpha}{2} \cdot \|\tilde{w}_n\|_2^2 + |([w_n,w_n],\tilde{F}_0)| \cdot \|\Delta v_0(\tilde{w}_n)\|^2 \leq 1 + \frac{M_n}{|([w_n,w_n],\tilde{F}_0)|} \leq 1.$$

Thus we can suppose that there exists $\tilde{w} \in V$ such that $\tilde{w}_n \rightharpoonup \tilde{w}$ weakly in V and hence $\tilde{w}_n \to \tilde{w}$ strongly in $H^{2-\varepsilon}(\Omega)$ for any $\varepsilon > 0$. Therefore it follows from (1.4.16) that $[\tilde{w}_n,\tilde{w}_n] \to [\tilde{w},\tilde{w}]$ strongly in $H^{-1-\varepsilon}(\Omega)$. We also have that $\|\Delta v_0(\tilde{w}_n)\|^2 \to 0$. Consequently $[\tilde{w},\tilde{w}] = 0$. Thus

$$\frac{\|[w_n,w_n]\|_{-1-\varepsilon}}{|([w_n,w_n],\tilde{F}_0)|} \to 0 \quad \text{as } n \to \infty.$$

or

$$\left|\left(\frac{[w_n,w_n]}{\|[w_n,w_n]\|_{-1-\varepsilon}},\tilde{F}_0\right)\right| \to \infty \quad \text{as } n \to \infty.$$

However (1.5.14) implies that $\tilde{F}_0 \in H_0^{1+\varepsilon}(\Omega)$ for $\varepsilon < \frac{1}{2}$. Therefore

$$\left|\left(\frac{[w_n,w_n]}{\|[w_n,w_n]\|_{-1-\varepsilon}},\tilde{F}_0\right)\right| \leq \|\tilde{F}_0\|_{1+\varepsilon} < \infty.$$

Thus we arrive at contradiction. Hence $\Psi(w)$ is bounded from below on V. □

By Lemma 1.5.10 we obtain the coercivity estimate

$$(w,\mathscr{A}w+\tilde{B}w) \geq \frac{\alpha}{2} \cdot \|w\|_2^2 - C, \quad v \in V.$$

As in the proof of Proposition 1.5.6 it is also easy to see that $\mathscr{A}w+\tilde{B}w$ is proper. Thus the first part of the theorem is proved.

The second part can be established in the same way as in Theorem 1.5.7. □

1.5.11. Remark.

- There are examples which show that problem (1.5.1)–(1.5.3) can possess several weak solutions (see, e.g., [85]).
- If we assume that $F_0 \in H^3(\Omega)\cap H_0^2(\Omega)$ then by symmetry of the von Karman bracket we have

  $$([w,w],F_0) = -(\Delta^2 v_0(w),F_0) = -(\Delta v_0(w),\Delta F_0), \quad w \in V.$$

  Therefore it follows from (1.5.15) that

  $$(w,\mathscr{A}w+\tilde{B}w) \geq \alpha \cdot \|w\|_2^2 - \frac{1}{4}\cdot \|\Delta \tilde{F}_0\|^2, \quad w \in V.$$

  Thus coercivity is very simple for the case $F_0 \in H^3(\Omega)\cap H_0^2(\Omega)$. However, from the mechanical point of view our weaker assumptions concerning F_0 mean that no external stresses are applied to the shell/plate at its free edge and some stresses are allowed on the clamped and simply supported parts. In the case $F_0 \in H^3(\Omega)\cap H_0^2(\Omega)$ we have no external stresses on the whole boundary. We refer to [85] for a more detailed discussion.
- In Theorem 1.5.9 instead of positivity of $(L(u),u)$ we can assume that the relation $(L(u),u) \geq -\varepsilon\|u\|^2$ holds with $\varepsilon > 0$ small enough.

■

The following observation is important in our subsequent considerations.

1.5.12. Remark. If L is a self-adjoint operator in $L^2(\Omega)$, then in the both cases (see Theorems 1.5.7 and 1.5.9) every weak solution is an extreme point of the functional

$$\Pi(u) = \frac{1}{2}\left(a(u,u) + \frac{1}{2}\|\Delta v(u)\|^2 + (-[u,F_0]+L(u),u) - 2([f,F_0]+p,u)\right)$$

on the space $V = \left\{u \in H^2(\Omega) \,:\, u|_{\Gamma_1\cup\Gamma_2} = 0,\ \nabla u|_{\Gamma_1} = 0\right\}$. Here $a(u,v)$ is defined by (1.5.6) and $v(u)$ is determined from (1.5.3). The functional $\Pi(u)$ can be interpreted as the potential energy of the shell/plate. The coercivity property means that the energy $\Pi(u)$ is bounded from below on the space V under the hypotheses of Theorem 1.5.7 or 1.5.9. We also note that functional $\Pi(u)$ can be written as

$$\Pi(u) = \frac{1}{2}a(u,u) + \frac{1}{4}\|\Delta v(u)\|^2 + \Pi_1(u),$$

where

$$\Pi_1(u) = \frac{1}{2}\left((-[u,F_0]+L(u),u) - 2([f,F_0]+p,u)\right).$$

Using Lemma 1.5.4 and also the argument given in the proof of Theorem 1.5.9 we can prove that for any $\eta > 0$ there exists C_η such that

$$|\Pi_1(u)| \leq \eta \left(a(u,u) + \|\Delta v(u)\|^2\right) + C_\eta, \tag{1.5.16}$$

provided the hypotheses of Theorem 1.5.7 or 1.5.9 hold. ∎

1.5.3 Modified mixed boundary conditions

The restrictions in Theorem 1.5.9 are boundary conditions (1.5.14) and the positivity of the form $(L(u),u)$. The main reason for these restrictions is that we have no analogues of Lemma 1.5.4 for the case when Γ_3 is nonempty. Here, the main issue is to guarantee the uniqueness property $[u,u] = 0 \to u = 0$. There are general theorems that provide this kind of conclusion for the "free" case as well but under the additional regularity of second-order derivatives on the boundary (see, e.g., [25]). These additional conditions are lost in the limit process which is the main tool in the contradiction argument used for the proof of Lemma 1.5.4. However, by slightly modifying boundary conditions (i.e., adding a superlinear term), we may conclude the assertion of this lemma, but for a modified functional. This modification corresponds to the modified boundary conditions for plates on a portion of the boundary Γ_3.

1.5.13. Lemma. *We consider the functional*

$$\hat{\Psi}(u;A,\delta) \equiv \Psi(u;A,\delta) + \int_{\Gamma_3} \beta(x) u^4 d\Gamma,$$

where the functional Ψ *is the same as in Lemma* 1.5.4 *and* $\beta \in L_\infty(\Gamma_3)$, $\beta(x) > 0$ *almost everywhere. Then the conclusion of Lemma* 1.5.4 *holds for* $\hat{\Psi}(u;A,\delta)$ *and for any function* $u \in H^2(\Omega)$ *such that* $u = 0$ *on* $\Gamma_1 \cup \Gamma_2$*:*

$$\inf\left\{\hat{\Psi}(u;A,\delta)\ :\ u \in H^2(\Omega),\ u|_{\Gamma_1 \cup \Gamma_2} = 0\right\} = c(A,\delta) > -\infty$$

for any $A > 0$ *and* $0 < \delta \leq 2$.

Proof. We use the same contradiction argument as in Lemma 1.5.4.

By contradiction, assume that there exists a sequence $\{u_n\} \subset H^2(\Omega)$ such that $u_n = 0$ on $\Gamma_1 \cup \Gamma_2$ and $M_n \equiv \hat{\Psi}(u_n;A,\delta) \to -\infty$ as $n \to \infty$. We can assume that $M_n < 0$ for all n, which then leads to $\|u_n\|_{2-\delta} \to \infty$. Let $w_n = u_n/\|u_n\|_{2-\delta}$. Then we have $\|w_n\|_{2-\delta} = 1$ and

$$\|u_n\|^2_{2-\delta} \cdot \left(\|w_n\|_2^2 + \|u_n\|^2_{2-\delta} \int_{\Gamma_3} \beta |w_n|^4 ds + \frac{\|\Delta v_0(u_n)\|^2}{\|u_n\|^2_{2-\delta}} - A \right) = M_n \to -\infty,$$

where $v_0(u)$ denotes the solution to (1.5.3) with $f \equiv 0$. Consequently

$$\|w_n\|_2^2 + \|u_n\|_{2-\delta}^2 \cdot \int_{\Gamma_3} \beta |w_n|^4 d\Gamma + \|u_n\|_{2-\delta}^2 \cdot \|\Delta v_0(w_n)\|^2 = A + \frac{M_n}{\|u_n\|_{2-\delta}^2} < A.$$

This implies that the sequence $\{w_n\}$ contains a subsequence which converges weakly in $H^2(\Omega)$. Thus we can suppose that $w_n \rightharpoonup w$ weakly in $H^2(\Omega)$ for some $w \in H^2(\Omega)$ and $w = 0$ on $\Gamma_1 \cup \Gamma_2$.

We also have

$$\|\Delta v_0(w_n)\|^2 + \int_{\Gamma_3} \beta |w_n|^4 d\Gamma \le \frac{A}{\|u_n\|_{2-\delta}^2} \to 0 \quad \text{as } n \to \infty. \tag{1.5.17}$$

Because $w_n \to w$ strongly in $H^{2-\varepsilon}(\Omega)$ for any $\varepsilon > 0$, we also conclude that

$$\|\Delta v_0(w)\|^2 + \int_{\Gamma_3} \beta |w|^4 d\Gamma = 0.$$

Because $\beta(x) > 0$ a.e., we have that $w = 0$ on Γ_3 and hence $w \equiv 0$ on Γ. The remaining part of the argument is now the same as in Lemma 1.5.4. □

1.5.14. Remark. We note that the motivation for selecting the new functional $\hat{\Psi}$ in Lemma 1.5.13 is the modification of free and nonlinear boundary conditions on Γ_3 given in the last line of (1.5.18). The nonlinear term u^3 corresponds to the presence of a static boundary damping. The potential energy for the plate problem with these modified boundary conditions corresponds to the functional $\hat{\Psi}$. From the mathematical point of view, the same effect could be achieved with any superlinear and dissipative static damping. ■

We consider problem (1.5.1) and (1.5.3) with the following boundary conditions

$$\left.\begin{aligned} &u|_{\Gamma_1} = \nabla u|_{\Gamma_1} = 0, && \text{(clamped)} \\ &u|_{\Gamma_2} = [\Delta u + (1-\mu)B_1 u]\big|_{\Gamma_2} = 0, && \text{(simply supported)} \\ &\left.\begin{aligned} &[\Delta u + (1-\mu)B_1 u]\big|_{\Gamma_3} = 0 \\ &\left[\frac{\partial}{\partial n}\Delta u + (1-\mu)B_2 u - \nu u - \beta u^3\right]\Big|_{\Gamma_3} = 0, \end{aligned}\right] && \text{(free modified)} \end{aligned}\right\} \tag{1.5.18}$$

In addition to the conditions listed above we assume that $\beta \in L_\infty(\Gamma_3)$ and $\beta(x) > 0$ almost everywhere.

1.5.15. Definition. A function u from the space V given by (1.5.4) is said to be a *weak solution* to the problem (1.5.1), (1.5.3), and (1.5.18) iff

$$a(u,w) + \int_\Omega (-[u+f, v(u)+F_0] + L(u) - p) \cdot w dx + \int_{\Gamma_3} \beta u^3 w d\Gamma = 0 \tag{1.5.19}$$

for any $w \in V$. Here as above $a(u,v)$ is defined by (1.5.6) and $v(u)$ is determined from (1.5.3). ■

We have the following theorem.

1.5.16. Theorem. *Let $p \in V'$, $f \in V$, $F_0 \in H^3(\Omega)$, and $L(\cdot)$ be a linear bounded operator from $H^{2-\sigma}(\Omega)$ into $H^{-\sigma}(\Omega)$ for all $0 \le \sigma \le \sigma_0$ with some $\sigma_0 > 0$. Then*

- *Problem* (1.5.1), (1.5.3), *and* (1.5.18) *has a weak solution. This solution can be obtained as a strong limit of a subsequence of Galerkin approximations. The set of all weak solutions is compact in V. Furthermore there exists an open dense set $\mathscr{R}$ in V' such that for any $p \in \mathscr{R}$ problem* (1.5.1), (1.5.3), *and* (1.5.18) *has a finite number of weak solutions.*
- *If $p \in L^2(\Omega)$ and $F_0 \in W^2_\infty(\Omega)$, then any weak solution belongs to*

$$W(\Omega,\Delta^2) = \left\{ u \in H^2(\Omega) \;:\; \Delta^2 u \in L^2(\Omega) \;\; and \;\; u \;\; satisfies \; (1.5.18) \right\}.$$

Moreover there exists an open dense set $\mathscr{R}_0$ in $L^2(\Omega)$ such that for any $p \in \mathscr{R}_0$ problem (1.5.1), (1.5.3), *and* (1.5.18) *has a finite number of solutions in the space $W(\Omega,\Delta^2)$. If $\overline{\Gamma_i} \cap \overline{\Gamma_j} = \emptyset$ for every $i \neq j$, then*

$$W(\Omega,\Delta^2) = \left\{ u \in H^4(\Omega) : \; u \;\; satisfies \; (1.5.18) \right\}.$$

Proof. As above we can write problem (1.5.1), (1.5.3), and (1.5.18) in the form

$$\mathscr{A} u + Bu + B_{\Gamma_3} u = p, \quad u \in V,$$

where $\mathscr{A} : V \mapsto V'$ and B are given by (1.5.7) and (1.5.8) and $B_{\Gamma_3} : V \mapsto V'$ is defined by the formula

$$(w, B_{\Gamma_3} u) = \int_{\Gamma_3} \beta u^3 w \, d\Gamma.$$

It is easy to see that B_{Γ_3} is strongly continuous. Therefore from Propositions 1.5.3 and 1.2.8 we have that $\mathscr{A} + B + B_{\Gamma_3}$ is a pseudomonotone bounded continuous operator from V into V' satisfying condition $(S)_+$. Using Proposition 1.2.11 we obtain that $\mathscr{A} + B + B_{\Gamma_3}$ is Fredholm with index zero.

As in the proof of Proposition 1.5.6 it follows from Lemma 1.5.13 that the operator $\mathscr{A} + B + B_{\Gamma_3}$ is coercive. It is also clear that $\mathscr{A} + B + B_{\Gamma_3}$ is proper.

Therefore we can apply Theorems 1.2.9 and 1.2.13 to complete the proof in the same way as it was done for Theorems 1.5.7 and 1.5.9. □

1.5.17. Remark. Even in the case $L \equiv 0$, problem (1.5.1), (1.5.3), and (1.5.18) is not potential for general F_0: the weak solutions cannot be obtained as extreme points of a functional describing potential energy. The point is that the relation

$$\int_\Omega [u, F_0] v \, dx = \int_\Omega [v, F_0] u \, dx$$

is not generally true for u and v from the space V given by (1.5.4). This is the main difference in comparison with the cases described in Theorems 1.5.7 and 1.5.9 (see Remark 1.5.12). ■

Chapter 2
Evolutionary Equations

2.1 Overview

This chapter provides preliminary material dealing with evolutionary abstract equations that serve as a prototype for von Karman evolutions discussed in subsequent chapters. Special attention is paid to the existence and uniqueness of solutions to nonlinear evolutionary equations. The methods used are based on monotone operator theory and their adaptation to nonmonotone problems. In what follows we provide a brief exposition of nonlinear semigroup theory and related concepts in maximal-monotone operator theory. We restrict our attention to single-valued operators, although many results stated below remain true in the multivalued setting.

We also state several results pertinent to linear plate equations in a form convenient for applications in subsequent chapters.

2.2 Accretive operators in Hilbert spaces

Let H be a Hilbert space. We consider (nonlinear) operators A in H defined on a subset $\mathscr{D}(A)$ of H with the range $R(A) = \{Ax : x \in \mathscr{D}(A)\} \subseteq H$.

We begin with several definitions.

2.2.1. Definition. An operator $A : \mathscr{D}(A) \subseteq H \mapsto H$ is called *accretive* iff we have

$$(Ax_1 - Ax_2, x_1 - x_2)_H \geq 0 \quad \text{for any } x_1, x_2 \in \mathscr{D}(A).$$

An accretive operator A is said to be *maximal accretive (m-accretive)* if the relation

$$(Av - u^*, v - u)_H \geq 0$$

for some $u, u^* \in H$ and for all $v \in \mathscr{D}(A)$ implies that $u \in \mathscr{D}(A)$ and $u^* = Au$. ■

I. Chueshov, I. Lasiecka, *Von Karman Evolution Equations*,
Springer Monographs in Mathematics, DOI 10.1007/978-0-387-87712-9_2,

We note that the accretive operator is a particular case of a monotone operator (cf. Definition 1.2.1).

A convenient characterization of m-accretive operators is given by the following well-known result; see, for example, [34] or [260, Chapter IV].

2.2.2. Theorem. *An accretive operator $A : \mathscr{D}(A) \subseteq H \mapsto H$ on the Hilbert space H is m-accretive if and only if $R(\lambda I + A) = H$ for every (equivalently for some) $\lambda > 0$. Here $R(B)$ denotes the range of an operator B.*

The following perturbation type of result is frequently used in the context of PDEs (see, e.g., [260, Section IV.2]).

2.2.3. Proposition. *Let $A : \mathscr{D}(A) \subseteq H \mapsto H$ be an m-accretive operator on the Hilbert space H. Then*

- *If $B : H \mapsto H$ is accretive and Lipschitz; that is,*

$$\|B(u_1) - B(u_2)\| \leq L\|u_1 - u_2\|, \quad u_1, u_2 \in H,$$

then $A + B$ is m-accretive.
- *If $B : H \mapsto H$ is Lipschitz, then for any $\omega \geq L$ the operator $A + B + \omega I$ is m-accretive.*

The following result presents some continuity properties of m-accretive operators in the spirit of Proposition 1.2.6.

2.2.4. Proposition. *Let $A : \mathscr{D}(A) \subseteq H \mapsto H$ be an m-accretive operator on the Hilbert space H. Let $\{v_n\}$ be a sequence in $\mathscr{D}(A)$. Assume that there exist elements v and f from H such that $v_n \to v$ and $Av_n \to f$ weakly in H and*

$$(Av_n - Av_m, v_n - v_m)_H \to 0 \quad \text{when } n, m \to \infty.$$

Then $v \in \mathscr{D}(A)$, $f = Av$ and $\lim_{n\to\infty}(Av_n, v_n)_H = (Av, v)_H$.

Proof. We apply the second part of Proposition 1.2.6 with $V = V' = H$. □

In the context of monotone operators, the notion of coercivity is fundamental (cf. Definition 1.2.4).

2.2.5. Definition. We say that an operator $A : \mathscr{D}(A) \subseteq H \mapsto H$ is *coercive* iff

$$\lim_{\|u\|_H \to \infty} \frac{(Au, u)_H}{\|u\|_H} = \infty, \quad u \in \mathscr{D}(A).$$

■

The following assertion holds (see, e.g., [260, Section IV.2, p. 170]).

2.2.6. Proposition. *Let A be m-accretive. If $\lim_{\|u\|_H \to \infty} \|A(u)\|_H = \infty$, then A is surjective; that is $R(A) = H$. If A is coercive then $R(A) = H$.*

2.2.7. Example. Any nonnegative self-adjoint operator in a Hilbert space is an m-accretive operator. ■

2.3 Abstract differential equations

Let A be an m-accretive operator on a Hilbert space H. We consider the following abstract differential equation:

$$\frac{du}{dt} + Au = f; \quad 0 < t < T, \quad u|_{t=0} = u_0 \in H, \tag{2.3.1}$$

where $0 < T \leq \infty$.

2.3.1. Definition. A *strong solution* of (2.3.1) is a continuous function u from $[0,T)$ into H that is absolutely continuous on each subinterval $[a,b]$, $0 < a < b < T$, hence u is differentiable a.e. with $du/dt \in L_1(a,b;H)$, and for a.e. $t > 0$ we have that $u(t) \in \mathscr{D}(A)$ and (2.3.1) holds. ■

Here, as in Section 1.1.4, we denote by $L_p(a,b;H)$ with $1 \leq p \leq \infty$ the Banach space of (equivalence classes of) Bochner measurable functions $f : [a,b] \mapsto H$ such that $\|f(\cdot)\|_H \in L_p(a,b)$. We also recall (see Section 1.1.4) that we denote by $C(a,b;H) \equiv C([a,b];H)$ the space of strongly continuous functions on $[a,b]$ with values in the space H. A similar meaning has the notations $C([a,b);H)$ and $C((a,b];H)$. We also use the space

$$W_p^1(a,b;H) = \left\{ f \in C(a,b;H) \ : \ f' \in L_p(a,b;H) \right\},$$

where $f'(t)$ is the strong derivative (almost everywhere) of $f(t)$ with respect to t. We note that $W_1^1(a,b;H)$ coincides with the set of absolutely continuous functions from $[a,b]$ into H (see, e.g., [260, Theorem 1.7, p. 105]). Higher-order analogues $W_p^m(a,b;H)$ of the space $W_p^1(a,b;H)$ are defined in (1.1.21). More details and references concerning the spaces above can be found in Section 1.1.

A fundamental result that asserts an existence of strong solutions for m-accretive operators is the following result due to Kato (see [260, p. 180]).

2.3.2. Theorem. *Let A be m-accretive on the Hilbert space H. Assume that $u_0 \in \mathscr{D}(A)$ and $f : [0,T] \mapsto H$ is absolutely continuous. Then there is a unique strong solution of* (2.3.1). *In addition, u is Lipschitz continuous from $[0,T]$ into H and strongly right-differentiable in H for $t \geq 0$. Moreover $u(t) \in \mathscr{D}(A)$ for all $t \geq 0$ and $u' \in L_\infty(0,T;H)$.*

We note that in some cases strong solutions can be constructed by the Galerkin method (see, e.g., [121]).

In our considerations a weaker concept of solution, the so-called generalized solution is used frequently. This is defined by considering (strong) limits of strong solutions.

2.3.3. Definition. A *generalized solution* to (2.3.1) on a (closed) interval $[0,T]$ is a continuous function $u \in C(0,T;H)$ such that $u(0) = u_0$ and for which there exists a sequence of strong solutions u_n defined on $[0,T]$ to the problem

$$\frac{d}{dt}u_n + Au_n = f_n, \quad n = 1,2,\ldots$$

with $f_n \to f$ in $L_1(0,T;H)$ and $u_n \to u$ in $C(0,T;H)$. A function $u(t)$ from the class $C([0,T);H)$ is a generalized solution to problem (2.3.1) on a *semi-interval* $[0,T)$, iff u is a generalized solution to (2.3.1) on every closed interval $[0,T']$ with $T' < T$. ■

The existence result for generalized solutions is given below [260, p. 183].

2.3.4. Theorem. *Let A be m-accretive on the Hilbert space H, $f \in L_1(0,T;H)$ and $u_0 \in \overline{\mathscr{D}(A)}$, where $\overline{\mathscr{D}(A)}$ is the closure of $\mathscr{D}(A)$ in H. Then there exists a unique generalized solution to* (2.3.1)*. Moreover, any two generalized solutions $u_1(t)$ and $u_2(t)$ with data $\{u_{10}, f_1\}$ and $\{u_{20}, f_2\}$ satisfy the following stability estimates with $0 \leq s < t \leq T$*

$$\| u_1(t) - u_2(t) \|_H^2 \leq \| u_1(s) - u_2(s) \|_H^2 + 2\int_s^t (f_1(z) - f_2(z), u_1(z) - u_2(z))_H dz \qquad (2.3.2)$$

and

$$\| u_1(t) - u_2(t) \|_H \leq \| u_1(s) - u_2(s) \|_H + \int_s^t \| f_1(z) - f_2(z) \|_H \, dz. \qquad (2.3.3)$$

2.3.5. Remark. We note that the concept of *generalized* solution is only weakly associated with the differential equation, via the limit procedure. Thus, it may happen that the generalized solution does not really satisfy the differential equation, even in a weak sense. However, under additional regularity properties one may be able to show that generalized solutions satisfy a weak (variational) form of differential equation. Such solutions are called *weak solutions* and are also often referred as *variational* solutions. We return to this point later, when discussing more specific structures of abstract evolution equations. ■

We continue this section with several perturbation results.
Let $B : \overline{\mathscr{D}(A)} \mapsto H$ be a Lipschitz function such that

$$\| B(u) - B(v) \|_H \leq L \| u - v \|_H; \quad u, v \in \overline{\mathscr{D}(A)}. \qquad (2.3.4)$$

We consider the perturbed equation

$$\frac{d}{dt}u + (A+B)u = f, \quad u|_{t=0} = u_0. \qquad (2.3.5)$$

2.3.6. Theorem. *We assume that A is m-accretive and B satisfies* (2.3.4).

- **Strong solutions:** *For each $u_0 \in \mathscr{D}(A)$ and f absolutely continuous from $[0,T]$ into H there is a unique strong solution $u(t)$ to the problem* (2.3.5)*. The function $t \mapsto u(t)$ is Lipschitz continuous from $[0,T]$ into H and strongly right-differentiable in H with $u(t) \in \mathscr{D}(A)$ for any $t \geq 0$. Moreover,*

$$u_t \equiv \frac{du}{dt} \in L_\infty(0,T;H), \quad A(u) \in L_\infty(0,T;H) \qquad (2.3.6)$$

and the functions $t \mapsto u_t(t)$ *and* $t \mapsto A(u(t))$ *are weakly continuous and strongly right-continuous as mappings from* $[0,T]$ *into* H.

- **Generalized solutions:** *For each* $u_0 \in \overline{\mathscr{D}(A)}$ *and* $f \in L_1(0,T;H)$ *there is a unique generalized solution to problem* (2.3.5) *on* $[0,T]$.

Moreover, any two generalized solutions $u_1(t)$ *and* $u_2(t)$ *to problem* (2.3.5) *with data* $\{u_{10}, f_1\}$ *and* $\{u_{20}, f_2\}$ *satisfy the following stability estimates with* $0 \le s < t \le T$

$$\| u_1(t) - u_2(t) \|_H^2 \le e^{2L(t-s)} \| u_1(s) - u_2(s) \|_H^2 + 2\int_s^t e^{2L(t-z)}(f_1(z) - f_2(z), u_1(z) - u_2(z))_H dz \tag{2.3.7}$$

and

$$\| u_1(t) - u_2(t) \|_H \le e^{L(t-s)} \| u_1(s) - u_2(s) \|_H + \int_s^t e^{L(t-z)} \| f_1(z) - f_2(z) \|_H \, dz. \tag{2.3.8}$$

The above theorem is a standard generation theorem which can be found in many books on monotone operator theory [18, 34, 260]. The main ingredient of the proof is the fact that $A + B + LI$ is an m-accretive operator (see Proposition 2.2.3). The proof of estimates (2.3.7) and (2.3.8) can be found in [260, p. 183]. The regularity properties of strong solutions follow from the argument given in [260, p. 175]. Here we also rely on Proposition 2.2.4.

In what follows we give a generalization of the result to locally Lipschitz perturbations B. Such result is a main tool in proving global existence for semilinear PDEs with a priori bounds typically obtained by some sort of energy method.

2.3.7. Definition. *A locally Lipschitz operator* $B : \overline{\mathscr{D}(A)} \mapsto H$ is a function such that for any $K > 0$ there exists a positive constant $L(K)$ such that

$$\| B(u) - B(v) \|_H \le L(K) \| u - v \|_H \tag{2.3.9}$$

for all $u, v \in \overline{\mathscr{D}(A)}$ such that $\| u \|_H \le K$ and $\| v \|_H \le K$. ■

The following theorem can be derived from Theorem 2.3.6 via the suitable (well-known) procedure of truncating B in order to consider suitable translation of the accretive operator. This result is known to researchers and often applied in the context of PDEs. For the sake of completeness we give the proof that follows the line of argument presented in [62] and relies on the idea found in [177, Section 2.3].

2.3.8. Theorem. *Let* $A + \lambda I$ *be an m-accretive operator for some* $\lambda \ge 0$. *We assume that B is subject to* (2.3.9).

- **Local strong solutions:** *For each element* $u_0 \in \mathscr{D}(A)$ *and any function* $f : \mathbb{R}_+ \mapsto H$ *that is absolutely continuous on every finite interval* $[0,T]$ *there exists* $t_{\max} \le \infty$ *such that there is a unique strong solution* $u(t)$ *to the problem* (2.3.5) *defined on* $[0, t_{\max})$. *The function* $t \mapsto u(t)$ *is Lipschitz continuous and right-differentiable* *with $u(t) \in \mathscr{D}(A)$ for any $t \in [0, t_{\max})$. Moreover, relations (2.3.6) hold for any $T < t_{\max}$ and the the functions $t \mapsto u_t(t)$ and $t \mapsto A(u(t))$ are weakly continuous and strongly right-continuous as mappings from $[0, t_{\max})$ into H.*

- **Local generalized solutions:** *For each element u_0 from the closure $\overline{\mathscr{D}(A)}$ of $\mathscr{D}(A)$ in H and for every function $f \in L_1^{\mathrm{loc}}([0, +\infty); H)$ there is a unique generalized solution to the problem (2.3.5) defined on $[0, t_{\max})$.*
- *In both cases we have $\lim_{t \to t_{\max}} \| u(t) \|_H = \infty$, provided $t_{\max} < \infty$.*
- **Global solutions:** *Let $u(t)$ be a strong (or generalized) solution with some data $\{u_0, f\}$. Assume that there exist $T > 0$ and $C_T(u_0, f)$ such that*

$$\sup_{[0,T^*)} \|u(t)\|_H \le C_T(u_0, f) \tag{2.3.10}$$

for every interval $[0, T^*) \subset [0, T]$ of the existence of the solution $u(t)$. Then the solution $u(t)$ exists on $[0, T]$. Moreover, if for any $T > 0$ there exists $C_T(u_0, f)$ such that (2.3.10) holds, then the solution $u(t)$ can be extended on $[0, +\infty)$; that is, $t_{\max} = \infty$.

Proof. We can assume that $\lambda = 0$ (otherwise we redefine $A := A + \lambda I$ and $B := B - \lambda I$) Moreover, we can assume that $0 \in \mathscr{D}(A)$ and $A0 = 0$. The point is that if $u(t)$ is a (strong or generalized) solution to (2.3.5), then for any $w_* \in \mathscr{D}(A)$ the function $w(t) = u(t) - w_*$ solves (in the same sense as $u(t)$) the problem

$$\frac{d}{dt} w + A_* w + B_*(w) = f_*, \quad w|_{t=0} = u_0 - w_*,$$

where A_*, B_* and f_* are given by

$$A_* w = A(w_* + w) - A w_*, \quad w \in \mathscr{D}(A_*) = -w_* + \mathscr{D}(A),$$

$$B_*(w) = B(w_* + w), \quad w \in H, \quad f_* = f - A w_*.$$

It is clear that A_*, B_*, and f_* satisfy the conditions of Theorem 2.3.8 (with another Lipschitz constant $L(K)$) and, in addition, $0 \in \mathscr{D}(A_*)$ and $A_* 0 = 0$.

To apply Theorem 2.3.6 we need to consider some regularization of the operator B. We set

$$\tilde{B}u = \begin{cases} Bu & \text{if } \|u\| \le K, \\ B\left(K \frac{u}{\|u\|}\right) & \text{if } \|u\| > K. \end{cases}$$

Here and below $\|\cdot\| \equiv \|\cdot\|_H$ and $(\cdot,\cdot)$ is the corresponding inner product. We claim that the mapping $\tilde{B}$ satisfies $\|\tilde{B}u - \tilde{B}v\| \le L(K)\|u - v\|$; that is, $\tilde{B}$ is globally Lipschitz. In order to prove this we distinguish three cases.

(i) If $\|u\|, \|v\| \le K$, the conclusion follows because of $\tilde{B}u = Bu$, $\tilde{B}v = Bv$, and formula (2.3.9).

(ii) In the case $\|u\| \le K$ and $\|v\| > K$ we have

$$\|\tilde{B}u - \tilde{B}v\| \le L(K) \left\| u - K\frac{v}{\|v\|} \right\| .$$

We show that

$$\left\| u - K\frac{v}{\|v\|} \right\| \le \|u - v\|$$

This is done by the following computation which relies on the Cauchy–Schwartz inequality

$$\begin{aligned}
\|u-v\|^2 - \left\| u - K\frac{v}{\|v\|} \right\|^2 &= \|v\|^2 - K^2 - 2\left(1 - \frac{K}{\|v\|}\right)(u,v) \\
&\ge \|v\|^2 - K^2 - 2\left(1 - \frac{K}{\|v\|}\right)\|u\|\|v\| \\
&\ge \|v\|^2 - K^2 - 2\left(1 - \frac{K}{\|v\|}\right)K\|v\| \\
&= \|v\|^2 - 2K\|v\| + K^2 \ge 0.
\end{aligned}$$

(iii) In the case $\|u\| > K$, $\|v\| > K$ we have

$$\|\tilde{B}u - \tilde{B}v\| \le L(K) \left\| K\frac{u}{\|u\|} - K\frac{v}{\|v\|} \right\|$$

and we claim that

$$\left\| K\frac{u}{\|u\|} - K\frac{v}{\|v\|} \right\| \le \|u - v\|.$$

Again, using the Cauchy–Schwartz inequality we have

$$\begin{aligned}
&\|u-v\|^2 - K^2 \left\| \frac{u}{\|u\|} - \frac{v}{\|v\|} \right\|^2 \\
&= \|u\|^2 - 2K^2 + \|v\|^2 + 2\left(\frac{K^2}{\|u\|\|v\|} - 1\right)(u,v) \\
&\ge \|u\|^2 + \|v\|^2 - 2K^2 + 2\left(\frac{K^2}{\|u\|\|v\|} - 1\right)\|u\|\|v\| = (\|u\| - \|v\|)^2 \ge 0 .
\end{aligned}$$

This shows that $\tilde{B}$ is indeed globally Lipschitz.

According to Theorem 2.3.6, assuming $u_0 \in \mathscr{D}(A)$ and $f \in W_1^1(0,T;H)$ for all $T > 0$ the equation

$$u_t + Au + \tilde{B}u = f \tag{2.3.11}$$

has a unique strong solution u. The same theorem guarantees a unique generalized solution on $\mathbb{R}_+$ provided $u_0 \in \overline{\mathscr{D}(A)}$ and $f \in L_1(0,T;H)$ for all $T > 0$.

Let $u_0 \in \mathscr{D}(A)$ be arbitrary. We choose K such that $\|u_0\| < K$ and assume for now that u is a strong solution to (2.3.11). Because $A0 = 0$, the function $w \equiv 0$ is a strong solution to the problem

$$w_t + Aw + \tilde{B}w = B(0), \quad w(0) = 0.$$

Therefore from (2.3.8) we have that

$$\|u(t)\| \le e^{L(K)t} \left[\|u_0\| + \int_0^t e^{-L(K)s} (\|f(s)\| + \|B(0)\|)\, ds \right]$$

for all $t \ge 0$. This inequality also holds for generalized solutions to (2.3.11). Now we recall that $\|u_0\| < K$ and choose $T > 0$ such that

$$V(K,T) \equiv \|u_0\| + \int_0^T e^{-L(K)s} (\|f(s)\| + \|B(0)\|)\, ds < K.$$

Hence, for

$$t \le t^* = \min \left\{ T, \frac{1}{L(K)} \log \frac{K}{V(K,T)} \right\}$$

we obtain $\|u(t)\| \le K$. Consequently, $\tilde{B}(u(t)) = B(u(t))$ for $t \le t^*$ and this means that $u(t)$ solves equation (2.3.5) on that interval.

From what we have just proved it follows that if u is a solution to (2.3.5) on $[0,t^*]$ it can be extended to a solution on the interval $[0,t^*+\delta]$ for some $\delta > 0$. For this we use the same method as outlined above with the initial value t^* and with a larger K. Of course, δ depends on $\|u(t^*)\|$.

Let $[0,t_{\max})$ be the maximal interval of existence of the solution. If $t_{\max} < \infty$, then $\lim_{t \nearrow t_{\max}} \|u(t)\| = \infty$. Otherwise there exists a sequence $t_n \nearrow t_{\max}$ such that $\|u(t_n)\| \le C$. This would allow us to extend u as a solution to (2.3.5) to an interval $[0,t_n+\delta]$ with $\delta > 0$ independent of n. Hence u can be extended beyond $t_{\max}$ which contradicts the construction of $t_{\max}$.

To prove the global existence part we need only note that (2.3.10) allows us to extend a solution inside the interval $[0,T]$ with time steps independent of $T^* < T$.

To prove the uniqueness of generalized solutions to (2.3.5) we note that any two solutions $u_1(t)$ and $u_2(t)$ on a semi-interval $[0,T)$ also solve (2.3.11) on any interval $[0,T']$ with $T' < T$ and $K = 1 + \max_{[0,T']} \{\|u_1(t)\| + \|u_2(t)\|\}$. Thus the uniqueness statement follows from Theorem 2.3.6. □

2.3.9. Remark. Let $u_1(t)$ and $u_2(t)$ be two generalized solutions to problem (2.3.5) with data $\{u_{10}, f_1\}$ and $\{u_{20}, f_2\}$ defined on a common interval $[0,T)$ of the existence. Then under the hypotheses of Theorem 2.3.8 it follows from (2.3.8) that the stability estimate

$$\| u_1(t) - u_2(t) \|_H \le C \left\{ \| u_1(s) - u_2(s) \|_H + \int_s^t \| f_1(\tau) - f_2(\tau) \|_H \, d\tau \right\} \quad (2.3.12)$$

holds for any $0 \le t \le T' < T$, where the constant C depends on T' and the norms $\sup_{[0,T']} \|u_i(t)\|_H$, $i = 1,2$. ■

2.3.10. Remark. Assume that the hypotheses of Theorem 2.3.8 hold. It follows from the argument given above that if $f \in W_1^1(0,T;H)$ and $u_0 \in \overline{\mathscr{D}(A)}$, then for any generalized solution $u(t)$ to problem (2.3.5) on the interval $[0,T]$ there exists a sequence of strong solutions $u_n(t)$ to (2.3.5) with data $\{u_{0n}, f\}$ such that $u_n(t) \to u(t)$ in $C(0,T;H)$. Thus it is not necessary to approximate $f(t)$ when we construct generalized solutions for smooth $f(t)$. ∎

2.4 Second-order abstract equations

In this section we specialize general results of the previous section to more concrete models of the second order in time. We start with a general model that covers almost all of our applications, including boundary forced models arising in nonlinear plate dynamics.

2.4.1 General model

We treat the following second-order abstract equation in a Hilbert space $\mathscr{H}$,

$$\begin{aligned} &Mu_{tt}(t) + \mathscr{A}u(t) + \mathscr{A}Gg(G^*\mathscr{A}u_t(t)) + D_0h(u(t)) \\ &+D_0D_0^*u_t(t) + Du_t(t) = F(u(t), u_t(t)) + p(t), \quad t > 0, \end{aligned} \tag{2.4.1}$$

with the following initial data

$$u|_{t=0} = u_0, \quad u_t|_{t=0} = u_1. \tag{2.4.2}$$

We consider issues such as existence, uniqueness, and energy inequalities associated with (2.4.1). To this end, the following set of assumptions is imposed.

2.4.1. Assumption.

1. $\mathscr{A}$ is a closed, linear positive self-adjoint operator acting on a Hilbert space $\mathscr{H}$ with $\mathscr{D}(\mathscr{A}) \subset \mathscr{H}$. We denote by $|\cdot|$ and $(\cdot,\cdot)$ the norm of $\mathscr{H}$ and the scalar product in $\mathscr{H}$. We use the same symbol to denote the duality pairing between $\mathscr{D}(\mathscr{A}^{1/2})$ and $\mathscr{D}(\mathscr{A}^{1/2})'$.
2. Let V be another Hilbert space such that $\mathscr{D}(\mathscr{A}^{1/2}) \subset V \subset \mathscr{H} \subset V' \subset \mathscr{D}(\mathscr{A}^{1/2})'$, all injections being continuous and dense, $M \in L(V,V')$, the form (Mu,v) is symmetric on V and $(Mu,u) \geq \alpha_0|u|_V^2$, where $\alpha_0 > 0$ and $(\cdot,\cdot)$ is understood as a duality pairing between V and V'. Hence, $M^{-1} \in L(V',V)$. Setting $\bar{M} = M|_H$ with $\mathscr{D}(\bar{M}) = \{u \in V; Mu \in \mathscr{H}\}$ we have $\mathscr{D}(\bar{M}^{1/2}) = V$. In what follows we do not distinguish between M and $\bar{M}$. Moreover, we can assume that $|\cdot|_V = |M^{1/2}\cdot|$ and $(\cdot,\cdot)_V = (M\cdot,\cdot)$.
3. Let U be another Hilbert space and U_0 be a reflexive Banach space such that $U_0 \subseteq U \subseteq U_0'$. We denote by $\langle \cdot,\cdot \rangle$ the scalar product on U and the duality pairing between U_0 and U_0'. We assume that $g : U_0 \mapsto U_0'$ is a continuous mapping

such that $g(0) = 0$ and

$$\langle g(v_1) - g(v_2), v_1 - v_2 \rangle \geq 0 \quad \text{for all } \; v_1, v_2 \in U_0.$$

4. The linear operator $G : U_0' \mapsto \mathscr{H}$ satisfies: $\mathscr{A}^{1/2}G : U_0' \mapsto \mathscr{H}$ is bounded, or equivalently, $G^*\mathscr{A} : \mathscr{D}(\mathscr{A}^{1/2}) \mapsto U_0$ is bounded, where the adjoint operator G^* is defined by the relation $\langle G^*u, v\rangle = (u, Gv)$.
5. The nonlinear operator $F : \mathscr{D}(\mathscr{A}^{1/2}) \times V \mapsto V'$ is locally Lipschitz; that is,
$$|F(u_1,v_1) - F(u_2,v_2)|_{V'} \leq L(K)\left(|\mathscr{A}^{1/2}(u_1-u_2)| + |v_1 - v_2|_V\right) \qquad (2.4.3)$$
for all $(u_i; v_i) \in \mathscr{D}(\mathscr{A}^{1/2}) \times V$ such that $|\mathscr{A}^{1/2}u_i|, |v_i|_V \leq K$.
6. The operator $D : \mathscr{D}(\mathscr{A}^{1/2}) \mapsto [\mathscr{D}(\mathscr{A}^{1/2})]'$ is monotone and hemicontinuous[1] (see Definitions 1.2.1 and 1.2.3).
7. Let Z be a given Hilbert space. The linear operator $D_0 : Z \mapsto \left[\mathscr{D}(\mathscr{A}^{1/2})\right]'$ is bounded and nonlinear function $h(u)$ is Lipschitz from $\mathscr{D}(\mathscr{A}^{1/2})$ into Z; that is
$$|h(u_1) - h(u_2)|_Z \leq L|\mathscr{A}^{1/2}(u_1 - u_2)|, \quad u_1, u_2 \in \mathscr{D}(\mathscr{A}^{1/2}). \qquad (2.4.4)$$
8. $p(t) \in L_1(0,T;V')$ for every $T > 0$.

2.4.2. Remark.

1. Because $\mathscr{A}$ is a positive self-adjoint operator, one can use interpolation theory (see, e.g., [222] or [275]) to construct the scale $\{\mathscr{H}_s : s \in \mathbb{R}\}$ of Hilbert spaces such that (a) $\mathscr{H}_{s_1} \subset \mathscr{H}_{s_2}$ for $s_1 > s_2$; (b) $\mathscr{H}_0 = \mathscr{H}$ and $\mathscr{H}_s = \mathscr{D}(\mathscr{A}^s)$ for $s > 0$; (c) $(\mathscr{H}_s)' = \mathscr{H}_{-s}$. The operator $\mathscr{A}$ can be extended to a family of operators that map $\mathscr{H}_s$ into $\mathscr{H}_{s-1}$ for any $s \in \mathbb{R}$. We denote this extension by the same symbol $\mathscr{A}$. We can also define the powers $\mathscr{A}^\gamma$ for any $\gamma \in \mathbb{R}$ as operators from $\mathscr{H}_s$ into $\mathscr{H}_{s-\gamma}$ possessing the property $\mathscr{A}^{\gamma_1+\gamma_2} = \mathscr{A}^{\gamma_1}\mathscr{A}^{\gamma_2}$. In Assumption 2.4.1(4) and below we understand the operator $\mathscr{A}$ in this (extended) sense.

2. In concrete applications, the term $\mathscr{A}Gg(G^*\mathscr{A}u_t)$ models boundary dissipation. In fact the operator G typically represents a suitable Green's map. A distinct feature of this class of problems is that the composition operator $\mathscr{A}G$ is never in the energy space $\mathscr{H}$. This corresponds to the fact that boundary dissipation problems are very intrinsically unbounded. We also note that in applications below the operator g is a Nemytskij-type operator (see Example 1.2.2) and the assumption that $g : U_0 \mapsto U_0'$ is continuous forces certain growth conditions imposed on the corresponding scalar-valued function $g(x,\xi)$.

3. The interior dissipation is modeled by the damping operator D. We note that the presence of the operator D is redundant in the model (2.4.1). Indeed, this term can be incorporated into the term $\mathscr{A}Gg_0G^*\mathscr{A}$ by taking $G \equiv \mathscr{A}^{-1/2}$ and $g_0 \equiv \mathscr{A}^{-1/2}D\mathscr{A}^{-1/2}$. We can also include $D_0D_0^*u_t$ into the damping term Du_t and, at least formally, $D_0h(u)$ can be included into the nonlinearity $F(u)$. However, we prefer to single out these terms, mostly for the clarity of the exposition. In several

[1] Thus by Proposition 1.2.5 D is maximal monotone.

applications these terms play a rather special role in describing the coupling in the structure. In particular, the term $D_0h(u)$ results from nonlinear boundary conditions and, if $g=0$, an additional component $D_0D_0^*u_t$ is necessary for uniqueness of solutions. As we show in Chapter 4 global Lipschitz condition (2.4.4) for $h(u)$ can be relaxed. ∎

Let us rewrite problem (2.4.1) as a first order equation. In order to accomplish this we introduce the following operator $A : H \mapsto H$, where $H \equiv \mathscr{D}(\mathscr{A}^{1/2}) \times V$, defined by

$$A = \begin{pmatrix} 0 & -I \\ M^{-1}[\mathscr{A} + D_0h(\cdot)] & M^{-1}[\mathscr{A}GgG^*\mathscr{A} + D + D_0D_0^*] \end{pmatrix}, \tag{2.4.5}$$

where $\mathscr{D}(A)$ consists of elements $u = (x;y) \in \mathscr{D}(\mathscr{A}^{1/2}) \times \mathscr{D}(\mathscr{A}^{1/2})$ possessing property

$$\mathscr{A}(x + Gg(G^*\mathscr{A}y)) + D_0h(x) + Dy + D_0D_0^*y \in V'.$$

With the above notation we easily see that the original evolution problem (2.4.1) is equivalent to the equation

$$\begin{cases} \dfrac{d}{dt}U(t) + AU(t) = \begin{pmatrix} 0 \\ M^{-1}F(u(t),u_t(t)) \end{pmatrix} + \begin{pmatrix} 0 \\ M^{-1}p(t) \end{pmatrix} \\ U(0) = U_0 \equiv (u_0;u_1), \end{cases} \tag{2.4.6}$$

where $U(t) = (u(t);u_t(t))$. This structure (2.4.6) of problem (2.4.1) and the considerations of the previous section lead to the following definitions.

2.4.3. Definition. A function $u(t)$ is said to be a *strong solution* to problem (2.4.1) and (2.4.2) on a semi-interval $[0,T)$, iff

- $u(t) \in C([0,T);\mathscr{D}(\mathscr{A}^{1/2})) \cap C^1([0,T);V)$.
- $u \in W_1^1(a,b;\mathscr{D}(\mathscr{A}^{1/2}))$ and $u_t \in W_1^1(a,b;V)$ for any $0 < a < b < T$.
- $\mathscr{A}[u(t) + Gg(G^*\mathscr{A}u_t(t))] + Du_t(t) + D_0h(u(t)) + D_0D_0^*u_t(t) \in V'$ for almost all $t \in [0,T]$.
- Equation (2.4.1) is satisfied in V' for almost all $t \in [0,T]$.
- Initial data (2.4.2) hold.

This function $u(t)$ is a strong solution on the interval $[0,T]$, if, in addition, we have that $(u(t);u_t(t))$ is continuous at $t=T$; that is, $u(t) \in C(0,T;\mathscr{D}(\mathscr{A}^{1/2})) \cap C^1(0,T;V)$. ∎

2.4.4. Definition. A function $u(t)$ is said to be a *generalized solution* to problem (2.4.1) and (2.4.2) on the interval $[0,T]$, iff

- $u(t) \in C(0,T;\mathscr{D}(\mathscr{A}^{1/2})) \cap C^1(0,T;V)$.
- Initial data (2.4.2) hold.

- There exist sequences of functions $\{p_n(t)\} \subset L_1(0,T;V)$ and strong solutions $\{u_n(t)\}$ to problem (2.4.1) and (2.4.2) defined on $[0,T]$ with p_n instead of p and $(u_{0n};u_{1n})$ instead of $(u_0;u_1)$ such that

$$\lim_{n\to\infty} \max_{t\in[0,T]} \left\{ |M^{1/2}(\partial_t u(t) - \partial_t u_n(t))| + |\mathscr{A}^{1/2}(u(t) - u_n(t))| \right\} = 0 \qquad (2.4.7)$$

and

$$\lim_{n\to\infty} \int_0^T |p(t) - p_n(t)|_{V'} dt = 0. \qquad (2.4.8)$$

This function $u(t)$ is a generalized solution on a semi-interval $[0,T)$, if $u(t)$ is a generalized solution on each subinterval $[0,T'] \subset [0,T)$. ■

Our main result in this section is the following theorem.

2.4.5. Theorem. *Under Assumption* 2.4.1 *with the reference to problem* (2.4.1) *and* (2.4.2) *the following statements hold.*

- **Local strong solutions:** *For every* $p \in W_1^1(0,T;V')$ *and* $u_0 \in \mathscr{D}(\mathscr{A}^{1/2})$ *and* $u_1 \in \mathscr{D}(\mathscr{A}^{1/2})$ *such that*

$$\mathscr{A}[u_0 + Gg(G^*\mathscr{A}u_1)] + Du_1 + D_0 h(u_0) + D_0 D_0^* u_1 \in V' \qquad (2.4.9)$$

there exist $t_{\max} > 0$ *and a unique strong solution such that*

$$(u;u_t) \in C([0,t_{\max});\mathscr{D}(\mathscr{A}^{1/2}) \times V),$$

$$(u_t;u_{tt}) \in C_r([0,t_{\max});\mathscr{D}(\mathscr{A}^{1/2}) \times V) \cap L_\infty^{\mathrm{loc}}([0,t_{\max});\mathscr{D}(\mathscr{A}^{1/2}) \times V),$$

$$\mathscr{A}[u + Gg(G^*\mathscr{A}u_t)] + Du_t + D_0 h(u) + D_0 D_0^* u_t \in C_r([0,t_{\max});V').$$

Here and below we denote by C_r *the space of functions that are continuous from the right. Moreover, the function* $t \mapsto (u_t(t);u_{tt}(t))$ *is weakly continuous in* $\mathscr{D}(\mathscr{A}^{1/2}) \times V$.
- **Local generalized solutions:** *Let* $(u_0;u_1) \in \overline{\mathscr{D}(A)} \subseteq \mathscr{D}(\mathscr{A}^{1/2}) \times V$, *where* $\overline{\mathscr{D}(A)}$ *is the closure of* $\mathscr{D}(A)$ *in* $\mathscr{D}(\mathscr{A}^{1/2}) \times V$. *Then there exist* $t_{\max} > 0$ *and a unique generalized solution such that*

$$(u;u_t) \in C([0,t_{\max});\mathscr{D}(\mathscr{A}^{1/2}) \times V).$$

- **Global solutions:** *If, in addition, strong (or generalized) solutions satisfy*

$$\sup_{t\in[0,t_*)} \left\{ |\mathscr{A}^{1/2}u(t)| + |u_t(t)|_V \right\} \le M(t_*,u_0,u_1)$$

for every existence semi-interval $[0,t_*)$, *then the local solutions referred to above are global; that is,* $t_{\max} = \infty$.
- **Stability estimate:** *Let* $u^1(t)$ *and* $u^2(t)$ *be two generalized solutions to problem* (2.4.1) *with data* $\{u_0^1, u_1^1, p_1\}$ *and* $\{u_0^2, u_1^2, p_2\}$ *defined on a common interval*

$[0,T)$ of the existence. Then the estimate

$$|\mathscr{A}^{1/2}[u^1(t)-u^2(t)]|+|u_t^1(t)-u_t^2(t)|_V \tag{2.4.10}$$
$$\leq C_1\left(|\mathscr{A}^{1/2}[u^1(s)-u^2(s)]|+|u_t^1(s)-u_t^2(s)|_V\right)+C_2\int_s^t|p_1(\tau)-p_2(\tau)|_{V'}d\tau$$

holds for any $0\leq t\leq T'<T$, where the constants C_1 and C_2 may depend on T' and $\sup_{t\in[0,T']}\{|\mathscr{A}^{1/2}u^i(t)|+|u_t^i(t)|_V\}$, $i=1,2$.

Proof. The result of Theorem 2.4.5 is a particular case of the theorem proved in [179] by applying the contraction mapping principle. However, the proof can be simplified by directly using Theorem 2.3.8.

In view of this, the conclusion stated in Theorem 2.4.5 follows once we prove

- $A+\lambda I$ is an m-accretive operator on $H=\mathscr{D}(\mathscr{A}^{1/2})\times V$ for some $\lambda\geq 0$.
- $B(U)=-\begin{pmatrix}0\\ M^{-1}F(x,y)\end{pmatrix}$ is locally Lipschitz on H, $U=(x;y)\in H$.

2.4.6. Lemma. *Under Assumption 2.4.1, there exist a constant $\lambda>0$ such that the operator $A+\lambda I$ is m-accretive, where A is defined in (2.4.5).*

The proof of this lemma relies on the following propositions. We begin with the assertion that states the property of accretivity for $A+\lambda I$.

2.4.7. Proposition. *For any $\lambda\geq L^2/2$, where L is a Lipschitz constant associated with h in (2.4.4), the operator $A+\lambda I$ is accretive.*

Proof. We take arbitrary elements $u=(u_1;u_2)$, $w=(w_1;w_2)\in\mathscr{D}(A)$. Let $\xi=(\xi_1;\xi_2)=A(u)$ and $\eta=(\eta_1;\eta_2)=A(w)$. Thus, in particular $\xi_1=-u_2,\eta_1=-w_2$. Moreover

$$\xi_2=M^{-1}[\mathscr{A}u_1+\mathscr{A}G\beta_{u_2}+Du_2+D_0h(u_1)+D_0D_0^*u_2]$$

and

$$\eta_2=M^{-1}[\mathscr{A}w_1+\mathscr{A}G\beta_{w_2}+Dw_2+D_0h(w_1)+D_0D_0^*w_2],$$

where $\beta_{u_2}=g(G^*\mathscr{A}u_2)$ and $\beta_{w_2}=g(G^*\mathscr{A}w_2)$.

Because

$$(A(u)-A(w),u-w)_H=(\xi-\eta,u-w)_H$$
$$=(\mathscr{A}^{1/2}(\xi_1-\eta_1),\mathscr{A}^{1/2}(u_1-w_1))+(M^{1/2}(\xi_2-\eta_2),M^{1/2}(u_2-w_2)),$$

we obtain that

$$(A(u)-A(w),u-w)_H$$
$$=(\mathscr{A}^{1/2}(\xi_1-\eta_1),\mathscr{A}^{1/2}(u_1-w_1))$$
$$+(M^{1/2}M^{-1}(\mathscr{A}(u_1+G\beta_{u_2}-w_1-G\beta_{w_2})+Du_2-Dw_2$$

$$
\begin{aligned}
&+D_0h(u_1)-D_0h(w_1)+D_0D_0^*(u_2-w_2)),M^{1/2}(u_2-w_2))\\
&=(\mathscr{A}^{1/2}(\xi_1-\eta_1),\mathscr{A}^{1/2}(u_1-w_1))+(\mathscr{A}(u_1-w_1+G(\beta_{u_2}-\beta_{w_2}))\\
&\quad+Du_2-Dw_2+D_0(h(u_1)-h(w_1))+D_0D_0^*(u_2-w_2),u_2-w_2)\\
&=(\mathscr{A}G(\beta_{u_2}-\beta_{w_2})+Du_2-Dw_2,u_2-w_2)\\
&\quad+(h(u_1)-h(w_1),D_0^*(u_2-w_2)_Z+|D_0^*(u_2-w_2)|_Z^2\\
&\geq\frac{1}{2}|D_0^*(u_2-w_2)|_Z^2-\frac{1}{2}L^2|\mathscr{A}^{1/2}(u_1-u_2)|^2\geq-\frac{1}{2}L^2|\mathscr{A}^{1/2}(u_1-u_2)|^2,
\end{aligned}
$$

where we have used monotonicity assumptions of D and g and we recall the constant L denotes the Lipschitz constant associated with $h(u)$. The accretivity of $A+\lambda I$ follows now by taking $\lambda\geq\frac{1}{2}L^2$. □

We introduce next the operator $T_\lambda:\mathscr{H}\mapsto\mathscr{H}$ which is defined by

$$
T_\lambda v=\mathscr{A}^{1/2}Gg(G^*\mathscr{A}^{1/2}v)+K_\lambda v, \tag{2.4.11}
$$

where

$$
K_\lambda v=\mathscr{A}^{-1/2}\left(D\mathscr{A}^{-1/2}v+D_0D_0^*\mathscr{A}^{-1/2}v\right)+\frac{1}{2\lambda}v+\mathscr{A}^{-1/2}D_0h\left(\mathscr{A}^{-1/2}\frac{a+v}{\lambda}\right)
$$

and a is an element in $\mathscr{H}$.

2.4.8. Proposition. *For $\lambda>0$ sufficiently large the operator T_λ is m-accretive in $\mathscr{H}$ with the domain $\mathscr{D}(T_\lambda)=\mathscr{H}$.*

Proof. In order to show accretivity of both operators T_λ and K_λ, the argument is very similar to the proof of accretivity of $A+\lambda I$. The details are omitted. We take advantage of the Lipschitz condition satisfied by h which is then calibrated by taking large λ.

As for *maximal accretivity* of T_λ, we argue as follows. First of all we notice that by Assumption 2.4.1(3,4) the operator

$$
\mathscr{A}^{1/2}GgG^*\mathscr{A}^{1/2}:\mathscr{H}\mapsto\mathscr{H}
$$

is accretive. Because the mappings $\mathscr{A}^{1/2}G$, $G^*\mathscr{A}^{1/2}$ and g are continuous, by the first assertion of Proposition 1.2.5 $\mathscr{A}^{1/2}GgG^*\mathscr{A}^{1/2}$ is m-accretive. Therefore the operator T_λ can be viewed as a sum of the m-accretive operator $\mathscr{A}^{1/2}GgG^*\mathscr{A}^{1/2}$ and the operator K_λ. By Proposition 1.2.5(5) it is sufficient to show that K_λ is m-accretive. By Assumption 2.4.1(6) the operator $\mathscr{A}^{-1/2}D\mathscr{A}^{-1/2}$ is m-accretive. Now we note that $K_\lambda-\mathscr{A}^{-1/2}D\mathscr{A}^{-1/2}$ is Lipschitz and accretive on $\mathscr{H}$ for $\lambda>0$ large enough. Therefore by Proposition 2.2.3(1) K_λ is m-accretive, as desired. □

2.4.9. Proposition. *$A+\lambda I$ is maximal on $H=\mathscr{D}(\mathscr{A}^{1/2})\times V$, for λ sufficiently large.*

Proof. By Theorem 2.2.2 and Proposition 2.4.7 it suffices to show that

$$R(A+\lambda I)=H \quad \text{for some } \lambda>0.$$

The above entails to solving the following system of equations: given $f_0 \in \mathscr{D}(\mathscr{A}^{1/2})$, $f_1 \in V$ find $(x;y) \in \mathscr{D}(A)$ such that

$$\begin{aligned} -y+\lambda x &= f_0, \\ \mathscr{A}x+D_0h(x)+\mathscr{A}Gg(G^*\mathscr{A}y)+D_0D_0^*y+Dy+\lambda My &= Mf_1. \end{aligned} \quad (2.4.12)$$

Eliminating x yields

$$\begin{aligned} &\frac{1}{\lambda}\mathscr{A}y+D_0h\left(\frac{f_0+y}{\lambda}\right)+\mathscr{A}Gg(G^*\mathscr{A}y) \\ &+D_0D_0^*y+Dy+\lambda My=Mf_1-\frac{1}{\lambda}\mathscr{A}f_0, \end{aligned} \quad (2.4.13)$$

which can be written as

$$\frac{1}{2\lambda}\mathscr{A}y+\mathscr{A}^{1/2}T_\lambda\mathscr{A}^{1/2}y+\lambda My=Mf_1-\frac{1}{\lambda}\mathscr{A}f_0 \in [\mathscr{D}(\mathscr{A}^{1/2})]',$$

where T_λ is given by (2.4.11) with $a=\mathscr{A}^{1/2}f_0$. If we denote $v=\mathscr{A}^{1/2}y$, then we obtain the relation

$$S_\lambda v \equiv \frac{1}{2\lambda}v+T_\lambda v+\lambda\mathscr{A}^{-1/2}M\mathscr{A}^{-1/2}v=\mathscr{A}^{-1/2}\left(Mf_1-\frac{1}{\lambda}\mathscr{A}f_0\right)\in\mathscr{H}. \quad (2.4.14)$$

By Proposition 2.4.8 T_λ is m-accretive in $\mathscr{H}$. Because $\mathscr{D}(\mathscr{A}^{1/2})\subset V$, it is also clear that $\mathscr{A}^{-1/2}M\mathscr{A}^{-1/2}$ is a bounded linear positive operator in $\mathscr{H}$. Consequently by the first assertion of Proposition 2.2.3 the operator S_λ is m-accretive and coercive in $\mathscr{H}$. Thus, by Proposition 2.2.6 we conclude $R(S_\lambda)=\mathscr{H}$. Therefore there exists $v\in\mathscr{H}$ which solves (2.4.14). Consequently $y=\mathscr{A}^{-1/2}v\in\mathscr{D}(\mathscr{A}^{1/2})$ is solution to (2.4.13) for the variable $y\in\mathscr{D}(\mathscr{A}^{1/2})$. Going back to the first equation in (2.4.12) we obtain $x\in\mathscr{D}(\mathscr{A}^{1/2})$. It is now straightforward to verify that $(x;y)\in\mathscr{D}(A)$. Indeed, the above follows from the structure of the second equation in (2.4.12) and the definition of the domain of the operator A. The proof of the maximality property has been completed. □

The result stated in Lemma 2.4.6 follows from Propositions 2.4.7 and 2.4.9.

To complete the proof of the Theorem 2.4.5 we appeal to Theorem 2.3.8 after noticing that the operator

$$B(U)=-\begin{pmatrix} 0 \\ M^{-1}F(x,y)\end{pmatrix}, \quad U=(x;y)\in H,$$

is locally Lipschitz on H. Indeed, the above follows from the local Lipschitz property of $M^{-1/2}F$ acting between $\mathscr{D}(\mathscr{A}^{1/2})$ and $\mathscr{H}$ and the definition of the norm in V. All the statements in Theorem 2.4.5 are direct consequences of statements in

Theorem 2.3.8 specialized to the system in question. The estimate (2.4.10) easily follows from relation (2.3.12) given in Remark 2.3.9. □

2.4.10. Remark. Assume that mappings D and D_0 have additional regularity properties. Namely we assume that

$$D : \mathscr{D}(\mathscr{A}^{1/2}) \mapsto V' \text{ and } D_0 : Z \mapsto V'$$

are continuous mappings. In this case the compatibility condition (2.4.9) can be written in the form

$$u_0 + Gg(G^*\mathscr{A}u_1) \in W \equiv \left\{ v \in \mathscr{D}(\mathscr{A}^{1/2}) : \mathscr{A}v \in V' \right\} \tag{2.4.15}$$

and the strong solution $u(t)$ possesses the property

$$\mathscr{A}[u + Gg(G^*\mathscr{A}u_t)] \in C_r([0,t_{\max});V').$$

We note that (2.4.15) holds if $u_0 \in W$ and $u_1 \in \mathscr{D}(\mathscr{A}^{1/2}) \cap \operatorname{Ker}\,[G^*\mathscr{A}]$. Consequently, if we assume that the set $\mathscr{D}(\mathscr{A}^{1/2}) \cap \operatorname{Ker}\,[G^*\mathscr{A}]$ is dense in V (which holds for many applications of boundary operators), then the generalized solutions in the second part of Theorem 2.4.5 exist for any initial data $(u_0;u_1)$ from $H = \mathscr{D}(\mathscr{A}^{1/2}) \times V$. The point is that the closure of $\mathscr{D}(A)$ is dense in H in this case. ■

The following assertion gives sufficient conditions for a generalized solution to satisfy (2.4.1) in a weak (variational) sense.

2.4.11. Proposition. *Additionally to Assumption* 2.4.1 *assume that the mappings*

$$v(t) \mapsto \mathscr{A}Gg(G^*\mathscr{A}v(t)), \quad v(t) \mapsto Dv(t) \text{ and } v(t) \mapsto D_0D_0^*v(t) \tag{2.4.16}$$

are continuous from $C(0,T;V)$ into $L_1([0,T];[\mathscr{D}(\mathscr{A}^{1/2})]')$ endowed with weak topology. Then any generalized solution to problem (2.4.1) *and* (2.4.2) *is also weak; that is, the relation*

$$\begin{aligned}(Mu_t(t),\phi) = (Mu_1,\phi) - \int_0^t &\big[(\mathscr{A}u(\tau),\phi) + (\mathscr{A}Gg(G^*\mathscr{A}u_t(\tau)),\phi)\\ &+(Du_t(\tau) + D_0h(u(\tau)) + D_0D_0^*u_t(\tau),\phi)\\ &-(F(u(\tau),u_t(\tau)) + p(\tau),\phi)\big]d\tau\end{aligned} \tag{2.4.17}$$

holds for every $\phi \in \mathscr{D}(\mathscr{A}^{1/2})$ and for almost all $t \in [0,T]$.

Proof. Because any strong solution satisfies (2.4.17), we can use relation (2.4.7) and continuity of the mappings (2.4.16) to pass with the limit in (2.4.17). □

The first assumption required by (2.4.16) is clearly restrictive. In typical applications to boundary damping problems, the boundedness of the operator $G^*\mathscr{A}$ on V

is too strong an assumption. In order to offset this we impose another—structural type—condition on function g that allows us to model a large variety of boundary dissipative mechanisms. To this end we further specify the nature of the monotone mapping $g(u)$.

2.4.12. Assumption. Let $g(z) = \partial\Phi(z)$ where $\partial\Phi$ is a gradient (Gateaux differential) of a convex and continuous function $\Phi : \mathscr{U}_0 \subset L_2(0,T;U) \to \mathbb{R}_+$ with $\Phi(0) = 0$, where $\mathscr{U}_0$ is a reflexive Banach space such that $\mathscr{U}_0 \subset L_2(0,T;U) \subset \mathscr{U}_0'$ with dense and continuous injections. We denote by $\langle\cdot,\cdot\rangle$ the corresponding duality pairing between $\mathscr{U}_0$ and $\mathscr{U}_0'$ and assume that

- **Boundedness:** $g = \partial\Phi : \mathscr{U}_0 \to \mathscr{U}_0'$ is bounded and continuous.
- **Coercivity:** $|z|_{\mathscr{U}_0} \le C(\Phi(z))$ where $C(s)$ is a locally bounded function on $\mathbb{R}_+$.

By convexity of Φ we have that

$$\Phi(v) \le \Phi(u) + \langle\partial\Phi(v), v-u\rangle \quad \text{for all } v,u \in \mathscr{U}_0.$$

This implies that $g = \partial\Phi$ is a monotone operator. Thus Proposition 1.2.5 $g = \partial\Phi$ is continuous and m-monotone.

2.4.13. Remark. In typical applications to boundary damping the space $\mathscr{U}_0$ is identified with a suitable L_q-space of the type $L_q(0,T;U_0)$. For instance, let g be the Nemytskij operator $u(x) \mapsto |u(x)|^p u(x)$ from $U_0 = L_{p+1}(\Gamma)$ into $U_0' = L_{(p+1)/p}(\Gamma)$. Then $\mathscr{U}_0 = L_{p+1}([0,T]\times\Gamma)$ and $\Phi(v) = (p+1)^{-1}\int_0^T\int_\Gamma |v(x,t)|^{p+1}dxdt$. ∎

2.4.14. Proposition. *Additionally to Assumption* 2.4.1 *assume that Assumption* 2.4.12 *is satisfied and*

$$v(t) \mapsto Dv(t) \;\; \text{and} \;\; v(t) \mapsto D_0D_0^*v(t) \tag{2.4.18}$$

are continuous from $C(0,T;V)$ *into* $L_1(0,T;[\mathscr{D}(\mathscr{A}^{1/2})]')$ *endowed with weak topology. Then:*

- *any generalized solution to problem* (2.4.1) *and* (2.4.2) *is also weak; that is, the relation* (2.4.17) *holds for every* $\phi \in \mathscr{D}(\mathscr{A}^{1/2})$ *and for almost all* $t \in [0,T]$.
- *Each generalized solution satisfies* $G^*\mathscr{A}u_t \in \mathscr{U}_0$.
- *The following stability estimate*

$$\begin{aligned} &|\mathscr{A}^{1/2}[u^1(t)-u^2(t)]|^2 + |u_t^1(t)-u_t^2(t)|_V^2 \\ &\quad + \int_s^t \langle g(G^*\mathscr{A}u_t^1) - g(G^*\mathscr{A}u_t^2), G^*\mathscr{A}(u_t^1-u_t^2)\rangle d\tau \\ &\le C_1|\mathscr{A}^{1/2}[u^1(s)-u^2(s)]|^2 + |u_t^1(s)-u_t^2(s)|_V^2 + C_2\left[\int_s^t |p_1(\tau)-p_2(\tau)|_{V'}d\tau\right]^2 \end{aligned} \tag{2.4.19}$$

holds for any two generalized solutions u^1 *and* u^2*, where* $0 \le s < t \le T$ *and the constant* C_1 *and* C_2 *depend on* T *and the energy bounds for solutions* u^1 *and* u^2.

Proof. Taking into consideration the arguments in Proposition 2.4.11, the first statement in Proposition 2.4.14 is proved as soon as we justify passage with the limit on strong solutions

$$\int_0^t \langle \mathscr{A} G g(G^* \mathscr{A} u_t^n(\tau)), \phi \rangle d\tau \to \int_0^t \langle g(G^* \mathscr{A} u_t(\tau)), G^* \mathscr{A} \phi \rangle d\tau \quad \text{as } n \to \infty \tag{2.4.20}$$

almost everywhere in t and for all $\phi \in \mathscr{D}(\mathscr{A}^{1/2})$. For this we make use of maximal monotonicity of the operator $\partial\Phi : \mathscr{U}_0 \to \mathscr{U}_0'$. To wit, we first note that the stability inequality in (2.4.19) is valid for any couple u^1 and u^2 of strong solutions. Therefore for any sequence $\{u^n\}$ of strong solutions corresponding to the forcing terms $p^n \to p$ in $L_1((0,T),V')$ and initial data $(u^n(0); u_t^n(0)) \to (u_0; u_1)$ in H we have that

$$\int_0^T \langle g(G^* \mathscr{A} u_t^n) - g(G^* \mathscr{A} u_t^m), G^* \mathscr{A}(u_t^n - u_t^m) \rangle dt \tag{2.4.21}$$
$$\leq C_T \left(|\mathscr{A}^{1/2}(u^n(0) - u^m(0))|^+ |u_t^n(0) - u_t^m(0)|_V^2 + + |p^n - p^m|^2_{L_1(0,T;V')} \right)$$

for every $m, n \geq 0$. In particular,

$$\int_0^T \langle g(G^* \mathscr{A} u_t^n), G^* \mathscr{A}(u_t^n) \rangle dt \leq C_T \left(|\mathscr{A}^{1/2} u(0)|, |u_t(0)|_V, |p|_{L_1(0,T;V')} \right), \tag{2.4.22}$$

which, by convexity of Φ implies

$$\begin{aligned} \Phi(G^* \mathscr{A} u_t^n)) &\leq \int_0^T \langle \partial\Phi(G^* \mathscr{A} u_t^n), G^* \mathscr{A} u_t^n \rangle dt \\ &\leq C_T \left(|\mathscr{A}^{1/2} u(0)|, |u_t(0)|_V, |p|_{L_1(0,T;V')} \right) = C_T. \end{aligned}$$

From Assumption 2.4.12 we infer $|G^* \mathscr{A} u_t^n|_{\mathscr{U}_0} \leq C_T$ and $|\partial\Phi(G^* \mathscr{A} u_t^n)|_{\mathscr{U}_0'} \leq C_T$ with the corresponding convergence:

$$u_t^n \to u_t \ \text{ in } L_2(0,T,V),$$
$$G^* \mathscr{A} u_t^n \to G^* \mathscr{A} u_t, \text{ weakly in } \mathscr{U}_0, \quad \partial\Phi(G^* \mathscr{A} u_t^n) \to l, \text{ weakly in } \mathscr{U}_0'. \tag{2.4.23}$$

We identify the limit l by proving that $l = \partial\Phi(G^* \mathscr{A} u_t)$. For this we notice first that the stability estimate in (2.4.21) implies

$$\lim_{n,m\to\infty} \int_0^T \langle \partial\Phi(G^* \mathscr{A} u_t^n) - \partial\Phi(G^* \mathscr{A} u_t^m), G^* \mathscr{A}(u_t^n - u_t^m) \rangle dt \to 0. \tag{2.4.24}$$

Thus, in conclusion, the equality $l = \partial\Phi(G^* \mathscr{A} u_t)$ follows now from (i) m-monotonicity of $\partial\Phi(z) : \mathscr{U}_0 \to \mathscr{U}_0'$; (ii) weak convergence in (2.4.23); and (iii) condition in (2.4.24) by application of Proposition 1.2.6. Thus

$$\partial\Phi(G^* \mathscr{A} u_t^n) \to \partial\Phi(G^* \mathscr{A} u_t), \ \text{weakly in } \mathscr{U}_0',$$

$$\int_0^T \langle \partial\Phi(G^*\mathscr{A}u_t^n), G^*\mathscr{A}u_t^n\rangle dt \to \int_0^T \langle \partial\Phi(G^*\mathscr{A}u_t), G^*\mathscr{A}u_t\rangle dt. \quad (2.4.25)$$

Because $G^*\mathscr{A}$ is bounded $\mathscr{D}(\mathscr{A}^{1/2}) \to U$, the first assertion in (2.4.25) allows us to conclude (2.4.20). This, in turn, justifies the weak form of variational equality in (2.4.17).

The second assertion in (2.4.25) allows us to pass with the limit on the stability estimate with the retention of the term

$$\int_0^T \langle \partial\Phi(G^*\mathscr{A}u_t), G^*\mathscr{A}u_t\rangle dt = \int_0^T \langle g(G^*\mathscr{A}u_t), G^*\mathscr{A}u_t\rangle dt$$

and obtain (2.4.19) for generalized solutions. □

2.4.2 Simplified nonlinear model

In this section we consider the following special case ($h=0, D_0=0$) of problem (2.4.1),

$$\begin{cases} Mu_{tt}(t) + \mathscr{A}u(t) + \mathscr{A}Gg(G^*\mathscr{A}u_t) + Du_t(t) = F(u(t), u_t(t)) + p(t), \\ u|_{t=0} = u_0, \quad u_t|_{t=0} = u_1, \end{cases} \quad (2.4.26)$$

Here our hypotheses concerning M, $\mathscr{A}$, and $p(t)$ are the same as in Assumption 2.4.1. The hypotheses concerning F are stronger in comparison than the ones given in Assumption 2.4.1(5). The case when the hypothesis on g also includes Assumption 2.4.12 is also considered. Additional conditions imposed on the nonlinear map F are aimed at guaranteeing global solvability of equations. For the reader's convenience we collect below all the hypotheses used in this section.

2.4.15. Assumption.

1. $\mathscr{A}$ is a closed, linear positive self-adjoint operator acting on a Hilbert space $\mathscr{H}$ with $\mathscr{D}(\mathscr{A}) \subset \mathscr{H}$. As above denote by $|\cdot|$ the norm of $\mathscr{H}$ and by $(\cdot,\cdot)$ the corresponding scalar product (and also the corresponding duality pairing).
2. Let $M : V \mapsto V'$ be a linear operator in a Hilbert space such that $\mathscr{D}(\mathscr{A}^{1/2}) \subset V \subset \mathscr{H} \subset V' \subset \mathscr{D}(\mathscr{A}^{1/2})'$, all injections being continuous and dense. Moreover the form (Mu, v) is symmetric on V and $(Mu,u) \geq \alpha_0|u|_V^2$, where $\alpha_0 > 0$ and $(\cdot,\cdot)$ is understood as a duality pairing between V and V'.
3. The nonlinear operator $F : \mathscr{D}(\mathscr{A}^{1/2}) \times V \mapsto V'$ is locally Lipschitz, i.e.

$$|F(u_1,v_1) - F(u_2,v_2)|_{V'} \leq L(K)\left(|\mathscr{A}^{1/2}(u_1-u_2)| + |v_1-v_2|_V\right) \quad (2.4.27)$$

for all $(u_i; v_i) \in \mathscr{D}(\mathscr{A}^{1/2}) \times V$ such that $|\mathscr{A}^{1/2}u_i|, |v_i|_V \leq K$. Moreover, we assume that F has the form

$$F(u,v) = -\Pi'(u) + F^*(u,v), \tag{2.4.28}$$

where $\Pi(u)$ is a C^1-functional on $\mathscr{D}(\mathscr{A}^{1/2})$, $\Pi'(u)$ stands for Fréchet derivative whose value on an element w is given by the inner product $(\Pi'(u),w)$ for $u,w \in \mathscr{D}(\mathscr{A}^{1/2})$, and $F^*(u,v)$ is a (nonlinear) Lipschitz mapping; that is, there exists $L^* > 0$ such that

$$|F^*(u_1,v_1) - F^*(u_2,v_2)|_{V'} \le L^* \left(|\mathscr{A}^{1/2}(u_1-u_2)| + |v_1 - v_2|_V \right) \tag{2.4.29}$$

for all $(u_i;v_i) \in \mathscr{D}(\mathscr{A}^{1/2}) \times V$. We also assume that

$$\Pi(u) = \Pi_0(u) + \Pi_1(u),$$

where $\Pi_0(u) \ge 0$, $\Pi_0(u)$ is bounded on bounded sets from $\mathscr{D}(\mathscr{A}^{1/2})$ and $\Pi_1(u)$ possesses the property[2]: $\forall \eta > 0, \exists C_\eta > 0$ such that

$$|\Pi_1(u)| \le \eta \cdot \left(|\mathscr{A}^{1/2}u|^2 + \Pi_0(u) \right) + C_\eta\,, \quad u \in \mathscr{D}(\mathscr{A}^{1/2})\,. \tag{2.4.30}$$

4. The operator $D : \mathscr{D}(\mathscr{A}^{1/2}) \mapsto \left[\mathscr{D}(\mathscr{A}^{1/2})\right]'$ is a monotone hemicontinuous operator:
$$(Du - Dv, u-v) \ge 0 \quad \text{for all} \quad u,v \in \mathscr{D}(\mathscr{A}^{1/2}) \tag{2.4.31}$$
and $\lambda \mapsto (D(u+\lambda v),v)$ is a continuous function from $\mathbb{R}$ into itself.
5. The operator G is the same as in point 4 of Assumption 2.4.1; that is, $G : U_0' \mapsto \mathscr{H}$ is linear operator such that $G^*\mathscr{A} : \mathscr{D}(\mathscr{A}^{\frac{1}{2}}) \mapsto U_0$ is bounded (the case $G \equiv 0$ is allowed). The function g satisfies Assumption 2.4.1(3), i.e. $g : U_0 \mapsto U_0'$ is a continuous mapping such that $g(0) = 0$ and $\langle g(v_1) - g(v_2), v_1 - v_2 \rangle \ge 0$ for all $v_1, v_2 \in U_0$. Here $U_0 \subset U \subset U_0'$ is a Gelfand triple with U being pivot.

Model (2.4.26)is used below to study von Karman evolution equations with both internal and boundary damping. The hypotheses above are motivated by the structure of the potential energy of the plate (cf. Remark 1.5.12). In particular, relation (2.4.30) corresponds to estimate (1.5.16). The presence of the term F^* in (2.4.28) exemplifies the operator L in (1.5.1). In some applications F^* may be interpreted as a nonconservative force. We refer to [75] for a discussion of problems in the framework of (2.4.26), however, under another set of hypotheses concerning F and for $g \equiv 0$, $G \equiv 0$.

Under the above assumption the following well-posedness result for problem (2.4.26) is established.

2.4.16. Theorem. *Let $T > 0$ be arbitrary. Under Assumption 2.4.15 the following statements hold.*

- **Strong solutions:** *For every $p \in W_1^1(0,T;V')$ and $(u_0;u_1) \in \mathscr{D}(\mathscr{A}^{1/2}) \times \mathscr{D}(\mathscr{A}^{1/2})$ such that $\mathscr{A}(u_0 + Gg(G^*\mathscr{A}u_1)) + D(u_1) \in V'$, there exists a unique strong solu-*

[2] Relation (2.4.30) reflects the fact that potential energy $\Pi(u)$ is bounded from below.

tion on the interval $[0,T]$ *such that*

$$(u;u_t) \in C(0,T;\mathscr{D}(\mathscr{A}^{1/2}) \times V)$$

and the function $t \mapsto (u_t(t);u_{tt}(t))$ *is weakly continuous in* $\mathscr{D}(\mathscr{A}^{1/2}) \times V$. *Moreover,*

$$(u_t;u_{tt}) \in C_r(0,T;\mathscr{D}(\mathscr{A}^{1/2}) \times V) \cap L_\infty(0,T;\mathscr{D}(\mathscr{A}^{1/2}) \times V), \tag{2.4.32}$$

and

$$\mathscr{A}(u(t) + Gg(G^*\mathscr{A}u_t(t)) + Du_t(t) \in C_r(0,T;V') \cap L_\infty(0,T;V'). \tag{2.4.33}$$

As above C_r *here denotes the space of right-continuous functions. This solution satisfies the energy relation*

$$\begin{aligned} &\mathscr{E}(u(t),u_t(t)) + \int_0^t [(Du_t(\tau),u_t(\tau)) + \langle g(G^*\mathscr{A}u_t(\tau)), G^*\mathscr{A}u_t(\tau)\rangle]\,d\tau \\ &= \mathscr{E}(u_0,u_1) + \int_0^t (F^*(u(\tau),u_t(\tau)) + p(\tau),u_t(\tau))d\tau\,, \end{aligned} \tag{2.4.34}$$

where $\mathscr{E}(u_0,u_1) = E(u_0,u_1) + \Pi_1(u_0)$ *with*

$$E(u_0,u_1) = E_0(u_0,u_1) + \Pi_0(u_0) \equiv \frac{1}{2}((Mu_1,u_1) + (\mathscr{A}u_0,u_0)) + \Pi_0(u_0)\,.$$

- **Generalized solutions:** *Assume that the set*

$$\mathscr{L} = \left\{(u_0;u_1) \in \mathscr{D}(\mathscr{A}^{1/2}) \times \mathscr{D}(\mathscr{A}^{1/2}) : \mathscr{A}[u_0 + Gg(G^*\mathscr{A}u_1)] + Du_1 \in V'\right\} \tag{2.4.35}$$

is dense in the space $\mathscr{D}(\mathscr{A}^{1/2}) \times V$ *(see also Remark* 2.4.17 *below). Then for every* $p \in L_1(0,T;V')$, $u_0 \in \mathscr{D}(\mathscr{A}^{1/2})$, *and* $u_1 \in V$ *there exists a unique generalized solution* $u(t)$ *such that* $(u;u_t) \in C(0,T;\mathscr{D}(\mathscr{A}^{1/2}) \times V)$.

Both strong and generalized solutions satisfy the energy inequality

$$\mathscr{E}(u(t),u_t(t)) \le \mathscr{E}(u_0,u_1) + \int_0^t (F^*(u(\tau),u_t(\tau)) + p(\tau),u_t(\tau))d\tau \tag{2.4.36}$$

and also the estimate

$$E(u(t),u_t(t)) \le c_0\left(1 + E(u_0,u_1) + \left[\int_0^t |p(\tau)|_{V'}d\tau\right]^2\right)e^{c_1 t} \tag{2.4.37}$$

for all $t \in [0,T]$ *with some constants* c_0 *and* c_1.

If we assume in addition that Assumption 2.4.12 *is in force, then we have that* $G^*\mathscr{A}u_t(t) \in \mathscr{U}_0$. *In this case both strong and generalized solutions satisfy the energy inequality*

$$\mathscr{E}(u(t),u_t(t))+\int_0^t \langle g(G^*\mathscr{A}u_t(\tau)),G^*\mathscr{A}u_t(\tau)\rangle d\tau$$
$$\leq \mathscr{E}(u_0,u_1)+\int_0^t (F^*(u(\tau),u_t(\tau))+p(\tau),u_t(\tau))d\tau\,. \qquad (2.4.38)$$

2.4.17. Remark. One can see that the following requirement

$$\mathscr{W}\equiv \mathrm{Ker}\;[G^*\mathscr{A}]\cap\left\{v\in\mathscr{D}(\mathscr{A}^{1/2})\,:\,D(v)\in V'\right\}\quad \text{is dense in } V. \qquad (2.4.39)$$

is a sufficient condition for the density of the set $\mathscr{L}$ given by (2.4.35). Indeed, under condition (2.4.39) the set $\mathscr{D}(\mathscr{A})\times\mathscr{W}$ is dense in $\mathscr{D}(\mathscr{A}^{1/2})\times V$ and belongs to $\mathscr{L}$. Thus in Theorem 2.4.16 we can replace the density of the set $\mathscr{L}$ by the condition in (2.4.39). This latter condition is convenient in cases when the operator G is Green's map corresponding to some boundary value problems (see Chapters 3 and 4). ■

Proof. By Theorem 2.4.5 strong and generalized solutions exist on some semi-interval $[0,t_{max})$.

Let $u(t)$ be a strong solution. In this case (2.4.26) holds as an equality in V' for almost all $t\in[0,t_{\max})$. Therefore, multiplying (2.4.26) by u_t in $\mathscr{H}$ after some standard calculations we obtain

$$E_0(u(t),u_t(t))+\int_0^t\left[(Du_t(\tau),u_t(\tau))+\langle g(G^*\mathscr{A}u_t(\tau)),G^*\mathscr{A}u_t(\tau)\rangle\right]d\tau$$
$$=E_0(u_0,u_1)+\int_0^t (F(u(\tau),u_t(\tau))+p(\tau),u_t(\tau))d\tau \qquad (2.4.40)$$

for $t\in[0,t_{\max})$. It is easy to see from (2.4.28) that

$$(F(u(t),u_t(t)),u_t(t))=-\frac{d}{dt}\Pi(u(t))+(F^*(u(t),u_t(t)),u_t(t)),\quad t\in[0,t_{\max}).$$

Therefore from (2.4.40) we obtain energy relation (2.4.34) valid for the strong solution $u(t)$ and for $t\in[0,t_{\max})$. We note that (2.4.30), with appropriate choice of η implies that there exists a constant $c>0$ such that

$$\frac{1}{2}E(u_0,u_1)-c\leq\mathscr{E}(u_0,u_1)\leq 2E(u_0,u_1)+c,\quad u_0\in\mathscr{D}(\mathscr{A}^{1/2}),\quad u_1\in V\,. \quad (2.4.41)$$

Denote

$$\widetilde{E}(t)=1+\max\left\{E(u(\tau),u_t(\tau))\,:\,\tau\in[0,t]\right\}.$$

Using (2.4.34) and (2.4.41) we obtain that

$$\widetilde{E}(t)\leq c(1+E(u_0,u_1))+2\int_0^t |(F^*(u(\tau),u_t(\tau))+p(\tau),u_t(\tau))|d\tau$$

with some constant $c>0$. From (2.4.29) we also have that

$$(F^*(u(\tau),u_t(\tau)),u_t(\tau))\leq c_0E(u(\tau),u_t(\tau))+c_1\leq c_2\widetilde{E}(\tau)$$

with some constants c_i. Therefore

$$\widetilde{E}(t) \leq c_1\,(1+E(u_0,u_1)) + c_2\int_0^t \widetilde{E}(\tau))d\tau + 2\left(2\widetilde{E}(t)\right)^{1/2}\cdot\int_0^t |p(\tau)|_{V'}d\tau.$$

This implies

$$\widetilde{E}(t) \leq c_0\left(1+E(u_0,u_1)+\left[\int_0^t |p(\tau)|_{V'}d\tau\right]^2\right) + c_1\int_0^t \widetilde{E}(\tau)d\tau$$

with some constants c_0 and c_1. Hence by Gronwall's lemma we have

$$\widetilde{E}(t) \leq c_0\left(1+E(u_0,u_1)+\left[\int_0^t |p(\tau)|_{V'}d\tau\right]^2\right)e^{c_1 t}, \quad t\in[0,t_{\max}).$$

After rescaling of constants we obtain

$$E(u(t),u_t(t)) \leq c_0\left(1+E(u_0,u_1)+\left[\int_0^t |p(\tau)|_{V'}d\tau\right]^2\right)e^{c_1 t}, \quad t\in[0,t_{\max}),$$

with some constants c_0 and c_1. Using (2.4.7) and and the continuity of $E(u,u_t)$ it is easy to find that the last relation is also true for generalized solutions. Thus the third part of Theorem 2.4.5 implies the global existence (and uniqueness) of both strong and generalized solutions.

We emphasize that generalized solutions exist for *all* initial data $(u_0;u_1)\in\mathscr{D}(\mathscr{A}^{1/2})\times V$. This is so because the set $\mathscr{L}$, given by (2.4.35), coincides with the domain $\mathscr{D}(A)$ of the corresponding accretive operator (see (2.4.5)) in the proof of Theorem 2.4.5.

The inequality in (2.4.36) is obvious for strong solutions. For generalized solutions it easily follows via the limiting process.

If Assumption 2.4.12 is in force, then the inclusion $G^*\mathscr{A}u_t(t)\in\mathscr{U}_0$ and the energy inequality in (2.4.38) follow from the argument given in the proof of Proposition 2.4.14. □

2.4.18. Remark. It follows from (2.4.10) that for any two generalized solutions $u^1(t)$ and $u^2(t)$ to problem (2.4.26) with initial data $(u_0^i;u_1^i)$ and with the same function $p(t)$ the estimate

$$|\mathscr{A}^{1/2}[u^1(t)-u^2(t)]|+|u_t^1(t)-u_t^2(t)|_V \leq C\left(|\mathscr{A}^{1/2}[u_0^1-u_0^2]|+|u_1^1-u_1^2|_V\right)$$

holds for any $0<t<T$, where the constants C may depend on T and the energies $E_0(u_0^i,u_1^i)$, $i=1,2$. In particular, if $p\in W_1^1(0,T;V')$, then for any generalized solution $u(t)$ we can choose a sequence of initial data and the corresponding strong solutions $u_n(t)$ such that

$$\lim_{n\to\infty} \sup_{t\in[0,T]} \left\{ |\mathscr{A}^{1/2}[u(t) - u_n(t)]| + |u_t(t) - u_{nt}(t)|_V \right\} = 0.$$

Thus in this case to construct a generalized solution we can approximate the initial data only (cf. Remark 2.3.10). ∎

2.4.19. Remark. In the case when $F^* \equiv 0$ and $p(t) \equiv p$ (autonomous case) the full energy $\mathscr{E}(u(t), u_t(t)) - (p, u(t))$ is not increasing along trajectories. However, in general, the dynamical system may not possess this property, as is most often the case when the term F^* is present in the model. ∎

The following assertion shows that under some additional conditions the energy inequality (2.4.38) valid for generalized solutions can be stated in a stronger form.

2.4.20. Proposition. *In addition to Assumptions* 2.4.15 *and* 2.4.12, *assume that there exists a convex function* $\varphi : V \mapsto \mathbb{R} \cup \{+\infty\}$ *such that*

- φ *is lower semicontinuous on* V; *that is,*

$$\{v_n \to v \ \ in\ V\} \Longrightarrow \liminf_{n\to\infty} \varphi(v_n) \geq \varphi(v);$$

- $\varphi(v) \leq (Dv, v)$ *for any* $v \in \mathscr{D}(\mathscr{A}^{1/2})$.

Then any generalized solution $u(t)$ *to problem* (2.4.26) *satisfies the energy inequality of the form*

$$\begin{aligned} &\mathscr{E}(u(t), u_t(t)) + \int_s^t [\varphi(u_t(\tau)) + \langle g(G^*\mathscr{A}u_t(\tau)), G^*\mathscr{A}u_t(\tau)\rangle] d\tau \\ &\leq \mathscr{E}(u(s), u_t(s)) + \int_s^t (F^*(u(\tau), u_t(\tau)) + p(\tau), u_t(\tau)) d\tau \qquad (2.4.42) \end{aligned}$$

for all $0 \leq s \leq t \leq T$.

Proof. Let $u(t)$ be a generalized solution and $\{u^n(t)\}$ be a sequence of strong solutions such that (2.4.7) holds. We use the energy relation

$$\begin{aligned} &\mathscr{E}(u^n(t), u^n_t(t)) + \int_s^t [(Du^n_t(\tau), u^n_t(\tau)) + \langle g(G^*\mathscr{A}u^n_t(\tau)), G^*\mathscr{A}u^n_t(\tau)\rangle] d\tau \\ &= \mathscr{E}(u^n(s), u^n_t(s)) + \int_s^t (F^*(u^n(\tau), u^n_t(\tau)) + p^n(\tau), u^n_t(\tau)) d\tau. \qquad (2.4.43) \end{aligned}$$

Because the energy $\mathscr{E}(u, u_t)$ is continuous on $\mathscr{D}(\mathscr{A}^{1/2}) \times V$, relation (2.4.7) implies that

$$\lim_{n\to\infty} \mathscr{E}(u^n(t), u^n_t(t)) = \mathscr{E}(u(t), u_t(t)) \quad \text{for every } t \in [0, T].$$

Using Lipschitz property (2.4.29) of F^* and relation (2.4.7), again, it is easy to see that

$$\lim_{n\to\infty} \int_s^t (F^*(u^n(\tau), u^n_t(\tau)), u^n_t(\tau)) d\tau = \int_s^t (F^*(u(\tau), u_t(\tau)), u_t(\tau)) d\tau\,.$$

By (2.4.7) and (2.4.8) we also have that

$$\lim_{n\to\infty}\int_s^t (p^n(\tau),u_t^n(\tau))d\tau = \int_s^t (p(\tau),u_t(\tau))d\tau\,.$$

Because φ is lower semicontinuous, by Fatou's lemma we have that

$$\liminf_{n\to\infty}\int_s^t (Du_t^n(\tau),u_t^n(\tau))d\tau \ge \int_s^t \liminf_{n\to\infty}\varphi(u_t^n(\tau))d\tau \ge \int_s^t \varphi(u_t(\tau))d\tau\,. \quad (2.4.44)$$

It is also follows from the argument given in the proof of Proposition 2.4.14 (see the second relation in (2.4.25)) that

$$\lim_{n\to\infty}\int_s^t \langle g(G^*\mathscr{A}u_t^n),G^*\mathscr{A}u_t^n\rangle d\tau = \int_s^t \langle g(G^*\mathscr{A}u_t),G^*\mathscr{A}u_t\rangle d\tau.$$

Thus, relation (2.4.42) follows from (2.4.43). □

The assertion stated below gives a condition under which generalized solutions satisfy energy identity (2.4.34) and also a weak (variational) form of equation (2.4.26). The conditions required are rather demanding, and therefore of limited applicability. However, for the sake of completeness we provide a concise formulation with a brief proof.

2.4.21. Proposition. *Let the hypotheses of Theorem* 2.4.16 *(along with Assumption* 2.4.12*) hold. Assume additionally that the operator D maps V into V' and is a monotone hemicontinuous operator bounded on bounded sets; that is,*

$$\sup\{|D(v)|_{V'}\;:\;v\in V,\,|v|_V\le\rho\}<\infty \quad \textit{for any}\;\; \rho>0. \qquad (2.4.45)$$

Then generalized solutions satisfy the energy relation (2.4.34)*. Moreover, every generalized solution is also weak: the relation*

$$\begin{aligned}(Mu_t(t),\psi) = (Mu_1,\psi) - \int_0^t \big((\mathscr{A}u(\tau),\psi)+(Du_t(\tau),\psi) \qquad\qquad (2.4.46)\\ + \langle g(G^*\mathscr{A}u_t(\tau)),G^*\mathscr{A}\phi\rangle - (F(u(\tau),u_t(\tau))+p(\tau),\psi)\big)d\tau\end{aligned}$$

holds for every $\psi\in\mathscr{D}(\mathscr{A}^{1/2})$ and for almost all $t\in[0,T]$.

Proof. The statement that generalized solutions are also weak solutions easily follows from Proposition 2.4.14.

Let us prove the energy equality (2.4.34) for generalized (weak) solutions.

As the argument given in the proof of Proposition 2.4.20 shows, to obtain (2.4.34) we need only check that

$$\lim_{n\to\infty}\int_0^t (Du_t^n(\tau),u_t^n(\tau))d\tau = \int_0^t (Du_t(\tau),u_t(\tau))d\tau\,, \qquad (2.4.47)$$

where $u(t)$ is a generalized solution and $\{u^n(t)\}$ is a sequence of strong solutions such that (2.4.7) holds.

By Proposition 1.2.5(4) $D : V \mapsto V'$ is demicontinuous; that is, $Du^n(t) \to Du(t)$ weakly in V' for every t and hence

$$\lim_{n\to\infty} (Du_t^n(\tau), u_t^n(\tau)) = (Du_t(\tau), u_t(\tau)), \quad \tau \in [0,T].$$

Thus (2.4.45) and the Lebesgue dominated convergence theorem imply (2.4.47). □

2.4.22. Remark. Assume that the operator D has the form $D = D_1 + D_2$, where the operator D_1 possesses the properties described in the statement of Proposition 2.4.20 with some convex lower semicontinuous function φ and the operator D_2 enjoys the requirements of Proposition 2.4.21. Then the same arguments as in Propositions 2.4.20 and 2.4.21 show that any generalized solution satisfies the energy inequality of the form

$$\mathscr{E}(u(t), u_t(t)) + \int_s^t [\varphi(u_t(\tau)) + (D_2 u_t(\tau), u_t(\tau)) + \langle g(G^*\mathscr{A} u_t(\tau)), G^*\mathscr{A} u_t(\tau)\rangle]\, d\tau$$
$$\leq \mathscr{E}(u(s), u_t(s)) + \int_s^t (F^*(u(\tau), u_t(\tau)) + p(\tau), u_t(\tau)) d\tau.$$

We can also suggest other versions of additional assumptions concerning D that guarantee the validity of energy relation (2.4.34) and variational equality (2.4.46) for generalized solutions. For instance, one can formulate conditions concerning D in the spirit of Assumption 2.4.12 with reference to the argument given in Proposition 2.4.14. However, for the sake of simplicity, we prefer to consider them in subsequent chapters in the context of concrete models. ■

2.4.3 *Linear nonhomogeneous problem*

In this section we consider the following linear second order abstract equation,

$$\begin{aligned} &Mu_{tt}(t) + \mathscr{A} u(t) + \mathscr{A} G[gG^*\mathscr{A} u_t(t) + \psi(t)] \\ &\quad + D_0 h(t) + D_0 D_0^* u_t(t) + Du_t(t) = p(t), \quad t > 0, \end{aligned} \tag{2.4.48}$$

with the following initial data

$$u|_{t=0} = u_0, \quad u_t|_{t=0} = u_1, \tag{2.4.49}$$

where the functions ψ, h and p are given forcing terms. We consider problem (2.4.48) under the following standing hypothesis.

2.4.23. Assumption.

- Operators M, $\mathscr{A}$, G, and D_0 satisfy the requirements listed in Assumption 2.4.1.

- The operators

$$D : \mathscr{D}(\mathscr{A}^{1/2}) \mapsto \left[\mathscr{D}(\mathscr{A}^{1/2})\right]' \quad \text{and} \quad g : U \mapsto U$$

are linear monotone operators.
- The functions ψ, h, and p possess the properties

$$\psi(t) \in L_1^{loc}(\mathbb{R}_+;U), \quad h(t) \in L_1^{loc}(\mathbb{R}_+;Z), \quad p(t) \in L_1^{loc}(\mathbb{R}_+;V'). \tag{2.4.50}$$

If $\psi \equiv 0$ and $h \equiv 0$, then we can apply our main Theorem 2.4.5 to obtain an existence and uniqueness result for problem (2.4.48) and (2.4.49). We also can reduce the case to Theorem 2.4.5 under the condition

$$\mathscr{A} G\psi(t) + D_0 h(t) \in L_1^{loc}(\mathbb{R}_+;V'). \tag{2.4.51}$$

However, the condition in (2.4.51) is too restrictive, due to potential incompatibility between the domain of $\mathscr{A}$ and the range of G. For this reason we consider a more general setup.

We begin with the following assertion on global existence of strong solutions.

2.4.24. Proposition (Strong solutions). *Let Assumption* 2.4.23 *be valid and*

$$\psi \in W_1^2(0,T;U), \quad -D_0 h + p \in W_1^1(0,T;V') \tag{2.4.52}$$

for every $T > 0$. Assume that $u_0 \in \mathscr{D}(\mathscr{A}^{1/2})$ and $u_1 \in \mathscr{D}(\mathscr{A}^{1/2})$ are such that

$$\mathscr{A}[u_0 + Gg(G^*\mathscr{A} u_1) + G\psi(0)] + Du_1 + D_0 D_0^* u_1 \in V'. \tag{2.4.53}$$

Then problem (2.4.48) *and* (2.4.49) *possesses a unique strong solution on $\mathbb{R}_+$ such that*

$$(u;u_t) \in C(\mathbb{R}_+;\mathscr{D}(\mathscr{A}^{1/2}) \times V), \tag{2.4.54}$$

$$(u_t;u_{tt}) \in C_r(\mathbb{R}_+;\mathscr{D}(\mathscr{A}^{1/2}) \times V) \cap L_\infty^{loc}(\mathbb{R}_+;\mathscr{D}(\mathscr{A}^{1/2}) \times V), \tag{2.4.55}$$

$$\mathscr{A}[u + Gg(G^*\mathscr{A} u_t) + G\psi] + Du_t + D_0 D_0^* u_t \in C_r(\mathbb{R}_+;V'). \tag{2.4.56}$$

Here, as above, we denote by C_r the space of functions that are continuous from the right. Moreover, the function $t \mapsto (u_t(t);u_{tt}(t))$ is weakly continuous in $\mathscr{D}(\mathscr{A}^{1/2}) \times V$ and we have the following energy relation

$$E_0(u(t),u_t(t)) + \int_0^t \left[(Du_t,u_t) + \langle g(G^*\mathscr{A} u_t), G^*\mathscr{A} u_t\rangle + |D_0^* u_t|_Z^2\right] d\tau$$

$$= E_0(u_0,u_1) + \int_0^t \left[-\langle \psi, G^*\mathscr{A} u_t\rangle - (h, D_0^* u_t)_Z + (p,u_t)\right] d\tau, \tag{2.4.57}$$

where, as above, $E_0(u_0,u_1) = \frac{1}{2}\left[(Mu_1,u_1) + (\mathscr{A} u_0,u_0)\right]$.

Proof. When $\psi = 0$, the conclusion of Proposition 2.4.24 follows directly from the abstract results of the previous section. Indeed, Theorem 2.4.5 implies the existence of a unique local strong solution $u(t)$ that satisfies energy relation (2.4.57) on every existence interval. By (2.4.52) this relation implies that

$$E_0(u(t),u_t(t)) \le E_0(u_0,u_1) + \int_0^t |-D_0 h(\tau)+p(\tau)|_{V'} |u_t(\tau)|_V d\tau$$

for every $t \ge 0$, which makes it possible to establish an appropriate a priori estimate and use the third statement of Theorem 2.4.5 to obtain global existence of strong solutions.

When $\psi \ne 0$, the situation is a bit more subtle. Theorem 2.4.5 does not apply directly. This is due to the fact that an element $\mathscr{A} G\psi$ may not be in V', no matter how smooth ψ is. In fact, this is a typical situation with boundary inputs, where incompatibility between the domain of $\mathscr{A}$ and the range of G is very strong. In order to cope with the issue we use an equivalent formulation of the problem via semigroup theory. To this end we introduce the operator

$$A \equiv \begin{pmatrix} 0 & -I \\ M^{-1}\mathscr{A} & M^{-1}[\mathscr{A} Gg(G^*\mathscr{A}) + D + D_0 D_0^*] \end{pmatrix}, \tag{2.4.58}$$

where $A : D(A) \subset H \mapsto H$, $H = \mathscr{D}(\mathscr{A}^{1/2}) \times V$, and

$$D(A) = \left\{ U = (u_1; u_2) \in H \;\middle|\; \begin{array}{l} u_1 \in \mathscr{D}(\mathscr{A}^{1/2}),\ u_2 \in \mathscr{D}(\mathscr{A}^{1/2}), \\ \mathscr{A} u_1 + \mathscr{A} Gg(G^*\mathscr{A} u_2) + Du_2 + D_0 D_0^* u_2 \in V' \end{array} \right\}.$$

We note that this operator has the form (2.4.5) with $h \equiv 0$. Therefore by Lemma 2.4.6 (see also Proposition 2.4.7 which is valid in our case with $L = 0$) the operator A is an m-accretive linear operator in H. Thus by Proposition 4.6 in [241] A is densely defined and by the Lumer–Phillips theorem (see, e.g., Theorem 4.3 in [241]) the operator $-A$ is a generator of a contraction semigroup e^{-At} on $H = D(\mathscr{A}^{1/2}) \times V$.

Note that

$$\begin{pmatrix} 0 \\ M^{-1}\mathscr{A} G\psi \end{pmatrix} = A \begin{pmatrix} G\psi \\ 0 \end{pmatrix} \in W_1^2(0,T;[\mathscr{D}(A)]')$$

due to the fact that $(G\psi;0) \in W_1^2(0,T;H)$. In addition, on the strength of the hypothesis in (2.4.52) we have that

$$\begin{pmatrix} 0 \\ M^{-1}p - M^{-1}D_0 h \end{pmatrix} \in W_1^1(0,T;H).$$

Thus, we are in a position to write down semigroup representation of the solution $U = (u; u_t)$ to the nonhomogeneous problem (2.4.48):

$$U(t) = e^{-At}U_0 + \int_0^t Ae^{-A(t-s)} \begin{pmatrix} -G\psi(s) \\ 0 \end{pmatrix} ds + \int_0^t e^{-A(t-s)} \begin{pmatrix} 0 \\ M^{-1}(p - D_0 h) \end{pmatrix} ds.$$

Integrating the first integral by parts (with values in $[\mathscr{D}(A)]'$) yields

$$U(t) = e^{-At}U_0 - e^{-A(t-s)} \begin{pmatrix} G\psi(s) \\ 0 \end{pmatrix} \Bigg|_{s=0}^{s=t} + \int_0^t e^{-A(t-s)} \begin{pmatrix} G\psi_t(s) \\ 0 \end{pmatrix} ds$$
$$+ \int_0^t e^{-A(t-s)} \begin{pmatrix} 0 \\ M^{-1}(p - D_0 h) \end{pmatrix} ds.$$

Introducing the notation

$$\hat{U}(t) \equiv \begin{pmatrix} \hat{u}_1(t) \\ \hat{u}_2(t) \end{pmatrix} = U(t) + \begin{pmatrix} G\psi(t) \\ 0 \end{pmatrix} = \begin{pmatrix} u(t) + G\psi(t) \\ u_t(t) \end{pmatrix},$$

we arrive at the problem

$$\hat{U}_t + A\hat{U} = \mathscr{F}(t) \equiv \begin{pmatrix} G\psi_t \\ M^{-1}p - M^{-1}D_0 h \end{pmatrix}, \quad \hat{U}(0) = \begin{pmatrix} u_0 + G\psi(0) \\ u_1 \end{pmatrix}. \quad (2.4.59)$$

Thus, by Theorem 2.3.2 problem (2.4.59) admits strong solutions $\hat{U}(t)$ provided $\hat{U}(0) \in \mathscr{D}(A)$ and $\mathscr{F}(t) \in W_1^1(0,T;H)$. In our case these properties are guaranteed by the compatibility conditions in (2.4.52) and (2.4.53).

Applying the standard energy method to equation (2.4.59) yields

$$\frac{1}{2}|\hat{U}(t)|_H^2 - \frac{1}{2}|\hat{U}(0)|_H^2 + \int_0^t \langle g(G^*\mathscr{A}\hat{u}_2), G^*\mathscr{A}\hat{u}_2\rangle ds \quad (2.4.60)$$
$$+ \int_0^t |D_0^*\hat{u}_2|^2 ds + \int_0^t (D\hat{u}_2, \hat{u}_2) ds = \int_0^t (G\psi_t, \mathscr{A}\hat{u}_1) ds + \int_0^t (p - D_0 h, \hat{u}_2) ds.$$

By using the relations $\hat{u}_1 = u + G\psi$ and $\hat{u}_2 = u_t$ we have that

$$|\hat{U}(t)|_H^2 = |U(t)|_H^2 + 2(\mathscr{A}^{1/2}G\psi(t), \mathscr{A}^{1/2}u(t)) + |\mathscr{A}^{1/2}G\psi(t)|^2$$

and

$$\int_0^t (G\psi_t, \mathscr{A}\hat{u}_1) ds = \left[\frac{1}{2}|\mathscr{A}^{1/2}G\psi(s)|^2 + (G\psi(s), \mathscr{A}u(s))\right]_{s=0}^{s=t} - \int_0^t (G\psi, \mathscr{A}u_t) ds.$$

Thus relation (2.4.60) is equivalent to

$$\frac{1}{2}|U(t)|_H^2 + \int_0^t \langle g(G^*\mathscr{A}u_t), G^*\mathscr{A}u_t\rangle ds + \int_0^t (Du_t, u_t) ds + \int_0^t |D_0^* u_t|^2 ds$$
$$= \frac{1}{2}|U(0)|_H^2 - \int_0^t \langle \psi, G^*\mathscr{A}u_t\rangle ds + \int_0^t (p - D_0 h, u_t) ds,$$

which is precisely the energy identity in (2.4.57). □

We note that the term $D_0 h$ can be written as $D_0 h = \mathscr{A}\mathscr{A}^{-1}D_0 h = \mathscr{A}G_1 h$, where $G_1 = \mathscr{A}^{-1}D_0 : Z \mapsto \mathscr{D}(\mathscr{A}^{1/2})$ is bounded. Thus, this term can be treated in exactly the same manner as the term $\mathscr{A}G\psi$. In view of this observation, we can restate Proposition 2.4.24 in the following form.

2.4.25. Proposition (Strong solutions). *Let Assumption* 2.4.23 *be valid and*

$$\psi \in W_1^2(0,T;U), \quad h \in W_1^2(0,T;Z), \quad p(t) \in W_1^1(0,T;V') \tag{2.4.61}$$

for every $T > 0$. *Assume that* $u_0 \in \mathscr{D}(\mathscr{A}^{1/2})$ *and* $u_1 \in \mathscr{D}(\mathscr{A}^{1/2})$ *are such that*

$$\mathscr{A}[u_0 + Gg(G^*\mathscr{A}u_1) + G\psi(0)] + Du_1 + D_0h(0) + D_0D_0^*u_1 \in V'. \tag{2.4.62}$$

Then problem (2.4.48) *and* (2.4.49) *possesses a unique strong solution on* $\mathbb{R}_+$ *possessing the properties* (2.4.54) *and* (2.4.55) *and also satisfying the relation*

$$\mathscr{A}[u + Gg(G^*\mathscr{A}u_t) + G\psi] + Du_t + D_0h + D_0D_0^*u_t \in C_r(\mathbb{R}_+;V'). \tag{2.4.63}$$

Moreover, the function $t \mapsto (u_t(t);u_{tt}(t))$ *is weakly continuous in* $\mathscr{D}(\mathscr{A}^{1/2}) \times V$ *and energy relation* (2.4.57) *holds.*

We note that in contrast to Proposition 2.4.24, Proposition 2.4.25 assumes more smoothness of the function $h(t)$ but it does not assume its compatibility with the force $p(t)$.

To study the regularity properties of a nonlinear problem with nonlinear boundary conditions we need a more general assertion that deals with generalized and weak solutions of linear problem (2.4.48).

The generalized solution to (2.4.48) and (2.4.49) is understood as a limit of a sequence of strong solutions u^n with convergent data $\{u_0^n, u_1^n, \psi^n, h^n, p^n\}$. We also recall the notion of weak solutions.

2.4.26. Definition. A function $u(t)$ is said to be a weak solution to problem (2.4.48) and (2.4.49) on an interval $[0,T]$ if $(u;u_t) \in L_2(0,T;\mathscr{D}(\mathscr{A}^{1/2}) \times V)$, $u(0) = u_0$, and $u(t)$ satisfies the relation

$$\begin{aligned} &-\int_0^T [(Mu_t,\phi_t) + (Du,\phi_t) + \langle gG^*\mathscr{A}u, G^*\mathscr{A}\phi_t\rangle + (D_0^*u, D_0^*\phi_t)_Z]\,d\tau \\ &+\int_0^T (\mathscr{A}u,\phi)d\tau + \int_0^T [\langle\psi, G^*\mathscr{A}\phi\rangle + (h, D_0^*\phi)_Z - (p,\phi)]\,d\tau \\ &= (Mu_1 + Du_0, \phi(0)) + \langle gG^*\mathscr{A}u_0, G^*\mathscr{A}\phi(0)\rangle + (D_0^*u_0, D_0^*\phi(0))_Z \end{aligned} \tag{2.4.64}$$

for any $\phi \in W_2^1(0,T;\mathscr{D}(\mathscr{A}^{1/2}))$ such that $\phi(T) = 0$. ∎

2.4.27. Remark. Another equivalent formulation of weak solutions (used in subsequent chapters) involves static test functions ϕ. Instead of (2.4.64) one may take

$$\begin{aligned} &(Mu_t(t),\phi) + (Du(t),\phi) + \langle gG^*\mathscr{A}u(t), G^*\mathscr{A}\phi\rangle + (D_0^*u(t), D_0^*\phi)_Z \\ &+\int_0^t (\mathscr{A}u,\phi)d\tau + \int_0^t [\langle\psi, G^*\mathscr{A}\phi\rangle + (h, D_0^*\phi)_Z - (p,\phi)]\,d\tau \\ &= (Mu_1 + Du_0, \phi) + \langle gG^*\mathscr{A}u_0, G^*\mathscr{A}\phi\rangle + (D_0^*u_0, D_0^*\phi)_Z \end{aligned} \tag{2.4.65}$$

for any $\phi \in \mathscr{D}(\mathscr{A}^{1/2})$. ∎

For existence of generalized and weak solutions we also need the following hypotheses.

2.4.28. Assumption.

- If $\psi \neq 0$, then $\langle gu, u\rangle \geq \alpha |u|_U^2$ for all $u \in U$ and for some $\alpha > 0$.
- The functions ψ, h, and p possess the properties

$$\psi(t) \in L_2(0,T;U), \quad h(t) \in L_2(0,T;Z), \quad p(t) \in L_1(0,T;V') \tag{2.4.66}$$

for every $T > 0$.

2.4.29. Theorem (Generalized and weak solutions). *Let Assumptions* 2.4.23 *and* 2.4.28 *be valid. Then for any initial data* $(u_0; u_1) \in (\mathscr{A}^{1/2}) \times V$ *problem* (2.4.48) *and* (2.4.49) *possesses a unique generalized solution such that*

$$(u; u_t) \in C(0,T; \mathscr{D}(\mathscr{A}^{1/2}) \times V) \tag{2.4.67}$$

and

$$\begin{gathered}\partial_t\left[D_{\text{sym}}^{1/2} u\right] \in L_2^{\text{loc}}([0,+\infty), \mathscr{H}), \\ \partial_t\left[G^*\mathscr{A} u\right] \in L_2^{\text{loc}}([0,+\infty), U), \quad \partial_t\left[D_0^* u\right] \in L_2^{\text{loc}}([0,+\infty), Z),\end{gathered} \tag{2.4.68}$$

where D_{sym} *is the self-adjoint nonnegative operator generated by bilinear form* $(D_{\text{sym}}\phi, \psi) = \frac{1}{2}[(D\phi, \psi) + (\phi, D\psi)]$ *for* $\phi, \psi \in \mathscr{D}(\mathscr{A}^{1/2})$. *Moreover, the solution* u *is also weak and satisfies energy relation* (2.4.57), *where we use the identifications* $D_{\text{sym}}^{1/2} u_t = \partial_t\left[D_{\text{sym}}^{1/2} u\right]$, $G^*\mathscr{A} u_t = \partial_t\left[G^*\mathscr{A} u\right]$, *and* $D_0^* u_t = \partial_t\left[D_0^* u\right]$.

Weak solutions to (2.4.48) *are unique. They are also generalized. In particular, any weak solution possesses properties* (2.4.67) *and* (2.4.68) *and satisfies energy relation* (2.4.57).

Proof. We start with the following observation that follows from energy equality (2.4.57): for any strong solution $u(t)$ (given by Proposition 2.4.24 or Proposition 2.4.25) by the standard Gronwall type argument (cf. the proof of the inequality in (2.4.37)) we have the following estimates

$$\begin{aligned} E_0(u(t), u_t(t)) &+ \int_0^t (Du_t, u_t)ds + \alpha \int_0^t |G^*\mathscr{A} u_t|_U^2 ds + \int_0^t |D_0^* u_t|_Z^2 ds \\ &\leq C_1 E_0(u_0, u_1) + C_2 \int_0^t \left[|\psi|_U^2 + |h|_Z^2\right] ds + C_3 \left[\int_0^t |p|_{V'} ds\right]^2. \end{aligned} \tag{2.4.69}$$

The above inequality, controlled by L_2 norms of the forcing terms ψ and h and by L_1 norms of p is a key in proving the existence of generalized or weak solutions.

Let $\psi^n \in W_1^2(0,T;U)$ and $h^n \in W_1^2(0,T;Z)$ be such that $\psi_n(0) = 0$, $h_n(0) = 0$, and

$$\lim_{n\to\infty}\left\{\int_0^T|\psi(t)-\psi^n(t)|_U^2dt+\int_0^T|h(t)-h^n(t)|_Z^2dt\right\}=0. \tag{2.4.70}$$

We choose p^n such that

$$p^n\in W_1^1(0,T;V'),\quad \lim_{n\to\infty}\int_0^T|p(t)-p^n(t)|_{V'}dt=0.$$

and $(u_0^n;u_1^n)\in\mathscr{D}(\mathscr{A}^{1/2})\times\mathscr{D}(\mathscr{A}^{1/2})$, such that $(u_0^n;u_1^n)\to(u_0;u_1)$ in $\mathscr{D}(\mathscr{A}^{1/2})\times V$ and

$$\mathscr{A}[u_0^n+Gg(G^*\mathscr{A}u_1^n)]+Du_1^n+D_0D_0^*u_1^n\in V'.$$

This choice of $(u_0^n;u_1^n)$ is possible for every $(u_0;u_1)\in H=\mathscr{D}(\mathscr{A}^{1/2})\times V$ because the domain $\mathscr{D}(A)$ of the operator A given by (2.4.58) is dense in H.

We denote by u^n the corresponding strong solution (which exists due to Proposition 2.4.25). Let $u^{n,m}=u^n-u^m$. Because $u^{n,m}$ is a strong solution to some linear problem of the form (2.4.48), it follows from (2.4.69) that

$$\begin{aligned}&E_0(u^{n,m}(t),u_t^{n,m}(t))+\int_0^t\left[|D_{\rm sym}^{1/2}u_t^{n,m}|^2+\alpha|G^*\mathscr{A}u_t^{n,m}|_U^2+|D_0^*u_t^{n,m}|_Z^2\right]d\tau\\&\le c_1E_0(u_0^{n,m},u_1^{n,m})+c_2\int_0^t\left[|\psi^n-\psi^m|_U^2+|h^n-h^m|_Z^2\right]d\tau\\&\quad+c_3\left[\int_0^t|p^n-p^m|_{V'}d\tau\right]^2.\end{aligned}$$

Therefore

$$\lim_{n,m\to\infty}\max_{[0,T]}E_0(u^{n,m}(t),u_t^{n,m}(t))=0$$

and

$$\lim_{n,m\to\infty}\int_0^T\left[|D_{\rm sym}^{1/2}u_t^{n,m}|^2+\alpha|G^*\mathscr{A}u_t^{n,m}|_U^2+|D_0^*u_t^{n,m}|_Z^2\right]d\tau=0.$$

Therefore there exists a function $u(t)$ possessing properties (2.4.67) and (2.4.68) such that

$$\lim_{n,m\to\infty}\max_{[0,T]}\left\{|M^{1/2}(u_t^n-u_t)|^2+|\mathscr{A}^{1/2}(u^n-u)|^2\right\}=0$$

and

$$\lim_{n\to\infty}\int_0^T\left[|D_{\rm sym}^{1/2}(u_t^n-u_t)|^2+|G^*\mathscr{A}(u_t^n-u_t)|_U^2+|D_0^*(u_t^n-u_t)|_Z^2\right]d\tau=0.$$

Thus $u(t)$ is a generalized solution to problem (2.4.48) and (2.4.49). Moreover, the convergence above makes it possible to prove that $u(t)$ satisfies energy equality (2.4.57) and variational relation (2.4.64); that is, $u(t)$ is also a weak solution.

To prove the uniqueness of weak solutions we note that the difference $u(t)$ of two weak solutions satisfies the relation

$$\int_0^T [(Mu_t, \phi_t) + (Du, \phi_t) + \langle gG^* \mathscr{A} u, G^* \mathscr{A} \phi_t \rangle + (D_0^* u, D_0^* \phi_t)_Z - (\mathscr{A} u, \phi)] \, d\tau = 0$$

for any $\phi \in W_2^1(0,T; \mathscr{D}(\mathscr{A}^{1/2}))$ such that $\phi(T) = 0$. Therefore, if we choose

$$\phi(t) \equiv \phi^s(t) = \begin{cases} -\int_t^s u(\tau) d\tau, & 0 \le t \le s; \\ 0, & s < t \le T, \end{cases}$$

for $s \in [0,T]$, then using the same idea as in [222, Chapter 3] we can prove that $u(t) = 0$ for almost every $t \in [0,T]$. This completes the proof of Theorem 2.4.29.

□

2.4.30. Remark. We also note that the same a priori bound as in (2.4.69) can be obtained for models of the form (2.4.48) with nonlinear damping operators D and g under monotonicity assumptions imposed on D and g as in Assumption 2.4.1. In the case of separable spaces this can be easily achieved by using the Galerkin method along with weak convergence. This method produces the result on the existence of weak solutions under the following (additional) set of assumptions.

- The operator $g: U \mapsto U$ possesses the property $\langle g(u), u \rangle \ge \alpha |u|_U^2$ for some $\alpha > 0$ and is weakly continuous as a mapping from $L_2(0,T;U)$ into itself.
- The operator D maps $L_\infty(0,T;V)$ into $L_2(0,T;[\mathscr{D}(\mathscr{A}^{1/2})]')$ continuously with respect to $*$-weak topology.
- The functions ψ, h, and p possess properties (2.4.66).

In this case weak solutions exist in the space $C_w(0,T;H)$ and possesse the additional property: $G^* \mathscr{A} u_t \in L_2(0,T;U)$. The argument relies on the energy inequality

$$E_0(u(t), u_t(t)) + \int_0^t |G^* \mathscr{A} u_t|_U^2 d\tau \le E_0(u_0, u_1) + C \int_0^t \left[|\psi(t)|_U^2 + |h(t)|_Z^2 + |p(t)|_{V'} |u_t|_V \right] d\tau.$$

■

As a particular case of problem (2.4.48) we now consider the following problem which may be seen as an "abstract harmonic oscillator."

$$\begin{cases} Mu_{tt}(t) + \mathscr{A} u(t) + Du_t(t) = p(t), \\ u|_{t=0} = u_0, \quad u_t|_{t=0} = u_1, \end{cases} \tag{2.4.71}$$

where M and $\mathscr{A}$ satisfy Assumption 2.4.1(1,2), $p(t) \in L_1(0,T;V')$, and D is a monotone linear operator from V into V'.

2.4.31. Theorem. *Let Assumption* 2.4.1(1,2) *hold. Assume that D is a monotone linear operator from V into V'. Then the following statements hold.*

- **Strong solutions:** *Let* $p \in W_1^1(0,T;V')$ *and* $(u_0;u_1) \in W \times \mathscr{D}(\mathscr{A}^{1/2})$, *where*

$$W = \left\{ u \in \mathscr{D}(\mathscr{A}^{1/2}) \,:\, \mathscr{A}u \in V' \right\}. \tag{2.4.72}$$

Then there exists a unique strong solution to problem (2.4.71) *on the interval* $[0,T]$ *such that*

$$(u;u_t) \in C(0,T;\mathscr{D}(\mathscr{A}^{1/2}) \times V), \tag{2.4.73}$$

and also

$$u_{tt} \in C(0,T;V), \quad u_t \in C(0,T;\mathscr{D}(\mathscr{A}^{1/2})), \quad \mathscr{A}u(t) \in C(0,T;V'). \tag{2.4.74}$$

- **Generalized (weak) solutions:** *For every* $p \in L_1(0,T;V')$ *and* $u_0 \in \mathscr{D}(\mathscr{A}^{1/2})$, $u_1 \in V$ *there exists a unique generalized solution to problem* (2.4.71) *with property* (2.4.73). *Moreover every generalized solution is weak and vice versa.*

Both strong and generalized solutions satisfy the following energy relation

$$E_0(u(t),u_t(t)) + \int_0^t (Du_t(\tau),u_t(\tau))d\tau = E_0(u_0,u_1) + \int_0^t (p(\tau),u_t(\tau))d\tau\,, \tag{2.4.75}$$

where $E(u_0,u_1) = \frac{1}{2}\left((Mu_1,u_1) + (\mathscr{A}u_0,u_0)\right)$.

Proof. It follows from Theorem 2.4.5 that there exists a strong solution $u(t)$ on some interval $[0,T') \subset [0,T]$ possessing property (2.4.73) with T' instead of T and such that

$$u_{tt} \in C_r([0,T');V) \text{ and } \mathscr{A}u(t) \in C_r([0,T');V'). \tag{2.4.76}$$

It is easy to see that this solution satisfies the energy relation (2.4.75) for $t \in [0,T')$. This relation implies that

$$E_0(u(t),u_t(t)) \le E_0(u_0,u_1) + \sqrt{2} \max_{\tau\in[0,t]} \left[E_0(u(\tau),u_t(\tau))\right]^{1/2} \cdot \int_0^t |p(\tau)|_{V'} d\tau$$

and therefore

$$E_0(u(t),u_t(t)) \le 2\left(E_0(u_0,u_1) + \left(\int_0^t |p(\tau)|_{V'} d\tau\right)^2\right), \quad t > 0.$$

Thus by the third assertion of Theorem 2.4.5, there exists a unique strong solution on the interval $[0,T]$ and possessing property (2.4.76) with $T = T'$.

Similarly, Theorem 2.4.5 implies the existence of generalized solutions. The energy relation—valid for generalized solutions—follows from (2.4.75) for strong solutions and from the approximation property (2.4.7) for generalized solutions. By Proposition 2.4.21 every generalized solution is weak. Thus, to prove that any weak solution is generalized it is sufficient to establish uniqueness of weak solutions. This can be done in the same way as in the proof of Theorem 2.4.29.

To obtain the improvement (2.4.74) of the property (2.4.76) we note that for any strong solution $u(t)$ the function $z(t) = u_t(t)$ is a weak solution to the problem

$$\begin{cases} Mz_{tt}(t) + \mathscr{A}z(t) + Dz_t(t) = p_t(t), \\ z|_{t=0} = u_1 \in \mathscr{D}(\mathscr{A}^{1/2}), \;\; z_t|_{t=0} = M^{-1}(-\mathscr{A}u_0 - Du_1 + p(0)) \in V. \end{cases}$$

As above, any weak solution to this problem is generalized. Therefore by the definition of generalized solutions we obtain that $(u_t; u_{tt}) \in C(0,T; \mathscr{D}(\mathscr{A}^{1/2}) \times V)$. Directly from (2.4.71) we obtain that $\mathscr{A}u \in C(0,T;V')$. Thus (2.4.74) holds. □

We conclude this section with a short analysis of the following linear, nonautonomous problem which arises naturally in the context of linearization,

$$\begin{cases} Mu_{tt}(t) + \mathscr{A}u(t) + \mathscr{A}Gg(G^*\mathscr{A}u_t) + D(t)u_t(t) = F(t)u(t) + p(t), \\ u|_{t=0} = u_0, \quad u_t|_{t=0} = u_1. \end{cases} \tag{2.4.77}$$

Regularity results for nonautonomous processes generated by (2.4.77) are used in the context of studying higher regularity of solutions.

We impose the following set of hypotheses.

2.4.32. Assumption.

- The spaces V, $\mathscr{H}$, and U are separable and the operators M and $\mathscr{A}$ satisfy the hypotheses in Assumption 2.4.1(1,2).
- $D(t) = D_0(t) + D_1(t)$, where[3] (i) $D_0(t)$ is a family of linear nonnegative self-adjoint operators in $\mathscr{H}$, and (ii) $D_0(t) : \mathscr{D}(\mathscr{A}^{1/2}) \mapsto V'$ and $D_1(t) : V \mapsto V'$ are measurable families of linear bounded operators for each $t \in \mathbb{R}_+$. Moreover

$$t \mapsto d_0(t) \equiv \|D_0(t)\|_{\mathscr{L}(\mathscr{D}(\mathscr{A}^{1/2}),V')} \quad \text{and} \quad t \mapsto d_1(t) \equiv \|D_1(t)\|_{\mathscr{L}(V,V')}$$

  are locally integrable functions on $[0,\infty)$.
- $F(t) : \mathscr{D}(\mathscr{A}^{1/2}) \mapsto V'$ is a linear bounded operator for each $t \in \mathbb{R}_+$ such that $t \mapsto F(t)$ is measurable and the function $t \mapsto \|F(t)\|_{\mathscr{L}(\mathscr{D}(\mathscr{A}^{1/2}),V')}$ is locally integrable on $[0,\infty)$.
- $p \in L_1^{\mathrm{loc}}([0,\infty); V')$.
- The operator G satisfies Assumption 2.4.1(4) with some Gelfand triple $U_0 \subset U \subset U_0'$, and either $g = 0$ or else $g \in \mathscr{L}(U)$, and $\langle g(u), u \rangle \geq \alpha |u|_U^2, \alpha > 0$.

We introduce the notion of a weak solution for problem (2.4.77) in the following way.

2.4.33. Definition. A function $u(t)$ is said to be a weak solution for problem (2.4.77) on the interval $[0,T]$ if

- $(u; u_t) \in L_\infty(0,T; \mathscr{D}(\mathscr{A}^{1/2}) \times V)$.

[3] We emphasize that we do not assume that $D(t)$ is a monotone operator.

- $D_0(t)^{1/2}u_t \in L_2(0,T;\mathscr{H})$ and $G^*\mathscr{A}u_t \in L_2(0,T;U)$ when $g \neq 0$.
- We have that $u(0) = u_0$ and $u(t)$ satisfies the relation

$$\begin{aligned} & -\int_0^T [(Mu_t,\phi_t) + \langle gG^*\mathscr{A}u, G^*\mathscr{A}\phi_t\rangle]\,dt \\ & + \int_0^T (u_t,(D_0(t)+D_1^*(t))\phi)d\tau + \int_0^T (\mathscr{A}u,\phi)dt \\ = {} & (Mu_1,\phi(0)) + \langle gG^*\mathscr{A}u_0, G^*\mathscr{A}\phi(0)\rangle + \int_0^T (F(t)u+p,\phi)dt \end{aligned} \tag{2.4.78}$$

for any $\phi \in W_2^1(0,T;\mathscr{D}(\mathscr{A}^{1/2}))$ such that $\phi(T) = 0$.

■

We note that in the case when $D_0(t) = $const, $D_1 \equiv 0$, $F \equiv 0$ this definition due to Theorem 2.4.29 is equivalent to Definition 2.4.26.

For further reference, we collect some (additional) properties of weak solutions. This is done in the lemma below.

2.4.34. Lemma. *Under Assumption* 2.4.32 *any weak solution* $u(t)$ *to problem* (2.4.77) *possesses the properties:*

- $Mu_{tt} \in L_1(0,T;[\mathscr{D}(\mathscr{A}^{1/2})]')$, *that also implies the following.*
- *The function* $t \mapsto (u(t);u_t(t))$ *is weakly continuous in* $H = \mathscr{D}(\mathscr{A}^{1/2}) \times V$.
- *The equation in* (2.4.77) *is satisfied for almost all* $t \in [0,T]$ *as an equality in* $[\mathscr{D}(\mathscr{A}^{1/2})]'$.

Proof. Inasmuch as

$$|D_0(t)^{1/2}\phi|^2 = (D_0(t)\phi,\phi) \le \|D_0(t)\|_{\mathscr{L}(\mathscr{D}(\mathscr{A}^{1/2}),V')}|\mathscr{A}^{1/2}\phi||\phi|_V \le Cd_0(t)|\mathscr{A}^{1/2}\phi|^2,$$

we have that $D_0(t)^{1/2} : \mathscr{D}(\mathscr{A}^{1/2}) \mapsto \mathscr{H}$ and thus $D_0(t)^{1/2} = [D_0(t)^{1/2}]^*$ maps $\mathscr{H}$ into $[\mathscr{D}(\mathscr{A}^{1/2})]'$ and the relation

$$\|D_0(t)^{1/2}\|_{\mathscr{L}(\mathscr{D}(\mathscr{A}^{1/2}),\mathscr{H})} = \|D_0(t)^{1/2}\|_{\mathscr{L}(\mathscr{H},[\mathscr{D}(\mathscr{A}^{1/2})]')} \le C\sqrt{d_0(t)}$$

holds. Therefore $|D(t)u_t(t)|_{[\mathscr{D}(\mathscr{A}^{1/2})]'} \le C\sqrt{d_0(t)}|D_0(t)^{1/2}u_t(t)| \in L_1(0,T)$. A similar argument for other terms from equation (2.4.77) and variational relation (2.4.78) makes it possible to prove that $Mu_{tt} \in L_1(0,T;[\mathscr{D}(\mathscr{A}^{1/2})]')$. It is standard to show that this last relation implies the second and the third statement of Lemma 2.4.34.

□

2.4.35. Theorem. *Let Assumption* 2.4.32 *be in force. Then*

- *For any* $(u_0;u_1) \in \mathscr{D}(\mathscr{A}^{1/2}) \times V$ *problem* (2.4.77) *has a weak solution* $u(t)$ *such that for every* $t \in [0,T]$ *the following inequality*

$$E_0(u(t),u_t(t)) \le C_T \left(E_0(u_0,u_1) + \left[\int_0^t |p(\tau)|_{V'}d\tau \right]^2 \right) \tag{2.4.79}$$

holds, where, as above, $E_0(u_0,u_1) = \frac{1}{2}[(Mu_1,u_1)+(\mathscr{A}u_0,u_0)]$.

- *If, in addition, we assume that there exist nonnegative self-adjoint operator* $\tilde{D} : \mathscr{D}(\tilde{D}) \subset \mathscr{H} \mapsto \mathscr{H}$, *and the constants* $\beta_0, \beta_1 > 0$ *such that*

$$\beta_0\tilde{D} \le D_0(t) < \beta_1\tilde{D} \quad \textit{for all } t\in[0,T], \tag{2.4.80}$$

then weak solution $u(t)$ *is unique, possesses property*

$$(u;u_t) \in C(0,T;\mathscr{D}(\mathscr{A}^{1/2})\times V) \textit{ for every } T>0, \tag{2.4.81}$$

and satisfies the following energy identity

$$\begin{aligned} E_0(u(t),u_t(t)) &+ \int_0^t \left[|D_0(\tau)^{1/2}u_t|^2 + \langle g(G^*\mathscr{A}u_t),G^*\mathscr{A}u_t\rangle\right]d\tau \\ &= E_0(u_0,u_1) + \int_0^t [(-D_1(\tau)u_t + F(\tau)u + p, u_t)]\,d\tau. \end{aligned} \tag{2.4.82}$$

Proof. One applies the standard compactness method based on Galerkin approximations (see, e.g., [222, Chapter 3]) to prove the existence of weak solutions.

We first define Galerkin approximations in the following way.

Let $\{e_k\}_{k=1}^\infty$ be a basis in the Hilbert space $\mathscr{D}(\mathscr{A}^{1/2})$. A function $u^n(t)$ of the form

$$u^n(t) = \sum_{k=1}^n g_k(t)e_k, \quad n=1,2,\dots$$

is said to be an*n-order Galerkin approximate solution* to problem (2.4.77) on the interval $[0,T]$, iff $g_k(t)\in W_\infty^2(0,T;\mathbb{R})$ are scalar functions and $u^n(t)$ satisfies the relations

$$\begin{cases} \frac{d^2}{dt^2}(Mu^n(t),e_k) + (\mathscr{A}u^n(t),e_k) + (D(t)u_t^n(t),e_k) + \langle g(G^*\mathscr{A}u_t^n),G^*\mathscr{A}e_k\rangle \\ \qquad\qquad - (F(t)u^n(t)+p(t),e_k), \quad k=1,2,\dots,n, \\ u^n|_{t=0} = u_{n0}, \quad u_t^n|_{t=0} = u_{n1}, \end{cases} \tag{2.4.83}$$

where $u_{n0},u_{n1}\in \mathrm{Span}\{e_k : k=1,2,\dots,n\}$ are chosen such that $u_{n0}\to u_0$ in $\mathscr{D}(\mathscr{A}^{1/2})$ and $u_{n1}\to u_1$ in V.

In what follows we assume that $g\ne 0$ for definiteness (the case $g\equiv 0$ is simpler).

It is easy to see that the approximate solution $u^n(t)$ satisfies the following energy inequality

$$\begin{aligned} E_0(u^n(t),u_t^n(t)) &+ \int_0^t (D_0(\tau)u_t^n(\tau),u_t^n(\tau))d\tau + \alpha\int_0^t |G^*\mathscr{A}u_t^n(\tau)|_U^2 d\tau \\ &\le E_0(u_{n0},u_{n1}) + \int_0^t (-D_1(\tau)u_t^n(\tau) + F(\tau)u^n(\tau) + p(\tau), u_t^n(\tau))d\tau, \end{aligned}$$

where $E(u_0,u_1) = \frac{1}{2}((Mu_1,u_1) + (\mathscr{A}u_0,u_0))$. Assumption 2.4.32 allows us to infer that

$$E_0(u^n(t),u_t^n(t)) \le C_T \left(E_0(u_{n0},u_{n1}) + \left[\int_0^t |p(\tau)|_{V'} d\tau \right]^2 \right) \le C_T \quad (2.4.84)$$

for all $t \in [0,T]$, $n = 1,2,\dots$, and also the a priori bounds:

$$\int_0^T |D_0(\tau)^{1/2} u_t^n|^2 d\tau \le C_T, \quad \int_0^T |G^* \mathscr{A} u_t^n|_U^2 d\tau \le C_T.$$

Therefore there exists a subsequence $\{n_k\}$ and a function $u(t)$ from the space $L_\infty(0,T;\mathscr{D}(\mathscr{A}^{1/2}))$ such that $u_t(t) \in L_\infty(0,T;V)$ and as $k \to \infty$ we have that

$$\begin{array}{lll} u^{n_k} \to u & *\text{-weakly in } L_\infty(0,T;\mathscr{D}(\mathscr{A}^{1/2})), \\ u_t^{n_k} \to u_t & *\text{-weakly in } L_\infty(0,T;V), \\ D_0^{1/2} u_t^{n_k} \to d & \text{weakly in } L_2(0,T;\mathscr{H}), \\ G^* \mathscr{A} u_t^{n_k} \to l & \text{weakly in } L_2(0,T;U). \end{array}$$

We obviously have that

$$\int_0^T \langle G^* \mathscr{A} u_t^n(\tau), \phi_0 \rangle r(\tau) d\tau = -\int_0^T \langle u^n(\tau), \mathscr{A} G \phi_0 \rangle r_t(\tau) d\tau$$

for any $\phi_0 \in U$ and $r(t) \in C_0^\infty(0,T)$. This implies that

$$\int_0^T \langle l(\tau), \phi_0 \rangle r(\tau) d\tau = -\int_0^T \langle G^* \mathscr{A} u(\tau), \phi_0 \rangle r_t(\tau) d\tau.$$

Thus the element l can be identified with the correct limit $G^* \mathscr{A} u_t$. Because

$$\int_0^T (D_0^{1/2} u_t^{n_k}, \phi) d\tau = \int_0^T (u_t^{n_k}, D_0^{1/2} \phi) d\tau \to \int_0^T (u_t, D_0^{1/2} \phi) d\tau$$

for any $\phi \in L_2(0,T;\mathscr{D}(\mathscr{A}^{1/2}))$ as $k \to \infty$, the element d can be easily identified with $D_0^{1/2} u_t$. One can also see that this limiting function u is a weak solution to problem (2.4.77) on the interval $[0,T]$ and that relation (2.4.79) follows from (2.4.84). This completes the proof of the first part of Theorem 2.4.35.

To prove the second part we first establish the following assertion.

2.4.36. Proposition. *Let Assumption* 2.4.32 *be in force and* (2.4.80) *hold. Then any weak solution $u(t)$ to problem* (2.4.77) *satisfies the energy identity in* (2.4.82).

Proof. We use the same idea as in [164]. Again, we concentrate on the case $g \ne 0$.

We first extend solution $u(t)$ on the whole real line. For this we consider the following autonomous problem

$$Mw_{tt} + \mathscr{A}[w + Gg(G^* \mathscr{A} w_t)] + \tilde{D} w_t = 0,\ t > 0, \quad w|_{t=0} = w_0, \quad w_t|_{t=0} = w_1,$$

By Theorem 2.4.29 for any $(w_0;w_1) \in H = \mathscr{D}(\mathscr{A}^{1/2}) \times V$ this problem has a unique generalized solution $w(t) \equiv w(t;w_0,w_1)$ that possesses the properties

- $t \mapsto (w(t);w_t(t))$ is strongly continuous in H.
- $\int_t^{t+1} \left[|\tilde{D}^{1/2}w_t|^2 + |G^*\mathscr{A}w_t|_U^2\right] d\tau \le C$ for all $t \ge 0$.

Now we define the extension of $u(t)$ by the formula

$$\tilde{u}(t) = \begin{cases} w(-t;u(0),u_t(0)), & \text{if } t < 0; \\ u(t), & \text{if } t \in [0,T]; \\ w(t-T;u(T),u_t(T)), & \text{if } t > T. \end{cases}$$

It is clear from Lemma 2.4.34 that this function $\tilde{u}(t)$ possesses the properties

- $(\tilde{u};\tilde{u}_t) \in L_\infty(a,b;\mathscr{D}(\mathscr{A}^{1/2}) \times V)$ and $M\tilde{u}_{tt} \in L_1(0,T;[\mathscr{D}(\mathscr{A}^{1/2})]')$ for all $a < b$.
- The function $t \mapsto (\tilde{u}(t);\tilde{u}_t(t))$ is weakly continuous on $[0,T]$ and strongly continuous on $\mathbb{R} \setminus (0,T)$ as a function with values in in $H = \mathscr{D}(\mathscr{A}^{1/2}) \times V$.
- We have that

$$\int_t^{t+1} \left[|\tilde{D}^{1/2}\tilde{u}_t|^2 + |G^*\mathscr{A}\tilde{u}_t|_U^2\right] d\tau \le C \quad \text{for all } t \in \mathbb{R}. \tag{2.4.85}$$

Now as in [164] we multiply (2.4.77) in $\mathscr{H}$ by $\tilde{u}^h(t) = (2h)^{-1}(\tilde{u}(t+h) - \tilde{u}(t-h))$ and then integrate over interval $[0,T]$. Our goal is to make the limit transition in each term of the equality obtained.

Using integration by parts and an appropriate change of variables we have that

$$\int_0^T (Mu_{tt},\tilde{u}^h)dt = (Mu_t,\tilde{u}^h)\Big|_{t=0}^{t=T} - \int_0^T (Mu_t,\tilde{u}_t^h)dt = \mathscr{I}_h^1 - \mathscr{I}_h^2(T) + \mathscr{I}_h^2(0),$$

where

$$\mathscr{I}_h^1 = \frac{1}{2h}\int_{-h}^{h} \left[(M^{1/2}u_t(T),M^{1/2}\tilde{u}_t(T+\tau)) - (M^{1/2}u_t(0),M^{1/2}\tilde{u}_t(\tau))\right] d\tau,$$

$$\mathscr{I}_h^2(T) = \frac{1}{2h}\int_{T-h}^{T} (M^{1/2}u_t(t),M^{1/2}\tilde{u}_t(t+h))dt,$$

$$\mathscr{I}_h^2(0) = \frac{1}{2h}\int_{-h}^{0} (M^{1/2}u_t(t+h),M^{1/2}\tilde{u}_t(t))dt.$$

The scalar function $(M^{1/2}u_t(T),M^{1/2}\tilde{u}(T+\tau)) - (M^{1/2}u_t(0),M^{1/2}\tilde{u}(\tau))$ is continuous with respect to τ, therefore we obviously have that

$$\lim_{h\to 0} \mathscr{I}_h^1 = |M^{1/2}u_t(T)|^2 - |M^{1/2}u_t(0)|^2.$$

Because $M^{1/2}\tilde{u}_t(t)$ is strongly continuous outside of $(0,T)$ (and weakly in $[0,T]$), we have that

$$\lim_{h\to 0}\left\{\sup_{T-h\le t\le T}\left|(M^{1/2}u_t(t),M^{1/2}\tilde{u}_t(t+h))-|M^{1/2}u_t(T)|^2\right|\right\}=0.$$

This implies that $\lim_{h\to 0}\mathscr{I}_h^2(T)=\frac{1}{2}|M^{1/2}u_t(T)|^2$. A similar argument yields that $\mathscr{I}_h^2(0)\to\frac{1}{2}|M^{1/2}u_t(0)|^2$ as $h\to 0$. Therefore we obtain that

$$\lim_{h\to 0}\int_0^T(Mu_{tt},\tilde{u}^h)dt=\frac{1}{2}\left(|M^{1/2}u_t(T)|^2-|M^{1/2}u_t(0)|^2\right).$$

Applying similar calculations we also have that

$$\lim_{h\to 0}\int_0^T(\mathscr{A}u,\tilde{u}^h)dt=\frac{1}{2}\left(|\mathscr{A}^{1/2}u(T)|^2-|\mathscr{A}^{1/2}u_t(0)|^2\right).$$

Now we make the limit transition in the damping term involving $D_0(t)$. It follows from (2.4.80) and (2.4.85) that

$$\int_0^T|D_0(t)^{1/2}\tilde{u}^h|^2dt\le C\quad\text{for all }\ 0<h<1.$$

Thus there exists element $k(t)\in L_2(0,T;\mathscr{H})$ such that $D_0^{1/2}(t)\tilde{u}^h(t)\to k(t)$ weakly in $L_2(0,T;\mathscr{H})$ as $h\to 0$ along a subsequence. On smooth functions ϕ we have that

$$\int_0^T(D_0^{1/2}\tilde{u}^h,\phi)dt=\int_0^T(\tilde{u}^h,D_0^{1/2}\phi)dt.$$

It is also clear that $\tilde{u}^h(t)\to\tilde{u}_t(t)$ weakly in V for every t that $|\tilde{u}^h(t)|$ is uniformly bounded with respect to $t\in[0,T]$ and $h\in(0,1)$. Therefore the Lebesgue dominated convergence theorem yields that

$$D_0(t)^{1/2}\tilde{u}^h(t)\to D_0(t)^{1/2}\tilde{u}_t(t)\quad\text{weakly in }\ L_2(0,T;\mathscr{H}).$$

Thus we have that

$$\lim_{h\to 0}\int_0^T(D_0(t)u_t(t),\tilde{u}^h(t))dt=\int_0^T|D_0^{1/2}(t)u_t(t)|^2dt.$$

In a similar way we can conclude that

$$\lim_{h\to 0}\int_0^T\langle\mathscr{A}GgG^*\mathscr{A}u_t(t),\tilde{u}^h(t)\rangle dt=\int_0^T\langle gG^*\mathscr{A}u_t(t),G^*\mathscr{A}u_t(t)\rangle dt.$$

The limit transition in the term

$$\int_0^T(-D_1(t)u_t(t)+F(t)u(t),\tilde{u}^h(t))dt$$

is obvious due to the fact that $\tilde{u}^h(t) \to \tilde{u}_t(t)$ weakly in V for every t that $|\tilde{u}^h(t)|_V$ is uniformly bounded with respect to $t \in [0,T]$ and $h \in (0,1)$. This completes the proof of Proposition 2.4.36. □

Completion of the proof of Theorem 2.4.35. We first note that the energy identity (2.4.82) via Gronwall's type argument implies the estimate (2.4.79) for *every* weak solution to problem (2.4.77). In particular this fact implies the uniqueness of these solutions under conditions of the second part of the theorem.

It is also clear from identity (2.4.82) that the scalar function

$$t \mapsto E_0(u(t), u_t(t)) \equiv \frac{1}{2}\left(|M^{1/2}u_t(t)|^2 + |\mathscr{A}^{1/2}u(t)|^2\right)$$

is continuous. Therefore the weak continuity of $(u(t); u_t(t))$ in H implies its strong continuity; that is, the function $u(t)$ possesses properties (2.4.81). Thus the proof of Theorem 2.4.35 is complete. □

2.4.4 *On higher regularity of solutions*

One of the typical approaches to construction of regular solutions relies on the study of the equation that can be obtained by differentiating (with respect to t) the original equation (see, e.g., [123] and the references therein). For instance, if we deal with the problem of the form

$$\begin{cases} Mu_{tt}(t) + \mathscr{A}u(t) + \mathscr{A}Gg(G^*\mathscr{A}u_t(t)) \\ \qquad\qquad + Du_t(t) + D_0h(u(t)) + D_0D_0^*u_t = F(u(t)), \quad t > 0, \\ u|_{t=0} = u_0, \quad u_t|_{t=0} = u_1, \end{cases} \tag{2.4.86}$$

then the formal differentiation gives us a nonautonomous linear differential equation with respect to $z = u_t$ of the form

$$\begin{cases} Mz_{tt} + \mathscr{A}z + D'(u_t(t))z_t + \mathscr{A}Gg'(G^*\mathscr{A}u_t(t))G^*\mathscr{A}z_t \\ \qquad\qquad + D_0h'(u(t))z + D_0D_0^*z_t = F'(u(t))z, \\ z|_{t=0} = z_0 \equiv u_1, \quad z_t|_{t=0} = z_1 \equiv u_{tt}(0), \end{cases} \tag{2.4.87}$$

where $D'(v)$, $g'(v)$, $h'(u)$, and $F'(u)$ denote the corresponding Fréchet derivatives that are linear operators in appropriate spaces for every fixed v and u. The initial velocity $z_1 \equiv u_{tt}(0)$ is calculated from (2.4.86):

$$z_1 = -M^{-1}\left[\mathscr{A}u_0 + \mathscr{A}Gg(G^*\mathscr{A}u_1) + D(u_1) + D_0h(u_0) + D_0D_0^*u_1 - F(u_0)\right].$$

The equation in (2.4.87) is linear in z with the coefficients depending on $u(t)$.

In order to obtain information on the additional regularity of solutions to a problem of the form (2.4.87), further regularity hypotheses imposed on the nonlinear operators in (2.4.86) are needed, in addition to the set of assumptions given in Assumption 2.4.1. We do not pursue full generality of the approach, but rather we consider simplified model (2.4.26) with $g \equiv 0$ and with nonlinearity F of the form $F(u,v) = F(u) + H(v)$ which depends both on displacement u and velocity v. More precisely, the model we discuss in this section is the following:

$$\begin{cases} Mu_{tt}(t) + \mathscr{A}u(t) + D(u_t(t)) = F(u(t)) + H(u_t(t)) + p(t), \\ u|_{t=0} = u_0, \quad u_t|_{t=0} = u_1, \end{cases} \tag{2.4.88}$$

under the set of requirements imposed in Assumption 2.4.15. In the first step we slightly improve the regularity property (2.4.32) of strong solutions. The corresponding result is formulated below.

2.4.37. Proposition. *In addition to the hypotheses of Theorem* 2.4.16 *assume that the spaces V and $\mathscr{H}$ are separable, the operators D and F and H are Fréchet differentiable, and the following mappings are continuous.*

- $D'(v), H'(v) : V \mapsto V'$ *for every* $v \in V$.
- $F'(v) : \mathscr{D}(\mathscr{A}^{1/2}) \mapsto V'$ *for every* $v \in \mathscr{D}(\mathscr{A}^{1/2})$.

Moreover we assume that

$$\|D'(v)\|_{\mathscr{L}(V,V')} + \|H'(v)\|_{\mathscr{L}(V,V')} + \|F'(w)\|_{\mathscr{L}(\mathscr{D}(\mathscr{A}^{1/2}),V')} \le C_r$$

for every $v \in \mathscr{D}(\mathscr{A}^{1/2})$ and $w \in W$ such that $|\mathscr{A}^{1/2}v| < r$ and $|\mathscr{A}w|_{V'} < r$, where W is defined by (2.4.72). *If $p \in W_1^1(0,T;V')$ and $(u_0;u_1) \in W \times \mathscr{D}(\mathscr{A}^{1/2})$, then the strong solution $u(t)$ to* (2.4.88) *possesses the following regularity property*

$$u_{tt} \in C(0,T;V), \quad u_t \in C(0,T;\mathscr{D}(\mathscr{A}^{1/2})) \text{ and } \mathscr{A}u(t) \in C(0,T;V').$$

Proof. If $u(t)$ is a strong solution to (2.4.88), then it is easy to see that $z(t) = u_t(t)$ is a weak solution to the problem

$$\begin{cases} Mz_{tt}(t) + \mathscr{A}z(t) = \widetilde{p}(t), \quad z|_{t=0} = u_1 \in \mathscr{D}(\mathscr{A}^{1/2}), \\ z_t|_{t=0} = M^{-1}(-\mathscr{A}u_0 - Du_1 + H(u_1) + F(u_0) + p(0)) \in V, \end{cases}$$

where time derivatives are considered in the sense of distributions and

$$\widetilde{p}(t) \equiv \left[-D'(u_t(t)) + H'(u_t(t))\right] u_{tt}(t) + F'(u(t))u_t(t) + p_t(t).$$

It follows from (2.4.32), (2.4.33) and from the hypotheses concerning the Frechet derivatives D', H' and F' that $\widetilde{p} \in L_1(0,T;V')$. Therefore we can apply Theorem 2.4.31 to conclude that

$$(u_t; u_{tt}) \equiv (z; z_t) \in C(0,T; \mathscr{D}(\mathscr{A}^{1/2}) \times V).$$

Therefore directly from (2.4.88) we obtain that $\mathscr{A}u(t) \in C(0,T;V')$. □

We note that the conditions on D and H imposed by Proposition 2.4.37 are rather restrictive. If $V = \mathscr{H} = L_2(\Omega)$ and D is the Nemytskij operator, then the property $D'(v) : \mathscr{H} \mapsto \mathscr{H}$ means in fact that D is an affine operator. Thus, abstract formulation of this sort forces assumptions that may be too strong. In fact, in the study of von Karman evolutions we are able to use the structure of nonlinear terms and the resulting cancelations, in order to overcome the limitations dictated by a more general abstract framework; see Sections 3.2.3 and 4.1.5, for instance.

One could also formulate higher-order regularity theorems. This requires an additional set of assumptions imposed on nonlinear quantities (and the initial data). In order to keep the exposition focused, we do not do this at the abstract level for the nonlinear case. However, later when dealing with a more specific context, we show that the method used for the proof of Proposition 2.4.37 applies with straightforward modifications and leads to regularity (differentiability) of solutions of higher order. We also refer the reader to the paper [123], where these ideas are presented for second-order nonlinear equations with linear internal damping. A typical result on higher regularity is presented in the following assertion which deals with the linear problem (2.4.71).

2.4.38. Theorem. *Let M and $\mathscr{A}$ satisfy Assumption* 2.4.1(1,2). *Assume that D is a monotone linear operator from V into V' and $p \in W_1^m(0,T,V')$ for some $m \geq 1$. Then weak solutions $u(t)$ to the problem*

$$\begin{cases} Mu_{tt}(t) + \mathscr{A}u(t) + Du_t(t) = p(t), \\ u|_{t=0} = u_0, \quad u_t|_{t=0} = u_1, \end{cases} \tag{2.4.89}$$

enjoy the following additional regularity properties,

$$\begin{cases} u^{(k)}(t) \in C(0,T;W) \ \textit{for} \ k = 0,1,2,\dots,m-1, \\ u^{(m)}(t) \in C(0,T;\mathscr{D}(\mathscr{A}^{1/2})), \quad u^{(m+1)}(t) \in C(0,T;V), \end{cases} \tag{2.4.90}$$

where W is defined by (2.4.72), *if and only if the following compatibility conditions*

$$\begin{cases} u^{(k)}(0) \in W \ \textit{for} \ k = 0,1,2,\dots,m-1, \\ u^{(m)}(0) \in \mathscr{D}(\mathscr{A}^{1/2}), \quad u^{(m+1)}(0) \in V, \end{cases} \tag{2.4.91}$$

hold, where the values $u^{(k)}(0)$ are defined by the recurrence relations

$$\begin{cases} u^{(0)}(0) = u_0, \quad u^{(1)}(0) = u_1, \\ u^{(k)}(0) = M^{-1}\left(-\mathscr{A}u^{(k-2)}(0) - Du^{(k-1)}(0) + p^{(k-2)}(0)\right), \quad k \geq 2. \end{cases} \tag{2.4.92}$$

Proof. We follow the line of argument given in [123].

It is clear that (2.4.90) implies (2.4.91).

Assume now that (2.4.91) holds and consider the problem

$$\begin{cases} Mv_{tt}(t) + \mathscr{A}v(t) + Dv_t(t) = p^{(k)}(t), \\ u|_{t=0} = v_{0k} \equiv u^{(k)}(0), \quad u_t|_{t=0} = v_{1k} \equiv u^{(k+1)}(0), \end{cases} \tag{2.4.93}$$

where $k \in \{1,\dots,m\}$. By (2.4.91) $(v_{0k}, v_{1k}) \in W \times \mathscr{D}(\mathscr{A}^{1/2})$ for $1 \le k \le m-1$ and $(v_{0m}; v_{1m}) \in \mathscr{D}(\mathscr{A}^{1/2}) \times V$. Therefore Theorem 2.4.31 implies that problem (2.4.93) has solution $v_k(t)$ from $C^1(0,T;\mathscr{D}(\mathscr{A}^{1/2})) \cap C(0,T;W)$ for $1 \le k \le m-1$ and from $C^1(0,T;V) \cap C(0,T;\mathscr{D}(\mathscr{A}^{1/2}))$ for $k = m$. Problem (2.4.89) is linear, therefore using uniqueness of a weak solution by induction in k we find that $u^{(k)}(t) = v_k(t)$ for every $1 \le k \le m$. □

2.5 Linear plate models

We collect here some results pertaining to linear plate equations with various boundary conditions. Some of the estimates derived here are used later in the context of nonlinear plate theory. Our main emphasis is on "special trace" regularity results. By "special," we mean trace results that do not follow from the interior regularity followed by trace theory. These results are sometimes referred to as "hidden" trace regularity.

We begin with equations equipped with homogeneous boundary conditions.

Below we use the notation $\|\cdot\|_{s,\mathscr{O}}$ for the norm in the Sobolev space $H^s(\mathscr{O})$. We also denote by $\|\cdot\|_{\mathscr{O}} \equiv \|\cdot\|_{s,\mathscr{O}}$ and $(\cdot,\cdot)_{\mathscr{O}}$ the norm and the inner product in $L_2(\Omega)$ The subscript $\mathscr{O}$ in the norms and inner products can be omitted when apparent from context.

2.5.1 *Homogeneous boundary conditions*

Three different types of boundary conditions are considered: clamped, hinged (simply supported) and free.

2.5.1.1 Clamped boundary conditions

We first consider clamped boundary conditions. This is to say, we study the following problem,

$$\begin{cases} u_{tt} - \alpha\Delta u_{tt} + \Delta^2 u = f(x,t) & \text{in } Q \equiv \Omega \times (0,T) \\ u = \frac{\partial}{\partial n} u = 0 & \text{on } \Sigma = \Gamma \times (0,T) \\ u|_{t=0} = u_0(x), \quad u_t|_{t=0} = u_1(x). \end{cases} \quad (2.5.1)$$

We can rewrite this problem in the form (2.4.71) with $\mathscr{H} = L_2(\Omega)$ and $V = V_\alpha$, where $V_\alpha = H_0^1(\Omega)$ for $\alpha > 0$ and $V_\alpha = L_2(\Omega)$ for $\alpha = 0$. In this case $D = 0$, $M = I - \alpha\Delta$, $\mathscr{D}(M) = (H^2 \cap H_0^1)(\Omega)$ when $\alpha > 0$, and $\mathscr{A} = \Delta^2$ with the domain $\mathscr{D}(\mathscr{A}) = (H^4 \cap H_0^2)(\Omega)$. We also have that $\mathscr{D}(\mathscr{A}^{1/2}) = H_0^2(\Omega)$.

Applying Theorem 2.4.31 we obtain the following assertion.

2.5.1. Proposition. *Assume that $f(x,t) \in L_1(0,T;V_\alpha')$ and $(u_0;u_1)$ belongs to the space $H_0^2(\Omega) \times V_\alpha$. Then problem (2.5.1) has a unique generalized solution such that $(u;u_t) \in C(0,T;H_0^2(\Omega) \times V_\alpha)$ and the energy equality*

$$E_0(u(t),u_t(t)) = E_0(u_0,u_1) + \int_0^t \int_\Omega f(\tau,x) u_t(\tau,x)\, dx d\tau \quad (2.5.2)$$

holds. Here

$$E_0(u_0,u_1) = \frac{1}{2} \int_\Omega \left(|u_1(x)|^2 + \alpha|\nabla u_1(x)|^2 + |\Delta u_0(x)|^2 \right) dx. \quad (2.5.3)$$

Any generalized solution is also weak and vice versa.

Proof. This is a direct corollary of the second part of Theorem 2.4.31. □

The following assertion gives additional regularity of weak (generalized) solutions. Below we use the notation $w^{(k)}(t) = \partial_t^k w(x,t)$ and define values $u^{(k)}(0)$ by the recurrence relations

$$\begin{cases} u^{(0)}(0) = u_0, \quad u^{(1)}(0) = u_1, \\ u^{(k)}(0) = (I - \alpha\Delta_D)^{-1} \left(\Delta^2 u^{(k-2)}(0) + f^{(k-2)}(0) \right), \quad k \geq 2, \end{cases} \quad (2.5.4)$$

where Δ_D is the Laplace operator with the Dirichlet boundary conditions.

2.5.2. Proposition. *Let $m \geq 1$.*

- **Case $\alpha > 0$.** *Assume that $f \in W_1^m(0,T;H^{-1}(\Omega))$ and*

$$f^{(k)}(x,t) \in C(0,T;H^{m-2-k}(\Omega)), \quad k = 0,1,\dots,m-2, \quad (2.5.5)$$

and the compatibility conditions

$$u^{(k)}(0) \in (H^3 \cap H_0^2)(\Omega), \quad k = 0,1,\dots,m-1,$$

$$u^{(m)}(0) \in H_0^2(\Omega), \quad u^{(m+1)}(0) \in H_0^1(\Omega),$$

hold. Then the weak solution $u(t)$ to problem (2.5.1) *belongs to the class*

$$\left\{v(t)\,:\,v^{(k)}(t)\in C(0,T;H^{m+2-k}(\Omega)),\ k=0,\ldots,m+1\right\}. \tag{2.5.6}$$

Moreover for every $t\in[0,T]$ *we have* $u^{(k)}(t)\in H_0^2(\Omega)$ *for* $k=0,\ldots,m$ *and* $u^{(m+1)}(t)\in H_0^1(\Omega)$.

- **Case** $\alpha=0$. *Assume that* $f\in W_1^m(0,T;L_2(\Omega))$ *and*

$$f^{(k)}(x,t)\in C(0,T;H^{2(m-k-1)}(\Omega)),\quad k=0,1,\ldots,m-2, \tag{2.5.7}$$

and the compatibility conditions

$$u^{(k)}(0)\in(H^4\cap H_0^2)(\Omega),\quad k=0,1,\ldots,m-1,$$

$$u^{(m)}(0)\in H_0^2(\Omega),\quad u^{(m+1)}(0)\in L_2(\Omega),$$

hold. Then the weak solution $u(t)$ *to problem* (2.5.1) *belongs to the class*

$$\left\{v(t)\,:\,v^{(k)}(t)\in C(0,T;H^{2(m+1-k)}(\Omega)),\ k=0,\ldots,m+1\right\}. \tag{2.5.8}$$

Moreover for every $t\in[0,T]$ *we have* $u^{(k)}(t)\in H_0^2(\Omega)$ *for* $k=0,\ldots,m$.

Proof. Let $\alpha>0$. First we apply Theorem 2.4.38. In the case considered we have that $V=H_0^1(\Omega)$, $V'=H^{-1}(\Omega)$ and $f\in W_1^m(0,T;V')$. We also have that

$$W\equiv\left\{u\in\mathscr{D}(\mathscr{A}^{1/2})\,:\,\mathscr{A}u\in V'\right\}=(H^3\cap H_0^2)(\Omega)$$

in the present case. Therefore Theorem 2.4.38 implies that

$$u^{(m+1)}(t)\in C(0,T;H_0^1(\Omega)),\quad u^{(m)}(t)\in C(0,T;H_0^2(\Omega)),$$

$$u^{(k-1)}(t)\in C(0,T;(H^3\cap H_0^2)(\Omega)),\quad k=1,\ldots,m.$$

In particular,

$$u^{(m-j)}(t)\in C(0,T;H^{j+2}(\Omega)), \tag{2.5.9}$$

for $j=-1,0,1$. To conclude the proof it remains to check that this relation holds for every $j\le m$. From (2.5.1) we obtain that

$$\Delta^2u^{(m-j)}(t)=-(I-\alpha\Delta)u^{(m-[j-2])}+f^{(m-j)},\quad j=2,\ldots,m. \tag{2.5.10}$$

By (2.5.5) we have that $f^{(m-j)}(t)\in C(0,T;H^{j-2}(\Omega))$ for $j=2,\ldots,m$. Therefore, induction in j leads to (2.5.9) for $j=2,\ldots,m$. Thus the assertion for $\alpha>0$ is proved.

In the case $\alpha=0$ we have

$$W=\left\{u\in\mathscr{D}(\mathscr{A}^{1/2})\,:\,\mathscr{A}u\in V'\right\}=(H^4\cap H_0^2)(\Omega).$$

As above, application of Theorem 2.4.38 implies that

$$u^{(m-j)}(t) \in C(0,T;H^{2(j+1)}(\Omega)),$$

for $j=-1,0,1$. By (2.5.7) we have that $f^{(m-j)}(t) \in C(0,T;H^{2(j-1)}(\Omega))$ for $j=2,\dots,m$. Consequently, by induction in j, the relation (2.5.10) with $\alpha=0$ leads to the desired result. □

In our further considerations the trace estimate for $\Delta u|_\Gamma$ is essential. A variant of this estimate has been used in the context of studying stability of unforced plates [9]. We need a "weighted" form of this trace estimate which is given below. It asserts existence of Δu on the boundary for energy (weak) solutions.

2.5.3. Theorem. *Let u be a weak solution to the linear problem*

$$\begin{cases} u_{tt} - \alpha\Delta u_{tt} + \Delta^2 u = f(x,t), \quad x\in\Omega, t>0, \\ u = \frac{\partial}{\partial n} u = 0 \text{ on } \Gamma, t>0, \quad u\big|_{t=0} = u_0, u_t\big|_{t=0} = u_1. \end{cases} \tag{2.5.11}$$

Let $h(x)$ be a smooth vector field on $\overline{\Omega}$ such that $h|_\Gamma = n$, on the boundary Γ, where n is the outward normal to the boundary.[4] *Then for any $\omega \geq 0$ we have*

$$c_0\int_{\Sigma_t} e^{-\omega(t-s)}|\Delta u|^2 dxds \leq c_1(1+\omega)\int_0^t e^{-\omega(t-s)}E_0(s)ds + c_2\left[E_0(0)e^{-\omega t} + E_0(t)\right] + \left|\int_0^t e^{-\omega(t-s)}\int_\Omega f(x,s)h\nabla u dxds\right|, \tag{2.5.12}$$

where $\Sigma_t \equiv \Gamma\times(0,t)$ and $E_0(t)\equiv E_0(u(t),u_t(t))$ is given by (2.5.3). All the constants c_i are uniform in $0\leq\alpha<\alpha_0$ and $\omega\geq 0$.

Proof. We follow the same line of argument as in [76]. For computations we consider smooth solutions that are guaranteed by Proposition 2.5.2. The final conclusion is achieved by the passage to the limit.

We multiply both sides of equation (2.5.11) by $h\nabla u e^{\omega t}$ and integrate over $Q_t = \Omega\times(0,t)$. This gives

$$\int_0^t (Mu_{tt} + \Delta^2 u - f, h\nabla u)e^{\omega s}ds = 0, \tag{2.5.13}$$

where we use notation $M = 1-\alpha\Delta$. Here and below we omit the index Ω in the $L_2(\Omega)$ inner product. We use partial integrations which gives us the following relation

$$\int_0^t (Mu_{tt}, h\nabla u e^{\omega s})ds = -\int_0^t (Mu_t, h\nabla u_t)e^{\omega s}ds + \Psi_0(t), \tag{2.5.14}$$

where

[4] The vector field $h(x)$ can be easily constructed for domains that are smooth. In the case of rectangles (or more generally Lipschitz domains) one uses vector fields that are transverse to the boundaries with normals forming an acute angle to the boundary. Because this technical detail is not essential to our analysis, we treat in detail the case of perpendicular vectors.

$$|\Psi_0(t)| \le c_1 \left[E_0(t)e^{\omega t} + E_0(0)\right] + c_2\omega \int_0^t E_0(s)e^{\omega s}ds$$

with the constants c_1 and c_2 independent of $\alpha \in [0, \alpha_0]$ and $\omega \ge 0$.

The critical term is biharmonic. Application of Green's formula and the fact that $u = 0$ and $\nabla u = 0$ on Γ give:

$$(\Delta^2 u, h\nabla u) = (\Delta u, \Delta(h\nabla u)) - \left(\Delta u, \frac{\partial}{\partial n}(h\nabla u)\right)_\Gamma. \tag{2.5.15}$$

On the other hand direct calculations imply

$$(\Delta u, \Delta(h\nabla u)) = \psi_0 + (\Delta u, h\cdot\nabla\Delta u), \tag{2.5.16}$$

where $|\psi_0| \le C\|u\|^2_{2,\Omega}$. By applying divergence theorem to the last term above yields:

$$\begin{aligned}(\Delta u, h\cdot\nabla\Delta u) &= \frac{1}{2}\int_\Omega \operatorname{div}\{h(\Delta u)^2\}dx - \frac{1}{2}\int_\Omega \operatorname{div}\{h\}|\Delta u|^2 dx\\ &= \frac{1}{2}\int_\Gamma (h,n)|\Delta u|^2 dx - \frac{1}{2}\int_\Omega \operatorname{div}\{h\}|\Delta u|^2 dx\\ &= \frac{1}{2}\int_\Gamma |\Delta u|^2 dx - \frac{1}{2}\int_\Omega \operatorname{div}\{h\}|\Delta u|^2 dx.\end{aligned} \tag{2.5.17}$$

On the other hand, using the boundary conditions for u and the representation for $\Delta u\big|_\Gamma$ given by (1.3.8) we have that

$$\left(\Delta u, \frac{\partial}{\partial n}(h\nabla u)\right)_\Gamma \equiv \int_\Gamma \Delta u\cdot\frac{\partial}{\partial n}(h\nabla u)d\Gamma = \int_\Gamma |\Delta u|^2 d\Gamma.$$

Therefore combining (2.5.15)–(2.5.17) yields

$$(\Delta^2 u, h\nabla u) = -\frac{1}{2}\|\Delta u\|^2_\Gamma + \psi_1,$$

where $|\psi_1| \le C\|\Delta u\|^2$. Hence we obtain from (2.5.13) and (2.5.14) that

$$\frac{1}{2}\int_{\Sigma_t} |\Delta u|^2 e^{\omega s}dxds = -\int_0^t (Mu_t, h\nabla u_t)e^{\omega s}ds + \Psi(t), \tag{2.5.18}$$

where

$$\begin{aligned}|\Psi(t)| \le c_1 \left[E_0(t)e^{\omega t} + E_0(0)\right] + c_2(1+\omega)\int_0^t E_0(s)e^{\omega s}ds\\ + \left|\int_0^t e^{\omega s}\int_\Omega f(x,s)h\nabla u dxds\right|\end{aligned}$$

with the constants c_1 and c_2 independent of $\alpha \in [0, \alpha_0]$ and $\omega \ge 0$. Thus to conclude the proof we need to estimate the term $(Mv, h\nabla v)$ for any smooth v such that $v = 0$

and $\nabla v = 0$ on Γ. We obviously have that

$$(Mv, h\nabla v) = (v, h\nabla v) - \alpha(\Delta v, h\nabla v).$$

The first term can be rewritten in the form

$$\begin{aligned}(v, h\nabla v) &= \frac{1}{2}\int_\Omega h\nabla|v|^2 dx = \frac{1}{2}\int_\Omega \operatorname{div}\{h|v|^2\}dx - \frac{1}{2}\int_\Omega \operatorname{div}\{h\}|v|^2 dx \\ &= \frac{1}{2}\int_\Gamma (h,n)|v|^2 dx - \frac{1}{2}\int_\Omega \operatorname{div}\{h\}|v|^2 dx = -\frac{1}{2}\int_\Omega \operatorname{div}\{h\}|v|^2 dx.\end{aligned}$$

As for the second term, after simple calculation we have that

$$\Delta v \cdot h\nabla v = \operatorname{div}\left\{\nabla v \cdot h\nabla v - \frac{1}{2}h|\nabla v|^2\right\} + \frac{1}{2}\operatorname{div}\{h\}|\nabla v|^2 - \sum_{i=1,2}\frac{\partial v}{\partial x_i}\left(\frac{\partial h}{\partial x_i}, \nabla v\right).$$

Therefore applying the divergence theorem and using the boundary conditions for ∇v yield that

$$\left|\int_\Omega \Delta v h \nabla v dx\right| \leq c\int_\Omega |\nabla v|^2 dx.$$

Consequently the relations above give us the estimate

$$\left|\int_0^t (Mu_t, h\nabla u_t)e^{\omega s} ds\right| \leq c\int_0^t \|M^{1/2}u_t\|_{0,\Omega}^2 e^{\omega s} ds$$

which, due to (2.5.18), implies the desired inequality (2.5.12) claimed in Theorem 2.5.3. □

As a corollary of Theorem 2.5.3 we obtain the following assertion.

2.5.4. Theorem. *Let $\alpha \geq 0$ and $u(t)$ be a weak solution to* (2.5.1). *Then*

$$\int_\Sigma |\Delta u|^2 d\Sigma \leq c_T \sup_{t\in[0,T]} \left[\|u(t)\|_{2,\Omega}^2 + \|u_t(t)\|_{0,\Omega}^2 + \alpha\|u_t(t)\|_{1,\Omega}^2\right] + c\left[\int_0^T \|f\|_{-1,\Omega} dt\right]^2.$$

Consequently by the energy relation in (2.5.2), *when $\alpha > 0$*

$$\int_\Sigma |\Delta u|^2 d\Sigma \leq c_T \left[\|u(0)\|_{2,\Omega}^2 + \|u_t(0)\|_{0,\Omega}^2 + \alpha\|u_t(0)\|_{1,\Omega}^2\right] + c\left[\int_0^T \|f\|_{-1,\Omega} dt\right]^2,$$

and when $\alpha = 0$

$$\int_\Sigma |\Delta u|^2 d\Sigma \leq c_T \left[\|u(0)\|_{2,\Omega}^2 + \|u_t(0)\|_{0,\Omega}^2\right] + c\left[\int_0^T \|f\|_{0,\Omega} dt\right]^2.$$

We note that the result of Theorem 2.5.4 provides $\frac{1}{2}$ derivative more than the usual trace theorems. Indeed, for finite energy solutions in $H^2(\Omega)$, the trace theorem, which can only be applied formally, will give at most $\Delta u \in H^{-1/2}(\Gamma)$. The result of Theorem 2.5.4 can also be obtained from the corresponding regularity result proved for the wave equation in [193]. In fact, by using the same multiplier $h\nabla u$, explicit proof of inequality in Theorem 2.5.4 is given in [178].

Assuming more regularity of solutions one obtains higher regularity of traces.

2.5.5. Theorem. *Let $\alpha \geq 0$ and $u(t)$ be a solution to* (2.5.1) *from the class*

$$C(0,T;H^3(\Omega)) \cap C^1(0,T;H_0^2(\Omega)).$$

Then

$$\int_\Sigma |\nabla \Delta u|^2 d\Sigma \leq c \sup_{t\in[0,T]} \left[\|u(t)\|_{3,\Omega}^2 + \|u_t(t)\|_{1,\Omega}^2 + \alpha \|u_t(t)\|_{2,\Omega}^2 \right] + c \left[\int_0^T \|f\|_{0,\Omega} dt \right]^2.$$

The result in this theorem is given in [211]. The proof is carried as follows. First regularity in Theorem 2.5.4 is applied to tangential derivatives of the solution. This gives the correct regularity of $(\partial/\partial\tau)\Delta u|_\Gamma$. Then applying multiplier $h\nabla\Delta u$ (with the same h as in Theorem 2.5.3) one reconstructs the regularity of $(\partial/\partial n)\Delta u$ on the boundary. The details are given in [216].

In what follows we also consider models with mixed boundary conditions. These involve combinations of clamped boundary conditions defined on Γ_0 with free (or hinged) boundary conditions on Γ_1, with $\Gamma = \Gamma_0 \cup \Gamma_1$. It is assumed that the intersection of the closures of these two parts of the boundary is empty. In such cases, the trace regularity announced in Theorems 2.5.3 and 2.5.4 is the same, but applied only to the clamped part of the boundary Γ_0. This is due to the fact that the argument used in the proofs of these theorems is completely local hence applicable to each part of the boundary separately. Indeed, the construction of a C^1 vector field h parallel to the normal direction to Γ_0 is required to vanish in the neighborhood of Γ_1. This is possible to accomplish, due to the fact that both parts of the boundary are separated (otherwise one may run into questions of the regularity of h). The above procedure leads to a "hidden" regularity result obtained for the following problem,

$$\begin{cases} u_{tt} - \alpha\Delta u_{tt} + \Delta^2 u = f(x,t) \text{ in } Q \equiv \Omega \times (0,T), \\ u = \frac{\partial}{\partial n} u = 0 \text{ on } \Sigma_0 = \Gamma_0 \times (0,T), \\ \text{Hinged or Free boundary conditions on } \Sigma_1 = \Gamma_1 \times (0,T). \\ u|_{t=0} = u_0(x), \quad u_t|_{t=0} = u_1(x). \end{cases} \tag{2.5.19}$$

We refer to (2.5.20) (resp., (2.5.22)) for the exact description of the hinged (resp., free) boundary conditions in the homogeneous case (see (2.5.45) and (2.5.55) for the case of nonhomogeneous boundary conditions on Σ_1). We also refer to Propositions 2.5.7, 2.5.14, and 2.5.16 below for well-posedness results concerning problem

(2.5.19) in the case of homogeneous boundary conditions. Concerning the nonhomogeneous, see Theorems 2.5.24, 2.5.25, and 2.5.28.

The following theorem is an analog of Theorem 2.5.4 for the mixed boundary conditions considered.

2.5.6. Theorem. *Assume that $\overline{\Gamma_0} \cap \overline{\Gamma_1} = \emptyset$. Let u be a weak solution to equation (2.5.19) with $\alpha \geq 0$ which lies in the class $C(0,T;H^2(\Omega)) \cap C^1(0,T;H_\alpha(\Omega))$, where $H_\alpha(\Omega) = H^1(\Omega)$ when $\alpha > 0$ and $H_\alpha(\Omega) = L_2(\Omega)$ if $\alpha = 0$. Then the following trace estimate takes place,*

$$\int_{\Sigma_0} |\Delta u|^2 d\Sigma \leq c \sup_{t\in[0,T]} \left[\|u(t)\|_{2,\Omega}^2 + \|u_t(t)\|_{0,\Omega}^2 + \alpha \|u_t(t)\|_{1,\Omega}^2 \right]$$
$$+ c \left[\int_0^T \|f\|_{-1,\Omega} dt \right]^2 .$$

2.5.1.2 Hinged–clamped boundary conditions

We consider a mixture of hinged–clamped boundary conditions. Let Γ_0 and Γ_1 be two disjoint parts of the boundary Γ. We study the following problem.

$$\begin{cases} u_{tt} - \alpha \Delta u_{tt} + \Delta^2 u = f(x,t) \text{ in } Q \equiv \Omega \times (0,T), \\ u = \nabla u = 0 \text{ on } \Sigma_0 = \Gamma_0 \times (0,T), \\ u = \Delta u = 0 \text{ on } \Sigma_1 = \Gamma_1 \times (0,T), \\ u|_{t=0} = u_0(x), \quad u_t|_{t=0} = u_1(x). \end{cases} \tag{2.5.20}$$

As above we can rewrite this problem in the form (2.4.71). We set:

$$\mathscr{H} = L_2(\Omega), \quad \mathscr{A} u = \Delta^2 u \text{ for } u \in \mathscr{D}(\mathscr{A}),$$

where the domain of $\mathscr{A}$ is given by

$$\mathscr{D}(\mathscr{A}) = \left\{ u \in H^4(\Omega) \cap H^2_{\Gamma_0}(\Omega),\ u = \Delta u = 0 \text{ on } \Gamma_1 \right\},$$
$$\mathscr{D}(\mathscr{A}^{1/2}) = H_0^1(\Omega) \cap H^2_{\Gamma_0}(\Omega).$$

Here above

$$H^2_{\Gamma_0}(\Omega) = \left\{ u \in H^2(\Omega),\ u = 0,\ \nabla u = 0 \text{ on } \Gamma_0 \right\}. \tag{2.5.21}$$

We set $M = I - \alpha\Delta$ on $\mathscr{D}(M) = H_0^1(\Omega) \cap H^2(\Omega)$ and introduce the space $V_\alpha = \mathscr{D}(M^{1/2})$ which coincides with $H_0^1(\Omega)$, when $\alpha > 0$ and $L_2(\Omega)$ when $\alpha = 0$.

Applying the second part of Theorem 2.4.31 we obtain the following assertion on the existence and uniqueness weak solutions.

2.5.7. Proposition. *Assume that*

$$f(x,t) \in L_1(0,T;V'_\alpha) \text{ and } (u_0;u_1) \in \mathscr{D}(\mathscr{A}^{1/2}) \times V_\alpha.$$

Then problem (2.5.20) *has a unique generalized solution such that*

$$(u;u_t) \in C(0,T;(H^2_{\Gamma_0} \cap H^1_0)(\Omega) \times V_\alpha)$$

and the energy equality (2.5.2) *holds. Any generalized solution is also weak and vice versa.*

As for the clamped case (see Proposition 2.5.2) we can apply Theorem 2.4.38 to obtain additional regularity of weak solutions. In Proposition 2.5.8, for the sake of clarity, we state the corresponding result for the case of the hinged boundary conditions ($\Gamma_0 = \emptyset$, $\Gamma_1 = \Gamma$). The statement in a more general case ($\Gamma_0 \neq \emptyset$, $\Gamma_1 \neq \emptyset$) is quite similar to Propositions 2.5.8 and 2.5.2.

2.5.8. Proposition. *Let $\Gamma_0 = \emptyset$ and $m \geq 1$.*

- **Case $\alpha > 0$.** *Let $f \in W_1^m(0,T;H^{-1}(\Omega))$ and* (2.5.5) *be valid. Assume that the compatibility conditions*

$$u^{(k)}(0) \in (H^3 \cap H^1_0)(\Omega), \quad \Delta u^{(k)}(0) \in H^1_0(\Omega), \quad k = 0,1,\dots,m-1,$$

$$u^{(m)}(0) \in (H^2 \cap H^1_0)(\Omega), \quad u^{(m+1)}(0) \in H^1_0(\Omega),$$

hold. Then the weak solution $u(t)$ to problem (2.5.20) *belongs to the class* (2.5.6). *Moreover for every $t \in [0,T]$ we have $u^{(k)}(t)\big|_{\partial\Omega} = 0$ for $k = 0,\dots,m+1$ and $\Delta u^{(k)}(t)\big|_{\partial\Omega} = 0$ for $k = 0,\dots,m-1$.*
- **Case $\alpha = 0$.** *Assume that the function $f \in W_1^m(0,T;L_2(\Omega))$ possesses the properties* (2.5.7) *and the compatibility conditions*

$$u^{(k)}(0) \in (H^4 \cap H^1_0)(\Omega), \ \Delta u^{(k)}(0) \in (H^2 \cap H^1_0)(\Omega), \ k = 0,\dots,m-1,$$

$$u^{(m)}(0) \in (H^2 \cap H^1_0)(\Omega), \quad u^{(m+1)}(0) \in L_2(\Omega),$$

hold. Then the weak solution $u(t)$ to problem (2.5.20) *belongs to the class* (2.5.8). *Moreover for every $t \in [0,T]$ we have $u^{(k)}(t)\big|_{\partial\Omega} = 0$ for $k = 0,\dots,m$ and $\Delta u^{(k)}(t)\big|_{\partial\Omega} = 0$ for $k = 0,\dots,m-1$.*

Proof. We apply Theorem 2.4.38 and the arguments similar to those given in the proof of Proposition 2.5.2. We note that in the case considered

$$W = \left\{ u \in H^{r(\alpha)}(\Omega) \ : \ u = \Delta u = 0 \text{ on } \partial\Omega \right\},$$

where $r(\alpha) = 3$ if $\alpha > 0$ and $r(\alpha) = 4$ for $\alpha = 0$. □

We turn now to the issue of *boundary regularity* of solutions to plate equations and consider boundary regularity of solutions to (2.5.20). We begin by first considering more regular solutions (i.e. those belonging to $H^3(\Omega)$ at least).

2.5.9. Theorem. *Let* $\alpha \geq 0$ *and* $u(t)$ *be a solution to* (2.5.20) *from the class* $C(0,T;H^3(\Omega)) \cap C^1(0,T;H^2(\Omega))$. *Then*

$$\int_\Sigma |\frac{\partial}{\partial n} \Delta u|^2 d\Sigma + \int_\Sigma |\frac{\partial}{\partial n} u_t|^2 d\Sigma$$
$$\leq c \sup_{t\in[0,T]} \left[\|u(t)\|^2_{3,\Omega} + \|u_t(t)\|^2_{1,\Omega} + \alpha\|u_t(t)\|^2_{2,\Omega}\right] + c\left[\int_0^T \|f\|_{0,\Omega} dt\right]^2.$$

The regularity on the clamped part of the boundary (i.e., Γ_0) follows from Theorem 2.5.5. Proof of the regularity on Γ_1 follows by applying multiplier $\nu\nabla\Delta u$ [178]. Detailed calculations are given in [216], vol. II, Theorem 10.7.3.1, p. 993, ($\alpha > 0$) and Theorem 10.8.10.1, p. 1022, ($\alpha = 0$).

2.5.10. Remark. When the initial data are trivial ($u_0 = 0$, $u_1 = 0$) one obtains the estimate in terms of the forcing term only. For this, it suffices to apply Duhamel's principle to the semigroup solution. As a consequence, one obtains that

$$\int_\Sigma |\frac{\partial}{\partial n} \Delta u|^2 d\Sigma + \int_\Sigma |\frac{\partial}{\partial n} u_t|^2 d\Sigma \leq c\left[\int_0^T \|f\|_{0,\Omega} dt\right]^2 \quad \text{for} \quad \alpha > 0,$$
$$\int_\Sigma |\frac{\partial}{\partial n} \Delta u|^2 d\Sigma + \int_\Sigma |\frac{\partial}{\partial n} u_t|^2 d\Sigma \leq c\left[\int_0^T \|f\|_{1,\Omega} dt\right]^2 \quad \text{for} \quad \alpha = 0,$$

where $f \in L_1(0,T;H_0^1(\Omega))$. We refer to Sections 10.7 and 10.8 in [216] for some details. ■

Change of variables $\bar{u} \equiv \Delta_D^{-1} u$, where Δ_D is the Laplace operator equipped with Dirichlet boundary conditions, leads to the following lower-level regularity result.

2.5.11. Theorem. *Let* $\alpha \geq 0$ *and* $u(t)$ *be a weak solution to* (2.5.20). *Then*

$$\int_\Sigma |\frac{\partial}{\partial n} u|^2 d\Sigma \leq c \sup_{t\in[0,T]} \left[\|u(t)\|^2_{1,\Omega} + \|u_t(t)\|^2_{-1,\Omega} + \alpha\|u_t(t)\|^2_{0,\Omega}\right]$$
$$+ c\left[\int_0^T \|\Delta_D^{-1} f\|_{0,\Omega} dt\right]^2.$$

2.5.12. Remark. The global estimate for zero initial data ($u_0 = 0$, $u_1 = 0$) reads

$$\int_\Sigma |\frac{\partial}{\partial n} u|^2 d\Sigma \leq c\left[\int_0^T \|\Delta_D^{-1} f\|_{0,\Omega} dt\right]^2 \quad \text{when} \quad \alpha > 0,$$
$$\int_\Sigma |\frac{\partial}{\partial n} u|^2 d\Sigma \leq c\left[\int_0^T \|f\|_{-1,\Omega} dt\right]^2 \quad \text{for} \quad \alpha = 0.$$

■

We note that Theorem 2.5.9 provides, again, $\frac{1}{2}$ derivative more than predicted by the trace theory. This is certainly true for both terms on the right side of inequality and $\alpha = 0$. In the case $\alpha > 0$ the regularity of $(\partial/\partial n)u_t$ is redundant, as it follows from a priori interior regularity. This raises the following question. Can we do better when $\alpha > 0$? And the answer is positive as shown by the theorem below.

2.5.13. Theorem. *Let $\alpha > 0$. Then*

$$\int_\Sigma |\frac{\partial}{\partial n} u_t|^2 d\Sigma \le c \sup_{t\in[0,T]} \left[\|u(t)\|_{2,\Omega}^2 + \|u_t(t)\|_{1,\Omega}^2\right] + c\left[\int_0^T \|f\|_{-1,\Omega} dt\right]^2.$$

The proof of this theorem is rather technical and it follows from reduction of the dynamics to the wave equation [210, Theorem 1.3].

2.5.1.3 Free–clamped boundary conditions

We consider

$$\begin{cases} u_{tt} - \alpha\Delta u_{tt} + \Delta^2 u = f(x,t) \text{ in } Q \equiv \Omega \times (0,T), \\ u = \nabla u = 0 \text{ on } \Sigma_0 = \Gamma_0 \times (0,T), \\ \Delta u + (1-\mu)B_1 u = 0 \text{ on } \Sigma_1 = \Gamma_1 \times (0,T), \\ \frac{\partial}{\partial n}\Delta u + (1-\mu)B_2 u - \alpha\frac{\partial}{\partial n} u_{tt} - \nu_1 u = 0 \text{ on } \Sigma_1, \\ u|_{t=0} = u_0(x), \;\; u_t|_{t=0} = u_1(x), \end{cases} \tag{2.5.22}$$

where B_1 and B_2 are defined by (1.3.20), $0 < \mu < 1$, and Γ_0 , Γ_1 are two disjoint parts of the boundary such that $\Gamma_0 \cup \Gamma_1 = \Gamma$. We assume that $\nu_1 \ge 0$ ($\nu_1 > 0$ if $\Gamma_0 = \emptyset$).

To reformulate problem (2.5.22) in an abstract form (2.4.71) we take into account the following formal observation.

Let us multiply the equation in (2.5.22) by smooth function ϕ such that $\phi = \nabla\phi = 0$ on Γ_0 and integrate over Ω. We obtain

$$\int_\Omega u_{tt}\phi dx - \alpha\int_\Omega \Delta u_{tt}\phi dx + \int_\Omega \Delta^2 u\phi dx = \int_\Omega f\phi dx.$$

Integration by parts, formula (1.3.3), and the boundary conditions in (2.5.22) allow us to obtain the following relation

$$\int_\Omega u_{tt}\phi dx + \alpha\int_\Omega \nabla u_{tt}\nabla\phi dx + a(u,\phi) = \int_\Omega f\phi dx, \tag{2.5.23}$$

for all $\phi \in H^2_{\Gamma_0}(\Omega)$, where $H^2_{\Gamma_0}(\Omega)$ is defined by (2.5.21) and

$$a(u,\phi) = a_0(u,\phi) + \nu_1\int_{\Gamma_1} u\phi d\Gamma \tag{2.5.24}$$

with $a_0(u,\phi)$ given by (1.3.4):

$$a_0(u,v) = \mu\int_\Omega \Delta u\Delta\phi\, dx + (1-\mu)\int_\Omega (u_{x_1x_1}\phi_{x_1x_1} + 2u_{x_1x_2}\phi_{x_1x_2} + u_{x_2x_2}\phi_{x_2x_2})\, dx.$$

The form $a(u,w)$ considered on $H^2_{\Gamma_0}(\Omega)$ is generated by the operator $\mathscr{A}$ given in $L_2(\Omega)$ by the relations (see Section 1.3.4)

$$\mathscr{A}u = \Delta^2 u,\ \mathscr{D}(\mathscr{A}) = \left\{ u \in H^4(\Omega) \left| \begin{array}{l} [\Delta u + (1-\mu)B_1 u]\Big|_{\Gamma_1} = 0, \\ \left[\frac{\partial}{\partial n}\Delta u + (1-\mu)B_2 u - \nu_1 u\right]\Big|_{\Gamma_1} = 0 \\ u = \nabla u = 0 \text{ on } \Gamma_0 \end{array} \right. \right\}. \tag{2.5.25}$$

Therefore

$$a(u,w) = (\mathscr{A}^{1/2}u, \mathscr{A}^{1/2}w)_\Omega = (\mathscr{A}u, w)_{\mathscr{D}(\mathscr{A}^{1/2}),[\mathscr{D}(\mathscr{A}^{1/2})]'}$$

with $\mathscr{D}(\mathscr{A}^{1/2}) = H^2_{\Gamma_0}(\Omega)$. Similarly,

$$\int_\Omega u_{tt}\phi dx + \alpha \int_\Omega \nabla u_{tt} \nabla \phi dx = (M^{1/2}u_{tt}, M^{1/2}\phi)_\Omega = (Mu_{tt}, \phi)_{H^1(\Omega),[H^1(\Omega)]'},$$

where $M = I - \alpha\Delta_{N,D}$ and $\Delta_{N,D}$ is the Laplace operator equipped with Neumann boundary conditions on Γ_1 and the Dirichlet on Γ_0. Thus (2.5.23) can be written in the form

$$(Mu_{tt}, \phi)_{H^1_{\Gamma_0}(\Omega),[H^1_{\Gamma_0}(\Omega)]'} + (\mathscr{A}u, \phi)_{\mathscr{D}(\mathscr{A}^{1/2}),[\mathscr{D}(\mathscr{A}^{1/2})]'} = \int_\Omega f\phi dx,$$

where

$$H^1_{\Gamma_0}(\Omega) = \left\{ u \in H^1(\Omega) :\ u = 0 \text{ on } \Gamma_0 \right\}, \tag{2.5.26}$$

and problem (2.5.22) can be presented in the form (2.4.71) with $\mathscr{H} \equiv L_2(\Omega)$, $V_\alpha \equiv H^1_{\Gamma_0}(\Omega)$ for $\alpha > 0$, $D \equiv 0$, and $\mathscr{A}$ and M described above.

Application of Theorem 2.4.31 gives the following assertion.

2.5.14. Proposition. *Let $\alpha > 0$. Then the following statements hold.*

- **Strong solutions:** *For $f \in W^1_1(0,T;[H^1_{\Gamma_0}(\Omega)]')$ and $(u_0;u_1) \in W \times H^2_{\Gamma_0}(\Omega)$, where*
$$W = \left\{ u \in H^2_{\Gamma_0}(\Omega)\ :\ \mathscr{A}u \in [H^1_{\Gamma_0}(\Omega)]' \right\}, \tag{2.5.27}$$
there exists a unique strong solution on the interval $[0,T]$ such that
$$\begin{aligned} (u;u_t) &\in C(0,T;H^2_{\Gamma_0}(\Omega) \times H^1_{\Gamma_0}(\Omega)), \\ u_{tt} &\in C(0,T;H^1_{\Gamma_0}(\Omega)) \text{ and } \mathscr{A}u(t) \in C(0,T;[H^1_{\Gamma_0}(\Omega)]'). \end{aligned} \tag{2.5.28}$$
- **Generalized (weak) solutions:** *Let $f \in L_1(0,T;[H^1_{\Gamma_0}(\Omega)]')$ and $u_0 \in H^2_{\Gamma_0}(\Omega)$, $u_1 \in H^1_{\Gamma_0}(\Omega)$. Then there exists a unique generalized solution such that* (2.5.28) *holds. Moreover every generalized solution is weak and vice versa.*

Both strong and generalized solutions satisfy the following energy relation

$$E_0^{(\alpha)}(u(t),u_t(t)) = E_0^{(\alpha)}(u_0,u_1) + \int_0^t \int_\Omega f(\tau,x)u_t(\tau,x)\,dx d\tau. \tag{2.5.29}$$

Here

$$E_0^{(\alpha)}(u_0,u_1) = \frac{1}{2}\int_\Omega \left(|u_1(x)|^2 + \alpha|\nabla u_1(x)|^2\right) dx + \frac{1}{2}a(u_0,u_0),$$

where $a(u,v)$ is given by (2.5.24).

2.5.15. Remark. Because $\mathscr{D}(\mathscr{A}^{1/4}) = \mathscr{D}(M^{1/2}) = V = H^1_{\Gamma_0}(\Omega)$, one can see that $W = \mathscr{D}(\mathscr{A}^{3/4})$, where W is given by (2.5.27). Therefore the elliptic regularity theory (see, e.g., [216, Proposition 3A.1, p. 283] and the references therein) implies

$$W = \left\{u \in (H^3 \cap H^2_{\Gamma_0})(\Omega)),\ \Delta u + (1-\mu)B_1 u = 0 \text{ on } \Gamma_1\right\} \subset H^3(\Omega).$$

Thus a strong solution to problem (2.5.22) with $\alpha > 0$ possesses the property

$$u(t) \in C(0,T;H^3(\Omega)) \cap C^1(0,T;H^2(\Omega))$$

and satisfies the boundary conditions $u = 0$, $\nabla u = 0$ on Γ_0 and $\Delta u + (1-\mu)B_1 u = 0$ on Γ_1.

As above (cf. Theorem 2.4.38 and Propositions 2.5.2 and 2.5.8) we can obtain assertions concerning the regularity of weak solutions. In particular, if

$$f(t) \in W_1^2(0,T;(H^1_{\Gamma_0}(\Omega))') \cap C(0,T;L_2(\Omega))$$

and the compatibility conditions

$$u_0, u_1 \in W,\quad u^{(2)}(0) = M^{-1}(-\mathscr{A}u_0 + f(0)) \in \mathscr{D}(\mathscr{A}^{1/2}) = H^2_{\Gamma_0}(\Omega),$$

$$u^{(3)}(0) = M^{-1}(-\mathscr{A}u_1 + f'(0)) \in V = H^1_{\Gamma_0}(\Omega),$$

hold, then a weak solution $u(t)$ to problem (2.5.22) with $\alpha > 0$ possesses the property

$$u(t) \in C^k(0,T;H^{4-k}(\Omega)),\quad k = 0,1,2,3. \tag{2.5.30}$$

Indeed, by Theorem 2.4.38 we have that (2.5.30) holds for $k = 1,2,3$. Using relation (2.5.23) and integrating by parts we obtain that

$$(\mathscr{A}u,\phi)_{H^2_{\Gamma_0}(\Omega),[H^2_{\Gamma_0}(\Omega)]'} + \alpha\int_{\Gamma_1} \frac{\partial u_{tt}}{\partial n}\phi d\Gamma = \int_\Omega (f - u_{tt} + \alpha\Delta u_{tt})\phi dx$$

for any $\phi \in H^2_{\Gamma_0}(\Omega)$. Let G_2 be a Green operator given by $v \equiv G_2 g$, where v solves the problem

$$\Delta^2 v = 0 \text{ in } \Omega,\quad v = \frac{\partial}{\partial n}v = 0 \text{ on } \Gamma_0,$$

$$\Delta v + (1-\mu)B_1 v = 0 \text{ and } \frac{\partial}{\partial n}\Delta v + (1-\mu)B_2 v - \nu_1 v = g \text{ on } \Gamma_1.$$

Then using (1.3.3) we obtain for all $\phi \in H^2_{\Gamma_0}(\Omega)$,

$$(\mathscr{A} G_2 \frac{\partial u_{tt}}{\partial n}, \phi)_{H^2(\Omega),[H^2(\Omega)]'} \equiv a\left(G_2 \frac{\partial u_{tt}}{\partial n}, \phi\right) = -\int_{\Gamma_1} \frac{\partial u_{tt}}{\partial n} \phi d\Gamma.$$

Thus we obtain

$$(\mathscr{A}\left(u - \alpha G_2 \frac{\partial u_{tt}}{\partial n}\right), \phi)_{H^2_{\Gamma_0}(\Omega),[H^2_{\Gamma_0}(\Omega)]'} = \int_\Omega (f - u_{tt} + \alpha \Delta u_{tt}) \phi dx$$

for any $\phi \in H^2_{\Gamma_0}(\Omega)$. This implies that

$$\mathscr{A}\left(u - \alpha G_2 \frac{\partial u_{tt}}{\partial n}\right) \in C(0,T;L_2(\Omega)). \tag{2.5.31}$$

Therefore by elliptic regularity theory we have that $u - \alpha G_2(\partial u_{tt}/\partial n)$ belongs to $C(0,T;H^4(\Omega))$. Relation (2.5.30) for $k = 2$ and the trace theorem yield that $\partial u_{tt}/\partial n \in H^{1/2}(\partial\Omega)$. Therefore, because $G_2 : H^{1/2}(\Gamma_1) \mapsto H^4(\Omega)$ is a bounded operator (see (1.3.23) in the case when $\Gamma_0 = \emptyset$), it follows that $\alpha G_2(\partial u_{tt}/\partial n) \in C(0,T;H^4(\Omega))$. Consequently $u \in C(0,T;H^4(\Omega))$. Thus we obtain (2.5.30).

We also note that relation (2.5.31) and the structure of the operator $\mathscr{A}$ imply that the solution $u(t)$ satisfies the boundary conditions in (2.5.22). ■

In the case $\alpha = 0$ we have the following assertion.

2.5.16. Proposition. *Let $\alpha = 0$. Then the following statements hold.*

- **Strong solutions:** *For every $f \in W_1^1(0,T;L_2(\Omega))$ and $(u_0,u_1) \in W \times H^2_{\Gamma_0}(\Omega)$, where*

$$W = \left\{u \in H^2_{\Gamma_0}(\Omega) \, : \, \mathscr{A}u \in L_2(\Omega)\right\} \equiv \mathscr{D}(\mathscr{A}) \tag{2.5.32}$$

  with $\mathscr{D}(\mathscr{A})$ given by (2.5.25)*, there exists a unique strong solution on the interval $[0,T]$ such that*

$$\begin{aligned}(u;u_t) &\in C(0,T;H^2_{\Gamma_0}(\Omega)\times L_2(\Omega)),\\ u_{tt} &\in C(0,T;L_2(\Omega)) \text{ and } \mathscr{A}u(t) \in C(0,T;L_2(\Omega)).\end{aligned} \tag{2.5.33}$$

- **Generalized (weak) solutions:** *For every $f \in L_1(0,T;L_2(\Omega))$ and $u_0 \in H^2_{\Gamma_0}(\Omega)$, $u_1 \in L_2(\Omega)$ there exists a unique generalized solution such that* (2.5.33) *holds. Moreover every generalized solution is weak and vice versa.*

Both strong and generalized solutions satisfy the energy relation in (2.5.29) *with $\alpha = 0$.*

2.5.17. Remark. In the case $\alpha = 0$ we have $W \subset H^4(\Omega)$ and as above (cf. the second parts of Propositions 2.5.2 and 2.5.8) we can prove the existence of more regular solutions. ■

Now we consider boundary regularity of solutions to (2.5.22). In comparison with clamped and hinged cases the situation is here different. Although the biharmonic operator with free boundary conditions satisfies standard elliptic regularity assumptions, the corresponding dynamic plate equation does not comply with the dynamic-type regularity hypotheses (see [212] and the references therein). Thus, "hidden regularity" for the dynamic problem with free boundary conditions is not expected. Getting the results at the level of formal application of trace theory is already an achievement. Indeed, the following regularity results have been proved by microlocal analysis methods [212].

2.5.18. Theorem. *Let $u(t)$ be a weak solution of* (2.5.22). *Then*

$$\int_0^T \alpha|\frac{\partial}{\partial n}u_t|^2_{H^{-1/2}(\Gamma_1)}dt \le c \sup_{t\in[0,T]} \left[\|u(t)\|^2_{2,\Omega} + \alpha\|u_t(t)\|^2_{1,\Omega}\right]$$
$$+ c\left[\int_0^T \|f\|_{[H^1(\Omega)]'}dt\right]^2.$$

The above inequality is given in Theorem 1.1 [212] when free boundary conditions are considered ($\Gamma_0 = \emptyset$). The analysis is purely local (in a layer near the boundary) and both portions of the boundary Γ_0 and Γ_1 are disjoint, thus the same result holds for the clamped–free plate, as stated above.

2.5.2 Nonhomogeneous boundary conditions. Regularity theory

Here we are concerned with regularity of solutions due to nonhomogeneous boundary terms. It is known, that such regularity (at the optimal level) does not follow from standard interior regularity followed by semigroup or variational methods. It turns out that this regularity results via duality from "hidden trace" regularity presented in Section 2.5.1. In what follows we treat separately three different types of boundary conditions: clamped, hinged, and free. For all three cases we provide both a variational and semigroup framework that leads to representation of weak solutions driven by inhomogeneity on the boundary.

2.5.2.1 Clamped boundary conditions

We consider clamped nonhomogeneous boundary conditions. This is to say, we study the following problem,

$$\begin{cases} u_{tt} - \alpha\Delta u_{tt} + \Delta^2 u = f(x,t) \text{ in } Q \equiv \Omega\times(0,T), \\ u = 0, \;\; \frac{\partial}{\partial n}u = g(x,t) \text{ on } \Sigma = \Gamma\times(0,T), \\ u|_{t=0} = u_t|_{t=0} = 0 \text{ in } \Omega. \end{cases} \tag{2.5.34}$$

Weak solutions to boundary value problems driven by "rough" boundary data can be defined in a variety of ways. The most typical methods are via transposition or via "boundary variation of parameter formula." In order to explain the method we consider the case when $\alpha = 0$. Case $\alpha > 0$, with clamped nonhomogeneous boundary conditions, is more complex and involves the use of factor spaces [215].

Abstract setup: In the case $\alpha = 0$ we are looking for a solution on the space $H \equiv L_2(\Omega) \times H^{-2}(\Omega)$. By $\mathscr{A}$ we denote, as usual, the biharmonic operator $\mathscr{A}u = \Delta^2 u$ for $u \subset \mathscr{D}(\mathscr{A})$, where

$$\mathscr{D}(\mathscr{A}) = H^4(\Omega) \cap H_0^2(\Omega) \text{ and also } \mathscr{D}(\mathscr{A}^{1/2}) = H_0^2(\Omega).$$

In order to introduce a "boundary" model defining solutions to (2.5.34) we define, following [216], Green's map $G : L_2(\Gamma) \mapsto L_2(\Omega)$ which is a biharmonic extension of the Neumann boundary data into the interior of Ω. This map $v \equiv Gg$ is given by

$$\Delta^2 v = 0 \text{ in } \Omega, \quad v = 0 \text{ and } \frac{\partial}{\partial n} v = g \text{ on } \Gamma. \tag{2.5.35}$$

From elliptic theory we have that $G : H^s(\Gamma) \to H^{s+3/2}(\Omega)$ is a bounded operator for all real s. We also note that for sufficiently smooth u satisfying boundary conditions in (2.5.34) and for $g \in H^{5/2}(\Gamma)$ one obtains $u - Gg \in \mathscr{D}(\mathscr{A})$. For $u \in H^2(\Omega) \cap H_0^1(\Omega)$ and $g \in H^{1/2}(\Gamma)$ we always have $u - Gg \in \mathscr{D}(\mathscr{A}^{1/2})$. This last assertion follows from the fact that $(\partial/\partial n)(u - Gg) = 0$ on Γ. With the above notation the original equation (with $\alpha = 0$) can be rewritten as an abstract equation

$$u_{tt} + \mathscr{A}(u - Gg) = f(\cdot, t) \text{ on } (0,T) \text{ in } [\mathscr{D}(\mathscr{A})]' \tag{2.5.36}$$

or, equivalently, in a weak (variational) form as

$$(u_{tt}, \phi) + (u, \mathscr{A}\phi) - \langle g, G^*\mathscr{A}\phi\rangle = (f(t), \phi) \text{ on } (0,T),\ \forall \phi \in \mathscr{D}(\mathscr{A}), \tag{2.5.37}$$

where we use notation $\langle \cdot, \cdot \rangle$ for $(\cdot, \cdot)_\Gamma$. Green's formula leads to the equality

$$G^*\mathscr{A}\phi = -\Delta\phi|_\Gamma,$$

which when applied to (2.5.37) yields

$$(u_{tt}, \phi) + (u, \mathscr{A}\phi) + \langle g, \Delta\phi\rangle = (f(t), \phi) \text{ on } (0,T),\ \forall \phi \in \mathscr{D}(\mathscr{A}). \tag{2.5.38}$$

The above formulation allow us to define the solution to the nonhomogeneous on the boundary problem via "transposition." To wit, let $h \in L_1(0,T; L_2(\Omega))$ and let $\psi_0 \in H_0^2(\Omega)$, $\psi_1 \in L_2(\Omega)$ be arbitrary given elements. We consider the adjoint equation

$$\psi_{tt} + \mathscr{A}\psi = h,\ 0 \le t < T, \quad \psi(T) = \psi_0,\ \psi_t(T) = \psi_1. \tag{2.5.39}$$

By Theorem 2.4.38 and hidden regularity Theorem 2.5.4 (applied after the change of time $t := T - t$) we infer that the map

$$(\psi_0; \psi_1; h) \to (\psi; \psi_t; \Delta\psi|_\Sigma) \tag{2.5.40}$$

is bounded from the space $H_0^2(\Omega) \times L_2(\Omega) \times L_1(0,T; L_2(\Omega))$ into the space

$$C(0,T;H_0^2(\Omega))\times C(0,T;L_2(\Omega))\times L_2(\Sigma).$$

We apply (2.5.38) with $\phi=\psi$, where ψ satisfies the adjoint equation (2.5.39) corresponding to sufficiently smooth h, so that by virtue of classical semigroup theory the corresponding $\psi(t)\in\mathscr{D}(\mathscr{A})$ is sufficiently smooth as well. Integration by parts in time after accounting for zero initial–terminal conditions gives:

$$(u_t(T),\psi_0)-(u(T),\psi_1)+\int_0^T(u,h)dt+\int_0^T\langle g,\Delta\psi(\psi_0,\psi_1,h)\rangle dt \quad (2.5.41)$$
$$=\int_0^T(f(t),\psi(\psi_0,\psi_1,h))dt.$$

Here we denote by $\psi(t)=\psi(t;\psi_0,\psi_1,h)$ the solution to problem (2.5.39). The above formula can be used for a definition of weak solution u corresponding to the boundary data $g\in L_2(\Sigma)$ and the forcing term $f\in L_1(0,T;H^{-2}(\Omega))$. Indeed, taking $\psi_0=0$, $\psi_1=0$, we arrive at the following definition.

2.5.19. Definition. The function $u\in L_\infty(0,T;L_2(\Omega))$ is said to be a weak (via transposition) solution to problem (2.5.34) with $\alpha=0$ if

$$\int_0^T(u,h)dt+\int_0^T\langle g,\Delta\psi_*(h)\rangle dt=\int_0^T(f(t),\psi_*(h))dt \quad (2.5.42)$$

for all $h\in L_1(0,T;L_2(\Omega))$, where $\psi_*(t,h)=\psi(t,0,0,h)$. ■

The Riesz representation theorem followed by the standard density argument implies the following regularity of weak solution u.

2.5.20. Theorem. *Let $\alpha=0$, $g\in L_2(\Sigma)$, and $f\in L_1(0,T;H^{-2}(\Omega))$; then weak solution corresponding to* (2.5.34) *and defined by* (2.5.42) *exists and has the following regularity*

$$u\in C(0,T;L_2(\Omega)),\quad u_t\in C(0,T;H^{-2}(\Omega)).$$

Proof. The proof is standard and follows from the classical transposition method [222, 221]. Regularity of the map given by (2.5.40) implies that

$$h\mapsto-\int_0^T\langle g,\Delta\psi_*(h)\rangle dt+\int_0^T(f(t),\psi_*(h))dt$$

is a bounded linear functional on $L_1(0,T;L_2(\Omega))$. Hence there exists element u from $L_\infty(0,T;L_2(\Omega))$ satisfying (2.5.42). If in (2.5.42) we choose $h(t)=\chi''(t)\phi+\chi(t)\mathscr{A}\phi$ with $\chi\in C_0^\infty(0,T;\mathbb{R})$ and $\phi\in\mathscr{D}(\mathscr{A})$, then after simple calculations we arrive at relation (2.5.38) which implies that $u_{tt}\in L_1(0,T;[\mathscr{D}(\mathscr{A})]')$. In particular this yields that the scalar function $\mapsto(u(t),\phi)$ is C^1. Therefore using the density argument we can obtain from (2.5.38) relation (2.5.41) written on the interval $[0,t]$ instead of $[0,T]$. By choosing $\psi_1=0$, $h=0$ this makes it possible via Riesz representation to obtain that $u_t\in L_\infty(0,T;H^{-2}(\Omega))$. The standard density argument improves L_∞ to C yielding $u_t\in C(0,T;\mathscr{D}(\mathscr{A}^{1/2})')=C(0,T;H^{-2}(\Omega))$ and $u\in C(0,T;L_2(\Omega))$, as desired. □

We now present an alternative approach to the definition of a weak solution by using a *semigroup approach* and an extended variation of constants formula. To this end we follow the approach taken in [216]. Let A denote the generator of the semigroup associated with the homogeneous plate dynamics considered on the space

$$H = L_2(\Omega) \times [\mathscr{D}(\mathscr{A})^{1/2})]'.$$

This is to say:

$$A(u;v) = (v; -\mathscr{A}u), \quad \mathscr{D}_H(A) = \mathscr{D}(\mathscr{A}^{1/2}) \times L_2(\Omega) = H_0^2(\Omega) \times L_2(\Omega).$$

Because the semigroup commutes with the generator, we can treat e^{At} as acting on a scale of spaces involving suitable domains of powers of $\mathscr{A}$ including $H_{-1} \equiv [\mathscr{D}(\mathscr{A})^{1/2})]' \times [\mathscr{D}(\mathscr{A})]'$ (see [216]). This allows us to define a solution to (2.5.36) via the following semigroup formula,

$$U(t) = L(g)(t) + K(f)(t), \quad U(t) = (u(t); u_t(t)),$$

where

$$L(g)(t) \equiv \int_0^t e^{A(t-s)} \begin{pmatrix} 0 \\ \mathscr{A}Gg(s) \end{pmatrix} ds, \quad K(f)(t) \equiv \int_0^t e^{A(t-s)} \begin{pmatrix} 0 \\ f(s) \end{pmatrix} ds. \quad (2.5.43)$$

The standard semigroup property along with $[\mathscr{D}(\mathscr{A}^{1/2})]' = H^{-2}(\Omega)$ implies that

$$K \in \mathscr{L}\left(L_1(0,T; H^{-2}(\Omega)) \mapsto C(0,T;H)\right).$$

Considering extension of the semigroup e^{At} as acting on a larger extended space $H_{-1} = [\mathscr{D}(\mathscr{A})^{1/2})]' \times [\mathscr{D}(\mathscr{A})]'$ and noting that $Gg \in L_2(\Omega)$, we also obtain (conservatively) that $L \in \mathscr{L}(L_2(\Sigma) \mapsto C(0,T;H_{-1}))$. This regularity of the map L can be much improved, as shown below.

2.5.21. Theorem. *The map L defined in* (2.5.43) *satisfies*

$$L \in \mathscr{L}(L_2(\Sigma) \to C(0,T;H)).$$

Proof. We first note that the operator L is both closed and densely defined as a mapping from $L_2(\Sigma)$ into $C([0,T],H)$. This statement follows from the following facts (which were shown in [216, vol. II, Remark 7.1.3, p. 646]).

- $A^{-1}L$ is bounded $L_2(\Sigma) \mapsto C([0,T],H)$ and A is closed on H.
- $L : H_0^1(0,T; L_2(\Gamma)) \to C([0,T]; H)$ is continuous.

The first property implies closedness of L; the second one implies the density (note $H_0^1(0,T;L_2(\Gamma))$ is dense in $L_2(\Sigma)$). Computing the adjoint of $L_t(g) = L(g)(t)$ for each point $t \in [0,T]$ (with respect to topologies $L_2(\Sigma_t) \mapsto H$, $\Sigma_t \equiv \Gamma \times (0,t)$) yields

$$L_t^*(\Phi)(s) = -\Delta\mathscr{A}^{-1}\psi_2(s)\big|_{\Sigma_t}, \quad \Psi(s) = e^{-A(t-s)}\Phi, \ s \in [0,t], \quad (2.5.44)$$

where $\Phi = (\phi_1;\phi_2) \in H$, $\Psi(s) = (\psi_1(s);\psi_2(s))$ (here we have also used that the operator A is skew-symmetric; i.e., $A^* = -A$). Noting that $\psi_2(s)$ satisfies

$$\psi_{2ss} + \mathscr{A}\psi_2 = 0 \quad \text{for } s \in (0,t)$$

with terminal conditions $\psi_2(t) = \phi_2 \in H^{-2}(\Omega)$ and $\psi_{2s}(t) = -\mathscr{A}\psi_1(t) = -\mathscr{A}\phi_1$, we obtain that $z(s) \equiv \mathscr{A}^{-1}\psi_2(s)$ satisfies

$$z_{ss} + \mathscr{A}z = 0 \quad \text{on } (0,t)$$

with terminal conditions $z(t) = \mathscr{A}^{-1}\phi_2 \in H_0^2(\Omega)$ and $z_s(t) = -\phi_1 \in L_2(\Omega)$. Applying the trace estimate given in Theorem 2.5.4 to the z equation, yields

$$\int_{\Sigma_t} |\Delta z|^2 d\Sigma \le C\|\Phi\|_H^2.$$

The above along with (2.5.44) implies

$$L_t^* : H \to L_2(\Sigma_t) \text{ is bounded uniformly for all } t \in [0,T]$$

hence $L : L_2(\Sigma) \to L_\infty(0,T;H)$ is bounded as well. The standard density argument boosts L_∞ to C. □

The above considerations lead to the following assertion.

2.5.22. Theorem. *Assume that* $\alpha = 0$. *Weak solution* $u(t)$ *to problem* (2.5.34) *with* $g \in L_2(\Sigma)$, *defined either via transposition or via semigroup formula satisfies* $(u;u_t) \in C(0,T;H)$ *and obeys the following estimate,*

$$\|u(t)\|_{0,\Omega} + \|u_t(t)\|_{-2,\Omega} \le c\left(\|g\|_{L_2(\Sigma)} + \|f\|_{L^1(0,T;H^{-2}(\Omega))}\right).$$

2.5.23. Remark.

1. We note that the result of Theorem 2.5.22 provides $\frac{1}{2}$ derivative more than the standard theory. The key element leading to this improvement is the trace inequality in Theorem 2.5.4.

2. Assuming higher regularity of the boundary data, one can obtain more regular solutions. This can be easily accomplished by applying the same argument to time derivatives of the equation.

3. In the case $\alpha > 0$ with clamped-type nonhomogeneous boundary conditions, the situation is more complicated and involves introduction of special factor spaces. This is because of the use of duality which leads to rather exotic Sobolev-type spaces [214, 215]. ■

2.5.2.2 Clamped–hinged boundary conditions

We assume that the boundary Γ of the domain Ω consists of two disjoint parts Γ_0 and Γ_1 and consider clamped–hinged nonhomogeneous boundary conditions; that is, we study the following problem,

$$\begin{cases} u_{tt}-\alpha\Delta u_{tt}+\Delta^2 u=f(x,t) & \text{in } Q\equiv\Omega\times(0,T),\\ u=\nabla u=0 & \text{on } \Sigma_0=\Gamma_0\times(0,T),\\ u=0,\ \Delta u=g(x,t) & \text{on } \Sigma_1=\Gamma_1\times(0,T),\\ u|_{t=0}=u_t|_{t=0}=0 & \text{in } \Omega. \end{cases} \tag{2.5.45}$$

Assume that $f\in L_2(0,T;H^{-1}(\Omega))$ and $g\in L_2(\Sigma_1)$. We recast the model in an abstract form based on introduction of suitable Green's maps.
Abstract setup: *Biharmonic generator* $\mathscr{A}:\mathscr{D}(\mathscr{A})\subset L_2(\Omega)\to L_2(\Omega)$ given by $\mathscr{A}u=\Delta^2 u$ for $u\in\mathscr{D}(\mathscr{A})$, where

$$\mathscr{D}(\mathscr{A})\equiv\{u\in H^4(\Omega)\cap H_0^1(\Omega)\cap H^2_{\Gamma_0}(\Omega),\ \ \Delta u=0\ \text{ on }\ \Gamma_1\}.$$

Clearly $\mathscr{D}(\mathscr{A}^{1/2})=H^2_{\Gamma_0}(\Omega)\cap H_0^1(\Omega)$. Here $H^2_{\Gamma_0}(\Omega)$ is given by (2.5.21).
Mass operator $M=(I-\alpha\Delta):\mathscr{D}(M)\mapsto L_2(\Omega)$ with $\mathscr{D}(M)=H^2(\Omega)\cap H_0^1(\Omega)$.
Green's map G. In order to introduce the "boundary" model defining solutions to (2.5.45) we define, following [216] (see also Chapter 1 for the case $\Gamma_0=\emptyset$), Green's map $G:L_2(\Gamma_1)\mapsto L_2(\Omega)$ which is a biharmonic extension of the clamped–hinged boundary data into the interior of Ω. This map $v\equiv Gg$ is given by

$$\Delta^2 v=0\ \text{ in }\ \Omega,\quad v=0\ \text{ on }\ \Gamma,\ \ \frac{\partial}{\partial n}v=0\ \text{ on }\ \Gamma_0\ \text{ and }\ \Delta v=g\ \text{ on }\ \Gamma_1. \tag{2.5.46}$$

From elliptic theory we have that $G:H^s(\Gamma_1)\to H^{s+5/2}(\Omega)$ is a bounded operator for all $s\in\mathbb{R}$. Hence, in particular, $G:L_2(\Gamma_1)\to\mathscr{D}(\mathscr{A}^{1/2})$ is bounded. We also note that if $g\in H^{3/2}(\Gamma_1)$ then for sufficiently smooth u satisfying boundary conditions in (2.5.45) one obtains $u-Gg\in\mathscr{D}(\mathscr{A})$. For $u\in H^2_{\Gamma_0}(\Omega)\cap H_0^1(\Omega)$ and $g\in L_2(\Gamma_1)$ we always have $u-Gg\in\mathscr{D}(\mathscr{A}^{1/2})$. With the above notation the original equation can be rewritten as an abstract equation

$$Mu_{tt}+\mathscr{A}(u-Gg)=f(\cdot,t)\quad\text{on}\ \ (0,T)\ \text{ in }\ [\mathscr{D}(\mathscr{A})]' \tag{2.5.47}$$

or, equivalently, in a weak (variational) form as

$$(Mu_{tt},\phi)+(u,\mathscr{A}\phi)-\langle g,G^*\mathscr{A}\phi\rangle_{\Gamma_1}=(f(t),\phi)\ \text{ on }(0,T),\ \forall\phi\in\mathscr{D}(\mathscr{A}), \tag{2.5.48}$$

where $\langle\cdot,\cdot\rangle_{\Gamma_1}$ is the inner product in $L_2(\Gamma_1)$. Green's formula (see (1.3.16) in the case $\Gamma_0=\emptyset$)

$$G^*\mathscr{A}\phi=\frac{\partial}{\partial n}\phi|_{\Gamma_1},\quad \phi\in\mathscr{D}(\mathscr{A}),$$

applied to (2.5.48) yields the following variational definition of the solution:

$$(Mu_{tt},\phi)+(u,\mathscr{A}\phi)-\langle g,\frac{\partial}{\partial n}\phi\rangle_{\Gamma_1}=(f(t),\phi)\quad\text{on}\ \ (0,T),\ \forall\phi\in\mathscr{D}(\mathscr{A}). \tag{2.5.49}$$

As in the clamped case, one applies the "transposition" method in order to obtain an equivalent definition of the solution. This leads to

$$(Mu_t(T),\psi(T))-(Mu(T),\psi_t(T))+\int_0^T(u(t),h) - \int_0^T\langle g,\frac{\partial}{\partial n}\psi(h)\rangle_{\Gamma_1}=\int_0^T(f,\psi(h)), \tag{2.5.50}$$

where $\psi=\psi(t;h)$ satisfies the equation $M\psi_{tt}+\mathscr{A}\psi=h$, $0\le t<T$, with preassigned terminal data $\psi(T)$ and $\psi_t(T)$.

For $\alpha=0$ we take $h\in L_1(0,T;H^{-1}(\Omega))$, $\psi(T)\in H_0^1(\Omega)$, $\psi_t(T)\in H^{-1}(\Omega)$. By Theorem 2.5.11 and Remark 2.5.12, $(\partial/\partial n)\psi\in L_2(\Sigma_1)$, hence, in the same way as in the case of problem (2.5.34) with clamped boundary conditions, the Riesz representation theorem applied to (2.5.50) implies that $u\in L_\infty(0,T;H_0^1(\Omega))$ and $u_t\in L_\infty(0,T;H^{-1}(\Omega))$. Thus, after lifting L_∞ to C, $g\in L_2(\Sigma_1)$ provides solution $(u;u_t)\in C(0,T;H_0^1(\Omega)\times H^{-1}(\Omega))$.

For $\alpha>0$ the same argument applied with data $h\in L_1(0,T;[\mathscr{D}(\mathscr{A}^{1/2})]')$, $\psi(T)\in H_0^1(\Omega)$, and $\psi_t(T)\in L_2(\Omega)$ via Theorem 2.5.11 and Remark 2.5.12 implies that $(\partial/\partial n)\psi\in L_2(\Sigma)$. Hence $u\in L_\infty(0,T;\mathscr{D}(\mathscr{A}^{1/2}))$, $u_t\in L_\infty(0,T;\mathscr{D}(M^{1/2}))$, and consequently $g\in L_2(\Sigma_1)$ provides solution $(u;u_t)\in C(0,T;H^2(\Omega)\times H^1(\Omega))$.

The same conclusion can be arrived at if one uses a semigroup framework to represent the solution with nonhomogeneous boundary data. To see this, we introduce the space $H\equiv\mathscr{D}(\mathscr{A}^{1/2})\times\mathscr{D}(M^{1/2})$ and the matrix operator $A:\mathscr{D}_H(A)\subset H\mapsto H$,

$$\begin{aligned} A(u;v)&=(v;-M^{-1}\mathscr{A}u),\\ \mathscr{D}_H(A)&=\{(u;v)\in H\,:\,v\in\mathscr{D}(\mathscr{A}^{1/2}),M^{-1/2}\mathscr{A}u\in L_2(\Omega)\}.\end{aligned}$$

The operator A generates a strongly continuous semigroup of contractions $e^{At}:H\mapsto H$. This allows us to construct the solution:

$$U(t)=L(g)(t)+K(f)(t),\quad U(t)=(u(t);u_t(t)), \tag{2.5.51}$$

where the operators K and L are given by

$$L(g)(t)\equiv\int_0^t e^{A(t-s)}\begin{pmatrix}0\\ M^{-1}\mathscr{A}Gg(s)\end{pmatrix}ds \tag{2.5.52}$$

and

$$K(f)(t)\equiv\int_0^t e^{A(t-s)}\begin{pmatrix}0\\ M^{-1}f(s)\end{pmatrix}ds. \tag{2.5.53}$$

Note that $A^{-1}(0;M^{-1}\mathscr{A}Gg)=-(Gg;0)\in H$ with $g\in L_2(\Gamma_1)$. Hence, by regularity of the Green's map $G:L_2(\Gamma_1)\mapsto\mathscr{D}(\mathscr{A}^{1/2})$, we have that $A^{-1}L$ maps $L_2(\Sigma_1)$ into $C(0,T;H)$. Consequently $L\in\mathscr{L}(L_2(\Sigma_1)\mapsto C(0,T;[\mathscr{D}(A)]'))$ which provides a good definition of the map L on extended (dual) spaces. Similarly

$$A^{-1}K \in \mathscr{L}(L_1(0,T;[\mathscr{D}(\mathscr{A}^{1/2})]') \mapsto C(0,T;H).$$

Thus for every $g \in L_2(\Sigma_1)$ and $f \in L_1(0,T;[\mathscr{D}(\mathscr{A}^{1/2})]')$, solution $U(t)$ given by (2.5.51) is a well-defined element in $C(0,T;[\mathscr{D}(A)]')$. However, the above regularity is clearly not sharp and more can be said about smoothness properties of weak solutions with L_2 boundary inputs. This is accomplished by using a "hidden" regularity.

To state our main results we distinguish two cases: $\alpha = 0$ and $\alpha > 0$.

2.5.24. Theorem. *Assume $\alpha > 0$. Then*

$$\|u(t)\|_{2,\Omega} + \|u_t(t)\|_{1,\Omega} \le c\left(\|g\|_{L_2(\Sigma_1)} + \|f\|_{L_1(0,T;H^{-1}(\Omega))}\right), \quad t \in [0,T].$$

Moreover

$$\|\frac{\partial}{\partial n} u_t\|_{L_2(\Sigma_1)} \le c\left(\|g\|_{L_2(\Sigma_1)} + \|f\|_{L_1(0,T;H^{-1}(\Omega))}\right).$$

When $\alpha = 0$ the solution is, not surprisingly, less regular.

2.5.25. Theorem. *Assume $\alpha = 0$. Then for every $t \in [0,T]$ we have*

$$\|u(t)\|_{1,\Omega} + \|u_t(t)\|_{-1,\Omega} + \|\frac{\partial}{\partial n} u\|_{L_2(\Sigma_1)} \le c\left(\|g\|_{L_2(\Sigma_1)} + \|f\|_{L_1(0,T;H^{-1}(\Omega))}\right).$$

The first statement in Theorem 2.5.24 follows by reduction to the wave equation and applying [193]. The details of this argument are given in [216, Theorem 10.7.3.2]. The second statement is proved in [210]. As for the interior regularity in Theorem 2.5.25 it follows via duality from Theorem 2.5.9. The boundary regularity of $(\partial/\partial n)u$ follows by applying multiplier techniques to the corresponding boundary value problem. The details are given in [216, Theorem 10.8.10.1, p. 1023].

It is worth noticing that the proofs of both theorems are based on duality with respect to the "hidden" regularity of boundary traces corresponding to homogeneous problems. Theorems 2.5.9 and 2.5.11 are the basis for the proof of Theorem 2.5.25, and Theorem 2.5.13 implies the result of Theorem 2.5.24. In order to give a flavor of the method, we prove the first statement in Theorem 2.5.24 (the proof of the second statement is more technical [210] and requires microlocalization).

2.5.26. Theorem. *Let $\alpha > 0$. The map L defined by* (2.5.52) *satisfies*

$$L \in \mathscr{L}(L_2(\Sigma_1) \mapsto C(0,T;H)) \ \ with \ \ H = \mathscr{D}(\mathscr{A}^{1/2}) \times \mathscr{D}(M^{1/2}).$$

Proof. We use the same idea as in the proof of Theorem 2.5.21. We first note that by the same argument as in the proof of Theorem 2.5.21 the operator L is both closed and densely defined $L_2(\Sigma_1) \mapsto C(0,T;H)$. Computing the adjoint of $L_t(g) = L(g)(t)$ for each point $t \in [0,T]$ (with respect to topologies $L_2(\Sigma_t^1) \mapsto H$ with $\Sigma_t^1 \equiv \Gamma_1 \times (0,t)$) yields

$$L_t^*(\Phi)(s) = \frac{\partial}{\partial n}\psi_2(s)\big|_{\Sigma_t^1}, \quad \Psi(s) = e^{-A(t-s)}\Phi, \ \ s \in [0,t],$$

where $\Phi = (\phi_1;\phi_2) \in H$, $\Psi(s) = (\psi_1(s);\psi_2(s))$. Noting that $\psi_2(s) = \partial_s \psi_1(s)$ and $\psi_1(s)$ satisfies

$$M\psi_{1ss}(s) + \mathscr{A}\psi_1(s) = 0 \ \text{ for } \ s \in (0,t)$$

with terminal conditions

$$\psi_1(t) = \phi_1 \in \mathscr{D}(\mathscr{A}^{1/2}) \subset H^2(\Omega), \quad \psi_{1s}(t) = \phi_2 \in V = \mathscr{D}(M^{1/2}) \sim H_0^1(\Omega),$$

we reduce the problem to the boundary estimate for

$$L_t^*(\Phi)(s) = \frac{\partial}{\partial n}\frac{d}{ds}\psi_1(s)\Big|_{\Sigma_t^1}. \tag{2.5.54}$$

With $z(s) \equiv \psi_1(s)$, we are led to the problem

$$Mz_{ss} + \mathscr{A}z = 0 \ \text{ on } \ (0,t)$$

with terminal conditions $z(t) = z_0 \in H^2_{\Gamma_0}(\Omega) \cap H_0^1(\Omega)$ and $z_s(t) = \phi_1 \in H_0^1(\Omega)$. Applying the trace estimate from Theorem 2.5.13 along with a standard semigroup estimate to the z equation, yields

$$\int_{\Sigma_t^1} |\frac{\partial}{\partial n} z_s|^2 d\Sigma \le C\|\Phi\|_H^2.$$

The above inequality along with (2.5.54) implies that

$$L_t^* : H \mapsto L_2(\Sigma_t^1) \ \text{ is bounded uniformly for all } t \in [0,T],$$

hence $L : L_2(\Sigma_1) \mapsto L_\infty(0,T;H)$ is bounded as well. The standard density argument boosts L_∞ to C. □

2.5.27. Remark. The proof of Theorem 2.5.25 is simpler that the foregoing. It requires, however, a different scale of spaces. One should take $H \equiv \mathscr{D}(\mathscr{A}^{1/4}) \times [\mathscr{D}(\mathscr{A}^{1/4})]'$ (as in the transposition method). The trace result needed for the proof is that of Theorem 2.5.11. ■

2.5.2.3 Clamped–free boundary conditions

Let Γ_0 and Γ_1 be two disjoint parts constituting the boundary Γ. We consider mixed boundary conditions by imposing a combination of clamped and free boundary conditions supported on two disjoint parts of the boundary: we deal with the model

$$\begin{cases} u_{tt} - \alpha\Delta u_{tt} + \Delta^2 u = f(x,t) & \text{in } Q \equiv \Omega \times (0,T), \\ u = \nabla u = 0 & \text{on } \Sigma_0 = \Gamma_0 \times (0,T) \\ \Delta u + (1-\mu)B_1 u = g_1(x,t) & \text{on } \Sigma_1 = \Gamma_1 \times (0,T), \\ \frac{\partial}{\partial n}\Delta u + (1-\mu)B_2 u - \alpha\frac{\partial}{\partial n}u_{tt} - \nu_1 u = g_2(x,t) & \text{on } \Sigma_1, \\ u|_{t=0} = u_t|_{t=0} = 0 & \text{in } \Omega, \end{cases} \tag{2.5.55}$$

where B_1 and B_2 are defined by (1.3.20) and $0 < \mu < 1$. We assume that $\nu_1 \geq 0$ ($\nu_1 > 0$ if $\Gamma_0 = \emptyset$) and

$$f \in L_2(0,T;[H^1(\Omega)]'), \quad g_1 \in L_2(0,T;H^{1/2}(\Gamma_1)), \quad g_2 \in L_2(0,T;H^{-1/2}(\Gamma_1)).$$

We study the regularity properties of solutions to (2.5.55) driven by a boundary input g. Here the boundary conditions do not satisfy the Lopatinski condition. Therefore, we do not expect much of hidden regularity.

Abstract setup: *Biharmonic generator* $\mathscr{A} : \mathscr{D}(\mathscr{A}) \subset L_2(\Omega) \mapsto L_2(\Omega)$ given by $\mathscr{A}u = \Delta^2 u$ for u from $\mathscr{D}(\mathscr{A})$ defined in (2.5.25). We note $\mathscr{D}(\mathscr{A}^{1/2}) = H^2_{\Gamma_0}(\Omega)$.

Mass operator $M : \mathscr{D}(\mathscr{A}^{1/4}) \mapsto [\mathscr{D}(\mathscr{A}^{1/4})]'$ is defined variationally as

$$(Mu,\phi) = (u,\phi) + \alpha(\nabla u, \nabla\phi), \quad \text{for all } \phi \in \mathscr{D}(\mathscr{A}^{1/4}) \sim H^1_{\Gamma_0}(\Omega).$$

We note that for $\alpha > 0$ the operator M can be identified with the action of $M = I - \alpha\Delta$ on the domain

$$\mathscr{D}(M) = H^1_{\Gamma_0}(\Omega) \cap \left\{ u \in H^2(\Omega) : \frac{\partial}{\partial n} u = 0 \text{ on } \Gamma_1 \right\}.$$

In this case $\mathscr{D}(M^{1/2}) = \mathscr{D}(\mathscr{A}^{1/4}) \sim H^1_{\Gamma_0}(\Omega)$. In the case $\alpha = 0$ we have $M = I$.

Green's maps G_i. In order to introduce the "boundary" model defining solutions to (2.5.55) we define, following [216] (see also (1.3.21) and (1.3.22) in Chapter 1), Green's map $G_i : L_2(\Gamma_1) \mapsto L_2(\Omega)$ which is a biharmonic extension of the boundary data on Γ_1 into the interior of Ω. The map $v \equiv G_1 g$ is given by

$$\Delta^2 v = 0 \text{ in } \Omega, \quad v = \frac{\partial}{\partial n} v = 0 \text{ on } \Gamma_0, \tag{2.5.56}$$

$$\Delta v + (1-\mu)B_1 v = g \text{ and } \frac{\partial}{\partial n}\Delta v + (1-\mu)B_2 v - \nu_1 v = 0 \text{ on } \Gamma_1.$$

Similarly, $v \equiv G_2 g$ is given by

$$\Delta^2 v = 0 \text{ in } \Omega, \quad v = \frac{\partial}{\partial n} v = 0 \text{ on } \Gamma_0, \tag{2.5.57}$$

$$\Delta v + (1-\mu)B_1 v = 0 \text{ and } \frac{\partial}{\partial n}\Delta v + (1-\mu)B_2 v - \nu_1 v = g \text{ on } \Gamma_1.$$

From the elliptic theory for all real s we have that $G_1 : H^s(\Gamma_1) \to H^{s+5/2}(\Omega)$ and $G_2 : H^s(\Gamma_1) \to H^{s+7/2}(\Omega)$ are bounded operators. In particular,

$$G_i : L_2(\Gamma_1) \to \mathscr{D}(\mathscr{A}^{1/2}), \quad i = 1,2,$$

are bounded operators. As in the clamped or hinged case we also note that for sufficiently smooth g_i and u satisfying boundary conditions in (2.5.55) one obtains $u - G_1 g_1 - G_2(g_2 + \alpha(\partial/\partial n)u_{tt}) \in \mathscr{D}(\mathscr{A})$. For $u \in H^2_{\Gamma_0}(\Omega)$ and $g \in L_2(\Gamma_1)$ we always have $u - G_i g \in \mathscr{D}(\mathscr{A}^{1/2})$ for $i = 1,2$. With the above notation the original equation can be rewritten as

$$(I - \alpha\Delta)u_{tt} + \mathscr{A}\left(u - G_1 g_1 - G_2 g_2 - \alpha G_2 \frac{\partial}{\partial n} u_{tt}\right) = f(\cdot,t) \quad \text{on } (0,T) \quad (2.5.58)$$

in $[\mathscr{D}(\mathscr{A})]'$ or, equivalently, in a weak (variational) form as

$$\begin{aligned}(M^{1/2}u_{tt}, M^{1/2}\phi) - \alpha\langle\frac{\partial}{\partial n}u_{tt},\phi\rangle_{\Gamma_1} + (u,\mathscr{A}\phi) - \langle g_1, G_1^*\mathscr{A}\phi\rangle_{\Gamma_1} \\ -\langle g_2, G_2^*\mathscr{A}\phi\rangle_{\Gamma_1} - \alpha\langle\frac{\partial}{\partial n}u_{tt}, G_2^*\mathscr{A}\phi\rangle_{\Gamma_1} = (f(t),\phi) \text{ on } (0,T) \quad (2.5.59)\end{aligned}$$

for all $\phi \in \mathscr{D}(\mathscr{A})$. As before we denote by $\langle\cdot,\cdot\rangle_{\Gamma_1}$ the corresponding integral on the boundary. Green's formula (see (1.3.3) and also (1.3.24) and (1.3.25)) implies

$$G_1^*\mathscr{A}\phi = \frac{\partial}{\partial n}\phi|_{\Gamma_1} \text{ and } G_2^*\mathscr{A}\phi = -\phi|_{\Gamma_1}, \quad \phi \in \mathscr{D}(\mathscr{A}).$$

The above identifications applied to (2.5.59) (after cancellation the terms involving $(\partial/\partial n)u_{tt}$) yield that

$$(M^{1/2}u_{tt}, M^{1/2}\phi) + (u,\mathscr{A}\phi) - \langle g_1, \frac{\partial}{\partial n}\phi\rangle_{\Gamma_1} + \langle g_2,\phi\rangle_{\Gamma_1} = (f(t),\phi) \quad (2.5.60)$$

on $(0,T)$ for all $\phi \in \mathscr{D}(\mathscr{A})$. This relation provides variational formulation of a solution to the plate problem with nonhomogeneous free boundary conditions. By noting that $(\mathscr{A}^{1/2}u, \mathscr{A}^{1/2}\phi) = a(u,\phi)$ for all $u,\phi \in \mathscr{D}(\mathscr{A}^{1/2})$, where $a(u,\phi)$ is given by (2.5.24), we can rewrite (2.5.60) with $u \in H^2_{\Gamma_0}(\Omega)$ as

$$(u_{tt},\phi) + \alpha(\nabla u_{tt},\nabla\phi) + a(u,\phi) - \langle g_1, \frac{\partial}{\partial n}\phi\rangle_{\Gamma_1} + \langle g_2,\phi\rangle_{\Gamma_1} = (f(t),\phi) \quad (2.5.61)$$

on $(0,T)$ for all $\phi \in H^2_{\Gamma_0}(\Omega)$. The above form can also be arrived at by formal integration by parts of the original equation with a test function $\phi \in H^2_{\Gamma_0}(\Omega)$.

Alternatively to variational formulation one can use the semigroup framework. For this, we proceed similarly as in the hinged case. Let $H \equiv \mathscr{D}(\mathscr{A}^{1/2}) \times \mathscr{D}(M^{1/2}) \sim H^2_{\Gamma_0}(\Omega) \times V_\alpha$, where $V_\alpha = H^1_{\Gamma_0}(\Omega)$ for $\alpha > 0$ and $V_\alpha = L_2(\Omega)$ for $\alpha = 0$. The operator $A : \mathscr{D}_H(A) \subset H \to H$ is defined by

$$A(u;v) = (v; -M^{-1}\mathscr{A}u)$$

with $\mathscr{D}_H(A) = \{(u;v) \in H, v \in \mathscr{D}(\mathscr{A}^{1/2}), M^{-1/2}\mathscr{A} u \in L_2(\Omega)\}$. The operator A generates a strongly continuous semigroup of contractions $e^{At} : H \to H$. As above this allows us to define the operators

$$U(t) = L_1(g_1)(t) + L_2(g_2)(t) + K(f)(t), \quad U(t) = (u(t); u_t(t)), \tag{2.5.62}$$

where

$$L_i(g)(t) \equiv \int_0^t e^{A(t-s)} \begin{pmatrix} 0 \\ M^{-1}\mathscr{A} G_i g(s) \end{pmatrix} ds, \quad i = 1,2 \tag{2.5.63}$$

and

$$K(f)(t) \equiv \int_0^t e^{A(t-s)} \begin{pmatrix} 0 \\ M^{-1} f(s) \end{pmatrix} ds. \tag{2.5.64}$$

Because $A^{-1}(0; M^{-1}\mathscr{A} Gg) = -(Gg; 0) \in \mathscr{D}(\mathscr{A}^{1/2}) \times V_\alpha = H$, the operators L_i are bounded from $L_2(\Sigma_1)$ into $C(0,T; [\mathscr{D}(A)]')$. As above this observation allows us to define by (2.5.62) the solution $U(t) \in C(0,T; [\mathscr{D}(A)]')$, due to the boundary inputs $g_i \in L_2(\Sigma)$. However, in what follows we provide a stronger notion (regularity) of solution.

Before stating the main results for problem (2.5.55) we note that the variational formulation and standard energy method allows us to deduce that for $\alpha > 0$

$$g_2 \in L_2(0,T; H^{-1/2}(\Gamma_1)) \ \text{ implies } \ L_2(g_2) \in C(0,T; H).$$

However, the energy method does not apply to deduce the regularity due to g_1 (unless time differentiability of g_1 is assumed). For the latter task we use the improved trace regularity result given by Theorem 2.5.18. This leads to the following theorem.

2.5.28. Theorem. *Let $\alpha > 0$. Then for every $t \in [0,T]$ we have*

$$\begin{aligned} &\|u(t)\|_{2,\Omega} + \|u_t(t)\|_{1,\Omega} \\ &\le c_T \left(\|g_1\|_{L_2(0,T;H^{1/2}(\Gamma_1))} + \|g_2\|_{L_2(0,T;H^{-1/2}(\Gamma_1))} + \|f\|_{L^2(0,T;[H^1(\Omega)]')} \right). \end{aligned}$$

In view of the comment above, the proof of the inequality in Theorem 2.5.28 follows from the next lemma.

2.5.29. Lemma. *Let $\alpha > 0$. The map L_1 defined by* (2.5.63) *satisfies*

$$L_1 \in \mathscr{L}\left(L_2(0,T; H^{1/2}(\Gamma_1)) \mapsto C(0,T; H) \right) \ \text{ with } \ H = H^2_{\Gamma_0}(\Omega) \times H^1_{\Gamma_0}(\Omega).$$

Proof. The argument is similar to that given in the proof of Theorem 2.5.21 (see also Theorem 2.5.26). We use the identification $L = L_1$. Fix $t \in [0,T]$. Computing the adjoint of $L_t(g) = L(g)(t)$ as an operator from $L_2(\Sigma_t^1)$ into H (where we recall $\Sigma_t^1 \equiv \Gamma_1 \times (0,t)$) yields

$$L_t^*(\Phi)(s) = \frac{\partial}{\partial n} \psi_2(s) \Big|_{\Sigma_t^1} \ \text{ and } \ \Psi(s) = e^{-A(t-s)}\Phi, \ s \in [0,t],$$

where $\Phi = (\phi_1;\phi_2) \in H$, $\Psi(s) = (\psi_1(s);\psi_2(s))$. Noting that $z(s) \equiv \psi_1(s)$ satisfies

$$Mz_{ss} + \mathscr{A}z = 0 \text{ on } (0,t) \tag{2.5.65}$$

with terminal conditions $z(t) = \phi_1 \in H^2_{\Gamma_0}(\Omega)$ and $z_s(t) = \phi_2 \in H^1_0(\Omega)$, we reduce the problem to the boundary estimate for

$$L_t^*(\Phi)(s) = \frac{\partial}{\partial n} z_s(s)\Big|_{\Sigma_t^1}. \tag{2.5.66}$$

Applying the trace estimate from Theorem 2.5.18 for solutions to (2.5.65) with time $s := t - s$ along with a standard semigroup estimate to the z equation, yields

$$\int_0^t |\frac{\partial}{\partial n} z_s|^2_{-1/2,\Gamma_1} d\Sigma \leq C\|\Phi\|^2_H.$$

The above along with (2.5.66) implies

$$L_t^* : H \mapsto L_2(0,t;H^{-1/2}(\Gamma_1)) \text{ is bounded uniformly for all } t \in [0,T].$$

Hence $L : L_2(0,T;H^{1/2}(\Gamma_1)) \mapsto L_\infty(0,T;H)$ is bounded as well. Now we apply the standard density argument to conclude the proof. □

The above list of regularity results is not exhaustive. One can shift up and down topologies and so on. It is important to understand the methods that lead to the desired improvement of regularity. Then such methods can be applied to the problem in hand (see [193, 201] for instance).

Chapter 3
Von Karman Models with Rotational Forces

Chapter 3 (resp., 4) treats evolutionary von Karman equations with (resp., without) rotational inertia forces. By way of introduction, we begin by describing a canonical case of von Karman equations equipped with standard clamped homogeneous boundary conditions:

$$\begin{cases} (1-\alpha\Delta)u_{tt} + d_0(x)g_0(u_t) - \alpha \mathrm{div}(d(x)g(\nabla u_t)) \\ \qquad\qquad + \Delta^2 u - [u+f, v+F_0] + Lu = p(x), \quad x\in\Omega, t>0, \\ u|_{\partial\Omega} = \left.\dfrac{\partial u}{\partial n}\right|_{\partial\Omega} = 0, \quad u|_{t=0} = u_0(x), \quad u_t|_{t=0} = u_1(x), \end{cases} \tag{3.0.1}$$

where $v = v(u)$ is a solution of the problem

$$\Delta^2 v + [u+2f, u] = 0, \quad v|_{\partial\Omega} = \left.\frac{\partial v}{\partial n}\right|_{\partial\Omega} = 0. \tag{3.0.2}$$

Here Ω is a smooth bounded domain in $\mathbb{R}^2$, $[u,v]$ is a von Karman bracket defined by (1.4.1), and F_0 is a given function determined by the in-plane mechanical loads. The function f determines the initial form of the shell. The case $f\equiv 0$ corresponds to plate theory. The parameter $\alpha\geq 0$ takes into account the rotational inertial momenta of the elements of the shell/plate. Here L is a linear bounded operator from $H^2(\Omega)$ into $L_2(\Omega)$. Usually L is a first-order differential operator with smooth coefficients. Terms $g_0(u_t)$ and $g(\nabla u_t)$ represent mechanical (potentially nonlinear) damping in the system with damping densities denoted by $d_0(x)$ and $d(x)$, which are assumed nonnegative in Ω.

The study of various variants of this system has received a great deal of attention in the past. The majority of work in this field considered the dynamics subjected to linear damping and a given transverse force $p(x)$ (see, e.g., the monographs [173], [220], [237]).

In addition to clamped boundary conditions we also consider a variety of other boundary conditions such as hinged and free (see (1.3.14) and (1.3.19)), and also nonlinear absorbing boundary conditions corresponding to hinged and free bound-

I. Chueshov, I. Lasiecka, *Von Karman Evolution Equations*,
Springer Monographs in Mathematics, DOI 10.1007/978-0-387-87712-9_3,

ary conditions. The presence of the damping in these boundary conditions has a very clear physical interpretation as damping via moments, torques, or shears. A specific form of boundary conditions is given later.

It has been known for a long time (see [220, 237]) that the treatment of well-posedness of solutions to dynamic von Karman equations depends on whether the value of the parameter is strictly positive, $\alpha > 0$, or just $\alpha = 0$. In fact, in the former case $\alpha > 0$ provides a certain regularizing effect on the velocity component which helps critically in dealing with the nonlinear von Karman bracket and makes the analysis simpler. When $\alpha = 0$, instead, this smoothing mechanism is no longer in place and full Hadamard well-posedness has been an open problem [220]. It is the development of regularity properties for the von Karman bracket (see Theorem 1.4.3) that has changed the situation (for a detailed discussion we refer to Chapter 4).

Our main goal in this chapter is to provide the statements of well-posedness and regularity of solutions corresponding to various evolutionary von Karman models in (3.0.1) which include rotational inertia; that is, in the case when $\alpha > 0$.

In order to accomplish this we apply the theory of monotone operators (see Chapter 2) which needs to be suitably adjusted in order to handle the problems described above. We consider separately the models with internal and boundary dissipation.

3.1 Well-posedness for models with internal dissipation

We start with von Karman evolutions accounting for rotational inertial forces along with a nonlinear internal dissipation:

$$\begin{cases} (1-\alpha\Delta)u_{tt} + d_0(x)g_0(u_t) - \alpha \mathrm{div}(d(x)g(\nabla u_t)) \\ \qquad\qquad + \Delta^2 u - [u+f, v+F_0] + Lu = p(x), \quad x \in \Omega, t > 0, \\ u|_{t=0} = u_0(x), \quad u_t|_{t=0} = u_1(x), \end{cases} \tag{3.1.1}$$

where $v = v(u)$ is a solution of the problem

$$\Delta^2 v + [u+2f, u] = 0, \quad v|_{\partial\Omega} = \frac{\partial v}{\partial n}\Big|_{\partial\Omega} = 0. \tag{3.1.2}$$

We assume the following hypotheses imposed on the given functions in (3.1.1).

3.1.1. Assumption.

- The function $g_0 : \mathbb{R} \mapsto \mathbb{R}$, $g_0(0) = 0$, is continuous and there exists $a_0 \geq 0$ such that $g_0(\sigma) + a_0\sigma$ is monotone nondecreasing.
- The function $g = (g_1; g_2)$ maps $\mathbb{R}^2$ into $\mathbb{R}^2$ and is a continuous, monotone, nondecreasing mapping in $\mathbb{R}^2$; that is,

$$\sum_{i=1,2} [g_i(s_1, s_2) - g_i(s_1^*, s_2^*)]\,(s_i - s_i^*) \geq 0, \quad (s_1; s_2),\ (s_1^*; s_2^*) \in \mathbb{R}^2.$$

Moreover g is of polynomial growth at infinity: we have estimate

$$|g(s)| \le C(1+|s|^p), \quad s=(s_1;s_2)\in \mathbb{R}^2, \ |s|=\sqrt{s_1^2+s_2^2}.$$

with some constants $C>0$ and $p\ge 1$. We also assume (without loss of generality) that $g(0)=0$.
- The functions $d_0(x)$ and $d(x)$ are nonnegative bounded measurable functions.
- $f\in H^2(\Omega), \ F_0\in H^2(\Omega), \ p\in H^{-1}(\Omega)$.
- L is a linear bounded operator from $H^2(\Omega)$ into $H^{-1}(\Omega)$.

A stronger version of Assumption 3.1.1 is needed later. This is formulated below.

3.1.2. Assumption. In addition to Assumption 3.1.1 we assume that the estimates

$$|g_0(\sigma)|\le C(1+|\sigma|^q) \text{ and } |g(s)|\le C(1+|s|)$$

hold with some constants $C>0$ and $q\ge 1$.

Boundary conditions associated with equations (3.1.1) and (3.1.2) are of clamped, hinged, or free type. Later, we also consider a combination of several different types of boundary conditions. However, for the time being, with an eye on more focused presentation, we restrict ourselves to the following three basic types of boundary conditions,

1. **[clamped]**: $u=\nabla u=0$ on $\Gamma\equiv\partial\Omega$,
2. **[hinged]**: $u=\Delta u=0$ on Γ,
3. **[free]**: $\Delta u+(1-\mu)B_1u=0$ on Γ, and

$$\frac{\partial}{\partial n}\Delta u+(1-\mu)B_2u-\alpha\frac{\partial}{\partial n}u_{tt}-\alpha d(x)g(\nabla u_t)\cdot n=\nu_1 u+\beta(x)u^3 \text{ on } \Gamma,$$

where B_1 and B_2 are given by (1.3.20), $\nu_1>0$, $\beta(x)\ge 0$.

We note that free boundary conditions are intrinsically nonlinear even in the case $\beta=0$ (due to the term $\alpha d(x)g(\nabla u_t)\cdot n$). In addition, the presence of terms such as

$$-\alpha\left[\frac{\partial}{\partial n}u_{tt}+dg(\nabla u_t)\cdot n\right]$$

is typical for free (natural) boundary conditions. Indeed, the derivation based on physical principles leads to a variational form that corresponds (after integration by parts) to the strong form of the free boundary condition defined above (see [173, Chapter 2] where several plate models with free boundary conditions are derived).

We denote by $\|\cdot\|_s\equiv\|\cdot\|_{s,\Omega}$ the norm in the Sobolev space $H^s(\Omega)$ and by $\|\cdot\|\equiv\|\cdot\|_\Omega$ and $(\cdot,\cdot)\equiv(\cdot,\cdot)_\Omega$ the norm and the inner product in $L_2(\Omega)$. We use the index Ω in the notations for the norms and inner products, when some ambiguity may arise.

3.1.1 Clamped boundary condition

We start with the simplest version of problem (3.1.1) when clamped boundary conditions are imposed:

$$u = \nabla u = 0 \quad \text{on} \quad \Gamma \equiv \partial\Omega. \tag{3.1.3}$$

To describe the energy of the model we introduce the following functional,

$$\mathscr{E}(t) \equiv \mathscr{E}(u,u_t) = \frac{1}{2}\int_\Omega \left[\alpha|\nabla u_t|^2 + |u_t|^2 + |\Delta u|^2\right] dx + \Pi(u), \tag{3.1.4}$$

where

$$\Pi(u) = \frac{1}{4}\int_\Omega \left[|\Delta v(u)|^2 - 2[u+2f,F_0]u - 4up\right] dx \tag{3.1.5}$$

and $v(u) \in H_0^2(\Omega)$ is determined from (3.1.2). The functional $\Pi(u)$ can be presented in the form

$$\Pi(u) = \frac{1}{4}\int_\Omega |\Delta v(u)|^2 dx + \Pi_1(u),$$

where

$$\Pi_1(u) = -\frac{1}{2}\int_\Omega ([u+2f,F_0]u + 2up)\, dx. \tag{3.1.6}$$

In what follows we also use another energy variable E which consists of the positive part of $\mathscr{E}$; that is,

$$E(t) \equiv E(u,u_t) = \frac{1}{2}\int_\Omega \left[\alpha|\nabla u_t|^2 + |u_t|^2 + |\Delta u|^2 + \frac{1}{2}|\Delta v(u)|^2\right] dx. \tag{3.1.7}$$

We have the obvious relation

$$\mathscr{E}(u,u_t) = E(u,u_t) + \Pi_1(u).$$

One can see that $|\Pi_1(u)| \le \delta\|u\|_2^2 + C_\delta\left[1 + \|u\|_1^2\right]$ for every $\delta > 0$. Therefore it follows from Lemma 1.5.4 that the potential energy $\frac{1}{2}\|\Delta u\|^2 + \Pi(u)$ is bounded from below on $H_0^2(\Omega)$. Therefore, the energy functional $\mathscr{E}(u,u_t)$ is bounded from below with respect to the topology of the energy space $H_0^2(\Omega)\times H_0^1(\Omega)$. Moreover, for any $\eta > 0$ there exists C_η such that

$$|\Pi_1(u)| \le \eta \int_\Omega \left[|\Delta u|^2 + \frac{1}{2}|\Delta v(u)|^2\right] dx + C_\eta, \tag{3.1.8}$$

and hence there exist positive constants c_0, c_1 and a positive constant K (depending on $||F_0||_2, ||f||_2, ||p||_{-1}$) such that

$$c_0\mathscr{E}(u,u_t) - K \le E(u,u_t) \le c_1\mathscr{E}(u,u_t) + K \tag{3.1.9}$$

for any $(u;u_t) \in H_0^2(\Omega) \times H_0^1(\Omega)$. Thus, in view of inequalities (3.1.9) physical energy $\mathscr{E}(u,u_t)$ is topologically equivalent to $E(u,u_t)$. We also note (see Proposition 1.4.2) that under condition (3.1.3) the term $\Pi_1(u)$ can be written in the following form,

$$\Pi_1(u) = -\frac{1}{2}\int_\Omega [u,u]F_0 dx - \int_\Omega u([f,F_0]+p)\,dx. \qquad (3.1.10)$$

In what follows we introduce the concepts of *generalized* and *strong* solutions. A specification of Definition 2.4.3 along with Remarks 2.3.10 and 2.4.18 allows us to give the following definition.

3.1.3. Definition. A function $u(t) \in C(0,T;H_0^2(\Omega)) \cap C^1(0,T;H_0^1(\Omega))$ possessing the properties $u(x,0) = u_0(x)$ and $u_t(x,0) = u_1(x)$ is said to be

1. A *strong solution* to problem (3.1.1) and (3.1.2) with the clamped boundary conditions (3.1.3) on the interval $[0,T]$, iff
   - $u(t)$ is an absolutely continuous function with values in $H_0^2(\Omega)$ and $u_t \in L_1(a,b;H_0^2(\Omega))$ for any $0<a<b<T$.
   - u_t is an absolutely continuous function with values in $H_0^1(\Omega)$ and $u_{tt} \in L_1(a,b;H_0^1(\Omega))$ for any $0<a<b<T$.
   - $u(t) \in H^3(\Omega) \cap H_0^2(\Omega)$ for almost all $t \in [0,T]$.
   - Equation (3.1.1) is satisfied (as an equality in $H^{-1}(\Omega)$) for almost all $t \in [0,T]$ with $v = v(u)$ defined according (3.1.2).
2. A *generalized solution* to problem (3.1.1)–(3.1.3) on the interval $[0,T]$, iff there exists a sequence $\{u_n(t)\}$ of strong solutions to (3.1.1)–(3.1.3) with initial data $(u_{0n};u_{1n})$ such that

$$\lim_{n\to\infty}\max_{t\in[0,T]}\{\|\partial_t u(t) - \partial_t u_n(t))\|_1 + \|u(t)-u_n(t)\|_2\} = 0. \qquad (3.1.11)$$

We recall that generalized solutions are defined in line with the semigroup concept of strong solutions (see, e.g., [241] and also the discussion in Chapter 2). ∎

Our main result in this section is the following theorem.

3.1.4. Theorem. *Under Assumptions* 3.1.1 *with reference to* (3.1.1), *subject to the clamped boundary conditions* (3.1.3), *the following statements are valid with any* $T>0$*:*

- **Generalized solutions:** *For all initial data* $u_0 \in H_0^2(\Omega)$, $u_1 \in H_0^1(\Omega)$ *there exists unique generalized solution* $u(t)$ *such that*

$$u \in C(0,T;H_0^2(\Omega)), \quad u_t \in C(0,T;H_0^1(\Omega)). \qquad (3.1.12)$$

  Moreover

$$d_0 u_t g_0(u_t),\ d\nabla u_t g(\nabla u_t) \in L_1(\Omega\times[0,T]), \qquad (3.1.13)$$

  and the following energy inequality holds for $s \le t$,

$$\mathscr{E}(u(t),u_t(t)) + \int_s^t \int_\Omega [d_0 g_0(u_t)u_t + \alpha d g(\nabla u_t)\nabla u_t]\, dx d\tau$$
$$\le \mathscr{E}(u(s),u_t(s)) - \int_s^t (Lu,u_t)_\Omega d\tau. \tag{3.1.14}$$

Thus, if $L \equiv 0$ and $g_0(s)s \ge 0$ for all $s \in \mathbb{R}$, then the energy $\mathscr{E}(u(t),u_t(t))$ of the system is nonincreasing; that is, $\mathscr{E}(u(t),u_t(t)) \le \mathscr{E}(u(s),u_t(s))$ for all $s \le t$. Assuming, in addition, that Assumption 3.1.2 holds, we have the energy identity:

$$\mathscr{E}(u(t),u_t(t)) + \int_s^t \int_\Omega [d_0 g_0(u_t)u_t + \alpha d g(\nabla u_t)\nabla u_t]\, dx d\tau$$
$$= \mathscr{E}(u(s),u_t(s)) - \int_0^t \int_\Omega Luu_t dx d\tau \tag{3.1.15}$$

and also the following regularity

$$u_{tt} \in L_2(0,T;L_2(\Omega)). \tag{3.1.16}$$

- **Strong solutions:** *Assuming $u_0 \in H_0^2(\Omega) \cap H^3(\Omega), u_1 \in H_0^2(\Omega)$ there exists a unique strong solution such that*

$$\left.\begin{array}{l} u \in C_r(0,T;(H^3 \cap H_0^2)(\Omega)) \cap L_\infty(0,T;(H^3 \cap H_0^2)(\Omega)), \\ u_t \in C_r(0,T;H_0^2(\Omega)) \cap L_\infty(0,T;H_0^2(\Omega)), \\ u_{tt} \in C_r(0,T;H_0^1(\Omega)) \cap L_\infty(0,T;H_0^1(\Omega)), \end{array}\right\} \tag{3.1.17}$$

where $C_r(0,T;X)$ is the space of strongly right-continuous functions with values in X. Strong solutions satisfy the energy identity in (3.1.15).
If $g_0(\sigma)$ and $g(s)$ are differentiable and

$$\sup_{|\sigma|\le R} |g_0'(\sigma)| \le C_R \text{ for every } R \text{ and } |\nabla_s g(s)| \le C \text{ for all } s \in \mathbb{R}^2 \tag{3.1.18}$$

with some constants $C_R > 0$ and $C > 0$, then

$$u \in C(0,T;(H^3 \cap H_0^2)(\Omega)), \quad u_t \in C(0,T;H_0^2(\Omega)), \quad u_{tt} \in C(0,T;H_0^1(\Omega)). \tag{3.1.19}$$

3.1.5. Remark. The energy identity in (3.1.15) for generalized solutions can be also established under weaker (than in Assumption 3.1.2) hypotheses imposed on rotational damping function $g : \mathbb{R}^2 \mapsto \mathbb{R}^2$. For instance, we can assume, instead of a linear growth condition, that the function g possesses the properties

$$(g(s),s)_{\mathbb{R}^2} \ge m|s|^{q_*} \text{ and } |g(s)| \le M|s|^{q_*-1} \text{ for } s \in \mathbb{R}^2,\ |s| \ge s_0, \tag{3.1.20}$$

where m, M, and s_0 are positive constants, $q^* \geq 1$. In this case instead of (3.1.16) we have that

$$u_{tt} \in L_{q^*/(q^*-1)}(0,T;L_2(\Omega)). \tag{3.1.21}$$

We do not provide detailed proofs of these facts and refer the reader to Section 3.2 where the corresponding argument is presented in a more demanding case of boundary damping.

Similarly, the growth condition imposed on function g in (3.1.18) can be relaxed by requiring that the derivative of g satisfies the coercivity condition in line with relations (3.1.20). Because this is not an essential point for further development, for the sake of clarity of exposition we adhere to a more restrictive hypothesis imposed by (3.1.18). ■

3.1.1.1 Proof of Theorem 3.1.4—Generalized and strong solutions

In order to prove Theorem 3.1.4 we first rewrite the von Karman equations as a second-order abstract equation and then we apply Theorem 2.4.16.

In order to accomplish this we introduce the following spaces and operators.

- $\mathscr{H} \equiv L_2(\Omega)$, $\mathscr{A}u \equiv \Delta^2 u$, $u \in \mathscr{D}(\mathscr{A})\,:\, \mathscr{D}(\mathscr{A}) \equiv H_0^2(\Omega) \cap H^4(\Omega)$.
- $V \equiv H_0^1(\Omega)$, $Mu \equiv u - \alpha\Delta u$, $u \in \mathscr{D}(M)\,:\, \mathscr{D}(M) \equiv H_0^1(\Omega) \cap H^2(\Omega)$.
- Hence $V' = H^{-1}(\Omega)$, $\mathscr{D}(\mathscr{A}^{1/2}) = H_0^2(\Omega)$, $[\mathscr{D}(\mathscr{A}^{1/2})]' = H^{-2}(\Omega)$.
- $F(u,v) = \widetilde{F}(u) + a_0 d_0(x)v$ with $\widetilde{F}(u) \equiv [u+f, v(u)+F_0] - Lu + p$, where $v(u) \in H_0^2(\Omega)$ solves the elliptic problem (3.1.2).
- $D(v) \equiv d_0(x)\,[g_0(v) + a_0 v] - \alpha \mathrm{div}(d(x)g(\nabla v))$ for $v \in \mathscr{D}(\mathscr{A}^{1/2}) = H_0^2(\Omega)$.

With the above notation, the abstract form of the equation (3.1.1)–(3.1.3) becomes

$$Mu_{tt} + \mathscr{A}u + Du_t = F(u,u_t) \tag{3.1.22}$$

and thus at least formally we arrive at (2.4.26) with $G = 0$.

Our first step is to verify that Assumption 2.4.15 in Theorem 2.4.16 is satisfied with the Green's map $G = 0$.

3.1.6. Lemma. *Operators $\mathscr{A}, M, F$, and D introduced above comply with Assumption* 2.4.15.

Proof.

1. *Hypotheses* 1 *and* 2 *in Assumption* 2.4.15: We note that with the definition of $\mathscr{A}$ as above, $\mathscr{A}$ is closed, positive, and densely defined on $\mathscr{H}$ with $\mathscr{D}(\mathscr{A}^{1/2}) = H_0^2(\Omega)$. Moreover, we have that

$$V = H_0^1(\Omega) = \mathscr{D}(M^{1/2}), \quad V' = H^{-1}(\Omega) \text{ and } [\mathscr{D}(\mathscr{A}^{1/2})]' = H^{-2}(\Omega).$$

So we have dense continuous injections: $\mathscr{D}(\mathscr{A}^{1/2}) \subset V \subset \mathscr{H} \subset V' \subset [\mathscr{D}(\mathscr{A}^{1/2})]'$. This shows that the requirements 1 and 2 in Assumption 2.4.15 are satisfied.

2. *Hypothesis* 3 *in Assumption* 2.4.15: This hypothesis requires showing that the nonlinear term $F(u,v) = \widetilde{F}(u) + a_0 d_0(x)v$ is locally Lipschitz from $H_0^2(\Omega) \times H_0^1(\Omega)$ into $H^{-1}(\Omega)$. It suffices to consider the term $\widetilde{F}(u)$ only.

For any $u_1, u_2 \in H^2(\Omega)$ we have that

$$\widetilde{F}(u_1) - \widetilde{F}(u_2) = [u, F_0] + [f, v(u_1) - v(u_2)] + [u_1, v(u_1)] - [u_1, v(u_2)] - Lu,$$

where $u = u_1 - u_2$. It follows from relation (1.4.17) in Theorem 1.4.3 that

$$\|[u, F_0]\|_{-1} \le C\|u\|_2 \|F_0\|_2.$$

Because by Corollary 1.4.5 (see (1.4.24))

$$\|v(u_1) - v(u_2)\|_{W^2_\infty(\Omega)} \le C\left(\|u_1\|_2 + \|u_2\|_2 + \|f\|_2\right)\|u_1 - u_2\|_2,$$

we have that

$$\|[f, v(u_1) - v(u_2)]\| \le C\left(\|u_1\|_2 + \|u_2\|_2 + \|f\|_2\right)\|f\|_2 \|u_1 - u_2\|_2.$$

By Assumption 3.1.1 we also have that $\|Lu\|_{-1} \le \|u\|_2$. Thus in order to prove that $\widetilde{F}$ is locally Lipschitz from $H^2(\Omega)$ into $H^{-1}(\Omega)$, it suffices to verify this property for the nonlinear term $[u, v(u)]$. The key result used is Corollary 1.4.5. In fact, from (1.4.26) with $\delta = 1$ of Corollary 1.4.5 we obtain that

$$\|[u_1, v(u_1)] - [u_2, v(u_2)]\|_{-1} \le C\left(\|u_1\|_2^2 + \|u_2\|_2^2 + \|f\|_2^2\right)\|u_1 - u_2\|_1. \quad (3.1.23)$$

Thus

$$\|\widetilde{F}(u_1) - \widetilde{F}(u_2)\|_{-1} \le C\left(1 + \|F_0\|_2 + \|f\|_2^2 + \|u_1\|_2^2 + \|u_2\|_2^2\right)\|u_1 - u_2\|_2, \quad (3.1.24)$$

and therefore $\widetilde{F} : \mathscr{D}(\mathscr{A}^{1/2}) \times V \mapsto V'$ and (2.4.27) holds.

It is also clear that F has the form (2.4.28) with $F^*(u,v) = -Lu + a_0 d_0 v$ and $\Pi(u) = \Pi_0(u) + \Pi_1(u)$, where $\Pi_0(u) = \frac{1}{4}\|\Delta v(u)\|^2$ and $\Pi_1(u)$ is given in (3.1.6). The estimate (2.4.30) is the same as (3.1.8). Obviously, by Assumption 3.1.1 $F^*(u,v) = -Lu + a_0 d_0 v$ satisfies (2.4.29). Thus requirement 3 in Assumption 2.4.15 is satisfied.

3. *Hypothesis* 4 *in Assumption* 2.4.15: In order to assert continuity of the operator D from $H_0^2(\Omega)$ into $H^{-2}(\Omega)$ we need to exploit the growth condition of functions g (see Assumption 3.1.1) along with Sobolev's embeddings. This is done as follows.

Because $H_0^2(\Omega) \subset C(\overline{\Omega})$, the Nemytskij operator $u \mapsto g_0(u) + a_0 d_0 u \in C(\overline{\Omega})$ is continuous from $H_0^2(\Omega)$ into $C(\overline{\Omega})$ hence trivially in $H^{-2}(\Omega)$. As for the second component associated with the function g, we have

$$\partial_{x_i} u \in H^1(\Omega) \subset L_p(\Omega) \quad \text{for all} \quad 1 \le p < \infty \quad \text{and} \quad u \in H^2(\Omega).$$

Hence for any polynomially growing at infinity function g we have the hemicontinuity of the mapping $u \mapsto g(\nabla u)$ from $H_0^2(\Omega)$ into $[L_2(\Omega)]^2$. Therefore, the mapping

$u \mapsto \operatorname{div}[d(x)g(\nabla u)] \in H^{-1}(\Omega)$ is hemicontinuous from $H_0^2(\Omega)$ into $H^{-1}(\Omega) \subset H^{-2}(\Omega)$. These observations allow us to conclude hemicontinuity of the operator D from $H_0^2(\Omega)$ into $H^{-2}(\Omega)$.

Monotonicity of D follows from the relation

$$(D(u_1)-D(u_2),u_1-u_2) = \int_\Omega d_0(x)(g_0(u_1)-g_0(u_2)+a_0(u_1-u_2))(u_1-u_2)dx$$
$$+\alpha\int_\Omega d(x)(g(\nabla u_1)-g(\nabla u_2))\nabla(u_1-u_2)dx \geq 0, \tag{3.1.25}$$

where the last conclusion follows from the monotonicity imposed on the functions g and g_0 in Assumption 3.1.1. The proof of Lemma 3.1.6 is thus completed. □

3.1.7. Remark. Using the same argument as in the proof of Lemma 3.1.8 below one can show that D maps $H_0^2(\Omega)$ into $H^{-1}(\Omega)$ continuously. However, this property is not needed for application of Theorem 2.4.16. ■

Lemma 3.1.6 asserts the validity of our standing Assumption 2.4.15 with $G = 0$. This allows us to apply the result of Theorem 2.4.16. Because $D(v) \in H^{-1}(\Omega)$ for all $v \in H_0^2(\Omega)$, the compatibility condition $\mathscr{A}u_0 + D(u_1) \in V'$ holds if $u_1 \in H_0^2(\Omega)$ and $u_0 \in W \equiv \{u \in \mathscr{D}(\mathscr{A}^{1/2}) : \mathscr{A}u \in V'\} = H^3(\Omega)\cap H_0^2(\Omega)$. Therefore Theorem 2.4.16 (along with Remark 2.4.17) implies all assertions of Theorem 3.1.4 except for (i) energy inequality (3.1.14), (ii) the regularity properties (3.1.16) and (3.1.19), and (iii) energy relation (3.1.15) for generalized solutions in the case when Assumption 3.1.2 holds.

In order to prove (3.1.14) we note that from energy equality (3.1.15), which is satisfied for all strong solutions, it follows that

$$cE(u(t),u_t(t)) + \int_s^t\int_\Omega \psi_N\left(d_0[g_0(u_t)+a_0u_t]u_t + \alpha d g(\nabla u_t)\nabla u_t\right)dxd\tau$$
$$\leq a_0\int_s^t\int_\Omega d_0|u_t|^2dxd\tau + \mathscr{E}(u(s),u_t(s)) - \int_s^t (Lu,u_t)_\Omega d\tau \tag{3.1.26}$$

for any strong solution, where $\psi_N(s) = \min\{s,N\}$ and N is a positive number. From the definition of generalized solutions any sequence $\{u^n\}$ of strong solutions converging to a generalized solution u contains a subsequence $\{u^{n_k}\}$ such that

$$u_t^{n_k}(x,t) \to u_t(x,t), \quad \nabla u_t^{n_k}(x,t) \to \nabla u_t(x,t) \ \text{ a.e. in } \Omega\times[0,T].$$

Therefore (3.1.26) holds for every generalized solution u and for $N = 1,2,\ldots$. It is clear that $\psi_N(s) \leq \psi_{N+1}(s)$. Therefore the Levi–Lebesgue theorem (see Theorem B.4.2 in Appendix B) on monotone convergence implies the validity of energy inequality (3.1.14) satisfied with every generalized solution.

Relation (3.1.15) for generalized solutions follows from Proposition 2.4.21 and the following assertion.

3.1.8. Lemma. *Let Assumption* 3.1.2 *be valid. Then the operator D defined by the formula*

$$D(u) \equiv d_0(x)\,[g_0(u)+a_0u] - \alpha\,\mathrm{div}(d(x)g(\nabla u))$$

is an m-monotone continuous mapping from $H_0^1(\Omega)$ into $H^{-1}(\Omega)$ and the relation

$$\sup\left\{\|Dv\|_{-1,\Omega}\;:\; v\in H_0^1(\Omega),\; \|v\|_{1,\Omega}\le\rho\right\}<\infty \tag{3.1.27}$$

holds for any $\rho>0$.

Proof. We first prove the continuity of mapping D. It is sufficient to prove the continuity of the mappings $u\mapsto g_0(u)$ and $u\mapsto g_i(\nabla u)$ from $H_0^1(\Omega)$ into $L_2(\Omega)$.

Let's contradict and argue that the mapping $u\mapsto g_0(u)$ is not continuous as specified above. Then for some $u\in H_0^1(\Omega)$ there exists a sequence $\{u_n\}\subset H_0^1(\Omega)$ such that $\|u_n-u\|_1\to 0$ and

$$\int_\Omega |g_0(u_n(x))-g_0(u(x))|^2dx\ge\delta>0,\quad n=1,2,\dots. \tag{3.1.28}$$

We can choose a subsequence $\{u_{n_k}\}$ such that $u_{n_k}\to u$ almost everywhere in Ω and thus $g_0(u_{n_k}(x))\to g_0(u(x))$ for almost all $x\in\Omega$. In addition

$$|g_0(u_{n_k}(x))-g_0(u(x))|^2\le C\left(1+|u_{n_k}(x)|^{2q}+|u(x)|^{2q}\right)$$

and because $\|u_{n_k}-u\|_1\to 0$ as $k\to\infty$, the embedding $H_1(\Omega)\subset L_p(\Omega)$ for $1\le p<\infty$ implies that

$$\int_\Omega |u_{n_k}(x)|^{2q}dx\to\int_\Omega |u(x)|^{2q}dx,\quad k\to\infty.$$

The Lebesgue dominated convergence theorem (see Theorem B.4.3 in Appendix B) implies now that

$$\lim_{k\to\infty}\int_\Omega |g_0(u_{n_k}(x))-g_0(u(x))|^2dx=0,$$

which contradicts (3.1.28). Thus $u\mapsto g_0(u)$ is continuous from $H_0^1(\Omega)$ into $L_2(\Omega)$.

A similar argument applies to the mapping $u\mapsto g_i(\nabla u)$ (we recall that by Assumption 3.1.2 $g_i(s)$ is linearly bounded). Thus D is continuous from $H_0^1(\Omega)$ into $H^{-1}(\Omega)$.

Monotonicity of D follows from (3.1.25). Because D is continuous, D is m-monotone by Proposition 1.2.5. □

Now we can apply Lemma 3.1.8 and the result of Proposition 2.4.21 to obtain the energy relation (3.1.15) for generalized solutions.

To establish improved regularity in (3.1.19) we consider $w:=u_t$ as a weak solution of a nonautonomous problem of the form

$$(1-\alpha\Delta)w_{tt}+\Delta^2w=F(x,t),\quad w\big|_\Gamma=0,\;\nabla w\big|_\Gamma=0,$$

with $F\in L_2(0,T;H^{-1}(\Omega))$, which is obtained by formal differentiation of (3.1.1) in t. Then we apply Proposition 2.5.1.

The regularity of second time derivatives of generalized solutions in (3.1.16) is based on the following trace regularity.

3.1.9. Proposition (Trace regularity). *Under Assumption* 3.1.2 *we have the estimate*

$$\int_{\Sigma_T} |\Delta u|^2 d\Sigma \le C_T(E(0)) \tag{3.1.29}$$

for generalized solutions. Here $\Sigma_T = \partial\Omega \times [0,T]$.

Proof. This estimate relies on multipliers/energy calculations similar to those performed in the linear case (see Theorem 2.5.4).

It follows from (3.1.9) and (3.1.14) that

$$E(u(t),u_t(t)) \le c_1(1+E(u_0,u_1))e^{c_2 t}, \quad t>0,$$

It is also clear that

$$\|F(u)\|_{-1} \le C(1+\|u\|_2^3) \le c_1(1+E(u_0,u_1))^{3/2}e^{c_2 t}, \quad t>0,$$

Thus considering $u(t)$ as a solution to the linear problem (2.5.1) with

$$f(x,t) = -d_0(x)g_0(u_t) + \alpha\,\mathrm{div}(d(x)g(\nabla u_t)) + [u+f, v+F_0] - L(u) + p(x),$$

we can apply Theorem 2.5.4 and obtain (3.1.29). □

In order to prove the regularity of the second-order time derivative, we recall from (3.1.22) that

$$u_{tt} = M^{-1}[-\mathscr{A}u - D(u_t) + F(u,u_t)].$$

From properties of D and F under Assumption 3.1.2 we have

$$\|M^{-1}D(u_t)\|_1 \le C_T(E(0)), \quad \|M^{-1}F(u,u_t)\|_1 \le C_T(E(0)). \tag{3.1.30}$$

On the other hand, by integrating by parts and using Green's formula we also obtain the following equality valid with any function $w \in L_2(\Omega)$:

$$\begin{aligned}\int_\Omega M^{-1}\mathscr{A}uwdx &= \int_\Omega \mathscr{A}uM^{-1}wdx = \int_\Omega \Delta^2 uM^{-1}wdx \\ &= \int_\Omega \Delta u\Delta M^{-1}wdx - \int_\Gamma \Delta u\frac{\partial}{\partial n}M^{-1}wd\Gamma. \end{aligned} \tag{3.1.31}$$

By using the trace regularity stated in Proposition 3.1.9 along with elliptic regularity

$$\|\Delta M^{-1}w\|_0 \le \|w\|_0, \quad \left\|\frac{\partial}{\partial n}M^{-1}w\right\|_{1/2,\Gamma} \le C\|w\|_0,$$

we obtain

$$\int_0^T \int_\Omega |M^{-1}\mathscr{A}u|^2 dxdt \le C_T(E(0)).$$

This proves the desired regularity of second time derivatives of generalized solutions, thus completing the proof of Theorem 3.1.4.

3.1.10. Remark. It is remarkable that in the case of von Karman evolutions, intrinsically related to hyperbolic dynamics with finite speed of propagation $\alpha > 0$, there are a number of regularity properties on the boundary that do not follow from interior regularity. In fact, in the case of clamped boundary conditions the regularity result (3.1.29) is valid for weak (finite energy) solutions. We note that this regularity provides $\frac{1}{2}$ derivative more on the boundary than the trace theorem (which can be applied only formally) would predict. ■

3.1.1.2 Weak Solutions

We note that Theorem 3.1.4 stated above asserts global existence and uniqueness of both generalized and strong solutions. By Definition 3.1.3 strong solutions satisfy a differential equation in a pointwise manner. However, this fails for generalized solutions that are only related to strong solutions via the corresponding limits. This raises an interesting question whether generalized solutions are related to a differential equation. A natural idea to pursue is to ask whether these solutions satisfy variational form of the equation, which are often called "weak solutions".

3.1.11. Definition. We say that u is a *weak solution* to equation (3.1.1) on an interval $[0,T]$ with the clamped boundary conditions (3.1.3) iff

$$u \in L_\infty(0,T;H_0^2(\Omega)) \cap W_\infty^1(0,T;H_0^1(\Omega)),$$

and also (i) $u|_{t=0} = u_0$; (ii) the functions

$$t \mapsto (dg(\nabla u_t), \nabla\phi)_\Omega \;\text{ and }\; t \mapsto (d_0 g_0(u_t), \phi)_\Omega \text{ are integrable} \tag{3.1.32}$$

for every $\phi \in H_0^2(\Omega)$; and (iii) the following variational relation

$$\begin{aligned}&(u_t(t),\phi)_\Omega + \alpha(\nabla u_t(t),\nabla\phi)_\Omega - (u_1,\phi)_\Omega - \alpha(\nabla u_1,\nabla\phi)_\Omega \\ &+ \int_0^t \big[(\Delta u, \Delta\phi)_\Omega + (d_0 g_0(u_t),\phi)_\Omega + \alpha(dg(\nabla u_t),\nabla\phi)_\Omega \\ &+ (-[v(u)+F_0, u+f] + Lu - p, \phi)_\Omega\big]\, d\tau = 0\end{aligned} \tag{3.1.33}$$

holds for any $\phi \in H_0^2(\Omega)$, where $v(u) \in H_0^2(\Omega)$ is defined by (3.1.2). ■

3.1.12. Theorem. *Assume that the rotational damping function* $g : \mathbb{R}^2 \mapsto \mathbb{R}^2$ *satisfies Assumption* 3.1.2 *or else possesses the property*

$$|g(s_1,s_2)| \le C\big(1 + [s_1 g_1(s_1,s_2) + s_2 g_2(s_1,s_2)]^\eta\big) \tag{3.1.34}$$

for some $C \ge 0$ *and* $\eta < 1$. *Then under Assumption 3.1.1 generalized solutions of the von Karman system* (3.1.1) *with the clamped boundary conditions* (3.1.3) *are also weak.*

If g_0 and g satisfy Assumption 3.1.2, then weak solutions are unique. This implies that every weak solution with initial data $(u_0;u_1) \in H_0^2(\Omega) \times H_0^1(\Omega)$ is also generalized.

3.1.13. Remark. We note that condition (3.1.34) holds true if in addition to the hypotheses in Assumption 3.1.1 g has the structure $g(s_1,s_2) = (g_1(s_1);g_2(s_2))$. Indeed, in this case by the polynomial growth conditions we have that

$$|g_i(s_i)| \le |g_i(s_i)|^{\delta}|g_i(s_i)|^{1-\delta} \le C_\delta |s_i|^{\delta p}|g_i(s_i)|^{1-\delta}, \quad |s_i| \ge 1,\ i=1,2.$$

Thus, if we choose $\delta = (1+p)^{-1}$, then we obtain that

$$|g_i(s_i)| \le C\left(1 + [s_i g_i(s_i)]^{p/(1+p)}\right), \quad s_i \in \mathbb{R},\ i=1,2.$$

This implies (3.1.34) with $\eta = p(1+p)^{-1}$. ■

3.1.14. Remark. The result stated in first part of Theorem 3.1.12 remains true if we assume that the rotational damping function g satisfies (3.1.20). Moreover, after some minor modification (which takes into account the coercivity given by (3.1.20)) of the definition of the weak solutions the uniqueness of weak solutions remains in force for this case. The point is that the coercivity condition imposed in (3.1.20) allows us to establish the appropriate energy inequality for the difference of two weak solutions. This fact alone allows us to prove uniqueness. Our main focus is on generalized solutions, therefore we omit these technicalities. ■

Proof. The variational form specified in Definition 3.1.11 is verified for strong solutions. Using the limit definition of generalized solutions along with weak continuity of the nonlinear terms involving Airy stress function, one can pass with the limit on these terms without any difficulty.

Thus, the only remaining issue is the passage through the limit on nonlinear damping terms. It is sufficient to prove that the convergence in (3.1.11) implies (at least along some subsequence) the convergence

$$\int_Q g_0(u_t^n)\phi dQ \to \int_Q g_0(u_t)\phi dQ, \quad n\to\infty, \tag{3.1.35}$$

and

$$\int_Q dg(\nabla u_t^n)\nabla\phi dQ \to \int_Q dg(\nabla u_t)\nabla\phi dQ, \quad n\to\infty, \tag{3.1.36}$$

for all $\phi \in L_\infty(0,T;H_0^2(\Omega))$, where $Q = (0,T)\times\Omega$.

To prove (3.1.35) we note that by (3.1.11) we can assume that $u_t^n \to u_t$ almost everywhere in Q. Thus $g_0(u_t^n) \to g_0(u_t)$ a.e. in Q. Because $g_0(s)+\alpha_0 s$ is monotone nondecreasing, it follows the existence of constants $C_1, C_2 \ge 0$ such that

$$|g_0(s)| \le C_1 + C_2 s^2 + s g_0(s), \quad s\in\mathbb{R}. \tag{3.1.37}$$

Thus (3.1.14) along with (3.1.37) imply $\sup_n \int_Q d_0(x)|g_0(u_t^n)|dQ < \infty$. Therefore, by Fatou's lemma (see Theorem B.4.1 in the appendix)

$$d_0(x)|g_0(u_t)| \in L_1(Q). \tag{3.1.38}$$

We claim that the sequence $d_0(x)g_0(u_t^n) \to d_0(x)g_0(u_t)$ in $L_1(Q)$; that is,

$$\lim_{n\to\infty} \int_Q d_0(x)\,|g_0(u_t^n) - g_0(u_t)|\,dQ = 0. \tag{3.1.39}$$

Let $E \subset Q$, $A = \{(t;x) \in E, |u_t^n| \geq \lambda\}$ and $B = E \setminus A$. We obviously have that

$$\int_E d_0(x)\,|g_0(u_t^n)|\,dQ \leq \int_B d_0\,|g_0(u_t^n)|\,dQ + \int_A d_0\,|g_0(u_t^n)|\,dQ$$
$$\leq b_\lambda \operatorname{mes}(E) + \frac{1}{\lambda}\int_Q d_0|g_0(u_t^n)u_t^n|dQ \leq b_\lambda \operatorname{mes}(E) + \frac{C}{\lambda}$$

for any $\lambda > 0$, where b_λ and C are positive constants. Consequently

$$\int_Q d_0(x)\,|g_0(u_t^n) - g_0(u_t)|\,dQ \leq b_\lambda \operatorname{mes}(E) + \frac{C}{\lambda}$$
$$+ \int_E d_0(x)\,|g_0(u_t)|\,dQ + \int_{Q\setminus E} d_0(x)\,|g_0(u_t^n) - g_0(u_t)|\,dQ. \tag{3.1.40}$$

By Egorov's theorem (Theorem B.4.4 in the appendix) for any $\varepsilon > 0$ there is $E \subset Q$ such that $\operatorname{mes}(E) \leq \varepsilon$ and $d_0(x)\,|g_0(u_t^n) - g_0(u_t)| \to 0$ uniformly on $Q \setminus E$. Therefore from (3.1.40) and (3.1.38) we have that

$$\limsup_{n\to\infty} \int_Q d_0(x)\,|g_0(u_t^n) - g_0(u_t)|\,dQ \leq m_\lambda(\varepsilon) + C\lambda^{-1}$$

for any $\lambda > 0$, where $m_\lambda(\varepsilon) \to 0$ as $\varepsilon \to 0$ for every fixed $\lambda > 0$. This implies (3.1.39) and hence by the embedding $H_0^2(\Omega) \subset C(\overline{\Omega})$ (3.1.35) holds.

As for (3.1.36) under Assumption 3.1.2, we use the monotonicity method. Indeed, Lemma 3.1.8 applied with $d_0(x) = 0$ gives us that $v \mapsto \operatorname{div}\{dg(\nabla v)\}$ is an m-monotone continuous mapping from $H_0^1(\Omega)$ into $H^{-1}(\Omega)$. This allows for the correct identification of the limit, hence proving (3.1.36).

In the case when (3.1.34) is satisfied, (3.1.36) is proved by using a similar argument as the one used for $g_0(u_t)$. In order to prove the convergence

$$\lim_{n\to\infty} \int_Q d(x)\,|g_i(\nabla u_t^n) - g_i(\nabla u_t)|^r\,dQ = 0 \quad \text{for } 1 < r < \frac{1}{\eta},$$

we simply use condition (3.1.34) that replaces (3.1.37) and also the relation

$$\int_E d(x)\,|g(\nabla u_t^n)|^r\,dQ \leq C[\operatorname{mes}(E)]^{1-r\eta}\left[\int_E d(x)\,(1 + g(\nabla u_t^n)\nabla u_t^n)\,dQ\right]^{r\eta}$$

That follows from the Hölder inequality. Sobolev's embedding of $H_0^2(\Omega)$ into $W_p^1(\Omega)$ for all $1 < p < \infty$, concludes the proof of (3.1.36). Thus, under either assumption any generalized solution is also weak.

In order to prove *uniqueness of weak solutions* we argue by contradiction. Let there exist two solutions u^1 and u^2. Their difference $u = u^1 - u^2$ is a weak solution to the linear problem

$$\begin{cases} u_{tt} - \alpha\Delta u_{tt} + \Delta^2 u = f(x,t), \text{ in } Q \equiv \Omega \times (0,T) \\ u = \frac{\partial}{\partial n} u = 0 \text{ on } \Sigma = \Gamma \times (0,T), \quad u|_{t=0} = u_t|_{t=0} = 0, \end{cases}$$

where $f(x,t) \equiv G(x,t) + F(x,t)$ with

$$G(x,t) = -d_0\left(g_0(u_t^1) - g_0(u_t^2) + a_0(u_t^1 - u_t^2)\right) + \alpha \mathrm{div}[d(g(\nabla u_t^1) - g(\nabla u_t^2)]$$

and

$$F(x,t) = a_0 d_0(u_t^1 - u_t^2) + [u^1 + f, v(u^1) + F_0] - [u^2 + f, v(u^2) + F_0] - L(u^1 - u^2).$$

A simple calculation involving Assumption 3.1.2 shows that

$$g(\nabla u_t^i) \in L_\infty(0,T;L_2(\Omega)) \text{ and } g_0(u_t^i) \in L_\infty(0,T;L_2(\Omega)), \quad i = 1,2.$$

Hence $G \in L_\infty(0,T;H^{-1}(\Omega))$. In addition

$$\|F(t)\|_{-1} \le C_T \cdot (\|u_t^1 - u_t^2\|_1 + (\|u^1 - u^2)\|_2), \quad t \in [0,T], \tag{3.1.41}$$

where the constant also depends on the values $\sup_{\tau\in[0,T]}\left\{|u_t^j(\tau)\|_1 + (\|u^j(\tau))\|_2\right\}$, $j = 1,2$. Now using the energy relation (2.5.2) for the linear problem (2.5.1) (which is legitimate on the strength of $f \in L_1(0,T;H^{-1}(\Omega))$, we obtain

$$E_0(u,u_t) + \int_0^t\int_\Omega d_0(x)\left[g_0(u_t^1) - g_0(u_t^2) + a_0(u_t^1 - u_t^2)\right] u_t dxds$$
$$+\alpha\int_0^t\int_\Omega d(x)[g(\nabla u_t^1) \quad g(\nabla u_t^2)]\nabla(u_t^1 \quad u_t^2)dxds - \int_0^t\int_\Omega F(x,s)u_t(x,s)dxds$$

The estimate in (3.1.41) and the monotonicity of the damping functions followed by Gronwall's lemma applied to this inequality imply $u \equiv 0$. □

3.1.2 Hinged boundary conditions

Now we consider problem (3.1.1) and (3.1.2) with hinged boundary conditions

$$u = \Delta u = 0 \quad \text{on } \Gamma \equiv \partial\Omega. \tag{3.1.42}$$

Similar to the case of clamped boundary conditions we give the following definition.

3.1.15. Definition. A function

$$u(t) \in C(0,T;H^2(\Omega)\cap H_0^1(\Omega))\cap C^1(0,T;H_0^1(\Omega))$$

possessing the properties $u(x,0) = u_0(x)$ and $u_t(x,0) = u_1(x)$ is said to be

1. A *strong solution* to problem (3.1.1) and (3.1.2) with hinged boundary conditions (3.1.42) on the interval $[0,T]$, iff
   - $u(t)$ is an absolutely continuous function with values in $H^2(\Omega)\cap H_0^1(\Omega)$ and $u_t \in L_1(a,b;H^2(\Omega)\cap H_0^1(\Omega))$ for any $0 < a < b < T$.
   - u_t is an absolutely continuous function with values in $H_0^1(\Omega)$ and $u_{tt} \in L_1(a,b;H_0^1(\Omega))$ for any $0 < a < b < T$.
   - $u(t) \in \{v \in H^3(\Omega)\cap H_0^1(\Omega) : \Delta v|_{\partial\Omega} = 0\}$ for almost all $t \in [0,T]$.
   - Equation (3.1.1) is satisfied in $H^{-1}(\Omega)$ for almost all $t \in [0,T]$ with $v = v(u)$ defined by (3.1.2);
2. A *generalized solution* to problem (3.1.1), (3.1.2), and (3.1.42) on the interval $[0,T]$, iff there exists a sequence $\{u_n(t)\}$ of strong solutions to (3.1.1), (3.1.2), and (3.1.42) with the initial data $(u_{0n};u_{1n})$ such that (3.1.11) holds.

■

For hinged boundary conditions we have a similar well-posedness result to that in the case of clamped homogeneous boundary conditions (cf. Theorem 3.1.4). A precise statement is given below.

3.1.16. Theorem. *Under the Assumption* 3.1.1 *with reference to* (3.1.1)*, subject to hinged boundary conditions* (3.1.42)*, the following statements are valid with any* $T > 0$.

- **Generalized solutions:** *For all initial data* $u_0 \in H_0^1(\Omega)\cap H^2(\Omega)$, $u_1 \in H_0^1(\Omega)$ *there exists a unique generalized solution* $u(t)$ *such that*

$$u \in C(0,T;H_0^1(\Omega)\cap H^2(\Omega)), \quad u_t \in C(0,T;H_0^1(\Omega)). \tag{3.1.43}$$

  Moreover, relations (3.1.13) *and energy inequality* (3.1.14) *hold for all generalized solutions, with the energy* $\mathscr{E}$ *defined by* (3.1.4)*. Assuming, in addition Assumption* 3.1.2,[1] *we also obtain that*

$$u_{tt} \in C(0,T;L_2(\Omega)), \tag{3.1.44}$$

  and that the energy identity in (3.1.15) *is valid.*
- **Strong solutions:** *Assume that*

$$u_0 \in W \equiv \{v \in (H^3\cap H_0^1)(\Omega) : \Delta v|_{\partial\Omega} = 0\} \quad \textit{and} \quad u_1 \in (H^2\cap H_0^1)(\Omega).$$

  Then there exists a unique strong solution such that

[1] As in the clamped case this assumption can be relaxed; see Remark 3.1.5.

$$\left. \begin{aligned} &u \in C_r(0,T;W) \cap L_\infty(0,T;W), \\ &u_t \in C_r(0,T;(H^2 \cap H_0^1)(\Omega)) \cap L_\infty(0,T;(H^2 \cap H_0^1)(\Omega)), \\ &u_{tt} \in C_r(0,T;H_0^1(\Omega)) \cap L_\infty(0,T;H_0^1(\Omega)), \end{aligned} \right\} \tag{3.1.45}$$

where as above $C_r(0,T;X)$ is the space of strongly right-continuous functions with values in X. Strong solutions satisfy the energy identity in (3.1.15).
If $g_0(\sigma)$ and $g(s)$ are differentiable and satisfy (3.1.18), *then*

$$u \in C(0,T;W), \quad u_t \in C(0,T;H^2 \cap H_0^1)(\Omega)), \quad u_{tt} \in C(0,T;H_0^1(\Omega)). \tag{3.1.46}$$

3.1.17. Remark. If we compare (3.1.16) and (3.1.44) we see that the regularity of u_{tt} for a generalized solution is lower in the case of clamped boundary conditions. In fact, this has to do with certain behavior of traces on the boundary that depends critically on boundary conditions imposed (cf. also the case of the free boundary conditions; see (3.1.64) in Theorem 3.1.22 below). ■

3.1.2.1 Proof of Theorem 3.1.16

The proof parallels the arguments given for Theorem 3.1.4. We indicate the points where differences occur.

Our first step is to rewrite the von Karman equation as a second-order abstract equation. This allows us application of Theorem 2.4.16. To this end, we introduce the following spaces and operators.

- $\mathscr{H} \equiv L_2(\Omega), \ V \equiv H_0^1(\Omega)$.
- $\mathscr{A} u \equiv \Delta^2 u, \ u \in \mathscr{D}(\mathscr{A}) : \mathscr{D}(\mathscr{A}) \equiv \{u \in H^4(\Omega), \ u = \Delta u = 0 \text{ on } \Gamma\}$.
- $Mu \equiv u - \alpha \Delta u, \ u \in \mathscr{D}(M)$, where $\mathscr{D}(M) \equiv H_0^1(\Omega) \cap H^2(\Omega)$.
- Hence $V' = H^{-1}(\Omega), \ \mathscr{D}(\mathscr{A}^{1/2}) = H^2(\Omega) \cap H_0^1(\Omega)$.
- $F(u,v) = \widetilde{F}(u) + a_0 d_0(x) v$ with $\widetilde{F}(u) \equiv [u+f, v(u) + F_0] - Lu + p$, where $v(u)$ satisfies the elliptic equation in (3.0.2).
- $D(v) \equiv d_0(x)\,[g_0(v) + a_0 v] - \alpha \operatorname{div}(d(x) g(\nabla v))$ for $v \in \mathscr{D}(\mathscr{A}^{1/2})$.

With the above notation the abstract form of the equation becomes

$$Mu_{tt} + \mathscr{A} u + Du_t = F(u, u_t) \tag{3.1.47}$$

and the only difference in comparison with the case of clamped boundary conditions is a different structure of the positive operator $\mathscr{A}$. In the hinged case we have

$$\mathscr{A} = A_D^2 \quad \text{and} \quad M = I + \alpha A_D, \tag{3.1.48}$$

where A_D is a negative Laplace operator subject to Dirichlet boundary conditions. However, the arguments are identical to those in the proof of Theorem 3.1.4. To obtain representation (2.4.28) for $F(u,v)$ in the hinged case with the same Π and

F^* as for the clamped case we use the relation

$$\int_\Omega [F_0, u] w dx = \int_\Omega [F_0, w] u dx, \quad u, w \in H^2(\Omega) \cap H_0^1(\Omega),$$

which follows from (1.4.9) and (1.4.10) with $\Gamma_1 = \Gamma_3 = \emptyset$ and $\Gamma_2 = \partial\Omega$. We note that relation (3.1.10) is not true for general F_0 in the case of boundary conditions (3.1.42).

Thus the only difference in the proof is at the level of proving regularity of the second time derivative in (3.1.44) under Assumption 3.1.2. In fact, here the argument is simpler than in the clamped case. This is due to the fact that additional boundary regularity is not needed. Indeed, in order to prove the regularity of the second time derivatives, we recall (3.1.47) to obtain that

$$u_{tt} = M^{-1}[-\mathscr{A} u - D(u_t) + F(u, u_t)]. \tag{3.1.49}$$

The mapping $(u; v) \mapsto F(u, v)$ is continuous from $H^2(\Omega) \times H_0^1(\Omega)$ into the space $H^{-1}(\Omega)$, therefore the function $t \mapsto F(u(t), u_t(t))$ is continuous in $H^{-1}(\Omega)$ for any generalized solution $u(t)$. The damping term $D(u_t(t))$ is handled via Lemma 3.1.8, as in the clamped case. Using (3.1.48) we also easily obtain that $M^{-1}A_D$ is a bounded operator in $L_2(\Omega)$. Thus $M^{-1}\mathscr{A} u(t) \in C(0, T; L^2(\Omega))$. Consequently relation (3.1.49) implies (3.1.44). Thus Theorem 3.1.16 is proved.

3.1.2.2 Weak solutions

Weak solutions are considered next.

3.1.18. Definition. We say that u is a *weak solution* to equation (3.1.1) with hinged boundary conditions (3.1.42) iff

$$u \in L_\infty(0, T; (H^2 \cap H_0^1)(\Omega)) \cap W_\infty^1(0, T; H_0^1(\Omega)),$$

$u|_{t=0} = u_0$, and the integrability condition (3.1.32) and the variational relation (3.1.33) hold for any $\phi \in H^2(\Omega) \cap H_0^1(\Omega)$. ■

Similarly to Theorem 3.1.12 one obtains the following result.

3.1.19. Theorem. *Under the hypotheses of Theorem* 3.1.12, *generalized solutions of von Karman system* (3.1.1) *with hinged boundary conditions* (3.1.42) *are also weak. If, g_0 and g satisfy Assumption* 3.1.2, *then weak solutions are unique. This, in turn, implies that every weak solution with initial data* $(u_0; u_1) \in (H^2 \cap H_0^1)(\Omega) \times H_0^1(\Omega)$ *is also generalized.*

3.1.20. Remark.

1. We note that the statements formulated in Remark 3.1.14 remain in force in the case of hinged boundary conditions.

2. As in the clamped case (cf. Remark 3.1.10)) we have the additional regularity properties on the boundary that do not follow from the interior regularity. For the hinged case the following result is valid for weak solutions,

$$\int_\Sigma |\nabla u_t|^2 d\Gamma dt \le C(E(0)).$$

Here $\Sigma = \partial\Omega \times [0,T]$ and the function $C(\cdot)$ also depends on T. This result easily follows from Theorem 2.5.13 (cf. idea of the proof of Proposition 3.1.9). We note again that the regularity stated above provides $\frac{1}{2}$ derivative more on the boundary than the trace theorem (which can be applied only formally) would predict. ∎

3.1.3 Boundary conditions of the free type

We consider problem (3.1.1) with the following boundary conditions

$$\begin{aligned} &\Delta u + (1-\mu)B_1 u = 0, \\ &\frac{\partial}{\partial n}\Delta u + (1-\mu)B_2 u - \alpha\frac{\partial}{\partial n}u_{tt} - \alpha d(x) g(\nabla u_t)\cdot n = \nu_1 u + \beta(x)u^3, \end{aligned} \tag{3.1.50}$$

on $\Gamma \equiv \partial\Omega$, where the boundary operators B_1 and B_2 are given by (1.3.20), $\nu_1 > 0$ is a parameter, and $\beta(x) \ge 0$ is a bounded measurable function.

In order to capture boundary conditions given by (3.1.50) we introduce the energy functional $\mathscr{E}$ which has a different form than that used in the hinged and clamped cases. This is due to the fact that (i) free boundary conditions lead to another (elliptic) quadratic form and (ii) the mapping $u \mapsto [F_0, u]$ is not potential on $H^2(\Omega)$ (see Remark 1.5.17 and also Remark 3.1.23 below).

We define the energy by the formula

$$\mathscr{E}(t) \equiv \mathscr{E}(u,u_t) = \frac{1}{2}\int_\Omega \left[\alpha|\nabla u_t|^2 + |u_t|^2\right]dx + \frac{1}{2}a(u,u) + \Pi(u), \tag{3.1.51}$$

where

$$a(u,w) = a_0(u,w) + \nu_1\int_\Gamma uw\,d\Gamma \tag{3.1.52}$$

with $a_0(u,w)$ defined by (1.3.4) and

$$\Pi(u) = \int_\Omega \left[\frac{1}{4}|\Delta v(u)|^2 - [f,F_0]u - up\right]dx + \frac{1}{4}\int_\Gamma \beta u^4 d\Gamma. \tag{3.1.53}$$

Here $v(u) \in H_0^2(\Omega)$ is determined from (3.1.2). The functional $\Pi(u)$ can be written in the form

$$\Pi(u) = \Pi_0(u) + \Pi_1(u), \tag{3.1.54}$$

where

$$\Pi_0(u) = \frac{1}{4}\int_\Omega |\Delta v(u)|^2 dx + \frac{1}{4}\int_\Gamma \beta u^4 d\Gamma \geq 0 \tag{3.1.55}$$

and

$$\Pi_1(u) = -\int_\Omega ([f, F_0]u + up)\, dx. \tag{3.1.56}$$

In what follows we also use the notation

$$E(t) \equiv E(u, u_t) = \frac{1}{2}\int_\Omega \left[\alpha|\nabla u_t|^2 + |u_t|^2\right] dx + \frac{1}{2}a(u,u) + \Pi_0(u) \tag{3.1.57}$$

for the positive part of the energy $\mathscr{E}$. Obviously, $\mathscr{E}(u, u_t) = E(u, u_t) + \Pi_1(u)$ and due to the relation

$$|\Pi_1(u)| \leq \eta a(u,u) + C_\eta \quad \text{for any } \eta > 0, \tag{3.1.58}$$

there exist positive constants c_0, c_1, and K (depending on F_0, p, f) such that

$$c_0\mathscr{E}(u, u_t) - K \leq E(u, u_t) \leq c_1\mathscr{E}(u, u_t) + K \tag{3.1.59}$$

for any $(u; u_t) \in H^2(\Omega) \times H^1(\Omega)$.

In order to introduce notions of strong and generalized solutions to problem (3.1.1) and (3.1.50) we first analyze the functional analytic setup, which is somewhat different from those of the clamped and hinged cases. The situation here is reminiscent of the semigroup treatment given for linear problems (Section 2.5).

We let

- $\mathscr{H} \equiv L_2(\Omega)$, $\mathscr{A}u \equiv \Delta^2 u, u \in \mathscr{D}(\mathscr{A})$, where

$$\mathscr{D}(\mathscr{A}) = \left\{ u \in H^4(\Omega) \;\middle|\; \begin{array}{l} [\Delta u + (1-\mu)B_1 u]\Big|_{\partial\Omega} = 0, \\ \left[\frac{\partial}{\partial n}\Delta u + (1-\mu)B_2 u - \nu_1 u\right]\Big|_{\partial\Omega} = 0 \end{array} \right\}.$$

- $V \equiv H^1(\Omega)$, $M \equiv I + \alpha A_N$ where $A_N u = -\Delta u$, and

$$\mathscr{D}(A_N) \equiv \left\{ u \in H^2(\Omega) : \; \frac{\partial}{\partial n} u = 0 \text{ on } \Gamma \right\}.$$

- Hence $V' = [H^1(\Omega)]'$, $\mathscr{D}(\mathscr{A}^{1/2}) = H^2(\Omega)$, $[\mathscr{D}(\mathscr{A}^{1/2})]' = [H^2(\Omega)]'$.
- $F(u,v) = \widetilde{F}(u) + a_0 d_0(x)v$ with $\widetilde{F}(u) \equiv [u+f, v(u)+F_0] - Lu + p - B_\Gamma u$, where $v(u)$ satisfies the elliptic equation in (3.1.2) and $B_\Gamma : \mathscr{D}(\mathscr{A}^{1/2}) \to V'$ introduced in the proof of Theorem 1.5.16 is given by

$$(B_\Gamma(u), \phi)_{V,V'} = \int_\Gamma \beta u^3 \phi d\Gamma, \quad \phi \in H^1(\Omega). \tag{3.1.60}$$

- $D(v) \equiv d_0(x)\left[g_0(v) + a_0 v\right] + \alpha D_1(v)$, where D_1 is defined variationally

$$\int_{\Omega} D_1(v)\phi dx = \int_{\Omega} d(x)g(\nabla v)\nabla\phi dx \quad \text{for all } \phi \in H^1(\Omega).$$

The original PDE model can be now formulated as an abstract evolution problem of the second order:

$$Mu_{tt} + \mathscr{A}u + Du_t = F(u,u_t). \tag{3.1.61}$$

To see this, it suffices to perform the following line of computations valid for smooth solutions,

$$\begin{aligned}
\mathscr{I} &:= -\alpha\int_{\Omega}\Delta u_{tt} w dx - \alpha\int_{\Omega}\operatorname{div}\left[d(x)g(\nabla u_t)\right] w dx + \int_{\Omega}\Delta^2 u w dx \\
&= \alpha\int_{\Omega}\nabla u_{tt}\nabla w dx - \alpha\int_{\Gamma}\frac{\partial}{\partial n}u_{tt} w d\Gamma - \alpha\int_{\Gamma} d(x)g(\nabla u_t)\cdot n w d\Gamma \\
&\quad + \alpha\int_{\Omega} d(x)g(\nabla u_t)\nabla w dx + a_0(u,w) \\
&\quad + \int_{\Gamma}\left[\left(\frac{\partial}{\partial n}\Delta u + (1-\mu)B_2 u\right) w - (\Delta u + (1-\mu)B_1 u)\frac{\partial}{\partial n} w\right] d\Gamma.
\end{aligned}$$

We have used here the Gauss–Ostrogradsky formula and Green's formula (1.3.3) for Δ^2. Noting the cancellation in the boundary conditions and the equivalence

$$a(u,w) = (\mathscr{A}^{1/2}u, \mathscr{A}^{1/2}w)_{\mathscr{H}},$$

where $a(u,w)$ is given by (3.1.52), we obtain that

$$\begin{aligned}
\mathscr{I} &= \alpha\int_{\Omega}\nabla u_{tt}\nabla w dx + \alpha\int_{\Omega} d(x)g(\nabla u_t)\nabla w dx + \int_{\Omega}\mathscr{A}^{1/2}u\mathscr{A}^{1/2}w dx + \int_{\Gamma}\beta u^3 w d\Gamma \\
&= \alpha(A_N^{1/2}u_{tt}, A_N^{1/2}w)_{\mathscr{H}} + \alpha(D_1(u),w)_{H^1,[H^1]'} + (\mathscr{A}^{1/2}u, \mathscr{A}^{1/2}w)_{\mathscr{H}} + \int_{\Gamma}\beta u^3 w d\Gamma \\
&= (\alpha A_N u_{tt} + \alpha D_1(u) + B_\Gamma(u), w)_{V,V'} + (\mathscr{A}u, w)_{\mathscr{D}(\mathscr{A}^{1/2}),[\mathscr{D}(\mathscr{A}^{1/2})]'}.
\end{aligned}$$

Thus smooth solutions to (3.1.1) and (3.1.50) satisfy (3.1.61), where the latter should be understood as an equality in the space $[H^2(\Omega)]'$.

Similar to the clamped and hinged cases we give the following definition.

3.1.21. Definition. A function $u(t) \in C(0,T;H^2(\Omega)) \cap C^1(0,T;H^1(\Omega))$ possessing the properties $u(x,0) = u_0(x)$ and $u_t(x,0) = u_1(x)$ is said to be

1. A *strong solution* to problem (3.1.1) and (3.1.2) with the free-type boundary conditions (3.1.50) on the interval $[0,T]$, iff
   - $u(t)$ is an absolutely continuous function with values in $H^2(\Omega)$ and $u_t \in L_1(a,b;H^2(\Omega))$ for any $0 < a < b < T$.
   - u_t is an absolutely continuous function with values in $H^1(\Omega)$ and $u_{tt} \in L_1(a,b;H^1(\Omega))$ for any $0 < a < b < T$.
   - $u(t) \in W \equiv \left\{u \in H^3(\Omega) : (\Delta u + (1-\mu)B_1 u)|_{\partial\Omega} = 0\right\}$ for almost all $t \in [0,T]$.
   - Equality (3.1.61) holds in $V' = [H^1(\Omega)]'$ for almost all $t \in [0,T]$.

2. A *generalized solution* to problem (3.1.1), (3.1.2), and (3.1.50) on the interval $[0,T]$, iff there exists a sequence $\{u_n(t)\}$ of strong solutions to (3.1.1), (3.1.2), and (3.1.50) with initial conditions given by $(u_{0n};u_{1n})$ and such that (3.1.11) holds.

∎

The theorem stated below supplies existence and uniqueness of strong and generalized solutions.

3.1.22. Theorem. *Let $p \in [H^1(\Omega)]'$, $F_0 \in H^{5/2+\delta}$ for some $\delta > 0$ and L be a linear bounded operator from $H^2(\Omega)$ into $[H^1(\Omega)]'$. Then under Assumption* 3.1.1 *with reference to* (3.1.1), *subject to free boundary conditions* (3.1.50), *the following statements are valid with any $T > 0$.*

- **Generalized solutions:** *For all initial data $u_0 \in H^2(\Omega), u_1 \in H^1(\Omega)$ there exists a unique generalized solution $u(t)$ such that*

$$u \in C(0,T;H^2(\Omega)), \quad u_t \in C(0,T;H^1(\Omega)). \tag{3.1.62}$$

Moreover, the following energy inequality holds for all generalized solutions and for all $s \le t$:

$$\begin{aligned} &\mathcal{E}(u(t),u_t(t)) + \int_s^t \int_\Omega [d_0 g_0(u_t)u_t + \alpha d g(\nabla u_t)\nabla u_t] dx d\tau \\ &\le \mathcal{E}(u(s),u_t(s)) - \int_s^t \int_\Omega ([u,F_0] + Lu) u_t dx d\tau. \end{aligned} \tag{3.1.63}$$

If, in addition, Assumption 3.1.2 *is valid,*[2] *then*

$$u_{tt} \in C(0,T;L_2(\Omega)) \tag{3.1.64}$$

and the following energy identity

$$\begin{aligned} &\mathcal{E}(u(t),u_t(t)) + \int_s^t \int_\Omega [d_0 g_0(u_t)u_t + \alpha d g(\nabla u_t)\nabla u_t] dx d\tau \\ &= \mathcal{E}(u(s),u_t(s)) - \int_s^t \int_\Omega ([u,F_0] + Lu) u_t dx d\tau \end{aligned} \tag{3.1.65}$$

holds.

- **Strong solutions:** *Under the assumption that $u_0 \in H^3(\Omega), u_1 \in H^2(\Omega)$, and, moreover,*

$$\Delta u_0 + (1-\mu)B_1 u_0 = 0 \text{ on } \Gamma,$$

there exists a unique strong solution $u(t)$ such that

---

[2] As in the clamped and hinged cases this hypothesis can be relaxed in the line with Remark 3.1.5 such that both the energy identity and the regularity in (3.1.21) hold for generalized solutions.

$$\left.\begin{aligned} &u \in C_r(0,T;W) \cap L_\infty([0,T];W), \\ &u_t \in C_r(0,T;H^2(\Omega)) \cap L_\infty(0,T;H^2(\Omega)), \\ &u_{tt} \in C_r(0,T;H^1(\Omega)) \cap L_\infty(0,T;H^1(\Omega)), \end{aligned}\right\} \tag{3.1.66}$$

where $W = \{u \in H^3(\Omega) : \Delta u + (1-\mu)B_1 u = 0 \text{ on } \Gamma\}$ *and* $C_r(0,T;X)$ *is the space of strongly right-continuous functions with values in* X. *Strong solutions satisfy energy identity in* (3.1.65).

If $g_0(\sigma)$ *and* $g(s)$ *are differentiable and satisfy* (3.1.18), *then*

$$u \in C(0,T;W), \quad u_t \in C(0,T;H^2(\Omega)), \quad u_{tt} \in C(0,T;H^1(\Omega)). \tag{3.1.67}$$

3.1.23. Remark. In contrast with the clamped and hinged cases the term $[u,F_0]$ is not conservative (i.e., the mapping $u \mapsto [u,F_0]$ is not of a potential type) in general. This explains why we cannot include this term in the energy expression. However, if we assume additionally that $F_0 \equiv 0$ and $\nabla F_0 \equiv 0$ on $\partial\Omega$, then by the symmetry of the von Karman bracket (see Proposition 1.4.2) the following relation holds for any generalized solution,

$$\int_0^t ([u,F_0],u_t)_\Omega d\tau = \frac{1}{2}([u(t),F_0],u(t)]) - \frac{1}{2}([u_0,F_0],u_0]).$$

Therefore, under the condition $F_0 \in H_0^2(\Omega)$ the energy relations (3.1.63) and (3.1.65) can be written in the same form as for the clamped and hinged cases. ∎

3.1.3.1 Proof of Theorem 3.1.22

As in the clamped and hinged case we apply Theorem 2.4.16 with $G = 0$. To this end, we establish the validity of the following lemma:

3.1.24. Lemma. *Operators* $\mathscr{A}, M, F$, *and* D *introduced above comply with Assumption* 2.4.15. *The nonlinear term* F *has the form*

$$F(u,v) = -\Pi'(u) + F^*(u),$$

where $\Pi(u)$ *is given by* (3.1.54) *and* $F^*(u,v) = [u,F_0] - Lu + a_0 d_0 v$.

Proof. It is almost the same as in the clamped case (see Lemma 3.1.6). The only difference is that now $\mathscr{D}(\mathscr{A}^{1/2}) = H^2(\Omega)$ and $V = H^1(\Omega)$ without any boundary conditions. Thus, one needs to be more careful in treating dual spaces, which no longer are spaces of distributions. However, this fact has no major implications in our context. Indeed, to obtain (2.4.27) we use formula (1.4.26) with $\delta < \frac{1}{2}$ and the relation $H^{-1/2+\varepsilon}(\Omega) \subset [H^1(\Omega)]' = V'$ which yields the estimate

$$||[u_1,v(u_1)] - [u_2,v(u_2)]||_{[H^1(\Omega)]'} \leq C(||u_1||_2^2 + ||u_2||_2^2 + ||f||_2^2)||u_1 - u_2||_2.$$

The verification of regularity properties of term B_Γ, which appears in the definition of $F(u)$, is also straightforward. Indeed, the map $u \mapsto \beta u^3|_\Gamma$ is locally Lipschitz from $H^2(\Omega) \to L_2(\Gamma)$ This follows from Sobolev's embeddings and the trace theory (accounting for the two-dimensionality of the domain). We also obviously have that $\phi|_\Gamma \in L_2(\Gamma)$ for $\phi \in V = H^1(\Omega)$, hence B_Γ is locally Lipschitz from $\mathscr{D}(\mathscr{A}^{1/2})$ into V', as desired.

We note, that in the above argument acritical role is played by the fact that rotational forces are present in the model (i.e., $\alpha > 0$). If $\alpha = 0$ then the above regularity does not hold because the corresponding V coincides with $L_2(\Omega)$. How to deal with the case $\alpha = 0$ is discussed in Chapter 4.

In order to verify the regularity of the term corresponding to F_0, we use formula (1.4.16) applied with $j = 0$, $\theta = 1/2 - \varepsilon$ and obtain

$$||[u, F_0]||_{-1/2+\varepsilon} \leq C||u||_2||F_0||_{5/2+\varepsilon}$$

which provides the desired estimate (2.4.29) for $F^*(u, v) = [u, F_0] - Lu + a_0 d_0 v$. □

The corresponding version of Lemma 3.1.8 remains true for the free-type boundary case. Thus the energy relation (3.1.65) is valid for generalized solutions under Assumption 3.1.2.

Regarding the regularity (3.1.64) of the second time derivative, this is straightforward. We have from (3.1.61)

$$u_{tt} = M^{-1}[-\mathscr{A}u - D(u_t) + F(u, u_t)].$$

Because $\mathscr{D}(\mathscr{A}^{1/2}) = H^2(\Omega)$, from standard elliptic estimates we have that the operator $M^{-1}\mathscr{A}^{1/2}$ is bounded in $\mathscr{H} = L_2(\Omega)$. Indeed, we have that

$$(M^{-1}\mathscr{A}^{1/2}u, v) = (\mathscr{A}^{1/2}u, M^{-1}v) = (u, \mathscr{A}^{1/2}M^{-1}v)$$

for any $u \in H^2(\Omega)$ and $v \in L_2(\Omega)$. Inasmuch as

$$M^{-1} : L_2(\Omega) \mapsto \mathscr{D}(M) \subset H^2(\Omega) \text{ and } \mathscr{A}^{1/2} : H^2(\Omega) \mapsto L_2(\Omega)$$

continuously, we obtain that $|(M^{-1}\mathscr{A}^{1/2}u, v)| \leq C\|u\|\|v\|$. This implies the boundedness of $M^{-1}\mathscr{A}^{1/2}$ in $\mathscr{H} = L_2(\Omega)$. Thus we can apply the same argument as in the hinged case. This gives $u_{tt} \in C(0, T; L_2(\Omega))$. The proof of Theorem 3.1.22 is complete.

3.1.3.2 Weak solutions

We next consider weak solutions.

3.1.25. Definition. We say that u is a *weak solution* to equation (3.1.1) with the free-type boundary conditions (3.1.50) iff

$$u \in L_\infty(0,T;H^2(\Omega)) \cap W^1_\infty(0,T;H^1(\Omega))$$

and also (i) $u|_{t=0} = u_0$, (ii) the integrability condition in (3.1.32) holds for every $\phi \in H^2(\Omega)$, and (iii) the following variational relation

$$\begin{aligned}(u_t(t),\phi)_\Omega &+ \alpha(\nabla u_t(t),\nabla\phi)_\Omega - (u_1,\phi)_\Omega - \alpha(\nabla u_1,\nabla\phi)_\Omega \\ &+ \int_0^t \big[a(u,\phi) + (d_0 g_0(u_t),\phi)_\Omega + \alpha(dg(\nabla u_t),\nabla\phi)_\Omega \\ &+ (-[v(u)+F_0,u+f] + Lu - p,\phi)_\Omega + (\beta u^3,\phi)_\Gamma\big]\, d\tau = 0 \end{aligned} \quad (3.1.68)$$

holds for any $\phi \in H^2(\Omega)$, where $v(u) \in H_0^2(\Omega)$ is defined from (3.1.2) and the form $a(u,\phi)$ is given by (3.1.52). ∎

A counterpart of Theorems 3.1.12 and 3.1.19 is the following.

3.1.26. Theorem. *Under the hypotheses of Theorem* 3.1.12 *generalized solutions of von Karman system* (3.1.1) *with the free-type boundary conditions* (3.1.50) *are also weak. If g_0 and g satisfy Assumption* 3.1.2, *weak solutions are unique. Hence every weak solution with initial data $(u_0;u_1) \in H^2(\Omega)\times H^1(\Omega)$ is also generalized.*

In the proof of the second part of this theorem we use properties of linear problem (2.5.22) stated in Proposition 2.5.14. We also note that, the same as in the clamped and hinged cases, Remark 3.1.14 remains valid for the free boundary conditions.

3.1.4 Regular solutions

In this section we give criteria of additional smoothness of solutions. We rely on the approach presented in [123]. To avoid a bit of complicated combinatorial calculations we consider the case of linear rotational damping only. However we keep *nonlinear* viscous damping which is represented by the general monotone polynomially bounded function $g_0(u_t)$. This case is representative of the main difficulties and, at the same time, simple enough to convey with clarity the main steps of the approach. More precisely, we consider the following equation

$$\begin{cases} (1-\alpha\cdot\Delta)u_{tt} + d_0(x)g_0(u_t) - \alpha\,\mathrm{div}\,[d(x)\nabla u_t] \\ \qquad + \Delta^2 u - [u+f, v+F_0] + Lu = p(x), \quad x\in\Omega,\ t>0, \\ u|_{t=0} = u_0(x), \quad \partial_t u|_{t=0} = u_1(x), \end{cases} \quad (3.1.69)$$

where, as usual, the function $v = v(u)$ is a solution of the problem

$$\Delta^2 v + [u+2f,u] = 0, \quad v|_{\partial\Omega} = \left.\frac{\partial v}{\partial n}\right|_{\partial\Omega} = 0. \quad (3.1.70)$$

We concentrate on free-type boundary conditions:

$$\Delta u + (1-\mu)B_1 u = 0,$$
$$\frac{\partial}{\partial n}\Delta u + (1-\mu)B_2 u - \alpha\frac{\partial}{\partial n}u_{tt} - \alpha d\frac{\partial}{\partial n}u_t = \nu_1 u + \beta(x)u^3, \tag{3.1.71}$$

on $\Gamma \equiv \partial\Omega$, where B_1 and B_2 are given by (1.3.20), ν_1 is a positive parameter, and β is a nonnegative bounded measurable function. The cases of clamped and hinged boundary conditions are simpler and can be treated in a similar way.

Our main hypothesis in this section is the following one.

3.1.27. Assumption.

- The function $g_0 : \mathbb{R} \mapsto \mathbb{R}$ belongs to $C^\infty(\mathbb{R})$ and possesses the properties (i) $g_0(0) = 0$, (ii) there exists $a_0 \geq 0$ such that $g_0(\sigma) + a_0\sigma$ is monotone nondecreasing, and (iii) the derivatives $g_0^{(k)}$ satisfy the relations

$$|g_0^{(k)}(\sigma)| \leq C_k(1+|\sigma|^{q_k}), \quad k = 1,2,\ldots, \tag{3.1.72}$$

  for some constants $C_k > 0$ and $q_k \geq 1$.
- The densities $d_0(x)$ and $d(x)$ are nonnegative.
- $f \in H^2(\Omega)$, $p \in \left[H^1(\Omega)\right]'$ and $F_0 \in H^{5/2+\delta}$ for some $\delta > 0$.
- L is a linear bounded operator from $H^2(\Omega)$ into $\left[H^1(\Omega)\right]'$.

Problem (3.1.69)–(3.1.71) can be written in the form (cf. Section 3.1.3):

$$Mu_{tt} + \mathscr{A}u + \widetilde{D}u_t = \widetilde{F}(u), \tag{3.1.73}$$

where M and $\mathscr{A}$ are the same as in Section 3.1.3. The operators

$$\widetilde{D} : H^1(\Omega) \mapsto [H^1(\Omega)]' \text{ and } \widetilde{F} : H^2(\Omega) \mapsto [H^1(\Omega)]'$$

are defined by the relations

$$\int_\Omega \widetilde{D}v\phi dx = \int_\Omega (d_0 g_0(v)\phi + \alpha d(x)\nabla v\nabla\phi)\,dx, \quad \forall\phi \in H^1(\Omega),$$

and $\widetilde{F}(u) = [u+f, v(u)+F_0] - Lu + p - B_\Gamma u$, where B_Γ is given by (3.1.60).

Let us define the values $u^{(k)}(0)$ by the recurrence relations

$$\begin{cases} u^{(0)}(0) = u_0, \quad u^{(1)}(0) = u_1, \\ u^{(k)}(0) = M^{-1}\left(-\mathscr{A}u^{(k-2)}(0) + \frac{d^{k-2}}{dt^{k-2}}\left[-\widetilde{D}(u_t) + \widetilde{F}(u)\right]\Big|_{t=0}\right), \end{cases} \tag{3.1.74}$$

where $k \geq 2$. We also denote by $\mathscr{L}_m$, $m \geq 1$, the class of functions that possess the following properties

$$\begin{cases} u^{(k)}(t) \in C(0,T;W) \text{ for } k = 0,1,2,\ldots,m-1, \\ u^{(m)}(t) \in C(0,T;H^2(\Omega)), \quad u^{(m+1)}(t) \in C(0,T;H^1(\Omega)), \end{cases} \tag{3.1.75}$$

for any $T > 0$, where $u^{(k)}(t) = \partial_t^k u(t)$ and

$$W = \{u \in H^2(\Omega) : \mathscr{A}u \in [H^1(\Omega)]'\}$$
$$= \{u \in H^3(\Omega) : (\Delta u + (1-\mu)B_1 u)|_{\partial\Omega} = 0\}.$$

3.1.28. Theorem. *Assume that Assumption* 3.1.27 *holds. Then a generalized solution to problem* (3.1.69)–(3.1.71) *belongs to* $\mathscr{L}_m$ *for* $m \geq 1$ *if and only if the following compatibility conditions*

$$\begin{cases} u^{(k)}(0) \in W \ \ for \ \ k = 0, 1, 2, \ldots, m-1, \\ u^{(m)}(0) \in H^2(\Omega), \quad u^{(m+1)}(0) \in H^1(\Omega), \end{cases} \tag{3.1.76}$$

hold, where the values $u^{(k)}(0)$ *are defined by the recurrence relations* (3.1.74).

Proof. The idea of the proof is the usual one when studying regularity of weak solutions. It uses induction and the "boot-strap" argument. Knowing the existence of weak solutions, one considers distributional time derivatives of these solutions. In order to apply the existence theorem to these time derivatives, one needs to read off the initial value for the second order-time derivatives. It is at this step that compatibility conditions enter the game. This step leads to obtaining regularity of time derivatives. The improvement on the space variable follows by reading off regularity of the elliptic equation.

In what follows we present the rigorous development of this idea which is executed inductively allowing us to reach an arbitrary level of regularity (assuming sufficient regularity of the initial data).

It is clear that any solution $u(t)$ from $\mathscr{L}_m$ possesses property (3.1.76). Thus we need only prove that compatibility conditions (3.1.76) imply that $u(t)$ belongs to the class $\mathscr{L}_m$. To prove it we rely on the idea presented in [123] (see also Theorem 2.4.38 and Propositions 2.5.2 and 2.5.8 in Chapter 2) and use induction in m.

For $m = 1$ the assertion is contained in the second part of Theorem 3.1.22 (see (3.1.67)) that also guarantees that the solution $u(t)$ is strong. Suppose that (3.1.76) holds for some $m > 1$ and that a (strong) solution $u(t)$ belongs to the class $\mathscr{L}_{n-1}$ for $2 \leq n \leq m$. Let us prove that $u(t) \in \mathscr{L}_n$.

We start with calculations of the n-order time derivative of the function $\widetilde{F}(u(t))$. The main idea is to separate the term with the highest-order derivative $u^{(n)}(t)$ which is the most singular. The formal rule of differentiation yields

$$\frac{d^n}{dt^n}\widetilde{F}(u(t)) = \langle \widetilde{F}'(u(t)), u^{(n)}(t)\rangle + G_n(t) + B_n(t).$$

Here $G_1 \equiv B_1 \equiv 0$ and for $n \geq 2$ the quantity $G_n(t) \equiv G_n(u(t), u'(t), \ldots, u^{(n-1)}(t))$ is a linear combination of functions of the form $[v(w_1, w_2), w_3]$, where w_j are either $u + f$, or one of the derivatives $u^{(k)}(t)$, $1 \leq k \leq n-1$. The value $v = v(w_1, w_2) \in H_0^2(\Omega)$ solves the problem

$$\Delta^2 v + [w_1, w_2] = 0, \quad v|_{\partial\Omega} = \frac{\partial v}{\partial n}\Big|_{\partial\Omega} = 0.$$

Similarly, for $n \geq 2$ the quantity $B_n(t) \equiv B_n(u(t), u'(t), \dots, u^{(n-1)}(t))$ is a linear combination of elements h from $[H^1(\Omega)]'$ which are defined by the relations

$$\int_\Omega h\phi dx = \int_\Gamma \beta u^{(j_1)}(t) u^{(j_2)}(t) u^{(j_3)}(t)\phi d\Gamma, \quad j_1 + j_2 + j_3 = n, \ 0 \leq j_i \leq n-1,$$

for any $\phi \in H^1(\Omega)$. The dependence on $u^{(n-1)}$ in G_n and B_n can be linear only in the case when $n > 2$.

By the induction hypothesis $u \in \mathscr{L}_{n-1}$. Thus $u^{(n-1)} \in C(0,T; H^2(\Omega))$ and $u^{(k)} \in C(0,T;W)$ for $k \leq n-2$. Therefore properties of the von Karman bracket allow us to assert that $G_n(t) + B_n(t)$ is a strongly continuous function with values in $V' = [H^1(\Omega)]'$.

In a similar way we have that

$$\frac{d^n}{dt^n}\widetilde{D}(u_t(t)) = \langle \widetilde{D}'(u_t(t)), u^{(n+1)}(t)\rangle + \hat{D}_n(t),$$

where $\langle \widetilde{D}'(u_t), \psi\rangle$ is defined by the relation

$$\int_\Omega \langle \widetilde{D}'(u_t), \psi\rangle \phi dx = \int_\Omega \left(d_0 g_0'(u_t)\psi\phi + \alpha d(x)\nabla\psi\nabla\phi\right) dx, \ \ \forall \phi \in H^1(\Omega),$$

$\hat{D}_1(t) \equiv 0$ and $\hat{D}_n(t)$ is the sum of elements of the form

$$\text{const} \cdot g_0^{(r)}(u_t) \cdot u^{(k_1)} \cdot \ldots \cdot u^{(k_r)}$$

with $2 \leq r \leq n$, $2 \leq k_j \leq n$ for every $j = 1, \dots, r$, and $\sum k_j = n + r$. By the induction hypothesis we have that $u^{(k)} \in C(0,T; H^2(\Omega))$ for every $k \leq n-1$ and $u^{(n)} \in C(0,T; H^1(\Omega))$. Therefore, owing to linear dependence on $u^{(n)}$ for $n > 2$, it is easy to see that $t \mapsto \hat{D}_n(t)$ is a continuous function with values in $V' = [H^1(\Omega)]'$.

Let us prove that the function $w(t) = u^{(n)}(t)$ belongs to the space

$$C(0,T; H^2(\Omega)) \cap C^1(0,T; H^1(\Omega)) \tag{3.1.77}$$

for any $T > 0$. To prove this we observe that the function $w(t)$ is a weak solution of the linear nonautonomous problem

$$\begin{cases} Mw_{tt} + \langle \widetilde{D}'(u_t), w_t\rangle + \mathscr{A} w - \langle \widetilde{F}'(u(t)), w(t)\rangle = G_n(t) + B_n(t) - \hat{D}_n(t), \\ w|_{t=0} = u^{(n)}(0), \quad w_t|_{t=0} = u^{(n+1)}(0). \end{cases} \tag{3.1.78}$$

By Theorem 2.4.35 (which applies with $D_0(t) = 0$ and $g = 0$) any solution to problem (3.1.78) belongs to the class (3.1.77). Thus we obtain that

$$u^{(n)}(t) \in C(0,T;H^2(\Omega)) \text{ and } u^{(n+1)}(t) \in C(0,T;H^1(\Omega))$$

for any $T > 0$. We obviously have that

$$\mathscr{A} u^{(n-1)}(t) = -Mu^{(n+1)}(t) - \langle \widetilde{D}'(u_t), u^{(n)}(t)\rangle + \langle F'(u(t)), u^{(n-1)}(t)\rangle + C_n(t),$$

where $C_n(t) = G_{n-1}(t) + B_{n-1}(t) - \hat{D}_{n-1}(t)$. Because $\mathscr{A}^{-1} : V' \mapsto W$, this implies that $u^{(n-1)}(t)$ lies in $C(0,T;W)$. Thus $u(t) \in \mathscr{L}_n$. □

3.1.29. Remark. As for linear problems (see Propositions 2.5.2 and 2.5.8 and also Remark 2.5.15) elliptic regularity theory allows us to boost further regularity of solutions in the spatial direction. Indeed, one can derive from Theorem 3.1.28 existence of spatially smooth solutions provided that (i) the coefficients $\{p, f, F_0\}$ are sufficiently smooth, and (ii) operator L is sufficiently regular. We do not engage in providing a detailed proof of this result but rather refer to Chapter 4 for the corresponding statements and arguments in the case when the rotational inertia is neglected ($\alpha = 0$). ■

3.2 Well-posedness in the case of nonlinear boundary dissipation

We consider von Karman evolutions with rotational forces and with dissipative boundary conditions

$$\begin{cases} (1-\alpha\Delta)u_{tt} + d_0(x)b(u_t) + \Delta^2 u = [u+f, v+F_0] - Lu + p(x),\ x \in \Omega, t > 0, \\ u|_{t=0} = u_0(x), \quad u_t|_{t=0} = u_1(x), \end{cases} \tag{3.2.1}$$

where $v = v(u)$ is a solution of the problem

$$\Delta^2 v + [u+2f, u] = 0, \quad v|_{\partial\Omega} = \frac{\partial v}{\partial n}\Big|_{\partial\Omega} = 0. \tag{3.2.2}$$

We consider a combination of several different types of boundary conditions. In order to have a more focused presentation, we restrict ourselves to the following canonical types of nonlinear boundary conditions that are defined on two disjoint portions Γ_0 and Γ_1 of the boundary $\Gamma = \Gamma_0 \cup \Gamma_1$,

1. **[clamped–hinged]**: $u = \nabla u = 0$ on Γ_0, $u = 0, \Delta u = -g_1((\partial/\partial n)u_t)$ on Γ_1;
2. **[clamped–free]**: $u = \nabla u = 0$ on Γ_0 and

$$\Delta u + (1-\mu)B_1 u = -g_1(\tfrac{\partial}{\partial n} u_t) \quad \text{on } \Gamma_1,$$

$$\tfrac{\partial}{\partial n}\Delta u + (1-\mu)B_2 u = \alpha \tfrac{\partial}{\partial n} u_{tt} + \nu_1 u + \beta u^3 + g_0(u_t) - \tfrac{\partial}{\partial \tau} g_2(\tfrac{\partial}{\partial \tau} u_t) \quad \text{on } \Gamma_1,$$

where B_1 and B_2 are given by (1.3.20), $n = (n_1; n_2)$ is the outer normal, $\tau = (-n_2; n_1)$ is the unit tangent vector, $\nu_1 \geq 0$, $\beta \in L_\infty(\Gamma_1)$ is nonnegative.

As in Section 3.1 we can also include in (3.2.1) the rotational damping of the form $\alpha \,\mathrm{div}\,\{d(x)g(\nabla u_t)\}$. However, because our goal in this section is to consider effects of boundary damping, for the sake of simplification we omit these internal rotational damping terms.

We assume the following hypotheses.

3.2.1. Assumption.

- The function $b : \mathbb{R} \mapsto \mathbb{R}$ is continuous and there exists $b_0 \geq 0$ such that $b(s) + b_0 \cdot s$ is nondecreasing, $b(0) = 0$. The function $d_0(x)$ is a nonnegative bounded measurable function.
- The function $g_0 : \mathbb{R} \mapsto \mathbb{R}$ is continuous, $g_0(0) = 0$, and there exists $a_0 \geq 0$ such that $g_0(s) + a_0 \cdot s$ is nondecreasing. Moreover $g_0(s)$ satisfies the estimate

$$|g_0(s)| \leq C(1 + |s|^{p_0}), \quad s \in \mathbb{R}, \ \text{for some } C > 0 \text{ and } p_0 \geq 1. \tag{3.2.3}$$

- The functions g_1 and g_2 are monotone nondecreasing and continuous from $\mathbb{R}$ into itself and polynomially bounded; that is, the estimate

$$|g_i(s)| \leq C(1 + |s|^p), \quad s \in \mathbb{R}, \ i = 1, 2,$$

holds with some $C > 0$ and $p \geq 1$. We also assume that $g_i(0) = 0$ for $i = 1, 2$. Moreover, we assume that g_2 is locally Lipschitz;[3] that is, for any $r > 0$ there exists C_r such that

$$|g_2(s) - g_2(\sigma)| \leq C_r |s - \sigma|, \quad s, \sigma \in [-r, r] \subset \mathbb{R}. \tag{3.2.4}$$

- $f \in H^2(\Omega)$, $F_0 \in H^2(\Omega)$, $p \in [H^1(\Omega)]'$.
- L is a linear bounded operator from $H^2(\Omega)$ into $[H^1(\Omega)]'$.

For some results a stronger version of Assumption 3.2.1 is required. This is given below.

3.2.2. Assumption. Assumption 3.2.1 is satisfied and the estimates

$$|b(s)| \leq C(1 + |s|^q), \quad s \in \mathbb{R},$$

and

$$m s^{q_*} \leq g_i(s) s \leq M s^{q_*}, \quad |s| \geq s_0, \ i = 1, 2, \tag{3.2.5}$$

hold with some constants $C > 0$, $q \geq 1$, $q_* > 1$, $s_0 > 0$, and $0 < m < M$.

---

[3] This Lipschitz requirement makes assumptions on g_1 and g_2 not symmetric. We need (3.2.4) to check the density condition of the form (2.4.35) which guarantees the existence of generalized solutions for all initial data with finite energy (see the proof of Theorem 3.2.12).

3.2.1 Clamped–hinged boundary conditions

We first consider problem (3.2.1) with the following combination of boundary conditions

$$u = \nabla u = 0 \text{ on } \Gamma_0; \quad u = 0, \ \Delta u = -g_1\left(\frac{\partial}{\partial n} u_t\right) \text{ on } \Gamma_1. \tag{3.2.6}$$

As in the previous section, the following functional describes the energy of the model,

$$\mathscr{E}(t) \equiv \mathscr{E}(u, u_t) = \frac{1}{2}\int_\Omega \left[\alpha|\nabla u_t|^2 + |u_t|^2 + |\Delta u|^2\right] dx + \Pi(u), \tag{3.2.7}$$

where

$$\Pi(u) = \frac{1}{4}\int_\Omega \left[|\Delta v(u)|^2 - 2[u + 2f, F_0]u - 4up\right] dx \tag{3.2.8}$$

and $v(u) \in H_0^2(\Omega)$ is determined from (3.2.2).

As in Section 3.1 we infer that there exist positive constants c_0, c_1, K such that

$$c_0\mathscr{E}(u, u_t) - K \le E(u, u_t) \le c_1\mathscr{E}(u, u_t) + K, \tag{3.2.9}$$

where we denote by $E(u, u_t)$ the positive part of the energy $\mathscr{E}$; that is,

$$E(t) \equiv E(u, u_t) = \frac{1}{2}\int_\Omega \left[\alpha|\nabla u_t|^2 + |u_t|^2 + |\Delta u|^2 + \frac{1}{2}|\Delta v(u)|^2\right] dx. \tag{3.2.10}$$

3.2.1.1 Abstract framework

In order to introduce notions of strong and generalized solutions to problem (3.2.1) and (3.2.6) a functional analytic setup is introduced first. This is in line with the abstract framework presented in Section 2.4 of Chapter 2. To this end the following spaces and operators are introduced.

- $\mathscr{H} \equiv L_2(\Omega)$, $V \equiv H_0^1(\Omega)$, $U \equiv L_2(\Gamma_1)$, $U_0 \equiv H^{1/2}(\Gamma_1)$, $U_0' = H^{-1/2}(\Gamma_1)$.
- $\mathscr{A}u \equiv \Delta^2 u$, $u \in \mathscr{D}(\mathscr{A})$, where

$$\mathscr{D}(\mathscr{A}) \equiv \left\{u \in H_0^1(\Omega) \cap H^4(\Omega) \ : \ \Delta u = 0 \text{ on } \Gamma_1, \ \frac{\partial}{\partial n} u = 0 \text{ on } \Gamma_0\right\}.$$

- $Mu \equiv u - \alpha\Delta u$, $u \in \mathscr{D}(M) \equiv H_0^1(\Omega) \cap H^2(\Omega)$.
- Hence $V' = H^{-1}(\Omega)$ and

$$\mathscr{D}(\mathscr{A}^{1/2}) = \left\{v \in H_0^1(\Omega) \cap H^2(\Omega) \ : \ \frac{\partial}{\partial n} v = 0 \text{ on } \Gamma_0\right\} \equiv H_0^1(\Omega) \cap H_{\Gamma_0}^2(\Omega),$$

where

$$H^2_{\Gamma_0}(\Omega) \equiv \left\{ u \in H^2(\Omega) : \quad u = 0, \;\; \nabla u = 0 \text{ on } \Gamma_0 \right\} . \tag{3.2.11}$$

- $F(u,v) = \widetilde{F}(u) + b_0 v$ with $\widetilde{F}(u) \equiv [u+f, v(u)+F_0] - Lu + p$, where $v(u)$ satisfies the elliptic equation in (3.2.2) and b_0 is the constant from the first requirement in Assumption 3.2.1.
- $D(v) \equiv d_0(x)\,[b(v) + b_0 v]$ for$v \in \mathscr{D}(\mathscr{A}^{1/2})$.
- $G : L_2(\Gamma_1) \to L_2(\Omega)$ denotes biharmonic extension (Green's map) of the boundary values defined on Γ_1. That is

$$Gv \equiv u \;\text{ iff }\; \Delta^2 u = 0 \;\text{ in } \Omega, \quad u = 0 \text{ on } \Gamma, \quad \frac{\partial}{\partial n} u = 0 \;\text{ on } \Gamma_0, \quad \Delta u = v \;\text{ on } \Gamma_1.$$

- The mapping $g : U_0 \mapsto U_0'$ is determined by the function $g_1(s)$ according to the formula

$$(g(v), w)_{U_0, U_0'} = \int_{\Gamma_1} g_1(v) w d\Gamma, \quad v, w \in U_0.$$

We notice that by classical elliptic theory [222]

$$G \in \mathscr{L}(H^s(\Gamma_1)) \mapsto H^{s+5/2}(\Omega)), \;\; s \in \mathbb{R}. \tag{3.2.12}$$

By Green's formula (1.3.17) in he same way as in the proof of relation (1.3.16) we obtain with an arbitrary $u \in \mathscr{D}(\mathscr{A})$ and $v \in L_2(\Gamma_1)$

$$\int_{\Gamma_1} G^* \mathscr{A} u v d\Gamma_1 = \int_\Omega \mathscr{A} u G v dx = \int_\Omega \Delta^2 u G v dx = \int_{\Gamma_1} \frac{\partial}{\partial n} u v d\Gamma.$$

Hence $G^* \mathscr{A} u = (\partial/\partial n) u|_{\Gamma_1}$ for all $u \in \mathscr{D}(\mathscr{A})$. From (3.2.12) and characterization of fractional power of elliptic operators [129] we have the continuity of the mapping

$$G : L_2(\Gamma_1) \mapsto H^{5/2}(\Omega) \subset \mathscr{D}(\mathscr{A}^{5/8-\varepsilon}).$$

Hence $G^* : [\mathscr{D}(\mathscr{A}^{5/8-\varepsilon})]' \mapsto L_2(\Gamma_1)$, and thus $G^* \mathscr{A} : \mathscr{D}(\mathscr{A}^{3/8+\varepsilon}) \mapsto L_2(\Gamma_1)$ is also bounded. The above regularity and the density of domains of fractional powers of the generator $\mathscr{A}$ allows us to extend the domain of $G^* \mathscr{A}$. Indeed, we have

$$G^* \mathscr{A} u = \frac{\partial}{\partial n} u\Big|_{\Gamma_1}, \; \forall u \in \mathscr{D}(\mathscr{A}^{(3/8)+\varepsilon}) \subset H^{(3/2)+4\varepsilon}(\Omega) \cap H^1_0(\Omega). \tag{3.2.13}$$

Let $u(t)$ be a smooth solution to problem (3.2.1) and (3.2.6). Then for any function w from the space $\mathscr{D}(\mathscr{A}^{1/2}) = H^2_{\Gamma_0}(\Omega) \cap H^1_0(\Omega)$, we find that

$$(\Delta^2 u, w)_\Omega = (\Delta u, \Delta w)_\Omega - \int_{\Gamma_1} \Delta u \frac{\partial}{\partial n} w d\Gamma .$$

Therefore, using (3.2.13) and the boundary condition for Δu we obtain that

$$((I-\alpha\Delta)u_{tt}+\Delta^2 u, w)_\Omega = (Mu_{tt},w)_{V,V'} + (\mathscr{A}u,w)_{\mathscr{D}(\mathscr{A}^{1/2}),[\mathscr{D}(\mathscr{A}^{1/2})]'}$$
$$+ (\mathscr{A}Gg(G^*\mathscr{A}u_t),w)_{\mathscr{D}(\mathscr{A}^{1/2}),[\mathscr{D}(\mathscr{A}^{1/2})]'}$$

and the abstract model for the underlying PDE becomes

$$Mu_{tt}+\mathscr{A}u+Du_t+\mathscr{A}Gg(G^*\mathscr{A}u_t) = F(u,u_t); \tag{3.2.14}$$

that is, problem (3.2.1) with boundary conditions (3.2.6) can be written in the form (2.4.26) with $p(t)=0$. Problem (3.2.14) can be also written in the equivalent form as follows

$$Mu_{tt}+Du_t+\mathscr{A}\left[u+Gg_1\left(\frac{\partial}{\partial n}u_t\right)\right] = F(u,u_t), \tag{3.2.15}$$

where this latter representation underscores the membership of $u+Gg_1((\partial/\partial n)u_t)$ is the domain of $\mathscr{A}$, provided u is smooth enough.

3.2.1.2 Strong and generalized solutions

The discussion carried above leads to the following definition.

3.2.3. Definition. A function

$$u \in C(0,T;H_0^1(\Omega)\cap H^2(\Omega))\cap C^1(0,T;H_0^1(\Omega)) \tag{3.2.16}$$

with the properties $\nabla u|_{\Gamma_0}=0$, $u(x,0)=u_0(x)$ and $u_t(x,0)=u_1(x)$ is said to be

1. A *strong solution* to problem (3.2.1) with the clamped–hinged boundary conditions (3.2.6) on the interval $[0,T]$, iff
   - $u(t)$ is an absolutely continuous function with values in $H^2(\Omega)\cap H_0^1(\Omega)$ and $u_t \in L_1(a,b;H^2_{\Gamma_0}(\Omega)\cap H_0^1(\Omega))$ for any $0<a<b<T$, where $H^2_{\Gamma_0}(\Omega)$ is given by (3.2.11).
   - u_t is an absolutely continuous function with values in $H_0^1(\Omega)$ and $u_{tt}\in L_1(a,b;H_0^1(\Omega))$ for any $0<a<b<T$,
   - $u(t)+Gg_1((\partial/\partial n)u_t(t))\in W$ for almost all $t\in[0,T]$, where

   $$W=\left\{u\in H^3(\Omega)\cap H_0^1(\Omega) : \frac{\partial}{\partial n}u|_{\Gamma_0}=0,\ \Delta u|_{\Gamma_1}=0\right\}. \tag{3.2.17}$$

   - Equality (3.2.15) holds in $H^{-1}(\Omega)$ for almost all $t\in[0,T]$.
2. A *generalized solution* to problem (3.2.1) and (3.2.6) on the interval $[0,T]$, iff there exists a sequence $\{u_n(t)\}$ of strong solutions to (3.2.1) and (3.2.6) with initial data given by $(u_{0n};u_{1n})$ and such that (3.1.11) holds.

■

Applying the results from Chapter 2 we obtain the following assertion.

3.2.4. Theorem. *Under Assumptions* 3.2.1 *with reference to* (3.2.1), *subject to clamped–hinged boundary conditions* (3.2.6), *the following statements are valid with any* $T > 0$.

- **Generalized solutions:** *Let* $u_0 \in H_0^1(\Omega) \cap H^2(\Omega)$, $u_1 \in H_0^1(\Omega)$, $\nabla u_0 = 0$ *on* Γ_0.

  1. *There exists a unique generalized solution* $u(t)$ *such that the following energy inequality*

  $$\mathscr{E}(u(t), u_t(t)) + \int_0^t \int_\Omega d_0 b(u_t) u_t dx d\tau \le \mathscr{E}(u_0, u_1) - \int_0^t (Lu, u_t)_\Omega d\tau \quad (3.2.18)$$

  holds, where the energy $\mathscr{E}$ *is given by* (3.2.7). *Thus, if* $Lu = 0$ *and* $sb(s) \ge 0$ *for all* $s \in \mathbb{R}$, *then* $\mathscr{E}(u(t), u_t(t)) \le \mathscr{E}(u_0, u_1)$ *for all* $t > 0$.
  2. *If* $g_1(s)$ *possesses the property*

  $$m|s|^{1+\delta} \le g_1(s)s, \quad |s| \ge s_0, \quad (3.2.19)$$

  for some constants $m > 0$, $\delta > 0$, *and* $s_0 \ge 0$, *then*

  $$\frac{\partial}{\partial n} u_t \equiv \partial_t \left[\frac{\partial}{\partial n} u \right] \in L_{1+\delta}([0,T] \times \Gamma_1)$$

  and the energy inequality can be written in the form

  $$\mathscr{E}(u(t), u_t(t)) + \int_0^t \left[\int_\Omega d_0 b(u_t) u_t dx + \int_{\Gamma_1} \psi\left(\frac{\partial}{\partial n} u_t \right) d\Gamma \right] d\tau$$
  $$\le \mathscr{E}(u_0, u_1) - \int_0^t \int_\Omega L u u_t dx d\tau \quad (3.2.20)$$

  for every continuous convex function ψ *such that*

  $$0 \le \psi(s) \le g_1(s)s, \quad \text{for all } s \in \mathbb{R}. \quad (3.2.21)$$

  3. *If* $b(s)$ *and* $g_1(s)$ *satisfy Assumption* 3.2.2, *then*

  $$\frac{\partial}{\partial n} u_t \in L_{q_*}([0,T] \times \Gamma_1), \; g_1\left(\frac{\partial}{\partial n} u_t \right) \in L_{q_*/(q_*-1)}([0,T] \times \Gamma_1), \quad (3.2.22)$$

  and the energy identity

  $$\mathscr{E}(u(t), u_t(t)) + \int_0^t \left[\int_\Omega d_0 b(u_t) u_t dx + \int_{\Gamma_1} g_1\left(\frac{\partial}{\partial n} u_t \right) \frac{\partial}{\partial n} u_t d\Gamma \right] d\tau$$
  $$= \mathscr{E}(u_0, u_1) - \int_0^t \int_\Omega L u u_t dx d\tau \quad (3.2.23)$$

  holds. Moreover, $u_{tt} \in L_{q_*/(q_*-1)}(0, T, L_2(\Omega))$ *provided* $q_* \ge 2$ *in* (3.2.5).

- **Strong solutions:** *If* $u_0 \in H^3(\Omega) \cap H_0^1(\Omega)$ *and* $u_1 \in H_0^1(\Omega) \cap H^2(\Omega)$ *satisfy the compatibility condition*

$$\Delta u_0 = -g_1(u_1) \;\; on\; \Gamma_1, \quad \nabla u_0 = 0 \;\; on\; \Gamma_0, \quad \nabla u_1 = 0 \;\; on\; \Gamma_0,$$

then there exists a unique strong solution such that

$$\left.\begin{aligned} & u + Gg_1\left(\frac{\partial}{\partial n} u_t\right) \in C_r(0,T;W) \cap L_\infty(0,T;W), \\ & u_t \in C_r(0,T;(H^2_{\Gamma_0} \cap H_0^1)(\Omega)) \cap L_\infty(0,T;(H^2_{\Gamma_0} \cap H_0^1)(\Omega)), \\ & u_{tt} \in C_r(0,T;H_0^1(\Omega)) \cap L_\infty(0,T;H_0^1(\Omega)), \end{aligned}\right\} \tag{3.2.24}$$

where W *is given by* (3.2.17) *and* $C_r(0,T;X)$ *is the space of strongly right-continuous functions. Strong solutions satisfy the energy identity in* (3.2.23).

3.2.5. Remark. The properties of u_t stated in (3.2.24) for strong solutions imply $(\partial/\partial n)u_t \in L_\infty(0,T;H^{1/2}(\Gamma_1))$. Therefore, if g_1 *is globally Lipschitz*, then $g_1((\partial/\partial n)u_t) \in L_\infty(0,T;H^{1/2}(\Gamma_1))$. In this case by (3.2.12) $Gg((\partial/\partial n)u_t)$ lies in $L_\infty(0,T;H^3(\Omega))$. Thus it follows from the first line in (3.2.24) that strong solutions u possess the regularity: $u \in L_\infty(0,T;H^3(\Omega))$ and, similarly, $u \in C_r(0,T;H^3(\Omega))$. ■

In order to prove Theorem 3.2.4 we proceed as before: by applying Theorem 2.4.16.

We first verify that Assumption 2.4.15 in Theorem 2.4.16 is satisfied.

3.2.6. Lemma. *Operators* $\mathscr{A}, M, D, G, F$, *and* g *introduced above comply with the requirements in Assumption* 2.4.15.

Proof. The verification of the requirements 1–4 in Assumption 2.4.15 is the same as in the case of internal damping, so it is not repeated. We refer to the arguments given in the proof of Lemma 3.1.6.

It suffices therefore to verify the validity of Assumption 2.4.15(5) (which is the same as requirements 3 and 4 in Assumption 2.4.1).

By the identification (3.2.13) of $G^*\mathscr{A}$ we easily infer that $G^*\mathscr{A} : \mathscr{D}(\mathscr{A}^{1/2}) \mapsto U_0 = H^{1/2}(\Gamma_1)$ is bounded. Indeed, the above follows from the classical trace theorem [222] which asserts $(\partial/\partial n) : H^2(\Omega) \mapsto H^{1/2}(\Gamma)$ is bounded. Thus, the requirement in (4) of Assumption 2.4.1 has been verified. In fact, in this case we also have the surjectivity of $G^*\mathscr{A}$, by virtue of the same trace theorem. Finally, we need to verify requirement 3 of Assumption 2.4.1; that is, the continuity and monotonicity of Nemytskij operator $g_1(s)$, as acting between the spaces $H^{1/2}(\Gamma_1)$ and $H^{-1/2}(\Gamma_1)$. Because by Assumption 3.2.1 the function $g_1(s)$ is assumed of a polynomial growth at infinity, the above continuity follows from standard nonlinear analysis and classical Sobolev embedding $H^{1/2}(\Gamma_1) \subset L_p(\Gamma_1)$ for any $1 \le p < \infty$. We refer to similar arguments given in the proof of Lemma 3.1.8. The monotonicity of the mapping g_1 is obvious. □

3.2.1.3 Proof of Theorem 3.2.4.

We split the proof into several steps.

Step 1: Global existence and uniqueness. Once our standing Assumption 2.4.15 holds, we are in a position to apply the result of Theorem 2.4.16. This theorem provides global existence and uniqueness of strong solutions that satisfy the energy relation of the form (3.2.23). The smoothness properties in (3.2.24) follow from (2.4.32) and (2.4.33). Moreover, one can see from (3.2.13) that Ker $[G^*\mathscr{A}] \supset C_0^\infty(\Omega)$ in our case. Therefore the set $\mathscr{W}$ defined in (2.4.39) contains $C_0^\infty(\Omega)$ which is dense in $H_0^1(\Omega)$. Thus Theorem 2.4.16 (along with Remark 2.4.17) implies global existence and uniqueness of generalized solutions.

Because the energy is bounded from below (see (3.2.9)) and monotonicity of $b(s) + b_0 s$ and $g_1(s)$ is assumed, energy identity (3.2.23) yields the a priori bound for strong solutions. This is to say,

$$E(t) \le C(E(0)) + C\int_0^t \left(||u||_2^2 + ||u_t||_1^2\right) d\tau$$

for any $t \in [0,T]$, where $E(t)$ is the positive part of the energy given by (3.2.10). Thus, from Gronwall's inequality we conclude the following a priori bound

$$||u(t)||_2 + ||u_t(t)||_1 + ||v(u)||_2 \le C_T, \quad t \in [0,T], \tag{3.2.25}$$

where the constant C_T depends on $||u_0||_2, ||u_1||_1, ||f||_2, ||F_0||_2$, and $||p||_{-1}$. Inequality (3.2.25) is stable with respect to the norm generated by the energy,therefore the same a priori bound holds for generalized solutions.

We also note that other properties of solutions can be derived as consequences of Theorem 2.4.16. However, we opt for a more direct treatment relying on the structure of the problem under consideration.

Step 2: Energy inequality and energy identity for generalized solutions.

Energy inequality. In order to obtain energy relations for generalized solutions, we need to perform a limit process on strong solutions. Let u_0 and u_1 be initial data of finite energy and let $(u(t); u_t(t))$ be a corresponding solution. From the definition of generalized solutions we know that such solutions are strong limits of strong solutions denoted by $u_n(t)$. We thus have

$$u_n \to u \ \text{ in } C(0,T;H^2(\Omega)), \quad u_{nt} \to u_t \ \text{ in } C(0,T;H_0^1(\Omega)). \tag{3.2.26}$$

Denoting by $\mathscr{E}_n(t)$ the energy corresponding to strong solutions $u_n(t)$, we can rewrite energy identity (3.2.23) in the form

$$\mathscr{E}_n(t) + \int_0^t \int_\Omega d_0\, [b(u_{nt}) + b_0 u_{nt}]\, u_{nt}\, dx d\tau + \int_0^t \int_{\Gamma_1} g_1\left(\frac{\partial}{\partial n} u_{nt}\right) \frac{\partial}{\partial n} u_{nt}\, dx d\tau$$
$$= \mathscr{E}_n(0) + b_0 \int_0^t \int_\Omega d_0 u_{nt}^2\, dx dt - \int_0^t (L u_n, u_{nt})_\Omega d\tau.$$

This implies that

$$\mathscr{E}_n(t)+\int_0^t\int_\Omega d_0\chi_N\left([b(u_{nt})+b_0u_{nt}]\,u_{nt}\right)dxdt+\int_0^t\int_{\Gamma_1}\psi\left(\frac{\partial}{\partial n}u_{nt}\right)dxd\tau$$
$$\le \mathscr{E}_n(0)+b_0\int_0^t\int_\Omega d_0u_{nt}^2dxdt-\int_0^t(Lu_n,u_{nt})_\Omega d\tau, \tag{3.2.27}$$

where $\chi_N(s)=\min\{N,s[b(s)+b_0s]\}$ and ψ possesses property (3.2.21). It follows from (3.2.19) that

$$\int_0^T\int_{\Gamma_1}\left|\frac{\partial}{\partial n}u_{nt}\right|^{1+\delta}dxd\tau\le C_1+C_2\int_0^T\int_{\Gamma_1}g_1\left(\frac{\partial}{\partial n}u_{nt}\right)\frac{\partial}{\partial n}u_{nt}dxd\tau\le C_T.$$

Therefore $\{(\partial/\partial n)u_{nt}\}$ is weakly compact in $L_{1+\delta}((0,T)\times\Gamma_1)$. Thus using the fact that

$$v\mapsto\int_0^T\int_{\Gamma_1}\psi(v)dxd\tau,\quad v\in L_{1+\delta}([0,T]\times\Gamma_1),$$

is (weakly) lower semicontinuous function (see [260, Proposition 8.1, p. 85]) and the fact that the energy function $\mathscr{E}_n(t)$ is continuous with respect to topology in (3.2.26), we can pass to the limit in (3.2.27) as $n\to\infty$ (at least along a subsequence) to obtain the relation

$$\mathscr{E}(t)+\int_0^t\int_\Omega d_0\chi_N\left([b(u_t)+b_0u_t]\,u_t\right)dxdt+\int_0^t\int_{\Gamma_1}\psi\left(\frac{\partial}{\partial n}u_t\right)dxd\tau$$
$$\le \mathscr{E}(0)+b_0\int_0^t\int_\Omega d_0u_t^2dxdt-\int_0^t(Lu,u_t)_\Omega d\tau.$$

It is clear that $\chi_N(s)\le\chi_{N+1}(s)$. Therefore by the Levi–Lebesgue theorem on monotone convergence (see Theorem B.4.2) we obtain the desired energy inequality (3.2.20) for every generalized solution under condition (3.2.19). If (3.2.19) fails, then clearly a weaker form of energy inequality (3.2.18) holds.

Energy equality. In order to obtain energy identity we assume that Assumption 3.2.2 holds.

By Sobolev's embeddings and polynomial growth (at infinity) of the function $b(s)$ we have that $v\mapsto d_0(b(v)+b_0v)$ is a monotone hemicontinuous operator from $H^1(\Omega)$ into $H^{-1}(\Omega)$ (cf. Lemma 3.1.8). By Proposition 1.2.5 we also have that b is demicontinuous. Therefore from (3.2.26) we have $d_0b(u_{nt}(t))\to d_0b(u_t)$ in $H^{-1}(\Omega)$ weakly. It is also clear that $||d_0b(u_{nt})||_{-1}\le C$ for $t\in[0,T]$. Thus using (3.2.26) we obtain that

$$\int_0^t\int_\Omega d_0(x)b(u_{nt})u_{nt}dxdt\to\int_0^t\int_\Omega d_0(x)b(u_t)u_tdxdt\quad\text{for any } t\in[0,T]. \tag{3.2.28}$$

In order to pass with the limit on the boundary term, the argument is more complicated. This is due to the fact that the a priori interior regularity provides no information on the behavior of normal derivatives of velocity on the boundary. For this reason we need additional assumption (3.2.5) on the bound from below of g_1.

The corresponding result is formulated below.

3.2.7. Lemma. *Let $h(s)$ be a monotone nondecreasing continuous function on $\mathbb{R}$ such that*

$$ms^q \le sh(s) \le Ms^q, \quad |s| \ge s_0, \tag{3.2.29}$$

for some $s_0 \ge 0$ and $q > 1$, where m and M are positive constants. Assume that $\{w_n\}$ is a sequence in $L_q([0,T]\times\Gamma_1)$ possessing the properties

$$\int_0^T \int_{\Gamma_1} w_n(x,t)h(w_n(x,t))d\Gamma dt \le C \tag{3.2.30}$$

and

$$\int_0^T \int_{\Gamma_1} (w_n(x,t)-w_k(x,t))(h(w_n(x,t))-h(w_k(x,t)))d\Gamma dt \to 0 \tag{3.2.31}$$

whenever $n,k\to\infty$. Then there exist a subsequence $\{n_l\}$ and a function $w(x,t)$ from $L_q([0,T]\times\Gamma_1)$ such that $h(w)\in L_p([0,T]\times\Gamma_1)$ for $p=q/(q-1)$ and

$$w_{n_l} \to w \ \text{ in } \ L_q([0,T]\times\Gamma_1), \quad h(w_{n_l}) \to h(w) \ \text{ in } \ L_p([0,T]\times\Gamma_1) \tag{3.2.32}$$

in the sense of weak convergence, and

$$\lim_{l\to\infty} \int_0^t \int_{\Gamma_1} w_{n_l}(x,\tau)h(w_{n_l}(x,\tau))d\Gamma d\tau = \int_0^t \int_{\Gamma_1} w(x,\tau)h(w(x,\tau))d\Gamma d\tau. \tag{3.2.33}$$

Proof. The result stated in the lemma follows from Lemma 1.3 [18]. For the sake of completeness we provide the details of the proof. Let $\Sigma_1 = [0,T]\times\Gamma_1$ and $\Sigma_1^n = \Sigma_1 \cap \{(x;t) : |w_n(x,t)| \ge s_0\}$. It follows from (3.2.30) that

$$\int_{\Sigma_1} |w_n|^q d\Gamma dt \le \frac{1}{m}\int_{\Sigma_1^n} w_n h(w_n) d\Gamma dt + s_0^q \text{mes}(\Sigma_1 \setminus \Sigma_1^n) \le C. \tag{3.2.34}$$

It follows from (3.2.29) that

$$|h(s)| \le C(1+|s|)^{q-1}, \quad s\in\mathbb{R}. \tag{3.2.35}$$

Therefore by (3.2.34) we have that $\int_{\Sigma_1} |h(w_n)|^p d\Gamma dt \le C$ with $p=q/(q-1)$. This and (3.2.34) imply weak convergence on a subsequence:

$$w_{n_l}(x,t) \to w(x,t) \ \text{ in } \ L_q(\Sigma_1), \quad h(w_{n_l}(x,t)) \to h^*(x,t) \ \text{ in } \ L_p(\Sigma_1), \tag{3.2.36}$$

where $w\in L_q(\Sigma_1)$ and $h^*\in L_p(\Sigma_1)$.

It is clear that $w\mapsto h(w)$ is a monotone operator from $L_q(\Sigma_1)$ into $L_p(\Sigma_1) = [L_q(\Sigma_1)]'$. Using the continuity of h, relation (3.2.35), and the Lebesgue dominated convergence theorem we obtain that

$$\lim_{\lambda\to 0}\int_{\Sigma_1} h(u+\lambda w)vd\Gamma d\tau = \int_{\Sigma_1} h(u)vd\Gamma d\tau$$

for any $u,w,v \in L_2(\Sigma_1)$. Thus $w \mapsto h(w)$ is hemicontinuous and hence by Proposition 1.2.5 it is a maximal monotone operator. Now we can apply the second part of Proposition 1.2.6 and derive from (3.2.31) and (3.2.36) that $h^* = h(w)$ and relation (3.2.33) holds. □

We apply Lemma 3.2.7 to our situation.

We consider the difference of two strong solutions: say $u_n(t)$ and $u_m(t)$. By standard calculations, using uniform (in n and m) estimates of the form (3.2.25) and locally Lipschitz property of nonlinearity $F(u,v)$, for these solutions we obtain the following energy relation:

$$\begin{aligned}
&||u_n(t)-u_m(t)||_2^2 + ||u_{nt}(t)-u_{mt}(t)||_1^2 \\
&+\int_0^t\int_\Omega [d_0(b(u_{nt})-b(u_{mt})+b_0(u_{nt}-u_{mt}))(u_{nt}-u_{mt})]d\Gamma ds \\
&+\int_0^t\int_{\Gamma_1}\left[(g_1(\frac{\partial}{\partial n}u_{nt})-g_1(\frac{\partial}{\partial n}u_{mt}))(\frac{\partial}{\partial n}u_{nt}-\frac{\partial}{\partial n}u_{mt})\right]d\Gamma ds \\
&\le C_T\int_0^t (||u_n(t)-u_m(t)||_2^2 + ||u_{nt}(t)-u_{mt}(t)||_1^2)\,ds \\
&\quad + ||u_n(0)-u_m(0)||_2^2 + ||u_{nt}(0)-u_{mt}(0)||_1^2 .
\end{aligned}$$

Therefore using the stability estimate of the form (2.4.10) (see also Remark 2.4.18) and (uniform) a priori bounds of strong solutions in (3.2.25) along with monotonicity of $b(s)+b_0 s$ we conclude that

$$\begin{aligned}
&||u_n(t)-u_m(t)||_2^2 + ||u_{nt}(t)-u_{mt}(t)||_1^2 \\
&+\int_0^t\int_{\Gamma_1} [(g_1(\frac{\partial}{\partial n}u_{nt})-g_1(\frac{\partial}{\partial n}u_{mt}))(\frac{\partial}{\partial n}u_{nt}-\frac{\partial}{\partial n}u_{mt})]d\Gamma ds \\
&\le C_T\left[||u_n(0)-u_m(0)||_2^2 + ||u_{nt}(0)-u_{mt}(0)||_1^2\right] \to 0,
\end{aligned} \tag{3.2.37}$$

whenever

$$||u_n(0)-u_0||_2 \to 0, \quad ||u_{nt}(0)-u_1||_1 \to 0.$$

Thus, in addition to having strong limits (3.2.26) we also have:

$$\int_0^t\int_{\Gamma_1}\left[g_1(\frac{\partial}{\partial n}u_{nt})-g_1(\frac{\partial}{\partial n}u_{mt})\right]\left[\frac{\partial}{\partial n}u_{nt}-\frac{\partial}{\partial n}u_{mt}\right]d\Gamma ds \to 0. \tag{3.2.38}$$

From the energy inequality we have that

$$\int_0^t\int_{\Gamma_1} g_1(\frac{\partial}{\partial n}u_{nt})\frac{\partial}{\partial n}u_{nt}d\Gamma ds \le C_T, \quad t\in[0,T].$$

Therefore, by Lemma 3.2.7 we obtain that on a subsequence:

$$\frac{\partial}{\partial n} u_{n_l t} \to w \text{ in } L_{q_*}(\Sigma_1), \quad g_1(\frac{\partial}{\partial n} u_{n_l t}) \to g_1(w) \text{ in } L_{q_*/(q_*-1)}(\Sigma_1) \tag{3.2.39}$$

for some $w \in L_{q_*}(\Sigma_1)$ in the sense of weak convergence (here as above we denote $\Sigma_1 \equiv [0,T] \times \Gamma_1$) and also

$$\lim_{l\to\infty} \int_0^t \int_{\Gamma_1} \frac{\partial}{\partial n} u_{n_l t} g_1(\frac{\partial}{\partial n} u_{n_l t}) d\Gamma d\tau = \int_0^t \int_{\Gamma_1} w g_1(w) d\Gamma d\tau.$$

However, by (3.2.26) we have that $(\partial/\partial n)u_n \to (\partial/\partial n)u$ in $L_2(0,T;L_2(\Gamma_1))$. Thus we can easily identify the limit w as being equal to $(\partial/\partial n)u_t$ in the sense of generalized functions. This implies the validity of (3.2.22), and also

$$\int_0^t \int_{\Gamma_1} g_1(\frac{\partial}{\partial n} u_{n_l t}) \frac{\partial}{\partial n} u_{n_l t} d\Gamma ds \to \int_0^t \int_{\Gamma_1} g_1(\frac{\partial}{\partial n} u_t) \frac{\partial}{\partial n} u_t d\Gamma ds, \quad l \to \infty.$$

The above convergence along with (3.2.28) allow us to pass with the limit on the energy identity written first for strong solutions. The limit process yields this identity valid for all generalized solutions.

Step 3: Regularity of the second time derivative. Let Assumption 3.2.2 hold with some $q_* \geq 2$ in (3.2.5) and $u(t)$ be a strong solution. From (3.2.14) we have

$$u_{tt} = -M^{-1}\left[D(u_t) + \mathscr{A}\left[u + Gg(G^* \mathscr{A} u_t)\right] - F(u, u_t)\right].$$

From properties of D, F we have

$$||M^{-1}D(u_t)|| \leq C_T, \quad ||M^{-1}F(u,u_t)||_1 \leq C_T. \tag{3.2.40}$$

On the other hand, by integrating by parts and using Green's formula we also obtain the following inequality valid with any function $w \in L_2(\Omega)$:

$$\begin{aligned} \int_\Omega M^{-1}\left[\mathscr{A}\left[u + Gg\left(\frac{\partial}{\partial n} u_t\right)\right]\right] w dx &= \int_\Omega \mathscr{A}\left[u + Gg\left(\frac{\partial}{\partial n} u_t\right)\right] M^{-1} w dx \quad (3.2.41) \\ &= \int_\Omega \Delta^2 u M^{-1} w dx \\ &= \int_\Omega \Delta u \Delta M^{-1} w dx - \int_\Gamma \Delta u \frac{\partial}{\partial n}\left(M^{-1} w\right) d\Gamma. \end{aligned}$$

By applying boundary conditions, using growth at infinity of $g_1(s)$ (see (3.2.5)) and the standard trace regularity estimate we obtain

$$\begin{aligned} \left| \int_{\Sigma_1} \Delta u \frac{\partial}{\partial n}\left(M^{-1} w\right) d\Gamma dt \right| &\leq \int_{\Sigma_1} |g_1(\frac{\partial}{\partial n} u_t)| \left| \frac{\partial}{\partial n}\left(M^{-1} w\right) \right| d\Gamma dt \\ &\leq C \|g_1(\frac{\partial}{\partial n} u_t)\|_{L_{q*/(q*-1)}(\Sigma_1)} \left\| \frac{\partial}{\partial n}\left(M^{-1} w\right) \right\|_{L_{q*}(\Sigma_1)} \\ &\leq C(E(0)) \|w\|_{L_{q*}(0,T;L_2(\Omega))} \end{aligned}$$

for any $w \in L_{q_*}(0,T;L_2(\Omega))$. On the other hand, by Proposition 3.1.9 applicable to clamped portion of the boundary Γ_0 (see also Theorem 2.5.6 and Theorem 2.5.4 in Chapter 2) we also infer

$$\left| \int_{\Sigma_0} \Delta u \frac{\partial}{\partial n} M^{-1} w d\Gamma dt \right| \leq C \|\Delta u\|_{L_2(\Sigma_0)} \|w\|_{L_2((0,T)\times\Omega)} \leq C_T \|w\|_{L_2((0,T)\times\Omega)}.$$

Because $q_* \geq 2$, the above estimate together with (3.2.40), (3.2.41) yields

$$\int_0^T \|u_{tt}\|^{q_*/(q_*-1)} dt \leq C_T(E(0)).$$

The last estimate remains true after limit transition for generalized solutions. This proves the desired regularity of second time derivatives of generalized solutions, thus completing the proof of Theorem 3.2.4.

3.2.1.4 Weak solutions

In what follows we assert that under an additional condition imposed on the damping, generalized solutions satisfy weak form of differential equality.

3.2.8. Theorem. *Assume in addition to Assumption* 3.2.1 *that the function* $g_1(s)$ *either possesses the property*

$$(s_1 - s_2)(g_1(s_1) - g_1(s_2)) \geq c_0 |s_1 - s_2|^r, \quad s_1, s_2 \in \mathbb{R}, \tag{3.2.42}$$

for some $c_0 > 0$ *and* $r \geq 1$ *or else inequality* (3.2.5) *in Assumption* 3.2.2 *is satisfied. Then generalized solutions of von Karman system* (3.2.1) *with clamped–hinged boundary conditions* (3.2.6) *are also weak solutions. This means that (i) any generalized solution* $u(t)$ *satisfies the integrability condition:* $t \mapsto (d_0 b(u_t), \phi)_\Omega$ *and* $t \mapsto (g_1((\partial/\partial n)u_t), (\partial/\partial n)\phi)_{\Gamma_1}$ *are locally integrable on* $\mathbb{R}_+$ *for any* $\phi \in H^2(\Omega)$, *and (ii) the following variational equality*

$$\begin{aligned}
&(u_t(t), \phi)_\Omega + \alpha(\nabla u_t(t), \nabla\phi)_\Omega - (u_1, \phi)_\Omega - \alpha(\nabla u_1, \nabla\phi)_\Omega \\
&+ \int_0^t \left[(\Delta u, \Delta\phi)_\Omega + (d_0 b(u_t), \phi)_\Omega + (g_1(\frac{\partial}{\partial n} u_t), \frac{\partial}{\partial n}\phi)_{\Gamma_1} \right] d\tau \\
&+ \int_0^t \left[(-[v(u) + F_0, u + f] + Lu - p, \phi)_\Omega \right] d\tau = 0
\end{aligned}$$

holds for any $\phi \in H_0^1(\Omega) \cap H^2(\Omega)$ *such that* $\nabla\phi = 0$ *on* Γ_0, *where* $v = v(u)$ *is defined from* (3.2.2) *and, as above,* $(\cdot,\cdot)_{\mathscr{O}}$ *denotes the inner product in* $L_2(\mathscr{O})$.

3.2.9. Remark. As an example of a damping satisfying (3.2.42) consider $g(s) = g_0|s|^{p-1}s$ with $p \geq 1$. Indeed, one can see that the function

$$G_a(s) \equiv p \int_{-a+s}^{a+s} |\sigma|^{p-1} d\sigma \equiv |a+s|^{p-1}(a+s) - |-a+s|^{p-1}(-a+s)$$

for all $s \geq 0$ satisfies the relations: $G'_a(s) \geq 0$ if $a \geq 0$ and $G'_a(s) \leq 0$ when $a < 0$. This implies that $aG_a(s) \geq aG_a(0)$ for all $a \in \mathbb{R}$ and $s \geq 0$. Therefore choosing $s = (s_1 + s_2)/2$ and $a = (s_1 - s_2)/2$ we obtain (3.2.42) with $c_0 = g_0 2^{-p+1}$ and $r = p+1$ in the case when $s_1 + s_2 \geq 0$. The case $s_1 + s_2 < 0$ can be reduced to the previous one by changing variables $s_1 := -s_1$ and $s_2 := -s_2$. ■

Proof. As before, the variational form of the definition is established by taking first strong solutions and integrating by parts. The final conclusion follows after passing with the limit on strong solutions. This is accomplished by using: (i) weak continuity of the von Karman bracket, and (ii) monotonicity of the dissipative terms g_i. The main difference in comparison with Theorem 3.1.12 is the limit transition in the term

$$\int_0^t \left(g_1(\frac{\partial}{\partial n} u_t), \frac{\partial}{\partial n} \phi \right)_{\Gamma_1} d\tau.$$

To accomplish this task note that (3.2.37) and (3.2.42) imply convergence to zero of the limit

$$\lim_{n,m\to\infty} \int_0^t \int_{\Gamma_1} \left| \frac{\partial}{\partial n} u_{nt} - \frac{\partial}{\partial n} u_{mt} \right|^r d\Gamma d\tau = 0$$

for any sequence $\{u_n\}$ of strong solutions that converges to u in the sense of relation (3.1.11). This implies that $(\partial/\partial n)u_t \in L_r([0,T] \times \Gamma_1)$ and there exists a subsequence $\{n_k\}$ such that

$$\frac{\partial}{\partial n} u_{n_k t} \to \frac{\partial}{\partial n} u_t \quad \text{strongly in} \quad L_r([0,T] \times \Gamma_1)$$

and also almost everywhere in $[0,T] \times \Gamma_1$. Therefore we can apply the same argument as in the case of interior damping (see the proof of Theorem 3.1.12).

In the case when (3.2.5) holds, the limit transition is accomplished via the argument given in Lemma 3.2.7 which, in turn, is based on monotonicity. □

3.2.2 Clamped–free boundary conditions

We consider next clamped–free boundary conditions imposed on the edge of the plate:

$$\begin{cases} u = \nabla u = 0 \text{ on } \Gamma_0, \\ \Delta u + (1-\mu)B_1 u = -g_1(\frac{\partial u_t}{\partial n}) \text{ on } \Gamma_1, \\ \frac{\partial}{\partial n}\Delta u + (1-\mu)B_2 u \quad \alpha \frac{\partial u_{tt}}{\partial n} \\ \qquad\qquad = \nu_1 u + \beta u^3 + g_0(u_t) - \frac{\partial}{\partial \tau} g_2(\frac{\partial u_t}{\partial \tau}) \text{ on } \Gamma_1, \end{cases} \tag{3.2.43}$$

where B_1 and B_2 are given by (1.3.20), $\nu_1 \geq 0$ ($\nu_1 > 0$ if $\Gamma_0 = \emptyset$), and $\beta \in L_\infty(\Gamma_1)$ is a nonnegative function. In this case we define the energy by the formula

$$\mathscr{E}(t) \equiv \mathscr{E}(u,u_t) = \frac{1}{2}\int_\Omega \left[\alpha|\nabla u_t|^2 + |u_t|^2\right] dx + \frac{1}{2}a(u,u) + \Pi(u), \tag{3.2.44}$$

where

$$a(u,w) = a_0(u,w) + \nu_1 \int_{\Gamma_1} uw \, d\Gamma$$

with $a_0(u,w)$ defined by (1.3.4) and

$$\Pi(u) = \int_\Omega \left[\frac{1}{4}|\Delta v(u)|^2 - [f,F_0]u - up\right] dx + \frac{1}{4}\int_{\Gamma_1} \beta u^4 d\Gamma. \tag{3.2.45}$$

Here $v(u) \in H_0^2(\Omega)$ is determined from (3.2.2).

We note that in the case of clamped–free boundary conditions the term $[u,F_0]$ in (3.2.1) is not potential for general $F_0 \in H^2(\Omega)$. Therefore, in contrast with the clamped–hinged case (see (3.2.7)), the term $[u,F_0]$ does not appear in the definition of the energy functional (see also Remark 3.1.23). As a consequence, the energy relation (see (3.2.72) will have to accommodates for a nondissipative term that depends quadratically on the solution.

The functional $\Pi(u)$ can be written in the form

$$\Pi(u) = \Pi_0(u) + \Pi_1(u), \tag{3.2.46}$$

where

$$\Pi_0(u) = \frac{1}{4}\int_\Omega |\Delta v(u)|^2 dx + \frac{1}{4}\int_{\Gamma_1} \beta u^4 d\Gamma \geq 0 \tag{3.2.47}$$

and

$$\Pi_1(u) = -\int_\Omega ([f,F_0]u + up)\, dx. \tag{3.2.48}$$

As above we also use another energy variable that consists of a positive part of $\mathscr{E}$:

$$E(t) \equiv E(u,u_t) = \frac{1}{2}\int_\Omega \left[\alpha|\nabla u_t|^2 + |u_t|^2\right] dx + \frac{1}{2}a(u,u) + \Pi_0(u). \tag{3.2.49}$$

Obviously we have the relation $\mathscr{E}(u,u_t) = E(u,u_t) + \Pi_1(u)$. As in Section 3.1.3 using (3.1.58) we infer that there exist positive constants c_0, c_1, and K such that

$$c_0\mathscr{E}(u,u_t) - K \le E(u,u_t) \le c_1\mathscr{E}(u,u_t) + K, \quad (u;u_t) \in H^2(\Omega)\times H^1(\Omega). \quad (3.2.50)$$

3.2.2.1 Functional analytic framework

As in the clamped–hinged case we first analyze the corresponding functional analytic setup. For this the following spaces and operators are introduced.

- $\mathscr{H} \equiv L_2(\Omega)$, $V = H^1_{\Gamma_0}(\Omega) \equiv \{u \in H^1(\Omega) : u|_{\Gamma_0} = 0\}$, $U \equiv L_2(\Gamma_1)\times L_2(\Gamma_1)$, $U_0 \equiv H^{1/2}(\Gamma_1)\times H^{1/2}(\Gamma_1)$, $U_0' = H^{-1/2}(\Gamma_1)\times H^{-1/2}(\Gamma_1)$.
- $\mathscr{A}u \equiv \Delta^2 u, u \in \mathscr{D}(\mathscr{A})$, where

$$\mathscr{D}(\mathscr{A}) \equiv \left\{ u \in H^2(\Omega) \;\middle|\; \begin{array}{l} \Delta^2 u \in L_2(\Omega), \quad u = \nabla u = 0 \text{ on } \Gamma_0, \\ \frac{\partial}{\partial n}\Delta u + (1-\mu)B_2 u - \nu_1 u = 0 \text{ on } \Gamma_1, \\ \Delta u + (1-\mu)B_1 u = 0 \text{ on } \Gamma_1 \end{array} \right\}. \quad (3.2.51)$$

  Because Γ_0 and Γ_1 are disjoint, elliptic regularity implies that the domain $\mathscr{D}(\mathscr{A})$ consists of the functions from $H^4(\Omega)$ satisfying the boundary conditions listed in (3.2.51) (see Remark 1.3.16). Thus

$$\mathscr{D}(\mathscr{A}) \equiv \left\{ u \in H^4(\Omega) \;\middle|\; \begin{array}{l} u = \nabla u = 0 \text{ on } \Gamma_0, \\ \frac{\partial}{\partial n}\Delta u + (1-\mu)B_2 u - \nu_1 u = 0 \text{ on } \Gamma_1, \\ \Delta u + (1-\mu)B_1 u = 0 \text{ on } \Gamma_1 \end{array} \right\}.$$

- $M \equiv I + \alpha A_N$, where $A_N u = -\Delta u$ with the domain $\mathscr{D}(A_N) \equiv \{u \in L_2(\Omega), u = 0, \text{ on } \Gamma_0, (\partial/\partial n)u = 0 \text{ on } \Gamma_1, \Delta u \in L_2(\Omega)\}$. Because $\Gamma_0 \cap \Gamma_1 = \emptyset$, we have

$$\mathscr{D}(A_N) \equiv \left\{ u \in H^2(\Omega) \;\middle|\; u = 0 \text{ on } \Gamma_0, \; \frac{\partial}{\partial n} u = 0 \text{ on } \Gamma_1 \right\}. \quad (3.2.52)$$

- Hence $V' = [H^1_{\Gamma_0}(\Omega)]'$ and $\mathscr{D}(\mathscr{A}^{1/2}) = H^2_{\Gamma_0}(\Omega)$, where the subscript Γ_0 refers to zero Dirichlet–Neumann boundary conditions on Γ_0 (see (3.2.11)). Moreover, we also have that $\mathscr{D}(\mathscr{A}^{1/4}) = \mathscr{D}(M^{1/2}) = V = H^1_{\Gamma_0}(\Omega)$.

To define the corresponding nonlinear mappings F and D we first introduce the operators $Q_{\Gamma_1}, D_{\Gamma_1} : V \mapsto V'$ and $B_{\Gamma_1} : \mathscr{D}(\mathscr{A}^{1/2}) \mapsto V'$ by the formulas

$$(Q_{\Gamma_1}u,\phi) = a_0\int_{\Gamma_1} u\phi d\Gamma_1, \quad (D_{\Gamma_1}u,\phi) = \int_{\Gamma_1}[g_0(u) + a_0 u]\phi d\Gamma_1, \quad (3.2.53)$$

$$(B_{\Gamma_1}u,\phi) = \int_{\Gamma_1}\beta u^3\phi d\Gamma_1 \quad \text{for any } \phi \in H^1_{\Gamma_0}(\Omega).$$

Continuing with the abstract setup, we define:

- $F(u,v) = \widetilde{F}(u) + b_0 v + Q_{\Gamma_1}(v)$ with $\widetilde{F}(u) \equiv [u+f, v(u)+F_0] - Lu + p - B_{\Gamma_1} u$, where $v(u)$ solves elliptic problem (3.2.2).
- $D(u) \equiv d_0(x)\,[b(u) + b_0 u] + D_{\Gamma_1} u$.
- $G : U \to \mathscr{H}$, $G(g_1, g_2) \equiv G_1 g_1 + G_2(\partial/\partial\tau) g_2$, where G_i, $i = 1,2$, denote biharmonic extensions of the boundary values defined on Γ_1. More specifically:
  $G_1 v \equiv u$, iff $\Delta^2 u = 0$, in $\Omega, u = 0, \nabla u = 0$ on Γ_0 and

$$\Delta u + (1-\mu)B_1 u = v \text{ on } \Gamma_1, \quad \frac{\partial}{\partial n}\Delta u + (1-\mu)B_2 u = \nu_1 u \text{ on } \Gamma_1;$$

  $G_2 v \equiv u$, iff $\Delta^2 u = 0$, in $\Omega, u = 0, \nabla u = 0$ on Γ_0 and

$$\Delta u + (1-\mu)B_1 u = 0 \text{ on } \Gamma_1, \quad \frac{\partial}{\partial n}\Delta u + (1-\mu)B_2 u = \nu_1 u + v \text{ on } \Gamma_1.$$

The mapping $g : U_0 \mapsto U_0'$ has the form

$$g : (v_1; v_2) \mapsto (g_1(v_1); g_2(v_2)), \quad (v_1; v_2) \in U_0.$$

We now rewrite problem (3.2.1) with boundary conditions (3.2.43) in the abstract form. For this, we notice first that by classical elliptic theory [222]

$$G_1 \in \mathscr{L}(H^s(\Gamma_1) \mapsto H^{s+5/2}(\Omega)) \text{ and } G_2 \in \mathscr{L}(H^s(\Gamma_1) \mapsto H^{s+7/2}(\Omega)) \tag{3.2.54}$$

for every $s \in \mathbb{R}$. Here, again, the regularity of Green's maps owns to the fact that the two portions of the boundary Γ_0 and Γ_1 are disjoint.

In the next step we compute the adjoints of G_i^* (recall that the duality is always understood with respect to L_2 topologies). In order to accomplish this we use Green's formula. By Green's formula (1.3.26), we obtain with an arbitrary $u \in \mathscr{D}(\mathscr{A})$ and $v \in L_2(\Gamma_1)$ that

$$\int_{\Gamma_1} G_1^* \mathscr{A} u v d\Gamma = \int_\Omega \mathscr{A} u G_1 v dx = \int_\Omega \Delta^2 u G_1 v dx - \int_{\Gamma_1} \frac{\partial}{\partial n} u v d\Gamma,$$

where we have used properties of biharmonic extension G_1 together with the fact that elements u satisfy appropriate boundary conditions determined by the membership in $\mathscr{D}(\mathscr{A})$ (see also (1.3.24) for the case $\Gamma_0 = \emptyset$). Hence $G_1^* \mathscr{A} u = (\partial/\partial n) u|_{\Gamma_1}$ for every $u \in \mathscr{D}(\mathscr{A})$. By continuity of $G_1 : L_2(\Gamma_1) \mapsto \mathscr{D}(\mathscr{A}^{5/8-\varepsilon})$ we have that $G_1^* : [\mathscr{D}(\mathscr{A}^{5/8-\varepsilon})]' \mapsto L_2(\Gamma_1)$. Hence the operator $G_1^* \mathscr{A} : \mathscr{D}(\mathscr{A}^{3/8+\varepsilon}) \mapsto L_2(\Gamma_1)$ is also bounded.

The above regularity and the density of domains of fractional powers of the generator $\mathscr{A}$ allows us to extend the domain of $G_1^* \mathscr{A}$. Indeed, we obtain:

$$G_1^* \mathscr{A} u = \frac{\partial}{\partial n} u\Big|_{\Gamma_1}, \quad \forall u \in \mathscr{D}(\mathscr{A}^{3/8+\varepsilon}) \subset H^{3/2+4\varepsilon}(\Omega). \tag{3.2.55}$$

By the above identification of $G_1^* \mathscr{A}$ we easily infer that

$$G_1^* \mathscr{A} : \mathscr{D}(\mathscr{A}^{1/2}) \mapsto H^{1/2}(\Gamma_1) \quad \text{is bounded and surjective.} \tag{3.2.56}$$

Indeed, the above follows from the classical trace theorem [222] which asserts that $(\partial/\partial n) : H^2(\Omega) \mapsto H^{1/2}(\Gamma)$ is bounded.

A similar argument applies to the operator G_2. In fact, by the Green's formula stated in (1.3.26) after accounting for appropriate boundary conditions we obtain (cf. (1.3.25) for the case $\Gamma_0 = \emptyset$):

$$\int_{\Gamma_1} G_2^* \mathscr{A} uvd\Gamma = \int_{\Omega} \mathscr{A} uG_2 vdx = -\int_{\Gamma_1} uvd\Gamma. \tag{3.2.57}$$

Therefore

$$G_2^* \mathscr{A} u = -u|_{\Gamma_1}, \ \forall u \in \mathscr{D}(\mathscr{A}), \tag{3.2.58}$$

which identification can be extended by density to all $u \in H^{1/2+\varepsilon}(\Omega)$. By the standard trace theorem [222]) we obtain that

$$G_2^* \mathscr{A} : H^2_{\Gamma_0}(\Omega) \mapsto H^{3/2}(\Gamma_1) \ \text{ is bounded and surjective.} \tag{3.2.59}$$

Now it is straightforward to verify that

$$G^*(u) = (G_1^* u; -\frac{\partial}{\partial \tau} G_2^* u), \tag{3.2.60}$$

hence

$$G^* \mathscr{A} u = (\frac{\partial}{\partial n} u; \frac{\partial}{\partial \tau} u) = \nabla u|_{\Gamma_1}, \quad \forall u \in \mathscr{D}(\mathscr{A}^{1/2}). \tag{3.2.61}$$

Let $u(t)$ be a smooth solution to problem (3.2.1) and (3.2.43). Then for any $w \in \mathscr{D}(\mathscr{A}^{1/2}) \equiv H^2_{\Gamma_0}(\Omega)$ from Green's formula (1.3.3) and boundary conditions (3.2.43) we find that

$$\begin{aligned} &((I - \alpha\Delta)u_{tt} + \Delta^2 u, w) = (Mu_{tt}, w)_{V,V'} + a(u,w) \\ &+ \int_{\Gamma_1} [\beta u^3 + g_0(u_t)]\, wd\Gamma + \int_{\Gamma_1} \frac{\partial}{\partial n} w g_1(\frac{\partial}{\partial n} u_t) d\Gamma - \int_{\Gamma_1} w \frac{\partial}{\partial \tau} g_2(\frac{\partial}{\partial \tau} u_t) d\Gamma. \end{aligned}$$

Thus using (3.2.61) we obtain

$$\begin{aligned} &((I - \alpha\Delta)u_{tt} + \Delta^2 u, w) = (Mu_{tt}, w)_{V,V'} + (\mathscr{A} u, w)_{\mathscr{D}(\mathscr{A}^{1/2}),[\mathscr{D}(\mathscr{A}^{1/2})]'} \\ &+ (\mathscr{A} Gg(G^* \mathscr{A} u_t), w)_{\mathscr{D}(\mathscr{A}^{1/2}),[\mathscr{D}(\mathscr{A}^{1/2})]'} + \int_{\Gamma_1} [\beta u^3 + g_0(u_t)]\, wd\Gamma. \end{aligned}$$

Consequently with the above notation the abstract model for problem (3.2.1) and (3.2.43) takes the form

$$Mu_{tt} + \mathscr{A} u + Du_t + \mathscr{A} Gg(G^* \mathscr{A} u_t) = F(u, u_t) \tag{3.2.62}$$

or equivalently

$$(I+\alpha A_N)u_{tt}+Du_t+\mathscr{A}\left[u+G_1g_1(\frac{\partial}{\partial n}u_t)+G_2\frac{\partial}{\partial \tau}(g_2(\frac{\partial}{\partial \tau}u_t))\right]=F(u,u_t), \tag{3.2.63}$$

where we have used characterization (3.2.61).

3.2.10. Remark. This last representation in (3.2.63), even for smooth solutions, should be interpreted in the dual space $[H^1_{\Gamma_0}(\Omega)]'$, or even in $[H^2_{\Gamma_0}(\Omega)]'$. This is in contrast with the clamped–hinged case where (3.2.15) has pointwise meaning for smooth solutions. In the clamped–free case the presence of dynamic term in the boundary conditions causes that $u+G_1g_1((\partial/\partial n)u_t)+G_2(\partial/\partial\tau)(g_2((\partial/\partial\tau)u_t))$ (resp., u_{tt}) is not in the domain of $\mathscr{A}$ (resp., A_N) even for smooth solutions. This fact adds some subtlety to PDE calculations performed later. ■

3.2.2.2 Generalized and strong solutions

Now we are in position to introduce concepts of strong and generalized solutions as pertinent to the clamped–free evolution problem.

3.2.11. Definition. A function

$$u\in C(0,T;H^2_{\Gamma_0}(\Omega))\cap C^1(0,T;H^1_{\Gamma_0}(\Omega)) \tag{3.2.64}$$

possessing the properties $u(x,0)=u_0$ and $u_t(x,0)=u_1$ is said to be

1. A *strong solution* to problem (3.2.1) with the clamped–free boundary conditions (3.2.43) on the interval $[0,T]$, iff
   - $u(t)$ is an absolutely continuous function with values in $H^2_{\Gamma_0}(\Omega)$ and $u_t\in L_1(a,b;H^2_{\Gamma_0}(\Omega))$ for any $0<a<b<T$.
   - u_t is an absolutely continuous function with values in $H^1_{\Gamma_0}(\Omega)$ and $u_{tt}\in L_1(a,b;H^1_{\Gamma_0}(\Omega))$ for any $0<a<b<T$.
   - $u(t)+G_1g_1((\partial/\partial n)u_t)+G_2(\partial/\partial\tau)(g_2((\partial/\partial\tau)u_t))\in W$ for almost all $t\in[0,T]$, where

   $$W=\left\{u\in(H^3\cap H^2_{\Gamma_0})(\Omega)\,:\,\Delta u+(1-\mu)B_1u\big|_{\Gamma_1}=0\right\}. \tag{3.2.65}$$

   - Equality (3.2.63) holds in $\left[H^1_{\Gamma_0}(\Omega)\right]'$ for almost all $t\in[0,T]$.
2. A *generalized solution* to problem (3.2.1) and (3.2.43) on the interval $[0,T]$, iff there exists a sequence $\{u_n(t)\}$ of strong solutions to (3.2.1) and (3.2.43) with the initial data $(u_{0n};u_{1n})$ and such that $u_n\to u$ in the space $C(0,T;H^2_{\Gamma_0}(\Omega))\cap C^1(0,T;H^1_{\Gamma_0}(\Omega))$.

■

As before, applying the results from Chapter 2 leads to the following result.

3.2.12. Theorem. *Let $F_0 \in H^{5/2+\delta}(\Omega)$ for some $\delta > 0$. Then under Assumption 3.2.1 with reference to (3.2.1), subject to clamped–free boundary conditions (3.2.43), the following statements are valid with any $T > 0$.*

- **Generalized solutions.** *Assume that $u_0 \in H^2_{\Gamma_0}(\Omega)$ and $u_1 \in H^1_{\Gamma_0}(\Omega)$.*

  1. *For all initial data u_0 and u_1 there exists unique generalized solution $u(t)$ such that the following energy inequality holds,*

$$\mathscr{E}(t) + \int_0^t \left[\int_\Omega d_0 b(u_t) u_t dx + \int_{\Gamma_1} g_0(u_t) u_t d\Gamma \right] d\tau$$
$$\leq \mathscr{E}(0) + \int_0^t ([u, F_0] - Lu, u_t)_\Omega dx ds. \tag{3.2.66}$$

  2. *If $g_i(s)$ possesses the property*

$$ms^{1+\delta} \leq g_i(s)s, \quad |s| \geq s_0,\ i = 1,2, \tag{3.2.67}$$

  for some constants $m > 0$, $\delta > 0$, and $s_0 \geq 0$, then

$$\frac{\partial}{\partial n} u_t\Big|_{\Gamma_1}, \quad \frac{\partial}{\partial \tau} u_t\Big|_{\Gamma_1} \in L_{1+\delta}([0,T] \times \Gamma_1) \tag{3.2.68}$$

  and the energy inequality can be written in the form

$$\mathscr{E}(u(t), u_t(t)) + \int_0^t \left[\int_\Omega d_0 b(u_t) u_t dx + \int_{\Gamma_1} g_0(u_t) u_t d\Gamma \right] d\tau$$
$$+ \int_0^t \int_{\Gamma_1} \left[\psi_1 \left(\frac{\partial}{\partial n} u_t \right) + \psi_2 \left(\frac{\partial}{\partial \tau} u_t \right) \right] d\Gamma d\tau$$
$$\leq \mathscr{E}(u_0, u_1) + \int_0^t \int_\Omega ([u, F_0] - Lu) u_t dx d\tau \tag{3.2.69}$$

  for every pair of continuous convex functions ψ_1 and ψ_2 such that

$$0 \leq \psi_i(s) \leq g_i(s)s, \quad \text{for all} \quad s \in \mathbb{R}. \tag{3.2.70}$$

  3. *If, Assumption 3.2.2 holds, then*

$$\frac{\partial u_t}{\partial n}, \frac{\partial u_t}{\partial \tau} \in L_{q_*}(\Sigma_1), \quad g_1\left(\frac{\partial u_t}{\partial n}\right), g_2\left(\frac{\partial u_t}{\partial \tau}\right) \in L_{q_*/(q_*-1)}(\Sigma_1), \tag{3.2.71}$$

  where $\Sigma_1 = [0,T] \times \Gamma_1$, and the following energy identity holds,

$$\mathscr{E}(t) + \int_0^t \left[\int_\Omega d_0 b(u_t) u_t dx + \int_{\Gamma_1} g_0(u_t) u_t d\Gamma \right] d\tau \tag{3.2.72}$$
$$+ \int_0^t \int_{\Gamma_1} g(\nabla u_t) \nabla u_t d\Gamma ds = \mathscr{E}(0) + \int_0^t \int_\Omega ([u, F_0] - Lu) u_t dx ds,$$

where we denote

$$g(\nabla u_t)\nabla u_t = g_1\left(\frac{\partial}{\partial n}u_t\right)\frac{\partial}{\partial n}u_t + g_2\left(\frac{\partial}{\partial \tau}u_t\right)\frac{\partial}{\partial \tau}u_t \,. \tag{3.2.73}$$

Moreover, $u_{tt} \in L_{q_*/(q_*-1)}(0,T,L_2(\Omega))$ *when* $q_* \geq 2$, *provided either* Γ_0 *is empty, or else the damping functions* g_1 *and* g_2 *are globally Lipschitz (in the latter case* $q_* = 2$*).*

- **Strong solutions.** *We assume that* $u_0 \in H^3(\Omega) \cap H^2_{\Gamma_0}(\Omega)$, $u_1 \in H^2_{\Gamma_0}(\Omega)$, *and, moreover,*

$$\Delta u_0 + (1-\mu)B_1 u_0 = -g_1(\frac{\partial}{\partial n}u_1) \quad on\ \Gamma_1.$$

Then there exists unique strong solution such that

$$\left.\begin{array}{l} w \in C_r(0,T;W)\cap L_\infty(0,T;W), \;\; u \in L_\infty(0,T;H^{5/2}(\Omega)), \\ u_t \in C_r(0,T;H^2_{\Gamma_0}(\Omega))\cap L_\infty(0,T,H^2_{\Gamma_0}(\Omega)), \\ u_{tt} \in C_r(0,T;H^1_{\Gamma_0}(\Omega))\cap L_\infty(0,T;H^1_{\Gamma_0}(\Omega)), \end{array}\right\} \tag{3.2.74}$$

where the space W *is given by (3.2.65) and*

$$w \equiv w(t) = u(t) + G_1 g_1\left(\frac{\partial}{\partial n}u_t(t)\right) + G_2\frac{\partial}{\partial \tau}g_2\left(\frac{\partial}{\partial \tau}u_t(t)\right).$$

Strong solutions satisfy the energy identity in (3.2.72).

3.2.13. Remark. It follows from (3.2.66) that the energy $\mathscr{E}(t)$ of a generalized solution $u(t)$ is monotone nonincreasing provided $L \equiv 0$, $F_0 \equiv 0$, $sb(s) \geq 0$, and $sg_0(s) \geq 0$. However, if we assume that $F_0 \in H^2_{\Gamma_1}(\Omega)$, then (1.4.14) implies that

$$\int_\Omega [u,F_0]u_t dx = \int_\Omega [u,u_t]F_0 dx = \frac{1}{2}\frac{d}{dt}\int_\Omega [u,u]F_0 dx$$

for any strong solution $u(t)$. Thus the energy relation (3.2.72) implies that the functional

$$\widetilde{\mathscr{E}}(t) = \widetilde{\mathscr{E}}(t)(u(t),u_t(t)) = \mathscr{E}(t)(u(t),u_t(t)) - \frac{1}{2}\int_\Omega [u(t),u(t)]F_0 dx$$

is nonincreasing for any strong solution $u(t)$ provided $F_0 \in (H^2_{\Gamma_1} \cap H^{5/2+\delta})(\Omega)$. Passing with a limit on strong solutions the same property remains valid for generalized solutions. It is also clear that in this case the energy relation (3.2.72) can be rewritten with the new energy $\widetilde{\mathscr{E}}(t)$ and without the integral term involving F_0. ∎

3.2.14. Remark. As in the proof of Theorem 3.2.12, under additional hypotheses that g_i *are globally Lipschitz functions*, one can improve the regularity of strong solution u stated in the first line of (3.2.74). In this case we have that the strong solution $u(t)$ belongs to $L_\infty(0,T;H^3(\Omega)) \cap C_r(0,T;H^3(\Omega))$. For the same effect in the clamped–hinged case see Remark 3.2.5. ∎

In order to prove Theorem 3.2.12 we proceed as before, by starting with verification of Assumption 2.4.15 in Theorem 2.4.16, in the context of problem (3.2.62).

3.2.15. Lemma. *Under Assumption* 3.2.1 *operators* $\mathscr{A}, M, D, G, F$, *and* g *introduced above comply with the requirements in Assumption* 2.4.15.

Proof. The verification of the requirements 1 and 2 is the same as before (see the argument given in Lemma 3.1.6 and Lemma 3.2.6).

As for the requirement 3, the only difference with the case of free boundary condition with purely internal damping (see Lemma 3.1.24) is that now we need to prove that the operator Q_{Γ_1} given in (3.2.53) is Lipschitz from $H^1_{\Gamma_0}(\Omega)$ into $\left[H^1_{\Gamma_0}(\Omega)\right]'$. The latter is due to the linear structure of Q_{Γ_1}.

To check Assumption 2.4.15(4) it is sufficient to establish that the operator D_{Γ_1} defined in (3.2.53) is a monotone hemicontinuous operator from $H^1_{\Gamma_0}(\Omega)$ into $\left[H^1_{\Gamma_0}(\Omega)\right]'$. Indeed, by (3.2.3) using embedding $H^{1/2}(\Gamma_1) \subset L_q(\Gamma_1)$ for every $1 \le q < \infty$ and also the trace theorem we have that

$$|(D_{\Gamma_1} v, \phi)| \le \|g_0(v) + a_0 v\|_{\Gamma_1} \|\phi\|_{\Gamma_1} \le C\left(1 + \|v\|^{p_0}_{L_{2p_0}(\Gamma_1)}\right) \|\phi\|_{\Gamma_1}$$
$$\le C\left(1 + \|v\|^{p_0}_{1/2,\Gamma_1}\right) \|\phi\|_{1/2,\Gamma_1} \le C\left(1 + \|v\|^{p_0}_{1,\Omega}\right) \|\phi\|_{1,\Omega}$$

for any $v, \phi \in H^1_{\Gamma_0}(\Omega)$. This implies that D_{Γ_1} is well-defined as a bounded operator from $H^1_{\Gamma_0}(\Omega)$ into $\left[H^1_{\Gamma_0}(\Omega)\right]'$. Obviously, D_{Γ_1} is monotone. The continuity of the mapping

$$\lambda \mapsto \int_{\Gamma_1} [g_0(u + \lambda v) + a_0(u + \lambda v)] v d\Gamma, \quad u, v \in H^1_{\Gamma_0}(\Omega),$$

follows from the polynomial bound (3.2.3) and continuity of g_0. Thus D_{Γ_1} is hemicontinuous. Thus requirements 4 is established.

The requirement 5 in Assumption 2.4.15 concerning G reduces to the fact that the operator $G^* \mathscr{A}^{1/2} : \mathscr{H} \mapsto U_0$ is bounded, where G^* is given by (3.2.60). By (3.2.61) the above translates into

$$G^* \mathscr{A}^{1/2} u = \left(\frac{\partial}{\partial n} \mathscr{A}^{-1/2} u; \frac{\partial}{\partial \tau} \mathscr{A}^{-1/2} u\right)$$

is bounded from $L_2(\Omega)$ into $H^{1/2}(\Gamma_1) \times H^{1/2}(\Gamma_1)$. The last assertion follows from the fact that $\mathscr{D}(\mathscr{A}^{1/2}) \subset H^2(\Omega)$ and the trace theorem which implies that the maps

$$\frac{\partial}{\partial n} : H^2(\Omega) \mapsto H^{1/2}(\Gamma) \text{ and } \frac{\partial}{\partial \tau} : H^2(\Omega) \mapsto H^{1/2}(\Gamma)$$

are bounded. This proves that G satisfies Assumption 2.4.15(5)

Finally, we need to verify requirement 5 concerning g. Because the functions g_1 and g_2 are assumed of a polynomial growth at infinity, the same argument as in Lemma 3.2.6 (see also the proof of Lemma 3.1.8) leads to the continuity of the Nemytskij operator $(g_1(v_1);g_2(v_2))$, as acting between the spaces $H^{1/2}(\Gamma_1)\times H^{1/2}(\Gamma_1)$ and $H^{-1/2}(\Gamma_1)\times H^{-1/2}(\Gamma_1)$. Monotonicity of this mapping g is straightforward. □

3.2.2.3 Proof of Theorem 3.2.12

As in case of clamped–hinged boundary conditions (see Theorem 3.2.4) we split the proof into several steps.

Step 1: Global existence and uniqueness. Validity of Assumption 2.4.15 allows us to apply the result of Theorem 2.4.16. This theorem provides global existence and uniqueness of strong solutions that satisfy energy relation (3.2.72). Due to Remark 2.5.15 the space W has the form (2.5.27). Therefore the smoothness properties in (3.2.74) follow from (2.4.32) and (2.4.33). The relation $u\in L_\infty(0,T;H^{5/2}(\Omega))$ in (3.2.74) follows from: $\nabla u_t|_{\Gamma_1}\in L_\infty(0,T;H^{1/2}(\Gamma_1))$, hence on the strength of polynomial growth conditions in Assumption 3.2.1),

$$g_1(\frac{\partial}{\partial n}u_t),\ \ g_2(\frac{\partial}{\partial \tau}u_t)\in L_\infty(0,T;L_2(\Gamma_1)),$$

and by elliptic regularity (3.2.54)

$$G_1g_1(\frac{\partial}{\partial n}u_t)\in L_\infty(0,T;H^{5/2}(\Omega))\ \text{ and }\ G_2\frac{\partial}{\partial \tau}g_2(\frac{\partial}{\partial \tau}u_t)\in L_\infty(0,T;H^{5/2}(\Omega)).$$

The above argument taken together with the statement in (3.2.74) concerning w implies $H^{5/2}(\Omega)$ regularity of $u(t)$. In the case when g_i are globally Lipschitz we can conclude that $g(\nabla u_t)\in L_\infty([0,T];H^{1/2}(\Gamma_1))$ and thus justify additional $\frac{1}{2}$ derivative of u, as stated in Remark 3.2.14.

Moreover, Theorem 2.4.16 implies the global existence and uniqueness of generalized solutions for *all* initial data $(u_0;u_1)$ from $H^2_{\Gamma_0}(\Omega)\times H^1_{\Gamma_0}(\Omega)$. For this we need to assert the validity of the density condition in (2.4.35) only. To this end we introduce the set

$$\mathscr{L}_0=W\times V_*=W\times\left\{v\in C^\infty(\overline{\Omega})\ :\ v|_{\Gamma_0}=0,\ \frac{\partial}{\partial n}v\Big|_{\Gamma_0\cup\Gamma_1}=0\right\},$$

where W is given by (3.2.65). It is obvious that $\mathscr{L}_0$ is dense in $H=H^2_{\Gamma_0}(\Omega)\times H^1_{\Gamma_0}(\Omega)$. Thus we need to show that $\mathscr{L}_0\subset\mathscr{L}$, where $\mathscr{L}$ is defined in (2.4.35). To see this, we note that $G_1g_1((\partial/\partial n)u_1)=0$ for $u_1\in V_*$ and hence one needs to check that $G_2\left[(\partial/\partial\tau)g_2((\partial/\partial\tau)u_1)\right]\in W$ for any $u_1\in V_*$. Because ∇u_1 belongs to $C^\infty(\Gamma)$ for this case, using the Lipschitz continuity of g_2 (by (3.2.4) in Assumption 3.2.1) one can see that $g_2((\partial/\partial\tau)u_1) \in H^{1/2}(\Gamma_1)$. Therefore (3.2.54) implies the desired conclusion.

In the same way as in the clamped–hinged case (Theorem 3.2.4) from the energy identity (3.2.72) for strong solutions we obtain the a priori bound for strong solutions. This is to say, with T any positive number we have

$$||u(t)||_2 + ||u_t(t)||_1 + ||v(u)||_2 \le C_T \quad \text{for all } t \le T, \tag{3.2.75}$$

where C_T may depend on the bounds for $||u_0||_2, ||u_1||_1, ||f||_2, ||F_0||_{5/2+\delta}$ and $||p||_{-1}$ only. Inequality (3.2.75) is stable with respect to the norm generated by the energy, thus the same a priori bound holds for generalized solutions.

Step 2: Energy inequality and energy identity for generalized solutions.

Energy inequality. In order to obtain energy relations for generalized solutions, we need to pass with the limit on strong solutions. Let $(u_0; u_1)$ be initial data of finite energy and let $u(t)$ be the corresponding solution. From the definition of generalized solutions we know that they are strong limits of strong solutions denoted by $u_n(t)$. We thus have

$$u_n \to u \ \text{ in } C(0,T;H^2(\Omega)), \quad u_{nt} \to u_t \ \text{ in } C(0,T;H^1(\Omega)). \tag{3.2.76}$$

Denoting by $\mathscr{E}_n(t)$ the energy corresponding to strong solutions, we have

$$\mathscr{E}_n(t) + \int_0^t \left[\int_\Omega d_0[b(u_{nt}) + b_0 u_{nt}]u_{nt} dx + \int_{\Gamma_1} [g_0(u_{nt}) + a_0 u_{nt}]u_{nt} d\Gamma \right] d\tau$$
$$+ \int_0^t \int_{\Gamma_1} g(\nabla u_{nt}) \nabla u_{nt} d\Gamma d\tau = \mathscr{E}_n(0) + \int_0^t \int_\Omega ([u_n, F_0] - Lu_n) u_{nt}) dx d\tau$$
$$+ \int_0^t \left[b_0 \int_\Omega d_0 u_{nt}^2 dx + a_0 \int_{\Gamma_1} u_{nt}^2 d\Gamma \right] d\tau.$$

Now we can use the same method as in the clamped–hinged case (see the proof of Theorem 3.2.4) to obtain both inequalities (3.2.66) and (3.2.69) and also relation (3.2.68) under condition (3.2.67).

Energy equality. Assume that Assumption 3.2.2 holds. Then the same monotonicity/continuity argument as in the clamped–hinged case used for justifying (3.2.28) allows us to obtain

$$\int_0^t \int_\Omega d_0(x) b(u_{nt}) u_{nt} dx dt \to \int_0^t \int_\Omega d_0(x) b(u_t) u_t dx dt \tag{3.2.77}$$

and

$$\int_0^t \int_{\Gamma_1} g_0(u_{nt}) u_{nt} dx dt \to \int_0^t \int_{\Gamma_1} g_0(u_t) u_t dx dt, \tag{3.2.78}$$

where $u(t)$ is a generalized solution and $\{u_n(t)\}$ are strong solutions such that (3.2.76) holds.

In order to pass with the limit on the boundary terms g_1 and g_2 we again apply the same type of argument as in the case of clamped–hinged boundary conditions, where the latter is based on monotonicity and Lemma 3.2.7. For the sake of completeness we repeat this argument.

By standard calculations we obtain the following energy inequality.

$$
\begin{aligned}
&||u_n(t) - u_m(t)||_2^2 + ||u_{nt}(t) - u_{mt}(t)||_1^2 \\
&+ \int_0^t \int_{\Gamma_1} \left(g_1(\frac{\partial}{\partial n} u_{nt}) - g_1(\frac{\partial}{\partial n} u_{mt}) \right) \left(\frac{\partial}{\partial n} u_{nt} - \frac{\partial}{\partial n} u_{mt} \right) d\Gamma d\tau \\
&+ \int_0^t \int_{\Gamma_1} \left(g_2(\frac{\partial}{\partial \tau} u_{nt}) - g_2(\frac{\partial}{\partial \tau} u_{mt}) \right) \left(\frac{\partial}{\partial \tau} u_{nt} - \frac{\partial}{\partial \tau} u_{mt} \right) d\Gamma d\tau \\
\leq\ & C\left(||u_n(0) - u_m(0)||_2^2 + ||u_{nt}(0) - u_{mt}(0)||_1^2 \right) \\
&+ C_T(r) \int_0^t \left(||u_n(t) - u_m(t)||_2^2 + ||u_{nt}(t) - u_{mt}(t)||_1^2 \right) d\tau
\end{aligned}
$$

for $||u_n(t)||_2 + ||u_{nt}(t)||_1 \leq r$. Stability estimate (2.4.10) and a priori bounds of strong solutions in (3.2.75) yield for $t \in [0,T]$,

$$
\begin{aligned}
&||u_n(t) - u_m(t)||_2^2 + ||u_{nt}(t) - u_{mt}(t)||_1^2 \\
&+ \int_0^t \int_{\Gamma_1} \left(g_1(\frac{\partial}{\partial n} u_{nt}) - g_1(\frac{\partial}{\partial n} u_{mt}) \right) \left(\frac{\partial}{\partial n} u_{nt} - \frac{\partial}{\partial n} u_{mt} \right) d\Gamma dt \\
&+ \int_0^t \int_{\Gamma_1} \left(g_2(\frac{\partial}{\partial \tau} u_{nt}) - g_2(\frac{\partial}{\partial \tau} u_{mt}) \right) \left(\frac{\partial}{\partial \tau} u_{nt} - \frac{\partial}{\partial \tau} u_{mt} \right) d\Gamma dt \\
\leq\ & C_T \left(||u_n(0) - u_m(0)||_2^2 + ||u_{nt}(0) - u_{mt}(0)||_1^2 \right) \to 0,
\end{aligned}
$$

whenever $||u_n(0) - u_0||_2 \to 0$ and $||u_{nt}(0) - u_1||_1 \to 0$. Thus, in addition to having strong limits (3.2.76) we also have:

$$\int_0^t \int_{\Gamma_1} \left(g_1(\frac{\partial}{\partial n} u_{nt}) - g_1(\frac{\partial}{\partial n} u_{mt}) \right) \left(\frac{\partial}{\partial n} u_{nt} - \frac{\partial}{\partial n} u_{mt} \right) d\Gamma dt \to 0,$$

and

$$\int_0^t \int_{\Gamma_1} \left(g_2(\frac{\partial}{\partial \tau} u_{nt}) - g_2(\frac{\partial}{\partial \tau} u_{mt}) \right) \left(\frac{\partial}{\partial \tau} u_{nt} - \frac{\partial}{\partial \tau} u_{mt} \right) d\Gamma dt \to 0.$$

From the energy equality (3.2.72) for strong solutions we obtain that

$$\int_0^t \int_{\Gamma_1} g_1(\frac{\partial}{\partial n} u_{nt}) \frac{\partial}{\partial n} u_{nt} d\Gamma dt + \int_0^t \int_{\Gamma_1} g_2(\frac{\partial}{\partial \tau} u_{nt}) \frac{\partial}{\partial \tau} u_{nt} d\Gamma dt \leq C(E(0)).$$

Therefore by (3.2.5) from Lemma 3.2.7 we have that there exist a subsequence $\{n_l\}$ and functions w_1 and w_2 from $L_{q_*}(\Sigma_1)$, where $\Sigma_1 = [0,T] \times \Gamma_1$, such that

$$\frac{\partial}{\partial n} u_{n_l t} \to w_1, \quad \frac{\partial}{\partial \tau} u_{n_l t} \to w_2 \quad \text{weakly in } L_{q_*}(\Sigma_1),$$

$$g_1(\frac{\partial}{\partial n} u_{n_l t}) \to g_1(w_1), \quad g_2(\frac{\partial}{\partial \tau} u_{n_l t}) \to g_2(w_2) \quad \text{weakly in } L_{q_*/(q_*-1)}(\Sigma_1)$$

and

$$\int_0^t \int_{\Gamma_1} g(\nabla u_{n_l t}) \nabla u_{n_l t} d\Gamma dt \to \int_0^t \int_{\Gamma_1} (g_1(w_1)w_1 + g_2(w_2)w_2)\, d\Gamma dt,$$

where we use notation (3.2.73). By (3.2.76) $\nabla u_n \to \nabla u$ in $L_2(\Sigma_1)$. Therefore it is standard to prove that $w_1 = (\partial/\partial n)u_t$ and $w_2 = (\partial/\partial\tau)u_t$ in the sense of generalized functions. Thus (3.2.71) holds and

$$\int_0^t \int_{\Gamma_1} g(\nabla u_{n_l t}) \nabla u_{n_l t} d\Gamma dt \to \int_0^t \int_{\Gamma_1} g(\nabla u_t) \nabla u_t d\Gamma dt,$$

when $l \to \infty$. The above convergence along with (3.2.77) and (3.2.78) allows us to pass with the limit on energy identity written first for strong solutions. The limit process yields this identity valid for all generalized solutions.

Step 3: Regularity of the second time derivative. By (3.2.63) we write

$$u_{tt} = -M^{-1}\left[\mathscr{A}\left[u + G_1 g_1(\frac{\partial}{\partial n}u_t)\right] + \mathscr{A}G_2\frac{\partial}{\partial\tau}(g_2(\frac{\partial}{\partial\tau}u_t)) + D(u_t) - F(u,u_t)\right]. \tag{3.2.79}$$

Let $q_* = 2$. From the properties of D and F we easily infer

$$||M^{-1}D(u_t)||_1 \le C_T(E(0)), \quad ||M^{-1}F(u,u_t)||_1 \le C_T(E(0)). \tag{3.2.80}$$

The influence of the boundary damping $g_2((\partial/\partial\tau)u_t)$ is analyzed next. By using relation (3.2.5) with $q_* = 2$ for the function g_2 we obtain

$$||M^{-1}\mathscr{A}G_2\frac{\partial}{\partial\tau}(g_2(\frac{\partial}{\partial\tau}u_t))|| \le C(1 + ||\nabla u_t||_{0,\Gamma_1}). \tag{3.2.81}$$

Indeed, to see the above we first note that $\mathscr{A}G_2(\partial/\partial\tau)g \in W = \mathscr{D}(\mathscr{A}^{3/4})$ when $g \in H^{1/2}(\Gamma_1)$ (see the properties of the Green's operator G_2 and also Remark 2.5.15) and that $M^{-1}w \in V = \mathscr{D}(\mathscr{A}^{1/4})$ for $w \in L_2(\Omega)$. Therefore, using representation (3.2.58) we can write with any $w \in L_2(\Omega)$, $g \in H^{1/2}(\Gamma_1)$,

$$\left|\int_\Omega wM^{-1}\mathscr{A}G_2\frac{\partial}{\partial\tau}g dx\right| = \left|\int_{\Gamma_1}\frac{\partial}{\partial\tau}gG_2^*\mathscr{A}M^{-1}wd\Gamma\right| = \left|\int_{\Gamma_1}g\frac{\partial}{\partial\tau}M^{-1}wd\Gamma\right|.$$

Thus, by elliptic regularity and the trace theory

$$\left|\int_\Omega wM^{-1}\mathscr{A}G_2\frac{\partial}{\partial\tau}g dx\right| \le ||g||_{-1/2,\Gamma_1}||\frac{\partial}{\partial\tau}M^{-1}w||_{1/2,\Gamma_1} \le C||g||_{-1/2,\Gamma_1}||w||. \tag{3.2.82}$$

From (3.2.5) with $q_* = 2$ we have that $|g_2(s)| \le C(1+|s|)$. Therefore inequality in (3.2.82) easily leads to (3.2.81).

By using the condition $g_i(s)s \ge m|s|^2$ for $|s| \ge s_0$ along with the energy equality (3.2.72) we also conclude that

$$\int_0^T \int_{\Gamma_1} |\nabla u_t|^2 d\Gamma dt \equiv \int_0^T \int_{\Gamma_1} \left[|\frac{\partial}{\partial n} u_t|^2 + |\frac{\partial}{\partial \tau} u_t|^2 \right] d\Gamma dt \le C_T(E(0))$$

which combined with (3.2.81) gives

$$\int_0^T ||M^{-1}\mathscr{A} G_2 \frac{\partial}{\partial \tau}(g_2(\frac{\partial}{\partial \tau} u_t))||^2 dt \le C_T(E(0)). \tag{3.2.83}$$

From this point on we consider separately two cases.

1. The case when Γ_0 is empty: In this case $\mathscr{D}(M) \subset \mathscr{D}(\mathscr{A}^{1/2})$ and thus $M^{-1}\mathscr{A}^{1/2}$ is a bounded operator in $L_2(\Omega)$. Hence for any generalized solution

$$||M^{-1}\mathscr{A} u|| \le ||M^{-1}\mathscr{A}^{1/2}||_{\mathscr{L}(L_2(\Omega))} \|\mathscr{A}^{1/2} u\| \le C(E(0)). \tag{3.2.84}$$

This relation also implies that

$$\begin{aligned} ||M^{-1}\mathscr{A} G_1 g_1(\frac{\partial}{\partial n} u_t)|| &\le C\|G_1 g_1(\frac{\partial}{\partial n} u_t)\|_{2,\Omega} \\ &\le C\|g_1(\frac{\partial}{\partial n} u_t)\|_{-1/2,\Gamma_1} \le C(1+||\nabla u_t||_{0,\Gamma_1}) \end{aligned}$$

Returning to the formula (3.2.79) and exploiting the estimates (3.2.80), (3.2.84), and also (3.2.83) yield the final conclusion (when $q_* = 2$ and $\Gamma_0 = \emptyset$):

$$\int_0^T ||u_{tt}||^2 dt \le C_T(E(0)).$$

If $q^* > 2$, we use the same argument as in the clamped–hinged case.

2. *The case when g_1 and g_2 are globally Lipschitz:* In this case we may assume that Γ_0 is nonempty and we need to account for the boundary conditions on the clamped portion of the boundary, which in turn makes the domain $\mathscr{A}^{1/2}$ more restrictive imposing boundary conditions on Γ_0. The inequality (3.2.84) is no longer valid and there is an additional influence of Δu supported on the clamped part of the boundary. This term is not bounded directly by the initial energy. In order to cope with the difficulty we evoke Proposition 3.1.9 applied to Γ_0 portion of the boundary. To proceed, we first establish the following representation for strong solutions u and any $w \in L_2(\Omega)$,

$$(\mathscr{A}(u + G_1 g_1(\frac{\partial}{\partial n} u_t)), M^{-1} w) = a(u, M^{-1} w) - \langle \Delta u, \frac{\partial}{\partial n} M^{-1} w \rangle_{\Gamma_0} \tag{3.2.85}$$

Let u be a strong solution. Then, because g_1 and g_2 are globally Lipschitz functions, by Remark 3.2.14 we conclude that $\hat{u} \equiv u + G_1 g_1((\partial/\partial n) u_t)$ lies in $H^3(\Omega)$ and satisfies the boundary conditions:

$$\hat{u} = \nabla \hat{u} = 0 \ \text{ on } \Gamma_0, \quad \Delta \hat{u} + (1-\mu) B_1 \hat{u} = 0 \ \text{ on } \Gamma_1, \tag{3.2.86}$$

which implies $\hat{u} \in \mathscr{D}(\mathscr{A}^{3/4})$. Thus the element $\hat{u}$ can be approximated in $H^3(\Omega)$ topology by a sequence $u_n \in \mathscr{D}(\mathscr{A})$. On the other hand, the following formula

$$(\mathscr{A}u, M^{-1}w) = a(u, M^{-1}w) - \langle \Delta u, \frac{\partial}{\partial n} M^{-1}w \rangle_{\Gamma_0} \tag{3.2.87}$$

is valid for all $u \in \mathscr{D}(\mathscr{A})$ and $w \in L_2(\Omega)$. Indeed, this follows from standard Green's formula (1.3.3) and from Proposition 1.3.5 which guarantees that $B_1 u = 0$ on Γ_0. Because (3.2.87) is stable with respect to $H^3(\Omega)$ norms, it can be extended by density to hold for $\hat{u}$. This gives:

$$\begin{aligned}(\mathscr{A}(u + G_1 g_1(\frac{\partial}{\partial n} u_t)), M^{-1}w) &= a(u + G_1 g_1(\frac{\partial}{\partial n} u_t), M^{-1}w) \\ &\quad - \langle \Delta u + \Delta G_1 g_1(\frac{\partial}{\partial n} u_t), \frac{\partial}{\partial n} M^{-1}w \rangle_{\Gamma_0}.\end{aligned} \tag{3.2.88}$$

Direct calculations, involving Green's formula (1.3.3) and the definition of the map G_1 (note that functions gave enough of regularity) show that

$$a(G_1 g, \phi) - \langle \Delta G_1 g, \frac{\partial}{\partial n} \phi \rangle_{\Gamma_0} = 0 \quad \text{for all } g \in H^{1/2}(\Gamma_1) \text{ and } \phi \in D(M).$$

This, along with (3.2.88) and after setting $\phi = M^{-1}w \in D(M)$ and $g = g_1((\partial/\partial n)u_t)$ (which belongs to $H^{1/2}(\Gamma_1)$ due to the global Lipschitz property of g_1) gives the representation in (3.2.85).

In order to estimate the boundary integral in (3.2.85) we exploit Theorem 2.5.6 which provides the needed trace estimate

$$\int_{\Sigma_0} |\Delta u|^2 d\Sigma_0 \le C_T(E(0)) + C_T \int_0^T ||F(u, u_t) + D(u_t)||^2_{-1,\Omega} dt \le C_T((E(0)) \tag{3.2.89}$$

where in the last step we have used (3.2.80). Combining (3.2.85) with (3.2.89) and noting that $||(\partial/\partial n)M^{-1}w|| \le C||w||$, $M^{-1} \in \mathscr{L}(L_2(\Omega) \to H^2(\Omega))$ gives:

$$\int_0^T ||M^{-1}\mathscr{A}(u + G_1 g_1(\frac{\partial}{\partial n} u_t))||^2 dt \le C_T(E(0)). \tag{3.2.90}$$

Returning to (3.2.79) we see that the property $u_{tt} \in L_2((0,T) \times \Omega)$ follows now by combining (3.2.80), (3.2.83), and (3.2.90). This completes the proof of Theorem 3.2.12.

3.2.16. Remark. In principle, to establish L_2 regularity of the derivative u_{tt} in the case when $\Gamma_0 \neq \emptyset$ we can also use the so-called localization method which does not require that both g_1 and g_2 are globally Lipschitz functions. Below we sketch the main idea behind this method.

We first decompose solution u into two parts $u = z + v$ with z supported near Γ_0 and v supported near Γ_1. This can be achieved via localization, owning to the fact that both parts of the boundary are separated. Let $\phi(x) \in C^\infty(\Omega)$ be a localizer such that $\phi(x) = 1$ in the neighborhood of Γ_1 and zero near Γ_0. We define $v = u\phi$ and $z = u(1-\phi)$.

It is straightforward to verify that the new variable v satisfies zero free boundary conditions on Γ_0 and free boundary conditions given by (3.2.43) on Γ_1. Thus this is a "free boundary condition" on both sides of the boundary In addition the variable v satisfies the equation:

$$v_{tt} = -M^{-1}\left[\mathscr{A}v + \mathscr{A}Gg(\nabla u_t) + \phi D(u_t) - \phi F(u,u_t) + f_v\right],$$

with $f_v \equiv -\{\Delta^2,\phi\}u + \alpha\{\Delta,\phi\}u_{tt}$, where the brackets $\{\cdot,\cdot\}$ stand for the commutators of differential operators, $\{A,B\} = AB - BA$, and the operators M, $\mathscr{A}$, G are appropriately modified to account for zero free (rather than clamped) boundary conditions imposed on Γ_0. We also note that the boundary term $g(\nabla u_t)$ is not affected by the localization, because $\phi = 1$ near Γ_1. Thus, for the variable v we can apply the same analysis as in the case of Γ_0 empty set.

The variable z is supported near the boundary Γ_0 and vanishes near Γ_1. Thus z satisfies clamped zero boundary conditions on the entire boundary $\Gamma = \Gamma_0 \cup \Gamma_1$ ($z = \nabla z = 0$ on Γ). The equation satisfied by z is the following

$$z_{tt} = -M^{-1}\left[\mathscr{A}z + (1-\phi)D(u_t) - (1-\phi)F(u,u_t) + f_z\right],$$

with $f_z \equiv -\{\Delta^2, 1-\phi\}u + \alpha\{\Delta, 1-\phi\}u_{tt}$, where the operators $M.\mathscr{A}$ correspond to clamped boundary conditions imposed on entire Γ. Thus we can apply the same argument as in the proof of Theorem 3.1.4 (see also Remark 3.1.5) to obtain to the relation $\|M^{-1}\mathscr{A}z\| \le C\left(\|z\|_2 + \|\Delta z\|_{\Gamma_0}\right)$. To estimate the boundary term we can use Theorem 2.5.4; see also Proposition 3.1.9.

One can see that the main difficulty in the application of localization method is related to the terms $\alpha M^{-1}\{\Delta,\phi\}u_{tt}$ and $\alpha M^{-1}\{\Delta, 1-\phi\}u_{tt}$ in the forces f_v and f_z. It is easy to show that these terms can be bounded by the H^{-1}-norm of the value ψu_{tt} for some $\psi \in C_0^\infty(\Omega)$ (this function ψ is produced by the commutators). Thus we can reduce the original problem concerning L_2 norm of u_{tt} to the question of estimating ψu_{tt} in the negative Sobolev space $H^{-1}(\Omega)$. The latter task can be accomplished by performing another localization in the interior of the domain and away from the both parts of the boundary (so the boundary conditions do not enter the analysis). However we do not perform detailed calculations because in further considerations we use the corresponding result in the case of globally Lipschitz damping functions g_1 and g_2 only. ∎

3.2.2.4 Weak solutions

As before, (cf. Theorem 3.2.8) , the following result emerges for weak solutions.

3.2.17. Theorem. *In addition to Assumption* 3.2.1 *assume that the functions* $g_1(s)$ *and* $g_2(s)$ *either possess the property*

$$(s_1 - s_2)(g_i(s_1) - g_i(s_2)) \geq c_0|s_1 - s_2|^r, \quad s_1, s_2 \in \mathbb{R},\ i = 1,2, \tag{3.2.91}$$

for some $c_0 > 0$ *and* $r \geq 1$*, or else satisfy Assumption 3.2.2. Then, the generalized solutions of von Karman equation (3.2.1) with clamped–free boundary conditions (3.2.43) are also weak solutions; that is, the functions* $t \mapsto (d_0 b(u_t), \phi)_\Omega$, $t \mapsto (g_0(u_t), \phi)_{\Gamma_1}$*, and* $t \mapsto (g(\nabla u_t), \nabla\phi)_{\Gamma_1}$ *are locally integrable on* $\mathbb{R}_+$ *for* $\phi \in H^2(\Omega)$ *and the following variational relation*

$$\begin{aligned}
&(u_t(t), \phi)_\Omega - \alpha(\nabla u_t(t), \nabla\phi)_\Omega - (u_1, \phi)_\Omega + \alpha(\nabla u_1, \nabla\phi)_\Omega \\
&+ \int_0^t [a(u,\phi) + (d_0 b(u_t), \phi)_\Omega + (g_0(u_t), \phi)_{\Gamma_1}]\, ds \\
&+ \int_0^t \left[(g_1(\frac{\partial}{\partial n} u_t), \frac{\partial}{\partial n}\phi)_{\Gamma_1} + (g_2(\frac{\partial}{\partial \tau} u_t), \frac{\partial}{\partial \tau}\phi)_{\Gamma_1}\right] ds \\
&+ \int_0^t [(-[v(u) + F_0, u + f] + Lu - p, \phi)_\Omega + (\beta u^3, \phi)_{\Gamma_1}]\, ds = 0
\end{aligned}$$

holds for any $\phi \in H^2_{\Gamma_0}(\Omega)$ *where* $v = v(u)$ *is defined from* (3.2.2).

The proof of this theorem goes along the same lines as in the hinged-clamped case; see Theorem 3.2.8. For comments concerning the requirement in (3.2.91) we refer to Remark 3.2.9.

3.2.3 Regular solutions

Under the additional differentiability properties of the nonlinear damping functions and additional regularity properties of initial data we can prove existence of regular-classical solutions.

For simplicity of presentation we consider the clamped–hinged case only and restrict ourselves by the case of linear boundary dissipation. Thus, we consider problem (3.2.1) with the following boundary conditions

$$u = \nabla u = 0 \ \text{ on } \Gamma_0, \quad u = 0,\ \Delta u = -g_1 \frac{\partial}{\partial n} u_t \ \text{ on } \Gamma_1, \tag{3.2.92}$$

where $g_1 \geq 0$ is a constant. Our main hypothesis in this section is the following one.

3.2.18. Assumption.

- Assumption 3.2.1 concerning b, f, F_0, p, and L holds.
- $b \in C^\infty(\mathbb{R})$ and its derivatives satisfy the relations

$$|b^{(k)}(\sigma)| \leq C_k(1 + |\sigma|^{q_k}), \quad k = 1,2,\ldots, \tag{3.2.93}$$

  for some constants $C_k > 0$ and $q_k \geq 1$.

Problem (3.2.1) with boundary conditions (3.2.92) can be written in the form (cf. equation (3.2.15) in Section 3.2.1):

$$Mu_{tt} + \widetilde{D}u_t + \mathscr{A}\left[u + g_1 \cdot G\left(\frac{\partial}{\partial n} u_t\right)\right] = \widetilde{F}(u), \tag{3.2.94}$$

where M, $\mathscr{A}$, G and $\widetilde{F}$ are the same as in Section 3.2.1 and $\widetilde{D}v = d_0(x)b(v)$.

Let us define the values $u^{(k)}(0)$ by the recurrence relations

$$\begin{cases} u^{(0)}(0) = u_0, \quad u^{(1)}(0) = u_1, \\ u^{(k)}(0) = -M^{-1}\mathscr{A}\left[u^{(k-2)}(0) + g_1 \cdot G\left(\frac{\partial}{\partial n} u^{(k-1)}\right)\right] \\ \qquad + M^{-1}\frac{d^{k-2}}{dt^{k-2}}\left[-\widetilde{D}(u_t) + \widetilde{F}(u)\right]\Big|_{t=0}, \end{cases} \tag{3.2.95}$$

where $k \geq 2$. We also denote by $\mathscr{L}_m$, $m > 1$, the class of functions that possess the following properties

$$\begin{cases} u^{(k)} + g_1 \cdot G\left(\frac{\partial}{\partial n} u^{(k+1)}\right) \in C(0,T;W), \quad k = 0,1,\ldots,m-1, \\ u^{(m)} \in C(0,T;(H^2_{\Gamma_0} \cap H^1_0)(\Omega)), \quad u^{(m+1)} \in C(0,T;H^1_0(\Omega)), \end{cases} \tag{3.2.96}$$

for any $T > 0$, where $H^2_{\Gamma_0}(\Omega)$ is given by (3.2.11) and W is defined by (3.2.17).

3.2.19. Theorem. *Assume that Assumption* 3.2.18 *holds. Then a generalized solution to problem* (3.2.1) *with boundary conditions* (3.2.92) *belongs to* $\mathscr{L}_m$ *for* $m \geq 1$ *if and only if the following compatibility conditions*

$$\begin{cases} u^{(k)}(0) + g_1 \cdot G\left(\frac{\partial}{\partial n} u^{(k+1)}(0)\right) \in W \quad \text{for} \quad k = 0,1,2,\ldots,m-1, \\ u^{(m)}(0) \in H^2_{\Gamma_0}(\Omega) \cap H^1_0(\Omega), \quad u^{(m+1)}(0) \in H^1_0(\Omega), \end{cases} \tag{3.2.97}$$

hold, where the values $u^{(k)}(0)$ are defined by the recurrence relations (3.2.95).

Proof. We use the same method as in the proof of Theorem 3.1.28 by relying on the properties of the corresponding linearized problem (see Theorems 2.4.29 and 2.4.35). In order to focus reader's attention on the technicalities related to the presence of the boundary damping, we present only the first iteration of the argument: the improvement of regularity by one time derivative (this corresponds to the case $m = 1$).[4]

Thus we are led to study the equation (3.2.94). If (3.2.97) holds for $m = 1$, then $u(t)$ is a strong solution Therefore $w \equiv u_t$ satisfies

---

[4] This result does not follow from the second part of Theorem 3.2.4 which gives us the right-continuity property (3.2.24) only

$$Mw_{tt} + d_0(x)b'(u_t)w_t + \mathscr{A}(w + g_1 G(\frac{\partial}{\partial n} w_t)) = \langle \tilde{F}'(u), w \rangle. \tag{3.2.98}$$

The above equation is nonautonomous, but linear in the variable w. We are thus in a position to apply linear theory. The operator $F(t) \equiv \tilde{F}'(u(t))$ complies with the requirements of Assumption 2.4.32. It is also clear that the operator $D_1(t)v \equiv d_0(x)b'(u_t)v$ is bounded from $H_0^1(\Omega)$ into $L_2(\Omega)$. Thus the damping operator in our case satisfies Assumption 2.4.32 with $D_0 \equiv 0$. Therefore, by Theorem 2.4.35 in order to conclude that

$$w \in C(0,T; H^2_{\Gamma_0}(\Omega) \cap H_0^1(\Omega)), \;\; w_t \in C(0,T; H_0^1(\Omega))$$

one needs to verify that $w(0) = u_t(0) \in \mathscr{D}(\mathscr{A}^{1/2}) = H^2_{\Gamma_0}(\Omega) \cap H_0^1(\Omega)$ and $w_t(0) = u_{tt}(0) \in H_0^1(\Omega)$, where

$$u_{tt}(0) = M^{-1}\left[-\tilde{D}u_t(0) - \mathscr{A}\left(u(0) + g_1 G(\frac{\partial}{\partial n} u_t(0)\right) + \tilde{F}(u(0)) \right].$$

The above is satisfied because the conditions in (3.2.97) holds for $m = 1$. Thus

$$u_t \in C(0,T; H^2_{\Gamma_0}(\Omega) \cap H_0^1(\Omega)), \;\; u_{tt} \in C(0,T; H_0^1(\Omega))$$

and hence it follows directly from (3.2.94) that $u + g_1 G(\frac{\partial}{\partial n} u_t) \in C(0,T; W)$. Thus Theorem 3.2.19 is proved for $m = 1$. In order to obtain the result for arbitrary $m \geq 1$ one should reiterate this procedure in the same way as it was done in the proof of Theorem 3.1.28. □

In conclusion we note that using regularity properties of the mapping $\{f; h\} \mapsto u$ generated by the elliptic problem

$$\Delta^2 u = f \in H^{-1}(\Omega),$$
$$u = \nabla u = 0 \;\text{ on } \Gamma_0, \quad u = 0, \; \Delta u = h \in H^{1/2}(\Gamma_1) \;\text{ on } \Gamma_1,$$

one can show that if $u \in \mathscr{L}_1$, then $u \in C([0,T], H^3(\Omega))$. As in the linear case (see, e.g., Proposition 2.5.2) this regularity can be easily reiterated by taking the subsequent time derivatives. This procedure leads to additional spatial regularity of solutions to the problem (3.2.1) and (3.2.92).

3.3 Other models with rotational inertia

The methods and ideas developed in previous sections of this chapter can be also applied to different types of perturbations/modifications of the von Karman model with rotational inertia as long as additional terms satisfy appropriate local Lipschitz conditions and suitable a priori bounds. Our goal in this section is to present several such (physically motivated) models which are based on von Karman evolution equations accounting for rotational terms. These include: retarded models, models with a memory and quasi-static models. Because our main focus in the book is the standard (basic) von Karman equations, our presentation is short and does not pretend to be a survey of all approaches related to the models described below.

3.3.1 Models with delay

We first consider the following von Karman system with (linear) retarded terms,

$$(1-\alpha\Delta)\partial_t^2 u+(d_1-\alpha d_2\Delta)\partial_t u+\Delta^2 u$$
$$= [u+f,v+F_0]-Lu+p(u^t;t), \quad x\in\Omega, t>0, \tag{3.3.1}$$

where, as above, $v=v(u)$ is a solution of the problem

$$\Delta^2 v+[u+2f,u]=0, \quad v|_{\partial\Omega}=\frac{\partial v}{\partial n}\Big|_{\partial\Omega}=0. \tag{3.3.2}$$

We assume that the plate is clamped; that is,

$$u|_{\partial\Omega}=\frac{\partial u}{\partial n}\Big|_{\partial\Omega}=0. \tag{3.3.3}$$

We assume that f and F_0 are given functions specified below and $d_1,d_2\geq 0$ are constants. If the plate is located in a potential linearized flow of gas then the aerodynamic pressure of the flow can be taken into account (see [54, 168] and the references therein and also Chapter 6 below) by assuming that

$$p(u^t;t)=p_0(x)+q(u^t;t), \tag{3.3.4}$$

where $p_0(x)\in L^2(\Omega)$, u^t is an element from $L^2(-r,0;H^2(\Omega))$ defined by the formula $u^t(s)=u(t+s)$, $s\in(-r,0)$, and $q(\cdot;t)$ is a continuous linear mapping from $L^2(-r,0;H^2(\Omega))$ into $H^{-1}(\Omega)$ possessing the property

$$\| q(u^t,t)\|_{-1,\Omega}^2\leq C\int_{t-r}^{t}\| u(\tau)\|_{2,\Omega}^2\,d\tau. \tag{3.3.5}$$

The parameter $r>0$ is the time of retardation and it depends on velocity of the unperturbed flow and on the size of the domain Ω. The retarded character of the problem requires the initial conditions of functional type

$$u\big|_{t\in(-r,0)}=\varphi(x,t), \quad u|_{t=0}=u_0(x), \quad \partial_t u|_{t=0}=u_1(x). \tag{3.3.6}$$

We assume that

$$\varphi(x,t)\in L_2(-r,0;H_0^2(\Omega)), \quad u_0(x)\in H_0^2(\Omega), \quad u_1(x)\in H_0^1(\Omega). \tag{3.3.7}$$

We also suppose that $L : H_0^2(\Omega) \mapsto H^{-1}(\Omega)$ is a linear bounded operator and the functions f, F_0 and p_0 in (3.3.1) and (3.3.4) possess properties

$$f(x) \in H^2(\Omega), \quad F_0(x) \in H^2(\Omega), \quad p_0(x) \in L_2(\Omega). \tag{3.3.8}$$

As in Section 3.1 we define a *weak solution* (on the interval $[0,T]$) to problem (3.3.1)–(3.3.6) as a function

$$u(t) \in L_2(-r,T;H_0^2(\Omega)) \cap L_\infty(0,T;H_0^2(\Omega)) \cap W_\infty^1(0,T;H_0^1(\Omega))$$

such that $u\big|_{t\in(-r,0)} = \varphi(x,t)$, $u|_{t=0} = u_0(x)$ and variational relation (3.1.33) holds with $p = p(u^t;t)$ and the corresponding changes in the damping terms.

Using the Galerkin method we obtain the following assertion (which was originally proved in [54]).

3.3.1. Theorem. *Under the assumptions above problem* (3.3.1)–(3.3.4) *with the initial data* (3.3.7) *has a unique weak solution on any interval* $[0,T]$. *This solution belongs to the class* $C(0,T;H_0^2(\Omega)) \cap C^1(0,T;H_0^1(\Omega))$ *and satisfies the energy relation*

$$\begin{aligned} &\mathscr{E}(u(t),u_t(t)) + \int_s^t \int_\Omega \left[d_1 u_t^2 + \alpha d_2 |\nabla u_t|^2\right] dx d\tau \\ &= \mathscr{E}(u(s),u_t(s)) + \int_s^t \int_\Omega (-Lu + p(u^\tau;\tau)) u_t dx d\tau. \end{aligned}$$

Here

$$\mathscr{E}(u,u_t) = \frac{1}{2}\int_\Omega \left[|u_t|^2 + \alpha|\nabla u_t|^2 + |\Delta u|^2 + \frac{1}{2}|\Delta v(u)|^2 - [u+2f,F_0]u \right] dx,$$

where $v(u) \in H_0^2(\Omega)$ *is determined from* (3.3.2). *Moreover, there exist constants* $C > 0$ *and* $a \geq 0$ *such that*

$$E(u(t),u_t(t)) \leq C\left(1 + E(u_0,u_1) + \int_{-r}^0 \|\varphi(\tau)\|_2^2 d\tau\right) \cdot e^{at}, \quad t > 0, \tag{3.3.9}$$

where

$$E(u,u_t) = \frac{1}{2}\int_\Omega \left[|u_t|^2 + \alpha|\nabla u_t|^2 + |\Delta u|^2 + \frac{1}{2}|\Delta v(u)|^2 \right] dx. \tag{3.3.10}$$

Proof. Let $\{e_k\}$ be the basis in $H_0^1(\Omega)$ consisting of eigenfunctions of the spectral boundary value problem

$$\Delta^2 v = \lambda(1 - \alpha\Delta)v, \qquad v|_{\partial\Omega} = \frac{\partial v}{\partial n}\Big|_{\partial\Omega} = 0. \tag{3.3.11}$$

We consider Galerkin m-order approximate solutions of the form

$$u_m(t) = \sum_{k=1}^{m} g_k(t) e_k,$$

where the scalar functions $g_k(t)$ are chosen such that

$$\begin{aligned} &((1-\alpha\Delta)\partial_t^2 u_m + (d_1 - \alpha d_2 \Delta)\partial_t u_m, e_j) + (\Delta u_m, \Delta e_j) \\ &= ([u_m + f, v(u_m) + F_0] - Lu_m + p(u_m^t; t), e_j) \quad j = 1, 2, \ldots, m,\ t > 0, \end{aligned}$$

and

$$\begin{aligned} &(u_m(0), e_j) = (u_0, e_j),\ (\partial_t u_m(0), e_j) = (u_1, e_j), \\ &(u_m(s), e_j) = (\varphi(s), e_j),\ s \in (-r, 0),\ j = 1, 2, \ldots, m. \end{aligned}$$

A local theorem of existence of approximate solutions can be obtained if we rewrite the corresponding ODE system with delay for g_k in integral form and use the method of successive approximations (see, e.g., [133]).

Using the symmetry of the von Karman bracket it is easy to obtain the energy relation for approximate solutions:

$$\begin{aligned} &\mathscr{E}(u_m(t), \partial_t u_m(t)) + 2\int_0^t \left[d_1 \|\partial_t u_m\|^2 + \alpha d_2 \|\nabla \partial_t u_m\|^2\right] d\tau \\ &= \mathscr{E}(u_m(0), \partial_t u_m(0)) + \int_0^t (-Lu + p(u_m^\tau; \tau), \partial_t u_m) d\tau. \end{aligned}$$

The above along with (3.3.5) and the relation between $\mathscr{E}$ and E given by (3.1.9) lead to the estimate

$$\begin{aligned} E(u_m(t), \partial_t u_m(t)) \le C_1 &\left(1 + E(u_0, u_1) + \int_{-r}^{0} \|\varphi(\tau)\|_2^2 d\tau\right) \\ &+ \int_0^t E(u_m(\tau), \partial_t u_m(\tau)) d\tau, \end{aligned} \tag{3.3.12}$$

where $E(u, u_t)$ is defined by (3.3.10). Therefore by Gronwall's lemma we find

$$\|\partial_t u_m\|^2 + \alpha \|\nabla \partial_t u_m\|^2 + \|\Delta u_m\|^2 + \|\Delta v(u_m)\|^2 < C_T, \tag{3.3.13}$$

for any t from an interval $[0, T)$ of the existence of the approximate solution. Therefore it follows from [133, Theorem 2.3.2] that the approximate solution u_m exists on $\mathbb{R}_+$. Relation (3.3.13) also implies that the sequence of approximate solutions $\{(u_m; \partial_t u_m)\}$ is $*$-weakly compact in $L_\infty(0, T; H_0^2(\Omega) \times H_0^1(\Omega))$. Now applying the standard method (see, e.g., [220]) we can prove the existence of a weak solution $u(t)$ to problem (3.3.1)–(3.3.4) with the initial data (3.3.7).

Considering now $u(t)$ as a weak solution to the linear problem of the form

$$(1-\alpha\Delta)\partial_t^2 u + \Delta^2 u = F(t),\ u|_{\partial\Omega} = \left.\frac{\partial u}{\partial n}\right|_{\partial\Omega} = 0, \tag{3.3.14}$$

with

$$F(t) = -(d_1 - \alpha d_2 \Delta)\partial_t u + [u+f, v+F_0] - L(u) + p(u^t;t) \in L_\infty(0,T;H^{-1}(\Omega))$$

and using Proposition 2.5.1 it is easy to prove the continuity properties of the weak solution and to establish the energy relation. Uniqueness of weak solutions to problem (3.3.1)–(3.3.4) and (3.3.7) follows from the energy relation for linear problem (3.3.14). Estimate (3.3.9) follows from (3.3.12). □

3.3.2 *Models with memory*

We consider von Karman equations with long-range memory term. Usually these terms model viscoelastic materials; that is, materials whose stresses at any instant depend on the complete history of strains that the material has undergone (see, e.g., [173]). We consider the following equation

$$(1-\alpha\Delta)\partial_t^2 u + \Delta^2(u + G[u](t)) = [u+f, v+F_0] - Lu + p(x) \qquad (3.3.15)$$

for $x \in \Omega$ and $t > 0$, where, as above, $v = v(u)$ solves (3.3.2). The viscoelastic operator G has either the form (see, e.g., [173]):

$$G_\infty[w](t) = -\int_{-\infty}^{t} g(t-s)w(x,s)ds$$

or the form (cf. [230]):

$$G_0[w](t) = -\int_0^t g(t-s)w(x,s)ds.$$

Here $g(s)$ is a scalar function possessing the properties

$$g, g', g'' \in L^1(0,\infty), \quad g(t) \geq 0,\ g'(t) \leq 0, \quad \beta = 1 - \int_0^\infty g(\tau)d\tau > 0.$$

The standard example of this function g is $g = \gamma_1 e^{-\gamma_2 s}$ with $0 < \gamma_1 < \gamma_2$.

The structure of the viscoelastic operator G_∞ requires us to consider equation (3.3.15) as a problem with infinite delay. However, sometimes it is convenient to represent the value $G_\infty[\Delta^2 u](t)$ in the form

$$G_\infty[\Delta^2 u](t) = G_0[\Delta^2 u](t) + p(x,t),$$

where $p(x,t) = -\int_{-\infty}^0 g(t-s)\Delta^2 u(x,s)ds$. Since we can consider $p(x,t)$ as a given additional transverse load, it is possible to put G_0 and additional (nonautonomous) load $p(x,t)$ into (3.3.15) instead of G_∞. In this case it is not necessary to assume that the initial data for problem (3.3.15) are given for all $t \in (-\infty, 0]$. Moreover $G_0 \equiv G_\infty$ in the case $u(x,t)|_{t\in(-\infty,0)} \equiv 0$.

In order to have a more focused presentation we consider equation (3.3.15) with the following clamped–free type boundary conditions defined on two disjoint portions Γ_0, Γ_1 of the boundary $\Gamma = \Gamma_0 \cup \Gamma_1$:

$$u = \nabla u = 0 \ \text{ on } \Gamma_0 \tag{3.3.16}$$

$$\Delta(u + G[u]) + (1-\mu)B_1(u + G[u]) = 0 \ \text{ on } \Gamma_1, \tag{3.3.17}$$

$$\frac{\partial}{\partial n}\Delta(u + G[u]) + (1-\mu)B_2(u + G[u]) - \alpha\frac{\partial}{\partial n}u_{tt} = 0, \ \text{ on } \Gamma_1, \tag{3.3.18}$$

where B_1 and B_2 are given by (1.3.20).

We consider the case $G \equiv G_\infty$. As above for given initial data

$$u(x,t)|_{t\in(-\infty,0)} = w(x,t), \quad u|_{t=0} = u_0(x), \quad \partial_t u|_{t=0} = u_1(x), \tag{3.3.19}$$

we can define a *weak solution* $u(t)$ on interval $[0,T]$ to problem (3.3.15)–(3.3.19) as an element from

$$L_\infty(-\infty,T;H^2_{\Gamma_0}(\Omega)) \cap W^1_\infty(0,T;H^1_{\Gamma_0}(\Omega))$$

such that $u\big|_{t\in(-\infty,0)} = w(x,t)$, $u|_{t=0} = u_0(x)$ and the corresponding variational relation holds. Here and below $H^2_{\Gamma_0}(\Omega)$ is defined by (3.2.11) and

$$H^1_{\Gamma_0}(\Omega) = \{u \in H^1(\Omega),\ u = 0 \text{ on } \Gamma_0\}.$$

One can prove the following existence and uniqueness theorem.

3.3.2. Theorem. *Let* (3.3.8) *be valid and*

$$w(x,t) \in L_\infty(-\infty,0;H^2_{\Gamma_0}(\Omega)) \quad u_0(x) \in H^2_{\Gamma_0}(\Omega), \quad u_1(x) \in H^1_{\Gamma_0}(\Omega).$$

We also suppose $L : H^2_{\Gamma_0}(\Omega) \mapsto \left[H^1_{\Gamma_0}(\Omega)\right]'$ *is a linear bounded operator. Then problem* (3.3.15)–(3.3.19) *has a unique weak solution on any interval* $[0,T]$*. This solution belongs to the class* $C(0,T;H^2_{\Gamma_0}(\Omega)) \cap C^1(0,T;H^1_{\Gamma_0}(\Omega))$.

This theorem is proved in [230] by using the compactness method in the case $w \equiv 0$, $f \equiv F_0 \equiv 0$ and $L \equiv 0$. However the same method can be applied for the case considered above. Under some compatibility conditions it is also possible to prove the existence of smooth solutions. We refer to [230] for details.

3.3.3 Quasi-static model

In the case when the inertia forces are small in comparison with the resisting forces of a medium, it is natural to consider the problem of the oscillations of a plate in a quasi-static formulation. For the clamped plate with linear damping this leads to the system:

$$\begin{cases} (1-\alpha\Delta)\partial_t u + \Delta^2 u - [u+f, v+F_0] + Lu = p(x), \\ u|_{\partial\Omega} = \left.\frac{\partial u}{\partial n}\right|_{\partial\Omega} = 0, \quad u|_{t=0} = u_0(x), \end{cases} \tag{3.3.20}$$

where $v = v(u)$, the Airy's stress function, solves the problem (3.3.2).

By *weak solution* of problem (3.3.20) on the interval $[0,T]$ we mean the function

$$u(t) \in L_\infty(0,T;H_0^1(\Omega)) \cap L_2(0,T;H_0^2(\Omega))$$

which is weakly continuous in $H_0^1(\Omega)$ and satisfies relations (3.3.20) in the sense of generalized functions.

3.3.3. Theorem. *Let* (3.3.8) *be valid and* $u_0 \in H_0^1(\Omega)$. *We also suppose that* $L : H_0^2(\Omega) \mapsto H^{-1}(\Omega)$ *is a linear bounded operator. Then problem* (3.3.20) *on the interval* $[0,T]$ *has a unique weak solution* $u(t)$ *that belongs to the class*

$$C^1([0,T];H_0^1(\Omega)) \cap C((0,T];H_0^2(\Omega)).$$

Moreover, the solution $u(t)$ *possesses the properties.*

- *There exist positive constants* C *and* β *such that*

$$\|u(t)\|^2 + \alpha\|\nabla u(t)\|^2 \le (\|u_0\|^2 + \alpha\|\nabla u_0\|^2)\, e^{-\beta t} + C\left(1 - e^{-\beta t}\right), \quad t \ge 0. \tag{3.3.21}$$

- *For any positive* R *and* T *there exists a constant* $C_{R,T}$ *such that*

$$\|u(t)\|_2 \le C_{R,T} t^{-1/2} \quad \text{for all } t \in (0,T],\ \|u_0\|_1 \le R. \tag{3.3.22}$$

Proof. Let $\{e_k\}$ be the basis in $H_0^1(\Omega)$ consisting of eigenfunctions of the spectral boundary value problem (3.3.11). We consider Galerkin m-order approximate solutions of the form

$$u_m(t) = \sum_{k=1}^{m} g_k(t) e_k,$$

where the scalar functions $g_k(t)$ are chosen such that

$$((1-\alpha\Delta)\partial_t u_m, e_j) + (\Delta u_m, \Delta e_j) = ([u_m + f, v(u_m) + F_0] - Lu_m + p, e_j), \quad t > 0,$$

and $(u_m(0), e_j) = (u_0, e_j)$ for $j = 1, 2, \dots, m$.

A local theorem of existence of approximate solutions follows from the standard argument. Using the von Karman bracket properties it is easy to obtain the following relation,

$$\frac{d}{dt}(\|\partial_t u_m(t)\|^2 + \alpha\|\nabla\partial_t u_m(t)\|^2) + \|\Delta u_m(t)\|^2 + a_0\|\Delta v(u_m(t))\|^2 \le C,$$

where a_0 and C are positive constants. This relation allows us as to prove the compactness of approximate solutions. The subsequent arguments are the usual ones and can be found in [49]. □

Chapter 4
Von Karman Equations Without Rotational Inertia

This chapter treats well-posedness of solutions to von Karman models that do not account for rotational terms. The absence of rotational inertia terms raises several questions related to generation of well-posed flows, making the subject more challenging. This is due to the lack of regularizing effect on the velocity s expressed by the term $\alpha\Delta u_{tt}$ in (3.1.1). In fact, in the absence of this term the issue of uniqueness of finite energy solutions has been an open problem in the literature. It turns out that the key to solvability of this problem is sharp regularity of Airy's stress function (see Corollary 1.4.5). In fact, this sharp regularity property allows us to use abstract results formulated in Chapter 2. The proof of well-posedness for the model without rotational forces follows from the abstract results in Theorems 2.4.5 and 2.4.16.

4.1 Models with interior dissipation

We study von Karman evolutions without rotational forces and with nonlinear internal dissipation. For $x \in \Omega$ and $t > 0$ we consider:

$$\begin{cases} u_{tt} + d_0(x)g_0(u_t) + \Delta^2 u - [u+f, v+F_0] + Lu = p(x), \\ u|_{t=0} = u_0(x), \quad u_t|_{t=0} = u_1(x), \end{cases} \tag{4.1.1}$$

where $v = v(u)$ the Airy's stress function is a solution of the problem

$$\Delta^2 v + [u+2f, u] = 0, \quad v|_{\partial\Omega} = \frac{\partial v}{\partial n}\Big|_{\partial\Omega} = 0. \tag{4.1.2}$$

We study this problem under the following hypotheses.

4.1.1. Assumption.

- The function $g_0 : \mathbb{R}^1 \mapsto \mathbb{R}^1$ is continuous and there exists $a_0 \geq 0$ such that $g_0(s) + a_0 s$ is increasing. The function $d_0(x)$ is a nonnegative bounded measurable function. We also assume (without loss of generality) that $g_0(0) = 0$.

I. Chueshov, I. Lasiecka, *Von Karman Evolution Equations*,
Springer Monographs in Mathematics, DOI 10.1007/978-0-387-87712-9_4,

- We suppose that for some $\delta > 0$ we have

$$f \in H^2(\Omega), \quad F_0 \in H^{3+\delta}(\Omega), \quad p \in L_2(\Omega). \tag{4.1.3}$$

- L is a linear bounded operator from $H^2(\Omega)$ into $L_2(\Omega)$.

Boundary conditions associated with problem (4.1.1) and (4.1.2) are of clamped, hinged, or free type:

1. **[clamped]**: $u = \nabla u = 0$ on $\Gamma \equiv \partial\Omega$.
2. **[hinged]**: $u = \Delta u = 0$ on Γ.
3. **[free]**: $\Delta u + (1-\mu)B_1 u = 0$ on Γ, and

$$\frac{\partial}{\partial n}\Delta u + (1-\mu)B_2 u = \nu_1 u + \beta(x)(u^3 + cu_t) \quad \text{on } \Gamma,$$

where B_1 and B_2 are given by (1.3.20), $\nu_1 > 0$ and $\beta(x) \geq 0$ is a bounded measurable function, the constant c is positive. The term $\beta(x)cu_t$ represents frictional force acting on the boundary.

4.1.2. Remark. Boundary conditions introduced above can be easily generalized by incorporating mixed boundary conditions acting on two disjoints parts of the boundary Γ_0 and Γ_1 (these were already considered in Chapters 2 and 3). By doing so one could treat various combinations of boundary conditions such as clamped–free, hinged–clamped or hinged–free boundary conditions. As long as the partition of the boundary is disjoint (so the issues of the loss of elliptic regularity will not enter the picture), the technical treatment of the corresponding mixed problem is a natural generalization of a single-type boundary value problem. In order to keep the exposition focused, we limit ourselves in this section to single-type boundary conditions. Only later in this chapter, when considering boundary damping, we formulate the problem in the context of a mixed boundary value problem. ∎

Historically, the existence and uniqueness of solutions to von Karman equations (particularly for models without rotational inertia) has been studied in the context of strong solutions. Indeed, global existence and uniqueness of strong solutions has been known since [51], for the case of linear dissipation and clamped boundary conditions. Similar and related results were later proved in [166, 278]. Also, existence of weak solutions for models with linear dissipation has been shown in [279] (see also [220] and [237]) by means of the Faedo–Galerkin method. The problem that was open for a long time is that of *uniqueness* of weak solutions [279, 220], as well as of *continuous dependence on initial data*. This problem has been settled by proving additional regularity of Airy's stress function; see Corollary 1.4.5 in Chapter 1. In fact, rather unexpected regularity of Airy's function allows us (as we show in the next sections) to use the general abstract framework of Chapter 2 in order to establish the well-posedness. On the other hand, in view of recent developments in the area, there is also another way of asserting well-posedness of rotation-free solutions to von Karman equations. Indeed, uniqueness of weak solutions was shown in [29] by exploiting Sedenko's method (see [255, 256]). This method, however, leads to energy estimates at the lower topological level. Hence, it does not provide continuous dependence on the initial data. On the other hand, as shown in [164], once the energy identity for weak solutions is established, a soft argument and the already obtained uniqueness also implies continuous dependence of finite energy weak solutions with respect to the initial data. In conclusion, the result in [29] supplemented with the method in [164] gives an alternative way of proving full Hadamard well-posedness of weak solutions. This method does not depend on sharp regularity of the Airy stress function. We refer to Appendix A.3 for a detailed discussion of Sedenko's method in the context of von Karman evolutions.

4.1.1 Clamped boundary conditions

We start with the most natural initial boundary value problem associated with (4.1.1) and equipped with clamped boundary conditions:

$$u = \nabla u = 0 \quad \text{on } \Gamma \equiv \partial\Omega. \tag{4.1.4}$$

To describe the energy of the model we introduce the following functional,

$$\mathscr{E}(t) \equiv \mathscr{E}(u,u_t) = \frac{1}{2}\int_\Omega \left(|u_t|^2 + |\Delta u|^2\right) dx + \Pi(u), \tag{4.1.5}$$

where

$$\Pi(u) = \frac{1}{4}\int_\Omega \left[|\Delta v(u)|^2 - 2[u+2f,F_0]u - 4up\right] dx$$

and $v(u) \in H_0^2(\Omega)$ is determined from (4.1.2). The functional $\Pi(u)$ is related to the potential energy of the system and can be presented in the form

$$\Pi(u) = \frac{1}{4}\int_\Omega |\Delta v(u)|^2 dx + \Pi_1(u), \tag{4.1.6}$$

where

$$\Pi_1(u) = -\frac{1}{2}\int_\Omega \left([u+2f,F_0]u + 2up\right) dx.$$

As in Chapter 3 we also use another energy variable that consists of a positive part of $\mathscr{E}$:

$$E(t) \equiv E(u,u_t) = \frac{1}{2}\int_\Omega \left[|u_t|^2 + |\Delta u|^2 + \frac{1}{2}|\Delta v(u)|^2\right] dx. \tag{4.1.7}$$

Obviously we have the relation $\mathscr{E}(u,u_t) = E(u,u_t) + \Pi_1(u)$ and, as in Section 3.1.1 we have that for any $\eta > 0$ there exists C_η such that

$$|\Pi_1(u)| \le \eta \int_\Omega \left[|\Delta u|^2 + \frac{1}{2}|\Delta v(u)|^2\right] dx + C_\eta, \quad u \in (H^2 \cap H_0^1)(\Omega). \tag{4.1.8}$$

Hence there exist positive constants c_0, c_1, and K such that

$$c_0\mathscr{E}(u,u_t) - K \le E(u,u_t) \le c_1\mathscr{E}(u,u_t) + K, \quad (u;u_t) \in (H^2 \cap H_0^1)(\Omega) \times L_2(\Omega). \tag{4.1.9}$$

In view of inequalities (4.1.9) the physical energy $\mathscr{E}$ is topologically equivalent to E. On the other hand, it follows from (4.1.7) and regularity of the Airy function $v(u)$ (see Corollary 1.4.5) that the topology generated by $E(u,u_t)$ corresponds to that of $H^2(\Omega) \times L_2(\Omega)$. Thus, the velocity has one less derivative than in the case of rotational forces included in the model.

Similar to the case with rotational forces (cf. Section 3.1.1) we introduce notions of strong and generalized solutions.

4.1.3. Definition. A function

$$u(t) \in C(0,T;H_0^2(\Omega)) \cap C^1(0,T;L_2(\Omega)),$$

possessing the properties $u(x,0) = u_0(x)$ and $u_t(x,0) = u_1(x)$ is said to be

1. A *strong solution* to problem (4.1.1) and (4.1.2) with the clamped boundary conditions (4.1.4) on the interval $[0,T]$, iff
   - $u(t)$ is an absolutely continuous function with values in $H_0^2(\Omega)$ and $u_t \in L_1(a,b;H_0^2(\Omega))$ for any $0 < a < b < T$.
   - u_t is an absolutely continuous function with values in $L_2(\Omega)$ and $u_{tt} \in L_1(a,b;L_2(\Omega))$ for any $0 < a < b < T$.
   - $u(t) \in H^4(\Omega) \cap H_0^2(\Omega)$ for almost all $t \in [0,T]$.
   - Equation (4.1.1) is satisfied in $L_2(\Omega)$ for almost all $t \in [0,T]$ with $v = v(u)$ defined by (4.1.2).
2. A *generalized solution* to problem (4.1.1), (4.1.2), and (4.1.4) on the interval $[0,T]$, iff there exists a sequence $\{u_n(t)\}$ of strong solutions to (4.1.1), (4.1.2), and (4.1.4) with $(u_{0n};u_{1n})$ instead of $(u_0;u_1)$ such that

$$\lim_{n\to\infty} \max_{t\in[0,T]} \left\{ \|\partial_t u(t) - \partial_t u_n(t))\|_\Omega + \|u(t) - u_n(t)\|_{2,\Omega} \right\} = 0. \tag{4.1.10}$$

■

We have the following theorem.

4.1.4. Theorem. *Under Assumption* 4.1.1 *with reference to* (4.1.1)*, subject to clamped boundary conditions* (4.1.4)*, the following statements are valid with any* $T > 0$.

- **Generalized solutions:** *For all initial data* $u_0 \in H_0^2(\Omega), u_1 \in L_2(\Omega)$ *there exists a unique generalized solution* $u(t)$ *such that the following energy inequality holds for* $s \le t$,

$$\mathscr{E}(u(t),u_t(t)) + \int_s^t \int_\Omega d_0 g_0(u_t) u_t dx d\tau \le \mathscr{E}(u(s),u_t(s)) - \int_s^t (Lu,u_t)_\Omega d\tau. \tag{4.1.11}$$

Thus, if $L \equiv 0$ and $sg_0(s) \geq 0$ for all $s \in \mathbb{R}$ the energy of the system is nonincreasing: $\mathscr{E}(u(t), u_t(t)) \leq \mathscr{E}(u(s), u_t(s))$, $t > s$.
Assuming, in addition, that $g_0(s)s \leq Ms^2$ for $|s| \geq 1$ we also obtain the energy identity

$$\mathscr{E}(u(t), u_t(t)) + \int_s^t \int_\Omega d_0 g_0(u_t) u_t dx d\tau = \mathscr{E}(u(s), u_t(s)) - \int_0^t \int_\Omega L u u_t dx d\tau. \tag{4.1.12}$$

- **Strong solutions:** *Assume that $u_0 \in H^4(\Omega) \cap H_0^2(\Omega)$ and $u_1 \in H_0^2(\Omega)$. Then, for any $T > 0$, there exists a unique strong solution such that*

$$\left. \begin{aligned} & u \in C_r(0,T;(H^4 \cap H_0^2)(\Omega)) \cap L_\infty(0,T;(H^4 \cap H_0^2)(\Omega)), \\ & u_t \in C_r(0,T;H_0^2(\Omega)) \cap L_\infty(0,T;H_0^2(\Omega)), \\ & u_{tt} \in C_r(0,T;L_2(\Omega)) \cap L_\infty(0,T;L_2(\Omega)), \end{aligned} \right\} \tag{4.1.13}$$

where $C_r(0,T;X)$ is the space of strongly right-continuous functions with values in X. Strong solutions satisfy the energy identity in (4.1.12). *If $g_0(s)$ is continuously differentiable, then*

$$u \in C(0,T;(H^4 \cap H_0^2)(\Omega)), \quad u_t \in C(0,T;H_0^2(\Omega)), \quad u_{tt} \in C(0,T;L_2(\Omega)). \tag{4.1.14}$$

4.1.5. Remark. In order to derive energy equality for generalized solutions, a linear bound at infinity imposed on the damping is assumed. This restriction can be significantly relaxed, provided more information on the structure of the damping is given. For instance, damping that is polynomially bounded from below and above, in the sense that $m|s|^p \leq g_0(s)s \leq M|s|^p$ for $p > 1$ and $|s| \geq s_0$, allows for a limit passage on the monotone term, a critical step for the derivation of energy identity. The argument relies on the corresponding analogue of Lemma 3.2.7 and is similar to ones given in the proofs of Theorems 3.2.4 and 3.2.12 devoted to the boundary dissipation case in the presence of rotational inertia. ∎

4.1.6. Remark. Exceptional regularity of traces of solutions on the boundary also takes place for models without rotational inertia. In fact, in the case of clamped boundary conditions, it follows from Theorem 2.5.4 and from the energy inequality (4.1.11) that

$$\int_\Sigma |\Delta u|^2 d\Sigma \leq C(E(0), \|f\|_2, \|F_0\|_{3+\delta}, \|p\|)$$

provided $g_0(u_t) \in L_1(0,T;H^{-1}(\Omega))$. This latter condition is guaranteed by suitable growth conditions imposed on the damping g. We note that the regularity stated above provides $\frac{1}{2}$ derivative more on the boundary than the classical trace theorem would (even if applied only formally). ∎

Proof of Theorem 4.1.4

In order to prove Theorem 4.1.4 we rewrite the von Karman equations as a second-order abstract equation and we apply Theorem 2.4.16. In order to accomplish this we introduce the following spaces and operators,

- $\mathscr{H} \equiv L_2(\Omega)$.
- $\mathscr{A}u \equiv \Delta^2 u,\ u \in \mathscr{D}(\mathscr{A}) \equiv H_0^2(\Omega) \cap H^4(\Omega),\ \mathscr{D}(\mathscr{A}^{1/2}) = H_0^2(\Omega)$.
- $F(u,v) = \widetilde{F}(u) + d_0 a_0 v$ with $\widetilde{F}(u) \equiv [u+f, v(u)+F_0] - Lu + p$, where $v(u)$ satisfies the elliptic equation in (4.1.2).
- $D(u) \equiv d_0(x)\,[g_0(u) + a_0 u]$.

With the above notation, the abstract form of the equation becomes:

$$u_{tt} + \mathscr{A}u + D(u_t) = F(u, u_t).$$

Thus we arrive at the simplified model (2.4.26) with $M \equiv I$, $G \equiv 0$, and $g \equiv 0$.
We verify the validity of Assumption 2.4.15 in abstract Theorem 2.4.16.

4.1.7. Lemma. *The operators $\mathscr{A}, D$, and F introduced above comply with the requirements in Assumption* 2.4.15 *with $M = I$ and $V = \mathscr{H} = L_2(\Omega)$.*

Proof. 1. *Hypotheses* 1 *and* 2 *in Assumption* 2.4.15: We note that with the definition of $\mathscr{A}$ as above, $\mathscr{A}$ is closed, positive, densely defined on $\mathscr{H}$ with $\mathscr{D}(\mathscr{A}^{1/2}) = H_0^2(\Omega)$. Moreover, because $V = L_2(\Omega) = \mathscr{D}(M^{1/2}) = V'$ and $[\mathscr{D}(\mathscr{A}^{1/2}]' = H^{-2}(\Omega)$, we have dense continuous injections:

$$\mathscr{D}(\mathscr{A}^{1/2}) \subset V = \mathscr{H} = V' \subset [\mathscr{D}(\mathscr{A}^{1/2})]'.$$

This shows requirements 1 and 2 in Assumption 2.4.15 are satisfied.

2. *Hypothesis* 3 *in Assumption* 2.4.15: This hypothesis requires that the nonlinear term $\widetilde{F}(u)$ is locally Lipschitz from $H_0^2(\Omega)$ into $L_2(\Omega)$. It is at this point where sharp regularity of the Airy stress function plays a major role. Note that in the case $\alpha > 0$ (rotational forces accounted) a much weaker condition was needed. In this latter case it was enough to show Lipschitz regularity with values in $H^{-1}(\Omega)$ (see Chapter 3).

By using regularity assumed in (4.1.3) we verify first the required regularity for the terms involving f and F_0. By Corollary 1.4.5

$$||v(u)||_{W_\infty^2(\Omega)} \le C[||u||_{2,\Omega}^2 + ||u||_{2,\Omega}||f||_{2,\Omega}]$$

hence $||[f, v(u)]||_\Omega \le C[||u||_{2,\Omega}^2 + ||u||_{2,\Omega}||f||_{2,\Omega}]||f||_{2,\Omega}$. Similarly, we obtain

$$||[u, F_0]||_\Omega \le C||u||_{2,\Omega}||F_0||_{W_\infty^2(\Omega)} \le C||u||_{2,\Omega}||F_0||_{3+\delta,\Omega},$$

where we have used Sobolev's embedding (1.1.5). The regularity of the term $Lu - p$ is straightforward. Thus $F : H^2(\Omega) \mapsto L^2(\Omega)$.

We verify next the Lipschitz property for the nonlinear term $[u,v(u)]$. The key result used is in Corollary 1.4.5. In fact, from (1.4.26) of Corollary 1.4.5 applied with $\delta = 0$,

$$||[u_1,v(u_1)]-[u_2,v(u_2)]||_\Omega \le C(||u_1||^2_{2,\Omega}+||u_2||^2_{2,\Omega}+||f||^2_{2,\Omega})||u_1-u_2||_{2,\Omega},$$

which inequality gives the desired local Lipschitz property for the nonlinear term.

It is also clear that F has the form (2.4.28) with $F^*(u,v) = -Lu + a_0 d_0 v$ and $\Pi(u)$ given by (4.1.6). The estimate (2.4.30) is the same as (4.1.8). Obviously, by Assumption 4.1.1 $F^*(u,v)$ satisfies (2.4.29). Thus requirement 3 in Assumption 2.4.15 is satisfied.

3. *Hypothesis* 4 *in Assumption* 2.4.15: Here the argument is identical to that in the proof of Lemma 3.1.6. For the reader's convenience we repeat the steps.

In order to assert continuity of $D(u) : H_0^2(\Omega) \mapsto H^{-2}(\Omega)$ we exploit Sobolev's embedding. For $u \in H_0^2(\Omega) \subset C(\overline{\Omega})$, we have $g_0(u) \in C(\overline{\Omega}) \subset L_2(\Omega) \subset H^{-2}(\Omega)$ with obvious continuity of the operator $u \mapsto g_0(u)$. This argument proves continuity of the map $D : H_0^2(\Omega) \to H^{-2}(\Omega)$. As for monotonicity, we simply write:

$$\begin{aligned}&(D(u_1)-D(u_2),u_1-u_2)_{H_0^2,H^{-2}}\\ &= \int_\Omega d_0(x)(g_0(u_1)-g_0(u_2)+a_0(u_1-u_2))(u_1-u_2)dx \ge 0\end{aligned}$$

and the conclusion follows from the monotonicity property imposed on g_0. □

We have shown that our standing Assumption 2.4.15 holds. Moreover, the set $\mathscr{W}$ defined in (2.4.39) is $\mathscr{D}(\mathscr{A}^{1/2}) = H_0^2(\Omega)$ and hence is dense in $V = L_2(\Omega)$. Thus due to Remark 2.4.17 the density condition holds, so we are in a position to apply the result of Theorem 2.4.16. This theorem implies all the assertions of Theorem 3.1.4 except for: (i) energy inequality (4.1.11), (ii) the additional regularity of strong solutions in (4.1.14), and (iii) the energy relation (4.1.12) for generalized solutions. We concentrate on these three items now.

Energy inequality (4.1.11) can be established in the same way as in Theorem 3.1.4 by regularizing the damping product $(a_0 u_t + g(u_t))u_t$ with the help of the function $\psi_N(s) = \min\{s,N\}$.

The regularity of strong solutions can be asserted quickly by noticing that M is an identity operator on $L_2(\Omega)$ and $\mathscr{D}(\mathscr{A}^{1/2}) = H_0^2(\Omega)$. Moreover, to establish relations in (4.1.14) we also use the obvious fact that $w(t) = u_t(t)$ solves some linear problem which satisfies hypotheses of Theorem 2.4.35.

As for energy relation (4.1.12) for generalized solutions it follows from Proposition 2.4.21 and the next assertion.

4.1.8. Lemma. *Let Assumption* 4.1.1 *be valid. Assume additionally that*

$$sg_0(s) \le Ms^2 \quad \textit{for all } |s| \ge 1,$$

where M is a positive constant. Then the operator D defined by the formula $D(u) \equiv d_0(x)[g_0(u) + a_0 u]$ is an m-accretive hemicontinuous mapping from $L_2(\Omega)$

into itself and the relation

$$\sup\{\|Dv\|_\Omega \,:\, v\in L_2(\Omega),\ \|v\|_\Omega \le \rho\} < \infty \tag{4.1.15}$$

holds for any $\rho > 0$.

Proof. The argument is the same as in the proof of Lemma 3.1.8. □

The proof of Theorem 4.1.4 is thus completed.

4.1.2 Hinged boundary conditions

Now we consider problem (4.1.1) and (4.1.2) in the case of hinged boundary conditions

$$u = \Delta u = 0 \quad \text{on } \Gamma \equiv \partial\Omega. \tag{4.1.16}$$

4.1.9. Definition. A function

$$u \in C([0,T]; H_0^1(\Omega)\cap H^2(\Omega))\cap C^1([0,T];L_2(\Omega)) \tag{4.1.17}$$

possessing the properties $u(x,0) = u_0$ and $u_t(x,0) = u_1$ is said to be

1. A *strong solution* to problem (4.1.1) and (4.1.2) with the hinged boundary conditions (4.1.16) on the interval $[0,T]$, iff
   - $u(t)$ is an absolutely continuous function with values in $H^2(\Omega)\cap H_0^1(\Omega)$ and $u_t \in L_1(a,b;H^2(\Omega)\cap H_0^1(\Omega))$ for any $0 < a < b < T$.
   - u_t is an absolutely continuous function with values in $L_2(\Omega)$ and $u_{tt} \in L_1(a,b;L_2(\Omega))$ for any $0 < a < b < T$.
   - $u(t) \in W$ for almost all $t \in [0,T]$, where

   $$W = \left\{w \in H^4(\Omega)\cap H_0^1(\Omega) \,:\, \Delta w|_{\partial\Omega} = 0\right\}. \tag{4.1.18}$$

   - Equation (4.1.1) is satisfied in $L_2(\Omega)$ for almost all $t \in [0,T]$ with $v = v(u)$ defined according (4.1.2).
2. A *generalized solution* to problem (4.1.1), (4.1.2), and (4.1.16) on the interval $[0,T]$, iff there exists a sequence $\{u_n(t)\}$ of strong solutions to (4.1.1), (4.1.2), and (4.1.16) with $(u_{0n};u_{1n})$ instead of $(u_0;u_1)$ such that (4.1.10) holds.

■

In this case we have a similar well-posedness result as for clamped homogeneous boundary conditions (cf. Theorem 4.1.4). A precise statement is given below.

4.1.10. Theorem. *Under Assumptions* 4.1.1 *with reference to* (4.1.1), *subject to hinged boundary conditions* (4.1.16), *the following statements are valid with any* $T > 0$.

- **Generalized solutions.** *For all initial data* $u_0 \in H_0^1(\Omega) \cap H^2(\Omega), u_1 \in L_2(\Omega)$ *there exists a unique generalized solution* $u(t)$ *such that the following energy inequality holds for all generalized solutions and* $s \le t$,

$$\mathscr{E}(u(t), u_t(t)) + \int_s^t \int_\Omega d_0 g_0(u_t) u_t dx d\tau \le \mathscr{E}(u(s), u_t(s)) \quad \int_s^t (Lu, u_t)_\Omega d\tau,$$

where the energy $\mathscr{E}$ *is defined by* (4.1.5). *Thus, if* $L \equiv 0$ *and* $s g_0(s) \ge 0$ *for all* $s \in \mathbb{R}$, *then* $\mathscr{E}(u(t), u_t(t)) \le \mathscr{E}(u(s), u_t(s))$ *for* $t > s$.
Assuming, in addition, that $g_0(s)s \le Ms^2$ *for* $|s| \ge 1$,[1] *we also obtain the energy identity*

$$\mathscr{E}(u(t), u_t(t)) + \int_s^t \int_\Omega d_0 g_0(u_t) u_t dx d\tau = \mathscr{E}(u(s), u_t(s)) - \int_s^t \int_\Omega L u u_t dx d\tau. \tag{4.1.19}$$

- **Strong solutions.** *Assume that* $(u_0; u_1) \in W \times (H^2 \cap H_0^1)(\Omega)$, *where* W *is given by* (4.1.18). *Then there exists a unique strong solution such that*

$$\left. \begin{aligned} & u \in C_r(0, T; W) \cap L_\infty(0, T; W), \\ & u_t \in C_r(0, T; (H^2 \cap H_0^1)(\Omega)) \cap L_\infty(0, T; (H^2 \cap H_0^1)(\Omega)), \\ & u_{tt} \in C_r(0, T; L_2(\Omega)) \cap L_\infty(0, T; L_2(\Omega)), \end{aligned} \right\}$$

where as above $C_r(0, T; X)$ *is the space of strongly right-continuous functions. Strong solutions satisfy the energy identity in* (4.1.19). *If* $g_0(s)$ *is* C^1, *then*

$$u \in C([0, T]; W), \quad u_t \in C([0, T]; (H^2 \cap H_0^1)(\Omega)), \quad u_{tt} \in C(0, T; L_2(\Omega)).$$

4.1.11. Remark. Similarly, as in the clamped case, one can obtain the exceptional regularity of traces of solutions on the boundary $\Sigma = \Gamma \times (0, T)$:

$$\|\frac{\partial}{\partial n} u_t\|_\Sigma + \|\frac{\partial}{\partial n} \Delta u\|_\Sigma \le C \sup_{t \in [0, T]} (\|u(t)\|_{3, \Omega} + \|u_t(t)\|_{1, \Omega}) + C_T(E(0), f, F_0, p)$$

provided, for instance, the damping function g_0 is linearly bounded. The above regularity result can be deduced from Theorem 2.5.9. Tangential rescaling of this estimate (i.e., application of tangential operator of anisotropic order 1 to both sides of the equation) leads to lower level estimates (in terms of the energy of solution $E(t)$). This step, more technical, is not pursued any further. ∎

[1] As in the clamped case this assumption can be relaxed; see Remark 4.1.5.

Proof of Theorem 4.1.10

The proof of this theorem parallels the arguments given for Theorem 4.1.4. We indicate the points where differences occur. Our first step is to rewrite the von Karman equation as a second-order abstract equation and then to apply Theorem 2.4.5. In order to accomplish this we introduce the following spaces and operators.

- $\mathscr{H} \equiv L_2(\Omega)$.
- $\mathscr{A} u \equiv \Delta^2 u, u \in \mathscr{D}(\mathscr{A}) \equiv \{u \in H^4(\Omega),\ u = \Delta u = 0 \text{ on } \Gamma\}$.
- $\mathscr{D}(\mathscr{A}^{1/2}) = H^2(\Omega) \cap H_0^1(\Omega)$.
- $F(u,v) = \widetilde{F}(u) + d_0 a_0 v$, $\widetilde{F}(u) \equiv [u+f, v(u)+F_0] - Lu + p$, $v(u)$ solves (4.1.2).
- $D(u) \equiv d_0(x)\,[g_0(u) + a_0 u]$.

With the above notation the abstract form of problem (4.1.1) and (4.1.16) becomes

$$u_{tt} + \mathscr{A} u + D(u_t) = F(u, u_t),$$

and the arguments needed for the proof are identical to these in Theorem 4.1.4. This, in particular includes verification of Assumption 2.4.15 in Theorem 2.4.16.

4.1.3 Free boundary conditions

We consider now problem (4.1.1) equipped with the following boundary conditions which are referred to as "free,"

$$\Delta u + (1-\mu)B_1 u = 0, \quad \frac{\partial}{\partial n}\Delta u + (1-\mu)B_2 u = \nu_1 u + \beta(x)(u^3 + c u_t), \quad (4.1.20)$$

on $\Gamma \equiv \partial\Omega$, where B_1 and B_2 are given by (1.3.20) and ν_1 and c are positive parameters, and $\beta \in L_\infty(\Gamma)$, $\beta(x) \geq 0$.

The energy functional $\mathscr{E}$ in the free case has a different form. It is defined by the formula

$$\mathscr{E}(t) \equiv \mathscr{E}(u, u_t) = \frac{1}{2}\int_\Omega |u_t|^2\,dx + \frac{1}{2}a(u,u) + \Pi(u),$$

where

$$a(u,w) = a_0(u,w) + \nu_1 \int_\Gamma u w\, d\Gamma$$

with $a_0(u,w)$ defined by (1.3.4) and

$$\Pi(u) = \int_\Omega \left[\frac{1}{4}|\Delta v(u)|^2 - [f, F_0]u - up\right] dx + \frac{1}{4}\int_\Gamma \beta u^4 d\Gamma.$$

Here $v(u) \in H_0^2(\Omega)$ is determined from (4.1.2). In what follows we also use another energy variable which consists of a positive part of $\mathscr{E}$. That is:

$$E(t) \equiv E(u, u_t) = \frac{1}{2}\int_\Omega |u_t|^2 dx + \frac{1}{2}a(u,u) + \frac{1}{4}\int_\Omega |\Delta v(u)|^2 dx + \frac{1}{4}\int_\Gamma \beta u^4 d\Gamma.$$

As in Section 3.1 (cf. (3.1.59)) we can conclude that there exist positive constants c_0, c_1, and K such that

$$c_0\mathscr{E}(u,u_t) - K \le E(u,u_t) \le c_1\mathscr{E}(u,u_t) + K, \quad (u;u_t) \in H^2(\Omega)\times L_2(\Omega). \tag{4.1.21}$$

In the free case, the following functional analytic setup applies:

- $\mathscr{H} \equiv L_2(\Omega)$, $Z \equiv L_2(\Gamma)$.
- $\mathscr{A}u \equiv \Delta^2 u$, $u \in \mathscr{D}(\mathscr{A})$, where $\mathscr{D}(\mathscr{A}^{1/2}) = H^2(\Omega)$ and

$$\mathscr{D}(\mathscr{A}) = \left\{ u \in H^4(\Omega) \;\middle|\; \begin{array}{l} [\Delta u + (1-\mu)B_1 u]\Big|_{\partial\Omega} = 0, \\ \left[\frac{\partial}{\partial n}\Delta u + (1-\mu)B_2 u - \nu_1 u\right]\Big|_{\partial\Omega} = 0 \end{array} \right\}. \tag{4.1.22}$$

- $F(u,v) = \widetilde{F}(u) + d_0 a_0 v$ with $\widetilde{F}(u) \equiv [u+f, v(u)+F_0] - Lu + p$, $v(u)$ solves (4.1.2).
- $D(u) \equiv d_0(x)\,[g_0(u) + a_0 u]$.
- If $\beta \neq 0$, then the effect of a nonlinear boundary condition can be built into operator $D_0 h(u)$, which coincides with operator B_{Γ_3} introduced in the proof of Theorem 1.5.16. To see this we define the linear operator $D_0 : L_2(\Gamma) \to [\mathscr{D}(\mathscr{A}^{1/2})]'$ by the following variational formula,

$$\int_\Omega D_0 u\phi dx = \int_\Gamma \sqrt{c\beta} u\phi d\Gamma, \quad \forall \phi \in H^2(\Omega). \tag{4.1.23}$$

It is clear that the adjoint (dual) operator $D_0^* : \mathscr{D}(\mathscr{A}^{1/2}) \mapsto L_2(\Gamma)$ is the trace type operator: $D_0^*\phi = \sqrt{c\beta}\phi|_\Gamma$. The nonlinear function $h : H^2(\Omega) \to L_2(\Gamma)$ is then defined as $h(u) \equiv \sqrt{c^{-1}\beta} u^3|_\Gamma$.

We show that the original PDE model is equivalent to the operator model:

$$u_{tt} + \mathscr{A}u + D_0 h(u) + D_0 D_0^* u_t + D(u_t) = F(u,u_t). \tag{4.1.24}$$

To see this, we recall (see 1.3.3)) the following form of Green's formula applied to strong solutions

$$\begin{aligned}\int_\Omega \Delta^2 uw dx &= a_0(u,w) \\ &+ \int_\Gamma \left[\left(\frac{\partial}{\partial n}\Delta u + (1-\mu)B_2 u\right) w - (\Delta u + (1-\mu)B_1 u)\frac{\partial}{\partial n} w\right] d\Gamma.\end{aligned}$$

Noting the equivalence $a(u,w) = (\mathscr{A}^{1/2}u, \mathscr{A}^{1/2}w)_{\mathscr{H}}$ and cancellation in boundary conditions, we find that

$$\begin{aligned}\int_\Omega \mathscr{A}^{1/2}u\mathscr{A}^{1/2}w dx + \int_\Gamma \beta[u^3 + cu_t]w d\Gamma &= (\mathscr{A}u, w)_{\mathscr{D}(\mathscr{A}^{1/2}),[\mathscr{D}(\mathscr{A}^{1/2}]'} \\ + (D_0 h(u), w)_{\mathscr{D}(\mathscr{A}^{1/2}),[\mathscr{D}(\mathscr{A}^{1/2}]'} &+ (D_0^* u_t, D_0^* w)_\Gamma.\end{aligned}$$

If we introduce Green's operator $G_2 : L_2(\Gamma) \mapsto L_2(\Omega)$ by the formula

$$G_2 v \equiv u \ \text{ iff } \begin{cases} \Delta^2 u = 0, & x \in \Omega, \\ \Delta u + (1-\mu) B_1 u = 0, & x \in \Gamma, \\ \frac{\partial}{\partial n}\Delta u + (1-\mu) B_2 u - \nu_1 u = v, & x \in \Gamma, \end{cases} \tag{4.1.25}$$

then as in Section 3.2.2 we find (cf. (3.2.58) and also (1.3.25)) that

$$D_0^* u = \sqrt{c\beta}\, u|_\Gamma = -\sqrt{c\beta}\, G_2^* \mathscr{A} u, \quad u \in H^{1/2+\varepsilon}(\Omega).$$

Thus $D_0 w = -\mathscr{A} G_2 \left[\sqrt{c\beta} w\right]$ and equation (4.1.24) can be also written in the following equivalent form

$$u_{tt} + D(u_t) + \mathscr{A}\left[u - G_2\left(c\beta u_t\big|_\Gamma + \beta u^3\big|_\Gamma\right)\right] = F(u, u_t). \tag{4.1.26}$$

This relation is our motivation for the following definition.

4.1.12. Definition. A function

$$u \in C(0,T; H^2(\Omega)) \cap C^1(0,T; L_2(\Omega)) \tag{4.1.27}$$

satisfying conditions $u(x,0) = u_0$ and $u_t(x,0) = u_1$ is said to be

1. A *strong solution* to problem (4.1.1) and (4.1.2) with the free boundary conditions (4.1.20) on the interval $[0,T]$, iff
   - $u(t)$ is an absolutely continuous function with values in $H^2(\Omega)$ and $u_t \in L_1(a,b; H^2(\Omega))$ for any $0 < a < b < T$.
   - u_t is an absolutely continuous function with values in $L_2(\Omega)$ and $u_{tt} \in L_1(a,b; L_2(\Omega))$ for any $0 < a < b < T$.
   - $u - G_2\left(c\beta u_t\big|_\Gamma + \beta u^3\big|_\Gamma\right) \in \mathscr{D}(\mathscr{A})$ for almost all $t \in [0,T]$, where $\mathscr{D}(\mathscr{A})$ is given by (4.1.22) and Green's operator G_2 is defined by (4.1.25).
   - Equation (4.1.26) is satisfied in $L_2(\Omega)$ for almost all $t \in [0,T]$.
2. A *generalized solution* to problem (4.1.1), (4.1.2), and (4.1.20) on the interval $[0,T]$, iff there exists a sequence $\{u_n(t)\}$ of strong solutions to (4.1.1), (4.1.2), and (4.1.20) with $(u_{0n}; u_{1n})$ instead of $(u_0; u_1)$ such that (4.1.10) holds.

■

4.1.13. Theorem. *Under the Assumptions* 4.1.1 *with reference to* (4.1.1), *subject to free boundary conditions* (4.1.20), *the following statements are valid with any* $T > 0$.

- **Generalized solutions**. *For all initial data* $u_0 \in H^2(\Omega), u_1 \in L_2(\Omega)$ *there exists a unique generalized solution* $u(t)$ *such that* $\sqrt{c\beta} u_t\big|_\Gamma \in L_2([0,T] \times \Gamma)$ *and the following energy inequality holds for all generalized solutions and* $s \le t$,

$$\mathscr{E}(u(t), u_t(t)) + \int_s^t \left[\int_\Omega d_0 g_0(u_t) u_t dx + \int_\Gamma c\beta u_t^2 d\Gamma\right] d\tau$$

$$\leq \mathscr{E}(u(s),u_t(s))+\int_s^t\int_\Omega([u,F_0]-Lu)u_t dxd\tau.$$

If, in addition $g_0(s)s \leq Ms^2$ *for* $|s|\geq 1$, [2] *then we have the energy identity:*

$$\mathscr{E}(u(t),u_t(t))+\int_s^t\left[\int_\Omega d_0 g_0(u_t)u_t dx+\int_\Gamma c\beta u_t^2 d\Gamma\right]d\tau$$
$$=\mathscr{E}(u(s),u_t(s))+\int_s^t\int_\Omega([u,F_0]-Lu)u_t dxd\tau. \quad (4.1.28)$$

- **Strong solutions.** *Assume that* $u_0\in H^4(\Omega), u_1\in H^2(\Omega)$ *and, moreover,*

$$\Delta u_0+(1-\mu)B_1u_0 = 0 \ \ on\ \Gamma,$$
$$\frac{\partial}{\partial n}\Delta u_0+(1-\nu)B_2u_0 = \nu_1 u_0+\beta(u_0^3+cu_1) \ \ on\ \Gamma. \quad (4.1.29)$$

Then, there exists a unique strong solution such that

$$\left.\begin{aligned}
&u-G_2\left(c\beta u_t\big|_\Gamma+\beta u^3\big|_\Gamma\right)\in C_r(0,T;\mathscr{D}(\mathscr{A}))\cap L_\infty(0,T;\mathscr{D}(\mathscr{A})),\\
&u_t\in C_r(0,T;H^2(\Omega))\cap L_\infty(0,T;H^2(\Omega)),\\
&u_{tt}\in C_r(0,T;L_2(\Omega))\cap L_\infty(0,T;L_2(\Omega)),
\end{aligned}\right\} \quad (4.1.30)$$

where $\mathscr{D}(\mathscr{A})$ *is given by* (4.1.22) *and* $C_r(0,T;X)$ *is the space of strongly right-continuous functions. Strong solutions satisfy the energy identity in* (4.1.28).

4.1.14. Remark. As we can see from the energy relation in (4.1.28) nonconservative forces in the model are presented by two terms $[u,F_0]$ and Lu. In general the force $[u,F_0]$ has no potential in the configuration space $H^2(\Omega)$; see also Remarks 3.1.23 and 1.5.17. ■

4.1.15. Remark. By the second line in (4.1.30) and by (4.1.27) from the standard trace theorem we have

$$cu_t\big|_\Gamma+u^3\big|_\Gamma\in C_r(0,T;H^{3/2}(\Gamma))\cap L_\infty(0,T;H^{3/2}(\Gamma)) \ \text{ for any strong solution } u.$$

Therefore it follows from the first line in (4.1.30) and from the regularity of Green's map G_2 given by (1.3.23) that in the case sufficiently smooth $\beta(x)$ any strong solution $u(t)$ to problem (4.1.1), and (4.1.2) and (4.1.20) (in addition to (4.1.30)) possesses the property $u(t)\in C_r(0,T;H^4(\Omega))\cap L_\infty(0,T;H^4(\Omega))$ and satisfies the boundary conditions in (4.1.20). ■

In comparison with the clamped or hinged case equation (4.1.24) has additional (boundary) terms that prevent from direct applicability of Theorem 2.4.16. We instead verify the validity of Assumption 2.4.1 in abstract Theorem 2.4.5.

[2] As in the clamped and hinged cases this assumption can be relaxed; see Remark 4.1.5.

4.1.16. Lemma. *The operators $\mathscr{A}, D, F$ introduced above comply with the requirements* 1,2 *and* 5,6 *in Assumption* 2.4.1 *with $M = I$ and $V = \mathscr{H} = L_2(\Omega)$.*

Proof. We apply the same arguments as in Lemma 4.1.7. The only difference is that now $\mathscr{D}(\mathscr{A}^{1/2}) = H^2(\Omega)$ without any boundary conditions and that $F^*(u,v) = [u, F_0] - Lu + a_0 d_0 v$. However, these facts do not alter the arguments. □

We note that in contrast with the case when rotational inertia are taken into account ($\alpha > 0$) the term on the boundary βu^3 provides unbounded perturbation that cannot be modeled by the generic term $F(u)$ in the case $\alpha = 0$. The presence of linear boundary damping βu_t has been found beneficial in offsetting the influence of the cubic term.

4.1.17. Lemma. *The operator D_0 introduced in* (4.1.23) *complies with Requirement* 7 *in Assumption* 2.4.1 *with $Z = L_2(\Gamma)$.*

Proof. The argument is straightforward. Indeed, the operator D_0 is bounded from $Z = L_2(\Gamma)$ into $[\mathscr{D}(\mathscr{A}^{1/2})]'$. The above follows by observing that the adjoint (dual) operator D_0^* given by the formula $D_0^* u = \sqrt{c\beta} u|_\Gamma$ is bounded from $H^2(\Omega)$ into $L_2(\Gamma)$, where the latter follows (abundantly) from the trace theorem. □

We note, however, that the function $h(u) = \sqrt{c^{-1}\beta} u^3|_\Gamma$ is not globally Lipschitz from $H^2(\Omega)$ into $L_2(\Gamma)$ and thus Requirement 7 in Assumption 2.4.1 does not hold for this h. Therefore application of Theorem 2.4.5 requires, as a preliminary step, some regularization of the term $h(u)$. We do this using the same idea as in the proof of Theorem 2.3.8.

Proof of Theorem 4.1.13.

Step 1: Regularization: Let

$$h_K(s) = \begin{cases} s^3, & \text{for } |s| \le K, \\ K^2 s, & \text{for } |s| \ge K. \end{cases} \tag{4.1.31}$$

It is clear that $h_K : \mathbb{R} \mapsto \mathbb{R}$ is a globally Lipschitz function and its antiderivative has the form

$$H_K(s) \equiv \int_0^s h_K(\sigma) d\sigma = \frac{1}{4} \begin{cases} s^4, & \text{for } |s| \le K, \\ 2K^2 s^2 - K^2, & \text{for } |s| \ge K. \end{cases}$$

Because $H^2(\Omega) \subset C(\overline{\Omega})$, the mapping $u \mapsto \sqrt{\beta} h_K(u)$ is globally Lipschitz from $H^2(\Omega)$ into $L_2(\Gamma)$. Moreover, we have that

$$h_K(u(x)) = [u(x)]^3 \quad \text{for all } x \in \overline{\Omega} \tag{4.1.32}$$

for any function $u \in H^2(\Omega)$ such that $\|u\|_2 \le c_0^{-1} K$, where

$$c_0 = \sup \left\{ \max_{x \in \overline{\Omega}} |u(x)| \; : \; u \in H^2(\Omega), \; \|u\|_{2,\Omega} \le 1 \right\}$$

is the embedding constant of $H^2(\Omega)$ into $C(\overline{\Omega})$.

Now we take arbitrary $R > 0$ and for the initial data $u_0 \in H^2(\Omega)$ and $u_1 \in L_2(\Omega)$ possessing the property

$$\|u_0\|_{2,\Omega}^2 + \|u_1\|_{\Omega}^2 \leq R^2, \tag{4.1.33}$$

instead of equation (4.1.24) we consider the following problem

$$u_{tt} + \mathscr{A}u + D_0\left[\sqrt{c^{-1}}\beta h_K(u)\right] + D_0 D_0^* u_t + D(u_t) = F(u, u_t), \tag{4.1.34}$$

where we choose the parameter K such that $K \geq 2c_0 R$. This choice of K guarantees relation (4.1.32) for the initial data considered (in particular, compatibility condition (4.1.29) is the same for the both problems (4.1.24) and (4.1.34)).

Step 2: Local solutions to problem (4.1.34). Once our standing Assumption 2.4.1 holds for problem (4.1.34), we are in a position to apply the result of Theorem 2.4.5. The first two statements in this theorem provide local (in time) existence and uniqueness of generalized solutions and also strong (local) solutions with additional regularity given by (4.1.30) (which is valid on every existence interval). We also note that local generalized solutions exist for all initial data $(u_0; u_1)$ from $H = H^2(\Omega) \times L_2(\Omega)$ because the set

$$\mathscr{L} = \mathscr{D}^* \times C_0^\infty(\Omega) \;\text{ with }\; \mathscr{D}^* = \left\{w \in H^2(\Omega) : \mathscr{A}\left(w - G_2\left[\beta h_K(w)\right]\right) \in L_2(\Omega)\right\} \tag{4.1.35}$$

is dense in H and belongs to the domain of the corresponding nonlinear operator A given by (2.4.5).

Step 3: A priori bounds and global existence for problem (4.1.34). The regularity guaranteed for strong solutions (see (4.1.30) and also Remark 4.1.15) allows us to perform classical integration by parts and to derive for strong solutions the energy identity (4.1.28) stated in Theorem 4.1.13 with $\mathscr{E}_K(u, u_t)$ instead of $\mathscr{E}(u, u_t)$, where

$$\mathscr{E}_K(u, u_t) = E_K(u, u_t) - \int_\Omega ([f, F_0]u + up)\, dx$$

with

$$E_K(u, u_t) = \frac{1}{2}\int_\Omega |u_t|^2 dx + \frac{1}{2}a(u, u) + \frac{1}{4}\int_\Omega |\Delta v(u)|^2 dx + \frac{1}{4}\int_\Gamma \beta H_K(u) d\Gamma.$$

By using the fact that the energy is bounded from below (see (4.1.21)) along with monotonicity of $g_0(s) + a_0 s$ we can obtain the a priori bound for strong solutions This is to say, with T any positive number we have that

$$E_K(u(t), u_t(t)) \leq C_1(R) + C_2 \int_0^t \left[||u||_{2,\Omega}^2 + ||u_t||_\Omega^2\right] dt$$

for all $t \in [0, T]$, where the constants $C_1(R)$ and C_2 do not depend on K. Thus, from Gronwall's inequality we conclude the following a priori bound

$$||u(t)||_{2,\Omega} + ||u_t(t)||_\Omega + ||v(u)||_{2,\Omega} \leq C_T(R), \quad 0 \leq t \leq T, \tag{4.1.36}$$

where the constant $C_T(R)$ does not depend on K (it may depend on $||f||_{2,\Omega}$, $||p||_\Omega$ and $||F_0||_{3+\delta,\Omega}$). Because inequality (4.1.36) is stable with respect to the norm generated by the energy, the same a priori bound holds for generalized solutions. By Theorem 2.4.5 this proves global (in time) existence of generalized and strong solutions to problem (4.1.34).

Step 4: Global existence for the original problem. Let $[0,T]$ be an arbitrary interval. By the previous argument on this interval we have the existence of both strong and generalized solutions to problem (4.1.34). These solutions admit estimate (4.1.36) for any initial data possessing the property (4.1.33). Thus by (4.1.32) for every $K > c_0 \max\{2R, C_T(R)\}$ these solutions also solve the original problem (4.1.34). Thus the strong and generalized solutions to problem (4.1.24) exist on any interval $[0,T]$. Clearly these solutions are unique due to the fact that any solution to (4.1.24) solves regularized problem (4.1.34) for K large enough.

Step 5: Energy inequality and identity for generalized solutions. In order to obtain energy relations for generalized solutions, we need to perform limit process on strong solutions. Let $(u_0;u_1)$ be the initial data of finite energy and let $(u(t);u_t(t))$ be a corresponding solution. From the definition of generalized solutions we know that such solutions are strong limits of strong solutions denoted by $u_n(t)$; we thus have

$$u_n \to u \text{ in } C(0,T;H^2(\Omega)), \quad u_{nt} \to u_t \text{ in } C(0,T;L_2(\Omega)) \tag{4.1.37}$$

Denoting by $\mathscr{E}_n(t)$ the energy corresponding to strong solutions, we have

$$\mathscr{E}_n(t) + \int_s^t \left[\int_\Omega d_0[g_0(u_{nt}) + a_0 u_{nt}]u_{nt}dx + \int_\Gamma c\beta u_{nt}^2 d\Gamma \right] d\tau$$
$$\le \mathscr{E}_n(s) + \int_s^t ([u_n, F_0] - Lu_n + a_0 d_0 u_{nt}, u_{nt})_\Omega d\tau.$$

The energy function $\mathscr{E}_n(t)$ is continuous with respect to topology in (4.1.37). Thus using the same idea as before (with introduction of the functions $\psi_N(s) = \min\{s,N\}$; see the proof of the corresponding fact in Theorem 3.1.4) we can pass with the limit $n \to \infty$ on the inequality above. This leads to the desired energy inequality valid for all generalized solutions.

In order to obtain energy identity we restrict the growth of nonlinearity by assuming that g_0 is linearly bounded at infinity. We consider the difference of two strong solutions, say $u_n(t)$ and $u_m(t)$. As in Section 3.2 by standard calculations using (i) local Lipschitz property for nonlinearity $F(u,v)$, (ii) a priori bounds of strong solutions in (4.1.36), and (iii) monotonicity of $g_0(s) + a_0 s$ and finally Gronwall's inequality we conclude that

$$||u_n(t) - u_m(t)||_{2,\Omega}^2 + ||u_{nt}(t) - u_{mt}(t)||_\Omega^2 + \int_0^t \int_\Gamma c\beta |u_{nt} - u_{mt}|^2 d\Gamma d\tau$$
$$\le C_T [||u_n(0) - u_m(0)||_{2,\Omega}^2 + ||u_{nt}(0) - u_{mt}(0)||_\Omega^2 \to 0,$$

whenever $||u_n(0)-u_0||_2 \to 0$ and $||u_{nt}(0)-u_1|| \to 0$. Thus in addition to having strong limits (4.1.37) we also have:

$$\int_0^t \int_\Gamma c\beta |u_{nt}-u_{mt}|^2 d\Gamma d\tau \to 0, \quad n,m\to\infty.$$

Thus there exists $v \in L_2([0,T]\times\Gamma)$ such that

$$\int_0^t \int_\Gamma |\sqrt{c\beta}u_{nt}-v|^2 d\Gamma d\tau \to 0, \quad n\to\infty.$$

We can identify $v(t)$ with $\sqrt{c\beta}u_t|_\Gamma$. Therefore using also the demicontinuity of g_0 which follows from Lemma 4.1.8 and Proposition 1.2.5 we can pass with the limit on energy identity written first for strong solutions. The limit process yields this identity valid for all generalized solutions. The proof of Theorem 4.1.13 is thus completed.

4.1.4 Weak solutions

In analogy to models with rotational forces we can also consider variational (weak) solutions. We consider all three types of boundary conditions.

4.1.18. Definition.

- We say that u is a *weak solution* to equation (4.1.1) with the *clamped* boundary conditions (4.1.4) iff $u \in L_\infty(0,T;H_0^2(\Omega)) \cap W_\infty^1(0,T;L_2(\Omega))$, $u|_{t=0}=u_0$,

$$\text{the function} \quad t \mapsto (d_0 g_0(u_t),\phi)_\Omega \quad \text{is integrable} \tag{4.1.38}$$

  for $\phi \in H_0^2(\Omega)$, and the following variational relation holds

$$(u_t(t),\phi)_\Omega - (u_1,\phi)_\Omega + \int_0^t (\Delta u, \Delta\phi)_\Omega d\tau \tag{4.1.39}$$
$$+\int_0^t [(d_0 g_0(u_t) - [v(u)+F_0,u+f] + Lu - p,\phi)_\Omega]d\tau = 0$$

  for all $\phi \in H_0^2(\Omega)$, where $v=v(u)$ solves the elliptic problem (4.1.2).
- A function u is a *weak solution* to equation (4.1.1) with the *hinged* boundary conditions (4.1.16) iff $u \in L_\infty(0,T;(H^2\cap H_0^1)(\Omega)) \cap W_\infty^1(0,T;L_2(\Omega))$, $u|_{t=0}=u_0$ and the relations (4.1.38) and (4.1.39) hold for all $\phi \in H^2(\Omega)\cap H_0^1(\Omega)$.
- We say that u is a *weak solution* to equation (4.1.1) with the *free* boundary conditions (4.1.20) iff $u \in L_\infty(0,T;H^2(\Omega)) \cap W_\infty^1(0,T;L_2(\Omega))$, $u|_{t=0}=u_0$, (4.1.38) is satisfied with $\phi \in H^2(\Omega)$, and the following variational relation holds

$$(u_t(t),\phi)_\Omega - (u_1,\phi)_\Omega$$

$$+\int_0^t [a(u,\phi)+(d_0 g_0(u_t)-[v(u)+F_0,u+f]+Lu-p,\phi)_\Omega]\,d\tau$$

$$+\int_0^t (\beta u^3,\phi)_\Gamma d\tau+(c\beta[u(t)-u_0],\phi)_\Gamma=0$$

for all $\phi\in H^2(\Omega)$, where $v=v(u)\in H_0^2(\Omega)$ solves problem (4.1.2).

■

4.1.19. Theorem. *Let Assumptions* 4.1.1 *be valid. Then any generalized solution of von Karman system* (4.1.1) *with either clamped* (4.1.4), *hinged* (4.1.16), *or free* (4.1.20) *boundary conditions is also a weak solution.*

If additionally g_0 *is of linear growth; that is,* $|g_0(s)|\le M_1(1+|s|)$ *for* $s\in\mathbb{R}$ *with some constant* $M_1>0$, *then weak solutions of von Karman system* (4.1.1) *with either clamped* (4.1.4), *hinged* (4.1.16), *or free* (4.1.20) *boundary conditions are unique.*[3] *Thus any weak solution is also generalized in this case.*

Proof. The variational form specified in Definition 4.1.18 is verified for strong solutions. Using the limit definition of generalized solutions along with weak continuity of the nonlinear terms involving Airy stress function, we can pass with the limit by routine argument. The passage through the limit on nonlinear damping terms is accomplished in the same way as in Theorem 3.1.12.

To prove uniqueness of weak solutions in the case of clamped or hinged boundary conditions we use the same method as in Theorem 3.1.12.

We assume that there exist two solutions u^1 and u^2. Their difference $u=u^1-u^2$ is a weak solution to the linear problem

$$\begin{cases} u_{tt}+\Delta^2 u=f(x,t), \text{ in } Q\equiv\Omega\times(0,T)\\ u|_{t=0}=u_t|_{t=0}=0,\end{cases}$$

where u is subjected to either clamped (4.1.4) or hinged (4.1.16) boundary conditions and

$$f(x,t)=G(x,t)+F(x,t), \tag{4.1.40}$$

where $G(x,t)=-d_0(x)\left(g_0(u_t^1)-g_0(u_t^2)\right)$ and

$$F(x,t)=[u^1+f,v(u^1)+F_0]-[u^2+f,v(u^2)+F_0]-L(u^1-u^2).$$

Because g_0 is of linear growth, we have that $G\in L_\infty(0,T;L_2(\Omega))$. A calculation similar the one given in the proof of Lemma 4.1.7 shows that

$$\|F(t)\|_\Omega\le C_T\cdot(\|u_t^1-u_t^2\|_\Omega+\|u^1-u^2\|_{2,\Omega}),\quad t\in[0,T], \tag{4.1.41}$$

where C_T also depends on the values $\sup_{\tau\in[0,T]}\left\{|u_t^j(\tau)\|_\Omega+(\|u^j(\tau))\|_{2,\Omega}\right\}$ for $j=1,2$. Now using the energy relation (2.5.2) for the linear problems (2.5.1) and

[3] The uniqueness statement can be also proved in the case when $ms^p\le sg_0(s)\le Ms^p$ for some $p\ge 2$ an for all $|s|\ge s_0$, where $s_0>0$; cf. Remark 3.1.14 in the rotational inertia case.

(2.5.20) (cf. Propositions 2.5.1 and 2.5.7), Gronwall's lemma, and monotonicity of $g_0(s) + a_0 s$ we easily find that $u \equiv 0$.

The case of the "free" boundary condition is more involved, as the role of the nonlinear boundary term is critical.

If $u = u^1 - u^2$ is the difference of two weak solutions to problem (4.1.1) with free-type boundary conditions (4.1.20), then u is a weak solution to the linear problem

$$\begin{cases} u_{tt} + \Delta^2 u = f(x,t), \text{ in } Q \equiv \Omega \times (0,T) \\ \Delta u + (1-\mu)B_1 u = 0, \\ \frac{\partial}{\partial n}\Delta u + (1-\mu)B_2 u = \nu_1 u + c\beta u_t + h(x,t), \\ u|_{t=0} = u_t|_{t=0} = 0, \end{cases}$$

where $f(x,t)$ is given by (4.1.40) and $h(x,t) = \beta((u^1)^3 - (u^2)^3)$. This problem can be written in the form (2.4.48):

$$u_{tt} + \mathscr{A} u + D_0 h(t) + D_0 D_0^* u_t = f(t),$$

where D_0 is defined by (4.1.23). Therefore from energy relation (2.4.57) given by Theorem 2.4.29 for weak solutions and also from monotonicity of $g_0(s) + a_0 s$ we obtain that

$$\|u_t(t)\|_\Omega^2 + \|u(t)\|_{2,\Omega}^2 \le \int_0^t \left(\|h(\tau)\|_\Gamma^2 + \|F(\tau)\|_\Omega^2 + \|u_t(\tau)\|_\Omega^2\right) d\tau.$$

The trace theorem implies:

$$\|h(t)\|_{3/2,\Gamma} \le C \cdot \|u^1 - u^2\|_{2,\Omega}(\|u^1\|_{2,\Omega}^2 + \|u^2\|_{2,\Omega}^2).$$

Therefore using (4.1.41) we can conclude that

$$\|u_t(t)\|_\Omega^2 + \|u(t)\|_{2,\Omega}^2 \le C\int_0^t \left(\|u(\tau)\|_{2,\Omega}^2 + \|u_t(\tau)\|_\Omega^2\right) d\tau.$$

This implies that $u \equiv 0$. □

4.1.20. Remark. In the problems above weak solutions can be constructed by the Galerkin method (see, e.g., [220] or [237]). For example, in the clamped case we can proceed as follows.

Let $\{e_k\}$ be a basis in the space $H_0^2(\Omega)$. We define an *n-order Galerkin approximate solution* to problem (4.1.1) with clamped boundary conditions (4.1.4) on the interval $[0,T]$ as a function $u^n(t)$ of the form

$$u^n(t) = \sum_{i=1}^n g_k(t) e_k, \quad n = 1, 2, \ldots,$$

where $g_k(t) \in W_\infty^2(0,T;\mathbb{R})$ are scalar functions such that $u^n(t)$ satisfies the relations

$$\begin{cases} \frac{d^2}{dt^2}(u^n(t), e_k)_\Omega + (d_0 g_0(u_t^n(t)), e_k)_\Omega + (\Delta u^n(t), \Delta e_k)_\Omega \\ \qquad - ([u^n(t) + f, v(u^n(t)) + F_0], e_k)_\Omega + (Lu^n(t), e_k)_\Omega \\ \qquad = (p, e_k)_\Omega \ \text{ for every } \ k = 1, 2, \dots, n, \\ u^n|_{t=0} = u_{n0}, \quad u_t^n|_{t=0} = u_{n1}, \end{cases} \tag{4.1.42}$$

where $v(u^n)$ is determined by u^n according (4.1.2) and $u_{n0}, u_{n1} \in \text{Span}\{e_k : k = 1, 2, \dots, n\}$ are chosen such that $u_{n0} \to u_0$ in $H_0^2(\Omega)$ and $u_{n1} \to u_1$ in $L_2(\Omega)$.

Multiplying (4.1.42) by $\dot{g}_k(t)$ and taking the sum for k from 1 to n we obtain

$$\frac{1}{2}\frac{d}{dt}\left\{\|(u^n(t)\|_\Omega^2 + \|\Delta u^n(t)\|_\Omega^2 - 2(p, u^n(t))_\Omega\right\} + (d_0 g_0(u_t^n(t)), u_t^n(t))_\Omega$$

$$-([u^n(t) + f, v(u^n(t)) + F_0], u_t^n(t))_\Omega + (Lu^n(t), u_t^n(t))_\Omega = 0.$$

Using the symmetry properties of the von Karman bracket and integrating over interval $[0,t]$ we obtain the energy equality for approximate solutions:

$$\mathcal{E}(u^n(t), u_t^n(t)) + d_0 \int_0^t (g_0(u_t^n), u_t^n)_\Omega d\tau = \mathcal{E}(u_{n0}, u_{n1}) - \int_0^t (Lu^n, u_t^n)_\Omega d\tau,$$

where the energy $\mathcal{E}(u, u_t)$ is given by (4.1.5). Now using (4.1.9) it is easy to obtain the a priori estimate:

$$\|u_t^n(t)\|_\Omega^2 + \|\Delta u^n(t)\|_\Omega^2 + \|\Delta v(u^n(t))\|_\Omega^2 \le C_T, \quad t \in [0, T].$$

This estimate implies that there exist a function $u(t)$ such that

$$u(t) \in L_\infty(0, T; H_0^2(\Omega)), \quad u_t(t) \in L_\infty(0, T; L_2(\Omega))$$

and a sequence $\{n_l\}$ such that

$$\begin{aligned} u^{n_l}(t) &\to u(t) \quad *\text{-weakly in } L_\infty(0, T; H_0^2(\Omega)), \\ u_t^{n_l}(t) &\to u_t(t) \quad *\text{-weakly in } L_\infty(0, T; L_2(\Omega)), \end{aligned}$$

when $l \to \infty$. If $g_0(s)$ is an increasing continuous function such that $g_0(0) = 0$ and there exists $m, M > 0$ and $p \ge 2$ such that

$$m|s|^p \le s g_0(s) \le M|s|^p \quad \text{for all } |s| \ge 1,$$

then we can also use the compactness/monotonicity argument (see [220, Chapter 2]). The properties above allow us to make (along a sequence) the limit transition $n \to \infty$ in (4.1.42) for every fixed k and to prove the existence of weak solutions. We refer to [220, Chapter 1] and [237] for details in the case of linear damping. ∎

4.1.5 Regular solutions

We consider the clamped case only. Hinged and free boundary conditions can be treated in a similar way (cf. Section 3.1.4). Thus, we consider the following equation

$$\begin{cases} u_{tt} + d_0 g_0(u_t) + \Delta^2 u - [u+f, v+F_0] + L(u) = p(x), \quad x \in \Omega, t > 0, \\ u|_{\partial\Omega} = \frac{\partial}{\partial n} u|_{\partial\Omega} = 0, \\ u|_{t=0} = u_0(x), \quad u_t|_{t=0} = u_1(x). \end{cases} \tag{4.1.43}$$

As above the function $v = v(u) \in H_0^2(\Omega)$ is a solution of the problem

$$\Delta^2 v + [u+2f, u] = 0, \quad v|_{\partial\Omega} = \left.\frac{\partial v}{\partial n}\right|_{\partial\Omega} = 0. \tag{4.1.44}$$

In addition to Assumption 4.1.1 we assume that

$$g_0 \in C^m(\mathbb{R}) \text{ and } f \in H^{3+\delta}(\Omega) \text{ for some } m \geq 1 \text{ and } \delta > 0. \tag{4.1.45}$$

As we already know (see, e.g.,Theorem 3.1.28), higher regularity of solutions requires compatibility conditions for initial and boundary data. For this we define the values $u^{(k)}(0)$ by the recurrence relations

$$\begin{aligned} &u^{(0)}(0) = u_0, \quad u^{(1)}(0) = u_1, \\ &u^{(k)}(0) = -\Delta^2 u^{(k-2)}(0) + \frac{d^{k-2}}{dt^{k-2}} \{-d_0 g_0(u_t) + F(u)\}\Big|_{t=0}, \end{aligned} \tag{4.1.46}$$

where $k \geq 2$ and $F(u) \equiv [u+f, v(u)+F_0] - Lu + p$. We also denote by $\mathscr{L}_m$ the class of functions that possess the following properties

$$\begin{cases} u^{(k)}(t) \in C(0,T; H^4(\Omega) \cap H_0^2(\Omega)) \quad \text{for} \quad k = 0,1,2,\ldots,m-1, \\ u^{(m)}(t) \in C(0,T; H_0^2(\Omega)), \quad u^{(m+1)}(t) \in C(0,T; L_2(\Omega)) \end{cases} \tag{4.1.47}$$

for any $T > 0$, where $u^{(k)}$ denotes the derivative with respect to t of the order k.

4.1.21. Theorem. *Let Assumption* 4.1.1 *be in force. Assume that g_0 and f possess properties listed in* (4.1.45) *for some $m \geq 1$ and $\delta > 0$. Then a generalized solution to problem* (4.1.43) *and* (4.1.44) *belongs to $\mathscr{L}_m$ if and only if the following compatibility conditions*

$$\begin{cases} u^{(k)}(0) \in H^4(\Omega) \cap H_0^2(\Omega) \quad \textit{for} \quad k = 0,1,2,\ldots,m-1, \\ u^{(m)}(0) \in H_0^2(\Omega), \quad u^{(m+1)}(0) \in L_2(\Omega), \end{cases} \tag{4.1.48}$$

hold, where the values $u^{(k)}(0)$ are defined by the recurrence relations (4.1.46).

To prove this theorem we need the following property of the Fréchet derivative of the mapping $F(u)$.

4.1.22. Lemma. *Assume that $u \in H^4(\Omega) \cap H_0^2(\Omega)$, $\|u\|_{4,\Omega} \le R$ and $w \in H_0^2(\Omega)$. Denote by $\langle F'(u); w\rangle$ the value of the Fréchet derivative of the mapping $F(u)$ on element w. Then*

$$\langle F'(u); w\rangle = [w, v(u) + F_0] - Lw + \widetilde{F}(u,w),$$

where $\widetilde{F}(u,w)$ can be estimated as

$$\|\widetilde{F}(u,w)\|_\Omega \le C_R \|w\|_{1,\Omega}. \tag{4.1.49}$$

Proof. It can be easily seen that $\widetilde{F}(u,w) = 2[u+f, v(u+f,w)]$, where $v = v(w_1,w_2)$ is determined from $w_1, w_2 \in H^2(\Omega)$ as the solution to the problem

$$\Delta^2 v + [w_1, w_2] = 0, \quad v|_{\partial\Omega} = \frac{\partial v}{\partial n}\Big|_{\partial\Omega} = 0. \tag{4.1.50}$$

The structure of the von Karman bracket and the embedding $H^{3+\delta}(\Omega) \subset C^2(\overline{\Omega})$ yield

$$\|[u+f, v(u+f,w)]\|_\Omega \le C_R \|v(u+f,w)\|_{2,\Omega}.$$

Because (see (1.3.11)) the Green operator $G = (\Delta_D^2)^{-1}$ of the Dirichlet problem for the biharmonic operator possesses the estimate

$$\|(\Delta_D^2)^{-1} h\|_{s+4,\Omega} \le C\|h\|_{s,\Omega}, \quad h \in H^s(\Omega),\ s \ge -2, \tag{4.1.51}$$

using (1.4.17) we obtain that

$$\|v(u+f,w)\|_{2,\Omega} \le C\|[u+f,w]\|_{-2,\Omega} \le C\|u+f\|_{2,\Omega}\|w\|_{1,\Omega}.$$

This yields (4.1.49). □

We also need the following von Karman bracket property.

4.1.23. Lemma. *The following estimate holds,*

$$\|[w_1, v(w_2,w_3)]\|_\Omega \le C\|w_1\|_{2+\delta,\Omega}\|w_2\|_{2,\Omega}\|w_3\|_{2-\delta,\Omega}, \quad 0 \le \delta < 1, \tag{4.1.52}$$

where $v(w_1,w_2)$ is defined by (4.1.50) and elements w_j lie in the corresponding Sobolev spaces.

Proof. It easily follows from Theorem 1.4.3 and Corollary 1.4.4. □

Proof of Theorem 4.1.21. We follow the line of argument given in [55]. It is clear that any solution $u(t)$ from $\mathscr{L}_m$ possesses property (4.1.48). Thus we need only prove that compatibility conditions (4.1.48) imply that $u(t) \in \mathscr{L}_m$. To prove it we rely on the idea presented in [123] and use induction in m.

For $m = 1$ the assertion coincides with the second part of Theorem 4.1.4. Suppose that (4.1.48) holds for some $m > 1$ and that a strong solution $u(t)$ belongs to the class $\mathscr{L}_{n-1}$ for $2 \le n \le m$. Let us prove $u(t) \in \mathscr{L}_n$.

The formal rule of differentiation yields

$$\frac{d^n}{dt^n} F(u(t)) = \langle F'(u(t)); u^{(n)}(t) \rangle + G_n(t).$$

Here $G_1 \equiv 0$ and for $n \ge 2$ the quantity

$$G_n(t) \equiv G_n(u(t), u'(t), \dots, u^{(n-1)}(t))$$

is a linear combination of functions of the form $[v(w_1, w_2), w_3]$, where w_j are either $u + f$ or one of the derivatives $u^{(k)}(t)$, $1 \le k \le n-1$, and the value $v(w_1, w_2)$ is defined by (4.1.50). Moreover, $G_n(t)$ depends linearly on $u^{(n-1)}(t)$ when $n > 2$. These properties, the induction hypothesis, and Lemma 4.1.23, allow us to assert that $G_n(t)$ is a strongly continuous function with values in $L_2(\Omega)$ possessing the property $\|G_n(t)\| \le C_T$ for any $t \in [0, T]$.

In a similar way we have that

$$\frac{d^n}{dt^n} g_0(u_t(t)) = g_0'(u_t(t)) u^{(n+1)}(t) + \hat{D}_n(t),$$

where $\hat{D}_1(t) \equiv 0$ and $\hat{D}_n(t) = c_n g_0''(u_t) u^{(n)}(t) u^{(2)}(t) + D_n^*(t)$ for $n \ge 2$ with some constant c_n. Here $D_2^*(t) \equiv 0$ and for $n \ge 3$ we have that $D_n^*(t)$ is a sum of elements of the form

$$c_{r, k_1, \dots, k_r} \cdot g_0^{(r)}(u_t) \cdot u^{(k_1)} \cdot \dots \cdot u^{(k_r)}, \quad 2 \le r \le n,$$

with $2 \le k_j \le n-1$ for every $j = 1, \dots, r$. By the induction hypothesis we have $u \in \mathscr{L}_{n-1}$. Therefore we have that $u^{(k)} \in C(0, T; H_0^2(\Omega))$ for every $k \le n-1$. Thus $D_n^*(t) \in C(0, T; L_2(\Omega))$.

Let us prove that

$$w(t) = u^{(n)}(t) \in C(0, T; H_0^2(\Omega)) \cap C^1(0, T; L_2(\Omega)) \quad \text{for any } T > 0. \tag{4.1.53}$$

To do this we observe that the function $w(t)$ is a solution (in the sense of distributions) of the problem

$$\begin{cases} w_{tt} + c(t) w_t + \Delta_D^2 w + b(t) w - \langle F'(u(t)); w(t) \rangle = Q_n(t), \\ w|_{t=0} = u^{(n)}(0), \quad w_t|_{t=0} = u^{(n+1)}(0), \end{cases} \tag{4.1.54}$$

which is obtained by formal n-order differentiation of (4.1.43) with respect to t. Here

$$c(t) := c(t, x) = d_0 g_0'(u_t(t, x)) \in C([0, T] \times \bar{\Omega})$$

is the nonnegative function,

$$b(t) := b(t,x) = c_n d_0 g_0''(u_t(t,x))u_{tt}(t,x) \in C(0,T;L_2(\Omega))$$

and $Q_n(t) = G_n(t) - d_0 D_n^*(t) \in C(0,T;L_2(\Omega))$.

4.1.24. Remark. If $u \in \mathscr{L}_{n-1}$, then the term $b(t)w = c_n d_0 g_0''(u_t)u^{(n)}u_{tt}$ can be included in $Q_n(t)$ provided $n > 2$, but not in the case $n = 2$. This is why we keep the term $b(t)w$ in equation (4.1.54). ■

Now we continue with the proof of (4.1.53).

The arguments presented below have a formal character. They can be made rigorous if the Galerkin approximation of the problem given by (4.1.54) and of a similar problem for $\tilde{w} = u^{(n-1)}$ are considered.

Multiplying (4.1.54) by w_t in $L_2(\Omega)$ and using Lemma 4.1.22, we have

$$\frac{1}{2}\frac{d}{dt}\left\{\|w_t\|_\Omega^2 + \|\Delta w\|_\Omega^2\right\} + \int_\Omega c(t)|w_t|^2 dx - ([w, v(u(t)) + F_0], w_t)_\Omega$$
$$\leq -\int_\Omega b(t)ww_t dx - (Lw, w_t) + C(1 + \|w\|_{1,\Omega})\|w_t\|_\Omega.$$

It is clear that

$$2([w, v(u(t)) + F_0], w_t) = \frac{d}{dt}([w,w], v(u) + F_0) - 2([w, v(u+f, u_t)], w).$$

Therefore Lemma 4.1.23 and the induction hypothesis yield

$$\frac{d}{dt}\left\{\|w_t\|_\Omega^2 + \|\Delta w\|_\Omega^2 - ([w,w], v(u) + F_0)_\Omega\right\} \leq C(1 + \|w_t\|_\Omega^2 + \|w\|_{2,\Omega}^2). \quad (4.1.55)$$

From Theorem 1.4.3 and Corollary 1.4.5 we have

$$|([w,w], v(u) + F_0)_\Omega| \leq C(1 + \|u\|_{2,\Omega}^2)\|w\|_\Omega\|\Delta w\|_\Omega \leq C\|\Delta w\|_\Omega,$$

where in the last inequality we use the fact that $\|u\|_{2,\Omega}$ and $\|w\|_\Omega = \|u^{(n)}\|_\Omega$ are bounded according to the induction hypothesis. Therefore using Gronwall's lemma and the compatibility conditions (4.1.48) from (4.1.55) we can extract an a priori estimate of the form

$$\|w_t(t)\|_\Omega^2 + \|\Delta w(t)\|_\Omega^2 \leq C_T, \quad t \in [0,T],$$

for any $T > 0$. Thus $w = u^{(n)}$ is a weak solution to (4.1.54). Now we can apply Theorem 2.4.35 with $D_0 \equiv 0$ to conclude that any solution to problem (4.1.54) possesses the property (4.1.53). Thus we obtain that

$$u^{(n)}(t) \in C(0,T;H_0^2(\Omega)) \text{ and } u^{(n+1)}(t) \in C(0,T;L_2(\Omega))$$

for any $T > 0$. Because

$$\Delta^2 u^{(n-1)}(t) = -u^{(n+1)}(t) - d_0\left(g_0'(u_t)u^{(n)}(t) + \hat{D}_{n-1}(t)\right)$$

$$+\langle F'(u(t));u^{(n-1)}(t)\rangle + G_{n-1}(t),$$

the elliptic regularity implies that $u^{(n-1)}(t)$ lies in $C(0,T;H^4(\Omega)\cap H_0^2(\Omega))$. Thus $u(t)\in\mathscr{L}_n$. This completes the proof of Theorem 4.1.21.

Theorem 4.1.21 and the elliptic regularity theory lead to the following assertion.

4.1.25. Theorem. *Let $m\geq 1$. In addition to Assumption* 4.1.1 *and relations in* (4.1.45) *in the case when $m\geq 2$ we assume that $L: H^{k+2}(\Omega)\mapsto H^k(\Omega)$ for $k=0,\dots,2m-2$, $g_0\in C^{2m-2}(\mathbb{R})$,*

$$d_0(x), p(x)\in H^{2m-2}(\Omega),\quad f(x)\in H^{2m}(\Omega),\quad F_0(x)\in H^{2m+1}(\Omega),$$

and the compatibility conditions (4.1.48) *hold. Then strong solution $u(t)$ to problem* (4.1.43) *and* (4.1.44) *possesses the property*

$$u^{(j)}(t)\in C(0,T;H^{2(m+1-j)}(\Omega)),\quad j=0,1,\dots,m+1,$$

for any $T>0$. In particular, if $m\geq 2$, then the solution $u(t)$ is classical.

The proof of this theorem can be obtained in the same way as in [123] for wave equations with linear damping (see also Proposition 2.5.2 in the case of linear plate models and [61] in the case the Berger plates). However, we find it convenient to use the inequality given below (see also [55]).

4.1.26. Lemma. *Suppose that $w_j(x)\in H^{k+2}(\Omega)$ for some integer $k\geq 0$ and for $j=1,2,3$. Let $v(w_1,w_2)$ be defined by* (4.1.50). *Then the quantity $[w_1,v(w_2,w_3)]$ lies in space $H^k(\Omega)$ and*

$$\|[w_1,v(w_2,w_3)]\|_{k,\Omega}\leq C\|w_1\|_{k+2,\Omega}\|w_2\|_{k+2,\Omega}\|w_3\|_{k+2,\Omega}. \tag{4.1.56}$$

Proof. For $k=0$ this is the statement of Lemma 4.1.23 with $\delta=0$

Consider the case $k\geq 0$. It sufficient to prove (4.1.56) only for smooth functions $w_j(x)$. Let $D^l v$ be any partial derivative of v of order l. It is clear that a quantity $D^l(D^2wD^2v)$ is a linear combination of terms of the form

$$D^{2+l-j}w\cdot D^{j+2}v,\quad j=0,1,\dots,l.$$

From (1.4.6) with $s=0$ we have

$$\|D^{2+l}wD^2v\|_\Omega\leq C\|w\|_{2+l,\Omega}\|v\|_{3+\theta,\Omega},\qquad \theta>0,$$

and (1.4.2) yields

$$\|D^{2+l-j}wD^{j+2}v\|_\Omega\leq C\|w\|_{2+l-j+1-\theta,\Omega}\|v\|_{j+2+\theta,\Omega},\quad j=1,\dots,l,\ 0<\theta<1$$

These inequalities and the structure of the von Karman bracket make it possible to assert that

$$\|[w,v]\|_{l,\Omega}\leq C\|w\|_{2+l,\Omega}\|v\|_{2+l+\theta,\Omega}\ \text{ for every } l=1,2,\dots \text{ and } \theta>0. \tag{4.1.57}$$

From this and (4.1.51) we obtain that

$$\|[w_1, v(w_2,w_3)]\|_{k,\Omega} \le C\|w_1\|_{k+2,\Omega}\|[w_2,w_3]\|_{k-2+\theta,\Omega}.$$

To conclude the proof of (4.1.56) for $k=1$ we use (1.4.16) with $j=0$, $\beta = 1-\theta$ instead of θ. In the case $k \ge 2$ we use relation $\|[w_2,w_3]\|_{k-2+\theta,\Omega} \le C\|[w_2,w_3]\|_{k-1,\Omega}$ and apply (4.1.57) with $l=k-1$ and $\theta=1$. □

Proof of Theorem 4.1.25. For $m=1$ the statement of this theorem is contained in Theorem 4.1.4. Thus we can assume that $m \ge 2$ and apply the following form of inductive argument.

Let $2 \le k \le m$. Assume that

$$u^{(j)}(t) \in C(0,T;H^{2(k-j)}(\Omega)), \quad 0 \le j \le k. \tag{4.1.58}$$

We need to prove this relation for $k := k+1$; that is, to prove that

$$u^{(j)}(t) \in C(0,T;H^{2(k+1-j)}(\Omega)) \quad \text{for all} \quad 0 \le j \le k+1. \tag{4.1.59}$$

By (4.1.47) and Theorem 4.1.21 relation (4.1.59) holds for $j = k-1, k, k+1$. Consider the case $j = k-2$. From (4.1.43) we have that

$$\Delta^2 u^{(k-2)} = -u^{(k)} + \frac{d^{k-2}}{dt^{k-2}}\{-d_0 g_0(u_t) + F(u)\}$$

and thus for $k \ge 3$ we obtain that

$$\Delta^2 u^{(k-2)} = -u^{(k)} - d_0 g_0'(u_t)u^{(k-1)} - d_0\hat{D}_{k-2}(t) + \langle F'(u), u^{(k-2)}\rangle + G_{k-2}(t),$$

where $\hat{D}_{k-1}(t)$ and $G_{k-2}(t)$ are defined in the proof of Theorem 4.1.21.

It follows from (4.1.58) that

$$u^{(j)}(t) \in C(0,T;H^4(\Omega)) \ \text{for}\ 0 \le j \le k-2. \tag{4.1.60}$$

Therefore from Lemma 4.1.26 one can see

$$\langle F'(u), u^{(k-2)}\rangle + G_{k-2}(t) \in C(0,T;H^2(\Omega)).$$

Because $d_0\hat{D}_{k-2}(t)$ contains time derivatives $u^{(j)}$ of order less than or equal to $k-2$, by (4.1.60) we conclude that $d_0\hat{D}_{k-2}(t) \in C(0,T;H^2(\Omega))$. At last, by (4.1.59) for $j = k, k-1$ we also have that

$$-u^{(k)} - d_0 g_0'(u_t)u^{(k-1)} \in C(0,T;H^2(\Omega)).$$

Thus $\Delta^2 u^{(k-2)} \in C(0,T;H^2(\Omega))$ and therefore by elliptic regularity theory (see, e.g., [222]) we conclude that $u^{(k-2)} \in C(0,T;H^6(\Omega))$; that is, (4.1.59) holds for $j = k-2$. In similar way, step by step, we establish (4.1.59) for all $0 \le j \le k+1$.

This completes the proof of Theorem 4.1.25.

We also refer to [27] where the well-posedness of problem (4.1.43) with $f = F_0 = 0$ and $L \equiv 0$ was studied for initial data from the interpolation spaces between $(H^4 \cap H_0^2)(\Omega) \times H_0^2(\Omega)$ and $H_0^2(\Omega) \times L_2(\Omega)$.

4.1.6 On a model with delay

In this section we briefly consider the nonrotational version of von Karman evolution equations with retarded terms. For the case when the rotational forces are included we refer to Section 3.3.1. Thus we consider the following von Karman system with a linear retarded term:

$$u_{tt} + du_t + \Delta^2 u = [u + f, v + F_0] - Lu + p(u^t;t), \quad x \in \Omega, t > 0, \tag{4.1.61}$$

where, as above, $v = v(u) \in H_0^2(\Omega)$ is the Airy stress function that solves the elliptic problem in (4.1.2), and $d \geq 0$ is a constant. Let the plate be clamped; that is, the displacement function u satisfies the boundary conditions in (4.1.4).

We assume that $f \in H^2(\Omega)$ and $F_0 \in H^{3+\delta}(\Omega)$ are given functions and

$$p(u^t;t) = p_0(x) + q(u^t;t),$$

where $p_0(x) \in L^2(\Omega)$ is given, u^t denotes an element from $L^2(-r,0;H^2(\Omega))$ defined by the formula $u^t(s) = u(t+s)$, $s \in (-r,0)$, and $q(\cdot;t)$ is a continuous linear mapping from $L^2(-r,0;H^2(\Omega))$ into $L_2(\Omega)$ possessing the property

$$\| q(u^t,t) \|_\Omega^2 \leq C \int_{t-r}^{t} \| u(\tau) \|_{2,\Omega}^2 \, d\tau.$$

As in Section 3.3.1, this choice of the retarded term $p(u^t;t)$ corresponds to the situation when the plate is located in a potential linearized flow of gas (see [54, 168] and the references therein and also Chapter 6 below). The term $p(u^t;t)$ describes the aerodynamic pressure of the flow on the plate. The parameter $r > 0$ is the time of retardation and it depends on the velocity of the unperturbed flow and on the size of the domain Ω. The retarded character of the problem requires the initial conditions to be defined functionally as

$$u\big|_{t\in(-r,0)} = \varphi(x,t), \quad u|_{t=0} = u_0(x), \quad \partial_t u|_{t=0} = u_1(x). \tag{4.1.62}$$

We assume that

$$\varphi(x,t) \in L_2(-r,0;H_0^2(\Omega)), \quad u_0(x) \in H_0^2(\Omega), \quad u_1(x) \in L_2(\Omega).$$

We also suppose that $L : H_0^2(\Omega) \mapsto L_2(\Omega)$ is a linear bounded operator.

As above (see also Section 3.3.1) we can define a *weak solution* (on the interval $[0,T]$) to problem (4.1.61) and (4.1.62) with the clamped boundary condition (4.1.4) as a function

$$u(t) \in L_2(-r,T;H_0^2(\Omega)) \cap L_\infty(0,T;H_0^2(\Omega)) \cap W_\infty^1(0,T;L_2(\Omega))$$

such that $u\big|_{t\in(-r,0)} = \varphi(x,t)$, $u|_{t=0} = u_0(x)$ and variational relation (4.1.39) holds with $p = p(u^t;t)$ and $d_0 g_0(s) = ds$.

Using the Galerkin method we can prove the following assertion.

4.1.27. Theorem. *Under the assumptions above problem* (4.1.61) *with the clamped boundary condition* (4.1.4) *and the initial data* (4.1.62) *has a unique weak solution on any interval* $[0,T]$. *This solution belongs to the class*

$$C(0,T;H_0^2(\Omega)) \cap C^1(0,T;L_2(\Omega))$$

and satisfies the energy relation

$$\mathcal{E}(t) + 2\int_s^t \int_\Omega du_t^2 dx d\tau = \mathcal{E}(s) + \int_s^t \int_\Omega (-Lu + p(u^\tau;\tau))u_t dx d\tau.$$

Here

$$\mathcal{E}(t) \equiv \mathcal{E}(u(t),u_t(t)) = \frac{1}{2}\int_\Omega \left[|u_t|^2 + |\Delta u|^2 + \frac{1}{2}|\Delta v(u)|^2 - [u+2f,F_0]u \right] dx,$$

where $v(u) \in H_0^2(\Omega)$ *is the Airy stress function determined from* (4.1.2). *Moreover, there exist constants* $C > 0$ *and* $a \geq 0$ *such that*

$$E(u(t),u_t(t)) \leq C\left(1 + E(u_0,u_1) + \int_{-r}^0 \|\varphi(\tau)\|_{2,\Omega}^2 d\tau \right) \cdot e^{at}, \quad t > 0,$$

where $E(u,u_t)$ *is a positive part of the energy* $\mathcal{E}(u,u_t)$ *given by* (4.1.7).

Proof. Our argument relies on the compactness method via Galerkin approximations and involves sharp estimates of the von Karman bracket given in Section 1.4. The corresponding a priori estimate can be derived in the same manner as in the proof of Theorem 3.3.1. We also refer to [33] and [80] for similar considerations in the case of Berger plate. □

4.2 Models with nonlinear boundary dissipation

We now consider von Karman evolutions with dissipative boundary conditions, in the framework of Chapter 2. The equations are given below.

$$\begin{cases} u_{tt} + d_0(x)g_0(u_t) + \Delta^2 u = [u+f, v+F_0] - Lu + p(x), & x \in \Omega, t > 0, \\ u|_{t=0} = u_0(x), \quad u_t|_{t=0} = u_1(x), \end{cases} \tag{4.2.1}$$

where $v = v(u)$ is a solution of the problem

$$\Delta^2 v + [u+2f, u] = 0, \quad v|_{\partial\Omega} = \frac{\partial v}{\partial n}\bigg|_{\partial\Omega} = 0. \tag{4.2.2}$$

As in Chapter 3 we restrict ourselves to the following basic, physically significant, types of boundary conditions defined on two disjoint portions Γ_0 and Γ_1 of the boundary $\Gamma = \Gamma_0 \cup \Gamma_1$:

1. **[clamped–hinged]**: $u = \nabla u = 0$ on Γ_0 and $u = 0$, $\Delta u = g_1((\partial/\partial n)u_t)$ on Γ_1;
2. **[clamped–free]**: $u = \nabla u = 0$ on Γ_0, $\Delta u + (1-\mu)B_1 u = -g_1((\partial/\partial n)u_t)$ on Γ_1, and
$$\frac{\partial}{\partial n}\Delta u + (1-\mu)B_2 u = \nu_1 u + \beta(x)[u^3 + cu_t] + g_2(u_t) \text{ on } \Gamma_1,$$
where B_1 and B_2 are defined by (1.3.20), $0<\mu<1$, $\nu_1>0$ ($\nu_1 \ge 0$ if $\Gamma_0 \neq \emptyset$) and $c>0$ are parameters and $\beta = \beta(x)$ is bounded on Ω and nonnegative function. The term βu_t could be included in the damping $g_2(u_t)$. However, even in the case β =const we prefer to keep the damping in this split form for more direct applications of abstract results from Chapter 2.

We assume the following hypotheses.

4.2.1. Assumption.

- The function $g_0 : \mathbb{R}^1 \mapsto \mathbb{R}^1$ is continuous and there exists $a_0 \ge 0$ such that $g_0(s) + a_0 s$ is increasing. The function $d_0(x)$ is a nonnegative bounded measurable function. We also assume (without loss of generality) that $g_0(0) = 0$.
- The functions g_1 and g_2 are monotone increasing, continuous from $\mathbb{R}$ into itself, and g_1 satisfies polynomial growth condition:
$$|g_1(s)| \le C(1+|s|^p), \quad s \in \mathbb{R},$$
with some constants $C>0$ and $p \ge 1$. We also assume that $g_i(0) = 0$ for $i = 1,2$.
- $f \in H^2(\Omega)$, $F_0 \in H^{3+\delta}(\Omega)$, $p \in L_2(\Omega)$, $\beta(x)$ lies in $L_\infty(\Omega)$ and is nonnegative.
- L is a linear bounded operator from $H^2(\Omega)$ into $L_2(\Omega)$.

For some of our results we also need a stronger version of Assumption 4.2.1, which is stated below.

4.2.2. Assumption. Assumption 4.2.1 is satisfied and the estimates

$$|g_0(s)| \le C(1+|s|),\ s \in \mathbb{R}, \quad \text{and} \quad ms^2 \le g_i(s)s \le Ms^2,\ |s| \ge 1,\ i = 1,2,$$

hold with some positive constants C, m and M.

4.2.1 Clamped–hinged boundary conditions

We consider problem (4.2.1) with the following combination of clamped and hinged boundary conditions

$$u = \nabla u = 0 \text{ on } \Gamma_0 \quad \text{and} \quad u = 0,\, \Delta u = -g_1(\frac{\partial}{\partial n} u_t) \text{ on } \Gamma_1, \tag{4.2.3}$$

where $\Gamma_0 \cup \Gamma_1 = \Gamma = \partial\Omega$. As before, we always assume that Γ_0 and Γ_1 are disjoint. This assumption is made in order to avoid potential difficulties with singularities of solutions to elliptic problems with mixed boundary conditions.

The energy functionals $\mathscr{E}$ and E for this case are the same as in Section 4.1 (cf. (4.1.5) and (4.1.7)):

$$\mathscr{E}(t) \equiv \mathscr{E}(u,u_t) = \frac{1}{2}\int_\Omega \left[|u_t|^2 + |\Delta u|^2 + \frac{1}{2}|\Delta v(u)|^2 - [u+2f, F_0]u - 2up \right] dx$$

and

$$E(t) \equiv E(u,u_t) = \frac{1}{2}\int_\Omega \left[|u_t|^2 + |\Delta u|^2 + \frac{1}{2}|\Delta v(u)|^2 \right] dx$$

As before we also have that

$$c\mathscr{E}(t) - K \le E(t) \le C\mathscr{E}(t) + K \tag{4.2.4}$$

with positive constant c, C, and K.

In this case the following functional analytic setup applies,

- $\mathscr{H} \equiv L_2(\Omega)$, $V \equiv L_2(\Omega)$, $U \equiv L_2(\Gamma_1)$, $U_0 \equiv H^{1/2}(\Gamma_1), U_0' = H^{-1/2}(\Gamma_1)$.
- $\mathscr{A} u \equiv \Delta^2 u, u \in \mathscr{D}(\mathscr{A})$, where

$$\mathscr{D}(\mathscr{A}) \equiv \left\{ u \in H_0^1(\Omega) \cap H^4(\Omega),\ \Delta u = 0, \text{ on } \Gamma_1, \frac{\partial}{\partial n} u = 0 \text{ on } \Gamma_0 \right\}. \tag{4.2.5}$$

- $Mu \equiv u$, $\mathscr{D}(M) = L_2(\Omega)$.
- Hence $V' = V = L_2(\Omega)$ and

$$\mathscr{D}(\mathscr{A}^{1/2}) = \left\{ v \in H_0^1(\Omega) \cap H^2(\Omega) \ :\ \frac{\partial}{\partial n} v = 0 \text{ on } \Gamma_0 \right\} \equiv H_0^1(\Omega) \cap H_{\Gamma_0}^2(\Omega),$$

  where $H_{\Gamma_0}^2(\Omega)$ is given by (3.2.11).
- $F(u,v) = \widetilde{F}(u) + a_0 d_0 v$ with $\widetilde{F}(u) \equiv [u+f, v(u)+F_0] - Lu + p$, where $v(u)$ solves (4.2.2).
- $D(u) \equiv d_0(x)\,[g_0(u) + a_0 u]$.
- As in Section 3.2.1, $G : L_2(\Gamma_1) \mapsto L_2(\Omega)$ denotes the biharmonic extension of the boundary values defined on Γ_1. That is,

$$Gv \equiv u \ \text{ iff } \ \begin{cases} \Delta^2 u = 0, \text{ in } \Omega, \\ u = 0 \text{ on } \Gamma,\ \frac{\partial}{\partial n} u = 0 \text{ on } \Gamma_0,\ \Delta u = v \text{ on } \Gamma_1. \end{cases} \tag{4.2.6}$$

  The mapping $g : U_0 \mapsto U_0'$ is determined by the function $g_1(s)$ according to the formula

$$(g(v), w)_{U_0, U_0'} = \int_{\Gamma_1} g_1(v) w d\Gamma, \quad v, w \in U_0.$$

The only difference in comparison with the clamped–hinged case with rotational forces (see Section 3.2.1) is that $M = I$ and $V = L_2(\Omega)$. Thus with the above notation the abstract model for the underlying PDE becomes

$$u_{tt} + \mathscr{A}u + D(u_t) + \mathscr{A}Gg(G^*\mathscr{A}u_t) = F(u,u_t). \tag{4.2.7}$$

Using the characterization of $G^*\mathscr{A} = \partial/\partial n$ which is given in (3.2.13) we can rewrite (4.2.7) in the following equivalent form

$$u_{tt} + Du_t + \mathscr{A}\left(u + Gg_1(\frac{\partial}{\partial n}u_t)\right) = F(u,u_t). \tag{4.2.8}$$

Thus the problem (4.2.1) and (4.2.3) is presented in the form (2.4.26).

In order to proceed, we next recall definitions of strong and generalized solutions.

4.2.3. Definition. A function

$$u \in C(0,T; H_0^1(\Omega) \cap H^2_{\Gamma_0}(\Omega)) \cap C^1(0,T,L_2(\Omega))$$

possessing the properties $u(x,0) = u_0(x)$ and $u_t(x,0) = u_1(x)$ is said to be

1. A *strong solution* to problem (4.2.1) with the clamped–hinged boundary conditions (4.2.3) on the interval $[0,T]$, iff
   - $u(t)$ is an absolutely continuous function with values in $H^2_{\Gamma_0}(\Omega) \cap H_0^1(\Omega)$ and $u_t \in L_1(a,b; H^2_{\Gamma_0}(\Omega) \cap H_0^1(\Omega))$ for any $0 < a < b < T$.
   - u_t is an absolutely continuous function with values in $L_2(\Omega)$ and $u_{tt} \in L_1(a,b; L_2(\Omega))$ for any $0 < a < b < T$.
   - $u(t) + Gg_1((\partial/\partial n)u_t(t)) \in \mathscr{D}(\mathscr{A})$ for almost all $t \in [0,T]$, where $\mathscr{D}(\mathscr{A})$ is given by (4.2.5) and G is defined in (4.2.6).
   - Equality (4.2.8) holds in $L_2(\Omega)$ for almost all $t \in [0,T]$.
2. A *generalized solution* to problem (4.2.1) and (4.2.3) on the interval $[0,T]$, iff there exists a sequence $\{u_n(t)\}$ of strong solutions to (4.2.1) and (4.2.3) with $(u_{0n}; u_{1n})$ instead of $(u_0; u_1)$ such that (4.1.10) holds.

■

4.2.4. Theorem. *Under Assumption* 4.2.1 *with reference to* (4.2.1), *subject to clamped–hinged boundary conditions* (4.2.3), *the following statements are valid with any* $T > 0$.

- **Generalized solutions:** *Assume that the initial data possess the properties* $u_0 \in H_0^1(\Omega) \cap H^2(\Omega), u_1 \in L_2(\Omega)$ *and also* $\nabla u_0 = 0$ *on* Γ_0.
  1. *Then there exists a unique generalized solution* $u(t)$ *such that the following energy inequality holds for all generalized solutions*

$$\mathscr{E}(t) + \int_s^t \int_\Omega d_0 g_0(u_t)u_t dx \le \mathscr{E}(s) - \int_s^t (Lu,u_t)_\Omega d\tau \tag{4.2.9}$$

for $t \geq s \geq 0$. *Thus, if* $L \equiv 0$ *and* $\sigma g_0(\sigma) \geq 0$ *for all* $\sigma \in \mathbb{R}$, *then* $\mathscr{E}(t) \leq \mathscr{E}(s)$ *for* $t \geq s \geq 0$.

2. *Assume that* $g_1(s)$ *possesses the property*

$$m|s|^{1+\delta} \leq g_1(s)s, \quad |s| \geq s_0, \tag{4.2.10}$$

for some constants $m > 0$, $\delta > 0$ *and* $s_0 \geq 0$. *Then*

$$\frac{\partial}{\partial n} u_t \equiv \partial_t \left[\frac{\partial}{\partial n} u \right] \in L_{1+\delta}([0,T] \times \Gamma_1) \tag{4.2.11}$$

and the energy inequality can be written in the form

$$\begin{aligned} &\mathscr{E}(u(t), u_t(t)) + \int_s^t \left[\int_\Omega d_0 g_0(u_t) u_t dx + \int_{\Gamma_1} \psi\left(\frac{\partial}{\partial n} u_t\right) d\Gamma \right] d\tau \\ &\leq \mathscr{E}(u(s), u_t(s)) - \int_s^t \int_\Omega L u u_t dx d\tau \end{aligned} \tag{4.2.12}$$

for every continuous convex function ψ *such that*

$$0 \leq \psi(s) \leq g_1(s)s \ \textit{for all}\ s \in \mathbb{R}. \tag{4.2.13}$$

3. *Assuming, in addition, that Assumption* 4.2.2 *holds, we also obtain the energy identity*

$$\begin{aligned} &\mathscr{E}(u(t), u_t(t)) + \int_s^t \left[\int_\Omega d_0 g_0(u_t) u_t dx + \int_{\Gamma_1} g_1\left(\frac{\partial}{\partial n} u_t\right) \frac{\partial}{\partial n} u_t d\Gamma \right] d\tau \\ &= \mathscr{E}(u(s), u_t(s)) - \int_s^t \int_\Omega L u u_t dx d\tau. \end{aligned} \tag{4.2.14}$$

- **Strong solutions:** *In addition to regularity of initial conditions assumed in the first part of the theorem, we assume the following compatibility condition (which we understand in the variational sense):* $u_0, u_1 \in H_0^1(\Omega) \cap H_{\Gamma_0}^2(\Omega)$ *are such that* $\Delta^2 u_0 \in L_2(\Omega)$ *and* $\Delta u_0 = -g_1(u_1)$ *on* Γ_1. *Then, there exists a unique strong solution such that*

$$\left.\begin{aligned} &u + G g_1\left(\tfrac{\partial}{\partial n} u_t\right) \in C_r(0,T;\mathscr{D}(\mathscr{A})) \cap L_\infty(0,T;\mathscr{D}(\mathscr{A})), \\ &u_t \in C_r(0,T;(H_{\Gamma_0}^2 \cap H_0^1)(\Omega)) \cap L_\infty(0,T;(H_{\Gamma_0}^2 \cap H_0^1)(\Omega)), \\ &u_{tt} \in C_r(0,T;L_2(\Omega)) \cap L_\infty(0,T;L_2(\Omega)), \end{aligned}\right\} \tag{4.2.15}$$

where $\mathscr{D}(\mathscr{A})$ *is given by* (4.2.5), G *is defined in* (4.2.6) *and as above* $C_r(0,T;X)$ *is the space of strongly right-continuous functions with values in* X. *Strong solutions satisfy the energy identity in* (4.2.14).

4.2.5. Remark. If one assumes a more precise structure of nonlinear terms g_i (e.g., g_i have polynomial bounds from below and above) then conditions of linear bounds assumed In Assumption 4.2.2 for derivation of energy equality for generalized solutions can be substantially relaxed. For instance, using the same idea as in the proofs of Theorems 3.2.4 and 3.2.12 we can assume (cf. (3.2.5) in Assumption 3.2.2) that

$$ms^{q_*} \le g_i(s)s \le m's^{q_*}, \quad |s| \ge s_0, \; i = 0,1,$$

with some constants $q_* > 1$, $s_0 > 0$, $m, m' > 0$. ■

In order to prove Theorem 4.2.4 we proceed in the same manner as before: we verify that Assumption 2.4.15 in Theorem 2.4.16 is satisfied.

4.2.6. Lemma. *Operators $\mathscr{A}, D, G, F$ introduced above comply with the requirements in Assumption 2.4.15 with $M \equiv I$.*

Proof. The verification of Assumptions 1–4 is the same as in the case of internal damping, so it is not repeated (see the argument given in the proof of Lemma 4.1.7). As for Assumption 2.4.15(5) (which is the requirements 3 and 4 in Assumption 2.4.1) we refer to the corresponding argument given in the proof of Lemma 3.2.6. We just recall here that the key role in the argument is played by the following facts: (i) $H^{1/2}(\Gamma_1) \subset L_q(\Gamma_1)$ for every $1 \le q < \infty$, (ii) $(\partial/\partial n) : H^2(\Omega) \mapsto H^{1/2}(\Gamma_1)$ is continuous, and (iii) $g_1 : L_q(\Gamma_1) \to L_{q/p}(\Gamma_1) \subset H^{-1/2}(\Gamma_1)$ is continuous and monotone for $q > p$. The latter fact can be proved by the same arguments as in Lemma 3.1.8. □

Proof of Theorem 4.2.4

Step 1: Global existence and uniqueness. We first note that from $G^*\mathscr{A} = \partial/\partial n$, we obtain that $Ker[G^*\mathscr{A}] \cap \mathscr{D}(\mathscr{A}^{1/2})$ contains $H_0^2(\Omega)$, which in turn is dense in $L_2(\Omega)$. Hence the density requirement in (2.4.39) (and thus in (2.4.35)) is amply satisfied. Therefore Lemma 4.2.6 allows for application of Theorem 2.4.16 which provides global existence of generalized and strong solutions. By Theorem 2.4.16 strong solutions satisfy the energy relation in (4.2.14) and possess the regularity stated in (4.2.15). Moreover, as in the proof of Theorem 3.2.4 from the energy identity (4.2.14) via Gronwall's lemma we can conclude that

$$||u(t)||_{2,\Omega} + ||u_t(t)||_\Omega + ||v(u)||_{2,\Omega} \le C_T, \quad \forall t \le T, \tag{4.2.16}$$

where the constant C_T depends on $||u_0||_{2,\Omega}, ||u_1||_\Omega, ||f||_{2,\Omega}, ||F_0||_{3+\delta,\Omega}$, and $||p_0||_\Omega$. As above, the same bound holds for generalized solutions.

Step 3: Energy inequality and identity for generalized solutions. We use the same idea as before (cf., e.g., the proofs of Theorems 3.2.4 and 4.1.4).

Energy inequality. Let $(u_0; u_1)$ be initial data of finite energy and let $u(t)$ be a corresponding generalized solution. Denote by $u_n(t)$ strong solutions such that

$$u_n \to u \text{ in } C(0,T;H^2(\Omega)), \quad u_{tn} \to u_t \text{ in } C(0,T;L_2(\Omega)). \tag{4.2.17}$$

By the same argument as in Theorem 3.2.4 we obtain the desired energy inequality (4.2.9).

From the coercivity assumption in (4.2.10) and from the energy relation (4.2.14) for strong solutions one can easily derive energy inequality (4.2.12) for generalized solutions along with regularity property in (4.2.11).

Energy equality. This step is more technical and requires more regularity imposed on nonlinear dissipation. Recall (see Remark 4.2.5) that these assumptions can be relaxed, provided more specific structure of nonlinearity is known. In order to obtain energy identity we assume that the dampings are of linear growth at infinity (such assumption is necessary in studying long-time behavior). More precisely we assume that Assumption 4.2.2 holds.

By Lemma 4.1.8 $D(v) = d_0\,[g_0(v) + a_0 v]$ is an m-accretive hemicontinuous operator in $L_2(\Omega)$ and hence by Proposition 1.2.5 it is demicontinuous. Therefore it follows from (4.2.17) that

$$\int_0^t \int_\Omega g_0(u_{nt}) u_{nt}\, dx dt \to \int_0^t \int_\Omega g_0(u_t) u_t\, dx dt. \tag{4.2.18}$$

In order to pass with the limit on the boundary term we use the same arguments as in the proof of of Theorem 3.2.4. We recall them shortly in this case.

We consider the difference of two strong solutions, say $u_n(t)$ and $u_m(t)$. In the same way as in the proof of Theorem 3.2.4 from the locally Lipschitz property of $F(u,v)$ using (i) a priori bounds of strong solutions in (4.2.16) along with (ii) monotonicity properties of the damping functions g_0 and g_1 we find that

$$\begin{aligned} &||u_n(t) - u_m(t)||_{2,\Omega}^2 + ||u_{nt}(t) - u_{mt}(t)||_\Omega^2 \\ &\quad + 2\int_0^t \int_{\Gamma_1} \left(g_1(\frac{\partial}{\partial n} u_{nt}) - g_1(\frac{\partial}{\partial n} u_{mt}) \right) \left(\frac{\partial}{\partial n} u_{nt} - \frac{\partial}{\partial n} u_{mt} \right) d\Gamma dt \\ &\le C_T \int_0^t [||u_n(t) - u_m(t)||_{2,\Omega}^2 + ||u_{nt}(t) - u_{mt}(t)||_\Omega^2] ds \\ &\quad + ||u_n(0) - u_m(0)||_{2,\Omega}^2 + ||u_{nt}(0) - u_{mt}(0)||_\Omega^2. \end{aligned}$$

Using Gronwall's inequality we conclude that

$$\begin{aligned} &||u_n(t) - u_m(t)||_{2,\Omega}^2 + ||u_{nt}(t) - u_{mt}(t)||_\Omega^2 \\ &\quad + \int_0^t \int_{\Gamma_1} \left(g_1(\frac{\partial}{\partial n} u_{nt}) - g_1(\frac{\partial}{\partial n} u_{mt}) \right) \left(\frac{\partial}{\partial n} u_{nt} - \frac{\partial}{\partial n} u_{mt} \right) d\Gamma dt \\ &\le C_T [||u_n(0) - u_m(0)||_{2,\Omega}^2 + ||u_{nt}(0) - u_{mt}(0)||_\Omega^2. \end{aligned}$$

Thus, in addition to having strong limits (4.2.17) we also have:

$$\int_0^t \int_{\Gamma_1} \left(g_1(\frac{\partial}{\partial n} u_{nt}) - g_1(\frac{\partial}{\partial n} u_{mt}) \right) \left(\frac{\partial}{\partial n} u_{nt} - \frac{\partial}{\partial n} u_{mt} \right) d\Gamma dt \to 0. \tag{4.2.19}$$

Is is also clear that from energy inequality and from (4.2.4) that

$$\int_0^t \int_{\Gamma_1} g_1(\frac{\partial}{\partial n} u_{nt}) \frac{\partial}{\partial n} u_{nt} d\Gamma dt \le C(E(0)).$$

Therefore, as in the proof of Theorem 3.2.4 applying Lemma 3.2.7 we obtain that (4.2.11) holds with $\delta = 1$ and there exists a subsequence $\{n_l\}$ such that

$$\int_0^t \int_{\Gamma_1} g_1(\frac{\partial}{\partial n} u_{n_l t}) \frac{\partial}{\partial n} u_{n_l t} d\Gamma dt \to \int_0^t \int_{\Gamma_1} g_1(\frac{\partial}{\partial n} u_t) \frac{\partial}{\partial n} u_t d\Gamma dt. \tag{4.2.20}$$

Moreover (see (3.2.32)) we have that

$$g_1(\frac{\partial}{\partial n} u_{n_l t}) \to g_1(\frac{\partial}{\partial n} u_t), \quad \frac{\partial}{\partial n} u_{n_l t} \to \frac{\partial}{\partial n} u_t \quad \text{weakly in } L_2((0,t) \times \Gamma_1).$$

The convergences (4.2.18) and (4.2.20) allow us to pass with the limit on energy identity written first for strong solutions. The limit process yields this identity valid for all generalized solutions. This completes the proof of Theorem 4.2.4.

Under some conditions generalized solutions do satisfy a weak form of differential equality.

4.2.7. Theorem. *Assume in addition to Assumption* 4.2.1 *that (i) either $g_1(s)$ satisfies Assumption* 4.2.2, *or else, (ii) the function $g_1(s)$ possesses the property*[4]

$$(s_1 - s_2)(g_1(s_1) - g_1(s_2)) \ge c_0 |s_1 - s_2|^r, \quad s_1, s_2 \in \mathbb{R}, \tag{4.2.21}$$

for some $c_0 > 0$ and $r \ge 1$. Then generalized solutions of the von Karman system (4.2.1) *with clamped–hinged boundary conditions* (4.2.3) *are also weak solutions. This means that for any generalized solutions $u(t)$ the functions $t \mapsto (d_0 g_0(u_t), \phi)_\Omega$ and $t \mapsto (g_1((\partial/\partial n)u_t), (\partial/\partial n)\phi)_{\Gamma_1}$ are integrable for $\phi \in H_0^1(\Omega) \cap H^2(\Omega)$, $\nabla\phi\big|_{\Gamma_0} = 0$, and the following variational relation*

$$\begin{aligned} &(u_t(t), \phi)_\Omega - (u_1, \phi)_\Omega \\ &+ \int_0^t \left[(\Delta u, \Delta \phi)_\Omega + (d_0 g_0(u_t), \phi)_\Omega + (g_1(\frac{\partial}{\partial n} u_t), \frac{\partial}{\partial n} \phi)_{\Gamma_1} \right] d\tau \\ &+ \int_0^t [(-[v(u) + F_0, u + f] + Lu - p, \phi)_\Omega] \, d\tau = 0 \end{aligned} \tag{4.2.22}$$

holds for any $\phi \in H_0^1(\Omega) \cap H^2(\Omega)$ such that $\nabla\phi = 0$ on Γ_0, where $v = v(u)$ is defined from (4.2.2) *and, as above, $(\cdot,\cdot)_{\mathscr{O}}$ denotes the inner product in $L_2(\mathscr{O})$.*

Proof. In the case (i) we can use the same argument as in the proof of energy equality for generalized solutions. We first write (4.2.22) for strong solutions and then using (4.2.17) and weak convergence $g_1((\partial/\partial n)u_{n_l t}) \to g_1((\partial/\partial n)u_t)$ make limit transition.

---

[4] See Remark 3.2.9 concerning the validity of this hypothesis.

In case (ii) we argue similarly and in the same way as in the proof of Theorem 3.2.8 by noting that relations (4.2.19) and (4.2.21) imply strong convergence $(\partial/\partial n)u_{nt} \to (\partial/\partial n)u_t$ in $L_r((0,T)\times\Gamma_1)$. □

4.2.2 Clamped–free boundary conditions

Now we consider the case of clamped–free boundary conditions:

$$\begin{cases} u = \nabla u = 0 \text{ on } \Gamma_0, \\ \Delta u + (1-\mu)B_1 u = -g_1(\frac{\partial u_t}{\partial n}) \text{ on } \Gamma_1, \\ \frac{\partial}{\partial n}\Delta u + (1-\mu)B_2 u = \nu_1 u + \beta(x)(u^3 + cu_t) + g_2(u_t) \text{ on } \Gamma_1, \end{cases} \tag{4.2.23}$$

where B_1 and B_2 are given by (1.3.20), $\nu_1 > 0$ ($\nu_1 \geq 0$ if $\Gamma_0 \neq \emptyset$), β is a nonnegative bounded measurable function and c is a positive constant.

In this case the energy has the form

$$\mathscr{E}(t) \equiv \frac{1}{2}\int_\Omega \left[|u_t|^2 + \frac{1}{2}|\Delta v(u)|^2 - 2[f,F_0]u - 2up\right]dx + \frac{1}{2}a(u,u) + \frac{1}{4}\int_{\Gamma_1}\beta u^4 d\Gamma,$$

where $a(u,w) = a_0(u,w) + \nu_1\int_{\Gamma_1} uw\,d\Gamma$ with $a_0(u,w)$ defined by (1.3.4). Again in what follows we denote by E the positive part of $\mathscr{E}$,

$$E(t) \equiv \frac{1}{2}\int_\Omega \left[|u_t|^2 + \frac{1}{2}|\Delta v(u)|^2\right]dx + \frac{1}{2}a(u,u) + \frac{1}{4}\int_{\Gamma_1}\beta u^4 d\Gamma.$$

As above there exist positive constant c, C, K such that

$$c\mathscr{E}(t) - K \leq E(t) \leq C\mathscr{E}(t) + K. \tag{4.2.24}$$

For functional analytic setup we introduce the following spaces and operators,

- $\mathscr{H} = V \equiv L_2(\Omega)$, $U \equiv L_2(\Gamma_1)\times L_2(\Gamma_1)$, $U_0 \equiv H^{1/2}(\Gamma_1)\times H^{3/2}(\Gamma_1), U_0' = H^{-1/2}(\Gamma_1)\times H^{-3/2}(\Gamma_1), Z \equiv L_2(\Gamma_1)$.
- $\mathscr{A}u \equiv \Delta^2 u, u \in \mathscr{D}(\mathscr{A})$, where

$$\mathscr{D}(\mathscr{A}) \equiv \left\{ u \in H^4(\Omega) \;\middle|\; \begin{array}{l} u = \nabla u = 0 \text{ on } \Gamma_0, \\ \frac{\partial}{\partial n}\Delta u + (1-\mu)B_2 u - \nu_1 u = 0 \text{ on } \Gamma_1, \\ \Delta u + (1-\mu)B_1 u = 0 \text{ on } \Gamma_1 \end{array} \right\}.$$

- Hence $V' = L_2(\Omega)$ and $\mathscr{D}(\mathscr{A}^{1/2}) = H^2_{\Gamma_0}(\Omega)$, where

$$H^2_{\Gamma_0}(\Omega) \equiv \left\{ u \in H^2(\Omega) : \; u = 0, \; \frac{\partial}{\partial n}u = 0 \text{ on } \Gamma_0 \right\}.$$

- $F(u,v) = \widetilde{F}(u) + a_0 d_0 v$ with $F(u) \equiv [u+f, v(u)+F_0] - Lu + p$, where $v(u)$ solves (4.2.2).
- $D(u) \equiv d_0(x)\,[g_0(u) + a_0 u]$.
- If $\beta \neq 0$, then, as in Section 4.1.3 the effect of the nonlinear boundary condition can be built into operator $D_0 h(u)$. Indeed, we define $D_0 : L_2(\Gamma_1) \to [\mathscr{D}(\mathscr{A}^{1/2})]'$

$$\int_\Omega D_0(u)\phi dx = \int_{\Gamma_1} \sqrt{c\beta} u\phi d\Gamma, \quad \forall \phi \in H^2_{\Gamma_0}(\Omega).$$

The operator $h : \mathscr{D}(\mathscr{A}^{1/2}) = H^2_{\Gamma_0}(\Omega) \to L_2(\Gamma_1)$ is given by $h(u) \equiv \sqrt{\beta c^{-1}} u^3\big|_{\Gamma_1}$.
- $G : U_0' \to \mathscr{H}, \ \ G(g_1, g_2) \equiv G_1 g_1 - G_2 g_2$ where G_i, $i = 1,2$ are the same as in Section 3.2.2:

$$G_1 v \equiv u \ \text{ iff } \begin{cases} \Delta^2 u = 0, \text{ in } \Omega, \quad u = 0, \nabla u = 0 \text{ on } \Gamma_0, \\ \Delta u + (1-\mu) B_1 u = v \text{ on } \Gamma_1, \\ \frac{\partial}{\partial n} \Delta u + (1-\mu) B_2 u = \nu_1 u \text{ on } \Gamma_1, \end{cases}$$

and

$$G_2 v \equiv u \ \text{ iff } \begin{cases} \Delta^2 u = 0, \text{ in } \Omega, \quad u = 0, \nabla u = 0 \text{ on } \Gamma_0, \\ \Delta u + (1-\mu) B_1 u = 0 \text{ on } \Gamma_1, \\ \frac{\partial}{\partial n} \Delta u + (1-\mu) B_2 u = \nu_1 u + v \text{ on } \Gamma_1. \end{cases}$$

The mapping $g : U_0 \mapsto U_0'$ has the form

$$g : (v_1; v_2) \mapsto (g_1(v_1); g_2(v_2)), \quad (v_1; v_2) \in U_0.$$

By calculation given in Section 3.2.2 (see also Proposition 1.3.13) we have that the mappings

$$G_1 : L_2(\Gamma_1) \mapsto \mathscr{D}(\mathscr{A}^{5/8-\varepsilon}), \qquad G_2 : L_2(\Gamma_1) \mapsto \mathscr{D}(\mathscr{A}^{7/8-\varepsilon}),$$

$$G_1^* \mathscr{A} : \mathscr{D}(\mathscr{A}^{1/2}) \mapsto H^{1/2}(\Gamma_1), \quad G_2^* \mathscr{A} : \mathscr{D}(\mathscr{A}^{1/2}) \mapsto H^{3/2}(\Gamma_1),$$

are bounded and for every $u \in \mathscr{D}(\mathscr{A}^{3/8+\varepsilon}) \subset H^{3/2+4\varepsilon}(\Omega)$ we have that

$$G^* \mathscr{A} u = (G_1^* \mathscr{A} u; -G_2^* \mathscr{A} u) = \left(\frac{\partial}{\partial n} u|_{\Gamma_1}; u|_{\Gamma_1} \right). \tag{4.2.25}$$

Using Green's formula (1.3.3) and boundary conditions (4.2.23) we obtain that

$$\int_\Omega \Delta^2 u w dx = a(u,w) + \int_{\Gamma_1} [\beta(u^3 + cu_t) + g_2(u_t)] w d\Gamma + \int_{\Gamma_1} \frac{\partial}{\partial n} w g_1(\frac{\partial}{\partial n} u_t) d\Gamma$$

for any $w \in \mathscr{D}(\mathscr{A}^{1/2})$, where $u(t)$ is a smooth solution to problem (4.2.1) and (4.2.23). Therefore

$$\int_\Omega \Delta^2 u w dx = (\mathscr{A} u + D_0 D_0^* u_t + D_0 h(u) + \mathscr{A} G g(G^* \mathscr{A} u_t), w)_{\mathscr{D}(\mathscr{A}^{1/2}), [\mathscr{D}(\mathscr{A}^{1/2})]'}.$$

Thus with the above notation the abstract model takes the form

$$u_{tt} + \mathscr{A}u + D_0 h(u) + D_0 D_0^* u_t + D(u_t) + \mathscr{A} G g(G^* \mathscr{A} u_t) = F(u, u_t)$$

or equivalently

$$u_{tt} + Du_t + \mathscr{A}\left[u + G_1 g_1(\frac{\partial}{\partial n} u_t) - G_2\left(g_2(u_t|_{\Gamma_1}) + \beta c u_t|_{\Gamma_1} + \beta u^3|_{\Gamma_1}\right)\right] = F(u, u_t). \tag{4.2.26}$$

In order to proceed, we restate the concepts of strong and generalized solutions.

4.2.8. Definition. A function

$$u \in C(0, T; H^2_{\Gamma_0}(\Omega)) \cap C^1(0, T; L_2(\Omega))$$

possessing the properties $u(x, 0) = u_0$ and $u_t(x, 0) = u_1$ is said to be

1. A *strong solution* to problem (4.2.1) with the clamped–free boundary conditions (4.2.23) on the interval $[0, T]$, iff
   - $u(t)$ is an absolutely continuous function with values in $H^2_{\Gamma_0}(\Omega)$ and $u_t \in L_1(a, b; H^2_{\Gamma_0}(\Omega))$ for any $0 < a < b < T$.
   - u_t is an absolutely continuous function with values in $L_2(\Omega)$ and $u_{tt} \in L_1(a, b; L_2(\Omega))$ for any $0 < a < b < T$.
   - $u + G_1 g_1((\partial/\partial n) u_t) - G_2(g_2(u_t|_{\Gamma_1}) + \beta c u_t|_{\Gamma_1} + \beta u^3|_{\Gamma_1}) \in \mathscr{D}(\mathscr{A})$ for almost all $t \in [0, T]$.
   - Equality (4.2.26) holds in $L_2(\Omega)$ for almost all $t \in [0, T]$.
2. A *generalized solution* to problem (4.2.1) and (4.2.23) on the interval $[0, T]$, iff there exists a sequence $\{u_n(t)\}$ of strong solutions to (4.2.1) and (4.2.23) with $(u_{0n}; u_{1n})$ instead of $(u_0; u_1)$ such that $u_n \to u$ in the space $C(0, T; H^2_{\Gamma_0}(\Omega)) \cap C^1(0, T; L_2(\Omega))$.

■

As above, applying the results from Chapter 2 we obtain the following assertion.

4.2.9. Theorem. *Under Assumption* 4.2.1 *with reference to* (4.2.1)*, subject to clamped–free boundary conditions* (4.2.23)*, the following statements are valid with any* $T > 0$.

- **Generalized solutions:** *Let* $u_0 \in H^2_{\Gamma_0}(\Omega)$ *and* $u_1 \in L_2(\Omega)$.
  1. *Then there exists a unique generalized solution* $u(t)$ *such that the following energy inequality holds for every* $t > 0$,

$$\mathscr{E}(t) + \int_0^t \int_\Omega d_0 g_0(u_t) u_t \, dx d\tau \le \mathscr{E}(0) + \int_0^t ([u, F_0] - Lu, u_t) d\tau. \tag{4.2.27}$$

  2. *Assume that* $g_i(s)$ *possesses the property*

$$m|s|^{1+\delta_i} \le g_i(s)s, \quad |s| \ge s_0,\ i = 1,2, \tag{4.2.28}$$

for some constants $m > 0$ and $s_0 \ge 0$, and $\delta_i > 0$. Then

$$u_t \in L_{1+\delta_2}([0,T] \times \Gamma_1), \quad \frac{\partial}{\partial n} u_t \in L_{1+\delta_1}([0,T] \times \Gamma_1) \tag{4.2.29}$$

and the energy inequality can be written in the form

$$\begin{aligned} &\mathscr{E}(t) + \int_0^t \int_\Omega d_0 g_0(u_t) u_t dx d\tau \\ &+ \int_0^t \int_{\Gamma_1} \left[\psi_1\left(\frac{\partial}{\partial n} u_t\right) + \psi_2(u_t) + \beta c u_t^2 \right] d\Gamma d\tau \\ &\le \mathscr{E}(0) + \int_0^t \int_\Omega ([u, F_0] - Lu) u_t dx d\tau \end{aligned} \tag{4.2.30}$$

for every pair of continuous convex functions ψ_1 and ψ_2 such that

$$0 \le \psi_i(s) \le g_i(s)s, \quad \text{for all } s \in \mathbb{R}. \tag{4.2.31}$$

3. *If, in addition Assumption 4.2.2 is valid,*[5] *then the following energy identity holds:*

$$\begin{aligned} &\mathscr{E}(t) + \int_0^t \int_\Omega d_0 g_0(u_t) u_t dx d\tau \\ &+ \int_0^t \int_{\Gamma_1} \left[g_1\left(\frac{\partial}{\partial n} u_t\right) \frac{\partial}{\partial n} u_t + g_2(u_t) u_t + \beta c u_t^2 \right] d\Gamma d\tau \\ &= \mathscr{E}(0) + \int_0^t \int_\Omega ([u, F_0] - Lu) u_t dx d\tau. \end{aligned} \tag{4.2.32}$$

- **Strong solutions:** *We assume that $u_0, u_1 \in H^2_{\Gamma_0}(\Omega)$ and, moreover, in the variational sense we have that $\Delta^2 u_0 \in L_2(\Omega)$ and*

$$\Delta u_0 + (1-\mu) B_1 u = -g_1\left(\frac{\partial}{\partial n} u_1\right) \text{ on } \Gamma_1,$$

$$\frac{\partial}{\partial n} \Delta u_0 + (1-\mu) B_1 u_0 = \nu_1 u_0 + \beta[u_0^3 + c u_1] + g_2(u_1) \text{ on } \Gamma_1.$$

Then, there exists a unique strong solution such that

[5] This hypothesis can be relaxed in the line with Remark 4.2.5.

$$\left.\begin{aligned} &w \in C_r(0,T;\mathscr{D}(\mathscr{A})) \cap L_\infty(0,T;\mathscr{D}(\mathscr{A})), \\ &u_t \in C_r(0,T;H^2_{\Gamma_0}(\Omega)) \cap L_\infty(0,T;H^2_{\Gamma_0}(\Omega)), \\ &u_{tt} \in C_r(0,T;L_2(\Omega)) \cap L_\infty(0,T;L_2(\Omega)), \end{aligned}\right\} \tag{4.2.33}$$

where

$$w = u(t) + G_1 g_1\left(\frac{\partial}{\partial n} u_t(t)\right) - G_2\left(g_2(u_t(t)|_{\Gamma_1}) + \beta c u_t(t)|_{\Gamma_1} + \beta u(t)^3|_{\Gamma_1}\right)$$

and $C_r(0,T;X)$ *is the set of strongly right-continuous functions. Strong solutions satisfy the energy identity in* (4.2.32).

4.2.10. Remark. The model considered in (4.2.23) can be slightly generalized in order to incorporate a more general form of free boundary conditions

$$\frac{\partial}{\partial n}\Delta u + (1-\mu)B_2 u = \nu_1 u + \beta(x)(u^3 + c(u_t)) + g_2(u_t) \text{ on } \Gamma_1, \tag{4.2.34}$$

where $c(s)$ is a continuous monotone function subject to the same growth conditions as g_2 and such that $c(s)s \geq m|s|^2$, $|s| >> 1$. Although the model considered above does not directly fit the abstract setting developed earlier and used for the proof of Theorem 4.2.9, rather obvious modifications of that setting (see [74] for a similar modification in the case of the wave equation) allow us to extend the validity of the results of Theorem 4.2.9 to hold for models with free boundary conditions given in (4.2.34). Moreover, we can also allow that the damping function $g_2(s)$ may depend on $x \in \Gamma_1$ in a regular way (with bounds that are uniform in $x \in \Gamma_1$). ■

In order to prove Theorem 4.2.9 we proceed in the same manner as before: we verify that Assumption 2.4.1 in Theorem 2.4.5 is satisfied. However, as in Theorem 4.1.13 we need to involve a regularization of the term $h(u)$.

4.2.11. Lemma. *Operators* $\mathscr{A}, D, G, F, D_0$ *introduced above comply with the requirements 1–7 in Assumption* 2.4.1 *with* $M \equiv I$.

Proof. Verification of the requirements 1,2,5,6 is the same as in the case of internal damping (cf. Lemma 4.1.7), so it is not repeated.

Let us verify Requirement 3. Because the function $g_1(s)$ is polynomially bounded and $g_2(s)$ is continuous, the continuity of the Nemytskij operator

$$(g_1, g_2) \,:\, U_0 \equiv H^{1/2}(\Gamma_1) \times H^{3/2}(\Gamma_1) \mapsto U_0' \equiv H^{-1/2}(\Gamma_1) \times H^{-3/2}(\Gamma_1)$$

follows from standard nonlinear analysis and classical Sobolev embedding

$$H^{1/2}(\Gamma_1) \subset L_p(\Gamma_1) \text{ for any } 1 \leq p < \infty \text{ and } H^{3/2}(\Gamma_1) \subset C(\Gamma_1).$$

The monotonicity of the mapping $(g_1; g_2)$ is obvious.

The requirement 4 in Assumption 2.4.1 reduces to the fact that

$$G^*\mathscr{A}^{1/2} : \mathscr{H} \to U_0 \equiv H^{1/2}(\Gamma_1) \times H^{3/2}(\Gamma_1)$$

is bounded. By (4.2.25) the above translates into

$$u \mapsto G^*\mathscr{A}^{1/2}u = (\frac{\partial}{\partial n}\mathscr{A}^{-1/2}u, \mathscr{A}^{-1/2}u)$$

is bounded from $L_2(\Omega)$ into $H^{1/2}(\Gamma_1) \times H^{3/2}(\Gamma_1)$. This last assertion follows from the fact that $\mathscr{D}(\mathscr{A}^{1/2}) = H^2_{\Gamma_0}(\Omega)$ and the trace theorem which implies that the maps

$$\frac{\partial}{\partial n} : H^2(\Omega) \to H^{1/2}(\Gamma), \quad \cdot|_\Gamma : H^2(\Omega) \to H^{3/2}(\Gamma)$$

are bounded. This proves validity of Requirement 4 in Assumption 2.4.1.

To check Requirement 7 concerning D_0 of Assumption 2.4.1 we refer to the argument used in Lemma 4.1.17. □

Proof of Theorem 4.2.9.

Step 1: Regularized problem. As in the proof of Theorem 4.1.13 we consider the regularized problem

$$u_{tt} + \mathscr{A}u + D_0[\beta h_K(u)] + D_0 D_0^* u_t + D(u_t) + \mathscr{A}Gg(G^*\mathscr{A}u_t) = F(u, u_t), \quad (4.2.35)$$

where the regularization h_K of the function h is given by (4.1.31). With the same notations as in the proof of Theorem 4.1.13 for arbitrary $R > 0$ we choose K such that $K \geq 2c_0 R$ and take the initial data possessing property (4.1.33).

Step 2: Local solutions to regularized problem. Applying Lemma 4.2.11 and Theorem 2.4.5 we obtain local (in time) existence of generalized solutions and also strong (local) solutions to problem (4.2.35) possessing the regularity (4.2.33). Local generalized solutions exist for all initial data $(u_0; u_1)$ from $H = H_0^2(\Omega) \times L_2(\Omega)$. The point is that by (4.2.25) we have that $\mathrm{Ker}\,[G^*\mathscr{A}] = H_0^2(\Omega) \supset C_0^\infty(\Omega)$. Therefore as in the proof of Theorem 4.1.13 one can show that that the set $\mathscr{L}$ given by (4.1.35) with $\mathscr{D}^* \cap H^2_{\Gamma_0}(\Omega)$ instead of $\mathscr{D}^*$ is dense in H and belongs to the domain of the corresponding nonlinear operator A given by (2.4.5).

Step 3: A priori bounds and global existence for (4.2.35). The energy identity stated in Theorem 4.2.9 follows, as before, by standard integration by parts which is fully justified for strong solutions. Indeed, the regularity guaranteed for strong solutions allows us to perform classical integration by parts.

By using property (4.2.24) we obtain

$$E_K(t) + \int_0^t \int_\Omega d_0\left[g_0(u_t) + a_0 u_t\right] u_t dx dt$$
$$+ \int_0^T \int_{\Gamma_1} \left[g_1\left(\frac{\partial}{\partial n} u_t\right) \frac{\partial}{\partial n} u_t + g_2(u_t) u_t + c\beta u_t^2 \right] d\Gamma dt$$

$$\leq C(E(0)) + C\int_0^t \left(||u||_{2,\Omega}^2 + ||u_t||_{\Omega}^2\right) dt$$

for $t \in [0,T]$, where $[0,T]$ is any interval of the existence of solutions to (4.2.35) and E_K is the corresponding regularized positive part of the energy. Thus, from the monotonicity properties of g_i and Gronwall's inequality we conclude the following a priori bound

$$||u(t)||_{2,\Omega} + ||u_t(t)||_{\Omega} \leq C_T(R) \equiv C_T(R, ||f||_{2,\Omega}, ||F_0||_{3+\delta,\Omega}, ||p_0||_{\Omega}) \quad (4.2.36)$$

for all $t \in [0,T]$, where $C_T(R)$ does not depend on the regularizing parameter K. Because inequality (4.2.36) is stable with respect to the norm generated by the energy, the same a priori bound holds for generalized solutions. This, in turn, proves global (in time) existence of generalized and strong solutions to problem (4.2.35).

Step 4: Existence and uniqueness of solutions to the original problem. The argument is the same as in the proof of Theorem 4.1.13. The regularity property in (4.2.29) follows from the coercivity assumption in (4.2.28) and the energy relation written above.

Step 5: Energy relations for generalized solutions.

Energy inequality. The energy inequalities (4.2.27) and (4.2.30) for a generalized solution $u(t)$ follow from the energy identity for strong solutions $u_n(t)$ which approximate $u(t)$ in the sense that

$$u_n \to u \ \text{ in } C([0,T]; H^2(\Omega)), \quad u_{nt} \to u_t \ \text{ in } C([0,T], L_2(\Omega)). \quad (4.2.37)$$

The remaining arguments are the same as in the proof of Theorem 3.2.12.

Energy equality. In order to obtain energy identity we assume that the boundary dissipation is of linear growth at infinity (see Assumption 4.2.2). As above (see the proof of Theorem 4.2.4, for instance) in addition to having strong limits (4.2.37) we also have:

$$\int_0^t \int_{\Gamma_1} \left[g_1\left(\frac{\partial}{\partial n} u_{nt}\right) - g_1\left(\frac{\partial}{\partial n} u_{mt}\right)\right]\left(\frac{\partial}{\partial n} u_{nt} - \frac{\partial}{\partial n} u_{mt}\right) d\Gamma d\tau$$
$$+ \int_0^t \int_{\Gamma_1} [g_2(u_{nt}) - g_2(u_{mt})](u_{nt} - u_{mt}) d\Gamma d\tau \to 0. \quad (4.2.38)$$

From the energy identity for strong solutions we also have that

$$\int_0^t \int_{\Gamma_1} g_1\left(\frac{\partial}{\partial n} u_{nt}\right) \frac{\partial}{\partial n} u_{nt} d\Gamma dt + \int_0^t \int_{\Gamma_1} g_2(u_{nt}) u_{nt} d\Gamma d \leq C.$$

Therefore we can apply Lemma 3.2.7 to conclude that (4.2.29) holds with $\delta_1 = \delta_2 = 1$ and to obtain weak convergence on a subsequence:

$$\left(\frac{\partial}{\partial n} u_{nt}; u_{nt}\right) \to \left(\frac{\partial}{\partial n} u_t; u_t\right) \ \text{ weakly in } \ L_2(\Sigma_1) \times L_2(\Sigma_1)$$

and

$$\left(g_1(\frac{\partial}{\partial n}u_{nt}); g_2(u_{nt})\right) \to \left(g_1(\frac{\partial}{\partial n}u_t); g_2(u_t)\right) \quad \text{weakly in } L_2(\Sigma_1) \times L_2(\Sigma_1).$$

As above we use notation $\Sigma_1 = [0,T] \times \Gamma_1$. Moreover, we have

$$\lim_{n\to\infty} \int_0^t \int_{\Gamma_1} \left[g_1\left(\frac{\partial}{\partial n}u_{nt}\right) \frac{\partial}{\partial n}u_{nt} + g_2(u_{nt})u_{nt} \right] d\Gamma d\tau$$
$$= \int_0^t \int_{\Gamma_1} [g_1(\frac{\partial}{\partial n}u_t)\frac{\partial}{\partial n}u_t + g_2(u_t)u_t] d\Gamma d\tau. \tag{4.2.39}$$

By (4.2.37) and linear growth (at infinity) of $g_0(s)$ as in the previous section we can prove that

$$\int_0^t \int_\Omega g_0(u_{nt})u_{nt}dxdt \to \int_0^t \int_\Omega g_0(u_t)u_t dxdt. \tag{4.2.40}$$

The above convergence (4.2.39) and (4.2.40) allows us to pass with the limit on the energy identity written first for strong solutions. The limit process yields this identity valid for all generalized solutions.

This completes the proof of Theorem 4.2.9.

Finally we show that generalized solutions are also weak.

4.2.12. Theorem. *Assume in addition to Assumption* 4.2.1 *that (i) either $g_1(s)$ and $g_2(s)$ satisfy Assumption* 4.2.2, *or else, (ii) the functions $g_1(s)$ and $g_2(s)$ possess the property*

$$(s_1 - s_2)(g_i(s_1) - g_i(s_2)) \geq c_0|s_1 - s_2|^r, \quad s_1, s_2 \in \mathbb{R},\ i = 1,2,$$

for some $c_0 > 0$ and $r \geq 1$. Then the generalized solutions of von Karman system (4.2.1) *and* (4.2.2) *the clamped–free boundary conditions* (4.2.23) *are also weak: any generalized solution $u(t)$ (i) possesses the property that the functions $t \mapsto (d_0 g_0(u_t), \phi)_\Omega$, $t \mapsto (g_1((\partial/\partial n)u_t), (\partial/\partial n)\phi)_{\Gamma_1}$, and $t \mapsto (g_2(u_t), \phi)_{\Gamma_1}$ are integrable, and (ii) satisfies the following variational relation*

$$(u_t(t), \phi)_\Omega - (u_1, \phi)_\Omega + \int_0^t [a(u,\phi) + (d_0 g_0(u_t), \phi)_\Omega]\, ds \tag{4.2.41}$$
$$+ \int_0^t \left[(g_1(\frac{\partial}{\partial n}u_t), \frac{\partial}{\partial n}\phi)_{\Gamma_1} + (g_2(u_t), \phi)_{\Gamma_1} \right] ds$$
$$+ \int_0^t [(-[v(u) + F_0, u + f] + Lu - p, \phi)_\Omega]\, ds + \int_0^t (\beta[u^3 + cu_t], \phi)_{\Gamma_1}\, ds = 0$$

for any $\phi \in H^2_{\Gamma_0}(\Omega)$, where $v = v(u)$ is defined from (4.2.2).

Proof. We first write (4.2.41) for strong solutions. Then we apply the same argument as above (see the comments in the proof of Theorem 4.2.7 and also Theorem 3.2.8, for instance) to perform the limit transition. □

4.2.13. Remark. In concluding this section we note that one could also consider the von Karman evolutions with boundary conditions which contain memory terms on free part of the boundary. One of the possible formulations of the problem is the following,

$$\partial_t^2 u + \Delta^2 u - [u+f, v+F_0] = p(x), \quad x \in \Omega,\ t > 0,$$

where $v = v(u) \in H_0^2(\Omega)$ the Airy stress function (which solves (4.2.2)). It is also assumed that the boundary $\Gamma = \partial\Omega$ consists of two disjoint portions Γ_0 and Γ_1 ($\Gamma = \Gamma_0 \cup \Gamma_1$) and the following boundary conditions (of the clamped–free type) are imposed,

$$u = \nabla u = 0 \ \text{ on } \Gamma_0,$$

$$\frac{\partial}{\partial n} u + \int_0^t g_1(t-s)\left[\Delta u(s) + (1-\mu)B_1 u(s) + \rho_1 \frac{\partial}{\partial n} u(s)\right] ds = 0 \ \text{ on } \Gamma_1,$$

$$u - \int_0^t g_2(t-s)\left[\frac{\partial}{\partial n}\Delta u(s) + (1-\mu)B_2 u(s) - \rho_2 u(s)\right] ds = 0 \ \text{ on } \Gamma_1,$$

where B_1 and B_2 is defined by (1.3.20). We equip the problem with the initial data

$$u|_{t=0} = u_0(x) \in H^2_{\Gamma_0}(\Omega), \quad \partial_t u|_{t=0} = u_1(x) \in L_2(\Omega).$$

Well-posedness of this problem was studied in [231] in the case when $f \equiv 0$, $F_0 \equiv 0$, and $\Gamma_0 = \emptyset$. In the general case the argument is similar. ■

4.3 Quasi-static model with clamped boundary condition

In this section we consider problem (4.1.1) and (4.1.2) with clamped boundary conditions (4.1.4) in quasi-static formulation.The model describes the case when the inertia forces are small in comparison with the resisting forces of a medium. We mainly follow the presentation given in [56].

We have the following system:

$$\begin{cases} \partial_t u + \Delta^2 u - [u+f, v+F_0] + Lu = p(x), \\ u|_{\partial\Omega} = \left.\frac{\partial u}{\partial n}\right|_{\partial\Omega} = 0, \quad u|_{t=0} = u_0(x), \end{cases} \tag{4.3.1}$$

where $v = v(u)$, the Airy stress function, is a solution of the problem

$$\Delta^2 v + [u+2f, u] = 0, \quad v|_{\partial\Omega} = \left.\frac{\partial v}{\partial n}\right|_{\partial\Omega} = 0. \tag{4.3.2}$$

By *weak solution* of problem (4.3.1) and (4.3.2) on the interval $[0,T]$ we mean a function $u(t) \in L_\infty(0,T;L_2(\Omega)) \cap L_2(0,T;H_0^2(\Omega))$ which is weakly continuous in $L_2(\Omega)$ and satisfies relations (4.3.1) in the sense of distributions.

4.3.1. Theorem. *Assume that* $f(x) \in H^2(\Omega)$, $F_0(x) \in H^{3+\delta}(\Omega)$, $p(x) \in L_2(\Omega)$. *We also suppose* $L : H^2(\Omega) \mapsto L_2(\Omega)$ *is a bounded linear operator. Then for any* $u_0 \in L_2(\Omega)$ *problem* (4.3.1) *and* (4.3.2) *on any interval* $[0,T]$ *has a weak solution* $u(t)$ *that possesses the properties*

$$\|u(t)\|_\Omega^2 + \int_0^t \left(\|\Delta u\|_\Omega^2 + \|\Delta v(u)\|_\Omega^2\right) d\tau \le \|u_0\|_\Omega^2 + Ct, \quad t > 0, \tag{4.3.3}$$

and there exists $\omega > 0$ *such that*

$$\|u(t)\|_\Omega^2 \le \|u_0\|_\Omega^2 e^{-\omega t} + C, \quad t > 0. \tag{4.3.4}$$

Proof. Let $\{e_k\}$ be the basis of eigenfunctions of the biharmonic operator with the Dirichlet boundary conditions:

$$\Delta^2 e_k = \lambda_k e_k, \quad e_k|_{\partial\Omega} = \frac{\partial e_k}{\partial n}\Big|_{\partial\Omega} = 0, \quad k = 1, 2, \dots .$$

We define the *n-order Galerkin approximate solution* to problem (4.3.1) on the interval $[0,T]$ as a function $u^n(t)$ of the form

$$u^n(t) = \sum_{i=1}^{n} g_k(t) e_k, \quad n = 1, 2, \dots,$$

where $g_k(t) \in W^1_\infty(0,T;\mathbb{R})$ are scalar functions such that $u^n(t)$ satisfies the relations

$$\begin{cases} \frac{d}{dt}(u^n(t), e_k)_\Omega + (\Delta u^n(t), \Delta e_k)_\Omega - ([u^n(t) + f, v(u^n(t)) + F_0], e_k)_\Omega \\ \qquad + (Lu^n(t), e_k)_\Omega = (p, e_k)_\Omega \ \text{ for every } k = 1, 2, \dots, n, \\ u^n|_{t=0} = u_{n0}, \end{cases} \tag{4.3.5}$$

where $v(u^n))$ is determined by u^n according to (4.3.2) and $u_{n0} \in \mathrm{Span}\{e_k : k = 1, 2, \dots, n\}$ are chosen such that $u_{n0} \to u_0$ in $L_2(\Omega)$.

We obviously have that

$$\begin{aligned} ([u^n(t) + f, v(u^n(t)) + F_0], u^n)_\Omega &= -\|\Delta v(u^n(t))\|_\Omega^2 - (f, [u^n(t), v(u^n(t))])_\Omega \\ &\quad + ([u^n(t) + f, u^n(t)], F_0)_\Omega. \end{aligned}$$

It follows from Theorem 1.4.3 that

$$|(f, [u^n, v(u^n)])|_\Omega + |([u^n + f, u^n], F_0)_\Omega| \le \frac{1}{4}\|\Delta v(u^n)\|_\Omega^2 + C_1\|u^n\|_{1,\Omega}^2 + C_2.$$

Therefore using Lemma 1.5.4 we find that

$$\frac{d}{dt}\|u^n(t)\|_\Omega^2 + \|\Delta u^n(t)\|_\Omega^2 + \|\Delta v(u^n(t))\|_\Omega^2 \le C. \tag{4.3.6}$$

This implies

$$\|u^n(t)\|_\Omega^2 + \int_0^t \left(\|\Delta u^n\|_\Omega^2 + \|\Delta v(u^n)\|_\Omega^2\right) d\tau \le \|u_{n0}\|_\Omega^2 + Ct. \tag{4.3.7}$$

It follows from (4.3.5) that

$$(\partial_t u^n(t), h)_\Omega = -(\Delta u^n(t), \Delta h)_\Omega + ([u^n(t) + f, v(u^n(t)) + F_0], P_n h)_\Omega$$
$$- (Lu^n(t), P_n h)_\Omega + (p, P_n h)_\Omega$$

for any $h \in H_0^2(\Omega)$, where P_n is the orthoprojector on Span$\{e_1, \dots, e_n\}$. Therefore using Theorem 1.4.3 it is easy to see that

$$\|\partial_t u^n(t)\|_{-2,\Omega} \le C\left(1 + \|\Delta u^n(t)\|_\Omega + \|u^n(t) + f\|_{1,\Omega}\|v(u^n(t))\|_{2,\Omega}\right).$$

Hence, (4.3.7) and interpolation inequalities imply that $\int_0^T \|\partial_t u^n(t)\|_{-2,\Omega}^{4/3} d\tau \le C_T$, where C_T does not depend on n. This estimate and also (4.3.7) and Corollary 1.4.5 allow us by the compactness method to prove the existence of a weak solution. Inequality (4.3.3) follows from (4.3.7) and semicontinuity of norms with respect to weak convergence. Relation (4.3.4) can be easily obtained from (4.3.6) which implies

$$\|u^n(t)\|_\Omega^2 \le \|u^n(0)\|_\Omega^2 e^{-\omega t} + C/\omega, \quad t > 0, \tag{4.3.8}$$

for some $\omega > 0$. □

Unfortunately the uniqueness of weak solutions to problem (4.3.1) and (4.3.2) is not known. However the uniqueness does hold for strong solutions.

4.3.2. Definition. A weak solution $u(t)$ to problem (4.3.1) is said to be *strong*, if it belongs to the class $L_\infty(0,T; H_0^2(\Omega)) \cap L_2(0,T; H^4(\Omega) \cap H_0^2(\Omega))$. ■

It follows directly from equation (4.3.1) that strong solutions possess the property $u_t \in L_2(0,T; L_2(\Omega))$. Therefore by the interpolation argument (see, e.g., [222]) any strong solution, if it exists, lies in the space $C(0,T; H_0^2(\Omega))$.

4.3.3. Theorem. *Assume that hypotheses of Theorem* 4.3.1 *hold and L is a bounded operator from $H^{2-\eta}(\Omega)$ into $L_2(\Omega)$ for some $\eta > 0$. Then for any $u_0 \in H_0^2(\Omega)$ problem* (4.3.1) *and* (4.3.2) *on any interval $[0,T]$ has a unique strong solution $u(t)$. Moreover, this solution for every $t > 0$ possesses the properties*

$$\|\Delta u(t)\|_\Omega^2 + \|\Delta v(u(t))\|_\Omega^2 \le A_1\left(\|\Delta u_0\|_\Omega^2 + \|\Delta v(u_0)\|_\Omega^2\right) e^{-\gamma t} + A_2, \tag{4.3.9}$$

$$\int_0^t \|u_t(\tau)\|_\Omega^2 d\tau \le A_1\left(\|\Delta u_0\|_\Omega^2 + \|\Delta v(u_0)\|_\Omega^2\right) + A_2 + A_3 t, \tag{4.3.10}$$

and

$$\int_0^t \|u(\tau)\|_{4,\Omega}^2 d\tau \le A_0(1+t)\left(1 + \|\Delta u_0\|_\Omega^2 + \|\Delta v(u_0)\|_\Omega^2\right), \tag{4.3.11}$$

where A_i are positive constants.

Proof. Let $u^n(t)$ be the approximate solution considered in the proof of Theorem 4.3.1. Multiply (4.3.5) by $\dot{g}_k(t)$ and take a sum from 1 to n. As a result we obtain

$$\frac{d}{dt}\Pi_*(u^n(t)) + \|u_t^n(t)\|_\Omega^2 + (Lu^n(t), u_t^n(t))_\Omega = 0. \tag{4.3.12}$$

where $\Pi_*(u) = \frac{1}{2}\|\Delta u\|^2 + \Pi(u)$ with $\Pi(u)$ given by (4.1.6). By (4.1.8) we have

$$a_1\left(\|\Delta u\|_\Omega^2 + \|\Delta v(u)\|_\Omega^2\right) - b_1 \le \Pi_*(u) \le a_2\left(\|\Delta u\|_\Omega^2 + \|\Delta v(u)\|_\Omega^2\right) + b_1, \tag{4.3.13}$$

for any $u \in H_0^2(\Omega)$ with positive constants a_i and b_i. We also have that

$$|(Lu^n, u_t^n)_\Omega| \le \delta\|u^n\|_{2,\Omega}^2 + \frac{1}{2}\|u_t^n\|_\Omega^2 + C_\delta\|u^n\|_\Omega^2 \quad \text{for every } \delta > 0.$$

Therefore, if we add relation (4.3.6) to (4.3.12), then it is easy to find that

$$\begin{aligned}&\frac{d}{dt}\left(\|u^n(t)\|_\Omega^2 + \Pi_*(u^n(t))\right)\\ &+\gamma_*\left(\|u^n(t)\|_\Omega^2 + \Pi_*(u^n(t))\right) + \frac{1}{2}\|u_t^n(t)\|_\Omega^2 \le B\left(1 + \|u^n(t)\|_\Omega^2\right),\end{aligned}$$

where γ_* and B are positive numbers that do not depend on n. This implies that

$$\begin{aligned}\|u^n(t)\|_\Omega^2 + \Pi_*(u^n(t)) \le &\left[\|u^n(0)\|_\Omega^2 + \Pi_*(u^n(0))\right]e^{-\gamma_* t}\\ &+\frac{B}{\gamma_*} + B\int_0^t e^{-\gamma_*(t-\tau)}\|u^n(\tau)\|_\Omega^2 d\tau.\end{aligned}$$

Therefore using (4.3.8) we obtain

$$\|u^n(t)\|_\Omega^2 + \Pi_*(u^n(t)) \le C_1\left[\|u^n(0)\|_\Omega^2 + \Pi_*(u^n(0))\right]e^{-\gamma t} + C_2 \tag{4.3.14}$$

for some $\gamma > 0$ and

$$\int_0^t \|u_t^n(\tau)\|_\Omega^2 d\tau \le C_1\left[\|u^n(0)\|_\Omega^2 + \Pi_*(u^n(0))\right] + C_2 t. \tag{4.3.15}$$

Estimates (4.3.13)–(4.3.15) allow us to prove the existence of a weak solution $u(t)$ possessing the properties (4.3.9), and (4.3.10). It follows from (4.3.1), (4.3.9) and (4.3.10) that this solution satisfies the estimate (4.3.11). This makes it possible to prove that $u(t)$ is a strong solution.

Now we prove uniqueness of strong solutions. Let $u_1(t)$ and $u_2(t)$ be strong solutions to problem (4.3.1). Then we can consider the function $w(t) \equiv u_1(t) - u_2(t)$ as a solution to some linear problem. Therefore one can see that

$$\|w(t)\|_\Omega^2 + 2\int_0^t \|\Delta w(\tau)\|_\Omega^2 d\tau = \|w(0)\|_\Omega^2 + 2\int_0^t [M(\tau) - (Lw, w)_\Omega]d\tau, \tag{4.3.16}$$

where $M(\tau) = ([u_1 + f, v(u_1) + F_0] - ([u_2 + f, v(u_2) + F_0], w)_\Omega$. It is easy to see that

$$M(\tau) = ([w, \frac{1}{2}(v(u_1) + v(u_2)) + F_0], w)_\Omega + \frac{1}{2}([u_1 + u_2 + 2f, v(u_1) - v(u_2), w)_\Omega.$$

Because

$$v(u_1) + v(u_2) = -(\Delta_D^2)^{-1}\left([w,w] + 2[u_1,u_2] + 2[u_1 + u_2, f]\right)$$

and

$$v(u_1) - v(u_2) = -(\Delta_D^2)^{-1}\left([w, u_1 + u_2 + 2f]\right),$$

we obtain that

$$M(\tau) \le ([F_0 + \psi(u_1,u_2), w], w).$$

Here $\psi(u_1,u_2) = -(\Delta_D^2)^{-1}\left([u_1,u_2] + [u_1 + u_2, f]\right)$. Therefore we can prove that

$$M(\tau, u_1, u_2) - (Lw, w) \le C(1 + \|\psi(u_1,u_2)\|_2^4)\|w\|^2 + \|\Delta w\|^2.$$

Consequently from (4.3.16) we can find that

$$\|w(t)\|_\Omega \le \|w(0)\|_\Omega \exp\left\{c\int_0^t (1 + \|\psi(u_1,u_2)\|_2^4) d\tau\right\}. \tag{4.3.17}$$

This relation implies the uniqueness of strong solutions. □

4.3.4. Remark. Using relation (4.3.17) and the estimates for the von Karman bracket given in Theorem 1.4.3 we can also prove that if $u_1(t)$ is a weak solution and $u_2(t)$ is a strong solution with the same initial data, then $u_1(t) \equiv u_2(t)$. ■

In conclusion of this section we refer to the paper [65] studying a quasi-static model with a displacement-dependent coefficient before u_t in (4.3.1).

Chapter 5
Thermoelastic Plates

In this chapter we study well-posedness and regularity issues associated with thermoelastic plates. The PDE model of thermoelasticity consists of coupled second-order (plate) and first-order (heat) equations. For this reason, treatment of thermoelastic models does not follow from the corresponding treatment of abstract second-order equations. A new functional framework needs to be developed.

In this chapter we undertake a study of solutions associated with von Karman evolution equations subject to thermal dissipation. We consider models *with* and *without* the rotational inertia terms.

5.1 PDE model

Let Ω be a bounded domain in $\mathbb{R}^2$ with the boundary $\partial\Omega = \Gamma$, Δ denotes the Laplace operator, and F_0 and p are given functions with regularity specified later. As before we set

$$[u,v] = \partial_{x_1}^2 u \cdot \partial_{x_2}^2 v + \partial_{x_2}^2 u \cdot \partial_{x_1}^2 v - 2 \cdot \partial_{x_1x_2}^2 u \cdot \partial_{x_1x_2}^2 v\,, \tag{5.1.1}$$

which expression defines the von Karman bracket.

In what follows we describe the model under consideration which is a thermoelastic von Karman plate subjected to an external and internal forcing. The corresponding equations (see, e.g., [173, 178] and the references therein) have the following form

$$\begin{cases} u_{tt} - \alpha\Delta u_{tt} + \mu\Delta\theta + \Delta^2 u - [u, v + F_0] = p(x), \quad x \in \Omega, t > 0, \\ \kappa\theta_t - \eta\Delta\theta - \mu\Delta u_t = 0, \quad x \in \Omega, t > 0, \end{cases} \tag{5.1.2}$$

where Airy's stress function $v = v(u)$ is a solution to the problem

I. Chueshov, I. Lasiecka, *Von Karman Evolution Equations*,
Springer Monographs in Mathematics, DOI 10.1007/978-0-387-87712-9_5,

$$\Delta^2 v + [u,u] = 0, \quad v|_{\partial\Omega} = \frac{\partial v}{\partial n}\Big|_{\partial\Omega} = 0. \tag{5.1.3}$$

The temperature θ satisfies the Dirichlet boundary condition of the form:

$$\theta = 0 \text{ on } \Gamma. \tag{5.1.4}$$

The boundary conditions imposed on the displacement u are either "clamped":

$$u = \frac{\partial}{\partial n} u = 0 \text{ on } \Gamma, \tag{5.1.5}$$

where n is the outer normal vector, or else "hinged":

$$u = \Delta u = 0 \text{ on } \Gamma. \tag{5.1.6}$$

The parameters μ and η are positive and α and κ are nonnegative. The case $\alpha > 0$ corresponds to taking into account rotational inertia of filaments of the plate. The parameter κ has the meaning of heat/thermal capacity. In the case of $\kappa = 0$ equations (5.1.2) can be decoupled. Indeed, substituting $\Delta\theta$ from the second equation, the first equation becomes just a scalar equation of structurally damped plate with the damping term equal to $-\mu^2\eta^{-1}\Delta u_t$, instead of $\mu\Delta\theta$. For the sake of some simplifications in contrast with the model in (3.0.1) and (3.0.2) (or in (4.1.1) and (4.1.2)) we assume that the function f which describes the initial form of the plate is zero.

In addition to clamped and hinged boundary conditions, one can also consider "free" boundary conditions which are given by:

$$\begin{aligned} \Delta u + (1-\hat{\mu})B_1 u + \mu\theta &= 0 \text{ on } \Gamma, \\ \frac{\partial}{\partial n}\Delta u + (1-\hat{\mu})B_2 u - \alpha\frac{\partial}{\partial n}u_{tt} + \mu\frac{\partial}{\partial n}\theta &= \nu_1 u + \beta u^3 \text{ on } \Gamma, \\ \frac{\partial}{\partial n}\theta + b\theta &= 0 \text{ on } \Gamma, \end{aligned}$$

where $\nu_1, \beta \geq 0$, $b > 0$, $\hat{\mu} \in (0, \frac{1}{2})$, and boundary operators B_1 and B_2 are given by (1.3.20). However, in order to focus the attention we concentrate on the first two types of boundary conditions: clamped and hinged. The main results remain valid in the free case, however, the details of the proof are more technical.

We note that the considered model is a coupled system of nonlinear plate and linear heat equations, rather than more frequently studied thermoelastic system coupling wave and heat equations, [91, 152] and the references therein. The characteristics of these two models, particularly with respect to stability analysis, are very different. From physical point of view the main peculiarities of the model in (5.1.2) are (i) the possibility of large deflections of the plate and (ii) small changes of the temperature near the reference temperature of the plate (which is reasonable in the absence of phase transitions). We refer to [173, 178] for further discussions and references.

5.2 Abstract formulation

In what follows we assume that the domain Ω is either smooth or rectangular. Let $H^s(\Omega)$ denote, as before, the L_2-based Sobolev space of the order s and by $H_0^s(\Omega)$ the closure of $C_0^\infty(\Omega)$ in $H^s(\Omega)$. We also recall the following notation.

$$||u|| \equiv \|u\|_{L_2(\Omega)}, \quad (u,v) \equiv (u,v)_{L_2(\Omega)}, \quad \|u\|_s \equiv \|u\|_{H^s(\Omega)}.$$

In the space $\mathscr{H} = L_2(\Omega)$ we define the operator A_D which stands for the minus Laplacian equipped with zero Dirichlet boundary data. This is to say

$$A_D u = -\Delta u, \quad u \in \mathscr{D}(A_D) = H^2(\Omega) \cap H_0^1(\Omega)$$

and consider the operator $M_\alpha = I + \alpha A_D$. It is well known that both operators A_D and M_α are positive self-adjoint operators on $\mathscr{H}$. We also introduce a biharmonic operator

$$\mathscr{A} u = \Delta^2 u, \quad u \in \mathscr{D}(\mathscr{A}) = \begin{cases} H^4(\Omega) \cap H_0^2(\Omega), & \text{(clamped b.c.);} \\ \{u \in H^4(\Omega) : u = \Delta u = 0 \text{ on } \Gamma\}, & \text{(hinged b.c.)} \end{cases}$$

In the "commutative" case of hinged boundary conditions (5.1.6) one has $\mathscr{A} = A_D^2$ which provides a lot of symmetry for the problem. Indeed, all the operators A_D, M_α, and $\mathscr{A}$ do commute. This feature substantially simplifies the analysis with respect to clamped case (5.1.5), where the latter requires several additional technical estimates that account for the lack of commutativity.

We also introduce nonlinear mapping $F(\cdot)$ by the formula

$$F(u) = [u, v(u) + F_0] + p(x), \quad u \in H^2(\Omega), \tag{5.2.1}$$

where $v(u) \in H_0^2(\Omega)$ is determined by u via (5.1.3).

With the above notation, equations in (5.1.2) with the boundary conditions considered can be written in the form

$$\begin{cases} M_\alpha u_{tt} - \mu A_D \theta + \mathscr{A} u = F(u), \\ \kappa \theta_t + \eta A_D \theta + \mu A_D u_t = 0. \end{cases} \tag{5.2.2}$$

We equip (5.2.2) with initial data

$$u|_{t=0} = u_0, \quad u_t|_{t=0} = u_1, \quad \theta|_{t=0} = \theta_0. \tag{5.2.3}$$

In the case $\kappa = 0$ the problem has the form

$$M_\alpha u_{tt} + \mu^2 \eta^{-1} A_D u_t + \mathscr{A} u = F(u), \quad u|_{t=0} = u_0,\ u_t|_{t=0} = u_1. \tag{5.2.4}$$

5.3 Linear problem

We begin by discussing properties of a linear thermoelastic system. This leads to consideration of the system

$$\begin{cases} M_\alpha u_{tt} - \mu A_D \theta + \mathscr{A} u = 0, \\ \kappa \theta_t + \eta A_D \theta + \mu A_D u_t = 0. \end{cases} \tag{5.3.1}$$

We introduce appropriate phase (energy) spaces $H_{\alpha,\kappa}$ that capture dependence on the varying parameters $0 \le \alpha \le m_\alpha$ and $0 \le \kappa \le m_\kappa$ for some (fixed) positive constants m_γ and m_κ (below we take $m_\alpha = m_\kappa = 1$ for simplicity).

For every couple $(\alpha;\kappa) \in \Lambda \equiv \{0 \le \alpha \le 1, 0 \le \kappa \le 1\}$ we introduce the Hilbert space

$$H_{\alpha,\kappa} = \begin{cases} \mathscr{D}(\mathscr{A}^{1/2}) \times V_\alpha \times \mathscr{H}, & \text{for } \kappa > 0; \\ \mathscr{D}(\mathscr{A}^{1/2}) \times V_\alpha, & \text{for } \kappa = 0, \end{cases} \tag{5.3.2}$$

where $\mathscr{H} = L_2(\Omega)$, $\mathscr{D}(\mathscr{A}^{1/2}) = H_0^2(\Omega)$ in the clamped case and $\mathscr{D}(\mathscr{A}^{1/2}) = \mathscr{D}(A_D) = (H^2 \cap H_0^1)(\Omega)$ in the hinged case, and $V_\alpha \equiv D(M_\alpha^{1/2})$ which is $H_0^1(\Omega)$ for $\alpha > 0$ and $L_2(\Omega)$ for $\alpha = 0$. We equip the space $H_{\alpha,\kappa}$ with the norm

$$|U|_{\alpha,\kappa}^2 = \|\mathscr{A}^{1/2} u_0\|^2 + \|M_\alpha{}^{1/2} u_1\|^2 + \kappa \|\theta\|^2, \quad U = (u_0; u_1; \theta),$$

for $(\alpha;\kappa) \in \Lambda$ (in the case $\kappa = 0$ the last term should be omitted). In both cases of boundary conditions we have that $\mathscr{D}(\mathscr{A}^{1/2}) \subseteq \mathscr{D}(A_D)$ and $\|\mathscr{A}^{1/2} u_0\| = \|A_D u_0\|$ for $u \in \mathscr{D}(\mathscr{A}^{1/2})$. We use this observation when dealing with the norm $|\cdot|_{\alpha,\kappa}$. We also point out that the corresponding inner product in $H_{\alpha,\kappa}$ has the form

$$(U,\hat{U})_{\alpha,\kappa} = (\mathscr{A}^{1/2} u_0, \mathscr{A}^{1/2} \hat{u}_0) + (M_\alpha{}^{1/2} u_1, M_\alpha{}^{1/2} \hat{u}_1) + \kappa(\theta,\hat{\theta})$$

for $U = (u_0; u_1; \theta)$ and $\hat{U} = (\hat{u}_0; \hat{u}_1; \hat{\theta})$.

In the space $H_{\alpha,\kappa}$ problem (5.3.1) can be written as an abstract first-order evolution of the form

$$\frac{dU(t)}{dt} + A_{\alpha,\kappa} U(t) = 0, \tag{5.3.3}$$

where in the case $\kappa > 0$ we have $U \equiv U(t) = (u(t); u_t(t); \theta(t))$ and the operator $A_{\alpha,\kappa} : \mathscr{D}(A_{\alpha,\kappa}) \subset H_{\alpha,\kappa} \mapsto H_{\alpha,\kappa}$ given by

$$A_{\alpha,\kappa} = \begin{pmatrix} 0 & -I & 0 \\ M_\alpha{}^{-1}\mathscr{A} & 0 & -\mu M_\alpha{}^{-1} A_D \\ 0 & \mu \kappa^{-1} A_D & \eta \kappa^{-1} A_D \end{pmatrix} \tag{5.3.4}$$

with the domain

$$\mathscr{D}(A_{\alpha,\kappa}) = \left\{ (u;v;\theta) \in \mathscr{D}(\mathscr{A}^{1/2}) \times \mathscr{D}(\mathscr{A}^{1/2}) \times \mathscr{D}(A_D) \ : \ M_\alpha{}^{-1/2} \mathscr{A} u \in \mathscr{H} \right\}.$$

In the degenerate case $\kappa = 0$ we have $U(t) = (u(t); u_t(t))$ and the operator $A_{\alpha,0}$ is given by

$$A_{\alpha,0} = \begin{pmatrix} 0 & -I \\ M_\alpha{}^{-1}\mathscr{A} & \mu^2\eta^{-1}M_\alpha{}^{-1}A_D \end{pmatrix} \tag{5.3.5}$$

on the domain

$$\mathscr{D}(A_{\alpha,0}) = \left\{(u;v) \in \mathscr{D}(\mathscr{A}^{1/2}) \times \mathscr{D}(\mathscr{A}^{1/2}) :\ M_\alpha{}^{-1/2}\mathscr{A}u \in \mathscr{H}\right\}.$$

5.3.1 Generation of strongly continuous semigroup

We have the following result concerning well-posedness of the problem (5.3.1).

5.3.1. Proposition. *For any $(\alpha;\kappa) \in \Lambda$ the system defined in* (5.3.1) *(or in* (5.3.3)*) generates a strongly continuous linear semigroup $e^{-A_{\alpha,\kappa}t}$ of contractions in $H_{\alpha,\kappa}$. Moreover, the semigroup $e^{-A_{\alpha,\kappa}t}$ is exponentially stable,*

$$|e^{-A_{\alpha,\kappa}t}|_{\mathscr{L}(H_{\alpha,\kappa})} \le Ce^{-\omega t}, \quad t \ge 0,$$

where the positive constants C and ω do not depend on $(\alpha;\kappa) \in \Lambda$.

Proof. In order to prove the generation result it is sufficient to show that the operator $A_{\alpha,\kappa}$ and its adjoint $A^*_{\alpha,\kappa}$ are both accretive (see [241, Corollary 4.4, p. 15]).

Case $\kappa > 0$: One can see that

$$(A_{\alpha,\kappa}U,\hat{U})_{\alpha,\kappa} = -(\mathscr{A}v,\hat{u}) + (\mathscr{A}u - \mu A_D\theta,\hat{v}) + \mu(A_Dv,\hat{\theta}) + \eta(A_D\theta,\hat{\theta}) \tag{5.3.6}$$

for any $U = (u;v;\theta) \in \mathscr{D}(A_{\alpha,\kappa})$ and $\hat{U} = (\hat{u};\hat{v};\hat{\theta}) \in H_{\alpha,\kappa}$. Therefore

$$(A_{\alpha,\kappa}U,U)_{\alpha,\kappa} = \eta\|A_D^{1/2}\theta\|^2 \ge 0 \ \text{ for } \ U = (u;v;\theta) \in \mathscr{D}(A_{\alpha,\kappa});$$

that is, $A_{\alpha,\kappa}$ is accretive.

It also follows from (5.3.6) that the adjoint operator $A^*_{\alpha,\kappa}$ has the form

$$A^*_{\alpha,\kappa} = \begin{pmatrix} 0 & I & 0 \\ -M_\alpha{}^{-1}\mathscr{A} & 0 & \mu M_\alpha{}^{-1}A_D \\ 0 & -\mu\kappa^{-1}A_D & \eta\kappa^{-1}A_D \end{pmatrix} \tag{5.3.7}$$

and $\mathscr{D}(A^*_{\alpha,\kappa}) = \mathscr{D}(A_{\alpha,\kappa})$. The same calculations as above reveal that $A^*_{\alpha,\kappa}$ is also accretive:

$$(A^*_{\alpha,\kappa}U,U)_{\alpha,\kappa} = \eta\|A_D^{1/2}\theta\|^2 \ge 0 \ \text{ for } \ U - (u;v;\theta) \in D(A^*_{\alpha,\kappa}) = \mathscr{D}(A_{\alpha,\kappa}).$$

Thus, semigroup generation theory (see [241, Corollary 4.4, p. 15]) applies leading to the conclusion that both $A_{\alpha,\kappa}$ and $A^*_{\alpha,\kappa}$ generate on $H_{\alpha,\kappa}$ strongly continuous semigroup of contractions $e^{-A_{\alpha,\kappa}t}$ and $e^{-A^*_{\alpha,\kappa}t}$.

Exponential stability of the semigroup, with the uniform control (in α) of the constants, has been proved in [8, 9]. For a complete presentation (including other boundary conditions, such as free) see Proposition 3.11.1.3 [216, p. 229]. To obtain the bounds for the stability constants C and ω that are uniform in both parameters α and κ we can use the same argument as in [8, 9]; see also the proof of Theorem 11.3.1 in Section 11.3 for a similar argument with control of dependence on both α and κ for a nonlinear model.

Case $\kappa = 0$: In this case

$$(A_{\alpha,0}U,\hat{U})_{\alpha,0} = -(\mathscr{A}v,\hat{u}) + (\mathscr{A}u + \mu^2\eta^{-1}A_D v,\hat{v})$$

for $U = (u;v) \in \mathscr{D}(A_{\alpha,0})$ and $\hat{U} = (\hat{u};\hat{v}) \in H_{\alpha,0}$. This implies that

$$A^*_{\alpha,0} = \begin{pmatrix} 0 & I \\ -M_\alpha{}^{-1}\mathscr{A} & \mu^2\eta^{-1}M_\alpha{}^{-1}A_D \end{pmatrix} \quad \text{with } \mathscr{D}(A^*_{\alpha,0}) = \mathscr{D}(A_{\alpha,0})$$

and

$$(A_{\alpha,0}U,U)_{\alpha,0} = (A^*_{\alpha,0}U,U)_{\alpha,0} = \mu^2\eta^{-1}\|A_D^{1/2}v\|^2 \geq 0$$

for $U = (u;v) \in \mathscr{D}(A_{\alpha,0}) = \mathscr{D}(A^*_{\alpha,0})$. Therefore we can apply the result from [241] mentioned above in the case $\kappa = 0$.

As for (uniform) exponential stability in this case, for solution $U(t) = e^{-A_{\alpha,0}t}U_0 \equiv (u(t);u_t(t))$ we can use (the standard) Lyapunov function of the form

$$V(t) \equiv V(u;u_t) = \frac{1}{2}\left(\|M_\alpha{}^{1/2}u_t\|^2 + \|\mathscr{A}^{1/2}u\|^2\right) + \varepsilon\left((M_\alpha u_t,u) + \frac{1}{2}(A_D u,u)\right)$$

with sufficiently small $\varepsilon > 0$. After simple calculations involving equation (5.2.4) with $F(u) \equiv 0$ one can see that $V'(t) + \gamma V(t) \leq 0$ with the positive constant γ independent of α. This implies exponential stability in the case $\kappa = 0$ with uniform (in α) control of the constants. □

We consider next the nonhomogeneous thermoelastic problem

$$\begin{cases} M_\alpha u_{tt} - \mu A_D\theta + \mathscr{A}u = f(t), \\ \kappa\theta_t + \eta A_D\theta + \mu A_D u_t = g(t). \end{cases} \tag{5.3.8}$$

along with the initial data

$$u|_{t=0} = u_0,\ u_t|_{t=0} = u_1, \quad \theta|_{t=0} = \theta_0, \tag{5.3.9}$$

and assume that $f \in L_1(\mathbb{R}_+;V'_\alpha)$ and $g \in L_1(\mathbb{R}_+;\mathscr{H})$ Here we denote by V'_α the dual (with respect to $\mathscr{H}$) to V_α. In the space $H_{\alpha,\kappa}$ problem (5.3.8) and (5.3.9) can be written in the form

$$\frac{dU(t)}{dt} + A_{\alpha,\kappa}U(t) = F(t), \quad U(0) = U_0, \tag{5.3.10}$$

where in the case $\kappa > 0$ we have $U \equiv U(t) = (u(t);u_t(t);\theta(t))$, $U_0 = (u_0;u_1;\theta_0)$, and $F(t) = (0;M_\alpha^{-1}f(t);g(t))$.

By Proposition 5.3.1 $A_{\alpha,\kappa}$ generates a strongly continuous semigroup, which then allows us to apply the theory developed in [241] (the methods presented in Section 2.3 are also applicable) to obtain the following result (which we state for the case $\kappa > 0$ only; modifications for the case $\kappa = 0$ are obvious).

5.3.2. Proposition. *Let $\kappa > 0$. With the reference to problem* (5.3.8) *and* (5.3.9) *we have the following assertions.*

- **Strong solutions:** *Let $f(t) \in W_1^1(\mathbb{R}_+;V'_\alpha)$ and $g(t) \in W_1^1(\mathbb{R}_+;\mathscr{H})$. Assume that $U_0 = (u_0;u_1;\theta_0) \in \mathscr{D}(A_{\alpha,\kappa})$. Then there exists a unique strong (in the sense of Definition* 2.3.1*) solution $U(t) = (u(t);u_t(t);\theta(t))$. This solution satisfies the energy balance equation*

$$|U(t)|^2_{\alpha,\kappa} + 2\eta \int_s^t ||A_D^{1/2}\theta(\tau)||^2 d\tau \tag{5.3.11}$$
$$= |U_0|^2_{\alpha,\kappa} + 2\int_0^t [(f(\tau),u_t(\tau)) + (g(\tau),\theta(\tau))]d\tau, \quad t > 0.$$

- **Generalized solutions:** *Let $f(t) \in L_1(\mathbb{R}_+;V'_\alpha)$ and $g(t) \in L_1(\mathbb{R}_+;\mathscr{H})$. Assume that $U_0 = (u_0;u_1;\theta_0) \in H_{\alpha,\kappa}$. Then there exists a unique generalized (in the sense of Definition* 2.3.3*) solution $U(t) = (u(t);u_t(t);\theta(t))$ such that $\theta(t) \in L_2^{\mathrm{loc}}(\mathbb{R}_+;\mathscr{D}(A_D^{1/2}))$ and the energy relation in* (5.3.11) *is valid. This solution is also mild; that is, it can be presented in the form*

$$U(t) = e^{-A_{\alpha,\kappa}t}U(0) + \int_0^t e^{-A_{\alpha,\kappa}(t-\tau)}F(\tau)d\tau,$$

and also satisfies the corresponding variational inequality (i.e., $U(t)$ is also weak).

Proof. Because $A_{\alpha,\kappa}$ is a generator of a strongly continuous semigroup (see Proposition 5.3.1), to obtain the existence and uniqueness of a strong solution we can apply [241, Corollary 2.10, p. 109]. Because for strong solutions equation (5.3.8) holds in $\mathscr{H}$ for almost all $t > 0$, we can use the standard procedure to obtain energy relation (5.3.11). It suffices to multiply the first equation in (5.3.8) by u_t, the second by θ, integrate by parts, and account for cancellation of the thermal coupling.

The existence of generalized solutions and their properties follows from the considerations given in [241, Section 4.2]. They can be also derived from the general abstract Theorem 2.3.8 □

5.3.2 *Analyticity of the semigroup for the model without rotational inertia*

In the case when $\alpha = 0$ the semigroup $e^{-A_{\alpha,\kappa}t}$ has an additional smoothing effect for positive times. More specifically, we have

5.3.3. Theorem. *Let* $\alpha = 0$*. Then the contraction semigroup* $e^{-A_{0,\kappa}t}$ *given by Proposition* 5.3.1 *and defined by equation* (5.3.3) *is analytic on* $H_{0,\kappa}$.

For the definition and properties of an analytic semigroup see [241] or [159], for instance.

The result in Theorem 5.3.3 has been known for several years, [224, 216] and references therein (see also [43] for the case $\kappa = 0$). We reproduce the proof below. In fact, we first provide a more general statement of the analyticity result formulated for abstract operators such that the problem under consideration is a special case.

Consider abstract systems of the form

$$\begin{cases} u_{tt} - \mu\mathscr{B}\theta + \mathscr{A}u = 0, \\ \theta_t + \mathscr{B}\theta + \mu\mathscr{B}u_t = 0, \end{cases} \tag{5.3.12}$$

and

$$u_{tt} + \mu^2\mathscr{B}u_t + \mathscr{A}u = 0, \tag{5.3.13}$$

where $\mu > 0$ is a parameter. We note that problem (5.3.1) with $\alpha = 0$ and $\kappa > 0$ can be written in the form (5.3.12) if we introduce new variables $\tilde{u}$ and $\tilde{\theta}$ such that $u(t) = \tilde{u}(\sqrt{\kappa}t)$ and $\theta(t) = \tilde{\theta}(\sqrt{\kappa}t)$. In this case we arrive at (5.3.12) with $\mathscr{B} := \eta\kappa^{-3/2}A_D$, $\mu := \mu\eta^{-1}\kappa^{1/2}$ and $\mathscr{A} := \kappa^{-1}\mathscr{A}$. It is also obvious that (5.3.1) with $\alpha = \kappa = 0$ is a special case of (5.3.13).

We impose the following set of assumptions.

5.3.4. Assumption.

- [i] $\mathscr{A} : \mathscr{D}(\mathscr{A}) \subset \mathscr{H} \to \mathscr{H}$ and $\mathscr{B} : \mathscr{D}(\mathscr{B}) \subset \mathscr{H} \to \mathscr{H}$ are two strictly positive and self-adjoint operators in a Hilbert space $\mathscr{H}$ (with the norm $|\cdot|_{\mathscr{H}}$).
- [ii] $\mathscr{D}(\mathscr{A}^{1/2}) \subset \mathscr{D}(\mathscr{B})$.
- [iii] There exists a constant $c > 0$ such that for all $x \in \mathscr{D}(\mathscr{A}^{1/2})$ we have:

$$c|\mathscr{A}^{1/4}x|_{\mathscr{H}} \le |\mathscr{B}^{1/2}x|_{\mathscr{H}}. \tag{5.3.14}$$

5.3.5. Remark. We note that the second bullet in Assumption 5.3.4 implies that $\mathscr{D}(\mathscr{A}^{1/4}) \subset \mathscr{D}(\mathscr{B}^{1/2})$. Therefore, by the closed graph theorem one has the reverse of the (5.3.14) inequality. Hence we have the two-sided bound:

$$c_1|\mathscr{B}^{1/2}x|_{\mathscr{H}} \le |\mathscr{A}^{1/4}x|_{\mathscr{H}} \le c_2|\mathscr{B}^{1/2}x|_{\mathscr{H}}, \quad x \in \mathscr{D}(\mathscr{A}^{1/4}), \tag{5.3.15}$$

for some positive constants c_1 and c_2. One can also see from this relation that the operator $\mathscr{B}^{1/2}\mathscr{A}^{-1/4}$ and thus its adjoint $\mathscr{A}^{-1/4}\mathscr{B}^{1/2}$ are both bounded on $\mathscr{H}$. The same is true concerning $\mathscr{A}^{1/4}\mathscr{B}^{-1/2}$ and $\mathscr{B}^{-1/2}\mathscr{A}^{1/4}$. ∎

Under these hypotheses in the same way as in Proposition 5.3.1 one can show that problems (5.3.12) and (5.3.13) generate strongly continuous contraction semigroups e^{-At} and e^{-A_1t} in the spaces $X \equiv \mathscr{D}(\mathscr{A}^{1/2}) \times \mathscr{H} \times \mathscr{H}$ and $Y \equiv \mathscr{D}(\mathscr{A}^{1/2}) \times \mathscr{H}$, respectively. The infinitesimal generators A and A_1 have the form

$$A \equiv \begin{pmatrix} 0 & -I & 0 \\ \mathscr{A} & 0 & -\mu\mathscr{B} \\ 0 & \mu\mathscr{B} & \mathscr{B} \end{pmatrix} \quad \text{and} \quad A_1 = \begin{pmatrix} 0 & -I \\ \mathscr{A} & \mu^2\mathscr{B} \end{pmatrix}$$

and defined on their natural domains

$$\mathscr{D}(A) = \mathscr{D}(\mathscr{A}) \times \mathscr{D}(\mathscr{A}^{1/2}) \times \mathscr{D}(\mathscr{B}) \quad \text{and} \quad \mathscr{D}(A_1) = \mathscr{D}(\mathscr{A}) \times \mathscr{D}(\mathscr{A}^{1/2}).$$

The corresponding accretivity properties have the form

$$(AU,U)_X = |\mathscr{B}^{1/2}\theta|^2_{\mathscr{H}} \geq 0, \quad U = (u;v;\theta) \in \mathscr{D}(A), \tag{5.3.16}$$

and

$$(A_1U,U)_Y = \mu^2|\mathscr{B}^{1/2}v|^2_{\mathscr{H}} \geq 0, \quad U = (u;v) \in \mathscr{D}(A_1).$$

To prove analyticity of the semigroups e^{-At} and e^{-A_1t} we use the following criterion.

5.3.6. Proposition. *Let $T(t) = e^{-At}$ be a strongly continuous semigroup of contractions in a Banach space X with the infinitesimal generator A such that the resolvent set $\rho(A)$ contains 0 (i.e., there exists A^{-1} as a bounded operator). Then $T(t)$ is an analytic semigroup if and only if there exists a constant $C > 0$ such that*

$$\|R(\lambda,-A)\|_{\mathscr{L}(X)} \leq C|\lambda|^{-1} \quad \textit{for all } \lambda \textit{ with } Re\lambda > 0, \tag{5.3.17}$$

where $R(\lambda,-A) = (\lambda I + A)^{-1}$ is the resolvent of the operator $-A$.

Proof. Because $T(t)$ is a contraction semigroup, by the Hille–Yosida theorem (see [241, Corollary 3.6, p. 11]) we have that

$$\|R(\lambda,-A)\|_{\mathscr{L}(X)} \leq \frac{1}{Re\lambda} \quad \text{for } Re\lambda > 0. \tag{5.3.18}$$

Therefore (5.3.17) is equivalent to the requirement

$$\|R(\lambda,-A)\|_{\mathscr{L}(X)} \leq \frac{C}{|Im\lambda|} \quad \text{for } Re\lambda > 0,\ Im\lambda \neq 0$$

for some constant $C > 0$. This follows from the obvious relation

$$\min\left\{\frac{1}{Re\lambda}, \frac{1}{|Im\lambda|}\right\} \leq \frac{1}{\max\{Re\lambda, |Im\lambda|\}} \leq \frac{\sqrt{2}}{|\lambda|} < \frac{\sqrt{2}}{|Im\lambda|}.$$

Consequently we can apply [241, Theorem 5.2, p. 61] to conclude the proof. □

We first consider the case of the semigroup $e^{-A_1 t}$ generated by the strongly damped problem (5.3.13) (which corresponds to the case $\kappa = 0$). The following proposition is also a preliminary step in the proof of analyticity in the case $\kappa > 0$.

5.3.7. Proposition. *Let Assumption* 5.3.4 *hold true. Then, the operator* A_1 *generates a strongly continuous and analytic contraction semigroup on* $Y = \mathscr{D}(\mathscr{A}^{1/2}) \times \mathscr{H}$ *and the following resolvent estimate holds*

$$\|R(\lambda, -A_1)\|_{\mathscr{L}(Y)} \le C|\lambda|^{-1} \quad \textit{for all } \lambda \textit{ with } Re\,\lambda > 0. \tag{5.3.19}$$

Proof. Proposition 5.3.7 is a special case of a more general result proved in [43]. For the reader's convenience, we sketch the main steps of the proof.

Without loss of generality we can assume that $\mu = 1$ in (5.3.13).

We apply Proposition 5.3.6. One can see that

$$A_1^{-1}(f_1; f_2) = (\mathscr{A}^{-1}(\mathscr{B} f_1 + f_2); -f_1), \quad (f_1; f_2) \in Y = \mathscr{D}(\mathscr{A}^{1/2}) \times \mathscr{H},$$

which implies that $0 \in \rho(A_1)$. Thus our goal now is to prove the estimate in (5.3.19). For this we first introduce the operator

$$V_{\mathscr{B}} = V_{\mathscr{B}}(\lambda) \equiv \lambda^2 I + \mathscr{A} + \lambda \mathscr{B}.$$

The operator $V_{\mathscr{B}}(\lambda)$ is clearly injective for all $Re\,\lambda > 0$. Indeed, this follows from

$$Re\,(\lambda^{-1} V_{\mathscr{B}}(\lambda)x, x)_{\mathscr{H}} = Re\,\lambda |x|_{\mathscr{H}}^2 + \frac{Re\,\lambda}{|\lambda|^2} |\mathscr{A}^{1/2} x|_{\mathscr{H}}^2 + |\mathscr{B}^{1/2} x|_{\mathscr{H}}^2 \tag{5.3.20}$$

which equality implies injectivity of both $V_{\mathscr{B}}(\lambda)$ and its adjoint for $Re\,\lambda > 0$. In fact, more is true. The operator $V_{\mathscr{B}}(\lambda)$ is bounded invertible on $\mathscr{H}$. Indeed, because $V_{\mathscr{B}}(\lambda)^* = V_{\mathscr{B}}(\bar{\lambda})$ and by (5.3.20) the null space of the latter contains only zero element, the range of $V_{\mathscr{B}}(\lambda)$ is dense. Thus, invertibility follows once we show that the range of $V_{\mathscr{B}}(\lambda)$ is closed. For this, it suffices to show the following bound is valid for all $Re\,\lambda > 0$,

$$|V_{\mathscr{B}}(\lambda)x|_{\mathscr{H}} \ge c|x|_{\mathscr{H}}. \tag{5.3.21}$$

In order to establish (5.3.21) it is convenient to write

$$|(V_{\mathscr{B}}(\lambda)x, x)_{\mathscr{H}}|^2 = |Im\,(V_{\mathscr{B}}(\lambda)x, x)_{\mathscr{H}}|^2 + |Re\,(V_{\mathscr{B}}(\lambda)x, x)_{\mathscr{H}}|^2,$$

where with $Re\,\lambda \ge 0$ and $\lambda = \alpha + i\beta$ we have

$$\begin{aligned} |Im\,(V_{\mathscr{B}}(\lambda)x, x)_{\mathscr{H}}|^2 &= \left|2\alpha\beta |x|_{\mathscr{H}}^2 + \beta(\mathscr{B}x, x)_{\mathscr{H}}\right|^2 = |\beta|^2 \left[2\alpha |x|_{\mathscr{H}}^2 + |\mathscr{B}^{1/2} x|_{\mathscr{H}}^2\right]^2 \\ &\ge |Im\,\lambda|^2 (\mathscr{B}x, x)_{\mathscr{H}}^2 = |Im\,\lambda|^2 |\mathscr{B}^{1/2} x|_{\mathscr{H}}^4. \end{aligned}$$

Because by the Cauchy–Schwarz inequality

$$|V_{\mathscr{B}}(\lambda)x|_{\mathscr{H}} |x|_{\mathscr{H}} \ge |(V_{\mathscr{B}}(\lambda)x, x)| \ge |\beta| |\mathscr{B}^{1/2} x|_{\mathscr{H}}^2 \ge c_1 |\beta| |x|_{\mathscr{H}}^2,$$

for $\beta = Im\lambda \neq 0$ (5.3.21) is satisfied. For $Im\lambda = 0$ the operator $V_{\mathscr{B}}$ is self-adjoint and strictly positive. Thus the invertibility of $V_{\mathscr{B}}(\lambda)$ is straightforward.

Now we return to the proof of the estimate in (5.3.19). By direct computation one shows

$$R(\lambda, -A_1) = \begin{bmatrix} R_1(\lambda) & R_2(\lambda) \\ R_3(\lambda) & R_4(\lambda) \end{bmatrix} \equiv \begin{bmatrix} V_{\mathscr{B}}^{-1}(\lambda I + \mathscr{B}) & V_{\mathscr{B}}^{-1} \\ -V_{\mathscr{B}}^{-1}\mathscr{A} & \lambda V_{\mathscr{B}}^{-1} \end{bmatrix}. \tag{5.3.22}$$

The following bounds, uniform in $Re\lambda > 0$ are the consequences of (5.3.15) as shown in [43].

5.3.8. Lemma **([43]).** *There exists a constant $M > 0$ such that*

$$\begin{aligned} |\mathscr{A}^{1/2} V_{\mathscr{B}}^{-1} \mathscr{A}^{1/2}|_{\mathscr{L}(\mathscr{H})} &\leq M, \\ |\lambda \mathscr{A}^{1/2} V_{\mathscr{B}}^{-1}|_{\mathscr{L}(\mathscr{H})} &\leq M, \\ |\lambda^2 V_{\mathscr{B}}^{-1}|_{\mathscr{L}(\mathscr{H})} &\leq M, \end{aligned} \tag{5.3.23}$$

where the estimates are uniform in $Re\lambda > 0$.

Proof. We briefly sketch the main idea behind the estimates leading to the result of the lemma. The proof carried in [43] consists of two steps: (i) prove the estimate with $\mathscr{B} = \mathscr{A}^{1/2}$, and (ii) use the perturbation argument for the treatment of $\mathscr{B}$ subject to the two-sided bounds (5.3.15) (the latter follows from Assumption 5.3.4).

We carry the calculations for the case $\mathscr{B} = \mathscr{A}^{1/2}$. Denote $V(\lambda) \equiv V_{\mathscr{A}^{1/2}}$.

We start with

$$\begin{aligned} |V(\lambda)\lambda^{-1}\mathscr{A}^{-1/2}x|_{\mathscr{H}}^2 &= |x|_{\mathscr{H}}^2 + |\lambda\mathscr{A}^{-1/2}x + \lambda^{-1}\mathscr{A}^{1/2}x|_{\mathscr{H}}^2 \\ &\quad + 2Re\,(\lambda\mathscr{A}^{-1/2}x + \lambda^{-1}\mathscr{A}^{1/2}x, x)_{\mathscr{H}} \geq |x|_{\mathscr{H}}^2, \end{aligned} \tag{5.3.24}$$

where we have used the fact that the real part of the mixed product is nonnegative. Thus, (5.3.24) proves the second inequality in the lemma with $\mathscr{B} = \mathscr{A}^{1/2}$.

Writing next

$$V(\lambda)\lambda^{-2} = I + \lambda^{-1}\mathscr{A}^{1/2} + \lambda^{-2}\mathscr{A} = I + \lambda^{-1}\mathscr{A}^{1/2} + (\lambda^{-1}\mathscr{A}^{1/2})^2,$$

we obtain

$$\begin{aligned} |V(\lambda)\lambda^{-2}x|_{\mathscr{H}}^2 &= |\lambda^{-1}\mathscr{A}^{1/2}x|_{\mathscr{H}}^2 + |\lambda^{-2}\mathscr{A}x|_{\mathscr{H}}^2 + |x|_{\mathscr{H}}^2 + 2Re\,\lambda^{-2}|\mathscr{A}^{1/2}x|_{\mathscr{H}}^2 \\ &\quad + 2Re\,\lambda^{-1}\left[(\mathscr{A}^{1/2}x, x)_{\mathscr{H}} + |\lambda^{-1}\mathscr{A}^{3/4}x|_{\mathscr{H}}^2\right]. \end{aligned}$$

We note that the last term in the above expression is positive for $Re\lambda > 0$. Therefore, because $\mathscr{A}$ is self-adjoint, we have

$$\begin{aligned} |V(\lambda)\lambda^{-2}x|_{\mathscr{H}}^2 &\geq ||\lambda|^{-1}\mathscr{A}^{1/2}x|_{\mathscr{H}}^2 + ||\lambda|^{-2}\mathscr{A}x|_{\mathscr{H}}^2 + |x|_{\mathscr{H}}^2 - 2|\lambda|^{-2}|\mathscr{A}^{1/2}x|_{\mathscr{H}}^2 \\ &\geq \left(\left[|\lambda|^{-4}\mathscr{A}^2 - |\lambda|^{-2}\mathscr{A} + I\right]x, x\right)_{\mathscr{H}} \geq \frac{3}{4}|x|_{\mathscr{H}}^2, \end{aligned}$$

which proves the last inequality in Lemma 5.3.8. Finally, for the first inequality we compute

$$V(\lambda)\mathscr{A}^{-1} = I + \lambda\mathscr{A}^{-1/2} + (\lambda\mathscr{A}^{-1/2})^2$$

and after some calculations

$$\begin{aligned}|V(\lambda)\mathscr{A}^{-1}x|^2_{\mathscr{H}} &= |x|^2_{\mathscr{H}} + |(\lambda\mathscr{A}^{-1/2})^2x|^2_{\mathscr{H}} + |\lambda\mathscr{A}^{-1/2}x|^2_{\mathscr{H}} \\ &\quad + 2Re\,((\lambda\mathscr{A}^{-1/2})^2x,x)_{\mathscr{H}} + P,\end{aligned}$$

where $P \equiv 2Re\,\lambda\left[|\mathscr{A}^{-1/4}x|^2_{\mathscr{H}} + |\lambda\mathscr{A}^{-3/4}x|^2_{\mathscr{H}}\right]$ is a positive term. Therefore, as above we obtain that

$$|V(\lambda)\mathscr{A}^{-1}x|^2_{\mathscr{H}} \geq \left(\left[|\lambda|^4\mathscr{A}^{-2} - |\lambda|^2\mathscr{A}^{-1} + I\right]x,x\right)_{\mathscr{H}} \geq \frac{3}{4}|x|^2_{\mathscr{H}},$$

and (note V commutes with $\mathscr{A}^{\alpha}$) $|\mathscr{A}^{-1/2}V(\lambda)\mathscr{A}^{-1/2}x|^2_{\mathscr{H}} \geq c|x|^2_{\mathscr{H}}$ completing the proof of Lemma 5.3.8 for $\mathscr{B} = \mathscr{A}^{1/2}$.

The actual proof with more general $\mathscr{B}$ is quite involved and can be traced back from [43]. □

To complete the proof of Proposition 5.3.7, we note that bounds in Lemma 5.3.8 along with the representation of the resolvent in (5.3.22) lead to the sought-after estimate in (5.3.19). □

Now we are in position to consider the semigroup generated by (5.3.12).

5.3.9. Proposition. *Under Assumption* 5.3.4 *the operator A generates an analytic contraction semigroup e^{-At} on $X = \mathscr{D}(\mathscr{A}^{1/2}) \times \mathscr{H} \times \mathscr{H}$.*

Proof. The result of this proposition is a special case of Theorem 2.2 [207]. The arguments presented below are from [207].

Step 1: We already know that A generates a strongly continuous semigroup e^{-At} of contractions on X. A direct calculation yields

$$A^{-1}F = (\mathscr{A}^{-1}[\mu^2\mathscr{B}f_1 + f_2 + \mu f_3]; -f_1; \mu f_1 + \mathscr{B}^{-1}f_3), \quad F = (f_1; f_2; f_3) \in X.$$

Therefore by Proposition 5.3.6 in order to prove Proposition 5.3.9 it suffices to establish the resolvent estimate (5.3.17). To accomplish this, we construct a suitable perturbation argument.

Step 2: Let $U(t) = e^{-At}U_0$, with $U_0 = (u_0; v_0; \theta_0) \in X$. Then one can see from (5.3.12) that $U(t) = (u(t); v(t); \theta(t))$ with $v(t) = u_t(t)$ satisfies the equation

$$\frac{d}{dt}\begin{pmatrix} u(t) \\ v(t) \end{pmatrix} + A_1\begin{pmatrix} u(t) \\ v(t) \end{pmatrix} = \begin{pmatrix} 0 \\ -\mu\theta_t(t) \end{pmatrix}.$$

Therefore

$$\begin{pmatrix} u(t) \\ v(t) \end{pmatrix} = e^{-A_1t}\begin{pmatrix} u_0 \\ v_0 \end{pmatrix} + \int_0^t e^{-A_1(t-\tau)}\begin{pmatrix} 0 \\ -\mu\theta_t(\tau) \end{pmatrix}d\tau.$$

Taking Laplace's transform

$$U(t) \mapsto \hat{U}(\lambda) \equiv \int_0^\infty e^{-\lambda t} U(t) dt \equiv (\hat{u}(\lambda); \hat{v}(\lambda); \hat{\theta}(\lambda)),$$

so that $\hat{U}(\lambda) = R(\lambda, -A)U_0$, we obtain

$$\hat{w}(\lambda) \equiv \begin{pmatrix} \hat{u}(\lambda) \\ \hat{v}(\lambda) \end{pmatrix} = R(\lambda, -A_1) \begin{pmatrix} u_0 \\ v_0 + \mu\theta_0 \end{pmatrix} - R(\lambda, -A_1) \begin{pmatrix} 0 \\ \mu\lambda\hat{\theta}(\lambda) \end{pmatrix}. \quad (5.3.25)$$

Step 3: One can see from (5.3.16) that

$$Re\,(U_0, \hat{U}(\lambda))_X = Re\,((\lambda + A)\hat{U}(\lambda), \hat{U}(\lambda))_X = (Re\,\lambda)|\hat{U}(\lambda)|_X^2 + |\mathscr{B}^{1/2}\hat{\theta}(\lambda)|_{\mathscr{H}}^2.$$

Therefore

$$|\mathscr{B}^{1/2}\hat{\theta}(\lambda)|_{\mathscr{H}}^2 \le Re\,(U_0, \hat{U}(\lambda))_X = Re\,(w_0, \hat{w}(\lambda))_Y + Re\,(\theta_0, \hat{\theta}(\lambda))_{\mathscr{H}}, \quad (5.3.26)$$

where $w_0 = (u_0; v_0)$. On the other hand

$$\begin{aligned} &Re\,(\mathscr{B}^{-1/2}\theta_0, \mathscr{B}^{1/2}\hat{\theta}(\lambda))_{\mathscr{H}} + Re\,(w_0, \hat{w}(\lambda))_Y \\ &\le \varepsilon|\mathscr{B}^{1/2}\hat{\theta}(\lambda)|_{\mathscr{H}}^2 + C_\varepsilon|\mathscr{B}^{-1/2}\theta_0|_{\mathscr{H}}^2 + |(w_0, \hat{w}(\lambda))_Y| \end{aligned}$$

Selecting small ε and combining with (5.3.26) yields

$$|\mathscr{B}^{1/2}\hat{\theta}(\lambda)|_{\mathscr{H}}^2 \le C\left(|\mathscr{B}^{-1/2}\theta_0|_{\mathscr{H}}^2 + |(\hat{w}(\lambda), w_0)_Y|\right) \quad (5.3.27)$$

Step 4: The following properties are the consequences (see [207, Lemma 3.2]) of Assumption 5.3.4: $\mathscr{D}(A_1^{1/2}) = \mathscr{D}(\mathscr{A}^{3/4}) \times \mathscr{D}(\mathscr{A}^{1/4})$ and

$$\mathscr{D}(A^{1/2}) = \mathscr{D}(\mathscr{A}^{3/4}) \times \mathscr{D}(\mathscr{A}^{1/4}) \times \mathscr{D}(\mathscr{B}^{1/2}) = \mathscr{D}(A_1^{1/2}) \times \mathscr{D}(\mathscr{B}^{1/2}).$$

Applying $A_1^{1/2}$ to both sides of equation (5.3.25) and exploiting analyticity of $e^{-A_1 t}$ along with the estimate (5.3.19), Assumption 5.3.4 gives

$$|A_1^{1/2}\hat{w}(\lambda)|_Y \le C|A_1^{-1/2}w_0|_Y + C|\mathscr{B}^{-1/2}\theta_0|_{\mathscr{H}}. \quad (5.3.28)$$

Indeed, by virtue of the estimate (5.3.19) $|A_1 R(\lambda, -A_1)|_{\mathscr{L}(Y)} \le M$. Thus, we have

$$|A_1^{1/2}R(\lambda, -A_1)w_0|_Y \le C|A_1^{-1/2}w_0|_Y$$

and

$$\begin{aligned} \left|A_1^{1/2}R(\lambda, -A_1)\begin{pmatrix} 0 \\ \theta_0 \end{pmatrix}\right|_Y &\le C\left|A_1^{-1/2}\begin{pmatrix} 0 \\ \theta_0 \end{pmatrix}\right|_Y \le C\left|A_1^{1/2}\begin{pmatrix} \mathscr{A}^{-1}\theta_0 \\ 0 \end{pmatrix}\right|_Y \\ &\le C|\mathscr{A}^{3/4}\mathscr{A}^{-1}\theta_0|_{\mathscr{H}} \le C|\mathscr{A}^{-1/4}\theta_0|_{\mathscr{H}} \le C|\mathscr{B}^{-1/2}\theta_0|_{\mathscr{H}}, \end{aligned}$$

where we have used the relation $A_1(\mathscr{A}^{-1}\theta_0;0) = (0;\theta_0)$, the characterization of $D(A_1^{1/2})$ given above and also Remark 5.3.5. Finally, by (5.3.19) we have that

$$\left| A_1^{1/2} R(\lambda, -A_1) \begin{pmatrix} 0 \\ \lambda\hat{\theta}(\lambda) \end{pmatrix} \right|_Y \le C \left| A_1^{1/2} \begin{pmatrix} 0 \\ \hat{\theta}(\lambda) \end{pmatrix} \right|_Y$$
$$\le C|\mathscr{A}^{1/4}\hat{\theta}(\lambda)|_{\mathscr{H}} \le C|\mathscr{B}^{1/2}\hat{\theta}(\lambda)|_{\mathscr{H}}.$$

Combining the above with (5.3.27) yields the conclusion in (5.3.28).

Step 5: Combining (5.3.27) and (5.3.28) gives:

$$|A_1^{1/2}\hat{w}(\lambda)|_Y^2 + |\mathscr{B}^{1/2}\hat{\theta}(\lambda)|_{\mathscr{H}}^2 \le C\left(|A_1^{-1/2} w_0|_Y^2 + |\mathscr{B}^{-1/2}\theta_0|_{\mathscr{H}}^2 \right).$$

Therefore recalling that $\mathscr{D}(A^{1/2}) = \mathscr{D}(A_1^{1/2}) \times \mathscr{D}(\mathscr{B}^{1/2})$ we obtain the combined inequality

$$|A^{1/2}\hat{U}(\lambda)|_X^2 = |A^{1/2}R(\lambda,-A)U_0|_X^2 \le C|A_1^{-1/2}w_0|_Y^2 + C|\mathscr{B}^{-1/2}\theta_0|_{\mathscr{H}}^2 \le C|A^{-1/2}U_0|_X^2$$

and from the resolvent identity after rescaling we conclude:

$$|\lambda R(\lambda,-A)U_0|_X^2 \le |U_0|_X^2 + |AR(\lambda,-A)U_0|_X^2 \le C|U_0|_X^2.$$

Proposition 5.3.6 now concludes the proof of Proposition 5.3.9. □

Completion of the proof of Theorem 5.3.3.

In the case $\kappa > 0$ we apply abstract Proposition 5.3.9 with appropriate changes of variables. For the case $\kappa = 0$ we use Proposition 5.3.7. The key part is to verify Assumption 5.3.4. It is here where the boundary conditions play critical role.

Hinged case:

- We obviously have that $\mathscr{A} = A_D^2$. This implies that Assumption 5.3.4 holds $\mathscr{B}$ proportional A_D.

Clamped case:

- Because $\mathscr{D}(\mathscr{A}^{1/2}) = H_0^2(\Omega) \subset \mathscr{D}(A_D) = H^2(\Omega) \cap H_0^1(\Omega) = \mathscr{D}(\mathscr{B})$, the second condition is verified.
- $\mathscr{D}(\mathscr{A}^{1/4}) = H_0^1(\Omega) = \mathscr{D}(A_D^{1/2})$, so the inequality in the third bullet holds via the closed graph theorem.

Thus, Assumption 5.3.4 is verified; hence Theorem 5.3.3 is proved.

5.3.10. Remark. In the case of the free mechanical and Dirichlet thermal boundary conditions the situation is more complicated. Although the result of Theorem 5.3.3 does hold, this cannot be deduced from Proposition 5.3.9. It requires an independent proof [208]. The main obstacle in applying abstract Proposition 5.3.9 is the second requirement in Assumption 5.3.4. Indeed, $\mathscr{D}(\mathscr{A}^{1/2}) = H^2(\Omega)$ which is not contained in $\mathscr{D}(\mathscr{B}) = H^2(\Omega) \cap H_0^1(\Omega)$. The difficulty is at the level of boundary conditions that require special treatment. We also refer to [216, Chapter 3] for more results concerning linear thermoelastic plates with different boundary conditions.

■

The analyticity of linear semigroup $e^{-A_{0,\kappa}t}$ proved in Theorem 5.3.3 and the fact that $A_{0,\kappa}^{-1}$ exists as a bounded operator makes it possible to apply [241, Theorem 6.13, p. 74] in order to obtain the following assertion.

5.3.11. Proposition. *The semigroup $e^{-A_{0,\kappa}t}$ generated by (5.3.1) for $\alpha = 0$ possesses the following properties.*

- $e^{-A_{0,\kappa}t} : H_{0,\kappa} \mapsto \mathscr{D}(A_{0,\kappa}^{\beta})$ *for every $\beta \geq 0$ and $t > 0$.*
- *For every $\beta \geq 0$ there exist $M = M_{\kappa,\beta} > 0$ and $\delta = \delta_\kappa > 0$ such that*

$$\|A_{0,\kappa}^{\beta} e^{-A_{0,\kappa}t}\|_{\mathscr{L}(H_{0,\kappa})} \leq M t^{-\beta} e^{-\delta t}, \quad t > 0. \tag{5.3.29}$$

As a consequence, we have the following "smoothing" effect

$$|e^{-A_{0,\kappa}t} U|_{H^3(\Omega)\times H^1(\Omega)\times H^1(\Omega)} \leq C|A_{0,\kappa}^{1/2} e^{-A_{0,\kappa}t} U|_{H_{0,\kappa}} \leq C t^{-1/2} |U|_{H_{0,\kappa}} \tag{5.3.30}$$

for the case $\kappa > 0$. The similar relation also holds in the case $\kappa = 0$.

Proof. The first (abstract) part is a consequence of [241, Theorem 6.13, p. 74]. The relation in (5.3.30) follows from the characterization

$$\mathscr{D}(A_{0,\kappa}^{1/2}) \sim \left[H^3(\Omega) \cap \mathscr{D}(\mathscr{A}^{1/2})\right] \times H_0^1(\Omega) \times H_0^1(\Omega), \quad \kappa > 0.$$

□

5.4 Generation of a nonlinear semigroup

Our aim here is to present well-posedness of a continuous semiflow corresponding to the models (5.2.2) and (5.2.4) considered on the phase space $H_{\alpha,\kappa}$ given by (5.3.2). By this we mean existence, uniqueness, and continuous dependence of solutions with respect to initial data and $t > 0$.

The following well-posedness result follows from regularity of the von Karman bracket (5.1.1) (see Section 1.4) along with properties of the corresponding linear semigroup (see Section 5.3).

5.4.1. Theorem (Well-posedness of finite energy solutions). *Assume that $F_0 \in W_\infty^2(\Omega)$ and $p \in L_2(\Omega)$. Let $(\alpha;\kappa) \in \Lambda \equiv \{0 \leq \alpha \leq 1, 0 \leq \kappa \leq 1\}$. Then*

- **Case $\kappa > 0$:** *for all initial data $U_0 = (u_0; u_1; \theta_0) \in H_{\alpha,\kappa}$ problem (5.2.2) and (5.2.3) possesses a unique (generalized) solution*

$$U(t) \equiv (u(t); u_t(t); \theta(t)) \in C([0,\infty); H_{\alpha,\kappa})$$

which depends continuously on the initial data. This solution is also weak; that is, satisfies the appropriate variational equation. Moreover, the energy balance equality

$$\mathcal{E}_{\alpha,\kappa}(u(t),u_t(t),\theta(t)) + \eta \int_s^t ||A_D^{1/2}\theta(\tau)||^2 d\tau = \mathcal{E}_{\alpha,\kappa}(u(s),u_t(s),\theta(s)) \quad (5.4.1)$$

holds for all $t \geq s \geq 0$, *where* $\mathcal{E}_{\alpha,\kappa}(u,u_t,\theta)$ *is the energy functional for the model* (5.2.2) *which is given by the formula*

$$\mathcal{E}_{\alpha,\kappa}(u,u_t,\theta) = E_{\alpha,\kappa}(u,u_t,\theta) - \frac{1}{2}\int_\Omega ([F_0,u]u + 2pu)dx \quad (5.4.2)$$

with

$$E_{\alpha,\kappa}(u,u_t,\theta) = \frac{1}{2}\left[||A_D u||^2 + ||M_\alpha{}^{1/2}u_t||^2 + \frac{1}{2}||\Delta v(u)||^2 + \kappa||\theta||^2 \right]. \quad (5.4.3)$$

Moreover, when $\alpha = 0$, *we have that*

$$U(t) = (u(t); u_t(t); \theta(t)) \in C\left((0,T]; H^3(\Omega)\times H^1(\Omega)\times H^1(\Omega)\right)$$

for every $T > 0$, *and*

$$\|u(t)\|_3^2 + \|u_t(t)\|_1^2 + \|\theta(t)\|_1^2 \leq \frac{C_\kappa(T,R)}{t}, \quad t \in (0,T],\ |U|_{0,\kappa} \leq R. \quad (5.4.4)$$

- **Case** $\kappa = 0$**:** *for every* $U_0 = (u_0; u_1) \in H_{\alpha,0}$ *problem* (5.2.4) *possesses a unique (generalized and weak) solution* $U(t) \equiv (u(t); u_t(t)) \in C(\mathbb{R}_+, H_{\alpha,0})$ *which depends continuously on the initial data and satisfies the energy equality*

$$\mathcal{E}_{\alpha,0}(u(t),u_t(t)) + \frac{\mu^2}{\eta}\int_s^t ||A_D^{1/2}u_t(\tau)||^2 d\tau = \mathcal{E}_{\alpha,0}(u(s),u_t(s)) \quad (5.4.5)$$

for all $t \geq s \geq 0$, *where* $\mathcal{E}_{\alpha,0}(u,u_t)$ *is the energy function for the model* (5.2.4) *which is given by the formula*

$$\mathcal{E}_{\alpha,0}(u,u_t) = E_{\alpha,0}(u,u_t) - \frac{1}{2}\int_\Omega ([F_0,u]u + 2pu)dx \quad (5.4.6)$$

with

$$E_{\alpha,0}(u,u_t) = \frac{1}{2}\left[||A_D u||^2 + ||M_\alpha{}^{1/2}u_t||^2 + \frac{1}{2}||\Delta v(u)||^2 \right]$$

Moreover, when $\alpha = 0$, *we also have that*

$$U(t) = (u(t); u_t(t)) \in C\left((0,T]; H^3(\Omega)\times H^1(\Omega)\right)$$

for every $T > 0$, and

$$\|u(t)\|_3^2 + \|u_t(t)\|_1^2 \le \frac{C(T,R)}{t} \quad \textit{for} \quad t \in (0,T],\ |U|_{0,0} \le R.$$

The proof of the well-posedness theorem, along with energy relations, is fairly routine. The main idea is to consider the nonlinear evolution as a locally Lipschitz perturbation of a contraction linear semigroup on $H_{\alpha,\kappa}$. This is possible due to sharp regularity of Airy's stress function $v(u)$. Indeed, the nonlinear term $F(u)$ is locally Lipschitz, on the strength of the regularity result given in Section 1.4. Thus, abstract results (see [241, Chapter 6] and also [62, Theorem 7.2] or Theorem 2.3.8 in Chapter 2) on generation of nonlinear semigroups apply, in order to conclude existence of nonlinear semigroups $S_t^{\alpha,\kappa}$ on $H_{\alpha,\kappa}$.

Proof of Theorem 5.4.1 (well-posedness). We consider the more challenging case $\kappa > 0$. The starting point is the linear part of the dynamics that can be represented by the operator $A_{\alpha,\kappa} : H_{\alpha,\kappa} \to H_{\alpha,\kappa}$ given by (5.3.4) and equipped with the natural domain $\mathscr{D}(A_{\alpha,\kappa})$. By Proposition 5.3.1 $A_{\alpha,\kappa}$ generates a strongly continuous semigroup $e^{-A_{\alpha,\kappa}t}$ of contractions on $H_{\alpha,\kappa}$. Now we rewrite nonlinear problem (5.2.2) and (5.2.3) as a first-order problem of the form

$$\frac{d}{dt}Y(t) + A_{\alpha,\kappa}Y(t) = \mathscr{B}(Y(t)),\ t > 0, \quad Y(0) = Y_0 = (u_0; u_1; \theta_0)$$

where $Y(t) \equiv (u(t); u_t(t); \theta(t))$ and $\mathscr{B}(Y) = \left(0; M_\alpha^{-1}F(u); 0\right)$ with $F(u)$ given by (5.2.1). It follows from Corollary 1.4.5 that the nonlinear term $\mathscr{B}(Y)$ is locally Lipschitz on $H_{\alpha,\kappa}$. Indeed, for $Y_i = (u_i; v_i; \theta_i)$, $i = 1,2$, we have

$$\begin{aligned} |\mathscr{B}(Y_1) - \mathscr{B}(Y_2)|_{\alpha,\kappa} &= \|M_\alpha^{-1/2}(F(u_1) - F(u_2))\| \le \|F(u_1) - F(u_2)\| \\ &\le C(\|[u_1, v(u_1)] - [u_2, v(u_2)]\| + \|[u_1 - u_2, F_0]\|). \end{aligned}$$

Therefore relation (1.4.26) with $\delta = 0$ in Corollary 1.4.5 implies that

$$|\mathscr{B}(Y_1) - \mathscr{B}(Y_2)|_{\alpha,\kappa} \le C\left(\|u_1\|_2^2 + \|u_1\|_2^2 + \|F_0\|_{W_\infty^2(\Omega)}\right)\|u_1 - u_2\|_2, \tag{5.4.7}$$

where C does not depend on α and κ. This means that $\mathscr{B}(Y)$ is locally Lipschitz on $H_{\alpha,\kappa}$.

Step 1: Local solutions. Thus, we can apply Theorem 2.3.8 from Chapter 2, which implies local existence of generalized solutions $Y(t) = (u(t); u_t(t); \theta(t))$ for any initial data $Y_0 = (u_0; u_1; \theta_0) \in H_{\alpha,\kappa}$. If $Y_0 \in \mathscr{D}(A_{\alpha,\kappa})$, then solution $Y(t)$ is strong and possesses the property

$$Y(t) \in C_r([0,T); \mathscr{D}(A_{\alpha,\kappa})) \subset C_r\left([0,T); W_\alpha \times H^2(\Omega) \times H^2(\Omega)\right)$$

for some $T > 0$, where C_r is the corresponding space of right-continuous functions and

$$W_\alpha = \left\{u \in \mathscr{D}(\mathscr{A}^{1/2}) : M_\alpha^{-1/2}\mathscr{A}u \in L_2(\Omega)\right\} \subset H^3(\Omega).$$

The function $t \mapsto Y(t)$ is also weakly continuous in $W_\alpha \times H^2(\Omega) \times H^2(\Omega)$.

Step 2: Energy equality and global solutions. The smoothness properties of strong solutions allows us to obtain by the standard energy method, the energy relation stated in (5.4.1) on the existence interval. It suffices to multiply the first equation in (5.1.2) by u_t, the second by θ, integrate by parts, and account for cancellation of the thermal coupling. This energy relation and also the inequalities of the form

$$c_1 E_{\alpha,\kappa}(u,u_t,\theta) - C_{F_0,p} \le \mathscr{E}_{\alpha,\kappa}(u,u_t,\theta) \le c_2 E_{\alpha,\kappa}(u,u_t,\theta) + C_{F_0,p}$$

imply a priori bounds for solutions in $H_{\alpha,\kappa}$ that are uniform in the parameters α and κ:

$$||A_D u(t)||^2 + ||M_\alpha{}^{1/2} u_t(t)||^2 + \kappa ||\theta(t)||^2 \le C(r, |F_0|_{W^2_\infty}, ||p||)\,. \tag{5.4.8}$$

under the condition $||A_D u_0||, ||M_\alpha{}^{1/2} u_1||, \sqrt{\kappa}||\theta_0|| \le r$. Moreover, the constant C in (5.4.8) does not depend on t (and on the size of the existence interval). Therefore we apply Theorem 2.3.8 again to show that local (in time) solutions are global and globally bounded on $\mathbb{R}_+$ with the values in $\mathscr{H}_{\alpha,\kappa}$.

In order to obtain energy relations valid for generalized solutions (strong limits in H_α of regular solutions), it suffices to write down energy relations for the difference of two regular solutions Y_n and Y_m. Because the nonlinear terms are locally Lipschitz, we obtain additional strong convergence

$$\theta_n \to \theta \ \text{ in } L_2\left(0,T;\mathscr{D}(A_D^{1/2})\right).$$

This allows us to pass with the limit on energy relations which then leads to energy relations valid for all generalized solutions, the fact stated in (5.4.1). This argument can be also used to show that any generalized solution is also weak and vice versa.

Step 3: Additional smoothness. In order to prove the inequality in (5.4.4), we evoke analyticity of linear semigroup $e^{A_{-\alpha,\kappa}t}$ which holds for $\alpha = 0$, as proved in Theorem 5.3.3, and for which Proposition 5.3.11 is valid. Indeed, using (5.3.30) and the variation of constants formula

$$Y(t) = e^{-A_{0,\kappa}t}Y(0) + \int_0^t e^{-A_{0,\kappa}(t-\tau)}\mathscr{B}(Y(\tau))d\tau$$

by the standard method (see, e.g., [61, Chapter 2]) we can obtain (5.4.4). □

The solutions to problems (5.2.2) and (5.2.4) generate a family of dynamical systems with the phase spaces $H_{\alpha,\kappa}$ given by (5.3.2). The evolution operator $S_t^{\alpha,\kappa}$ for $\kappa > 0$ is given by the formula $S_t^{\alpha,\kappa}(u_0;u_1;\theta_0) = (u(t);u_t(t);\theta(t))$, where $u(t)$ and $\theta(t)$ solve (5.2.2) with initial data (5.2.3), and for $\kappa = 0$ is defined by the relation $S_t^{\alpha,0}(u_0;u_1) = (u(t);u_t(t))$, where $u(t)$ is a solution to (5.2.4). So, in all cases considered we have a well-defined semiflow on the space $H_{\alpha,\kappa}$. When $\alpha > 0$, the corresponding semiflow is predominantly hyperbolic. When $\alpha = 0$ the semiflow is parabolic-like.

5.5 Regularity of the semiflow

In this section we study regularity properties of the flow associated with evolution operator $S_t^{\alpha,\kappa}$. This includes properties such as Lipschitz continuity, differentiability, and higher energy regularity. We pay special attention to the dependence of the corresponding bounds on parameters $(\alpha;\kappa)\in\Lambda=\{0\le\alpha\le 1, 0\le\kappa\le 1\}$.

5.5.1. Proposition (Lipschitz property). *Let $U_1,U_2\in H_{\alpha,\kappa}$, and $|U_i|_{\alpha,\kappa}\le R$. Then*

$$|S_t^{\alpha,\kappa}U_1 - S_t^{\alpha,\kappa}U_2|_{\alpha,\kappa}\le e^{a_R t}|U_1-U_2|_{\alpha,\kappa},\quad t>0, \tag{5.5.1}$$

where the constant $a_R>0$ does not depend on $(\alpha,\kappa)\in\Lambda$.

Proof. Let $S_t^{\alpha,\kappa}U_i=(u^i(t);u_t^i(t);\theta^i(t))$. Then the difference

$$(z(t);\xi(t))\equiv(u^1(t)-u^2(t);\theta^1(t)-\theta^2(t))$$

of two solutions satisfies the linear problem

$$\begin{cases} M_\alpha z_{tt}-\mu A_D\xi+\mathscr{A}z=F(t), & z|_{t=0}=z_0,\ z_t|_{t=0}=z_1, \\ \kappa\xi_t+\eta A_D\xi+\mu A_D z_t=0, & \xi|_{t=0}=\xi_0, \end{cases} \tag{5.5.2}$$

where $F(t)\equiv F(u^1)-F(u^2)\equiv[v(u^1)-v(u^2),u^1]+[v(u^2),z]+[F_0,z]$. Therefore it follows from energy relation (5.3.11) in Proposition 5.3.2 that

$$|S_t^{\alpha,\kappa}U_1-S_t^{\alpha,\kappa}U_2|_{\alpha,\kappa}^2\le|U_1-U_2|_{\alpha,\kappa}^2+2\int_0^t(F(\tau),z_t(\tau))d\tau.$$

One can see that under the hypotheses of Proposition 5.5.1 we have from (5.4.8) that $|S_t^{\alpha,\kappa}U_i|_{\alpha,\kappa}\le C_R$ for all $t\ge 0$. Therefore it follows from (5.4.7) that

$$|(F(\tau),z_t(\tau))|\le C_R|S_\tau^{\alpha,\kappa}U_1-S_\tau^{\alpha,\kappa}U_2|_{\alpha,\kappa}^2,\quad \tau\ge 0.$$

Consequently Gronwall's lemma leads to the result. □

5.5.2. Proposition (Differentiability). *The mapping $U\mapsto S_t^{\alpha,\kappa}U$ is Fréchet differentiable on $H_{\alpha,\kappa}$ for every $t\ge 0$. Moreover the following properties hold.*

- *The Fréchet derivative $D\left[S_t^{\alpha,\kappa}U_0\right]: H_{\alpha,\kappa}\mapsto H_{\alpha,\kappa}$ is a mapping of the form*

$$D\left[S_t^{\alpha,\kappa}U_0\right]W_0=W(t)\equiv(w(t);w_t(t);\xi(t)),\quad W_0=(w_0;w_1;\xi_0), \tag{5.5.3}$$

where $(w(t);w_t(t);\xi(t))\in C([0,\infty),H_{\alpha,\kappa})$ is a unique solution to the problem

$$\begin{cases} M_\alpha w_{tt}-\mu A_D\xi+\mathscr{A}w=\langle F'(u(t));w\rangle, & w|_{t=0}=w_0,\ w_t|_{t=0}=w_1, \\ \kappa\xi_t+\eta A_D\xi+\mu A_D w_t=0, & \xi|_{t=0}=\xi_0. \end{cases} \tag{5.5.4}$$

Here $u(t)$ is the first (plate) component of $S_t^{\alpha,\kappa}U_0 \equiv (u(t);u_t(t);\theta(t))$ and $F'(u)$ is the Fréchet derivative of $F(u)$; that is, $F'(u)$ is a mapping of the form

$$\langle F'(u(t));w\rangle = [w,v(u)+F_0]+2[u,v(u,w)], \tag{5.5.5}$$

where $v(u)$ is determined from (5.1.3) and $v(w_1,w_2)\in H_0^2(\Omega)$ solves the problem

$$\Delta^2 v(w_1,w_2) = -[w_1,w_2] \text{ in } \Omega, \quad \frac{\partial}{\partial n}v(w_1,w_2) = v(w_1,w_2) = 0 \text{ on } \Gamma. \tag{5.5.6}$$

- *The Fréchet derivative $D\left[S_t^{\alpha,\kappa}U\right]$ possesses the properties*

$$\left|D\left[S_t^{\alpha,\kappa}U\right]\right|_{\mathscr{L}(H_{\alpha,\kappa})} \le e^{a_R t}, \quad |U|_{\alpha,\kappa} \le R, \tag{5.5.7}$$

and

$$\left|D\left[S_t^{\alpha,\kappa}U_1\right] - D\left[S_t^{\alpha,\kappa}U_2\right]\right|_{\mathscr{L}(H_{\alpha,\kappa})} \le b_R e^{a_R t}|U_1-U_2|_{\alpha,\kappa}, \quad t>0, \tag{5.5.8}$$

for any $U_1,U_2 \in H_{\alpha,\kappa}$ such that $|U_i|_{\alpha,\kappa} \le R$. Here the positive constants a_R and b_R do not depend on $(\alpha,\kappa)\in\Lambda$.

Proof. We first prove by the standard energy method that for every element $W_0 = (w_0;w_1;\xi_0)$ from the space $H_{\alpha,\kappa}$ the linear problem in (5.5.4) possesses a unique weak solution $(w(t);\xi(t))$ such that

$$W(t) \equiv (w(t);w_t(t);\xi(t)) \in C([0,\infty),H_{\alpha,\kappa}).$$

and

$$|W(t)|_{\alpha,\kappa} \le e^{a_R t}|W_0|_{\alpha,\kappa}, \quad t\ge 0, \tag{5.5.9}$$

provided $\|u(t)\|_2 \le C_R$ for all $t\ge 0$ (below the constants a_R and C_R may change from line to line).

We next consider the difference

$$Y(t) = (z(t);z_t(t);\zeta(t)) \equiv S_t^{\alpha,\kappa}[U_0+W_0] - S_t^{\alpha,\kappa}[U_0] - W(t).$$

It is clear that the couple $(z(t);\zeta(t))$ solves linear problem (5.5.2) with zero initial data and with

$$\begin{aligned} F(t) &= F(u^*(t)) - F(u(t)) - \langle F'(u(t));w(t)\rangle \\ &= \int_0^1 \langle F'(u(t)+\lambda(u^*(t)-u(t)) - F'(u(t)),w(t)\rangle d\lambda, \end{aligned}$$

where $u(t)$ (resp., $u^*(t)$) is the first (plate) component of the expression $S_t^{\alpha,\kappa}[U_0]$ (resp., $S_t^{\alpha,\kappa}[U_0+W_0]$). Using regularity properties of the von Karman bracket and relations (5.5.1) and (5.5.9) one can see that

$$\|F(t)\| \le C_R\|u^*(t)-u(t)\|_2\|w(t)\|_2 \le C_R e^{a_R t}|W_0|^2_{\alpha,\kappa}.$$

Therefore the energy type argument applied to (5.5.2) in the case considered yields that

$$|Y(t)|^2_{\alpha,\kappa} \le C_R e^{a_R t}|W_0|^4_{\alpha,\kappa} + \int_0^t |Y(\tau)|^2_{\alpha,\kappa} d\tau, \quad t \ge 0.$$

Gronwall's lemma then implies that $|Y(t)|^2_{\alpha,\kappa} \le C_R e^{a_R t}|W_0|^4_{\alpha,\kappa}$. By the definition of the Frechet derivative this implies relation (5.5.3). The inequality in (5.5.7) follows from (5.5.9). As for (5.5.8), it can be derived by the same method as in the proof of Proposition 5.5.1. □

More regular solutions are considered next. The existence of such is asserted below.

5.5.3. Proposition (Strong solutions). *Let $W_\alpha = \{u \in \mathscr{D}(\mathscr{A}^{1/2}) : \mathscr{A}u \in V'_\alpha\}$, where V'_α is dual to $V_\alpha = \mathscr{D}(M_\alpha^{1/2})$. We equip W_α with the norm $\|\mathscr{A}\cdot\|_{V'_\alpha}$. For the initial data such that*

$$u_0 \in W_\alpha, \; u_1 \in \mathscr{D}(\mathscr{A}^{1/2}), \; \theta_0 \in \mathscr{D}(A_D), \tag{5.5.10}$$

the corresponding solutions $(u(t);\theta(t))$ to problem (5.2.2) *and* (5.2.3) *have the following regularity,*

$$(u(t);u_t(t);u_{tt}(t);\theta(t);\theta_t(t)) \in C\left(\mathbb{R}_+; W_\alpha \times \mathscr{D}(\mathscr{A}^{1/2}) \times V_\alpha \times \mathscr{D}(A_D) \times L_2(\Omega)\right).$$

We note that, because $\mathscr{D}(\mathscr{A}^{1/4}) \sim H_0^1(\Omega) \sim \mathscr{D}(M_\alpha^{1/2})$ in the case $\alpha > 0$ [129] and thus by the closed graph theorem $M_\alpha^{-1/2}\mathscr{A}^{1/4}$ is an isomorphism on $L_2(\Omega)$, we have that $W_\alpha = \mathscr{D}(\mathscr{A}^{3/4}) \subset H^3(\Omega)$ and $V_\alpha = H_0^1(\Omega)$. In the case $\alpha = 0$ we obviously have that $W_\alpha = \mathscr{D}(\mathscr{A}) \subset H^4(\Omega)$ and $V_\alpha = L_2(\Omega)$. We also note that similar regularity results also hold in the case $\kappa = 0$.

Proof. The proof is standard and relies on analysis of time derivatives of solutions. Indeed, one can see that a couple $(w(t);\xi(t)) \equiv (u_t(t);\theta_t(t))$ satisfies (5.5.4) with initial data

$$w_0 = u_1, \quad w_1 = M_\alpha^{-1}\left[\mu A_D\theta_0 - \mathscr{A}u_0 + F(u_0)\right], \quad \xi_0 = -\frac{\eta}{\kappa}A_D\theta_0 - \frac{\mu}{\kappa}A_D u_1.$$

By (5.5.10) we have that $(w_0;w_1;\xi_0) \in H_{\alpha,\kappa}$. Therefore the same argument as in Proposition 5.5.2 yields

$$(u_{tt}(t);u_t(t);\theta_t(t)) \in C(\mathbb{R}_+;\mathscr{H}_{\alpha,\kappa}) \equiv C\left(\mathbb{R}_+;\mathscr{D}(\mathscr{A}^{1/2}) \times V_\alpha \times L_2(\Omega)\right).$$

From the above regularity and from (5.2.2) we obtain that

$$(M_\alpha^{-1/2}\mathscr{A}u(t);\theta(t)) \in C(\mathbb{R}_+;L_2(\Omega) \times \mathscr{D}(A_D)).$$

This completes the proof of Proposition 5.5.3. □

We note that the existence of strong solutions stated in Proposition 5.5.3 can also be derived from Theorem 2.3.8. However, this theorem does not guarantee the desired smoothness and therefore the additional argument (presented just above) is needed.

5.6 Backward uniqueness of the semiflow

Backward uniqueness for a thermoelastic nonlinear plate, beside being of interest on its own, arises as an issue in the context of studying properties of attractors. Indeed, it becomes a tool in proving certain characteristics of the attractors. Because the thermoelastic dynamics is represented by a continuous semiflow—and not a flow—the issue of backward uniqueness is far from obvious. When $\alpha = 0$ the analyticity of the underlying linear semigroup provides a tool (see, e.g., [139, Section 7.3]) for the backward unique continuation. Thus, evolutions with analytic (in time) coefficients are tractable by this method. In the case when $\alpha > 0$, the problem is more subtle due to the parabolic–hyperbolic mixing of the dynamics. In fact, even for linear thermoelastic plates with time-independent coefficients, this property has been shown only recently [218] by using complex analysis methods.

Backward uniqueness, quantitatively, means that two trajectories coinciding at a given time $t > 0$ must coincide also at any earlier time. Precise formulation of the corresponding backward uniqueness result is given below.

5.6.1. Theorem (Backward uniqueness). *Let $\kappa > 0$ and $\alpha > 0$. Assume that $p \in L_2(\Omega)$ and $F_0 \in W^2_\infty(\Omega)$. Then the following statements hold.*

- *Let $(u^1(t);\theta^1(t))$ and $(u^2(t);\theta^2(t))$ be two solutions of equations* (5.2.2) *on an interval $[0,T]$ such that*

$$U^i(t) \equiv (u^i(t);u^i_t(t);\theta^i(t)) \in C(0,T;H_{\alpha,\kappa}), \quad i = 1,2.$$

If $U^1(T) = U^2(T)$, then $U^1(t) = U^2(t)$ for every $t \in [0,T]$.
- *Let $u(t) \in C(0,T;\mathscr{D}(\mathscr{A}^{1/2})$ and $(w(t);\xi(t))$ be a solution to the linear (nonautonomous) equation* (5.5.4) *such that*

$$W(t) \equiv (w(t);w_t(t);\xi(t)) \in C(0,T;H_{\alpha,\kappa}). \tag{5.6.1}$$

If $W(T) = 0$, then $W(t) = 0$ for every $t \in [0,T]$.

5.6.2. Remark. A similar backward uniqueness theorem remains also true in the case $\kappa = 0$ and $\alpha > 0$; that is, for problem (5.2.4). The point is that in this case the dynamics is time reversible. ∎

5.6.3. Remark. When $\alpha > 0$, the proof of backward uniqueness (given below) depends on partial "hyperbolicity" of (5.2.2). This property is no longer true when $\alpha = 0$. In this latter case, one needs to rely on parabolic or analytic characteristics of the model. The difficulty, in the nonlinear problems, is that the forcing term coefficients—$F(u)$ in (5.2.2) (resulting from nonlinear terms)—produce neither terms which are analytic in time nor low order perturbation of a positive operator. Thus, the standard methods (see [139, Section 7.3] and the literature cited therein) have limited applicability to the problem in question. There are also some developments of the "parabolic" methods mentioned which are suitable to treat backward uniqueness without reliance on analyticity (see, e.g., [122, 169] and the references therein). However, these methods also require some sort of symmetry of the linear operator. For thermal plates this does not hold in general, except for special cases when the boundary conditions are simply supported. For this latter case, we show in Proposition 5.6.8 how to adapt parabolic methods in order to obtain backward uniqueness also for the case $\alpha = 0$ but with hinged boundary conditions.

■

The proof of Theorem 5.6.1 is based on adaptation of the technique presented in [165], where linear and unforced thermal plates with space– and time-dependent coefficients are considered.

In order to prove the first part of Theorem 5.6.1 we consider the difference of two solutions:

$$W(t) := U^1(t) - U^2(t) \equiv (w(t); w_t(t); \xi(t)) \in C(0,T; H_{\alpha,\kappa}).$$

It is clear that $(w(t); w_t(t); \xi(t))$ solves the linear nonautonomous equations

$$\begin{cases} M_\alpha w_{tt} - \mu A_D \xi + \mathscr{A} w = L(t)w, \\ \kappa \xi_t + \eta A_D \xi + \mu A_D w_t = 0. \end{cases} \tag{5.6.2}$$

Here the linear operator $L(t)$ defined on $\mathscr{D}(\mathscr{A}^{1/2})$ has the form

$$L(t)w = [v(u^1(t) + u^2(t), w), u^1(t)] + [v(u^2(t), u^2(t)) + F_0, w] \tag{5.6.3}$$

for $w \in \mathscr{D}(\mathscr{A}^{1/2})$, where $v(w_1, w_2) \in H_0^2(\Omega)$ is defined by (5.5.6). Thus to prove the first part of Theorem 5.6.1 we need to show that for any solution $(w(t); \xi(t))$ to equations (5.6.2) possessing property (5.6.1) the relation $W(T) = 0$ implies that $W(t) = 0$ for every $t \in [0,T]$.

The statement of the second part of Theorem 5.6.1 is also the backward uniqueness property for the linear problem (5.6.2) but with the operator $L(t)$ of the form

$$L(t)w = \langle F'(u(t)); w\rangle = 2[v(u(t), w), u(t)] + [v(u(t), u(t)) + F_0, w].$$

One can see that in both cases the operator $L(t)$ possesses the properties: (i) $t \mapsto L(t)w$ is a strongly continuous function with values in $L_2(\Omega)$, and (ii) there exists a constant $C_L > 0$ such that

$$\|L(t)w\| \le C_L \|A_D w\| \quad \text{for all } t \in [0,T],\ w \in \mathscr{D}(A_D). \tag{5.6.4}$$

Thus the proof of Theorem 5.6.1 follows from the following general assertion.

5.6.4. Proposition (Backward uniqueness 2). *Let $\kappa > 0$, $\alpha > 0$. Assume that $L(t)$; $\mathscr{D}(A_D) \mapsto L_2(\Omega)$ is a (measurable) family of linear operators satisfying* (5.6.4) *and $(w(t); \xi(t))$ is a solution to the linear (nonautonomous) equation* (5.6.2) *on the interval $[0,T]$ such that* (5.6.1) *holds. If $w(T) = w_t(T) = \xi(T) = 0$, then $w(t) = \xi(t) = 0$ for every $t \in [0,T]$.*

Proof. We adapt the idea from [165].

Let $\delta \in (0,T)$ be arbitrary and $\chi : \mathbb{R} \mapsto [0,1]$ be a C^∞-smooth function such that (i) $\chi(t) = 0$ for $t \leq 0$ and (ii) $\chi(t) = 1$ for $t \geq \delta$. Let $(w;\xi)$ be a couple of functions satisfying the hypotheses of Proposition 5.6.4. We introduce a new couple $(\widetilde{w};\widetilde{\xi})$ by the formulas

$$\widetilde{w}(t) = \chi(t)w(t) \text{ and } \widetilde{\xi}(t) = \chi(t)\xi(t). \tag{5.6.5}$$

It is clear that the couple $(\widetilde{w}(t);\widetilde{\xi}(t))$ satisfies the relations

$$\widetilde{w}(0) = \widetilde{w}_t(0) = \widetilde{\xi}(0) = 0, \quad \widetilde{w}(T) = \widetilde{w}_t(T) = \widetilde{\xi}(T) = 0, \tag{5.6.6}$$

and solves the equations

$$\begin{cases} M_\alpha \widetilde{w}_{tt} - \mu A_D \widetilde{\xi} + \mathscr{A}\widetilde{w} = L(t)\widetilde{w} + f(t), \\ \kappa \widetilde{\xi}_t + \eta A_D \widetilde{\xi} + \mu A_D \widetilde{w}_t = g(t), \end{cases} \tag{5.6.7}$$

on the interval $[0,T]$, where

$$f(t) = 2\chi_t(t)M_\alpha w_t(t) + \chi_{tt}(t)M_\alpha w(t), \quad g(t) = \chi_t(t)\left[\kappa\xi(t) + \mu A_D w(t)\right]. \tag{5.6.8}$$

5.6.5. Lemma. *Let $\widetilde{E}(t) = \frac{1}{2}\left[||A_D\widetilde{w}(t)||^2 + ||{M_\alpha}^{1/2}\widetilde{w}_t(t)||^2\right]$. Then*

$$\widetilde{E}(t) \leq C_1 \int_t^T \Psi(s)ds + C_2||\widetilde{\xi}(t)||_{-1}^2, \quad t \in [0,T], \tag{5.6.9}$$

where

$$\Psi(t) = \widetilde{E}(t) + ||\widetilde{\xi}(t)||^2 + ||f(t)||_{-1}^2 + ||g(t)||_{-1}^2, \quad t \in [0,T]. \tag{5.6.10}$$

Proof. Applying the standard energy arguments to the first equation in (5.6.7) and using (5.6.6) we find that

$$\widetilde{E}(t) = -\int_t^T \left(\mu A_D\widetilde{\xi}(s) + L(s)\widetilde{w}(s) + f(s), \widetilde{w}_t(s)\right) ds$$

It follows from the second equation in (5.6.7) that $\eta A_D\widetilde{\xi} = -\kappa\widetilde{\xi}_t - \mu A_D\widetilde{w}_t + g(t)$. Thus we have that

$$-(A_D\widetilde{\xi},\widetilde{w}_t) = \eta^{-1}(\kappa\widetilde{\xi}_t + \mu A_D\widetilde{w}_t - g(t),\widetilde{w}_t)$$

$$= \eta^{-1}\mu(A_D\widetilde{w}_t,\widetilde{w}_t) - \eta^{-1}(g(t),\widetilde{w}_t) + \eta^{-1}\kappa\frac{d}{dt}(\widetilde{\xi},\widetilde{w}_t) - \eta^{-1}\kappa(\widetilde{\xi},\widetilde{w}_{tt}).$$

Therefore one can see that for any $\varepsilon > 0$

$$\widetilde{E}(t) \le C_1 \int_t^T \left[\Psi(s) + \varepsilon\|\widetilde{w}_{tt}(s)\|^2 + C_\varepsilon\|\widetilde{\xi}(s)\|^2\right] ds + C_2\|\widetilde{\xi}(t)\|_{-1}^2, \quad t \in [0,T],$$

where $\Psi(t)$ is given by (5.6.10). It follows from the first equation in (5.6.7) that

$$\|\widetilde{w}_{tt}(s)\|^2 \le C_1\left[\|M_\alpha{}^{-1}\mathscr{A}\widetilde{w}(s)\|^2 + \widetilde{E}(s) + \|\widetilde{\xi}(s)\|^2 + \|f(s)\|_{-1}^2\right].$$

Thus we obtain

$$\widetilde{E}(t) \le C_1 \int_t^T \left[\Psi(s) + \varepsilon\|M_\alpha{}^{-1}\mathscr{A}\widetilde{w}(s)\|^2 + C_\varepsilon\|\widetilde{\xi}(s)\|^2\right] ds + C_2\|\widetilde{\xi}(t)\|_{-1}^2 \quad (5.6.11)$$

for any $t \in [0,T]$ and $\varepsilon \in (0,1]$. Therefore to obtain relation (5.6.9) we need to estimate the term $\|M_\alpha{}^{-1}\mathscr{A}\widetilde{w}(s)\|^2$. In the case of the hinged boundary conditions (5.1.6) we obviously have that

$$\|M_\alpha{}^{-1}\mathscr{A}\widetilde{w}(s)\|^2 \le C\|A_D\widetilde{w}(s)\|^2 \le 2C\widetilde{E}(s), \quad s \in [0,T],$$

which implies the conclusion. In the clamped case we use Theorem 2.5.3 via the following argument. Green's formula and the trace theorem imply

$$\begin{aligned}(M_\alpha{}^{-1}\mathscr{A}\widetilde{w},z) &= (\mathscr{A}\widetilde{w},M_\alpha{}^{-1}z)\\ &\le C\left[\|\widetilde{w}\|_2\|M_\alpha{}^{-1}z\|_2 + \|\Delta\widetilde{w}\|_{L_2(\Gamma)}\left\|\frac{\partial}{\partial n}M_\alpha{}^{-1}z\right\|_{L_2(\Gamma)}\right]\\ &\le C\left[\|\widetilde{w}\|_2 + \|\Delta\widetilde{w}\|_{L_2(\Gamma)}\right]\|z\|\end{aligned}$$

for any $z \in L_2(\Omega)$. Therefore

$$\|M_\alpha{}^{-1}\mathscr{A}\widetilde{w}(s)\|^2 \le C\left[\widetilde{E}(s) + \|\Delta\widetilde{w}(s)\|_{L_2(\Gamma)}^2\right], \quad s \in [0,T].$$

It follows from Theorem 2.5.3 with $\omega = 0$ for the interval $[t,T]$ that

$$\begin{aligned}\int_t^T \|\Delta\widetilde{w}(s)\|_{L_2(\Gamma)}^2 ds &\le c_1\int_t^T \widetilde{E}(s)ds\\ &\quad + c_2\left[\widetilde{E}(t) + \left|\int_t^T(\mu A_D\widetilde{\xi} + L(s)\widetilde{w} + f(s),h\nabla\widetilde{w})ds\right|\right] \quad (5.6.12)\\ &\le c_1\int_t^T\left[\widetilde{E}(s) + \|f(s)\|_{-1}^2\right]ds + c_2\left[\widetilde{E}(t) + \left|\int_t^T(\mu A_D\widetilde{\xi},h\nabla\widetilde{w})ds\right|\right].\end{aligned}$$

As above, it follows from the second equation in (5.6.7) that

$$
\begin{aligned}
& -(A_D\widetilde{\xi}, h\nabla\widetilde{w}) = \eta^{-1}(\kappa\widetilde{\xi}_t + \mu A_D\widetilde{w}_t - g(t), h\nabla\widetilde{w}) \\
&= \eta^{-1}\left[\mu(A_D\widetilde{w}_t, h\nabla\widetilde{w}) - (g(t), h\nabla\widetilde{w}) + \kappa\frac{d}{dt}(\widetilde{\xi}, h\nabla\widetilde{w}) - \kappa(\widetilde{\xi}, h\nabla\widetilde{w}_t)\right].
\end{aligned}
$$

Therefore one can see that

$$
\left|\int_t^T (\mu A_D\widetilde{\xi}, h\nabla\widetilde{w})ds\right| \le C_1\left[\widetilde{E}(t) + \|\widetilde{\xi}(t)\|_{-1}^2\right] + C_2\int_t^T \Psi(s)ds.
$$

Thus from (5.6.12) we have that

$$
\int_t^T \|M_\alpha{}^{-1}\mathscr{A}\widetilde{w}(s)\|^2 ds \le C_1\left[\widetilde{E}(t) + \|\widetilde{\xi}(t)\|_{-1}^2\right] + C_2\int_t^T \Psi(s)ds
$$

in the case of the clamped boundary conditions. Consequently relation (5.6.9) follows from (5.6.11).

□

The following assertion provides the estimates for the thermal component $\widetilde{\xi}$.

5.6.6. Lemma. *The following relation*

$$
\int_t^T \|\widetilde{\xi}(s)\|^2 ds \le C_1\|\widetilde{\xi}(t)\|_{-1}^2 + C_2\int_t^T \left[\widetilde{E}(s) + \|g(s)\|_{-1}^2\right] ds \tag{5.6.13}
$$

holds for any $t \in [0,T]$. Let $h(t) = \tau e^t$ for some positive τ. Then we have

$$
\int_0^T e^{2h(s)}\|\widetilde{\xi}(s)\|_{-1}^2 ds \le \frac{C}{\tau}\int_0^T e^{2h(s)}\left[\widetilde{E}(s) + \|g(s)\|_{-1}^2\right] ds, \tag{5.6.14}
$$

where the constant C does not depend on τ.

Proof. Multiplying the second equation in (5.6.7) by $A_D^{-1}\widetilde{\xi}(t)$ and integrating from t to T with the use of (5.6.6) we obtain:

$$
2\eta\int_t^T \|\widetilde{\xi}(s)\|^2 ds = \kappa\|A_D^{-1/2}\widetilde{\xi}(t)\|^2 + 2\int_t^T (-\mu A_D\widetilde{w}_t(s) + g(s), A_D^{-1}\widetilde{\xi}(s))ds.
$$

This implies (5.6.13).

To prove (5.6.14) we note that the function $\xi^*(t) = e^{h(t)}\widetilde{\xi}(t)$ solves the equation

$$
\kappa\xi_t^* + \eta A_D\xi^* - \kappa h_t\xi^* = g^*(t) \equiv e^h\left[-\mu A_D\widetilde{w}_t + g(t)\right].
$$

Therefore

$$
\begin{aligned}
\|A_D^{-1/2}g^*(t)\|^2 &= \|A_D^{-1/2}[\kappa\xi_t^* + \eta A_D\xi^* - \kappa h_t\xi^*]\|^2 \\
&\ge 2(\kappa\xi_t^*, \eta\xi^* - \kappa h_t A_D^{-1}\xi^*) \\
&= \kappa\frac{d}{dt}\left[\eta\|\xi^*\|^2 - \kappa h_t\|A_D^{-1/2}\xi^*\|^2\right] + \kappa^2 h_{tt}\|A_D^{-1/2}\xi^*\|^2.
\end{aligned}
$$

Using (5.6.6) after integration we have that

$$\int_0^T \|A_D^{-1/2} g^*(s)\|^2 ds \geq \kappa \int_0^T h_{tt} \|A_D^{-1/2} \xi^*\|^2 ds.$$

Because $h_{tt}(t) = \tau e^t \geq \tau$ for $t \geq 0$, we can easily obtain (5.6.14). □

Now using Lemmas 5.6.5 and 5.6.6 we establish the following assertion.

5.6.7. Lemma. *Let $h(t) = \tau e^t$ with $\tau > 0$. Then there exists $\tau_0 > 0$ such that for every $\tau \geq \tau_0$ we have that*

$$\int_0^T e^{2h(s)} \widetilde{E}(s) ds \leq \frac{C}{\tau} \int_0^T e^{2h(s)} \left[\|f(s)\|_{-1}^2 + \|g(s)\|_{-1}^2 \right] ds, \tag{5.6.15}$$

where the constant C does not depend on τ. Here $\widetilde{E}(t)$ is the same as in Lemma 5.6.5 and the functions f and g are given by (5.6.8).

Proof. We first note that (5.6.9) and (5.6.13) imply that

$$\widetilde{E}(t) \leq C_1 \int_t^T \Psi_0(s) ds + C_2 \|\widetilde{\xi}(t)\|_{-1}^2, \quad t \in [0,T], \tag{5.6.16}$$

where

$$\Psi_0(t) = \widetilde{E}(t) + \|f(t)\|_{-1}^2 + \|g(t)\|_{-1}^2, \quad t \in [0,T]. \tag{5.6.17}$$

A simple calculation based on integration by parts yields

$$\begin{aligned} I_T := \int_0^T e^{2h(t)} \int_t^T \Psi_0(s) ds dt &= \int_0^T \frac{1}{2h_t(t)} \left[\int_t^T \Psi_0(s) ds \right] \frac{d}{dt} e^{2h(t)} dt \\ &\leq - \int_0^T e^{2h(t)} \frac{d}{dt} \left[\frac{1}{2h_t(t)} \int_t^T \Psi_0(s) ds \right] dt \\ &= \int_0^T e^{2h(t)} \frac{h_{tt}(t)}{2[h_t(t)]^2} \int_t^T \Psi_0(s) ds dt + \int_0^T e^{2h(t)} \frac{1}{2h_t(t)} \Psi_0(t) dt. \end{aligned}$$

Because

$$\frac{h_{tt}(t)}{[h_t(t)]^2} = \frac{e^{-t}}{\tau} \leq \frac{1}{\tau} \quad \text{and} \quad \frac{1}{h_t(t)} = \frac{e^{-t}}{\tau} \leq \frac{1}{\tau} \quad \text{for all } t \geq 0,$$

we obtain

$$I_T \leq \frac{1}{2\tau} I_T + \frac{1}{2\tau} \int_0^T e^{2h(t)} \Psi_0(t) dt.$$

Therefore we have that

$$\int_0^T e^{2h(t)} \int_t^T \Psi_0(s) ds dt \leq \frac{1}{\tau} \int_0^T e^{2h(t)} \Psi_0(t) dt$$

for all $\tau \geq 1$. Thus, it follows from (5.6.16) and (5.6.14) that

$$\int_0^T e^{2h(s)}\widetilde{E}(s)ds \le \frac{C}{\tau}\int_0^T e^{2h(s)}\left[\widetilde{E}(s) + \|f(s)\|_{-1}^2 + \|g(s)\|_{-1}^2\right] ds$$

for all $\tau \ge 1$, where the constant C does not depend on τ. This implies (5.6.15) for τ large enough. □

Completion of the proof of Proposition 5.6.4. Because $f(t)$ and $g(t)$ given by (5.6.8) are supported on $(0,\delta)$, Lemma 5.6.7 implies

$$e^{2h(\delta)}\int_\delta^T \widetilde{E}(s)ds \le \int_0^T e^{2h(s)}\widetilde{E}(s)ds \le \frac{C}{\tau}e^{2h(\delta)}\int_0^\delta \left[\|f(s)\|_{-1}^2 + \|g(s)\|_{-1}^2\right] ds$$

for all $\tau \ge \tau_0$. Therefore, it follows from the representations in (5.6.5) and (5.6.8) that

$$\int_\delta^T \left[||A_D w||^2 + ||M_\alpha{}^{1/2} w_t||^2\right] ds \le \frac{C_\delta}{\tau}\int_0^\delta \left[||M_\alpha{}^{1/2} w_t||^2 + ||A_D w||^2 + ||\xi||^2\right] ds$$

for all $\tau \ge \tau_0$. In the limit $\tau \to \infty$ we obtain $w(t) = w_t(t) = 0$ for all $\delta \le t \le T$. It also follows from (5.6.2) that $\xi(t) = 0$ for all $\delta \le t \le T$. Inasmuch as $\delta > 0$ is arbitrary, this concludes the proof of Proposition 5.6.4. □

Now we turn to the nonrotational case when $\alpha = 0$. In this case, the result of Proposition 5.6.4 holds whenever the operator $L(t)w$, for $w \in D(A_D)$ has analytic dependence in time and obeys the estimate (5.6.4). Indeed, this conclusion follows from the analytic continuation principle, which can be applied on the strength of analyticity of the semigroup associated with the linear operator $A_{0,\kappa}$; see Theorem 5.3.3.

Another instance when backward uniqueness can be proved for a thermoelastic system of the form (5.6.2) with $\alpha = 0$ is the commutative case when $\mathscr{A} = aA_D^2$ with $a > 0$. In fact, this follows from a more general result stated below.

We define a product space $H \equiv \mathscr{H} \times \mathscr{H} \times \mathscr{H}$.

5.6.8. Proposition. *Let $\mathscr{M}$ be a 3×3 real coefficient matrix with distinct eigenvalues λ_i such that Re $\lambda_i > 0$, $i = 1,2,3$. Assume that a family of (measurable) linear operators $\mathscr{L}(t) : H \mapsto H$ satisfies $|\mathscr{L}(t)Y|_H \le c|Y|_H$. Consider the evolution on H given by*

$$Y_t(t) + A_D\mathscr{M}Y(t) = \mathscr{L}(t)Y(t), \quad t > 0. \tag{5.6.18}$$

Let $Y(t)$ be a solution to (5.6.18) and let $Y(T) = 0$ for some $T > 0$. Then, backward uniqueness holds; that is, $Y(t) \equiv 0$ for all $t \le T$.

Proof. Because $\mathscr{M}$ is diagonalizable and invertible, we can write $\mathscr{M} = P^{-1}\Lambda P$, where Λ is a diagonal matrix with the eigenvalues having positive real parts and P is a similarity transformation. We introduce a new variable $\hat{Y} \equiv PY$, so that the original evolution becomes

$$\hat{Y}_t + A_D\Lambda\hat{Y} = P\mathscr{L}(t)P^{-1}\hat{Y}. \tag{5.6.19}$$

Now we note that both the operator $A_D \Lambda$ and $P\mathscr{L}(t)P^{-1}$ satisfy (component-wise) the hypotheses imposed in Theorem 1.1 [122]. Thus, the method of the proof in [122] applies and yields $\hat{Y}(T) = 0 \Rightarrow \hat{Y}(t) \equiv 0$. Clearly, the same conclusion is valid for $Y(t)$. □

As a consequence, we obtain the following corollary.

5.6.9. Corollary. *The results stated in Theorem* 5.6.1 *and Proposition* 5.6.4 *remain true with* $\alpha = 0$ *when* $\mathscr{A} = aA_D^2$, *for any positive constant a.*

Proof. Proposition 5.6.8 immediately implies the backward uniqueness property for the system (5.6.2) with $\mathscr{A} = aA_D^2$, $\alpha = 0$. Indeed, it suffices to rewrite the system in the variable $Y(t) = (A_D w(t); w_t(t); \theta(t))$. It follows from (5.3.4) that in this case

$$\mathscr{M} \equiv \begin{pmatrix} 0 & -1 & 0 \\ a & 0 & -\mu \\ 0 & \mu\kappa^{-1} & \eta\kappa^{-1} \end{pmatrix}.$$

Then the matrix $\mathscr{M}$ has one positive eigenvalue and two complex conjugate eigenvalues with positive real part, see [216, Section 3D, p. 311]. Thus, applying Proposition 5.6.8 with

$$\mathscr{L}(t) \equiv \begin{pmatrix} 0 & 0\,0 \\ L(t)A_D^{-1} & 0\,0 \\ 0 & 0\,0 \end{pmatrix}$$

and accounting for the assumption (5.6.4) we obtain the statement of Proposition 5.6.4 in the case $\alpha = 0$, which also implies the statement of Theorem 5.6.1 for the case considered. □

5.6.10. Remark. The backward uniqueness property for nonrotational Berger thermoelastic plates with hinged boundary conditions was proved by a similar method, in [242]; see also [126]. ■

5.7 Stationary solutions

In this short section we describe some properties of stationary solutions associated with the thermoelastic system given by (5.1.2). The importance of stationary solutions stems from the fact that they characterize long time behavior of dynamics. We assume that $\kappa > 0$ and introduce the set of stationary points of $S_t^{\alpha,\kappa}$ denoted by $\mathscr{N}$ (as we show this set does not depend on α and κ):

$$\mathscr{N} = \left\{ V \in H_{\alpha,\kappa} \, : \, S_t^{\alpha,\kappa} V = V \text{ for all } t \geq 0 \right\}.$$

One can see that every stationary point V has the form $V = (u;0;0)$ where $u = u(x) \in H^2(\Omega)$ is a weak (variational) solution to the problem

$$\Delta^2 u = [v(u) + F_0, u] + p \quad \text{in } \Omega,$$

with the corresponding boundary condition (either (5.1.5) or (5.1.6)), where the function $v(u)$ solves (5.1.3). In particular, the stationary points do not depend on the parameters α, κ, μ, and η. One can also see that

$$\mathscr{N} \subset \{U \in H_{\alpha,\kappa} : |U|_{\alpha,\kappa} \le R_0\},$$

where R_0 depends on $\|F_0\|_{W^2_\infty(\Omega)}$ and $\|p\|_{L_2(\Omega)}$ only. The same is true in the case when $\kappa = 0$. In Chapter 11 we use this fact to prove some uniform estimates for the attractor.

It follows from the corresponding energy relation that the full energy $\mathscr{E}_{\alpha,\kappa}$ given by (5.4.2) (by (5.4.6) in the case $\kappa = 0$) is nonincreasing. Therefore the set

$$\mathscr{E}_R^{\alpha,\kappa} = \left\{U = (u_0; u_1; \theta_0) \in H_{\alpha,\kappa} \,:\, \mathscr{E}_{\alpha,\kappa}(u_0, u_1, \theta_0) \le R^2\right\} \tag{5.7.1}$$

is forward invariant for every $R > 0$; that is, $S_t^{\alpha,\kappa}\mathscr{E}_R^{\alpha,\kappa} \subset \mathscr{E}_R^{\alpha,\kappa}$ for $t \ge 0$. One can also see, because of the topological equivalence between the norm induced by the energy and the topology of $H_{\alpha,\kappa}$, that there exists $R_0^* \ge R_0$ which depends on $\|F_0\|_{W^2_\infty(\Omega)}$ and $\|p\|_{L_2(\Omega)}$ only and such that $\mathscr{N} \subset \mathscr{E}_{R_0^*}^{\alpha,\kappa}$. This observation regarding the set $\mathscr{N}$ of stationary solutions is important in the study of long-time dynamics and is used in Chapter 11.

Chapter 6
Structural Acoustic Problems and Plates in a Potential Flow of Gas

6.1 Introduction

In this chapter we discuss several models that involve coupled PDE structures. The coupling is between two second-order evolution equations and takes place on the lower-dimensional manifold interface between the two media. More specifically, we consider

- *Structural acoustic interaction.* These are governed by wave equations, describing acoustic pressure, and nonlinear plate equations, describing oscillations of the structure. The coupling occurs via boundary traces defined on a part of the boundary interface between the structure and acoustic environment occupying a bounded domain $\mathscr{O} \subset \mathbb{R}^3$.
- *Flow-structure interaction.* These are governed by wave equations defined on $\mathbb{R}^3_+$, describing flow of the gas, and plate equations, modeling oscillations in the structure in both the subsonic and supersonic flow of gas. The interaction takes place on the interface separating two environments.

Thus we deal with the problem of interaction of a clamped plate with an enclosed acoustic field or flow of gas. The dynamics of a plate are described by the von Karman evolution equations. The chamber can be finite or not. In the latter case (Case 2 above) we naturally arrive at the problem of nonlinear oscillations of a clamped plate in a flow of gas. In order to describe the influence of the gas we make use of the theory of potential flows that leads to the equations in (6.1.3) and (6.1.5).

To describe the structure of the model let us consider von Karman equations with clamped boundary conditions of the form:

$$\begin{cases} u_{tt} - \alpha\Delta u_{tt} + \Delta^2 u - [u, v + F_0] = p(x,t), \quad x \in \Omega,\ t > 0, \\ u|_{\partial\Omega} = \dfrac{\partial u}{\partial n}\Big|_{\partial\Omega} = 0, \\ u|_{t=0} = u_0(x), \quad \partial_t u|_{t=0} = u_1(x), \end{cases} \tag{6.1.1}$$

I. Chueshov, I. Lasiecka, *Von Karman Evolution Equations*,
Springer Monographs in Mathematics, DOI 10.1007/978-0-387-87712-9_6,

where as usual $v = v(u)$ is the Airy stress function. We assume that the domain $\Omega \subset \mathbb{R}^2$ embedded in $\mathbb{R}^3$ such that any point $x \in \Omega \subset \mathbb{R}^3$ has a form $(x_1, x_2, 0)$ and that this embedded domain Ω belongs to the boundary of a domain $\mathcal{O} \subset \mathbb{R}^3_+$.

Acoustic structure interaction. If the plate dynamics interacts with an enclosed acoustic field filling a bounded domain $\mathcal{O} \subset \mathbb{R}^3$, then the external force p in (6.1.1) is of the following form

$$p(x,t) = p_0(x) + \nu \cdot \partial_t \phi|_{|x_3=0}, \quad x \in \Omega, \tag{6.1.2}$$

(see, e.g., [28]), where $p_0(x) \in L^2(\Omega)$ and the parameter $\nu > 0$ characterizes the intensity of the interaction between the gas chamber and the plate. The acoustic velocity potential $\phi(x,t) = \phi(x_1, x_2, x_3; t)$ of the gas satisfies the equations:

$$\begin{cases} \phi_{tt} = \Delta \phi, \quad x \in \mathcal{O}, \\ \dfrac{\partial \phi}{\partial x_3} = \partial_t u(x_1, x_2, t) \ \text{ for } \ (x_1, x_2) \in \Omega, \ x_3 = 0, \\ \dfrac{\partial \phi}{\partial n} = 0 \ \text{ for } \ x \in \mathcal{O} \setminus \Omega, \\ \phi_{|t=0} = \phi_0(x), \quad \partial_t \phi_{|t=0} = \phi_1(x). \end{cases} \tag{6.1.3}$$

We study some natural generalizations of problem (6.1.1)–(6.1.3) in Section 6.2. In Section 6.3 instead of (6.1.1) we also consider thermoelastic plates described by equations of the form (5.1.2).

Flow Structure Interaction. Another kind of problem arises when

$$\mathcal{O} = \mathbb{R}^3_+ \equiv \{(x_1, x_2, x_3) \in \mathbb{R}^3 : x_3 > 0\}.$$

In this case it is natural to consider the problem from an aerodynamical point of view. If we assume that the gas flows above the plate in the direction of the x_1-axis, then the aerodynamical pressure of the flow on the plate is given by,

$$p(x,t) = p_0(x) + \nu \cdot (\partial_t + U \cdot \partial_{x_1})(\phi|_{|x_3=0}), \quad x \in \Omega, \tag{6.1.4}$$

(see, e.g., [28] and [99]) with $p_0(x) \in L^2(\Omega)$, the parameter $\nu > 0$ characterizes the intensity of the interaction between the flow and the plate, the parameter $U \geq 0$ ($U \neq 1$) represents the nonperturbed flow velocity and the velocity potential $\phi(x,t) = \phi(x_1, x_2, x_3; t)$ of the perturbed flow satisfies the equations:

$$\begin{cases} (\partial_t + U \partial_{x_1})^2 \phi = \Delta \phi, \quad x \in \mathbb{R}^3_+, \\ \left.\dfrac{\partial \phi}{\partial x_3}\right|_{x_3 = 0} = \begin{cases} (\partial_t + U \partial_{x_1}) u(x_1, x_2, t), & (x_1, x_2) \in \Omega, \\ 0, & (x_1, x_2) \notin \Omega, \end{cases} \\ \phi_{|t=0} = \phi_0(x), \qquad \partial_t \phi_{|t=0} = \phi_1(x). \end{cases} \tag{6.1.5}$$

In Sections 6.4–6.6 we present results on the well-posedness of problem (6.1.1) and (6.1.4), (6.1.5) for the following values of parameters: (i) $\alpha \geq 0$ and $0 < U < 1$ (see Section 6.5) and (ii) $\alpha > 0$, $U > 0$, $U \neq 1$ (see Section 6.6). Thus in the case $\alpha > 0$ and $0 < U < 1$ we present two different approaches. The case $\{\alpha = 0, U > 1\}$ remains beyond the scope of our considerations (see Section 6.4 for more comments).

6.1.1. Remark. One could consider other boundary conditions associated with the plate equation, such as hinged or partially clamped and hinged. However, in order to focus our treatment we concentrate on clamped boundary conditions which are very representative, physically most relevant, and mathematically challenging.

Motivated by physical considerations, we have assumed that the domain $\mathcal{O}$ is bounded in the case of structural acoustic interaction, and $\mathcal{O} = \mathbb{R}^3_+$, in the case of flow structure interaction. However, most of the results presented remain valid for more general domains that do not need to be bounded. We do not pursue these generalizations. ∎

Although both models have a lot in common in terms of the structure of PDEs involved, there are number of mathematical differences and subtleties. For this reason we treat these problems separately. In what follows, we present abstract models for the respective problems and we recast these problems within the framework of previous abstract models. The present chapter deals with well-posedness and regularity of solutions to the corresponding PDE systems. Chapter 12, instead, discusses long-time behavior.

Notations: As above $H^s(D)$ is the Sobolev space of order s on D and $H^s_0(D)$ is the closure in $H^s(D)$ of the set C_0^∞ of smooth functions with compact support in D. In this chapter D is either $\mathcal{O} \subseteq \mathbb{R}^3_+$ or $\Omega \subset \mathbb{R}^2$ which is naturally embedded in $\mathbb{R}^3$ according to the formula $(x_1;x_2) \mapsto (x_1;x_2;0)$.

We denote by $\|\cdot\|_{s,D}$ the norm in $H^s(D)$ and by $\|\cdot\|_D$ and $(\cdot,\cdot)_D$ the norm and the inner product in $L^2(D)$. For the sake of simplicity we sometimes write $\|\cdot\|_s$ and $\|\cdot\|$ instead of $\|\cdot\|_{s,D}$ and $\|\cdot\|_D$, if this does not lead to confusion.

In order to unify the treatment for both rotational and nonrotational cases we use the topology induced by the operator

$$M_\alpha u \equiv (I - \alpha\Delta_D)u, \quad u \in \mathscr{D}(M_\alpha) \equiv \mathscr{D}(\Delta_D) = H^2(\Omega) \cap H^1_0(\Omega). \tag{6.1.6}$$

where $\Delta_D : \mathscr{D}(\Delta_D) \subset L_2(\Omega) \to L_2(\Omega)$ denotes the Laplace operator with zero Dirichlet boundary conditions ($M_\alpha = I$ when $\alpha = 0$). According to the values of α, one has

$$V_\alpha := \mathscr{D}(M_\alpha^{1/2}) \equiv \begin{cases} H^1_0(\Omega), & \alpha > 0, \\ L_2(\Omega), & \alpha = 0. \end{cases} \tag{6.1.7}$$

The inner product and the norm on $V_\alpha = \mathscr{D}(M_\alpha^{1/2})$ are given by the formulas

$$(u,v)_{V_\alpha} = (u,v)_\Omega + \alpha(\nabla u, \nabla v)_\Omega \quad \text{and} \quad \|u\|^2_{V_\alpha} = \|u\|^2_\Omega + \alpha\|\nabla u\|^2_\Omega. \tag{6.1.8}$$

Thus V_α is topologically equivalent to $H_0^1(\Omega)$ when $\alpha > 0$ and $L_2(\Omega)$ when $\alpha = 0$. Later we also need the dual space V'_α (where duality is with respect to the pivot space $L_2(\Omega)$), and we have $V_\alpha \subseteq L_2(\Omega)) \subseteq V'_\alpha$.

6.2 Structural acoustic problem

6.2.1 Description of the model

The mathematical model under consideration consists of a semilinear wave equation defined on a bounded domain $\mathscr{O}$, which is strongly coupled with the von Karman plate equation acting only on a part of the boundary of $\mathscr{O}$. This kind of model, referred to in the literature as structural acoustic interactions, arises in the context of modeling gas pressure in an acoustic chamber surrounded by a combination of hard (rigid) and flexible walls. The pressure of a gas in the chamber is described by the solution to a wave equation, and vibrations of the flexible wall are described by the solution to a plate equation.

More precisely, let $\mathscr{O} \subset \mathbb{R}^3$ be a smooth bounded domain, with the boundary $\partial\mathscr{O} = \overline{\Omega \cup \Omega_*}$ consisting of two open (in induced topology) connected disjoint parts Ω and Ω_* of positive measure; Ω is *flat* and is referred to as the elastic wall, whose dynamics is described by the von Karman plate equation. The acoustic medium in the chamber $\mathscr{O}$ is described by a semilinear wave equation. Thus, we consider the following PDE system (which generalizes (6.1.1)–(6.1.3) written for the variables[1] $z \equiv -\phi$ and u),

$$\begin{cases} z_{tt} + g(z_t) - \Delta z + f(z) = 0 \text{ in } \mathscr{O} \times (0,T), \\ \dfrac{\partial z}{\partial n} = 0 & \text{on } \Omega_* \times (0,T), \\ \dfrac{\partial z}{\partial n} = \gamma\kappa u_t & \text{on } \Omega \times (0,T), \\ z(0,\cdot) = z^0, \quad z_t(0,\cdot) = z^1 \text{ in } \mathscr{O}, \end{cases} \tag{6.2.1}$$

and

$$\begin{cases} u_{tt} - \alpha\Delta u_{tt} + b_0(u_t) - \alpha\left[(b_1(u_{x_1t}))_{x_1} + (b_1(u_{x_2t}))_{x_2}\right] \\ \qquad + \Delta^2 u - [u, v + F_0] + \beta\kappa z_t|_\Omega - p_0(x) = 0 \text{ in } \Omega \times (0,T), \\ u = \dfrac{\partial u}{\partial n} = 0 & \text{on } \partial\Omega \times (0,T), \\ u(0,\cdot) = u^0(x), \quad u_t(0,\cdot) = u^1(x) & \text{in } \Omega, \end{cases} \tag{6.2.2}$$

[1] It is convenient to choose another sign of the velocity potential in structural acoustic problems in bounded domains; see [188] and the references therein.

where the Airy stress function $v = v(u)$ is a solution of the problem

$$\Delta^2 v + [u,u] = 0, \quad v|_{\partial\Omega} = \frac{\partial v}{\partial n}\bigg|_{\partial\Omega} = 0. \tag{6.2.3}$$

In the above system, $g(s)$, $b_0(s)$, and $b_1(s)$ are nondecreasing functions describing the dissipation effects in the model, and the term $f(z)$ represents a nonlinear force acting on the wave component; n is the outer normal vector, β and γ are positive constants; the parameter $\kappa \geq 0$ has been introduced in order to cover also the case of noninteracting wave and plate equations ($\kappa = 0$). The rotational term $\alpha\Delta u_{tt}$ represents moments of inertia. We consider both cases $\alpha > 0$ and $\alpha = 0$.

The part Ω_* of the boundary describes a rigid (hard) wall, whereas Ω is a flexible wall where the coupling with the plate equation takes place. The boundary term $\beta\kappa z_t|_\Omega$ describes back pressure exercised by the acoustic medium on the wall. The function F_0 describes in-plane forces applied to the plate, and $p_0 \in L_2(\Omega)$ accounts for transversal forces. The described interactive system involves a coupling between three- and two-dimensional manifolds, and as such is of hybrid type [223].

Structural acoustic models are well known in both the physical and mathematical literature and go back to the canonical models considered in [19, 238, 146]. More recently, these models were studied in the context of control theory, where problems such as the active control of pressure and vibrations by means of actuators placed on the flexible wall, or stabilization and controllability of the overall structure become issues of focal interest. There is a very large literature devoted to this topic; the reader is referred to the monograph [188], which also provides a rather comprehensive overview of related works. More recent contributions include [14, 15, 39, 131, 185], where questions of exact controllability or uniform stability are dealt with for interactions of wave/Kirchhoff plates [15], wave/Reissner–Mindlin plates [131], or wave/shell models [39], respectively.

In this section we aim to prove the existence and uniqueness theorem for problem (6.2.1) and (6.2.2) under rather general conditions on the nonlinear functions g, f, b_0, and b_1; see Assumption 6.2.1. In particular, it is not assumed that the damping functions g and b are (i) differentiable, nor (ii) strictly increasing.

Our main result in this section is Theorem 6.2.5 on well-posedness of the PDE problem (6.2.1) and (6.2.2).

6.2.2 Basic assumption

We impose the following basic assumptions on the nonlinear functions g and f which affect the wave component of the system and on the damping functions b_0 and b_1 in the plate equation.

6.2.1. Assumption. • $g \in C(\mathbb{R})$ is a nondecreasing function, $g(0) = 0$, and there exists a constant $C > 0$ such that

$$|g(s)| \le C(1+|s|^p), \quad s \in \mathbb{R}, \quad \text{for some} \quad 1 \le p \le 5. \tag{6.2.4}$$

- $f \in Lip_{\rm loc}(\mathbb{R})$, and there exists a positive constant M such that

$$|f(s_1) - f(s_2)| \le M(1+|s_1|^q + |s_2|^q)|s_1 - s_2| \quad \text{for } s_1, s_2 \in \mathbb{R}, \tag{6.2.5}$$

  where $q \le 2$. Moreover, the following dissipativity condition holds true,

$$\mu := \frac{1}{2} \liminf_{|s|\to\infty} \frac{f(s)}{s} > 0. \tag{6.2.6}$$

- $b_0 \in C(\mathbb{R})$ is a nondecreasing function such that $b_0(0) = 0$ and $b_1 \in C(\mathbb{R})$, $b_1(0) = 0$, nondecreasing and such that

$$|b_1(s)| \le C(1+|s|^{p_1}), \quad s \in \mathbb{R}, \quad \text{for some} \quad p_1 \ge 1.$$

- $p_0 \in L_2(\Omega)$, $F_0 \in H^{3+\delta}(\Omega)$ for some $\delta > 0$. [2]

Note that the values $p = 5$ and $q = 3$ correspond to "critical" values for the damping and the source for wave dynamics (see [75] and the references therein).

6.2.3 Abstract formulation

We find it convenient to represent the PDE system (6.2.1) and (6.2.2) in an abstract-semigroup form. In fact, we show that the resulting formulation is a special case of a general nonlinear structural acoustic model given in [188, Section 2.6]. In order to accomplish this we introduce the following spaces and operators.

Let $A : \mathscr{D}(A) \subset L_2(\mathscr{O}) \to L_2(\mathscr{O})$ be a positive self-adjoint operator defined by

$$Ah = -\Delta h + \mu h, \quad \mathscr{D}(A) = \left\{ h \in H^2(\mathscr{O}) : \frac{\partial h}{\partial n}\Big|_{\partial\mathscr{O}} = 0 \right\}; \tag{6.2.7}$$

where $\mu > 0$ is given by (6.2.6). Next, let N_0 be the Neumann map from $L_2(\Omega)$ to $L_2(\mathscr{O})$, defined by

$$\psi = N_0\varphi \Longleftrightarrow \left\{ (-\Delta + \mu)\psi = 0 \text{ in } \mathscr{O}; \ \frac{\partial\psi}{\partial n}\Big|_{\Omega} = \varphi, \ \frac{\partial\psi}{\partial n}\Big|_{\Omega_*} = 0 \right\}. \tag{6.2.8}$$

It is well known (see, e.g., [216, Chapter 3]) that

$$N_0 \text{ continuous } : L_2(\Omega) \to H^{3/2}(\mathscr{O}) \subset \mathscr{D}(A^{3/4-\varepsilon}), \quad \varepsilon > 0.$$

In particular, we have that

---

[2] In the case $\alpha > 0$ this requirement can be relaxed.

$$A^{3/4-\varepsilon}N_0 \text{ continuous} : L_2(\Omega) \to L_2(\mathscr{O}). \tag{6.2.9}$$

By Green's formula, the following trace result holds true (see, e.g., [216]):

$$N_0^* A h = h|_\Omega \quad \text{for } h \in \mathscr{D}(A), \tag{6.2.10}$$

where $N_0^* : L_2(\mathscr{O}) \to L_2(\Omega)$ is the adjoint of N_0. The validity of (6.2.10) may be extended to all $h \in H^1(\mathscr{O}) \equiv \mathscr{D}(A^{1/2})$, as $\mathscr{D}(A)$ is dense in $\mathscr{D}(A^{1/2})$ and $h|_\Omega$ is bounded on $H^1(\mathscr{O})$.

Regarding the plate model, let $\mathscr{A} : \mathscr{D}(\mathscr{A}) \subset L_2(\Omega) \to L_2(\Omega)$ be the positive selfadjoint operator defined by

$$\mathscr{A} w = \Delta^2 w, \quad \mathscr{D}(\mathscr{A}) = H^4(\Omega) \cap H_0^2(\Omega).$$

Finally we consider the stiffness (inertia) operator M_α, which is given by (6.1.6).

By using the above dynamic operators, the coupled PDE problem (6.2.1) and (6.2.2) can be rewritten as the following abstract second order system,

$$z_{tt} + A(z - \gamma\kappa N_0 u_t) + D(z_t) + F_1(z) = 0, \tag{6.2.11}$$

$$M_\alpha u_{tt} + \mathscr{A} u + B(u_t) + \beta\kappa N_0^* A z_t + F_2(u) = 0, \tag{6.2.12}$$

$$z(0) = z^0, \quad z_t(0) = z^1; \quad u(0) = u^0, \quad u_t(0) = u^1, \tag{6.2.13}$$

where we have introduced the operators

$$D(h) := g(h), \quad F_1(z) = f(z) - \mu z, \tag{6.2.14}$$

in (6.2.11), whereas

$$B(w) := b_0(w) - \alpha \operatorname{div} b_1(\nabla w), \quad F_2(u) = -[u, v(u) + F_0] - p_0$$

in (6.2.12), where we denote $b_1(\nabla w) = (b_1(w_{x_1}); b_1(w_{x_2}))$.

Regarding the nonlinear force terms we have that

$$F_1(z) = \Phi'(z) \text{ with } \Phi(z) = \int_{\mathscr{O}} \int_0^{z(x)} (f(\xi) - \mu\xi)\, d\xi\, dx, \tag{6.2.15}$$

where $'$ stands for the Fréchet derivative in an appropriate space. It readily follows from (6.2.6) that

$$\Phi(z) \geq \delta_f \|z\|_{\mathscr{O}}^2 - M_f, \quad z \in H^1(\mathscr{O}), \tag{6.2.16}$$

for some positive constants δ_f and M_f. Similarly, we have that

$$F_2(u) = \Pi'(u) \quad \text{with} \quad \Pi(u) = \frac{1}{4}\|\Delta v(u)\|_\Omega^2 - \frac{1}{2}([u,u], F_0)_\Omega - (p_0, v)_\Omega. \tag{6.2.17}$$

The phase spaces Y_1 for the wave component $(z; z_t)$ and Y_2 for the plate component $(u; u_t)$ of system (6.2.11)–(6.2.13) are given, respectively, by:

$$Y_1 := \mathscr{D}(A^{1/2}) \times L_2(\mathscr{O}) \equiv H^1(\mathscr{O}) \times L_2(\mathscr{O});$$

$$Y_2 := \mathscr{D}(\mathscr{A}^{1/2}) \times V_\alpha \equiv H_0^2(\Omega) \times \mathscr{D}(M_\alpha^{1/2}),$$

where $V_\alpha = \mathscr{D}(M_\alpha^{1/2})$ is given in (6.1.7). The state space for problem (6.2.11)–(6.2.13) is then

$$Y = Y_1 \times Y_2 = \mathscr{D}(A^{1/2}) \times L_2(\mathscr{O}) \times \mathscr{D}(\mathscr{A}^{1/2}) \times V_\alpha, \tag{6.2.18}$$

which is supplemented with the following norm

$$||y||_Y^2 = ||(z_1;z_2;v_1;v_2)||_Y^2 := \beta||(z_1;z_2)||_{Y_1}^2 + \gamma||(v_1;v_2)||_{Y_2}^2 \tag{6.2.19}$$

and with the corresponding inner product.

6.2.2. Remark. Notice that an important consequence of Assumption 6.2.1 and of criticality of the parameter q in (6.2.5) is that the nonlinear operator F_1 is locally Lipschitz continuous from $H^1(\mathscr{O})$ into $L_2(\mathscr{O})$. Similarly, Airy stress function regularity implies that F_2 is locally Lipschitz from $H_0^2(\Omega)$ into $L_2(\Omega)$. This is to say:

$$\begin{aligned} ||F_1(z_1) - F_1(z_2))||_{\mathscr{O}} &\le C(\rho)||z_1 - z_2||_{1,\mathscr{O}}, \quad ||z_i||_{1,\mathscr{O}} \le \rho < \infty, \\ ||F_2(u_1) - F_1(u_2)||_\Omega &\le C(\rho)||u_1 - u_2||_{2,\Omega}, \quad ||u_i||_{2,\Omega} \le \rho < \infty, \end{aligned} \tag{6.2.20}$$

where $C(\rho)$ denotes a function bounded for bounded arguments. It is important to emphasize that although the operators are bounded on the respective spaces, they are not compact. This fact justifies the notion of "criticality" for the parameter q and for the nonlinear terms F_1 and F_2. ■

The natural (nonlinear) energy functions associated with the solutions to the uncoupled wave and plate models are given, respectively, by

$$\mathscr{E}_z(z(t), z_t(t)) := E_z^0(z(t), z_t(t)) + \Phi(z(t)),$$

$$\mathscr{E}_u(u(t), u_t(t)) := E_u^0(u(t), u_t(t)) + \Pi(u(t)),$$

where we have set

$$E_z^0(z(t), z_t(t)) := \frac{1}{2}\left\{||A^{1/2}z(t)||_{\mathscr{O}}^2 + ||z_t(t)||_{\mathscr{O}}^2\right\} := E_z^0(t), \tag{6.2.21}$$

$$E_u^0(u(t), u_t(t)) := \frac{1}{2}\left\{||\mathscr{A}^{1/2}u(t)||_\Omega^2 + ||M_\alpha^{1/2}u_t(t)||_\Omega^2\right\} := E_u^0(t). \tag{6.2.22}$$

Because both energy functionals $\mathscr{E}_z$ and $\mathscr{E}_u$ may be negative, it is convenient to introduce the following positive energy functions

$$E_z(z, z_t) := E_z^0(z, z_t) + \Phi(z) + M_f = \mathscr{E}_z(z, z_t) + M_f, \tag{6.2.23}$$

where M_f is the constant in (6.2.16), and

$$E_u(u,u_t) := E_u^0(u,u_t) + \frac{1}{4}||\Delta v(u)||^2_\Omega. \tag{6.2.24}$$

Finally, we introduce the total energy $\mathscr{E}(t) = \mathscr{E}(z(t), z_t(t), u(t), u_t(t))$ of the system, namely

$$\mathscr{E}(t) = \mathscr{E}(z(t), z_t(t), u(t), u_t(t)) := \beta\mathscr{E}_z(z,z_t) + \gamma\mathscr{E}_u(u,u_t), \tag{6.2.25}$$

whose positive part is given by

$$E(t) = E(z,z_t,u,u_t) := \beta E_z(z,z_t) + \gamma E_u(u,u_t). \tag{6.2.26}$$

It is easy to see from the structure of the energy functionals and in view of (6.2.16) and (6.2.17) that for any $\beta, \gamma > 0$ there exist positive constants c, C, and M_0 such that

$$cE(z,z_t,u,u_t) - M_0 \leq \mathscr{E}(z,z_t,u,u_t) \leq CE(z,z_t,u,u_t) + M_0, \tag{6.2.27}$$

where $\mathscr{E}$ and E are the energies defined in (6.2.25) and (6.2.26). For the proof of the corresponding estimate for the plate component we refer to Section 3.1.1 (see (3.1.9)).

6.2.4 Well-posedness

In this section we study well-posedness of problem (6.2.1) and (6.2.2). Because the corresponding abstract system (6.2.11)–(6.2.13) is a special case of a general abstract model studied in [188, Section 2.6], local and global existence results for the solutions can be deduced from Theorem 2.6.1 and Theorem 2.6.2 in [188]. However, for the sake of completeness and because [188] is focused on nonlinear *boundary* damping, which requires additional technicalities, we provide a self-contained proof, tailored for the specific problem under investigation (see also [38, Appendix A] and [36]).

In order to make our statements precise, following the same idea as in Chapters 3 and 4 we introduce the concepts of strong and generalized solutions.

6.2.3. Definition. A pair of functions

$$(z(t);u(t)) \in C(0,T;\mathscr{D}(A^{1/2})) \times \mathscr{D}(\mathscr{A}^{1/2})) \cap C^1(0,T;L_2(\mathscr{O}) \times \mathscr{D}(M_\alpha^{1/2}))$$

that satisfy the initial conditions (6.2.13) is said to be

(S) A *strong solution* to problem (6.2.11)–(6.2.13) on the interval $[0,T]$, iff

- For any $0 < a < b < T$ one has

$$(z_t(t);u_t(t)) \in L_1(a,b;\mathscr{D}(A^{1/2}) \times \mathscr{D}(\mathscr{A}^{1/2}))$$

and

$$(z_{tt}(t);u_{tt}(t)) \in L_1(a,b;L_2(\mathscr{O}) \times \mathscr{D}(M_\alpha^{1/2})).$$

- $A[z(t) - \gamma\kappa N_0 u_t(t)] + D(z_t(t)) \in L_2(\mathscr{O})$ and $\mathscr{A}u(t) \in V'_\alpha \equiv \left[\mathscr{D}(M_\alpha^{1/2})\right]'$ for almost all $t \in [0,T]$.
- Equations (6.2.11) and (6.2.12) are satisfied in $L_2(\mathscr{O}) \times V'_\alpha$ for almost all $t \in [0,T]$.

(G) A *generalized solution* to problem (6.2.11)–(6.2.13) on the interval $[0,T]$, iff there exists a sequence of strong solutions $\{(z_n(t);u_n(t))\}_n$ to (6.2.11)–(6.2.13) with initial data $(z_n^0;z_n^1;u_n^0;u_n^1)$ (in place of $(z^0;z^1;u^0;u^1)$) such that

$$\lim_{n\to\infty}\max_{t\in[0,T]}\left\{\|\partial_t z(t) - \partial_t z_n(t))\|_{\mathscr{O}} + \|A^{1/2}(z(t) - z_n(t))\|_{\mathscr{O}}\right\} = 0$$

and

$$\lim_{n\to\infty}\max_{t\in[0,T]}\left\{\|M_\alpha^{1/2}(\partial_t u(t) - \partial_t u_n(t))\|_\Omega + \|\mathscr{A}^{1/2}(u(t) - u_n(t))\|_\Omega\right\} = 0.$$

■

6.2.4. Remark. The definition of strong solution does imply that $u(t) \in (H^3 \cap H_0^2)(\Omega)$ in the case $\alpha > 0$ and $u(t) \in (H^4 \cap H_0^2)(\Omega)$ when $\alpha = 0$. We can also obtain additional space regularity of the variable $z(t)$. Indeed, this follows from the elliptic theory applied to the condition $A[(z(t) - \gamma\kappa N_0 u_t(t)] + D(z_t(t)) \in L_2(\mathscr{O})$. The amount of additional regularity inherited by $z(t)$ depends on the nonlinearity of the damping $D(z_t)$. This topic being tangential to present considerations is not discussed. We refer to [74] and to the references therein for some details. ■

The main result describing well-posedness for problem (6.2.1) and (6.2.2) is the following assertion.

6.2.5. Theorem. *Under Assumption* 6.2.1 *for every* $\alpha \geq 0$ *system* (6.2.1) *and* (6.2.2) *is well-posed on*

$$Y = H^1(\mathscr{O}) \times L_2(\mathscr{O}) \times H_0^2(\Omega) \times \mathscr{D}(M_\alpha^{1/2});$$

that is, for any $(z^0;z^1;u^0;u^1) =: y_0 \in Y$ *there exists a unique generalized solution* $y(t) = (z(t);z_t(t);u(t);u_t(t))$ *on* $\mathbb{R}_+$ *that depends continuously on initial data. This solution satisfies the energy inequality*

$$\mathscr{E}(t) + \beta\int_s^t (D(z_t),z_t)_{\mathscr{O}}\,d\tau + \gamma\int_s^t (B(u_t),u_t)_\Omega\,d\tau \leq \mathscr{E}(s), \qquad 0 \leq s \leq t, \quad (6.2.28)$$

with the total energy $\mathscr{E}(t)$ *given by* (6.2.25), *where*

$$(B(u_t),u_t)_\Omega = (b_0(u_t),u_t)_\Omega + \alpha(b_1(\nabla u_t),\nabla u_t)_\Omega.$$

Moreover, if initial data y_0 *are such that* $z^0, z^1 \in \mathscr{D}(A^{1/2})$, $u^0, u^1 \in \mathscr{D}(\mathscr{A}^{1/2})$, *and*

$$A[z^0 - \gamma\kappa N_0 u^1] + D(z^1) \in L_2(\mathscr{O}), \quad M_\alpha^{-1/2}\mathscr{A} u^0 \in L_2(\Omega),$$

then there exists a unique strong solution $y(t)$ *such that*

$$(z_{tt}; u_{tt}) \in C_r(\mathbb{R}_+; L_2(\mathscr{O}) \times \mathscr{D}(M_\alpha^{1/2})) \cap L_\infty^{loc}(\mathbb{R}_+; L_2(\mathscr{O}) \times \mathscr{D}(M_\alpha^{1/2})), \tag{6.2.29}$$

where C_r *stands for the space of right-continuous functions. This solution satisfies the* energy identity:

$$\mathscr{E}(t) + \beta \int_s^t (D(z_t), z_t)_{\mathscr{O}} d\tau + \gamma \int_s^t (B(u_t), u_t)_\Omega d\tau = \mathscr{E}(s), \quad 0 \le s \le t. \tag{6.2.30}$$

Both strong and generalized solutions satisfy the inequality

$$\mathscr{E}(t) \equiv \mathscr{E}(z(t), z_t(t), u(t), u_t(t)) \le \mathscr{E}(z(s), z_t(s), u(s), u_t(s)) \equiv \mathscr{E}(s) \tag{6.2.31}$$

for $t \ge s$*, which implies, in particular, (see* (6.2.27)*)*

$$E(z(t), z_t(t), u(t), u_t(t)) \le C\left(1 + E(z^0, z^1, u^0, u^1)\right) \ \ for\ t \ge 0. \tag{6.2.32}$$

Proof. As in Section 2.4 we use monotone operator theory and rely on Theorem 2.3.8.

Step 1. Existence and uniqueness of a local solution. We consider the abstract second-order system (6.2.11)–(6.2.13) corresponding to the PDE model (6.2.1) and (6.2.2), and introduce the overall dynamics operator $T : \mathscr{D}(T) \subset Y \to Y$; that is,

$$T \begin{bmatrix} z_1 \\ z_2 \\ u_1 \\ u_2 \end{bmatrix} = \begin{bmatrix} -z_2 \\ Az_1 + D(z_2) + z_2 - \gamma\kappa A N_0 u_2 \\ -u_2 \\ M_\alpha^{-1}(\mathscr{A} u_1 + \beta\kappa N_0^* A z_2 + B(u_2) + u_2) \end{bmatrix}, \tag{6.2.33}$$

whose domain is

$$\mathscr{D}(T) = \Big\{(z_1; z_2; u_1; u_2) \in \mathscr{D}(A^{1/2}) \times \mathscr{D}(A^{1/2}) \times \mathscr{D}(\mathscr{A}^{1/2}) \times \mathscr{D}(\mathscr{A}^{1/2}) \\ : A[z_1 - \gamma\kappa N_0 u_2] + D(z_2) \in L_2(\mathscr{O}), M_\alpha^{-1/2}\mathscr{A} u_1 \in L_2(\Omega)\Big\}$$

which is dense in Y. Setting $y = (z; z_t; u; u_t)$, it is elementary to derive from the second-order system (6.2.11)–(6.2.13) the first-order formulation of the original PDE problem (6.2.1) and (6.2.2), namely

$$y' + Ty = C(y), \quad y(0) = y_0, \tag{6.2.34}$$

with T defined by (6.2.33) and C given by

$$C(z_1; z_2; u_1; u_2) = \left(0; -F_1(z_1) + z_2; 0; M_\alpha^{-1}(-F_2(u_1) + u_2)\right).$$

As a consequence of (6.2.20), the nonlinear term C is locally Lipschitz on the phase space Y. Thus, according to Theorem 2.3.8 in Chapter 2 a local existence result follows once we prove that T is a maximal accretive operator.
Accretivity. If $y=(z_1;z_2;u_1;u_2)$, $\tilde{y}=(\tilde{z}_1;\tilde{z}_2;\tilde{u}_1;\tilde{u}_2)\in\mathscr{D}(T)$, then

$$
\begin{aligned}
&(Ty-T\tilde{y},y-\tilde{y})_Y\\
&=-\beta\gamma\kappa(AN_0(u_2-\tilde{u}_2),z_2-\tilde{z}_2)+\beta(g(z_2)-g(\tilde{z}_2)+z_2-\tilde{z}_2,z_2-\tilde{z}_2)\\
&\quad+\gamma\big(B(u_2)-B(\tilde{u}_2)+u_2-\tilde{u}_2,u_2-\tilde{u}_2\big)+\beta\gamma\kappa\left(N_0^*A(z_2-\tilde{z}_2),u_2-\tilde{u}_2\right)\\
&\geq\beta\left(g(z_2)-g(\tilde{z}_2),z_2-\tilde{z}_2\right)+\gamma\big(b_0(u_2)-b_0(\tilde{u}_2),v_2-\tilde{u}_2\big)\\
&\quad+\gamma\alpha\left(b_1(\nabla u_2)-b_1(\nabla\tilde{u}_2),\nabla u_2-\nabla\tilde{u}_2\right)\geq 0,
\end{aligned}
$$

where we have used (6.2.18) and (6.2.19) to compute the inner product, and the monotonicity properties of g and b_i to obtain the latest inequality.

Maximality. By Theorem 2.2.2 (see also [260, p. 18]), in order to show that T is maximal, we only need to prove that $\mathscr{R}(I+T)=Y$. We can do it in the same way as in [10, 36].

Given $h=(\varphi_1;\varphi_2;\psi_1;\psi_2)\in Y$, we seek to solve the equation $(I+T)y=h$, which explicitly reads as

$$
\begin{cases}
z_1-z_2=\varphi_1\\
2z_2+Az_1+g(z_2)-\gamma\kappa AN_0u_2=\varphi_2\\
u_1-u_2=\psi_1\\
u_2+M_\alpha^{-1}(u_2+\beta\kappa N_0^*Az_2+\mathscr{A}u_1+B(u_2))=\psi_2.
\end{cases}
\tag{6.2.35}
$$

Eliminating z_1 and u_1 we arrive at the equation

$$
(L+G)\begin{bmatrix}z_2\\u_2\end{bmatrix}=\begin{bmatrix}\varphi_2-A\varphi_1\\\psi_2-M_\alpha^{-1}\mathscr{A}\psi_1\end{bmatrix},
\tag{6.2.36}
$$

with L and G given by

$$
L=\begin{pmatrix}I+A & -\gamma\kappa AN_0\\ \beta\kappa M_\alpha^{-1}N_0^*A & M_\alpha^{-1}(I+\mathscr{A})\end{pmatrix},\quad G\begin{bmatrix}z_2\\u_2\end{bmatrix}=\begin{bmatrix}g(z_2)+z_2\\M_\alpha^{-1}B(u_2)+u_2\end{bmatrix}.
\tag{6.2.37}
$$

Let us observe that the right hand side of (6.2.36) belongs to the dual space $\mathscr{Y}'$ of $\mathscr{Y}:=\mathscr{D}(A^{1/2})\times\mathscr{D}(\mathscr{A}^{1/2})$ with respect to the duality $\langle\cdot,\cdot\rangle_{\mathscr{Y}',\mathscr{Y}}$ given by

$$
\langle(z,u),(h,w)\rangle_{\mathscr{Y}',\mathscr{Y}}=\beta(z,h)_{\mathscr{O}}+\gamma(M_\alpha u,w)_\Omega\,.
$$

Moreover, we have in particular that $L\in\mathscr{L}(\mathscr{Y},\mathscr{Y}')$ and $G:\mathscr{Y}\to\mathscr{Y}'$ is a monotone hemicontinuous operator by Assumption 6.2.1. The above decomposition is useful to show the following result.

6.2.6. Lemma. *Let L and G be the operators defined in* (6.2.37). *Then* $\mathscr{R}(L+G) \equiv \mathscr{Y}'$.

Proof. By [18, Corollary 1.3, p. 48] (see also the last bullet in Proposition 1.2.5), it is necessary to verify the following properties: (i) L is a monotone hemicontinuous operator from $\mathscr{Y}$ to $\mathscr{Y}'$, (ii) G is a maximal monotone operator from $\mathscr{Y}$ to $\mathscr{Y}'$, and (iii) $L+G$ is coercive.

We first compute

$$\left\langle L\begin{bmatrix} z_2 \\ u_2 \end{bmatrix}, \begin{bmatrix} z_2 \\ u_2 \end{bmatrix} \right\rangle_{\mathscr{Y}',\mathscr{Y}} = \left\langle \begin{bmatrix} z_2 + Az_2 - \gamma\kappa A N_0 u_2 \\ M_\alpha^{-1}(\beta\kappa N_0^* A z_2 + u_2 + \mathscr{A} u_2) \end{bmatrix}, \begin{bmatrix} z_2 \\ u_2 \end{bmatrix} \right\rangle_{\mathscr{Y}',\mathscr{Y}}$$

$$= \beta\|z_2\|^2 + \beta\|A^{1/2}z_2\|^2 - \beta\gamma\kappa(A^{1/2}N_0u_2, A^{1/2}z_2) + \gamma\beta\kappa(N_0^*Az_2, u_2)$$

$$+ \gamma\|u_2\|^2 + \gamma\|\mathscr{A}^{1/2}u_2\|^2 \geq \beta\|A^{1/2}z_2\|^2 + \gamma\|\mathscr{A}^{1/2}u_2\|^2 = \left\|(z_2; u_2)\right\|_{\mathscr{Y}}^2.$$

This shows that L is coercive, hence as a linear operator it is a monotone hemicontinuous operator, and thus (i) holds true. Moreover, because G is monotone, then $L+G$ is coercive, as well, so that (iii) is satisfied. It remains to be shown that G is maximal monotone as a mapping from $\mathscr{Y}$ to $\mathscr{Y}'$. Because $g_1(s) := g(s) + s$ is increasing, then $g_1 = \partial\Phi(\cdot)$, as a mapping from $\mathscr{D}(A^{1/2})$ to $[\mathscr{D}(A^{1/2})]'$, where Φ is some proper, convex, lower semicontinuous functional on $\mathscr{D}(A^{1/2})$ and $\partial\Phi$ denotes the subgradient of Φ. The same property holds for $b_0(\cdot)$ and $\operatorname{div} b_1(\nabla\cdot)$ (in proper spaces). Thus, we can invoke [18, Theorem 2.1, p. 54] to obtain that G is a maximal monotone operator from $\mathscr{Y}$ to $\mathscr{Y}'$, as desired. □

Solving the subsystem (6.2.36) yields $(z_2; u_2) \in \mathscr{Y}$, and returning to system (6.2.35) we then have $z_1 = z_2 + \varphi_1 \in \mathscr{D}(A^{1/2})$, $u_1 = u_2 + \psi_1 \in \mathscr{D}(\mathscr{A}^{1/2})$. Using the first and third relations in (6.2.35) we can conclude that $y = (z_1; z_2; u_1; u_2) \in \mathscr{D}(T)$ and satisfies $(I+T)y = h$; hence T is m-accretive.

We now see that equation (6.2.34) is a locally Lipschitz perturbation of an evolution equation with m-accretive generator. Therefore, by Theorem 2.3.8 for initial data $y^0 = (z^0; z^1; u^0; u^1) \in \mathscr{D}(T)$, there exists a unique strong solution $y(t) = (z; z_t; u; u_t)$ to (6.2.1) and (6.2.2) on an appropriate interval $(0, t_{\max})$ possessing properties (6.2.29) on the existence interval. Instead, $y^0 \in Y$ produces a unique generalized solution $y(t) \in C([0, t_{\max}), Y)$. In either case $t_{\max}$ depends on $\|y^0\|_Y$; furthermore, $t_{\max} < \infty$ implies $\lim_{t \nearrow t_{\max}} \|y(t)\|_Y = +\infty$.

Step 2. Energy inequalities/equalities. To derive the energy identity (6.2.30) for strong solutions on the existence time interval a standard procedure applies (see, e.g., [36] or Lemma 12.2.10 in Chapter 12), provided boundedness of the damping operator $D(z_t)$ and $B(u_t)$ as a map from $\mathscr{D}(A^{1/2})$ (resp., $\mathscr{D}(\mathscr{A}^{1/2})$) into the duals $[\mathscr{D}(A^{1/2})]'$ (resp., $[\mathscr{D}(\mathscr{A}^{1/2}))]'$) is ascertained. The latter follows from criticality of the parameter p (in the case of D) and from Sobolev's embeddings $\mathscr{D}(\mathscr{A}^{1/2}) \subset C(\overline{\Omega})$ and $\nabla\mathscr{D}(\mathscr{A}^{1/2}) \subset L_p(\Omega)$ for every $p \geq 1$ (in the case of B). In fact, the damping B is smoother than required, because $B(\mathscr{D}(\mathscr{A}^{1/2})) \subset H^{-1}(\Omega)$.

To establish the energy inequality (6.2.28) for generalized solutions, we only need to justify the limit transition (from strong to generalized solutions) in the damping terms. This can be done exploiting the properties of L_2-convergence and appropriate approximations of the damping functions g and b (the argument is the same as in the proof of Theorem 3.1.4; see also [73]). Finally, the crucial property (6.2.31) that the energy is nonincreasing immediately follows from the energy inequality (6.2.28). The upper bound (6.2.32) is obtained combining (6.2.31) with (6.2.27).

Step 3. Global existence. It follows from (6.2.32) that the solution cannot blow up in finite time. Therefore the same argument as in Theorem 2.3.8 in Chapter 2 shows global existence for both strong and generalized solutions. □

6.2.7. Remark. The existence of generalized solutions established in Theorem 6.2.5 is obtained by using the theory of nonlinear semigroups. As such, these solutions are defined as strong limits of regular (strong) solutions, as in the part (**G**) of Definition 6.2.3. This does not automatically imply that generalized solutions satisfy a variational equality. However, the regularity of g and f enable us to compute appropriate limits and to obtain the variational form stated below. Indeed, using the same argument as in Chapters 3 and 4 (see Theorems 3.1.12 and 4.1.19) one can prove that any generalized solution $y(t) = (z(t); z_t(t); u(t); u_t(t))$ to problem (6.2.11)–(6.2.13)is also weak: it satisfies the following (variational) relations,

$$\frac{d}{dt}(z_t, \phi)_{\mathscr{O}} + (\nabla z, \nabla \phi)_{\mathscr{O}} + (g(z_t), \phi)_{\mathscr{O}} - \gamma\kappa(u_t, \phi)_\Omega + (f(z), \phi)_{\mathscr{O}} = 0, \quad (6.2.38)$$

$$\frac{d}{dt}(M_\alpha u_t + \beta\kappa z, \psi)_\Omega + (\Delta u, \Delta\psi)_\Omega + (B(u_t), \psi)_\Omega$$
$$-([u, v(u) + F_0], \psi)_\Omega = (p_0, \psi)_\Omega, \quad (6.2.39)$$

for any $\phi \in H^1(\mathscr{O})$ and $\psi \in H^2(\Omega) \cap H_0^1(\Omega)$ in the sense of distributions. ■

Theorem 6.2.5 makes it possible to define a dynamical system (Y, S_t) with the phase space Y given by (6.2.18) and with the evolution operator $S_t : Y \to Y$ given by the relation

$$S_t y = (z(t); z_t(t); u(t); u_t(t)), \quad y = (z^0; z^1; u^0; u^1), \quad (6.2.40)$$

where $(z(t); u(t))$ is a generalized solution to (6.2.11)–(6.2.13). Moreover, the monotonicity of the damping operators D and B, combined with the estimate in (6.2.20) and the boundedness property given by (6.2.32) imply (by a pretty routine argument) that the semiflow S_t is locally Lipschitz on Y; that is, there exist $a > 0$ and $\omega(\rho) > 0$ such that

$$||S_t y_1 - S_t y_2||_Y \le a e^{\omega(\rho)t} ||y_1 - y_2||_Y, \quad ||y_i||_Y \le \rho, \, t \ge 0. \quad (6.2.41)$$

6.3 Coupled wave and thermoelastic plate equations

In this section we consider a nonlinear structural acoustic model that includes thermal effects and does not contain any mechanical dissipation in the plate component. Accordingly, the PDE system displays an additional coupling of the elastic equation with a heat equation. It is our aim to prove well-posedness; that is, existence and uniqueness of the solution, as well as continuous dependence on the initial data for the corresponding model. We follow the line of argument presented in [37] for the case of the Berger plates.

6.3.1 Description of the model

The PDE system under investigation is described as follows. Let, as in Section 6.2 $\mathscr{O} \subset \mathbb{R}^3$ be an open bounded domain with boundary $\partial\mathscr{O} = \overline{\Omega \cup \Omega_*}$ comprising two open (in the induced topology), connected, disjoint parts Ω and Ω_* of positive measure. We assume that either $\mathscr{O}$ is sufficiently smooth (e.g., $\partial\mathscr{O} \in C^2$) or else $\mathscr{O}$ is convex. Ω is *flat* and is referred to as the *elastic wall*, whose dynamics is described by the von Karman equations. The acoustic medium in the chamber $\mathscr{O}$ is described by a semilinear wave equation in the variable z, whereas u denotes the vertical displacement of the plate, This leads to the following coupled PDE system,

$$\begin{cases} z_{tt} + g(z_t) - \Delta z + f(z) = 0 \ \text{ in } \ \mathscr{O} \times (0,T), \\ \dfrac{\partial z}{\partial n} = 0 \ \text{ on } \ \Omega_* \times (0,T), \quad \dfrac{\partial z}{\partial n} = \gamma\kappa v_t \ \text{ on } \ \Omega \times (0,T), \\ u_{tt} - \alpha\Delta u_{tt} + \Delta^2 u - [u, v + F_0] + \beta\kappa z_t|_\Omega + \Delta\theta = p_0 \ \text{ in } \ \Omega \times (0,T), \\ u = \Delta u = 0 \ \text{ on } \ \partial\Omega \times (0,T), \\ \theta_t - \Delta\theta - \Delta v_t = 0 \ \text{ in } \ \Omega \times (0,T), \quad \theta = 0 \ \text{ on } \ \partial\Omega \times (0,T), \end{cases} \tag{6.3.1}$$

where $v = v(u) \in H_0^2(\Omega)$ is Airy's stress function given by (6.2.3). We supplement the problem (6.3.1) with initial data

$$\begin{cases} z(0,\cdot) = z^0, \quad z_t(0,\cdot) = z^1 & \text{in } \mathscr{O}, \\ u(0,\cdot) = u^0, \quad u_t(0,\cdot) = u^1, \quad \theta(0,\cdot) = \theta^0 & \text{in } \Omega. \end{cases} \tag{6.3.2}$$

As in Section 6.2, $g(s)$ is a nondecreasing function describing the dissipation that may affect the wave component of the system, and the term $f(z)$ represents a nonlinear force; n is the outer normal vector, and β and γ are positive constants; the parameter $0 \le \kappa \le 1$ has been introduced in order to cover also the case of non-interacting wave and plate equations ($\kappa = 0$). The boundary term $\beta\kappa z_t|_\Omega$ describes the acoustic pressure. The parameter $\alpha \in [0,1]$ describes the rotational inertia of the plate filaments. In contrast with Section 6.2 for the sake of diversity we consider here a hinged boundary condition for the plate.

Concerning functions $g \in C(\mathbb{R})$ and $f \in Lip_{\rm loc}(\mathbb{R})$ and the loads p_0 and F_0 we assume that Assumption 6.2.1 is in force.

6.3.2 Abstract formulation

As in Section 6.2 to study the dynamics of the PDE problem (6.3.1) and (6.3.2), it is useful to recast it as an abstract evolution in an appropriate Hilbert space. The operators and spaces needed for this abstract setup are the following. Let $A : \mathscr{D}(A) \subset L_2(\mathscr{O}) \mapsto L_2(\mathscr{O})$ be the positive selfadjoint operator defined by (6.2.7) and N_0 be the Neumann map from $L_2(\Omega)$ to $L_2(\mathscr{O})$ defined by (6.2.8).

Regarding the thermoelastic plate model, it is convenient to introduce the positive selfadjoint operator $\mathscr{A} : \mathscr{D}(\mathscr{A}) \subset L_2(\Omega) \mapsto L_2(\Omega)$ defined by

$$\mathscr{A} w = -\Delta w, \quad \mathscr{D}(\mathscr{A}) = H^2(\Omega) \cap H_0^1(\Omega).$$

We use this operator $\mathscr{A}$ in the abstract versions of both the plate and heat equations. Finally, we define the inertia operator $M_\alpha = I + \alpha\mathscr{A}$, with obvious domain and introduce the space $V_\alpha = \mathscr{D}(M_\alpha^{1/2}) = \mathscr{D}(\mathscr{A}^{1/2})$ which structure is described in (6.1.7). We denote by V_α' its dual with respect to the pivot space $L_2(\Omega)$.

With the above dynamic operators, the initial/boundary value problem (6.3.1) and (6.3.2) can be rewritten as the following abstract second order system:

$$z_{tt} + A\left(z - \gamma\kappa N_0 u_t\right) + D(z_t) + F_1(z) = 0, \tag{6.3.3}$$

$$M_\alpha u_{tt} + \mathscr{A}^2 u + \beta\kappa N_0^* A z_t - \mathscr{A}\theta + F_2(u) = 0, \tag{6.3.4}$$

$$\theta_t + \mathscr{A}\theta + \mathscr{A} u_t = 0, \tag{6.3.5}$$

$$z(0) = z^0, \quad z_t(0) = z^1; \quad u(0) = u^0, \quad u_t(0) = u^1, \quad \theta(0) = \theta^0, \tag{6.3.6}$$

where we have introduced the Nemytskij operators $D(h)$ and $F_1(z)$ given by (6.2.14) in (6.3.3), whereas $F_2(u) = -[u, v(u) + F_0] - p_0$ in (6.3.4). Regarding the properties of the nonlinear force terms we also refer to (6.2.15)–(6.2.17) and (6.2.20).

The state spaces Y_1 for the wave component $(z; z_t)$ and Y_2 for the plate component $(u; u_t)$ of system (6.3.3)–(6.3.6) are given by

$$Y_1 := \mathscr{D}(A^{1/2}) \times L_2(\mathscr{O}) \equiv H^1(\mathscr{O}) \times L_2(\mathscr{O}),$$
$$Y_2 := \mathscr{D}(\mathscr{A}) \times V_\alpha \equiv \left[H^2 \cap H_0^1\right](\Omega) \times \mathscr{D}(M_\alpha^{1/2}),$$

with respective norms

$$||(z_1; z_2)||_{Y_1}^2 = \|A^{1/2} z_1\|_{\mathscr{O}}^2 + \|z_2\|_{\mathscr{O}}^2, \quad ||(v_1; v_2)||_{Y_2}^2 = \|\mathscr{A} v_1\|_\Omega^2 + \|M_\alpha^{1/2} v_2\|_\Omega^2,$$

and V_α as in (6.1.7). The natural state space for the thermal component θ is $Y_3 = L_2(\Omega)$. The phase space for problem (6.3.3)–(6.3.6) is then

$$Y = Y_1 \times Y_2 \times Y_3 = \mathscr{D}(A^{1/2}) \times L^2(\mathscr{O}) \times \mathscr{D}(\mathscr{A}) \times V_\alpha \times L_2(\Omega), \tag{6.3.7}$$

endowed with the norm

$$||y||_Y^2 = ||(z_1; z_2; v_1; v_2; \theta)||_Y^2 := \beta||(z_1; z_2)||_{Y_1}^2 + \gamma\left(||(v_1; v_2)||_{Y_2}^2 + \|\theta\|_\Omega^2\right)$$

(and obvious corresponding inner product).

The natural (nonlinear) energy functions associated with the solutions to the uncoupled wave and plate models are given, respectively, by

$$\mathscr{E}_z(z(t), z_t(t)) := E_z^0(z(t), z_t(t)) + \Phi(z(t)),$$

$$\mathscr{E}_v(u(t), u_t(t)) := E_u^0(u(t), u_t(t)) + \Pi(u(t)),$$

where the potentials $\Phi(z)$ and $\Pi(u)$ given by (6.2.15) and (6.2.17). We also set

$$E_z^0(t) \equiv E_z^0(z(t), z_t(t)) = \frac{1}{2}\left\{\|A^{1/2}z(t)\|_\mathscr{O}^2 + \|z_t(t)\|_\mathscr{O}^2\right\}, \tag{6.3.8}$$

$$E_u^0(t) \equiv E_u^0(u(t), u_t(t)) = \frac{1}{2}\left\{\|\mathscr{A}u(t)\|_\Omega^2 + \|M_\alpha^{1/2}u_t(t)\|_\Omega^2\right\}. \tag{6.3.9}$$

As in Section 6.2 it is convenient to introduce the positive energy functions: $E_z(z, z_t)$ by (6.2.23) for the wave component and $E_u(u, u_t)$ by (6.2.24) for the plate component. The thermal energy is described by $\mathscr{E}_\theta(t) := \mathscr{E}_\theta(\theta(t)) = \frac{1}{2}\|\theta(t)\|_\Omega^2$. Now we introduce the total energy $\mathscr{E}(t) = \mathscr{E}(z(t), z_t(t), v(t), v_t(t), \theta(t))$ of the system, namely

$$\mathscr{E}(t) = \mathscr{E}(z(t), z_t(t), u(t), u_t(t), \theta(t)) := \beta\mathscr{E}_z(z, z_t) + \gamma\Big(\mathscr{E}_u(u, u_t) + \frac{1}{2}\|\theta\|_\Omega^2\Big), \tag{6.3.10}$$

whose positive part is given by

$$E(t) = E(z, z_t, u, u_t, \theta) := \beta E_z(z, z_t) + \gamma\Big(E_u(u, u_t) + \frac{1}{2}\|\theta\|_\Omega^2\Big). \tag{6.3.11}$$

It follows from (6.2.27) that for any $\gamma, \beta > 0$ there exist constants $c, C, M_0 > 0$ such that

$$cE(z, z_t, v, v_t, \theta) - M_0 \le \mathscr{E}(z, z_t, v, v_t, \theta) \le CE(z, z_t, v, v_t, \theta) + M_0, \tag{6.3.12}$$

where $\mathscr{E}$ and E are the energies defined in (6.3.10) and (6.3.11).

6.3.3 Well-posedness

As above to study well-posedness of problem (6.3.1) and (6.3.2), we may view the corresponding system (6.3.3)–(6.3.6) as an abstract first order equation and apply Theorem 2.3.8 from Chapter 2 (see also [188, Remark 2.6.2] focused on a structural acoustic model with thermal effects).

We start with definitions of strong and generalized solutions.

6.3.1. Definition. A triplet of functions $(z(t);u(t);\theta(t))$ that satisfy the initial conditions (6.3.6) and such that

$$(z(t);u(t)) \in C(0,T;\mathscr{D}(A^{1/2}) \times \mathscr{D}(\mathscr{A})) \cap C^1(0,T;L_2(\mathscr{O}) \times V_\alpha)$$

and $\theta \in C(0,T;L_2(\Omega))$ is said to be

(S) A *strong solution* to problem (6.3.3)–(6.3.6) on the interval $[0,T]$, iff

- For any $0 < a < b < T$ one has that

$$(z_t(t);u_t(t)) \in L_1(a,b;\mathscr{D}(A^{1/2}) \times \mathscr{D}(\mathscr{A}^{1/2})), \quad \theta_t \in L_1(a,b;L_2(\Omega))$$

  and $(z_{tt}(t);u_{tt}(t)) \in L_1(a,b;L_2(\mathscr{O}) \times V_\alpha)$.
- $A[z(t) - \gamma\kappa N_0 u_t(t)] + D(z_t(t)) \in L_2(\mathscr{O})$, $\mathscr{A}^2 u(t) \in V'_\alpha$ and $\theta(t) \in \mathscr{D}(\mathscr{A})$ for almost all $t \in [0,T]$.
- Equations (6.3.3)–(6.3.5) are satisfied in $L_2(\mathscr{O}) \times V'_\alpha \times L_2(\Omega)$ for almost all $t \in [0,T]$.

(G) A *generalized solution* to problem (6.3.3)–(6.3.6) on the interval $[0,T]$, iff there exists a sequence $\{z_n(t);u_n(t);\theta_n(t)\}_n$ of strong solutions to (6.3.3)–(6.3.6), with initial data $(z_n^0;z_n^1;u_n^0;u_n^1;\theta_n^0)$ (in place of $(z^0;z^1;v^0;v^1;\theta^0)$), such that

$$\lim_{n\to\infty} \max_{t\in[0,T]} \left\{ \|\partial_t z(t) - \partial_t z_n(t)\|_{\mathscr{O}} + \|A^{1/2}(z(t) - z_n(t))\|_{\mathscr{O}} \right\} = 0,$$

$$\lim_{n\to\infty} \max_{t\in[0,T]} \left\{ \|M_\alpha^{1/2}(\partial_t u(t) - \partial_t u_n(t))\|_\Omega + \|\mathscr{A}(u(t) - u_n(t))\|_\Omega \right\} = 0.$$

and

$$\lim_{n\to\infty} \max_{t\in[0,T]} \|\theta(t) - \theta_n(t)\|_\Omega = 0.$$

■

In the statement of well-posedness of problem (6.3.3)–(6.3.6), we also need the function space defined by

$$W_\alpha = \left\{ u \in \mathscr{D}(\mathscr{A}) : \mathscr{A}^2 u \in V'_\alpha \right\}, \tag{6.3.13}$$

where V'_α denotes the dual space of V_α in (6.1.7). It is readily verified that

$$W_\alpha = \begin{cases} \mathscr{D}(\mathscr{A}^{3/2}) \equiv \{u \in H^3(\Omega) : \; u = \Delta u = 0 \text{ on } \partial\Omega\} & \text{if } \alpha > 0; \\ \mathscr{D}(\mathscr{A}^{2}) \equiv \{u \in H^4(\Omega) : \; u = \Delta u = 0 \text{ on } \partial\Omega\} & \text{if } \alpha = 0. \end{cases} \tag{6.3.14}$$

6.3.2. Theorem. *Let $\alpha \geq 0$. Then under Assumption* 6.2.1 *(concerning damping g, forcing f, and external loads F_0 and p_0) the PDE system* (6.3.1) *is well-posed on*

$$Y = H^1(\mathscr{O}) \times L_2(\mathscr{O}) \times [H^2(\Omega) \cap H_0^1(\Omega)] \times V_\alpha \times L_2(\Omega);$$

that is, for any $(z^0; z^1; u^0; u^1; \theta^0) =: y_0 \in Y$ there exists a unique generalized solution $y(t) = (z(t); z_t(t); u(t); u_t(t); \theta(t))$ on $\mathbb{R}_+$ that depends continuously on initial data. This solution satisfies the energy inequality

$$\mathscr{E}(t) + \beta \int_s^t (D(z_t), z_t)_{\mathscr{O}}\, d\tau + \gamma \int_s^t \|\mathscr{A}^{1/2}\theta\|_\Omega^2\, d\tau \leq \mathscr{E}(s), \quad 0 \leq s \leq t, \tag{6.3.15}$$

with the total energy $\mathscr{E}(t)$ given by (6.3.10)*. Moreover, if initial data are such that*

$$z^0, z^1 \in \mathscr{D}(A^{1/2}), \quad u^0 \in W_\alpha,\, u^1 \in \mathscr{D}(\mathscr{A}), \quad \theta^0 \in \mathscr{D}(\mathscr{A})$$

and $A[z^0 - \gamma\kappa N_0 u^1] + D(z^1) \in L_2(\mathscr{O})$, then there exists a unique strong solution $y(t)$ satisfying the energy identity:

$$\mathscr{E}(t) + \beta \int_s^t (D(z_t), z_t)_{\mathscr{O}}\, d\tau + \gamma \int_s^t \|\mathscr{A}^{1/2}\theta\|_\Omega^2\, d\tau = \mathscr{E}(s), \quad 0 \leq s \leq t. \tag{6.3.16}$$

The strong solution possesses the property

$$(z_{tt}; u_{tt}; \theta_t) \in (C_r \cap L_\infty^{loc})(\mathbb{R}_+; L_2(\mathscr{O}) \times \mathscr{D}(M_\alpha^{1/2}) \times L_2(\Omega)).$$

Both strong and generalized solutions satisfy the inequality

$$\mathscr{E}(t) \leq \mathscr{E}(s) \;\; \text{for } t \geq s, \tag{6.3.17}$$

and thus (see (6.3.12)*)*

$$E(z(t), z_t(t), u(t), u_t(t), \theta(t)) \leq C\left(1 + E(z^0, z^1, u^0, u^1, \theta^0)\right) \;\; \text{for } t \geq 0. \tag{6.3.18}$$

Sketch of the proof of Theorem 6.3.2. The proof follows the arguments used several times before, so we provide only a brief sketch. Consider the abstract system (6.3.3)–(6.3.6) corresponding to the PDE model (6.3.1) and give its first-order abstract formulation in the variable $y = (z; z_t; u; u_t; \theta)$. For this we introduce the operator $L : \mathscr{D}(L) \subset Y \to Y$ defined by

$$L\begin{bmatrix} z_1 \\ z_2 \\ u_1 \\ u_2 \\ \theta \end{bmatrix} = \begin{bmatrix} -z_2 \\ Az_1 + D(z_2) + z_2 - \gamma\kappa A N_0 u_2 \\ -u_2 \\ M_\alpha^{-1}\left[\mathscr{A}^2 u_1 + \beta\kappa N_0^* A z_2 + u_2 - \mathscr{A}\theta\right] \\ \mathscr{A}\theta + \mathscr{A}u_2 \end{bmatrix}, \tag{6.3.19}$$

with domain

$$\mathscr{D}(L) = \left\{(z_1; z_2; u_1; u_2; \theta) \in \mathscr{D}_* : A[z_1 - \gamma\kappa N_0 u_2] + D(z_2) \in L_2(\mathscr{O})\right\}, \tag{6.3.20}$$

where $\mathscr{D}_* \equiv \mathscr{D}(A^{1/2}) \times \mathscr{D}(A^{1/2}) \times W_\alpha \times \mathscr{D}(\mathscr{A}) \times \mathscr{D}(\mathscr{A})$ The domain $\mathscr{D}(L)$ is dense in Y. Then, we see that system (6.3.3)–(6.3.6) can be rewritten as

$$y' + Ly = C(y), \quad y(0) = y_0, \tag{6.3.21}$$

with L defined by (6.3.19) and C given by

$$C(z_1; z_2; u_1; u_2; \theta) = \left(0; -F_1(z_1) + z_2; 0; M_\alpha{}^{-1}[-F_2(u_1) + u_2]; 0\right).$$

It is not difficult to show that L is a maximal monotone operator, and hence the proof is omitted; a detailed proof in the isothermal case is found in Section 6.2 (see also [38, Appendix A], [36] and [188]). Moreover, as in Section 6.2 by (6.2.20), the nonlinear term C is locally Lipschitz on the phase space Y. Thus, the nonlinear equation (6.3.21) is a locally Lipschitz perturbation of a system driven by an m-monotone operator and one may invoke Theorem 2.3.8, which yields the local existence (and uniqueness) for both strong and generalized solutions.

The next step consists in establishing the energy inequality (6.3.15) and the identity (6.3.16) on the solutions' existence interval. Indeed, the equality (6.3.16) pertaining to strong solutions is easily shown using a standard argument. That generalized solutions satisfy the inequality (6.3.15) follows by a limit procedure, in view of (6.3.16) and Definition 6.3.1(G). Moreover, (6.3.15) implies (6.3.17) and (6.3.18). Finally, the latter estimate makes it possible to establish a global existence result and therefore to conclude the proof. □

6.3.3. Remark. As in Remark 6.2.7 one can prove that any generalized solution $y(t) = (z(t); z_t(t); u(t); u_t(t); \theta(t))$ to problem (6.3.3)–(6.3.6) is also weak; that is, it satisfies the following system of (variational) equations,

$$\frac{d}{dt}(z_t, \phi)_{\mathscr{O}} + (\nabla z, \nabla\phi)_{\mathscr{O}} + (g(z_t), \phi)_{\mathscr{O}} - \gamma\kappa(u_t, \phi)_\Omega + (f(z), \phi)_{\mathscr{O}} = 0, \tag{6.3.22}$$

$$\frac{d}{dt}(M_\alpha u_t + \beta\kappa z, \psi)_\Omega + (\Delta u, \Delta\psi)_\Omega - (\theta, \Delta\psi)_\Omega \\ - ([u, v(u) + F_0], \psi)_\Omega = (p_0, \psi)_\Omega, \tag{6.3.23}$$

$$\frac{d}{dt}(\theta, \chi)_\Omega + (\nabla\theta, \nabla\chi)_\Omega + (\nabla v_t, \nabla\chi)_\Omega = 0, \tag{6.3.24}$$

for any $\phi \in H^1(\mathcal{O})$, $\psi \in H^2(\Omega) \cap H_0^1(\Omega)$, and $\chi \in H_0^1(\Omega)$. ■

Theorem 6.3.2 enables us to define a dynamical system (Y, S_t) with the phase space Y given by (6.3.7) and with the evolution operator $S_t : Y \to Y$ given by the relation

$$S_t y_0 = (z(t); z_t(t); u(t); u_t(t); \theta(t)), \quad y_0 = (z^0; z^1; u^0; u^1; \theta^0), \tag{6.3.25}$$

where $(z(t); u(t); \theta(t))$ is a generalized solution to (6.3.3)–(6.3.6). Moreover, as in Section 6.2 (see relation (6.2.41)), we can conclude that there exist $a > 0$ and $b(\rho) > 0$ such that

$$||S_t y_1 - S_t y_2||_Y \le a e^{b(\rho)t} ||y_1 - y_2||_Y, \quad \forall \, ||y_i||_Y \le \rho, \, t \ge 0, \tag{6.3.26}$$

which means that the semiflow S_t is locally Lipschitz on Y.

6.4 Plates in a flow of gas: Description of the model

In this section we consider a coupled PDE system arising in flow–structure interactions. The structure is described by a plate equation with a damping:

$$(1 - \alpha\Delta)\partial_t^2 u + \Delta^2 u + b_0(u_t) - \alpha \mathrm{div}\,[b_1(\nabla u_t)] - [u, v + F_0] = p(x,t), \quad x \in \Omega, t > 0, \tag{6.4.1}$$

$$u|_{\partial\Omega} = \frac{\partial u}{\partial n}|_{\partial\Omega} = 0, \quad u|_{t=0} = u_0(x), \quad \partial_t u|_{t=0} = u_1(x), \tag{6.4.2}$$

where $v = v(u)$, Airy's stress function, is a solution of the problem

$$\Delta^2 v + [u, u] = 0, \quad v|_{\partial\Omega} = \frac{\partial v}{\partial n}|_{\partial\Omega} = 0. \tag{6.4.3}$$

Here, as above F_0 is a given in-plane force and $b_0(s)$ and $b_1(s)$ are non decreasing functions, $\alpha \ge 0$; we also use the notation $b_1(\nabla u) = (b_1(u_{x_1 t}); b_1(u_{x_2 t}))$.

We consider interaction between the plate and linearized potential flow of gas. If the gas flows above the plate in the direction of the x_1-axis, then aerodynamical pressure of the flow on the plate is given by:

$$p(x,t) = p_0(x) + \nu \cdot (\partial_t + U\partial_{x_1})(\phi|_{|x_3=0}), \quad x \in \Omega, \tag{6.4.4}$$

(see, e.g., [28], [99]) with $p_0(x) \in L^2(\Omega)$; the parameter $\nu > 0$ characterizes the intensity of the interaction between the flow and the plate, the parameter $U \ge 0$ ($U \ne 1$) represents the nonperturbed flow velocity, and the velocity potential $\phi(x,t) = \phi(x_1, x_2, x_3; t)$ of the perturbed flow satisfies the equations:

$$(\partial_t + U\partial_{x_1})^2 \phi = \Delta\phi, \quad x \in \mathbb{R}^3_+ \equiv \{(x_1, x_2, x_3) \in \mathbb{R}^3 : x_3 > 0\}, \tag{6.4.5}$$

$$\left.\frac{\partial\phi}{\partial x_3}\right|_{x_3=0} = \begin{cases} (\partial_t + U\partial_{x_1})u(x_1,x_2,t), & (x_1,x_2)\in\Omega, \\ 0, & (x_1,x_2)\notin\Omega, \end{cases} \tag{6.4.6}$$

$$\phi_{|t=0} = \phi_0(x), \qquad \partial_t\phi_{|t=0} = \phi_1(x). \tag{6.4.7}$$

We first consider the *subsonic* case: $0 \le U < 1$, for both rotational ($\alpha > 0$) and nonrotational ($\alpha = 0$) situations. This case can be recast within an abstract theory, as in the case of a structural acoustic model. This, in turn, leads to a unified treatment of the well-posedness for the problem addressed.

For the *supersonic* case $U > 1$, the analysis is more special and requires a more detailed structure of fundamental solutions along with more refined trace results pertaining to hyperbolic flows [235]. In the supersonic case, we conduct this analysis for rotational case only. The reason for this restriction is the following. As in the case of structural acoustic problems, the interaction between the plate and the wave equation takes place via the hyperbolic Neumann–Dirichlet map. This corresponds to the failure of the (dynamic) Lopatinski condition which results in the loss of "$\frac{1}{2}$ derivative" for the regularity of back pressure acting on the plate. It is known by now, that "sharp trace hyperbolic" regularity recovers this regularity, provided that one has extra $\frac{1}{2}$ derivative, either in time or tangential direction. The latter happens only in the rotational case, where velocities of the plate are in $H^1(\Omega)$. In the subsonic case, this additional regularity of the velocity of the plate equation is not essential, because the coupled structure obeys dissipative law with the energy function bounded from below. In the supersonic case, this is no longer the case. As a consequence, one needs to look at the "decoupled" equations. For this, boundary regularity becomes of the essence. As a final comment, we note that similar phenomena of "sharp" trace regularity was used in [15] in studying controllability of structural acoustic interaction.

Although the approach of Section 6.6 to the problem (6.4.1)–(6.4.7) allows us to cover (in the rotational case) both subsonic ($0 \le U < 1$) and supersonic ($U > 1$) cases simultaneously, the method presented does not allow us to consider fully nonlinear (monotone) damping imposed on the plate equation. For this reason we prefer to give a separate treatment for the subsonic case. This treatment has an advantage of being more general without any reliance on the structure of fundamental solutions to the flow equation and, in addition, it allows us to consider arbitrary monotone damping implanted on the plate structure.

We note that for a subsonic flow ($0 < U < 1$) the problem (6.4.1)–(6.4.7) with absence of damping terms $b_0(u_t) - \alpha\,\mathrm{div}\,\{b_1(\nabla u_t)\}$ was considered in [52] for $\alpha \ge 0$ (see also [30] where $\alpha = 0$ and interaction with elastic base is also included). The papers [31] and [32] were devoted to the development of an approach to problem (6.4.1)–(6.4.7) for $\alpha > 0$ which allows one to cover both subsonic ($0 \le U < 1$) and supersonic ($U > 1$) cases simultaneously.

The problem of nonlinear oscillations of a plate in a supersonic gas flow ($U > 1$) in various formulations was studied in [54, 80, 30, 253] (see also Section 3.3.1). We also refer to the papers [252, 254] devoted to the problem of dynamics of a clamped von Karman plate in a gas flow in the presence of thermal effects.

We introduce below definitions of strong, weak, and generalized solutions for the model considered.

6.4.1. Definition. The pair of functions $(u(x_1,x_2,t);\phi(x_1,x_2,x_3,t))$ such that

$$u(x,t) \in C(0,T;H_0^2(\Omega))\cap C^1(0,T;V_\alpha), \tag{6.4.8}$$
$$\phi(x,t) \in C(0,T;H^1(\mathbb{R}^3_+))\cap C^1(0,T;L^2(\mathbb{R}^3_+)) \tag{6.4.9}$$

is said to be a *strong* solution of problem (6.4.1)–(6.4.7) on the interval $[0,T]$, if

- $(\phi_t;u_t) \in L_1(a,b;H^1(\mathbb{R}^3_+)\times H_0^2(\Omega))$ for any $(a,b)\subset[0,T]$.
- $(\phi_{tt};u_{tt}) \in L_1(a,b;L_2(\mathbb{R}^3_+)\times V_\alpha)$ for any $(a,b)\subset[0,T]$.
- $\phi(t) \in H^2(\mathbb{R}^3_+)$ and $\Delta^2 u(t) + b_0(u_t(t)) - \alpha\,\mathrm{div}\, b_1(\nabla u_t(t)) \in V'_\alpha$ for almost all $t\in[0,T]$, where V_α is given by (6.1.7).
- Equation (6.4.1) holds in V'_α for almost all $t>0$, whereas equations (6.4.3) (resp., (6.4.5)) are satisfied for almost all $t>0$, $x\in\Omega$ (resp., $x\in\mathbb{R}^3_+$);
- The boundary conditions in (6.4.2) (resp., (6.4.6)) hold for almost all $t\in[0,T]$ and $x\in\partial\Omega$ (resp., $x\in\mathbb{R}^2$);
- The initial conditions are satisfied pointwise; that is,

$$\phi(0)=\phi_0, \quad \phi_t(0)=\phi_1, \quad u(0)=u_0, \quad u_t(0)=u_1.$$

■

Generalized solutions are defined as strong limits of strong solutions.

6.4.2. Definition. A pair of functions $(u(x_1,x_2,t);\phi(x_1,x_2,x_3,t))$ is a *generalized* solution of problem (6.4.1)–(6.4.7) on an interval $[0,T]$, if (6.4.8) and (6.4.9) are satisfied and there exists a sequence of strong solutions $(\phi_n(t);u_n(t))$ such that

$$\lim_{n\to\infty}\max_{t\in[0,T]}\left\{\|\partial_t\phi-\partial_t\phi_n(t)\|_{\mathbb{R}^3_+}+\|\phi(t)-\phi_n(t)\|_{1,\mathbb{R}^3_+}\right\}=0$$

and

$$\lim_{n\to\infty}\max_{t\in[0,T]}\left\{\|\partial_t u(t)-\partial_t u_n(t)\|_{V_\alpha}+\|u(t)-u_n(t)\|_{2,\Omega}\right\}=0,$$

where $\|\cdot\|_{V_\alpha}$ is given by (6.1.8). ■

6.4.3. Definition. A pair of functions $(u(x_1,x_2,t);\phi(x_1,x_2,x_3,t))$ is said to be a *weak* solution of problem (6.4.1)–(6.4.7) on an interval $[0,T]$, if

$$u(x,t)\in\mathscr{W}_T\equiv\left\{u(x,t)\in L_\infty(0,T;H_0^2(\Omega)),\quad \partial_t u(x,t)\in L_\infty(0,T;V_\alpha)\right\},$$
$$\phi(x,t)\in\mathscr{V}_T\equiv\left\{\phi(x,t)\in L_\infty(0,T;H^1(\mathbb{R}^3_+)),\quad \partial_t\phi(x,t)\in L_\infty(0,T;L^2(\mathbb{R}^3_+))\right\};$$

the initial data $u(x,0)=u_0(x)$, $\phi(x,0)=\phi_0$ hold, we have that

$$b_0((\partial_t u(t))\in L_1(0,T;H^{-2}(\Omega)),\quad b_1(\partial_{tx_i}u(t))\in L_1(0,T;H^{-1}(\Omega)),\ i=1,2;$$

and the following variational relations hold,

$$
\begin{aligned}
&-\int_0^T ((1-\alpha\Delta)\partial_t u(t), \partial_t w(t))_\Omega dt \\
&+\int_0^T [(b_0(\partial_t u(t)), w(t))_\Omega + \alpha(b_1(\nabla\partial_t u(t)), \nabla w(t))_\Omega + (\Delta u(t), \Delta w(t))_\Omega]\, dt \\
&= \int_0^T ([u(t), v(u(t)) + F_0] + p_0, w(t))_\Omega dt + ((1-\alpha\Delta)u_1 - \nu\gamma[\phi_0], w(0))_\Omega \\
&-\nu \int_0^T (\gamma[\phi(t)], \partial_t w(t) + U\partial_{x_1} w(t))_\Omega dt
\end{aligned}
\tag{6.4.10}
$$

and

$$
\begin{aligned}
&-\int_0^T ((\partial_t + U\partial_{x_1})\phi, (\partial_t + U\partial_{x_1})\psi)_{\mathbb{R}^3_+} dt + \int_0^T (\nabla\phi, \nabla\psi)_{\mathbb{R}^3_+} dt \\
&= (\phi_1 + U\partial_{x_1}\phi_0, \psi(0))_{\mathbb{R}^3_+} - \int_0^T ((\partial_t + U\partial_{x_1})u, \gamma[\psi])_\Omega dt,
\end{aligned}
\tag{6.4.11}
$$

for all test functions $w \in \mathscr{W}_T$ such that $w(T) = 0$ and $\psi \in \mathscr{V}_T$ such that $\psi(T) = 0$. Here and below we denote by $\gamma[\phi]$ the trace of the function $\phi(x) \in H^1(\mathbb{R}^3_+)$ on the plane $\{x : x_3 = 0\}$. ■

In what follows we present several approaches that are available for studying well-posedness of flow–structure interaction. We first present the Galerkin approach for construction of strong solutions that provides good results for the subsonic case which may be rotational or not: $\alpha \geq 0$. Then we use these strong solutions to obtain via limit transition generalized solutions which are weak under some natural constraints (see Assumption 6.5.1). In the supersonic case, a method based on the representation formula describing flow equations in combination with the Galerkin method for a plate component is very successful. This method leads to weak solutions for both subsonic and supersonic cases but under the restriction that $\alpha > 0$.

6.5 Plates in a flow of gas: Subsonic case

In this section we prove well-posedness of the system (6.4.1)–(6.4.7) for the subsonic ($U < 1$) case and without any restrictions on $\alpha \geq 0$. So we deal with both cases of presence ($\alpha > 0$) and absence ($\alpha = 0$) of rotational inertia. We also note that instead of $\mathbb{R}^3_+$ we can consider more general geometric configurations $\mathscr{O}$; however, in order to avoid additional technical details we confine our attention to a positive half-space $\mathbb{R}^3_+$.

We impose the following basic assumptions on the nonlinear damping functions b_0 and b_1 and on the forces F_0 and p.

6.5.1. Assumption. • $b_0, b_1 \in C^1(\mathbb{R})$ are nondecreasing and such that $b_i(0) = 0$ for $i = 0, 1$. Moreover, $b_1(s)$ is polynomially bounded:

$$|b_1(s)| \leq C(1 + |s|^p) \text{ for some } p \geq 0.$$

- $p_0 \in L_2(\Omega)$, $F_0 \in H^{3+\delta}(\Omega)$ for some $\delta > 0$ (in the case $\alpha > 0$ this requirement can be relaxed).

6.5.1 The statement of the main results

Our main result in this section is the following theorem.

6.5.2. Theorem. *Let $0 \le U < 1$ and $\alpha \ge 0$. We assume that Assumption 6.5.1 is in force.*

Part I: Strong solutions. *Assume that $u_1 \in H_0^2(\Omega)$ and*

$$u_0 \in W_\alpha \equiv \left\{ v \in H_0^2(\Omega) \; : \; \Delta^2 v \in V'_\alpha \equiv [\mathscr{D}(M_\alpha^{1/2})]' \right\},^3$$

where V'_α is dual to V_α given by (6.1.7). Moreover we assume that $\phi_0 \in H^2(\mathbb{R}^3_+)$ and $\phi_1 \in H^1(\mathbb{R}^3_+)$ with the following compatibility condition holds:

$$\partial_{x_3}\phi_0 = \begin{cases} u_1 + U\partial_{x_1}u_0, & (x_1;x_2) \in \Omega, \\ 0, & (x_1;x_2) \notin \Omega. \end{cases} \tag{6.5.1}$$

Then the problem (6.4.1)–(6.4.7) has a unique strong solution for any interval $[0,T]$. This solution possesses the properties

$$(\phi;\phi_t;\phi_{tt}) \in L_\infty(0,T;H^2(\mathbb{R}^3_+) \times H^1(\mathbb{R}^3_+) \times L^2(\mathbb{R}^3_+)), \tag{6.5.2}$$

$$(u;u_t;u_{tt}) \in L_\infty(0,T;W_\alpha \times H_0^2(\Omega) \times V_\alpha) \tag{6.5.3}$$

and satisfies the energy equality:

$$\mathscr{E}(t) + \int_s^t [(b_0(u_t), u_t))_\Omega + \alpha(b_1(\nabla u_t), \nabla u_t)_\Omega] d\tau = \mathscr{E}(s) \;\; for \;\; t > s, \tag{6.5.4}$$

where

$$\mathscr{E}(t) = E_{pl}(u(t), \partial_t u(t)) + E_{fl}(\phi(t), \partial_t \phi(t)) + E_{int}(u(t), \phi(t)), \tag{6.5.5}$$

Here the energy of the plate has the form

$$E_{pl}(u(t), \partial_t u(t)) = \frac{1}{2} \parallel \partial_t u(t) \parallel_\Omega^2 + \frac{\alpha}{2} \parallel \nabla \partial_t u(t) \parallel_\Omega^2 + \frac{1}{2} \|\Delta u(t)\|_\Omega^2 + \Pi(u(t))$$

with

$$\Pi(u) = \frac{1}{4} \parallel \Delta v(u) \parallel_\Omega^2 - \frac{1}{2}([u,u], F_0)_\Omega - (p_0, u)_\Omega$$

---

[3] One can see that $W_\alpha = (H^3 \cap H_0^2)(\Omega)$ for $\alpha > 0$ and $W_\alpha = (H^4 \cap H_0^2)(\Omega)$ when $\alpha = 0$.

$$\equiv \Pi_0(u) - \frac{1}{2}([u,u],F_0)_\Omega - (p_0,u)_\Omega. \tag{6.5.6}$$

The flow energy $E_{fl}(\phi,\partial_t\phi)$ and the energy $E_{int}(u,\phi)$ of the plate–flow interaction are defined as follows,

$$E_{fl}(\phi,\partial_t\phi) = \frac{\nu}{2}\left\{\| \partial_t\phi \|^2_{\mathbb{R}^3_+} + \| \nabla\phi \|^2_{\mathbb{R}^3_+} - U^2 \| \partial_{x_1}\phi \|^2_{\mathbb{R}^3_+}\right\}, \tag{6.5.7}$$

$$E_{int}(u,\phi) = \nu U(\gamma[\phi],\partial_{x_1}u)_\Omega. \tag{6.5.8}$$

Here and below $\gamma[\phi] = \phi_{|x_3=0}$.

Part II: Generalized and weak solutions. *We assume*

$$u_0 \in H_0^2(\Omega), \quad u_1 \in V_\alpha, \quad \phi_0 \in H^1(\mathbb{R}^3_+), \quad \phi_1 \in L^2(\mathbb{R}^3_+).$$

Then the problem (6.4.1)–(6.4.7) *has a unique generalized solution in any interval* $[0,T]$. *This solution satisfies the energy inequality*

$$\mathscr{E}(t) + \int_s^t [(b_0(u_t),u_t))_\Omega + \alpha(b_1(\nabla u_t),\nabla u_t)_\Omega]d\tau \le \mathscr{E}(s) \ \textit{for} \ t > s. \tag{6.5.9}$$

Every generalized solution is also weak[4].

6.5.3. Remark (Potential generalizations). Theorem 6.5.2 (with obvious changes in the energy equalities) remains true for a more general plate (shallow shell) equation (as in (3.1.1) and (3.1.2) in Chapter 3) which includes a function f to take into account the initial form of the shell. Similarly, we could incorporate nonlinear damping, both on the plate and flow. The assumptions that need to be imposed on the damping terms are the same as in the case of each scalar equation (see the requirements in Assumption 6.2.1). In order to focus our attention on the model itself, we do not add damping terms to the wave component. The framework considered above is not exhaustive. One could easily introduce various other generalizations, for instance:

- Different boundary conditions associated with the plate equation (hinged, free or combination thereof).
- Boundary dissipation added to the plate model.
- Presence of dissipative sources attached to the flow (and also plate) equation.
- Conditions imposed in Assumption 6.5.1 on the dissipation b_0 and b_1 can be relaxed. This is in line with the treatment of nonlinear damping in plate theory presented in Chapter 3 ($\alpha > 0$) and Chapter 4 ($\alpha = 0$).
- Abstract form of the problem with general locally Lipschitz nonlinearities $F_1(\phi)$ and $F_2(u)$ in both equations (see (6.5.16)–(6.5.18) below).

[4] Concerning uniqueness of weak solutions we refer to Remark 6.5.9 below.

The above-described problems have already been treated in the context of a single equation. However, combining these features within the coupled model should not present any major difficulties. ■

The subsequent sections are devoted to the proof of Theorem 6.5.2. The outline of the rest is as follows. Section 6.5.2 provides preliminary material. Section 6.5.3 contains the construction of Galerkin approximations and studies their properties. Section 6.5.4 completes the proof of Theorem 6.5.2 for strong solutions via limit transition in Galerkin approximations. In Section 6.5.5 we deal with generalized and weak solutions. In Section 6.5.6 we consider stationary solutions.

6.5.2 Preliminaries and abstract setting

We find it convenient to represent the PDE system (6.4.1)–(6.4.7) in an abstract-semigroup form. In fact, the resulting formulation is a special case of a general nonlinear structural acoustic model given in [188, Section 2.6]. In order to accomplish this we introduce the following spaces and operators.

Let $A : \mathscr{D}(A) \subset L_2(\mathbb{R}^3_+) \to L_2(\mathbb{R}^3_+)$ be the self-adjoint operator defined by

$$Ah = -\Delta h + U^2 \partial^2_{x_1} + \mu h, \quad \mathscr{D}(A) = \left\{ h \in H^2(\mathbb{R}^3_+) : \left. \frac{\partial h}{\partial x_3} \right|_{x_3=0} = 0 \right\},$$

where $\mu > 0$. Because $0 \le U < 1$, A is positive and $\mathscr{D}(A^{1/2}) = H^1(\mathbb{R}^3_+)$. Next we introduce the Neumann map N_0 from $L_2(\Omega)$ to $L_2(\mathbb{R}^3_+)$, defined by

$$\psi = N_0\varphi \Longleftrightarrow \left\{ (-\Delta + U^2\partial^2_{x_1} + \mu)\psi = 0 \text{ in } \mathbb{R}^3_+ ; \; \left. \frac{\partial \psi}{\partial n} \right|_{\Omega} = \varphi, \; \left. \frac{\partial \psi}{\partial n} \right|_{\Omega_*} = 0 \right\}. \tag{6.5.10}$$

where $\Omega_* = \partial\mathbb{R}^3_+ \setminus \Omega$. It is well known (see, e.g., [222]) that the operator N_0 in the case $\mathscr{O} = \mathbb{R}^3_+$ possesses the same properties as the corresponding operator in the case of a bounded domain (see the definition in (6.2.8)). Namely we have that

$$N_0 \text{ continuous } : L_2(\Omega) \to H^{3/2}(\mathbb{R}^3_+) \subset \mathscr{D}(A^{3/4-\varepsilon}), \quad \varepsilon > 0,$$

and

$$A^{3/4-\varepsilon} N_0 \text{ continuous} : L^2(\Omega) \to L^2(\mathbb{R}^3_+). \tag{6.5.11}$$

Moreover, as in Section 6.2.3 by Green's formula we have that

$$N_0^* A h = h|_{\Omega} \quad \text{for } h \in \mathscr{D}(A), \tag{6.5.12}$$

where $N_0^* : L_2(\mathbb{R}^3_+) \to L_2(\Omega)$ is the adjoint of N_0. The validity of (6.5.12) may be extended to all $h \in H^1(\mathbb{R}^3_+) \equiv \mathscr{D}(A^{1/2})$, as $\mathscr{D}(A)$ is dense in $\mathscr{D}(A^{1/2})$ and the trace operator is bounded on $H^1(\mathbb{R}^3_+)$. Denote:

$$F_1(\phi) \equiv -\mu\phi, \quad D(\phi) \equiv 2U\phi_{x_1} \tag{6.5.13}$$

The plate equation is modeled abstractly as usual by introducing the operators:

$$\mathscr{A}w = \Delta^2 w, \quad \mathscr{D}(\mathscr{A}) = H^4(\Omega) \cap H_0^2(\Omega),$$

$$B(w,z) := b_0(w) - \alpha \operatorname{div} b_1(z), \quad F_2(u) := -[u, v(u) + F_0] - p_0. \tag{6.5.14}$$

Regarding the nonlinear force terms we have

$$F_2(u) = \Pi'(u) \text{ with } \Pi(u) = \frac{1}{4}||\Delta v(u)||_\Omega^2 - \frac{1}{2}([u,u], F_0) - (p_0, u). \tag{6.5.15}$$

With the above notation, the abstract model for flow-structure interaction is the following.

$$\phi_{tt} + A(\phi + N_0(u_t + U u_{x_1})) + D(\phi_t) + F_1(\phi) = 0, \tag{6.5.16}$$

$$M_\alpha u_{tt} + \mathscr{A}u + B(u_t, \nabla u_t) - \nu N_0^* A(\phi_t + U\phi_{x_1}) + F_2(u) = 0, \tag{6.5.17}$$

$$\phi(0) = \phi_0, \phi_t(0) = \phi_1, u(0) = u_0, u_t(0) = u_1. \tag{6.5.18}$$

The state space for problem (6.5.16)–(6.5.18) is

$$Y = Y_1 \times Y_2 = \mathscr{D}(A^{1/2}) \times L^2(\mathbb{R}^3_+) \times \mathscr{D}(\mathscr{A}^{1/2}) \times \mathscr{D}(M_\alpha^{1/2}), \tag{6.5.19}$$

where the phase spaces Y_1 for the flow component $(\phi;\phi_t)$ and Y_2 for the plate component $(u;u_t)$ are given, respectively, by:

$$Y_1 := \mathscr{D}(A^{1/2}) \times L_2(\mathbb{R}^3_+) \equiv H^1(\mathbb{R}^3_+) \times L_2(\mathbb{R}^3_+);$$

$$Y_2 := \mathscr{D}(\mathscr{A}^{1/2}) \times \mathscr{D}(M_\alpha^{1/2}) \equiv [H^2(\Omega) \cap H_0^1(\Omega)] \times V_\alpha.$$

The norm in the space Y is defined by the relation

$$||(z_1;z_2;v_1;v_2)||_Y^2 := \nu\left(||A^{1/2}z_1||^2_{\mathbb{R}^3_+} + ||z_2||^2_{\mathbb{R}^3_+}\right) + ||\mathscr{A}^{1/2}v_1||_\Omega^2 + ||M_\alpha^{1/2}v_2||_\Omega^2$$

and generates the corresponding inner product.

A critical property that is responsible for Hadamard well-posedness of the plate–flow interaction is Lipschitz continuity of the nonlinear term F_2, a property that results from Airy sharp regularity; see the second relation in (6.2.20) and the discussion in Remark 6.2.2.

The full energy of the system may be negative, thus it is important to assert that it is bounded from below. This last property follows from the following properties of functionals $\Pi(u)$ and $E_{int}(u,\phi)$.

6.5.4. Lemma. *The potential energy $\Pi_*(u) = \frac{1}{2}||\Delta u||_\Omega^2 + \Pi(u)$ of the plate is a continuous function of $u \in H_0^2(\Omega)$. It is bounded from below for any $F_0(x) \in H^2(\Omega)$.*

Proof. It follows from Lemma 1.5.4 (see also Remark 1.5.12). □

6.5.5. Lemma. *The interaction plate–flow energy $E_{int}(u,\phi)$ satisfies*

$$| E_{int}(u,\phi) | \le \delta \| \nabla\phi \|^2_{\mathbb{R}^3_+} + \frac{C}{\delta} \| u \|^2_{1,\Omega} \tag{6.5.20}$$

for any $\delta > 0$ and for all $\phi \in H^1(\mathbb{R}^3_+)$ and $u \in H^1(\Omega)$.

Proof. Using the Hardy inequality

$$\frac{1}{4}\int_{\mathbb{R}^3} | x |^{-2} \cdot | \phi(x) |^2 \, dx \le \int_{\mathbb{R}^3} | \nabla\phi(x) |^2 \, dx,$$

it is easy to see that

$$\int_{Q_R} | \phi(x) |^2 \, dx \le C_R \cdot \int_{\mathbb{R}^3_+} | \nabla\phi(x) |^2 \, dx, \tag{6.5.21}$$

where $Q_R = \{x \in \mathbb{R}^3_+ : | x | \le R\}$ and $\phi \in H^1(\mathbb{R}^3_+)$. Therefore using the trace theorem we obtain

$$\| r_\Omega \gamma[\phi] \|_\Omega \le C \| \nabla\phi \|_{\mathbb{R}^3_+}, \quad \phi \in H^1(\mathbb{R}^3_+), \tag{6.5.22}$$

where r_Ω is the restriction operator from $L_2(\mathbb{R}^2)$ into $L_2(\Omega)$ $(r_\Omega u = u(x)|_{x\in\Omega})$. Inequality (6.5.20) then follows from (6.5.22). □

Now as in Section 6.2 (see (6.2.27)) we easily arrive at the following assertion.

6.5.6. Corollary. *There exist positive constants c, C, and $M = M_{F_0,p_0}$ such that*

$$cE(t) - M \le \mathscr{E}(t) \le CE(t) + M,$$

where $\mathscr{E}(t)$ is given by (6.5.5) and

$$E(t) = \frac{1}{2}\left(||u_t||^2_\Omega + \alpha||\nabla u_t||^2_\Omega + ||\Delta u||^2\right) + \Pi_0(u) + E_{fl}(t).$$

This Corollary 6.5.6 makes it possible to derive from Theorem 6.5.2 the following properties of solutions to (6.4.1)–(6.4.7).

6.5.7. Proposition. *Let the hypotheses of Theorem 6.5.2 be in force. Then any generalized solution $(u(t);\phi(t))$ to problem (6.4.1)–(6.4.7) on $\mathbb{R}_+$ possesses the property*

$$\sup_{t\ge 0}\Big\{ \| \partial_t u(t) \|^2_\Omega + \alpha \| \nabla\partial_t u(t) \|^2_\Omega + \| \Delta u(t) \|^2_\Omega + \| \partial_t\phi(t) \|^2_{\mathbb{R}^3_+} + \| \nabla\phi(t) \|^2_{\mathbb{R}^3_+} \Big\} < +\infty.$$

Proof. Because by Theorem 6.5.2 any generalized solution satisfies the energy inequality (6.5.9), we have that $\mathscr{E}(t) \le \mathscr{E}(s)$. Thus Corollary 6.5.6 applied to the latter inequality yields the estimate stated above. □

6.5.3 Galerkin approximations

The proof of Theorem 6.5.2 is based on the Galerkin method. Let $\{e_k\}$ be the basis in $L^2(\Omega)$ of eigenvectors corresponding to the biharmonic operator Δ^2 with Dirichlet conditions on $\partial\Omega$ and let $\{\chi_k\}$ be a basis in the space $V = H^2(\mathbb{R}^3_+)$. We denote

$$H_N \equiv \operatorname{span}\{e_i,\ i = 1,2\ldots N\} \quad \text{and} \quad V_N \equiv \operatorname{span}\{\chi_i,\ i = 1,2\ldots N\}.$$

We define the N-order approximate solution of the problem (6.4.1)–(6.4.7) as a pair of functions

$$u_N(t) = \sum_{k=1}^{N} g_k(t)e_k \quad \text{and} \quad \phi_N(t) = \sum_{k=1}^{N} \eta_k(t)\chi_k$$

that satisfy the following equations for all basis test functions $e_j \in H_N$, $\chi_j \in V_N$,

$$\begin{aligned}
&(\partial_t^2 u_N, e_j)_\Omega + \alpha(\nabla\partial_t^2 u_N, \nabla e_j)_\Omega \\
&+ (b_0(\partial_t u_N), e_j)_\Omega + \alpha(b_1(\nabla\partial_t u_N), \nabla e_j)_\Omega + (\Delta u_N, \Delta e_j)_\Omega \\
&= ([u_N, v(u_N) + F_0] + p_0, e_j)_\Omega + \nu((\partial_t + U\partial_{x_1})\gamma[\phi_N], e_j)_\Omega,
\end{aligned} \tag{6.5.23}$$

and

$$\begin{aligned}
&(\partial_t^2 \phi_N, \chi_j)_{\mathbb{R}^3_+} - (2U\partial_t\phi_N + U^2\partial_{x_1}\phi_N, \partial_{x_1}\chi_j)_{\mathbb{R}^3_+} + (\nabla\phi_N, \nabla\chi_j)_{\mathbb{R}^3_+} \\
&+ ((\partial_t + U\partial_{x_1})u_N, \gamma[\chi_j])_\Omega = 0,
\end{aligned} \tag{6.5.24}$$

where $j = 1,2,\ldots N$, $\gamma[\psi] = \psi_{|x_3=0}$. We choose initial conditions such that

$$\begin{aligned}
&\| u_N(0) - u_0 \|_{2,\Omega} + \| M_\alpha^{1/2}[\partial_t u_N(0) - u_1] \|_{0,\Omega} \\
&+ \| \phi_N(0) - \phi_0 \|_{1,\mathbb{R}^3_+} + \| \partial_t\phi_N(0) - \phi_1 \|_{0,\mathbb{R}^3_+} \to 0,
\end{aligned} \tag{6.5.25}$$

when N goes to infinity.

The following assertion deals with Galerkin approximate solutions for the fixed N and provides preliminary backgrounds for a priori estimates.

6.5.8. Proposition. *Let Assumption* 6.5.1 *be in force. Then for every initial datum*[5] $y_N^0 = (\phi_N(0); \partial_t\phi_N(0); u_N(0); \partial_t u_N(0))$ *from* $V_N \times V_N \times H_N \times H_N$ *and for any interval* $[0,T]$ *there exists a unique Galerkin N-order approximate solution* $(\phi_N(t); u_N(t))$

[5] Because this proposition deals with fixed N, we do not assume the convergence property in (6.5.25).

that belongs to the class $C^2(0,T;V_N \times H_N)$. This solution possesses the following properties.

- *The energy relation*

$$\mathscr{E}_N(t) + \int_s^t [(b_0(\partial_t u_N), \partial_t u_N))_\Omega + \alpha(b_1(\nabla\partial_t u_N), \nabla\partial_t u_N)_\Omega] d\tau = \mathscr{E}_N(s) \quad (6.5.26)$$

holds for $t > s$, where

$$\mathscr{E}_N(t) = E_{pl}(u_N(t), \partial_t u_N(t)) + E_{fl}(\phi_N(t), \partial_t \phi_N(t)) + E_{int}(u_N(t), \phi_N(t)).$$

- *For any $R > 0$ there exists C_R independent of N such that*

$$\sup_{t\in[0,T]} \Big\{ \| \partial_t u_N(t) \|_\Omega^2 + \alpha \| \nabla\partial_t u_N(t) \|_\Omega^2 \quad (6.5.27)$$
$$+ \| \Delta u_N(t) \|_\Omega^2 + \| \partial_t \phi_N(t) \|_{\mathbb{R}^3_+}^2 + \| \nabla\phi_N(t) \|_{\mathbb{R}^3_+}^2 \Big\} < C_R$$

provided $\|y_N^0\|_Y \le R$.

- *Let $(\phi_N; u_N)$ and $(\tilde{\phi}_N; \tilde{u}_N)$ be Galerkin approximations of the same order N that correspond to different initial data y_N^0 and $\tilde{y}_N^0$ such that $\|y_N^0\|_Y, \|\tilde{y}_N^0\|_Y \le R$. Then there exist constants $a > 0$ and b_R independent of N such that*

$$\|y_N(t) - \tilde{y}_N(t)\|_Y \le a e^{b_R t} \|y_N(0) - \tilde{y}_N(0)\|_Y, \quad t > 0, \quad (6.5.28)$$

where $y_N(t) \equiv (\phi_N; \partial_t\phi_N; u_N; \partial_t u_N)$ and $\tilde{y}_N(t) \equiv (\tilde{\phi}_N; \partial_t\tilde{\phi}_N; \tilde{u}_N; \partial_t\tilde{u}_N)$.

- *For y_N^0 such that $\|\tilde{y}_N^0\|_Y \le R$ we also have that*

$$\|\partial_t y_N(t)\|_Y \le a e^{b_R t} \|\partial_t y_N(0)\|_Y, \quad t > 0, \quad (6.5.29)$$

which can be written in the form

$$\sup_{t\in[0,T]} \Big\{ \| \partial_{tt} u_N(t) \|_\Omega^2 + \alpha \| \nabla\partial_{tt} u_N(t) \|_\Omega^2 + \| \partial_t u_N(t) \|_{2,\Omega}^2 \Big\} \quad (6.5.30)$$
$$+ \sup_{t\in[0,T]} \Big\{ \| \partial_{tt} \phi_N(t) \|_{\mathbb{R}^3_+}^2 + \| \partial_t \phi_N(t) \|_{1,\mathbb{R}^3_+}^2 \Big\}$$
$$\le C(T,R) \Big\{ \| \partial_{tt} u_N(0) \|_\Omega^2 + \alpha \| \nabla\partial_{tt} u_N(0) \|_\Omega^2 + \| \partial_t u_N(0) \|_{2,\Omega}^2$$
$$+ \| \partial_{tt} \phi_N(0) \|_{\mathbb{R}^3_+}^2 + \| \partial_t \phi_N(0) \|_{1,\mathbb{R}^3_+}^2 \Big\},$$

where the constant $C(T,R)$ is independent of N.

Proof. Because b_0 and b_1 are C^1 functions, the approximate solutions exist, at least locally. Standard energetic calculations show that energy dissipative law (6.5.26) is valid for $u_N(t)$ and $\phi_N(t)$ on any existence interval. Relation (6.5.26) implies that $\mathscr{E}_N(t) \le \mathscr{E}_N(s)$ for $t \ge s$. Therefore (6.5.27) follows from Corollary 6.5.6.

In particular, estimate (6.5.27) implies the global existence of Galerkin approximations for every fixed N.

To prove (6.5.28) we note that the difference

$$y_N(t) - \tilde{y}_N(t) \equiv (\psi_N(t); \partial_t \psi_N(t); w_N(t); \partial_t w_N(t))$$

satisfies relations

$$\begin{aligned}
&(\partial_t^2 w_N, e_j)_\Omega + \alpha(\nabla \partial_t^2 w_N, \nabla e_j)_\Omega + (b_0(\partial_t u_N) - b_0(\partial_t \tilde{u}_N), e_j)_\Omega \\
&+ \alpha(b_1(\nabla \partial_t u_N) - b_1(\nabla \partial_t \tilde{u}_N), \nabla e_j)_\Omega + (\Delta w_N, \Delta e_j)_\Omega \\
&= (F_N(t), e_j)_\Omega + \nu((\partial_t + U\partial_{x_1})\gamma[\psi_N], e_j)_\Omega,
\end{aligned}$$

where $F_N(t) \equiv [u_N, v(u_N) + F_0] - [\tilde{u}_N, v(\tilde{u}_N) + F_0]$, and

$$\begin{aligned}
&(\partial_t^2 \psi_N, \chi_j)_{\mathbb{R}^3_+} - (2U\partial_t \psi_N + U^2 \partial_{x_1} \psi_N, \partial_{x_1} \chi_j)_{\mathbb{R}^3_+} + (\nabla \psi_N, \nabla \chi_j)_{\mathbb{R}^3_+} \\
&+ ((\partial_t + U\partial_{x_1}) w_N, \gamma[\chi_j])_\Omega = 0,
\end{aligned}$$

where $j = 1, 2, \dots N$. Again by the standard energetic calculations this implies

$$\mathscr{E}_N^0(t) \le \mathscr{E}_N^0(s) + \int_s^t (F_N(\tau), \partial_t w_N(\tau))_\Omega d\tau \quad \text{for } t > s, \tag{6.5.31}$$

where

$$\mathscr{E}_N^0(t) = E_{pl}^0(w_N(t), \partial_t w_N(t)) + E_{fl}(\psi_N(t), \partial_t \psi_N(t)) + E_{int}(w_N(t), \psi_N(t)).$$

Here the energies E_{fl} and E_{int} are defined by (6.5.7) and (6.5.8) and E_{pl}^0 is given by

$$E_{pl}^0(u, \partial_t u) = \frac{1}{2}\left(\| \partial_t u \|_\Omega^2 + \alpha \| \nabla \partial_t u \|_\Omega^2 + ||\Delta u||_\Omega^2\right).$$

In the same way as in Chapters 3 and 4 using the uniform (in N) estimate in (6.5.27) one can see that

$$|(F_N(\tau), \partial_t w_N(\tau))_\Omega| \le C_R \left(||\Delta w_N(\tau)|_\Omega^2 + \|\partial_t w_N(\tau)\|_\Omega^2\right) \le C_R E_{pl}^0(w_N(\tau), \partial_t w_N(\tau))$$

with the constant C_R independent of N. Lemma 6.5.5, along with the interpolation inequality $||w||_{1,\Omega} \le \eta ||\Delta w||_\Omega + C_\eta ||w||_\Omega$ for all $\eta > 0$, implies that for any small $\delta > 0$ there exists $C_\delta > 0$ such that

$$|E_{int}(w_N(t), \phi_N(t))| \le \delta \left(||\nabla \psi_N(t)||_{\mathbb{R}^3_+}^2 + ||\Delta w_N(t)||_\Omega^2\right) + C_\delta ||w_N(t)||_\Omega^2.$$

Therefore denoting

$$E_N^0(t) = E_{pl}^0(w_N(t), \partial_t w_N(t)) + E_{fl}(\psi_N(t), \partial_t \psi_N(t))$$

and noting that

$$||w_N(t)||^2_\Omega \le ||w_N(s)||^2_\Omega + 2\int_s^t ||w_N(\tau)||_\Omega ||\partial_t w_N(\tau)||_\Omega d\tau$$

we obtain that

$$|E_{int}(w_N(t),\phi_N(t))| \le \delta E^0_N(t) + C_\delta \left[E^0_N(s) + \int_s^t E^0_N(\tau)d\tau\right], \quad t > s, \qquad (6.5.32)$$

for every small $\delta > 0$ with the constant C_δ independent of N and R. Consequently, it follows from (6.5.31) that

$$E^0_N(t) \le c_0 E^0_N(s) + c_R \int_s^t E^0_N(\tau)d\tau,$$

which implies that

$$E^0_N(t) \le c_0 E^0_N(s) e^{c_R(t-s)}, \quad t > s. \qquad (6.5.33)$$

Inasmuch as

$$a_0 E^0_N(t) \le \|y_N(t) - \tilde{y}_N(t)\|^2_Y \le a_1 \left(E^0_N(t) + ||\psi_N(t)||^2_{\mathbb{R}^3_+}\right)$$

with the constants a_0 and a_1 independent of N and R, and

$$||\psi_N(t)||_{\mathbb{R}^3_+} \le ||\psi_N(s)||_{\mathbb{R}^3_+} + \int_s^t ||\partial_t \psi_N(\tau)||_{\mathbb{R}^3_+} d\tau,$$

we obtain (6.5.28) from (6.5.33).

Relation (6.5.29) easily follows from (6.5.28). Indeed, using (6.5.28) with $\tilde{y}_N(t) = y_N(t+h)$ for $0 < h < 1$ we obtain the estimate

$$\|y_N(t+h) - y_N(t)\|_Y \le a e^{b_R t} \|y_N(h) - y_N(0)\|_Y, \quad t > 0,$$

which is uniform in N and h. Thus dividing by h after the limit transition $h \to 0$ we obtain (6.5.29) and also (6.5.30). This completes the proof of Proposition 6.5.8. □

6.5.4 Strong solutions—Proof of Part I of Theorem 6.5.2

Let initial conditions satisfy the following compatibility conditions,

$$\phi_0 \in H^2(\mathbb{R}^3_+),\ \phi_1 \in H^1(\mathbb{R}^3_+),\ u_0, u_1 \in H^2_0(\Omega),\ M_\alpha^{-1/2}\mathscr{A}u_0 \in L_2(\Omega) \qquad (6.5.34)$$

along with the compatibility condition on the boundary Ω given in (6.5.1). In order to obtain strong solutions via the Galerkin method we need appropriate a priori estimates. By Proposition 6.5.8 we have the uniform (in N) bound (6.5.27) and also estimate (6.5.30) which right-hand part depends on N via approximate initial data. Thus to obtain an a priori estimate from (6.5.30) we need to determine the approximate initial conditions such that the values $\partial_t w_N(0) = \partial_t^2 u_N(0)$ and $\partial_t \psi_N(0) = \partial_t^2 \phi_N(0)$ admit bounds that are independent of N. This requires a careful definition of initial conditions chosen for the approximation. At this point compatibility relations imposed on the initial data become critical. We proceed as follows.

Step 1: Let $u_{1,N} \in H_N$ and $\phi_{1,N} \in V_N$, such that

$$||u_{N,1} - u_1||_{2,\Omega} + ||\phi_{N,1} - \phi_1||_{1,\mathbb{R}^3_+} \to 0, \quad N \to \infty$$

The main issue is to find suitable approximations of $u_{N,0}$, $\phi_{N,0}$ that are compatible with the equation. To accomplish this we use elliptic projections. This leads to: find $\phi_{N,0}$ and $u_{N,0}$ that satisfy the following elliptic system,

$$\begin{aligned} & -(U^2\partial_{x_1}\phi_{N,0}, \partial_{x_1}\chi_j)_{\mathbb{R}^3_+} + (\nabla\phi_{N,0}, \nabla\chi_j)_{\mathbb{R}^3_+} \\ & + (u_{N,1} + U\partial_{x_1}u_{N,0}, \gamma[\chi_j])_\Omega + \lambda(\phi_{N,0}, \chi_j)_{\mathbb{R}^3_+} \\ = & -(U^2\partial_{x_1}\phi_0, \partial_{x_1}\chi_j)_{\mathbb{R}^3_+} + (\nabla\phi_0, \nabla\chi_j)_{\mathbb{R}^3_+} + (u_1 + U\partial_{x_1}u_0, \gamma[\chi_j])_\Omega + \lambda(\phi_0, \chi_j)_{\mathbb{R}^3_+} \\ = & -(\Delta\phi_0 - U^2\partial_{x_1}^2\phi_0 - \lambda\phi_0, \chi_j)_{\mathbb{R}^3_+} \end{aligned} \tag{6.5.35}$$

(the last equality invokes compatibility condition imposed on the initial data) and

$$\begin{aligned} & (\Delta u_{N,0}, \Delta e_j)_\Omega - \nu(\gamma[\phi_{N,1}] + U\partial_{x_1}\gamma[\phi_{N,0}], e_j)_\Omega + \lambda(u_{N,0}, e_j)_\Omega \\ = & (\Delta u_0, \Delta e_j)_\Omega - \nu(\gamma[\phi_1] + U\partial_{x_1}\gamma[\phi_0], e_j)_\Omega + \lambda(u_0, e_j)_\Omega, \end{aligned} \tag{6.5.36}$$

where $\chi_j \in V_N$ and $e_j \in H_N$ are the corresponding basis elements, $j = 1, 2, \ldots, N$. The parameter $\lambda > 0$ is chosen large enough and fixed[6] such that (6.5.35) and (6.5.36) have a unique solution $(\phi_{N,0}; u_{N,0})$. Due to the uniform estimate (which follows from (6.5.35) and (6.5.36) in the standard way)

$$\begin{aligned} & ||A^{1/2}\phi_{N,0}||^2_{\mathbb{R}^3_+} + ||\mathscr{A}^{1/2}u_{N,0}||^2_\Omega \le C_1\left[||u_{1,N}||^2_\Omega + ||\phi_{N,1}||^2_{1,\mathbb{R}^3_+}\right] \\ & + C_2\left[||u_1||^2_\Omega + ||\phi_1||^2_{1,\mathbb{R}^3_+} + ||\phi_0||^2_{2,\mathbb{R}^3_+} + ||\mathscr{A}^{1/2}u_0||^2_\Omega\right] \le C, \end{aligned}$$

we claim there exists unique solution $(u_{N,0}; \phi_{N,0})$ to system (6.5.35) and (6.5.36), which converges to $(u_0; \phi_0)$ in the $H_0^2(\Omega) \times H^1(\mathbb{R}^3_+)$ norm.

Step 2: The above "elliptic" approximation allows us to determine the second time derivatives for the approximating solutions. Indeed, using (6.5.35) from (6.5.24) we have that

$$\begin{aligned} (\partial_t^2\phi_N(0), \chi)_{\mathbb{R}^3_+} & = (2U\phi_{N,1} + U^2\partial_{x_1}\phi_{N,0}, \partial_{x_1}\chi)_{\mathbb{R}^3_+} - (\nabla\phi_{N,0}, \nabla\chi)_{\mathbb{R}^3_+} \\ & \quad - (u_{N,1} + U\partial_{x_1}u_{N,0}, \gamma[\chi])_\Omega \\ & = -2U(\partial_{x_1}\phi_{N,1}, \chi)_{\mathbb{R}^3_+} + (\Delta\phi_0 - U^2\partial_{x_1}^2\phi_0 - \lambda[\phi_0 - \phi_{N,0}], \chi)_{\mathbb{R}^3_+} \end{aligned} \tag{6.5.37}$$

[6] Because this parameter plays no role in the subsequent argument, the dependence of the constants on λ is neglected.

for all $\chi \in V_N$, Similarly, from (6.5.23) and (6.5.36) we have

$$\begin{aligned}(M_\alpha \partial_t^2 u_N(0), e)_\Omega = &-(b_0(u_{N,1}), e)_\Omega - \alpha(b_1(\nabla u_{N,1}), \nabla e)_\Omega \\ &-(\Delta u_{N,0}, \Delta e)_\Omega + ([u_{N,0}, v(u_{N,0}) + F_0] + p_0, e)_\Omega \\ &+ \nu(\gamma[\phi_{N,1}] + U\partial_{x_1}\gamma[\phi_{N,0}], e)_\Omega \\ = &-(b_0(u_{N,1}), e)_\Omega - \alpha(b_1(\nabla u_{N,1}), \nabla e)_\Omega \\ &-(\Delta^2 u_0 + \lambda[u_0 - u_{N,0}], e)_\Omega \\ &+([u_{N,0}, v(u_{N,0}) + F_0] + p_0, e)_\Omega + \nu(\gamma[\phi_1] + U\partial_{x_1})\gamma[\phi_0], e)_\Omega\end{aligned} \tag{6.5.38}$$

for every $e \in H_N$. The relations in (6.5.37) and (6.5.38) along with the boundedness of expressions on the right sides of (6.5.37), (6.5.38) for all $\chi \in L_2(\mathbb{R}^3_+)$ and $e \in V_\alpha$, allow us to determine

$$\partial_t^2 \phi_N(0) \in L_2(\mathbb{R}^3_+), \quad \partial_t^2 u_N(0) \in \mathscr{D}(M_\alpha^{1/2}),$$

which are bounded uniformly in N by the increasing function of the norm of the initial data in the space

$$D \equiv \left\{ y^0 = (\phi_0; \phi_1; u_0; u_1) \in Y \;\middle|\; \begin{array}{l} u_0, u_1 \in H_0^2(\Omega),\ \Delta^2 u_0 \in V'_\alpha, \\ \phi_0 \in H^2(\mathbb{R}^3_+),\ \phi_1 \in H^1(\mathbb{R}^3_+),\ \text{(6.5.1) holds} \end{array} \right\}.$$

Step 3: Thus from (6.5.27) and (6.5.30) we obtain the following a priori estimates

$$\|M_\alpha^{1/2} \partial_t^2 u_N(t)\|_\Omega^2 + \|\partial_t u_N(t)\|_{2,\Omega}^2 + \|u_N(t)\|_{2,\Omega}^2 \le C(T,R), \quad t \in [0,T], \tag{6.5.39}$$

and

$$\|\partial_t^2 \phi_N(t)\|_{\mathbb{R}^3_+}^2 + \|\partial_t \phi_N(t)\|_{1,\mathbb{R}^3_+}^2 + \|\phi_N(t)\|_{1,\mathbb{R}^3_+}^2 \le C(T,R), \quad t \in [0,T], \tag{6.5.40}$$

provided $|y^0|_D \le R$, where R is an arbitrary (fixed) number (below we suppress dependence on R). To obtain improved space regularity estimates for approximate solutions, we go back to the variational form defining Galerkin approximate solutions (see (6.5.23) and (6.5.24)) and we rewrite this by using additional information on the regularity of time derivatives:

$$\begin{aligned}&(M_\alpha \partial_t^2 u_N, e)_\Omega + (b_0(\partial_t u_N), e)_\Omega + \alpha(b_1(\partial_t \nabla u_N), \nabla e)_\Omega + (\mathscr{A}^{3/4} u_N, \mathscr{A}^{1/4} e)_\Omega \\ &= ([u_{N,0}, v(u_{N,0}) + F_0] + p_0 + \nu(\partial_t \gamma[\phi_N] + U\partial_{x_1}\gamma[\phi_N], e)_\Omega,\end{aligned} \tag{6.5.41}$$

for any $e \in H_N$, and

$$(\partial_t^2 \phi_N + 2U\partial_t\partial_{x_1}\phi_N - \mu\phi_N, \chi)_{\mathbb{R}^3_+} + (A^{1/2}[\phi_N + N_0(\partial_t + U\partial_{x_1})u_N], A^{1/2}\chi)_{\mathbb{R}^3_+} = 0, \tag{6.5.42}$$

for any $\chi \in V_N$. In the calculations above we have used the identifications

$$(\Delta u_N, \Delta e)_\Omega = (u_N, \mathscr{A} e)_\Omega$$

and

$$\begin{aligned}(\nabla \phi_N, \nabla \chi)_{\mathbb{R}^3_+} - (U^2 \partial_{x_1} \phi_N, \partial_{x_1} \chi)_{\mathbb{R}^3_+} + \mu(\phi_N, \chi)_{\mathbb{R}^3_+} + ((\partial_t + U\partial_{x_1}) u_N, \gamma[\chi])_\Omega \\ = -(A^{1/2}[\phi_N + N_0(\partial_t + U\partial_{x_1}) u_N], A^{1/2}\chi)_{\mathbb{R}^3_+},\end{aligned}$$

which incorporate boundary conditions imposed by the Galerkin scheme. By exploiting the additional regularity in (6.5.39) and (6.5.40) we obtain from (6.5.41) and (6.5.42) that

$$\begin{aligned}&|(\mathscr{A}^{3/4} u_N(t), \mathscr{A}^{1/4} e)_\Omega| \le C\|e\|_{1,\Omega}, \qquad \forall e \in H_N, \\ &|(A[\phi_N(t) + N_0(\partial_t + U\partial_{x_1}) u_N(t)], \chi)_{\mathbb{R}^3_+}| \le C\|\chi\|_{\mathbb{R}^3_+}, \quad \forall \chi \in V_N. \end{aligned} \tag{6.5.43}$$

By density argument, the estimates in (6.5.43) imply

$$\sup_{t\in[0,T]} \left\{ \|u_N(t)\|_{3,\Omega} + \|\phi_N(t) + N_0(\partial_t + U\partial_{x_1}) u_N(t)\|_{2,\mathbb{R}^3_+} \right\} \le C_T. \tag{6.5.44}$$

Because $\|(\partial_t + U\partial_{x_1}) u_N\|_{1,\Omega} \le C_T$ for all $t \in [0,T]$, we have that

$$\sup_{t\in[0,T]} \|N_0(\partial_t + U\partial_{x_1}) u_N\|_{2,\mathbb{R}^3_+} \le C_T$$

and it follows from (6.5.44) that

$$\sup_{t\in[0,T]} \|\phi_N(t)\|_{2,\mathbb{R}^3_+} \le C_T. \tag{6.5.45}$$

Going back to (6.5.41) we obtain that

$$|(\mathscr{A} u_N(t), e)_\Omega| \le C\|M_\alpha^{1/2} e\|_\Omega, \quad e \in H_N.$$

Thus, we arrive at the following space regularity a priori estimate

$$\|\phi_N(t)\|_{2,\mathbb{R}^3_+} + \|M_\alpha^{-1/2} \mathscr{A} u_N(t)\|_\Omega \le C_T, \quad 0 \le t \le T. \tag{6.5.46}$$

Now using a priori estimates in (6.5.39), (6.5.40), and (6.5.46) we can pass to the limit $N \to \infty$ and establish the first part of Theorem 6.5.2.

6.5.5 Generalized and weak solutions—Proof of Part II of Theorem 6.5.2

To obtain the existence of generalized solutions we note that (6.5.28) implies that

$$\|y(t) - \tilde{y}(t)\|_Y \le a e^{b_R t} \|y(0) - \tilde{y}(0)\|_Y, \quad t > 0 \tag{6.5.47}$$

for any couple $y(t) \equiv (\phi; \partial_t \phi; u; \partial_t u)$ and $\tilde{y}(t) \equiv (\tilde{\phi}; \partial_t \tilde{\phi}; \tilde{u}; \partial_t \tilde{u})$ of strong solutions with initial data y^0 and $\tilde{y}^0$ such that $\|y^0\|_Y, \|\tilde{y}^0\|_Y \le R$. Because the initial data y^0 for which there exist strong solutions are dense in Y, we can use (6.5.47) to obtain the existence (and uniqueness) of a generalized solution by approximating the corresponding initial data.

To prove that every generalized solution is also weak we use the same argument as in Theorem 3.1.12 for the case $\alpha > 0$ (see also Remark 3.1.13 after this theorem noting that (3.1.34) is satisfied due to the polynomial bound imposed on $b_1(s)$) and Theorem 4.1.19 when $\alpha = 0$.

The proof of Theorem 6.5.2 is thus completed.

6.5.9. Remark. The uniqueness of weak solutions to problem (6.4.1)–(6.4.7) can be easily established under the following additional conditions on the damping terms b_0 and b_1.

- *Case* $\alpha > 0$: $b_0(s)$ is polynomially bounded and b_1 is linearly bounded:

$$|b_0(s)| \le C(1 + |s|^q) \quad \text{for some} \quad q \ge 0, \quad |b_1(s)| \le C(1 + |s|); \tag{6.5.48}$$

- *Case* $\alpha = 0$: $b_0(s)$ is linearly bounded: $q = 1$ in (6.5.48).

To obtain this result we first need to consider the following linear problem,

$$\phi_{tt} + A(\phi + N_0(u_t + U u_{x_1})) + 2U \phi_{t x_1} - \mu \phi + F = 0, \tag{6.5.49}$$

$$M_\alpha u_{tt} + \mathscr{A} u - \nu N_0^* A(\phi_t + U \phi_{x_1}) + G = 0, \tag{6.5.50}$$

$$\phi(0) = \phi_0, \quad \phi_t(0) = \phi_1, \quad u(0) = u_0, \quad u_t(0) = u_1, \tag{6.5.51}$$

where we have used the same notations as in (6.5.16)–(6.5.18) and the given functions F, G satisfy

$$F \in L_1(0,T; L_2(\mathbb{R}^3_+)), \quad G \in L_1(0,T; [\mathscr{D}(M_\alpha^{1/2})]').$$

By the same method[7] as in the nonlinear case we can construct first strong and then generalized and weak solutions to (6.5.49)–(6.5.51). Then using the same idea as in the proof of Theorem 2.4.35 we can prove that every weak solution $y(t) = (\phi(t); \phi_t(t); u(t); u_t(t))$ to (6.5.49)–(6.5.51) (i) satisfies the corresponding energy relation, (ii) is strongly continuous in Y, and (iii) satisfies the inequality

$$\|y(t)\|_Y \le C e^{\lambda t} \left[\|y(0)\|_Y + \int_0^t \left(\|F(\tau)\|_{\mathbb{R}^3_+} + \|M_\alpha^{-1/2} G(\tau)\|_\Omega \right) d\tau \right] \tag{6.5.52}$$

[7] Note that the a priori bounds for E_{int} established above (see Lemma 6.5.5 and relation (6.5.32)) do not depend on the structure of nonlinear terms in plate equations. Thus, they are applicable in the linear problem (6.5.49)–(6.5.51).

for all $t \geq 0$, where $C, \lambda > 0$ are constants. Note that the difficulty in proving the above-listed properties lies in the fact that the elements ϕ_{tx_1}, $AN_0(u_t + Uu_{x_1})$, $N_0^* A\phi_t$ are not well defined in the phase space Y. However, the approximation method used in the proof of Theorem 2.4.35 allows us to exploit crucial cancellations and yields the desired conclusions via the limit process.

The relation in (6.5.52) makes it possible to obtain the uniqueness of weak solutions to the original nonlinear problem in (6.4.1)–(6.4.7) in the same way as in Theorem 3.1.12 ($\alpha > 0$) and Theorem 4.1.19 ($\alpha = 0$).

We also note that the conditions on the damping functions b_0 and b_1 given above can be relaxed in the way described in Remark 3.1.14. ∎

6.5.6 Stationary solutions

Now we consider stationary solutions to the problem (6.4.1)–(6.4.7) in the case of subsonic flow. It is convenient to introduce the following spaces

$$W_1 \equiv W_1(\mathbb{R}^3_+) = \{\phi(x) \in L_2^{\mathrm{loc}}(\mathbb{R}^3_+) : \nabla\phi(x) \in L_2(\mathbb{R}^3_+)\}$$

with the norm $\|\cdot\|_{W_1} = \|\nabla\cdot\|_{\mathbb{R}^3_+}$ and

$$W_2 \equiv W_2(\mathbb{R}^3_+) = \{\phi(x) \in W_1(\mathbb{R}^3_+) : \partial_{x_ix_j}\phi(x) \in L_2(\mathbb{R}^3_+)\}$$

with the norm $\|\cdot\|^2_{W_2} = \|\cdot\|^2_{W_1} + \sum_{i,j}\|\partial_{x_ix_j}\phi\|^2_{\mathbb{R}^3_+}$. It is clear that $W_k(\mathbb{R}^3_+) \cap L_2(\mathbb{R}^3_+) = H^k(\mathbb{R}^3_+)$ and $H^k(\mathbb{R}^3_+)$ is dense in $W_k(\mathbb{R}^3_+)$, $k = 1, 2$. We note that inequality (6.5.21) gives that the spaces $W_1(\mathbb{R}^3_+)$ and $W_2(\mathbb{R}^3_+)$ are the Hilbert spaces with the inner products

$$(\cdot,\cdot)_{W_1} = (\nabla\cdot,\nabla\cdot)_{\mathbb{R}^3_+} \quad \text{and} \quad (\cdot,\cdot)_{W_2} = (\cdot,\cdot)_{W_1} + \sum_{i,j}(\partial_{x_ix_j}\cdot,\partial_{x_ix_j}\cdot)_{\mathbb{R}^3_+}.$$

6.5.10. Theorem. *Let $0 \leq U < 1$. Then there exist a pair of functions $(u(x);\phi(x))$ from $(H^4 \cap H_0^2)(\Omega) \times W_2(\mathbb{R}^3_+)$ satisfying the equations*

$$\Delta^2 u - [u, v + F_0] = p_0(x) + \nu \cdot U \cdot \partial_{x_1}(\phi|_{x_3=0}), \quad x \in \Omega, \tag{6.5.53}$$

$$u|_{\partial\Omega} = \frac{\partial u}{\partial n}|_{\partial\Omega} = 0, \tag{6.5.54}$$

where the value $v = v(u)$ is defined by u as a solution of problem (6.4.3), and the velocity potential $\phi(x) = \phi(x_1, x_2, x_3)$ is defined as a solution of the problem:

$$\Delta\phi - U^2 \cdot \partial_{x_1}^2\phi = 0, \quad x \in \mathbb{R}^3_+, \tag{6.5.55}$$

$$\left.\frac{\partial \phi}{\partial x_3}\right|_{x_3=0} = \begin{cases} U\partial_{x_1} u(x_1,x_2), & (x_1,x_2) \in \Omega, \\ 0 & (x_1,x_2) \notin \Omega. \end{cases} \tag{6.5.56}$$

The pair $(u(x);\phi(x)) \in (H^4 \cap H_0^2)(\Omega) \times W_2(\mathbb{R}_+^3)$ *is the solution to problem* (6.5.53)–(6.5.56) *if and only if* $(u(x);\phi(x))$ *is an extreme point for the potential energy functional:*

$$D(u,\phi) = \frac{1}{2}\|\Delta u\|^2 + \Pi(u) + \frac{\nu}{2}\left(\| \nabla\phi \|^2_{\mathbb{R}_+^3} - U^2 \| \partial_{x_1}\phi \|^2_{\mathbb{R}_+^3}\right) + E_{int}(u,\phi)$$

in the space $H_0^2(\Omega) \times W_1(\mathbb{R}_+^3)$*, where* $\Pi(u)$ *and* $E_{int}(u,\phi)$ *are given by* (6.5.6) *and* (6.5.8)*. The latter fact is equivalent to the statement that* $(u(x);\phi(x))$ *is a weak solution to problem* (6.5.53)–(6.5.56)*, which is defined as a pair* $(u;\phi) \in H_0^2(\Omega) \times W_1(\mathbb{R}_+^3)$ *such that*

$$(\Delta u, \Delta w)_\Omega - ([u, v(u) + F_0], w)_\Omega + \nu U(\gamma[\phi], \partial_{x_1} w)_\Omega = (p_0, w)_\Omega \tag{6.5.57}$$

and

$$(\nabla\phi, \nabla\psi)_{\mathbb{R}_+^3} - U^2(\partial_{x_1}\phi, \partial_{x_1}\psi)_{\mathbb{R}_+^3} + U(\partial_{x_1} u, \gamma[\psi])_\Omega = 0 \tag{6.5.58}$$

for any $w \in H_0^2(\Omega)$ *and* $\psi \in W_1(\mathbb{R}_+^3)$.

The following remark explains why problem (6.5.53)–(6.5.56), may have several solutions.

6.5.11. Remark. Assume that $p_0(x) \equiv 0$ and $F_0(x) = -\lambda(x_1^2 + x_2^2)$. Then there exist $\lambda_0 > 0$ such that for $\lambda > \lambda_0$ we have at least three pairs of stationary solutions: $(0;0)$, $(u;\phi)$, and $(-u;-\phi)$. Really, in this case we have $D(0,0) = 0$. Therefore due to Theorem 6.5.10 it suffices to show that $D(u,0) \equiv \Pi(u) < 0$ for some $u \in H_0^2(\Omega)$. This fact follows from the considerations given in [85]. Thus with this choice of p_0 and F, problem (6.5.53)–(6.5.56) has three solutions at least. We also note that for generic $p_0(x)$ and $F_0(x)$ the number of stationary solutions is finite. The corresponding argument is the same as in Chapter 1 (see Theorem 1.5.7 for instance). ■

Proof of Theorem 6.5.10.

Let $\{e_k\}$ and $\{\omega_k\}$ be bases in $H_0^2(\Omega)$ and $W_1(\mathbb{R}_+^3)$, respectively. We define the N-order approximate solution of stationary problem (6.5.53)–(6.5.56) as the pair of functions

$$u_N(x) = \sum_{k=1}^N g_k e_k(x) \text{ and } \phi_N(x) = \sum_{k=1}^N \eta_k \omega_k(x)$$

which satisfies the following equations,

$$(\Delta u_N, \Delta e_j)_\Omega - ([u_N, v(u_N) + F_0], e_j)_\Omega + \nu U(\gamma[\phi_N], \partial_{x_1} e_j)_\Omega = (p_0, e_j)_\Omega \tag{6.5.59}$$

and

$$(\nabla\phi_N, \nabla\omega_j)_{\mathbb{R}^3_+} - U^2(\partial_{x_1}\phi_N, \partial_{x_1}\omega_j)_{\mathbb{R}^3_+} + U(\partial_{x_1}u_N, \gamma[\omega_j])_\Omega = 0, \qquad (6.5.60)$$

where $j = 1,2,\dots,N, \gamma[\psi] = \psi_{|x_3=0}$ and the value $v_N = v(u_N)$ is found from u_N according to (6.4.3). It is easy to see that one of the solutions to the problem (6.5.59) and (6.5.60) can be obtained as a solution to the minimization problem $\min D(u_N, \phi_N)$. The potential energy functional $D(u, \phi)$ is smooth on the space $H_0^2(\Omega) \times W_1$ and possesses the property

$$D(u,\phi) \geq c_0 \cdot \{\| \Delta u \|_\Omega^2 + \| \nabla\phi \|_{\mathbb{R}^3_+}^2\} - C \qquad (6.5.61)$$

with some constants $c_0 > 0$ and $C \geq 0$. The above follows from the superlinear structure of potential function $\Pi(u)$ which is specific to von Karman evolutions. In view of the above, there exists at least one N-order approximate solution $(u_N; \phi_N)$ of the problem (6.5.59) and (6.5.60) satisfying the equality

$$D(u_N, \phi_N) = \min\{D(u,\phi) \, : \{u,\phi\} \in \mathscr{L}_N\}, \qquad (6.5.62)$$

where $\mathscr{L}_N = \text{Lin}\,\{e_1,\dots,e_N\} \times \text{Lin}\,\{\omega_1,\dots,\omega_N\}$. From (6.5.61) and (6.5.62) we also have the a priori estimate

$$\| \Delta u_N \|_\Omega^2 + \| \nabla\phi_N \|_{\mathbb{R}^3_+}^2 \leq c_0^{-1} \cdot (D(0,0) + C).$$

Therefore using this estimate we can make a limit transition and prove the existence of a pair $\{u,\phi\} \in H_0^2(\Omega) \times W_1(\mathbb{R}^3_+)$ such that (6.5.57) and (6.5.58) hold for any $w \in H_0^2(\Omega)$ and $\psi \in W_1(\mathbb{R}^3_+)$. It is clear that (6.5.57) and (6.5.58) are the Euler–Lagrange equations for the functional $D(u,\phi)$. So any pair $(u;\phi) \in H_0^2(\Omega) \times W_1(\mathbb{R}^3_+)$ satisfying (6.5.57) and (6.5.58) is an extreme point for $D(u,\phi)$. Because $\partial_{x_1}u \in H_0^1(\Omega)$, we can consider (6.5.58) as a variational formulation of the Neumann problem (6.5.55) and (6.5.56) with a given boundary function from $H^1(\mathbb{R}^2)$. Consequently the standard regularity theory for elliptic boundary problems gives that $\phi \in W_2(\mathbb{R}^3_+)$ and $\gamma[\phi] \in H^{3/2}(\Omega)$. Therefore we have $p_0(x) + \nu U\partial_{x_1}\gamma[\phi] \in L^2(\Omega)$. We can now apply the regularity theorem and obtain that $u \in (H^4 \cap H_0^2)(\Omega)$. Thus any weak solution $(u;\phi) \in H_0^2(\Omega) \times W_1(\mathbb{R}^3_+)$ (defined by (6.5.57) and (6.5.58)) belongs to $(H^4 \cap H_0^2)(\Omega) \times W_2(\mathbb{R}^3_+)$. This completes the proof of Theorem 6.5.10.

6.6 Plates with rotational inertia in both subsonic and supersonic gas flow cases

In this section we prove the well-posedness of the system introduced in (6.4.1)–(6.4.7) for both cases subsonic $U < 1$ and supersonic $0 < U, U \neq 1$ simultaneously in the presence of rotational inertia of filaments of the plate. In this section we mainly follow the presentation given in [31, 32].

6.6.1 The statement of the main results

Our main result un this section is the following theorem.

6.6.1. Theorem. *Assume that $\alpha > 0$. Let $U \geq 0$ and $U \neq 1$. We also assume that $b_0(s) = b_1(s) = 0$ and*

$$u_0 \in H_0^2(\Omega), \quad u_1 \in H_0^1(\Omega), \quad \phi_0 \in H^1(\mathbb{R}^3_+), \quad \phi_1 \in L^2(\mathbb{R}^3_+).$$

Let $F_0(x) \in H^4(\Omega)$ and $p_0(x) \in L^2(\Omega)$. Then the problem (6.4.1)–(6.4.7) has a unique weak solution in any interval $[0,T]$. This solution has the following properties,

$$u(x,t) \in C(0,T;H_0^2(\Omega)) \cap C^1(0,T;H_0^1(\Omega)) \tag{6.6.1}$$

and

$$\phi(x,t) \in C(0,T;H^1(\mathbb{R}^3_+)) \cap C^1(0,T;L^2(\mathbb{R}^3_+)) \tag{6.6.2}$$

and it satisfies the equalities

$$\mathscr{E}^{(1)}(t) = \mathscr{E}^{(1)}(0), \tag{6.6.3}$$

$$\mathscr{E}^{(2)}(t) = \mathscr{E}^{(2)}(0) + \nu U \int_0^t d\tau \int_\Omega \partial_{x_1}[\partial_t u + U\partial_{x_1} u] \cdot \gamma[\phi]\, dx_1 dx_2, \tag{6.6.4}$$

where we have set

$$\mathscr{E}^{(1)}(t) = E_{pl}(u(t), \partial_t u(t)) + \nu E_{fl}^{(1)}(\phi(t), \partial_t \phi(t)) + E_{int}(u(t), \phi(t)),$$

$$\mathscr{E}^{(2)}(t) = E_{pl}(u(t), \partial_t u(t)) + \nu E_{fl}^{(2)}(\phi(t), \partial_t \phi(t)) + E_{int}(u(t), \phi(t)).$$

Here $E_{pl}(u, \partial_t u)$ denotes the energy of the plate, which has the form

$$E_{pl}(u, \partial_t u) = \frac{1}{2}\left(\| \partial_t u \|_{1,\Omega}^2 + \| \Delta u \|_\Omega^2\right) + \Pi(u) \tag{6.6.5}$$

with

$$\Pi(u) = \frac{1}{4} \| \Delta v(u) \|_\Omega^2 - \frac{1}{2}([u,u], F_0)_\Omega - (p_0, u)_\Omega;$$

and $E_{fl}^{(j)}(\phi, \partial_t \phi)$ is defined by

$$E_{fl}^{(1)}(\phi, \partial_t \phi) = \frac{1}{2}\left(\| \partial_t \phi \|_{\mathbb{R}^3_+}^2 + \| \nabla \phi \|_{\mathbb{R}^3_+}^2 - U^2 \| \partial_{x_1} \phi \|_{\mathbb{R}^3_+}^2\right), \tag{6.6.6}$$

$$E_{fl}^{(2)}(\phi, \partial_t \phi) = \frac{1}{2}\left(\| (\partial_t + U\partial_{x_1})\phi \|_{\mathbb{R}^3_+}^2 + \| \nabla \phi \|_{\mathbb{R}^3_+}^2\right). \tag{6.6.7}$$

The energy $E_{int}(u, \phi)$ of the plate–flow interaction is defined by

$$E_{int}(u, \phi) = \nu U (\gamma[\phi], \partial_{x_1} u)_\Omega,$$

where as above $\gamma[\phi] = \phi_{|x_3=0}$.

6.6.2. Remark. Theorem 6.6.1 (with obvious changes in the energy equalities (6.6.3) and (6.6.4)) remain true when instead of (6.4.1) we consider the following equation,

$$(1-\alpha\Delta)\partial_t^2 u + (\varepsilon_1 - \varepsilon_2\Delta)\partial_t u + \Delta^2 u - [u+f, \hat{v}(u) + F_0] = p(x,t) \qquad (6.6.8)$$

for $x \in \Omega, t > 0$, where $\varepsilon_1, \varepsilon_2 \geq 0$ and $f(x) \in H^2(\Omega)$ are given. The value $\hat{v}(u)$ is defined as a solution of the problem

$$\Delta^2 \hat{v} + [u + 2f, u] = 0, \quad v|_{\partial\Omega} = \frac{\partial v}{\partial n}\Big|_{\partial\Omega} = 0. \qquad (6.6.9)$$

Equations (6.6.8) and (6.6.9) arise when one studies the dynamics of the shallow elastic shell, taking into account the resistance of the medium to the rotation of elements of the shell (see Chapter 3). We can also incorporate nonlinear damping terms into the plate (shell) equation of the model. However, for the sake of clarity we do not pursue these generalizations. ■

The proof of Theorem 6.6.1 is based on the properties of the solutions of (6.4.5)–(6.4.7) for the given function $u(x,t)$ (see the next section) and also, as for problem (6.4.1)–(6.4.3) with given $p(x,t)$, on the properties of the von Karman bracket proved in Chapter 1 (see Theorem 1.4.3).

6.6.3. Remark. The result of Theorem 6.6.1 holds for both the subsonic and supersonic case. However, as we could see in Section 6.5 in the subsonic case a more general approach is possible that relies on an abstract setting of the model without reliance on the structure of the fundamental solution. In addition, this approach also provides results for the nonrotational model as well as a model with nonlinear damping. This motivates why the subsonic model had a separate treatment. On the other hand the approach presented here makes it possible to cover both subsonic and supersonic flows simultaneously in the unified way for the case $\alpha > 0$. This is why the result was presented in a full generality covering both supersonic and subsonic models. ■

6.6.2 Flow potentials with given boundary conditions

This section is devoted to the study of problem (6.4.5)–(6.4.7). In order to build preparatory background for the unified proof of well-posedness of solutions in the case $\alpha > 0$ (Theorem 6.6.1) regularity of solutions to the flow equation driven by nonhomogeneous Neumann boundary conditions is of critical importance. This is discussed below. We establish energy-type equalities and derive integral representation for the solution $\phi(x,t)$.

Thus we consider the following linear hyperbolic problem in the half-space $\mathbb{R}^3_+ \equiv \{(x_1,x_2,x_3) \in \mathbb{R}^3 : x_3 > 0\}$:

$$(\partial_t + U\partial_{x_1})^2\phi = \Delta\phi, \quad x \in \mathbb{R}^3_+, \quad t > 0, \tag{6.6.10}$$

$$\left.\frac{\partial \phi}{\partial x_3}\right|_{x_3=0} = h(x_1,x_2,t), \tag{6.6.11}$$

$$\phi_{|t=0} = \phi_0(x), \quad \partial_t\phi_{|t=0} = \phi_1(x) \tag{6.6.12}$$

where $h(x,t)$ is a given function. As above we assume $U \geq 0$ and $U \neq 1$. Applying Theorem 2 of [235] to our case we obtain the following assertion.

6.6.4. Theorem. *Let*

$$\phi_0 \in H^1(\mathbb{R}^3_+), \quad \phi_1 \in L_2(\mathbb{R}^3_+), \quad h(x,t) \in C(\mathbb{R}_+; H^{1/2}(\mathbb{R}^2)).$$

Then there exists a unique solution $\phi(x,t)$ to (6.6.10)–(6.6.12) *such that*

$$\phi(x,t) \in C(\mathbb{R}_+; H^1(\mathbb{R}^3_+)), \quad \partial_t\phi(x,t) \in C(\mathbb{R}_+; L_2(\mathbb{R}^3_+)). \tag{6.6.13}$$

This solution possesses the following property for some $\sigma > 0$:

$$\| \partial_t\phi(t) \|^2_{\mathbb{R}^3_+} + \| \phi(t) \|^2_{1,\mathbb{R}^3_+} + \int_0^t \| \gamma[\phi(\tau)] \|^2_{1/2,\mathbb{R}^2} d\tau$$
$$\leq Ce^{\sigma t}\left(\| \phi_1 \|^2_{\mathbb{R}^3_+} + \| \phi_0 \|^2_{1,\mathbb{R}^3_+} + \int_0^t \| h(\tau) \|^2_{1/2,\mathbb{R}^2} d\tau \right). \tag{6.6.14}$$

Furthermore if

$$\phi_0 \in H^{k+1}(\mathbb{R}^3_+), \quad \phi_1 \in H^k(\mathbb{R}^3_+), \quad \partial_t^j h(x,t) \in C(\mathbb{R}_+; H^{k-j+1/2}(\mathbb{R}^2)) \tag{6.6.15}$$

for some $k \geq 1$ and for $j = 0,1,\dots,k$ and if the following compatibility conditions are fulfilled,

$$\left.\frac{\partial \phi_j(0)}{\partial x_3}\right|_{x_3=0} = \partial_t^j h(x_1,x_2,0), \quad j = 0,1,\dots,k-1, \tag{6.6.16}$$

(the value $\phi_j(x,0) \equiv [\partial_t^j\phi](x,0)$ is determined by ϕ_0 and ϕ_1 inductively from (6.6.10)*), then the solution belongs to the space $C^j(\mathbb{R}_+; H^{k+1-j}(\mathbb{R}^3_+))$ for $j = 0,1,\dots,k+1$ and for some $\sigma_k^* > 0$ we have*

$$\| \partial_t\phi(t) \|^2_{k,\mathbb{R}^3_+} + \| \phi(t) \|^2_{k+1,\mathbb{R}^3_+} + \int_0^t \sum_{i+j\leq k} \| \partial_x^j\partial_t^i\gamma[\phi(\tau)] \|^2_{1/2+k-i-j,\mathbb{R}^2} d\tau$$
$$\leq Ce^{\sigma_k^* t}\left[\| \phi_1 \|^2_{k,\mathbb{R}^3_+} + \| \phi_0 \|^2_{k+1,\mathbb{R}^3_+} + \int_0^t \sum_{j\leq k} \| \partial_t^j h(\tau) \|^2_{k-j+1/2,\mathbb{R}^2} d\tau \right]. \tag{6.6.17}$$

6.6.5. Remark. Because $C(\mathbb{R}_+;H^{1/2}(\mathbb{R}^2))$ is dense in $L_2^{\mathrm{loc}}(\mathbb{R}_+;H^{1/2}(\mathbb{R}^2))$, we can guarantee the existence of a unique solution of (6.6.10)–(6.6.12) possessing the properties (6.6.13) and (6.6.14) if we have

$$\phi_0 \in H^1(\mathbb{R}^3_+), \quad \phi_1 \in L_2(\mathbb{R}^3_+), \quad h(x,t) \in L_2^{\mathrm{loc}}(\mathbb{R}_+;H^{1/2}(\mathbb{R}^2)). \tag{6.6.18}$$

It is also easy to see that the solution $\phi(x,t)$ is a classical solution if conditions (6.6.15) and (6.6.16) are satisfied for $k \geq 3$. ■

6.6.6. Remark. Theorem 6.6.4 is a special case of Theorem 2 of [235] which states that the solutions of the wave equation driven by nonhomogeneous Neumann data in $L_2(0,T;H^{1/2}(\mathbb{R}^2))$ are of finite energy, (i.e., $H^1(\mathbb{R}^3_+)\times L_2(\mathbb{R}^3_+)$). It is known by now that $L_2(0,T,L_2(\mathbb{R}^2))$ boundary data do not produce finite energy solutions, unless the dimension of the domain is equal to one. The loss of differentiability ($\frac{1}{2}$ derivative) occurs in the characteristic sector where Fourier's variables corresponding to time and space tangential localization are comparable. This is due to uniform (dynamical) Lopatinski condition not being satisfied. Thus, improving the regularity of the boundary data only in the space tangential direction (by $\frac{1}{2}$ derivative), provides an overall benefit for the entire region of Fourier–Laplace variables. More recent results [199, 268], show that it is not necessary to feed the entire $\frac{1}{2}$ derivative. The same effect can be obtained by increasing the space tangential regularity of boundary data by only $\frac{1}{3}$ derivative. ■

Below we say for simplicity that a solution $\phi(x,t)$ to (6.6.10)–(6.6.12) is k-smooth if

$$\phi(x,t) \in C^j(\mathbb{R}_+;H^{k+1-j}(\mathbb{R}^3_+)) \quad \text{for } j=0,1,\dots,k+1. \tag{6.6.19}$$

We need the following lemmas on approximation of 0-smooth and 1-smooth solutions.

6.6.7. Lemma. *Assume that* (6.6.18) *holds and* $\phi(x,t)$ *is the corresponding 0-smooth solution. Then for any* $m \geq 1$ *there exists a sequence* $\{\phi_N(x,t)\}$ *of* m*-smooth solutions such that*

$$\begin{aligned}&\sup_{t\in[0,T]} \| \partial_t(\phi(t)-\phi_N(t)) \|^2_{\mathbb{R}^3_+} + \sup_{t\in[0,T]} \| \phi(t)-\phi_N(t) \|^2_{1,\mathbb{R}^3_+} \\ &+ \int_0^T \| h(\tau)-\gamma[\partial_{x_3}\phi_N(\tau)] \|^2_{1/2,\mathbb{R}^2}\, d\tau \to 0\end{aligned} \tag{6.6.20}$$

for any $T>0$*, when* $N\to\infty$.

Proof. It is clear that there exists a sequence $\{\bar{h}_N(x,t)\}$ of smooth functions such that

$$\lim_{N\to\infty}\int_0^T \| h(\tau)-\bar{h}_N(\tau) \|^2_{1/2,\mathbb{R}^2}\, d\tau = 0 \tag{6.6.21}$$

for any $T>0$ and $(\partial_t^j \bar{h}_N)(0)=0$ for $j=0,1,\dots,m-1$. Assume that $\phi_0^N(x)$ and $\phi_1^N(x)$ are smooth approximations of ϕ_0 and ϕ_1 in $H^1(\mathbb{R}^3_+)$ and $L_2(\mathbb{R}^3_+)$, respectively. Let $\phi_j^N(x)$ be defined by $\phi_0^N(x)$ and $\phi_1^N(x)$ from (6.6.10) inductively. Consider problem (6.6.10)–(6.6.12) with $\phi_0 = \phi_0^N(x)$, $\phi_1 = \phi_1^N(x)$, and

$$h(t) = h_N(t) \equiv \bar{h}_N(t) + \beta(A_N t) \cdot \sum_{i=0}^{m-1} \gamma[\partial_{x_3}\phi_i^N] \cdot \frac{t^i}{i!}, \tag{6.6.22}$$

where

$$A_N = N \cdot \left(1 + \sum_{i=0}^{m-1} \| \phi_i^N \|^2_{2,\mathbb{R}^3_+}\right)$$

and $\beta(t)$ is a C^∞-function such that $\beta(t) = 1$ for $0 \le t \le 1$ and $\beta(t) = 0$ for $t \ge 2$. It is clear that for these $\phi_0^N(x)$, $\phi_1^N(x)$, and $h_N(x,t)$ conditions (6.6.15) and (6.6.16) with $k = m$ are valid. Therefore there exists an m-smooth solution $\phi_N(x,t)$ with these data. Property (6.6.20) follows from (6.6.14), (6.6.21), and (6.6.22). □

6.6.8. Lemma. *Assume that*

$$\begin{gathered}\phi_0 \in H^2(\mathbb{R}^3_+), \quad \phi_1 \in H^1(\mathbb{R}^3_+), \\ \partial_t^j h(x,t) \in L_2(\mathbb{R}_+; H^{1-j+1/2}(\mathbb{R}^2)), \quad j = 0,1,\end{gathered} \tag{6.6.23}$$

and (6.6.16) *with* $k = 1$ *hold. Then problem* (6.6.10)–(6.6.12) *has a unique* 1*-smooth solution* $\phi(x,t)$ *with the property* (6.6.17) *for* $k = 1$ *and for any* $m \ge 2$ *there exists a sequence* $\{\phi_N(x,t)\}$ *of* m*-smooth solutions such that*

$$\begin{aligned}&\sup_{t\in[0,T]} \| \partial_t(\phi(t) - \phi_N(t)) \|^2_{1,\mathbb{R}^3_+} + \sup_{t\in[0,T]} \| \phi(t) - \phi_N(t) \|^2_{2,\mathbb{R}^3_+} \\ &+ \int_0^T \| \partial_t(h(\tau) - \gamma[\partial_{x_3}\phi_N(\tau)]) \|^2_{1/2,\mathbb{R}^2} d\tau + \int_0^T \| h(\tau) - \gamma[\partial_{x_3}\phi_N(\tau)] \|^2_{3/2,\mathbb{R}^2} d\tau \to 0\end{aligned} \tag{6.6.24}$$

for any $T > 0$, *when* $N \to \infty$.

Proof. Assume that $h(x,t)$ possesses the property (6.6.15) with $k = 1$. Then there exists a sequence $\{\bar{h}_N(x,t)\}$ of functions in $C^m(\mathbb{R}_+; H^{m+1}(\mathbb{R}^2))$ such that

$$\sup_{t\in[0,T]} \| h(t) - \bar{h}_N(t) \|^2_{3/2,\mathbb{R}^2} + \int_0^T \| \partial_t(h(\tau) - \bar{h}_N(\tau)) \|^2_{1/2,\mathbb{R}^2} d\tau \to 0 \tag{6.6.25}$$

for any $T > 0$ when $N \to \infty$ and $(\partial_t^j \bar{h}_N)(0) = 0$ for $j = 1, \dots, m-1$. Let $\phi_0 \in H^3(\mathbb{R}^3_+)$, $\phi_1 \in H^2(\mathbb{R}^3_+)$ and $\phi_0^N(x)$ and $\phi_1^N(x)$ be smooth approximations of ϕ_0 and ϕ_1 in $H^3(\mathbb{R}^3_+)$ and $H^2(\mathbb{R}^3_+)$ respectively. As above let $\phi_j^N(x)$ be defined by $\phi_0^N(x)$ and $\phi_1^N(x)$ from (6.6.10) inductively. Let us consider

$$h_N(t) = \bar{h}_N(t) + \beta(B_N t) \cdot \sum_{i=1}^{m-1} g_i^N \cdot \frac{t^i}{i!} + g_0^N - \bar{h}_N(0),$$

where $\beta(t)$ is as in the proof of Lemma 6.6.7 and $g_i^N = \gamma[\partial_{x_3}\phi_i^N]$. Because

$$\|g_0^N - \gamma[\partial_{x_3}\phi_0]\|_{3/2} \to 0$$

and in view of the compatibility condition we can choose constants B_N such that (6.6.25) hold with h_N instead of $\bar{h}_N$. Therefore Theorem 6.6.4 implies that the solutions $\phi_N(x,t)$ of problem (6.6.10)–(6.6.12) with $\phi_0 = \phi_0^N(x)$, $\phi_1 = \phi_1^N(x)$, and $h = h_N(x,t)$ possess the property (6.6.24). It is clear $\phi_N(x,t)$ are m-smooth solutions.

Now assume $\phi_0 \in H^2(\mathbb{R}^3_+)$ and $\phi_1 \in H^1(\mathbb{R}^3_+)$. Let $p : H^{3/2}(\mathbb{R}^2) \to H^3(\mathbb{R}^3_+)$ be an extension operator such that $\gamma[\partial_{x_3} p(h)] = h$. Due to the compatibility condition we can write $\phi_0 = p(h(0)) + \psi$, where ψ belongs to the space

$$V = \{\phi \in H^2(\mathbb{R}^3_+) : \partial_{x_3}\phi_{|x_3=0} = 0\}.$$

Because $V \cap H^3(\mathbb{R}^3_+)$ is dense in V we can approximate $\phi_0 \in H^2(\mathbb{R}^3_+)$ by functions $\phi_0^N \in H^3(\mathbb{R}^3_+)$ satisfying the same compatibility condition. Thus we can obtain (6.6.24) under conditions (6.6.15) and (6.6.16) with $k = 1$. Finally we can approximate the data $\{\phi_0, \phi_1, h\}$ with the properties (6.6.23) and (6.6.16) with $k = 1$ by a sequence $\{\phi_0^N, \phi_1^N, h_N\}$ possessing the properties (6.6.15) and (6.6.16) with $k = 1$ such that

$$\int_0^T \{\| h(\tau) - \bar{h}_N(\tau) \|^2_{3/2,\mathbb{R}^2} + \| \partial_t (h(\tau) - \bar{h}_N(\tau)) \|^2_{1/2,\mathbb{R}^2}\} d\tau \to 0$$

and also

$$\sup_{t\in[0,T]} \| h(t) - h_N(t) \|_{1/2,\mathbb{R}^2_+} \to 0 \;\text{ and }\; \phi_0^N = (\phi_0 - p(h(0))) + p(h_N(0)),$$

where the extension operator p defined above is chosen so that it is also continuous from $H^{1/2}(\mathbb{R}^2)$ into $H^2(\mathbb{R}^3_+)$. □

Using Lemma 6.6.8 we obtain the following assertion.

6.6.9. Theorem. *Assume that the conditions* (6.6.23) *and* (6.6.16) *with* $k = 1$ *hold. Then the solution* $\phi(x,t)$ *of the problem* (6.6.10)–(6.6.12) *satisfies the energy equalities:*

$$E_{fl}^{(1)}(\phi(t), \partial_t\phi(t)) = E_{fl}^{(1)}(\phi_0, \phi_1) - \int_0^t \int_{\mathbb{R}^2} h(x,\tau) \cdot \partial_t \gamma[\phi(x,\tau)] dx d\tau \qquad (6.6.26)$$

and

$$E_{fl}^{(2)}(\phi(t), \partial_t\phi(t)) = E_{fl}^{(2)}(\phi_0, \phi_1) - \int_0^t \int_{\mathbb{R}^2} h(x,\tau) \cdot \{(\partial_t + U\partial_{x_1})\gamma[\phi(x,\tau)]\} dx d\tau, \qquad (6.6.27)$$

where $E_{fl}^{(1)}(\phi, \partial_t\phi)$ *and* $E_{fl}^{(2)}(\phi, \partial_t\phi)$ *are defined by* (6.6.19) *and* (6.6.20).

Proof. It is clear that for classical solutions $\phi(x,t)$ we have

$$\begin{aligned}
&\frac{1}{2}\partial_t\{(\partial_t\phi(x,t))^2 + (\nabla\phi(x,t))^2 - U^2(\partial_{x_1}\phi(x,t))^2\} \\
&= \text{div}\,\{\partial_t\phi(x,t)\cdot\nabla\phi(x,t)\} - U\partial_{x_1}\{(\partial_t\phi(x,t) + U\partial_{x_1}\phi(x,t))\cdot\partial_t\phi(x,t)\}
\end{aligned}$$

and

$$\begin{aligned}
&\frac{1}{2}\partial_t\{(\partial_t\phi(x,t) + U\partial_{x_1}\phi(x,t))^2 + (\nabla\phi(x,t))^2\} \\
&= \text{div}\,\{(\partial_t\phi(x,t) + U\partial_{x_1}\phi(x,t))\cdot\nabla\phi(x,t)\} \\
&\quad - \frac{U}{2}\partial_{x_1}\{(\partial_t\phi(x,t) + U\partial_{x_1}\phi(x,t))^2 + (\nabla\phi(x,t))^2\}.
\end{aligned}$$

Integrating these equalities over $(0,t)\times[\{|x|\le\rho\}\cap\mathbb{R}^3_+]$ leads to (6.6.26) and (6.6.27) valid for classical solutions. Passage with the limit on these identities leads to the same relations satisfied for 1-smooth solutions. □

6.6.10. Theorem. *Let $\phi_0 \equiv \phi_1 \equiv 0$. Assume that*

$$h(x,t) \in L^p_{loc}(\mathbb{R}_+; H^1(\mathbb{R}^2)) \text{ for some } p > 2.$$

Then the weak solution $\phi(x,t)$ to (6.6.10)–(6.6.12) has the form:

$$\phi(x,t) = -\frac{\chi(t-x_3)}{2\pi}\int_{x_3}^{t} ds \int_0^{2\pi} d\theta\, h(x_1 - k_1(\theta,s,x_3), x_2 - k_2(\theta,s,x_3), t-s). \tag{6.6.28}$$

Here and below $\chi(t)$ is the Heaviside function ($\chi(t)=1$ for $t\ge 0$ and $\chi(t)=0$ when $t<0$) and

$$k_1(\theta,s,x_3) = Us + \sqrt{s^2-x_3^2}\sin\theta, \quad k_2(\theta,s,x_3) = \sqrt{s^2-x_3^2}\cos\theta.$$

In this case we also have

$$\begin{aligned}
(\partial_t + U\partial_{x_1})\gamma[\phi] &= -h(x_1,x_2,t) \\
&\quad + \frac{1}{2\pi}\int_0^t ds\int_0^{2\pi} d\theta\,[M_\theta h](x_1-(U+\sin\theta)s, x_2 - s\cos\theta, t-s),
\end{aligned} \tag{6.6.29}$$

where $M_\theta = \sin\theta\cdot\partial_{x_1} + \cos\theta\cdot\partial_{x_2}$.

Proof. Let us first assume that $h(x,t)\in C_0^\infty(\mathbb{R}^2\times\mathbb{R}_+)$. Using the Fourier–Laplace method it is easy to see that in this case the following formula gives a classical solution to (6.6.10) and (6.6.11)

$$\phi(x,t) = -\frac{1}{2\pi}\int_{\mathbb{R}^2} d\eta \int_{-\infty}^{\infty} d\sigma e^{(\delta+i\sigma)t}\cdot e^{i\langle x,\eta\rangle}\cdot F(x_3,\delta+i\sigma,\eta)\cdot\hat{h}(\eta,\delta+i\sigma)$$

with

$$F(x_3,\tau,\eta)=\frac{\exp\{-x_3\sqrt{(\tau+iU\eta_1)^2+\eta^2}\}}{\sqrt{(\tau+iU\eta_1)^2+\eta^2}},\quad \tau=\delta+i\sigma,\quad \delta=Re\tau>0,$$

where the square root is chosen such that $Re\sqrt{z^2+\eta^2}\geq Rez/\sqrt{2}>0$ and

$$\hat{h}(\eta,\tau)=\frac{1}{(2\pi)^2}\int_{\mathbb{R}^2}dx\int_0^{+\infty}dte^{-\tau t}\cdot e^{-i\langle x,\eta\rangle}\cdot h(x,t),\quad Re\tau>0.$$

A simple calculation shows that this solution possesses the property (6.6.16) for any $k>0$. Now using the known formula

$$\frac{\exp\{-\alpha\sqrt{p^2+1}\}}{\sqrt{p^2+1}}=\int_\alpha^\infty \mathscr{J}_0(\sqrt{t^2-\alpha^2})\cdot e^{-pt}\,dt,$$

where $\mathscr{J}_0(z)$ is the zero-order Bessel function, and the theorem on the convolution for the Fourier–Laplace transformation we have

$$\phi(x,t)=-\chi(t-x_3)\int_{\mathbb{R}^2}d\eta\int_{x_3}^t dse^{i\langle x,\eta\rangle}\cdot e^{-iU\eta_1 s}\mathscr{J}_0(|\eta|\sqrt{s^2-x_3^2})\cdot\bar{h}(\eta,t-s),$$

where

$$\bar{h}(\eta,t)=\frac{1}{(2\pi)^2}\int_{\mathbb{R}^2}dxe^{-i\langle x,\eta\rangle}\cdot h(x,t).$$

Because

$$2\pi a\int_{\mathbb{R}^2}d\eta e^{i\langle y,\eta\rangle}\cdot\mathscr{J}_0(a|\eta|)\cdot\bar{h}(\eta)=\int_{|x|=a}h(y-x)dS_x$$

we have

$$\phi(x,t)=-\frac{1}{2\pi}\chi(t-x_3)\cdot\int_{x_3}^t\frac{ds}{\sqrt{s^2-x_3^2}}\int_{K(s,x_3)}dS_\xi\, h(x_1-Us-\xi_1,x_2-\xi_2,t-s),$$

where $K(s,x_3)=\{\xi\in\mathbb{R}^2:|\xi|=\sqrt{s^2-x_3^2}\}$. Using polar coordinates we obtain (6.6.28) in the case of a smooth functions $h(x,t)$. In order to prove (6.6.29) in this case we note that (6.6.28) implies

$$\gamma[\phi](x_1,x_2,t)=-\frac{1}{2\pi}\cdot\int_0^t ds\int_0^{2\pi}d\theta\, h(x_1-(U+\sin\theta)s,x_2-s\cos\theta,t-s).$$

Therefore applying the operator $\partial_t+U\partial_{x_1}$ to this formula and using the equality

$$\begin{aligned}\partial_t h=&-\frac{d}{ds}h(x_1-(U+\sin\theta)s,x_2-s\cos\theta,t-s)\\&-(U+\sin\theta)\partial_{x_1}h-\cos\theta\partial_{x_2}h,\end{aligned}\tag{6.6.30}$$

we obtain (6.6.29) for $h(x,t) \in C_0^\infty(\mathbb{R}^2 \times \mathbb{R}_+)$. To conclude the proof of the theorem we should make the limit transition. We can accomplish this by relying on the following assertion.

6.6.11. Lemma. *Assume that $h(x,t) \in C_0^\infty(\mathbb{R}^2 \times \mathbb{R}_+)$ and $\phi(x,t)$ is defined by* (6.6.28). *Then*

$$\| \partial_t \phi(t) \|^2_{\mathbb{R}^3_+} + \| \phi(t) \|^2_{1,\mathbb{R}^3_+}$$
$$\leq C_1 \int_0^t \| h(\tau) \|^2_{\mathbb{R}^2} d\tau + C_2(r) t^{3-2/r} \left(\int_0^t \| \nabla h(\tau) \|^r_{\mathbb{R}^2} d\tau \right)^{2/r} \quad (6.6.31)$$

for any $r > 2$ and

$$\| (\partial_t + U \cdot \partial_{x_1}) \gamma[\phi(t)] \|_{\mathbb{R}^2} \leq \| h(t) \|_{\mathbb{R}^2} + \int_0^t \| \nabla h(\tau) \|_{\mathbb{R}^2} d\tau. \quad (6.6.32)$$

Proof. The inequality (6.6.32) is an obvious corollary of (6.6.29).
Let us prove (6.6.31). One can see from (6.6.28) that

$$\|\phi(.,x_3,t)\|_{\mathbb{R}^2} \leq \chi(t-x_3) \cdot \int_{x_3}^t \|h(t-s)\|_{\mathbb{R}^2} ds,$$

$$\|\partial_{x_j} \phi(.,x_3,t)\|_{\mathbb{R}^2} \leq \chi(t-x_3) \cdot \int_{x_3}^t \|\partial_{x_j} h(t-s)\|_{\mathbb{R}^2} ds \quad \text{for} \quad j = 1,2,$$

and

$$\|\partial_{x_3} \phi(.,x_3,t)\|_{\mathbb{R}^2} \leq \chi(t-x_3) \left(\|h(t-x_3)\|_{\mathbb{R}^2} + \int_{x_3}^t \frac{x_3 ds}{\sqrt{s^2 - x_3^2}} \|\nabla h(t-s)\|_{\mathbb{R}^2} \right).$$

Using a formula similar to (6.6.30) we also have

$$\|\partial_t \phi(.,x_3,t)\|_{\mathbb{R}^2} \leq \chi(t-x_3) \left[\|h(t-x_3)\|_{\mathbb{R}^2} + C \int_{x_3}^t \frac{s ds}{\sqrt{s^2 - x_3^2}} \|\nabla h(t-s)\|_{\mathbb{R}^2} \right].$$

Thus we have

$$\left(\|\phi(.,x_3,t)\|^2_{1,\mathbb{R}^2} + \|\partial_{x_3} \phi(.,x_3,t)\|^2_{\mathbb{R}^2} + \|\partial_t \phi(.,x_3,t)\|^2_{\mathbb{R}^2} \right)^{1/2} \quad (6.6.33)$$
$$\leq C \cdot \chi(t-x_3) \left(\|h(t-x_3)\|_{\mathbb{R}^2} + \int_{x_3}^t \frac{s ds}{\sqrt{s^2 - x_3^2}} \|\nabla h(t-s)\|_{\mathbb{R}^2} \right).$$

Because

$$\left(\int_{x_3}^{t}\left(\frac{s}{\sqrt{s^2-x_3^2}}\right)^q ds\right)^{1/q} \le C_q\cdot\left((t-x_3)^{1/q}+x_3^{1/2}(t-x_3)^{1/q-1/2}\right)$$

for $0<q<2$, we have

$$\left(\int_{x_3}^{t}\frac{s\,ds}{\sqrt{s^2-x_3^2}}\|\nabla h(t-s)\|_{\mathbb{R}^2}\right)^2 \le C_r\cdot(t^{2-2/r}+x_3t^{1-2/r})\cdot\left(\int_0^t\|\nabla h(s)\|_{\mathbb{R}^2}^r\,ds\right)^{2/r}$$

for any $r>2$. Therefore from (6.6.33) we obtain (6.6.31). □

Now we establish the following Kirchhoff type of integral representation of the solutions of problem (6.6.10)–(6.6.12) when $h(x,t)\equiv 0$.

6.6.12. Theorem. *Let $h(x,t)\equiv 0$ and $\phi_0\in H^1(\mathbb{R}^3_+)$ and $\phi_1\in L^2(\mathbb{R}^3_+)$. Then the weak solution $\phi(x,t)$ to (6.6.10)–(6.6.12) is given by the following formula,*

$$\phi(x,t)=\frac{1}{4\pi}\cdot\int_S dS_\xi\,[\hat\phi_0-t(\xi-U\mathbf{e}_1,\nabla\hat\phi_0)+t\hat\phi_1](x-(\xi+U\mathbf{e}_1)t), \tag{6.6.34}$$

where S is the unit sphere in $\mathbb{R}^3$, dS_ξ its surface measure, $\mathbf{e}_1=(1,0,0)\in\mathbb{R}^3$ and $\hat\phi_0\in H^1(\mathbb{R}^3)$ and $\hat\phi_1\in L^2(\mathbb{R}^3)$ are defined by the formula

$$\hat\phi_j(x)=\begin{cases}\hat\phi_j(x_1,x_2,x_3), & x_3\ge 0,\\ \hat\phi_j(x_1,x_2,-x_3), & x_3<0.\end{cases}$$

Proof. It is sufficient to prove (6.6.34) for the problem

$$(\partial_t+U\partial_{x_1})^2\phi=\Delta\phi,\quad x\in\mathbb{R}^3,\quad t>0, \tag{6.6.35}$$
$$\phi_{|t=0}=\hat\phi_0(x),\qquad \partial_t\phi_{|t=0}=\hat\phi_1(x). \tag{6.6.36}$$

We first note that the smooth solution $\phi(x,t)$ to (6.6.35) and (6.6.36) is given by

$$\phi(x,t)=\psi(\hat\phi_1;x,t)+(\partial_t+2U\partial_{x_1})\psi(\hat\phi_0;x,t), \tag{6.6.37}$$

where $\psi(f;x,t)$ is the solution to (6.6.35) and (6.6.36) with $\hat\phi_0\equiv 0$ and $\hat\phi_1=f$. Using the Fourier method one easily sees that

$$\psi(f;x,t)=\frac{1}{(2\pi)^{3/2}}\int_{\mathbb{R}^3}e^{-iU\xi_1t}\frac{\sin|\xi|t}{|\xi|}\cdot e^{ix\xi}\hat f(\xi)\,d\xi,$$

where $\hat f$ is the Fourier transformation of f. After a simple calculation, the theorem on the convolution gives

$$\psi(f;x,t)=\frac{t}{4\pi}\cdot\int_S dS_\xi\,f(x-(\xi+U\mathbf{e}_1)t). \tag{6.6.38}$$

Now (6.6.37) and (6.6.38) imply (6.6.34) for smooth functions $\hat{\phi}_0$ and $\hat{\phi}_1$. It is also clear that the solution $\phi(x,t)$ given by (6.6.34) satisfies the estimate

$$\|\phi(t)\|_{\mathbb{R}^3_+} \le C\cdot(1+t)\cdot\left(\|\phi_1\|_{\mathbb{R}^3_+} + \|\phi_0\|_{1,\mathbb{R}^3_+}\right).$$

Lemma 6.6.7 and this estimate imply (6.6.34) for 0-smooth (weak) solutions of (6.6.10)–(6.6.12) with $h \equiv 0$. □

6.6.3 Construction of approximate solutions

We now construct approximate solutions and prove a priori estimates for these.

Let $\{e_k\}$ be an orthonormal basis in $H_0^1(\Omega)$ consisting of eigenvectors of the problem

$$\Delta^2 w = \lambda(1-\alpha\Delta)w, \quad w|_{\partial\Omega} = \left.\frac{\partial w}{\partial n}\right|_{\partial\Omega} = 0. \tag{6.6.39}$$

and $0 < \lambda_1 \le \lambda_2 \le \cdots$ the corresponding eigenvalues. The spectral relation in (6.6.39) can be written in the form $M_\alpha^{-1}\mathscr{A}w = \lambda w$, where M_α and $\mathscr{A}$ are the same as above (see, e.g., Section 6.2.3). Let P_N be an orthoprojector onto Span $\{e_1,\dots,e_n\}$.

We define approximate solutions by the formula

$$u_N(t) = \sum_{k=1}^N g_k(t)e_k \in \mathscr{L}_T^N \equiv C^1(0,T;P_N H_0^1(\Omega))$$

such that the following equations in $H_0^1(\Omega)$ hold,

$$u_N(t) = u_N(0) + \int_0^t \partial_t u_N(\tau)\,d\tau \tag{6.6.40}$$

and

$$\begin{aligned} \partial_t u_N(t) &= \partial_t u_N(0) \\ &\quad + \int_0^t \{-M_\alpha^{-1}\mathscr{A}u_N(\tau) + P_N M_\alpha^{-1}([u_N(\tau), v(u_N(\tau)) + F_0] + p_0)\}\,d\tau \\ &\quad + \nu P_N M_\alpha^{-1} r_\Omega(\gamma[\phi_N(t)] - \gamma[\phi_0]) + \nu U \int_0^t P_N M_\alpha^{-1} r_\Omega(\partial_{x_1}\gamma[\phi_N(\tau)])\,d\tau \end{aligned} \tag{6.6.41}$$

for all $0 \le t < T$. Here (i) $u_N(0) = P_N u_0$ and $\partial_t u_N(0) = P_N u_1$; (ii) $v(u_N)$ is defined by u_N from (6.4.3); and (iii) $\phi_N(x,t)$ is the solution of problem (6.6.10)–(6.6.12) with $h(x,t) = h_N(x,t)$, where

$$h_N(x,t) = \begin{cases} (\partial_t + U\chi_N(x)\partial_{x_1})u_N(x,t), & x = (x_1;x_2) \in \Omega, \\ 0, & x = (x_1;x_2) \notin \Omega. \end{cases} \tag{6.6.42}$$

Here $\chi_N(x) \in C_0^\infty(\Omega)$ chosen such that $0 \le \chi_N(x) \le 1$, $\chi_N(x) \to 1$ almost surely and $|\nabla\chi_N(x)| \cdot \mathrm{dist}(x, \partial\Omega) \le C$ for any $x \in \Omega$ with C independent of N. We also denote by r_Ω the restriction operator from $L_2(\mathbb{R}^2)$ into $L_2(\Omega)$ ($r_\Omega u = u(x)|_{x\in\Omega}$).

Because $h_N(x,t) \in C(0,T;H^2(\mathbb{R}^2))$ for $u_N \in \mathscr{L}_T^N$, from Theorem 6.6.4 we have $\phi_N(x,t) \in C(0,T;H^1(\mathbb{R}^3_+))$ when $\phi_0 \in H^1(\mathbb{R}^3_+)$, $\phi_1 \in L^2(\mathbb{R}^3_+)$. Therefore $r_\Omega\gamma[\phi_N] \in H^{1/2}(\Omega)$ and the right-hand side of (6.6.41) is well defined.

6.6.13. Theorem. *Let (i) $U \ge 0$, $U \ne 1$, $\alpha > 0$; (ii) $F_0 \in H^4(\Omega)$ and $p_0 \in L_2(\Omega)$. Assume that*

$$u_0 \in H_0^2(\Omega), \quad u_1 \in H_0^1(\Omega), \quad \phi_0 \in H^1(\mathbb{R}^3_+), \quad \phi_1 \in L_2(\mathbb{R}^3_+).$$

Then for any interval $[0,T)$ there exists a unique approximate solution of order N $(u_N(t);\phi_N(t))$ of problem (6.4.1)–(6.4.7). *This solution possesses the property*

$$\| \partial_t u_N \|_{1,\Omega}^2 + \| \Delta u_N \|_\Omega^2 + \| \partial_t \phi_N \|_{\mathbb{R}^3_+}^2 + \| \nabla\phi_N \|_{\mathbb{R}^3_+}^2 \le C_T, \quad t \in [0,T], \tag{6.6.43}$$

where C_T is independent of N, and satisfies the equalities

$$\mathscr{E}_N^{(1)}(t) = \mathscr{E}_N^{(1)}(0) - \nu U \int_0^t d\tau \int_\Omega (1-\chi_N)\partial_{x_1}\partial_t u_N \cdot \gamma[\phi_N]\, dx_1 dx_2 \tag{6.6.44}$$

and

$$\mathscr{E}_N^{(2)}(t) = \mathscr{E}_N^{(2)}(0) + \nu U \int_0^t d\tau \int_\Omega [\chi_N \partial_{x_1}\partial_t u_N + U\partial_{x_1}(\chi_N\partial_{x_1} u_N)] \cdot \gamma[\phi_N]\, dx_1 dx_2, \tag{6.6.45}$$

where

$$\mathscr{E}_N^{(1)}(t) = E_{pl}(u_N(t), \partial_t u_N(t)) + \nu E_{fl}^{(1)}(\phi_N(t), \partial_t\phi_N(t)) + E_{int}^N(u_N(t), \phi_N(t))$$

and

$$\mathscr{E}_N^{(2)}(t) = E_{pl}(u_N(t), \partial_t u_N(t)) + \nu E_{fl}^{(2)}(\phi_N(t), \partial_t\phi_N(t)) + E_{int}^N(u_N(t), \phi_N(t)).$$

Here the energy $E_{pl}(u,\partial_t u)$ and the values $E_{fl}^{(j)}(\phi,\partial_t\phi)$ are defined by (6.6.5)–(6.6.7) *and $E_{int}^N(u_N,\phi_N) = \nu U(\gamma[\phi_N], \chi_N \cdot \partial_{x_1} u_N)_\Omega$.*

We first prove the local existence of approximate solutions.

6.6.14. Lemma. *Let $w_0, w_1 \in P_N H_0^1(\Omega)$, $\phi_0 \in H^1(\mathbb{R}^3_+)$, $\phi_1 \in L_2(\mathbb{R}^3_+)$, and*

$$\| w_1 \|_{1,\Omega}^2 + \| w_0 \|_{2,\Omega}^2 + \| \phi_1 \|_{\mathbb{R}^3_+}^2 + \| \phi_0 \|_{1,\mathbb{R}^3_+}^2 \le R^2, \tag{6.6.46}$$

for some $R > 0$. Then there exist $T_0 = T_0(N,R)$ such that the problem (6.6.40) *and* (6.6.41) *with $u_N(0) = w_0$ and $\partial_t u_N(0) = w_1$ has unique solution $(u_N(t);\phi_N(t))$ on the interval $[0,T_0)$.*

Proof. Let us consider the space

$$\mathscr{M}_T^N = \{(v_0; v_1) : v_0 \in C(0,T; P_N H_0^2(\Omega)),\, v_1 \in C(0,T; P_N H_0^1(\Omega))\}$$

with the norm $|(v_0;v_1)|_{\mathscr{M}_T^N} = \max_{[0,T]} \| v_0(t) \|_{2,\Omega} + \max_{[0,T]} \| v_1(t) \|_{1,\Omega}$.

Denote by $\mathscr{B}$ the mapping from $\mathscr{M}_T^N$ into itself defined by

$$\mathscr{B}(v_0;v_1) = (b_0(v_0;v_1); b_1(v_0;v_1)),$$

where $b_0(v_0;v_1) = w_0 + \int_0^t v_1(\tau)\, d\tau$ and

$$b_1(v_0;v_1) = w_1 + \int_0^t \{-M_\alpha^{-1}\mathscr{A} v_0(\tau) + P_N M_\alpha^{-1}([v_0(\tau), v^*(v_0(\tau)) + F_0] + p_0)\}\, d\tau$$
$$+ \nu P_N M_\alpha^{-1} r_\Omega(\gamma[\phi(t)] - \gamma[\phi_0]) + \nu U \int_0^t P_N M_\alpha^{-1} r_\Omega(\partial_{x_1}\gamma[\phi(\tau)])\, d\tau.$$

Here $v = v^*(v_0)$ is defined by v_0 from (6.4.3) with $u = v_0$ and $\phi(x,t)$ is the solution of problem (6.6.10)–(6.6.12) with

$$h(x,t) = \begin{cases} v_1(x_1,x_2,t) + U\chi_N(x_1,x_2)\partial_{x_1} v_0(x_1,x_2,t), & (x_1,x_2) \in \Omega, \\ 0, & (x_1,x_2) \notin \Omega. \end{cases}$$

Because $v_1, \partial_{x_1} v_0 \in C(0,T; H_0^1(\Omega))$, we have that $h \in C(0,T; H_0^1(\mathbb{R}^2))$.

It is sufficient to prove that there exists $T_0 = T_0(N,R)$ such that the mapping $\mathscr{B}$ has a fixed point in the ball

$$\mathscr{K}(w_0;w_1) = \{(v_0;v_1) \in \mathscr{M}_T^N : |(v_0;v_1) - (w_0;w_1)|_{\mathscr{M}_T^N} \le 1\}, \quad T < T_0.$$

Assume that $(v_0;v_1) \in \mathscr{K}(w_0;w_1)$; then (6.6.46) implies

$$\max_{[0,T]} \| v_0(t) \|_{2,\Omega} + \max_{[0,T]} \| v_1(t) \|_{1,\Omega} \le 1 + 2R.$$

Therefore

$$\|b_0(v_0;v_1)(t) - w_0\|_{2,\Omega} \le c\sqrt{\lambda_N}(1+2R)T, \quad t \in [0,T]. \tag{6.6.47}$$

Using the von Karman bracket properties (see Theorem 1.4.3) and the properties of the operators A and J we also have

$$\|b_1(v_0;v_1)(t) - w_1\|_{1,\Omega} \le (1+\lambda_N)C_R T$$
$$+ C_N\left[\|\gamma[\phi(t)] - \gamma[\phi_0]\|_{\mathbb{R}^2} + \sqrt{T}\left(\int_0^t \|\gamma[\phi(\tau)])\|_{\mathbb{R}^2}^2 d\tau\right)^{1/2}\right]$$

for $t \in [0,T]$. The interpolation and the trace theorems give

$$\|\gamma[\phi(t)] - \gamma[\phi(s)]\|_{\mathbb{R}^2} \le \|\phi(t) - \phi(s)\|_{1/2+\delta,\mathbb{R}^3_+}$$

$$\le C\|\phi(t)-\phi(s)\|_{\mathbb{R}^3_+}^{1/2-\delta}\cdot\left(\|\phi(t)\|_{1,\mathbb{R}^3_+}+\|\phi(s)\|_{1,\mathbb{R}^3_+}\right)^{1/2+\delta}$$
$$\le C|t-s|^{1/2-\delta}\cdot\left(\max_{[0,T]}\|\partial_t\phi(t)\|_{\mathbb{R}^3_+}\right)^{1/2-\delta}\cdot\left(\|\phi(t)\|_{1,\mathbb{R}^3_+}+\|\phi(s)\|_{1,\mathbb{R}^3_+}\right)^{1/2+\delta}$$

for any $t,s\in[0,T)$ and for some $\delta<\frac{1}{2}$, so Theorem 6.6.4 implies

$$\|b_1(v_0;v_1)(t)-w_1\|_{1,\Omega}\le(1+\lambda_N)C_RT$$
$$+C_NT^{1/2-\delta}(1+T^\delta)\exp(\sigma T/2)\left(R^2+\int_0^T\|h(\tau)\|^2_{1/2,\Omega}d\tau\right)^{1/2} \quad (6.6.48)$$

It follows from Theorem 11.8 of [222] that

$$\|\chi_N v\|_{1,\Omega}\le C\|v\|_{1,\Omega} \quad \text{for any } v\in H^1_0(\Omega), \quad (6.6.49)$$

Therefore the properties of $h(\tau)$ and (6.6.48) give

$$\|b_1(v_0;v_1)(t)-w_1\|_{1,\Omega}\le C_{R,N}\left(T+T^{1/2-\delta}(1+T)\exp(\sigma T/2)\right). \quad (6.6.50)$$

It follows from (6.6.47) and (6.6.50) that $\mathscr{B}$ maps $\mathscr{K}(w_0;w_1)$ into itself provided $T\le T_0(N,R)$. A similar argument shows that $\mathscr{B}$ is a contraction in this ball for $T_0(N,R)$ small enough. □

Let p be the continuous linear extension operator from $H^1_0(\Omega)$ into $H^2(\mathbb{R}^3_+)$ such that

$$\left.\frac{\partial p[h]}{\partial x_3}\right|_{x_3=0}=\begin{cases}h(x_1,x_2), & (x_1,x_2)\in\Omega,\\ 0, & (x_1,x_2)\notin\Omega.\end{cases}$$

6.6.15. Lemma. *Let $(u_N(t);\phi_N(t))$ be an approximate solution of problem* (6.4.1)–(6.4.7) *on some interval $[0,T_*)$. Assume that*

$$\phi_1\in H^1(\mathbb{R}^3_+) \quad \text{and} \quad \phi_0(x)=p[h_N(0)](x)+\psi(x), \quad (6.6.51)$$

where $h_N(x,t)$ defined by (6.6.42) *and $\psi(x)\in V=\{\phi\in H^2(\mathbb{R}^3_+):\partial_{x_3}\phi_{|x_3=0}=0\}$. Then for this approximate solution satisfies* (6.6.44) *and* (6.6.45) *for all $t\in[0,T_*)$.*

Proof. Let us represent $\phi_N(t)$ as a sum of two functions $\phi_N(t)=\phi_N^*(t)+\phi_N^{**}(t)$, where $\phi_N^*(x,t)$ is the solution of problem (6.6.10)–(6.6.12) with $\phi_0=\phi_1=0$ and $h(x,t)=h_N(x,t)-h_N(x,0)$, $\phi_N^{**}(x,t)$ is the solution of the problem (6.6.10)–(6.6.12) with the given ϕ_0, ϕ_1 and $h(x,t)=h_N(x,0)$. Theorem 6.6.10 implies that

$$r_\Omega(\partial_t+U\partial_{x_1})\gamma[\phi_N^*]\in C(0,T_*;L^2(\Omega)).$$

It also follows from Theorem 6.6.4 that

$$\phi_N^{**}(t)\in C(0,T_*;H^2(\mathbb{R}^3_+)) \quad \text{and} \quad \partial_t\phi_N^{**}(t)\in C(0,T_*;H^1(\mathbb{R}^3_+)).$$

Consequently $r_\Omega(\partial_t + U\partial_{x_1})\gamma[\phi_N^{**}] \in C(0,T_*;L^2(\Omega))$. Therefore we can differentiate (6.6.41) with respect t and obtain that

$$\partial_t^2 u_N(t) = -M_\alpha^{-1}\mathscr{A} u_N(t) + P_N M_\alpha^{-1}([u_N(t), v(u_N(t)) + F_0] + p_0) + \nu P_N M_\alpha^{-1} r_\Omega(\gamma[(\partial_t + U\partial_{x_1})\phi_N(t)] \quad (6.6.52)$$

belongs to $C(0,T_*;P_N H_0^1(\Omega)) \subset C(0,T_*;H_0^2(\Omega))$. We also obviously have that $\partial_{x_1 t} u_N(t) \in C(0,T_*;H_0^1(\Omega))$. Thus we have properties (6.6.15) and (6.6.16) with $k=1$. Applying Theorem 6.6.9 we obtain the equalities

$$E_{fl}^{(1)}(\phi_N(t), \partial_t\phi_N(t)) - E_{fl}^{(1)}(\phi_0, \phi_1) = -\int_0^t \int_\Omega (\partial_t + U\chi_N(x)\partial_{x_1})u_N(x,\tau)) \cdot \partial_t\gamma[\phi_N(x,\tau)]dxd\tau \quad (6.6.53)$$

and

$$E_{fl}^{(2)}(\phi_N(t), \partial_t\phi_N(t)) - E_{fl}^{(2)}(\phi_0, \phi_1) = -\int_0^t \int_\Omega (\partial_t + U\chi_N(x)\partial_{x_1})u_N(x,\tau)) \cdot (\partial_t + U\partial_{x_1})\gamma[\phi_N(x,\tau)]\, dxd\tau. \quad (6.6.54)$$

If we multiply (6.6.52) in $H_0^1(\Omega)$ by $\partial_t u_N(t)$, we get

$$E_{pl}(u_N(t), \partial_t u_N(t)) - E_{pl}(u_N(0), \partial_t u_N(0)) = \nu \int_0^t \int_\Omega \partial_t u_N(x,\tau) \cdot (\partial_t + U\partial_{x_1})\gamma[\phi_N(x,\tau)]dxd\tau. \quad (6.6.55)$$

Comparing (6.6.53)–(6.6.55) we obtain (6.6.44) and (6.6.45) in this case. □

6.6.16. Lemma. *Assume that conditions of Lemma* 6.6.15 *are valid. Then for the approximate solution* $(u_N(t);\phi_N(t))$ *we have the estimate*

$$\| \partial_t u_N(t) \|_{1,\Omega}^2 + \| \Delta u_N(t) \|_\Omega^2 + \| \partial_t\phi_N(t) \|_{\mathbb{R}^3_+}^2 + \| \nabla\phi_N(t) \|_{\mathbb{R}^3_+}^2 \le C_1 e^{C_2 t}\left(1 + E_{pl}^0(u_N(0), \partial_t u_N(0)) + E_{fl}^{(2)}(\phi_0, \phi_1)\right) \quad (6.6.56)$$

for any $t \in [0,T_*)$ *with constants* C_1 *and* C_2 *independent of* N, *where energy* $E_{fl}^{(2)}(\phi_0,\phi_1)$ *is defined by* (6.6.7) *and*

$$E_{pl}^0(u, \partial_t u) = \frac{1}{2}\left(\| \partial_t u \|_{1,\Omega}^2 + \| \Delta u \|_\Omega^2 + \frac{1}{2} \| \Delta v(u) \|_\Omega^2 \right).$$

If $(u_{j,N}(t);\phi_{j,N}(t))$, $j=1,2$, *are two approximate solutions such that*

$$E_{pl}^0(u_{j,N}(0), \partial_t u_{j,N}(0)) + E_{fl}^{(2)}(\phi_{j,0}, \phi_{j,1}) \le R^2, \quad j=1,2,$$

then

$$
\begin{aligned}
&\| \partial_t(u_{1,N}(t) - u_{2,N}(t)) \|_{1,\Omega}^2 + \| \Delta(u_{1,N}(t) - u_{2,N}(t)) \|_{\Omega}^2 \\
&\le C(T_*,R)\Big\{ \| \partial_t u_{1,N}(0) - \partial_t u_{2,N}(0) \|_{1,\Omega}^2 + \| \Delta(u_{1,N}(0) - u_{2,N}(0)) \|_{\Omega}^2 \\
&\quad + \| \phi_{1,1} - \phi_{2,1} \|_{\mathbb{R}^3_+}^2 + \| \nabla(\phi_{1,0} - \phi_{2,0}) \|_{\mathbb{R}^3_+}^2 \Big\}
\end{aligned} \tag{6.6.57}
$$

and

$$
\begin{aligned}
&\| \partial_t(\phi_{1,N}(t) - \phi_{2,N}(t)) \|_{\mathbb{R}^3_+}^2 + \| \nabla(\phi_{1,N}(t) - \phi_{2,N}(t)) \|_{\mathbb{R}^3_+}^2 \\
&\le C(T_*,R)\Big\{ \| \partial_t u_{1,N}(0) - \partial_t u_{2,N}(0) \|_{1,\Omega}^2 + \| \Delta(u_{1,N}(0) - u_{2,N}(0)) \|_{\Omega}^2 \\
&\quad + \| \phi_{1,1} - \phi_{2,1} \|_{\mathbb{R}^3_+}^2 + \| \nabla(\phi_{1,0} - \phi_{2,0}) \|_{\mathbb{R}^3_+}^2 \Big\},
\end{aligned} \tag{6.6.58}
$$

where $C(T_*,R)$ is independent of N.

Proof. Lemmas 6.5.4 and 6.5.5 give

$$
\beta_1 \mathscr{E}^{(2,0)}(u_0,u_1;\phi_0,\phi_1) - c_1 \le \mathscr{E}^{(2)}(u_0,u_1;\phi_0,\phi_1) \le \beta_2 \mathscr{E}^{(2,0)}(u_0,u_1;\phi_0,\phi_1) + c_2,
$$

where β_j and c_j are positive constants and

$$
\mathscr{E}^{(2,0)}(u_0,u_1;\phi_0,\phi_1) = E^0_{pl}(u_0,u_1) + E^{(2)}_{fl}(\phi_0,\phi_1).
$$

Using (6.6.49) we have

$$
\|\chi_N \partial_{x_1} \partial_t u_N + U \partial_{x_1}(\chi_N \partial_{x_1} u_N)\|_{\Omega} \le C(\| \partial_t u_N \|_{1,\Omega} + \| \Delta u_N \|_{\Omega}).
$$

Thus we can obtain (6.6.56) using (6.5.22), (6.6.45) and Gronwall's lemma. To prove (6.6.57) and (6.6.58) we use standard arguments relying on relations (6.6.52), (6.6.56), and (6.6.14). □

We can now prove Theorem 6.6.13 for ϕ_0 and ϕ_1 possessing property (6.6.51). Indeed, assume that $[0,T_*)$ is the maximal interval of existence of approximate solution. Then using (6.6.52) and (6.6.56) we can conclude that $\|\partial_t^2 u_N(t)\|_{\Omega}$ is bounded on this interval. Therefore $u_N(t)$ and $\partial_t u_N(t)$ can be extended as a continuous functions on the segment $[0,T_*]$. It is also clear from Theorem 6.6.4 that $\phi_N(t)$ and $\partial_t \phi_N(t)$ are continuous functions on $[0,T_*]$ with values in $H^1(\mathbb{R}^3_+)$ and $L^2(\mathbb{R}^3_+)$. Therefore we can apply Lemma 6.6.14 with initial data

$$
\{u_N(t), \partial_t u_N(t); \phi_N(t), \partial_t \phi_N(t)\}|_{t=T_*}
$$

at time $t = T_*$, so we can continue the solution on a longer interval. Thus the approximate solution $(u_N;\phi_N)$ exists on any interval $[0,T]$ for initial data (ϕ_0,ϕ_1) satisfying (6.6.51). To conclude the proof of Theorem 6.6.13 for general initial data we use (6.6.57), (6.6.58), and the fact that the set $\{\phi \in H^2(\mathbb{R}^3_+) : \partial_{x_3}\phi_{|x_3=0} = h(x)\}$ is dense in $H^1(\mathbb{R}^3_+)$ for any $h(x) \in H^1_0(\mathbb{R}^2)$.

To take limits when $N \to \infty$ we also need the following estimate.

6.6.17. Lemma. *Assume that the conditions of Theorem* 6.6.13 *are valid. Then for the approximate solution* $(u_N(t);\phi_N(t))$ *we have the estimate*

$$\| \mathscr{A}^{-1/2} M_\alpha (\partial_t u_N(t) - \partial_t u_N(s)) \|_\Omega \le C_T |t-s|^{1/2-\delta}, \quad t,s \in [0,T] \qquad (6.6.59)$$

for some $\delta < 1/2$*, where* C_T *is independent of* N*. This means that* $\partial_t u_N$ *is (uniformly) Hölder continuous in the space* $\widetilde{L}_2(\Omega)$ *defined as the dual to* $\mathscr{D}(\mathscr{A}^{1/2})$ *with respect to* $V_\alpha = \mathscr{D}({M_\alpha}^{1/2})$ *taken as a pivot space (see [215]).*

Proof. It follows from (6.6.41) that we have

$$\partial_t u_N(t) = \partial_t u_N(s) - \int_s^t \left[M_\alpha^{-1}\mathscr{A} u_N(\tau) - P_N M_\alpha^{-1}([u_N(\tau), v(u_N(\tau)) + F_0] - p_0)\right] d\tau$$
$$+ \nu P_N M_\alpha^{-1} r_\Omega(\gamma[\phi_N(t)] - \gamma[\phi_N(s)]) + \nu U \int_s^t P_N M_\alpha^{-1} r_\Omega(\partial_{x_1}\gamma[\phi_N(\tau)])\, d\tau$$

for $0 \le s \le t \le T$. Therefore

$$|(\partial_t u_N(t) - \partial_t u_N(s), w)_1|$$
$$\le C_1 \|w\|_2 \cdot \int_s^t [\|u_N(\tau)\|_2 + \|[u_N(\tau), v(u_N(\tau)) + \theta] + p_0\|_{-2}]\, d\tau$$
$$+ C_2 \|w\|_1 \cdot \left[\|r_\Omega(\gamma[\phi_N(t)] - \gamma[\phi_N(s)])\| + \int_s^t \|r_\Omega(\partial_{x_1}\gamma[\phi_N(\tau)])\|\, d\tau\right]$$

for any $w \in H_0^2(\Omega)$. Therefore using Theorem 6.6.4 as in Lemma 6.6.14, we obtain

$$|(M_\alpha(\partial_t u_N(t) - \partial_t u_N(s)), w)| \le C|t-s|^{1/2-\delta} \cdot \|w\|_2,$$

which implies (6.6.59). □

6.6.4 Limit transition

In this subsection we conclude the proof of the main theorems.

It follows from (6.6.43) that there exist a subsequence $\{N_k\}$ and a function $u(t) \in \mathscr{W}_T$ such that

$$(u_{N_k}(t); \partial_t u_{N_k}(t)) \to (u(t); \partial_t u(t)) \quad *\text{-weakly in} \quad L_\infty(0,T; H_0^2(\Omega) \times H_0^1(\Omega)),$$

Lemma 6.6.17 and the Aubin-Doubinsky theorem (see Theorem A.2.4 and Theorem 5 in [263]) allow us to assert that

$$u_{N_k}(t) \to u(t) \quad \text{strongly in} \quad C(0,T; H_0^{2-\varepsilon}(\Omega)), \qquad (6.6.60)$$
$$\partial_t u_{N_k}(t) \to \partial_t u(t) \quad \text{strongly in} \quad C(0,T; H_0^{1-\varepsilon}(\Omega)) \qquad (6.6.61)$$

for any $0 < \varepsilon < 1$. For the sake of simplicity we suppose below $N_k \equiv N$. The interpolation of the Sobolev spaces and the estimate (6.6.43) give

$$\int_0^T \|\chi_N \partial_{x_1} u_N(\tau) - \partial_{x_1} u(\tau)\|^2_{1-\varepsilon,\Omega}\, d\tau \le C_T^{\varepsilon} \int_0^T \|\chi_N \partial_{x_1} u_N(\tau) - \partial_{x_1} u(\tau)\|^{2\varepsilon}_{\Omega}\, d\tau.$$

Thus it follows from (6.6.60) and (6.6.61) that we have a strong limit:

$$\partial_t u_N(t) + U\chi_N(x)\partial_{x_1} u_N(t) \to \partial_t u(t) + U\partial_{x_1} u(t) \tag{6.6.62}$$

in $L_2(0,T;H_0^{1-\varepsilon}(\Omega))$. Theorem 6.6.4 now implies that

$$\sup_{[0,T]} \| \partial_t(\phi_N(t) - \phi(t)) \|^2_{\mathbb{R}^3_+} + \sup_{[0,T]} \| \nabla(\phi_N(t) - \phi(t)) \|^2_{\mathbb{R}^3_+} \to 0, \quad N \to \infty, \tag{6.6.63}$$

where $\phi(t)$ is the solution of (6.4.4)–(6.4.7).

Let $w(t) \in P_M \mathscr{W}_T$ with $M \le N$ and $w(T) = 0$. Multiplying (6.6.41) by $\partial_t w(t)$ in $H_0^1(\Omega)$ and integrating with respect to t we get

$$\begin{aligned} &-\int_0^T ((1-\alpha\Delta)\partial_t u_N(t), \partial_t w(t))_\Omega dt \\ &+\int_0^T (\Delta u_N(t), \Delta w(t))_\Omega dt - \int_0^T ([u_N(t), v(u_N(t)) + \theta] + p_0, w(t))_\Omega dt \\ &= ((1-\alpha\Delta)\partial_t u_N(0) - \nu\gamma[\phi_0], w(0))_\Omega - \nu\int_0^T (\gamma[\phi_N(t)], \partial_t w(t) + U\partial_{x_1} w(t))_\Omega dt \end{aligned}$$

for any $w \in P_M \mathscr{W}_T$ such that $w(T) = 0$. The convergence properties of u_N and ϕ_N allow us to take limits when $N \to \infty$; this proves (6.4.10) for this $w(t)$. Letting $M \to \infty$ we obtain the existence of weak solutions.

Let us prove the uniqueness of these solutions. Assume that $(u_j;\phi_j)$ are two weak solutions of (6.4.1)–(6.4.7). Then $\phi(x,t) = \phi_1(x,t) - \phi_2(x,t)$ satisfies equations (6.6.10)–(6.6.12) with $\phi_0(x) = \phi_1(x) = 0$ and

$$h(x,t) = \begin{cases} (\partial_t + U\partial_{x_1})u(x_1,x_2,t), & (x_1,x_2) \in \Omega, \\ 0, & (x_1,x_2) \notin \Omega, \end{cases} \tag{6.6.64}$$

where $u = u_1 - u_2$. Because $h \in L_\infty(0,T;H^1(\mathbb{R}^2))$, Theorem 6.6.10 gives

$$r_\Omega(\partial_t + U\partial_{x_1})\gamma[\phi] \in L_\infty(0,T;L^2(\Omega)). \tag{6.6.65}$$

So we can consider $u(x,t)$ as a weak solution of problem (6.4.1)–(6.4.3) with a given transverse force $p(x,t)$ belonging to $L_\infty(0,T;L_2(\Omega))$. The standard argument together with Theorem 6.6.10 and the estimate (6.6.32) show that $u(x,t) \equiv 0$ and, therefore also $\phi(x,t) \equiv 0$.

The properties of weak convergence and Lemma 6.6.16 imply that weak solutions depend continuously on their initial data:

$$\begin{aligned}
&\| \partial_t (u_1(t) - u_2(t)) \|_{1,\Omega}^2 + \| \Delta (u_1(t) - u_2(t)) \|_\Omega^2 \\
&\le C_T \Big[\| u_{1,1} - u_{1,2} \|_{1,\Omega}^2 + \| \Delta (u_{0,1} - u_{0,2}) \|_\Omega^2 \\
&\quad + \| \phi_{1,1} - \phi_{1,2} \|_{\mathbb{R}^3_+}^2 + \| \nabla (\phi_{0,1} - \phi_{0,2}) \|_{\mathbb{R}^3_+}^2 \Big]
\end{aligned} \tag{6.6.66}$$

and

$$\begin{aligned}
&\| \partial_t (\phi_1(t) - \phi_2(t)) \|_{\mathbb{R}^3_+}^2 + \| \nabla (\phi_1(t) - \phi_2(t)) \|_{\mathbb{R}^3_+}^2 \\
&\le C_T \Big[\| u_{1,1} - u_{1,2} \|_{1,\Omega}^2 + \| \Delta (u_{0,1}(0) - u_{0,2}) \|_\Omega^2 \\
&\quad + \| \phi_{1,1} - \phi_{1,2} \|_{\mathbb{R}^3_+}^2 + \| \nabla (\phi_{0,1} - \phi_{0,2}) \|_{\mathbb{R}^3_+}^2 \Big],
\end{aligned} \tag{6.6.67}$$

where $(u_j(t); \phi_j(t))$ is the solution with initial data $(u_{0,j}; u_{1,j}; \phi_{0,j}; \phi_{1,j})$, $j = 1,2$. It is also clear that the solution $(u(t); \phi(t))$ possesses the properties (6.6.1) and (6.6.2) provided that

$$\phi_1 \in H^1(\mathbb{R}^3_+) \ \text{ and } \ \phi_0(x) = p[h](x) + \psi(x), \tag{6.6.68}$$

where $\psi(x) \in V = \{\phi \in H^2(\mathbb{R}^3_+) : \partial_{x_3} \phi_{|x_3=0} = 0\}$ and

$$h(x) = \begin{cases} u_1 + U\partial_{x_1} u_0, & (x_1, x_2) \in \Omega, \\ 0, & (x_1, x_2) \notin \Omega. \end{cases}$$

Therefore (6.6.66) and (6.6.67) imply (6.6.1) and (6.6.2) for any weak solution.

We prove the equalities (6.6.3) and (6.6.4) first in the case when (6.6.68) holds. In this case Theorems 6.6.4 gives (6.6.65). Therefore we can use the energy equality for problem (6.4.1)–(6.4.3) with the given $p(x,t)$ (see Section 3.1). We have

$$E_{pl}(u(t), \partial_t u(t)) - E_{pl}(u(0), \partial_t u(0)) = \nu \int_0^t \int_\Omega \partial_t u(x,t) (\partial_t + U\partial_{x_1}) \gamma[\phi(x,\tau)] dx d\tau. \tag{6.6.69}$$

Because $\phi(x,t) - \phi_N(x,t)$ is a solution to (6.6.10)–(6.6.12) with zero initial data, (6.6.29) and (6.6.62) imply

$$\int_0^T \| (\partial_t + U\partial_{x_1}) \gamma [\phi(t) - \phi_N(t)] \|_{-1+\varepsilon,\Omega}^2 \, dt \to 0 \ \text{ when } N \to \infty.$$

From this and (6.6.61) we conclude that the right-hand side of (6.6.55) tends to the right-hand side of (6.6.69), when $N \to \infty$. Consequently

$$E_{pl}(u(t), \partial_t u(t)) = \lim_{N \to \infty} E_{pl}(u_N(t), \partial_t u_N(t)). \tag{6.6.70}$$

Therefore in the case (6.6.68), relations (6.6.3) and (6.6.4) follow from (6.6.44), (6.6.45), (6.6.70) and (6.6.60)–(6.6.63). Now using (6.6.66) and (6.6.67) we obtain (6.6.3) and (6.6.4) in the general case.

6.6.5 Reduced retarded problem

The following result shows that problem (6.4.1)–(6.4.7) can be reduced to a retarded partial differential equation.

6.6.18. Theorem. *Assume that the conditions of Theorem* 6.6.1 *are valid and there exists* $R > 0$ *such that* $\phi_0(x) = \phi_1(x) = 0$ *for* $|x| \geq R$. *Let* $(u(x_1,x_2,t);\phi(x_1,x_2,x_3,t))$ *be a weak solution of problem* (6.4.1)–(6.4.7). *Then there exists a positive number* $t(R,U,\Omega)$ *such that for all* $t \geq t(R,U,\Omega)$ *the function* $u(x,t)$ *satisfies* (6.4.1) *with*

$$p(x,t) = p_0(x) - \nu(\partial_t u + U\partial_{x_1} u + q(u;x,t)) \tag{6.6.71}$$

with

$$q(u,x,t) = \frac{1}{2\pi} \cdot \int_0^{t_*} ds \int_0^{2\pi} d\theta [M_\theta^2 \hat{u}](x_1 - (U + \sin\theta)s, x_2 - s\cos\theta, t - s). \tag{6.6.72}$$

Here $\hat{u}$ *is the extension of* $u(x,t)$ *by zero outside of* Ω, $M_\theta = \sin\theta \cdot \partial_{x_1} + \cos\theta \cdot \partial_{x_2}$ *and*

$$t_* = \inf\{t : x(U,\theta,s) \notin \Omega \ \textit{for all}\ x \in \Omega,\ \theta \in [0,2\pi],\ s > t\}, \tag{6.6.73}$$

where $x(U,\theta,s) = (x_1 - (U + \sin\theta)s; x_2 - s\cos\theta) \in \mathbb{R}^2$ *and* $x = (x_1;x_2) \in \Omega \subset \mathbb{R}^2$.

Thus, after some time t, the function $u(x,t)$ can be treated as a solution of a retarded nonlinear partial differential equation (6.4.1) with $p(x,t)$ defined by (6.6.71).

6.6.19. Remark. Let us assume $\Omega \subset \{x \in \mathbb{R}^2 : |x - z| \leq \rho\}$ for some $z \in \mathbb{R}^2$ and $\rho > 0$. Because

$$(x_1 - z_1 - (U + \sin\theta)s)^2 + (x_2 - z_2 - s\cos\theta)^2 \geq -2\rho s(U+1) + (U-1)^2 s^2,$$

we have $(x_1 - (U + \sin\theta)s; x_2 - s\cos\theta) \notin \Omega$ provided that $s \geq t_*(U,\Omega)$, where

$$t_*(U,\Omega) = \rho \cdot \frac{1 + U + \sqrt{2 + 2U^2}}{(U-1)^2}.$$

Therefore we have estimate $t_* \leq t_*(U,\Omega)$ for all $U > 0$, $U \neq 1$. It is also easy to see that $t_* \leq l \cdot (U-1)^{-1}$, when $U > 1$. Here l is the size of Ω along the x_1-axis. ■

6.6.20. Remark. Let us suppose $U > 1$. Inasmuch as the integrand in (6.6.72) vanishes if $s \geq t_*$, we have

$$q(u;x,t) = \frac{1}{2\pi} \int_0^{\infty} ds \int_0^{2\pi} d\theta [M_\theta^2 \hat{u}](x_1 - (U + \sin\theta)s, x_2 - s\cos\theta, t - s)$$

for $t \geq t_*$. If we make the change of the variable $s \to \xi = x_1 - (U + \sin\theta)s$ and introduce a new angle θ^* by the formulas

$$\frac{\cos\theta}{U+\sin\theta} = \frac{1}{k}\cos\theta^* \quad \text{and} \quad \frac{1}{U+\sin\theta} = \frac{1}{k^2}\left(U-\sin\theta^*\right),$$

where $k^2 = U^2 - 1$, we can easily find that

$$q(u;x,t) = \frac{1}{2\pi k}\cdot\int_{-\infty}^{x_1} d\xi \int_0^{2\pi} d\theta [L_\theta^2 \hat{u}](\xi, x_2 - \kappa_1(\xi,x_1,\theta), t - \kappa_2(\xi,x_1,\theta)), \tag{6.6.74}$$

where the differential operator L_θ is defined by

$$L_\theta = \frac{U\sin\theta - 1}{U-\sin\theta}\cdot\frac{\partial}{\partial x_1} + \frac{k\cos\theta}{U-\sin\theta}\cdot\frac{\partial}{\partial x_2}$$

and

$$\kappa_1(\xi,x_1,\theta) = \frac{x_1-\xi}{k}\cos\theta, \quad \kappa_2(\xi,x_1,\theta) = \frac{x_1-\xi}{k^2}(U-\sin\theta).$$

This representation of q in (6.6.71) for the aerodynamical pressure was used earlier in the study of the long time behavior of plates and shallow shells (see [54] and also [80, 33]). The derivation of (6.6.74) in [54] relies on heuristic observations put forward in [168]. Thus Theorem 6.6.18 gives rigorous justification of the presentation for the aerodynamical pressure in (6.6.71) and (6.6.74). ∎

Proof of Theorem 6.6.18. Let $\phi(x,t) = \phi^*(x,t) + \phi^{**}(x,t)$, where $\phi^*(x,t)$ is the solution of (6.4.5)–(6.4.7) with $\phi_0 = \phi_1 = 0$ and $\phi^{**}(x,t)$ is the solution of (6.6.10)–(6.6.12) with $h \equiv 0$. If ϕ_0 and ϕ_1 satisfy the conditions of Theorem 6.6.18, it follows from (6.6.34) that we have $\gamma[\phi^{**}(t)] = 0$ in a neighborhood of $\Omega \subset \mathbb{R}^2$ for t large enough. Thus

$$r_\Omega(\partial_t + U\partial_{x_1})\gamma[\phi^{**}(t)] = 0 \ \text{ for } t \ge t_R \tag{6.6.75}$$

for some $t_R > 0$. Let us consider the value $r_\Omega(\partial_t + U\partial_{x_1})\gamma[\phi^*]$. Applying a formula similar to (6.6.30) to compute $\partial_t u$, we get

$$r_\Omega(\partial_t + U\partial_{x_1})\gamma[\phi^*] = -(\partial_t + U\partial_{x_1})u(x,t) - q_0(u_0;x,t) - q(u;x,t)$$

with

$$q_0(u_0;x,t) = \frac{1}{2\pi}\int_0^{2\pi} d\theta [M_\theta \hat{u}_0](x_1 - (U+\sin\theta)t, x_2 - t\cos\theta) \tag{6.6.76}$$

and

$$q(u;x,t) = \frac{1}{2\pi}\int_0^t ds \int_0^{2\pi} d\theta [M_\theta^2 \hat{u}](x_1 - (U+\sin\theta)s, x_2 - s\cos\theta, t-s), \tag{6.6.77}$$

where $\hat{w}$ is the extension of w by zero outside of Ω and $M_\theta = \sin\theta\cdot\partial_{x_1} + \cos\theta\cdot\partial_{x_2}$. Because $(x_1 - (U+\sin\theta)s; x_2 - s\cos\theta) \notin \Omega$ provided that $s \ge t_*$, where t_* is defined by (6.6.73), we have for $t \ge t_*$

$$r_\Omega(\partial_t + U\partial_{x_1})\gamma[\phi^*] = -(\partial_t + U\partial_{x_1})u(x,t) - q(u;x,t), \tag{6.6.78}$$

and

$$q(u;x,t) = \frac{1}{2\pi}\int_0^{t_*} ds \int_0^{2\pi} d\theta [M_\theta^2 \hat{u}](x_1 - (U + \sin\theta)s, x_2 - s\cos\theta, t - s). \quad (6.6.79)$$

Now (6.6.75), (6.6.78), and (6.6.79) imply (6.6.71) and (6.6.72). This completes the proof of Theorem 6.6.18.

6.6.21. Remark. We note that Theorem 6.6.18 (the same as Theorem 6.6.1, see Remark 6.6.2) remains true when instead of (6.4.1) and (6.4.3) we consider equations (6.6.8) and (6.6.9). More general damping functions can be also considered. ■

Theorem 6.6.18 allows us to formulate the problem of aeroelastic oscillations of the plate as a retarded model of the form (3.3.1)–(3.3.3). Moreover it is worth mentioning that from a mechanical point of view, Theorem 6.6.18 on the reduction of the problem (6.4.1)–(6.4.7) on nonlinear plate oscillations in a potential gas flow to the retarded partial differential equation allows us to suggest an effective method for description of aerodynamical forces and to simplify essentially an investigation of the stability problem for realistic plate constructions in a flow of gas. For details we refer to Sections 3.3.1 and 9.3.

6.6.22. Remark. In the applied literature (see, e.g., [28, 99] and references therein) many authors involve the "piston" theory based on the so-called plane sections law (see [147] and also [28]) for taking into account the interaction between the gas flow and the plate in the case of large supersonic speeds ($U \gg 1$). This law leads to the formula for the aerodynamical pressure of the form

$$p^*(x,t) = p_0(x) - \nu \cdot (\partial_t u(x,t) + U\partial_{x_1} u(x,t)). \quad (6.6.80)$$

On the other hand the equation (6.6.74) implies that for $U > 1$ we have

$$\| q(u,t) \|_\Omega^2 \leq \frac{c_0}{U} \cdot \int_{t-c_1/U}^{t} \| \Delta u(\tau) \|_\Omega^2 \, d\tau,$$

where the constants c_0 and c_1 depend on the size of the domain Ω only. Thus the formula (6.6.71) for the aerodynamical pressure in the limit of large U is in good agreement with the result (6.6.80) given by the semiempirical plane sections law. In this context, it is natural to raise a question of an asymptotic closedness (for large U) of solutions to the problem (6.4.1)–(6.4.3) with $p(x,t) = p^*(x,t)$ to solutions of problem (6.4.1)–(6.4.7) in the original formulation. The affirmative answer to this question would provide us with additional arguments in favor of validity of the plane sections law in the study of nonlinear dynamics of aeroelastic structures. ■

Part II
Long-Time Dynamics

Chapter 7
Attractors for Evolutionary Equations

This chapter provides a survey of quantitative theory pertinent to long-time behavior of infinite-dimensional dissipative systems. The results are presented in a convenient form for applications to the material presented in subsequent chapters. For other possible approaches to the topic we refer to the monographs [17, 61, 134, 139, 172, 259, 273].

The main focus is on questions such as existence of global attractors, and their structure, dimensionality, and smoothness. In this context, gradient systems play a prominent role as an important subclass of more general dissipative systems. With the aim of unifying specific criteria that lead to the desired properties of attractors (mentioned above), we single out a class of "quasi-stable" systems that enjoy the so-called stabilizability–observability inequality. This inequality, although often difficult to establish, once it is proved provides a string of consequences that describe various properties of attractors.

In addition to attractors, other long-time behavior objects such as inertial manifolds, exponential attractors, and determining functional sets are also considered in this chapter

7.1 Dissipative dynamical systems

By definition a dynamical system is a pair of objects (X, S_t) consisting of a complete metric space X and a family of continuous mappings $\{S_t : t \in \mathbb{R}_+\}$ of X into itself with the semigroup properties:

$$S_0 = I, \quad S_{t+\tau} = S_t \circ S_\tau.$$

We also assume that $y(t) = S_t y_0$ is continuous with respect to t for any $y_0 \in X$. Therewith X is called a *phase space* (or state space) and S_t is called an *evolution semigroup* (or evolution operator). Our canonical example is the following.

I. Chueshov, I. Lasiecka, *Von Karman Evolution Equations*,
Springer Monographs in Mathematics, DOI 10.1007/978-0-387-87712-9_7,

7.1.1. Example. Let X be a Banach space and $F : \mathscr{D}(F) \subseteq X \mapsto X$ be a (nonlinear) operator on X. Consider the equation

$$\frac{du(t)}{dt} = F(u(t)), \quad t \geq 0, \quad u(0) = u_0 \in X. \tag{7.1.1}$$

If this problem is well-posed (in the sense specified below), then we have a dynamical system (X, S_t) with S_t defined by $S_t u_0 = u(t, u_0)$, where $u(t, u_0)$ is the solution to problem (7.1.1). ■

This chapter deals with dissipative dynamical systems. From a physical point of view, dissipative systems are characterized by "relocation" and dissipation of the energy. By this we mean that the energy of higher modes is dissipated and relocated to low modes, which phenomenon leads to ultimately finite-dimensional structures. This process produces a formation of limit regimes that are stable in a suitable sense. The following definition describes several concepts that allow us to to quantitize both dissipation and relocation on a formal level.

7.1.2. Definition. Let (X, S_t) be a dynamical system.

- A closed set $B \subset X$ is said to be *absorbing* for (X, S_t) iff for any bounded set $D \subset X$ there exists $t_0(D)$ such that $S_t D \subset B$ for all $t \geq t_0(D)$.
- (X, S_t) is said to be *(bounded, or ultimately) dissipative* iff it possesses a bounded absorbing set B. If X is a Banach space, then a value $R > 0$ is said to be a *radius of dissipativity* of (X, S_t) iff $B \subset \{x \in X : \|x\|_X \leq R\}$
- The dynamical system (X, S_t) is said to be *point dissipative* iff there exists $B_0 \subset X$ such that for any $x \in X$ there is $t_0(x)$ such that $S_t x \in B_0$ for all $t \geq t_0(x)$.
- (X, S_t) is said to be *compact* iff it is dissipative and the absorbing set B is compact.
- (X, S_t) is said to be *asymptotically compact* iff there exists an attracting compact set K; that is, for any bounded set D we have

$$\lim_{t \to +\infty} d_X\{S_t D \,|\, K\} = 0, \tag{7.1.2}$$

  where $d_X\{A|B\} = \sup_{x \in A} \operatorname{dist}_X(x, B)$.
- (X, S_t) is said to be *asymptotically smooth* iff for any bounded set D such that $S_t D \subset D$ for $t > 0$ there exists a compact set K in the closure $\overline{D}$ of D, such that (7.1.2) holds.

■

If the phase space X is compact, then (X, S_t) is a compact dynamical system. If X is a finite-dimensional space, then any dissipative system is compact. It is also clear that every asymptotically compact system is dissipative and asymptotically smooth. We note that in most cases von Karman models described in the previous sections generate asymptotically smooth dynamical systems under appropriate conditions. This is the main reason why we discuss below asymptotic compactness/smoothness with details.

Now we recall several well-known notions from the theory of dynamical systems.

A set $D \subset X$ is said to be *forward* (or positively) *invariant* iff $S_t D \subseteq D$ for all $t \geq 0$. It is *backward* (or negatively) *invariant* iff $S_t D \supseteq D$ for all $t \geq 0$. The set D is said to be *invariant* iff it is both forward and backward invariant; that is, $S_t D = D$ for all $t \geq 0$.

Let $D \subset X$. The set

$$\gamma_D^t \equiv \bigcup_{\tau \geq t} S_\tau D$$

is called the *tail* (from the moment t) of the trajectories emanating from D. It is clear that $\gamma_D^t = \gamma_{S_t D}^0 \equiv \gamma_{S_t D}^+$. If $D = \{v\}$ is a single point set, then $\gamma_v^+ \equiv \gamma_D^0$ is said to be a *positive semitrajectory (or semiorbit)* emanating from v. A continuous curve $\gamma \equiv \{u(t) : t \in \mathbb{R}\}$ in X is said to be a *full trajectory* iff $S_t u(\tau) = u(t+\tau)$ for any $\tau \in \mathbb{R}$ and $t \geq 0$. Because S_t is not necessarily an invertible operator, a full trajectory may not exist. Semitrajectories are forward invariant sets. Full trajectories are invariant sets.

To describe the asymptotic behavior we use the concept of an ω-limit set. The set

$$\omega(D) \equiv \bigcap_{t>0} \gamma_D^t = \bigcap_{t>0} \overline{\bigcup_{\tau \geq t} S_\tau D} \tag{7.1.3}$$

is called the ω-*limit set* of the trajectories emanating from D (the bar over a set means the closure). It is equivalent to saying that $x \in \omega(D)$ if and only if there exist sequences $t_n \to +\infty$ and $x_n \in D$ such that $S_{t_n} x_n \to x$ as $n \to \infty$. It is clear that ω-limit sets (if they exist) are forward invariant.

If $\gamma = \{u(t) : t \in \mathbb{R}\}$ is a full trajectory, we can define both ω- and α-limit sets of γ by the formulas

$$\omega(\gamma) = \bigcap_{t>0} \overline{\bigcup \{u(\tau) : \tau \geq t\}} \quad \text{and} \quad \alpha(\gamma) = \bigcap_{t<0} \overline{\bigcup \{u(\tau) : \tau \leq t\}} \tag{7.1.4}$$

The following assertion provides conditions under which $\omega(D)$ is nonempty.

7.1.3. Proposition. *Assume that the system (X, S_t) is an asymptotically compact dynamical system with attracting compact set K. Then for any bounded set D the ω-limit set $\omega(D)$ is a nonempty compact invariant set.*

If (X, S_t) is asymptotically smooth and the tail γ_D^t is bounded for some $t \geq 0$, then ω-limit set $\omega(D)$ is also a nonempty compact invariant set.

For the proof see, for example, [61, 134, 172]).

The following proposition provides other (equivalent) descriptions of asymptotically compact systems. For the proof and discussion we refer to [273, Chapter 1] (see also [172] and [246]).

7.1.4. Proposition. *Assume that X is a Banach space and (X, S_t) is a dissipative dynamical system, Then the following assertions are equivalent.*

- *(X, S_t) is asymptotically compact.*
- *(X, S_t) is asymptotically smooth.*

- *There exists a decomposition $S_t = S_t^{(1)} + S_t^{(2)}$, where $S_t^{(1)}$ is uniformly compact for large t; that is, for any bounded set D there exists $t_0 = t_0(D)$ such that the set $\gamma^{(1)}(D;t_0) := \bigcup_{\tau \ge t_0} S_\tau^{(1)} D$ is relatively compact in X and $S_t^{(2)}$ is a continuous mapping in X such that*

$$r_D(t) = \sup\left\{ \|S_t^{(2)}x\|_X \; : \; x \in D \right\} \to 0 \quad as \quad t \to \infty \tag{7.1.5}$$

  for every bounded set D.
- *The Ladyzhenskaya condition is satisfied: for every bounded sequence $\{x_n\} \subset X$ and every sequence $t_n \to \infty$ the sequence $\{S_{t_n} x_n\}$ is relatively compact in X.*

It the system (X, S_t) is not dissipative, the equivalence described in Proposition 7.1.4 is no longer valid. However, it is clear that asymptotic compactness implies asymptotic smoothness. Moreover, the next two propositions show that the asymptotic smoothness condition is the weakest one among other compactness properties postulated by Proposition 7.1.4.

7.1.5. Proposition. *Let (X, S_t) be a dynamical system in a Banach space X. Assume that for each $t > 0$ there exists a decomposition $S_t = S_t^{(1)} + S_t^{(2)}$, where $S_t^{(2)}$ is a continuous mapping in X satisfying* (7.1.5) *and $S_t^{(1)}$ is completely continuous; that is, for each $t > 0$ the set $\{S_\tau^{(1)} B : 0 \le \tau \le t\}$ is bounded and $S_t^{(1)} B$ is a relatively compact set in X for all $t > 0$ large enough. Here B is an arbitrary bounded set in X. Then (X, S_t) is asymptotically smooth.*

Proof. See [134, Lemma 3.2.3]. □

Another useful condition of asymptotic smoothness is given in the following assertion.

7.1.6. Proposition. *Let (X, S_t) be a dynamical system in a Banach space X. Assume that the Ladyzhenskaya condition holds: for every bounded sequence $\{x_n\} \subset X$ and every sequence $t_n \to \infty$ the sequence $\{S_{t_n} x_n\}$ is relatively compact in X. Then (X, S_t) is asymptotically smooth.*

Proof. Let D be a bounded forward invariant set. It follows from the Ladyzhenskaya condition that $K = \omega(D)$ is a non-empty compact invariant set (see the argument given in [172] or in [61, Chapter 1]). The contradiction argument, applied in the same manner as in the proof of Theorem 1.5.1 [61], leads to (7.1.2). □

We conclude this section with several assertions that give other convenient criteria for asymptotic smoothness of a dynamical system. These are used later on in the context of von Karman systems with nonlinear damping.

The following assertion is a generalization of the results presented in [134] and [42].

7.1.7. Theorem. *Let (X, S_t) be a dynamical system on a Banach space X. Assume that for any bounded positively invariant set B in X there exist $T > 0$, a continuous nondecreasing function $g : \mathbb{R}_+ \mapsto \mathbb{R}_+$, and a pseudometric ρ_B^T on $C(0, T; X)$ such that* *(i)* $g(0) = 0$; $g(s) < s$, $s > 0$.

(ii) *The pseudometric* ρ_B^T *is precompact (with respect to the norm of* X*) in the following sense. Any sequence* $\{x_n\} \subset B$ *has a subsequence* $\{x_{n_k}\}$ *such that the sequence* $\{y_k\} \subset C(0,T;X)$ *of elements* $y_k(\tau) = S_\tau x_{n_k}$ *is Cauchy with respect to* ρ_B^T.

(iii) *The following estimate*

$$\|S_T y_1 - S_T y_2\| \le g\left(\|y_1 - y_2\| + \rho_B^T(\{S_\tau y_1\}, \{S_\tau y_2\})\right), \tag{7.1.6}$$

holds for every $y_1, y_2 \in B$, *where we denote by* $\{S_\tau y_i\}$ *the element in the space* $C(0,T;X)$ *given by function* $y_i(\tau) = S_\tau y_i$.

Then, (X, S_t) *is asymptotically smooth dynamical system.*

7.1.8. Remark. Instead of (7.1.6) one may also assume that

$$\|S_T y_1 - S_T y_2\| \le g(\|y_1 - y_2\|) + \rho_B^T(\{S_\tau y_1\}, \{S_\tau y_2\})$$

(pseudometric outside g). We also note that Theorem 7.1.7 remains valid in the case when X is a complete metric space (see [75, Chapter 2]). ■

Proof. We use the same idea as in [134, Lemma 2.3.6] (see also [69] and Theorem 2.4 in [75]) which is based on Kuratowski's α-measure of noncompactness. The latter is defined by the formula

$$\alpha(B) = \inf\{d : B \text{ has a finite cover of diameter } < d\}$$

on bounded sets of X. The α-measure has the following properties (see, e.g., [134] or [259, Lemma 22.2]).

(i) $\alpha(B) = 0$ if and only if B is precompact.
(ii) $\alpha(A \cup B) \le \max\{\alpha(A), \alpha(B)\}$.
(iii) $\alpha(A + B) \le \alpha(A) + \alpha(B)$.
(iv) $\alpha(\overline{\text{co}}\, B) = \alpha(B)$, where $\alpha(\overline{\text{co}}\, B)$ is the closed convex hull of B.
(v) If $B_1 \supset B_2 \supset B_3 \dots$ are nonempty closed sets in X such that $\alpha(B_n) \to 0$ as $n \to \infty$, then $\cap_{n \ge 1} B_n$ is nonempty and compact.

We first prove that

$$\alpha(S_T B) \le g(\alpha(B)). \tag{7.1.7}$$

For any $\varepsilon > 0$ there exist sets $F_1, \dots, F_n$ such that

$$B = F_1 \cup, \dots, \cup F_n, \quad \operatorname{diam} F_i < \alpha(B) + \varepsilon.$$

It follows from assumption (ii) that there exists a finite set $\mathcal{N} = \{x_i : i = 1, 2 \dots m\} \subset B$ such that for every $y \in B$ there is $x_i \in \mathcal{N}$ with the property $\rho_B^T(\{S_\tau y\}, \{S_\tau x_i\}) \le \varepsilon$. It means that

$$B = \cup_{i=1}^m C_i, \quad C_i = \{y \in B : \rho_B^T(\{S_\tau y\}, \{S_\tau x_i\}) \le \varepsilon\}.$$

Now we use representation

$$B = \cup_{i,j}(C_i \cap F_j) \text{ and } S_T B = \cup_{i,j}(S_T(C_i \cap F_j)). \tag{7.1.8}$$

Because $\operatorname{diam} F_j < \alpha(B) + \varepsilon$ and

$$\rho_B^T(\{S_\tau y_1\},\{S_\tau y_2\}) \le \rho_B^T(\{S_\tau y_1\},\{S_\tau x_i\}) + \rho_B^T(\{S_\tau x_i\},\{S_\tau y_2\})$$

for any $y_1, y_2 \in C_i \cap F_j$, it follows from (7.1.6) that

$$\|S_T y_1 - S_T y_2\| \le g([\alpha(B)+\varepsilon]+2\varepsilon)$$

for any $y_1, y_2 \in C_i \cap F_j$. Thus $\operatorname{diam}(S_T(C_j \cap F_i)) \le g(\alpha(B)+3\varepsilon)$. Therefore using (7.1.8) and the definition of α-measure we obtain (7.1.7).

Because $S_t B \subset B$, the following representation takes place,

$$\omega(B) = \bigcap_{k=1}^{\infty} B_k, \quad B_k \equiv \overline{S_{kT}B},$$

for the ω-limit set $\omega(B)$. It is also clear that $B_k \supset B_{k+1}$ for every k and

$$\alpha(B_k) \le \alpha(S_T B_{k-1}) \le g(\alpha(B_{k-1})) \quad k = 1,2,\ldots. \tag{7.1.9}$$

Because $g(s) < s$, the sequence $\{\alpha(B_k)\}$ is decreasing. Therefore there exists $\alpha_0 = \lim_{k\to\infty} \alpha(B_k)$. From (7.1.9) we have that $\alpha_0 \le g(\alpha_0)$ which is possible only if $\alpha_0 = 0$. Thus by property (v) of the α-measure, $\omega(B)$ is a nonempty compact set. The standard argument (see, e.g., [134, pp. 13,14]) allows us to conclude that $\omega(B)$ attracts B uniformly. Thus (X,S_t) is asymptotically smooth. □

Theorem 7.1.7 implies the following two propositions which are slight generalizations of the results presented in [134] and [42].

7.1.9. Proposition. *Let (X,S_t) be a dynamical system on a Banach space X. Assume that for any bounded positively invariant set B in X and for any $t \ge t_0 = t_0(B) \ge 0$ there exist a function $K_B(t)$ on $[t_0,+\infty)$ and a pseudometric ρ_B^t on $C(0,t;X)$ such that*

(i) $K_B(t) \ge 0$ and $\lim_{t\to\infty} K_B(t) = 0$.

(ii) The pseudometric ρ_B^t is precompact (with respect to the norm of X) in the following sense. Any sequence $\{x_n\} \subset B$ has a subsequence $\{x_{n_k}\}$ such that the sequence $\{y_k\} \subset C(0,t;X)$ of elements $y_k(\tau) = S_\tau x_{n_k}$ is Cauchy with respect to ρ_B^t.

(iii) The estimate

$$\|S_t y_1 - S_t y_2\| \le K_B(t)\cdot\|y_1 - y_2\| + \rho_B^t(\{S_\tau y_1\},\{S_\tau y_2\}), \quad t \ge t_0, \tag{7.1.10}$$

holds for every $y_1, y_2 \in B$, where we denote by $\{S_\tau y_i\}$ the element in $C(0,t;X)$ given by function $y_i(\tau) = S_\tau y_i$.

Then, (X, S_t) is an asymptotically smooth dynamical system.

Proof. We apply Theorem 7.1.7 with $g(s) = K_B(T) \cdot s$, where T is chosen such that $K_B(T) < 1$. □

Proposition 7.1.9 implies the following result which was proved earlier in [42].

7.1.10. Proposition. *Assume that a dynamical system (X, S_t) on a Banach space X possesses the following property. For any bounded positively invariant set B in X there exist functions $C_B(t) \geq 0$ and $K_B(t) \geq 0$ such that $\lim_{t\to\infty} K_B(t) = 0$, a time $t_0 = t_0(B)$, and a precompact pseudometric ρ on X such that*

$$\|S_t y_1 - S_t y_2\| \leq K_B(t) \cdot \|y_1 - y_2\| + C_B(t) \cdot \rho(y_1, y_2), \quad t \geq t_0, \tag{7.1.11}$$

for every $y_1, y_2 \in B$. Then (X, S_t) is an asymptotically smooth dynamical system.

We recall that a pseudometric ρ on a Banach space X is said to be *precompact* (with respect to the norm of X) if any bounded sequence (in the norm) has a subsequence which is Cauchy with respect to ρ.

Proof. It is clear that a pseudometric ρ_B^t defined on $C(0,t;X)$ by the formula

$$\rho_B^t(\{S_\tau y_1\}, \{S_\tau y_2\}) = C_B(t) \cdot \rho(y_1, y_2)$$

satisfies assumptions (ii) and (iii) in Proposition 7.1.9. □

Our next criterion of asymptotic smoothness relies on the idea presented in [160].

7.1.11. Theorem. *Let (X, S_t) be a dynamical system on a complete metric space X endowed with a metric d. Assume that for any bounded positively invariant set B in X and for any $\varepsilon > 0$ there exists $T \equiv T(\varepsilon, B)$ such that*

$$d(S_T y_1, S_T y_2) \leq \varepsilon + \Psi_{\varepsilon,B,T}(y_1, y_2), y_i \in B, \tag{7.1.12}$$

where $\Psi_{\varepsilon,B,T}(y_1, y_2)$ is a functional defined on $B \times B$ such that

$$\liminf_{m\to\infty} \liminf_{n\to\infty} \Psi_{\varepsilon,B,T}(y_n, y_m) = 0 \tag{7.1.13}$$

for every sequence $\{y_n\}$ from B. Then (X, S_t) is an asymptotically smooth dynamical system.

We note that the compactness criterion presented in Theorem 7.1.11 provides more flexibility, with respect to more standard methods such as given in Theorem 7.1.7, by allowing the taking of sequential limits (in n and m) rather than the simultaneous limits. This was an observation made for the first time in [160]. The result stated in Theorem 7.1.11 is an abstract version of Theorem 2 in [160] that can be derived from the arguments given in [160]. For the reader's convenience we give an independent and shorter proof of the same result.

Theorem 7.1.11 follows at once from the following assertion.

7.1.12. Proposition. *Let (X,S_t) be a dynamical system on a complete metric space (X,d). Assume that for any bounded positively invariant set B in X and for any $\varepsilon > 0$ there exists $T \equiv T(\varepsilon,B)$ such that*

$$\liminf_{m\to\infty}\liminf_{n\to\infty} d(S_T y_n, S_T y_m) \le \varepsilon \tag{7.1.14}$$

for every sequence $\{y_n\}$ from B. Then (X,S_t) is an asymptotically smooth dynamical system.

Proof. As in the proof of Theorem 7.1.7 it is sufficient to prove that

$$\lim_{t\to\infty} \alpha(S_t B) = 0,$$

where $\alpha(B)$ is the Kuratowski α-measure of noncompactness.

Because $S_{t_1}B \subset S_{t_2}B$ for $t_1 > t_2$, the function $\alpha(t) \equiv \alpha(S_t B)$ is nonincreasing. Therefore it is sufficient to prove that for any $\varepsilon > 0$ there exists $T > 0$ such that $\alpha(S_T B) \le \varepsilon$. If this is not true, then there is $\varepsilon_0 > 0$ such that $\alpha(S_T B) \ge 3\varepsilon_0$ for all $T > 0$. For this ε_0 we choose T_0 such that (7.1.14) holds. The relation $\alpha(S_{T_0}B) \ge 3\varepsilon_0$ implies that there exists an infinite sequence $\{y_n\}_{n=1}^{\infty}$ such that

$$d(S_{T_0}y_n, S_{T_0}y_m) \ge 2\varepsilon_0 \;\text{ for all } n \ne m,\; n,m = 1,2,\ldots$$

This contradicts (7.1.14). □

7.2 Global attractors

The main objects arising in the analysis of long-time behavior of infinite-dimensional dissipative dynamical systems are attractors. Their study allows us to answer a number of fundamental questions on the properties of limit regimes that can arise in the systems under consideration. At present, there are several general approaches and methods that allow us to prove the existence and finite-dimensionality of global attractors for a large class of dynamical systems generated by nonlinear partial differential equations (see, e.g., [17, 61, 134, 172, 273] and the references listed therein). Below we review some of them, with particular emphasis on methods applicable to second-order evolutions.

7.2.1. Definition. A bounded closed set $A \subset X$ is said to be a *global attractor* of the dynamical system (X,S_t) iff the following properties hold.

(i) A is an invariant set; that is, $S_t A = A$ for $t \ge 0$.
(ii) A is uniformly attracting; that is, for all bounded set $D \subset X$

$$\lim_{t\to+\infty} d_X\{S_t D | A\} = 0, \tag{7.2.1}$$

where $d_X\{A|B\} = \sup_{x\in A} \mathrm{dist}_X(x,B)$ is the Hausdorff semidistance.

■

The main result on the existence of global attractor is the following one.

7.2.2. Theorem. *Let (X,S_t) be an asymptotically compact dynamical system in a Banach space X with an attracting compact set K. Then (X,S_t) possesses a unique compact global attractor A such that $A \subset K$. This attractor is a connected set and has the form*

$$A = \omega(K) = \bigcap_{t>0} \overline{\bigcup_{\tau\geq t} S_\tau K}. \tag{7.2.2}$$

We also have the relation

$$A = \bigcap_{n\geq N} S_n K \ \ \textit{for all}\ N \in \mathbb{Z}_+. \tag{7.2.3}$$

Moreover, (i) a full trajectory $\gamma = \{u(t) : t \in \mathbb{R}\}$ belongs to the attractor if and only if γ is a bounded set and (ii) for any $x \in A$ there exist a full trajectory $\gamma = \{u(t) : t \subset \mathbb{R}\}$ such that $u(0) = x$ and $\gamma \subset A$. Thus the global attractor can be described as a set of all bounded full trajectories.

If B is a bounded absorbing set for (X,S_t), then $A = \omega(B)$ and

$$\lim_{t\to+\infty} \left(d_X\{S_t B\,|\,A\} + d_X\{A\,|\,S_t B\}\right) = 0, \tag{7.2.4}$$

where $d_X\{A|B\} = \sup_{x\in A} \mathrm{dist}_X(x,B)$. Thus A attracts bounded absorbing sets in the Hausdorff metric.

For the proof we refer to [17, 61, 273].

We note that property (7.2.4) means that for any $\varepsilon > 0$ and for any absorbing set B there exists $t_\varepsilon > 0$ such that $S_t B \subset \mathcal{O}_\varepsilon(A)$ and $A \subset \mathcal{O}_\varepsilon(S_t B)$ for all $t \geq t_\varepsilon$. Here $\mathcal{O}_\varepsilon(D)$ denotes the ε-vicinity of the set D.

Due to Proposition 7.1.4 Theorem 7.2.2 can restated in the following form.

7.2.3. Theorem. *Any dissipative asymptotically smooth dynamical system (X,S_t) in a Banach space X possesses a unique compact global attractor A. This attractor is a connected set and can be described as a set of all bounded full trajectories. Moreover $A = \omega(B)$ for any bounded absorbing set B of (X,S_t) and relation (7.2.4) holds.*

It is clear that if a system possesses a compact global attractor, then it is asymptotically compact. Thus Theorem 7.2.2 implies that (X,S_t) has a compact global attractor if and only if it is asymptotically compact. By Theorem 7.2.3 we also have that a dissipative system possesses a compact global attractor if and only if it is asymptotically smooth.

We can also prove the following assertion (see, e.g., [134, Theorem 2.4.6] or [246, Theorem 2.26]).

7.2.4. Theorem. *Let (X, S_t) be a dynamical system in a complete metric space X. Then (X, S_t) possesses a compact global attractor A if and only if*

(i) (X, S_t) is asymptotically smooth.
(ii) (X, S_t) is point dissipative.
(iii) For every bounded set $B \subset X$ there exist t_0 such that the tail $\gamma_B^{t_0}$ is bounded.

The attractor A has the representation

$$A = \bigcup \{\omega(B) \; : \; B \text{ is a bounded subset of } X\}. \tag{7.2.5}$$

Thus the properties (i)–(iii) above are equivalent to asymptotic compactness of the system.

This theorem and also Proposition 7.1.5 imply the following assertion.

7.2.5. Corollary. *Assume that (X, S_t) is a dissipative dynamical system in a Banach space X. Assume that for each $t > 0$ large enough the evolution operator S_t admits a decomposition $S_t = S_t^{(1)} + S_t^{(2)}$, where $S_t^{(2)}$ is a continuous mapping in X satisfying* (7.1.5) *and $S_t^{(1)}$ is compact; that is, $S_t^{(1)}B$ is a precompact set in X for all $t > 0$ large enough and for every bounded set $B \subset X$. Then (X, S_t) possesses a compact global attractor.*

In order to describe stability properties of attractors we need the following notions. A positively invariant set M is said to be *stable (in the Lyapunov sense)* iff for any vicinity $\mathcal{O}$ of M there exists a vicinity $M \subset \mathcal{O}' \subset \mathcal{O}$ such that $S_t\mathcal{O}' \subset \mathcal{O}$ for all $t \geq 0$. The set M is *asymptotically stable* iff it is stable and $S_t x \to M$ as $t \to \infty$ for every $x \in \mathcal{O}'$. This set is *uniformly asymptotically stable* iff it is stable and

$$\lim_{t \to +\infty} \sup_{x \in \mathcal{O}'} \operatorname{dist}_X(S_t x, M) = 0.$$

We have the following property of compact global attractors (see [17, 61]).

7.2.6. Theorem. *Let (X, S_t) be a dynamical system in a Banach space X possessing a compact global attractor A. Assume that there exist a bounded vicinity $\mathcal{O}$ of A such that the mapping $(t, x) \mapsto S_t x$ is continuous on $\mathbb{R}_+ \times \mathcal{O}$. Then A is uniformly asymptotically stable.*

We also have the following *reduction principle* (for the proof we refer to [61], for example).

7.2.7. Theorem. *Let (X, S_t) be a dissipative dynamical system in a Banach space X. Assume that there exists a positively invariant locally compact*[1] *set M possessing the property of uniform attraction: for any bounded set D we have*

$$\lim_{t \to +\infty} \sup_{x \in D} \operatorname{dist}_X(S_t x, M) = 0.$$

If A is a global attractor of the restriction (M, S_t) of the system (X, S_t) on M, then A is also a global attractor for (X, S_t).

---

[1] In the sense that every bounded subset of that set is relatively compact.

In certain situations this theorem enables us to decrease significantly the dimension of the phase space, which is important in the study of infinite-dimensional systems.

We next consider stability of attractors with respect to perturbations of a dynamical system. Assume that we have a family of dynamical systems (X, S_t^λ) with the same phase space X and with evolutionary operator S_t^λ depending on a parameter λ from a complete metric space Λ. The following assertion is proved by Kapitansky and Kostin [157] (see also [134] and [61]).

7.2.8. Theorem. *Assume that a dynamical system (X, S_t^λ) in a Banach space X possesses a compact global attractor A^λ for every $\lambda \in \Lambda$. Assume that the following conditions hold.*

(i) There exists a compact $K \subset X$ such that $A^\lambda \subset K$.
(ii) If $\lambda_k \to \lambda_0$, $x_k \to x_0$ and $x_k \in A^{\lambda_k}$, then $S_\tau^{\lambda_k} x_k \to S_\tau^{\lambda_0} x_0$ for some $\tau > 0$.

Then the family $\{A^\lambda\}$ of attractors is upper semicontinuous at the point λ_0; that is,

$$d_X\left\{A^{\lambda_k} \,|\, A^{\lambda_0}\right\} \equiv \sup\left\{\operatorname{dist}_X(x, A^{\lambda_0}) \,:\, x \in A^{\lambda_k}\right\} \to 0 \text{ as } \lambda_k \to \lambda_0. \qquad (7.2.6)$$

Moreover, the upper limit $A(\lambda_0, \Lambda)$ of the attractors A^λ at λ_0 defined by the formula

$$A(\lambda_0, \Lambda) = \bigcap_{\delta>0} \overline{\bigcup \{A^\lambda \,:\, \lambda \in \Lambda,\ 0 < \operatorname{dist}(\lambda, \lambda_0) < \delta\}}$$

is a nonempty compact invariant set lying in the attractor A^{λ_0}.

The situation with the continuity of attractors A^λ with respect to λ is more complicated. In general the family $\{A^\lambda\}$ is not lower semicontinuous at the point λ_0; that is, the property $d_X\left\{A^{\lambda_0} \,|\, A^{\lambda_k}\right\} \to 0$ as $\lambda_k \to \lambda_0$ does not hold. In order to prove lower semicontinuity under hypotheses of Theorem 7.2.8 some additional assumptions should be imposed (see [17]). However, the lower semicontinuity property is generic under simple compactness assumptions (see the discussion in the survey [246]).

The following concept is useful to describe long-time behavior of individual trajectories (see [172]).

7.2.9. Definition. A bounded closed set $A_{\min} \subset X$ is said to be a *global minimal attractor* of the dynamical system (X, S_t) iff the following properties hold.

(i) $A_{\min}$ is a positively invariant set, that is, $S_t A_{\min} \subseteq A_{\min}$ for $t \geq 0$;
(ii) $A_{\min}$ attracts every point x from X; that is,

$$\lim_{t \to +\infty} \operatorname{dist}_X(S_t x, A_{\min}) = 0 \text{ for any } x \in X;$$

(iii) $A_{\min}$ is minimal; that is, $A_{\min}$ has no proper subsets possessing (i) and (ii).

■

It is clear that if the system (X, S_t) admits global attractor A, then a global minimal attractor $A_{\min}$ also exists and $A_{\min} \subset A$.

Similar to Theorem 7.2.4 one can prove the following assertion.

7.2.10. Theorem. *Let (X,S_t) be a dynamical system in a Banach space X Then (X,S_t) possesses a global minimal attractor $A_{\min}$ if (X,S_t) is point dissipative and for every $x \in X$ there exist t_0 such that the tail $\gamma_x^{t_0}$ is relatively compact. Moreover the attractor $A_{\min}$ has the representation*

$$A_{\min} = \overline{\bigcup \{\omega(x) \,:\, x \in X\}}. \tag{7.2.7}$$

The most restrictive assumption guaranteeing the existence of a global attractor is asymptotic compactness of the corresponding dynamical system (see Theorem 7.2.2). However, this assumption can be circumvented if one limits considerations to weak topology. Thus, it is convenient to consider the property of uniform convergence (7.2.1) in a weak topology. This leads to the following definition.

7.2.11. Definition. Let X be a separable reflexive Banach space. A bounded weakly closed set $A \subset X$ is said to be a *weak global attractor* of the dynamical system (X,S_t) iff the following properties hold.

(i) A is an invariant set; that is, $S_t A = A$ for $t \geq 0$.
(ii) A is uniformly attracting in weak topology; that is, for any weak vicinity $\mathscr{O}$ of A and for bounded set $D \subset X$ there exists $t_0(D,\mathscr{O})$ such that $S_t D \subset \mathscr{O}$ for all $t \geq t_0(D,\mathscr{O})$.

■

It is clear that a global attractor, if it exists, is also weak. However, the following result on the existence of weak global attractors shows that the hypotheses of asymptotic compactness can be replaced by a weaker condition of weak closeability. Of course, the conclusion obtained is likewise weaker, strong attraction being replaced by weak attraction.

7.2.12. Theorem. *Let (X,S_t) be a dissipative dynamical system in a separable reflexive Banach space X with an absorbing set B. Assume that evolution operator S_t is weakly closed: for all $t > 0$ the weak convergences $u_n \rightharpoonup u$ and $S_t u_n \rightharpoonup v$ in X imply that $S_t u = v$. Then (X,S_t) possesses a unique weak global attractor A such that $A \subset B$. This attractor has the form*

$$A = \omega_w(B) = \bigcap_{t>0} \overline{\bigcup_{\tau \geq t} S_\tau B}^{\,w}, \tag{7.2.8}$$

where $\overline{D}^w$ denotes the closure of the set D in weak topology.

The proof relies on weak compactness of bounded sets in a separable reflexive Banach space. For the details we refer to [17] or [61].

We also note that sometimes it is convenient to use not only strong or weak convergences but also other topologies in the definition of global attractors. We refer to [17] for the theory of attractors involving two phase spaces with different topologies.

7.3 Dimension of global attractors

Finite-dimensionality is an important property of global attractors that can be established for many dynamical systems, including those arising in significant applications. There are several approaches that provide effective estimates for the dimension of attractors of dynamical systems generated by PDEs (see, e.g., [17, 172, 273]). Here we present two approaches that do not require C^1-smoothness of the evolutionary operator (as in [17, 273]). The reason for this focus is that dynamical systems of hyperbolic-like nature do not display smoothing effects, unlike parabolic equations. Therefore, the C^1 smoothness of the flows is most often beyond question, particularly in problems with a nonlinear dissipation. Instead, we consider more general locally Lipschitz flows, which can be studied by methods introduced by Ladyzhenskaya's theorem (see, e.g., [172]) on finite dimensionality of invariant sets. Another successful approach discussed in this section is related to squeezing property in the formulation considered in [244]. However, we wish to point out that the estimates of the dimension based on the theorems below usually tend to be conservative.

Fractal and Hausdorff dimensions are the most commonly used measures in the theory of infinite-dimensional dynamical systems.

7.3.1. Definition. Let M be a compact set in a metric space X.

- The *fractal (box-counting) dimension* $\dim_f M$ of M is defined by

$$\dim_f M = \limsup_{\varepsilon \to 0} \frac{\ln n(M,\varepsilon)}{\ln(1/\varepsilon)},$$

  where $n(M,\varepsilon)$ is the minimal number of closed balls of the radius ε which cover the set M.
- For positive d we define *d-dimensional Hausdorff measure* by the formula

$$\mu(M,d) = \lim_{\varepsilon \to 0} \mu(M,d,\varepsilon),$$

  where

$$\mu(M,d,\varepsilon) = \inf\left\{ \sum_j (r_j)^d \, : \, M \subset \bigcup_i B(x_j, r_j), \; r_j \le \varepsilon \right\}.$$

  Here $B(x_j, r_j)$ is the ball in X with center x_j and radius r_j. The *Hausdorff dimension* $\dim_H M$ of M is defined by the formula $\dim_H M = \inf\{d \, : \, \mu(M,d) = 0\}$.

■

One can show that the Hausdorff dimension does not exceed the fractal one:

$$\dim_H M \le \dim_f M. \tag{7.3.1}$$

Simple calculations show that (i) $\dim_H M = 0$ if $M = \{a_n\}$ is a sequence in $\mathbb{R}$, (ii) $\dim_f M = 1/2$ if $M = \{1/n\}$, and (iii) $\dim_H M = \dim_f M = \ln 2/\ln 3$ when M is the Cantor set obtained from the interval $[0,1]$ by sequential removal of the central thirds. We refer to [107] for details and for other properties of Hausdorff and fractal (box-counting) dimension.

We choose to deal with the fractal dimension of attractors for the following reasons: (i) fractal dimension is more convenient in calculations and (ii) it estimates the Hausdorff dimension from above (see (7.3.1)). The importance of the notion of finite fractal dimension is also illustrated by the following property (see [113]): if M compact set in X such that $\dim_f M < n/2$ for some $n \in \mathbb{N}$, then there exists an injective Lipschitz mapping $L : M \mapsto \mathbb{R}^n$ such that its inverse is Hölder continuous. This means that M can be placed in the graph of Hölder continuous mapping which maps a compact subset of $\mathbb{R}^n$ onto M.

We also note that the fractal dimension $\dim_f M$ of a compact set M can be represented by the formula

$$\dim_f M = \limsup_{\varepsilon \to 0} \frac{\ln N(M,\varepsilon)}{\ln(1/\varepsilon)}, \tag{7.3.2}$$

where $N(M,\varepsilon)$ is the minimal number of closed sets of the diameter 2ε that cover M.

The following version of Ladyzhenskaya's theorem has been proved in [61].

7.3.2. Theorem (Ladyzhenskaya). *Let M be a compact set in a Hilbert space H. Assume that V is a continuous mapping in H such that $V(M) \supseteq M$ and there exists a finite-dimensional projector P in H such that*

$$\|P(Vv_1 - Vv_2)\| \le l\|v_1 - v_2\|, \quad v_1, v_2 \in M, \tag{7.3.3}$$

and

$$\|(I-P)(Vv_1 - Vv_2)\| \le \delta\|v_1 - v_2\|, \quad v_1, v_2 \in M, \tag{7.3.4}$$

where $\delta < 1$. Then the fractal dimension $\dim_f M$ is finite and the estimate

$$\dim_f M \le \dim P \cdot \ln\frac{9l}{1-\delta}\left[\ln\frac{2}{1+\delta}\right]^{-1} \tag{7.3.5}$$

holds, provided that $l \ge 1-\delta$ (if $l < 1-\delta$ then M is a single-point set).

Roughly speaking, the assumptions (7.3.3) and (7.3.4) mean that the mapping V squeezes the set M along the space $(I-P)H$ although it does not stretch M too much along PH. The negative invariance of M gives us that $M \subseteq V^k M$ for all $k \in \mathbb{N}$. Thus the set M must be initially squeezed. This property is expressed by the assertion on finite dimensionality of M.

In problems with nonlinear damping the following criterion turns out to be useful (see [63, 68, 69], for instance).

7.3.3. Theorem. *Let H be a separable Hilbert space and M be a bounded closed set in H. Assume that there exists a mapping $V : M \mapsto H$ such that*

(i) $M \subseteq VM$.

(ii) V is Lipschitz on M; that is, there exists $L > 0$ *such that*

$$\|Vv_1 - Vv_2\| \le L\|v_1 - v_2\|, \quad v_1, v_2 \in M. \tag{7.3.6}$$

(iii) There exist compact seminorms $n_1(x)$ *and* $n_2(x)$ *on H such that*

$$\|Vv_1 - Vv_2\| \le \eta\|v_1 - v_2\| + K \cdot [n_1(v_1 - v_2) + n_2(Vv_1 - Vv_2)] \tag{7.3.7}$$

for any $v_1, v_2 \in M$*, where* $0 < \eta < 1$ *and* $K > 0$ *are constants (a seminorm* $n(x)$ *on H is said to be compact iff* $n(x_m) \to 0$ *for any sequence* $\{x_m\} \subset H$ *such that* $x_m \to 0$ *weakly in H).*

Then M is a compact set in H of a finite fractal dimension. Moreover, if the seminorms n_1 *and* n_2 *have the form* $n_i(v) = \|P_i v\|$*,* $i = 1, 2$*, where* P_1 *and* P_2 *are finite-dimensional orthoprojectors, then*

$$\dim_f M < (\dim P_1 + \dim P_2) \cdot \ln\left(1 + \frac{8(1+L)\sqrt{2}K}{1-\eta}\right) \cdot \left[\ln \frac{2}{1+\eta}\right]^{-1}. \tag{7.3.8}$$

7.3.4. Remark. We note (see [75]) that in the general case the estimate for the dimension has the form

$$\dim_f M \le \left[\ln \frac{2}{1+\eta}\right]^{-1} \cdot \ln m_0\left(\frac{4K(1+L^2)^{1/2}}{1-\eta}\right), \tag{7.3.9}$$

where $m_0(R)$ is the maximal number of pairs (x_i, y_i) in $H \times H$ possessing the properties

$$\|x_i\|^2 + \|y_i\|^2 \le R^2, \quad n_1(x_i - x_j) + n_2(y_i - y_j) > 1, \quad i \ne j.$$

We also note that a more general statement (for metric spaces) and some other versions and corollaries of Theorem 7.3.3 can be found in [75]. We also refer to [70] for an analysis of the dimension problem in the case of nonlinear relations of the type (7.3.7). ■

In the proof of Theorem 7.3.3 we follow the line of the argument given in [69] and rely on the following lemmas.

7.3.5. Lemma. *Assume that* $V : M \mapsto H$ *is a mapping such that* (7.3.6) *and* (7.3.7) *with some* $\eta > 0$ *hold. Then*

$$\alpha(VB) \le \eta \cdot \alpha(B) \quad \text{for any } B \subseteq M, \tag{7.3.10}$$

where $\alpha(B)$ *is the Kuratowski* α*-measure of noncompactness of the set B (for the definition see the proof of Theorem* 7.1.7*). Thus V is an* α*-contraction in the sense of* [134, p. 14].

Proof. We use the same idea as in Theorem 7.1.7. By the definition of $\alpha(B)$ (see the proof of Theorem 7.1.7), for any $\varepsilon > 0$ there exist sets $F_1, \ldots, F_n$ such that

$$B = F_1 \cup, \dots, \cup F_n, \quad \operatorname{diam} F_i < \alpha(B) + \varepsilon.$$

Let $\mathscr{N} = \{x_i : i = 1,2\dots m\} \subset B$ be a finite set such that for every $y \in B$ there is $i \in \{1,2,\dots,m\}$ with the property $n_1(y - x_i) + n_2(Vy - Vx_i) \le \varepsilon$. If there is no such set for some $\varepsilon > 0$, then there exists a sequence $\{z_n\} \subset B$ such that

$$n_1(z_n - z_m) + n_2(Vz_n - Vz_m) \ge \varepsilon \text{ for all } n \ne m. \tag{7.3.11}$$

We can assume that $z_n \to z$ and $Vz_n \to w$ weakly in H for some $z, w \in H$. Because the seminorms n_1 and n_2 are compact, this implies that $n_1(z_n - z) + n_2(Vz_n - w) \to 0$ when $n \to \infty$, which is impossible due to (7.3.11). Thus such a finite set $\mathscr{N}$ exists and

$$B = \cup_{i=1}^m C_i, \quad C_i = \{y \in B : n_1(y - x_i) + n_2(Vy - Vx_i) \le \varepsilon\}, \quad x_i \in \mathscr{N}.$$

By exploiting the representations $B = \cup_{i,j}(C_i \cap F_j)$ and $VB = \cup_{i,j}(V(C_i \cap F_j))$ one can see from (7.3.7) that $\operatorname{diam}(V(C_j \cap F_i)) \le \eta \cdot \alpha(B) + \varepsilon \cdot [2K + \eta]$. This implies (7.3.10). □

7.3.6. Lemma. *For any $\varepsilon > 0$ there exist a constant K_ε and finite-dimensional orthoprojectors P_1^ε and P_2^ε in H such that*

$$n_1(v) \le \varepsilon \|v\| + K_\varepsilon \|P_1^\varepsilon v\|, \quad v \in H, \tag{7.3.12}$$

and

$$n_2(Vv_1 - Vv_2) \le \varepsilon \|v_1 - v_2\| + K_\varepsilon \|P_2^\varepsilon(Vv_1 - Vv_2)\|, \quad v_1, v_2 \in M. \tag{7.3.13}$$

Proof. Assume that (7.3.12) is not true. Then there exists $\varepsilon_0 > 0$ and a sequence of orthoprojectors $\{P_m\}$ such that $P_m \to I$ strongly in H and

$$n_1(v_m) \ge \varepsilon_0 + c_m \|P_m v_m\|, \quad m = 1,2,\dots, \tag{7.3.14}$$

for some sequence $\{v_m\} \subset H$ with the property $\|v_m\| = 1$, where $c_m \to \infty$ as $m \to \infty$. From (7.3.14) we have that $\|P_m v_m\| \to 0$ as $m \to \infty$. We can also assume that $v_m \to v$ weakly in H for some $v \in H$. Because

$$P_m v = P_m(v - v_m) + P_m v_m \to 0 \text{ weakly in } H,$$

we conclude that $v = 0$. This implies that $n_1(v_m) \to 0$ as $m \to \infty$ which contradicts (7.3.14). Thus (7.3.12) holds.

To prove (7.3.13) we note that the same argument implies that

$$n_2(Vv_1 - Vv_2) \le \frac{\varepsilon}{L} \|Vv_1 - Vv_2\| + K_\varepsilon \|P_2^\varepsilon(Vv_1 - Vv_2)\|, \quad v_1, v_2 \in H.$$

Therefore using Lipschitz property (ii) we obtain (7.3.13). □

7.3.7. Remark. In most applications, the seminorms n_1 and n_2 are generated by norms of appropriate Sobolev spaces, say H_1 and H_2, such that injection $H \subset H_i, i = 1,2$ are compact and such that $|v|_{H_i} \le K_i\|v\|$, $i = 1,2$. In this case, the contradiction argument used in Lemma 7.3.6 can be avoided and explicit estimates for the constants K_ε can be given. Indeed, let ε be a given constant and $P_1^\varepsilon, P_2^\varepsilon$ denote orthonormal projections on H, with the properties $|I - P_i^\varepsilon|_{\mathscr{L}(H,H_i)} \le \varepsilon$ for $i = 1,2$. Then

$$n_1(v) = |v|_{H_1} \le |(I - P_1^\varepsilon)v|_{H_1} + |P_1^\varepsilon v|_{H_1} \le \varepsilon\|v\| + K_1\|P_1^\varepsilon v\|, \quad v \in M,$$

and similarly for $n_2(Vv_1 - Vv_2)$. Thus, in this case the constant K_ε does not depend on ε and is equal to $\max\{K_1, K_2\}$, where K_i denote norms of the canonical isomorphism $H \subset H_i$. ■

7.3.8. Lemma. *Let F be a closed bounded set in $\mathbb{R}^d$ equipped with Euclidean norm $|\cdot|$. Assume that $\operatorname{diam} F \le 2r$ for some $r > 0$. Then for any $\alpha > 0$ there exists a finite set $\{x_k : k = 1,\dots n_\alpha\} \subset F$ such that*

$$F \subset \bigcup_{k=1}^{n_\alpha} \{x \in \mathbb{R}^d : |x - x_k| \le \alpha r\} \text{ and } n_\alpha \le \left(1 + \frac{4}{\alpha}\right)^d.$$

Proof. Because F is compact in $\mathbb{R}^d$ there exists a set $\{x_k : k = 1,\dots,n_\alpha\} \subset F$ such that (i) for any $y \in F$ we can find x_i such that $|y - x_i| < \alpha r$ and (ii) $|x_j - x_i| \ge \alpha r$ for any $i \ne j$. Thus we need only prove the estimate for n_α. Consider the balls

$$B_k = \left\{x \in \mathbb{R}^d : |x - x_k| < \frac{\alpha r}{2}\right\}, \quad k = 1,\dots,n_\alpha.$$

These balls possess the properties $B_k \cap B_j = \emptyset$ for $k \ne j$, $k, j = 1,\dots n_\alpha$, and

$$B_k \subset \widetilde{B} \equiv \left\{x \in \mathbb{R}^d : |x - y_0| \le 2r + \frac{\alpha r}{2}\right\}, \quad k = 1,\dots,n_\alpha,$$

where y_0 is a point from F. Therefore $n_\alpha \cdot \mathrm{Vol}\,(B_1) = \sum_{k=1}^{n_\alpha} \mathrm{Vol}(B_k) \le \mathrm{Vol}(\widetilde{B})$. This implies the estimate for n_α. □

Proof of Theorem 7.3.3. Lemma 7.3.5 implies that $\alpha(M) \le \eta \cdot \alpha(M)$. Since $0 < \eta < 1$, this is possible only if $\alpha(M) = 0$. Thus M is compact.

Lemma 7.3.6 allows us to rewrite (7.3.7) in the following form

$$\|Vv_1 - Vv_2\| \le (\eta + \delta)\|v_1 - v_2\| + \widetilde{K}\left(\|P_1(v_1 - v_2)\|^2 + \|P_2(Vv_1 - Vv_2)\|^2\right)^{1/2} \tag{7.3.15}$$

for every $v_1, v_2 \in M$ and any $\delta > 0$, where the constant $\widetilde{K}$ and the orthoprojectors P_1 and P_2 may depend on δ.

Assume that $\{F_i : i = 1,\dots,N(M,\varepsilon)\}$ is the minimal covering of M by its closed subsets with a diameter less than 2ε. Let

$$\widetilde{F}_i = \{(P_1 y; P_2 V y) : y \in F_i\} \subset \widetilde{H} \equiv P_1 H \times P_2 H.$$

It is easy to see that $\operatorname{diam}\widetilde{F}_i \le (1+L)\operatorname{diam} F_i \le 2\varepsilon(1+L)$. Because $\widetilde{H}$ is finite-dimensional, the application of Lemma 7.3.8 to the set $\widetilde{F}_i$ with the parameter α replaced by $\alpha/(1+L)$ gives that there exists a finite set $\{x_k^i : k=1,\dots,n_{\alpha,i}\} \subset F_i$ such that

$$F_i \subset \bigcup_{k=1}^{n_{\alpha,i}} B_k^i,\ \ B_k^i \equiv \left\{v\in F_i : \|P_1(v-x_k^i)\|^2 + \|P_2(Vv - Vx_k^i)\|^2 \le (\alpha\varepsilon)^2\right\}$$

and

$$n_{\alpha,i} \le \left(1+\frac{4(1+L)}{\alpha}\right)^d,\quad d = d(\delta) = \dim P_1 + \dim P_2,\ \ i=1,\dots,N(M,\varepsilon).$$

If $y_1, y_2 \in B_k^i$, then from (7.3.15) we have

$$\|Vy_1 - Vy_2\| \le (\eta+\delta)\cdot 2\varepsilon + \widetilde{K}\cdot 2\alpha\varepsilon = 2\widetilde{\eta}\varepsilon,$$

where $\widetilde{\eta} = \widetilde{\eta}(\delta,\alpha) \equiv \eta + \delta + \alpha\widetilde{K}$. Thus $\operatorname{diam} VB_k^i \le 2\widetilde{\eta}\varepsilon$. Inasmuch as

$$M \subset VM \subset \bigcup_{i=1}^{N(M,\varepsilon)} \bigcup_{k=1}^{n_{\alpha,i}} VB_k^i,$$

we have that

$$N(M,\widetilde{\eta}\varepsilon) \le \left(1+\frac{4(1+L)}{\alpha}\right)^d \cdot N(M,\varepsilon).$$

Because $\eta < 1$, we can choose δ and α such that $\widetilde{\eta} < 1$ and after applying iteratively the above inequality with $\varepsilon = 1, \varepsilon = \widetilde{\eta}, \dots, \varepsilon = \widetilde{\eta}^{n-1}$ we obtain the inequality

$$N(M,\widetilde{\eta}^n) \le \left(1+\frac{4(1+L)}{\alpha}\right)^{dn} \cdot N(M,1),\quad n\in\mathbb{N}.$$

Now, for every $0<\varepsilon<1$ there exists $n(\varepsilon)$ such that $\widetilde{\eta}^n \le \varepsilon < \widetilde{\eta}^{n-1}$. Thus

$$N(M,\varepsilon) \le \left(1+\frac{4(1+L)}{\alpha}\right)^{dn(\varepsilon)} \cdot N(M,1)\ \text{ with } n(\varepsilon) \le 1 + \frac{\ln\varepsilon}{\ln\widetilde{\eta}}.$$

Finally, estimating the fractal dimension from the formula in (7.3.2) yields (see also [61, Section 1.8])

$$\dim_f M \le d\cdot\ln\left(1+\frac{4(1+L)}{\alpha}\right)\cdot\left[\ln\frac{1}{\eta+\delta+\alpha\widetilde{K}}\right]^{-1}, \tag{7.3.16}$$

where the parameters δ and α are chosen such that $\eta+\delta+\alpha\widetilde{K} < 1$ and $d \equiv \dim P_1 + \dim P_2$ depends on δ. Thus $\dim_f M < \infty$. Moreover, relation (7.3.16) implies estimate (7.3.8). The point is that in this case we can use (7.3.16) with $\delta = 0$ and $\widetilde{K} = \sqrt{2}K$ and if we choose $\alpha = (1-\eta)/(2\sqrt{2}K)$, we obtain (7.3.8). □

7.4 Fractal exponential attractors (inertial sets)

Theorems 7.3.2 and 7.3.3 pertain to negatively invariant sets M; this is to say $M \subseteq V(M)$. As for the positively invariant set, the finite-dimensionality is not guaranteed by conditions (7.3.3) and (7.3.4). However, as the following theorem states, they are attracted by finite-dimensional compacts with an exponential rate (for the proof we refer to [61]).

7.4.1. Theorem. *Let V be a continuous mapping defined on a compact set M of a Hilbert space H such that $V(M) \subseteq M$. Assume that there exists a finite-dimensional projector P in H such that* (7.3.3) *and* (7.3.4) *hold with $\delta < \frac{1}{2}$. Then for any $\theta \in (\delta, 1)$ there exists a positively invariant closed set $A_\theta \subset M$ such that*

$$\sup\left\{\operatorname{dist}(V^k u, A_\theta) \, : \, u \in M\right\} \le r\theta^k, \quad k = 1, 2, \ldots, \tag{7.4.1}$$

for some constant $r > 0$ and $\dim_f A_\theta \le C(l, \delta, \theta) \cdot \dim P$ where $C(l, \delta, \theta)$ is a constant (some bounds for $C(l, \delta, \theta)$ can be found in [61, Section 1.8]*).*

Note that the condition (7.4.1) means an exponential rate of attraction.

Combining the idea of the proof of Theorem 7.4.1 (see [61, Section 1.8]) with the method presented in the proof of Theorem 7.3.3 we can also obtain the following assertion which is a version of the result proved in [75] for metric spaces.

7.4.2. Theorem. *Let $V : M \mapsto M$ be a mapping defined on a closed bounded set M of a separable Hilbert space H. Assume that* (7.3.6) *holds and there exist compact seminorms n_1 and n_2 on H such that*

$$\|Vv_1 - Vv_2\| \le \eta\|v_1 - v_2\| + K \cdot [n_1(v_1 - v_2) + n_2(Vv_1 - Vv_2)] \tag{7.4.2}$$

holds for any $v_1, v_2 \in M$, where $0 < \eta < 1$ and $K > 0$ are constants. Then for any $\theta \in (\eta, 1)$ there exists a positively invariant compact set $A_\theta \subset M$ of finite fractal dimension satisfying (7.4.1)*. If $n_1(x) = \|P_1 x\|$ and $n_2(x) = \|P_2 x\|$, where P_1 and P_2 are finite-dimensional projectors in H, then*

$$\dim_f A_\theta \le \ln\left(1 + \frac{4\sqrt{2}(1+L)K}{\theta - \eta}\right)\left[\ln\frac{1}{\theta}\right]^{-1}(\dim P_1 + \dim P_2). \tag{7.4.3}$$

Proof. By Lemma 7.3.6 it is sufficient to consider the case when $n_1(x) = \|P_1 x\|$ and $n_2(x) = \|P_2 x\|$.

It follows from Lemma 7.3.5 that V is an α-contraction. Therefore, the set $M_0 = \cap_{n\ge 1} V^n M$ is a compact global attractor for the discrete dynamical system (M, V^k).

By Theorem 7.3.3, $\dim_f M_0 < \infty$. We construct a set A_θ as an extension of M_0. This construction relies on the following assertion.

7.4.3. Lemma. *Assume that $\theta > \eta$. Then there exists a collection of finite sets $\{E_m\}_{m=0}^\infty$ possessing the properties:*

(i) $E_m \subset V^m M$ for every $m = 0, 1, \dots,$ and

$$V^m M \subset \bigcup_{v \in E_m} \left(V^m M \cap B_{2\theta^m}(v)\right), \quad m = 0, 1, \dots, \tag{7.4.4}$$

where $B_\rho(v) = \{w \in H : \|w - v\| \le \rho\}$ is a ball with the center at v.
(ii) There exists a constant $N_0 > 0$ such that for every $m \ge 0$ we have

$$\operatorname{Card} E_m \le N(V^m M, \theta^m) \le N_0 \left(1 + \frac{4\sqrt{2}(1+L)K}{\theta - \eta}\right)^{md}, \tag{7.4.5}$$

where $d = \dim P_1 + \dim P_1$ and, as above, $N(M, \varepsilon)$ is the minimal number of closed sets of the diameter 2ε which cover M.

Proof. Let $E_m = \{a_i^m : i = 1, \dots, N_m\}$ be a maximal set in $V^m M$ possessing the property $\|a_i^m - a_j^m\| > 2\theta^m$, $i \ne j$. Then it is clear that (7.4.4) holds and hence we need only to establish relation (7.4.5).

The inequality

$$N_m \equiv \operatorname{Card} E_m \le N(V^m M, \theta^m), \quad m = 0, 1, \dots, \tag{7.4.6}$$

follows from the fact that two different elements from E_m cannot belong to the same set of the diameter $2\theta^m$.

By the same argument as in the proof of Theorem 7.3.3 we can conclude that for any $M^* \subseteq M$ we have that

$$N(VM^*, (\eta + \alpha\sqrt{2}K)\varepsilon) \le \left(1 + \frac{4(1+L)}{\alpha}\right)^d \cdot N(M^*, \varepsilon)$$

for every $\alpha > 0$. Therefore taking $M^* = V^{m-1}M$, $\alpha = (\theta - \eta)/(\sqrt{2}K)$ and $\varepsilon = \theta^{m-1}$ we obtain that

$$N(V^m M, \theta^m) \le \left(1 + \frac{4\sqrt{2}(1+L)K}{\theta - \eta}\right)^d \cdot N(V^{m-1}M, \theta^{m-1}), \quad m = 1, 2, \dots$$

By (7.4.6) this implies (7.4.5). □

Completion of the proof of Theorem 7.4.2. We prove that the set

$$A_\theta = M_0 \cup \{V^k E_m : k, m = 0, 1, 2, \dots\}$$

satisfies the conclusion of the theorem. It is easy to see that A_θ is a compact, positively invariant set. By (7.4.4) we have that

$$\mathrm{dist}_H(V^m y, A_\theta) \leq \mathrm{dist}_H(V^m y, E_m) \leq 2\theta^m, \quad m = 0, 1, 2, \ldots,$$

for every $y \in M$. This implies (7.4.1).

To estimate the fractal dimension of A_θ we use the idea presented in [75]. We first note that

$$A_\theta \subset V^n M \cup \{V^k E_m : k + m \leq n-1,\, k, m \geq 0\}$$

for every $n \geq 1$. Therefore

$$N(A_\theta, \varepsilon) \leq N(V^n M, \varepsilon) + \sum_{m=0}^{n-1} (n-m) \,\mathrm{Card}\, E_m$$

for every $n \geq 1$ and $\varepsilon > 0$. Consequently choosing $\varepsilon = \theta^n$ from Lemma 7.4.3 we have that

$$N(A_\theta, \theta^n) \leq N_0 \left(1 + \frac{4\sqrt{2}(1+L)K}{\theta - \eta}\right)^{nd} \left\{1 + \sum_{j=1}^{\infty} j \left(1 + \frac{4\sqrt{2}(1+L)K}{\theta - \eta}\right)^{-jd}\right\}$$

for every $n \geq 1$. As in the proof of Theorem 7.3.3 we take $0 < \varepsilon < 1$ and choose $n = n_\varepsilon$ such that $\theta^n \leq \varepsilon < \theta^{n-1}$. Thus

$$\ln N(A_\theta, \varepsilon) \leq \ln N(A_\theta, \theta_\varepsilon^n) \leq n_\varepsilon d \ln\left(1 + \frac{4\sqrt{2}(1+L)K}{\theta - \eta}\right) + C(K, L, \theta, \eta).$$

Because $n_\varepsilon \leq 1 + \ln(1/\varepsilon)\,[\ln(1/\theta)]^{-1}$, this implies (7.4.3). □

Under the hypotheses of either Theorem 7.4.1 or Theorem 7.4.2 the discrete dynamical system (M, V^k) possesses a compact global attractor M_0. This attractor uniformly attracts all the trajectories of the system (M, V^k) and by Theorem 7.3.2 (or by Theorem 7.3.3) $\dim_f M_0 < \infty$. Unfortunately, in general the rate of convergence to the attractor cannot be estimated. This rate may be very slow. However, Theorems 7.4.1 and 7.4.2 attest that the global attractor is contained in a finite-dimensional positively invariant set which attracts M uniformly and exponentially fast. Thus the dynamics of the system becomes finite-dimensional exponentially fast independent of the speed of convergence to the global attractor. Moreover, the reduction principle (see Theorem 7.2.7) is applicable in this case. Thus finite-dimensional positively invariant exponentially attracting sets can be useful for the description of qualitative behavior of infinite-dimensional systems. These sets are frequently called *inertial sets* or *fractal exponential attractors*. In some cases they turn out to be surfaces in the phase space (see Section 7.6 below). For details we refer to [102] and to the references therein.

Here we present only one result in this direction in a form convenient for future applications. We start with the following definition (see [102]).

7.4.4. Definition. A compact set $A_{\text{exp}} \subset X$ is said to be *inertial* (or a *fractal exponential attractor*) of the dynamical system (X, S_t) iff A is a positively invariant set of finite fractal dimension and for every bounded set $D \subset X$ there exist positive constants t_D, C_D and γ_D such that

$$d_X\{S_t D \,|\, A_{\text{exp}}\} \equiv \sup_{x \in D} \text{dist}_X(S_t x, A_{\text{exp}}) \le C_D \cdot e^{-\gamma_D(t-t_D)}, \quad t \ge t_D.$$

■

7.4.5. Remark. As we show below (see, e.g., Theorem 7.9.9) in some cases one can prove the existence of an exponential attractor whose dimension is finite in some extended space $\widetilde{X} \supset X$ only. We frequently call this exponentially attracting set a generalized fractal exponential attractor. ■

The following assertion is proved in [61, Chapter 1] (see also [101] for a similar approach to construction of exponential attractors).

7.4.6. Theorem. *Assume that a dynamical system (H, S_t) on a separable Hilbert space H possesses the properties:*

- *There exist a positively invariant compact set K and positive constants C and γ such that*
$$\sup\{\text{dist}_X(S_t x, K) \,:\, x \in D\} \le C \cdot e^{-\gamma(t-t_D)}$$
for every bounded set $D \subset X$ and for $t \ge t_D$.
- *There exist a neighborhood $\mathscr{O}$ of the compact K and numbers Δ_1 and α_1 such that*
$$\|S_t x_1 - S_t x_2\| \le \Delta_1 e^{\alpha_1 t} \|x_1 - x_2\|$$
provided that $S_t x_i$ belongs to the closure $\overline{\mathscr{O}}$ of $\mathscr{O}$ for all $t \ge 0$.
- *The mapping $t \mapsto S_t x$ is uniformly Hólder continuous on K; that is, there exist constants $C_K(T) > 0$ and $\eta \in (0,1]$ such that*
$$\|S_{t_1} x - S_{t_2} x\| \le C_K(T)|t_1 - t_2|^\eta, \quad t_1, t_2 \in [0,T],\ x \in K.$$
- *There exist a sequence of finite-dimensional projectors $\{P_n\}$ in H; positive constants Δ_2, α_2, and β; and a sequence of positive numbers $\{\rho_n\}$ tending to zero as $n \to \infty$ such that*
$$\|(1 - P_n)(S_t x_1 - S_t x_2)\| \le \Delta_2 e^{-\beta t}\left(1 + \rho_n e^{\alpha_2 t}\right) \|x_1 - x_2\| \text{ for any } x_1, x_2 \in K.$$

Then for any $\nu > \ln 2$ there exists an inertial set A_{exp} for (H, S_t) such that

$$\sup_{x \in D} \text{dist}_X(S_t x, A_{\text{exp}}) \le C(D, \nu) \cdot \exp\left\{-\frac{\gamma \cdot \nu}{\nu + \gamma + \alpha_1}(t - t_D)\right\}$$

for every bounded set $D \subset X$ and $t \ge t_D$. Moreover,

$$\dim_f A_{\exp} \leq C_0 \cdot (1 + \ln \|P_n\|) \dim P_n ,$$

where C_0 does not depends on ν and n and the number n is determined from the condition $\rho_n \leq (4\Delta_2)^{\alpha_2/\beta} \cdot \exp\{-\nu\alpha_2/\beta\}$.

7.5 Gradient systems

The study of the structure of the global attractors is an important problem from the point of view of applications. There are no universal approaches solving this problem. It is well known that even in finite-dimensional cases an attractor can possess extremely complicated structure. However, some sets that belong to the attractor can be easily pointed out. For example, every stationary point ($S_t x = x$ for all $t > 0$) belongs to the attractor of the system. Theorem 7.2.2 shows that any bounded full trajectory also lies in the global attractor.

7.5.1. Definition. Let $\mathcal{N}$ be the set of stationary points of the dynamical system (X, S_t):

$$\mathcal{N} = \{v \in X \,:\, S_t v = v \text{ for all } t \geq 0\}.$$

We define the *unstable manifold* $\mathcal{M}^u(\mathcal{N})$ emanating from the set $\mathcal{N}$ as a set of all $y \in X$ such that there exists a full trajectory $\gamma = \{u(t) \,:\, t \in \mathbb{R}\}$ with the properties

$$u(0) = y \text{ and } \lim_{t \to -\infty} \operatorname{dist}_X(u(t), \mathcal{N}) = 0.$$

■

It is clear that $\mathcal{M}^u(\mathcal{N})$ is an invariant set. It is also easy to prove the following assertion (see, e.g., [17], [61], or [273]).

7.5.2. Proposition. *Let $\mathcal{N}$ be the set of stationary points of the dynamical system (X, S_t) possessing a global attractor A. Then $\mathcal{M}^u(\mathcal{N}) \subset A$.*

For gradient systems it is possible to prove that the unstable manifold coincides with the attractor; that is, $\mathcal{M}^u(\mathcal{N}) = A$.

We give the following definition.

7.5.3. Definition. Let $Y \subseteq X$ be a forward invariant set of a dynamical system (X, S_t).

- The continuous functional $\Phi(y)$ defined on Y is said to be the *Lyapunov function* for the dynamical system (X, S_t) on Y iff the function $t \mapsto \Phi(S_t y)$ is a nonincreasing function for any $y \in Y$.
- The Lyapunov function $\Phi(y)$ is said to be *strict* on Y iff the equation $\Phi(S_t y) = \Phi(y)$ for *all* $t > 0$ and for some $y \in Y$ implies that $S_t y = y$ for all $t > 0$; that is, y is a stationary point of (X, S_t).
- The dynamical system (X, S_t) is said to be *gradient* iff there exists a strict Lyapunov function for (X, S_t) on the whole phase space X.

■

7.5.4. Example. The system $(\mathbb{R}^d, S_t)$ generated the ordinary differential equation

$$\dot{x} = \nabla F(x), \quad x \in \mathbb{R}^d, \, t > 0,$$

possesses a strict Lyapunov function $\Phi(x) = -F(x)$ on $\mathbb{R}^d$ provided the equation $\nabla F(x) = 0$ has isolated roots. ■

We have the following result on the existence of a global attractor for systems with a Lyapunov function.

7.5.5. Proposition. *Let a dynamical system (X, S_t) be point dissipative and asymptotically smooth. Assume that there exists a Lyapunov function $\Phi(x)$ for (X, S_t) on X such that (i) it is bounded from above on any bounded subset of X and (ii) the set $\Phi_R = \{x : \Phi(x) \le R\}$ is bounded for every R. Then (X, S_t) possesses a compact global attractor.*

Proof. Any bounded set B lies in the set Φ_R for some $R = R_B$. It is clear that $S_t \Phi_R \subseteq \Phi_R$ for any R. Therefore $\gamma_B^0 \subset \Phi_R$ and hence γ_B^0 is bounded. Thus we can apply Theorem 7.2.4. □

7.5.1 Geometric structure of the attractor

We have the following result on the structure of a global attractor (for the proof we refer to many sources, including [17, 61, 134, 139, 172, 273]).

7.5.6. Theorem. *Let a dynamical system (X, S_t) possess a compact global attractor A. Assume that there exists a strict Lyapunov function on A. Then $A = \mathscr{M}^u(\mathscr{N})$. Moreover, the global attractor A consists of full trajectories $\gamma = \{u(t) : t \in \mathbb{R}\}$ such that*

$$\lim_{t \to -\infty} \operatorname{dist}_X(u(t), \mathscr{N}) = 0 \quad \text{and} \quad \lim_{t \to +\infty} \operatorname{dist}_X(u(t), \mathscr{N}) = 0. \tag{7.5.1}$$

The following assertion is critical for von Karman models with conservative loads. Its main advantage (in contrast with other statements on the existence of a global attractor) is that it does not assume any dissipativity properties of the system in explicit form.

7.5.7. Corollary. *Assume that (X, S_t) is a gradient asymptotically smooth dynamical system. Assume its Lyapunov function $\Phi(x)$ is bounded from above on any bounded subset of X and the set $\Phi_R = \{x : \Phi(x) \le R\}$ is bounded for every R. If the set $\mathscr{N}$ of stationary points of (X, S_t) is bounded, then (X, S_t) possesses a compact global attractor $A = \mathscr{M}^u(\mathscr{N})$.*

Proof. We choose R_0 such that $\mathscr{N} \subset \Phi_{R_0}$. By Theorem 7.2.4 the dynamical system (Φ_R, S_t) possesses a compact global attractor A_R for every R. If $R \ge R_0$, then by

Theorem 7.5.6 we have that $A_R = \mathscr{M}^u(\mathscr{N})$. Thus A_R does not depend on R for $R \geq R_0$ and it is global attractor for (X, S_t). □

7.5.8. Remark. It follows from the first equality in (7.5.1) that under the hypotheses of Corollary 7.5.7 the following relation is valid,

$$\sup\{\Phi(u) : u \in A\} \leq \sup\{\Phi(u) : u \in \mathscr{N}\}. \tag{7.5.2}$$

If the Lyapunov function $\Phi(u)$ is topologically equivalent to the norm of the phase space X, and the existence of a global attractor is guaranteed, inequality in (7.5.2) can be used in order to provide an upper bound for the size of the absorbing ball. In fact, later on, we apply this method in order to obtain uniform (with respect to the parameters of the problem) bounds for the attractor. ■

If the system (X, S_t) is not gradient but possesses a Lyapunov function (which is not strict), we cannot guarantee that $A = \mathscr{M}^u(\mathscr{N})$. However, we can prove the following assertion (see also [61, Theorem 6.2, Chapter 1] and [75, Theorem 2.30]).

7.5.9. Theorem. *Let (X, S_t) be an asymptotically smooth dynamical system in a Banach space X. Assume that there exists a Lyapunov function $\Phi(x)$ for (X, S_t) on X such that $\Phi(x)$ is bounded from above on any bounded subset of X and the set $\Phi_R = \{x : \Phi(x) \leq R\}$ is bounded for every R. Let $\mathscr{B}$ be the set of elements $x \in X$ such that there exists a full trajectory $\{u(t) : t \in \mathbb{R}\}$ with the properties $u(0) = x$ and $\Phi(u(t)) = \Phi(x)$ for all $t \in \mathbb{R}$. If $\mathscr{B}$ is bounded, then (X, S_t) possesses a compact global attractor and $A = \mathscr{M}^u(\mathscr{B})$.*

Proof. As in the proof of Corollary 7.5.7 we choose R_0 such that $\mathscr{B} \subset \Phi_{R_0}$. By Theorem 7.2.4 the dynamical system (Φ_R, S_t) possesses a compact global attractor A_R for every R. If $R \geq R_0$, then $\mathscr{B} \subset \Phi_R$ and therefore by Theorem 1.6.2 [61] $A_R = \mathscr{M}^u(\mathscr{B})$ for all $R \geq R_0$. Thus $A = \mathscr{M}^u(\mathscr{B})$ is a global attractor for (X, S_t). □

The following assertion describes the long-time behavior of individual trajectories (for the proof we refer to [17] or [61]). This property is often referred to as *strong stability* of the set of equilibria.

7.5.10. Theorem. *Assume that a gradient dynamical system (X, S_t) possesses a compact global attractor A. Then for any $x \in X$ we have*

$$\lim_{t\to+\infty} \mathrm{dist}_X(S_t x, \mathscr{N}) = 0; \tag{7.5.3}$$

that is, any trajectory stabilizes to the set $\mathscr{N}$ of stationary points. In particular, this means that the global minimal attractor $A_{\min}$ coincides with the set of the stationary points, $A_{\min} = \mathscr{N}$.

Assume that $\mathscr{N} = \{z_1, \ldots, z_n\}$ is a finite set. In this case $A = \cup_{i=1}^n \mathscr{M}^u(z_i)$, where $\mathscr{M}^u(z_i)$ is the unstable manifold of the stationary point z_i. This is to say, $\mathscr{M}^u(z_i)$ consists of all $y \in X$ such that there exist a full trajectory $\gamma = \{u(t) : t \in \mathbb{R}\}$ with the properties $u(0) = y$ and $u(t) \to z_i$ as $t \to -\infty$.

From Theorem 7.5.10 we obtain the following assertion.

7.5.11. Corollary. *Assume that a gradient dynamical system (X,S_t) possesses a compact global attractor A and $\mathcal{N}$ is a finite set. Then*

(i) The global attractor A consists of full trajectories $\gamma = \{u(t) : t \in \mathbb{R}\}$ connecting pairs of stationary points: any $u \in A$ belongs to some full trajectory γ and for any $\gamma \subset A$ there exists a pair $\{z, z^*\} \subset \mathcal{N}$ such that

$$u(t) \to z \text{ as } t \to -\infty \text{ and } u(t) \to z^* \text{ as } t \to +\infty.$$

(ii) For any $v \in X$ there exists a stationary point z such that $S_t v \to z$ as $t \to +\infty$.

7.5.12. Remark. Assume that the hypotheses of Corollary 7.5.11 hold. Introduce m_0 distinct values $\Phi_1 < \Phi_2 < \cdots < \Phi_{m_0}$ of the set $\{\Phi(x) : x \in \mathcal{N}\}$ and let

$$\mathcal{N}^j = \{x \in \mathcal{N} : \Phi(x) = \Phi_j\}, \quad j = 1, \dots, m_0.$$

Then the sets $\mathcal{N}^1, \dots, \mathcal{N}^{m_0}$ provide *Morse decomposition* of the attractor A; that is, (i) the subsets $\mathcal{N}^j$ are compact, invariant, and disjoint; and (ii) for any $x \in A \setminus \cup_j \mathcal{N}^j$ and every full trajectory $\gamma_x \subset A$ through x there exist $k > l$ such that $\alpha(\gamma_x) \in \mathcal{N}^k$ and $\omega(\gamma_x) \in \mathcal{N}^l$, where $\alpha(\gamma_x)$ and $\omega(\gamma_x)$ are α- and ω-limit sets (see (7.1.4)). ■

The following theorem describes some additional properties of global attractors for gradient systems. Before stating it, the following definition is introduced.

7.5.13. Definition. Let X be a Banach space. Assume that the evolution operator S_t of a dynamical system (X,S_t) is of class C^1; that is, $S_t u$ has a continuous Fréchet derivative with respect to $u \in X$ for each $t > 0$. An equilibrium point z of dynamical system (X,S_t) is said to be *hyperbolic* iff the Fréchet derivative $S' \equiv DS_1(z)$ of $S_t z$ at the moment $t = 1$ is a linear operator in X with the spectrum $\sigma(S')$ possessing the property

$$\sigma(S') \cap \{w \in \mathbb{C} : |w| = 1\} = \emptyset.$$

We also define the index $\mathrm{ind}\,(z)$ (of instability) of the equilibrium z as a dimension of the spectral subspace of the operator S' corresponding to the set $\sigma_+(S') \equiv \{z \in \sigma(S') : |z| > 1\}$. ■

The following assertion is proved in [17].

7.5.14. Theorem. *Assume that a gradient dynamical system (X,S_t) in a Banach space X with a strict Lyapunov function $\Phi(u)$ possesses the following properties.*

(i) It admits a compact global attractor A.
(ii) $S_t \in C^{1+\alpha}$ and there exists a vicinity $\mathcal{O} \supset A$ such that

$$\|DS_t(u) - DS_t(v)\|_{\mathcal{L}(X,X)} \le C_T \|u - v\|_X^\alpha, \quad u, v \in \mathcal{O},\ t \in [0,T].$$

(iii) $(t,u) \mapsto S_t u$ is continuous over $\mathbb{R}_+ \times A$.
(iv) The operators S_t are injective on A for any $t > 0$ and S_t^{-1} are continuous on A.
(v) The Fréchet derivatives $DS_t(u)$ of $S_t u$ at any point $u \in A$ have zero kernel.

(vi) *The set* $\mathscr{N} = \{z_1, \dots, z_n\}$ *of equilibrium points is finite and every point* $z_j \in \mathscr{N}$ *is hyperbolic.*

Let the indexation of equilibrium points be such that

$$\Phi(z_1) \le \Phi(z_2) \le \cdots \le \Phi(z_n)$$

and let $\mathscr{M}_k = \cup_{j=1}^{k} M^u(z_j)$, $\mathscr{M}_0 = \emptyset$, *where* $M^u(z_j)$ *denotes the unstable manifold emanating from* z_j. *Assume also that the function* $t \mapsto \Phi(S_t u)$ *is strictly decreasing for* $u \notin \mathscr{N}$.

Then $A = \mathscr{M}_n$ *and the following properties hold.*

(i) $M^u(z_i) \cap M^u(z_j) = \emptyset$ *when* $i \neq j$.
(ii) $\mathscr{M}_k$ *is a compact invariant set.*
(iii) $\partial M^u(z_i) \equiv \overline{M^u(z_i)} \setminus M^u(z_i)$ *is an invariant set and* $\partial M^u(z_i) \subset \mathscr{M}_{i-1}$.
(iv) *For any compact set* $K \subset M^u(z_i) \setminus \{z_i\}$ *we have*

$$\lim_{t \to +\infty} \max\{\mathrm{dist}_X(S_t k, \mathscr{M}_{i-1}) : k \in K\} = 0.$$

(v) *Every set* $M^u(z_i)$ *is a* C^1*-manifold of finite dimension* d_i, *this manifold is diffeomorphic to* $\mathbb{R}^{d_i}$, *and the embedding* $M^u(z_i) \subset X$ *is of class* C^1 *in a vicinity of of any point* $v \in M^u(z_i)$. *Moreover* $d_i = \dim E_+(z_i)$, *where* $E_+(z_i)$ *is the spectral subspace of the operator* $S' = DS_1(z_i)$ *which corresponds to the set* $\{\lambda : |\lambda| > 1\}$.

We note that the injectivity property for S_t assumed in (iv) depends on the backward uniqueness property of S_t on the attractor.

7.5.2 Rate of convergence to global attractors

In many cases it is important to know how fast the trajectories starting from bounded sets converge to global attractors. For gradient systems this rate is related to the rates of convergence of individual trajectories to equilibria.

The result stated below provides conditions sufficient for the exponential rate of stabilization to the attractor along with some additional properties of the attractor (see, e.g., [17], [134] and also Theorems 4.7 and 4.8 in the survey [246]).

7.5.15. Theorem. *Let* X *be a Banach space and the hypotheses of Corollary 7.5.7 be in force. Assume that (i) an evolution operator* S_t *is* C^1, *(ii) the set* $\mathscr{N}$ *of equilibrium points is finite and all equilibria are hyperbolic, and (iii) there exists a Lyapunov* $\Phi(x)$ *function such that* $\Phi(S_t x) < \Phi(x)$ *for all* $x \in X$, $x \notin \mathscr{N}$ *and for all* $t > 0$. *Then*

- *For any* $y \in X$ *there exists* $e \in \mathscr{N}$ *such that*

$$\|S_t y - e\|_X \le C_y e^{-\omega t}, \quad t > 0.$$

Moreover, for any bounded set B in X we have that

$$\sup\{\mathrm{dist}(S_t y, A) : y \in B\} \le C_B e^{-\omega t}, \quad t > 0. \tag{7.5.4}$$

Here A is a global attractor, C_y, C_B, and ω are positive constants, and ω in (7.5.4) depends on the minimum, over $e \in \mathscr{N}$, of the distance of the spectrum of $D[S_1 e]$ to the unit circle in $\mathbb{C}$.

- *If we assume in addition that (i) S_1 is injective on the attractor and (ii) the linear map $D[S_1 y]$ is injective for every $y \in A$, then for each $e \in \mathscr{N}$ the unstable manifold $\mathscr{M}^u(e)$ is an embedded C^1-submanifold of X of finite dimension* $\mathrm{ind}(e)$, *which implies that* $\dim_H A = \max_{e \in \mathscr{N}} \mathrm{ind}(e)$.

We note that the proof of this result (see [17] or [134]) relies on geometric consideration of behavior of trajectories in a vicinity of equilibria points. The critical assumption for this is that the evolution S_t is C^1 and that equilibria are finite and hyperbolic. The above assumptions allow us to reduce the problem of convergence in the vicinity of equilibria to a linear problem.

Our main goal in this section is to present a result (see Theorem 7.5.17 below) which gives an estimate of the rate of convergence to the global attractor under the assumption that similar estimates are known in small vicinities of stationary points. To our best knowledge the first result in this direction has been obtained by Kostin [167] for the case of discrete dynamical systems. Inspired by the technique developed in [167] we obtain a similar result for time continuous systems in [75].

Our main assumption in this section is the following one.

7.5.16. Assumption. (X, S_t) is a dynamical system on a Banach space X possessing the properties:

- There exists a compact global attractor $A = \mathscr{M}^u(\mathscr{N})$, where $\mathscr{N}$ is the set of all equilibria (we do not assume that $\mathscr{N}$ is finite).
- There exists a strict Lyapunov function on X such that (i) $\Phi(x)$ is bounded from above on any bounded subset of X, (ii) the set $\Phi_R = \{x : \Phi(x) < R\}$ is bounded for every R, and (iii) the set $\{\Phi(x) : x \in \mathscr{N}\}$ is finite and $\Phi_1 < \Phi_2 < \cdots < \Phi_m$ are its m distinct values.
- There exist constants $c_R \ge 1$ and $L_R \ge 0$ such that

$$\|S_t y_1 - S_t y_2\|_X \le c_R e^{L_R t} \|y_1 - y_2\|_X \quad \text{for any } y_1, y_2 \in \Phi_R.$$

- For every set $\mathscr{N}^j = \{x \in \mathscr{N} : \Phi(x) = \Phi_j\}$, $j = 1, \dots, m$, there exist a vicinity $\mathscr{O}_j$ of $\mathscr{N}^j$ and a decreasing continuous function $\psi_j : \mathbb{R}_+ \mapsto \mathbb{R}_+$ such that the property $S_t z \in \mathscr{O}_j$ for all $t \in [0, T]$ implies that

$$\mathrm{dist}(S_t z, \mathscr{M}^u(\mathscr{N}^j)) \le \psi_j(t), \quad t \in [0, T].$$

We note that under Assumption 7.5.16 the sets $\mathscr{N}^j$ provide Morse's decomposition of the attractor A.

Let $\varepsilon > 0$ be chosen such that

$$\varepsilon < \min_i \{\Phi_{i+1} - \Phi_i\} \quad \text{and} \quad \varepsilon < \frac{1}{2}\min_i \left\{ \Phi_i - \sup_{u \in A_i \setminus \mathscr{O}_i} \Phi(u) \right\}, \tag{7.5.5}$$

where we have introduced the subsets

$$A_n = \bigcup_{j=1}^{n} \mathscr{M}^u(\mathscr{N}^j), \quad n = 1, 2, \ldots, m,$$

of the attractor A. It is clear that $A = A_m$.

Under Assumption 7.5.16 in the space X we consider the following sets

$$\mathscr{B}_i = \{u \in X : \Phi(u) < \Phi_i - \varepsilon\}, \quad i = 2, \ldots, m,$$

and $\mathscr{B}_{m+1} = \{u \in X : \Phi(u) < R_*\}$, where R_* is chosen such that the global attractor A belongs to $\mathscr{B}_{m+1}$. It is clear that every set $\mathscr{B}_i$ is forward invariant. One can show that for every $n = 1, \ldots, m$ the set A_n is a global attractor for the restriction $(\mathscr{B}_{n+1}, S_t)$ of (X, S_t) on $\mathscr{B}_{n+1}$.

7.5.17. Theorem. *Under Assumption* 7.5.16 *there exist numbers* $0 < T_1 < T_2 < \cdots < T_m < \infty$ *such that*

$$\sup_{y \in \mathscr{B}_{k+1}} \operatorname{dist}(S_{T_k+t}y, A_k) \le \Psi_k(t), \quad k = 1, 2, \ldots, m,\ t \ge 0,$$

where $\Psi_1(t) = \psi_1(t)$ *and*

$$\Psi_{k+1}(t) = c_{R_*} e^{L_{R_*} T_k} \cdot \left[\psi_{k+1}(t) + \sup_{s \in [0,t]} \min\left\{ \psi_{k+1}(s) e^{L_{R_*}(t-s)} ; \Psi_k(t-s) \right\} \right]$$

for every $k = 1, \ldots, m-1$. *Moreover, for any bounded set* B *in* X *there exists* $t_B \ge 0$ *such that*

$$\sup_{y \in B} \operatorname{dist}(S_t y, A) \le \Psi_m(t - t_B), \quad t \ge t_B,$$

where A *is the global attractor for the system* (X, S_t).

Proof. We refer to [75, Chapter 2]. □

The following assertion give a condition for the exponential rate of attraction of bounded sets.

7.5.18. Corollary. *Assume that Assumption* 7.5.16 *holds with the function* $\psi_j(t) = c_j e^{-\gamma_j t}$, *where* c_j *and* γ_j *are positive constants,* $j = 1, \ldots, m$. *Then there exists* $\gamma_0 > 0$ *such that for any bounded set* B *we can find positive constants* C_B *and* t_B *such that*

$$\sup_{y \in B} \operatorname{dist}(S_t y, A) \le C_B e^{-\gamma_0 t}, \quad t \ge t_B.$$

Proof. It is sufficient to show that every function $\Psi_k(t)$ from the statement of Theorem 7.5.17 admits the estimate $\Psi_k(t) \le C_k e^{-\beta_k t}$ for every $k = 1, \dots, m$, with positive C_k and β_k. This can be done by induction in k (see [75, Chapter 2] for details). □

7.6 General idea about inertial manifolds

If an infinite-dimensional system possesses a global attractor of finite dimension, then there is, at least theoretically, a possibility to reduce the study of its asymptotic regimes to the investigation of properties of a finite-dimensional system. However the structure of the attractor very often is not known, so the reduction of the flow to finite-dimensional subspace is often problematic. This motivates introduction of other concepts such as inertial manifolds. To this end, methods such as integral manifolds (see, e.g., [92, 139, 233]) and the reduction principle prove very useful. They lead to constructions (see [117]) of the inertial manifold associated with an infinite-dimensional dynamical system (see, e.g., [61, 90, 273] and references therein). These manifolds are finite-dimensional invariant surfaces, which contain global attractor and attract trajectories exponentially fast. Moreover, there is a possibility to reduce the study of limit regimes of the original infinite-dimensional system to solving similar problem for a class of ordinary differential equations. The theory of inertial manifolds has been developed and widely studied for deterministic systems by many authors (see, e.g., [45, 50, 61, 90, 236, 117, 227, 273] and the references therein). A typical condition required by all known results concerning existence of inertial manifolds is some sort of gap condition imposed on the spectrum of the linearized problem.

7.6.1. Definition. Let (X, S_t) be a dynamical system in a Banach space X. Let $\mathscr{M}$ be a finite-dimensional surface in the space X of the following structure

$$\mathscr{M} = \{p + \Phi(p) \; : \; p \in \mathscr{P}X, \; \Phi : \mathscr{P}X \to (I - \mathscr{P})X\}, \tag{7.6.1}$$

where $\mathscr{P}$ is a finite-dimensional projector and $\Phi(\cdot)$ is a continuous mapping satisfying the Lipschitz condition

$$\|\Phi(p_1) - \Phi(p_2)\| \le L \cdot \|p_1 - p_2\|$$

for all $p_j \in \mathscr{P}X$, where L is a positive constant. Here and below $\|\cdot\|$ denotes the norm in X. Then $\mathscr{M}$ is said to be an *inertial manifold* (IM) for the dynamical system (X, S_t) iff

- The surface $\mathscr{M}$ is invariant; that is $S_t u \in \mathscr{M}$ for any $u \in \mathscr{M}$ and $t \ge 0$.
- $\mathscr{M}$ is exponentially attracting; that is, for any bounded set $B \subset X$ there exists $C_B > 0$ and $\eta_B > 0$ such that

$$\sup\{\mathrm{dist}_X(S_t y, \mathscr{M}) \; : \; y \in B\} \le C_B e^{-\eta_B t} \quad \text{for all } t \ge 0.$$

■

We also need the concept (see, e.g., [90]) of asymptotic completeness of an inertial manifold $\mathscr{M}$ which means that given any orbit of a dynamical system, we can find another orbit lying in $\mathscr{M}$ that produces the same limit behavior at $t \to \infty$. If the system (X, S_t) is generated by some nonlinear PDE the asymptotic completeness property makes it possible to reduce the description of limit regimes of (X, S_t) to the study of long-time dynamics of the corresponding inertial form which is a finite-dimensional ODE (see, e.g., [90] and [61] and the references therein).

7.6.2. Definition. The inertial manifold $\mathscr{M}$ is said to be *exponentially (asymptotically) complete* iff for any $u \in X$ there exist $u^* \in \mathscr{M}$ and numbers $s \geq \tau \geq 0$ such that $S_t u^* \in \mathscr{M}$ for all $t \geq 0$ and

$$\|S_t u - S_{t-\tau} u^*\| \leq C \cdot e^{-\eta t} \quad \text{for all } t \geq s,$$

where C and η are positive constants. ■

We consider an evolution differential equation in a Banach space X of the type

$$\frac{du}{dt} + \mathscr{A} u = \mathscr{B}(u), \quad u|_{t=0} = u_0. \tag{7.6.2}$$

We assume that $\mathscr{A}$ is a generator of the C_0-semigroup $e^{-\mathscr{A} t}$ in X and $\mathscr{B}(\cdot)$ is a nonlinear mapping from X into X such that

$$\| \mathscr{B}(u) \| \leq M_0, \quad \| \mathscr{B}(u_1) - \mathscr{B}(u_2) \| \leq M_1 \cdot \| u_1 - u_2 \| . \tag{7.6.3}$$

Under the above assumptions it is well known by standard semigroup methods that for each initial datum $u_0 \in X$, there exists a unique global solution $u \in C(\mathbb{R}_+; X)$. Thus problem (7.6.2) generates a dynamical system (X, S_t) in the space X.

Assume that there exist orthoprojectors $\mathscr{P}$ and $\mathscr{Q} = I - \mathscr{P}$ such that $\dim \mathscr{P} < \infty$, $\mathscr{P}, \mathscr{Q}$ commute with $\mathscr{A}$ and

$$\|\mathscr{P} e^{\mathscr{A} t}\| \leq e^{\lambda^- t}, \ t > 0; \quad \|\mathscr{Q} e^{-\mathscr{A} t}\| \leq e^{-\lambda^+ t}, \ t > 0,$$

where $0 < \lambda^- < \lambda^+$.

The following theorem holds (see, e.g., [61, Chapter 3] and also Theorem 13.2.6 and Corollary 13.2.5 in Chapter 13).

7.6.3. Theorem. *Assume that the spectral gap condition*

$$\lambda^+ - \lambda^- \geq \frac{4M_1}{q} \tag{7.6.4}$$

holds for some $0 < q < 1$. *Then there exists a map* $\Phi : \mathscr{P}X \to \mathscr{Q}X$ *such that*

$$\|\Phi(p)\| \leq \frac{M_0}{\lambda^+}, \quad \|\Phi(p_1) - \Phi(p_2)\| \leq \frac{q}{2(1-q)} \|p_1 - p_2\| ,$$

and the set $\mathcal{M}$ *defined in* (7.6.1) *is a finite-dimensional invariant manifold for dynamics governed by* (7.6.2).

If the spectral gap condition (7.6.4) *holds for some* $0 < q < \frac{2}{3}$, *then the manifold* $\mathcal{M}$ *is exponentially complete; that is, for any* $u(t)$ *there exists a trajectory* $u^*(t) \in \mathcal{M}$ *such that*

$$\|u(t) - u^*(t)\| \le C_q \cdot e^{-\gamma t} \cdot \|\mathcal{Q}u(0) - \Phi(\mathcal{P}u(0))\|, \quad t > 0, \quad \gamma = \frac{\lambda^- + \lambda^+}{2}.$$

We refer to Chapter 13 for more details concerning inertial manifolds with applications to von Karman evolution equations.

7.7 Approximate inertial manifolds

The assumptions (see (7.6.3) and (7.6.4)) that guarantee the existence of inertial manifold (IM) are rather strong. That is why the concept of approximate IM has been introduced [114]. This manifold is a finite-dimensional smooth surface in a phase space, whose small vicinity attracts all the trajectories. We note also that this concept leads to new approaches to numerical investigation of long-time behavior of solutions (see, e.g., [228] and [96]). The approximate IM have been constructed for a wide class of parabolic partial differential equations (see [95, 114, 153, 274] and the references therein). But all these constructions rely on a regularizing effect for parabolic equations (the solution for the moment $t > 0$ is more smooth than it's at the initial value).

In this section we describe the construction of approximate inertial manifolds for second order in time evolution equations. We follow [57, 58] mainly. The main result of this section is presented in Theorem 7.7.1.

7.7.1 The main assumptions

We consider a dissipative evolution differential equation of the second order in time in the Hilbert space $\mathcal{H}$ of the type:

$$\partial_t^2 u + \gamma \partial_t u + \mathcal{A} u = B(u), \quad u_{|t=0} = u_0, \quad \partial_t u_{|t=0} = u_1, \tag{7.7.1}$$

where γ is a positive number, $\mathcal{A}$ is a positive self-adjoint operator with a discrete spectrum, $B(\cdot)$ is a nonlinear mapping from the domain $\mathcal{D}(\mathcal{A}^{1/2})$ of the operator $\mathcal{A}^{1/2}$ into $\mathcal{H}$ such that for some integer $m \ge 2$ the mapping $B(u)$ belongs to C^m, and

$$\|\langle B^{(k)}(u); w_1, \ldots, w_k \rangle\| \le C_\rho \prod_{j=1}^{k} \|\mathcal{A}^{1/2} w_j\|, \tag{7.7.2}$$

$$\|\langle B^{(k)}(u)-B^{(k)}(u^*);w_1,\dots,w_k\rangle\| \le C_\rho \|\mathscr{A}^{1/2}(u-u^*)\| \prod_{j=1}^{k} \|\mathscr{A}^{1/2}w_j\| \quad (7.7.3)$$

for every $\rho > 0$, where $k = 0,\dots,m$, $\|\cdot\|$ is the norm in space $\mathscr{H}$, $\|\mathscr{A}^{1/2}u\| \le \rho$, $\|\mathscr{A}^{1/2}u^*\| \le \rho$, $w_j \in \mathscr{D}(\mathscr{A}^{1/2})$. Here $B^{(k)}(u)$ is the k-order Fréchet derivative of $B(u)$ and $\langle B^{(k)}(u);w_1,\dots,w_k\rangle$ is the value of $B^{(k)}(u)$ on elements $w_1,\dots,w_k$. We assume that the problem (7.7.1) is well-posed on two scales of Hilbert spaces: in $\mathscr{D}(\mathscr{A}^{1/2})\times\mathscr{H}$ and $\mathscr{D}(\mathscr{A})\times\mathscr{D}(\mathscr{A}^{1/2})$.

Let $m \ge 1$ be an integer and let $R > 0$ be a positive number. We denote by $\mathscr{L}_{m,R}$ the class of solutions of the problem (7.7.1) that possess the following regularity and dissipativity properties:

(i) for $k = 0,1,2,\dots,m-1$ and for all $T > 0$ we have that $u^{(k)}(t) \in C(0,T;\mathscr{D}(\mathscr{A}))$ and also

$$u^{(m)}(t) \in C(0,T;\mathscr{D}(\mathscr{A}^{1/2})), \quad u^{(m+1)}(t) \in C(0,T;\mathscr{H}),$$

where $C(0,T;V)$ denotes the space of strongly continuous functions on $[0,T]$ with values in V, here and below $u^{(k)}(t) = \partial_t^k u(t)$.

(ii) For any $u(t) \in \mathscr{L}_{m,R}$ there exists $t^* > 0$ such that

$$\|u^{(k+1)}(t)\|^2 + \|\mathscr{A}^{1/2}u^{(k)}(t)\|^2 + \|\mathscr{A}u^{(k-1)}(t)\|^2 \le R^2, \quad (7.7.4)$$

where $k = 0,1,\dots,m$ and $t \ge t^*$.

In fact these classes $\mathscr{L}_{m,R}$ were investigated in [123] with applications to semilinear wave equations with linear damping (see Chapters 3 and 4 for similar classes in the case of von Karman equations).

7.7.2 Construction of approximate inertial manifolds

We fix an integer N and as above denote $P = P_N$ the projector in $\mathscr{H}$ onto the space spanned by the first N eigenvectors of $\mathscr{A}$, $Q \equiv I - P$. Applying P and Q to equation (7.7.1) we obtain the following coupled system of equations for $p(t) = Pu(t)$ and $q(t) = Qu(t)$,

$$\begin{aligned} \partial_t^2 p + \gamma\partial_t p + \mathscr{A}p &= PB(p+q), \\ \partial_t^2 q + \gamma\partial_t q + \mathscr{A}q &= QB(p+q). \end{aligned} \quad (7.7.5)$$

Assume that the system (7.7.5) has an invariant manifold in the phase space $\mathscr{D}(\mathscr{A}^{1/2})\times\mathscr{H}$ of the type:

$$M = \{(p+h(p,\dot p);\dot p + l(p,\dot p)) : p,\dot p \in PH\}, \quad (7.7.6)$$

where h and l are smooth mappings from $P\mathscr{H}\times P\mathscr{H}$ into $Q\mathscr{D}(\mathscr{A})$. Then substituting $q(t) = h(p(t),\partial_t p(t))$ and $\partial_t q(t) = l(p(t),\partial_t p(t))$ in (7.7.5) we can obtain the following equality,

$$\langle \delta_p l; \dot{p} \rangle + \langle \delta_{\dot{p}} l; -\gamma \dot{p} - \mathscr{A} p + PB(p + h(p, \dot{p})) \rangle$$
$$+ \gamma l(p, \dot{p}) + \mathscr{A} h(p, \dot{p}) = QB(p + h(p, \dot{p})).$$

And the compatibility condition $l(p(t), \partial_t p(t)) = \partial_t h(p(t), \partial_t p(t))$ implies that

$$l(p, \dot{p}) = \langle \delta_p h; \dot{p} \rangle + \langle \delta_{\dot{p}} h; -\gamma \dot{p} - \mathscr{A} p + PB(p + h(p, \dot{p})) \rangle.$$

Here and below $\delta_p f$ and $\delta_{\dot{p}} f$ are the Frechet derivatives of $f(p, \dot{p})$ with respect to p and $\dot{p}$; $\langle \delta_p f; w \rangle$ and $\langle \delta_{\dot{p}} f; w \rangle$ are the corresponding values on element w.

Using these formal equations we can suggest the following iterative process for the determination of a function family $\{h_k; l_k\}$ that defines the sequence of approximate inertial manifolds according to (7.7.6),

$$\begin{aligned} \mathscr{A} h_k(p, \dot{p}) &= QB(p + h_{k-1}(p, \dot{p})) - \gamma l_{\nu(k)}(p, \dot{p}) - \langle \delta_p l_{k-1}; \dot{p} \rangle \\ &\quad - \langle \delta_{\dot{p}} l_{k-1}; -\gamma \dot{p} - \mathscr{A} p + PB(p + h_{k-1}(p, \dot{p})) \rangle, \end{aligned} \tag{7.7.7}$$

where $k = 1, 2, 3, \ldots$ and integers $\nu(k)$ must be chosen such that $k - 1 \le \nu(k) \le k$. Here $l_k(p, \dot{p})$ is defined by the formula

$$l_k(p, \dot{p}) = \langle \delta_p h_{k-1}; \dot{p} \rangle + \langle \delta_{\dot{p}} h_{k-1}; -\gamma \dot{p} - \mathscr{A} p + PB(p + h_{k-1}(p, \dot{p})) \rangle \tag{7.7.8}$$

for $k = 1, 2, \ldots$. We also suppose

$$h_0(p, \dot{p}) \equiv l_0(p, \dot{p}) \equiv 0. \tag{7.7.9}$$

Now we introduce the induced trajectories $U_s(t) = (u_s(t); \bar{u}_s(t))$, where $s = 0, 1, 2, \ldots$ and

$$u_s(t) = p(t) + q_s(t), \quad \bar{u}_s(t) = \partial_t p(t) + \bar{q}_s(t). \tag{7.7.10}$$

Here $p(t) = Pu(t)$, $u(t)$ is the solution of the problem (7.7.1); $q_s(t)$ and $\bar{q}_s(t)$ are given by

$$q_s(t) = h_s(p(t), \partial_t p(t)), \quad \bar{q}_s(t) = l_s(p(t), \partial_t p(t)).$$

The main result is the following statement.

7.7.1. Theorem. *Let $u(t)$ be the solution of the problem (7.7.1) and $u(t) \in \mathscr{L}_{m,R}$, where $m \ge 2$. Suppose that $h_n(p, \dot{p})$ and $l_n(p, \dot{p})$ are defined according to (7.7.7) - (7.7.9). Then for $n \le m - 1$ and for t sufficiently large we have*

$$\|\mathscr{A} \partial_t^j (u(t) - u_n(t))\| + \|\mathscr{A}^{1/2} \partial_t^j (\partial_t u(t) - \bar{u}_n(t))\| \le C_{n,R} \cdot \lambda_{N+1}^{-n/2},$$

when $0 \le j \le m - n - 1$. Here $u_n(t)$ and $\bar{u}_n(t)$ are defined according to (7.7.10), λ_{N+1} is the $(N+1)$th eigenvalue of $\mathscr{A}$.

For the proof we refer to [57]; see also [55] for the case of the von Karman model.

Theorem 7.7.1 makes it possible to obtain the following result on localization of the global attractor to problem (7.7.1).

Consider the dynamical system (W, S_t) generated in the space $W = \mathscr{D}(\mathscr{A}) \times \mathscr{D}(\mathscr{A}^{1/2})$ by the formula $S_t U = (u(t); \partial_t u(t))$, where $u(t)$ is the solution to (7.7.1) with initial data $U = (u_0; u_1)$.

Theorem 7.7.1 implies the following localization result for the global attractor.

7.7.2. Theorem. *Assume that the global attractor A of the system $(\mathscr{H}, S_t)$ generated by (7.7.1) exists and any trajectory $U(t) = (u(t); \partial_t u(t))$ lying in A possesses the properties (7.7.4) for all $t \in \mathbb{R}$ and $k = 1, 2, \ldots$. Then for any $U(t) = (u(t); \partial_t u(t))$ from the attractor A we have the relation:*

$$\{\|\mathscr{A}\partial_t^j(u(t) - u_n(t))\|^2 + \|\mathscr{A}^{1/2}\partial_t^j(\partial_t u(t) - \bar{u}_n(t))\|^2\}^{1/2} \le C_{n,j} \cdot \lambda_{N+1}^{-n/2}$$

for $n = 1, 2, \ldots$; $j = 0, 1, \ldots$ and all $t \in \mathbb{R}$. Here $u_n(t)$ and $\bar{u}_n(t)$ are defined by (7.7.10). Furthermore

$$\sup\{\operatorname{dist}(U, M_n) : U \in A\} \le c_n \lambda_{N+1}^{-n/2}, \quad n = 1, 2, \ldots, \tag{7.7.11}$$

where M_n is the manifold of the form (7.7.6) with $h = h_n(p, p)$ and $l = l_n(p, p)$. Here $\operatorname{dist}(U, M_n)$ is the distance between U and M_n in the space $W = \mathscr{D}(\mathscr{A}) \times \mathscr{D}(\mathscr{A}^{1/2})$.

Relation (7.7.11) means that the global attractor lies in some (small) layer adjacent to the approximate inertial manifold M_n. Thus, in some sense, the surface M_n provides information about the "shape" of the attractor.

7.7.3 Nonlinear Galerkin method

As in the parabolic case (see, e.g., [228, 96, 240] and the references therein) the sequence of approximate inertial manifolds given by (7.7.7)–(7.7.9) can be used to construct new methods for numerical investigation of the long-time behavior of second-order evolution equations.

Indeed, the sequence $\{h_n(p, \dot{p})\}$ generates a family of approximate inertial forms of the problem (7.7.1):

$$\partial_t^2 p + \gamma \partial_t p + \mathscr{A} p = PB(p + h_n(p, \partial_t p)). \tag{7.7.12}$$

Every such form generates a finite-dimensional dynamical system in PW. If $n = 0$, then equation (7.7.12) becomes the usual Galerkin approximation of the problem. In view of Theorems 7.7.1 and 7.7.2 it is natural to use equations (7.7.12) with $n > 0$ for numerical investigation of the long-time behavior of the system. However, because $h_n(p, \dot{p})$ has values in the infinite-dimensional space QW, we must make an additional discretization. This can be made in the following manner.

As above we assume that $P = P_N$ is the same as in Theorem 7.7.1. We denote by $f_n(p, \dot{p})$ either $h_n(p, \dot{p})$ or $l_n(p, \dot{p})$, where h_n and l_n are defined by (7.7.7)–(7.7.9). Let

$$f_n^*(p, \dot{p}) \equiv f_{N,M,n}(p, \dot{p}) = \chi(R^{-1}(\| \mathscr{A}^{1/2} p \|) P_M f_n(p, \dot{p}),$$

where $M > N$ and the function $\chi(s)$ lies in $C^\infty(\mathbb{R}_+)$ and possesses the following properties: (i) $0 \le \chi(s) \le 1$; (ii) $\chi(s) = 1$ if $0 \le s \le 1$; (iii) $\chi(s) = 0$ if $s \ge 2$; R is a constant large enough.

We consider the following finite-dimensional equation in the space $P_N\mathscr{H}$,

$$\partial_t^2 p^* + \gamma\partial_t p^* + \mathscr{A} p^* = P_N B(p^* + h_n^*(p^*, \partial_t p^*)), \qquad (7.7.13)$$
$$p^*_{|t=0} = P_N u_0, \quad \partial_t p^*_{|t=0} = P_N u_1.$$

In order to determine the value h_n^* we must solve step-by-step some systems of linear algebraic equations in $(M-N)$-dimensional space. Therefore the usage of numerical schemes based on equation (7.7.13) has some computational advantages in comparison with the traditional Galerkin method in the space $P_M\mathscr{H}$. For parabolic systems similar numerical methods are known as nonlinear Galerkin methods [228] (see also [96, 153] and the references therein).

The following theorem allows us to estimate the accuracy of approximation of the exact solution of the problem (7.7.1) by the trajectory which is constructed according to (7.7.13).

7.7.3. Theorem. *Let $u(t)$ be the solution of the problem* (7.7.1) *possessing the properties* (7.7.4) *for some $m \ge 2$ and for all $t > 0$. Let*

$$u_n^*(t) = p^*(t) + h_n^*(p^*(t), \partial_t p^*(t)) \quad \textit{and} \quad \hat{u}_n^*(t) = p^*(t) + h_n^*(p^*(t), \partial_t p^*(t)),$$

where $p^*(t)$ is the solution of the problem (7.7.13). *Then for $n \le m-1$ the following estimate holds,*

$$\left(\|\mathscr{A}^{1/2}(u(t) - u_n^*(t))\|^2 + \|\partial_t u(t) - \hat{u}_n^*(t))\|^2\right)^{1/2}$$
$$\le \{\alpha_1 \cdot \lambda_{N+1}^{-(n+1)/2} + \alpha_2 \cdot \lambda_{M+1}^{-1/2}\} \cdot \exp(\beta t). \qquad (7.7.14)$$

Here the positive constants α_1, α_2, and β depend on the constant R_2 from (7.7.4)*; α_1 and α_2 also depend on m.*

In order to prove this theorem it is necessary to make use of the properties of the sequence $\{h_n\}$ and to compare the solution $p^*(t)$ of the problem (7.7.13) with the value $p(t) = P_N u(t)$ that satisfies the equation

$$\partial_t^2 p + \gamma\partial_t p + \mathscr{A} p = PB(p(t) + q(t)),$$

where $q(t) = (I - P_N)u(t)$ and $u(t)$ is the solution of the problem (7.7.1). We omit these rather standard arguments.

If in Theorem 7.7.3 we suppose $n = 0$ and $N = M$, then estimate (7.7.14) gives accuracy of the standard Galerkin method. Therefore, if parameters N, M, and n are correlated such that $\lambda_{M+1} \cong \lambda_{N+1}^{n+1}$, then the nonlinear Galerkin method has the same accuracy as the standard Galerkin method generated based on M basic functions. However, if we use the nonlinear method, then we should solve a system consisting of N ordinary differential equations and $M-N$ linear algebraic equations. In particular, to define h_n for $n = 1$ we need to solve the equation

$$\mathscr{A} h_1(p, \dot{p}) = (P_M - P_N)B(p)$$

and choose N and M such that $\lambda_{M+1} \cong \lambda_{N+1}^2$. Thus the nonlinear Galerkin method with $n = 1$ has certain advantages in comparison with the linear (standard) method. We also refer to [58] for some discussion of nonlinear Galerkin methods for dissipative perturbations of infinite-dimensional Hamiltonian systems.

7.8 General idea about determining functionals

Detailed study of the structure of the attractor is limited to few and rather special problems (see Section 7.5), therefore it becomes important to search for minimal (or close to minimal) sets of natural parameters of the problem that uniquely determine long-time behavior of the system. This problem was first discussed by Foias and Prodi [116] and by Ladyzhenskaya [170] for the 2D Navier–Stokes equations. They proved that the long-time behavior of the solutions is completely determined by the dynamics of the first N Fourier modes, provided N is sufficiently large. Later on, similar results were obtained for other parameters and equations (see, e.g., [88, 115, 118, 119, 171, 258] and the references quoted therein). In particular, the concepts of determining nodes [115, 118, 258] and determining local volume averages [119, 154, 155] were introduced in the above references. The relation between the problem of existence of finite number of determining parameters and some problems in interpolation theory has been discussed. The general concept of determining functionals was introduced (see [86, 87]). For further details we refer to the survey [60] and to the references quoted therein (see also [61, Chapter 5]).

7.8.1 Concept of a set of determining functionals

Let us consider a nonautonomous differential equation in a reflexive Banach space H of the type

$$\frac{du}{dt} = F(u,t), \quad t > 0, \quad u|_{t=0} = u_0. \tag{7.8.1}$$

Let's recall some of the notation employed below. $\mathscr{W}$ stands for a class of solutions to (7.8.1) defined on the semi-axis $\mathbb{R}_+ \equiv \{t : t \geq 0\}$ such that for any $u(t) \in \mathscr{W}$ there exists a moment of time $t_0 > 0$ such that

$$u(t) \in C(t_0, +\infty; H) \cap L_2^{\text{loc}}(t_0, +\infty; V), \tag{7.8.2}$$

where V is a reflexive Banach space continuously embedded into H. As above, here and below $C(a,b;X)$ is the space of strongly continuous functions on $[a,b]$ with values in X, $L_2^{\mathrm{loc}}(a,b;X)$ is defined analogously. The notations $\|\cdot\|_H$ and $\|\cdot\|_V$ are also used for norms in the spaces H and V, $\|\cdot\|_H \le \|\cdot\|_V$. The following definition is based on the property established in [116] for the Fourier modes of solutions to the 2D Navier–Stokes system with the periodic boundary conditions.

7.8.1. Definition. Let $\mathscr{L} = \{l_j : j = 1,\dots,N\}$ be a set of linear continuous functionals on V. Then $\mathscr{L}$ is said to be *a set of asymptotically* $(V,H,\mathscr{W})$*-determining functionals* (or elements) for problem (7.8.1), if for any two solutions $u,v \in \mathscr{W}$ the condition

$$\lim_{t\to\infty}\int_t^{t+1} |l_j(u(\tau)) - l_j(v(\tau))|^2 d\tau = 0 \ \text{ for } j = 1,\dots,N \tag{7.8.3}$$

implies that

$$\lim_{t\to\infty} \| u(t) - v(t) \|_H = 0. \tag{7.8.4}$$

■

The following theorem is the basis for many assertions known to date on the existence of finite sets of asymptotically determining functionals (for the proof we refer to [60] and [61, Chapter 5]).

7.8.2. Theorem. *Let* $\mathscr{L} = \{l_j : j = 1,\dots,N\}$ *be a family of linear continuous functionals on* V*. Suppose that there exists a continuous function* $\mathscr{V}(u,t)$ *on* $H \times \mathbb{R}_+$ *with values in* $\mathbb{R}_+$*, satisfying the following properties.*

- *There exist positive numbers* α *and* σ *such that*

$$\mathscr{V}(u,t) \ge \alpha \cdot \|u\|^\sigma \ \textit{for any } u \in H,\ t \in \mathbb{R}_+;$$

- *For any two solutions* $u(t), v(t) \in \mathscr{W}$ *to the problem* (7.8.1) *there exist (i) a point of time* $t_0 > 0$*, (ii) a function* $\psi(t)$ *that is locally integrable over the interval* $[t_0,\infty)$ *and such that*

$$\gamma_\psi^+ \equiv \liminf_{t\to\infty}\int_t^{t+a} \psi(\tau)\,d\tau > 0$$

and

$$\Gamma_\psi^+ \equiv \limsup_{t\to\infty}\int_t^{t+a} \max\{0,-\psi(\tau)\}\,d\tau < \infty$$

for some $a > 0$*, and also (iii) a positive constant* C *such that*

$$\begin{aligned} &\mathscr{V}(u(t)-v(t),t) + \textstyle\int_s^t \psi(\tau)\cdot\mathscr{V}(u(\tau)-v(\tau),\tau)\,d\tau \\ &\le \mathscr{V}(u(s)-v(s),s) + C\cdot \textstyle\int_s^t \max_{j=1,\dots,N} |l_j(u(\tau)) - l_j(v(\tau))|^2\,d\tau \end{aligned} \tag{7.8.5}$$

holds for all $t \ge s \ge t_0$.

Then $\mathscr{L}$ *is a set of asymptotically* $(V,H,\mathscr{W})$*-determining functionals for problem* (7.8.1).

Other approaches and definitions characterizing determining functionals are possible and have been used in the literature (see, e.g., [170, 86, 87] and the references therein). For instance, the definition below is an extension to a general dynamical system of the property proved by Ladyzhenskaya [170] for trajectories lying in the global attractor of the 2D Navier–Stokes equations.

7.8.3. Definition. Let $\overline{\mathscr{W}}$ be a class of solutions to problem (7.8.1) on the real axis $\mathbb{R}$ such that $\overline{\mathscr{W}} \subset L_2^{\rm loc}(\mathbb{R};V)$. A family $\mathscr{L} = \{l_j : j = 1,\dots,N\}$ of linear continuous functionals on V is said to be *a set of* $(V,\overline{\mathscr{W}})$*-determining functionals* (or elements) for problem (7.8.1), if for any two of its solutions $u, v \in \overline{\mathscr{W}}$ the condition

$$l_j(u(t)) = l_j(v(t)) \;\text{ for almost all }\; t \in \mathbb{R} \;\text{ and }\; j = 1,\dots,N,$$

implies $u(t) \equiv v(t)$. ■

It is easy to infer (see [60] and [61, Chapter 5]) the following version (reformulation) of Theorem 7.8.2.

7.8.4. Theorem. *Let $\mathscr{L} = \{l_j : j = 1,\dots,N\}$ be a family of linear continuous functionals on V. Let $\overline{\mathscr{W}}$ be a class of solutions to problem* (7.8.1) *on the real axis $\mathbb{R}$ such that $\overline{\mathscr{W}} \subset C(\mathbb{R};H) \cap L_2^{\rm loc}(\mathbb{R};V)$. Assume that there exists a continuous function $\mathscr{V}(u,t)$ on $H \times \mathbb{R}$ with values in $\mathbb{R}$ possessing the properties.*

- *There exist $\alpha, \sigma > 0$ such that $\mathscr{V}(u,t) \ge \alpha \cdot \|u\|^\sigma$ for all $u \in H$ and $t \in \mathbb{R}$.*
- *For any $u(t), v(t) \in \overline{\mathscr{W}}$ we have $\sup_{t\in\mathbb{R}} \mathscr{V}(u(t) - v(t), t) < \infty$.*
- *For any two solutions $u(t), v(t) \in \overline{\mathscr{W}}$ to problem* (7.8.1) *there exist (i) locally integrable over the axis $\mathbb{R}$ function $\psi(t)$ with the properties*

$$\gamma_\psi^- \equiv \liminf_{t\to-\infty} \int_t^{t+a} \psi(\tau)\, d\tau > 0$$

and

$$\Gamma_\psi^- = \limsup_{t\to-\infty} \int_t^{t+a} \max\{0, -\psi(\tau)\}\, d\tau < \infty$$

for some $a > 0$, and (ii) a positive constant C such that (7.8.5) *holds for all $t \ge s$.*

Then, $\mathscr{L}$ is a set of $(V,\overline{\mathscr{W}})$-determining functionals for the problem (7.8.1).

We note that the notions of determining functionals introduced above are rather general. As we show below in many cases it is convenient to adjust the corresponding definitions to the problem under consideration. In this connection we mention that Definitions 7.8.1 and 7.8.3 are given mainly for the sake of orientation.

7.8.2 Completeness defect of a set of functionals

For characterization of a set of determining elements we make use of the following concept of completeness defect which has been introduced in [59, 60].

7.8.5. Definition. Let V and H be reflexive Banach spaces and V is continuously and densely embedded into H. *The completeness defect* of a set $\mathscr{L}$ of linear functionals on V with respect to H is the value

$$\varepsilon_{\mathscr{L}}(V,H) = \sup\{\| w \|_H \colon w \in V,\, l(w) = 0,\, l \in \mathscr{L},\, \| w \|_V \le 1\}\,. \tag{7.8.6}$$

∎

We note that finite-dimensionality of the Span$\mathscr{L}$ of the set $\mathscr{L}$ is not assumed at this point. It is also obvious that $\varepsilon_{\mathscr{L}_1}(V,H) \ge \varepsilon_{\mathscr{L}_2}(V,H)$ provided Span$\mathscr{L}_1 \subset$ Span$\mathscr{L}_2$. In addition $\varepsilon_{\mathscr{L}}(V,H) = 0$ if and only if the class of functionals $\mathscr{L}$ is complete in V; that is, the following uniqueness condition holds: $l(w) = 0$ for all $l \in \mathscr{L}$ implies $w = 0$. We can generalize the notion of the completeness defect by considering some seminorms μ_V in (7.8.6) instead of the norm $\|\cdot\|_H$ (see the statement of Theorem 7.9.11 below).

The basic properties of completeness defect which we use in subsequent considerations are described in the following assertions (for the proofs we refer to [60] and [61, Chapter 5]).

7.8.6. Theorem. *Let $\varepsilon_{\mathscr{L}} = \varepsilon_{\mathscr{L}}(V,H)$ be the completeness defect of a set $\mathscr{L}$ of linear functionals on V with respect to H. Then there exists a positive constant $C_{\mathscr{L}}$ such that*

$$\|w\|_H \le C_{\mathscr{L}} \cdot \sup\{|l(w)| \,:\, l \in \bar{\mathscr{L}}, \| l \|_* \le 1\} + \varepsilon_{\mathscr{L}} \cdot \|w\|_V \quad \textit{for any } w \in V, \tag{7.8.7}$$

where $\bar{\mathscr{L}}$ is the closed linear span of the set $\mathscr{L}$ in V^*, and $\|\cdot\|_*$ is the norm in V^*.

Let W be a reflexive Banach space such that $V \subset W \subset H$ and all the embeddings are continuous and dense. Assume that the (interpolation) inequality

$$\|u\|_W \le a_\theta \|u\|_H^\theta \|u\|_V^{1-\theta}, \quad u \in V,$$

is valid with some constants $a_\theta > 0$ and $0 < \theta < 1$. Then for any set $\mathscr{L}$ of the linear functionals on W the following estimate holds,

$$\left[a_\theta^{-1} \varepsilon_{\mathscr{L}}(V,W)\right]^{1/\theta} \le \varepsilon_{\mathscr{L}}(V,H) \le \left[a_\theta \varepsilon_{\mathscr{L}}(W,H)\right]^{1/(1-\theta)}. \tag{7.8.8}$$

The next assertion establishes the connection between the completeness defect and some concepts (see, e.g., [4]) in approximation theory. The following definition is needed.

7.8.7. Definition. Let $V \subset H$ be the separable Hilbert spaces.

- Let R be a linear operator in H. The value

  $$e_V^H(R) = \sup\{\|u - Ru\|_H \,:\, \|u\|_V \le 1\}$$

  is said to be the *global approximation error* in H arising in the approximation of elements $v \in V$ by elements Rv.

- Let F be a subspace of H. The value

$$e_V^H(F) = \sup\{\mathrm{dist}_H(v,F) \ : \ \|v\|_V \leq 1\}$$

is said to be the *global error of approximation* in H of elements of V by elements of the subspace F.
- *The Kolmogorov N-width* of the embedding of V into H is defined by the formula

$$\kappa_N = \kappa_N(V,H) = \inf\{e_V^H(F) \ : \ F \in \mathscr{F}_N\},$$

where $\mathscr{F}_N$ is the family of all N-dimensional subspaces F of the space V.

In other words, the Kolmogorov N-width κ_N of the embedding of V into H is the minimal global error of approximation in H of elements of V by elements of some N-dimensional subspace. ■

7.8.8. Theorem. *Let V and H be the separable Hilbert spaces such that V is compactly and densely embedded into H. Let $\mathscr{L}$ be a set of linear functionals on V. Then we have the following relations*

$$\varepsilon_{\mathscr{L}}(V,H) = \min\{e_V^H(R) \ : \ R \in \mathscr{R}_{\mathscr{L}}\},$$

where $\mathscr{R}_{\mathscr{L}}$ is the family of linear bounded operators R that map V into H and such that $Rv = 0$ for all $v \in \mathscr{L}^\perp = \{v \in V \ : \ l(v) = 0, \ l \in \mathscr{L}\}$, and

$$\kappa_N(V,H) = \min\{\varepsilon_{\mathscr{L}}(V,H) : \mathscr{L} \subset V^*, \dim \mathrm{Lin}\, \mathscr{L} = N\} = \sqrt{\mu_{N+1}},$$

where $\{\mu_j\}$ are the non-increasing ordered eigenvalues of the operator K in V defined by the equality $(Ku,v)_V = (u,v)_H$ for $u,v \in V$.

7.8.9. Example (Modes). Let $\mathscr{A}$ be a positive operator with a discrete spectrum in a separable Hilbert space H; that is, there exists the orthonormal basis $\{e_k\}$ in H such that

$$\mathscr{A} e_k = \omega_k e_k, \quad 0 < \omega_1 \leq \omega_2 \leq \cdots, \quad \lim_{k\to\infty} \omega_k = \infty. \tag{7.8.9}$$

Let $\{\mathscr{F}_s\}_{s\in\mathbb{R}}$ be the scale of spaces generated by $\mathscr{A}$; that is, $\mathscr{F}_s = \mathscr{D}(\mathscr{A}^s)$ if $s \geq 0$ and $\mathscr{F}_s$ is completion of H with respect to the norm $\|\mathscr{A}^s \cdot\|$ when $s < 0$. Denote by $\mathscr{L}$ the set of functionals $\mathscr{L} = \{l_j(u) = (u,e_j)_H \ : \ j = 1,2,\ldots,N\}$. A simple calculation relying on Theorem 7.8.8 shows that $\varepsilon_{\mathscr{L}}(\mathscr{F}_s,\mathscr{F}_\sigma) = \kappa_N(\mathscr{F}_s,\mathscr{F}_\sigma) = \omega_{N+1}^{\sigma-s}$ for every $s > \sigma$. ■

7.8.3 *Estimates for completeness defect in Sobolev spaces*

Now we consider examples of two families of functionals on the Sobolev spaces $H^s(\Omega)$ that are important from the point of view of applications.

Let Ω be either a smooth domain or a parallelepiped in $\mathbb{R}^n$. Assume that Ω is divided into subdomains $\{\Omega_j \,:\, j = 1,2,\dots,N\}$ such that

$$\overline{\Omega} = \bigcup \{\overline{\Omega_j} \,:\, j = 1,2,\dots,N\}, \quad \Omega_j \bigcap \Omega_i = \emptyset, \ j \neq i,$$

where the bar denotes the closure of a set. Assume that $\lambda_j(x)$ is a function from $L_\infty(\Omega_j)$ such that

$$\operatorname{supp} \lambda_j \subset\subset \Omega_j, \quad \int_{\Omega_j} \lambda_j(x)dx = 1$$

and Ω_j is a star-like domain with respect to the support $\operatorname{supp} \lambda_j$. Thus we have the collection $\mathscr{T} = \{(\Omega_j, \lambda_j) \,:\, j = 1,2,\dots,N\}$.

7.8.10. Example (Generalized local volume averages). We define the set $\mathscr{L}$ of generalized local volume averages corresponding to the collection $\mathscr{T}$ as the family of functionals of the type

$$\mathscr{L} = \left\{ l_j(u) = \int_{\Omega_j} \lambda_j(x)u(x)\,dx, \ j = 1,2,\dots,N \right\}.$$

It follows from [60, Theorem 3.1] that there exist constants c_1 and c_2 depending on s, σ, and Ω such that for $\varepsilon_{\mathscr{L}}(s,\sigma) \equiv \varepsilon_{\mathscr{L}}(H^s(\Omega), H^\sigma(\Omega))$ the estimate

$$c_1 \cdot \left[\max_j d_j \right]^{s-\sigma} \le \varepsilon_{\mathscr{L}}(s,\sigma) \le c_2 \cdot \left[\max_j d_j \right]^{s-\sigma} \tag{7.8.10}$$

holds for every $0 \le \sigma \le s$, where $d_j = \operatorname{diam} \Omega_j \equiv \sup\{|x-y| \,:\, x,y \in \Omega_j\}$.

As a particular case we consider the situation when $\Omega = (0,l)^n$ is the cube in $\mathbb{R}^n$ with the edge of length l and construct the collection $\mathscr{T} = \{(\Omega_j, \lambda_j)\}$ assigning local volume averages in the following way. Let $K = (0,1)^n$ be the standard unit cube in $\mathbb{R}^n$ and ω be a measurable set in K with the positive Lebesgue measure, $\operatorname{mes} \omega > 0$. We define the function $\lambda(x)$ on K by the formula

$$\lambda(x) = \begin{cases} [\operatorname{mes} \omega]^{-1}, & x \in \omega, \\ 0, & x \in K \setminus \omega. \end{cases}$$

Assume that $\Omega_j = \left\{ x = (x_1,\dots,x_n) \,:\, j_i < h^{-1}x_i < j_i + 1, \, i = 1,\dots,n \right\}$, and

$$\lambda_j(x) = \frac{1}{h^n} \lambda \left(\frac{x}{h} - j \right), \quad x \in \Omega_j,$$

for any multi-index $j = (j_1,\dots,j_n)$, where $j_i = 0,1,\dots,N-1$, $h = l/N$. We obviously have that $\operatorname{diam} \Omega_j = \sqrt{n} \cdot h$ and hence in this case by (7.8.10) we have

$$c_1 \cdot h^{s-\sigma} \le \varepsilon_{\mathscr{L}}(H^s(\Omega), H^\sigma(\Omega)) \le c_2 \cdot h^{s-\sigma}, \quad 0 \le \sigma \le s, s \neq 0.$$

■

The completeness defect for another important class of functionals is considered in the following example.

7.8.11. Example (Nodes). Let the domain Ω be divided into subdomains $\{\Omega_j\}$ as described above. We choose the point x_j in each subdomain Ω_j and define the set of functionals on $H^m(\Omega)$, $m = [n/2]+1$ (we call them nodes):

$$\mathscr{L} = \{l_j(u) = u(x_j) \ : x_j \in \Omega_j, \ j = 1,2,\dots,N\}.$$

By Theorem 3.2 [60] estimate (7.8.10) remains true for $s \geq m$ and $0 \leq \sigma \leq s$. ∎

7.8.4 Existence of determining functionals

Now we formulate several results that involve the completeness defect to characterize sets of determining functionals.

7.8.4.1 Parabolic case

We first consider a question of the existence and the properties of determining functionals for the evolutions generated by an abstract parabolic equation defined on a separable Hilbert space $\mathscr{H}$. To wit, let's consider

$$u_t + \mathscr{A} u = B(u,t), \ t > 0, \quad u|_{t=0} = u_0 \in \mathscr{H}. \tag{7.8.11}$$

Here $\mathscr{A}$ is a positive operator with the discrete spectrum: there exists the orthonormal basis $\{e_i\}$ in the space $\mathscr{H}$ such that (7.8.9) holds. The function $B(u,t)$ is a continuous mapping from $\mathscr{D}(\mathscr{A}^{1/2}) \times \mathbb{R}$ into the space $\mathscr{D}(\mathscr{A}^{-\gamma})$ for some $0 \leq \gamma < 1/2$ and it possesses the property

$$\|\mathscr{A}^{-\gamma}(B(u_1,t) - B(u_2,t))\| \leq M(\rho)\|\mathscr{A}^{1/2}(u_1 - u_2)\| \tag{7.8.12}$$

for all $u_j \in \mathscr{D}(\mathscr{A}^{1/2})$ such that $\|\mathscr{A}^{1/2}u_j\| \leq \rho$, where ρ is an arbitrary positive number, $M(\rho) > 0$ is a constant, and $\|\cdot\|$ is the norm in $\mathscr{H}$. Here $\mathscr{D}(\mathscr{A}^{-\gamma})$ is a completion of $\mathscr{H}$ with respect to the norm $\|\mathscr{A}^{-\gamma}\cdot\|$. Assume that problem (7.8.11) is uniquely solvable within the class $\mathscr{W} = C(\mathbb{R}_+;\mathscr{H}) \cap C(\mathbb{R}_+ \setminus \{0\};\mathscr{D}(\mathscr{A}^{1/2}))$ and it is point dissipative; that is, there exists $R > 0$ such that

$$\|A^{1/2}u(t)\| \leq R, \quad \text{when } t \geq t_0(u), \quad \text{for every } u(t) \in \mathscr{W}. \tag{7.8.13}$$

7.8.12. Theorem. *Assume that conditions* (7.8.12) *and* (7.8.13) *are met. Let* $\mathscr{L} = \{l_j : j = 1,\dots,N\}$ *be a set of linear continuous functionals on the space* $V = \mathscr{D}(\mathscr{A}^{1/2})$ *with the norm* $\|\cdot\|_V = \|\mathscr{A}^{1/2}\cdot\|$. *Then* $\mathscr{L}$ *is a set of asymptotically* $(V,\mathscr{H},\mathscr{W})$*-determining functionals for the problem* (7.8.11) *provided the completeness defect* $\varepsilon_{\mathscr{L}}(V,\mathscr{H})$ *fulfills the inequality*

$$\varepsilon_{\mathscr{L}}(V,\mathscr{H}) < \varepsilon_0(\gamma,R) \equiv \sqrt{\frac{1+2\gamma}{1-2\gamma}} \cdot [(1+2\gamma)M(R)]^{-1/(1-2\gamma)},$$

where $M(\rho)$, γ, *and* R *are the same as in* (7.8.12) *and* (7.8.13).

Proof. The proof relies on Theorem 7.8.2 with the function $\mathscr{V}(u,t) = \|u\|^2$, where the properties of the completeness defect given in Theorem 7.8.6 are used. We refer to [60] for details. □

The following assertion shows that in some cases the determining functionals can be defined on certain model space.

7.8.13. Theorem. *Assume that the problem* (7.8.11) *is solvable within the class* $\mathscr{W}$ *and it is pointwise dissipative (see* (7.8.13)*). Let* Z_0 *and* Z_1 *denote Hilbert spaces with the norms* $\|\cdot\|_{Z_0}$ *and* $\|\cdot\|_{Z_1}$ *such that* $Z_1 \subset Z_0$. *Assume that there exists a linear operator* $J : V = \mathscr{D}(\mathscr{A}^{1/2}) \mapsto Z_1$ *such that* $\|Ju\|_{Z_1} \le K \cdot \|u\|_V$ *and (instead of* (7.8.12)*) the inequality*

$$\frac{1}{2}\|\mathscr{A}^{1/2}(u_1-u_2)\|^2 - (B(u_1,t)-B(u_2,t),u_1-u_2) \ge -C(R)\|J(u_1-u_2)\|^2_{Z_0}$$

fulfills for all $u_j \in \mathscr{D}(\mathscr{A}^{1/2})$ *possessing the properties* $\|\mathscr{A}^{1/2}u_j\| \le R$, *where* $R > 0$ *is the constant from* (7.8.13) *and* $C(R)$ *is a positive number. Here* $(\cdot,\cdot)$ *is a scalar product in* $\mathscr{H}$. *Let* $\mathscr{L} = \{l_i : i = 1,\dots,N\}$ *be a set of the linear continuous functionals on* Z_1. *Then*

$$J^*\mathscr{L} = \{l_i^J(w) = l_i(Jw) : l_i \in \mathscr{L}\}$$

is an asymptotically $(V,\mathscr{H},\mathscr{W})$*-determining set for the problem* (7.8.11) *provided* $\varepsilon_{\mathscr{L}}(Z_1,Z_0) < (2C(R)K^2)^{-1/2}$.

The standard situation covered by the framework of Theorem 7.8.13 is the case of boundary determining functionals; that is, the case when $Z_1 \subset Z_0$ are Sobolev type spaces on the boundary of nonlinear PDE in a bounded domain (see [60] and [61, Section 5.8] for details).

7.8.4.2 Second-order in time evolution equations

We now consider an abstract second-order evolution with linear damping defined on a separable Hilbert space $\mathscr{H}$:

$$u_{tt} + \gamma u_t + \mathscr{A}u + G(u,t) = 0, \;\; t > 0, \quad u|_{t=0} = u_0, \;\; u_t|_{t=0} = u_1. \tag{7.8.14}$$

Here as above, $\mathscr{A}$ is a positive operator with discrete spectrum and γ is a positive constant. We assume that the function $G(u,t)$ is the continuous mapping from $\mathscr{D}(\mathscr{A}^{1/2}) \times \mathbb{R}$ into the space $\mathscr{H}$ possessing the property

$$\|G(u_1,t) - G(u_2,t))\| \le M(\rho)\|\mathscr{A}^\theta(u_1 - u_2)\| \tag{7.8.15}$$

for some $0 \le \theta < \frac{1}{2}$ and for all $u_j \in \mathscr{D}(\mathscr{A}^{1/2})$ such that $\|\mathscr{A}^{1/2}u_j\| \le \rho$, where ρ is an arbitrary positive number, $M(\rho) > 0$ is a constant. Assume that the problem (7.8.14) is uniquely solvable within the class of functions $\mathscr{W} = C^1(\mathbb{R}_+;\mathscr{H}) \cap C(\mathbb{R}_+;\mathscr{D}(\mathscr{A}^{1/2}))$ and it is point dissipative in the following sense: there exists $R > 0$ such that

$$\|u_t(t)\|^2 + \|\mathscr{A}^{1/2}u(t)\|^2 \le R^2 \quad \text{for } t \ge t_0(u) \text{ for every } u \in \mathscr{W}. \tag{7.8.16}$$

7.8.14. Theorem. *Let $\mathscr{L} = \{l_j : j = 1,\dots,N\}$ be a set of the linear continuous functionals on the space $V = \mathscr{D}(\mathscr{A}^{1/2})$ endowed with the norm $\|\cdot\|_V = \|\mathscr{A}^{1/2}\cdot\|$. Assume that the completeness defect $\varepsilon = \varepsilon_{\mathscr{L}}(V,\mathscr{H})$ fulfills the inequality*

$$32\gamma^{-2}M(R)^2\varepsilon_{\mathscr{L}}^{2-4\theta} + \gamma^2\varepsilon_{\mathscr{L}}^2 < 8,$$

where $M(\rho)$, R, and γ are the same as in (7.8.14)–(7.8.16). *Then $\mathscr{L}$ is a set of asymptotically $(V, V \times \mathscr{H}, \mathscr{W})$-determining functionals for the problem* (7.8.14) *in the following sense: for any two of its solutions $u_1, u_2 \in \mathscr{W}$ the condition*

$$\lim_{t\to\infty} \int_t^{t+1} |l_j(u_1(\tau)) - l_j(u_2(\tau))|^2 d\tau = 0 \ \textit{for}\ \ j = 1,\dots,N,$$

implies

$$\lim_{t\to\infty} \{\|A^{1/2}(u_1(t) - u_2(t))\|^2 + \|\partial_t u_1(t) - \partial_t u_2(t)\|^2\} = 0.$$

Proof. We rely on Theorem 7.8.1 with

$$\mathscr{V}(u) = \frac{1}{2}\left(\|\partial_t u\|^2 + \|A^{1/2}u\|^2\right) + \varepsilon(u, \partial_t u),$$

where $\varepsilon > 0$ is small enough. See details in [60]. □

As we show later, this theorem can be applied to modified von Karman equations. More complicated models may require modifications of the Lyapunov-type function $\mathscr{V}$, in order to establish a property such as (7.8.5).

For other approaches to determining functionals which are more convenient for application to von Karman models with nonlinear damping we refer to Section 7.9 below and also to Section 8.9 in Chapter 8.

7.9 Stabilizability estimate and its consequences

In this section we consider a class of dissipative dynamical systems of a special form and study their dynamical properties. Our main motivation is related to nonlinear

PDE's of second order in time possibly interacting with parabolic equation. More precisely, we assume the following.

7.9.1. Assumption. Let X, Y, and Z be reflexive Banach spaces; X is compactly embedded in Y. We endow the space $H = X \times Y \times Z$ with the norm

$$|y|_H^2 = |u_0|_X^2 + |u_1|_Y^2 + |\theta|_Z^2, \quad y = (u_0; u_1; \theta_0).$$

The trivial case $Z = \{0\}$ is allowed. We assume that (H, S_t) is a dynamical system on $H = X \times Y \times Z$ with the evolution operator of the form

$$S_t y = (u(t); u_t(t); \theta(t)), \quad y = (u_0; u_1; \theta_0) \in H, \tag{7.9.1}$$

where the functions $u(t)$ and $\theta(t)$ possess the properties

$$u \in C(\mathbb{R}_+, X) \cap C^1(\mathbb{R}_+, Y), \quad \theta \in C(\mathbb{R}_+, Z).$$

The structure of the phase space H and the evolution operator S_t in Assumption 7.9.1 is motivated by the study of the system generated by equation of the second order in time in $X \times Y$ (e.g., as (2.4.1)) possibly interacting with some first-order evolution equation in space Z. This type of interaction arises in modeling of thermoelastic plates (see Chapter 5, for instance)

7.9.2. Definition. A dynamical system of the form (7.9.1) is said to be stable modulo compact terms (*quasi-stable*, for short) on a set $\mathscr{B} \subset H$ if there exist a compact seminorm $\mu_X(\cdot)$ on the space X and nonnegative scalar functions $a(t)$, $b(t)$, and $c(t)$ on $\mathbb{R}_+$ such that (i) $a(t)$ and $c(t)$ are locally bounded on $[0, \infty)$, (ii) $b(t) \in L_1(\mathbb{R}_+)$ possesses the property $\lim_{t\to\infty} b(t) = 0$, and (iii) for every $y_1, y_2 \in \mathscr{B}$ and $t > 0$ the following relations

$$|S_t y_1 - S_t y_2|_H^2 \le a(t) \cdot |y_1 - y_2|_H^2 \tag{7.9.2}$$

and

$$|S_t y_1 - S_t y_2|_H^2 \le b(t) \cdot |y_1 - y_2|_H^2 + c(t) \cdot \sup_{0\le s\le t} \left[\mu_X(u^1(s) - u^2(s))\right]^2 \tag{7.9.3}$$

hold. Here we denote $S_t y_i = (u^i(t); u_t^i(t); \theta^i(t))$, $i = 1, 2$. ∎

7.9.3. Remark. The definition of *quasi-stability* is rather natural from the point of view of long-time behavior. It pertains to decomposition of the flow into exponentially stable and compact part. This represents some sort of analogy with the "splitting" method [17, 273], however, the decomposition refers to the difference of two trajectories, rather than a single trajectory. The relation (7.9.3) is called a *stabilizability estimate* and, in the context of long-time dynamics, was originally introduced in [69] (see also [63, 68] and the discussion in [75]). To obtain such an estimate proves fairly technical (in critical problems) and requires rather subtle PDE tools to prove it. Illustrations of the method are given later in Section 8.5 for some abstract models and in Chapters 9–11 for a variety of von Karman models. We also refer to [37, 38, 63, 69, 71, 74, 75, 78, 79, 239] for similar considerations for other models.

The notion of quasi-stability introduced by Definition 7.9.2 requires a special structure of the semiflow and a special type of a (stabilizability) estimate. However, the idea behind this notion can be applied in many other cases (see, e.g., [68, 75, 77] and also the proof of Theorem 10.2.11 based on the stabilizability-type estimate established in Proposition 10.2.17 and the proof of Theorem 10.3.5 based on Proposition 10.3.10). Systems with delay/memory terms can be also included in this framework (see, e.g., [108, 109, 243, 253] and also the proof of Theorem 9.3.5). The same idea was recently applied in [65] for analysis of long-time dynamics in a parabolic-type model of the form

$$K(u)u_t + \mathscr{A}u + F(u) = 0$$

in a Hilbert space $\mathscr{H}$, where $\mathscr{A}$ is a positive linear operator with the dense domain $\mathscr{D}(\mathscr{A})$, $K(u)$ is a family of bounded positive operators in $\mathscr{H}$ for every $u \in \mathscr{D}(\mathscr{A}^{1/2})$, and $F : \mathscr{D}(\mathscr{A}^{1/2}) \mapsto \mathscr{H}$ is a (nonlinear) mapping. ■

In what follows our aim is to show that quasi-stable systems enjoy many nice properties that include (i) existence of global attractors that is both finite-dimensional and smooth, (ii) exponential attractors, and (iii) determining functionals, and so on.

7.9.4. Proposition. *Let Assumption 7.9.1 be in force. Assume that the dynamical system (H, S_t) is quasi-stable on every bounded forward invariant set $\mathscr{B}$ in H. Then, (H, S_t) is asymptotically smooth.*

Proof. Let

$$\widetilde{X} = \text{Closure}\left\{v \in X : |v|_{\widetilde{X}} \equiv \mu_X(v) + |v|_Y < \infty\right\}. \tag{7.9.4}$$

One can see that X is compactly embedded in the Banach space $\widetilde{X}$. Therefore by Theorem 1.1.8 we have that the space

$$W^1_{\infty,2}(0,T;X,Y) = \left\{f \in L_\infty(0,T;X) \,:\, f' \in L_2(0,T;Y)\right\}$$

is compactly embedded in $C(0,T;\widetilde{X})$. This implies that the pseudometric $\rho^t_{\mathscr{B}}$ in $C(0,t;H)$ defined by the formula

$$\rho^t_{\mathscr{B}}(\{S_\tau y_1\}, \{S_\tau y_2\}) = c(t) \sup_{\tau\in[0,t]} \mu_X(u^1(t) - u^2(t))$$

is precompact (with respect to H). Here we denote by $\{S_\tau y_i\}$ the element from $C(0,t;H)$ given by function $y_i(\tau) = S_\tau y_i \equiv (u^i(t); u^i_t(t); \theta^i(t))$. By (7.9.3) $\rho^t_{\mathscr{B}}$ satisfies the hypotheses of Proposition 7.1.9 with $K_B(t) = b(t)$. This implies the result. □

7.9.5. Corollary. *If the system (H, S_t) is dissipative and satisfies the hypotheses of Proposition 7.9.4, then it possesses a compact global attractor.*

Proof. By Proposition 7.9.4 the system (H, S_t) is asymptotically smooth. Thus the result follows from Theorem 7.2.3. □

7.9.1 *Finite dimension of global attractors*

We start with the following general assertion.

7.9.6. Theorem. *Let Assumption* 7.9.1 *be valid. Assume that the dynamical system* (H,S_t) *possesses a compact global attractor* A *and is quasi-stable on* A *(see Definition* 7.9.2*). Then the attractor* A *of has a finite fractal dimension* $\dim_f^H A$.

Proof. The idea of the proof is based on the method of "short" trajectories (see, e.g., [225, 226] and the references therein and also [75]).

We apply Theorem 7.3.3 in the space $H_T = H \times W_1(0,T)$ with an appropriate T. Here

$$W_1(0,T) = \left\{ z \in L_2(0,T;X) : |z|^2_{W_1(0,T)} \equiv \int_0^T \left(|z(t)|_X^2 + |z_t(t)|_Y^2\right) dt < \infty \right\}. \quad (7.9.5)$$

The norm in H_T is given by

$$\|U\|^2_{H_T} = |y|_H^2 + |z|^2_{W_1(0,T)}, \quad U = (y;z),\ y = (u_0;u_1;\theta_0). \quad (7.9.6)$$

Let $y_i = (u_0^i;u_1^i;\theta_0^i)$, $i=1,2$, be two elements from the attractor A. We denote

$$S_t y_i = (u^i(t);u_t^i(t);\theta^i(t)), \quad t \geq 0,\ i=1,2,$$

and $Z(t) = S_t y_1 - S_t y_2 \equiv (z(t);z_t(t);\xi(t))$, where

$$(z(t);z_t(t);\xi(t)) \equiv (u^1(t)-u^2(t);u_t^1(t)-u_t^2(t);\theta^1(t)-\theta^2(t)).$$

Integrating (7.9.3) from T to $2T$ with respect to t, we obtain that

$$\int_T^{2T} |S_t y_1 - S_t y_2|_H^2 dt \leq \widetilde{b}_T |y_1 - y_2|_H^2 + \widetilde{c}_T \sup_{0\leq s\leq 2T} [\mu_X(z(s))]^2, \quad (7.9.7)$$

where

$$\widetilde{b}_T = \int_T^{2T} b(t)dt \text{ and } \widetilde{c}_T = \int_T^{2T} c(t)dt.$$

It also follows from (7.9.3) that

$$|S_T y_1 - S_T y_2|_H^2 \leq b(T)\cdot |y_1 - y_2|_H^2 + c(T)\cdot \sup_{0\leq s\leq T} [\mu_X(z(s))]^2$$

and combining with (7.9.7) yields

$$|S_T y_1 - S_T y_2|_H^2 + \int_T^{2T} |S_t y_1 - S_t y_2|_H^2 dt \leq b_T |y_1 - y_2|_H^2 + c_T \sup_{0\leq s\leq 2T} [\mu_X(z(s))]^2, \quad (7.9.8)$$

where

$$b_T = b(T) + \int_T^{2T} b(t)dt \text{ and } c_T = c(T) + \int_T^{2T} c(t)dt. \tag{7.9.9}$$

Let A be the global attractor. Consider in the space H_T the set

$$A_T := \{U \equiv (u(0); u_t(0); \theta(0); u(t), t \in [0,T]) : (u(0); u_t(0); \theta(0)) \in A\},$$

where $u(t)$ is the first component of $S_t y(0)$ with $y(0) = (u(0); u_t(0); \theta(0))$, and define operator $V : A_T \mapsto H_T$ by the formula

$$V : (u(0); u_t(0); \theta(0); u(t)) \mapsto (S_T y(0); u(T+t)).$$

It is clear that V is Lipschitz on A_T and $VA_T = A_T$.

Because the space $\widetilde{X}$ given by (7.9.4) possesses the properties

$$X \subset \widetilde{X} \subset Y \quad \text{and} \quad X \subset \widetilde{X} \text{ is compact},$$

contradiction argument yields

$$[\mu_X(u)]^2 \le \varepsilon |u|_X^2 + C_\varepsilon |u|_Y^2 \quad \text{for any} \quad \varepsilon > 0. \tag{7.9.10}$$

Therefore it follows from (7.9.2) that

$$\sup_{0\le s\le 2T} [\mu_X(z(s))]^2 \le \frac{b_T}{c_T} |y_1 - y_2|_H^2 + C(a_T, b_T, c_T) \sup_{0\le s\le 2T} |z(s)|_Y^2,$$

where $a_T = \sup_{0\le s\le 2T} a(s)$ and b_T, c_T given by (7.9.9). Consequently, from (7.9.8) we obtain

$$\|VU_1 - VU_2\|_{H_T} \le \eta_T \|U_1 - U_2\|_{H_T} + K_T \cdot (n_T(U_1 - U_2) + n_T(VU_1 - VU_2)),$$

for any $U_1, U_2 \in A_T$, where $K_T > 0$ is a constant (depending on a_T, b_T, c_T, the embedding properties of X into Y, the seminorm μ_X), and

$$\eta_T^2 = 2b_T = 2b(T) + 2\int_T^{2T} b(t)dt. \tag{7.9.11}$$

The seminorm n_T has the form $n_T(U) := \sup_{0\le s\le T} |u(s)|_Y$. Because $W_1(0,T)$ is compactly embedded into $C(0,T;Y)$, $n_T(U)$ is a compact seminorm on H_T and we can choose $T > 1$ such that $\eta_T < 1$. We also have from (7.9.2) that

$$\|VU_1 - VU_2\|_{H_T} \le L_T \|U_1 - U_2\|_{\mathscr{H}_T} \quad \text{for} \quad U_1, U_2 \in A_T,$$

where $L_T^2 = a(T) + \int_T^{2T} a(t)dt$. Therefore we can apply Theorem 7.3.3 which implies that A_T is a compact set in H_T of finite fractal dimension.

Let $\mathscr{P} : H_T \mapsto H$ be the operator defined by the formula

$$\mathscr{P} : (u_0; u_1; \theta_0; z(t)) \mapsto (u_0; u_1; \theta_0).$$

$A = \mathscr{P}A_T$ and $\mathscr{P}$ is Lipshitz continuous, thus $\dim_f^H A \le \dim_f^{H_T} A_T < \infty$. Here $\dim_f^W$ stands for the fractal dimension of a set in the space W. This concludes the proof of Theorem 7.9.6. □

7.9.7. Remark. By Remark 7.3.4 the dimension $\dim_f^H A$ of the attractor admits the estimate

$$\dim_f^H A \le \left[\ln \frac{2}{1+\eta_T}\right]^{-1} \cdot \ln m_0 \left(\frac{4K_T(1+L_T^2)^{1/2}}{1-\eta_T}\right), \tag{7.9.12}$$

Here $m_0(R)$ is the maximal number of pairs $(x_i; y_i)$ in $H_T \times H_T$ possessing the properties

$$\|x_i\|_{H_T}^2 + \|y_i\|_{H_T}^2 \le R^2, \quad n_T(x_i - x_j) + n_T(y_i - y_j) > 1, \quad i \ne j.$$

It is clear that $m_0(R)$ can be estimated by the maximal number of pairs $(x_i; y_i)$ in $W_1(0,T) \times W_1(0,T)$ possessing the properties $\|x_i\|_{W_1(0,T)}^2 + \|y_i\|_{W_1(0,T)}^2 \le R^2$ and $n_T(x_i - x_j) + n_T(y_i - y_j) > 1$ for all $i \ne j$. Thus the bound in (7.9.12) depends on the functions a, b, c and seminorm μ_X in Definition 7.9.2 and also on the embedding properties of X into Y. We use this observation in further considerations. ■

7.9.2 *Regularity of trajectories from the attractor*

In this section we show how stabilizability estimates can be used in order to obtain additional regularity of trajectories lying on the global attractor. The theorem below provides regularity for time derivatives. The needed "space" regularity follows from the analysis of the respective PDE. It typically involves application of elliptic theory (see the corresponding results in Chapter 9).

7.9.8. Theorem. *Let Assumption* 7.9.1 *be valid. Assume that the dynamical system* (H, S_t) *possesses a compact global attractor A and is quasi-stable on the attractor A. Moreover, we assume that* (7.9.3) *holds with the function* $c(t)$ *possessing the property* $c_\infty = \sup_{t\in\mathbb{R}_+} c(t) < \infty$. *Then any full trajectory* $\{(u(t); u_t(t); \theta(t)) : t \in \mathbb{R}\}$ *that belongs to the global attractor enjoys the following regularity properties,*

$$u_t \in L_\infty(\mathbb{R}; X) \cap C(\mathbb{R}; Y), \quad u_{tt} \in L_\infty(\mathbb{R}; Y), \quad \theta_t \in L_\infty(\mathbb{R}; Z) \tag{7.9.13}$$

Moreover, there exists $R > 0$ *such that*

$$|u_t(t)|_X^2 + |u_{tt}(t)|_Y^2 + |\theta_t(t)|_Z^2 \le R^2, \quad t \in \mathbb{R}, \tag{7.9.14}$$

where R depends on the constant c_∞, *on the seminorm* μ_X *in Definition* 7.9.2, *and also on the embedding properties of X into Y.*

Proof. It follows from (7.9.3) that for any two trajectories

$$\gamma = \{U(t) \equiv (u(t); u_t(t); \theta(t)) : t \in \mathbb{R}\},$$
$$\gamma^* = \{U^*(t) \equiv (u^*(t); u_t^*(t); \theta^*(t)) : t \in \mathbb{R}\}$$

from the global attractor we have that

$$|Z(t)|_H^2 \leq b(t-s)|Z(s)|_H^2 + c(t-s) \sup_{s \leq \tau \leq t} [\mu_X(z(\tau))]^2 \tag{7.9.15}$$

for all $s \leq t$, $s,t \in \mathbb{R}$, where $Z(t) = U^*(t) - U(t)$ and $z(t) = u^*(t) - u(t)$. In the limit $s \to -\infty$ relation (7.9.15) gives us that

$$|Z(t)|_H^2 \leq c_\infty \sup_{-\infty \leq \tau \leq t} [\mu_X(z(\tau))]^2$$

for every $t \in \mathbb{R}$ and for every couple of trajectories γ and γ^*. Using relation (7.9.10) we can conclude that

$$\sup_{-\infty \leq \tau \leq t} |Z(\tau)|_H^2 < C \sup_{-\infty \leq \tau \leq t} |z(\tau)|_Y^2, \tag{7.9.16}$$

for every $t \in \mathbb{R}$ and for every couple of trajectories γ and γ^* from the attractor.

Now we fix the trajectory γ and for $0 < |h| < 1$ we consider the shifted trajectory $\gamma^* \equiv \gamma_h = \{y(t+h) : t \in \mathbb{R}\}$. Applying (7.9.16) for this pair of trajectories and using the fact that all terms (7.9.16) are quadratic with respect to Z we obtain that

$$\sup_{-\infty \leq \tau \leq t} \left\{ |u^h(\tau)|_X^2 + |u_t^h(\tau)|_Y^2 + |\theta_t^h(\tau)|_Z^2 \right\} \leq C \sup_{-\infty \leq \tau \leq t} |u^h(\tau)|_Y^2, \tag{7.9.17}$$

where $u^h(t) = h^{-1} \cdot [u(t+h) - u(t)]$ and $\theta^h(t) = h^{-1} \cdot [\theta(t+h) - \theta(t)]$. On the attractor we obviously have that

$$|u^h(t)|_Y \leq \frac{1}{h} \cdot \int_0^h |u_t(\tau + t)|_Y d\tau \leq C, \quad t \in \mathbb{R},$$

with uniformity in h. Therefore (7.9.17) implies that

$$|u^h(t)|_X^2 + |u_t^h(t)|_Y^2 + |\theta_t^h(t)|_Z^2 \leq C, \quad t \in \mathbb{R}.$$

Passing with the limit on h then yields relations (7.9.13) and (7.9.14). □

7.9.3 Fractal exponential attractors

For quasi-stable systems we have the following result pertaining to (generalized) fractal exponential attractors (see Definition 7.4.4 and Remark 7.4.5). For more details concerning fractal exponential attractors we refer to [102] and also to recent survey [232].

7.9.9. Theorem. *Let Assumption 7.9.1 be valid. Assume that the dynamical system (H,S_t) is dissipative and quasi-stable on some bounded absorbing set $\mathscr{B}$. We also assume that there exists a space $\widetilde{H} \supseteq H$ such that $t \mapsto S_t y$ is Hölder continuous in $\widetilde{H}$ for every $y \in \mathscr{B}$; that is, there exist $0 < \gamma \le 1$ and $C_{\mathscr{B},T} > 0$ such that*

$$|S_{t_1}y - S_{t_2}y|_{\widetilde{H}} \le C_{\mathscr{B},T}|t_1 - t_2|^{\gamma}, \quad t_1,t_2 \in [0,T],\ y \in \mathscr{B}. \tag{7.9.18}$$

Then the dynamical system (H,S_t) possesses a (generalized) fractal exponential attractor whose dimension is finite in the space $\widetilde{H}$.

Proof. We apply the same idea as in the proof of Theorem 7.9.6.

We can assume that absorbing set $\mathscr{B}$ is closed and forward invariant (otherwise, instead of $\mathscr{B}$, we consider $\mathscr{B}' = \overline{\cup_{t\ge t_0} S_t \mathscr{B}}$ for t_0 large enough, which lies in $\mathscr{B}$).

In the space $H_T = H \times W_1(0,T)$ equipped with the norm (7.9.6), where $W_1(0,T)$ is given by (7.9.5), we consider the set

$$\mathscr{B}_T := \{U \equiv (u(0);u_t(0);\theta(0);u(t),t \in [0,T]) \,:\, (u(0);u_t(0);\theta(0)) \in \mathscr{B}\},$$

where $u(t)$ is the first component of $S_t y(0)$ with $y(0) = (u(0);u_t(0);\theta(0))$ and define operator $V \,:\, \mathscr{B}_T \mapsto H_T$ by the formula

$$V \,:\, (u(0);u_t(0);\theta(0);u(t)) \mapsto (S_T y(0);u(T+t)).$$

It is clear that $\mathscr{B}_T$ is a closed bounded set in H_T which is forward invariant with respect to V.

It follows from (7.9.2) that

$$\|VU_1 - VU_2\|^2_{H_T} \le \left(a(T) + \int_T^{2T} a(t)dt\right) \|U_1 - U_2\|^2_{H_T}, \quad U_1,U_2 \in \mathscr{B}_T.$$

As in the proof of Theorem 7.9.6 we can obtain that

$$\|VU_1 - VU_2\|_{H_T} \le \eta_T \|U_1 - U_2\|_{H_T} + K_T \cdot (n_T(U_1 - U_2) + n_T(VU_1 - VU_2))$$

for any $U_1,U_2 \in \mathscr{B}_T$ and for some $T > 0$, where $K_T > 0$ is a constant, $n_T(U) := \sup_{0\le s\le T} |u(s)|_Y$ and $\eta_T < 1$ is given by (7.9.11). Therefore by Theorem 7.4.2 the mapping V possesses a fractal exponential attractor; that is, there exists a compact set $\mathscr{A}_T \subset \mathscr{B}_T$ and a number $0 < q < 1$ such that $\dim_f^{H_T} \mathscr{A}_T < \infty$, $V\mathscr{A}_T \subset \mathscr{A}_T$, and

$$\sup\left\{\text{dist}_{H_T}(V^k U, \mathscr{A}_T) \,:\, U \in \mathscr{B}_T\right\} \le Cq^k, \quad k = 1,2,\dots,$$

for some constant $C > 0$. In particular, this relation implies that

$$\sup\{\text{dist}_H(S_{kT}y, \mathscr{A}) \,:\, u \in \mathscr{B}\} \le Cq^k, \quad k = 1,2,\dots, \tag{7.9.19}$$

where $\mathscr{A}$ is the projection of $\mathscr{A}_T$ of the first components:

$$\mathscr{A} = \{(u(0);u_t(0);\theta(0)) \in \mathscr{B} \,:\, (u(0);u_t(0);u(t);\theta(0);t \in [0,T]) \in \mathscr{A}_T\}.$$

It is clear that $\mathscr{A}$ is a compact forward invariant set with respect to S_T; that is, $S_T\mathscr{A} \subset \mathscr{A}$. Moreover $\dim_f^H \mathscr{A} \leq \dim_f^{H_T} \mathscr{A}_T < \infty$. One can also see that

$$A_{\exp} = \cup\{S_t\mathscr{A} \,:\, t \in [0,T]\}$$

is a compact forward invariant set with respect to S_t; that is, $S_t A_{\exp} \subset A_{\exp}$. Moreover, it follows from (7.9.18) that $\dim_f^{\widetilde{H}} A_{\exp} \leq c\left[1+\dim_f^H \mathscr{A}\right] < \infty$. We also have from (7.9.19) and (7.9.2) that

$$\sup\left\{\mathrm{dist}_H(S_t y, A_{\exp}) \,:\, u \in \mathscr{B}\right\} \leq Ce^{-\gamma t}, \quad t \geq 0,$$

for some $\gamma > 0$. Thus $A_{\exp}$ is a (generalized) fractal exponential attractor. □

7.9.10. Remark. Hólder continuity (7.9.18) is needed in order to derive the finiteness of the fractal dimension $\dim_f^{\widetilde{H}} A_{\exp}$ from the finiteness of $\dim_f^H \mathscr{A}$. We do not know whether the same holds true without property (7.9.18) imposed in some vicinity of $A_{\exp}$. This is because $A_{\exp}$ is a *uncountable* union of (finite-dimensional) sets $S_t\mathscr{A}$. We also emphasize that fractal dimension depends on the topology. Indeed, [261] provides an example of a set with finite fractal dimension in one space and infinite fractal dimension in another (smaller) space. ■

7.9.4 Determining functionals

Another consequence of the stabilizability estimate is the following assertion pertaining to determining functionals (see Section 7.8 for general discussion of the theory of determining functionals).

7.9.11. Theorem. *Let Assumption* 7.9.1 *be valid. Assume that the dynamical system (H, S_t) is dissipative and quasi-stable on some bounded absorbing set $\mathscr{B}$. Let $\mathscr{L} = \{l_j : j = 1, \ldots, N\}$ be a set of linearly independent functionals on X and*

$$\varepsilon_{\mathscr{L}}(\mu_X) = \sup\{\mu_X(w) \,:\, w \in X, l(w) = 0, l \in \mathscr{L}, |w|_X \leq 1\}$$

be its completeness defect with respect to the seminorm μ_X. If there exists $\tau > 0$ such that

$$\eta_\tau \equiv b(\tau) + \varepsilon^2_{\mathscr{L}}(\mu_X)c(\tau) \cdot \sup_{s\in[0,\tau]} a(s) < 1 \tag{7.9.20}$$

then, the relation

$$\lim_{t\to\infty} l_j(u^1(s) - u^2(s)) = 0, \quad j = 1, 2, \ldots, N, \tag{7.9.21}$$

implies that $\lim_{t\to\infty} |S_t y_1 - S_t y_2|_H = 0$. *Here* $S_t y_i = (u^i(t); u^i_t(t); \theta^i(t))$, $i = 1, 2$.

Proof. We first note that the convergence in (7.9.21) is equivalent to the convergence

$$\Delta_{\mathscr{L}}(t) \equiv \sup_{s\in[t,t+\tau]} \max_j |l_j(u^1(s) - u^2(s))| = 0, \quad t \to \infty, \tag{7.9.22}$$

for every fixed $\tau > 0$ Assume now that $S_t y_i = (u^i(t); u^i_t(t); \theta^i(t)) \in \mathscr{B}$ for $t \geq t_0$, $i = 1, 2$. Then from (7.9.3) we have that

$$|S_{t+\tau}y_1 - S_{t+\tau}y_2|^2_H \leq b(\tau)|S_t y_1 - S_t y_2|^2_H + c(\tau) \sup_{t\leq s\leq t+\tau} \left[\mu_X(u^1(s) - u^2(s))\right]^2, \tag{7.9.23}$$

for any $t \geq t_0$. Similarly to (7.8.7) we have that

$$\mu_X(v) \leq \varepsilon_{\mathscr{L}}(\mu_X)|v|_X + C_{\mathscr{L}} \max_{j=1,\dots,N} |l_j(v)|, \quad \forall v \in X. \tag{7.9.24}$$

Indeed, let $\{e_j : j = 1, \dots, N\} \subset X$ be a biorthogonal system for $\mathscr{L}$ (i.e., $l_j(v_i) = 0$ if $j \neq i$ and $l_j(e_j) = 1$). In this case for any $v \in X$ the element $w = v - \sum_{i=1}^N l_i(v) e_i$ possesses the properties $l_j(w) = 0$ for $j = 1, \dots, N$. By the definition of $\varepsilon_{\mathscr{L}}(\mu_X)$ we have that $\mu_X(w) \leq \varepsilon_{\mathscr{L}}(\mu_X)|w|_X$. Therefore from the representation for w we obtain (7.9.24).

From (7.9.24) we have

$$[\mu_X(v)]^2 \leq (1+\delta)\varepsilon^2_{\mathscr{L}}(\mu_X)|v|^2_X + C_{\mathscr{L},\delta} \max_{j=1,\dots,N} |l_j(v)|^2, \quad \forall v \in X,$$

for each $\delta > 0$. By (7.9.2) this implies that

$$\sup_{t\leq s\leq t+\tau} \left[\mu_X(u^1(s) - u^2(s))\right]^2$$
$$\leq \left[(1+\delta)\varepsilon^2_{\mathscr{L}}(\mu_X) \sup_{s\in[0,\tau]} a(s)\right] |S_t y_1 - S_t y_2|^2_H + C_{\mathscr{L},\delta}\Delta^2_{\mathscr{L}}(t).$$

Consequently, (7.9.23) yields

$$|S_{t+\tau}y_1 - S_{t+\tau}y_2|^2_H \leq \eta |S_t y_1 - S_t y_2|^2_H + C_{\mathscr{L},\delta}\Delta^2_{\mathscr{L}}(t),$$

where $\eta = (1+\delta)\varepsilon^2_{\mathscr{L}}(\mu_X)c(\tau) \cdot \sup_{s\in[0,\tau]} a(s) + b(\tau)$. Under condition (7.9.20) we can choose $\delta > 0$ such that $\eta < 1$ and find that

$$|S_{t_0+n\tau}y_1 - S_{t_0+n\tau}y_2|^2_H \leq \eta^n \cdot |S_{t_0}y_1 - S_{t_0}y_2|^2_H + C \sum_{m=0}^{n-1} \eta^{n-m-1}\Delta^2_{\mathscr{L}}(t_0 + m\tau).$$

It is easy to see now that $\lim_{n\to\infty} |S_{t_0+n\tau}y_1 - S_{t_0+n\tau}y_2|^2_H = 0$ under conditions (7.9.20) and (7.9.22). Application of (7.9.2) completes the proof. □

Chapter 8
Long-Time Behavior of Second-Order Abstract Equations

The main aim in this chapter is to present general methods that can be used in the study of the long-time behavior of dynamical systems generated by damped second-order abstract equations. To accomplish this we use abstract results of Chapter 7 that describe long-time dynamics of more general dynamical systems. Particular assumptions imposed on the data of the problem (damping and sources) are motivated by applications to nonlinear plate equations, with particular emphasis on von Karman evolutions. In fact, in Chapter 9 we specialize these results to von Karman models with internal nonlinear damping and both subcritical and critical sources. However, abstract hypotheses formulated are tailored to accomodate larger classes of nonlinear dynamic plate equations. We refer to [75] for a more general approach covering a larger class of nonlinear second order evolution equations and to [103, 123, 124, 156] for the the case of the corresponding abstract semilinear problems with linear damping.

In this chapter we rely on the abstract results presented in Chapters 2 and 7.

8.1 Main assumptions

We start with the following second-order abstract equation:

$$\begin{cases} Mu_{tt}(t) + \mathscr{A}u(t) + kD(u_t(t)) = F(u(t)), \\ u|_{t=0} = u_0, \quad u_t|_{t=0} = u_1, \end{cases} \tag{8.1.1}$$

under the following set of assumptions (which is a special case of Assumptions 2.4.1 and 2.4.15).

8.1.1. Assumption.

(A) $\mathscr{A}$ is a closed, linear positive self-adjoint operator acting on a Hilbert space $\mathscr{H}$ with $\mathscr{D}(\mathscr{A}) \subset \mathscr{H}$. We denote by $|\cdot|$ the norm of $\mathscr{H}$; $(\cdot,\cdot)$ denotes a scalar product in $\mathscr{H}$. We use the same symbol to denote the duality pairing between $\mathscr{D}(\mathscr{A}^{1/2})$ and $\mathscr{D}(\mathscr{A}^{1/2})'$.

I. Chueshov, I. Lasiecka, *Von Karman Evolution Equations*,
Springer Monographs in Mathematics, DOI 10.1007/978-0-387-87712-9_8,

(M) Let V be another Hilbert space such that $\mathscr{D}(\mathscr{A}^{1/2}) \subset V \subset \mathscr{H} \subset V' \subset \mathscr{D}(\mathscr{A}^{1/2})'$, all injections being continuous and dense, $M \in L(V,V')$, the bilinear form (Mu,v) is symmetric, and $(Mu,u) \geq \alpha_0|u|_V^2$, where $\alpha_0 > 0$ and $(\cdot,\cdot)$ is understood as a duality pairing between V and V'. Hence, $M^{-1} \in L(V',V)$. Setting $\bar{M} = M|_{\mathscr{H}}$ with $\mathscr{D}(\bar{M}) = \{u \in V; Mu \in \mathscr{H}\}$ we have $D(\bar{M}^{1/2}) = V$ (below we do not distinguish the operators M and $\bar{M}$ and assume that $(\cdot,\cdot)_V = (M\cdot,\cdot)$ and $|\cdot|_V = |M^{1/2}\cdot|$).

(D) The operator $D : \mathscr{D}(\mathscr{A}^{1/2}) \to [\mathscr{D}(\mathscr{A}^{1/2})]'$ is assumed monotone and hemicontinuous with

$$D(0) = 0 \text{ and } (Du - Dv, u - v) \geq 0; \quad u, v \in \mathscr{D}(\mathscr{A}^{1/2}); \tag{8.1.2}$$

the intensity parameter k is positive.

(F) The nonlinear operator $F : \mathscr{D}(\mathscr{A}^{1/2}) \mapsto V'$ is locally Lipschitz; that is,

$$|F(u_1) - F(u_2)|_{V'} \leq L(\rho)|\mathscr{A}^{1/2}(u_1 - u_2)|, \quad \forall |\mathscr{A}^{1/2}u_i| \leq \rho\,. \tag{8.1.3}$$

We also assume that F has the form

$$F(u) = -\Pi'(u) + F^*(u), \tag{8.1.4}$$

where $\Pi(u)$ is a C^1-functional on $\mathscr{D}(\mathscr{A}^{1/2})$ with $\Pi'(u)$ Fréchet derivative[1] and $F^*(u)$ is a nonlinear mapping from $\mathscr{D}(\mathscr{A}^{1/2})$ into V' possessing the property

$$|F^*(u_1) - F^*(u_2)|_{V'}^2 \leq c_0|\mathscr{A}^{1/2}(u_1 - u_2)|^2, \quad u_1, u_2 \in \mathscr{D}(\mathscr{A}^{1/2}). \tag{8.1.5}$$

We assume that $\Pi(u) = \Pi_0(u) + \Pi_1(u)$, where $\Pi_0(u) \geq 0$, $\Pi_0(u)$ is bounded on bounded sets from $\mathscr{D}(\mathscr{A}^{1/2})$ and $\Pi_1(u)$ possesses the property

$$|\Pi_1(u)| \leq \eta \cdot \left(|\mathscr{A}^{1/2}u|^2 + \Pi_0(u)\right) + C_\eta\,, \quad u \in \mathscr{D}(\mathscr{A}^{1/2})\,, \tag{8.1.6}$$

for every $\eta > 0$.

8.1.2. Remark. The term $\Pi(u)$ represents the potential energy related to conservative (potential) forces and $F^*(u)$ corresponds to non-conservative forces (i.e., forces that cannot be represented by potential operators). The splitting of $\Pi(u)$ into two parts Π_0 and Π_1 is according to the sign of the energy function, with $\Pi_0(u)$ always being nonnegative. The inequality in (8.1.6), which in the case of von Karman equations is related to the uniqueness property of the von Karman bracket, is a critical ingredient in establishing the bound from below for the energy function. See also Remark 1.5.12. ∎

It was shown in Chapter 2 (see Theorem 2.4.16) that under Assumption 8.1.1 there exists a global generalized solution $u(t)$ to (8.1.1) from the class

---

[1] This is to say that the inequality $|\langle \Pi'(u); h\rangle| \leq C_r|\mathscr{A}^{1/2}h|$ takes place for $u \in \mathscr{D}(\mathscr{A}^{1/2})$ such that $|\mathscr{A}^{1/2}u| \leq r$.

$$C(\mathbb{R}_+,\mathscr{D}(\mathscr{A}^{1/2}))\cap C^1(\mathbb{R}_+,V).$$

This solution $u(t)$ can be approximated by strong solutions. For each strong solution $u(t)$ the following energy relation

$$\mathscr{E}(u(t),u_t(t))+k\int_0^t(D(u_t(\tau)),u_t(\tau))d\tau=\mathscr{E}(u_0,u_1)+\int_0^t(F^*(u(\tau)),u_t(\tau))d\tau \tag{8.1.7}$$

holds, where

$$\mathscr{E}(u_0,u_1)=E(u_0,u_1)+\Pi_1(u_0) \tag{8.1.8}$$

with

$$E(u_0,u_1)=\frac{1}{2}((Mu_1,u_1)+(\mathscr{A}u_0,u_0))+\Pi_0(u_0)\,. \tag{8.1.9}$$

We note that relation (8.1.6) implies that there exists a constant $c>0$ such that

$$\frac{1}{2}\cdot E(u_0,u_1)-c\le\mathscr{E}(u_0,u_1)\le 2\cdot E(u_0,u_1)+c,\quad u_0\in\mathscr{D}(\mathscr{A}^{1/2}),\ u_1\in V\,. \tag{8.1.10}$$

Any generalized solution satisfies (cf. (2.4.36)) the energy inequality

$$\mathscr{E}(u(t),u_t(t))\le\mathscr{E}(u_0,u_1)+\int_0^t(F^*(u(\tau)),u_t(\tau))d\tau. \tag{8.1.11}$$

Hence by (8.1.5) and (8.1.10) using Gronwall's lemma we find that

$$E(u(t),u_t(t))\le C(1+E(u_0,u_1))e^{at},\quad t\ge 0, \tag{8.1.12}$$

for any generalized solution, where C and a are positive constants. In the conservative case ($F^*\equiv 0$), we obviously have $a=0$ in (8.1.12).

Below we also need the following assertion.

8.1.3. Proposition. *Let Assumption* 8.1.1 *hold. Then for any $R>0$ we have the following estimate*

$$\begin{aligned}&|u_{1t}(t)-u_{2t}(t)|_V^2+|\mathscr{A}^{1/2}(u_1(t)-u_2(t))|^2\\&\le e^{L(R)t}\left(|u_{11}-u_{12}|_V^2+|\mathscr{A}^{1/2}(u_{01}-u_{02})|^2\right),\end{aligned} \tag{8.1.13}$$

where $u_1(t)$ and $u_2(t)$ are two solutions to problem (8.1.1) *with the initial data $(u_{01};u_{11})$ and $(u_{02};u_{12})$, respectively, such that $|\mathscr{A}^{1/2}(u_i(t)|\le R$ for all $t\ge 0$, $i=1,2$. Here $L(R)$ is the Lipschitz constant from* (8.1.3).

Proof. Let $u_1(t)$ and $u_2(t)$ be strong solutions to problem (8.1.1). From (8.1.1) for the difference $u(t)=u_1(t)-u_2(t)$ we have the relation

$$Mu_{tt}+k[Du_{1t}-Du_{2t}]+\mathscr{A}u=F(u_1)-F(u_2).$$

This implies that the following energy relation

$$E_0(u(t),u_t(t)) + k\int_0^t (Du_{1t}(\tau) - Du_{2t}(\tau), u_t(\tau))d\tau$$
$$= E_0(u(0),u_t(0)) + \int_0^t (F(u_1(\tau)) - F(u_1(\tau)), u_t(\tau))d\tau,$$

where $E_0(u,v) = \frac{1}{2}((Mv,v) + (\mathscr{A}u,u))$. Therefore using (8.1.2) and (8.1.3) we obtain that

$$E_0(u(t),u_t(t)) \le E_0(u(0),u_t(0)) + L(R)\int_0^t E_0(u(\tau),u_t(\tau))d\tau\,.$$

Gronwall's lemma and the limit transition to generalized solutions give the result.

□

Thus the results presented in Chapter 2 imply that problem (8.1.1) generates a dynamical system (H,S_t) in the space $H = \mathscr{D}(\mathscr{A}^{1/2}) \times V$ by the formula $S_t(u_0;u_1) = (u(t);u_t(t))$, where $u(t)$ is the solution to (8.1.1).

8.1.4. Remark. In the case of conservative forces ($F^* \equiv 0$), the energy inequality (8.1.11) implies that the energy $\mathscr{E}(u(t),u_t(t))$ is nonincreasing along trajectories and hence the set

$$\mathscr{E}_R = \left\{(u_0;u_1) \in H \,:\, \mathscr{E}(u_0,u_1) \le R^2\right\}$$

is forward invariant (i.e., $S_t\mathscr{E}_R \subseteq \mathscr{E}_R$). Moreover under some additional minor conditions on the damping operator D the dynamical system (H,S_t) is gradient (see Definition 7.5.3). Therefore, in order to show the existence of a global attractor it suffices to verify the hypotheses made Corollary 7.5.7. The case of nonconservative forces ($F^* \not\equiv 0$) is more complicated because the energy may behave in an almost arbitrary fashion. In this case we need apply more general Theorem 7.2.2 or Theorem 7.2.4 which require the dissipativity property. This leads to a more restrictive bound imposed on the nonlinear dissipation. ■

8.2 Dissipativity

The dissipativity property is fundamental to long-time behavior of solutions. The task of establishing dissipativity is particularly difficult for dynamical systems whose energy may not be nonincreasing. This may be due to the presence of restoring forces that are not conservative. It is our aim in this section to show that rather general class of abstract systems defined in (8.1.1) exhibits the dissipativity property under some rather natural additional assumptions. We pay main attention to the case of nonconservative forces. In the conservative case, as pointed out in Remark 8.1.4, the proof of dissipativity is not necessary.

It should also be noted that in applications to certain models not only dissipativity of the dynamical system is needed, but also control of the size of the absorbing set in terms of the parameters representing the damping.

To prove dissipativity of the system (H, S_t) generated by (8.1.1) we impose the following rather natural condition.

8.2.1. Assumption.

(D) • There exist constants $c_0 \geq 0$ and $c_1 > 0$ such that

$$(Mv,v) \leq c_0 + c_1(Dv,v), \quad v \in \mathscr{D}(\mathscr{A}^{1/2}). \tag{8.2.1}$$

• The operator D is linearly bounded in the following sense. There exist non-negative constants c_2 and c_3 such that

$$|Dv|_{[\mathscr{D}(\mathscr{A}^{1/2})]'} \leq c_2 + c_3|v|_V, \quad v \in \mathscr{D}(\mathscr{A}^{1/2}). \tag{8.2.2}$$

(F) • There exist positive constants $0 \leq \eta < 1$, c_4 and c_5 such that

$$(u, F(u)) \leq \eta|\mathscr{A}^{1/2}u|^2 - c_4\Pi_0(u) + c_5, \quad u \in \mathscr{D}(\mathscr{A}^{1/2}). \tag{8.2.3}$$

• For every $\eta > 0$ there exists $C_\eta > 0$ such that

$$|u|^2 \leq C_\eta + \eta\left(|\mathscr{A}^{1/2}u|^2 + \Pi_0(u)\right), \quad u \in \mathscr{D}(\mathscr{A}^{1/2}). \tag{8.2.4}$$

• The non-conservative term $F^*(u)$ satisfies the inequality

$$|F^*(u)|^2_{V'} \leq c_6 + c_7|\mathscr{A}^{1/2-\delta}u|^2 \quad \text{for some} \quad \delta > 0. \tag{8.2.5}$$

8.2.2. Remark. We note that in the case of von Karman evolutions the inequality in (8.2.4) is related to a "hidden" superlinearity of potential energy (see, e.g., Lemma 1.5.4 in Chapter 1). One can also see that relation (8.2.4) implies that for any positive $\delta > 0$ and any positive number η there exists a constant C_η such that

$$|\mathscr{A}^{1/2-\delta}u|^2 < \eta[|\mathscr{A}^{1/2}u|^2 + \Pi_0(u)] + C_\eta. \tag{8.2.6}$$

Moreover, we can suppose that $C_\eta = b(1/\eta)$, where $b(\xi)$ is a nondecreasing function of $\xi \in \mathbb{R}_+ \setminus \{0\}$. ∎

8.2.3. Theorem. *Under Assumptions* 8.1.1 *and* 8.2.1 *the system* (H, S_t) *generated by* (8.1.1) *in the space* $H = \mathscr{D}(\mathscr{A}^{1/2}) \times V$ *is (ultimately) dissipative. It possesses a forward invariant bounded absorbing set (whose size may depend on the damping parameter k).*

Assume, in addition, that Assumption 8.2.1*(D) holds with* $c_0 = c_2 = 0$ *in* (8.2.1) *and* (8.2.2)*; that is, there exist constants* $c_1 > 0$ *and* $c_3 > 0$ *such that*

$$(Mv,v) \leq c_1(Dv,v), \quad v \in \mathscr{D}(\mathscr{A}^{1/2}), \tag{8.2.7}$$

and

$$|Dv|_{[\mathscr{D}(\mathscr{A}^{1/2})]'} \leq c_3|v|_V, \quad v \in \mathscr{D}(\mathscr{A}^{1/2}). \tag{8.2.8}$$

Then we can choose an absorbing ball with radius that is independent of the damping parameter $k \in [k_0, +\infty)$ *for every fixed* $k_0 > 0$. *In this case a forward invariant bounded (uniformly with respect to* $k \geq k_0$*) absorbing set also exists.*

Proof. We refer to Theorem 3.10 [75, Chapter 3]. □

We note that Assumption 8.2.1 assumes linear bound on the non-conservative term $F^*(u)$ and on the damping. In some more special cases when the potential energy has strong coercivity properties, it is possible to relax these constraints at the expense of assuming large values of the damping parameter. The relevant result is formulated below.

8.2.4. Theorem. *Let Assumption* 8.1.1 *be valid. Assume that Assumption* 8.2.1 *holds with the following two relations replacing, respectively,* (8.2.2) *and* (8.2.5),

$$k|(Dv,u)| \leq \eta \left[|\mathscr{A}^{1/2}u|^2 + \Pi_0(u) \right] + C_{\eta,k} + C_1 \cdot (Dv,v) \quad v,u \in \mathscr{D}(\mathscr{A}^{1/2}), \quad (8.2.9)$$

where $\eta > 0$ *is arbitrary small and* C_1 *does not depend on k, and*

$$|F^*(u)|_{V'}^2 \leq C_2 + C_3 \left(|\mathscr{A}^{1/2}u|^2 + \Pi_0(u) \right), \quad u \in \mathscr{D}(\mathscr{A}^{1/2}). \quad (8.2.10)$$

Then there exists $k_* > 0$ *such that the system* (H, S_t) *generated by* (8.1.1) *in the space* $H = \mathscr{D}(\mathscr{A}^{1/2}) \times V$ *is ultimately dissipative for each* $k \geq k_*$. *It possesses a forward invariant bounded absorbing set (whose size may depend on the damping parameter k).*

Proof. See [75, Chapter 3, Theorem 3.11]. □

8.2.5. Remark. One can show (see [75, Chapter 3]) that Theorem 8.2.4 remains true if we replace (8.2.9) by the relation

$$|Dv|_{[\mathscr{D}(\mathscr{A}^{1/2-\delta})]'} \leq c_1 + c_2 |v|_V \quad \text{with } \delta > 0,$$

which is stronger than (8.2.2). However, in this case we avoid the linear bound and the compactness property for $F^*(u)$. ■

Now we briefly discuss the case when nonconservative force F^* is not present in the model; that is, we consider the following version of problem (8.1.1),

$$\begin{cases} Mu_{tt}(t) + \mathscr{A}u(t) + k \cdot D(u_t(t)) = F(u(t)) \equiv -\Pi'(u(t)), \\ u|_{t=0} = u_0 \in \mathscr{D}(\mathscr{A}^{1/2}), \quad u_t|_{t=0} = u_1 \in V = \mathscr{D}(M^{1/2}). \end{cases} \quad (8.2.11)$$

In addition to Assumption 8.1.1 we impose, instead of Assumption 8.2.1, the following hypotheses concerning D and F.

8.2.6. Assumption.

(D) • There exist constants $c_0 \geq 0$ and $c_1 > 0$ such that

$$(Mv,v) \leq c_0 + c_1(Dv,v) \quad \text{for any } v \in \mathscr{D}(\mathscr{A}^{1/2}). \quad (8.2.12)$$

- For any $\delta > 0$ there exists a nondecreasing function $K_\delta(s) > 0$ such that

$$|(Dv,u)| \le K_\delta(E_0(u,v)) \cdot (Dv,v) + \delta \cdot (1 + E_0(u,v)) \quad (8.2.13)$$

for any $u, v \in \mathscr{D}(\mathscr{A}^{1/2})$, where $E_0(u,v) = \frac{1}{2}((Mv,v) + (\mathscr{A}u,u))$.

(F) $F(u) = -\Pi'(u)$ and there exist $0 \le \eta < 1$ and $c_2 > 0$ such that

$$(u, F(u)) \le \eta|\mathscr{A}^{1/2}u|^2 + c_2, \quad u \in \mathscr{D}(\mathscr{A}^{1/2}).$$

8.2.7. Remark. Assumption (8.2.13) holds true provided that for any $\delta > 0$ we can find a nondecreasing function $\widetilde{K}_\delta(s) > 0$ such that

$$|Dv|^2_{[\mathscr{D}(\mathscr{A}^{1/2})]'} \le \delta + \widetilde{K}_\delta(|v|_V) \cdot (Dv,v), \quad u, v \in \mathscr{D}(\mathscr{A}^{1/2}).$$

We refer to [75, Remark 3.2] for details. ■

We have the following result on dissipativity in the case $F^* \equiv 0$.

8.2.8. Theorem. *Under Assumptions* 8.1.1 *and* 8.2.6 *the system* (H, S_t) *generated by* (8.2.11) *in the space* $H = \mathscr{D}(\mathscr{A}^{1/2}) \times V$ *is dissipative; that is, there exists* $R > 0$ *possessing the property: for any bounded set* B *from* H *there exists* $t_0 = t_0(B)$ *such that* $\|S_t y\|_H \le R$ *for all* $y \in B$ *and* $t \ge t_0$. *We can choose a radius* R *of an absorbing ball such that* R *does not depend on the damping parameter* k. *Moreover, there exists a forward invariant absorbing set* $\mathscr{B}_0$ *with the size that does not depend on* k.

Proof. The proof is based on Lyapunov's method; see [75, Chapter 3] for details. □

8.3 Existence of global attractors

In this section we prove several results on existence of global attractors for problem (8.1.1). We follow ideas presented in [75] which are applicable to a more general class of the second-order evolution equations.

8.3.1 Preliminary inequalities

This section provides technical preliminary inequalities which serve various purposes for the proofs of (i) asymptotic smoothness, (ii) regularity, and (iii) finite dimensionality of attractors. These inequalities are technical and perhaps not of major interest at this stage of reading. However, because these results provide a unifying thread for further development, we have included the corresponding results in this section.

8.3.1. Lemma (Equipartition energy inequality). *Under Assumption* 8.1.1 *there exists $T_0 > 0$ and a constant $c > 0$ independent of T such that for any pair w and v of strong solutions to* (8.1.1) *we have the following relation*

$$TE_z(T) + \int_0^T E_z(t)dt \le c\left\{ \int_0^T |M^{1/2} z_t(s)|^2 ds + k\int_0^T (D(t,z_t),z_t)dt \right. \\ \left. + k\int_0^T |(D(t,z_t),z)|\,dt + \Psi_T(w,v) \right\} \tag{8.3.1}$$

for every $T \ge T_0$, where $z(t) = w(t) - v(t)$ and we use the following notations:

$$\begin{aligned} E_z(t) &= E_0(z(t),z_t(t)) = \frac{1}{2}\left((Mz_t(t),z_t(t)) + (\mathscr{A}z(t),z(t))\right), \\ D(t,z_t) &= D(v_t(t)+z_t(t)) - D(v_t(t)), \\ \Psi_T(w,v) &= \left|\int_0^T (G_{w,v}(\tau),z_t(\tau))d\tau\right| + \left|\int_0^T (G_{w,v}(t),z(t))dt\right| \\ &\quad + \left|\int_0^T dt \int_t^T (G_{w,v}(\tau),z_t(\tau))d\tau\right| \end{aligned} \tag{8.3.2}$$

with $G_{w,v}(t)$ given by

$$G_{w,v}(t) = F(w(t)) - F(v(t)). \tag{8.3.3}$$

Proof. Our argument is the same as in [75]. The variable z satisfies the equation

$$Mz_{tt} + \mathscr{A}z + kD(t,z_t) = G_{w,v}(t) \tag{8.3.4}$$

and hence we have the following energy relation

$$E_z(T) + k\int_t^T (D(\tau,z_t),z_t)d\tau = E_z(t) + \int_t^T (G_{w,v}(\tau),z_t(\tau))d\tau, \quad t \in [0,T]. \tag{8.3.5}$$

Multiplying (8.3.4) by z and following with the integration over $\Omega \times [0,T]$ yields

$$\int_0^T E_z(t)dt \le c_0\left(E_z(T) + E_z(0)\right) + \int_0^T |M^{1/2} z_t(s)|^2 ds \\ + \frac{k}{2}\int_0^T |(D(s,z_t),z)|ds + \frac{1}{2}\int_0^T (G_{w,v}(s),z(s))ds. \tag{8.3.6}$$

From (8.3.5) we have

$$E_z(0) = E_z(T) + k\int_0^T (D(\tau,z_t),z_t)d\tau - \int_0^T (G_{w,v}(\tau),z_t(\tau))d\tau$$

and integrating (8.3.5) from 0 to T

$$TE_z(T) \le \int_0^T E_z(t)dt + \int_0^T dt \int_t^T (G_{w,v}(\tau),z_t(\tau))d\tau.$$

Therefore (8.3.1) follows from (8.3.6). □

In some sense the inequality in (8.3.1) represents equipartition of the energy of the difference of two solutions. The potential energy is reconstructed from the kinetic energy and the nonlinear quantities entering the equation.

For the estimates involving the damping operator we use the following inequality which is derived from energy inequality.

8.3.2. Lemma. *Let Assumption* 8.1.1 *be in force. Assume that w and v are strong solutions to* (8.1.1) *possessing property*

$$\max_{s\in[0,T]}\left\{|\mathscr{A}^{1/2}w(s)|^2+|M^{1/2}w_t(s)|^2+|\mathscr{A}^{1/2}v(s)|^2+|M^{1/2}v_t(s)|^2\right\}\le R^2 \quad (8.3.7)$$

for some $R>0$. Then, with the notations from the previous lemma, we have that

$$\frac{1}{2}\max_{t\in[0,T]}|\mathscr{A}^{1/2}z(t)|^2\le E_z(T)+k\int_0^T(D(t,z_t),z_t)dt+c_R\int_0^T E_z(t)dt, \quad (8.3.8)$$

and

$$\begin{aligned}&\int_0^T[(D(w_t),w_t)+(D(v_t),v_t)]\,|\mathscr{A}^{1/2}z|^2dt \quad (8.3.9)\\ &\le 2D_0^T\cdot\left[E_z(T)+k\int_0^T(D(t,z_t),z_t)dt+c_R\int_0^T E_z(t)dt\right],\end{aligned}$$

where $c_R>0$ does not depend on k and T,

$$D_0^T=D_0^T(w,v)\equiv\int_0^T[(D(w_t),w_t)+(D(v_t),v_t)]\,dt. \quad (8.3.10)$$

Proof. Lemma 8.3.1 implies that

$$|(G_{w,v}(t),z_t)|\le C_R\left(|\mathscr{A}^{1/2}z|+|M^{1/2}z_t|\right)|M^{1/2}z_t|\le C_R E_z(t)$$

under condition (8.3.7). Therefore, it follows from (8.3.5) that

$$\max_{t\in[0,T]}E_z(t)\le E_z(T)+k\int_0^T(D(\tau,z_t),z_t)d\tau+C_R\int_0^T E_z(\tau)d\tau.$$

Because $|\mathscr{A}^{1/2}z(t)|^2\le 2E_z(t)$, this implies (8.3.8) and (8.3.9). □

8.3.2 Main results on asymptotic smoothness

We start with the case of general nonlinear forces, including the nonconservative forces ($F^* \not\equiv 0$). In addition to basic Assumption 8.1.1, we impose the following hypotheses concerning D, Π, and F^*.

8.3.3. Assumption.

(D) *dissipation*: For any $\eta > 0$ there exist $C_\eta > 0$ such that

$$(Mv, v) \le \eta + C_\eta \cdot (D(u+v) - D(u), v) \quad \text{for any} \quad u, v \in \mathscr{D}(\mathscr{A}^{1/2}). \quad (8.3.11)$$

Moreover, there exist parameters $\kappa \in (0,2]$ and $0 < \delta < 1/2$ such that for every $\varepsilon > 0$ we have inequality

$$\begin{aligned} &|(D(u+v) - D(u), w)| \\ &\le C_1^\varepsilon(r) \cdot (D(u+v) - D(u), v) \\ &+ C_2(r)\,(1 + (D(u), u) + (D(u+v), u+v))\,[|\mathscr{A}^{1/2-\delta} w|^\kappa + \varepsilon |\mathscr{A}^{1/2} w|^2] \end{aligned} \quad (8.3.12)$$

for any $u, v, w \in \mathscr{D}(\mathscr{A}^{1/2})$ such that $|\mathscr{A}^{1/2} w| + |M^{1/2} u| + |M^{1/2} v| \le r$ with arbitrary $r > 0$, where $C_1^\varepsilon(r)$ and $C_2(r)$ are non-decreasing functions of r, $C_2(r)$ does not depend on ε.

(F) *forcing*:

(i) The potential energy functional $\Pi(u)$ is continuous on $\mathscr{D}(\mathscr{A}^{1/2-\delta})$ for some $\delta > 0$.
(ii) The mapping $u \mapsto \mathscr{A}^{-l}\Pi'(u)$ is continuous from $\mathscr{D}(\mathscr{A}^{1/2-\delta})$ into $\mathscr{H}$ for some $l, \delta > 0$.
(iii) There exist $0 < \tilde{\eta} \le 1/2$ such that

$$|(F^*(u) - F^*(\hat{u})|_{V'}^2 \le C(r) |\mathscr{A}^{1/2-\tilde{\eta}}(u - \hat{u})|^2 \quad (8.3.13)$$

for any u and $\hat{u}$ from $\mathscr{D}(\mathscr{A}^{1/2})$ such that $|\mathscr{A}^{1/2} u| \le r$ and $|\mathscr{A}^{1/2} \hat{u}| \le r$. Here $r > 0$ is arbitrary, and $C(r)$ is a nondecreasing function of r.

We note that Assumption 8.3.3 allows to consider conservative forcing terms at the critical level. Nonconservative forces F^* are assumed subcritical ($\tilde{\eta} > 0$ in (8.3.13)).

The following general result is obtained by appealing to the ideas originally presented in [160], and later generalized in [75, Chapter 3].

8.3.4. Theorem. *Let Assumption* 8.1.1 *and Assumption* 8.3.3 *hold. Assume that* $\mathscr{D}(\mathscr{A}^{1/2-\beta})$ *is compactly embedded in* V *for some* $\beta > 0$ *and the system* (H, S_t) *generated by problem* (8.1.1) *in* $H = \mathscr{D}(\mathscr{A}^{1/2}) \times V$ *is dissipative. Then the semiflow generated by problem* (8.1.1) *possesses a compact global attractor* A.

As a consequence of this theorem we have the following assertion.

8.3.5. Theorem. *Let Assumptions* 8.1.1 *and* 8.3.3 *be in force. Assume also that either Assumption* 8.2.1 *or else the hypotheses of Theorem* 8.2.4 *hold. Let* $\mathscr{D}(\mathscr{A}^{1/2-\beta})$ *be compactly embedded into* V *for some* $\beta > 0$. *Then the dynamical system* (H, S_t) *generated by problem* (8.1.1) *possesses a compact global attractor* A.

Proof. It follows either from Theorem 8.2.3 or from Theorem 8.2.4 that (H, S_t) is dissipative. Therefore we can apply Theorem 8.3.4. □

By appealing to Theorem 7.2.3, in order to prove Theorem 8.3.4 it suffices to establish the following proposition.

8.3.6. Proposition. *Assume that Assumptions* 8.1.1 *and* 8.3.3 *hold true and the embedding* $\mathscr{D}(\mathscr{A}^{1/2-\beta}) \subset V$ *is compact for some* $\beta > 0$. *Then, the semiflow* S_t *generated by problem* (8.1.1) *is asymptotically smooth.*

Proof. The proof relies on the inequality in Lemma 8.3.1 and employs an idea that is credited to [160] and developed in [75]. We apply Theorem 7.1.11 on a forward invariant set $\mathscr{B}$ which we consider as a metric space with the distance $d(y_1, y_2) = \|y_1 - y_2\|_H$, $y_1, y_2 \in \mathscr{B}$.

Let $w(t)$ and $v(t)$ be two solutions to (8.1.1) corresponding to initial data in a forward invariant bounded set $\mathscr{B}$:

$$(w(t); w_t(t)) \equiv S_t y_0, \quad (v(t); v_t(t)) \equiv S_t y_1, \quad y_0, y_1 \in \mathscr{B}. \tag{8.3.14}$$

In further calculations we can assume that $w(t)$ and $v(t)$ are strong solutions to (8.1.1). Thus $z(t) \equiv w(t) - v(t)$ satisfies the equation

$$Mz_{tt} + \mathscr{A} z + kD(t, z_t) = G_{w,v}(t), \quad (z(0); z_t(0)) = y_0 - y_1, \tag{8.3.15}$$

where we use the same notations as in (8.3.2) and (8.3.3):

$$D(t, z_t) = D(z_t(t) + v_t(t)) - D(v_t(t)), \quad G_{w,v}(t) = F(w(t)) - F(v(t)). \tag{8.3.16}$$

The function $z(t)$ satisfies the energy relation of the form (8.3.5) on every interval $[0, T]$.

Because $\mathscr{B}$ is a bounded forward invariant set, we have relation (8.3.7) for $T = \infty$ and some $R = R(\mathscr{B}) > 0$. Therefore from energy relation (8.1.7) and from (8.1.5) we always have that

$$D_0^T \equiv \int_0^T (D(w_t), w_t)dt + \int_0^T (D(v_t), v_t)dt \le k^{-1} C_{\mathscr{B}}(1+T), \quad \forall T > 0. \tag{8.3.17}$$

We start with the following consequence of Lemma 8.3.1.

8.3.7. Lemma. *Let* $T_0 > 0$ *be the same as in Lemma* 8.3.1. *Then for any* $\eta > 0$ *and* $T > T_0$ *there exist positive constants* $C^{(1)}_{\mathscr{B},\eta}$ *and* $C^{(2)}_{\mathscr{B},T}$ *such that*

$$E_z(T) \le \eta + \frac{C^{(1)}_{\mathscr{B},\eta}}{T} \cdot [1 + \Psi_T(w,v)] + C^{(2)}_{\mathscr{B},T} \sup_{t\in[0,T]} |\mathscr{A}^{1/2-\delta} z(t)|^{\kappa}, \tag{8.3.18}$$

where δ and κ are the constants from (8.3.12) *and $\Psi_T(w,v)$ is given by* (8.3.2).

Proof. Using (8.3.12) from Lemma 8.3.1 we have that

$$TE_z(T) \le c\left\{\int_0^T |M^{1/2}z_t|^2 dt + \Psi_T(w,v)\right\} + C_{\mathscr{B},\varepsilon}\int_0^T (D(t,z_t),z_t)dt$$
$$+C_{\mathscr{B}}\int_0^T [1+(D(w_t),w_t)+(D(v_t),v_t)]\left[|\mathscr{A}^{1/2-\delta}z|^\kappa + \varepsilon|\mathscr{A}^{1/2}z|^2\right]dt.$$

Since $|\mathscr{A}^{1/2}z(t)|^2 \le C_{\mathscr{B}}$ for all $t \ge 0$, by (8.3.17) we have that

$$\int_0^T [1+(D(w_t),w_t)+(D(v_t),v_t)]\left[|\mathscr{A}^{1/2-\delta}z|^\kappa + \varepsilon|\mathscr{A}^{1/2}z|^2\right]dt$$
$$\le \varepsilon C_{\mathscr{B}}(1+T) + C_{\mathscr{B},T}\sup_{t\in[0,T]}|\mathscr{A}^{1/2-\delta}z(t)|^\kappa.$$

Therefore relation (8.3.11) after rescaling $\varepsilon C_{\mathscr{B}} := \varepsilon$ yields that

$$TE_z(T) \le \varepsilon + (\eta+\varepsilon)\cdot T + C_{\mathscr{B},\eta,\varepsilon}\int_0^T (D(t,z_t),z_t)dt + c\Psi_T(w,v)$$
$$+C_{\mathscr{B},T}\sup_{t\in[0,T]}|\mathscr{A}^{1/2-\delta}z(t)|^\kappa \tag{8.3.19}$$

for any $\eta > 0$ and $\varepsilon > 0$. Because $E_z(0) \le C_{\mathscr{B}}$, using energy relation (8.3.5) with $t = 0$ we obtain

$$k\int_0^T (D(t,z_t),z_t)dt \le C_{\mathscr{B}} + \int_0^T (G_{w,v}(\tau),z_t(\tau))d\tau \le C_{\mathscr{B}} + \Psi_T(w,v). \tag{8.3.20}$$

Therefore substituting $\eta + \varepsilon := \eta$ in (8.3.19) we obtain (8.3.18). □

Using Lemma 8.3.7 we obtain the following assertion.

8.3.8. Lemma. *For any $\eta > 0$ and $T > T_0$, there exist positive constants $C^{(1)}_{\mathscr{B},\eta}$ and $C^{(2)}_{\mathscr{B},\eta,T}$ such that*

$$E_z(T) \le \eta + \frac{C^{(1)}_{\mathscr{B},\eta}}{T}\cdot[1+\Psi_*(w,v;T)], \tag{8.3.21}$$

where $z = w - v$ and

$$\Psi_*(w,v;T) = \left|\int_0^T (G_0(\tau),z_t(\tau))d\tau\right| + \left|\int_0^T dt\int_t^T (G_0(\tau),z_t(\tau))d\tau\right|$$
$$+C^{(2)}_{\mathscr{B},\eta,T}\sup_{t\in[0,T]}|\mathscr{A}^{1/2-\beta_*}z(t)|^{\kappa_*} \tag{8.3.22}$$

with positive β_* and κ_* and with $G_0(t)$ given by

$$G_0(t) \equiv G_0(w(t),v(t)) = -\Pi'(w(t)) + \Pi'(v(t)). \tag{8.3.23}$$

Proof. It follows from (8.3.16) and (8.3.13) that $G_{w,v}(t) = G_0(t) + G_1(t)$, where

$$|G_1(t)|_{V'} \le C_{\mathscr{B}} |\mathscr{A}^{1/2-\tilde{\eta}} z(t)|.$$

We also have that

$$\int_0^T (G_0(t), z(t))dt \le C_{\mathscr{B}} \int_0^T |z(t)|_V dt.$$

Therefore, because the function $\Psi_T(w,v)$ given by (8.3.2) is subadditive with respect to $G_{w,v}$, relation (8.3.21) follows from (8.3.18). This completes the proof of Lemma 8.3.8. □

To apply Theorem 7.1.11 it is sufficient to prove that

$$\liminf_{m\to\infty} \liminf_{n\to\infty} \Psi_*(w_n, w_m; T) = 0 \tag{8.3.24}$$

for every $T > 0$, where $\Psi_*(w,v;T)$ is given by (8.3.22) and $(w^n(t); w_t^n(t)) \equiv S_t y_0^n$ are solutions to (8.1.1) with initial data $\{y_0^n\}_{n=1}^\infty$ from $\mathscr{B}$. We can assume that

$$(w^n(t); w_t^n(t)) \to (w(t); w_t(t)) \quad \text{*-weakly in} \quad L_\infty(0,T; \mathscr{D}(\mathscr{A}^{1/2}) \times V). \tag{8.3.25}$$

Step 1: By the compactness of the embedding (see, e.g., Theorem 1.1.8)

$$C(0,T;\mathscr{D}(\mathscr{A}^{1/2})) \cap C^1(0,T;V) \subset C(0,T;\mathscr{D}(\mathscr{A}^{1/2-\beta_*})) \tag{8.3.26}$$

we obtain

$$\lim_{m,n\to\infty} \sup_{t\in[0,T]} |\mathscr{A}^{1/2-\beta_*}(w^n(t) - w^m(t))|^{\kappa_*} = 0. \tag{8.3.27}$$

Step 2: We claim that

$$\lim_{m\to\infty} \lim_{n\to\infty} \int_t^T (G_0^{n,m}(\tau), w_t^n(\tau) - w_t^m(\tau))d\tau = 0 \tag{8.3.28}$$

for any $t < T$, where $G_0^{n,m}(t) = G_0(w^n(t), w^m(t))$ and $G_0(w,v)$ is defined by (8.3.23). Indeed, because

$$\begin{aligned} &\int_t^T (G_0^{n,m}(\tau), w_t^n(\tau) \quad w_t^m(\tau))d\tau \\ &= -\Pi(w^n(T)) + \Pi(w^n(t)) - \Pi(w^m(T)) + \Pi(w^m(t)) + \mathscr{I}_t^T(n,m), \end{aligned}$$

where

$$\mathscr{I}_t^T(n,m) = \int_t^T \left[(\Pi'(w^n(\tau)), w_t^m(\tau)) + (\Pi'(w^m(\tau)), w_t^n(\tau))\right] d\tau,$$

by (F)(i) in Assumption 8.3.3 and (8.3.26) we have that

$$\lim_{m\to\infty}\lim_{n\to\infty}\int_t^T (G_0^{n,m}(\tau), w_t^n(\tau)-w_t^m(\tau))d\tau$$
$$= -2\Pi(w(T))+2\Pi(w(t))+\lim_{m\to\infty}\lim_{n\to\infty}\mathscr{I}_t^T(n,m). \tag{8.3.29}$$

It follows from Assumption 8.3.3(F)(ii), (8.3.25), and (8.3.26) that

$$\Pi'(w^n(t)) \to \Pi'(w(t)) \quad \text{*-weakly in} \quad L_\infty(0,T;V').$$

Therefore using the compactness of the embedding in (8.3.26) we have

$$\lim_{m\to\infty}\lim_{n\to\infty}\int_t^T (\Pi'(w^n(\tau)), w_t^m(\tau))d\tau = \lim_{m\to\infty}\int_t^T (\Pi'(w(\tau)), w_t^m(\tau))d\tau$$
$$= \int_t^T (\Pi'(w(\tau)), w_t(\tau))d\tau.$$

In a similar way we also have that

$$\lim_{m\to\infty}\lim_{n\to\infty}\int_t^T (\Pi'(w^m(\tau)), w_t^n(\tau))d\tau = \int_t^T (\Pi'(w(\tau)), w_t(\tau))d\tau.$$

Therefore

$$\lim_{m\to\infty}\lim_{n\to\infty}\mathscr{I}_t^T(n,m) = 2\int_t^T (\Pi'(w(\tau)), w_t(\tau))d\tau. \tag{8.3.30}$$

Hence after substituting (8.3.30) in (8.3.29) we get (8.3.28).

Concluding Step: It follows from (8.3.22), (8.3.27), and (8.3.28) and also from the Lebesgue dominated convergence theorem that (8.3.24) holds. Thus we can apply Theorem 7.1.11 to obtain the asymptotic smoothness of S_t and, hence, the statement of Proposition 8.3.6. The proof of Theorem 8.3.4 is thus completed. □

Conservative forces case. In the case when the nonlinear forces $F(u)$ in equation (8.1.1) *are conservative* ($F^*(u) \equiv 0$), the conditions imposed on the damping operator D in Assumption 8.3.3 can be simplified. More precisely, let us introduce the following hypotheses.

8.3.9. Assumption.

(D) *damping*:

(i) For any $\eta > 0$ there exist $C_\eta > 0$ such that

$$(Mv, v) \le \eta + C_\eta \cdot (D(v), v) \ \text{ for any } v \in \mathscr{D}(\mathscr{A}^{1/2}). \tag{8.3.31}$$

(ii) Relation (8.2.13) holds; that is, for any $\delta > 0$ there exists a nondecreasing function $K_\delta(s) > 0$ such that

$$|(Dv, u)| \le K_\delta(E_0(u,v)) \cdot (Dv, v) + \delta \cdot (1 + E_0(u,v)) \tag{8.3.32}$$

for any $u, v \in \mathscr{D}(\mathscr{A}^{1/2})$, where $E_0(u,v) = \frac{1}{2}((Mv,v)+(\mathscr{A}u,u))$.

(F) *forcing*:

(i) The potential energy functional $\Pi(u)$ is continuous on $\mathscr{D}(\mathscr{A}^{1/2-\delta})$ for some $\delta > 0$.
(ii) The mapping $u \mapsto \mathscr{A}^{-l}\Pi'(u)$ is continuous from $\mathscr{D}(\mathscr{A}^{1/2-\delta})$ into $\mathscr{H}$ for some $l, \delta > 0$.
(iii) The nonlinear forces $F(u)$ are conservative; that is, $F(u) = -\Pi'(u)$.

Under the above assumptions the following result holds.

8.3.10. Theorem. *Let Assumption* 8.1.1 *and Assumption* 8.3.9 *hold. Assume that* $\mathscr{D}(\mathscr{A}^{1/2-\beta})$ *is compactly embedded in* V *for some* $\beta > 0$ *and the nonlinear operator* $F(u)$ *possesses the following property. There exist* $0 \le \eta < 1$ *and a positive constant* c_2 *such that*

$$(u, F(u)) \le \eta |\mathscr{A}^{1/2}u|^2 + c_2, \quad u \in \mathscr{D}(\mathscr{A}^{1/2}) . \tag{8.3.33}$$

Then the system (H, S_t) *generated by problem* (8.1.1) *in* $H = \mathscr{D}(\mathscr{A}^{1/2}) \times V$ *possesses a compact global attractor* A.

Proof. To prove the existence of a compact global attractor we need to establish the following assertion.

8.3.11. Proposition. *Assume that Assumptions* 8.1.1 *and* 8.3.9 *hold true and the embedding* $\mathscr{D}(\mathscr{A}^{1/2-\beta}) \subset V$ *is compact for some* $\beta > 0$. *Then the semiflow* S_t *generated by problem* (8.1.1) *is asymptotically smooth.*

Proof. As above we apply Theorem 7.1.11 on the forward invariant set $\mathscr{B}$. For this we check that in the case of conservative forces it is possible to establish Lemma 8.3.7 and Lemma 8.3.8 under Assumption 8.3.9.

With the same notations as in the proof of Proposition 8.3.6, for any pair of strong solutions $(w(t); w_t(t))$ and $(v(t); v_t(t))$ with initial data from $\mathscr{B}$ from energy relation (8.1.7) with $F^* \equiv 0$ we always have that

$$D_0^T \equiv \int_0^T (D(w_t), w_t)dt + \int_0^T (D(v_t), v_t)dt \le C_{\mathscr{B}}, \tag{8.3.34}$$

where, in contrast to (8.3.17), the constant $C_{\mathscr{B}}$ does not depend on T. This fact is crucial in our argument.

We apply Lemma 8.3.1. For this we need to estimate the terms involving the damping operator. As above we assume that $w(t)$ and $v(t)$ are strong solutions to (8.1.1).

From (8.3.32) we have that

$$\begin{aligned} |(D(t,z_t),z)| &\le |(D(w_t),z)| + |(D(v_t),z)| \\ &\le C_{\mathscr{B}}^{\delta}[(D(w_t),w_t) + (D(v_t),v_t)] + \delta C_{\mathscr{B}} \end{aligned}$$

for any $\delta > 0$. Therefore by (8.3.34) after rescaling of δ we obtain that

$$\int_0^T |(D(t,z_t),z)|dt \le \delta T + C_{\mathscr{B}}^{\delta}$$

for any $\delta > 0$. In a similar way relation (8.3.31) implies that

$$\int_0^T |M^{1/2}z_t|^2 dt \le 2\int_0^T \left(|M^{1/2}w_t|^2 + |M^{1/2}v_t|^2\right) dt \le \eta T + C^{\eta}_{\mathscr{B}}$$

for any $\eta > 0$. Therefore using (8.3.20) and Lemma 8.3.1 we get the estimate

$$TE_z(T) \le (\delta+\eta)\cdot T + C_{\mathscr{B},\delta,\eta} + c\Psi_T(w,v), \quad T \ge T_0,$$

for any positive δ and η, where $\Psi_T(w,v)$ is given by (8.3.2). This implies relation (8.3.18) in Lemma 8.3.7. Thus we can obtain the conclusion of Lemma 8.3.8 with $\Psi_*(w,v;T)$ given by (8.3.22). Now we can apply Theorem 7.1.11. The argument is the same as in the proof of Proposition 8.3.6. □

To complete the proof of Theorem 8.3.10 we note that (8.3.33) makes it possible to check Assumption 8.2.6, Thus by Theorem 8.2.8 the system (H,S_t) is dissipative, Therefore the result follows from Proposition 8.3.11 and Theorem 7.2.3. □

Condition (8.3.33) is needed only to guarantee dissipativity. In the following theorem we show that this condition can be dispensed with by restricting the system on an invariant bounded set.

8.3.12. Theorem. *Let $F^* \equiv 0$ and Assumptions 8.1.1 and 8.3.9 hold. Assume that $\mathscr{D}(\mathscr{A}^{1/2-\beta})$ is compactly embedded into V for some $\beta > 0$. Let $(\mathscr{E}_R, S_t)$ be the restriction of the dynamical system (H,S_t) generated by problem (8.1.1) on the forward invariant set*

$$\mathscr{E}_R = \left\{(u_0;u_1) \in H \,:\, \mathscr{E}(u_0,u_1) \le R^2\right\},$$

where $\mathscr{E}(u_0,u_1)$ is the energy of (H,S_t) given by (8.1.8). Then

- *For any $R > 0$ the dynamical system $(\mathscr{E}_R, S_t)$ has a compact global attractor A_R.*
- *If in addition we assume that (i) the set*

$$\mathscr{N}_* = \left\{u \in \mathscr{D}(\mathscr{A}^{1/2}) \,:\, \mathscr{A}u = F(u)\right\}$$

is bounded in $\mathscr{D}(\mathscr{A}^{1/2})$, and (ii) the function $t \mapsto \mathscr{E}(u(t),u_t(t))$ is (strictly) decreasing[2] for every nonstationary solution $u(t)$ to (8.1.1). Then there exists $R_0 > 0$ such that the attractor A_R does not depend on the parameter R for $R \ge R_0$. Moreover, in this case $A \equiv A_{R_0}$ is a compact attractor for (H,S_t) and $A = \mathscr{M}^u(\mathscr{N})$, where $\mathscr{M}^u(\mathscr{N})$ is the unstable manifold (see Definition 7.5.1) emanating from the set $\mathscr{N} = \{(u;0) \,:\, u \in \mathscr{N}_*\}$ of stationary points of (H,S_t).

Proof. Because $\mathscr{E}_R$ is bounded and forward invariant, Proposition 8.3.11 implies that $(\mathscr{E}_R, S_t)$ is a dissipative asymptotically smooth dynamical system. Therefore by Theorem 7.2.4 we obtain the result in the first part of Theorem 8.3.12.

[2] We can relax this requirement by assuming that the property $\mathscr{E}(u(t),u_t(t)) = \text{const}$ for all $t \in \mathbb{R}_+$ implies that $u(t) \equiv u_* \in \mathscr{N}_*$.

We choose now $R_0 > 1$ such that the set $\mathscr{N}$ of equilibria lies in $\mathscr{E}_{R_0-1}$. It is clear that $\mathscr{E}(u_0,u_1)$ is a strict Lyapunov function for $(\mathscr{E}_R, S_t)$. Consequently by the same argument as in Corollary 7.5.7 we have that $\mathscr{A}_R = \mathscr{M}^u(\mathscr{N})$ and therefore $\mathscr{A}_R$ does not depend on R for $R \geq R_0$. □

8.4 Regular attractors. Rate of stabilization to equilibria

In this section we consider properties of global attractors in the case when nonconservative forces are absent ($F^* \equiv 0$ in representation (8.1.4)); that is, we deal with the problem

$$\begin{cases} Mu_{tt}(t) + \mathscr{A}u(t) + k\cdot D(u_t(t)) = -\Pi'(u(t)), \\ u|_{t=0} = u_0 \in \mathscr{D}(\mathscr{A}^{1/2}), \quad u_t|_{t=0} = u_1 \in V = \mathscr{D}(M^{1/2}). \end{cases}$$

In this situation, under some additional conditions we show that the corresponding dynamical system (H,S_t) is gradient and hence the attractor has a regular structure. In particular, any trajectory converges to the set $\mathscr{N}$ of equilibria. If $\mathscr{N}$ is finite and hyperbolic in some (weak) sense described below, we estimate the rate of this convergence. We note that the results on the rate of stabilization to equilibria which are available in the literature usually rely on the study of the linearization near an equilibrium point and the corresponding local unstable manifold (see, e.g., [17]). This approach requires a rather strong hyperbolicity condition (see Definition 7.5.13) which may fail in the case of nonlinear damping. In contrast, the method presented below is completely analytic and covers the case of a damping operator degenerated near zero.

We start with the following simple structural property.

8.4.1. Proposition. *Let Assumption* 8.1.1 *hold with $F^* \equiv 0$ in relation* (8.1.4). *Assume also that there exist a strictly increasing, concave function $H_0 \in C(\mathbb{R}^+)$ with the property $H_0(0) = 0$, such that*

$$H_0((D(v),v)) \geq (Mv,v) \ \ \textit{for any } v \in \mathscr{D}(\mathscr{A}^{1/2}). \tag{8.4.1}$$

Then the dynamical system (H,S_t) generated by (8.1.1) *in $H = \mathscr{D}(\mathscr{A}^{1/2}) \times V$ is gradient. Thus, if (H,S_t) possesses a compact global attractor A, then $A = \mathscr{M}^u(\mathscr{N})$; that is, A is the unstable manifold emanating from the set $\mathscr{N}$ of equilibria for* (8.1.1), *$\mathscr{N} = \{(u;0) : \mathscr{A}u = F(u)\}$. Moreover, the statements of Theorem* 7.5.10 *and Corollary* 7.5.11 *are in force. In particular, every trajectory $S_t y$ converges to the set $\mathscr{N}$ as $t \to \infty$.*

Proof. The energy $\mathscr{E}(u,u_t)$ given by (8.1.8) satisfies the relation

$$\mathscr{E}(u(t),u_t(t)) + k\int_0^t (Du_t(\tau),u_t(\tau))d\tau = \mathscr{E}(u_0,u_1)$$

on strong solutions. This implies that

$$\mathscr{E}(u(t),u_t(t))+k\int_0^t H_0^{-1}\left(|M^{1/2}u_t(\tau)|^2\right)d\tau \le \mathscr{E}(u_0,u_1). \tag{8.4.2}$$

Thus after limit transition we obtain that (8.4.2) remains true for any generalized solution. This property implies that the energy $\mathscr{E}(u_0,u_1)$ is a strict Lyapunov function for (H,S_t) (see Definition 7.5.3) and hence (H,S_t) is gradient. Therefore we can apply Theorems 7.5.6 and 7.5.10 and Corollary 7.5.11 to conclude the proof. □

8.4.2. Remark. We note that in a special case when $H_0(s)=k_0^{-1}s$ condition (8.4.1) is nothing else but a strong monotonicity requirement imposed on the damping operator D. However, in a more general case when H_0 is nonlinear, condition (8.4.1) allows us to obtain estimates with a damping whose behavior at the origin is not quantified (e.g., superlinear). In practical applications where $D(v)=g(v)$ and g is a monotone (scalar) function, we take $H_0(s)=s, |s|\ge 1$, and the behavior of $H_0(s)$ for $|s|\le 1$ is described by a suitable concave function depending on g (see [195] and also Proposition B.2.1 in Appendix B. We also note that coercivity condition (8.4.1) implies property (8.2.12) in Assumption 8.2.6 (see also (8.2.1)). ■

The main result of this section reads as follows.

8.4.3. Theorem. *Assume that Assumption* 8.1.1 *holds with* $F^*\equiv 0$ *in the representation* (8.1.4), $\mathscr{A}^{-1}$ *is compact, and the dynamical system* (H,S_t) *generated by* (8.1.1) *possesses a compact global attractor A. Assume additionally that*

- *Coercivity condition* (8.4.1) *holds and, also, the following inequality*

$$|(D(v),w)|\le C_1(r)\cdot(D(v),v)+C_2(r)\cdot|\mathscr{A}^{1/2}w|^2 \tag{8.4.3}$$

  holds for any $v,w\in\mathscr{D}(\mathscr{A}^{1/2})$ *such that* $|\mathscr{A}^{1/2}w|+|M^{1/2}v|\le r$ *with arbitrary* $r>0$, *where* $C_1(r)$ *and* $C_2(r)$ *are nondecreasing functions of* r.
- $F(u)$ *is Fréchet differentiable and its derivative* $F'(u)$ *possesses the properties: there exists* $\delta>0$ *such that*

$$|\langle F'(u);w\rangle|_{[\mathscr{D}(\mathscr{A}^{1/2})]'}\le C_R|\mathscr{A}^{1/2-\delta}w|, \tag{8.4.4}$$

  and

$$|\langle F'(u)-F'(v);w\rangle|_{[\mathscr{D}(\mathscr{A}^{1/2})]'}\le C_R|\mathscr{A}^{1/2-\delta}(u-v)|\cdot|\mathscr{A}^{1/2}w| \tag{8.4.5}$$

  for any $w,u,v\in\mathscr{D}(\mathscr{A}^{1/2})$ *such that* $|\mathscr{A}^{1/2}u|\le R$ *and* $|\mathscr{A}^{1/2}v|\le R$.
- *Any generalized solution* $u(t)$ *to* (8.1.1) *satisfies the energy inequality*

$$\mathscr{E}(u(t),u_t(t))+k\int_s^t(Du_t(\tau),u_t(\tau))d\tau\le\mathscr{E}(u(s),u_t(s)) \tag{8.4.6}$$

  for any $0\le s\le t<\infty$, *where the energy* $\mathscr{E}(u,u_t)$ *is given by* (8.1.8).

- *The set $\mathscr{N}$ of equilibrium points is finite and all equilibria are hyperbolic in the sense that the equation $\mathscr{A}u = \langle F'(w), u\rangle$ has only a trivial solution for each $(w;0) \in \mathscr{N}$.*

Then for any $y \in H$ there exists an equilibrium $y_0 \in \mathscr{N}$ such that

$$|S_t y - y_0|_H^2 \le C \cdot \sigma\left(\left[tT^{-1}\right]\right), \quad t > 0, \tag{8.4.7}$$

where C and T are positive constants depending on y, $[a]$ denotes the integer part of a, and $\sigma(t)$ satisfies the following ODE,

$$\frac{d\sigma}{dt} + Q(\sigma) = 0,\ t > 0, \quad \sigma(0) = C(y, y_0). \tag{8.4.8}$$

Here $C(y, y_0)$ is a constant depending on y and y_0, $Q(s) = s - (I + G_0)^{-1}(s)$ with $G_0(s) = c_1 (I + H_0)^{-1}(c_2 s)$, where c_1 and c_2 are positive numbers depending on y and y_0. In particular, if $H_0(s) = a_0 s$, then for any $y \in H$ there exist $\gamma > 0$, $C > 0$, and an equilibrium $y_0 \in \mathscr{N}$ such that

$$|S_t y - y_0|_H \le C e^{-\gamma t}, \quad t > 0. \tag{8.4.9}$$

Proof. We refer to [75, Section 4.3], see also [69] for the case $M \equiv I$ and Chapter 10 where the same idea is applied in the case of a Karman model with a boundary damping. □

We conclude this subsection with some comments concerning Theorem 8.4.3 which we put in three remarks below.

8.4.4. Remark. Relation (8.4.3) holds if we assume that (8.3.12) is satisfied for $u = 0$ with $\kappa = 2$. We also note that the substitution $\varepsilon \cdot w$ in (8.4.3) instead of w gives the relation

$$|(D(v), w)| \le C_1(r)\varepsilon^{-1} \cdot (D(v), v) + \varepsilon \cdot C_2(r) \cdot |\mathscr{A}^{1/2} w|^2, \quad 0 < \varepsilon < 1,$$

for any $v, w \in \mathscr{D}(\mathscr{A}^{1/2})$ such that $|\mathscr{A}^{1/2} w| + |M^{1/2} v| \le r$ with arbitrary $r > 0$. ■

8.4.5. Remark. The standard hyperbolicity condition (see Definition 7.5.13) requires that the spectrum of the linearization of the semiflow S_t around an equilibrium does not intersect the unit circumference in the complex plane. In Theorem 8.4.3 we do not assume the existence of this linearization. However, if the linearization exists, our hyperbolicity condition is equivalent to the requirement that $\lambda = 1$ does not belong to the spectrum.

We also note that one of crucial hypotheses of Theorem 8.4.3 is energy inequality (8.4.6) for generalized solutions. In general, the validity of (8.4.6) for all generalized solutions may require some additional assumptions concerning the damping operator D (see, e.g., Proposition 2.4.21 and Remark 2.4.22). However in the applications considered, the requirement in (8.4.6) can be derived under the same hypotheses which we need for the existence of a compact global attractor.

As mentioned in [75] we can state Theorem 8.4.3 in the conditional form: if some trajectory converges to an isolated equilibrium, then it converges with the rate prescribed in (8.4.7). Moreover, in the case of a single equilibrium ($\mathscr{N}$ is a single point set) there is no need to assume a priori convergence of a trajectory to some attractor. In this case Theorem 8.4.3 provides us with a rate of global stabilization to an equilibrium which, a posteriori, is a global attractor. ∎

8.4.6. Remark. One can see that any nonzero solution $\sigma(t)$ to problem (8.4.8) is decreasing and tends to zero when $t \to +\infty$. Moreover, the rate of the convergence function $\sigma(t)$ is determined by the behavior of the function $Q(s)$ (and hence $H_0(s)$) around zero. For an example, in the case when $H_0(s) \sim c_0 s^\alpha$ as $s \to +0$ with $0 < \alpha < 1$, one can see that $Q(s) \sim c_1 s^{1/\alpha}$ as $s \to +0$. This implies that

$$\sigma(t) \sim ct^{-\alpha/(1-\alpha)} \quad \text{as} \quad t \to +\infty.$$

We refer to Remark B.3.2 in Appendix B for details. ∎

8.5 Stabilizability and quasi-stability estimates

In this section we return to the general, potentially nonconservative case, and give conditions that guarantee the system (H, S_t) generated by (8.1.1) to be quasi-stable (see Definition 7.9.2). As we recall, quasi-stability is a very strong property that allows us to establish—including critical cases—several important properties of the attractor including finite-dimensionality and smoothness (see Section 7.9). Thus, the main task is to establish stabilizability estimate of the form (7.9.3), which eventually evolves into quasi-stability property described in Definition 7.9.2.

8.5.1 Basic theorem on quasi-stability

Our main hypothesis in this section is the following, rather technical, assumption which, however, may be guaranteed by several sets of sufficient simple conditions. In order not to curtail generality of the approach, we formulate this standing assumption in a weakest possible form.

8.5.1. Assumption. Let $\mathscr{B}$ be a forward invariant set for the system (H, S_t) generated by (8.1.1) and $S_t y_i = (u^i(t); u^i_t(t))$ with $y_i \in \mathscr{B}$, $i = 1, 2$. We assume that

(G) For any $\varepsilon > 0$ and $T > 0$ there exist $a_\varepsilon(\mathscr{B}, T)$ and $b_\varepsilon(\mathscr{B}, T)$ such that for $z(t) = u^1(t) - u^2(t)$ we have the following relation

$$\sup_{t\in[0,T]} \left| \int_{s+t}^{s+T} (G_{u^1,u^2}(\tau), z_t(\tau)) d\tau \right| \le \varepsilon \int_s^{s+T} E_z(\tau) d\tau \tag{8.5.1}$$

$$+a_\varepsilon(\mathscr{B},T)\int_s^{s+T} d(\tau;u^1,u^2)E_z(\tau)d\tau + b_\varepsilon(\mathscr{B},T)\sup_{0\le\tau\le T}|\mathscr{A}^\sigma z(s+\tau)|^2$$

for all $s\ge 0$, where $\sigma<1/2$. Here $G_{u^1,u^2}(t)$ and $E_z(t)$ are given by (8.3.3) and (8.3.2), and $d(\tau;u^1,u^2)$ is a nonnegative function such that

$$d_\infty = \int_0^\infty d(t;u^1,u^2)\,dt \le C_{\mathscr{B}} < \infty. \tag{8.5.2}$$

(D) There exist numbers $c_0>0$ such that

$$(D(u+v)-D(u),v)\ge c_0(Mv,v) \ \text{ for any } u,v\in\mathscr{D}(\mathscr{A}^{1/2}). \tag{8.5.3}$$

(S) There exists $\beta>0$ such that $\mathscr{D}(\mathscr{A}^{1/2-\beta})\subset V=\mathscr{D}(M^{1/2})$.

8.5.2. Remark. We provide the following sufficient condition for (8.5.1) which is frequently satisfied and easy to verify. Assume that $G_{u^1,u^2}(t)$ given by (8.3.3) admits the following representation

$$\left(G_{u^1,u^2}(t),u_t^1(t)-u_t^2(t)\right)=\frac{d}{dt}Q(t)+R(t), \tag{8.5.4}$$

where the scalar functions $Q(t)\equiv Q(u^1(t);u^2(t))$ and $R(t)\equiv R(y^1(t);y^2(t))$ for $y_1,y_2\in\mathscr{B}$ enjoy the following properties: (i) there exist a number $0\le\sigma<1/2$ and a locally bounded function $c(t)$ on $[0,\infty)$ such that

$$|Q(t)|\le c(t)|\mathscr{A}^\sigma(u^1(t)-u^2(t))|^2, \tag{8.5.5}$$

and (ii) for every $\varepsilon>0$ there exists $a_\varepsilon(T)>0$ and $b_\varepsilon(T)>0$ such that

$$\int_s^{s+T}|R(t)|dt\le\varepsilon\int_s^{s+T}E_z(t)dt \tag{8.5.6}$$
$$+a_\varepsilon(T)\int_s^{s+T}d(t;u^1,u^2)E_z(t)dt+b_\varepsilon(T)\sup_{0\le\tau\le T}|\mathscr{A}^\sigma z(s+\tau)|^2$$

for every $T>0$ and $s\ge 0$, where $z(t)=u^1(t)-u^2(t)$ and $E_z(t)$ is given by (8.3.2). Then it is clear that (8.5.1) holds. We refer to Assumption 8.5.7 given below and providing concrete conditions imposed on the semilinear source F that guarantee representation (8.5.4) with properties (8.5.5) and (8.5.6). Another sufficient condition for (8.5.1) with $d(\tau;u^1,u^2)\equiv 0$ is described in the proof of Theorem 8.5.6 in the subcritical case (see relation (8.5.20)). We also note that the condition in (8.5.1) describes a certain compensated compactness property enjoyed by the operator $\{u^1;u^2\}\mapsto G_{u^1,u^2}$. ■

Our main result in this section is the following theorem which establishes quasi-stability of a second-order evolution (8.1.1).

8.5.3. Theorem. *Let Assumption* 8.1.1 *and relation* (8.3.12) *with* $\kappa=2$ *be in force. Assume that Assumption* 8.5.1 *holds with some forward invariant set* $\mathscr{B}$. *Let* $y_1,y_2\in$

$\mathscr{B}$. Then for $S_t y_i = (u^i(t), u^i_t(t))$ we have the following relation

$$|S_t y_1 - S_t y_2|^2_H \le C_1(\mathscr{B}) e^{-\omega_{\mathscr{B}} t} |y_1 - y_2|^2_H \\ + C_2(\mathscr{B}) \max_{[0,t]} |(u^1(\tau) - u^2(\tau))|^2, \quad t > 0, \qquad (8.5.7)$$

where $C_1(\mathscr{B})$, $C_2(\mathscr{B})$, and $\omega_{\mathscr{B}}$ are positive constants.

Proof of Theorem 8.5.3. Without loss of generality we can assume that $u^1(t)$ and $u^2(t)$ are strong solutions. As above we denote $z(t) = u^1(t) - u^2(t)$.

We rely on the following lemmas.

8.5.4. Lemma. *Let Assumption* 8.1.1 *be in force and relation* (8.3.12) *with $\kappa = 2$ hold. Assume that w and v are strong solutions to* (8.1.1) *possessing the following property*

$$\max_{s \in [0,T]} \left\{ |\mathscr{A}^{1/2} w(s)|^2 + |M^{1/2} w_t(s)|^2 + |\mathscr{A}^{1/2} v(s)|^2 + |M^{1/2} v_t(s)|^2 \right\} \le R^2 \quad (8.5.8)$$

for some $R > 0$. Then for the difference $z(t) = w(t) - v(t)$ we have that

$$TE_z(T) + \int_0^T E_z(t)dt \le c_0 \int_0^T |M^{1/2} z_t|^2 dt \qquad (8.5.9)$$
$$+ C(R,T) \left[k \int_0^T (D(t,z_t), z_t)dt + (1+k) \max_{s \in [0,T]} |\mathscr{A}^{1/2-\delta} z(s)|^2 \right] + c_0 \Psi_T(w,v),$$

for $T \ge T_0$, where $c_0 > 0$ does not depend on R and T and the damping parameter k. If relation (8.3.12) *does not contain the term $\varepsilon |\mathscr{A}^{1/2} w|^2$, the constant $C(R,T)$ also does not depend on k. Here $E_z(t)$, $D(t,z_t)$, and $\Psi_T(w,v)$ are the same as in Lemma* 8.3.1*:*

$$\begin{aligned} E_z(t) &= E_0(z(t), z_t(t)) = \frac{1}{2} \left((Mz_t(t), z_t(t)) + (\mathscr{A} z(t), z(t)) \right), \\ D(t,z_t) &= D(v_t(t) + z_t(t)) - D(v_t(t)), \\ \Psi_T(w,v) &= \left| \int_0^T (G_{w,v}(\tau), z_t(\tau)) d\tau \right| + \left| \int_0^T (G_{w,v}(t), z(t)) dt \right| \\ &\quad + \left| \int_0^T dt \int_t^T (G_{w,v}(\tau), z_t(\tau)) d\tau \right| \end{aligned} \qquad (8.5.10)$$

with $G_{w,v}(t)$ given by (8.3.3)*; that is, $G_{w,v}(t) = F(w(t)) - F(v(t))$.*

Proof. It follows from the energy relation (8.3.5) and from (8.5.8) that

$$kD_0^T(w,v) \le C_R(1+T),$$

where $D_0^T(w,v)$ is given by (8.3.10). Therefore from Lemma 8.3.2 and from (8.3.12) with $\kappa = 2$ we have that

$$k\int_0^T |(D(t,z_t),z)|\,dt \le C(R,T)(1+k)\max_{s\in[0,T]} |\mathscr{A}^{1/2-\delta}z(s)|^2$$
$$+C_\varepsilon(R,T)k\int_0^T (D(t,z_t),z_t)dt + \varepsilon\left[E_z(T)+\int_0^T E_z(t)dt\right]$$

for any $\varepsilon>0$, where $C(R,T)$ and $C_\varepsilon(R,T)$ do not depend on k when the term $\varepsilon|\mathscr{A}^{1/2}w|^2$, is absent in (8.3.12). Therefore using inequality (8.3.1) in Lemma 8.3.1 we obtain (8.5.9). □

To complete the proof of Theorem 8.5.3 we need the following assertion.

8.5.5. Lemma. *Let Assumption* 8.1.1 *and Assumption* 8.5.1 *be in force. Assume that relation* (8.3.12) *with* $\kappa=2$ *hold. Let* $y^1(t)=S_ty_1=(u^1(t);u_t^1(t))$ *and* $y^2(t)=S_ty_2=(u^2(t);u_t^2(t))$ *with* y_1 *and* y_2 *taken from* $\mathscr{B}$. *Let* $T\ge 1+T_0$, *where* T_0 *is given by Lemma 8.5.4. Then there exists* $\gamma=\gamma(T)<1$ *such that*

$$E_z(s+T)\le \gamma E_z(s)+C_T\left[\sup_{0\le\tau\le T}|z(s+\tau)|^2+\int_s^{s+T} d(\tau;u^1,u^2)E_z(\tau)d\tau\right] \quad (8.5.11)$$

for all $s\ge 0$. *The constant* C_T *may depend on the parameter k and the size of* $\mathscr{B}$.

Proof. It follows from Lemma 8.5.4 and from (8.5.3) that

$$TE_z(s+T)+\int_s^{s+T} E_z(\tau)d\tau \quad (8.5.12)$$
$$\le C_T\left[k\int_s^{s+T}(D(t,z_t),z_t)dt+\max_{\tau\in[0,T]}|\mathscr{A}^{1/2-\delta}z(s+\tau)|^2\right]+c_0\Psi_T^s(u^1,u^2),$$

where

$$\Psi_T^s(u^1,u^2)=(1+T)\sup_{t\in[0,T]}\left|\int_{s+t}^{s+T}(G_{u^1,u^2}(\tau),z_t(\tau))d\tau\right|+\left|\int_s^{s+T}(G_{u^1,u^2}(t),z(t))dt\right|. \quad (8.5.13)$$

By Lemma 8.3.2 we have that

$$\max_{\tau\in[0,T]}|\mathscr{A}^{\rho}z(s+\tau)|^2\le C_\eta\max_{\tau\in[0,T]}|z(s+\tau)|^2+\eta\Big\{E_z(s+T) \quad (8.5.14)$$
$$+k\int_s^{s+T}(D(t,z_t),z_t)dt+c_{\mathscr{B}}\int_s^{s+T}E_z(t)dt\Big\},$$

for every $\eta>0$, where $0\le\rho<\frac{1}{2}$. Therefore (8.5.12) can be written in the form

$$TE_z(s+T)+\int_s^{s+T}E_z(\tau)d\tau \quad (8.5.15)$$
$$\le C_T\max_{\tau\in[0,T]}|z(s+\tau)|^2+C_Tk\int_s^{s+T}(D(t,z_t),z_t)dt+c_0\Psi_T^s(u^1,u^2)$$

for $T \geq 1 + T_0$. From energy relation (8.3.5) we also have that

$$k \int_s^{s+T} (D(t, z_t), z_t) dt \leq E_z(s) - E_z(s+T) + \Psi_T^s(u^1, u^2). \tag{8.5.16}$$

Consequently

$$\begin{aligned} & TE_z(s+T) + \int_s^{s+T} E_z(\tau) d\tau + k \int_s^{s+T} (D(t, z_t), z_t) dt \\ & \leq C_T \max_{\tau \in [0,T]} |z(s+\tau)|^2 + C_T (E_z(s) - E_z(s+T)) + c_T \Psi_T^s(u^1, u^2). \end{aligned} \tag{8.5.17}$$

Because

$$|(G_{u^1,u^2}(t), z(t))| \leq C_R |\mathscr{A}^{1/2} z| |z|_V \leq C_R |\mathscr{A}^{1/2} z| |\mathscr{A}^{1/2-\beta} z|,$$

by (8.5.1) we obtain that

$$\begin{aligned} \Psi_T^s(u^1, u^2) \leq \varepsilon & \int_s^{s+T} E_z(\tau) d\tau \\ & + \widetilde{a}_\varepsilon(R, T) \int_s^{s+T} d(\tau; u^1, u^2) E_z(\tau) d\tau + \widetilde{b}_\varepsilon(R, T) \sup_{0 \leq \tau \leq T} |\mathscr{A}^{\tilde{\sigma}} z(s+\tau)|^2 \end{aligned} \tag{8.5.18}$$

for every $\varepsilon > 0$, where $\tilde{\sigma} = \max\{\sigma, 1/2 - \beta\}$. Therefore using (8.5.14) with an appropriate choice of η and also (8.5.16) we obtain the following version of (8.5.18) with $\tilde{\sigma} = 0$,

$$\begin{aligned} \Psi_T^s(u^1, u^2) \leq \varepsilon & \left\{ E_z(s) + \int_s^{s+T} E_z(\tau) d\tau + k \int_s^{s+T} (D(t, z_t), z_t) dt \right\} \\ & + a_\varepsilon(R, T) \int_s^{s+T} d(\tau; u^1, u^2) E_z(\tau) d\tau + b_\varepsilon(R, T) \sup_{0 \leq \tau \leq T} |z(s+\tau)|^2 \end{aligned}$$

for every $\varepsilon > 0$. Substituting this relation (with appropriately chosen ε) in (8.5.17) we obtain (8.5.11). □

To conclude the proof of Theorem 8.5.3 we note that (8.5.11) yields that

$$E_z((m+1)T) \leq \gamma E_z(mT) + c_{\mathscr{B},T} b_m, \quad m = 0, 1, 2, \ldots,$$

with $0 < \gamma \equiv \gamma_{\mathscr{B},R} < 1$, where

$$b_m \equiv \sup_{\tau \in [mT, (m+1)T]} |z(\tau)|^2 + \int_{mT}^{(m+1)T} d(\tau; u^1, u^2) E_z(\tau) d\tau.$$

This yields

$$E_z(mT) \leq \gamma^m E_z(0) + c \sum_{l=1}^{m} \gamma^{m-l} b_{l-1}.$$

Because $\gamma < 1$, one can see that there exists $\omega > 0$ such that for all $t \geq 0$ we have

$$E_z(t) \leq C_1 e^{-\omega t} E_z(0) + C_2 \left\{ \sup_{\tau\in[0,t]} |z(\tau)|^2 + \int_0^t e^{-\omega(t-\tau)} d(\tau;u^1,u^2) E_z(\tau) d\tau \right\}.$$

Therefore, applying Gronwall's lemma to the function $E_z(t)e^{\omega t}$ we find that

$$E_z(t) \leq \left[C_1 E_z(0) e^{-\omega t} + C_2 \sup_{\tau\in[0,t]} |z(\tau)|^2 \right] \exp\left\{ C_2 \int_0^t d(\tau;u^1,u^2)\, d\tau \right\}.$$

Now using (8.5.2) we obtain estimate (8.5.7). This concludes the proof of Theorem 8.5.3.

8.5.2 *Sufficient conditions for quasi-stability*

As a simple consequence of Theorem 8.5.3 we have the following assertion.

8.5.6. Theorem. *Let Assumption* 8.1.1 *be valid and relation* 8.3.12 *hold with* $\varkappa = 2$. *In addition we assume that (i) there exists a number* $c_0 > 0$ *such that* (8.5.3) *holds, (ii)* $\mathscr{D}(\mathscr{A}^{1/2-\beta}) \subset V$ *for some* $\beta > 0$, *and (iii) there exists* $0 \leq \eta \leq 1/2$ *such that*

$$|F(u) - F(v)|_{V'} \leq C(\rho) |\mathscr{A}^{1/2-\eta}(u-v)| \tag{8.5.19}$$

for any u and v from $\mathscr{D}(\mathscr{A}^{1/2})$ *such that* $|\mathscr{A}^{1/2}u| \leq \rho$ *and* $|\mathscr{A}^{1/2}v| \leq \rho$. *Here* $\rho > 0$ *is arbitrary,* $C(\rho)$ *is a nondecreasing function of* $\rho > 0$. *Let* $\mathscr{B}$ *be a bounded forward invariant set for* (H, S_t). *Let* $u^1(t)$ *and* $u^2(t)$ *be two generalized solutions to* (8.1.1) *corresponding to initial data in the set* $\mathscr{B}$:

$$(u^1(t), u^1_t(t)) = S_t y_1, \quad (u^2(t), u^2_t(t)) = S_t y_2, \quad y_1, y_2 \in \mathscr{B}.$$

Then

- **Subcritical case** ($\eta > 0$): *There exist constants* $\omega_{\mathscr{B}} > 0$ *and* $C_i(\mathscr{B})$ *such that the stabilizability estimate in* (8.5.7) *holds.*
- **Critical case** ($\eta = 0$): *If in addition we assume that (i) relation* (8.3.12) *does not contain the term* $\varepsilon|\mathscr{A}^{1/2}w|^2$, *and (ii)* $R \equiv \sup\{|y|_H : y \in \mathscr{B}\}$ *does not depend on the damping parameter* $k \in [k_0, \infty)$ *for some* $k_0 > 0$, *then the stabilizability estimate in* (8.5.7) *holds for every* $k \geq k_*$, *where* $k_* \equiv k_*(R) \geq k_0$ *is large enough.*

Proof. In the case when $\eta > 0$ we obviously have that

$$\left| \int_t^T (F(z+v) - F(v), z_t) dt \right|$$
$$\leq \varepsilon \int_0^T E_0(z, z_t)(t) dt + C_\varepsilon(\mathscr{B}, T) \sup_{t\in[0,T]} |\mathscr{A}^{1/2-\eta} z(t)|^2. \tag{8.5.20}$$

Therefore (8.5.1) holds with $d(\tau,u_1,u_2)\equiv 0$ holds. Thus we can apply Theorem 8.5.3.

In the critical case we need some modification of the argument given in Theorem 8.5.3. Using (8.5.3) and (8.5.14) from Lemma 8.5.4 we have that

$$\begin{aligned} TE_z(s+T)+\int_s^{s+T}E_z(\tau)d\tau+k\int_s^{s+T}|M^{1/2}z_t(t)|^2dt \\ \le C(R,T)k\int_s^{s+T}(D(t,z_t),z_t)dt+C_{R,T,k}\max_{\tau\in[0,T]}|z(s+\tau)|^2+c_0\Psi_T^s(u^1,u^2), \end{aligned} \tag{8.5.21}$$

where $c_0>0$ does not depend on R and T, and $C(R,T)$ does not depend on the damping parameter $k\ge k_0>0$ and $\Psi_T^s(u^1,u^2)$ is given by (8.5.13). Therefore it follows from (8.5.16) that

$$\begin{aligned} TE_z(s+T)+\int_s^{s+T}E_z(\tau)d\tau+k\int_s^{s+T}|M^{1/2}z_t(t)|^2dt \\ \le C^1_{R,T}\left[E_z(s)-E_z(s+T)\right]+C_{R,T,k}\max_{\tau\in[0,T]}|z(s+\tau)|^2+C^2_{R,T}\Psi_T^s(u^1,u^2), \end{aligned} \tag{8.5.22}$$

where $C^1_{R,T}$ and $C^2_{R,T}$ do not depend on $k\ge k_0$. It is also clear that

$$\begin{aligned} C^2_{R,T}\Psi_T^s(u^1,u^2)\le \frac{C^*_{R,T}}{k}\int_s^{s+T}|\mathscr{A}^{1/2}z|^2dt+k\int_s^{s+T}|M^{1/2}z_t|^2dt \\ +C^{**}_{R,T,k}\max_{\tau\in[0,T]}|z(s+\tau)|^2, \end{aligned}$$

where $C^*_{R,T}$ does not depend on $k\ge k_0$. Consequently taking in (8.5.22) $k\ge k_*\equiv C^*_{R,T}$, we obtain that

$$E_z(s+T)\le C_1(R,T)\left[E_z(s)-E_z(s+T)\right]+C_2(R,T)\sup_{0\le\tau\le T}|z(s+\tau)|^2,\quad s>0,$$

for some fixed $T\ge T_0$ and $k\ge k_*$. This implies (8.5.11) with $d(\tau,u^1,u^2)\equiv 0$. The remaining part of the proof is now identical to that in the case of Theorem 8.5.3. □

The drawback of Theorem 8.5.6 is that the damping parameter k is assumed sufficiently large for the critical case ($\eta=0$). This restriction can be eliminated when nonconservative forces are absent. Indeed, in the case of *conservative forces* ($F^*(u)\equiv 0$) one can take advantage of the finiteness of the dissipation integral

$$\int_0^\infty (D(u_t(t)),u_t(t))dt \tag{8.5.23}$$

evaluated along trajectories in order to avoid the large damping hypothesis in the critical case (see the second part of Theorem 8.5.6). To exploit the beneficial effects of the damping, we formulate the following hypothesis imposed on the nonlinear term F.

8.5.7. Assumption. Assume that

- $F(u)$ does not contain nonconservative terms; that is, $F(u) = -\Pi'(u)$.
- The functional $\Pi : \mathscr{D}(\mathscr{A}^{1/2}) \mapsto \mathbb{R}$ is a Fréchet C^3-mapping.
- The second $\Pi^{(2)}(u)$ and the third $\Pi^{(3)}(u)$ Fréchet derivatives of $\Pi(u)$ satisfy the conditions

$$\left|\langle \Pi^{(2)}(u); v, v\rangle\right| \le C_\rho |\mathscr{A}^\sigma v|^2, \quad v \in \mathscr{D}(\mathscr{A}^{1/2}), \tag{8.5.24}$$

for some $\sigma < \frac{1}{2}$, and

$$\left|\langle \Pi^{(3)}(u); v_1, v_2, v_3\rangle\right| \le C_\rho |\mathscr{A}^{1/2} v_1||\mathscr{A}^{1/2} v_2||v_3|_V, \quad v_i \in \mathscr{D}(\mathscr{A}^{1/2}), \tag{8.5.25}$$

for all $u \in \mathscr{D}(\mathscr{A}^{1/2})$ such that $|\mathscr{A}^{1/2}u| \le \rho$, where $\rho > 0$ is arbitrary and C_ρ is a positive constant. Here above $\langle \Pi^{(k)}(u); v_1, \dots, v_k\rangle$ denotes the value of the derivative $\Pi^{(k)}(u)$ on elements $v_1, \dots, v_k$.

We recall that the Fréchet derivatives $\Pi^{(k)}(u)$ of the functional Π are symmetric k-linear continuous forms on $\mathscr{D}(\mathscr{A}^{1/2})$ (see, e.g., [41]). Moreover, if $\Pi \in C^3$, then $(F(u), v) \equiv -\langle \Pi'(u); v\rangle$ is C^2-functional for every fixed $v \in \mathscr{D}(\mathscr{A}^{1/2})$ and the following Taylor's expansion holds

$$(F(u+w) - F(u), v) = -\langle \Pi^{(2)}(u); w, v\rangle - \int_0^1 (1-\lambda)\langle \Pi^{(3)}(u+\lambda w); w, w, v\rangle d\lambda \tag{8.5.26}$$

for any $u, v \in \mathscr{D}(\mathscr{A}^{1/2})$ [41]. This structure of $F(u)$ is important for our subsequent considerations because it leads to a representation of the form (8.5.4). Indeed, assume that $u(t)$ and $z(t)$ belong to the class $C^1(a,b; \mathscr{D}(\mathscr{A}^{1/2}))$ for some interval $[a,b] \subseteq \mathbb{R}$. Then by the differentiation rule for composition of mappings [41] using the symmetry of the form $\Pi^{(2)}(u)$ we have that

$$\frac{d}{dt}\langle \Pi^{(2)}(u); z, z\rangle = \langle \Pi^{(3)}(u); u_t, z, z\rangle + 2\langle \Pi^{(2)}(u); z, z_t\rangle.$$

Therefore from (8.5.26) we obtain the representation

$$(F(u(t)+z(t)) - F(u(t)), z_t(t)) = \frac{d}{dt}Q(t) + R(t), \quad t \in [a,b] \subseteq \mathbb{R}, \tag{8.5.27}$$

with

$$Q(t) = -\frac{1}{2}\langle \Pi^{(2)}(u(t)); z(t), z(t)\rangle \tag{8.5.28}$$

and

$$R(t) = \frac{1}{2}\langle \Pi^{(3)}(u); u_t, z, z\rangle - \int_0^1 (1-\lambda)\langle \Pi^{(3)}(u+\lambda z); z, z, z_t\rangle d\lambda. \tag{8.5.29}$$

8.5.8. Remark. Instead of Assumption 8.5.7 one could also postulate the decomposition as in (8.5.27) with $Q(t)$ and $R(t)$ subject to (8.5.24) with $v = z(t)$ and (8.5.25) with $v_1 = v_2 = z(t)$ and $v_3 = z_t(t)$. ■

We also note that under Assumption 8.5.7 it is easy to see from (8.5.27) that for any $(u(t);u_t(t)),(z(t);z_t(t)) \in C(a,b;\mathscr{D}(\mathscr{A}^{1/2})\times V)$ such that

$$\sup_{t\in[a,b]}\left\{|\mathscr{A}^{1/2}u(t)|+|\mathscr{A}^{1/2}z(t)|\right\}\le R \tag{8.5.30}$$

we have the relation

$$\left|\int_s^t (G_{u+z,u}(\tau),z_t(\tau))d\tau\right| \le C_1(R)\sup_{s\le\tau\le t}|\mathscr{A}^\sigma z(\tau)|^2 \tag{8.5.31}$$
$$+C_2(R)\int_s^t (|u_t(\tau)|_V+|z_t(\tau)|_V)\,|\mathscr{A}^{1/2}z(\tau)|^2 d\tau$$

for all $a < s \le t < b$, where $\sigma < \frac{1}{2}$ and $G_{u^1,u^2}(t) = F(u^1(t)) - F(u^2(t))$ in the case considered. This property implies the following assertion, which constitutes the main ingredient of the proof of stabilizability estimate.

8.5.9. Proposition. *Let Assumption* 8.5.7 *be in force. Assume that vector functions* $(u^1(t);u^1_t(t))$ *and* $(u^2(t);u^2_t(t))$ *from* $C(\mathbb{R}_+;\mathscr{D}(\mathscr{A}^{1/2})\times V)$ *possess the property*

$$\max_{s\in\mathbb{R}_+}\left\{|\mathscr{A}^{1/2}u^1(s)|^2+|M^{1/2}u^1_t(s)|^2+|\mathscr{A}^{1/2}u^2(s)|^2+|M^{1/2}u^2_t(s)|^2\right\}\le R^2 \tag{8.5.32}$$

for some $R>0$. *Then for any* $\varepsilon>0$ *and* $T>0$ *there exist* $a_\varepsilon(R,T)$ *and* $b(R,T)$ *such that for* $z(t)=u^1(t)-u^2(t)$ *we have the following relation*

$$\sup_{t\in[0,T]}\left|\int_{s+t}^{s+T}(G_{u^1,u^2}(\tau),z_t(\tau))d\tau\right|\le\varepsilon\int_s^{s+T}E_z(\tau)d\tau \tag{8.5.33}$$
$$+a_\varepsilon(R,T)\int_s^{s+T}d(\tau;u^1,u^2)E_z(\tau)d\tau+b\sup_{0\le\tau\le T}|\mathscr{A}^\sigma z(s+\tau)|^2$$

for all $s\ge0$, *where* $\sigma<\frac{1}{2}$. *Here* $E_z(t)$ *and* $G_{u^1,u^2}(t)$ *are given by* (8.3.2) *and* (8.3.3), *and*

$$d(t;u^1,u^2)=|u^1_t(t)|^2_V+|u^2_t(t)|^2_V. \tag{8.5.34}$$

Now we are in position to state the following theorem which covers the case of critical nonlinearity and arbitrary damping parameter. The theorem below may be considered as the main result guaranteeing quasi-stability of the second order dynamical system.

8.5.10. Theorem. *Let Assumption* 8.1.1, *relations* (8.5.3) *and* (8.3.12) *with* $\kappa=2$ *be in force. Assume that* $\mathscr{D}(\mathscr{A}^{1/2-\beta})\subset V$ *for some* $\beta>0$ *and the nonlinear mapping* $F(u)$ *satisfies Assumption* 8.5.7. *Then the requirements in Assumption* 8.5.1 *hold* *true on any bounded forward invariant set $\mathscr{B}$ and thus the stabilizability estimate in* (8.5.7) *is valid.*

Proof. By (8.5.3) we have that

$$d(t;u^1,u^2) \equiv |u_t^1(t)|_V^2 + |u_t^2(t)|_V^2 \le c_0^{-1}\left[D(u_t^1(t),u_t^1(t)) + D(u_t^2(t),u_t^2(t))\right]. \tag{8.5.35}$$

Therefore it follows from (8.3.34) that relation (8.5.2) holds. Thus we can apply Theorem 8.5.3. □

The following assertion presents another set of assumptions that guarantees the validity of the stabilizability estimate, although it requires subcriticality of the source F.

8.5.11. Theorem. *Assume that Assumption* 8.1.1 *holds and the dynamical system* (H,S_t) *generated by* (8.1.1) *on a forward invariant set $\mathscr{B}$ possesses the properties:*

- *D is Fréchet differentiable as a mapping from $\mathscr{D}(\mathscr{A}^{1/2})$ into $\left[\mathscr{D}(\mathscr{A}^{1/2})\right]'$, the derivative $D'(u)$ generates a symmetric bilinear form on $\mathscr{D}(\mathscr{A}^{1/2})$, and there exist positive constants k_1 and k_2 such that*
$$\begin{aligned} k_1(Mw,w) &\le k(\langle D'(\lambda u + (1-\lambda)v);w\rangle,w) \\ &\le k_2\left[1 + (D(u),u) + (D(v),v)\right]|\mathscr{A}^{1/2}w|^2 \end{aligned} \tag{8.5.36}$$
for any $0 \le \lambda \le 1$ and $u,v,w \in \mathscr{D}(\mathscr{A}^{1/2})$.
- *There exist $0 < \eta \le \frac{1}{2}$ and $C_{\mathscr{B}} > 0$ such that*
$$|F(u) - F(v)|_{V'} \le C_{\mathscr{B}}|\mathscr{A}^{1/2-\eta}(u-v)| \tag{8.5.37}$$
for any $u,v \in Pr_{\mathscr{D}(\mathscr{A}^{1/2})}\mathscr{B} \equiv \left\{w \in \mathscr{D}(\mathscr{A}^{1/2}) : \exists v \in V : \ (w,v) \in \mathscr{B}\right\}$.

Then there exist positive constants $C_1(\mathscr{B})$, $C_2(\mathscr{B})$, and $\omega_{\mathscr{B}}$ such that for any $y_1,y_2 \in \mathscr{B}$ we have that

$$\begin{aligned} |S_t y_1 - S_t y_2|_H &\le C_1(\mathscr{B})e^{-\omega_{\mathscr{B}}t}|y_1 - y_2|_H \\ &\quad + C_2(\mathscr{B})\int_0^t e^{-\omega_{\mathscr{B}}(t-\tau)}|u^1(\tau) - u^2(\tau)|d\tau, \quad t > 0, \end{aligned} \tag{8.5.38}$$

where $S_t y_i = (u^i(t);u_t^i(t))$. In particular, the quasi-stability estimate in (8.5.7) *holds true.*

Proof. As before, we can assume that $u^1(t)$ and $u^2(t)$ are strong solutions. Let $z(t) = u^1(t) - u^2(t)$. Then $z(t)$ solves the equation

$$Mz_{tt} + kD(t)z_t + \mathscr{A}z = G_{u^1,u^2}(t), \tag{8.5.39}$$

where

$$D(t)w = \int_0^1 \langle D'(u_t^2(t) + \xi(u_t^1(t) - u_t^2(t)));w\rangle d\xi \tag{8.5.40}$$

and $G_{u^1,u^2}(t)$ given by (8.3.3); that is, $G_{u^1,u^2}(t) = F(u^1(t)) - F(u^2(t))$.

We need the following "exponential decay" estimate obtained for linear nonautonomous system, that is also used later. This is why in the proposition below we propose some properties of strong solutions as a hypothesis.

8.5.12. Proposition. *Let Assumption* 8.1.1*(A,M) hold. Assume that* $\{D(t)\}$ *is a family of self-adjoint operators in the space* $\mathscr{H}$ *that are bounded from* $\mathscr{D}(\mathscr{A}^{1/2})$ *into* $[\mathscr{D}(\mathscr{A}^{1/2})]'$ *and possesses the property*

$$k_1(Mw,w) \le k(Dw,w) \le k_2(1+d(t))|\mathscr{A}^{1/2}w|^2 \qquad (8.5.41)$$

for any $w \in \mathscr{D}(\mathscr{A}^{1/2})$, *where* k_1 *and* k_2 *are positive constants, and* $d \in L_1^{loc}(\mathbb{R})$ *is a nonnegative function possesses the property*

$$\sup_{s\in\mathbb{R}} \int_s^{s+T} d(\tau)d\tau \le d_T < \infty.$$

Let $f \in L_1^{\rm loc}([s,\infty),V')$. *Consider the following linear nonautonomous equation,*

$$Mv_{tt} + kD(t)v_t + \mathscr{A}v = f(t), \;\; t > s, \;\;\; v\big|_{t=s} = v_0, \; v_t\big|_{t=s} = v_1, \qquad (8.5.42)$$

Let $v(t) = v(t,s;v_0;v_1)$ *be a (weak or generalized) solution of* (8.5.42) *such that it can be approximated by either a sequence* $\{v^N\}$ *of strong (smooth) solutions or a sequence of Galerkin approximations* $\{v^N\}$ *with the properties:*

- $$\liminf_{N\to\infty} |\mathscr{A}^{1/2}v^N(s)|^2 \le |\mathscr{A}^{1/2}v_0|^2, \quad \liminf_{N\to\infty} |M^{1/2}v_t^N(s)|^2 \le |M^{1/2}v_1|^2.$$
- *For every* $T > 0$ *and* $s \in \mathbb{R}$*:*
$$(v^N;v_t^N) \to (v;v_t) \;\; *\text{-weakly} \;\; \text{in} \; L_\infty(s,s+T;\mathscr{D}(\mathscr{A}^{1/2})\times V).$$
- v^N *satisfies equation* (8.5.42) *with* $D(t)$ *replaced by* $D^N(t)$ *and* $f(t)$ *replaced by* $f^N(t)$, *where* $D^N(t)$ *satisfies the same condition as* $D(t)$ *and* $f^N(t) \to f(t)$ *in the space* $L_1(s,s+T;V')$ *for every* $T > 0$ *as* $N \to \infty$.

Let $y(t) = (v(t);v_t(t))$ *and* $y_0 = (v_0;v_1)$. *Then the following estimate*

$$|y(t)|_H \le Ce^{-\omega(t-s)}|y_0|_H + C\int_s^t e^{-\omega(t-\tau)}|f(\tau)|_{V'}d\tau, \quad t > s, \qquad (8.5.43)$$

holds, where ω *and* C *are positive constants depending on* k_1, k_2 *and the bound* d_T *for some* $T > 0$.

Proof. One can see that it is sufficient to establish (8.5.43) for every approximate solution $v^N(t)$ (below we suppress the superscript N). We begin by defining the energy function associated with the system (8.5.42). Let

$$E(t) \equiv \frac{1}{2}\left(|M^{1/2}v_t(t)|^2 + |\mathscr{A}^{1/2}v(t)|^2\right) = \frac{1}{2}|y(t)|_H^2.$$

The standard energy calculations for the solutions considered give:

$$E(t) + k\int_s^t (D(\tau)v_t(\tau), v_t(\tau))d\tau = E(s) + \int_s^t (f(\tau), v_t(\tau))d\tau. \quad (8.5.44)$$

From (8.5.41) we have

$$k(D(t)v_t, v_t) \geq k_1|M^{1/2}v_t|^2. \quad (8.5.45)$$

Multiplying the approximate equation for (8.5.42) with approximating solution v and integrating from s to $s+T$ give

$$\begin{aligned}
&-\int_s^{T+s} |M^{1/2}v_t|^2 dt + \int_s^{T+s} |\mathscr{A}^{1/2}v|^2 dt + k\int_s^{T+s} (D(t)v_t, v)dt \\
&\leq c_0[E(s) + E(T+s)] + \int_s^{T+s} (f(t), v(t))dt,
\end{aligned}$$

and by (8.5.45), (8.5.44) we have that

$$\begin{aligned}
\int_s^{T+s} E(t)dt + \sup_{s\leq t\leq T+s} E(t) &\leq C_1 k\int_s^{T+s} (D(t)v_t, v_t)dt \quad (8.5.46)\\
&\leq C_2 E(T+s) + C_3\left[\int_s^{T+s} |f(t)|_{V'}dt\right]^2 \\
&\quad + C_4 k\int_s^{T+s} |(D(t)v_t, v)|dt.
\end{aligned}$$

Recalling the right inequality in (8.5.41) we obtain

$$\begin{aligned}
k\int_s^{T+s} |(D(t)v_t, v)|dt &\leq C_\varepsilon k\int_s^{T+s} (D(t)v_t, v_t)dt + \varepsilon k\int_s^{T+s} (D(t)v, v)dt \\
&\leq C_\varepsilon k\int_s^{T+s} (D(t)v_t, v_t)dt + \varepsilon k_2(T+d_T)\sup_{s\leq t\leq T+s} |\mathscr{A}^{1/2}v(t)|^2.
\end{aligned}$$

Combining the above inequality with (8.5.46) and selecting ε suitably small give

$$\begin{aligned}
\int_s^{T+s} E(t)dt + \sup_{s\leq t\leq T+s} E(t) &\leq C_1 E(T+s) + C_T k\int_s^{T+s} (D(t)v_t, v_t)dt \\
&\quad + C_2\left[\int_s^{T+s} |f(t)|dt\right]^2. \quad (8.5.47)
\end{aligned}$$

It follows from (8.5.44) that

$$TE(T+s) \leq \int_s^{T+s} E(t)dt + T\int_s^{T+s} |(f(t), v_t(t))|dt.$$

Consequently,

$$TE(T+s) + \sup_{s\le t\le T+s} E(t) \le C_1 E(T+s) + C_T k \int_s^{T+s} (D(t)v_t, v_t)dt + C_2(1+T^2)\left[\int_s^{T+s} |f(t)|_{V'}dt\right]^2.$$

After taking $T \ge C_1 + 1$ and using

$$k\int_s^{T+s} (D(t)v_t, v_t)dt \le E(s) - E(T+s) + \varepsilon \sup_{s\le t\le T+s} E(t) + C_\varepsilon \left[\int_s^{T+s} |f(t)|_{V'}dt\right]^2$$

for any $\varepsilon > 0$ (which follows from (8.5.44)) yields

$$E(T+s) \le \frac{C_T}{C_T+1} E(s) + C_T \left[\int_s^{T+s} |f(t)|_{V'}dt\right]^2, \quad s \in \mathbb{R}.$$

Because $|y(s)|_H^2 = 2E(s)$, this implies that

$$|y(T+s)|_H \le \gamma |y(s)|_H + C\int_s^{T+s} |f(t)|_{V'}dt, \text{ where } \gamma < 1, \ s\in\mathbb{R}.$$

By the same procedure as in the proof of Theorem 8.5.3 we conclude that

$$|y(mT+\sigma)|_H \le \gamma^m |y(\sigma)|_H + C\sum_{l=1}^m \gamma^{m-l} \int_{\sigma+(l-1)T}^{\sigma+lT} |f(t)|_{V'}dt$$

for every $\sigma \in [s, T+s]$. From the energy relation (8.5.44) we obviously have that

$$|y(\sigma)|_H^2 \le 2 \sup_{s\le\sigma\le T+s} E(\sigma) \le 3E(s) + C_0 \left[\int_s^{T+s} |f(t)|_{V'}dt\right]^2.$$

Therefore

$$|y(mT+\sigma)|_H \le c_0\gamma^m |y(s)|_H + C_0\int_s^{T+s} |f(t)|_{V'}dt + C\int_\sigma^{\sigma+mT} e^{-((mT+\sigma)-t)\omega} |f(t)|_{V'}dt$$

for every $\sigma \in [s, T+s]$ and $m = 1, 2, \dots$ with $\omega = T^{-1}\ln(1/\gamma)$. This allows us to obtain (8.5.43) for approximate solutions $v = v^N$. The limit transition exploiting weak lower semicontinuity of the norm and hypotheses imposed on the approximations yields the desired conclusion. □

To conclude with the proof of Theorem 8.5.11 we note that $z(t)$ which solves (8.5.39) is a difference of two strong solutions of the initial nonlinear problem (8.1.1). Therefore we apply Proposition 8.5.12 with the operator $D(t)$ given by

(8.5.40). The assumptions imposed in Proposition 8.5.12 are satisfied due to the fact that u_t^i appearing in (8.5.40) correspond to strong solutions. Using (8.5.37) and Proposition 8.5.12 we get

$$|S_t y_1 - S_t y_2|_H \le C_{\mathscr{B}} e^{-\bar{\omega}_{\mathscr{B}} t}|y_1 - y_2|_H + C_{\mathscr{B}} \int_0^t e^{-\bar{\omega}_{\mathscr{B}}(t-\tau)}|\mathscr{A}^{1/2-\eta} z(\tau)| d\tau$$

with positive $C_{\mathscr{B}}$ and $\bar{\omega}_{\mathscr{B}}$. Using the interpolation we obtain that

$$|\mathscr{A}^{1/2-\eta} z| \le \varepsilon |\mathscr{A}^{1/2} z| + C_\varepsilon |z| \quad \text{for every } \varepsilon > 0.$$

Therefore the function $\Psi(t) = e^{\bar{\omega}_{\mathscr{B}} t}|S_t y_1 - S_t y_2|_H$ satisfies the inequality

$$\Psi(t) \le C_{\mathscr{B}}|y_1 - y_2|_H + C_{\varepsilon,\mathscr{B}} \int_0^t e^{\bar{\omega}_{\mathscr{B}}\tau}|z(\tau)| d\tau + \varepsilon \int_0^t \Psi(\tau) d\tau.$$

Thus using Gronwall's type argument with ε small enough we obtain (8.5.38). □

Theorems 8.5.6, 8.5.10, and 8.5.11 make it possible to apply the results of Section 7.9 to the dynamical system (H, S_t) generated by (8.1.1). This topic is considered in the next sections.

8.6 Finite dimension of global attractors

We start with the following general assertion which is a version of Theorem 7.9.6 adapted to problem (8.1.1) satisfying Assumption 8.1.1.

8.6.1. Theorem. *Let Assumption* 8.1.1 *be valid and the embedding* $\mathscr{D}(\mathscr{A}^{1/2}) \subset V$ *be compact. Assume that the system* (H, S_t) *generated by* (8.1.1))*possesses the global attractor* A *and there exist nonnegative scalar functions* $a(t)$, $b(t)$, *and* $c(t)$ *on* $\mathbb{R}_+$ *such that (i)* $a(t)$ *and* $c(t)$ *are locally bounded on* $[0,\infty)$, *(ii)* $b(t) \in L_1(\mathbb{R}_+)$ *possesses the property* $\lim_{t\to\infty} b(t) = 0$, *and (iii) for every* $y_1, y_2 \in A$ *and* $t > 0$ *the following relations*

$$|S_t y_1 - S_t y_2|_H^2 \le a(t) \cdot |y_1 - y_2|_H^2 \tag{8.6.1}$$

and

$$|S_t y_1 - S_t y_2|_H^2 \le b(t) \cdot |y_1 - y_2|_H^2 + c(t) \sup_{0 \le s \le t} |\mathscr{A}^\sigma (u^1(s) - u^2(s))|^2 \tag{8.6.2}$$

hold for some $\sigma \in [0, \frac{1}{2})$. *Here we denote* $S_t y_i = (u^i(t); u_t^i(t))$, $i = 1, 2$. *Then the attractor* A *has a finite fractal dimension.*

Proof. It is clear that the system (H, S_t) satisfies Assumption 7.9.1 with $X = \mathscr{D}(\mathscr{A}^{1/2})$, $Y = V$, and $Z = \{0\}$. By (8.6.1) and (8.6.2) this system is quasi-stable on the attractor (see Definition 7.9.2) with $\mu_X(v) = |\mathscr{A}^\sigma v|$, $\sigma \in [0, \frac{1}{2})$, which is a compact seminorm on $\mathscr{D}(\mathscr{A}^{1/2})$. Therefore the statement follows from Theorem 7.9.6. □

The theorem above provides a general tool for proving finiteness of dimensions of attractors. In what follows we use this result in order to establish finiteness of fractal dimension for attractors described in Theorem 8.3.5 and Theorem 8.3.12.

Our first result deals with possibly nonconservative forces and corresponds to Theorem 8.3.5. For this we need the following hypothesis.

8.6.2. Assumption.

- There exists a number $c_0 > 0$ such that (8.5.3) holds.
- There exist $0 \le \eta \le \frac{1}{2}$ such that

$$|F(u) - F(v)|_{V'} \le C(\rho)|\mathscr{A}^{1/2-\eta}(u-v)| \tag{8.6.3}$$

  for any u and v from $\mathscr{D}(\mathscr{A}^{1/2})$ such that $|\mathscr{A}^{1/2}u| \le \rho$ and $|\mathscr{A}^{1/2}v| \le \rho$. Here $\rho > 0$ is arbitrary, $C(\rho)$ is a nondecreasing function of $\rho > 0$.
- $\mathscr{D}(\mathscr{A}^{1/2-\beta})$ is compactly embedded into V for some $\beta > 0$.

8.6.3. Theorem. *Let Assumptions* 8.1.1, 8.2.1 *and* 8.6.2 *hold. Assume that Assumption* 8.3.3 *holds with* $\varkappa = 2$ *in* (8.3.12). *Then the global attractor A of the system* (H, S_t) *generated by* (8.1.1) *exists and has a finite fractal dimension provided the nonlinearity F is subcritical (i.e.,* $\eta > 0$ *in* (8.6.3)*). The same result remains true in the critical case (*$\eta = 0$*) provided (i) relations* (8.2.7) *and* (8.2.8) *hold, (ii) relation (8.3.12) does not contain the term* $\varepsilon|\mathscr{A}^{1/2}w|^2$*, and (iii) the damping parameter k is large enough.*

Proof. SUBCRITICAL CASE: By Theorem 8.2.3 the system (H, S_t) is dissipative. Estimate (8.6.1) on a bounded absorbing set follows from Proposition 8.1.3. To obtain (8.6.2) (with $\sigma = 0$) we use Theorem 8.5.6. Hence we can apply Proposition 7.9.4 with $X = \mathscr{D}(\mathscr{A}^{1/2})$, $Y = V$, $Z = \{0\}$, and $\mu_X(\cdot) = |\cdot|$ to obtain the existence of a global attractor. Its finite dimension now follows from Theorem 8.6.1.

CRITICAL CASE: The argument is the same. We only note that by (8.2.7) and (8.2.8) Theorem 8.2.3 implies the existence of a forward invariant absorbing set with the size independent of k. Therefore we can apply the second part of Theorem 8.5.6 to obtain stabilizability estimate (8.6.2) for k large enough. □

We conclude this section with the following theorem which supplies the main result pertaining to finite-dimensionality of attractors with the critical conservative forces and without assuming large values of the damping parameter.

8.6.4. Theorem. *Let Assumption* 8.1.1 *with* $F^*(u) \equiv 0$*, and relations* (8.5.3) *and* (8.3.12) *with* $\kappa = 2$*, and Assumption* 8.3.9 *be in force. Assume that* $\mathscr{D}(\mathscr{A}^{1/2-\beta})$ *is compactly embedded in V for some* $\beta > 0$*, the nonlinear mapping* $F(u) = -\Pi'(u)$ *satisfies Assumption* 8.5.7*, and the dissipativity condition in* (8.3.33) *holds. Then the system* (H, S_t) *generated by* (8.1.1) *possesses a compact global attractor A of a finite fractal dimension.*

Proof. The existence of a compact global attractor A follows from Theorem 8.3.10. The stabilizability estimate in (8.6.2) follows from Theorem 8.5.10. Thus we can apply Theorem 8.6.1 to prove that A has a finite dimension. □

8.7 Regularity of elements from attractors

In this section we address a question of additional smoothness of elements in the attractor. Relying on the stabilizability estimate we are able to prove, under some additional hypotheses, that elements in the attractor are more regular than indicated by the topology of the phase space.

Regularity of attractors of second-order in time evolution equations has been studied in the abstract setting in [123] for the case of *scalar linear* damping and $M = I$ (see also [55, 57] and also Chapter 9 for the case of von Karman models).

Our aim in this section is to present the results with nonlinear and possibly vectorial damping. We mainly follow the ideas presented in [75] and start with the following assertion which gives us the result in the case of the attractors constructed in Theorem 8.6.3 or in Theorem 8.6.4.

8.7.1. Theorem. *Let the hypotheses of either Theorem* 8.6.3 *or Theorem* 8.6.4 *be in force. Then the system* (H,S_t) *generated by* (8.1.1) *possesses a compact global attractor A of finite fractal dimension. Moreover, any full trajectory* $\{(u(t);u_t(t)) : t\in\mathbb{R}\}$ *that belongs to the global attractor enjoys the following regularity properties:*

$$\mathscr{A}u+kD(u_t)\in C_r(\mathbb{R};V'),\quad u_t\in C_r(\mathbb{R};\mathscr{D}(\mathscr{A}^{1/2})),\quad u_{tt}\in C_r(\mathbb{R};V),\tag{8.7.1}$$

where C_r *denotes the space right-continuous functions, and there exists* $R>0$ *such that*

$$|u_{tt}(t)|_V+|\mathscr{A}^{1/2}u_t(t)|+|\mathscr{A}^{1/2}u(t)|+|\mathscr{A}u(t)+kD(u_t(t))|_{V'}<R,\quad t\in\mathbb{R}.\tag{8.7.2}$$

In particular, the attractor A is a closed bounded set in $\mathscr{D}(\mathscr{A}^{1/2})\times\mathscr{D}(\mathscr{A}^{1/2})$*; more precisely, we have that*

$$A\subset\left\{(u;v)\in\mathscr{D}(\mathscr{A}^{1/2})\times\mathscr{D}(\mathscr{A}^{1/2})\;\middle|\;\begin{array}{l}w\equiv\mathscr{A}u+kD(v)\in V',\\ |w|_{V'}+|\mathscr{A}^{1/2}v|+|\mathscr{A}^{1/2}u|\le R\end{array}\right\}.\tag{8.7.3}$$

Proof. It follows from Theorem 8.5.6 (in the case of Theorem 8.6.3) and from Theorem 8.5.3 (in the case of Theorem 8.6.4) that the system (H,S_t) satisfies Assumption 7.9.1 with $X=\mathscr{D}(\mathscr{A}^{1/2})$, $Y=V$, and $Z=\{0\}$ and is quasi-stable (see Definition 7.9.2) on the attractor with $\mu_X(v)=|v|$, which is a compact seminorm on $\mathscr{D}(\mathscr{A}^{1/2})$. Moreover, the corresponding function $c(t)$ in (7.9.3) is a constant. Therefore we can apply Theorem 7.9.8 which yields the relation

$$|u_{tt}(t)|_V^2+|\mathscr{A}^{1/2}u_t(t)|^2\le C,\quad t\in\mathbb{R},\tag{8.7.4}$$

for any trajectory $\gamma = \{y(t) \equiv (u(t); u_t(t)) \,:\, t \in \mathbb{R}\}$ from the global attractor. Returning to the original equation we obtain that

$$|\mathscr{A} u(t) + kD(u_t(t))|_{V'} \le C|u_{tt}|_V + |F(u(t))|_{V'} \le C, \quad t \in \mathbb{R}. \tag{8.7.5}$$

Thus, we have that

$$(\mathscr{A} u + kD(u_t); u_t; u_{tt}) \in L_\infty(\mathbb{R}; V' \times \mathscr{D}(\mathscr{A}^{1/2}) \times V).$$

In particular $\mathscr{A} u(t) + kD(u_t(t)) \in V'$ for almost all $t \in \mathbb{R}$. Therefore by Theorem 2.4.16 $u(t)$ is a strong solution to the original equation and hence relations (8.7.1) hold. The estimate in (8.7.2) follows from (8.7.4) and (8.7.5). Relation (8.7.3) follows from (8.7.1) and (8.7.2). □

8.7.2. Corollary. *In addition to the hypotheses of Theorem* 8.7.1 *we assume that the damping operator D maps $\mathscr{D}(\mathscr{A}^{1/2})$ into V' and is bounded on bounded sets. Then the global attractor A is a closed bounded set in the space $W \times \mathscr{D}(\mathscr{A}^{1/2})$, where $W = \{u \in \mathscr{D}(\mathscr{A}^{1/2}) \,:\, \mathscr{A} u \in V'\}$.*

Proof. Under the additional assumption imposed on the damping operator D we have that $|D(u_t(t))|_{V'} \le C$ for all $t \in \mathbb{R}$ on the attractor. Therefore the conclusion follows from (8.7.2). □

The result provided below is based on the "stabilizablity estimate" obtained in Theorem 8.5.11.

8.7.3. Theorem. *Assume that Assumption* 8.1.1 *holds and the system (H, S_t) generated by* (8.1.1) *possesses a global attractor A. Additionally assume that*

- *The damping operator D maps $\mathscr{D}(\mathscr{A}^{1/2})$ into V' continuously and is bounded on bounded sets.*
- *D is Fréchet differentiable as a mapping from $\mathscr{D}(\mathscr{A}^{1/2})$ into $\left[\mathscr{D}(\mathscr{A}^{1/2})\right]'$, the derivative $D'(u)$ generates a symmetric bilinear form on $\mathscr{D}(\mathscr{A}^{1/2})$, and there exist positive constants k_1 and k_2 such that*

$$\begin{aligned} k_1(Mw, w) &\le k(\langle D'(\lambda u + (1-\lambda)v); w\rangle, w) \\ &\le k_2\left[1 + (D(u), u) + (D(v), v)\right] |\mathscr{A}^{1/2} w|^2 \end{aligned} \tag{8.7.6}$$

for any $0 \le \lambda \le 1$ and $u, v, w \in \mathscr{D}(\mathscr{A}^{1/2})$ such that $|u|_V, |v|_V \le 1 + R_0$, where R_0 is a dissipativity radius of the system.
- *The nonlinear mapping $F(u)$ is C^1 from $\mathscr{D}(\mathscr{A}^{1/2})$ into V' and*

$$|\langle F'(u), w\rangle|_{V'} \le C_\rho \cdot |\mathscr{A}^{1/2-\delta} w| \tag{8.7.7}$$

for every $\rho > 0$, where $|\mathscr{A}^{1/2} u| \le \rho$, $w \in \mathscr{D}(\mathscr{A}^{1/2})$ and $\delta \in (0, 1/2]$.

Then the global attractor is a bounded set in $W \times \mathscr{D}(\mathscr{A}^{1/2})$), where $W = \{u \in \mathscr{D}(\mathscr{A}^{1/2}) \,:\, \mathscr{A} u \in V'\}$ and any full trajectory $\{(u(t); u_t(t)) \,:\, t \in \mathbb{R}\}$ that belongs to the global attractor possesses the following regularity properties

$$u \in C_r(\mathbb{R};W), \quad u_t \in C_r(\mathbb{R};\mathscr{D}(\mathscr{A}^{1/2})), \quad u_{tt} \in C_r(\mathbb{R};V). \tag{8.7.8}$$

Moreover, there exists $R > 0$ such that

$$|u_{tt}(t)|_V^2 + |\mathscr{A}^{1/2}u_t(t)|^2 + |\mathscr{A}u(t)|_{V'}^2 \le R^2, \quad t \in \mathbb{R}. \tag{8.7.9}$$

Proof. We apply Theorem 8.5.11 which implies the stabilizability estimate in (8.5.7). Therefore, it follows from Theorem 7.9.8 that

$$|u_{tt}(t)|_V^2 + |\mathscr{A}^{1/2}u_t(t)|^2 + |\mathscr{A}^{1/2}u(t)|^2 \le C_R, \quad t \in \mathbb{R}. \tag{8.7.10}$$

Therefore using the fact that D is continuous from $\mathscr{D}(\mathscr{A})$ into V' and the same argument as in Theorem 8.7.1 we conclude the proof. □

Assuming more regularity on the damping D one can to obtain even higher regularity of trajectories lying in the attractor. However, the method necessitates rather strong limitations imposed on the nonlinear damping in the abstract framework. For this reason we do not pursue this line of argument, but rather limit ourselves to a linear damping where "infinite smoothness" of attractors can be shown.

8.7.4. Theorem. *Assume that Assumption* 8.1.1 *holds and the system (H,S_t) generated by* (8.1.1) *possesses a global attractor A. Additionally assume that*

- *The damping operator D is a linear positive self-adjoint operator in $\mathscr{H}$ bounded from $\mathscr{D}(\mathscr{A}^{1/2})$ into V' and possessing the property*

$$k_1(Mw,w) \le k(Dw,w) \le k_2|\mathscr{A}^{1/2}w|^2 \tag{8.7.11}$$

  for any $w \in \mathscr{D}(\mathscr{A}^{1/2})$, where k_1 and k_2 are positive constants.
- *Concerning the nonlinear mapping $F(u)$ we assume that $F(u)$ belongs to C^m as a mapping from $\mathscr{D}(\mathscr{A}^{1/2})$ into V' and*

$$|\langle F^{(k)}(u);w_1,\dots,w_k\rangle|_{V'} \le C_\rho \cdot \prod_{j=1}^{k} |\mathscr{A}^{1/2-\delta}w_j|, \tag{8.7.12}$$

  for every $\rho > 0$, where $k = 0,1,\dots,m$, $|\mathscr{A}^{1/2}u| \le \rho$, $w_j \in \mathscr{D}(\mathscr{A}^{1/2})$ and $\delta \in (0,\frac{1}{2}]$. We also use notations $F^{(k)}(u)$ for the k-order Fréchet derivative of $F(u)$ and $\langle F^{(k)}(u);w_1,\dots,w_k\rangle$ for the value of $F^{(k)}(u)$ on elements $w_1,\dots,w_k$.

Then any full trajectory $\{(u(t);u_t(t)) : t \in \mathbb{R}\}$ that belongs to the global attractor possesses the following regularity properties,

$$\begin{gathered} u^{(k)}(t) \in C(\mathbb{R};W), \quad k = 0,1,2,\dots,m-1, \\ u^{(m)}(t) \in C(\mathbb{R};\mathscr{D}(\mathscr{A}^{1/2})), \quad u^{(m+1)}(t) \in C(\mathbb{R};V), \end{gathered} \tag{8.7.13}$$

where $W = \{u \in \mathscr{D}(\mathscr{A}^{1/2}) : \mathscr{A}u \in V'\}$. Here and below $u^{(k)}(t) = \partial_t^k u(t)$. Moreover,

$$|\mathscr{A}u^{(k-1)}(t)|_{V'}^2 + |\mathscr{A}^{1/2}u^{(k)}(t)|^2 + |u^{(k+1)}(t)|_V^2 + \int_t^{t+1} |D^{1/2}u^{(k+1)}(\tau)|^2 d\tau \le R^2 \tag{8.7.14}$$

for every $t \in \mathbb{R}$, where $k = 1, \dots, m$.

The result stated above, in the special case of an abstract wave equation with linear scalar damping ($M = D = I$), has been proved in [123]. Thus Theorem 8.7.4 extends this result to more general vectorial structures. For similar results in the case of von Karman evolution both with linear and nonlinear damping we refer to Theorem 9.2.10, Remark 9.2.12, and Theorem 9.5.5 in Chapter 9.

Proof. We use induction in m.

CASE $m = 1$: By Theorem 8.7.3 we have that

$$u \in C_r(\mathbb{R}; W), \quad u_t \in C_r(\mathbb{R}; \mathscr{D}(\mathscr{A}^{1/2})), \quad u_{tt} \in C_r(\mathbb{R}; V),$$

where C_r stands for right-continuous functions, and relations (8.7.9) hold. Thus to obtain the statement of the theorem for $m = 1$ we need to show that the function $t \mapsto (u(t); u_t(t); u_{tt}(t))$ is continuous in $W \times \mathscr{D}(\mathscr{A}^{1/2}) \times V$ and

$$\int_a^{a+1} |D^{1/2}u_{tt}(\tau)|^2 d\tau \le C \text{ for all } a \in \mathbb{R}.$$

Let $v^h(t) = h^{-1}(u(t+h) - u(t))$. Then $v^h(t)$ is a strong solution to the linear problem

$$Mv_{tt}^h + kDv_t^h + \mathscr{A}v^h = F_h(t) \equiv h^{-1}[F(u(t+h)) - F(u(t))$$

and the following energy relation holds,

$$\begin{aligned} &\frac{1}{2}\left(|M^{1/2}v_t^h(t)\|^2 + |\mathscr{A}^{1/2}v^h(t)\|^2\right) + k\int_s^t |D^{1/2}v_t^h(\tau)|^2 d\tau \\ &= \frac{1}{2}\left(|M^{1/2}v_t^h(s)\|^2 + |\mathscr{A}^{1/2}v^h(s)\|^2\right) + \int_s^t (F_h(\tau), v_t^h(\tau))d\tau, \end{aligned}$$

which implies that

$$\int_s^t |D^{1/2}v_t^h(\tau)|^2 d\tau \le C_R(1 + |t-s|),$$

where C_R is independent of h. This makes it possible to show that $v(t) = u_t(t) = \lim_{h\to 0} v_h(t)$ is a weak solution to problem

$$Mv_{tt} + kDv_t + \mathscr{A}v = \langle F'(u(t)); v\rangle$$

satisfying $(v; v_t) \in L_\infty(\mathbb{R}; \mathscr{D}(\mathscr{A}^{1/2}) \times V)$ and $D^{1/2}v_t \in L_2^{loc}(\mathbb{R}; \mathscr{H})$. Therefore by Theorem 2.4.35 we can conclude that

$$(u_t; u_{tt}) \equiv (v; v_t) \in C(\mathbb{R}; \mathscr{D}(\mathscr{A}^{1/2}) \times V).$$

From the equation for u we also obtain that $u \in C(\mathbb{R}; W)$. Thus the relations in (8.7.13) and (8.7.14) hold for $m = 1$.

INDUCTIVE ARGUMENT: Assume that the hypotheses of the theorem are valid for $m = n+1$ and relations (8.7.13) and (8.7.14) hold true for $m = n$. Then $v(t) = u^{(n)}(t) \in C(\mathbb{R}; \mathscr{D}(\mathscr{A}^{1/2}))$ satisfies the equation

$$Mv_{tt} + kDv_t + \mathscr{A}v = \Phi(t) \equiv \Phi(v(t); u'(t), \dots, u^{(n-1)}(t)),$$

where

$$\Phi(v(t); u'(t), \dots, u^{(n-1)}(t)) = \frac{d^n}{dt^n} F(u(t)) \text{ with } v(t) = u^{(n)}(t).$$

Moreover

$$\int_t^{t+1} |D^{1/2} v_t(\tau)|^2 d\tau \le R, \quad t \in \mathbb{R}.$$

Therefore the function $v^h(t) = h^{-1}(v(t+h) - v(t))$, $h > 0$, possesses the same regularity properties as v and satisfies the equation

$$Mv_{tt}^h + kDv_t^h + \mathscr{A}v^h = \Phi^h(t) \equiv \frac{1}{h}(\Phi(t+h) - \Phi(t)).$$

This allows us to apply Theorem 2.4.35 and Proposition 8.5.12 to obtain that

$$|y^h(t)|_H \le C_1 e^{-\omega(t-s)} |y^h(s)|_H + C_2 \int_s^t e^{-\omega(t-\tau)} |\Phi^h(\tau)|_{V'} d\tau,$$

where $y^h(t) = (v^h(t); v_t^h(t))$. By the induction hypothesis for each fixed h the function $|y^h(t)|_H$ is bounded for all $t \in \mathbb{R}$ (with the bound possibly depending on h). Therefore sending $s \to -\infty$ we obtain with a fixed h the following representation:

$$|y^h(t)|_H \le C \int_{-\infty}^t e^{-\omega(t-\tau)} |\Phi^h(\tau)|_{V'} d\tau, \tag{8.7.15}$$

Using (8.7.12) and the induction hypothesis one can show that

$$|\Phi^h(t)|_{V'} \le \varepsilon |\mathscr{A}^{1/2} v^h|^2 + C_{R,\varepsilon}$$

with uniformity in h. Hence combining with (8.7.15) yields

$$|y^h(t)|_H \le C \int_{-\infty}^t e^{-\omega(t-\tau} |\Phi^h(\tau)|_{V'} d\tau \le \varepsilon \sup_{s \le t} |y^h(s)|_H + C_{\varepsilon,R}. \tag{8.7.16}$$

Taking ε suitably small gives $|y^h(t)|_H \le C_R$ for all $t \in \mathbb{R}$. Passing with the limit on h yields then

$$|v_{tt}(t)|_V^2 + |\mathscr{A}^{1/2} v_t(t)|^2 \le C_R, \quad t \in \mathbb{R}.$$

The energy relation in Theorem 2.4.35 also implies that

$$\int_t^{t+1} |D^{1/2}v_t^h(\tau)|^2 d\tau \le C_R, \quad t \in \mathbb{R}.$$

Thus in the limit we obtain

$$\int_t^{t+1} |D^{1/2}v_{tt}(\tau)|^2 d\tau \le C_R, \quad t \in \mathbb{R}.$$

This allows us to prove (8.7.13) and (8.7.14) for $m = n+1$ and conclude the proof of Theorem 8.7.4. □

Under some conditions it is also possible to transfer time regularity of trajectories lying in the attractor into spatial smoothness of its elements. For details we refer to [123], where the case $M = D = I$ is discussed. The case of von Karman evolutions is studied in Chapter 9 (see Corollary 9.2.11, Remark 9.2.12, and Corollary 9.5.8).

8.8 On "strong" attractors

Theorems presented in Section 8.7 assert an additional regularity of attractor. A natural question that can be asked in this context is that of existence of strong (i.e., corresponding to topology of strong solutions) attractors. Although this latter property is technically related to smoothness of attractors, the corresponding result does not follow from results of Section 8.7. In what follows we briefly discuss the issue of "strong attractors." We show that under some conditions strong solutions converge uniformly to the attractor A in a topology that is stronger than the topology of the space H. The key tool in this direction is furnished by the dissipativity property of strong solutions and also involves the stabilizability estimate (see (8.8.1) below).

8.8.1. Theorem (Dissipativity). *Let Assumption* 8.1.1 *be valid. Assume that the system* (H,S_t) *generated by* (8.1.1) *is dissipative, possesses a forward invariant absorbing set* $\mathscr{B}$, *and there exist nonnegative scalar function* $b(t)$ *on* $\mathbb{R}_+$ *and a constant* $c>0$ *such that (i)* $b(t) \in L_1(\mathbb{R}_+)$ *possesses the property* $\lim_{t\to\infty} b(t) = 0$ *and (ii) for every* $y_1, y_2 \in \mathscr{B}$ *and* $t>0$ *the following relation*

$$|S_t y_1 - S_t y_2|_H^2 \le b(t)\cdot |y_1 - y_2|_H^2 + c \cdot \sup_{0\le s\le t} |M^{1/2}(u^1(s) - u^2(s))|^2 \tag{8.8.1}$$

holds. Here we denote $S_t y_i = (u^i(t); u_t^i(t))$, $i = 1,2$. *For* $R>0$ *we define the set*

$$\mathscr{W}_R = \left\{ (u;v) \in \mathscr{D}(\mathscr{A}^{1/2}) \times \mathscr{D}(\mathscr{A}^{1/2}) \;\middle|\; \begin{array}{l} w \equiv \mathscr{A}u + kD(v) \in V', \\ |w|_{V'} + |\mathscr{A}^{1/2}v| + |\mathscr{A}^{1/2}u| \le R \end{array} \right\}. \tag{8.8.2}$$

Then there exists $R_0 > 0$ *such that for any* $R>0$ *we can find* $t_R > 0$ *such that*

$$\|u_{tt}(t)\|_V^2 + |\mathscr{A}^{1/2}u_t(t)|^2 + |\mathscr{A}u(t) + kD(u_t(t))|_{V'}^2 \le R_0^2 \quad \text{for all} \quad t \ge t_R \tag{8.8.3}$$

for any strong solution $u(t)$ *to problem (8.1.1) with initial data* $(u_0;u_1)$ *from* $\mathscr{W}_R$.

Proof. It follows from Proposition 8.1.3 and relation (8.1.12) that

$$|S_{t+h}y - S_t y|_H \le C(T,\rho)|S_h y - y|_H, \quad t \in [0,T],\ 0 < h < 1,\ |y|_H \le \rho.$$

Let $S_t y = (u(t); u_t(t))$ with $y \in \mathscr{W}_R$. After dividing on h and the limit transition $h \to 0$ we get that

$$|u_{tt}(t)|_V^2 + |\mathscr{A}^{1/2} u_t(t)|^2 \le C_T\left(|u_{tt}(0)|_V^2 + |\mathscr{A}^{1/2} u_t(0)|^2\right), \quad t \in [0,T].$$

This implies that for every $R > 0$ and $T > 0$ there exists R_T such that

$$S_t \mathscr{W}_R \subset \mathscr{W}_{R_T} \quad \text{for all } t \in [0,T].$$

Because (H, S_t) is dissipative, we can also choose $T = T_R$ such that $S_t \mathscr{W}_R \subset \mathscr{B}$ for all $t \ge T_R$. Thus for every $R > 0$ there exists R_* and $T = T_R$ such that

$$S_t \mathscr{W}_R \subset \mathscr{W}_{R_*} \quad \text{for } t \in [0, 1+T_R] \quad \text{and} \quad S_t \mathscr{W}_R \subset \mathscr{B} \quad \text{for } t \ge T_R. \tag{8.8.4}$$

It follows from (8.8.1) that

$$|S_{t+h}y - S_t y|_H^2 \le b(t-\tau)\cdot|S_{\tau+h}y - S_\tau y|_H^2 + c\cdot \sup_{\tau\le s\le t}|M^{1/2}(u(s+h)-u(s))|^2$$

for every $h > 0$ and $t > \tau \ge T_R$. Therefore, as above, after dividing by h^2 and taking the limit $h \to 0$ we obtain that

$$|u_{tt}(t)|_V^2 + |\mathscr{A}^{1/2} u_t(t)|^2 \le b(t-\tau)\left(|u_{tt}(\tau)|_V^2 + |\mathscr{A}^{1/2} u_t(\tau)|^2\right) + c \sup_{\tau\le s\le t}|u_t(s)|_V^2.$$

Taking $\tau = T_R$ and using (8.8.4) yields

$$|u_{tt}(t)|_V^2 + |\mathscr{A}^{1/2} u_t(t)|^2 \le b(t-T_R)C_{R_*} + C_{\mathscr{B}}, \quad t \ge T_R.$$

This implies the conclusion. □

8.8.2. Corollary. *In addition to the hypotheses of Theorem* 8.8.1 *we assume that the damping operator D is a continuous mapping from $\mathscr{D}(\mathscr{A}^{1/2})$ into V' and is bounded on bounded sets. Then problem* (8.1.1) *generates a dissipative dynamical system in the phase space*

$$H_{st} = W \times \mathscr{D}(\mathscr{A}^{1/2}),$$

where

$$W = \left\{u \in \mathscr{D}(\mathscr{A}^{1/2}) \,:\, \mathscr{A}u \in V'\right\}. \tag{8.8.5}$$

Proof. Because $|D(v)|_{V'} \le C_R$ for any v such that $|\mathscr{A}^{1/2}v| \le R$, it is clear from (8.8.3) that $|\mathscr{A}u(t)|_{V'}^2 \le C_{R_0}$ for $t \ge t_R$. This implies the desired conclusion. □

To prove the existence of a global attractor for the system (H_{st}, S_t) we need the following additional hypotheses.

8.8.3. Assumption.

(D) *damping*:

- The damping operator $D : \mathscr{D}(\mathscr{A}^{1/2}) \mapsto V'$ is locally Lipschitz and possesses the properties

$$|D(u_1) - D(u_2)|_{V'} \le C_R |\mathscr{A}^{1/2}(u_1 - u_2)|, \tag{8.8.6}$$

and

$$c_0 |M^{1/2}(u_1 - u_2)|^2 \le (D(u_1) - D(u_2), u_1 - u_2), \tag{8.8.7}$$

for any $u_1, u_2 \in \mathscr{D}(\mathscr{A}^{1/2})$ such that $|\mathscr{A}^{1/2}u_1|, |\mathscr{A}^{1/2}u_2| \le R$, where $R > 0$ is arbitrary and c_0 is a positive constant.

- The operator D is Fréchet differentiable[3] and (i) $\|D'(u)\|_{\mathscr{L}(V,V')} < \infty$ for each $u \in \mathscr{D}(\mathscr{A}^{1/2})$; (ii) the linear form $\Psi(w_1, w_2) = (\langle D'(u); w_1 \rangle, w_2)$ is symmetric on V; and (iii) for any $u_1, u_2 \in \mathscr{D}(\mathscr{A}^{1/2})$ and $v \in V$ such that $|\mathscr{A}^{1/2}u_1| + |\mathscr{A}^{1/2}u_2| + |v|_V \le R$ we have that

$$|\langle (D'(u_1) - D'(u_2)); v \rangle|_{V'} \le C_R |\mathscr{A}^{1/2-\delta}(u_1 - u_2)|, \tag{8.8.8}$$

where $R > 0$ is arbitrary and $\delta > 0$ is a constant.

(F) *force*: The operator $F : \mathscr{D}(\mathscr{A}^{1/2}) \mapsto V'$ is Frechet differentiable and

$$|\langle F'(w); v \rangle|_{V'} \le C_R |\mathscr{A}^{1/2-\delta} v|, \quad w \in W, \ |\mathscr{A} w|_{V'} \le R, \tag{8.8.9}$$

where W is given by (8.8.5), and

$$|\langle (F'(w_1) - F'(w_2)); v \rangle|_{V'} \le C_R |\mathscr{A}^{1-\delta}(w_1 - w_2)|, \tag{8.8.10}$$

for any $w_1, w_2 \in W$, $v \in \mathscr{D}(\mathscr{A}^{1/2})$ such that $|\mathscr{A} w_1|_{V'} + |\mathscr{A} w_2|_{V'} + |\mathscr{A}^{1/2} v| \le R$, where $R > 0$ is arbitrary and $\delta > 0$ is a constant.

8.8.4. Theorem (Strong Attractor). *Let Assumption* 8.8.3 *be in force. Assume that* $\mathscr{D}(\mathscr{A}^{1/2})$ *is compactly embedded in* V. *Then under the hypotheses of Theorem* 8.8.1 *the system* (H_{st}, S_t) *possesses a compact global attractor. Here* $H_{st} = W \times \mathscr{D}(\mathscr{A}^{1/2})$ *with* W *given by* (8.8.5).

Proof. Due to Corollary 8.8.2 we need to prove the asymptotic smoothness only. We use the Ceron–Lopes criteria given in Proposition 7.1.9.

It follows from the smoothness properties of D and F assumed in Assumption 8.8.3 that the function $v(t) = u_t(t)$ solves the problem

$$\begin{cases} M v_{tt}(t) + \mathscr{A} v(t) + k \langle D'(u_t(t)); v_t \rangle = \langle F'(u(t)); v \rangle, \\ v|_{t=0} = u_1 \in \mathscr{D}(\mathscr{A}^{1/2}), \quad v_t|_{t=0} = u_{tt}(0) \in V = \mathscr{D}(M^{1/2}). \end{cases}$$

---

[3] This requirement along with (subcritical) estimate (8.8.8) is rather restrictive even for application in the case of plates with rotational inertia; see Section 9.2.4 in Chapter 9. In this respect the result in Theorem 8.8.4 below, which is based on Assumption 8.8.3, is mainly of methodological interest.

Therefore the difference $w(t) = v^1(t) - v^2(t) \equiv u_t^1(t) - u_t^2(t)$ satisfies the equation

$$Mw_{tt}(t) + \mathscr{A} w(t) + k \cdot \langle D'(u_t^1(t)); w_t \rangle = \Phi(t),$$

where

$$\Phi(t) = \langle F'(u^1(t)); v^1 \rangle - \langle F'(u^2(t)); v^2 \rangle - k \langle D'(u_t^1(t)) - D'(u_t^2(t)); v_t^2 \rangle.$$

It follows from Assumption 8.8.3 that $D(t) := kD'(u_t^1(t))$ is a positive self-adjoint operator such that $\|D(t)\|_{\mathscr{L}(V,V')} \le C$ and

$$|\Phi(t)|_{V'} \le C\left(|\mathscr{A}^{1-\delta}(u^1(t) - u^2(t))| + |\mathscr{A}^{1/2-\delta}(u_t^1(t) - u_t^2(t))|\right) \tag{8.8.11}$$

in the absorbing ball for (H_{st}, S_t). Consequently by the second part of Theorem 2.4.35 the solution $w(t)$ can be constructed by the Galerkin method and thus from Proposition 8.5.12 we obtain that

$$\left\| \begin{bmatrix} w(t) \\ w_t(t) \end{bmatrix} \right\|_H \le Ce^{-\gamma t} \left\| \begin{bmatrix} w(0) \\ w_t(0) \end{bmatrix} \right\|_H + C \int_0^t e^{-\gamma(t-\tau)} |\Phi(\tau)|_{V'} d\tau, \quad \gamma > 0.$$

Therefore using relation (8.8.11) and Assumption 8.8.3 one can see that

$$\begin{aligned} |S_t y_1 - S_t y_2|_{H_{st}} \le\ & Ce^{-\gamma t} |y_1 - y_2|_{H_{st}} \\ & + C \int_0^t e^{-\gamma(t-\tau)} \Big(|\mathscr{A}^{1-\delta}(u^1(\tau) - u^2(\tau))| \\ & + |\mathscr{A}^{1/2-\delta}(u_t^1(\tau) - u_t^2(\tau))|\Big) d\tau. \end{aligned} \tag{8.8.12}$$

Thus we can apply Proposition 7.1.9 and conclude the proof. □

8.8.5. Remark. It follows from (8.8.12) and from Theorem 4.4 [75] that the global attractor has a finite dimension as a compact set in the space H_{st}. Moreover, in the case when the global attractor for (H, S_t) is bounded in H_{st}, this attractor is also an attractor for (H_{st}, S_t) provided the hypotheses of Theorem 8.8.4 hold true. ■

8.9 Determining functionals

As explained in Section 7.8 (see also the references quoted there) determining functionals is an important tool for the description of the long-time behavior of the dissipative dynamical system. In this section we develop the theory of determining functionals for the second-order evolution equation. As above we consider the following second-order abstract equation,

$$\begin{cases} Mu_{tt}(t) + \mathscr{A} u(t) + kD(u_t(t)) = F(u(t)), \\ u|_{t=0} = u_0, \quad u_t|_{t=0} = u_1. \end{cases} \tag{8.9.1}$$

8.9.1 An approach based on stabilizability estimate

We start with the following result based on Theorem 7.9.11.

8.9.1. Theorem. *Let the embedding $\mathscr{D}(\mathscr{A}^{1/2}) \subset V$ be compact. Assume that the system (H,S_t) generated by (8.9.1) possesses an absorbing set $\mathscr{B}$ and there exists nonnegative functions $a(t)$, $b(t)$, and $c(t)$ on $\mathbb{R}_+$ such that (i) $a(t)$ and $c(t)$ are locally bounded on $[0,\infty)$, (ii) $b(t) \in L_1(\mathbb{R}_+)$ and $\lim_{t\to\infty} b(t) = 0$, and (iii) for every $y_1, y_2 \in \mathscr{B}$ and $t > 0$ the following relations*

$$|S_t y_1 - S_t y_2|_H^2 \le a(t) \cdot |y_1 - y_2|_H^2 \tag{8.9.2}$$

and

$$|S_t y_1 - S_t y_2|_H^2 \le b(t) \cdot |y_1 - y_2|_H^2 + c(t) \cdot \sup_{0\le s\le t} |A^\sigma(u^1(s) - u^2(s))|^2 \tag{8.9.3}$$

hold for some $\sigma \in [0, \frac{1}{2})$. Here we denote $S_t y_i = (u^i(t); u^i_t(t))$, $i = 1,2$. Let $\mathscr{L} = \{l_j : j = 1,\dots,N\}$ be a set of functionals on $\mathscr{D}(\mathscr{A}^{1/2})$ with the completeness defect $\varepsilon_{\mathscr{L}} \equiv \varepsilon_{\mathscr{L}}(\mathscr{D}(\mathscr{A}^{1/2}), \mathscr{H})$ (see Definition 7.8.5). If there exists $\tau > 0$ such that

$$\eta_\tau \equiv \varepsilon_{\mathscr{L}}^{2-4\sigma} c(\tau) \cdot \sup_{s\in[0,\tau]} a(s) + b(\tau) < 1 \tag{8.9.4}$$

Then the relation

$$\lim_{t\to\infty} l_j(u^1(t) - u^2(t)) = 0, \quad j = 1,\dots,N, \tag{8.9.5}$$

implies that $\lim_{t\to\infty} |S_t y_1 - S_t y_2|_H = 0$. Here $S_t y_i = (u^i(t), u^i_t(t))$, $i = 1,2$.

Proof. We apply Theorem 7.9.11. It is obvious the system (H,S_t) satisfies Assumption 7.9.1 with $X = \mathscr{D}(\mathscr{A}^{1/2})$, $Y = V$, and $Z = \{0\}$. It is also clear from (8.9.2) and (8.9.3) that (H,S_t) is quasi-stable on the absorbing ball $\mathscr{B}$ (see Definition 7.9.2) with $\mu_X(v) = |\mathscr{A}^\sigma v|$, $\sigma \in [0, \frac{1}{2})$, which is a compact seminorm on $\mathscr{D}(\mathscr{A}^{1/2})$. By Theorem 7.8.6 we have that

$$\varepsilon_{\mathscr{L}}(\mu_X) = \varepsilon_{\mathscr{L}}(\mathscr{D}(\mathscr{A}^{1/2}), \mathscr{D}(\mathscr{A}^\sigma)) \le \left[\varepsilon_{\mathscr{L}}(\mathscr{D}(\mathscr{A}^{1/2}), \mathscr{H})\right]^{1-2\sigma},$$

where $\varepsilon_{\mathscr{L}}(\mu_X)$ is defined in the statement of Theorem 7.9.11. Therefore the application of Theorem 7.9.11 gives the desired conclusion. □

8.9.2. Corollary. *Let the hypotheses of either Theorem 8.6.3 or Theorem 8.6.4 be in force. Assume that $\mathscr{L} = \{l_j : j = 1,\dots,N\}$ is a set of functionals on $\mathscr{D}(\mathscr{A}^{1/2})$ with the completeness defect $\varepsilon_{\mathscr{L}} \equiv \varepsilon_{\mathscr{L}}(\mathscr{D}(\mathscr{A}^{1/2}), \mathscr{H})$. Then there exists $\varepsilon_0 > 0$ such that under the condition $\varepsilon_{\mathscr{L}} < \varepsilon_0$ relation (8.9.5) implies that $\lim_{t\to\infty} |S_t y_1 - S_t y_2|_H = 0$. Here $S_t y_i = (u^i(t), u^i_t(t))$, $i = 1,2$.*

Proof. In the case of the hypotheses of Theorem 8.6.3, by Theorem 8.5.6 relation (8.9.3) holds with $b(t) = Ce^{-\omega_{\mathscr{B}}t}$, $c(t) \equiv C(\mathscr{B})$, and $\sigma = 0$, where $\mathscr{B}$ is a forward invariant absorbing set (which exists by Theorem 8.2.3 or Theorem 8.2.4). Relation (8.9.2) with $a(t) = e^{L(\mathscr{B})t}$ follows from Proposition 8.1.3. Therefore, if $C(\mathscr{B})\varepsilon_{\mathscr{L}}^2$ is small enough, we can choose τ such that (8.9.4) holds for some $\tau > 0$. Thus we can apply Theorem 8.9.1.

In the case when the hypotheses of Theorem 8.6.4 are in force the argument is similar. It involves Theorem 8.5.10 to obtain (8.9.3). □

Our second result is based on a stabilizability estimate in the form given in relation (8.5.38) (with an integral compact part).

8.9.3. Theorem. *Assume that the system (H, S_t) generated by (8.9.1) is dissipative. Let the hypotheses of Theorem 8.5.11 be in force on a forward invariant absorbing set and $\mathscr{L} = \{l_j : j = 1, \dots, N\}$ be a set of the linear continuous functionals on the space $\mathscr{D}(A^{1/2})$ endowed with the norm $|A^{1/2}\cdot|$ with the completeness defect $\varepsilon_{\mathscr{L}} = \varepsilon_{\mathscr{L}}(\mathscr{D}(A^{1/2}), \mathscr{H})$. Then there exists $\varepsilon_0 > 0$ such that under the condition $\varepsilon_{\mathscr{L}} \le \varepsilon_0$ the set $\mathscr{L}$ is a set of asymptotically determining functionals for problem (8.9.1) in the sense: for any two of its solutions u_1 and u_2 the condition*

$$\lim_{t\to\infty} \int_t^{t+1} |l_j(u_1(\tau)) - l_j(u_2(\tau))| d\tau = 0 \;\; \textit{for } j = 1, \dots, N, \tag{8.9.6}$$

implies

$$\lim_{t\to\infty} \left\{ |\mathscr{A}^{1/2}(u_1(t) - u_2(t))|^2 + |\partial_t u_1(t) - \partial_t u_2(t)|_V^2 \right\} = 0. \tag{8.9.7}$$

Proof. Let $y_1 = (u_{01}; u_{11})$ and $y_2 = (u_{02}; u_{12})$ and $t_0 > 0$ be such that $S_t y_1$ and $S_t y_2$ belong to the absorbing set for all $t \ge t_0$. Then by Theorem 8.5.11 we have that

$$|S_t y_1 - S_t y_2|_H \le c_1 e^{-\omega(t-t_0)} |S_{t_0} y_1 - S_{t_0} y_2|_H + c_2 \int_{t_0}^t e^{-\omega(t-\tau)} |u^1(\tau) - u^2(\tau)| d\tau\,,$$

where $S_t y_1 = (u^1(t); u_t^1(t))$ and $S_t y_2 = (u^2(t); u_t^2(t))$ and the positive constants c_1, c_2, and ω depend on the absorbing set.

By Theorem 7.8.6 we have that

$$|v| \le \varepsilon_{\mathscr{L}} |\mathscr{A}^{1/2} v| + C_{\mathscr{L}} \max_j |l_j(v)| \quad \text{for } v \in \mathscr{D}(\mathscr{A}^{1/2}),$$

which implies that

$$|S_t y_1 - S_t y_2|_H \le c_1 e^{-\omega(t-t_0)} |S_{t_0} y_1 - S_{t_0} y_2|_H + \varepsilon_{\mathscr{L}} c_2 \int_{t_0}^t e^{-\omega(t-\tau)} |S_\tau y_1 - S_\tau y_2|_H d\tau$$

$$+ C(\mathscr{L}) \int_{t_0}^t e^{-\omega(t-\tau)} \max_j |l_j(u_1(\tau) - u_2(\tau))| d\tau\,.$$

Thus the function $\psi(t) = e^{\omega t}|S_t y_1 - S_t y_2|_H$ satisfies the inequality

$$\psi(t) \le c_1 \psi(t_0) + \varepsilon_{\mathscr{L}} c_2 \int_{t_0}^{t} \psi(\tau) d\tau + C(\mathscr{L}) \int_{t_0}^{t} e^{\omega\tau} \mathscr{N}(\tau) d\tau ,$$

where $\mathscr{N}(\tau) = \max_j |l_j(u_1(\tau) - u_2(\tau))|$. Using Gronwall's lemma we find that

$$\psi(t) \le c_1 \psi(t_0) e^{\varepsilon_{\mathscr{L}} c_2 (t-t_0)} + C(\mathscr{L}) e^{\omega t} \int_{t_0}^{t} e^{-\gamma(t-\tau)} \mathscr{N}(\tau) d\tau ,$$

where $\gamma = \omega - \varepsilon_{\mathscr{L}} c_2$. Therefore under the condition $\varepsilon_{\mathscr{L}} c_2 < \omega$ we have that

$$\limsup_{t\to\infty} |S_t y_1 - S_t y_2|_H \le C(\mathscr{L}) \limsup_{t\to\infty} \left\{ \int_{t_0}^{t} e^{-\gamma(t-\tau)} \mathscr{N}(\tau) d\tau \right\} .$$

From dissipativity of (H, S_t) we obtain that

$$\int_{t_0}^{t} e^{-\gamma(t-\tau)} \mathscr{N}(\tau) d\tau \le C_{\mathscr{L}} e^{-\gamma a} + \int_{t-a}^{t} \mathscr{N}(\tau) d\tau$$

for any $0 < a < t - t_0$. Therefore by (8.9.6) implies that

$$\limsup_{t\to\infty} |S_t y_1 - S_t y_2|_H \le C(\mathscr{L}, R) e^{-\gamma a}$$

for any $a > 0$. This relation implies the assertion of the theorem. □

8.9.4. Remark. In the case $D = kM$ we can also apply Theorem 7.8.14. The point is that in this case we can rewrite equation (8.9.1) in the space V (equipped with the inner product $(\cdot,\cdot)_V = (M\cdot,\cdot))$ in the form

$$u_{tt} + k u_t + \widetilde{\mathscr{A}} u = \widetilde{F}(u), \quad t > 0,$$

where $\widetilde{\mathscr{A}}$ is a positive self-adjoint operator in V defined by the relation

$$(\widetilde{\mathscr{A}} u, v)_V = (\mathscr{A} u, v), \quad u, v \in \mathscr{D}(\mathscr{A}^{1/2}),$$

and $\widetilde{F}$ is a nonlinear mapping from $\mathscr{D}(\mathscr{A}^{1/2}) = \mathscr{D}(\widetilde{\mathscr{A}}^{1/2})$ into V defined by the formula $(\widetilde{F}(u), v)_V = (F(u), v)$ for $u \in \mathscr{D}(\mathscr{A}^{1/2})$ and $v \in V$. ■

8.9.2 Energy approach

We now consider the case when only conservative forces are present ($F^* \equiv 0$) and, in contrast to the previous approach, we do not assume that the damping is linearly bounded from below (see (8.5.3)). This energy approach is based on ideas developed in [64, 67]. In addition to Assumption 8.1.1 it requires that $F^* \equiv 0$ and also the following hypotheses, responsible for long-time behavior of solutions to (8.9.1), are satisfied.

8.9.5. Assumption.

(D) There exists a continuous, increasing, concave function $H_0 : \mathbb{R}_+ \mapsto \mathbb{R}_+$, $H(0) = 0$, such that

$$(Mv, v) \le H_0((D(v), v)). \tag{8.9.8}$$

We also assume that

$$(D(v), u) \le C_1(|\mathscr{A}^{1/2}u|) \cdot (D(v), v) + c_2|\mathscr{A}^{1/2-\eta}u|^2, \tag{8.9.9}$$

for any $u, v \in \mathscr{D}(\mathscr{A}^{1/2})$, where $C_1(r)$ is nondecreasing function of $r > 0$, c_2 is a positive constant and $\eta \in (0, 1/2]$.

(F) There exist $\eta \in (0, \frac{1}{2}]$ and $b(r) > 0$ such that for every $r > 0$ we have

$$(F(w+u) - F(w+z\cdot u), u) \le (1-z)b(r)(\mathscr{A}^{1-2\eta}u, u)\,, \tag{8.9.10}$$

for any $z \in [0, 1]$, $u \in \mathscr{D}(\mathscr{A}^{1/2})$, $|\mathscr{A}^{1/2}u| \le r$ and w from the set $\mathscr{N} = \{u \in \mathscr{D}(\mathscr{A}^{1/2}) : \mathscr{A}u = F(u)\}$ of stationary solutions.

The requirement in (8.9.8) is the same as in (8.4.1) in Proposition 8.4.1 and Theorem 8.4.3. We refer to Remark 8.4.2 and to Appendix B for a discussion of this property.

Let $H_T(s) \equiv 3H_0(s/T)$, where $T > 0$. Because H_T is increasing, $(I + H_T)$ is invertible. Therefore the function $p(s) \equiv c(I + H_T)^{-1}(c_* s)$ is positive, continuous, and strictly increasing with $p(0) = 0$ for every $c, c_* > 0$. Finally we set $q(s) \equiv s - (I + p)^{-1}(s)$ for $s \ge 0$. It is clear that q is Lipschitz continuous, strictly increasing, positive and zero at the origin (see Proposition B.2.3 in Appendix B). With function q we associate the nonlinear differential equation:

$$\frac{d}{dt}\sigma(t) + q(\sigma(t)) = f(t),\ t > 0; \quad \sigma(0) = \sigma_0 \in \mathbb{R}_+. \tag{8.9.11}$$

It is clear that for any $f \in L_1(\mathbb{R}_+)$ there exists a unique global solution $\sigma \in C(\mathbb{R}_+)$. Moreover, $\sigma(t) \ge 0$ when $f(t) \ge 0$ and if $f(t) \to 0$ as $t \to \infty$, then $\lim_{t\to\infty} \sigma(t) = 0$. We refer to Proposition B.3.1 in Appendix B for details.

Our main result in this section is the following assertion.

8.9.6. Theorem. *Assume that Assumptions* 8.1.1 *and* 8.9.5 *hold and the dynamical system generated by* (8.9.1) *is dissipative*[4] *in the space* $\mathscr{D}(\mathscr{A}^{1/2}) \times \mathscr{H}$ *with a radius dissipativity R. Let* $u(t)$ *be a solution to problem* (8.9.1) *and* $w \in \mathscr{D}(\mathscr{A}^{1/2})$ *be a solution to the stationary problem* $\mathscr{A}u = F(u)$. *Assume that* $\mathscr{L} = \{l_j : j = 1, \dots, N\}$ *is a set of functionals on* $\mathscr{D}(\mathscr{A}^{1/2})$ *and the completeness defect* $\varepsilon_{\mathscr{L}} \equiv \varepsilon_{\mathscr{L}}(\mathscr{D}(\mathscr{A}^{1/2}), \mathscr{H})$ *possesses the property* $\varepsilon_{\mathscr{L}}^{4\eta}(b_0 + 2c_2) < 1$, *where* η, $b_0 = b(2R)$, *and* c_2 *are the constants from* (8.9.9) *and* (8.9.10). *Assume that*

[4] The dissipativity assumption can be omitted in the case when the constant $b(r)$ in (8.9.10) does not depend on r.

$$\lim_{t\to+\infty} l_j(u(t)) = l_j(w) \quad \textit{for all } j = 1,\dots,N, \tag{8.9.12}$$

and denote

$$h(t) = \sup_{s\geq t}\left\{\max_j |l_j(u(s)-w)|^2\right\}. \tag{8.9.13}$$

Then there exists $T \geq 2$ *such that*

$$|M^{1/2}u_t(t)|^2 + |\mathscr{A}^{1/2}(u(t)-w)|^2 \leq C_1\sigma\left(\left[\frac{t}{T}\right]\right) + C_2 h(t) \tag{8.9.14}$$

for all $t > 0$*, where* $\sigma(t)$ *satisfies the nonlinear ODE* (8.9.11) *(with parameters* c *and* c_* *defining* q *depending on the values of constants assumed in Assumptions* 8.1.1 *and* 8.9.5*) and with* $f(t) \equiv Th(T[t])$ *and* $\sigma(0)$ *depending on* u_0*,* u_1 *and* w*. Here* $C_1, C_2 > 0$ *are constants and* $[a]$ *is the notation for the integer part of* $a \in \mathbb{R}$*. Moreover, condition* (8.9.12) *implies that*

$$\lim_{t\to\infty}\left(|M^{1/2}u_t(t)|^2 + |\mathscr{A}^{1/2}(u(t)-w)|^2\right) = 0. \tag{8.9.15}$$

Proof. We follow the line of argument given in [67] and rely on some ideas developed in [64] for the wave equation with nonlinear dissipation.

Let $v(t) = u(t) - w$. Then $v(t)$ satisfies the following equation,

$$Mv_{tt} + kD(v_t) + \mathscr{A}v = F(w+v(t)) - F(w), \quad t > 0. \tag{8.9.16}$$

In the argument below we can assume that v is a strong solution. Moreover, we can assume that $|\mathscr{A}^{1/2}u(t)|, |\mathscr{A}^{1/2}w| \leq R$ and thus $|\mathscr{A}^{1/2}v(t)| \leq 2R$.

Multiplying (8.9.16) in $\mathscr{H}$ by v_t we obtain that

$$\frac{1}{2}\cdot\frac{d}{dt}\left(|M^{1/2}v_t(t)|^2 + |\mathscr{A}^{1/2}v(t)|^2\right) + k(D(v_t), v_t) = (F(u) - F(w), v_t). \tag{8.9.17}$$

It is not difficult to see that

$$(F(u) - F(w), v_t) = (F(u), u_t) - (F(w), v_t) = -\frac{d}{dt}\Phi(v(t))$$

with

$$\Phi(v) = \Pi(u) - \Pi(w) + (F(w), v) \equiv -\int_0^1 (F(w+\lambda v) - F(w), v)d\lambda.$$

Consequently from (8.9.17) we obtain the following energy type relation

$$\frac{d}{dt}\widetilde{E}(t) + k(D(v_t), v_t) = 0, \tag{8.9.18}$$

where

$$\widetilde{E}(t) = \frac{1}{2}\left(|M^{1/2}v_t(t)|^2 + |\mathscr{A}^{1/2}v(t)|^2\right) + \Phi(v(t)).$$

It follows from Assumption 8.9.5(F) with $u = \lambda v$ and $z = 0$ that

$$\Phi(v) \geq -\frac{b_0}{2}|\mathscr{A}^{1/2-\eta}v|^2 \tag{8.9.19}$$

with $b_0 = b(2R)$. Since

$$|\mathscr{A}^{1/2-\eta}v| \leq |\mathscr{A}^{1/2}v|^{1-2\eta} \cdot |v|^{2\eta}, \quad 0 \leq \eta \leq \frac{1}{2},$$

using (7.8.8) with $V = \mathscr{D}(\mathscr{A}^{1/2})$ and $H = \mathscr{H}$ and relation (7.8.7) we obtain that

$$|\mathscr{A}^{1/2-\eta}v|^2 \leq (1+\delta)\varepsilon_{\mathscr{L}}^{4\eta}|\mathscr{A}^{1/2}v|^2 + C_{\mathscr{L},\delta} \max_{j=1,\dots,N} |l_j(u)|^2, \tag{8.9.20}$$

for each $\delta > 0$. From (8.9.19) and (8.9.20) we get the following estimate.

8.9.7. Lemma.

$$\widetilde{E}(t) \geq \frac{1}{2}|M^{1/2}v_t|^2 + \left(\frac{1}{2} - \frac{b_0}{2}\varepsilon_{\mathscr{L}}^{4\eta}(1+\delta)\right)|A^{1/2}v|^2 - C_{\mathscr{L},\delta}\max_j |l_j(v)|^2. \tag{8.9.21}$$

Moreover, $\tilde{E}(t) \geq 0$ for all $t \geq 0$ provided $\varepsilon_{\mathscr{L}}^{4\eta} b_0 < 1$ and (8.9.12) holds.

Proof. The inequality in (8.9.21) is a consequence of inequalities (8.9.19) and (8.9.20). In order to prove positivity of $\tilde{E}$ we note that it follows from (8.9.18) that the function $\widetilde{E}(t)$ is nonincreasing. Inequality (8.9.21) implies that if $\varepsilon_{\mathscr{L}}^{4\eta} < b_0^{-1}$ and $\lim_{t\to+\infty} l_j(v(t)) = 0$ then $\lim_{t\to+\infty} \widetilde{E}(t) \geq 0$. Thus $\widetilde{E}(t) \geq 0$, $t > 0$, as desired. □

The following estimate is critical for the asymptotic behavior.

8.9.8. Lemma. *Let $T > T_0$, where $T_0 > 1$ is sufficiently large. Then*

$$p(\widetilde{E}(mT)) + \widetilde{E}(mT) \leq \widetilde{E}((m-1)T) + \mathscr{N}_{\mathscr{L}}(m,T)$$

for $m = 1, 2\dots$, where $\mathscr{N}_{\mathscr{L}}(m,T) = C_{\mathscr{L}} T \sup_{\tau\in[(m-1)T,mT]} \{\max_j |l_j(v(\tau))|^2\}$.

Proof. Multiplying equation (8.9.16) by v we find

$$\frac{d}{dt}(Mv_t, v) = |M^{1/2}v_t|^2 - |A^{1/2}v|^2 + (F(v+w) - F(w), v) - (D(v_t), v).$$

Because $-\frac{1}{2}|A^{1/2}v|^2 = -\widetilde{E}(t) + \frac{1}{2}|M^{1/2}v_t|^2 + \Phi(v)$, we have

$$\begin{aligned}\frac{d}{dt}(Mv_t, v) &= \frac{3}{2}|M^{1/2}v_t|^2 - \frac{1}{2}|A^{1/2}v|^2 - \widetilde{E}(t) \\ &\quad + \{\Phi(v) + (F(w+v) - F(w), v)\} - (D(v_t), v).\end{aligned}$$

By Assumption 8.9.5(F) we have

$$\Phi(v)+(F(w+v)-F(w),v)=\int_0^1 (F(w+v)-F(w+zv)),v)dz \le \frac{b_0}{2}|A^{1/2-\eta}v|^2.$$

Therefore using (8.9.20) we obtain the following inequality

$$\begin{aligned}\frac{d}{dt}(Mv_t,v) \le -\widetilde{E}(t) - \frac{1}{2}\left(1-\varepsilon_{\mathscr{L}}^{4\eta}(1+\delta)b_0\right)|A^{1/2}v|^2 \\ + C_{\mathscr{L},\delta}\max_j |l_j(v)|^2 + \frac{3}{2}|M^{1/2}v_t|^2 + (D(v_t),v).\end{aligned}$$

Integrating the last inequality from 0 to T with respect to t we obtain:

$$\begin{aligned}\int_0^T \widetilde{E}(s)ds \le -(Mv_t(T),v(T)) + (Mv_t(0),v(0)) + \frac{3}{2}\int_0^T |M^{1/2}v_t(s)|^2 ds \\ -\frac{1}{2}\left(1-\varepsilon_{\mathscr{L}}^{4\eta}(1+\delta)b_0\right)\int_0^T |A^{1/2}v(s)|^2 ds \\ + k\int_0^T (D(v_t(s)),v(s))ds + C_{\mathscr{L},\delta}\int_0^T \max_j |l_j(v(s))|^2 ds.\end{aligned} \tag{8.9.22}$$

From Lemma 8.9.7 we obtain

$$|A^{1/2}v(t)|^2 + |M^{1/2}v(t)|^2 \le C\widetilde{E}(t) + C_{\mathscr{L}}\max_j |l_j(v(t))|^2 \tag{8.9.23}$$

under the condition $\varepsilon_{\mathscr{L}}^{4\eta} b_0 < 1$. Thus, by direct computations

$$\begin{aligned}|(Mv_t(T),v(T))| + |(Mv_t(0),v(0))| \\ \le C\left[\widetilde{E}(T)+\widetilde{E}(0)\right] + C_{\mathscr{L}}\left(\max_j |l_j(v(T))|^2 + \max_j |l_j(v(0))|^2\right).\end{aligned} \tag{8.9.24}$$

By (8.9.9)

$$\int_0^T (D(v_t),v)ds \le C(R)\int_0^T (D(v_t),v_t)ds + c_2\int_0^T |\mathscr{A}^{1/2-\eta}v|^2 ds.$$

Therefore, using (8.9.20) we obtain

$$\begin{aligned}\int_0^T (D(v_t),v)ds \le C_0\int_0^T (D(v_t),v_t)ds + c_2\varepsilon_{\mathscr{L}}^{4\eta}(1+\delta)\int_0^T |A^{1/2}v(s)|^2 ds \\ + C_{\mathscr{L},\delta}\int_0^T \max_j |l_j(v(s))|^2 ds.\end{aligned} \tag{8.9.25}$$

From (8.9.8) and Jensen's inequality we also have

$$\int_0^T (Mv_t,v_t)ds \le \int_0^T H_0((D(v_t),v_t))ds \le \frac{T}{3}H_T\left(\int_0^T (D(v_t),v_t)ds\right). \tag{8.9.26}$$

Therefore using (8.9.24)–(8.9.26) from (8.9.22) we obtain

$$\int_0^T \widetilde{E}(s)ds \le C[\widetilde{E}(0)+\widetilde{E}(T)] + (CI+\frac{T}{2}H_T)\left(\int_0^T (D(v_t),v_t)ds\right) + \mathscr{N}_{\mathscr{L}}(1,T)$$

provided $\varepsilon_{\mathscr{L}}^{4\eta}(b_0+2c_2)<1$ and $T\ge 1$. Because $t\widetilde{E}(t)\le\int_0^t \widetilde{E}(s)ds$ (by (8.9.18) $\widetilde{E}(t)$ is a non increasing function), using the relation

$$\widetilde{E}(0) = \widetilde{E}(T) + k\int_0^T (D(v_t(s)),v_t(s))ds, \tag{8.9.27}$$

which follows from (8.9.18), we obtain the following inequality

$$T\widetilde{E}(T) \le C_1\widetilde{E}(T) + (C_0 I+\frac{T}{2}H_T)\left(\int_0^T (D(v_t),v_t)ds\right) + \mathscr{N}_{\mathscr{L}}(1,T)\,.$$

Therefore taking $T>\max[2C_1,2]$ and using (8.9.27) again, we obtain

$$\begin{aligned}\widetilde{E}(T) &\le C_T(I+H_T)\left(k^{-1}\left[\widetilde{E}(0)-\widetilde{E}(T)\right]\right) + \mathscr{N}_{\mathscr{L}}(1,T)\\ &\le C_T(I+H_T)\left(k^{-1}\left[\widetilde{E}(0)-\widetilde{E}(T)+\mathscr{N}_{\mathscr{L}}(1,T)\right]\right)\,.\end{aligned}$$

Using the concavity of H_T and recalling the definition of $p(s)$ we obtain

$$p(\widetilde{E}(T)) + \widetilde{E}(T) \le \widetilde{E}(0) + \mathscr{N}_{\mathscr{L}}(1,T)\,,$$

which gives the inequality in Lemma 8.9.8 for the case $m=1$, Reiterating this argument on each subinterval $((m-1)T,mT)$ gives the conclusion of Lemma 8.9.8. □

To complete the proof of Theorem 8.9.6 we use Proposition B.3.3 from Appendix B. We first note that $\mathscr{N}_{\mathscr{L}}(m,T)\le CTh((m-1)T)$, where $h(t)$ is given by (8.9.13). Therefore from Lemma 8.9.8 for $s_m:=\widetilde{E}(mT)$ we obtain that

$$s_{m+1}+p(s_{m+1})\le s_m+f_m,\quad m=0,1,2,\dots,$$

where $f_m\le\int_m^{m+1} f(\tau)d\tau$ with $f(t)=CTh(T[t])$. Therefore Proposition B.3.3 from Appendix B and monotonicity of $\widetilde{E}(t)$ imply that

$$\widetilde{E}(t)\le\widetilde{E}([tT^{-1}]T)\le\sigma([tT^{-1}]),\quad t\ge 0,$$

where $[\cdot]$ is the sign for the integer part and $\sigma(t)$ solves (8.9.11). Therefore using (8.9.23) we obtain the desired estimate in (8.9.14).

To prove the last statement we note that (8.9.12) implies that $h(t)\to 0$ as $t\to\infty$. Therefore $f(t)=CTh(T[t])\to 0$ as $t\to\infty$. Thus we can use the second part of Proposition B.3.1 to obtain (8.9.15). □

8.9.9. Remark. In the case of linear $H_0(s)$ the functions $p(s)$ and $q(s)$ are also linear. Therefore solving equation (8.9.11) we can derive from (8.9.14) the relation

$$|M^{1/2}u_t(t)|^2+|\mathscr{A}^{1/2}(u(t)-w)|^2 \le c_1 e^{-\widetilde{\gamma}t}+c_2\int_0^t e^{-\widetilde{\gamma}(t-\tau)}h(c_3[\tau])d\tau+c_4h(t)$$

for some $\widetilde{\gamma}>0$ and positive constants c_i, where $h(t)$ is given by (8.9.13). This formula describes the decay rates of the solution $u(t)$ to the equilibrium w. The corresponding decay rates are exponential when $h(t)=ce^{-\beta t}$ or polynomial if $h(t)=c(1+t)^{-\beta}$, $\beta>0$. We note that in general $h(t)$ may depend crucially on behavior of the solutions $u(t)$ and w. ■

Theorem 8.9.6 immediately implies the following assertion.

8.9.10. Corollary. *Let $w_1,w_2\in\mathscr{D}(\mathscr{A}^{1/2})$ be two stationary solutions to problem (8.9.1). Let $\mathscr{L}=\{l_j : j=1,\dots,N\}$ be a set of functionals on $\mathscr{D}(\mathscr{A}^{1/2})$ and $\varepsilon_{\mathscr{L}}^{4\eta}(b_0+2c_2)<1$, where η, b_0, and c_2 are the same as in Theorem* 8.9.6. *Then the condition $l_j(w_1)=l_j(w_2)$ for $j=1,\dots,N$ implies that $w_1=w_2$.*

The two corollaries formulated below deal with the case when the problem possesses precompact trajectories (this property holds if the system is asymptotically smooth, for instance).

Let $u(t)$ be a solution to the problem (8.9.1). We recall (see Section 7.1) that the set

$$\gamma_+(u_0,u_1)=\cup\{(u(t);u_t(t))\ :\ t\ge 0\}$$

in the space $H=\mathscr{D}(\mathscr{A}^{1/2})\times V$ is said to be the *semitrajectory* emanating from $(u_0;u_1)$ of the dynamical system generated by (8.9.1) in H. We also recall (see (7.1.4)) that the ω-limit set of the semitrajectory $\gamma_+(u_0,u_1)$ is defined by the formula

$$\omega(\gamma_+)\equiv\omega(u_0,u_1)=\cap_{\tau>0}\mathrm{Closure}_H\left\{\cup\{(u(t);u_t(t))\ :\ t\ge\tau\}\right\}.$$

8.9.11. Corollary. *Let $\mathscr{L}=\{l_j : j=1,\dots,N\}$ be a set of functionals on $\mathscr{D}(\mathscr{A}^{1/2})$ such that $\varepsilon_{\mathscr{L}}^{4\eta}(b_0+2c_2)<1$, where η, $b_0=b(2R)$, and c_2 are the constants from* (8.9.9) *and* (8.9.10). *Assume that $u(t)$ is a solution to problem* (8.9.1) *with relatively compact semitrajectory $\gamma_+=\gamma_+(u_0,u_1)$ and there exist the finite limits*

$$\lim_{t\to+\infty} l_j(u(t))\equiv l_j \quad \textit{for every}\quad j=1,\dots,N.$$

Then there exists a stationary solution $w\in\mathscr{D}(\mathscr{A}^{1/2})$ such that (8.9.15) *holds.*

Proof. The relative compactness of γ_+ implies that ω-limit set $\omega(\gamma_+)$ is a nonempty compact set in H. The energy relation in (8.4.2) implies that the functional $V(y)=\mathscr{E}(u_0,u_1)$, $y=(u_0;u_1)$, is a strict Lyapunov function on H for system (8.9.1) (see Definition 7.5.3) and therefore one can see that the ω-limit set $\omega(\gamma_+)$ lies in the set $\mathscr{N}\equiv\{(w;0)\in H : Aw=F(w)\}$ of equilibrium points to problem (8.9.1). From the convergence $l_j(u(t))\to l_j$ we have that $l_j(w)=l_j$ for all $(w;0)\in\omega(\gamma_+)\subset\mathscr{N}$. Consequently Corollary 8.9.10 implies that $\omega(\gamma_+)$ consists of a single point $(w;0)$ and therefore (8.9.15) holds. □

8.9.12. Corollary. *Let the assumptions of Corollary* 8.9.11 *be valid. Assume that* $u^{(1)}(t)$ *and* $u^{(2)}(t)$ *are solutions to equation* (8.9.1) *with relatively compact semi-trajectories* $\gamma_+^{(1)}$ *and* $\gamma_+^{(2)}$ *and*

$$\lim_{t\to+\infty}\left(l_j(u^{(1)}(t))-l_j(u^{(2)}(t))\right)=0,\quad j=1,\dots,N. \tag{8.9.28}$$

Then $\omega(\gamma_+^{(1)})\equiv\omega(\gamma_+^{(2)})$. *If the set* $\mathcal{N}$ *of equilibrium points is finite, then there exists a stationary solution* $w\in\mathcal{D}(\mathcal{A}^{1/2})$ *such that* (8.9.15) *holds for both solutions* $u_1(t)$ *and* $u_2(t)$.

Proof. Let $(z_1,0)\in\omega(\gamma_+^{(1)})\subset\mathcal{N}$. Then there exists a sequence $\{t_m\}$ such that $t_m\to+\infty$ and $u^{(1)}(t_m)\to z_1$ in the space $\mathcal{D}(\mathcal{A}^{1/2})$ when $m\to\infty$. Because $\gamma_+^{(2)}$ is a relatively compact set, we can choose a subsequence $\{t_{m_k}\}\subset\{t_m\}$ such that $u^{(2)}(t_{m_k})\to z_2$, where $(z_2;0)\in\omega(\gamma_+^{(2)})\subset\mathcal{N}$. Property (8.9.28) gives that $l_j(z_1)=l_j(z_2)$ for all $j=1,2,\dots,N$. Consequently Corollary 8.9.10 implies that $z_1=z_2$ and therefore we have $(z_1;0)\in\omega(\gamma_+^{(2)})$. This implies that $\omega(\gamma_+^{(1)})=\omega(\gamma_+^{(2)})$. If $\mathcal{N}$ is finite, then it is easy to see that $\omega(\gamma_+^{(1)})=\omega(\gamma_+^{(2)})$ consists of a single equilibrium point. This implies the assertion stated in the corollary. □

8.10 Exponential fractal attractors

In this section we present a result on the existence of fractal exponential attractors for the dynamical system (H,S_t) generated by the second-order evolution equation in (8.1.1).

We recall (see Definition 7.4.4) that a compact set $A_{\text{exp}}\subset H$ is said to be *inertial* (or a *fractal exponential attractor*) for the dynamical system (X,S_t) if (i) A_{exp} is a positively invariant set of finite fractal dimension (possibly, in the topology of some extended space) and (ii) for every bounded set $D\subset X$ there exist positive constants t_D, C_D and γ_D such that

$$d_X\{S_tD\,|\,A_{\text{exp}}\}\equiv\sup_{x\in D}\text{dist}_X(S_tx,A_{\text{exp}})\le C_D\cdot e^{-\gamma_D(t-t_D)},\quad t\ge t_D. \tag{8.10.1}$$

Using the stabilizability estimates (see Section 8.5) and the abstract result presented in Theorem 7.9.9 we can construct inertial sets for a class of systems of the form (8.1.1).

8.10.1. Theorem. *Let* $\mathcal{A}^{-1}$ *be compact and the hypotheses of either Theorem* 8.5.6 *or Theorem* 8.5.10 *be in force. Assume that the system* (H,S_t) *generated by* (8.1.1) *is dissipative and there exist numbers* $l\ge 0$ *and* $0<\alpha<1$ *such that*

$$|\mathcal{A}^{-l}D(v)|\le C(r)\,[1+(D(v),v)]^{\alpha},\quad v\in\mathcal{D}(\mathcal{A}^{1/2}),\ |v|_V\le r. \tag{8.10.2}$$

Then the system (H, S_t) *possesses a (generalized) fractal exponential attractor whose dimension is finite in the space* $\widetilde{H} = V \times W$*, where* W *is a completion of* V *with respect to the norm* $|\cdot|_W = |\mathscr{A}^{-l_*} M \cdot|$ *with* $l_* = \max\{l, 1/2\}$.

Proof. Applying either Theorem 8.5.6 or Theorem 8.5.10 we obtain the stabilizability estimate in (8.5.7) on an absorbing set $\mathscr{B}$.

It follows from (8.10.2) that

$$\int_0^T |\mathscr{A}^{-l_*} M u_{tt}(\tau)|^{1/\alpha} d\tau \leq C_{\mathscr{B}}(T)$$

for any strong solution $u(t)$ from the absorbing set $\mathscr{B}$. This implies (7.9.18) with $\gamma = 1 - \alpha$ in the space $\widetilde{H} = V \times W$. Therefore the result follows from Theorem 7.9.9. □

We note that the existence and properties of a fractal exponential attractor for wave equations with linear damping was studied earlier in [104, 105, 106]; see also the recent survey [232]. Thus, again, the result in Theorem 8.10.1 provides extensions of the respective results to problems with nonlinear dissipation. We also refer to [75] for other results on exponential attractors in the case of nonlinear damping.

8.11 Approximate inertial manifolds

We consider the case of linear damping with $D = kM$ only; that is, we consider the problem

$$M u_{tt} + kM u_t + \mathscr{A} u = F(u), \quad t > 0. \tag{8.11.1}$$

To construct approximate inertial manifolds for this problem we apply the results from Section 7.7.

Concerning the nonlinear mapping $F(u)$ we assume that $F(u)$ belongs to C^m as a mapping from $\mathscr{D}(\mathscr{A}^{1/2})$ into V' and

$$|\langle F^{(k)}(u); w_1, \ldots, w_k \rangle|_{V'} \leq C_\rho \cdot \prod_{j=1}^k |\mathscr{A}^{1/2} w_j|, \tag{8.11.2}$$

$$|\langle F^{(k)}(u) - F^{(k)}(u^*); w_1, \ldots, w_k \rangle|_{V'} \leq C_\rho \cdot |\mathscr{A}^{1/2}(u - u^*)| \cdot \prod_{j=1}^k |A^{1/2} w_j| \tag{8.11.3}$$

for every $\rho > 0$, where $k = 0, 1, \ldots, m$, $|\mathscr{A}^{1/2} u| \leq \rho$, $|\mathscr{A}^{1/2} u^*| \leq \rho$, $w_j \in \mathscr{D}(\mathscr{A}^{1/2})$. Here $F^{(k)}(u)$ is the k-order Fréchet derivative of $F(u)$ and $\langle F^{(k)}(u); w_1, \ldots, w_k \rangle$ is the value of $F^{(k)}(u)$ on elements $w_1, \ldots, w_k$; see [41] for some details concerning higher-order Fréchet derivatives.

As shown in Remark 8.9.4, equation (8.11.1) can be written in the space V (equipped with the inner product $(\cdot,\cdot)_V = (M\cdot,\cdot)$) in the form

$$u_{tt} + ku_t + \widetilde{\mathscr{A}} u = \widetilde{F}(u),\ t > 0, \tag{8.11.4}$$

where $\widetilde{\mathscr{A}}$ is a positive self-adjoint operator in V defined by the relation

$$(\widetilde{\mathscr{A}} u, v)_V = (\mathscr{A} u, v),\quad u, v \in \mathscr{D}(\mathscr{A}^{1/2}),$$

and $\widetilde{F}$ is a nonlinear mapping from $\mathscr{D}(\mathscr{A}^{1/2}) \equiv \mathscr{D}(\widetilde{\mathscr{A}}^{1/2})$ into V defined by the formula $(\widetilde{F}(u), v)_V = (F(u), v)$, $u \in \mathscr{D}(\mathscr{A}^{1/2})$, $v \in V$. These relations means that $\widetilde{\mathscr{A}} = M^{-1}\mathscr{A}$ with the domain $\mathscr{D}(\widetilde{\mathscr{A}}) \equiv W = \{u \in \mathscr{D}(\mathscr{A}^{1/2}) : \mathscr{A} u \in V'\}$ and $\widetilde{F}(u) = M^{-1}F(u)$ for $u \in \mathscr{D}(\mathscr{A}^{1/2})$.

If $\mathscr{D}(\mathscr{A}^{1/2})$ is compactly embedded in V, then there exists an orthonormal basis $\{e_n\}_{n\in\mathbb{N}}$ in V such that $e_n \in \mathscr{D}(\mathscr{A}^{1/2})$ for every $n \in \mathbb{N}$ and

$$\mathscr{A} e_n = \lambda_n M e_n,\quad 0 < \lambda_1 \le \lambda_2 \le \cdots,\quad \lim_{n\to\infty} \lambda_n = \infty. \tag{8.11.5}$$

It is clear that $\{e_n\}_{n\in\mathbb{N}}$ is the basis of eigenfunctions for the operator $\widetilde{\mathscr{A}}$.

Let P be the orthoprojector in V on the Span$\{e_k : k = 1, \ldots, N\}$ and $Q = I - P$. We also use the notations $\widetilde{P} = MPM^{-1}$ and $\widetilde{Q} = MQM^{-1}$. We construct approximate inertial manifolds in the following way. Let $h_n(p, \dot{p})$ and $l_n(p, \dot{p})$ be functions on $PV \times PV$ with values in QV defined by the formulas:

$$\begin{aligned}\mathscr{A} h_n(p, \dot{p}) &= \widetilde{Q}F(p + h_{n-1}(p, \dot{p})) - kMl_{n-1}(p, \dot{p}) - \langle \delta_p M l_{n-1}; \dot{p}\rangle \\ &\quad - \langle \delta_{\dot{p}} M l_{n-1}; -k\dot{p} + M^{-1}\Big(-\mathscr{A} p + \widetilde{P}F(p + h_{n-1}(p, \dot{p}))\Big)\rangle,\end{aligned} \tag{8.11.6}$$

where $n = 1, 2, \ldots$ Here $l_n(p, \dot{p})$ is defined by the formula

$$l_n(p, \dot{p}) = \langle \delta_p h_{n-1}; \dot{p}\rangle + \langle \delta_{\dot{p}} h_{n-1}; -k\dot{p} + M^{-1}\Big(-\mathscr{A} p + \widetilde{P}F(p + h_{n-1}(p, \dot{p}))\Big)\rangle, \tag{8.11.7}$$

where $n = 1, 2, \ldots$. Here $\delta_p f$ and $\delta \dot{p} f$ are the Frechet derivatives of $f(p, \dot{p})$ with respect to p and $\dot{p}$; $\langle \delta_p f; w\rangle$ and $\langle \delta_{\dot{p}} f; w\rangle$ are the corresponding values on element w. We also suppose $h_0(p, \dot{p}) \equiv l_0(p, \dot{p}) \equiv 0$.

Let $m \ge 1$ be an integer and let $R > 0$ be a positive number. We denote by $L_{m,R}$ the class of solutions of the problem (8.11.1) that possess the following regularity and dissipativity properties:

(i) We have $u^{(k)}(t) \in C(\mathbb{R}_+; W)$ for $k = 0, 1, 2, \ldots, m-1$ and

$$u^{(m)}(t) \in C(\mathbb{R}_+; \mathscr{D}(\mathscr{A}^{1/2})),\quad u^{(m+1)}(t) \in C(\mathbb{R}_+; V),$$

where $C(\mathbb{R}_+; X)$ denotes the space of strongly continuous functions with values in X, here and below $u^{(k)}(t) = \partial_t^k u(t)$.

(ii) For any $u(t) \in L_{m,R}$ there exists $t^* > 0$ such that

$$|u^{(k+1)}(t)|_V^2 + |\mathscr{A}^{1/2} u^{(k)}(t)|^2 + |\mathscr{A} u^{(k-1)}(t)|_{V'}^2 \le R^2, \tag{8.11.8}$$

where $k = 0, 1, \ldots, m$ and $t \geq t^*$.

Using ideas given in Theorem 2.4.38 and Theorem 8.7.4 we can provide conditions that guarantee that these classes $L_{m,R}$ are nonempty.

8.11.1. Theorem. *Assume that the embedding $\mathscr{D}(\mathscr{A}^{1/2}) \subset V$ is compact and conditions* (8.11.2) *and* (8.11.3) *hold. Let $u(t)$ be the solution of the problem* (8.11.1) *and $u(t) \in L_{m,R}$, where $m \geq 2$. Then for $n \leq m-1$ and for t sufficiently large we have*

$$|\mathscr{A}\partial_t^j(u(t) - u_n(t))|_{V'} \leq C_{n,R} \cdot \lambda_{N+1}^{-n/2}$$

and

$$|\mathscr{A}^{1/2}\partial_t^j(\partial_t u(t) - \bar{u}_n(t))| \leq C_{n,R} \cdot \lambda_{N+1}^{-n/2},$$

when $0 \leq j \leq m-n-1$. Here λ_{N+1} is the $(N+1)$th eigenvalue of the spectral problem (8.11.5) *and $u_n(t)$ and $\bar{u}_n(t)$ are defined by*

$$u_n(t) = p(t) + h_n(p(t), \partial_t p(t)), \quad \bar{u}_n(t) = \partial_t p(t) + l_n(p(t), \partial_t p(t)), \tag{8.11.9}$$

where $p(t) = Pu(t)$ and h_n, l_n are defined in (8.11.6) *and* (8.11.7).

Proof. This is a direct application of Theorem 7.7.1 to problem (8.11.4). □

We also have the following localization result which is a direct corollary of Theorems 7.7.2 and 8.7.4.

8.11.2. Theorem. *Let the hypotheses of Theorems* 8.7.4 *and* 8.11.1 *with arbitrary m be in force. Then for any trajectory $U(t) = (u(t); \partial_t u(t))$ from the global attractor A corresponding to problem* (8.11.1) *we have the relation:*

$$\left\{ |\mathscr{A}\partial_t^j(u(t) - u_n(t))|_{V'}^2 + |\mathscr{A}^{1/2}\partial_t^j(\partial_t u(t) - \bar{u}_n(t))|^2 \right\}^{1/2} \leq C_{n,R,j} \cdot \lambda_{N+1}^{-n/2}$$

for $n = 1, 2, \ldots$; $j = 0, 1, \ldots$ and all $t \in \mathbb{R}$. Here $u_n(t)$ and $\bar{u}_n(t)$ are defined by (8.11.9). *Furthermore*

$$\sup\{\mathrm{dist}(U, M_n) : U \in A\} \leq c_n \lambda_{N+1}^{-n/2}, \quad n = 1, 2, \ldots,$$

where

$$M_n = \{(p + h_n(p, \dot{p}); \dot{p} + l_n(p, \dot{p})) : p, \dot{p} \in PV\}$$

and dist(U, M_n) *is the distance between U and M_n in the space*

$$\left\{ u \in \mathscr{D}(\mathscr{A}^{1/2}) : \mathscr{A}u \in V' \right\} \times \mathscr{D}(\mathscr{A}^{1/2}).$$

Chapter 9
Plates with Internal Damping

In this chapter von Karman equations with nonlinear internal and fully supported in Ω damping are considered. The boundary conditions associated with the model are either clamped, hinged or else "free". In this latter case, the boundary conditions may involve naturally both dynamic and nonlinear terms. The well-posedness of solutions to the models considered follows from the results presented in Chapter 3, for models with rotational forces and in Chapter 4, for nonrotational models.

The main goal of this chapter is to study long-time behavior of dynamical systems generated by the evolutionary von Karman models. Special emphasis is placed on the role of nonlinearity of the damping, particularly when the latter is combined with criticality of the sources. We recast the corresponding models as the special cases of abstract equations discussed in Chapters 7 and 8. Although the general abstract treatment presented in these Chapters constitutes the foundation for the analysis, there are a number of more specific and technical issues associated with verification of abstract hypotheses for the concrete models. These require separate arguments which often rely on new developments in the theory of nonlinear plates.

9.1 Existence of global attractors for von Karman model with rotational forces

This section is devoted to the following problem.

$$\begin{cases} (1-\alpha\Delta)u_{tt} + d_0(x)g_0(u_t) - \alpha\,\mathrm{div}\,\{d(x)g(\nabla u_t)\} \\ \qquad\qquad + \Delta^2 u - [u+f, v+F_0] + Lu = p(x), \quad x\in\Omega, t>0, \\ u|_{t=0} = u_0(x), \quad u_t|_{t=0} = u_1(x), \end{cases} \tag{9.1.1}$$

where $v = v(u)$ is a solution of the equation

I. Chueshov, I. Lasiecka, *Von Karman Evolution Equations*,
Springer Monographs in Mathematics, DOI 10.1007/978-0-387-87712-9_9,

$$\Delta^2 v + [u + 2f, u] = 0, \quad v|_{\partial\Omega} = \frac{\partial v}{\partial n}\Big|_{\partial\Omega} = 0. \tag{9.1.2}$$

We recall that the case $L = 0$ corresponds to the gradient structure of the semiflow, and $L \neq 0$ is a nonconservative force. In this section we assume that the parameter α is positive and we restrict our consideration to the following canonical boundary conditions imposed on the displacement $u(x,t)$:

1. **[clamped]**:

$$u = \nabla u = 0 \text{ on } \Gamma = \partial\Omega. \tag{9.1.3}$$

2. **[hinged]**:

$$u = \Delta u = 0 \text{ on } \Gamma = \partial\Omega. \tag{9.1.4}$$

3. **[free]**:

$$\begin{cases} \Delta u + (1-\mu)B_1 u = 0 \text{ on } \Gamma, \\ \frac{\partial}{\partial n}\Delta u + (1-\mu)B_2 u - \alpha\frac{\partial}{\partial n}u_{tt} - \alpha d(x)g(\nabla u_t)\cdot n = \nu_1 u + \beta u^3 \text{ on } \Gamma. \end{cases} \tag{9.1.5}$$

Here ν_1 is a nonnegative constant, $0 < \mu < 1$, the boundary operators B_1 and B_2 are defined by (1.3.20), and $\beta \in L_\infty(\Gamma)$ is nonnegative.

9.1.1. Remark. Instead of hinged boundary conditions, one may also consider *simply supported* boundary conditions which are given by

$$u = 0, \quad \Delta u + (1-\mu)B_1 u = 0 \text{ on } \Gamma. \tag{9.1.6}$$

These boundary conditions lead to the self-adjoint structure of the biharmonic operator with the associated bilinear form given by (1.3.4); see Section 1.3.3. The functional framework (including phase space) for simply supported boundary conditions is the same as for the hinged boundary conditions. We also note that mixed boundary conditions that involve combinations of different boundary conditions imposed on Γ are discussed briefly later (see Section 9.1.4). ■

It was shown in Chapter 3 that problem (9.1.1) and (9.1.2) under each of the boundary conditions stated above admits the following abstract representation,

$$\begin{cases} Mu_{tt} + D(u_t) + \mathscr{A}u = F(u) \\ u(0) = u_0 \in \mathscr{D}(\mathscr{A}^{1/2}), \;\; u_t(0) = u_1 \in V = \mathscr{D}(M^{1/2}). \end{cases} \tag{9.1.7}$$

The mass operators M, the elastic operator $\mathscr{A}$, the damping operator D, and the von Karman nonlinear operator F depend on the type of boundary conditions considered. The operators M and D depend on the parameter α representing rotational inertia.

In this section we apply an abstract result from Chapter 8 to abstract equation (9.1.7) with the specified set of boundary conditions, considered in each case. Our basic assumption on the given functions in (9.1.1) is a special case of those given in Chapter 3 (cf. Assumptions 3.1.1)

9.1.2. Assumption.

- $g_0 \subset C^1(\mathbb{R})$ is a monotone nondecreasing function such that $g_0(0) = 0$.
- The function g has the form $g(s_1, s_2) = (g_1(s_1); g_2(s_2))$, where $(s_1; s_2) \in \mathbb{R}^2$ and $g_i \in C^1(\mathbb{R})$ is a monotone nondecreasing function such that $g_i(0) = 0$, $i = 1, 2$. Moreover, we assume that g_i are of polynomial growth at infinity; that is, we assume that
$$0 < m \le g_i'(s) \le M|s|^{p-1}, \quad |s| \ge 1,\ i = 1, 2, \tag{9.1.8}$$
with some constants m, M, and $p \ge 1$. In the case when $L \ne 0$ (non-conservative case), we assume that this inequality holds with $p = 1$ and also that g_0 possesses the same property.
- $d_0(x), d(x) \in L_\infty(\Omega)$ are nonnegative.
- $p(x) \in L_2(\Omega)$, $f(x) \in H^2(\Omega)$, and $F_0(x) \in H^3(\Omega)$.
- The mapping L is a linear bounded operator from $H^{2-\sigma}(\Omega)$ into $H^{-1}(\Omega)$ for some $\sigma > 0$ (into $[H^1(\Omega)]'$ in the case of free boundary conditions (9.1.5)).

The following assertion is a direct corollary of Theorems 3.1.4, 3.1.16, and 3.1.22.

9.1.3. Theorem. *Under Assumption* 9.1.2 *problem* (9.1.1) *and* (9.1.2) *with either clamped* (9.1.3), *or hinged* (9.1.4) *or free* (9.1.5) *boundary conditions for the displacement $u(x,t)$ generates a dynamical system (H, S_t) with the evolution operator S_t given by the formula*

$$S_t(u_0; u_1) = (u(t); u_t(t)), \quad (u_0; u_1) \in H, \ t \ge 0, \tag{9.1.9}$$

where $u(t)$ is a generalized solution to problem (9.1.1) *and* (9.1.2) *with the corresponding boundary conditions for u. The phase space H has the form*

- *$H = H_0^2(\Omega) \times H_0^1(\Omega)$ for the clamped boundary conditions* (9.1.3),
- *$H = (H^2(\Omega) \cap H_0^1(\Omega)) \times H_0^1(\Omega)$ in the hinged case* (9.1.4),
- *$H = H^2(\Omega) \times H^1(\Omega)$ in the free case* (9.1.5).

Our goal in this section is to study asymptotic properties of the evolution semigroup S_t given by (9.1.9). We apply the results presented in Chapter 8.

9.1.4. Remark. Assumptions imposed on $d_0(x)$ and g_0 may be substantially relaxed. This is due to the fact that in the case of rotational forces accounted for in the model, the damping created by g_0 is not essential. Here, the main mechanism of uniform dissipation of the energy are the damping terms represented by g_i. For instance, for the purpose of studying attractors in the clamped and hinged cases, one can take $d_0(x) \equiv 0$; that is, it is possible to neglect the term $g_0(u_t)$. Moreover, we can also assume that the damping function g_0 is not monotone (we refer to Theorem 6.8 in [75] for the corresponding result in the case of clamped boundary conditions).

Also, the assumption on the diagonal structure of g can be relaxed. However, in order to streamline the exposition, we prefer to think of g as a diagonal mapping of the form $g(s_1, s_2) = (g_1(s_1); g_2(s_2))$. ∎

9.1.1 Clamped boundary condition

We start with the simplest case of the clamped boundary condition given in (9.1.3). Our main result in this section is the following assertion.

9.1.5. Theorem. *Under Assumption* 9.1.2 *the dynamical system* (H, S_t) *generated by equations* (9.1.1)–(9.1.3) *with* $\alpha > 0$ *in the space* $H \equiv H_0^2(\Omega) \times H_0^1(\Omega)$ *possesses a global compact attractor A provided*

(i) *There exist positive constants* d_0 *and* d_1 *such that*

$$0 \le d_0(x) \le d_1, \quad 0 < d_0 \le d(x) \le d_1. \tag{9.1.10}$$

(ii) *In the case* $L \equiv 0$ *the damping functions* $g_i(s)$ *possess properties*

$$s^2 \le \eta + C_\eta s g_i(s) \text{ for every } \eta > 0,\ i = 1,2,\ s \in \mathbb{R}. \tag{9.1.11}$$

(iii) *In the case* $L \ne 0$ *the damping functions* $g_i(s)$ *are strictly increasing,* $i = 1,2$.

9.1.6. Remark. One can see that under Assumption 9.1.2 the inequality (9.1.11) is implied by the following property: for any $\delta > 0$ there exists $m_\delta > 0$ such that

$$|g_i(s)| \ge m_\delta |s| \text{ for } |s| \ge \delta. \tag{9.1.12}$$

On the other hand (9.1.11) is readily satisfied provided that (i) the function $g_i(s)$ is nondecreasing on $\mathbb{R}$ and strictly increasing in some (small) neighborhood of 0, and (ii) we have that $\liminf_{|s|\to\infty}(g_i(s)/s) > 0$, see Proposition B.1.1 in Appendix B. In particular, this means that our hypotheses allows the damping function g_i to be constants on some closed finite intervals that are away from zero. ∎

Here we provide a few comments on the relation of the result given in Theorem 9.1.5 with respect to the literature. Existence and finite-dimensionality of the global attractor, for the case of linear damping, was proved earlier in [46] (see also [47], where the structure of the attractor was described). The case of nonlinear damping was considered in [189] (see also[75]) and references therein. Theorem 9.1.5 allows us to treat strongly nonlinear damping without any restrictions on damping parameters.

Proof. We apply abstract results from Chapter 8. To this end we first describe the operators and spaces in the abstract representation (9.1.7) for the given case.

As in Chapter 3 we introduce the following spaces and operators.

- $\mathscr{H} \equiv L_2(\Omega), \ V \equiv H_0^1(\Omega)$.
- $\mathscr{A} u \equiv \Delta^2 u, u \in \mathscr{D}(\mathscr{A}) : \mathscr{D}(\mathscr{A}) \equiv H_0^2(\Omega) \cap H^4(\Omega)$.
- $Mu \equiv u - \alpha\Delta u, u \in \mathscr{D}(M) : \mathscr{D}(M) \equiv H_0^1(\Omega) \cap H^2(\Omega)$.
- Hence $V' = H^{-1}(\Omega), \ \mathscr{D}(\mathscr{A}^{1/2}) = H_0^2(\Omega), \ [\mathscr{D}(\mathscr{A}^{1/2})]' = H^{-2}(\Omega)$.
- $F(u) \equiv [u + f, v(u) + F_0] - Lu + p$, where $v(u)$ satisfies the elliptic equation in (9.1.2), $F^*(u) = -Lu$.

- $D(u) \equiv d_0(x)g_0(u) - \alpha \operatorname{div}\{d(x)g(\nabla u)\}$.

With the above notation problem (9.1.1)–(9.1.3) has the form (9.1.7) (see the proof of Theorem 3.1.4). Below we verify that the hypotheses in Assumptions 8.1.1, 8.2.1, 8.3.3, and 8.6.2 are satisfied.

We first note that Assumption 8.1.1 is a special case of Assumption 2.4.15, which implies the existence and uniqueness of generalized solutions to (9.1.7). Therefore by Lemma 3.1.6, Assumption 8.1.1 holds for our case.

9.1.1.1 Verification of Assumptions 8.2.1 and 8.3.3 imposed on the forcing F

The nonlinear term $F(u)$ can presented in the form (8.1.4) as

$$F(u) = -\Pi'(u) + F^*(u)$$

with

$$\Pi(u) = \frac{1}{4}\|\Delta v(u)\|^2 - \frac{1}{2}([u,F_0],u) - ([f,F_0]+p,u), \tag{9.1.13}$$

where $v(u) \in H_0^2(\Omega)$ is defined by (9.1.2), and $F^*(u) = -L(u)$. Thus we set

$$\Pi_0(u) = \frac{1}{4}\|\Delta v(u)\|^2 \quad \text{and} \quad \Pi_1(u) = -\frac{1}{2}([u,F_0],u) - ([f,F_0]+p,u). \tag{9.1.14}$$

We note that in the case considered we have also another representation

$$\Pi_1(u) = -\frac{1}{2}([u,u],F_0) - ([f,F_0]+p,u),$$

which follows from Proposition 1.4.2.

The following assertion is also useful in the hinged case.

9.1.7. Lemma. *Let the hypotheses concerning f, F_0, and p given in Assumption 9.1.2 be in force. Then for any $u \in H^2(\Omega)\cap H_0^1(\Omega)$ the following relations hold.*

$$\|F^*(u)\|_{-1} = \|Lu\|_{-1} \le C\|u\|_{2-\sigma} \ \textit{for some} \ 0<\sigma\le 1, \tag{9.1.15}$$

$$\|u\|^2 \le \eta\left(\|\Delta u\|^2 + \Pi_0(u)\right) + C_\eta \ \textit{for any} \ \eta>0, \tag{9.1.16}$$

$$|\Pi_1(u)| \le \eta\left(\|\Delta u\|^2 + \Pi_0(u)\right) + C_\eta \ \textit{for any} \ \eta>0, \tag{9.1.17}$$

$$(F(u),u) \le -\frac{1}{4}\|\Delta v(u)\|^2 + \eta\|\Delta u\|^2 + C_\eta \ \textit{for any} \ \eta>0. \tag{9.1.18}$$

We also have that

$$\|F(u_1)-F(u_2)\|_{-1} \le C_\rho\|u_1-u_2\|_{2-\sigma} \tag{9.1.19}$$

for some $\sigma>0$ and for any $u_1,u_2 \in H^2(\Omega)\cap H_0^1(\Omega)$ such that $\|u_i\|_2 \le \rho$, $i=1,2$, where $\rho>0$ is arbitrary and C_ρ is a nondecreasing function of ρ.

Furthermore, there exists $\delta > 0$ such that

$$|\Pi(u_1) - \Pi(u_2)| \le C\left(1 + \|u_1\|_{2-\delta}^3 + \|u_2\|_{2-\delta}^3\right) \|u_1 - u_2\|_{2-\delta}, \tag{9.1.20}$$

for any $u_1, u_2 \in H^2(\Omega) \cap H_0^1(\Omega)$.

Proof. Estimate (9.1.15) follows from the definition of $F^*(u)$ and from the properties of L described in Assumption 9.1.2.

Relation (9.1.16) follows from Lemma 1.5.4.

Using (1.4.17) with $j = 2$ and $\beta = 0$ we obtain that

$$\begin{aligned} |\Pi_1(u)| &\le C\|F_0\|_2 \cdot \|u\|_1 \cdot \|u\|_2 + \|u\|_2 \cdot \|[f, F_0] + p\|_{-2} \\ &\le \eta\|\Delta u\|^2 + C_\eta^{(1)}\|u\|^2 + C_\eta^{(2)} \end{aligned}$$

for any $\eta > 0$. Therefore (9.1.17) follows form (9.1.16).

A simple calculation using the symmetry of the von Karman bracket (see Proposition 1.4.2) gives

$$(F(u), u) = -\|\Delta v(u)\|^2 - ([f, u], v(u)) + ([u, F_0], u) + ([f, F_0] + p, u) - (Lu, u).$$

Therefore using the bounds for the von Karman bracket given in Theorem 1.4.3 and also the interpolation argument we obtain

$$(F(u), u) \le -\frac{1}{2}\|\Delta v(u)\|^2 + \eta\|\Delta u\|^2 + C_\eta^{(1)}\|u\|^2 + C_\eta^{(2)}$$

for any positive η. Now using (9.1.16) we get (9.1.18).

To prove (9.1.19) we note that

$$F(u_1) - F(u_2) = [u_1 + f, v(u_1) - v(u_2)] + [u_1 - u_2, v(u_2) + F_0] - L(u_1 - u_2).$$

Using Theorem 1.4.3 and Corollary 1.4.5 we obtain that

$$\|F(u_1) - F(u_2)\|_{-1} \le C(1 + \|u_1\|_2^2 + \|u_2\|_2^2)\|u_1 - u_2\|_{1+\delta} + \|L(u_1 - u_2)\|_{-1}$$

for every $\delta > 0$. Therefore we get (9.1.19) from (9.1.15).

To prove (9.1.20) we note that $[u_1, u_1] - [u_2, u_2] = [u_1 - u_2, u_1 + u_2]$ and

$$\begin{aligned} &\left|\|\Delta v(u_1)\|^2 - \|\Delta v(u_2)\|^2\right| \\ &\le \|\Delta(v(u_1) - v(u_2))\| \left(\|\Delta v(u_1)\| + \|\Delta v(u_2)\|\right) \\ &\le C\|[u_1 - u_2, u_1 + u_2 + 2f]\|_{-2} \left(\|[u_1, u_1 + 2f]\|_{-2} + \|u_2, u_2 + 2f]\|_{-2}\right). \end{aligned}$$

Therefore (9.1.20) follows from (1.4.17) with $j = 2$. □

Lemma 9.1.7 implies the validity of Assumptions 8.2.1(F) and 8.3.3(F) imposed on F. Indeed, (9.1.16) coincides with (8.2.4), and relation (9.1.18) is identical to (8.2.3) in the framework considered. Thus we obtain the following assertion.

9.1.8. Lemma. *The nonlinear mapping F satisfies the hypotheses postulated by Assumptions* 8.1.1, 8.2.1, *and* 8.3.3.

9.1.1.2 Verifications of Assumptions 8.2.1 and 8.3.3 imposed on the damping operator

We denote

$$D(v,u) \equiv \int_\Omega d_0(x)g_0(v)u dx + \alpha \sum_{i=1,2} \int_\Omega d(x)g_i(v_{x_i})u_{x_i} dx, \quad v,u \in H^1(\Omega). \quad (9.1.21)$$

The argument given in Chapter 3 (see Lemma 3.1.6) shows that Assumption 8.1.1(D) holds. Validity of Assumption 8.2.1(D) and Assumption 8.3.3(D) follows from the result stated in the lemma below.

9.1.9. Lemma. *Let Assumption* 9.1.2 *and relations* (9.1.10) *hold. Then*

- $$D(v,v) \geq c_0 \left(\|v\|^2 + \alpha\|\nabla v\|^2\right) - c_1, \quad v \in H_0^1(\Omega), \quad (9.1.22)$$

  where c_0 and c_1 are positive constants.
- *If $p = 1$ in* (9.1.8) *and $g_0'(s)$ is bounded (assumed when $L \neq 0$), then*

  $$\|D(v)\|_{-1} \leq c_2\|v\|_1, \quad v \in H^1(\Omega), \quad (9.1.23)$$

  and

  $$|D(u+v,w) - D(u,w)| \leq c_3\|w\|_1 \left[1 + D(u+v,u+v)^{1/2} + D(u,u)^{1/2}\right] \quad (9.1.24)$$

  for any $v,u \in H_0^1(\Omega)$ and $w \in H^1(\Omega)$, where $c_2 > 0$ and $c_3 > 0$ are constants.
- *If $1 \leq p < \infty$ (relevant when $L = 0$), then for any $\eta > 0$ we have that*

  $$|D(v,u)| \leq C_\eta \|u\|_s \cdot D(v,v) + \eta \|u\|_1, \quad v \in H^1(\Omega),\ u \in H^2(\Omega), \quad (9.1.25)$$

  for some $C_\eta > 0$, and

  $$|D(u+v,w) - D(u,w)| \leq C\|w\|_s \left[1 + D(u+v,u+v) + D(u,u)\right], \quad (9.1.26)$$

  for any $v,u \in H^1(\Omega)$ and $w \in H^2(\Omega)$, where $s = 2p/(1+p)$ if $p > 1$. When $p = 1$, the parameter $s > 1$ can be taken arbitrary.

Proof. By the Friedrichs inequality we have

$$\|v\|^2 + \alpha\|\nabla v\|^2 \leq C(1+\alpha)\|\nabla v\|^2, \quad v \in H_0^1(\Omega). \quad (9.1.27)$$

It follows from (9.1.8) that $sg_i(s) \geq m's^2 - c$ for all $s \in \mathbb{R}$ and for some $c > 0$ and $0 < m' < m$. We also have that $sg_0(s) \geq 0$. Therefore

$$D(v,v) \geq \alpha d_0 \sum_{i=1,2} \int_\Omega g_i(v_{x_i})v_{x_i}dx \geq \alpha d_0 m' \|\nabla v\|^2 - \alpha c|\Omega|.$$

Thus (9.1.22) follows from (9.1.27).

In the case $L \neq 0$ we have that

$$|g_0(v)| \leq a_1|v| \quad \text{and} \quad |g_1(v_1)|^2 + |g_2(v_2)|^2 \leq a_1^2(v_1^2 + v_2^2). \tag{9.1.28}$$

This implies that

$$|D(v,u)| \leq C \cdot (\|v\|\|u\| + \alpha\|\nabla v\|\|\nabla u\|) \leq C \cdot \|v\|_1\|u\|_1.$$

Therefore we obtain (9.1.23).

Similarly, from (9.1.28) we have

$$\begin{aligned} &|D(u+v,w) - D(u,w)| \\ &\leq C\int_\Omega (|u+v| + |u|)|w|dx + C\alpha \int_\Omega (|\nabla(u+v)| + |\nabla u|)|\nabla w|dx \\ &\leq C\left[\left(\|u+v\|^2 + \alpha^2\|\nabla(u+v)\|^2\right)^{1/2} + \left(\|u\|^2 + \alpha^2\|\nabla u\|^2\right)^{1/2}\right]\|w\|_1. \end{aligned}$$

Therefore (9.1.24) follows from (9.1.22).

Now we consider the case of an arbitrary finite p (of relevance when $L = 0$). We use the diagonal structure of the mapping $g = (g_1; g_2)$.

Let $\delta > 0$ be arbitrary. Denoting $\Omega_1 \equiv \{x \in \Omega : |v_{x_i}| \leq \delta\}$ and $\Omega_2 \equiv \{x \in \Omega : |v_{x_i}| \geq \delta\}$ for each $i = 1,2$ we have

$$\left|\int_\Omega dg_i(v_{x_i})u_{x_i}dx\right| \leq \int_{\Omega_1} |dg_i(v_{x_i})u_{x_i}|dx + \int_{\Omega_2} |dg_i(v_{x_i})u_{x_i}|dx. \tag{9.1.29}$$

It is clear that

$$\int_{\Omega_1} |dg_i(v_{x_i})u_{x_i}|dx \leq C\max_{|s|\leq\delta}|g_i(s)|\|u\|_1. \tag{9.1.30}$$

In the region Ω_2 we use Hölder's inequality applied with some $1 < r \leq 2$:

$$\int_{\Omega_2} d|g_i(v_{x_i})u_{x_i}|dx \leq d_1\left(\int_{\Omega_2} |g_i(v_{x_i})|^r dx\right)^{1/r}\left(\int_{\Omega_2} |u_{x_i}|^{\bar{r}}dx\right)^{1/\bar{r}},$$

where $r^{-1} + \bar{r}^{-1} = 1$. Because by (1.1.6) $H^{1-2/\bar{r}}(\Omega) \subset L_{\bar{r}}(\Omega)$, from polynomial growth condition (9.1.8) assumed on g_i we have

$$\int_{\Omega_2} |dg_i(v_{x_i})u_{x_i}|dx \leq C\left(\int_{\Omega_2} |g_i(v_{x_i})||v_{x_i}|^{p(r-1)}dx\right)^{1/r}\|u\|_{2-2/\bar{r}}.$$

If we choose $r = 1 + 1/p$, we obtain

$$\int_{\Omega_2} |dg_i(v_{x_i})u_{x_i}|dx \leq C\left(\int_{\Omega_2} |g_i(v_{x_i})||v_{x_i}|dx\right)^{p/(1+p)} \|u\|_{2-2/(1+p)}$$
$$\leq C_\delta \int_{\Omega_2} d|g_i(v_{x_i})||v_{x_i}|dx \cdot \|u\|_{2-2/(1+p)}, \quad (9.1.31)$$

where in the last inequality we use the fact that on Ω_2 we have

$$\left(\int_{\Omega_2} |g_i(v_{x_i})||v_{x_i}|dx\right)^{p/(1+p)} \leq \frac{C_\delta}{d_0}\int_{\Omega_2} d(x)|g_i(v_{x_i})||v_{x_i}|dx.$$

Combining (9.1.29)–(9.1.31) yields the inequality

$$\left|\int_{\Omega} dg_i(v_{x_i})u_{x_i}dx\right| \leq C_\delta\|u\|_{2p/(1+p)} \cdot \int_{\Omega} dg_i(v_{x_i})v_{x_i}dx + C\max_{|s|\leq\delta}|g_i(s)|\|u\|_1. \quad (9.1.32)$$

A similar, but simpler argument applies to the term g_0. Due to the Sobolev embedding $H^{1+\delta}(\Omega) \subset L_\infty(\Omega)$ (see, e.g., (1.1.5)), we need no growth conditions on $g_0(v)$. Indeed, on the set $\widetilde{\Omega}_2 \equiv \{x \in \Omega : |v(x)| \geq \delta\}$ we have

$$\left|\int_{\widetilde{\Omega}_2} d_0g_0(v)udx\right| \leq C\max_{\Omega}|u|\int_{\widetilde{\Omega}_2} d_0|g_0(v)|dx \leq C_{\delta,\beta}\|u\|_{1+\beta}\int_{\widetilde{\Omega}_2} d_0g_0(v)vdx \quad (9.1.33)$$

for every $\beta > 0$. Thus from (9.1.32) and (9.1.33) we we can easily obtain (9.1.25).

To verify relation (9.1.26) we note that

$$|D(u+v,w) - D(u,w)| \leq |D(u+v,w)| + |D(u,w)|.$$

Therefore we can apply (9.1.25) with $\eta = 1$ to obtain (9.1.26). □

To prove property (8.3.11) in Assumption 8.3.3 we observe the validity of the following inequality:

9.1.10. Lemma. *Let Assumption 9.1.2 hold. Assume that $g_i(s)$, $i = 1,2$, are increasing. Then for any $\varepsilon > 0$ there exist C_ε such that*

$$\|v\|^2 + \alpha\|\nabla v\|^2 \leq \varepsilon + C_\varepsilon\left[D(u+v,v) - D(u,v)\right], \quad (9.1.34)$$

for any $v \in H_0^1(\Omega)$ and $u \in H^1(\Omega)$, where the form $D(v,u)$ is given by (9.1.21). The same is true for $v \in H^1(\Omega)$ provided that in addition we assume that $g_0(s)$ is increasing and $g_0'(s) \geq m > 0$ for $|s| \geq 1$.

Proof. It follows from the definition of $D(v,u)$ and from Proposition B.1.2 in Appendix B. □

We are now in position to conclude the proof of Theorem 9.1.5. By Lemma 9.1.8 it is sufficient to check the hypotheses imposed on D in the corresponding assumptions.

Clearly Assumption 8.2.1(D) follows from (9.1.22) and (9.1.23). Therefore Theorem 8.2.3 implies dissipativity of the dynamical system (H,S_t) generated by problem (9.1.1)–(9.1.3) in the case $L \neq 0$.

Lemma 9.1.10 and relation (9.1.24) imply that the damping operator D satisfies Assumption 8.3.3 with $\varkappa = 1$. Thus Theorem 8.3.5 implies the existence of a compact global attractor in the case $L \neq 0$.

In the case $L = 0$ we note that by energy inequality (3.1.14) the system (H, S_t) is gradient. Because the set of stationary solutions is bounded (see Theorem 1.5.7), we can use Theorem 8.3.12. to prove the existence of a compact global attractor in the case $L = 0$. Assumption 8.3.9(D) follows from (9.1.11) and (9.1.25). □

9.1.11. Remark. Whether in the nonconservative case $L \neq 0$, the linear growth imposed on the damping g_i is necessary for the existence of the attractor in Theorem 9.1.5, is an open question. ■

9.1.2 Hinged or simply supported boundary conditions

We consider next equations (9.1.1) and (9.1.2) with hinged-type boundary conditions (9.1.4) or (9.1.6) for the displacement $u(x,t)$.

The following result is a counterpart of Theorem 9.1.5.

9.1.12. Theorem. *Let Assumption* 9.1.2 *hold and* (H, S_t) *be the dynamical system generated by problem* (9.1.1) *and* (9.1.2) *with hinged (resp., simply supported) boundary conditions* (9.1.4) *(resp.,* (9.1.6)*) in the space* $H = (H^2(\Omega) \cap H_0^1(\Omega)) \times H_0^1(\Omega)$. *Then the system* (H, S_t) *possesses a compact global attractor* A *under condition* (9.1.10) *provided*

(i) *The damping functions* $g_i(s)$ *possess properties* (9.1.11) *when* $L \equiv 0$.
(ii) *The damping functions* $g_i(s)$ *are strictly increasing when* $L \neq 0$.

Proof. Problem (9.1.1), (9.1.2), and (9.1.4) can be represented (see the proof of Theorem 3.1.16) in the abstract form (9.1.7) after introducing the following spaces and operators.

- $\mathscr{H} \equiv L_2(\Omega), \; V \equiv H_0^1(\Omega)$.
- $\mathscr{A} u \equiv \Delta^2 u, u \in \mathscr{D}(\mathscr{A}) : \mathscr{D}(\mathscr{A}) \equiv \{u \in H^4(\Omega), \; u = \Delta u = 0 \text{ on } \Gamma\}$.
- $Mu \equiv u - \alpha \Delta u, u \in \mathscr{D}(M)$, where $\mathscr{D}(M) \equiv H_0^1(\Omega) \cap H^2(\Omega)$.
- Hence $V' = H^{-1}(\Omega), \; \mathscr{D}(\mathscr{A}^{1/2}) = H^2(\Omega) \cap H_0^1(\Omega)$.
- $F(u) \equiv [u + f, v(u) + F_0] - Lu + p$, where $v(u)$ solves (9.1.2).
- $D(u) \equiv d_0(x) g_0(u) - \alpha \operatorname{div}(d(x) g(\nabla u))$.

The only difference with respect to the clamped case is the form of the domain of the operator $\mathscr{A}$. Therefore the arguments are essentially identical to those in the case of clamped boundary conditions. For the reader's convenience we summarize the main steps.

Step 1: The nonlinear term $F(u)$ can be represented in the form $F(u) = -\Pi'(u) + F^*(u)$ with $\Pi(u)$ given by (9.1.13) and $F^*(u) = -L(u)$. We define $\Pi_0(u)$ and $\Pi_1(u)$ by the same relations as in the clamped case (see (9.1.14)). In the hinged case, we also use symmetry of the von Karman bracket for functions in $H_0^1(\Omega)\cap H^2(\Omega)$ (see Proposition 1.4.2). In particular, we use the relation

$$\int_\Omega [u,F_0]w\,dx = \int_\Omega [w,F_0]u\,dx, \quad u,w \in H^2(\Omega)\cap H_0^1(\Omega), \tag{9.1.35}$$

which follows from (1.4.9) and (1.4.10) with $\Gamma_1 = \Gamma_3 = \varnothing$ and $\Gamma_2 = \partial\Omega$.

By Lemma 9.1.7 the nonlinearity $F(u)$ satisfies Assumptions 8.1.1, 8.2.1, and 8.3.3.

Step 2: The damping operators D are the same in both the clamped and hinged cases. Therefore Assumption 8.2.1(D) for D follows from Lemma 9.1.9. Thus Theorem 8.2.3 implies dissipativity of the dynamical system (H,S_t) generated by problem (9.1.1), (9.1.2), and (9.1.4) in the case $L \not\equiv 0$.

Step 3: The application of Lemma 9.1.10 and relations (9.1.24) and (9.1.26) imply that the damping operator D satisfies Assumption 8.3.3 with $\varkappa = 1$ for $L \not\equiv 0$.

The subsequent arguments are as in the clamped case. Thus by Theorem 8.3.5 or Theorem 8.3.12 the system (H,S_t) possesses a compact global attractor.

In the case of simply supported boundary conditions (9.1.6), the analysis is identical with only one difference in the definition of $\mathscr{D}(\mathscr{A})$ which now becomes:

$$\mathscr{D}(\mathscr{A}) = \left\{u \in H^4(\Omega) \,:\, u = \Delta u + (1-\mu)B_1 u = 0 \text{ on } \Gamma\right\}.$$

This does not affect the arguments above. In particular the symmetry property of the von Karman bracket which generally may depend on the boundary conditions imposed and which was used in (9.1.35) remains true (see Proposition 1.4.2). □

9.1.3 Free boundary conditions

We consider equations (9.1.1) and (9.1.2) with free boundary conditions (9.1.5) for the displacement $u(x,t)$. In contrast with the clamped and hinged cases, the free case requires additional assumption. Namely, we impose the following hypotheses.

9.1.13. Assumption.

- Functions $d_0(x)$ and $d(x)$ are strictly positive; that is,

$$0 < d_0 \le d_0(x), d(x) \le d_1 \quad \text{for all } x \in \Omega.$$

- We have that $\nu_1 > 0$, $\beta \subset L_\infty(\Gamma)$, $\beta(x) \ge 0$ a.e., and $F_0 \subset H^3(\Omega)\cap H_0^2(\Omega)$.
- In the case $L \equiv 0$, the damping functions $g_i(s)$ are required to have the following properties

$$s^2 \le \eta + C_\eta s g_i(s), \quad \text{for every } \eta > 0,\ i = 0,1,2. \tag{9.1.36}$$

- In the case $L \neq 0$, the damping functions $g_i(s)$ are strictly increasing for every $i = 0,1,2$.

9.1.14. Remark. The hypothesis $F_0 \in H^3(\Omega) \cap H_0^2(\Omega)$ is needed in order to include the term $[u, F_0]$ in the potential part of the nonlinearity. This is necessitated by the fact that solutions do not vanish on the boundary. Instead, with a new requirement it follows from Proposition 1.4.2 that

$$([u, F_0], w) = ([u, w], F_0) \quad \text{if } u, w \in H^2(\Omega) \text{ and } F_0 \in H_0^2(\Omega), \qquad (9.1.37)$$

as needed. We can also consider more general case when $F_0 = F_0^{(1)} + F_0^{(2)}$, where $F_0^{(1)} \in H^3(\Omega) \cap H_0^2(\Omega)$ and $F_0^{(2)} \in H^3(\Omega)$. However, this situation can be reduced to the previous one by redefining the operator L and the external force p in the following way: $L(u) := -[u, F_0^{(2)}] + L(u)$ and $p := [f, F_0^{(2)}] + p$. ∎

The main result of this section is the following assertion.

9.1.15. Theorem. *Let Assumptions* 9.1.2 *and* 9.1.13 *hold and* (H, S_t) *be the dynamical system generated by problem* (9.1.1) *and* (9.1.2) *with free boundary conditions* (9.1.5) *in the space* $H = H^2(\Omega) \times H^1(\Omega)$. *Then the system* (H, S_t) *possesses a compact global attractor* A.

Proof. The idea is the same as the clamped and hinged cases: we first rewrite the problem in the abstract form and then apply the results from Chapter 8.

In the free case, the functional analytic setup is somewhat different and the definitions of operators M and D are less direct (for details see Section 3.1.3). This is due to the fact that solutions do not vanish on the boundary. In order to describe the functional setup we introduce

- $\mathscr{H} \equiv L_2(\Omega)$, $V \equiv H^1(\Omega)$, $V' = [H^1(\Omega)]'$.
- $\mathscr{A} u \equiv \Delta^2 u, u \in \mathscr{D}(\mathscr{A})$, where $\mathscr{D}(\mathscr{A}^{1/2}) = H^2(\Omega)$ and

$$\mathscr{D}(\mathscr{A}) \equiv \left\{ u \in H^4(\Omega) \;\middle|\; \begin{array}{l} \frac{\partial}{\partial n}\Delta u + (1-\mu)B_2 u = \nu_1 u \text{ on } \Gamma, \\ \Delta u + (1-\mu)B_1 u = 0 \text{ on } \Gamma \end{array} \right\}$$

- $M \equiv I + \alpha A_N$, where $A_N u = -\Delta u$ with the domain

$$\mathscr{D}(A_N) \equiv \left\{ u \in H^2(\Omega) : \frac{\partial}{\partial n} u = 0 \text{ on } \Gamma \right\}.$$

- We set

$$F(u) \equiv [u + f, v(u) + F_0] - Lu + p - B_\Gamma u \equiv \widetilde{F}(u) - B_\Gamma u, \qquad (9.1.38)$$

where $v(u)$ solves (9.1.2) and the operator $B_\Gamma : H^1(\Omega) \mapsto [H^1(\Omega)]'$ is defined variationally:

$$\int_\Omega B_\Gamma(u)\phi dx = \int_\Gamma \beta u^3 \phi dx, \quad \forall \phi \in H^1(\Omega).$$

- $D(u) \equiv d_0(x) g_0(u) + \alpha D_1(u)$, where D_1 is also defined variationally:

$$\int_\Omega D_1(u)\phi dx = \int_\Omega d(x)g(\nabla u)\nabla\phi dx, \quad \forall\phi \in H^1(\Omega).$$

It is straightforward to verify by integrating by parts (see Section 3.1.3) that the original PDE given in (9.1.1) and (9.1.2) with the boundary conditions (9.1.5) is equivalent to the operator model (9.1.7).

With the above notation we need to verify that the assumptions from Chapter 8 are satisfied. As before, the arguments are similar to those in the clamped case, although the technical details are a little bit more involved. For the reader's convenience we summarize the main facts.

It is clear that Lemma 3.1.24 implies that Assumption 8.1.1 holds for M, $\mathscr{A}$, D, and F introduced above.

9.1.3.1 Verification of Assumptions 8.2.1 and 8.3.3 imposed on F

The nonlinear term $F(u)$ is represented in the form (8.1.4)

$$F(u) = -\Pi'(u) + F^*(u)$$

with

$$\Pi(u) = \frac{1}{4}\|\Delta v(u)\|^2 + \frac{1}{4}\int_\Gamma \beta u^4 d\Gamma - \frac{1}{2}([u,F_0],u) - ([f,F_0]+p,u), \tag{9.1.39}$$

where $v(u) \in H_0^2(\Omega)$ is defined by (9.1.2), and $F^*(u) = -L(u)$. We set:

$$\Pi_0(u) = \frac{1}{4}\|\Delta v(u)\|^2 + \frac{1}{4}\int_\Gamma \beta u^4 d\Gamma \text{ and } \Pi_1(u) = -\frac{1}{2}([u,F_0],u) - ([f,F_0]+p,u). \tag{9.1.40}$$

Lemma 1.5.13 implies in the free case that

$$\|u\|^2 \le \eta\left(\|u\|_2^2 + \Pi_0(u)\right) + C_\eta, \quad u \in H^2(\Omega), \tag{9.1.41}$$

so (8.2.4) is satisfied for any $\eta > 0$. Moreover we also have that

$$|\Pi_1(u)| \le \eta\left(\|u\|_2^2 + \Pi_0(u)\right) + C_\eta, \quad u \in H^2(\Omega),$$

for any $\eta > 0$. A simple calculation shows that

$$(F(u),u) = -\|\Delta v(u)\|^2 - \int_\Gamma \beta u^4 d\Gamma - ([f,u],v(u)) + ([u+f,F_0] - L(u) + p,u).$$

This implies $(F(u),u) \le -2\Pi_0(u) + \eta\|u\|_2^2 + C_\eta^{(1)}\|u\|^2 + C_\eta^{(2)}$. Therefore using the ellipticity of $\mathscr{A}$ and (9.1.41) we obtain that

$$(F(u),u) \le -\Pi_0(u) + \eta\|\mathscr{A}^{1/2}u\|_2^2 + C_\eta.$$

Thus we have (8.2.3). Together with (9.1.41) it gives us the validity of Assumption 8.2.1(F).

Our next task is to verify the validity of Assumption 8.3.3(F). Representation (9.1.38) implies

$$\|F(u)-F(w)\|_{V'} \le \|\widetilde{F}(u)-\widetilde{F}(w)\|_{[H^1(\Omega)]'} + |B_\Gamma(u)-B_\Gamma(w)\|_{[H^1(\Omega)]'}.$$

By Theorem 1.4.3 and Corollary 1.4.5 we have that

$$\|\widetilde{F}(u)-\widetilde{F}(w)\|_{[H^1(\Omega)]'} \le C(1+\|u\|_{2-\sigma}^2+\|w\|_{2-\sigma}^2)\|u-w\|_{2-\sigma}^2 \qquad (9.1.42)$$

for some $0<\sigma<\frac{1}{2}$. We also have

$$|(B_\Gamma(u)-B_\Gamma(w),\phi)| \le C(\max_\Omega |u| + \max_\Omega |w|)^2 \|u-w\|_{L_2(\Gamma)}\|\phi\|_{L_2(\Gamma)}$$

for any $\phi\in H^1(\Omega)$. Because $H^{1+\delta}(\Omega)\subset C(\overline{\Omega})$, $\delta>0$, we obtain

$$\|B_\Gamma(u)-B_\Gamma(w)\|_{[H^1(\Omega)]'} \le C(1+\|u\|_{2-\sigma}^2+\|w\|_{2-\sigma}^2)\|u-w\|_{2-\sigma}^2. \qquad (9.1.43)$$

Thus (9.1.42), (9.1.43), and the argument above makes it possible to conclude that the requirements (ii) and (iii) of Assumption 8.3.3(F) hold. To prove Requirement (i) in Assumption 8.3.3(F) we use the same argument as in the proof of relation (9.1.20) in Lemma 9.1.7.

We are now in position to conclude the proof of Theorem 9.1.15. For this, it suffices to check the hypotheses imposed on D in the corresponding assumptions. This is done below.

9.1.3.2 Verification of Assumptions 8.2.1 and 8.3.3 imposed on the damping operator

Although the analysis of the damping in the free case is similar to that in the clamped and hinged cases, there are a number of technical details which are different and more involved in the free case. This is due to different structure of the phase space, where the latter does not encode any boundary conditions imposed on the elements. In fact, as seen in the model, the boundary conditions produced by the variational principle are intrinsically nonlinear and by themselves force a certain amount of dissipation on the boundary.

To wit, let's recall the variational structure of the damping operator $D=d_0g_0+\alpha D_1$:

$$D(v,u)\equiv\int_\Omega d_0(x)g_0(v)u\,dx+\alpha\sum_{i=1,2}\int_\Omega d(x)g_i(v_{x_i})u_{x_i}dx, \quad v,u\in H^1(\Omega). \qquad (9.1.44)$$

Validity of Assumptions 8.2.1 and 8.3.3 follows from the result stated in the lemma below.

9.1.16. Lemma. *Let Assumptions* 9.1.2 *and* 9.1.13 *hold. Then*

- $$D(v,v) \geq c_0 \left(\|v\|^2 + \alpha\|\nabla v\|^2\right) - c_1, \quad v \in H^1(\Omega), \tag{9.1.45}$$

  where c_0 and c_1 are positive constants.
- *If $p = 1$ in* (9.1.8) *and $g_0'(s)$ is bounded (assumed when $L \neq 0$), then*

  $$\|D(v,u)\| \leq c_2\|u\|_1\|v\|_1, \quad v,u \in H^1(\Omega), \tag{9.1.46}$$

  and

  $$|D(u+v,w) - D(u,w)| \leq c_3\|w\|_1 \left[1 + D(u+v,u+v)^{1/2} + D(u,u)^{1/2}\right] \tag{9.1.47}$$

  for any $v,u \in H^1(\Omega)$ and $w \in H^1(\Omega)$, where $c_2 > 0$ and $c_3 > 0$ are constants.
- *If $1 \leq p < \infty$ (relevant when $L = 0$) then for any $\eta > 0$ we have*

  $$|D(v,u)| \leq C_\eta\|u\|_s \cdot D(v,v) + \eta\|u\|_1, \quad v \in H^1(\Omega),\ u \in H^2(\Omega), \tag{9.1.48}$$

  $$|D(u+v,w) - D(u,w)| \leq C\|w\|_s \left[1 + D(u+v,u+v) + D(u,u)\right], \tag{9.1.49}$$

  for any $v,u \in H^1(\Omega)$ and $w \in H^2(\Omega)$, where $s = 2p/(1+p)$ if $p > 1$ (if $p = 1$ then $s > 1$ is arbitrary). Constants C_η and C are positive.

Proof. The proof of the lemma follows the same line of argument as given in the clamped case. The main difference is due to the fact that elements do not vanish on the boundary, thus the Fridrichs inequality is no longer valid. However, this is compensated by the assumption that $d_0(x) \geq d_0 > 0$. Therefore, lower-order damping g_0 subject to the additional assumption (9.1.36) provides the needed bound in the first statement of the lemma in the case $L \equiv 0$. The remaining arguments in the lemma rely on differential Sobolev calculus that does not depend on the boundary conditions. □

With the result of Lemma 9.1.16 in hand, the proof of Theorem 9.1.15 can be completed as in clamped case. □

9.1.4 Mixed boundary conditions

Assume that the boundary $\partial\Omega$ is divided into three nonoverlapping parts Γ_i, each open relative to $\partial\Omega$ and such that $\partial\Omega = \cup_i \Gamma_i$. We allow a possibility where only some of these three (one or two) boundary conditions are imposed. This leads to consideration of equations (9.1.1) and (9.1.2) with the following boundary conditions

$$\left.\begin{aligned}
&u = \nabla u = 0 \text{ on } \Gamma_1, && \text{(clamped)}\\
&u = \Delta u + (1-\mu)B_1 u = 0 \text{ on } \Gamma_2, && \text{(simply supported)}\\
&\left.\begin{aligned}
&\Delta u + (1-\mu)B_1 u = 0 \text{ on } \Gamma_3,\\
&\frac{\partial}{\partial n}\Delta u + (1-\mu)B_2 u - \alpha\frac{\partial}{\partial n}w_{tt}\\
&\quad -\alpha d(x)g(\nabla u_t)\cdot n = \nu_1 u + \beta u^3 \text{ on } \Gamma_3.
\end{aligned}\right] && \text{(free)}
\end{aligned}\right\} \tag{9.1.50}$$

In this case the functional analytic setup is the following

- $\mathscr{H} \equiv L_2(\Omega)$, $V \equiv H^1_{\Gamma_1\cup\Gamma_2}(\Omega) = \{w \in H^1(\Omega) : u|_{\Gamma_1\cup\Gamma_2} = 0\}$.
- $\mathscr{A}u \equiv \Delta^2 u, u \in \mathscr{D}(\mathscr{A})$, where

$$\mathscr{D}(\mathscr{A}) \equiv \left\{ u \in H^4(\Omega) \;\middle|\; \begin{aligned} &u|_{\Gamma_1\cup\Gamma_2} = 0,\ \nabla u|_{\Gamma_1} = 0,\\ &\Delta u + (1-\mu)B_1 u = 0 \text{ on } \Gamma_2\cup\Gamma_3,\\ &\frac{\partial}{\partial n}\Delta u + (1-\mu)B_2 u = \nu_1 u \text{ on } \Gamma_3 \end{aligned} \right\}.$$

- $M \equiv I + \alpha A_N$, where $A_N u = -\Delta u$ with the domain

$$\mathscr{D}(A_N) \equiv \{u \in H^2(\Omega) : u = 0 \text{ on } \Gamma_1\cup\Gamma_2,\ \frac{\partial}{\partial n}u = 0 \text{ on } \Gamma_3\}.$$

- We set

$$F(u) \equiv [u+f, v(u)+F_0] - Lu + p - B_{\Gamma_3}u \equiv \widetilde{F}(u) - B_{\Gamma_3}u,$$

where $v(u)$ solves (9.1.2) and the operator $B_{\Gamma_3} : V \mapsto V'$ is defined variationally:

$$\int_\Omega B_{\Gamma_3}(u)\phi dx = \int_{\Gamma_3} \beta u^3 \phi dx, \quad \forall \phi \in V.$$

- $D(u) \equiv d_0(x)g_0(u) + \alpha D_1(u)$, where D_1 is also defined variationally:

$$\int_\Omega D_1(u)\phi dx = \int_\Omega d(x)g(\nabla u)\nabla\phi dx, \quad \forall \phi \in V.$$

One can easily assert—by integrating by parts—that the original PDE given in (9.1.1) and (9.1.2) with the boundary conditions (9.1.50) is equivalent to the operator model in (9.1.7).

Because the boundaries of the regions Γ_i are not overlapping, similar results on attractors, to the ones given above, can be established in a straightforward manner. Moreover, in the case $\Gamma_1\cup\Gamma_2 \neq \emptyset$ the nondegeneracy condition imposed in Assumption 9.1.13 on the damping function $g_0(s)$ can be relaxed.

9.2 Further properties of the attractor for von Karman model with rotational inertia

In this section we study properties of the global attractor for the dynamical system (H,S_t) generated by problem (9.1.1) and (9.1.2) with one of the boundary conditions (9.1.3), (9.1.4), or (9.1.5).

Below we assume the validity of the hypotheses of Section 9.1 which guarantee the existence of a compact global attractor. Thus the existence of a global attractor is our starting point and we look for additional properties of the said attractor.

9.2.1 Regular structure of the attractor

The precise structure of the global attractor A of the system (H,S_t) is, generally, not known. However, by Proposition 7.5.2 this attractor contains the unstable manifold $\mathscr{M}^u(\mathscr{N})$ (see the definition in Section 7.5) emanating from the set $\mathscr{N}$ of all stationary points of (H,S_t). The set $\mathscr{N}$ in our case consists of elements $(u;0)\in H$ such that u solves the problem

$$\Delta^2 u-[u+f,v+F_0]+Lu=p(x),\ \ x\in\Omega, \tag{9.2.1}$$

$$\Delta^2 v+[u+2f,u]=0,\ \ x\in\Omega,\quad v|_{\partial\Omega}=\frac{\partial v}{\partial n}\Big|_{\partial\Omega}=0, \tag{9.2.2}$$

where the displacement $u(x)$ satisfies either clamped (9.1.3), hinged (9.1.4), or free (9.1.5) boundary conditions. The latter for time-independent solutions has the form

$$\begin{aligned}&\Delta u+(1-\mu)B_1u=0\ \text{ on }\Gamma,\\ &\frac{\partial}{\partial n}\Delta u+(1-\mu)B_2u=\nu_1u+\beta u^3\ \text{ on }\Gamma.\end{aligned} \tag{9.2.3}$$

Here, as above, $\nu_1>0$ and $\beta\in L_\infty(\Gamma)$ is nonnegative, $0<\mu<1$, the boundary operators B_1 and B_2 are defined by (1.3.20).

Below we denote by $\mathscr{N}^*$ the set of all solutions to problem (9.2.1) and (9.2.2) with the corresponding boundary conditions (either (9.1.3), (9.1.4), or (9.2.3)).

In the case when the system is of gradient type, the set $\mathscr{N}$ is a carrier of information about the attractor. In our situation this requires nonconservative forces to be absent. Thus, if we assume that the (nonconservative) operator L is absent in (9.1.1) (and $F_0\in(H^3\cap H_0^2)(\Omega)$ in the case of free boundary conditions), then the energy

$$\mathscr{E}(u,u_t)=\frac{1}{2}\int_\Omega\left[\alpha|\nabla u_t|^2+|u_t|^2\right]dx+\Pi(u),$$

is a strict Lyapunov function for (H,S_t). Hence (H,S_t) is a gradient system (see Definition 7.5.3 in Section 7.5), and the application of Theorem 7.5.6 yields the following result.

9.2.1. Theorem. *Assume that* $L \equiv 0$ *in* (9.1.1) *and* $F_0 \in (H^3 \cap H_0^2)(\Omega)$ *in the case of free boundary conditions. Let* (H, S_t) *be the dynamical system generated by problem* (9.1.1) *and* (9.1.2) *with one of the boundary conditions* (9.1.3), (9.1.4), *or* (9.1.5). *Assume that* (H, S_t) *possesses a compact global attractor* A. *Then*

- $A = \mathscr{M}^u(\mathscr{N})$, *where* $\mathscr{N} = \{(w;0) : w \in \mathscr{N}^*\}$ *is the set of stationary point of* (H, S_t) *and* $\mathscr{M}^u(\mathscr{N})$ *is the the unstable manifold* $\mathscr{M}^u(\mathscr{N})$ *emanating from* $\mathscr{N}$ *which is defined as a set of all* $U \in H$ *such that there exists a full trajectory* $\gamma = \{U(t) = (u(t); u_t(t)) : t \in \mathbb{R}\}$ *with the properties*

$$U(0) = U \ \textit{and} \ \lim_{t \to -\infty} \operatorname{dist}_H(U(t), \mathscr{N}) = 0.$$

- *The global attractor* A *consists of full trajectories* $\gamma = \{(u(t); u_t(t)) : t \in \mathbb{R}\}$ *such that*

$$\lim_{t \to \pm\infty} \left\{ \|u_t(t)\|_1^2 + \inf_{w \in \mathscr{N}^*} \|u(t) - w\|_2^2 \right\} = 0.$$

- *Any generalized solution* $u(t)$ *to problem* (9.1.1) *and* (9.1.2) *with the corresponding boundary conditions stabilizes to the set of stationary points; that is,*

$$\lim_{t \to +\infty} \left\{ \|u_t(t)\|_1^2 + \inf_{w \in \mathscr{N}^*} \|u(t) - w\|_2^2 \right\} = 0. \tag{9.2.4}$$

This theorem along with generic-type results on the finiteness of the number of solutions to problem (9.2.1) and (9.2.2) given in Theorem 1.5.7 and Theorem 1.5.16 allow us to obtain the following result.

9.2.2. Corollary. *Under the hypotheses of Theorem* 9.2.1 *there exists an open dense set* $\mathscr{R}_0$ *in* $L_2(\Omega)$ *such that for every* $p \in \mathscr{R}_0$ *the set* $\mathscr{N}$ *of stationary points for* (H, S_t) *is finite. In this case* $A = \cup_{i=1}^N \mathscr{M}^u(z_i)$, *where* $z_i = (w_i; 0)$ *and* w_i *is a solution to problem* (9.2.1) *and* (9.2.2) *with the corresponding boundary conditions,* $i = 1, \ldots, N$. *Moreover,*

- *The global attractor* A *consists of full trajectories* $\gamma = \{(u(t); u_t(t)) : t \in \mathbb{R}\}$ *connecting pairs of stationary points; that is, any* $W \in A$ *belongs to some full trajectory* γ *and for any* $\gamma \subset A$ *there exists a pair* $\{w_-, w_+\} \subset \mathscr{N}^*$ *such that*

$$\lim_{t \to -\infty} \{\|u_t(t)\|_1^2 + \|u(t) - w_-\|_2^2\} = 0$$

  and

$$\lim_{t \to +\infty} \{\|u_t(t)\|_1^2 + \|u(t) - w_+\|_2^2\} = 0.$$

- *For any* $(u_0; u_1) \in H$ *there exists a stationary solution* $w \in \mathscr{N}^*$ *such that*

$$\lim_{t \to +\infty} \{\|u_t(t)\|_1^2 + \|u(t) - w\|_2^2\} = 0, \tag{9.2.5}$$

  where $u(t)$ *is a generalized solution to problem* (9.1.1) *and* (9.1.2) *with initial data* $(u_0; u_1)$ *and with the corresponding boundary conditions.*

Under additional assumptions we can prove that the rate of convergence in (9.2.5) is exponential. To state the result we first establish the following assertion.

9.2.3. Lemma. *Let Assumption* 9.1.2 *and relations in* (9.1.10) *hold. Assume that* $g_i(s)$, $i = 1,2$, *are increasing*[1] *(in the case of free boundary conditions the same condition is required for* $g_0(s)$ *and also* $d_0(x) \ge d_0 > 0$ *and* $g_0'(s) \ge m > 0$ *for* $|s|$ *large enough). Let the form* $D(v,u)$ *be given by* (9.1.21). *Then there exists a strictly increasing continuous concave function* $H_0 : \mathbb{R}_+ \mapsto \mathbb{R}_+$ *such that*

$$H_0(D(v,v)) \ge \|v\|^2 + \alpha\|\nabla v\|^2, \tag{9.2.6}$$

where $v \in (H_0^1 \cap H^2)(\Omega)$ *in the clamped and hinged cases and* $v \in H^2(\Omega)$ *in the case of free boundary conditions. If* $g_i'(s) \ge m > 0$ *for all* $s \in \mathbb{R}$ *(*$i = 1,2$ *in the clamped and hinged cases), then* (9.2.6) *holds with* $H_0(s) = h_0 \cdot s$.

Proof. In the case $g_i'(s) \ge m$ for all $s \in \mathbb{R}$ and $i = 0,1,2$ we obviously have the required relation with $H_0(s) = m^{-1}s$. In the general case we can take $H_0(s) = h(s) + m^{-1}s$, where function $h(s)$ a continuous, concave increasing function, $h(0) = 0$. The role of the function $h(s)$ is to account for the behavior of $g_i(s)$ for "small" values of s. The above follows from the construction in Proposition B.2.1 in the appendix (see also [62] and/or [195]). Indeed, it is shown there that under the assumptions that a given function g is strictly increasing, $g'(s) \ge m_1, |s| \ge 1$, and $g(0) = 0$ there exists a monotone concave function, $h : \mathbb{R} \to \mathbb{R}^+$, $h(0) = 0$, and such that with

$$s^2 \le \frac{1}{m_1}(sg(s) + h(sg(s))) \;\; \text{for all } s \in \mathbb{R}. \tag{9.2.7}$$

The above gives rise to the function $H_0(s)$ as stated above. Thus we can construct a strictly increasing continuous concave function $\widetilde{H}_0(s)$ such that $2s^2 \le \widetilde{H}_0(sg_i(s))$ for all $s \in \mathbb{R}$, where $i = 1,2$, and

$$s_1^2 + s_2^2 \le \widetilde{H}_0(s_1 g_i(s_1) + s_2 g_i(s_2)) \;\; \text{for all } \; s_1, s_2 \in \mathbb{R}. \tag{9.2.8}$$

Therefore Jensen's inequality implies (9.2.6) with an appropriate $H_0(s)$. □

Now we are ready to state the result on the rate of convergence to an equilibrium.

9.2.4. Theorem. *Assume that* $L \equiv 0$ *in* (9.1.1) *and the hypotheses of Section* 9.1 *that guarantee the existence of a compact global attractor A for the dynamical system* (H, S_t) *generated by problem* (9.1.1) *and* (9.1.2) *with one of the boundary conditions* (9.1.3), (9.1.4), *or* (9.1.5) *hold. Assume additionally that the damping functions* g_1 *and* g_2 *are increasing (in the case of free boundary conditions we also assume that* g_0 *possesses the same property,* $g_0'(s) \ge m > 0$ *for* $|s|$ *large enough, and* $F_0 \in (H^3 \cap H_0^2)(\Omega)$*).*

---

[1] We can slightly relax this condition by assuming that $g_i(s)$ are nondecreasing and $sg_i(s) > 0$ for all $s \ne 0$; see Proposition B.2.1

If the set of stationary solutions consists of finitely many isolated equilibrium points that are hyperbolic,[2] *then for any initial condition $y \in H$ there exists an equilibrium point $e = (w;0)$, $w \in H_0^2(\Omega)$, such that*

$$\|S_t y - e\|_H^2 \le C \cdot \sigma\left(\left[tT^{-1}\right]\right), \quad t > 0, \tag{9.2.9}$$

where C and T are positive constants depending on y and e, $[a]$ denotes the integer part of a and $\sigma(t)$ satisfies the following ODE,

$$\frac{d\sigma}{dt} + Q(\sigma) = 0,\ t > 0, \quad \sigma(0) = C(y,e). \tag{9.2.10}$$

Here $C(y,e)$ is a constant depending on y and e, $Q(s) = s - (I + G_0)^{-1}(s)$ with $G_0(s) = c_1 (I + H_0)^{-1}(c_2 s)$, where H_0 is defined in Lemma 9.2.3 and positive numbers c_1 and c_2 depend on y and e.

9.2.5. Remark.

1. As shown in [196, 94], the nonlinear function $Q(\sigma)$ is asymptotically (for small values of σ) comparable to H_0^{-1}. Thus, it is a behavior of H_0^{-1} that determines the decay rates produced by ODE (9.2.10). See also Remark B.3.2 in Appendix B.

2. Under condition (9.1.36) we can avoid the strict increasing requirement concerning $g_i(s)$ by noting that (9.1.36) implies that $sg_i(s) > 0$ for $s \ne 0$; see the proof of Lemma 9.2.3 and Proposition B.2.1. ∎

Proof. To prove (9.2.9) we apply Theorem 8.4.3. For this we need to check property (8.4.3) for the damping D and properties (8.4.4) and (8.4.5) for the force $F(u)$.

To obtain (8.4.3) we note that

$$\begin{aligned}
\left|\int_\Omega d g_i(v_{x_i}) u_{x_i} dx\right| &\le d_1 \int_\Omega \left| |g_i(v_{x_i}) v_{x_i}|^{1/2} \left|\frac{g_i(v_{x_i})}{v_{x_i}}\right|^{1/2} |u_{x_i}| dx \\
&\le \int_\Omega v_{x_i} g_i(v_{x_i}) dx + \frac{d_1^2}{4} \int_\Omega \left|\frac{g_i(v_{x_i})}{v_{x_i}}\right| |u_{x_i}|^2 dx \\
&\le \frac{1}{d_0} \int_\Omega d(x) v_{x_i} g_i(v_{x_i}) dx + \frac{d_1^2}{4} \|u\|_{L_{2\bar r}(\Omega)}^2 \left[\int_\Omega \left|\frac{g_i(v_{x_i})}{v_{x_i}}\right|^r dx\right]^{1/r},
\end{aligned}$$

where $r^{-1} + \bar r^{-1} = 1$, $r > 1$. We obviously have that

$$\int_\Omega \left|\frac{g_i(v_{x_i})}{v_{x_i}}\right|^r dx \le C_{r,\Omega} + \int_{\Omega_*} |g_i(v_{x_i})| \frac{|g_i(v_{x_i})|^{r-1}}{|v_{x_i}|^r} dx,$$

where $\Omega_* \equiv \{x \in \Omega : |v_{x_i}| \ge 1\}$. Thus by the growth conditions in (9.1.8) we obtain, after choosing $r = (p+1)(p-1)^{-1}$ (in the case when $p > 1$) that

[2] In the sense that the linearization of the problem (9.2.1) and (9.2.2) with the corresponding boundary conditions around every stationary solution has a trivial solution only.

$$\int_\Omega \left|\frac{g_i(v_{x_i})}{v_{x_i}}\right|^r dx \le C_1 + C_2 \int_\Omega v_{x_i} dg_i(v_{x_i})dx, \quad i = 1,2. \tag{9.2.11}$$

Similarly, because $g_0(v)v^{-1} \le C(1 + vg_0(v))$, we have that

$$\begin{aligned}\left|\int_\Omega d_0(x)g_0(v)udx\right| &\le \int_\Omega d_0 v g_0(v)dx + \frac{1}{4}\int_\Omega d_0 \left|\frac{g_0(v)}{v}\right| |u|^2 dx \\ &\le \int_\Omega d_0 v g_0(v)dx + \frac{1}{4}\max_{\overline{\Omega}} |u|^2 \int_\Omega d_0 \left|g_0(v)v^{-1}\right| dx \\ &\le \int_\Omega d_0 v g_0(v)dx + C\|u\|_2^2 \int_\Omega (1 + d_0 v g_0(v))dx. \quad (9.2.12)\end{aligned}$$

Property (8.4.3) easily follows from (9.2.11) and (9.2.12).

Now we consider the force $F(u)$. The Frechet derivative of $F(u)$ has the form

$$\langle F'(u); w\rangle = [w, v(u) + F_0] + [u + f, v(u,w)], \quad w \in H^2(\Omega),$$

where $v(u,w) \in H_0^2(\Omega)$ solves the problem

$$\Delta^2 v + 2[u + f, w] = 0, \quad x \in \Omega, \quad v|_{\partial\Omega} = \left.\frac{\partial v}{\partial n}\right|_{\partial\Omega} = 0.$$

Therefore properties (8.4.4) and (8.4.5) of the Frechét derivative of $F(u)$ easily follow from Theorem 1.4.3 and Corollary 1.4.5 concerning the von Karman bracket.

□

9.2.2 Finite dimension

9.2.6. Theorem. *Assume the hypotheses of Section* 9.1 *which guarantee the existence of a compact global attractor A for the dynamical system* (H, S_t) *generated by problem* (9.1.1) *and* (9.1.2) *with one of the boundary conditions* (9.1.3), (9.1.4), *or* (9.1.5) *hold. In addition, we assume that*

(i) $g_0(s)$ *is of polynomial growth at infinity; that is, there exist* $p_0 \ge 1$ *and* $M_0 > 0$ *such that*

$$0 \le m_0 \le g_0'(s) \le M_0[1 + |s|^{p_0 - 1}], \quad s \in \mathbb{R}, \tag{9.2.13}$$

where $m_0 > 0$ *in the case of free boundary conditions.*

(ii) *There exists* $0 \le \gamma < 1$ *such that the functions* g_i *satisfy the inequality*

$$0 < m \le g_i'(s) \le M[1 + sg_i(s)]^\gamma, \quad s \in \mathbb{R},\ i = 1,2, \tag{9.2.14}$$

Then the fractal dimension of the attractor $\mathscr{A}$ *is finite.*

9.2.7. Remark. As one can see from the argument below, instead of (9.2.13) we can assume that (9.2.14) holds for $i = 0$ with $\gamma = 1$.

It is also clear that the right inequality in (9.2.14) holds true if we assume that

$$g_i'(s) \le M'\left(1+|s|^{p-1}\right) \quad \text{and} \quad sg_i(s) \ge m''|s|^{(p-1)/\gamma} - M'' \quad \text{for all } s \in \mathbb{R}, \quad (9.2.15)$$

for some positive constants M', M'' and m''. We also note that the second requirement in (9.2.15) follows from (9.1.8) if $1 \le p \le 1+2\gamma$. In particular, the property $g_i'(s) \le M(1+sg_i(s))^\gamma$ for some $\gamma < 1$ holds true, if, in addition to (9.1.8), we assume that either $p < 3$ or else $p \ge 3$ and $g_i(s)s \ge m_1|s|^l$ for all $|s| \ge 1$ and for some $m_1 > 0$ and $l > p-1$. In addition, the condition (9.2.14) with $\gamma = 1$ allows also an exponential behavior for $g_i(s)$, for example, $|g_i(s)| \sim e^{\kappa|s|}$ as $|s| \to \infty$ for some $\kappa > 0$. ■

To prove the finiteness of fractal dimension of A we need the following Lemma.

9.2.8. Lemma. *Let* (9.1.10) *hold and* $d_0(x) \ge d_0 > 0$ *in the case of free boundary conditions. Let* $V = H_0^1(\Omega)$ *in the clamped and hinged cases and* $V = H^1(\Omega)$ *in the case of free boundary conditions. Then under Assumption* 9.1.2 *and additional hypotheses* (9.2.13) *and* (9.2.14) , *the form* $D(v,u)$ *given by* (9.1.21) *possesses the properties*

- *If* $p = 1$ *and* $g_0'(s)$ *is bounded (when* $L \ne 0$*), then*

$$|D(u+v,w) - D(u,w)| \le C_1[D(u+v,v) - D(u,v)] + C_2\|w\|_1^2 \qquad (9.2.16)$$

for any $v \in V$, $u,w \in H^1(\Omega)$.
- *If* $1 \le p < \infty$ *(when* $L = 0$*), then*

$$\begin{aligned} |D(u+v,w) - D(u,w)| &\le C_1[D(u+v,v) - D(u,v)] \qquad (9.2.17)\\ &+ C_2\|w\|_s^2\left[1+\|u\|_1^{2p_0-2} + \|v\|_1^{2p_0-2} + D(u+v,u+v) + D(u,u)\right] \end{aligned}$$

for any $v \in V$, $u \in H^1(\Omega)$, $w \in H^2(\Omega)$, *where* $s \in (1,2)$.

Proof. It is easy to see that in the case $L \ne 0$,

$$|D(u+v,w) - D(u,w)| \le C\|v\|_1\|w\|_1.$$

Thus from the lower bound for $g_i'(s)$ we obtain (9.2.16).

Let us prove (9.2.17). Property (9.2.14) implies the relation

$$m \le \frac{g_i(s_2) - g_i(s_1)}{s_2 - s_1} \le M(1+s_1g_i(s_1) + s_2g_i(s_2))^\gamma \quad \forall s_1 < s_2,\ i = 1,2,$$

where $0 \le \gamma < 1$. Therefore, for arbitrary $\delta > 0$ we have

$$\begin{aligned} \int_\Omega d|g_i((u+v)_{x_i}) - g_i(u_{x_i})||w_{x_i}|dx &\le \delta\int_\Omega d[g_i((u+v)_{x_i}) - g_i(u_{x_i})]\cdot|v_{x_i}|dx\\ &+ C_\delta\int_\Omega d(1+(u+v)_{x_i}g_i((u+v)_{x_i}) + u_{x_i}g_i((u)_{x_i}))^\gamma\cdot|w_{x_i}|^2dx \end{aligned}$$

$$\leq \delta[D(u+v,v)-D(u,v)]+C_\delta\|w\|_s^2\,[1+D(u+v,u+v)+D(u,u)]$$

with some $1 < s < 2$. Because $g_0(s)$ is of polynomial growth (see (9.2.13)), we also have

$$\int_\Omega d_0|g_0(u+v)-g_0(u)||w|dx \leq C\|w\|_s \cdot \int_\Omega \left(1+|u|^{p_0-1}+|v|^{p_0-1}\right)|v|dx$$
$$\leq C\|w\|_s\|v\|\,(1+\|u\|_1+\|v\|_1)^{p_0-1}, \quad s>1.$$

By (9.2.14) (using the Friedrichs inequality in the clamped and hinged cases) we obtain that

$$\|v\|^2 \leq C[D(u+v,v)-D(u,v)], \quad v\in V,\ u\in H_0^1(\Omega).$$

Therefore it is easy to obtain (9.2.17). □

Finite dimension of the attractor: Under the assumptions of Theorem 9.2.6 from Lemma 9.2.8 we have that (8.3.12) holds with $\varkappa = 2$. Therefore application of Theorem 8.6.3 establishes the finiteness of fractal dimension of the attractor under any boundary conditions considered.

9.2.3 Smoothness of elements from the attractor

The following assertion is a straightforward consequence of Theorem 8.7.1.

9.2.9. Theorem. *Assume the hypotheses of Section* 9.1 *that guarantee the existence of a compact global attractor A for the dynamical system* (H,S_t) *generated by problem* (9.1.1) *and* (9.1.2) *with one of the boundary conditions* (9.1.3), (9.1.4), *or* (9.1.5) *hold. If, in addition, we assume that (i)* $g_0(s)$ *is of polynomial growth at infinity;*[3] *that is, i.e. there exist* $p_0 \geq 1$ *and* $M_0 > 0$ *such that* (9.2.13) *holds, and (ii) there exists* $0 \leq \gamma < 1$ *such that the functions* g_i *satisfy the inequality* (9.2.14), *then any full trajectory* $\gamma = \{(u(t);u_t(t)) : t\in\mathbb{R}\}$ *from the attractor A of the dynamical system* (H,S_t) *generated by problem* (9.1.1) *and* (9.1.2) *with one of the boundary conditions* (9.1.3), (9.1.4), *or* (9.1.5), *possesses the properties*

$$u(t)\in C_r(\mathbb{R};W), \quad (u_t(t),u_{tt}(t))\in C_r(\mathbb{R};H), \tag{9.2.18}$$

where C_r *means right-continuous functions and*

- $H = H_0^2(\Omega)\times H_0^1(\Omega)$ *and* $W = H^3(\Omega)\cap H_0^2(\Omega)$ *for the clamped boundary conditions*(9.1.3).
- $H = (H^2\cap H_0^1)(\Omega)\times H_0^1(\Omega)$ *and* $W = \{w\in H^3(\Omega) : u|_{\partial\Omega} = \Delta u|_{\partial\Omega} = 0\}$ *in the hinged case* (9.1.4).

---

[3] Instead of a polynomial growth condition we can assume that $g_0(s)$ satisfies (9.2.14) with $\gamma = 1$, cf. Remark 9.2.7(1).

- $H = H^2(\Omega) \times H^1(\Omega)$ *and* $W = \{w \in H^3(\Omega) : \Delta u + (1-\mu)B_1 u|_\Gamma = 0\}$ *in the free case* (9.1.5).

Moreover,

$$\sup_{t \in \mathbb{R}} \left(\|u(t)\|_3^2 + \|u_t(t)\|_2^2 + \|u_{tt}(t)\|_1^2 \right) \leq C_A < \infty.$$

In particular we have that $A \subset H^3(\Omega) \times H^2(\Omega)$ *for each case.*

Proof. This is a consequence of Theorem 8.7.1. □

A natural question to ask is whether one can obtain even more regularity of the attractor, provided, of course, that the parameters defining the dynamical system are smooth enough. To answer this question one can use the method involving the reiteration of the procedure used for the first boost of regularity stated in the last part of Theorem 9.2.9. For simplicity of exposition, we reproduce the argument in the case of linear damping, where the regularity of the damping does not impede the iteration process. In fact, in that case we are able to show that the attractor may possess C^∞ smoothness. For possible generalizations to nonlinear damping we refer to Remark 9.2.12 below.

9.2.10. Theorem. *Assume that* $p \in L^2(\Omega)$, $f \in H^2(\Omega)$, *and* $F_0 \in H^3(\Omega)$ ($F_0 \in H^3(\Omega) \cap H_0^2(\Omega)$ *and* $\nu_1 > 0$ *in the case of free boundary conditions). Assume also* $d_0(x)$ *and* $d(x)$ *are continuous functions satisfying* (9.1.10) *and*

$$g_i(v) = g_i \cdot v, \quad i = 0,1,2, \; v \in \mathbb{R}, \tag{9.2.19}$$

where $g_i > 0$. *In the free case we also suppose that* $d_0(x) \geq d_0 > 0$. *Then the dynamical system* (H, S_t) *generated by problem* (9.1.1) *and* (9.1.2) *with one of the boundary conditions* (9.1.3), (9.1.4), *or* (9.1.5), *has a compact global attractor* A *and any full trajectory* $\gamma = \{(u(t); u_t(t)) : t \in \mathbb{R}\}$ *from the attractor* A *possesses the property*

$$u^{(k)}(t) \in C(\mathbb{R}; W), \quad k = 0,1,\ldots, \tag{9.2.20}$$

where W *is the same as in Theorem* 9.2.9, $u^{(0)}(t) = u(t)$, *and* $u^{(k)}(t) = \partial_t^k u(t)$ *for* $k = 1,2,\ldots$ *Moreover,*

$$\sup_{t \in \mathbb{R}} \|u^{(k)}(t)\|_3 \leq C_k(A) < \infty \quad \text{for each} \quad k = 0,1,\ldots.$$

Proof. We apply Theorem 8.7.4. It is clear that $F(u) \in C^\infty$ and relations (8.7.12) hold for every $m \geq 1$. □

In particular Theorem 9.2.10 means that the global attractor A belongs to the space $H^3(\Omega) \times H^3(\Omega)$. The following assertion provides conditions that guarantee $A \subset H^m(\Omega) \times H^m(\Omega)$ for $m \geq 4$.

9.2.11. Corollary. *Let the hypotheses of Theorem* 9.2.10 *be valid and* $m \geq 4$. *Assume that* $d_0(x)$ *and* $d(x)$ *are* C^∞*-functions and*

$$p + [f, F_0] \in H^{m-4}(\Omega), \quad f \in H^{m-2}(\Omega), \quad F_0 \in H^{m-2}(\Omega), \tag{9.2.21}$$

Let L be continuous from $H^{l-1}(\Omega)$ into $H^{l-4}(\Omega)$ for all $4 \le l \le m$. Then any full trajectory $\gamma = \{(u(t); u_t(t)) : t \in \mathbb{R}\}$ from the attractor A of the system (H, S_t) possesses the property

$$u^{(k)}(t) \in C(\mathbb{R}; H^m(\Omega)), \quad k = 0, 1, \ldots \tag{9.2.22}$$

In addition, the global attractor A is a bounded set in $H^m(\Omega) \times H^m(\Omega)$. If p, f, and F_0 are C^∞-functions, then $A \subset C^\infty(\overline{\Omega}) \times C^\infty(\overline{\Omega})$.

Proof. We rely on elliptic regularity of stationary von Karman equations, along with inductive argument. The idea is the same as in Section 3.1.4; see also [123]. For definiteness we consider the case of clamped boundary conditions (9.1.3).

From equation (9.1.1) we have that

$$\Delta^2 u^{(k)}(t) = -(1 - \alpha\Delta) u^{(k+2)}(t) - d_0 g_0 u^{(k+1)}(t) + \sum_{i=1,2} g_i \left(d u_{x_i}^{(k+1)}(t) \right)_{x_i}$$
$$+ \langle F'(u(t)); u^{(k)}(t) \rangle + G_k(t),$$

where

$$\langle F'(u); w \rangle = [w, v(u) + F_0] + 2[u + f, v(u + f, w)] - Lw.$$

We also have that $G_1 \equiv 0$ and $G_k(t) \equiv G_k(u(t), u'(t), \ldots, u^{(k-1)}(t))$ for $k \ge 2$ is a linear combination of functions of the form $[v(w_1, w_2), w_3]$, where w_j are either $u + f$ or one of the derivatives $u^{(i)}(t)$, $1 \le i \le k-1$, and the value $v = v(w_1, w_2) \in H_0^2(\Omega)$ is determined as the solution to the problem

$$\Delta^2 v + [w_1, w_2] = 0, \quad v|_{\partial\Omega} = \frac{\partial v}{\partial n}\bigg|_{\partial\Omega} = 0.$$

for given w_1 and w_2 from $H^2(\Omega)$.

Consequently, using Lemma 4.1.26 and relation (4.1.57) we can conclude that for $l \ge 2$ and $k \ge 1$ we have

$$\|\Delta^2 u^{(k)}(t)\|_{l-2} \le C\Big\{ \|u^{(k+2)}\|_l + \|u^{(k+1)}\|_l + \|u^{(k)}\|_l \left[\|u\|_l \|u + 2f\|_l + \|u + f\|_l^2 \right]$$
$$+ \|u^{(k)}\|_{l+1} \left[\|F_0\|_l + \|L\|_{\mathscr{L}(H^{l+1}, H^{l-2})} \right] \Big\} + \psi(\|u + f\|_l, \|u^{(j)}\|_l;\ j = 1, \ldots, k-1),$$

where ψ is a third-order polynomial of its arguments. For $l \ge 2$ and $k = 0$ we also have

$$\|\Delta^2 u(t)\|_{l-2} \le C\Big\{ \|u''\|_l + \|u'\|_l + \|u\|_l \|u + 2f\|_l \|u + f\|_l$$
$$+ \|u\|_{l+1} \left[\|F_0\|_l + \|L\|_{\mathscr{L}(H^{l+1}, H^{l-2})} \right] + \|p + [f, F_0]\|_{l-2} \Big\}.$$

Therefore by the elliptic regularity for all $l \ge 2$ and $k \ge 0$ we have that

$$\|u^{(k)}(t)\|_{l+2} \le \psi(\|f\|_l, \|u^{(j)}\|_l;\ j=0,1,2,\dots,k+2)$$
$$+C\Big\{\|u^{(k)}\|_{l+1}\Big[\|F_0\|_l + \|L\|_{\mathscr{L}(H^{l+1},H^{l-2})}\Big] + k\cdot\|p+[f,F_0]\|_{l-2}\Big\}.$$

By induction in l we conclude the proof of Corollary 9.2.11. □

9.2.12. Remark. The linear damping argument in Theorem 9.2.10 and Corollary 9.2.11 can be easily extended to the case when the rotational damping $\operatorname{div} g(\nabla u_t)$ is linear and the viscous damping $g_0(s)$ is a C^∞ (nonlinear) function satisfying Assumption 9.1.2 and such that

$$|g_0^{(k)}(s)| \le C_k(1+|s|^{p_{0,k}}), \quad k=0,1,2,\dots,$$

where $p_{0,k} < \infty$. In the free boundary condition case we also need the property $g_0'(s) \ge m > 0$. The corresponding argument also relies on the H^1-regularity of velocities u_t and ideas developed in the proof of Theorem 3.1.28. Moreover, in the case $L \equiv 0$ we can avoid the growth condition assumed above for $g_0^{(k)}(s)$ by assuming that $g_0(s)$ satisfies (9.2.14) with $\gamma = 1$. This can be accomplished by applying the method developed in Section 9.5.2 below where a more challenging case of nonrotational model is considered. The case of fully nonlinear damping, although phenomenologically similar, is technically more demanding and requires additional growth conditions imposed on the rotational damping. The reader may consult [194] in the case of a 2D wave model. ■

9.2.4 Strong attractors

The validity of quasi-stability property for the system allows us to deduce—for free—another important property of dynamical system. This is existence of strong attractors that is, attractors in a strong topology determined by the generators of the dynamical system. The corresponding result is formulated below.

9.2.13. Theorem. *Let the hypotheses of Theorem 9.2.9 be in force. Assume in addition that $g_0 \in C^2(\mathbb{R})$ with g_0'' polynomially bounded and the rotational damping is linear; that is, $g_i(s) = g_i \cdot s$ for $i = 1,2$. Then the global attractor is also strong: for any bounded set B from $W \times \widetilde{W}$ we have that*

$$\lim_{t\to\infty} \sup_{y\in B} \operatorname{dist}_{W\times\widetilde{W}}(S_t y, A) = 0,$$

where W ii the same as in Theorem 9.2.9 and $\widetilde{W} = H_0^2(\Omega)$ for clamped boundary conditions (9.1.3), $\widetilde{W} = (H^2 \cap H_0^1)(\Omega)$ in the hinged case (9.1.4), and $\widetilde{W} = H^2(\Omega)$ in the free case (9.1.5).

Proof. We apply Theorem 8.8.4. The hypotheses of Theorem 9.2.9 guarantees that the system (H, S_t) is quasi-stable; that is, satisfies an appropriate stabilizability estimate (see Theorem 8.5.6 in the subcritical case $\eta > 0$) Therefore we need only to check Assumption 8.8.3 which is simple in the case of linear rotational damping.

□

9.2.14. Remark. Whether the result of of Theorem 9.2.13 is valid with fully nonlinear damping, remains still an open question. However, in the case when rotational terms are absent, such result is proved in Section 9.5.3. We note that the nonrotational case is more complex with respect to nonlinear forcing (due to criticality) and the effect of nonlinear damping on nonrotational models is more gentle. In reiterating the argument for smoothness to higher derivatives, it is the nonlinear rotational damping that enters the system in nondissipative way. ■

9.2.5 Exponential attractor

An important property of the dynamical system is an existence of exponential attractors, which attract trajectories at the exponential rate. The quasi-stability property allows us to deduce an existence of exponential attractors. The corresponding result is formulated below.

9.2.15. Theorem. *Let Assumption* 9.1.2 *be valid and* (H,S_t) *be the dynamical system generated by problem* (9.1.1) *and* (9.1.2) *with one of the boundary conditions* (9.1.3), (9.1.4), *or* (9.1.5). *In addition we assume* (9.1.10) *in the clamped and hinged cases and Assumption* 9.1.13 *in the case of free boundary conditions. Then under conditions* (9.2.13) *and* (9.2.14) *the dynamical system* (H,S_t) *has a (generalized) fractal exponential attractor A (see Definition* 7.4.4 *and Remark* 7.4.5*) whose dimension is finite in the space* $H^1(\Omega) \times W'$, *where* W' *is a completion of either* $H_0^1(\Omega)$ *(in the clamped and hinged cases) or* $H^1(\Omega)$ *(in the case of the free b.c.) with respect to the norm* $\|\cdot\|_{W'} = \|(1-\alpha\Delta)\cdot\|_{-2}$.

Proof. We apply Theorem 8.10.1 which assumes the hypotheses of Theorem 8.5.6 and also relation (8.10.2). The latter is obviously true in the rotational (subcritical) case with $l = \frac{1}{2}$ in relation (8.10.2). □

9.2.16. Remark. The question of whether the global attractor is exponential is still open in the case of general damping. In the case when the damping is linear and $L \equiv 0$ we can apply Theorem 7.5.15 to obtain exponential convergence of bounded sets to the global attractor for the system (9.1.1) and (9.1.2). ■

9.2.6 Determining functionals

As in abstract theory (see Section 7.8) characterization of determining functionals is based on the notion of completeness defect (see Definition 7.8.5).

Our first result is based on Theorem 8.9.3 and deals with linearly bounded damping.

9.2.17. Theorem. *Let Assumption* 9.1.2 *be in force. Assume that*

$$0 \le m_i \le g_i'(s) \le M_i \ \ for\ i = 0,1,2, \quad with\ m_1, m_2 > 0$$

and $d_0(x)$ *and* $d(x)$ *satisfy* (9.1.10). *In free case we also suppose that* $m_0 > 0$ *and* $d_0(x) \ge d_0 > 0$. *Let* $\mathscr{L} = \{l_j : j = 1, \dots, N\}$ *be a set of the linear continuous functionals on the space* $H^2(\Omega)$ *with the completeness defect* $\varepsilon_{\mathscr{L}} = \varepsilon_{\mathscr{L}}(H^2(\Omega), L_2(\Omega))$. *Then there exists* $\varepsilon_0 > 0$ *such that under the condition* $\varepsilon_{\mathscr{L}} \le \varepsilon_0$ *the set* $\mathscr{L}$ *is a set of determining functionals for the dynamical system* (H, S_t) *generated by problem* (9.1.1) *and* (9.1.2) *with one of the boundary conditions* (9.1.3), (9.1.4), *or* (9.1.5); *that is, for any two generalized solutions* $u_1(t)$ *and* $u_2(t)$ *to problem* (9.1.1) *and* (9.1.2) *with the corresponding boundary conditions the relation*

$$\lim_{t\to\infty} \int_t^{t+1} |l(u_1(\tau)) - l(u_2(\tau))|^2 d\tau = 0 \ \ for\ every\ l \in \mathscr{L} \tag{9.2.23}$$

implies

$$\lim_{t\to\infty} \{ \|u_1(t) - u_2(t)\|_2^2 + \|\partial_t u_1(t) - \partial_t u_2(t)\|_1^2 \} = 0. \tag{9.2.24}$$

Proof. We apply Theorem 8.9.3. This theorem requires some boundedness condition concerning the damping operator; see (8.5.36). In our case this condition obviously holds true due to linear bounds imposed[4] on damping functions. □

The following assertion provides an example of an application of Theorem 9.2.17.

9.2.18. Corollary. *Under the hypotheses of Theorem* 9.2.17 *there exists a finite set* $\{x_k : k = 1, \dots, N\}$ *of nodes in* Ω *such that for any two generalized solutions* $u_1(t)$ *and* $u_2(t)$ *to problem* (9.1.1) *and* (9.1.2) *with one of the boundary conditions* (9.1.3), (9.1.4), *or* (9.1.5) *the relation in* (9.2.24) *holds provided*

$$\lim_{t\to\infty} \int_t^{t+1} |u_1(\tau, x_k) - u_2(\tau, x_k)|^2 d\tau = 0 \ \ for\ every\ k = 1, \dots, N. \tag{9.2.25}$$

Proof. It is clear $\mathscr{L} = \{l(u) = u(x_k) : k = 1, \dots, N,\}$ is a set of continuous functionals on $H^2(\Omega)$. By Theorem 3.2 from [60] (see also Example 7.8.11) we can choose $\{x_k\}$ in Ω such that $\varepsilon_{\mathscr{L}}(H^2(\Omega), L_2(\Omega)) \le c_\Omega N^{-1}$. The application of Theorem 9.2.17 gives the final result. □

For other examples of sets of functionals on $H^2(\Omega)$ that possess a small completeness defect we refer to Section 7.8.3.

In the case of more general nonlinear damping we can apply the corollaries of Theorem 8.9.1 and Theorem 8.9.6.

9.2.19. Theorem. *Let the hypotheses of Theorem* 9.2.6 *be valid and* (H, S_t) *be the dynamical system generated by problem (9.1.1) and (9.1.2) with one of the boundary conditions* (9.1.3), (9.1.4), *or* (9.1.5). *Assume that* $\mathscr{L} = \{l_j : j = 1, \dots, N\}$ *is a set of*

[4] We note that this requirement concerning $g_i(s)$ is not optimal and can be relaxed.

the linear continuous functionals on the space $H^2(\Omega)$ with the completeness defect $\varepsilon_{\mathscr{L}} = \varepsilon_{\mathscr{L}}(H^2(\Omega), L_2(\Omega))$. Then there exists $\varepsilon_0 > 0$ such that under the condition $\varepsilon_{\mathscr{L}} \le \varepsilon_0$ the set $\mathscr{L}$ is a set of determining functionals for (H, S_t) in the following sense: for any two generalized solutions $u_1(t)$ and $u_2(t)$ to problem (9.1.1) *and* (9.1.2) *with the corresponding boundary conditions the relation*

$$\lim_{t\to\infty} |l(u_1(t)) - l(u_2(t))| = 0 \ \text{for every}\ l \in \mathscr{L} \tag{9.2.26}$$

implies (9.2.24).

Proof. Lemma 9.2.8 implies that the hypotheses of Theorem 8.6.3 hold. Therefore we can apply Corollary 8.9.2. □

Another result is based on Theorem 8.9.6.

9.2.20. Theorem. *Let $L \equiv 0$ and Assumption* 9.1.2 *and relations* (9.1.10) *be valid. Assume that $g_i(s)$ are increasing for $i = 1, 2$. In the case of free boundary conditions we also suppose that Assumption* 9.1.13 *holds, $g_0(s)$ is increasing, and $d_0(x) \ge d_0 > 0$. Let $u(t)$ be a solution to problem* (9.1.1) *and* (9.1.2) *with one of the boundary conditions* (9.1.3), (9.1.4), *or* (9.1.5) *and $w \in H^2(\Omega)$ be a solution to stationary problem* (9.2.1) *and* (9.2.2) *with the corresponding boundary conditions. Assume that $\mathscr{L} = \{l_j : j = 1, \dots, N\}$ is a set of linear continuous functionals on the space $H^2(\Omega)$ with the completeness defect $\varepsilon_{\mathscr{L}} = \varepsilon_{\mathscr{L}}(H^2(\Omega), L_2(\Omega))$. Then there exists $\varepsilon_0 > 0$ such that under the condition $\varepsilon_{\mathscr{L}} \le \varepsilon_0$ the relation*

$$\lim_{t\to\infty} l(u(t)) = l(w) \ \text{for every}\ l \in \mathscr{L} \tag{9.2.27}$$

implies that

$$\lim_{t\to\infty} \left\{ \|u(t) - w\|_2^2 + \|\partial_t u(t)\|_1^2 \right\} = 0. \tag{9.2.28}$$

Proof. By relation (9.1.25) and Lemma 9.2.3, Assumption 8.9.5(D) holds with some $H(s)$. Thus we need only to check Assumption 8.9.5(F) for the mapping $F(u)$ given by the formula

$$F(u) = [u + f, v(u) + F_0] + p,$$

where as above $v = v(u)$ solves (9.1.2). We have that

$$\begin{aligned}
&(F(w+u) - F(w + z\cdot u), u) \\
&= ([w + u + f, v(w+u) + F_0] - [w + z\cdot u + f, v(w + z\cdot u) + F_0], u) \\
&= (1 - z)\left[([u,u], v(w+u)) + ([u, F_0], u)\right] \\
&\quad + ([w + z\cdot u + f, u], v(w+u) - v(w + z\cdot u))
\end{aligned}$$

for any $z \in [0,1]$ and $u, w \in H^2(\Omega)$. We can assume that $\|u\|_2 \le R$ and $\|w\|_2 \le R$, where R is the radius of the absorbing ball. Thus applying the estimates for the von Karman bracket given in Theorem 1.4.3 we obtain that

$$([u,u], v(w+u)) + ([u, F_0], u) \le C_R \|u\|_{3/2}^2$$

and

$$
\begin{aligned}
&([w+z\cdot u+f,u],v(w+u)-v(w+z\cdot u))\\
&\le C\|u\|_1\cdot\|w+z\cdot u+f\|_2\cdot\|v(w+u)-v(w+z\cdot u)\|_2\le C_R(1-z)\|u\|_1^2.
\end{aligned}
$$

Thus have that

$$
(F(w+u)-F(w+z\cdot u),u)\le C_R(1-z)\|u\|_{3/2}^2 \quad \text{for } z\in[0,1],\quad u,w\in H^2(\Omega),
$$

where C_R is a constant depending on the radius R of the absorbing ball. This implies Assumption 8.9.5(F) for the case considered. □

9.2.7 Approximate inertial manifolds

Here we consider the following problem with linear damping,

$$
\begin{cases}
(1-\alpha\Delta)u_{tt}+k(1-\alpha\Delta)u_t\\
\qquad +\Delta^2u-[u+f,v+F_0]+L(u)=p(x), \quad x\in\Omega,\, t>0,\\
u|_{t=0}=u_0(x), \quad u_t|_{t=0}=u_1(x),
\end{cases}
\tag{9.2.29}
$$

where $v=v(u)$ is a solution of the problem (9.1.2). We assume that the parameters α and k are positive and the displacement $u(x,t)$ is subjected to one of the boundary conditions (9.1.3), (9.1.4), or (9.1.5).

This problem can be written as a first-order equation of the form

$$
\frac{d}{dt}\begin{pmatrix}u\\ u_t\end{pmatrix}+\begin{pmatrix}0 & -1\\ M^{-1}\mathscr{A} & k\end{pmatrix}\begin{pmatrix}u\\ u_t\end{pmatrix}=\begin{pmatrix}0\\ M^{-1}F(u)\end{pmatrix}
\tag{9.2.30}
$$

with the corresponding choice of spaces $\mathscr{H}$ and V and operators M, $\mathscr{A}$ and $F(\cdot)$ (see their description in Section 9.1 for all types of the boundary conditions).

To construct an inertial manifold (see Definition 7.6.1) for equation (9.2.30) we can use Theorem 7.6.3. However, to check hypotheses of this theorem we need to establish some bounds for nonlinearity and also some spectral gap condition for the operator $M^{-1}\mathscr{A}$. In fact this spectral gap condition means that the distance between neighboring eigenvalues of $M^{-1}\mathscr{A}$ increases. We cannot establish this property for the system in (9.2.29) (see Chapter 13 for the corresponding analysis). However, as in Sections 7.7 and 8.11 we can construct approximate inertial manifolds for the case considered. Indeed, application of the results given in Section 8.11 allow us to obtain the following assertion easily.

9.2.21. Theorem. *Assume that* $p \in L^2(\Omega)$, $f \in H^2(\Omega)$, *and* $F_0 \in H^3(\Omega)$ *(*$F_0 \in H^3(\Omega) \cap H_0^2(\Omega)$ *and* $\nu_1 > 0$ *in the case of free boundary conditions). Then the statements of Theorems* 8.11.1 *and* 8.11.2 *hold for system* (9.2.29) *and* (9.1.2) *with the corresponding choice of spaces* $\mathscr{H}$ *and* V *and operators* M, $\mathscr{A}$ *and* $F(\cdot)$.

For further details concerning approximate inertial manifolds for von Karman models with rotational inertia we refer to the paper [57].

9.3 Attractors for other models with rotational inertia

In this section we prove the existence and study the properties of global attractors for other variants of von Karman evolutions. These include: (i) time-delayed models and (ii) quasi-static version of these equations (for well-posedness results for these models see Section 3.3).

9.3.1 Von Karman equations with retarded terms

We first consider the following von Karman system with (linear) retarded terms and restrict ourselves by the case of linear damping:

$$(1-\alpha\Delta)u_{tt} + (d_1 - \alpha d_2\Delta)u_t + \Delta^2 u = [u+f, v+F_0] - Lu + p(u^t;t), \quad x \in \Omega, t > 0, \tag{9.3.1}$$

where, as above, $v = v(u)$ is a solution of the fourth-order elliptic problem

$$\Delta^2 v + [u+2f, u] = 0, \quad v|_{\partial\Omega} = \frac{\partial v}{\partial n}\Big|_{\partial\Omega} = 0. \tag{9.3.2}$$

We assume for the sake of definiteness that the plate is clamped; that is,

$$u|_{\partial\Omega} = \frac{\partial u}{\partial n}\Big|_{\partial\Omega} = 0. \tag{9.3.3}$$

We also assume that

$$p(u^t;t) = p_0(x) + q(u^t;t), \tag{9.3.4}$$

where $p_0(x) \in L_2(\Omega)$, u^t is an element from $L_2(-t^*,0;H^2(\Omega))$ defined by the formula $u^t(s) = u(t+s)$, $s \in (-t^*,0)$, and $q(\cdot;t)$ is a continuous linear mapping from $L^2(-t^*,0;H^2(\Omega))$ into $L_2(\Omega)$. The parameter $t^* > 0$ is the time of retardation. The retarded character of the problem requires the initial conditions of the type

$$u\big|_{t\in(-t^*,0)} = \varphi(x,t), \quad u|_{t=0} = u_0(x), \quad \partial_t u|_{t=0} = u_1(x). \tag{9.3.5}$$

9.3.1. Remark. If the plate is located in a potential linearized flow of gas moving along the x_1-axis, then the aerodynamic pressure of the flow can be taken into account (see [54, 168] and the references therein and also Section 6.6.5) by assuming that

$$p(u^t;t) = p_0(x) - \nu\left(u_t + Uu_{x_1} + q(u^t;t)\right), \tag{9.3.6}$$

where $p_0(x) \in L_2(\Omega)$ and

$$q(u^t,x,t) = \frac{1}{2\pi}\cdot\int_0^{t^*} ds \int_0^{2\pi} d\theta [M_\theta^2 \hat{u}](x_1-(U+\sin\theta)s, x_2 - s\cos\theta, t-s). \tag{9.3.7}$$

Here $\hat{u}$ is the extension of $u(x,t)$ by zero outside of Ω, $M_\theta = \sin\theta\cdot\partial_{x_1} + \cos\theta\cdot\partial_{x_2}$, and

$$t^* = \inf\{t : x(U,\theta,s) \notin \Omega \quad \text{for all} \quad x\in\Omega,\ \theta\in[0,2\pi],\ s>t\},$$

where $x(U,\theta,s) = (x_1-(U+\sin\theta)s; x_2 - s\cos\theta) \in \mathbb{R}^2$ and $x = (x_1;x_2) \in \Omega \subset \mathbb{R}^2$. The delay parameter $t^* > 0$ depends on the velocity of the unperturbed flow U and on the size of the domain Ω. We refer to Remark 6.6.19 for bounds for this delay parameter $t^* = t^*(U,\Omega)$. ■

Below we consider problem (9.3.1)–(9.3.5) under the following set of hypotheses.

9.3.2. Assumption.

- The requirement of Assumption 9.1.2 concerning the operator L holds; that is, the mapping L is a linear bounded operator from $H^{2-\sigma}(\Omega)$ into $H^{-\sigma}(\Omega)$ for all $0 \le \sigma \le \sigma_0$ with some $\sigma_0 > 0$.
- The functions f, F_0, and p_0 in (9.3.1) and (9.3.4) possess properties

$$f(x) \in H^2(\Omega), \quad F_0(x) \in H^3(\Omega) \quad p_0(x) \in L_2(\Omega). \tag{9.3.8}$$

- $q(\cdot;t)$ is a continuous linear mapping from $L^2(-t^*,0;H^2(\Omega))$ into $H^{-1}(\Omega)$ possessing the property[5]

$$\| q(u^t,t) \|_{-1,\Omega}^2 \le C\int_{t-t^*}^t \| u(\tau) \|_{2-\sigma,\Omega}^2 \, d\tau \quad \text{for some} \quad \sigma > 0. \tag{9.3.9}$$

It was shown in Theorem 3.3.1 that under Assumption 9.3.2 for any initial data

$$\varphi(x,t) \in, L_2(-t^*,0;H_0^2(\Omega)), \quad u_0(x) \in H_0^2(\Omega), \quad u_1(x) \in H_0^1(\Omega) \tag{9.3.10}$$

problem (9.3.1)–(9.3.5) admits a unique weak solution $u(t)$ such that

$$u(t) \in L_2(-t^*,T;H_0^2(\Omega)) \cap C(0,T;H_0^2(\Omega)) \cap C^1(0,T;H_0^1(\Omega)) \tag{9.3.11}$$

for any $T > 0$. This weak solution satisfies the energy relation

$$\mathscr{E}(u(t),u_t(t)) + \int_s^t \int_\Omega \left[d_0 u_t^2 + \alpha d_1 |\nabla u_t|^2\right] dx d\tau$$

[5] By Lemma 12.4.2 this property holds in the case considered in Remark 9.3.1 with $\sigma = 1$.

$$= \mathscr{E}(u(s),u_t(s)) + \int_s^t \int_\Omega (-Lu + p(u^\tau;\tau))u_t dx d\tau.$$

Here

$$\mathscr{E}(u,u_t) = \frac{1}{2}\int_\Omega \left[\alpha|\nabla u_t|^2 + |u_t|^2 - [u+2f,F_0|u]\right] dx + \Pi_*(u) \tag{9.3.12}$$

with

$$\Pi_*(u) = \frac{1}{2}\int_\Omega \left[|\Delta u|^2 + \frac{1}{2}|\Delta v(u)|^2\right] dx, \tag{9.3.13}$$

where $v(u) \in H_0^2(\Omega)$ is determined from (9.3.2). Moreover, there exist constants $C > 0$ and $a \geq 0$ such that

$$E(u(t),u_t(t)) \leq C\left(1 + E(u_0,u_1) + \int_{-r}^0 \|\varphi(\tau)\|_2^2 d\tau\right) \cdot e^{at}, \quad t > 0, \tag{9.3.14}$$

where

$$E(u,u_t) = \frac{1}{2}\left[\|u_t\|^2 + \alpha\|\nabla u_t\|^2\right] + \Pi_*(u) \tag{9.3.15}$$

with $\Pi_*(u)$ given by (9.3.13).

9.3.3. Proposition. *Let $u^1(t)$ and $u^2(t)$ be two weak solutions to problem* (9.3.1)–(9.3.4) *with different initial data. Assume that*

$$\|u_t^j(t)\|^2 + \alpha\|\nabla u_t^j(t)\|^2 + \|\Delta u^j(t)\|^2 \leq R^2, \quad j = 1,2, \tag{9.3.16}$$

for some R and for all $t \in [0,T]$, where $T \leq \infty$. Then there exist $C_R > 0$ and $a_R \geq 0$ such that

$$\begin{aligned}&\|u_t^1(t) - u_t^2(t)\|^2 + \alpha\|\nabla(u_t^1(t) - u_t^2(t))\|^2 + \|\Delta(u^1(t) - u^2(t))\|^2 \\ &\leq C_R\Big(\|u_1^1 - u_1^2\|^2 + \alpha\|\nabla(u_1^1 - u_2^2)\|^2 + \|\Delta(u_0^1 - u_0^2)\|^2 \\ &\quad + \int_{-t^*}^0 \|\varphi^1(\tau) - \varphi^2(\tau)\|_2^2 d\tau\Big) e^{a_R t} \quad \textit{for all } t \in [0,T].\end{aligned}$$

Proof. The difference $u(t) = u^1(t) - u^2(t)$ is a solution to the problem

$$(1 - \alpha\Delta)u_{tt} + (d_1 - \alpha d_2\Delta)u_t + \Delta^2 u = F(t,u^1,u^2), \quad x \in \Omega, t > 0, \tag{9.3.17}$$

with the clamped boundary conditions (9.3.3), where

$$F(t,u^1,u^2) = [u, v(u^1) + F_0] + [u^2 + f, v(u^1) - v(u^2)] - L(u) + p(u^t;t). \tag{9.3.18}$$

Using relation (9.3.9), Theorem 1.4.3, and Corollary 1.4.5 we have the estimate

$$\|F(t,u^1,u^2)\|_{-1} \leq C\Big(\|u\|_{2-\sigma} \cdot \|v(u^1) + F_0\|_{2+\sigma} + \|u^2 + f\|_2 \cdot \|v(u^1) - v(u^2)\|_2$$

$$+\|u\|_{2-\sigma}+\left[\int_{t-t^*}^{t}\|u(\tau)\|_{2-\sigma}^2 d\tau\right]^{1/2}\Bigg)$$

for some positive σ. Therefore under condition (9.3.16) we obtain that

$$\|F(t,u^1,u^2)\|_{-1} \le C_R\|u(t)\|_{2-\sigma}+C\left[\int_{t-t^*}^{t}\|u(\tau)\|_{2-\sigma}^2 d\tau\right]^{1/2} \quad (9.3.19)$$

with $\sigma > 0$. In particular, this relation implies that

$$\int_0^t \|F(\tau,u^1,u^2)\|_{-1}^2 d\tau \le C_R(1+t^*)\int_0^t \|u(\tau)\|_2^2 d\tau + Ct^*\int_{-t^*}^{0}\|u(\tau)\|_2^2 d\tau.$$

Therefore the energy equality for problem (9.3.17) and Gronwall's lemma implies the conclusion desired. □

It follows from (9.3.11) and from Proposition 9.3.3 that problem (9.3.1)–(9.3.5) generates dynamical system $(\mathbf{H}, S_t)$ in the phase space

$$\mathbf{H} = H_0^2(\Omega)\times H_0^1(\Omega)\times L_2(-t^*,T;H_0^2(\Omega))$$

with

$$S_t(u_0;u_1,\varphi) = (u(t);u_t(t);u^t), \quad t>0,$$

Here $u(t)$ is a weak solution to (9.3.1)–(9.3.5) and u^t is an element from the space $L_2(-t^*,0;H^2(\Omega))$ defined by the formula $u^t(s) = u(t+s)$, $s\in(-t^*,0)$. Moreover, the evolution operator S_t possesses properties

$$\|S_t y_0\|_{\mathbf{H}} \le C_R e^{at} \quad \text{and} \quad \|S_t y_1 - S_t y_2\|_{\mathbf{H}} \le C_R e^{a_R t}\|y_1-y_2\|_{\mathbf{H}}, \quad (9.3.20)$$

for any $y_i\in\mathbf{H}$ such $\|y_i\|_{\mathbf{H}}\le R$, $i=0,1,2$, where R is arbitrary, and C_R, a_R, a are positive numbers.

9.3.4. Theorem. *The system $(\mathbf{H}, S_t)$ is dissipative. Moreover, there exists a positively invariant bounded absorbing set.*

Proof. For the trajectories of the system $(\mathbf{H}, S_t)$ we consider the following Lyapunov-type function

$$V(S_t y) = \mathscr{E}(u(t),u_t(t)) + \nu\left\{((1-\alpha\Delta)u_t,u)+\frac{1}{2}(d_1-\alpha d_2\Delta)u,u)\right\}$$
$$+\mu\int_0^{t^*} ds\int_{t-s}^{t}\Pi_*(u(\tau))d\tau,$$

where $S_t y \equiv y(t) = (u(t);u_t(t);u^t)$ for $t\ge 0$, $\mathscr{E}(u,u_t)$ and $\Pi_*(u)$ is given by (9.3.12) and (9.3.13), and ν and μ are positive numbers.

It follows from (3.1.9) that

$$c_0E(u(t),u_t(t)) - c \le V(S_t y) \le c_1 E(u(t),u_t(t)) + \mu t^* \int_{-t^*}^{0} \Pi_*(u(t+\tau))d\tau + c$$

for all $\nu \in (0,\nu_0)$, where $c_0, c_1, c > 0$ are constant and ν_0 is small enough. After some calculation with an appropriate choice of ν and μ we can find that

$$\frac{d}{dt}V(S_t y) + \gamma V(S_t y) \le C, \quad t > 0,$$

where γ is a positive constant. This implies that

$$V(S_t y) \le V(y)e^{-\gamma t} + \frac{C}{\gamma}\left(1 - e^{-\gamma t}\right), \quad t \ge 0.$$

Therefore the set $\mathscr{B} = \left\{y \in \mathbf{H} : V(y) \le R^2 \equiv 1 + C\gamma^{-1}\right\}$ is an absorbing positively invariant bounded set. Thus $(\mathbf{H}, S_t)$ is dissipative. □

9.3.5. Theorem. *The system $(\mathbf{H}, S_t)$ has a finite-dimensional global attractor.*

The proof of this theorem relies on the compactness criterion given in Proposition 7.1.9 and on the general result on dimension stated in Theorem 7.3.3. In order to apply these results we need the following inequality, which reflects the quasi-stability property.

9.3.6. Lemma. *Let $u^1(t)$ and $u^2(t)$ be two weak solutions to problem* (9.3.1)–(9.3.4) *with different initial data. Assume that* (9.3.16) *holds for some R and for all $t > 0$. Then there exist positive constants C, μ, and C_R such that*

$$\begin{aligned} &\|u_t^1(t) - u_t^2(t)\|_1^2 + \|u^1(t) - u^2(t)\|_2^2 \\ &\le C\left(\|u_1^1 - u_1^2\|_1^2 + \|u_0^1 - u_0^2\|_2^2 + \int_{-t^*}^{0} \|\varphi^1(\tau) - \varphi^2(\tau)\|_{2-\sigma}^2 d\tau\right)e^{-\mu t} \qquad (9.3.21) \\ &+ C_R \max_{0\le\tau\le t} \|u^1(t) - u^2(t)\|_{2-\sigma}^2 \quad \text{for any } t \ge t^*. \end{aligned}$$

Proof. Because the difference $u(t) = u^1(t) - u^2(t)$ solves problem (9.3.17) with the clamped boundary conditions (9.3.3), we can write the representation

$$z(t) = T_t z(0) + \int_0^t T_{t-s}\left(0; (I - \alpha\Delta)^{-1}F(s, u^1, u^2)\right) ds,$$

where $z(t) = (u(t), u_t(t)) \in H \equiv H_0^2(\Omega) \times H_0^1(\Omega)$ and T_t is the evolution operator in H of the linear problem (9.3.17) with $F \equiv 0$. One can show (see, e.g., Proposition 8.5.12 in the autonomous and homogeneous case: $D(t) = D(0)$, $f \equiv 0$) that

$$\|T_t h\|_H \le Ce^{-\mu_0 t}\|h\|_H, \ h = (h_1; h_2) \in H,$$

where C and μ_0 are positive. Therefore we obtain

$$\|z(t)\|_H \le Ce^{-\mu_0 t}\|z(0)\| + C\int_0^t e^{-\mu_0(t-s)}\|F(s,u^1,u^2)\|_{-1}ds. \tag{9.3.22}$$

From (9.3.19) we find

$$\int_0^t e^{-\mu_0(t-s)}\|F(s,u^1,u^2)\|_{-1}ds$$

$$\le \frac{C_R}{\mu_0}\max_{0\le\tau\le t}\|u(t)\|_{2-\sigma} + \frac{C}{\sqrt{\mu_0}}\left[\int_0^t e^{-\mu_0(t-s)}\int_{s-t^*}^s \|u(\tau)\|^2_{2-\sigma}d\tau ds\right]^{1/2}.$$

Using the formula

$$\int_0^t\int_{s-t^*}^s f(s,\tau)d\tau ds = \left(\int_{t^*}^t\int_{s-t^*}^s + \int_0^{t^*}\int_0^s\right) f(s,\tau)d\tau ds + \int_0^{t^*}\int_{s-t^*}^0 f(s,\tau)d\tau ds$$

with $f(s,\tau) = e^{-\mu_0(t-s)}\|u(\tau)\|^2_{2-\sigma}$, we find

$$\int_0^t e^{-\mu_0(t-s)}\int_{s-t^*}^s \|u(\tau)\|^2_{2-\sigma}d\tau ds$$

$$\le \frac{t^*}{\mu_0}\max_{0\le\tau\le t}\|u(t)\|^2_{2-\sigma} + \frac{e^{-\mu_0(t-t^*)}}{\mu_0}\cdot\int_{-t^*}^0 \|u(\tau)\|^2_{2-\sigma}d\tau.$$

Thus

$$\int_0^t e^{-\mu_0(t-s)}\|F(s,u^1,u^2)\|_{-1}ds$$

$$\le C_{R,t^*}\max_{0\le\tau\le t}\|u(t)\|_{2-\sigma} + Ce^{-\mu_0(t-t^*)/2}\cdot\left[\int_{-t^*}^0 \|u(\tau)\|^2_{2-\sigma}d\tau\right]^{1/2}.$$

Hence (9.3.21) follows from (9.3.22). □

9.3.1.1 Proof of Theorem 9.3.5

Step 1: *existence.* It follows from (9.3.21) that

$$\|S_t y_1 - S_t y_2\|_{\mathbf{H}} \le C_1 e^{-\mu t}\|y_1 - y_2\|_{\mathbf{H}} + C_2\max_{0\le\tau\le t}\|u^1(t) - u^2(t)\|_{2-\sigma} \tag{9.3.23}$$

for any $t \ge t^*$. Here C_1, C_2, and μ are positive numbers and

$$\left\{S_t y_i = (u^i(t); u^i_t(t); (u^i)^t) : t \ge 0\right\}, \quad i = 1,2,$$

are semitrajectories that belong to the absorbing set. It is easy to see that

$$\rho_t(\{S_\tau y_1\},\{S_\tau y_2\}) \equiv \max_{0\le\tau\le t} \|u^1(t)-u^2(t)\|_{2-\sigma}$$

is a precompact semimetric in the sense of Proposition 7.1.9. Therefore the application of this proposition gives that $(\mathbf{H},S_t)$ is an asymptotically smooth dynamical system. Therefore Theorem 7.2.3 implies the existence of a compact global attractor for $(\mathbf{H},S_t)$.

Step 2: *finite dimension.* Let $T \ge t^*$ and

$$W_1(0,T)=\left\{z\in L_2(0,T;H_0^2(\Omega))\ :\ |z|^2_{W_1(0,T)}\equiv\int_0^T \left(\|z(t)\|_2^2+\|z_t(t)\|_1^2\right)dt<\infty\right\}.$$

In the space $\mathbf{H}_T=\mathbf{H}\times W_1(0,T)$ we consider the set

$$A_T:=\left\{U\equiv(u(0);u_t(0);u^0;u(t),t\in[0,T])\ :\ y=(u(0);u_t(0);u^0)\in A\right\},$$

where A is the global attractor for $(\mathbf{H},S_t)$ and $u(t)$ is the solution to problem (9.3.1)–(9.3.5) with initial data $(u(0);u_t(0);u^0)$. Here, as above, u^t is the element from $L_2(-t^*,0;H_0^2(\Omega))$ defined by the formula $u^t(s)=u(t+s)$, $s\in(-t^*,0)$.

We define operator $V\ :\ A_T\mapsto\mathbf{H}_T$ by the formula

$$V\ :\ (u(0);u_t(0);u^0;u(t))\mapsto(u(T);u_t(T);u^T;u(T+t)).$$

By (9.3.20) V is Lipschitz on A_T. It is also clear that $VA_T=A_T$. Therefore, using relation (9.3.23) as in the proof of Theorem 7.9.6 we can easily check that with an appropriate choice of $T>0$ the mapping V satisfies the hypotheses of Theorem 7.3.3. Thus we can conclude that $\dim_f A<\infty$. □

9.3.7. Remark. Relying on Theorem 7.4.2 and using the idea given in the proof of Theorem 7.9.9 we can also establish the existence of fractal exponential attractors for $(\mathbf{H},S_t)$. ■

9.3.2 Quasi-static version of von Karman equations

We consider the von Karman model with a small mass parameter μ:

$$\begin{cases}\mu(1-\alpha\Delta)u_{tt}+\kappa(1-\alpha\Delta)u_t+\Delta^2u-[u+f,v+F_0]+Lu=p(x),\\ u|_{\partial\Omega}=\left.\dfrac{\partial u}{\partial n}\right|_{\partial\Omega}=0,\quad u|_{t=0}=u_0(x),\quad u_t|_{t=0}=u_1(x),\end{cases}\tag{9.3.24}$$

where $v=v(u)$ the Airy stress function is a solution of the problem

$$\Delta^2v+[u+2f,u]=0,\quad v|_{\partial\Omega}=\left.\frac{\partial v}{\partial n}\right|_{\partial\Omega}=0.\tag{9.3.25}$$

Here μ and κ are positive parameters. We note that the change of the time variable $t \mapsto t/\sqrt{\mu}$ reduces problem (9.3.24) to the following one,

$$\begin{cases} (1-\alpha\Delta)u_{tt} + \dfrac{\kappa}{\sqrt{\mu}}(1-\alpha\Delta)u_t + \Delta^2 u - [u+f, v+F_0] + Lu = p(x), \\ u|_{\partial\Omega} = \dfrac{\partial u}{\partial n}\Big|_{\partial\Omega} = 0, \quad u|_{t=0} = u_0(x), \quad u_t|_{t=0} = \sqrt{\mu}u_1(x). \end{cases}$$

Thus we can apply the all previous result on well-posedness and long-time dynamics established above to problem (9.3.24). We also emphasize that the case of small mass parameter μ corresponds to the case of large damping.

In the case when the inertia forces are small in comparison with the resisting forces of a medium ($\mu \ll \kappa$), it is natural to consider the problem of the oscillations of a plate in quasistatic formulation (putting formally $\mu = 0$):

$$\begin{cases} \kappa(1-\alpha\Delta)u_t + \Delta^2 u - [u+f, v+F_0] + Lu = p(x), \\ u|_{\partial\Omega} = \dfrac{\partial u}{\partial n}\Big|_{\partial\Omega} = 0, \quad u|_{t=0} = u_0(x), \end{cases} \tag{9.3.26}$$

where $v = v(u)$ solves (9.3.25).

The quasi-static model is of parabolic type with strong smoothing properties resulting from the analyticity of the underlined linear semigroup. In this section we prove the existence of the global attractor of problem (9.3.26) and establish its asymptotic relation with the attractor of the problem (9.3.24) with small μ.

9.3.8. Assumption.

The mapping L is a linear bounded operator from $H^{2-\sigma}(\Omega)$ into $H^{-\sigma}(\Omega)$ for all $0 \le \sigma \le \sigma_0$ with some $\sigma_0 > 0$.

- The functions f, F_0, and p in possesses properties

$$f(x) \in H^2(\Omega), \quad F_0(x) \in H^3(\Omega), \quad p(x) \in L_2(\Omega). \tag{9.3.27}$$

It was shown in Theorem 3.3.3 that under Assumption 9.3.8 for any initial data $u_0 \in H_0^1(\Omega)$ problem (9.3.26) has a unique weak solution $u(t)$ which belongs to the class

$$C^1([0,T]; H_0^1(\Omega)) \cap C((0,T]; H_0^2(\Omega))$$

for every interval $[0,T]$. Thus problem (9.3.26) generates dynamical system (V, S_t) in the space $V = H_0^1(\Omega)$ by the formula $S_t u_0 = u(t)$, $t > 0$, where $u(t)$ is a weak solution to (9.3.26).

9.3.2.1 Attractor for quasi-static problem (9.3.26)

The smoothing character of the dynamics in the quasi-static version is reflected by the following theorem on long-time dynamics below.

9.3.9. Theorem. *The dynamical system $(H_0^1(\Omega), S_t)$ generated by* (9.3.26) *is compact. Thus it has compact global attractor A. This attractor is a bounded set in the space $(H^3 \cap H_0^2)(\Omega)$ and has finite fractal dimension.*

Proof. It follows from (3.3.21) and (3.3.22) in Theorem 3.3.3 that the system $(H_0^1(\Omega), S_t)$ is compact (see Definition 7.1.2). Therefore by Theorem 7.2.2 it possesses a compact global attractor A. This attractor belongs to $H_0^2(\Omega)$.

The linear problem

$$\kappa(1-\alpha\Delta)u_t + \Delta^2 u = 0, \quad u|_{\partial\Omega} = \frac{\partial u}{\partial n}\Big|_{\partial\Omega} = 0, \tag{9.3.28}$$

generates an analytic exponentially stable linear semigroup T_t in $V = H_0^1(\Omega)$ such that

$$\|T_t h\|_{s'} \le Ce^{-\beta t}\left(1 + t^{-(s'-s)/2}\right)\|h\|_s, \quad 1 \le s \le s' \le 3. \tag{9.3.29}$$

Therefore the representation

$$u(t) = T_t u(0) + \int_0^t T_{t-\tau}(\kappa I - \kappa\alpha\Delta)^{-1} F(u(\tau))d\tau,$$

where $F(u) = [u+f, v+F_0] - L(u) + p$, allows us to obtain the estimate

$$\|u(t)\|_3 = Ct^{-1}\|u(0)\|_1 + C\int_0^t (t-\tau)^{-1+\delta}\|F(u(\tau)\|_{-1+2\delta}\,d\tau, \quad 0 < t \le 1,$$

with $0 < \delta < 1/2$. Because $\|u\|_2 \le C$ for an element u from the attractor, using properties of the von Karman bracket we conclude that $\|F(u(\tau)\|_{-1+2\delta} \le C$ and, therefore, $A \subset (H^3 \cap H_0^2)(\Omega)$.

To prove finiteness of the fractal dimension of the attractor we rely on Ladyzhenskaya's Theorem 7.3.2 with the projector $P = P_N$ on the span of the first N eigenfunctions of the spectral boundary value problem

$$\Delta^2 v = \lambda\kappa(1-\alpha\Delta)v, \qquad v|_{\partial\Omega} = \frac{\partial v}{\partial n}\Big|_{\partial\Omega} = 0, \tag{9.3.30}$$

with N large enough. For more details we refer to [49]. □

9.3.2.2 Upper semicontinuity of the attractor to problem (9.3.24)

Let (H, S_t^μ) be the dynamical system generated in the space $H = H_0^2(\Omega) \times H_0^1(\Omega)$ by problem (9.3.24). It follows from Theorem 9.1.5 that (H, S_t^μ) possesses a compact global attractor A_μ. By Theorem 9.2.9 this attractor is a bounded set in the space $W = (H^3 \cap H_0^2)(\Omega) \times H_0^2(\Omega)$.

We have the following result on the upper semicontinuity of the attractor A_μ.

9.3.10. Theorem. *Under Assumption* 9.3.8 *we have the relation*

$$\lim_{\mu \to 0} \sup \left\{ \operatorname{dist}_H(y, \hat{A}) \, : \, y \in A_\mu \right\} = 0, \tag{9.3.31}$$

where $\hat{A} = \left\{ (z_0; z_1) \, : \, z_0 \in A, \kappa z_1 = (I - \alpha\Delta)^{-1} \left(-\Delta^2 z_0 + F(z_0) \right) \right\}$ *and* A *is the global attractor for the dynamical system* $(H_0^1(\Omega), S_t)$ *generated by* (9.3.26). *Here, as above,* $F(u) = [u + f, v + F_0] - L(u) + p$.

Proof. Without loss of generality we assume that $\kappa = 1$ and redefine the norm in $H_0^1(\Omega)$ by the formula $\|\cdot\|_1^2 = \|\cdot\|^2 + \alpha\|\nabla \cdot\|^2$. The key to the result is estimates for the attractor A_μ which are uniform with respect to the parameter μ.

As in the proof of Theorem 9.1.5, relying on Theorem 8.2.3 in the case $L \neq 0$ and on Theorem 8.3.12 in the case $L \equiv 0$, we infer that there exist $R > 0$ and $\mu_0 > 0$ such that

$$S_t^\mu B \subset \left\{ y = (u_0; u_1) \in H \, : \, \mu\|u_1\|_1^2 + \|u_0\|_2^2 \leq R^2 \right\} \tag{9.3.32}$$

for all $t \geq t(B, \mu)$, where B is a bounded set in H and $0 < \mu \leq \mu_0$.

9.3.11. Lemma. *Assume that* $u(t)$ *is a solution to problem* (9.3.24) *such that* $\|\Delta u(t)\| \leq R$ *for* $t \geq 0$. *Then we have the estimate*

$$\frac{1}{2} \int_0^t \|u_t(\tau)\|_1^2 e^{-\beta(t-\tau)} d\tau \leq \left(\mu\|u_t(0)\|_1^2 + \|\Delta u(0)\|^2 \right) e^{-\beta t} + C(R, \beta),$$

where β *is a positive constant such that* $\beta\mu < \frac{1}{2}$.

Proof. It is clear $\|F(u(t))\|_{-1} \leq C_R$. Therefore from the energy relation for problem (9.3.24) we find that

$$\frac{1}{2} \frac{d}{dt} \left(\mu\|u_t(t)\|_1^2 + \|\Delta u(t)\|^2 \right) + \frac{1}{2}\|u_t(t)\|_1^2 \leq C_R.$$

We multiply this inequality by $2\exp(\beta t)$. Then for $\beta\mu < \frac{1}{2}$, on the strength of $\|\Delta u(t)\| \leq R$, we have

$$\frac{d}{dt} \left[e^{\beta t} \left(\mu\|u_t(t)\|_1^2 + \|\Delta u(t)\|^2 \right) \right] + \frac{1}{2}\|u_t(t)\|_1^2 e^{\beta t} \leq C_R e^{\beta t}(1 + \beta).$$

Integrating from 0 to t leads to the conclusion. □

9.3.12. Lemma. *Let* $u(t)$ *be a solution to problem* (9.3.24) *with initial data* $(u_0; u_1)$ *from the space* $W = (H^3 \cap H_0^2)(\Omega) \times H_0^2(\Omega)$ *such that for* $t \geq 0$ *one has* $\|u(t)\|_2 \leq R$. *Then for the function* $w(t) = u_t(t)$ *and* $\mu \in (0, \mu_0)$, *where* μ_0 *is sufficiently small, we have the estimate*

$$\mu\|w_t(t)\|_1^2 + \|\Delta w(t)\|^2 \leq C_1 \left(\mu\|w_t(0)\|_1^2 + \|\Delta w(0)\|^2 \right) e^{-\beta_0 t} + C_2 \tag{9.3.33}$$

where $\beta_0 > 0$, *and the numbers* C_1 *and* C_2 *do not depend on* $\mu \in (0, \mu_0)$.

Proof. We consider the function

$$W(t) = \frac{1}{2}\left(\mu\|w_t(t)\|_1^2 + \|\Delta w(t)\|^2\right) + \nu\left\{\mu(w_t(t), w(t))_1 + \frac{1}{2}\|w(t)\|_1^2\right\},$$

where $\nu > 0$. Obviously

$$\frac{\mu(1-\nu\mu)}{2}\|w_t\|_1^2 + \frac{1}{2}\|\Delta w\|^2 \le W(t) \le \frac{\mu(1+\nu\mu)}{2}\|w_t\|_1^2 + \frac{1+c_0\nu}{2}\|\Delta w\|^2. \tag{9.3.34}$$

Because the function $w(t)$ solves the equation obtained by differentiation of (9.3.24) with respect to t, using the estimate $\|\langle F'(u), w\rangle\|_{-1} \le C_R\|w\|_2$, it is easy to find that

$$\frac{d}{dt}W(t) \le -\frac{1-2\nu\mu}{2}\|w_t\|_1^2 - \frac{\nu - c_R}{2}\|\Delta w\|^2 + \frac{\nu c_R}{2}\|w\|_1^2.$$

Setting now $\nu = c_R + 2$, for sufficiently small μ_0, with the aid of (9.3.34) we obtain that

$$\frac{d}{dt}W(t) + \beta_0 W(t) \le C\|w(t)\|_1^2, \quad t \ge 0, \ \mu \in (0, \mu_0),$$

where the constants $\beta_0 > 0$ and C do not depend on μ. Consequently

$$W(t) \le W(0)e^{-\beta_0 t} + C\int_0^t \|u_t(\tau)\|_1^2 e^{-\beta_0(t-\tau)}d\tau.$$

Therefore estimate (9.3.33) follows from (9.3.34) and Lemma 9.3.11. □

Now we are in position to conclude the proof of Theorem 9.3.10.

Lemma 9.3.12 implies that for any full trajectory $y^\mu(t) = (u^\mu(t); u_t^\mu(t))$ from the attractor A_μ we have the estimate

$$\mu\|u_{tt}^\mu(t)\|_1^2 + \|\Delta u_t^\mu(t)\|^2 + \|u^\mu(t)\|_3^2 \le R_*^2, \quad t \in \mathbb{R}. \tag{9.3.35}$$

We prove next (9.3.31). Let $y^\mu = (u_0^\mu; u_1^\mu)$ be an element from A_μ such that

$$\mathrm{dist}_H\left(y^\mu, \hat{A}\right) = \sup\left\{\mathrm{dist}_H\left(y, \hat{A}\right) \,:\, y \in A^\mu\right\}.$$

This element y_μ exists because both A^μ and $\hat{A}$ are compact sets in H. We argue by contradiction in order to prove (9.3.31). If (9.3.31) does not hold, then

$$d(y^{\mu_k}) \equiv \mathrm{dist}_H\left(y^{\mu_k}, \hat{A}\right) \ge \delta > 0 \tag{9.3.36}$$

for some sequence $\{\mu_k\}$ such that $\mu_k \to 0$ as $k \to \infty$. Let $y^k(t) = (u^k(t); u_t^k(t))$ be a full trajectory from the attractor A_{μ_k} such that $y^k(0) = y^{\mu_k}$. It follows from (9.3.35) that the sequence $\{u^k(t)\}$ is compact in $C([a,b], H_0^2(\Omega))$ for any $a < b$. Therefore we can find a subsequence $\{u^{k_m}(t)\}$ and a function $u(t) \in C(\mathbb{R}, H_0^2(\Omega))$ such that

$$\lim_{m\to\infty}\max_{t\in[a,b]}\|u^{k_m}(t) - u(t)\|_2 = 0. \tag{9.3.37}$$

It also follows from (9.3.35) that $\mu\|u_{tt}^{k_m}(t)\|_1 \to 0$ as $m \to \infty$. Therefore, taking in (9.3.24) the limit as $\mu \to 0$ we obtain that the function $u(t)$ is a solution to problem (9.3.26) which is bounded on the entire time axis and, consequently, lies in the attractor of the system (V, S_t). Moreover, (9.3.35) and (9.3.37) imply that

$$d(y^{\mu_{k_m}}) \le \|y^{k_m}(0) - y_0\|_{H_0^2(\Omega)\times H_0^1(\Omega)} \to 0 \quad \text{as} \quad m \to \infty,$$

with $y_0 = (u(0); (I-\alpha\Delta)^{-1}\left(-\Delta^2 u(0) + F(u(0))\right)$. This contradicts (9.3.36). □

9.3.13. Remark. It follows from the uniform estimate in (9.3.35) and from the interpolation argument that upper semicontinuity property (9.3.31) is valid in a stronger topology of the space $H^{3-\delta}(\Omega)\times H^{2-\delta}(\Omega)$, where $\delta > 0$ is arbitrary small. ■

9.4 Global attractors for von Karman model without rotational inertia

In this section we consider von Karman equations without rotational forces ($\alpha = 0$). The model without rotational forces displays less regularity, hence the analysis of the nonlinear effects is more subtle. In fact, the results obtained often depend on a very structure of nonlinearity where certain estimates are derived by relying on a particular form of the von Karman bracket along with its cancellation properties.

To begin, we consider the following problem

$$\begin{cases} u_{tt} + kd_0(x)g_0(u_t) + \Delta^2 u - [u+f, v+F_0] + Lu = p(x), \quad x \in \Omega, t > 0, \\ u|_{t=0} = u_0(x), \quad \partial_t u|_{t=0} = u_1(x), \end{cases} \tag{9.4.1}$$

where, as above, $v = v(u)$ is a solution of the problem

$$\Delta^2 v + [u + 2f, u] = 0, \quad v|_{\partial\Omega} = \left.\frac{\partial v}{\partial n}\right|_{\partial\Omega} = 0. \tag{9.4.2}$$

As in Section 9.1 we restrict ourselves to the following basic three types of boundary conditions:

1. **[clamped]**:
$$u = \nabla u = 0 \ \text{ on } \Gamma; \tag{9.4.3}$$
2. **[hinged]**:
$$u = \Delta u = 0 \ \text{ on } \Gamma; \tag{9.4.4}$$
3. **[free]**:
$$\begin{aligned} &\Delta u + (1-\mu)B_1 u = 0 \ \text{ and} \\ &\frac{\partial}{\partial n}\Delta u + (1-\mu)B_2 u = \nu_1 u + \beta u^3 + \beta u_t \ \text{ on } \Gamma\,. \end{aligned} \tag{9.4.5}$$

Here $\nu_1 > 0$ is a parameter and β is a nonnegative measurable function such that $\beta(x) > 0$, a.e for $x \in \Gamma$, $0 < \mu < 1$. The boundary operators B_1 and B_2 are defined by (1.3.20).

One could also consider various combinations of mixed boundary conditions, as was done before in the rotational case. This generalization does not add new difficulties from the technical point of view.

The abstract version of the von Karman model without the rotational forces ($\alpha = 0$) has different forms depending on the type of boundary conditions (for details see Chapter 4). In the case of the clamped (9.4.3) or hinged conditions (9.4.4) we have formally the same (with $M \equiv I$) abstract form (9.1.7) as in the case of rotational forces. In the case of free boundary conditions (9.4.5) this abstract representation is the following

$$\begin{cases} u_{tt} + kD(u_t) + \mathscr{A} u + D_0 h(u) + D_0 D_0^* u_t = F(u), \\ u(0) = u_0 \in \mathscr{D}(\mathscr{A}^{1/2}), \;\; u_t(0) = u_1 \in \mathscr{H}. \end{cases} \tag{9.4.6}$$

In this section we use the results from Chapter 8 applied to the abstract model (either (9.1.7) or (9.4.6)) with appropriate specification of boundary conditions. We note that the main difference with respect to the rotational model studied in Sections 9.1 and 9.2 is the fact that the nonlinear term $F(u)$ is no longer compact with respect to the finite energy norm $H^2(\Omega) \times L_2(\Omega)$. This fact does not have any major bearing on the existence of an absorbing set, which depends heavily on the structure of the nonlinearity. It has, however, consequences on the theory of global attractors where the compactness property is critical. The arguments required for this are much more delicate.

A few comments explaining the presence of the operator D_0 in (9.4.6) are in order. The appearance of this term D_0, in the model (9.4.6) with free boundary conditions, may appear at first somewhat surprising. However, the purpose of this term is to model the effect of nonlinear and nonmonotone boundary conditions. The need for such is in the context of free boundary conditions, where the nonlinearity on the boundary guarantees that the energy is bounded from below (otherwise is not). In the case of rotational forces present in the model these nonlinear terms can be modeled by the term $F(u)$ which is allowed to be in a larger space V'. However, when $\alpha = 0$, the term F is required to be in $L_2(\Omega)$, thus even very mild distributions are not accomodated by the right-hand side of the equation. In view of this, the role of D_0 is precisely to model this special boundary effect.

Our basic hypotheses of this section are a slight enforcement of those in Chapter 4 (cf. Assumptions 4.1.1). They are collected in the following list.

9.4.1. Assumption.

- The function g_0 is assumed monotone nondecreasing and continuous. Moreover we assume that $g_0(0) = 0$. The function $d_0(x)$ is nonnegative bounded measurable functions, the damping coefficient k is positive.
- $f \in H^3(\Omega)$, $F_0 \in H^4(\Omega)$, $p \in L^2(\Omega)$.
- L is a linear bounded operator from $H^{2-\sigma}(\Omega)$ into $L_2(\Omega)$ for some $\sigma > 0$.

The following assertion is a direct corollary of Theorems 4.1.4, 4.1.10 and 4.1.13.

9.4.2. Theorem. *Under Assumption* 9.4.1 *problem* (9.4.1) *and* (9.4.2) *with either clamped* (9.4.3), *hinged* (9.4.4), *or free* (9.4.5) *boundary conditions for the displacement* $u(x,t)$ *generates a dynamical system* (H, S_t) *with the evolution operator* S_t *given by the formula*

$$S_t(u_0; u_1) = (u(t); u_t(t)), \quad (u_0; u_1) \in H, \ t \ge 0, \tag{9.4.7}$$

where $u(t)$ *is a generalized solution to problem* (9.4.1) *and* (9.4.2) *with the corresponding boundary conditions for* u. *The phase space* H *has the form*

- $H = H_0^2(\Omega) \times L_2(\Omega)$ *for the clamped boundary conditions* (9.4.3),
- $H = (H^2(\Omega) \cap H_0^1(\Omega)) \times L_2(\Omega)$ *in the hinged case* (9.4.4),
- $H = H^2(\Omega) \times L_2(\Omega)$ *in the free case* (9.4.5).

Our goal now is to study asymptotic properties of the evolution semigroup S_t given by (9.4.7). As in Section 9.1, to apply the results presented in Chapter 8 we need some additional hypotheses concerning $d_0(x)g_0(v)$. We collect them into two sets of assumptions.

9.4.3. Assumption.

- There exist positive constants d_0 and d_1 such that

$$d_0 \le d_0(x) \le d_1. \tag{9.4.8}$$

- For any $\varepsilon > 0$ there exists C_ε such that[6]

$$s^2 \le \varepsilon + C_\varepsilon s g_0(s) \quad \text{for all} \ s \in \mathbb{R}. \tag{9.4.9}$$

- In the case $L \ne 0$ we assume in addition that $g_0 \in C^1(\mathbb{R})$ is increasing and there exist $a_0 > 0$ and $a_1 > 0$ such that

$$g_0'(v) \ge a_0 \quad \text{and} \quad |g_0(v)| \le a_1 |v| \quad \text{for all} \ v \in \mathbb{R}. \tag{9.4.10}$$

The following hypothesis concerning the damping term is important for finite dimensionality of global attractors.

9.4.4. Assumption. There exist positive constants a_0 and a_1 such that

$$a_0 \le \frac{g_0(u+v) - g_0(u)}{v} \le a_1 \left(1 + (u+v)g_0(u+v) + u g_0(u)\right). \tag{9.4.11}$$

for any $u \in \mathbb{R}$ and $v > 0$.

9.4.5. Remark. If $g_0 \in C^1(\mathbb{R})$, then Assumption 9.4.4 holds if and only if

$$m_0 \le g_0'(s) \le m_1 \left(1 + s g_0(s)\right), \quad s \in \mathbb{R}, \tag{9.4.12}$$

---

[6] See Remark 9.1.6 and Proposition B.1.1 concerning the validity of this property.

where m_0 and m_1 are positive constants. We note that the right inequality in property (9.4.12) holds true if we assume that there exist $p > 1, M_1 > 0$ such that $g'(s) \le M_1[1+|s|^{p-1}]$, and when $p > 3$, $g(s)s \ge m_1|s|^{p-1}$ for $|s| \ge 1$ and for some $m_1 > 0$. However relation (9.4.12) also allows for exponential behavior for $g(s)$, for example, $|g(s)| \sim e^{\alpha|s|}$ as $|s| \to \infty$ for some $\alpha > 0$. See also Remark 9.2.7. ■

9.4.1 Clamped boundary condition

We begin with the case of the clamped boundary condition given in (9.4.3).

9.4.6. Theorem. *Assume that Assumptions* 9.4.1 *and* 9.4.3 *hold. Let* (H, S_t) *be the dynamical system generated by problem* (9.4.1) *and* (9.4.2) *with clamped boundary conditions* (9.4.3) *in the space* $H = H_0^2(\Omega) \times L_2(\Omega)$. *Then*

- *The system* (H, S_t) *possesses a compact global attractor A.*
- *If in addition Assumption* 9.4.4 *holds, then the attractor A has a finite fractal dimension, if either the damping coefficient k in* (9.4.1) *is large enough or else* $L \equiv 0$.

Proof. As shown in Section 4.1, in a Hilbert space $\mathscr{H}$ the problem considered can be written in the form

$$u_{tt} + kD(u_t) + \mathscr{A}u = F(u), \quad u(0) = u_0 \in \mathscr{D}(\mathscr{A}^{1/2}), \; u_t(0) = u_1 \in \mathscr{H}, \quad (9.4.13)$$

with the following spaces and operators.

- $\mathscr{H} = L_2(\Omega)$.
- $\mathscr{A}u \equiv \Delta^2 u$, $u \in \mathscr{D}(\mathscr{A}) \equiv H_0^2(\Omega) \cap H^4(\Omega)$, $[\mathscr{D}(\mathscr{A}^{1/2})]' = H^{-2}(\Omega)$.
- $F(u) \equiv [u + f, v(u) + F_0] - Lu + p$, where $v(u)$ solves (9.4.2).
- $D(u) \equiv d_0(x) g_0(u)$.

With the above notation we verify that the assumptions of Theorem 8.2.3 (dissipativity in the case $L \neq 0$), Proposition 8.3.6 (compactness), and Theorems 8.6.3 and Theorem 8.6.4 (finite dimension) are satisfied.

We first note that, because Assumption 8.1.1 is a special case of Assumption 2.4.15, by Lemma 4.1.7 Assumption 8.1.1 holds for our case. We recall that this Assumption 8.1.1 provides well-posedness for (9.4.13) and is present in all abstract assertions mentioned in the previous paragraph.

9.4.1.1 Verification of the assumptions imposed on F.

We recall (see Section 4.1) that $F(u)$ can be presented in the form (8.1.4):

$$F(u) = -\Pi'(u) + F^*(u) \quad (9.4.14)$$

with

$$\Pi(u) = \frac{1}{4}\|\Delta v(u)\|^2 - \frac{1}{2}([u,F_0],u) - ([f,F_0]+p,u), \tag{9.4.15}$$

where $v(u) \in H_0^2(\Omega)$ is defined by (9.4.2), and $F^*(u) = -L(u)$. Thus we set

$$\Pi_0(u) = \frac{1}{4}\|\Delta v(u)\|^2 \text{ and } \Pi_1(u) = -\frac{1}{2}([u,F_0],u) - ([f,F_0]+p,u). \tag{9.4.16}$$

9.4.7. Lemma. *The nonlinear mapping F given by* (9.4.14) *satisfies the hypotheses given in Assumptions* 8.1.1, 8.2.1, 8.3.3 *and in Assumption* 8.6.2 *with* $\eta = 0$.

Proof. It is clear that (8.2.5) and (8.3.13) hold for this $F^*(u)$. As for relations (8.2.3) and (8.2.4) in Assumption 8.2.1 they can be easily derived from Lemma 9.1.7 (see (9.1.16) and (9.1.17)). The properties of $\Pi(u)$ stated in Assumption 8.3.3(F) follow from (9.1.19) and (9.1.20).

Relation (8.6.3) with $\eta = 0$ in Assumption 8.6.2 is the same as (8.1.3) in Assumption 8.1.1 (the fact that η is not strictly positive in the model considered, has important implications). □

9.4.1.2 Verifications of the assumptions imposed on the damping D

Below we denote $(Dv,u) = \int_\Omega d_0(x)g_0(v(x))u(x)dx$ and start with the following lemmas.

9.4.8. Lemma. *Let* $g_0 \in C(\mathbb{R})$ *and Assumption* 9.4.3 *hold. Then*

- *If* $L \neq 0$, *then*

$$(Dv,v) = \int_\Omega d_0(x)g_0(v(x))v(x)dx \geq a_0 d_0\|v\|^2, \tag{9.4.17}$$

$$|(Dv,u)|^2 \leq (a_1 d_1)^2\|v\|^2\|u\|^2 \leq K\cdot(Dv,v)\|u\|^2, \quad v,u \in L_2(\Omega), \tag{9.4.18}$$

where $K = (a_1 d_1)^2/(a_0 d_0)$, *and*

$$|D(u+v,w) - D(u,w)| \leq a_1 d_1 \max_\Omega |w|\cdot[K_1 + D(u+v,u+v) + D(u,u)] \tag{9.4.19}$$

for any $v,u \in L_2(\Omega)$ *and* $w \in C(\Omega)$, *where* $K_1 = c_1(a_0 d_0)^{-1}$ *and* c_1 *depends on the size of* Ω *only.*

- *If* $L = 0$, *then for every* $\delta > 0$ *there exists* $C_\delta > 0$ *such that*

$$|(Dv,u)| \leq C_\delta \max_\Omega |u|\cdot D(v,v) + \delta\left(1+\|u\|^2\right) \tag{9.4.20}$$

for any $v \in H^2(\Omega)$, $u \in C(\Omega)$, *and*

$$|(D(u+v),w) - (Du,w)| \leq \frac{d_1}{d_0}\max_\Omega |w|\cdot[c_0 + D(u+v,u+v) + D(u,u)] \tag{9.4.21}$$

for any $v, u \in H^2(\Omega)$ *and* $w \in C(\Omega)$, *where* $c_0 = 2d_0a_2 \cdot Vol(\Omega)$ *with* $a_2 = \sup\{|g_0(v)| : |v| \leq 1\}$.

Proof. Property (9.4.17) follows directly from (9.4.8) and (9.4.10).
Relation (9.4.10) also implies that

$$\left|\int_\Omega d_0(x)g_0(v)u\,dx\right| \leq a_1 d_1 \int_\Omega |v|\cdot|u|\,dx \leq a_1 d_1 \|v\|\cdot\|u\|.$$

Therefore using (9.4.17) we obtain (9.4.18).
To prove (9.4.19) we note that (9.4.10) implies that

$$\begin{aligned}|(D(u+v)-D(u),w)| &\leq d_1 \max_\Omega |w| \cdot \int_\Omega (|g_0(u+v)|+|g_0(u)|)\,dx \\ &\leq a_1 d_1 \max_\Omega |w| \cdot \int_\Omega (|u+v|+|u|)\,dx \\ &\leq a_1 d_1 \max_\Omega |w| \cdot \left[\frac{\mathrm{Vol}(\Omega)}{2a_0 d_0} + a_0 d_0 \int_\Omega \left(|u+v|^2+|u|^2\right)dx\right].\end{aligned}$$

By (9.4.17) we obtain (9.4.19).
For the case $L=0$ we can afford a nonlinear growth of the dissipation. To handle this we split the region of the integration. Let $\Omega_1 \equiv \{x \in \Omega,\ |v| \leq \varepsilon\}$ and $\Omega_2 \equiv \{x \in \Omega,\ |v| > \varepsilon\}$. The parameter $\varepsilon > 0$ will be chosen later. We have

$$\int_\Omega d_0(x)g_0(v)u\,dx \leq d_1 \cdot \left[\int_{\Omega_1} |g_0(v)u|\,dx + \int_{\Omega_2} |g_0(v)u|\,dx\right]. \tag{9.4.22}$$

By the Cauchy–Schwarz inequality we obtain

$$\begin{aligned}d_1 \int_{\Omega_1} |g_0(v)u|\,dx &\leq \frac{d_1^2}{4\delta}\int_{\Omega_1} |g_0(v)|^2 dx + \delta \int_{\Omega_1} |u|^2 dx \\ &\leq \frac{d_1^2}{4\delta}\mathrm{Vol}(\Omega) \max_{|s|\leq\varepsilon} |g_0(s)|^2 + \delta\|u\|^2.\end{aligned}$$

If we choose $\varepsilon = \varepsilon_\delta$ such that $d_1^2 \mathrm{Vol}(\Omega) \max_{|s|\leq\varepsilon} |g_0(s)|^2 \leq 4\delta^2$, we obtain that

$$d_1 \int_{\Omega_1} |g_0(v)u|\,dx \leq \delta\left(1+\|u\|^2\right). \tag{9.4.23}$$

In the region Ω_2 we have that $\varepsilon_\delta^{-1} v \geq 1$ and thus

$$\int_{\Omega_2} |g_0(v)u|\,dx \leq \frac{1}{\varepsilon_\delta}\max_\Omega |u| \int_{\Omega_2} g_0(v)v\,dx \leq C_\delta \max_\Omega |u| \cdot (Dv,v). \tag{9.4.24}$$

Hence (9.4.22)–(9.4.24) imply (9.4.20).
Now we prove (9.4.21). We write

$$
\begin{aligned}
|(D(u+v)-D(u),w)| &\le d_1 \int_\Omega |g_0(u+v)-g_0(u)||w|dx \\
&\le d_1|w|_{C(\Omega)} \int_\Omega (|g_0(u+v)|+|g_0(u)|)\,dx. \quad (9.4.25)
\end{aligned}
$$

In the next step we split again the region of integration into two domains

$$
\Omega_1 \equiv \{x\in\Omega; |u(x)+v(x)|\le 1\} \text{ and } \Omega_2 \equiv \{x\in\Omega; |u(x)+v(x)|\ge 1\}.
$$

We have that

$$
\begin{aligned}
\int_\Omega |g_0(u+v)|dx &= \int_{\Omega_1} |g_0(u+v)|dx + \int_{\Omega_2} |g_0(u+v)|dx \\
&\le a_2 \mathrm{Vol}(\Omega) + \int_\Omega g_0(u+v)(u+v)dx. \qquad (9.4.26)
\end{aligned}
$$

The same argument applies to the term $g_0(u)$ giving

$$
\int_\Omega |g_0(u)|dx \le a_2\mathrm{Vol}(\Omega) + \int_\Omega g_0(u)udx. \qquad (9.4.27)
$$

Combining (9.4.25)–(9.4.27) we obtain (9.4.21). □

We note that an estimate of the form (9.4.19) (resp., (9.4.21)) can be easily derived from (9.4.18) (resp., (9.4.20)) with less precise constants. However we find it useful to give an independent argument which provides more accurate control of the constants with respect to the damping parameters. This is useful in estimating the size of absorbing set as being independent on the damping parameter k.

9.4.9. Lemma. *Under Assumptions 9.4.1 and 9.4.4 we have the following estimates*

$$
(D(u+v),v)-(Du,v) \ge d_0a_0\|v\|^2, \qquad (9.4.28)
$$

and, with arbitrary $\varepsilon>0$ and some constants b_1 and b_2,

$$
\begin{aligned}
|(D(u+v),w)-(Du,w)| \le{}& \frac{b_1}{\varepsilon}(D(u+v)-Du,v) \qquad (9.4.29) \\
&+ b_2\varepsilon \max_\Omega |w|^2 \cdot [1+D(u+v,u+v)+D(u,u)].
\end{aligned}
$$

Proof. Relation (9.4.28) is a direct consequence of the lower bound in (9.4.11).

As to the inequality (9.4.29), we carry the following computations:

$$
\begin{aligned}
&\left|\int_\Omega d_0(x)(g_0(u+v)-g_0(u))wdx\right| \qquad (9.4.30) \\
&\le \varepsilon \int_\Omega |g_0(u+v)-g_0(u)|\frac{|w|^2}{|v|}dx + \frac{C}{\varepsilon}\int_\Omega (g_0(u+v)-g_0(u))vdx
\end{aligned}
$$

for any $\varepsilon>0$. Using (9.4.11) we have

$$\int_\Omega |g_0(u+v)-g_0(u)|\frac{|w|^2}{|v|}dx \tag{9.4.31}$$

$$\leq a_1 \max_\Omega |w|^2 \left[1+\int_\Omega (g_0(u+v)(u+v)+g_0(u)u)dx\right].$$

Relations (9.4.30) and (9.4.31) imply (9.4.29). □

We are now in position to conclude the proof of Theorem 9.4.6.

Dissipativity (in the case $L \neq 0$: We need to check relations (8.2.7) and (8.2.8). They follow from (9.4.17) and (9.4.18). In this case, by Theorem 8.2.3 the dynamical system (H,S_t) generated by problem (9.4.1)–(9.4.3) is dissipative and the dissipativity radius does not depend on k.

Existence of attractor.

1. CASE $L \neq 0$: Relations (9.4.19) imply that the damping operator D satisfies (8.3.12) in Assumption 8.3.3(D) with $\varepsilon = 0$, $C_1^\varepsilon(r) = 0$, and $\varkappa = 1$. Inequality (8.3.11) is obvious due to (9.4.10). Thus assumption 8.3.3 holds and we can apply Theorem 8.3.5.

2. CASE $L = 0$: In this case by the energy inequality (4.1.11) the system (H,S_t) is gradient. Moreover the set of stationary solutions is bounded. Therefore we can apply Theorem 8.3.12.

Finite dimension of the attractor: Assumptions 9.4.1 and 9.4.4 and Lemma 9.4.9 imply that (8.3.12) holds with $\varkappa = 2$ and without the term $\varepsilon|\mathscr{A}^{1/2}w|^2$. Therefore we can apply Theorem 8.6.3 to prove the fractal dimension finiteness of the attractor in the case of large damping parameter k.

In the case when $L \equiv 0$ we can use Theorem 8.6.4. However for the reader's convenience we prefer to rely on a more general Theorem 8.6.1 by directly checking the stabilizability estimate in (8.6.2). For this we use the structure of the nonlinear term to obtain the main estimate (8.5.7) in Theorem 8.5.3. Therefore we need to check relation (8.5.1) for the case considered.

We denote by $U(t) = (u(t);u_t(t)) = S_t y_1$ and $W(t) = (w(t);w_t(t)) = S_t y_2$ the two solutions corresponding to initial conditions y_1 and y_2, respectively. We assume that y_i belong to a bounded forward invariant set $\mathscr{E}_R$ given by

$$\mathscr{E}_R = \left\{y = (u_0;u_1) \,:\, \mathscr{E}(u_0,u_1) \leq R^2\right\}.$$

Without loss of generality we can assume that $u(t)$ and $w(t)$ are strong solutions. Let $z(t) = u(t) - w(t)$.

To prove (8.5.1), the algebraic structure of von Karman bracket becomes critical. This allows us to present the difference $F(u) - F(w)$ in the following form

$$F(u) - F(w) = \widehat{\mathscr{R}}(u,w;z) + \mathscr{R}_*(u,w;z) \equiv \widehat{\mathscr{R}}(z) + \mathscr{R}_*(z),$$

where

$$\widehat{\mathscr{R}}(z) = [u,v(u,u)] - [w,v(w,w)] + [z,F_0] + 2[z,v(f,u)] \tag{9.4.32}$$

and

$$\mathscr{R}_*(z) = 2[w, v(f,z)] + [f, v(u+w+2f,z)].$$

Here we also denote by $v(u_1,u_2) \in H_0^2(\Omega)$ the solution to the problem

$$\Delta^2 v(u_1,u_2) = -[u_1,u_2] \text{ in } \Omega, \quad \frac{\partial}{\partial n} v(u_1,u_2) = v(u_1,u_2) = 0 \text{ on } \Gamma. \quad (9.4.33)$$

Because $f \in H^3(\Omega)$, one can see that

$$\|\mathscr{R}_*(z)\| \le C_R \|\mathscr{A}^{1/2-\eta} z\| \quad \text{with } \eta > 0.$$

Therefore to prove (8.5.1) we need to consider the corresponding integral with $\widehat{\mathscr{R}}(z)$ only.

A simple calculation relying on the symmetry properties of the von Karman bracket makes it possible to prove the following assertion.

9.4.10. Lemma. *Assume $u, w \in C(0,T;H^2(\Omega)) \cap C^1(0,T;L_2(\Omega))$. Let $z = u - v$ and $\mathscr{R}_0(z) = [u, v(u,u)] - [w, v(w,w)]$, where $v(u_1,u_2)$ is defined by (9.4.33).Then*

$$(\mathscr{R}_0(z), z_t) = \frac{1}{4}\frac{d}{dt} Q_0(z) + \frac{1}{2} P_0(z), \quad (9.4.34)$$

where

$$Q_0(z) = (v(u,u) + v(w,w), [z,z]) - \|\Delta v(u+w,z)\|^2$$

and

$$P_0(z) = -(u_t, [u, v(z,z)]) - (w_t, [w, v(z,z)]) - (u_t + w_t, [z, v(u+w,z)]). \quad (9.4.35)$$

Proof. It is sufficient to consider the case of smooth functions u and w (from $C^1(0,T;H^4(\Omega))$, for instance).

One can see that $[u, v(u,u)] - [w, v(w,w)] = [z, v(u,u)] + [w, v(u+w,z)]$. Therefore by the symmetry between u and w we have that

$$\mathscr{R}_0(z) = \frac{1}{2}[z, v(u,u) + v(w,w)] + \frac{1}{2}[u+w, v(u+w,z)].$$

By the symmetry properties of the von Karman bracket (see Proposition 1.4.2) this implies that

$$(\mathscr{R}_0(z), z_t) = \frac{1}{4}(v(u,u) + v(w,w), [z,z]_t) + \frac{1}{2}(v(u+w,z), [u+w, z_t]).$$

By the structure of the element $v(u_1,u_2)$ given by (9.4.33) one can see that

$$(\mathscr{R}_0(z), z_t) - \frac{1}{4}\frac{dQ_0}{dt} = -\frac{1}{2}(v(u,u_t) + v(w,w_t), [z,z]) - \frac{1}{2}(v(u+w,z), [u_t + w_t, z]).$$

Therefore the symmetry of the von Karman bracket implies (9.4.34). □

By Lemma 9.4.10 we have that

$$(\widehat{\mathscr{R}}(z), z_t) = \frac{1}{4}\frac{d}{dt}Q(z) + \frac{1}{2}P(z),$$

for $\widehat{\mathscr{R}}(z)$ given by (9.4.32), where $P(z) = P_0(z) - 2(u_t, [v(z,z), f])$ and

$$Q(z) = Q_0(z) + 4([z,z], v(f,u)] + 2(F_0, [z,z]).$$

Now we estimate values Q and P. It follows from (1.4.16) with $\theta = 2\delta$ and $\beta = \delta$ that

$$\|[u_1, u_2]\|_{-1-2\delta} \le C\|u_1\|_{2-\delta}\|u_2\|_{2-\delta}, \quad 0 < \delta < 1/2. \tag{9.4.36}$$

Therefore using the elliptic regularity of Δ^2 with the Dirichlet boundary conditions one can see that

$$|Q(z)| \le C_R\|z\|_{2-\eta}^2 \quad \text{for some } \eta > 0. \tag{9.4.37}$$

Now we estimate $P(z)$. We start with $P_0(z)$. The first term in $P_0(z)$ is estimated as

$$|(u_t, [u, v(z,z)])| \le C\|u_t\|\|[u, v(z,z)]\| \le C\|u_t\|\|u\|_2\|z\|_2^2 \le C_R\|u_t\|\|z\|_2^2. \tag{9.4.38}$$

Similarly

$$|(w_t, [w, v(z,z)])| \le C_R\|w_t\|\|z\|_2^2. \tag{9.4.39}$$

Now we consider the third term $(u_t + w_t, [z, v(u+w, z])$. Using (1.4.23) we can write

$$\|[z, v(u+w, z)]\| \le C\|z\|_2\|v[u+w, z]\|_{W_\infty^2} \le C_R\|z\|_2^2.$$

Thus we obtain that

$$|(u_t + w_t, [z, v(u+w, z])| \le C_R\left(\|u_t\| + \|w_t\|\right)\|z\|_2^2$$

and hence by (9.4.38) and (9.4.39) we have that $|P_0(z)| \le C_R\left(\|u_t\| + \|w_t\|\right)\|z\|_2^2$. In a similar way

$$|(u_t, [v(z,z), f])| \le C_R\|u_t\|\|z\|_2^2.$$

Therefore $|P(z)| \le C_R\left(\|u_t\| + \|w_t\|\right)\|z\|_2^2$. Now using (9.4.37) we conclude that there exists $\eta > 0$ such that

$$\left|\int_s^t (\widehat{\mathscr{R}}(z), z_t)d\tau\right| \le C_R \max_{\tau\in[s,t]} \|z(\tau)\|_{2-\eta}^2 + C_R\int_s^t \left(\|w_t\| + \|u_t\|\right)\|z\|_2^2 d\tau. \tag{9.4.40}$$

This implies that

$$\sup_{t\in[0,T]}\left|\int_{s+t}^{s+T} (\widehat{\mathscr{R}}(z), z_t)d\tau\right| \le C(R)\max_{\tau\in[0,T]}\|z(s+\tau)\|_{2-\eta}^2 + \varepsilon\int_s^{s+T}\|z\|_2^2 d\tau$$
$$+ C_\varepsilon(R)\int_s^{s+T}\left(\|u_t\|^2 + \|w_t\|^2\right)\|z\|_2^2 d\tau \tag{9.4.41}$$

for every $\varepsilon > 0$ and for some $\eta > 0$. In the case $L \equiv 0$ we have that

$$\int_0^\infty (\|u_t\|^2 + \|w_t\|^2)\,dt \le c_0 \int_0^\infty (D(u_t,u_t) + D(w_t,w_t))\,dt \le C_R < \infty,$$

thus we can conclude from (9.4.41) (see also Remark 8.5.2) that (8.5.1) is valid. Therefore we can apply Theorem 8.5.3 which provides us with a stabilizability estimate of the form

$$\|S_t y_1 - S_t y_2\|_H^2 \le C_R \|y_1 - y_2\|_H^2 e^{-\omega_R t} + C_R \max_{\tau\in[0,t]} \|u(\tau) - w(\tau)\|^2, \tag{9.4.42}$$

where $S_t y_1 = (u(t); u_t(t))$ and $S_t y_2 = (w(t); w_t(t))$ with $y_1, y_2 \in \mathscr{E}_R$. Then we apply Theorem 8.6.1 and conclude the proof of Theorem 9.4.6. □

9.4.11. Remark. The argument above shows that for any two functions $w(\tau)$ and $u(\tau)$ from the class $C(s,t;H^2(\Omega)) \cap C(s,t;L_2(\Omega))$ for some $s,t \in \mathbb{R}$, $s < t$, and such that

$$\|w(\tau)\|_2^2 + \|w_t(\tau)\|^2 \le R^2, \quad \|u(\tau)\|_2^2 + \|u_t(\tau)\|^2 \le R^2, \quad \tau \in [s,t],$$

relation (9.4.40) holds with $z = u - w$ and $\widehat{\mathscr{R}}(z) \equiv \mathscr{R}_0(z) = [u, v(u,u)] - [w, v(w,w)]$, where $v(u_1,u_2)$ is defined by (9.4.33) We use this observation in Chapter 10. ■

9.4.2 Hinged boundary conditions

Now we consider equations (9.4.1) and (9.4.2) with boundary conditions (9.4.4) for the displacement $u(x,t)$. As in the clamped case we have the following assertion stating the existence of the global attractor for models that do not contain nonconservative forces, or else the damping parameter is large.

9.4.12. Theorem. *Assume that Assumptions* 9.4.1 *and* 9.4.3 *hold. Let* (H,S_t) *be the dynamical system generated by problem* (9.4.1) *and* (9.4.2) *with hinged boundary conditions* (9.4.4) *in the space* $H = (H^2(\Omega) \cap H_0^1(\Omega)) \times L_2(\Omega)$. *Then*

- *The system* (H,S_t) *possesses a compact global attractor* A.
- *Let in addition Assumption* 9.4.4 *hold. The attractor* A *has a finite fractal dimension, if either the damping parameter* k *in equation* (9.4.1) *is large enough or else* $L \equiv 0$.

Proof. Problem (9.4.1), (9.4.2), and (9.4.4) can be presented (see the proof of Theorem 4.1.10) in the abstract form (9.4.13), if we introduce the following spaces and operators.

- $\mathscr{H} \equiv L_2(\Omega)$, $V \equiv L_2(\Omega)$.
- $\mathscr{A} u \equiv \Delta^2 u, u \in \mathscr{D}(\mathscr{A}) : \mathscr{D}(\mathscr{A}) \equiv \{u \in H^4(\Omega),\ u = \Delta u = 0 \text{ on } \Gamma\}$.
- $F(u) \equiv [u + f, v(u) + F_0] - Lu + p$, where $v(u) \in H_0^2(\Omega)$ satisfies (9.4.2).

- $D(u) \equiv d_0(x)g_0(u)$.

The only difference with the clamped case is the form of the domain of the operator $\mathscr{A}$ and the arguments are essentially identical to these in the case of clamped boundary conditions. We do not repeat them. We only note that (i) $F(u)$ has the same representation (9.4.14) as in the clamped case and to check the assumptions imposed on F, as in the proof of Theorem 9.1.12, we also use additional symmetry properties of von Karman bracket; (ii) the damping operators D are the same in both the clamped and hinged cases and the Lemmas 9.4.8 and 9.4.9 are valid with the same constants. We also have the stabilizability estimate (9.4.42) when $L \equiv 0$. Thus, the proof can be carried out as in the clamped case. □

9.4.3 Free boundary conditions

In this section we consider equations (9.4.1) and (9.4.2) with free boundary conditions (9.4.5) for the displacement $u(x,t)$. In contrast with the clamped and hinged cases the functional analytic setup is somewhat different. We have some extra terms in the case of the boundary conditions (9.4.5) (cf. (9.4.6) and (9.4.13)). Thus Theorems 8.2.3, 8.3.5 and 8.6.3 are not applied directly here.

As in the case of $\alpha > 0$ (cf. Assumption 9.1.13) we need additional hypotheses.

9.4.13. Assumption.

- Let $\nu_1 > 0$ and $\beta \in L_\infty(\Gamma)$, $\beta(x) \geq 0$ a.e.
- $F_0 \in H^4(\Omega) \cap H_0^2(\Omega)$.

We note that the hypothesis $F_0 = 0$ and $\nabla F_0 = 0$ on the free part of the boundary is needed to include the force $[u, F_0]$ in the potential energy. This hypothesis can be relaxed (see Remark 9.1.14).

Existence, uniqueness, and regularity of both weak and strong solutions follows from the results presented in Chapter 4, which, in turn, are based on the analysis of the abstract model of the form:

$$u_{tt} + kD(u_t) + \mathscr{A}u + D_0 D_0^* u_t + D_0 h(u) = F(u)$$

with the following choices of the operators and spaces.

- $\mathscr{H} = V = L_2(\Omega)$.
- $\mathscr{A}u = \Delta^2 u$, $D(\mathscr{A}) = \{u \in H^4(\Omega), \text{ with BC in } (9.4.5)\}$, $\mathscr{D}(\mathscr{A}^{1/2}) = H^2(\Omega)$.
- $D(v) = d_0(x)g_0(v)$.
- $F(u) = [u+f, v(u)+F_0] - Lu - p$, where $v(u)$ solves (9.4.2).
- $D_0 : L_2(\Gamma) \mapsto [\mathscr{D}(\mathscr{A}^{1/2})]'$ is given by $\int_\Omega D_0(u) v dx = \int_\Gamma \sqrt{\beta(x)} uv dx$ for every $v \in H^1(\Omega)$ and $D_0^* v = \sqrt{\beta} v\big|_\Gamma$ for $v \in H^1(\Omega)$.
- Nonlinear operator $h : \mathscr{D}(\mathscr{A}^{1/2}) \mapsto L_2(\Gamma)$ is defined by

$$h(u) = \sqrt{\beta(x)} u^3|_\Gamma = D_0^*[u^3] \quad \text{for } u \in H^2(\Omega).$$

The above model fits the framework of Chapter 4 which, in turn, provides existence, uniqueness, and regularity results necessary to define semiflow on the space $H = \mathscr{D}(\mathscr{A}^{1/2}) \times \mathscr{H}$. This leads to a well-defined dynamical system (H, S_t). In what follows we freely use notions of weak and strong solutions.

From the perspective of long-time behavior the presence of the terms D_0 and $h(u)$ complicates matters or, more precisely, prevents us from blindly jumping into the abstract model. Indeed, these two terms, unlike in the rotational model, do not satisfy the required hypotheses of boundedness in L_2. On the other hand, these terms are critical for dissipativity properties. For this reason, instead of appealing to abstract treatments, we present a separate treatment for this case.

We first note that any strong solution to problem (9.4.1), (9.4.2), (9.4.5) satisfies the following energy equality (see Theorem 4.1.13)

$$\mathscr{E}(u(t), u_t(t)) + \int_s^t \left[\int_\Omega k d_0 g_0(u_t) u_t dx + \int_\Gamma \beta(x) u_t^2 d\Gamma \right] d\tau$$
$$= \mathscr{E}(u(s), u_t(s)) - \int_s^t \int_\Omega L u u_t dx d\tau, \tag{9.4.43}$$

where

$$\mathscr{E}(t) \equiv \mathscr{E}(u, u_t) = \frac{1}{2} \int_\Omega |u_t|^2 dx + \frac{1}{2} a(u, u) + \Pi(u), \tag{9.4.44}$$

where $a(u,w) = a_0(u,w) + \nu_1 \int_\Gamma uw d\Gamma$ with $a_0(u,w)$ defined by (1.3.4) and

$$\Pi(u) = \frac{1}{4} \|\Delta v(u)\|^2 + \frac{1}{4} \int_\Gamma \beta u^4 dx - \frac{1}{2}([u,u], F_0) - ([f, F_0] + p, u), \tag{9.4.45}$$

where $v(u) \in H_0^2(\Omega)$ is defined by (9.1.2). In what follows we also use another energy variable which consists of a positive part of $\mathscr{E}$. We first set

$$\Pi_0(u) = \frac{1}{4} \|\Delta v(u)\|^2 + \frac{1}{4} \int_\Gamma \beta u^4 dx, \quad \Pi_1(u) = -\frac{1}{2}([u,u], F_0) - ([f, F_0] + p, u) \tag{9.4.46}$$

and

$$E(t) \equiv E(u, u_t) = \frac{1}{2} \int_\Omega |u_t|^2 dx + \frac{1}{2} a(u,u) + \Pi_0(u). \tag{9.4.47}$$

It is clear that $\mathscr{E}(u, u_t) = E(u, u_t) + \Pi_1(u)$. One can also see from Lemma 1.5.10 that

$$|\Pi_1(u)| \le \eta \left(a(u,u) + \Pi_0(u) \right) + C_\eta, \quad u \in H^2(\Omega),$$

for any $\eta > 0$. Therefore as in Section 4.1 (cf. (4.1.21)) we can conclude that there exist positive constants c_0, c_1, and K such that

$$c_0 \mathscr{E}(u, u_t) - K \le E(u, u_t) \le c_1 \mathscr{E}(u, u_t) + K \tag{9.4.48}$$

for all $(u; u_t) \in H^2(\Omega) \times H^1(\Omega)$.

Assume that $\beta(x) > 0$ a.e. Then the dissipativity of the dynamics (with both $L \neq 0$ and $L = 0$) follows from the previous arguments. The presence of the terms with β does not contribute to loss of dissipativity. Indeed, the nonlinear boundary term provides a fully conservative contribution, and the linear boundary damping adds dissipativity. The arguments totally parallel those described in Chapter 8. To see this, from (9.4.1) and from Green's formula we have that

$$\frac{d}{dt}(u_t,u) = \|u_t\|^2 - k\int_\Omega d_0 g_0(u_t)u dx - \int_\Gamma \beta u_t u d\Gamma + ([f,F_0]+p,u)$$
$$- \{a(u,u) + 4\Pi_0(u) + ([f,u],v(u)) - ([u,F_0],u) + (Lu,u)\}$$

for any strong solution u. This relation and Lemma 1.5.13 easily imply that

$$\frac{d}{dt}(u_t,u) \leq \|u_t\|^2 - k\int_\Omega d_0 g_0(u_t)u dx - \int_\Gamma \beta u_t u d\Gamma - \left\{\frac{1}{2}a(u,u) + 2\Pi_0(u)\right\} + C. \tag{9.4.49}$$

By using Lyapunov-type function $V(u,u_t) = \mathscr{E}(u,u_t) + \nu\cdot(u,u_t)$ and relations (9.4.43) and (9.4.49) we can obtain dissipativity estimates, independent of the size of the damping parameter, as in the clamped and hinged case. Thus, the system can be easily shown to be dissipative in both the nonconservative ($L \neq 0$) and conservative ($L = 0$) case under appropriate conditions concerning the damping function. We refer to [75] for more details at the abstract level.

The main result in this section is the following theorem.

9.4.14. Theorem. *Let Assumptions* 9.4.1, 9.4.3 *and* 9.4.13 *be in force. Assume that* (H,S_t) *is the dynamical system generated by problem* (9.4.1) *and* (9.4.2) *with free boundary conditions* (9.4.5) *in the space* $H = H^2(\Omega)\times L_2(\Omega)$. *Let either* $\beta(x) > 0$ *a.e. or else* $L \equiv 0$. *Then*

- *The system* (H,S_t) *possesses a compact global attractor* A.
- *Under the additional Assumption* 9.4.4, *the attractor* A *has a finite fractal dimension, provided that either damping parameter* k *is large enough or else* $L \equiv 0$.

From this point on the analysis is very similar to that developed in the clamped or hinged case. Asymptotic smoothness can be proved by using the same compensated compactness criterion, and the stabilizability estimate shows that the system is quasi-stable. However, the technical details, due to the more complex structure of boundary conditions are somewhat different. For this reason we provide a separate proof. Before we do this, few remarks are in order.

We first note that when $\beta = 0$ (and hence $L \equiv 0$) the problem fits directly in our abstract framework and generates a gradient system. Thus, verification of the abstract hypotheses (as in the cases of other boundary conditions) would suffice for proving the theorem. However, when $\beta > 0$, the treatment requires extra care. For this reason we provide complete arguments without resorting to an abstract framework.

We also mention that the presence of the boundary damping in the model may suggest that this damping should suffice to stabilize the plate. This is indeed true, but provided that $\inf_\Omega \beta$ is strictly positive (see Section 10.4 in Chapter 10). In the case of nonnegative β, boundary dissipation is not strong enough. Without assuming the lower bound on the parameter $\beta(x)$, the said uniform stability is false. Internal damping becomes the major carrier of stability.

Proof of Theorem 9.4.14.

The main goal is to establish some kind of compactness/stabilizability estimate valid for a difference of two solutions $z = u - w$. The nonlinearity is critical, therefore one must use some special features of the problem in order to control critical terms. We split our argument into several steps.

Preliminaries. Let $z \equiv z(t) = u(t) - w(t)$, where $U(t) = (u(t); u_t(t)) = S_t y_1$ and $W(t) = (w(t); w_t(t)) = S_t y_2$ are two (strong) solutions corresponding to the semiflow S_t and residing in some bounded, positively invariant set that belongs to the ball of the radius R in the space H. We can assume that R is independent of the damping parameter k. In particular $\|U(t)\|_H \le R$ and $\|W(t)\|_H \le R$.

The variable $(z; z_t)$ satisfies

$$z_{tt} + \Delta^2 z + k d_0 (g_0(z_t + w_t) - g_0(w_t)) = \mathscr{R}(z) \tag{9.4.50}$$

with free type boundary conditions generated by (9.4.5). Here $\mathscr{R}(z) \equiv F(u) - F(w)$. We denote

$$E_z(t) \equiv \frac{1}{2}\int_\Omega [|z_t|^2 + a(z,z)]dx,$$

$$D_s^t(z) \equiv k\int_s^t \int_\Omega d_0(x)(g_0(z_t + w_t) - g_0(w_t))z_t dx d\tau + \int_s^t \int_\Gamma \beta(x) z_t^2 d\Gamma d\tau, \tag{9.4.51}$$

$$lot(z) \equiv C_R \sup_{t\in[0,T]} \|z(t)\|_{2-\eta}, \quad lot_0(z) \equiv C_R \sup_{t\in[0,T]} \|z(t)\|^2. \tag{9.4.52}$$

By appealing to properties of the von Karman bracket and Airy stress function (see Theorem 1.4.3 and Corollary 1.4.5 in Chapter 1) we obtain that $\|\mathscr{R}(z(t))\| \le C_R\|z(t)\|_2$ for all ≥ 0. Moreover, the following energy relation holds true:

$$E_z(t) + D_s^t(z) = E_z(s) + \int_s^t \int_\Omega \mathscr{R}(z) z_t dx d\tau - \int_s^t \int_\Gamma \beta(u^2 + uw + w^2) z z_t d\Gamma d\tau. \tag{9.4.53}$$

Using

$$\int_s^t \int_\Gamma \beta(u^2 + uw + w^2) z z_t d\Gamma d\tau \le \delta\left[\int_s^t \int_\Gamma \beta z_t^2 d\Gamma d\tau + \int_s^t E_z(\tau) d\tau\right] + c_\delta T lot_0(z)$$

for $0 \le s \le t \le T$, one eliminates boundary terms in the energy inequality (9.4.53), obtaining

$$E_z(t) + \frac{1}{2}D_s^t(z) \le E_z(s) + \delta\int_s^t E_z(\tau)d\tau + \int_s^t \int_\Omega \mathscr{R}(z) z_t dx d\tau + c_\delta T lot_0(z) \tag{9.4.54}$$

and

$$E_z(s) \le E_z(t) + \delta \int_s^t E_z(\tau)d\tau + 2D_s^t(z) + \int_s^t \int_\Omega \mathscr{R}(z) z_t dx d\tau + c_\delta T\, lot_0(z) \quad (9.4.55)$$

for all $0 \le s \le t \le T$ with arbitrary small $\delta > 0$. By (9.4.43) single trajectories satisfy the bound (with $C_L = 0$ when $L \equiv 0$):

$$\mathscr{E}(t) + k \int_s^t \int_\Omega d_0 g_0(u_t) u_t + \int_s^t \int_\Gamma \beta u_t^2 \le \mathscr{E}(s) + C_L \int_s^t \|u_t\| \|u\|_{2-\eta}. \quad (9.4.56)$$

Step 1: Asymptotic smoothness. The main estimate leading to asymptotic smoothness is presented in the following proposition.

9.4.15. Proposition. *Under the hypotheses of the first part in Theorem* 9.4.14 *there exists* $\varepsilon_0 > 0$ *such that for all* $0 < \varepsilon < \varepsilon_0$ *and* $T > 1$, *there exist positive constants* $c_0, C_{R,\varepsilon}, C_T$ *such that*

$$\int_0^T E_z(t)dt \le \varepsilon T + c_0 D_0^T(z) + C_{R,\varepsilon} + C_T\, lot(z). \quad (9.4.57)$$

Proof. *Step 1—Reconstruction of potential energy:* Standard equipartition of energy argument (multiply equation (9.4.50) by z and integrate by parts), followed by the use of assumptions imposed on the damping, give

$$\int_0^T E_z(t)dt \le c_0 \left[\int_Q |z_t|^2 dQ + E_z(T) + E_z(0) \right] \quad (9.4.58)$$
$$+ k \left| \int_Q d_0 (g_0(z_t + w_t) - g_0(w_t)) z dQ \right| + \left| \int_\Sigma \beta(x) z_t z d\Sigma \right| + C_T\, lot_0(z),$$

where $Q = \Omega \times [0,T]$ and $\Sigma = \partial\Omega \times [0,T]$.

Step 2—Estimate for the nonlinear damping term: From (9.4.56) each single trajectory satisfies

$$\mathscr{E}(T) + k \int_0^T \int_\Omega d_0 g_0(u_t) u_t + \int_0^T \int_\Gamma \beta u_t^2 \le \mathscr{E}(0) + T C_R. \quad (9.4.59)$$

Hence, using (9.4.48) we obtain that[7]

$$\int_0^T \int_\Omega d_0 |(g_0(z_t + w_t) - g_0(w_t)) z| dQ$$
$$\le |z|_{C(Q)} \left[\int_0^T \int_\Omega d_0 \left(g_0(u_t) u_t + g_0(w_t) w_t \right) dQ + CT \right]$$

[7] Note the linear bound in the expression for $lot(z)$ below. The requirements in Assumptions 9.4.1 and 9.4.3 concerning the damping function prevent quadratic growth of lower-order terms yielding only linear growth.

$$\le \sup_{t\in[0,T]} \|z\|_{2-\eta} \left[\int_0^T \int_\Omega d_0(g_0(u_t)u_t + g_0(w_t)w_t) dQ + CT \right]$$
$$\le C \sup_{t\in[0,T]} \|z\|_{2-\eta}(E(0)+K+C_R T) \le C_T lot(z), \tag{9.4.60}$$

where we have also used embedding $H^{2-\eta}(\Omega) \subset C(\Omega)$ and the obvious relation $|g_0(s)| \le c_0 + s g_0(s)$ with $c_0 = \max_{|s|\le 1} |g_0(s)|$.

A similar estimate holds for the boundary damping:

$$\left| \int_0^T \int_\Gamma \beta(x) z_t z \right| = \frac{1}{2} \left| \int_\Gamma \beta(z^2(T)(x) - z^2(0)(x)) d\Gamma \right| d\Sigma \le C_T lot(z). \tag{9.4.61}$$

Step 3—Completion of the argument: From the assumption imposed on the damping in (9.4.9) in the case $L \equiv 0$ we have that

$$\int_s^t \int_\Omega z_t^2 dx d\tau \le 2 \int_s^t \int_\Omega (u_t^2 + w_t^2) dx d\tau$$
$$\le \varepsilon(t-s) + C_\varepsilon \int_s^t \int_\Omega d_0(x)(u_t g_0(u_t) + w_t g_0(w_t)) dx d\tau.$$

Therefore the energy relation in (9.4.56) yields that

$$\int_0^T \int_\Omega z_t^2 dx d\tau \le \varepsilon T + C_{R,\varepsilon} \quad \text{when} \quad L \equiv 0.$$

In the case when $L \not\equiv 0$ from Assumption 9.4.3 we have that

$$\int_s^t \int_\Omega z_t^2 dx d\tau \le c_0 \int_s^t \int_\Omega d_0(x)(g_0(u_t) - g_0(w_t)) z_t dx d\tau \le c_0 D_s^t(z).$$

Thus in the both cases of L we have that

$$\int_0^T \int_\Omega z_t^2 dx d\tau \le \varepsilon T + c_0 D_0^T(z) + C_{R,\varepsilon}. \tag{9.4.62}$$

Recalling $E_z(0) + E_z(T) \le C(R)$, the proof of Proposition 9.4.15 is completed upon collection of results in (9.4.58)–(9.4.62). □

From Proposition 9.4.15 and the energy inequality (9.4.54) we obtain the following.

9.4.16. Proposition. *Let $\mathscr{R}(z) \equiv F(u) - F(w)$. Then*

$$E_z(T) \le \varepsilon + \frac{C_\varepsilon(R)}{T} \left[1 + \left| \int_0^T (\mathscr{R}(z), z_t) \right| + \left| \int_0^T \int_s^T (\mathscr{R}(z), z_t) \right| \right] + C_\varepsilon lot(z).$$

Proof. The argument is basically the same as in the proof of Lemmas 8.3.7 and 8.3.8. We do not repeat it here. □

At this point we are in a position to connect with abstract Theorem 7.1.11. Let

$$\Psi_{R,T,\varepsilon}(U_0,W_0) \equiv C_{T,R,\varepsilon}\left[\left|\int_0^T(\mathscr{R}(z),z_t)\right| + \left|\int_0^T\int_s^T(\mathscr{R}(z),z_t)\right|\right] + C_T\, lot(z),$$

so $E_z(T) \le \varepsilon + \Psi_{\varepsilon,R,T}(U_0,W_0)$, for sufficiently large T, where the functional $\Psi_{\varepsilon,R,T}$ satisfies compensated compactness property in Theorem 7.1.11. For this latter claim, we provide the following line of argument which is an extension of the argument developed in [160] (see also argument in the proof of Proposition 8.3.6).

Let $w^n(t)$ be a sequence of solutions corresponding to initial data $y^n \equiv (w_0^n; w_1^n)$ from a bounded positively invariant set. By choosing a subsequence we can assume that

$$y^n(t) \equiv (w^n(t); w_t^n(t)) \to y(t) \equiv (w(t); w_t(t)) \tag{9.4.63}$$

$*$-weakly in $L_\infty(0,T;H)$ for some solution $(w(t); w_t(t)) \in L_\infty(0,T;H)$ and

$$\sup_{[0,T]}\left\{\|z^{n,m}(t)\|_{2-\eta}\right\} \to 0, \quad n,m\to\infty, \tag{9.4.64}$$

for some $\eta > 0$, where $z^{n,m}(t) \equiv w^n(t) - w^m(t)$.

By (9.4.64) we have that $lot(z^{n,m}) \to 0$. Therefore our proof is completed as soon as we show that

$$\lim_n \lim_m \hat{\Psi}_T(y^n, y^m) \to 0, \tag{9.4.65}$$

where

$$\hat{\Psi}_T(U_0,W_0) = \left|\int_0^T(\mathscr{R}(z),z_t)\right| + \left|\int_0^T\int_s^T(\mathscr{R}(z),z_t)\right|.$$

Because (for smooth solutions) we have

$$\begin{aligned}(\mathscr{R}(z),z_t) - (Lz,z_t) = \frac{1}{4}\frac{d}{dt}\left[-\|\Delta v(u)\|^2 - \|\Delta v(w)\|^2 + 2\int_\Omega [z,z]F_0 dx\right] \\ - ([v(w),w+f],u_t) - ([v(u),u+f],w_t),\end{aligned}$$

integrating in time one obtains

$$\begin{aligned}\int_t^T(\mathscr{R}(z),z_t)ds &= \frac{1}{4}\left[\|\Delta v(u)(t)\|^2 + \|\Delta v(w)(t)\|^2\right] \\ &\quad -\frac{1}{4}\left[\|\Delta v(u)(T)\|^2 + \|\Delta v(w)(T)\|^2\right] \\ &\quad +\frac{1}{2}\int_\Omega [z(T),z(T)]F_0 dx - \frac{1}{2}\int_\Omega [z(t),z(t)]F_0 dx \\ &\quad +\int_t^T(Lz,z_t)ds - \int_t^T[([v(w),w+f],u_t) + ([v(u),u+f],w_t)]ds.\end{aligned}$$

An important remark is that all the terms except the last one are compact on the finite energy space. This along with (9.4.63) and (9.4.64) implies

$$\lim_n \lim_m \int_t^T(\mathscr{R}(z^{n,m}), z_t^{n,m})ds = \frac{1}{2}\left[\|\Delta v(w)(t)\|^2 - \|\Delta v(w)(T)\|^2\right]$$

$$- \lim_n \lim_m \int_t^T [([v(w^n), w^n + f], w_t^m) + ([v(w^m), w^m + f], w_t^n)] \, ds. \quad (9.4.66)$$

As for the second term in (9.4.66), we use the convergence in (9.4.63) along with the relation $\|[v(w^n(t)), w^n(t) + f]\|_{0,\Omega} \le C_R$. The latter follows from the boundedness of the von Karman bracket (see Corollary 1.4.5). Thus (9.4.64) implies the convergence

$$[v(w^n), w^n + f] \to [v(w), w + f] \quad *\text{-weakly in } L_\infty(0, T; L_2(\Omega)) \text{ as } n \to \infty,$$

so we can conclude that

$$\lim_n \lim_m \int_t^T [([v(w^n), w^n + f], w_t^m) + ([v(w^m), w^m + f], w_t^n)] \, ds \quad (9.4.67)$$
$$= 2 \int_t^T ([v(w), w + f], w_t) ds = \frac{1}{2} \left[-\|\Delta v(w)(T)\|^2 + \|\Delta v(w)(t)\|^2 \right].$$

Combining (9.4.66) and (9.4.67) yields $\lim_n \lim_m \int_t^T (\mathscr{R}(z^{n,m}), z_t^{n,m}) ds = 0$, hence the functional $\Psi_{\varepsilon,R,T}$ satisfies the compensated compactness property postulated by Theorem 7.1.11. The proof of asymptotic smoothness is completed. Because the conservative ($L = 0$) system is a gradient system, and nonconservative system ($L \neq 0$) has an absorbing ball, with a radius independent of the damping parameter size, by Theorem 7.2.3 or Corollary 7.5.7 asymptotic smoothness also yields existence of compact attractors, hence implying the first part of Theorem 9.4.14.

Step 2: Finite dimensionality. In the case when the nonlinear force is conservative, our main goal is to establish the property of quasi-stability. In the nonconservative case, the result requires a large damping parameter. This latter case can be easily deduced from the corresponding treatments of the abstract model in the proof of Theorem 8.5.6. The large parameter technique is not sensitive with respect to addition of nonconservative forces (of course, at the expense of enlarging the parameter). Instead, for the model considered quasi-stability depends on the finiteness of the dissipation integral. The latter loses this property in the nonconservative case. For this reason we focus on quasi-stability for conservative models only (i.e., in the case $L \equiv 0$). Thus, our goal is to establish the following assertion.

9.4.17. Proposition. *Under the hypotheses of Theorem* 9.4.14 *(including Assumption* 9.4.4 *in the case* $L \equiv 0$*) the following stabilizability estimate*[8] *holds true: there exist constants* $\omega > 0$ *and* $C > 0$ *depending on* R *such that*

$$E_z(t) \le C E_z(0) e^{-\omega t} + lot_0(z), \quad t > 0, \quad (9.4.68)$$

where $lot_0(z) = C_R \max_{0 \le s \le T} \|z(s)\|^2$.

[8] Note that the lower-order term in (9.4.68) is quadratic, unlike Proposition 9.4.15. This is an important aspect of the estimate.

Proof. We use the same idea as in the proof of the second statement of Theorem 9.4.6 (see (9.4.40) and (9.4.41)) and start with the following lemma.[9]

9.4.18. Lemma. *Assume $w(\tau)$ and $u(\tau)$ are two functions from the class*

$$C(s,t;H^2(\Omega))\cap C^1(s,t;L_2(\Omega)),$$

for some $s,t\in\mathbb{R}$, $0\le t-s\le T_0$, such that

$$\|w(\tau)\|_2^2+\|w_t(\tau)\|^2\le R^2,\quad \|u(\tau)\|_2^2+\|u_t(\tau)\|^2\le R^2,\quad \tau\in[s,t].$$

Let $z(\tau)=u(\tau)-w(\tau)$ and $\mathscr{R}(z)=F(u)-F(w)$ as before. Then there exists $\eta>0$ such that

$$\left|\int_s^t(\mathscr{R}(z),z_t)d\tau\right|\le C_{R,T_0,\varepsilon}\max_{\tau\in[s,t]}\|z(\tau)\|_{2-\eta}^2 \tag{9.4.69}$$
$$+C_R\int_s^t(\|w_t\|+\|u_t\|)\,\|z\|_2^2d\tau+\varepsilon\int_s^t\|z_t\|^2d\tau.$$

Proof. The argument relies on the structure of the von Karman bracket and resulting cancellations. One can see that

$$F(u)-F(w)\equiv\mathscr{R}(z)=\widehat{\mathscr{R}}(z)+\mathscr{R}_*(z),$$

where

$$\widehat{\mathscr{R}}(z)=[u,v(u,u)]-[w,v(w,w)]+[z,F_0+2v(f,u)],$$
$$\mathscr{R}_*(z)=2[w,v(f,z)]+[f,v(u+w+2f,z)]-Lz.$$

Here we also denote by $v(u_1,u_2)\in H_0^2(\Omega)$ the solution to the problem

$$\Delta^2v(u_1,u_2)=-[u_1,u_2]\ \text{in}\ \Omega,\quad \frac{\partial}{\partial n}v(u_1,u_2)=v(u_1,u_2)=0\ \text{on}\ \Gamma. \tag{9.4.70}$$

Therefore using Lemma 9.4.10 we obtain that

$$(\mathscr{R}(z),z_t)=\frac14\frac{d}{dt}Q(z)+\frac12P(z)+(\mathscr{R}_*(z),z_t), \tag{9.4.71}$$

where

$$Q(z)=(v(u)+v(w),[z,z])-\|\Delta v(u+w,z)\|^2+2(F_0+2v(f,u),[z,z]),$$

and

$$P(z)=-(u_t,[u+2f,v(z,z)])-(w_t,[w,v(z,z)])-(u_t+w_t,[z,v(u+w,z)]). \tag{9.4.72}$$

[9] A similar assertion was proved earlier in the case when $f\equiv0$, $F_0\equiv0$, and $L\equiv0$; see Remark 9.4.11

To estimate the values Q and P we use the following estimate for the von Karman bracket

$$|[u_1,u_2]|_{-1-2\delta} \le C\|u_1\|_{2-\delta}\|u_2\|_{2-\delta}, \quad 0<\delta<1/2, \tag{9.4.73}$$

and also Airy's stress function property

$$\|v(u_1,u_2)\|_{W^2_\infty(\Omega)} \le C\|u_1\|_2\|u_2\|_2. \tag{9.4.74}$$

Using (9.4.73) and elliptic regularity of Δ^2 with the Dirichlet boundary conditions one can see that

$$|Q(z)| \le C_R\|z\|^2_{2-\eta} \quad \text{for some } \eta>0. \tag{9.4.75}$$

Now we estimate $P(z)$. Using (9.4.74) we have that

$$|(u_t,[u,v(z,z)])| \le C\|u_t\|\|[u,v(z,z)]\| \le C\|u_t\|\|u\|_2\|z\|^2_2 \le C_R\|u_t\|\|z\|^2_2, \tag{9.4.76}$$

and, similarly,

$$|(w_t,[w,v(z,z)])| \le C_R\|w_t\|\|z\|^2_2. \tag{9.4.77}$$

Now we consider the third term $(u_t+w_t,[z,v(u+w,z]))$. As above we can write

$$\|[z,v(u+w,z)]\| \le C\|z\|_2\|v[u+w,z]\|_{W^2_\infty(\Omega)} \le C_R\|z\|^2_2.$$

Thus we obtain

$$|(u_t+w_t,[z,v(u+w,z])| \le C_R\left(\|u_t\|+\|w_t\|\right)\|z\|^2_2$$

and hence by (9.4.76) and (9.4.77) we have that

$$|P(z)| \le C_R\left(\|u_t\|+\|w_t\|\right)\|z\|^2_2.$$

One can also see that

$$|(\mathscr{R}_*(z),z_t)| \le C_R\|z\|_{2-\eta}\|z_t\| \le \varepsilon\|z_t\|^2 + C_{R,\varepsilon}\|z\|^2_{2-\eta}.$$

Now using (9.4.75) we obtain (9.4.69). □

We return to energy relation (9.4.58).

By exploiting assumption (9.4.11) imposed on the damping and also relations (9.4.30) and (9.4.31) with $\varepsilon = C\delta^{-1}$, dissipation integrals are estimated as follows,

$$k\left|\int_0^T\int_\Omega d_0(g_0(z_t+w_t)-g_0(w_t))zdxdt\right| \le \delta D_0^T(z) + C_{\delta,R,T}\max_{[0,T]}\|z\|^2_{2-\eta} \tag{9.4.78}$$

and

$$\left|\int_0^T\int_\Gamma \beta z_t z d\Gamma dt\right| \le \delta D_0^T(z) + C_{\delta,R,T}\max_{[0,T]}\|z\|^2_{2-\eta} \tag{9.4.79}$$

for and $\delta > 0$ and for some $\eta > 0$, where $D_0^T(z)$ is defined in (9.4.51).

The above along with (9.4.58) and obvious inequality

$$\int_0^T \int_\Omega z_t^2 dxdt \le \frac{c}{k} D_0^T(z)$$

lead to the following *"reconstruction"* inequality:

$$\int_0^T E(t)dt \le c\left[k^{-1}D_0^T(z) + E(T) + E(0)\right] + C_{T,R} \max_{\tau\in[0,T]} \|z(\tau)\|_{2-\eta}^2, \tag{9.4.80}$$

where $c > 0$ is a constant independent of T and R.

Relation (9.4.80) states that the energy is reconstructed, modulo lower-order terms, from the damping. In order to infer the stabilizability inequality one should relate the dissipation term $D_0^T(z)$ to the difference $E(0) - E(T)$. As usual, this is done with the help of an energy identity. However, in doing so, we encounter the main difficulty which is the presence of a critical (noncompact) term $(\mathscr{R}(z), z_t)$. This difficulty can be circumvented when the damping coefficient is assumed large enough. In fact, when $L \neq 0$ the damping parameter is assumed large, hence the arguments leading from (9.4.80) to quasi-stability are reasonably straightforward, hence omitted.[10] However, the general case with an arbitrary damping parameter presents a challenge. To handle this, we use critically the "trick" introduced by Lemma 9.4.18. In order to handle critical terms, we apply Lemma 9.4.18 which introduces velocity terms $\|w_t\|$ and $\|u_t\|$ in the last integration in (9.4.69). These terms act as "small" parameters for large values of time. As we show below, this allows for absorption of noncompact contribution in energy relation (9.4.54). The details are given below.

From Lemma 9.4.18 we obtain the estimate for the $\mathscr{R}(z)z_t$ term:

$$\begin{aligned} \left|\int_0^T (\mathscr{R}(z), z_t)dt\right| &\le C_\varepsilon(R,T) \max_{[0,T]} \|z\|_{2-\eta}^2 + \varepsilon \int_0^T \left[\|z\|_2^2 + \|z_t\|^2\right] dt \\ &\quad + C_\varepsilon(R) \int_0^T \left(\|u_t\|^2 + \|w_t\|^2\right) \|z\|_2^2 dt \end{aligned} \tag{9.4.81}$$

for every $\varepsilon > 0$ and for some $\eta > 0$. Thus using the energy relation (9.4.54) we find from (9.4.81) that

$$\begin{aligned} E_z(T) + \frac{1}{2} D_s^T(z) &\le E_z(s) + \varepsilon \int_0^T [\|z\|_2^2 + \|z_t\|^2] ds \\ &\quad + C_\varepsilon(R) \int_0^T \left(\|u_t\|^2 + \|w_t\|^2\right) \|z\|_2^2 dt + C_\varepsilon(R,T) \max_{[0,T]} \|z\|_{2-\eta}^2. \end{aligned} \tag{9.4.82}$$

After integration with respect to s over the interval $[0,T]$ and choosing $\varepsilon = (2T)^{-1}$ we have

[10] See the proof of Lemma 3.59 in [75] for the abstract model and also the argument given in Theorem 8.5.6.

$$TE_z(T) + \int_0^T E_z(s)ds \le c_0 \int_0^T E_z(s)ds$$
$$+ C(R,T) \int_0^T \left(\|u_t\|^2 + \|w_t\|^2\right) \|z\|_2^2 dt + C(R,T) \max_{[0,T]} \|z\|_{2-\eta}^2.$$

Similarly, from (9.4.55) we have

$$E_z(0) \le E_z(T) + 2D_0^T(z) + \varepsilon \int_0^T E_z(s)ds$$
$$+ C_\varepsilon(R) \int_0^T \left(\|u_t\|^2 + \|w_t\|^2\right) \|z\|_2^2 dt + C_\varepsilon(R,T) \max_{[0,T]} \|z\|_{2-\eta}^2.$$

Thus by (9.4.80)

$$TE_z(T) + \int_0^T E_z(s)ds \le cD_0^T(z) + C(R,T) \int_0^T \left(\|u_t\|^2 + \|w_t\|^2\right) \|z\|_2^2 dt$$
$$+ C(R,T) \max_{\tau\in[0,T]} \|z(\tau)\|_{2-\eta}^2 \tag{9.4.83}$$

for all $T \ge T_0$ with some $T_0 \ge 1$. From (9.4.82) we also have that

$$\frac{1}{2} D_0^T(z) \le E_z(0) - E_z(T) + \varepsilon \int_0^T E_z(s)ds$$
$$C_\varepsilon(R) \int_0^T \left(\|u_t\|^2 + \|w_t\|^2\right) \|z\|_2^2 dt + C_\varepsilon(R,T) \max_{[0,T]} \|z\|_{2-\eta}^2$$

for any $\varepsilon > 0$. Therefore (9.4.83) implies

$$TE_z(T) + \int_0^T E_z(s)ds \le c\,(E_z(0) - E_z(T)) + C(R,T) \int_0^T \left(\|u_t\|^2 + \|w_t\|^2\right) \|z\|_2^2 dt$$
$$+ C(R,T) \max_{\tau\in[0,T]} \|z(\tau)\|_{2-\eta}^2.$$

By (9.4.54) we have that

$$\max_{\tau\in[0,T]} \|z(\tau)\|_2^2 \le c_0 \max_{\tau\in[0,T]} E_z(\tau) \le c_0 \left[E_z(0) + c_R \int_0^T E_z(s)ds + lot_0(z) \right],$$

where $lot_0(z) = C_R \max_{0\le s\le T} \|z(s)\|^2$. Therefore, because

$$\|z\|_{2-\eta}^2 \le \delta \|z\|_2^2 + c_\delta \|z\|^2 \quad \text{for every } \delta > 0,$$

after an appropriate choice of δ we obtain

$$(T+c)E_z(T) \le (c+c_0)E_z(0)$$
$$+ C(R,T) \int_0^T \left(\|u_t\|^2 + \|w_t\|^2\right) \|z\|_2^2 dt + C(R,T) lot_0(z)$$

for all $T \geq T_0$. Now, by already known abstract argument (see, e.g., the proof of Theorem 8.5.3) we obtain the estimate

$$E_z(t) \leq \left[C_1 E_z(0) e^{-\omega t} + C_2 \sup_{\tau \in [0,t]} \|z(\tau)\|^2 \right] \exp\left\{ C_2 \int_0^t \mathscr{K}(\tau)\, d\tau \right\}, \qquad (9.4.84)$$

for all $t \geq 0$, where $\mathscr{K}(\tau) = \|u_t(\tau)\|^2 + \|w_t(\tau)\|^2$. The key element now is the fact that $\mathscr{K}(t) \in L_1(\mathbb{R}_+)$, the property that reflects smallness of $w_t(t)$ and $u_t(t)$ for large time. This, in turn, follows from the lower bound in (9.4.11) (see also (9.4.12)) which implies

$$\int_0^t \mathscr{K}(\tau)\, d\tau \leq C \left[\int_0^t \int_\Omega d_0 g_0(u_t) u_t dx d\tau + \int_0^t \int_\Omega d_0 g_0(w_t) w_t dx d\tau \right]$$

for all $t \geq 0$, and consequently (inasmuch as $L = 0$)

$$\int_0^t \mathscr{K}(\tau)\, d\tau = \int_0^t [||u_t||^2 + ||w_t||^2] d\tau \leq C_R \quad \text{for all } t \geq 0. \qquad (9.4.85)$$

Therefore by (9.4.84) and (9.4.85) we obtain the estimate in (9.4.68). □

Proposition 9.4.17 allows us to apply the abstract Theorem 7.9.6 in Chapter 7, which asserts a finite fractal dimension of the attractor.

9.5 Further properties of the attractor for von Karman model without rotational inertia

In this section we study properties of the global attractor for the dynamical system (H, S_t) generated by problem (9.4.1) and (9.4.2) with one of the boundary conditions (9.4.3), or (9.4.4), or (9.4.5).

Below we assume the validity of the hypotheses of Section 9.4 that guarantee the existence of the compact global attractor.

9.5.1 Regular structure of the attractor

For the case considered here we have results similar to the case $\alpha > 0$ (see Section 9.2).

Consider the problem

$$\Delta^2 u - [u + f, v + F_0] + Lu = p(x), \quad x \in \Omega, \qquad (9.5.1)$$

$$\Delta^2 v + [u + 2f, u] = 0, \ x \in \Omega, \quad v|_\Gamma = \left.\frac{\partial v}{\partial n}\right|_\Gamma = 0, \qquad (9.5.2)$$

where the displacement $u(x)$ satisfies either clamped (9.4.3), hinged (9.4.4), or free (9.4.5) which for time-independent solutions has the form

$$\begin{aligned} &\Delta u + (1-\mu)B_1 u = 0 \ \text{ on } \Gamma, \\ &\frac{\partial}{\partial n}\Delta u + (1-\mu)B_2 u = \nu_1 u + \beta u^3 \ \text{ on } \Gamma. \end{aligned} \tag{9.5.3}$$

Here $\nu_1 > 0$ and $\beta \in L_\infty(\Gamma)$ is a nonnegative function, $0 < \mu < 1$; the boundary operators B_1 and B_2 are defined by (1.3.20).

We denote by $\mathcal{N}^*$ the set of all solutions to problem (9.5.1) and (9.5.2) with the corresponding boundary conditions (either (9.4.3), (9.4.4), or (9.5.3)).

We begin by listing properties that result from gradient structure of the attractor. For this, we assume that nonconservative forces are absent, (i.e., $L = 0$). Indeed, assuming that the operator L is absent in (9.1.1), then the energy

$$\mathcal{E}(u,u_t) = \frac{1}{2}\int_\Omega |u_t|^2 dx + \Pi(u),$$

is a strict Lyapunov function for (H,S_t) and hence (H,S_t) is a gradient system (see the definitions in Section 7.5). Therefore the application of Theorem 7.5.6 gives the following result.

9.5.1. Theorem. *Assume that* $L \equiv 0$ *in* (9.4.1). *Let* (H,S_t) *be the dynamical system generated by problem* (9.4.1) *and* (9.4.2) *with one of the boundary conditions* (9.4.3), (9.4.4), *or* (9.4.5). *Assume that* (H,S_t) *possesses a compact global attractor* A. *Then*

- $A = \mathcal{M}^u(\mathcal{N})$, *where* $\mathcal{N} = \{(w;0) : w \in \mathcal{N}^*\}$ *is the set of stationary points of* (H,S_t) *and* $\mathcal{M}^u(\mathcal{N})$ *is the the unstable manifold* $\mathcal{M}^u(\mathcal{N})$ *emanating from* $\mathcal{N}$.
- *The global attractor* A *consists of full trajectories* $\gamma = \{(u(t);u_t(t)) : t \in \mathbb{R}\}$ *such that*

$$\lim_{t\to\pm\infty}\left\{\|u_t(t)\|^2 + \inf_{w\in\mathcal{N}^*}\|u(t)-w\|_2^2\right\} = 0.$$

- *Any generalized solution* $u(t)$ *to problem* (9.4.1) *and* (9.4.2) *with the corresponding boundary conditions stabilizes to the set of stationary points; that is,*

$$\lim_{t\to+\infty}\left\{\|u_t(t)\|^2 + \inf_{w\in\mathcal{N}^*}\|u(t)-w\|_2^2\right\} = 0. \tag{9.5.4}$$

This theorem and also the results on the finiteness of the number of solutions to problem (9.5.1) and (9.5.2) given in Theorem 1.5.7 and Theorem 1.5.16 allow us to obtain the following assertion.

9.5.2. Corollary. *Under the hypotheses of Theorem* 9.5.1 *there exists an open dense set* $\mathcal{R}_0$ *in* $L_2(\Omega)$ *such that for every* $p \in \mathcal{R}_0$ *the set* $\mathcal{N}$ *of stationary points for* (H,S_t) *is finite. In this case* $A = \cup_{i=1}^N \mathcal{M}^u(z_i)$, *where* $z_i = (w_i;0)$ *and* w_i *is a solution to problem* (9.5.1) *and* (9.5.2) *with the corresponding boundary conditions,* $i = 1,\dots,N$. *Moreover,*

- *The global attractor A consists of full trajectories $\gamma = \{(u(t); u_t(t)) \, : \, t \in \mathbb{R}\}$ connecting pairs of stationary points; that is, any $W \in A$ belongs to some full trajectory γ and for any $\gamma \subset A$ there exists a pair $\{w_-, w_+\} \subset \mathscr{N}^*$ such that*

$$\lim_{t \to -\infty} \{\|u_t(t)\|^2 + \|u(t) - w_-\|_2^2\} = 0$$

and

$$\lim_{t \to +\infty} \{\|u_t(t)\|^2 + \|u(t) - w_+\|_2^2\} = 0.$$

- *For any $(u_0; u_1) \in H$ there exists a stationary solution $w \in \mathscr{N}^*$ such that*

$$\lim_{t \to +\infty} \{\|u_t(t)\|^2 + \|u(t) - w\|_2^2\} = 0, \tag{9.5.5}$$

where $u(t)$ is a generalized solution to problem (9.4.1) *and* (9.4.2) *with initial data $(u_0; u_1)$ and with the corresponding boundary conditions.*

Now we give a result on the rates of convergence of an individual trajectory to an equilibrium. In order to describe the decay rates, we need some notation. We first introduce a concave, strictly increasing, continuous function $h : \mathbb{R}_+ \mapsto \mathbb{R}_+$ which captures the behavior of $g(s)$ at the origin possessing the properties

$$h(0) = 0 \text{ and } s^2 + g^2(s) \le h(sg(s)) \text{ for } |s| \le 1. \tag{9.5.6}$$

Such a function can always be constructed due to the monotonicity of g; see [195] and also Proposition B.2.1 in Appendix B. Now we define

$$H_0(s) = h\left(\frac{s}{c_3}\right), \quad G_0(s) = c_1(I + H_0)^{-1}(c_2 s), \quad Q(s) = s - (I + G_0)^{-1}(s), \tag{9.5.7}$$

where c_1, c_2, and c_3 are positive constants. Clearly, $Q(s)$ is strictly monotone. Thus, the differential equation

$$\frac{d\sigma}{dt} + Q(\sigma) = 0, \quad t > 0, \quad \sigma(0) = \sigma_0 \in \mathbb{R}, \tag{9.5.8}$$

admits the global unique solution $\sigma(t)$ which, moreover, decays asymptotically to zero as $t \to \infty$. See Proposition B.3.1in Appendix B.

With these preparations we are ready to state our result.

9.5.3. Theorem (Rate of stabilization). *Let Assumptions* 9.4.1 *and* 9.4.3 *hold. (In the case of the free boundary conditions we assume that Assumption* 9.4.13 *is also true). Let (H, S_t) be the dynamical system generated by problem* (9.4.1) *and* (9.4.2) *with one of the boundary conditions* (9.4.3), (9.4.4), *or* (9.4.5). *In addition assume that (i) problem* (9.5.1) *and* (9.5.2) *with the corresponding boundary conditions has a finite number of solutions. Then for any $V \in H$ there exists a stationary point $E = (e; 0)$ such that $S_t V \to E$ as $t \to +\infty$. Moreover, if the equilibrium E is hyperbolic in the sense that the linearization of* (9.5.1) *and* (9.5.2) *around every solution has the trivial solution only, then there exist positive constants C and T depending on V*

and E such that we have the following rates of stabilization.

$$\|S_tV - E\|_H \le C\sigma([tT^{-1}]), \quad t > 0, \tag{9.5.9}$$

where $[a]$ *denotes the integer part of a and* $\sigma(t)$ *satisfies* (9.5.8) *with* $\sigma_0 = C(V,E)$, *where* $C(V,E) > 0$ *is a constant depending on* $V,E \in H$, *and* Q *is defined by* (9.5.7) *with the constants* c_i *depending on V and E. In particular if* $g'(0) > 0$, *then*

$$\|S_tV - E\|_H \le Ce^{-\omega t}$$

for some positive constants C and ω *depending on* $V,E \in H$.

Proof. This follows from Theorem 8.4.3 in the clamped and hinged cases. In the case of free-type boundary conditions (9.4.5) the argument is similar to the proof of Theorem 8.4.3 given in [75]. See also the proof of Theorem 10.4.10 below, where the same assertion is established in the case when boundary damping is the main factor responsible for stabilization. □

9.5.2 Smoothness of elements from the attractor

The additional smoothness of the attractor results from the quasi-stability property of the system. Indeed, as a consequence of the results stated in Section 8.7 we have the following.

9.5.4. Theorem. *Let Assumptions* 9.4.1 *and* 9.4.3 *hold (in the case of free-type boundary conditions we also assume* $\beta \in C^\infty(\Gamma)$ *and the validity of Assumption* 9.4.13. *Assume that either* $L \equiv 0$ *or else the damping parameter is large enough. If the damping function* g_0 *satisfies* (9.4.11), *then any full trajectory* $\gamma = \{(u(t);u_t(t)) : t \in \mathbb{R}\}$ *from the attractor A of the dynamical system* (H,S_t) *generated by problem* (9.4.1) *and* (9.4.2) *with one of the boundary conditions* (9.4.3), (9.4.4), *or* (9.4.5), *possesses the properties*

$$u(t) \in C_r(\mathbb{R};W), \quad (u_t(t);u_{tt}(t)) \in C_r(\mathbb{R};H), \tag{9.5.10}$$

where C_r *stands for right-continuous functions and*

- $H = H_0^2(\Omega) \times L_2(\Omega)$ *and* $W = H^4(\Omega) \cap H_0^2(\Omega)$ *for the clamped boundary conditions* (9.4.3).
- $H = (H^2 \cap H_0^1)(\Omega) \times L_2(\Omega)$ *and* $W = \{w \in H^4(\Omega) : w|_{\partial\Omega} = \Delta w|_{\partial\Omega} = 0\}$ *in the hinged case* (9.4.4).
- *In the case of free type boundary conditions* (9.4.5) *we have* $H = \{(w_1;w_2) : w_1 \in H^2(\Omega), w_2 \in L_2(\Omega)\}$ *and W consists of elements* $w_0 \in H^4(\Omega)$ *such that the pair* $(w_0;w_1)$ *satisfies the compatibility condition in* (4.1.29); *that is,*

$$\Delta w_0 + (1-\mu)B_1 w_0 = 0, \quad \frac{\partial}{\partial n}\Delta w_0 + (1-\nu)B_2 w_0 = \nu_1 w_0 + \beta(w_0^3 + w_1) \quad on\ \Gamma.$$

Moreover, $A \subset H^4(\Omega) \times H^2(\Omega)$ *for each case and*

$$\sup_{t\in\mathbb{R}} \left(\|u(t)\|_4^2 + \|u_t(t)\|_2^2 + \|u_{tt}(t)\|^2 \right) \leq C_A < \infty. \tag{9.5.11}$$

Proof. In the clamped or hinged case the result follows from the abstract Theorem 8.7.1 (see also Corollary 8.7.2). In the case of free boundary conditions we apply stabilizability estimate (9.4.68) from Proposition 9.4.17 and Theorem 7.9.8. For right-continuity in (9.5.10) in the free case we use Remark 4.1.15. □

As in the case when the rotational forces are present ($\alpha > 0$), a higher level of regularity of attractor can be obtained. The argument is based on boot-strapping regularity and an inductive procedure. Owing to the fact that in the nonrotational case von Karman nonlinearity is critical with respect to the dynamics (unlike the rotational case), the analysis of higher regularity of attractors benefits from the property of finiteness of the dissipation integral (which requires the absence of nonconservative loads). Under this assumption, the result of Theorem 9.5.4 can be improved in order to claim infinite time differentiability of elements from the attractor. The relevant result is formulated below.

9.5.5. Theorem. *In addition to the hypotheses of Theorem* 9.5.4 *we assume that* $L \equiv 0$ *and* $g_0(s) \in C^\infty(\mathbb{R})$. *Then the dynamical system* (H, S_t) *generated by problem* (9.4.1) *and* (9.4.2) *with one of the boundary conditions* (9.4.3), (9.4.4), *or* (9.4.5), *has a compact global attractor A and any full trajectory* $\gamma = \{(u(t); u_t(t)) : t \in \mathbb{R}\}$ *from the attractor A possesses the property*

$$u^{(n)}(t) \in C(\mathbb{R}; W), \quad n = 0, 1, \ldots, \tag{9.5.12}$$

where W is the same as in Theorem 9.5.4 *in the clamped and hinged cases and we choose* $W = H^4(\Omega)$ *in the case of free boundary conditions,* $u^{(0)}(t) = u(t)$ *and* $u^{(n)}(t) = \partial_t^n u(t)$ *for* $n = 1, 2, \ldots$ *Moreover,* $\sup_{t\in\mathbb{R}} \|u^{(n)}(t)\|_4 \leq C_n(A) < \infty$ *for each* $n = 0, 1, \ldots$.

One could also show that the elements from the attractor display an arbitrary level of space regularity, provided that the forces p and F_0 are sufficiently smooth. The corresponding result is formulated as Corollary 9.5.8 of the present theorem.

Proof. To focus the arguments we concentrate on the case of clamped boundary conditions (9.4.3) (in the hinged case the argument is the same, in the free case we need some obvious modifications due to the presence of nonlinear and nonhomogeneous terms in the boundary condition for displacement).

We start with the following preliminary assertion.

9.5.6. Lemma. *Let* $u \in H^2(\Omega)$ *and* $F'(u)$ *be the Fréchet derivative of F with* $L \equiv 0$*:*

$$\langle F'(u); w \rangle = [w, v(u + 2f, u) + F_0] + 2[u + f, v(u + f, w)], \tag{9.5.13}$$

where $v = v(w_1, w_2) \in H_0^2(\Omega)$ *is determined from* (9.4.70). *Then* $F'(u)$ *can be extended as a continuous linear operator from* $L_2(\Omega)$ *into* $H^{-2}(\Omega)$ *such that*

$$\|\langle F'(u);w\rangle\|_{-2} \le C(1+\|u\|_2^2)\|w\|. \tag{9.5.14}$$

Proof. Using the representation in (1.4.11) we have that

$$\|[w, v(u+2f,u)+F_0]\|_{-2} \le C\|w\|\|v(u+2f,u)+F_0\|_{W^2_\infty}.$$

Therefore from (1.4.23) we have $\|[w, v(u+2f,u)+F_0]\|_{-2} \le C(1+\|u\|_2^2)\|w\|$.
One can see that for every $\phi \in H_0^2(\Omega)$ we have

$$([u+f, v(u+f,w)],\phi) = ([u+f,\phi], v(u+f,w)) = ([v(u+f,\phi), u+f], w).$$

Therefore using (1.4.23) we obtain that

$$|([u+f, v(u+f,w)],\phi)| \le C\|u+f\|_2^2\|\phi\|_2\|w\|,$$

which implies $\|[u+f, v(u+f,w)]\|_{-2} \le C(1+\|u\|_2^2)\|w\|$ and makes it possible to conclude the proof of Lemma 9.5.6. □

By Theorem 9.5.4 any trajectory $(u(t);u_t(t))$ from the attractor provides a strong solution to problem

$$u_{tt} + kD(u_t) + \mathscr{A}u = F(u), \quad t \in \mathbb{R},$$

(cf.(9.4.13)), which we can consider as a problem on the whole real line $\mathbb{R}$. In particular, by (4.1.14) we have that

$$(u(t);u_t(t);u_{tt}(t)) \in C(\mathbb{R}; W \times H), \tag{9.5.15}$$

where W and H are the same as in Theorem 9.5.4. Moreover, one can see that $w(t) = u_t(t)$ is a weak solution for the nonautonomous equation

$$w_{tt} + kd_0 g_0'(u_t(t))w_t + \mathscr{A}w = \langle F'(u(t)), w\rangle, \quad t \in \mathbb{R}, \tag{9.5.16}$$

where $\langle F'(u(t));w\rangle$ given by (9.5.13) belongs $L_2(\Omega)$ (this follows from (9.5.11)). On the strength of (9.5.11) we also have that

$$||w(t)||_2^2 + ||w_t(t)||^2 \le C_A \quad \text{for all } t \in \mathbb{R}. \tag{9.5.17}$$

For further boosting of regularity of the attractor, $L_2(\mathbb{R})$ membership of functions in (9.5.17) is needed. Another way of saying this is we need to establish finiteness of time integrals over $\mathbb{R}$ of the energy corresponding to w. This is done in the following assertion which improves the boundedness property (9.5.11) of the trajectories $(u(t);u_t(t))$ from the attractor.

9.5.7. Proposition. *Let the hypotheses of Theorem* 9.5.5 *be in force. Then any trajectory* $\{(u(t);u_t(t)) : t \in \mathbb{R}\}$ *from the attractor possesses property* (9.5.15), *and, in addition to* (9.5.11), *we also have the finiteness of the following dissipation integral,*

$$\mathscr{I}_1[u] = \int_{-\infty}^{\infty} \int_{\Omega} \left[u_t g_0(u_t) + g_0'(u_t)|u_{tt}|^2 + |\Delta u_t|^2 \right] dxdt \le C_A < \infty, \tag{9.5.18}$$

which implies that

$$\int_{-\infty}^{\infty} \left(\|u_t(t)\|^2 + \|u_{tt}(t)\|^2 + \|\Delta u_t(t)\|^2 \right) dt \le C_A < \infty. \tag{9.5.19}$$

Proof. As above we concentrate on the clamped case. The finiteness of

$$\mathscr{I}_0[u] = \int_{-\infty}^{\infty} \int_{\Omega} u_t g_0(u_t) dxdt$$

follows from the energy relation in (4.1.11) with $L \equiv 0$ and from the boundedness of the energy on the attractor.

The finiteness of the remaining terms in (9.5.18) follows "morally" from the arguments leading to the stabilizability estimate in Section 8.5.2 in Chapter 8. However, in order to make this argument independent and self-contained we develop energetic estimates based on the appropriate construction of the Lyapunov function. This argument is then repeated and "boosted" to higher energies levels. The key to the construction of a suitable Lyapunov function is the following splitting property of the von Karman bracket,[11]

$$(\langle F'(u); w \rangle, w_t) = \frac{d}{dt} Q_0(t) + P(t), \tag{9.5.20}$$

where

$$\begin{aligned} Q_0(t) &= -\|\Delta v(u+f, w)\|^2 + \frac{1}{2}([w,w], v(u+2f, u) + F_0), \\ P(t) &= -3([w,w], v(u+f, w)) = -3([w, v(u+f, w)], w), \end{aligned}$$

where $w = u_t$ and $v = v(w_1, w_2) \in H_0^2(\Omega)$ is determined from (9.4.70). In fact, this decomposition is the key in handling the criticality of the von Karman forcing, present at the nonrotational level. Decomposition in (9.5.20) suggests the following choice of energy function

$$Q(t) = \frac{1}{2}\left[\|w_t\|^2 + \|\Delta w\|^2 \right] - Q_0(t) + \mu \|w\|^2,$$

where μ is selected large enough so that the functional $Q(t)$ is positive. More precisely:

$$a_0 \left(\|w_t(t)\|^2 + \|\Delta w(t)\|^2 \right) \le Q(t) \le a_1 \left(\|w_t(t)\|^2 + \|\Delta w(t)\|^2 \right) \tag{9.5.21}$$

---

[11] See Lemma 9.4.10 for the corresponding "difference" analogue that has been used for the proof of finite dimensionality of the attractor in Theorem 9.4.6 and quasi-stability estimate in Proposition 9.4.17.

for all $t \in \mathbb{R}$ with suitable constants a_0, a_1, and with $\mu = \mu(A)$ large enough. The above can be easily accomplished on the strength of a priori bounds (9.5.11) already established on the attractor. Indeed, because $||\Delta u(t)|| \le C_A$ on the attractor, by Theorem 1.4.3 and Corollary 1.4.4 we have

$$|([w,w], v(u+2f,u)+F_0)| \le C(1+\|\Delta u\|^2)\|w\|\|\Delta w\| \le C_A||w||^2 + \frac{1}{2}||\Delta w||^2 \tag{9.5.22}$$

which gives the desired lower bound in (9.5.21) for $\mu > C_A/2$. The upper bound is obvious.

Using the energy relation for (9.5.16) (see Theorem 2.4.35) and (9.5.20) one obtains

$$\frac{d}{dt}Q(t) + k\int_\Omega d_0 g_0'(u_t)|w_t|^2 dx = -3([w,w], v(u+f,w)) + 2\mu(w,w_t).$$

We introduce next a standard parametric Lyapunov-type function

$$W(t) = Q(t) + \varepsilon(w_t, w), \quad \varepsilon \in (0, \varepsilon_0)$$

where $\varepsilon_0 > 0$ is selected so that (see (9.5.21) for all $\varepsilon \in (0, \varepsilon_0)$ we have

$$\frac{1}{2}a_0\left(\|w_t(t)\|^2 + \|\Delta w(t)\|^2\right) \le W(t) \le 2a_1\left(\|w_t(t)\|^2 + \|\Delta w(t)\|^2\right). \tag{9.5.23}$$

By Theorem 1.4.3 and Corollary 1.4.4 we also have that

$$|([w,w], v(u+f,w))| \le C(1+\|\Delta u\|)\|w\|\|\Delta w\|^2. \tag{9.5.24}$$

$\|\Delta u(t)\|$ is bounded on the attractor, therefore from Lemma 9.5.6 we infer that

$$|(\langle F'(u(t)); w\rangle, w)| \le ||w||_2||\langle F'(u(t)), w\rangle||_{-2} \le C_A||w||||\Delta w||. \tag{9.5.25}$$

Standard calculations based on the relation $(d/dt)(w_t, w) = ||w_t||^2 + (w_{tt}, w)$ and the structure of the equation (9.5.16) lead to

$$\begin{aligned} \frac{dW}{dt} + k\int_\Omega d_0 g_0'(u_t)|w_t|^2 dx &\le C_1\|w\|\|\Delta w\|^2 + 2\mu\|w\|\|w_t\| \\ &+ \varepsilon\left[\|w_t\|^2 - k\int_\Omega d_0 g_0'(u_t) w_t w dx - \|\Delta w\|^2 + C_2\|w\|\|\Delta w\|\right]. \end{aligned}$$

Because $||u_t(t)||_{C(\Omega)} \le c||u_t(t)||_2 \le C_A$, we have that

$$\left|\int_\Omega d_0 g_0'(u_t) w_t w dx\right| \le \frac{1}{2}\int_\Omega d_0 g_0'(u_t)|w_t|^2 dx + C_{g_0,A}\int_\Omega |u_t|^2 dx. \tag{9.5.26}$$

It follows from the strong monotonicity of g_0 given by lower bound in (9.4.11) (or in (9.4.12)) that

$$\|w_t\|^2 \le c_0 \int_\Omega d_0 g_0'(u_t)|w_t|^2 dx. \tag{9.5.27}$$

Therefore for $\varepsilon > 0$ small enough there exist constants $\gamma_0, \gamma_1 > 0$ such that

$$\frac{dW}{dt} + \gamma_0 W + \gamma_1 \int_\Omega d_0 g_0'(u_t)|w_t|^2 dx \le c_0 \int_\Omega |u_t|^2 dx \, [1 + W(t)].$$

By (9.5.11) from (9.5.23) we have that $\sup_{t\in\mathbb{R}} W(t) \le C_A < \infty$. This implies that

$$\begin{aligned}\int_{-\infty}^{\infty} \left[\gamma_0 W(t) + \gamma_1 \int_\Omega d_0 g_0'(u_t)|w_t|^2 dx\right] dt &\le C_1 + C_2 \int_{-\infty}^{\infty} \|u_t\|^2 dt \\ &\le C_1 + C_2 \int_{-\infty}^{\infty} \int_\Omega d_0 u_t g_0(u_t) dx dt \le C_A,\end{aligned} \tag{9.5.28}$$

where the last term is already known to be finite. This completes the proof of Proposition 9.5.7. □

To continue with the proof of Theorem 9.5.5 we use induction. Indeed, by using induction in n we prove the property

$$(u^{(n-1)}(t); u^{(n)}(t); u^{(n+1)}(t)) \in C(\mathbb{R}; W \times H), \quad n = 1, 2, \ldots, \tag{9.5.29}$$

where W and H are the same as in Theorem 9.5.4, along with the corresponding uniform estimate

$$\|u^{(n-1)}(t)\|_4^2 + \|u^{(n)}(t)\|_2^2 + \|u^{(n+1)}(t)\|^2 \le R_n^2, \quad n = 1, 2, \ldots, \quad t \in \mathbb{R}, \tag{9.5.30}$$

and

$$\mathscr{I}_n[u] = \int_{-\infty}^{\infty} \int_\Omega \left[|u^{(n)}|^2 + g_0'(u_t)|u^{(n+1)}|^2 + |\Delta u^{(n)}|^2\right] dx dt \le C_n(A) < \infty. \tag{9.5.31}$$

For $n = 1$ this fact is already proved in Theorem 9.5.4 and in Proposition 9.5.7.

For the sake of clarity we start with the induction transition from $n = 1$ to $n = 2$. The general inductive step is explained later.

The main idea is the same as the one used for the derivation of the quasi-stability estimate in Proposition 9.4.17. Indeed, we derive the stabilizability estimate valid for a finite difference of two solutions, however—this time—the estimate is given at higher energetic levels; that is, for successive time derivatives of the solutions lying on the attractor. To accomplish this task we repeat and generalize the procedure used for the proof of stabilizability estimate.

Let

$$z(t) \equiv z^h(t) = h^{-1}[w(t+h) - w(t)], \quad \text{for } h > 0,$$

where as above $w = u_t$. It follows from (9.5.16) that $z = z^h$ satisfies the following equation:

$$z_{tt} + k d_0 g_0'(u_t(t+h)) z_t + \mathscr{A} z = \langle F'(u(t+h)); z\rangle + R_1(t,h) + R_2(t,h) \tag{9.5.32}$$

for all $t \in \mathbb{R}$. Here

$$R_1(t,h) = -kd_0\left[\int_0^1 g_0''(\lambda u_t(t+h)+(1-\lambda)u_t(t))d\lambda\right] w_t(t)z(t),$$

$$R_2(t,h) = [w(t), v(\hat{z}, u(t+h)+u(t)+2f)] \\ + 2[\hat{z}, v(u(t+h)+f, w(t))] + 2[u(t)+f, v(\hat{z}, w(t))],$$

where $\hat{z} \equiv \hat{z}^h(t) = h^{-1}[u(t+h)-u(t)]$ and, as above, $v = v(w_1,w_2) \in H_0^2(\Omega)$ is determined from (9.4.70).

Using the boundedness of $\|u_t(t)\|_2$ and $\|u(t)\|_2$ on the attractor along with Sobolev's embedding $H^2(\Omega) \subset C(\Omega)$, so that $g_0''(\lambda u_t(t+h,x)+(1-\lambda)u_t(t,x))$ is uniformly bounded on the attractor, we obtain

$$\|R_1(t,h)\| \le C_{g_0,A}\|w_t\|\|z\|_2 \quad \text{and} \quad \|R_2(t,h)\| \le C_A\|w\|_2\|\hat{z}\|_2, \tag{9.5.33}$$

where the constant $C_A, C_{g_0,A}$ do not depend on t and h.

In analogy with Proposition 9.5.7 we define Lyapunov function

$$W^h(t) = Q^h(t) + \varepsilon(z_t, z),$$

where

$$Q^h(t) = \frac{1}{2}\Big[\|z_t\|^2 + \|\Delta z\|^2 + 2\|\Delta v(u(t+h)+f,z)\|^2 \\ - ([z,z], v(u(t+h)+2f, u(t+h)) + F_0)\Big] + \mu\|z\|^2,$$

with suitably selected μ (large) and ε (small). The choice of the energy $Q^h(t)$ is dictated by the same considerations as in the previous case with the critical term $\langle F'(u(t+h)), z\rangle$ decomposed into a form similar to relation (9.5.20).

From (9.5.22) we obviously have that there exist $\varepsilon_0 > 0$ and $\mu \ge 0$ such that for every $\varepsilon \in [0,\varepsilon_0]$ we have

$$a_0\left(\|z_t(t)\|^2 + \|\Delta z(t)\|^2\right) \le W^h(t) \le a_1\left(\|z_t(t)\|^2 + \|\Delta z(t)\|^2\right),$$

where $a_0, a_1 > 0$ are constants independent of h. We also have that

$$\frac{dQ^h}{dt} = (z_{tt} + \mathscr{A}z - \langle F'(u(t+h)), z\rangle, z_t) - ([z,z], v(u(t+h)+f, u_t(t+h))) \\ -2([v(u(t+h)+f,z),z], u_t(t+h)) + 2\mu(z,z_t).$$

Using equation (9.5.32) gives

$$\frac{dQ^h}{dt} + k\int_\Omega d_0 g_0'(u_t(t+h))|z_t|^2 dx \\ = -([z,z], v(u(t+h)+f, u_t(t+h))) - 2([v(u(t+h)+f,z),z], u_t(t+h)) \\ + (R_1(t,h)+R_2(t,h), z_t) + 2\mu(z,z_t).$$

By appealing to the estimates for R_i terms given in (9.5.33) along with the estimates of the von Karman brackets we obtain that

$$\frac{dQ^h}{dt} + k\int_\Omega d_0 g_0'(u_t(t+h))|z_t|^2 dx$$
$$\leq 2\mu\|z\|\|z_t\| + C_1\|w_t\|\|\Delta z\|\|z_t\| + C_2\|\Delta u_t(t+h)\|\|\Delta z\|\|z\| + C_3\|\Delta w\|\|z_t\|\|\Delta\hat{z}\|.$$

By the estimate $||\Delta u_t(t+h)||\ ||\Delta z||\ ||z|| \leq \rho||\Delta z||^2 + C_{A,\rho}||z||^2$ for any $\rho > 0$ and relation (9.5.27) with $w_t = z_t$ we have that

$$\frac{dQ^h}{dt} + \frac{k}{2}\int_\Omega d_0 g_0'(u_t(t+h))|z_t|^2 dx$$
$$\leq \rho||\Delta z||^2 + C_\rho\|z\|^2 + C\left[\|w_t\|^2\|\Delta z\|^2 + \|\Delta w\|^2\|\Delta\hat{z}\|^2\right]$$

for any $\rho > 0$. The uniform bounds in (9.5.11) after straightforward calculations lead to the relations

$$\frac{d}{dt}(z, z_t) = ||z_t||^2 + (z, z_{tt})$$
$$\leq ||z_t||^2 - \frac{1}{2}||\Delta z||^2 + C_1\|z\|^2 + C_2\|\Delta\hat{z}\|^2 - k\int_\Omega d_0 g_0'(u_t(t+h))z_t z dx.$$

Using a priori bound $||u_t(t)||_{C(\Omega)} \leq C_A$ in the term $g_0'(u_t(t))$ one obtains

$$\frac{d}{dt}(z, z_t) \leq 2||z_t||^2 - \frac{1}{2}||\Delta z||^2 + C_1\|z\|^2 + C_2\|\Delta\hat{z}\|^2.$$

Thus combining the estimates above and selecting sufficiently small ρ and ε lead to the relation

$$\frac{dW^h}{dt} + \gamma_0 W^h + \gamma_1\int_\Omega d_0 g_0'(u_t(t+h))|z_t|^2 dx \leq c_1\left(\|z\|^2 + \|\Delta\hat{z}\|^2\right) + c_2 D(t)W^h(t) \tag{9.5.34}$$

with $D(t) = \|u_{tt}(t)\|^2$, where the positive constants γ_0, γ_1, c_1, c_2 do not depend on h. Thus we have that

$$W^h(t) \leq W^h(s)\exp\left\{-\gamma_0(t-s) + c_2\int_s^t D(\tau)d\tau\right\} \tag{9.5.35}$$
$$+ c_1\int_s^t \left(\|z(\tau)\|^2 + \|\Delta\hat{z}(\tau)\|^2\right)\exp\left\{-\gamma_0(t-\tau) + c_2\int_\tau^t D(r)dr\right\}d\tau$$

for every $t > s$. Because

$$\|z(t)\|^2 \leq h^{-1}\int_0^h \|u_{tt}(t+\tau)\|^2 d\tau \quad \text{and} \quad \||\Delta\hat{z}(t)\|^2 \leq h^{-1}\int_0^h \|\Delta u_t(t+\tau)\|^2 d\tau,$$

it follows from (9.5.31) for $n = 1$ that

$$\int_{-\infty}^{\infty} \|z(t)\|^2 dt \le \int_{-\infty}^{\infty} \|u_{tt}(t)\|^2 dt \equiv \int_{-\infty}^{\infty} D(t)dt \le C_A < \infty,$$
$$\int_{-\infty}^{\infty} \|\Delta\hat{z}(t)\|^2 dt \le \int_{-\infty}^{\infty} \|\Delta u_t(t)\|^2 dt \le C_A < \infty. \tag{9.5.36}$$

Therefore, because $\sup_{s\in\mathbb{R}} W^h(s)$ is finite for every fixed $h > 0$, we can pass to the limit $s \to -\infty$ in (9.5.35) and obtain that

$$W^h(t) \le c_1 \int_{-\infty}^{t} \left(\|z(\tau)\|^2 + \|\Delta\hat{z}(\tau)\|^2\right) \exp\left\{-\gamma_0(t-\tau) + c_2 \int_\tau^t D(r)dr\right\} d\tau \le C_A$$

for every $t \in \mathbb{R}$, where $\hat{z} = h^{-1}[u(t+h) - u(t)]$ and $z = \hat{z}_t$. Moreover, from (9.5.34) and (9.5.36) we also have that

$$\int_{-\infty}^{\infty} \left[W^h(t) + \int_\Omega d_0 g_0'(u_t(t+h))|z_t(t)|^2 dx\right] dt \le C_A.$$

After passing with the limit $h \to 0$ we obtain (9.5.30) and (9.5.31) for the case $n = 2$. The property (9.5.29) in this case follows by considering $u_{tt} = \lim_{h\to 0} z^h$ as a solution of the corresponding limiting equation, a linear (nonautonomous) equation satisfying the hypotheses of Theorem 2.4.35.

In order to proceed with subsequent induction steps we assume that (9.5.29)–(9.5.31) holds for all $1 \le n \le m$ (with $m \ge 2$). In this case $w(t) \equiv u^{(m)}(t) \in C(\mathbb{R}; H_0^2(\Omega)) \cap C^1(\mathbb{R}; L_2(\Omega))$ satisfies (in the variational sense) the equation

$$w_{tt} + kd_0 g_0'(u_t(t))w_t + \mathscr{A}w = \langle F'(u(t)); w\rangle - (m-1)kd_0 g_0''(u_t(t))u_{tt}w + R_m(t)$$

on the axis $\mathbb{R}$, where, as above, the Fréchet derivative $F'(u)$ is given by (9.5.13) and $R_m(t) = G_m(t) - kd_0 D_m(t)$. Here $G_m(t) \equiv G_n(u(t), u'(t), \ldots, u^{(m-1)}(t))$, $m \ge 2$, is a linear combination of functions of the form $[v(w_1, w_2), w_3]$, where w_j are either $u + f$ or one of the derivatives $u^{(k)}(t)$, $1 \le k \le m-1$, and the value $v = v(w_1, w_2)$ from $H_0^2(\Omega)$ is determined from (9.4.70). The value $D_m(t)$ is zero when $m = 2$ and for $m \ge 3$, $D_m(t)$ is a sum of elements of the form

$$c_{r,k_1,\ldots,k_r} \cdot g_0^{(r)}(u_t) \cdot u^{(k_1)} \cdot \ldots \cdot u^{(k_r)}, \quad 2 \le r \le m,$$

with $2 \le k_j \le m-1$ for every $j = 1, \ldots, r$.

One can see from the induction hypothesis that

$$h^{-2} \int_{-\infty}^{\infty} \|R_m(t+h) - R_m(t)\|^2 dt \le C_m(A) < \infty$$

uniformly in $0 < h < 1$. This observation makes it possible to apply the same argument as in the case of the transition from $n = 1$ to $n = 2$ and to conclude the proof of Theorem 9.5.5. □

The following assertion contains the result on spatial smoothness of the global attractor.

9.5.8. Corollary. *Let the hypotheses of Theorem* 9.5.5 *be valid. Assume that*

$$d_0 \in C^\infty(\Omega), \ \ p+[f,F_0] \in H^{2(m-2)}(\Omega), \ \ f \in H^{2m-2}(\Omega), \ \ F_0 \in H^{2m-1}(\Omega), \tag{9.5.37}$$

for some $m \geq 2$. Then any full trajectory $\gamma = \{(u(t);u_t(t)) \, : \, t \in \mathbb{R}\}$ from the attractor A of the system (H,S_t) possesses the properties

$$u^{(n)}(t) \in C(\mathbb{R};H^{2m}(\Omega)) \quad and \quad \sup_{t\in\mathbb{R}} \|u^{(n)}(t)\|_{2m} \leq C_{n,m}(A) < \infty \tag{9.5.38}$$

for $n = 0,1,\ldots$. In particular A is a bounded set in $H^{2m}(\Omega) \times H^{2m}(\Omega)$. If p, f, and F_0 are C^∞-functions, then the attractor A lies in $C^\infty(\overline{\Omega}) \times C^\infty(\overline{\Omega})$.

Proof. We rely on elliptic regularity of stationary von Karman equations. The idea the same as in the proof of Corollary 9.2.11; see also [123].

As in Theorem 9.5.5 we consider the case of clamped boundary conditions only. From equation (9.4.1) we have that

$$\begin{aligned} \Delta^2 u^{(n)}(t) &= -u^{(n+2)}(t) - kd_0 \sum g_0^{(r)}(u_t) u^{(k_1)} \ldots u^{(k_r)} \\ &\quad + \langle F'(u(t)); u^{(n)}(t) \rangle + G_n(t), \quad n = 1,2,\ldots, \end{aligned}$$

where $r \leq n$ and $k_j \leq n+1$ and $\langle F'(u);w\rangle$ is given by (9.5.13). As above, we also have that $G_1 \equiv 0$ and $G_n(t) \equiv G_n(u(t),u'(t),\ldots,u^{(n-1)}(t))$ for $n \geq 2$ is a linear combination of functions of the form $[v(w_1,w_2),w_3]$, where w_j are either $u+f$ or one of the derivatives $u^{(i)}(t)$, $1 \leq i \leq n-1$, and the value $v = v(w_1,w_2) \in H_0^2(\Omega)$ is determined as the solution to the problem (9.4.70) for given $w_1,w_2 \in H^2(\Omega)$.

Subsequently, using Lemma 4.1.26 and relation (4.1.57) as in the proof of Corollary 9.2.11 we can conclude that for $l \geq 2$ and $n \geq 0$ we have that

$$\|u^{(n)}(t)\|_{l+2} \leq \psi(\|f\|_l, \|F_0\|_{l+1}, \|u\|_l, \|u_t\|_l, \ldots, \|u^{(n+2)}\|_l) + Cn \cdot \|p+[f,F_0]\|_{l-2}$$

on the attractor, where ψ is a polynomial of the arguments listed. By induction in l we complete the proof of Corollary 9.5.8. □

9.5.3 Strong attractors

In the case of attractors described in Section 9.5.2, quasi-stability of the system under consideration allows us to establish attractivity properties of the attractor in a stronger topology.

In this section we restrict ourselves to the cases of clamped and hinged boundary conditions. Concerning the case of free-type boundary conditions, the reader is referred to Remark 9.5.12 below.

Let $H_{st} = W \times V$, where $W = H^4(\Omega) \cap H_0^2(\Omega)$ and $V = H_0^2(\Omega)$ in the case of the clamped boundary conditions (9.4.3), and $W = \{w \in H^4(\Omega) \, : \, u|_{\partial\Omega} = \Delta u|_{\partial\Omega} = 0\}$

and $V = (H^2 \cap H_0^1)(\Omega)$ in the hinged case (9.4.4). This Hilbert space H_{st} we endow with the norm $\|(w_0;w_1)\|^2_{H_{st}} = \|\Delta w_0\|^2 + \|w_1\|^2$.

By Theorem 4.1.4 (the clamped case) and Theorem 4.1.10 (the hinged case) the space H_{st} is a (maximal) set of initial data for strong solutions and thus H_{st} is positively invariant with respect to S_t (i.e., $S_t H_{st} \subset H_{st}$).

Our goal in this section is to study long-time dynamics of the semiflow S_t on the space H_{st}.

9.5.9. Proposition. *Let the hypotheses of Theorem* 9.5.4 *be in force. Then the system* (H_{st}, S_t) *generated by the strong solutions to problem* (9.4.1) *and* (9.4.2) *with either clamped* (9.4.3) *or hinged* (9.4.4) *conditions is dissipative.*

Proof. Under the hypotheses imposed we have the stabilizability estimate of the form (9.4.42). Thus we can apply Corollary 8.8.2. □

Our main result concerning long-time dynamics of strong solutions is the following theorem which asserts that the attractor A is attracting for strong orbits in the strong topology of H_{st}.

9.5.10. Theorem. *Under the conditions of Theorem* 9.5.5 *the global attractor A of the system* (H, S_t) *generated by the generalized solutions to problem* (9.4.1) *and* (9.4.2) *with either clamped* (9.4.3) *or hinged* (9.4.4) *conditions is also strong, i.e. A is a global attractor for the system* (H_{st}, S_t) *which means that A uniformly attracts bounded sets from* H_{st} *in the topology of the space* H_{st}*. Moreover, A has a finite dimension as a compact set in* H_{st}*.*

Proof. We use the same idea as in the proof of the abstract Theorem 8.8.4. and rely on the following stabilizability estimate established for a higher energy level represented by the topology in H_{st}.

9.5.11. Lemma (Higher-order quasi-stability). *Assume that* $S_t y_1 = (u^1(t); u_t^1(t))$ *and* $S_t y_2 = (u^2(t); u_t^2(t))$ *are two semitrajectories from the absorbing ball in the space* H_{st}*:*

$$\|u_{tt}^i(t)\|^2 + \|\Delta u_t^i(t)\|_2^2 + \|\Delta^2 u^i(t)\|^2 \le R^2, \quad t \ge 0,\ i = 1,2. \tag{9.5.39}$$

for some $R > 0$*. Then under the conditions of Theorem* 9.5.5 *we have that*

$$\|S_t y_1 - S_t y_2\|^2_{H_{st}} \le C_R^1 \|y_1 - y_2\|^2_{H_{st}} e^{-\gamma t} + C_R^2 \max_{\tau \in [0,t]} \|u^1(\tau) - u^2(\tau)\|^2. \tag{9.5.40}$$

This lemma means that the semiflow S_t is quasi-stable on an absorbing ball in H_{st}.

Proof. By the same argument as in the proof of Proposition 9.5.7 we conclude that

$$\int_0^\infty \int_\Omega \left[u_t^i g_0(u_t^i) + g_0'(u_t^i)|u_{tt}^i|^2 + |\Delta u_t^i|^2 \right] dxdt \le C_R < \infty, \quad i = 1,2. \tag{9.5.41}$$

As in Section 9.5.2, in particular see (9.5.16), we have that $w^i(t) = u_t^i(t)$ is a weak solution for nonautonomous equation

$$w^i_{tt} + kd_0 g_0'(u^i_t(t))w^i_t + \mathscr{A} w^i = \langle F'(u^i(t)); w^i \rangle,$$

on the semi-axis $\mathbb{R}_+$ with the given initial data, where $\langle F'(u); w \rangle$ is given by (9.5.13). Thus the difference $w = w^1 - w^2$ satisfies the equation

$$w_{tt} + kd_0 g_0'(u^1_t(t))w_t + \mathscr{A} w = \langle F'(u^1(t)); w \rangle + R_1(t) + R_2(t),$$

where

$$R_1(t) = -kd_0 \left[\int_0^1 g_0''(\lambda u^1_t + (1-\lambda)u^2_t) d\lambda \right] w^2_t w$$

and

$$R_2(t) = [w^2, v(w, u^1 + u^2 + 2f)] + 2[w, v(u^1 + f, w^2)] + 2[u^2 + f, v(u^1 - u^2, w^2)].$$

Here, as above, $v = v(w_1, w_2) \in H^2_0(\Omega)$ is determined from (9.4.70). By the same calculation as in Section 9.5.2 we have that

$$\|R_1(t)\| \le C_R \|w^2_t\| \|\Delta w\| \quad \text{and} \quad \|R_2(t)\| \le C_R \|w^2\|_2 \|\Delta w\| + C_R \|u^1(t) - u^2(t)\|_2.$$

Now as in Section 9.5.2 we introduce the Lyapunov function

$$W(t) = Q(t) + \varepsilon(w_t, w),$$

where

$$Q(t) = \frac{1}{2} \Big[\|w_t\|^2 + \|\Delta w\|^2 + 2\|\Delta v(u^1 + f, w)\|^2 - ([w,w], v(u^1 + 2f, u^1) + F_0) \Big] + \mu \|w\|^2.$$

As above, after an appropriate choice of ε and μ we obtain that

$$a_1 \left(\|w_t\|^2 + \|\Delta w\|^2 \right) \le W(t) \le a_2 \left(\|w_t\|^2 + \|\Delta w\|^2 \right)$$

and

$$\frac{dW}{dt} + \gamma W \le c_1 \left(\|w(t)\|^2 + \|u^1(t) - u^2(t)\|^2_2 \right) + c_2 D(t) W(t),$$

where $D(t) = \|u^2_{tt}(t)\|^2 + \|\Delta u^2_t(t)\|^2$. The positive constants a_1, a_2, γ, c_1, c_2 may depend on R. Because $w = u^1_t - u^2_t$, this implies that

$$W(t) \le W(0) \exp\left\{ -\gamma t + c_2 \int_0^t D(\tau) d\tau \right\} + c_1 \int_0^t \|S_\tau y_1 - S_\tau y_2\|^2_H \exp\left\{ -\gamma(t-\tau) + c_2 \int_\tau^t D(r) dr \right\} d\tau$$

for every $t > 0$. Due to the finiteness of the dissipation integral in (9.5.41) and stabilizability estimate (9.4.42) in the space H this implies (9.5.40) which is the result of Lemma 9.5.11. □

To conclude the proof of Theorem 9.5.10, in view of Lemma 9.5.11 it suffices to apply Corollary 7.9.5 and Theorem 7.9.6. □

9.5.12. Remark. Results similar to Proposition 9.5.9 and Theorem 9.5.10 can also be established in the case of free boundary conditions (9.4.5). In this case, instead of the Hilbert space H_{st} we need to consider, as a phase space, the set $\mathscr{L}_{st}$ which is defined as a closed set in $H^4(\Omega) \times H^2(\Omega)$ consisting of elements $(u_0; u_1)$ satisfying the compatibility condition in (4.1.29). By Theorem 4.1.13 $\mathscr{L}_{st}$ is a maximal set of initial data for strong solutions and thus $\mathscr{L}_{st}$ is positively invariant with respect to S_t: $S_t\mathscr{L}_{st} \subset \mathscr{L}_{st}$. We can apply the same argument as above to semiflow S_t on the closed metric space $\mathscr{L}_{st}$ to obtain the corresponding result in the case of free boundary conditions. ■

9.5.4 Exponential attractor

As in the rotational case (see Theorem 9.2.15) using the corresponding stabilizability estimate one can prove the existence of fractal exponential attractor.

9.5.13. Theorem. *Let Assumption* 9.4.1 *be valid and* (H, S_t) *be the dynamical system generated by problem* (9.4.1) *and* (9.4.2) *with one of the boundary conditions* (9.4.3), *or* (9.4.4), *or* (9.4.5). *Assume the validity of the hypotheses of Section* 9.4 *which guarantee the existence of a global finite-dimensional attractor. Assume in addition that*

$$|g_0(s)| \leq C(1 + sg_0(s))^\gamma, \quad s \in \mathbb{R}, \tag{9.5.42}$$

for some $0 \leq \gamma < 1$. *Then the system* (H, S_t) *has a (generalized) fractal exponential attractor* A *whose dimension is finite in the space* $\widetilde{H} = L_2(\Omega) \times H^{-2}(\Omega)$.

Proof. We apply Theorem 8.10.1. To check condition (8.10.2) we note that

$$\|\mathscr{A}^{-1/2}D(v)\| \leq \int_\Omega |g_0(v)|dx, \quad v \in H^2(\Omega).$$

Thus by (9.5.42) we have

$$\|\mathscr{A}^{-1/2}D(v)\| \leq C\int_\Omega (1 + vg_0(v))^\gamma dx \leq C_\gamma \left[\int_\Omega (1 + vg_0(v))\, dx\right]^\gamma,$$

which implies (8.10.2) with $l = \frac{1}{2}$. □

We note that as in the rotational case (see Remark 9.2.16) we do not know whether the global attractor is exponential under the conditions of Theorem 9.5.13.

9.5.5 Upper semicontinuity of the global attractor with respect to rotational inertia

In this section we show that the global attractor A_α of problem (9.1.1) and (9.1.2) is close to the attractor A of problem (9.4.1) and (9.4.2) when α is small enough (for other upper semicontinuity properties of global attractors for von Karman evolutions we refer to Sections 9.3.2 and 9.6.2).

For definiteness we consider the case of clamped plates only. We also assume that $L \equiv 0$; that is, nonconservative forces are absent (for the case $L \neq 0$ with linear damping we refer to [53]). More precisely, we consider the following problem

$$\begin{cases} (1-\alpha\Delta)u_{tt} + g_0(u_t) - \alpha \sum_{i=1,2} \left[g_i(u_{tx_i})\right]_{x_i} \\ \qquad\qquad + \Delta^2 u - [u+f, v+F_0] = p(x), \quad x \in \Omega, t > 0, \\ u = \nabla u = 0 \text{ on } \Gamma = \partial\Omega, \quad u|_{t=0} = u_0(x), \quad \partial_t u|_{t=0} = u_1(x), \end{cases} \tag{9.5.43}$$

where $v = v(u)$ is a solution of the elliptic problem (9.1.2).

9.5.14. Theorem. *Let $p \in L_2(\Omega)$, $f \in H^3(\Omega)$, and $F_0 \in H^4(\Omega)$. Assume that $g_i(s) \in C^1(\mathbb{R})$ are increasing functions such that (i) $g_i(0) = 0$; (ii) there exist $0 < m_0 \le m_1$ such that*

$$0 < m_0 \le g_i'(s) \le m_1(1 + s g_i(s))^{\gamma_i}, \quad i = 0,1,2,\ s \in \mathbb{R}, \tag{9.5.44}$$

where $\gamma_0 = 1$ and $\gamma_i < 1$ for $i = 1.2$; and (iii) the functions $g_1'(s)$ and $g_2'(s)$ are of polynomial growth; that is, (9.1.8) holds. Let A_α be a global attractor to problem (9.5.43) with $\alpha \ge 0$ (the case $\alpha = 0$ is included). Then for every $\beta \ge 0$ we have that

$$\lim_{\alpha\to\beta} \sup \left\{ \mathrm{dist}_{H_0^2(\Omega)\times L^2(\Omega)}(y, A_\beta) \ :\ y \in A_\alpha \right\} = 0. \tag{9.5.45}$$

Proof. We use the same procedure as in the proof of Theorem 9.3.10; see also [53].

By Theorem 9.2.9 and Theorem 9.5.4 the global attractors A_α exist for all $\alpha \ge 0$ and A_α is a bounded set in $(H^3 \cap H_0^2)(\Omega) \times H_0^2(\Omega)$ in the case $\alpha > 0$ and in $(H^4 \cap H_0^2)(\Omega) \times H_0^2(\Omega)$ for $\alpha = 0$. Thus to prove the theorem it is sufficient to obtain uniform (in α) estimates[12] for A_α; that is, to establish that there exists $R > 0$ independent of $\alpha \in [0, \alpha_0]$ such that

$$\|u^\alpha(t)\|_3^2 + \|\Delta u_t^\alpha(t)\|^2 + \|u_{tt}^\alpha(t)\|^2 + \alpha\|\nabla u_{tt}^\alpha(t)\|^2 \le R^2, \quad t \in \mathbb{R}, \tag{9.5.46}$$

for any full trajectory $y^\alpha(t) = (u^\alpha(t); u_t^\alpha(t))$ of the system (9.5.43) from the attractor A_α. Indeed, (9.5.46) implies that the family of trajectories $\{y^\alpha(t) : \alpha \to \beta \ge 0\}$

---

[12] The main source of possible nonuniformity in the estimate in (9.5.46) is related to behavior of solutions in the limit $\alpha \to 0$ and the case when $\beta = 0$ in (9.5.45) is the most important statement of Theorem 9.5.14.

is relatively compact in $C(a,b;H_0^2(\Omega)\times L^2(\Omega))$ for every interval $[a,b]\subset\mathbb{R}$. Therefore the same argument as in Theorem 9.3.10 gives (9.5.45).

To prove (9.5.46) we first note that the following uniform (in α) estimates

$$\|\Delta u^\alpha(t)\|^2+\|u_t^\alpha(t)\|^2+\alpha\|\nabla u_t^\alpha(t)\|^2\le R_0^2,\quad t\in\mathbb{R}, \tag{9.5.47}$$

and

$$\int_{-\infty}^{\infty}D_\alpha(t)dt\equiv\int_{-\infty}^{\infty}\int_\Omega[u_t^\alpha(t)g_0(u_t^\alpha(t))+\alpha g(\nabla u_t^\alpha(t))\nabla u_t^\alpha(t)]\,dxdt\le C_{R_0} \tag{9.5.48}$$

hold for any full trajectory $y^\alpha(t)=(u^\alpha(t);u_t^\alpha(t))$ from the attractor A_α. These relations follow from the facts that (i) the set $\mathscr{N}$ of equilibria independent of α and (ii) $\mathscr{A}_\alpha=\mathscr{M}_+(\mathscr{N})$, thus $\sup_{y\in\mathscr{A}_\alpha}\mathscr{E}(y)\le\sup_{y\in\mathscr{N}}\mathscr{E}(y)$, where $\mathscr{E}(y)$ is the full energy at the state y of the system generated by (9.5.43). To obtain (9.5.48) we also use the energy relation (3.1.14) on the attractor A_α.

Now we note that, due to the smoothness of elements from the attractor stated above the function $w(t)=u_t(t)$ is a solution to the following problem

$$(1-\alpha\Delta)w_{tt}+g_0'(u_t)w_t-\alpha\sum_{i=1,2}\left[g_i'(u_{tx_i})w_{tx_i}\right]_{x_i}+\Delta^2w=\langle F'(u);w\rangle,$$

where the derivative $F'(u)$ is given by (9.5.13). As in the proof of Proposition 9.5.7 (see (9.5.20)) using this equation for w one can see that the function

$$\begin{aligned}Q_\alpha(t)=\frac{1}{2}\Big[&\|w_t\|^2+\alpha\|\nabla w_t\|^2+\|\Delta w\|^2+2\|\Delta v(u+f,w)\|^2\\&-([w,w],v(u+2f,u)+F_0)\Big],\end{aligned}$$

where $v=v(w_1,w_2)\in H_0^2(\Omega)$ is determined from (9.4.70), satisfies the relation

$$\frac{d}{dt}Q_\alpha(t)+\int_\Omega\left[g_0'(u_t)|w_t|^2+\alpha\sum_{i=1,2}g_i'(u_{tx_i})|w_{tx_i}|^2\right]dx=-3([w,w],v(u+f,w)).$$

Now we introduce the Lyapunov-type function

$$W(t)=Q_\alpha(t)+\varepsilon((1-\alpha\Delta)w_t,w)+\mu\|w\|^2.$$

As in the proof of Proposition 9.5.7 by (9.5.22) we can conclude that there exist $\mu>0$ and $\alpha_0>0$ such that for ε small enough and for $\alpha\in[0,\alpha_0]$ we have that

$$a_0\left(\|w_t\|^2+\alpha\|\nabla w_t\|^2+\|\Delta w\|^2\right)\le W(t)\le a_1\left(\|w_t\|^2+\alpha\|\nabla w_t\|^2+\|\Delta w\|^2\right),$$

where the constants a_0 and a_1 do not depend on α but may depend on R_0.

By (9.5.44) we have that

$$
\begin{aligned}
&\left| \int_\Omega g_i'(u_{tx_i}) w_{tx_i} w_{x_i} dx \right| \\
&\le \frac{1}{4} \int_\Omega g_i'(u_{tx_i}) |w_{tx_i}|^2 dx + m_1 \left[\|w_{x_i}\|^2 + \int_\Omega [u_{tx_i} g_i(u_{tx_i})]^{\gamma_i} |w_{x_i}|^2 dx \right] \\
&\le \frac{1}{4} \int_\Omega g_i'(u_{tx_i}) |w_{tx_i}|^2 dx + m_1 \left[\|\nabla w\|^2 + \left(\int_\Omega u_{tx_i} g_i(u_{tx_i}) dx \right)^{\gamma_i} \|\nabla w\|^2_{L_{2/(1-\gamma_i)}} \right] \\
&\le \frac{1}{4} \int_\Omega g_i'(u_{tx_i}) |w_{tx_i}|^2 dx + \eta \|\Delta w\|^2 + C_\eta \left[\|w\|^2 + \int_\Omega \nabla u_t g(\nabla u_t) dx \|\Delta w\|^2 \right]
\end{aligned}
$$

for every $\eta > 0$. Therefore using the same calculations as in Proposition 9.5.7 one can see that there exist positive constants c_0, c_1 and c_2 independent of α such that

$$
\frac{d}{dt} W(t) + c_0 W(t) \le c_1 \|w(t)\|^2 + c_2 D_\alpha(t) W(t).
$$

Because $\|w(t)\|^2 \le m_0 D_\alpha(t)$, from (9.5.48) we have that $\sup_{t\in\mathbb{R}} W(t) \le C_{R_0} < \infty$ which allows us to conclude (9.5.46). □

9.5.15. Remark. One can see from (9.5.45) and from the uniform (with respect to α) estimate in (9.5.46) via interpolation argument that upper semicontinuity property (9.5.45) is valid in a stronger topology of the space $H^{3-\delta}(\Omega) \times H^{2-\delta}(\Omega)$ with $\delta > 0$ arbitrary small. ■

9.5.6 Determining functionals

In this section we prove existence of a wide collection of finite sets of functionals that completely determine the long-time behavior of strong solutions to the von Karman evolution equations. This collection contains finite sets of determining modes, nodes and local volume averages (see Examples 7.8.9, 7.8.10, and 7.8.11 in Section 7.8).

9.5.6.1 General case

We start with the case of nonlinear damping and with determining functionals taken from the energy space. Our result here is based on the stabilizability estimate for the system considered. We apply Theorem 8.9.1 for our case.

9.5.16. Theorem. *Let Assumptions* 9.4.1, 9.4.3 *and* 9.4.4 *be valid and* (H, S_t) *be the dynamical system generated by problem* (9.4.1) *and* (9.4.2) *with one of the boundary conditions* (9.4.3), *or* (9.4.4), *or* (9.4.5). *In addition we assume Assumption* 9.4.13 *in the case of free boundary conditions. Assume also that either* $L \equiv 0$ *or the damping parameter k is large enough. Let* $\mathscr{L} = \{l_j : j = 1, \dots, N\}$ *be a set of linear continuous functionals on the space* $H^2(\Omega)$ *with the completeness defect*

$\varepsilon_{\mathcal{L}} = \varepsilon_{\mathcal{L}}(H^2(\Omega), L_2(\Omega))$. *Then there exists* $\varepsilon_0 > 0$ *such that under the condition* $\varepsilon_{\mathcal{L}} \le \varepsilon_0$ *the set* $\mathcal{L}$ *is a set of determining functionals for* (H, S_t); *that is, for any two generalized solutions* $u_1(t)$ *and* $u_2(t)$ *to problem* (9.4.1) *and* (9.4.2) *with the corresponding boundary conditions the relation*

$$\lim_{t\to\infty} |l(u_1(t)) - l(u_2(t))|^2 = 0 \quad \textit{for every } l \in \mathcal{L} \tag{9.5.49}$$

implies

$$\lim_{t\to\infty} \left\{ \|u_1(t) - u_2(t)\|_2^2 + \|\partial_t u_1(t) - \partial_t u_2(t)\|^2 \right\} = 0. \tag{9.5.50}$$

We recall (see Definition 7.8.5) that the completeness defect $\varepsilon_{\mathcal{L}}$ of the set $\mathcal{L}$ with respect to the pair of spaces $H^2(\Omega)$ and $L_2(\Omega)$ is defined by the formula

$$\varepsilon_{\mathcal{L}} = \sup\{\| w \| : w \in H^2(\Omega),\ l_j(w) = 0,\ l_j \in \mathcal{L},\ \| w \|_2 \le 1\}. \tag{9.5.51}$$

Proof. We apply Theorem 8.9.1. The stabilizability estimate in (8.9.3) follows from the same argument as in the proof of finite-dimensionality of the attractor and it is based on verification of (8.3.12) with $\kappa = 2$ in the clamped and hinged cases (see also (9.4.42)). In the free boundary condition case we use the stabilizability estimate from Proposition 9.4.17. □

9.5.6.2 Linear damping and strong solutions

We can also provide conditions that guarantee the existence of determining functionals with respect to topology related to strong solutions; we prove that (9.5.49) implies the convergence in a stronger sense in comparison with (9.5.50). For the sake of some simplification as a model we consider von Karman evolution equations with clamped boundary conditions and with linear damping.[13]

Thus we deal with the following problem

$$\begin{cases} u_{tt} + ku_t + \Delta^2 u - [u+f, v+F_0] + Lu = p(x),\ x \in \Omega, t > 0, \\ u|_{\partial\Omega} = \left.\dfrac{\partial u}{\partial n}\right|_{\partial\Omega} = 0, \quad u|_{t=0} = u_0(x), \quad \partial_t u|_{t=0} = u_1(x), \end{cases} \tag{9.5.52}$$

where $v = v(u)$ is defined as a solution to the problem (9.4.2). We assume that L possesses the properties guaranteed by Assumption 9.4.1 and also that

$$p(x) \in L^2(\Omega), \quad f(x) \in H^3(\Omega), \quad F_0(x) \in H^4(\Omega).$$

Under these conditions (see Theorem 4.1.4) problem (9.5.52) has a (strong) solution with properties $u(t) \in C(\mathbb{R}; H_0^2(\Omega) \cap H^4(\Omega)) \cap C^1(\mathbb{R}; H_0^2(\Omega))$ provided $u_0 \in$

[13] In the case of nonlinear damping without nonconservative forces ($L \equiv 0$) we can obtain a result on determining functionals by relying on Theorem 8.9.1 and on the "strong" stabilizability estimate in (9.5.40).

$H_0^2(\Omega)\cap H^4(\Omega)$ and $u_1 \in H_0^2(\Omega)$. By a method similar to that applied in Theorem 8.7.1 (see also Theorem 8.8.1 and Proposition 9.5.9) one can prove that if either $L \equiv 0$ or else $k > 0$ is sufficiently large, then the corresponding dynamical system is dissipative in $H_{st} = H^4(\Omega)\cap H_0^2(\Omega)\times H_0^2(\Omega)$; that is, there exists $R > 0$ such that for any bounded set B in H_{st} we have

$$\| u_t(t) \|_2^2 + \| u(t) \|_4^2 \le R^2 \quad \text{when } t \ge t_B. \tag{9.5.53}$$

Here $u(t)$ is the solution to the problem (9.5.52) with initial data $(u_0;u_1)$ from B.

Our main result is the following.

9.5.17. Theorem. *Let L and $k > 0$ be such that* (9.5.53) *is in force and $\mathscr{L} = \{l_j : j = 1,\dots,N\}$ be a finite set of linearly independent functionals on $H_0^2(\Omega)$. Then there exists $\varepsilon_0 > 0$ depending on the dissipativity radius R and on the parameters of the problem* (9.5.52) *such that the inequality $\varepsilon_{\mathscr{L}} < \varepsilon_0$ implies that $\mathscr{L}$ is a strongly determining set for the problem* (9.5.52) *in the sense that for any two strong solutions $u_1(t)$ and $u_2(t)$ the condition*

$$\lim_{t\to\infty}\{l_j(u_1(t)) - l_j(u_2(t))\} = 0 \ \textit{for}\ j = 1,\dots,N \tag{9.5.54}$$

implies that

$$\lim_{t\to\infty}\{\| \partial_t(u_1(t) - u_2(t)) \|_2^2 + \| u_1(t) - u_2(t) \|_4^2\} = 0. \tag{9.5.55}$$

The results presented in Section 7.8.3 allow us to derive from Theorem 9.5.17 that modes, local volume averages, and nodes give examples of strongly determining functionals.

Our first step in the proof of Theorem 9.5.17 is the following assertion.

9.5.18. Theorem. *Let $u_1(t)$ and $u_2(t)$ be strong solutions to* (9.5.52) *with initial data $(u_{0j};u_{1j})$, $j = 1,2$. Assume that*

$$\| \partial_t u_j(t) \|_2^2 + \| u_j(t) \|_4^2 \le R^2, \quad t \ge 0, \quad j = 1,2. \tag{9.5.56}$$

Let $\mathscr{L} = \{l_j : j = 1,\dots,N\}$ be a finite set of linearly independent functionals on $H_0^2(\Omega)$ with the completeness defect $\varepsilon_{\mathscr{L}}$ defined by (9.5.51). *Then there exists ε_0 such that the inequality $\varepsilon_{\mathscr{L}} < \varepsilon_0$ implies that*

$$\begin{aligned}
&\| \partial_t(u_1(t) - u_2(t)) \|^2 + \| u_1(t) - u_2(t) \|_2^2 \\
&\le C_1 e^{-\beta t}\left(\| u_{11} - u_{12} \|^2 + \| u_{01} - u_{02} \|_2^2\right) \\
&\quad + C_2\int_0^t e^{-\beta(t-\tau)}[\mathscr{N}(u_1(\tau) - u_2(\tau))]^2 d\tau,
\end{aligned} \tag{9.5.57}$$

where $\mathscr{N}(w) = \max_{0\le j\le N}|l_j(w)|$ and β is a positive number.

From this theorem we obviously have that for $\varepsilon_{\mathscr{L}} < \varepsilon_0$, the condition (9.5.54) implies

$$\lim_{t\to\infty}\{\| \partial_t(u_1(t)-u_2(t)) \|^2 + \| u_1(t)-u_2(t) \|_2^2\} = 0.$$

Moreover, using dissipativity property (9.5.53) and the interpolation inequalities one can easily see that the condition (9.5.54) implies

$$\lim_{t\to\infty}\{\| \partial_t(u_1(t)-u_2(t)) \|_{2-\delta}^2 + \| u_1(t)-u_2(t) \|_{4-\delta}^2\} = 0 \tag{9.5.58}$$

for any $\delta > 0$ provided $\varepsilon_{\mathscr{L}} < \varepsilon_0$.

Proof of Theorem 9.5.18. we use the same idea as in Sections 9.5.3 and 9.5.5. Let $u(t) = u_1(t) - u_2(t)$ and $W(t) = W_0(t) + \nu W_1(t)$, where

$$W_0(t) = \frac{1}{2}\left(\| u_t \|^2 + \| \Delta u \|^2 - ([u,u], v(u_1+2f,u_1)+F_0) + \mu \| u \|^2\right),$$

where $v = v(w_1,w_2) \in H_0^2(\Omega)$ is given by (9.4.70), and $W_1(t) = (u,u_t) + (k/2) \| u \|^2$. As in Sections 9.5.3 and 9.5.5 we can choose a positive constant μ such that for all $0 < \nu \le k/2$:

$$\frac{1}{4}\left(\| u_t(t) \|^2 + \| \Delta u(t) \|^2\right) \le W(t) \le C_R \cdot \left(\| u_t(t) \|^2 + \| \Delta u(t) \|^2\right). \tag{9.5.59}$$

A simple calculation gives

$$\begin{aligned}\frac{d}{dt}W_0(t) &= -k \| u_t \|^2 + (\mu u - Lu, u_t) - \frac{1}{2}([u,u], v(u_1+f,u_{1t}))\\ &\quad + ([u_2+f, v(u_1+2f,u_1) - v(u_2+2f,u_2)], u_t),\end{aligned}$$

which makes it possible to obtain that

$$\frac{d}{dt}W_0(t) \le -\frac{k}{2} \| u_t \|^2 + \frac{\nu}{4}\|\Delta u\|^2 + C_R\|u\|^2. \tag{9.5.60}$$

Because

$$\begin{aligned}\frac{d}{dt}W_1(t) &= \| u_t \|^2 - \|\Delta u\|^2 + ([u,u], v(u_1+2f,u_1)+F_0)\\ &\quad + ([u_2+f, v(u_1+2f,u_1) - v(u_2+2f,u_2)], u) - (Lu,u),\end{aligned}$$

we also have that

$$\frac{d}{dt}W_1(t) \le \| u_t \|^2 - \frac{1}{2}\|\Delta u\|^2 + C_R\|u\|^2. \tag{9.5.61}$$

Thus, using (9.5.60) and (9.5.61) we can choose $\nu > 0$ small enough such that

$$\frac{d}{dt}W(t) \le -\delta\left(\| u_t \|^2 + \|\Delta u\|^2\right) + C_R\|u\|^2 \tag{9.5.62}$$

with some positive δ. It follows from Theorem 7.8.6 that

$$\|u\| \le C_{\mathscr{L}} \cdot \max_j |l_j(u)| + \varepsilon_{\mathscr{L}} \cdot \|u\|_2.$$

Consequently there exists $\varepsilon_0 > 0$ such that from (9.5.62) we have

$$\frac{d}{dt} W(t) \le -\frac{\delta}{2} \left(\| u_t \|^2 + \|\Delta u\|^2 \right) + C(R, \mathscr{L}) \cdot \left(\max_j |l_j(u)| \right)^2$$

provided $\varepsilon_{\mathscr{L}} < \varepsilon_0$. Now using (9.5.59) after simple calculations we get (9.5.57). This completes the proof of Theorem 9.5.18. □

9.5.19. Remark. Let $u_1(t)$ and $u_2(t)$ be strong solutions to (9.5.52) that are defined on the whole time axis. Assume that they possess the property (9.5.56) *for all* $t \in \mathbb{R}$. Using (9.5.57) one can see that the condition

$$l_j(u_1(t)) = l_j(u_2(t)) \quad \text{for all } t \in \mathbb{R} \text{ and for } l_j \in \mathscr{L}$$

implies the equality $u_1(t) = u_2(t)$ for all $t \in \mathbb{R}$ provided $\varepsilon_{\mathscr{L}} < \varepsilon_0$. Thus every solution that is bounded on the whole time axis (in the sense of (9.5.56)) is uniquely defined by its values on the functionals $l_j \in \mathscr{L}$ (cf. also Definition 7.8.3 and Theorem 7.8.4). This property of determining elements was exploited in [170]] for the definition of determining modes for the 2D Navier–Stokes equations. ■

Proof of Theorem 9.5.17. We need only to derive (9.5.55) from (9.5.58). Our argument below is similar to that used in the proof of Lemma 9.5.11.

Because $w(t) = \partial_t (u_1(t) - u_2(t))$ satisfies the equality

$$\partial_t^2 w + k\partial_t w + \Delta^2 w - [w, v(u_1 + 2f, u_1) + F_0] + \mu w = G(t), \quad t > 0,$$

where $G(t) = [\partial_t u_2, v(u_1 + 2f, u_1) - v(u_2 + 2f, u_2)] + F_1(t) - F_2(t)$ with

$$F_j(t) = 2[u_j(t) + f, v(u_j(t) + f, \partial_t u_j(t))] - L\partial_t u_j(t) + \mu \partial_t u_j(t)$$

and $v(w_1, w_2)$ is defined according to (9.4.70). Using (9.5.56) and Theorem 1.4.3 from (9.5.58) one can see that $\|G(t)\| \to 0$ as $t \to \infty$. Therefore, if we now consider the Lyapunov function

$$\begin{aligned} V(t) = W(w(t), \partial_t w(t)) &= \frac{1}{2} \left(\| \partial_t w(t) \|^2 + \| \Delta w(t) \|^2 \right) \\ &- \frac{1}{2} \left(([w(t), w(t)], v(u_1(t) + 2f, u_1(t)) + F_0) + \mu \| w(t) \|^2 \right) \\ &+ \nu \left((w(t), \partial_t w(t)) + \frac{k}{2} \| w(t) \|^2 \right) \end{aligned}$$

with appropriate μ and ν, then, as in the proof of Theorem 9.5.18, we have that

$$\frac{1}{4} \left(\| \partial_t w(t) \|^2 + \| \Delta w(t) \|^2 \right) \le V(t) \le C_R \left(\| \partial_t w(t) \|^2 + \| \Delta w(t) \|^2 \right)$$

and

$$\frac{d}{dt}V(t)+\omega V(t)\leq C_R\left(\| \partial_t(u_1(t)-u_2(t)) \|^2+\|G(t)\|^2\right)$$

with some positive ω. Therefore

$$\begin{aligned}\| \partial_t w(t) \|^2 \ + \| \Delta w(t) \|^2 &\leq C_1 e^{-\omega t}\left(\| w(0) \|^2+\| \Delta w(0) \|^2\right)\\ &+C_2\int_0^t e^{-\omega(t-\tau)}\left(\| \partial_t(u_1(\tau)-u_2(\tau)) \|^2+\|G(\tau)\|^2\right)d\tau.\end{aligned}$$

Consequently $\lim_{t\to\infty}\{\| \partial_t w(t) \|^2+\| \Delta w(t) \|^2\}=0$. Because $w=\partial_t(u_1-u_2)$, using (9.5.52) we obtain that $\|u_1(t)-u_2(t)\|_4\to 0$ when $t\to\infty$. This completes the proof of Theorem 9.5.17. □

9.6 Global attractor for quasi-static model

In this section we consider long-time behavior of problem (9.4.1) and (9.4.2) with small mass parameter. We restrict ourselves to clamped boundary conditions (9.4.3) and the case of linear damping. Thus we consider the following problem

$$\begin{cases}\mu u_{tt}+u_t+\Delta^2 u-[u+f,v+F_0]+Lu=p(x),\\ u|_{\partial\Omega}=\left.\frac{\partial u}{\partial n}\right|_{\partial\Omega}=0,\quad u|_{t=0}=u_0(x),\end{cases}\tag{9.6.1}$$

where $v=v(u)$, the Airy stress function, is a solution of the problem

$$\Delta^2 v+[u+2f,u]=0,\quad v|_{\partial\Omega}=\left.\frac{\partial v}{\partial n}\right|_{\partial\Omega}=0.\tag{9.6.2}$$

Here μ is a positive parameter. In the case when the inertia forces are small in comparison with the resisting (linear) forces of a medium ($\mu\ll 1$) we obtain the problem on oscillations of a plate in quasi-static formulation (putting formally $\mu=0$):

$$\begin{cases}\partial_t u+\Delta^2 u-[u+f,v+F_0]+Lu=p(x),\\ u|_{\partial\Omega}=\left.\frac{\partial u}{\partial n}\right|_{\partial\Omega}=0,\quad u|_{t=0}=u_0(x),\end{cases}\tag{9.6.3}$$

where $v=v(u)$ solves (9.6.2).

9.6.1. Assumption.

- The mapping L is a linear bounded operator from $H^\sigma(\Omega)$ into $H^{\sigma-2+\delta}(\Omega)$ for all $2-\delta\leq\sigma\leq 2$ with some $\delta>0$.
- The functions f, F_0, and p possess properties

$$f(x)\in H^3(\Omega),\quad F_0(x)\in H^4(\Omega),\quad p(x)\in H^1(\Omega).\tag{9.6.4}$$

Problem (9.6.3) was studied in Section 4.3. In particular were proved (see Theorem 4.3.3) the existence and uniqueness of solutions from the class

$$C(0,T;H_0^2(\Omega))\cap L_2(0,T;H^4(\Omega)\cap H_0^2(\Omega)) \text{ for any } T>0.$$

Moreover, problem (9.6.3) generates dynamical system $(H_0^2(\Omega),S_t)$ in the space $H_0^2(\Omega)$ by the formula $S_t u_0 = u(t),\ t>0$, where $u(t)$ is a solution to (9.6.3).

In this section we prove the existence of the global attractor of problem (9.6.3) and establish its relation to the attractor of the problem (9.6.1) with small μ. We also refer to Section 9.3.2 for similar results in the rotational case.

In subsequent considerations it is convenient to rewrite problem (9.6.3) in the abstract form

$$u_t + \mathscr{A}u = F(u), \quad u|_{t=0} = u_0, \tag{9.6.5}$$

in the space $\mathscr{H} = L_2(\Omega)$, where $\mathscr{A}u \equiv \Delta^2 u$, $u\in\mathscr{D}(\mathscr{A})\equiv H_0^2(\Omega)\cap H^4(\Omega)$, and, as above, $F(u)\equiv[u+f,v(u)+F_0]-Lu+p$ with $v(u)\in H_0^2(\Omega)$ solving (9.6.3).

9.6.1 The existence of attractor for quasi-static problem

To obtain the existence of the global attractor we prove that the system $(H_0^2(\Omega),S_t)$ is compact; that is, possesses a compact absorbing set. For this we need the following assertion.

9.6.2. Lemma. *Under Assumption* 9.6.1 *we have the estimate*[14]

$$\|\mathscr{A}^{1/2+\delta}u(t)\| \le C_R t^{-1/2}, \quad 0<t<1,\ 0<\delta<\frac{1}{8}, \tag{9.6.6}$$

for any solution $u(t)$ to (9.6.5) *with $u_0\in H_0^2(\Omega)$, $\|u_0\|_2\le R$.*

Proof. If we multiply (9.6.5) by $2t\mathscr{A}^{1+2\delta}u$ we obtain

$$\frac{d}{dt}\left[t\|\mathscr{A}^{1/2+\delta}u\|^2\right] - \|\mathscr{A}^{1/2+\delta}u\|^2 + 2t\|\mathscr{A}^{1+\delta}u\|^2 = 2t(F(u),\mathscr{A}^{1+2\delta}u).$$

This implies that

$$\frac{d}{dt}\left[t\|\mathscr{A}^{1/2+\delta}u\|^2\right] \le \|\mathscr{A}^{1/2+\delta}u\|^2 + 2t\|\mathscr{A}^{\delta}F(u)\|^2. \tag{9.6.7}$$

We obviously have that

---

[14] Using the analyticity of semigroup $e^{-\mathscr{A}t}$ and the variation constants formula we can easily obtain the estimate in (9.6.6) with $t^{-\delta}$ instead of $t^{-1/2}$. However we do not pursue this route because (i) it is sufficient for the existence of attractor to have smoothing property of trajectories after some short time with arbitrary behavior when $t\to 0$ and (ii) the presented argument provides an independent and self-contained tool for smoothing phenomena in parabolic-like problems without any need of referencing more advanced methods involving analyticity.

$$\|\mathscr{A}^{\delta}F(u)\| \le C(1+\|u\|_2+\|[u+f,v+F_0]\|_{4\delta}),\quad 0<\delta<\frac{1}{8}. \tag{9.6.8}$$

By (1.4.4) we have for $0<\sigma<1-4\delta$ that

$$\|[u+f,v+F_0]\|_{4\delta} \le C\|u+f\|_{2+4\delta+\sigma}\|v+F_0\|_{3-\sigma}.$$

Thus by Theorem 1.4.3 $\|[u+f,v+F_0]\|_{4\delta} \le C(1+\|u\|_3)(1+\|u\|_2^2)$. Therefore by (4.3.9) and (4.3.11) we can conclude that

$$\int_0^T \left(\|\mathscr{A}^{1/2+\delta}u\|^2+t\|\mathscr{A}^{\delta}F(u)\|^2\right)dt \le C(R,T)$$

when $\|u_0\|_2 \le R$. Therefore integrating (9.6.7) we obtain the conclusion. □

Now we are in position to prove the existence of a global attractor for the system generated the quasi-static version (9.6.3) of von Karman equations.

9.6.3. Theorem. *Under Assumption* 9.6.1 *the dynamical system* $(H_0^2(\Omega),S_t)$ *generated by* (9.6.3) *possesses a compact global attractor of finite fractal dimension.*

Proof. By (4.3.9) the system $(H_0^2(\Omega),S_t)$ is dissipative in $H_0^2(\Omega)$. It follows from Lemma 9.6.2 that $(H_0^2(\Omega),S_t)$ is also compact (there exists a compact absorbing set). Therefore by Theorem 7.2.3 a compact global attractor A for $(H_0^2(\Omega),S_t)$ exists. To prove its finite-dimensionality it is convenient to use Ladyzhenskaya's Theorem 7.3.2 with the projector $P=P_N$ (with appropriate N) which is spectral for $\mathscr{A}$. We omit these details because they are standard (see, e.g., [61]). □

9.6.2 Upper semicontinuity of the attractor to quasi-static problem

Now we prove an analogue of Theorem 9.3.10 in the case of absence of rotational damping.

Let (H,S_t^{μ}) be the dynamical system generated in the space $H=H_0^2(\Omega)\times L_2(\Omega)$ by problem (9.6.1). It follows from Theorem 9.4.6 that (H,S_t^{μ}) possesses a compact global attractor A_{μ}. By Theorem 9.5.4 this attractor is a bounded set in the space $W=(H^4\cap H_0^2)(\Omega)\times H_0^2(\Omega)$ (if $L\ne 0$ we should assume that μ is small enough, which corresponds to the case of a large damping parameter).

We have the following result on the upper semicontinuity of the attractor A_{μ}.

9.6.4. Theorem. *Under Assumption* 9.6.1 *we have the relation*

$$\lim_{\mu\to 0}\sup\left\{\mathrm{dist}_H(y,\hat{A})\,:\,y\in A_{\mu}\right\}=0, \tag{9.6.9}$$

where $\hat{A}=\left\{(z_0;z_1)\,:\,z_0\in A, z_1=-\Delta^2 z_0+F(z_1)\right\}$. *Here* A *is the global attractor for the dynamical system* $(H_0^2(\Omega),S_t)$ *generated by* (9.6.3).

As in Theorems 9.3.10 and 9.5.14 the main issue in the proof of this theorem is an appropriate uniform (in μ) estimate for solutions. For this we use the same idea as in Theorem 9.3.10 and rely on the following lemmas.

9.6.5. Lemma. *Assume that $u(t)$ is a solution to problem* (9.6.1) *possessing the property $\|\Delta u(t)\| \le R$ for $t \ge 0$. Then we have the estimate*

$$\frac{1}{2}\int_0^t \|u_t(\tau)\|^2 e^{-\beta(t-\tau)} d\tau \le \left(\mu\|u_t(0)\|^2 + \widetilde{\Pi}_0(u(0))\right) e^{-\beta t} + C(R,\beta),$$

where $\beta > 0$ is a constant such that $\beta\mu < \frac{1}{2}$ and $\widetilde{\Pi}_0(u) = \|\Delta u\|^2 + \frac{1}{2}\|\Delta v(u)\|^2$.

Proof. From the energy relation for problem (9.6.1) we find that

$$\frac{1}{2}\frac{d}{dt}\left(\mu\|u_t(t)\|^2 + \widetilde{\Pi}_0(u(t))\right) + \frac{1}{2}\|u_t(t)\|^2 \le C_R.$$

We multiply this inequality by $2\exp(\beta t)$. Then for $\beta\mu < \frac{1}{2}$, by virtue of the fact that $\|\Delta u(t)\| \le R$, we have

$$\frac{d}{dt}\left[e^{\beta t}\left(\mu\|u_t(t)\|^2 + \widetilde{\Pi}_0(u(t))\right)\right] + \frac{1}{2}\|u_t(t)\|^2 e^{\beta t} \le C_R e^{\beta t}(1+\beta).$$

Integrating from 0 to t we obtain the assertion of the lemma. □

9.6.6. Lemma. *Let $u(t)$ be a solution to problem* (9.6.1) *with initial data $(u_0; u_1)$ from the space $W = (H^4 \cap H_0^2)(\Omega) \times H_0^2(\Omega)$ such that for $t \ge 0$ one has $\|u(t)\|_2 \le R$. Then for the function $w(t) = u_t(t)$ and $\mu \in (0,\mu_0)$, where $\mu_0 = \mu_0(R)$ is sufficiently small, we have the estimate*

$$\mu\|w_t(t)\|^2 + \|\Delta w(t)\|^2 \le C_1\left(\mu\|w_t(0)\|^2 + \|\Delta w(0)\|^2\right) e^{-\beta_0 t} + C_2, \tag{9.6.10}$$

where $\beta_0 > 0$, and the numbers C_1 and C_2 do not depend on $\mu \in (0,\mu_0)$.

Proof. The function $w(t)$ solves the equation which is obtained by differentiation of (9.6.1) with respect to t:

$$\begin{cases} \mu w_{tt} + w_t + \mathscr{A} w = \langle F'(u(t)); w\rangle, \\ w(0) = u_1, \ \mu w_t(0) = -(u_1 + \mathscr{A} u_0) + F(u_0). \end{cases}$$

Here, as above,

$$\langle F'(u); w\rangle = [w, v(u+2f,u) + F_0] + 2[u+f, v(u+f,w)] - Lw,$$

where element $v(w_1,w_2)$ solves the problem (9.4.70). It follows from Corollaries 1.4.4 and 1.4.5 that

$$\|\langle F'(u), w\rangle\| \le C(1+\|u\|_2^2)\|w\|_2 \le C_R\|w\|_2. \tag{9.6.11}$$

Now we consider the function

$$W(t)=\frac{1}{2}\left(\mu\|w_t(t)\|^2+\|\Delta w(t)\|^2\right)+\nu\left\{\mu(w_t(t),w(t))+\frac{1}{2}\|w(t)\|^2\right\},$$

where $\nu>0$. Obviously

$$\frac{\mu(1-\nu\mu)}{2}\|w_t\|^2+\frac{1}{2}\|\Delta w\|^2\le W(t)\le\frac{\mu(1+\nu\mu)}{2}\|w_t\|^2+\frac{1+c_0\nu}{2}\|\Delta w\|^2. \tag{9.6.12}$$

Using (9.6.11) it is easy to find that

$$\frac{d}{dt}W(t)\le-\frac{1-2\nu\mu}{2}\|w_t(t)\|^2-\frac{\nu-c_R}{2}\|\Delta w\|^2+c_{R,\nu}\|w\|^2.$$

Setting now $\nu=c_R+2$, for sufficiently small μ_0, with the aid of (9.6.12) we obtain that

$$\frac{d}{dt}W(t)+\beta_0W(t)\le C\|w(t)\|^2,\quad t\ge0,\ \mu\in(0,\mu_0),$$

where the constants $\beta_0>0$ and C do not depend on μ. Consequently

$$W(t)\le W(0)e^{-\beta_0t}+C\int_0^t\|u_t(\tau)\|^2e^{-\beta_0(t-\tau)}d\tau.$$

Therefore estimate (9.6.10) follows from (9.6.12) and Lemma 9.6.5. □

To conclude the proof of Theorem 9.6.4 we note by the standard dissipativity argument (see, e.g., Theorem 8.2.3 and also [75, Section 3.1]) we have that

$$A_\mu\in\left\{(u_0;u_1)\in H_0^2(\Omega)\times L_2(\Omega):\mu\|u_1\|^2+\|\Delta u_0\|^2\le R^2\right\},$$

where R does not depend on μ. Therefore Lemma 9.6.6 implies that for any full trajectory $y^\mu(t)=(u^\mu(t);u_t^\mu(t))$ from the attractor A_μ we have the estimate

$$\mu\|u_{tt}^\mu(t)\|^2+\|\Delta u_t^\mu(t)\|^2+\|u^\mu(t)\|_4^2\le R_*^2,\quad t\in\mathbb{R}, \tag{9.6.13}$$

which is unform in $\mu\in(0,\mu_0)$. Further arguments are similar to those given in Section 9.3.2 for the case $\alpha>0$.

Chapter 10
Plates with Boundary Damping

10.1 Introduction: Overview

In this chapter we consider models where the main source of dissipation is generated by boundary damping. Typical physical examples include boundary friction applied to an edge of the plate. This type of problem—with geometrically constrained support of the damping—is mathematically more demanding. Indeed, the dissipation is active on a much more restrictive support, hence potentially less effective. However, recent developments in the area of boundary stabilization of plates provide new tools in order to carry out the appropriate estimates. These are based on a multipliers method and microlocal analysis estimates. We first consider models with the rotational inertia included. These problems are more tractable and simpler, due to the fact that the nonlinear term acts upon the system as a compact perturbation. Proving asymptotic smoothness in that case is an easier task. When the rotational inertia are absent in the model, the analysis is more involved and delicate, because it involves the need for compensating the lack of compactness.

Our analysis focuses on two types of physically significant boundary dissipation mechanisms: dissipation acting upon free and hinged boundary conditions. In order to focus our analysis on new difficulties resulting from the treatment of boundary dissipation, the curvature terms represented by the function f, which describes initial (free) form of the shell (cf. (3.0.1)), are not included in the model. This means that we deal here with plates (not with shells). Similarly the nonconservative terms such as operator L in (3.0.1) are not included in the present treatment.

For the sake of orientation we provide a brief overview of the type of models with boundary dissipation that are considered in this chapter.

We assume that Γ consists of two smooth disjoint parts Γ_0 and Γ_1 and study "weakly damped" von Karman evolution

$$u_{tt} - \alpha \Delta u_{tt} + \Delta^2 u + d_0(x) b(u_t) = [v(u) + F_0, u] + p \quad \text{in } \Omega \times (0, \infty) \qquad (10.1.1)$$

with Airy's stress function given by $v = v(u) \in H_0^2(\Omega)$, where $v(u)$ solves the problem

I. Chueshov, I. Lasiecka, *Von Karman Evolution Equations*,
Springer Monographs in Mathematics, DOI 10.1007/978-0-387-87712-9_10,

$$\Delta^2 v + [u,u] = 0, \quad u = \frac{\partial}{\partial n} u = 0 \text{ on } \Gamma = \partial\Omega, \tag{10.1.2}$$

and subject to clamped boundary conditions imposed on Γ_0

$$u = \nabla u = 0 \quad \text{on } \Gamma_0 \times (0,\infty).$$

Weak (or light) damping is represented by the term $d_0(x)b(u_t)$ with $d_0(x)$ possibly degenerating in Ω. Such a weak damping may provide strong stability of unforced solutions, but would not yield uniform stability (even in the linear case without any forcing). The role of weak damping, although physically natural, has the benefit of injecting gradient structure into the resulting dynamical system.

Thus, the main mechanism of dissipation is boundary damping supported on a portion of the boundary Γ_1. We consider two modes of boundary dissipation supported on Γ_1.

- **Shear torques moment dissipation** via the so-called "free" boundary conditions:

$$\Delta u + (1-\mu)B_1 u = -\alpha g_1(\frac{\partial}{\partial n} u_t), \tag{10.1.3}$$
$$\frac{\partial}{\partial n}\Delta u + (1-\mu)B_2 u - \alpha\frac{\partial}{\partial n} u_{tt} - v_1 u - \beta u^3 = -\alpha\frac{\partial}{\partial \tau} g_2(\frac{\partial}{\partial \tau} u_t) + g_0(u_t),$$

  where the boundary operators B_1 and B_2 are defined in (1.3.20).
- **Purely moment dissipation** via the so-called "hinged" or "simply supported" boundary conditions:

$$u = 0, \quad \Delta u = -g_1(\frac{\partial}{\partial n} u_t). \tag{10.1.4}$$

There are two major new issues arising in the context of boundary dissipation being the main dissipative mechanism. The first one is that of the role played by the geometry of the domain, and the second one is that of reduced number of boundary conditions used for dissipation of the energy. In fact, the two issues are somewhat related. Geometry may play a role in controlling or absorbing propagation of energy from the portion of the boundary that is not subject to dissipation.

Below we provide a brief overview of the main features of the problem studied by distinguishing the cases when rotational forces count or not in the model. Because the portion of the boundary (denoted by Γ_0) is not subjected to the dissipation, it is necessary (unless $\Gamma_0 = \emptyset$) to impose a geometric "ray trapping" condition on Γ_0. This amounts to asking if there exists $x_0 \in \mathbb{R}^2$ such that $(x - x_0)\cdot n \le 0$ for all $x \in \Gamma_0$.

- $\alpha > 0$ **and the dissipation acts via moments/shears (10.1.3):** In that case there is a need for dissipation in both boundary conditions: moments are stabilized by $g_1((\partial/\partial n)u_t)$ and shears (resp., torques) are stabilized by $g_0(u_t)$ (resp., $(\partial/\partial\tau)g_2((\partial/\partial\tau)u_t)$). The additional feature of this model is the presence of the cubic term $\beta(x)u^3$ in the boundary conditions. This term, controlled by $\beta(x) \ge 0$, is responsible for preventing the energy to be unbounded from below. It is a rather specific feature associated with "free" boundary conditions, which alone do not provide a good topological control for harmonic functions (unlike Dirichlet or Neumann boundary conditions). One could dispense with the necessity of introducing a cubic term on the boundary, but at the price of requiring that the internal in-plane forces F_0 vanish on the boundary. In this latter case one can take $\beta = 0$. Long-time behavior for this model is studied in Section 10.2.

- $\alpha > 0$ **and the dissipation acts via hinged boundary conditions (10.1.4):** In that case, it suffices to use dissipation only in one boundary condition, moment acting with $g_1((\partial/\partial n)u_t)$. Thus, dissipation of the energy is achieved by affecting one boundary condition only. This feature adds a number of technical difficulties which are elaborated below. The corresponding result and proofs are given in Section 10.3.
- $\alpha = 0$ **and the dissipation acts via free boundary conditions (10.1.3):** Here, again, it suffices to use dissipation in one boundary condition only: shears via $g_0(u_t)$ with moments and torques equal to zero ($g_1 = g_2 = 0$). The additional price to pay is the presence of a star-shaped geometric condition imposed on the active part of the boundary Γ_1. This is to say, we require that $(x - x_0) \cdot n \geq 0$ for $x \in \Gamma_1$. This last condition can be eliminated. Doing this, however, requires heavy use of microlocal analysis [185] which, in turn, relies on application of a more general method presented in [269, 185]. Long-time behavior of a nonrotational clamped–free plate is presented in Section 10.4.
- $\alpha = 0$ **and dissipation acts via hinged boundary conditions (10.1.4):** Here, again, the result presented requires the use of dissipation in one boundary condition only, by acting via moments prescribed by $g_1((\partial/\partial n)u_t)$. The mathematical treatment for this particular configuration is probably the most delicate. It compiles the difficulties due to (i) criticality of the nonlinear term, (ii) only one boundary condition activates the dissipation, and (iii) no geometric conditions are imposed on Γ_1. This problem is studied in Section 10.5.

Reducing the number of boundary conditions required for dissipation of the energy depends on rather technical estimates for the boundary traces of the corresponding solutions of von Karman evolutions. These are often referred as "sharp trace" regularity results. Such results are valid for solutions of evolutionary systems (rather than just for arbitrary functions), where the interaction on the boundary with evolution of solutions plays critical role.

We note that the first case discussed above, the rotational case with free boundary conditions, is the least regular from the point of view of linear dynamics. Indeed, this is a hyperbolic model with boundary conditions that do not satisfy (dynamical) Lopatinski conditions. These features are responsible for the need of using both boundary conditions as means for dissipating and propagating damping over the entire region.

Based on this brief overview, one can conclude that controlling long-time behavior from the boundary is a rather subtle issue that depends on the geometric properties of the domain.

In what follows we present specific results corresponding to each of the four cases listed above.

We also wish to add that although some of the abstract tools developed in Chapter 8 for the second-order equations are still applicable in the boundary case, we prefer to make the exposition independent and self-contained without strong reliance on the abstract framework of Chapter 8.

10.1.1. Remark. One could also address the issue of long-time behavior of von Karman evolutions with the dissipation that is localized on some subregion of Ω. In fact, the technical treatment of this problem is very close to that of boundary dissipation. The needed inequalities are similar in flavor and, in some cases, problems with partially localized dissipation can just be reduced to those with boundary dissipation (we refer to [78] and [79] for a discussion of similar relations in the case of the wave equation). ■

As in the previous chapters we use the notation $\|\cdot\|_{s,\mathcal{O}}$ for the Sobolev s-norm on the set $\mathcal{O}$; $\|\cdot\|_{\mathcal{O}} \equiv \|\cdot\|_{0,\mathcal{O}}$ and $(\cdot,\cdot)_{\mathcal{O}} \equiv (\cdot,\cdot)_{0,\mathcal{O}}$ denote the norm and inner product in $L_2(\mathcal{O})$. If $\mathcal{O} = \Omega$, we often write $\|\cdot\|$ and $(\cdot,\cdot)$ instead of $\|\cdot\|_{\Omega}$ and $(\cdot,\cdot)_{\Omega}$.

10.2 Global attractors for von Karman models with rotational forces and with dissipation in free boundary conditions

In the case of rotational forces included in the model, the analysis of asymptotic smoothness is simpler than in the case $\alpha = 0$. The main reason for simplification is the fact that the velocity is smoother by one derivative (when $\alpha > 0$) and, consequently, the effect of the von Karman nonlinear term provides compact perturbation to the model.

We start with the case of partially clamped plates subject to dissipation imposed on a free part of the boundary.

10.2.1 The model and the main result on the existence of compact attractors

We assume that $\Gamma = \partial\Omega$ consists of two smooth disjoint parts Γ_0 and Γ_1 and consider the following von Karman model with boundary dissipation

$$\begin{cases} u_{tt} - \alpha\Delta u_{tt} + \Delta^2 u + d_0(x)b(u_t) = [v(u)+F_0,u] + p \ \text{ in } \Omega, \\ u = \frac{\partial}{\partial n}u = 0 \ \text{ on } \Gamma_0, \\ \Delta u + (1-\mu)B_1 u = -\alpha g_1(\frac{\partial}{\partial n}u_t) \ \text{ on } \Gamma_1, \\ \frac{\partial}{\partial n}\Delta u + (1-\mu)B_2 u - \alpha\frac{\partial}{\partial n}u_{tt} \\ \qquad -\nu_1 u - \beta u^3 = -\alpha\frac{\partial}{\partial\tau}g_2(\frac{\partial}{\partial\tau}u_t) + g_0(u_t) \ \text{ on } \Gamma_1, \end{cases} \tag{10.2.1}$$

where $v = v(u) \in H_0^2(\Omega)$ solves the problem

$$\Delta^2 v + [u,u] = 0, \quad u = \frac{\partial}{\partial n} u = 0 \text{ on } \Gamma = \partial\Omega, \tag{10.2.2}$$

and the boundary operators B_1 and B_2 are defined in (1.3.20).

Below for any subset Γ_* of the boundary Γ we use the notation

$$H^s_{\Gamma_*}(\Omega) \equiv \left\{ u \in H^s(\Omega) : \left.\frac{\partial^k u}{\partial n^k}\right|_{\Gamma_*} = 0 \text{ for all } 0 \le k < s - \frac{1}{2},\ k \in \mathbb{N} \right\}. \tag{10.2.3}$$

Our main hypotheses concerning problem (10.2.1) are collected below.

10.2.1. Assumption.

- $\alpha > 0$ and $\nu_1 \ge 0$ ($\nu > 0$ if $\Gamma_0 = \emptyset$) are parameters, $d_0(x) \in L_\infty(\Omega)$ and $\beta(x) \in L_\infty(\Gamma_1)$ are nonnegative functions, and $p \in L_2(\Omega)$, $F_0 \in H^3(\Omega) \cap H^2_{\Gamma_1}(\Omega)$.
- $b, g_0 \in C(\mathbb{R})$ are locally Lipschitz and of polynomial growth; that is, there exist $q \ge 1$ and $C > 0$ such that

  $$|b(s_1) - b(s_2)| + |g_0(s_1) - g_0(s_2)| \le C\left(1 + |s_1|^{q-1} + |s_2|^{q-1}\right)|s_1 - s_2|$$

  for all $s_1, s_2 \in \mathbb{R}$, and possess the properties:
  (i) $g_0(0) = 0$ and $b(0) = 0$.
  (ii) $sb(s) > 0$ and $sg_0(s) \ge 0$ for $s \ne 0$.
  (iii) There exists $a_0 \ge 0$ such that the functions $b(s) + a_0 s$ and $g_0(s) + a_0 s$ are increasing;
- $g_1, g_2 \in C(\mathbb{R})$ are nondecreasing, with the property $g_i(0) = 0$, globally Lipshitz:

  $$|g_i(s_1) - g_i(s_2)| \le M|s_1 - s_2|, \quad s_1, s_2 \in \mathbb{R},\ i = 1,2, \tag{10.2.4}$$

  and for any $\varepsilon > 0$ there exists C_ε such that

  $$s^2 \le \varepsilon + C_\varepsilon s g_i(s) \text{ for all } s \in \mathbb{R},\ i = 1,2. \tag{10.2.5}$$

10.2.2. Remark. Note that the system considered is under the effect of internal damping of the form $d_0(x)b(u_t)$. However, in contrast with Section 9.1.3, we do not assume that the damping density $d_0(x)$ is uniformly positive (i.e., the relation $\inf_\Omega d_0(x) = 0$ is allowed). Moreover, we do not assume that the damping $b(u_t)$ (as well as the boundary damping $g_0(u_t)$) are monotone. It turns out that even when $d_0(x) \equiv 0$, some results on the attractor of local nature remain valid. Because such weak internal damping when acting alone does not suffice to dissipate the energy and stabilize the system, we interpret the term $d_0(x)b(u_t)$ as a *light* internal damping. Its presence is fully justified from the physical modeling point of view. We note that if we put in (10.2.1) a full interior damping of the form $d_0 b(u_t) - \alpha \text{div}\,\{d_1 b_1(\nabla u_t)\}$, then we could rely on the methods developed in Chapter 9. In this case the conditions concerning boundary dampings can be relaxed. We also note the property in (10.2.5) allows the damping functions g_1 and g_2 to be constants on some closed finite intervals away from zero (see Proposition B.1.1 in Appendix B and also Remark 9.1.6 for additional comments concerning property (10.2.5)). ■

We first refer to the result on well-posedness of semiflow associated with the system (10.2.1), which is a special case of Theorems 3.2.12 and 3.2.17 (one can see that Assumption 10.2.1 is a special case of Assumptions 3.2.1 and 3.2.2). For additional smoothness of strong solutions see also Remark 3.2.14.

10.2.3. Theorem (Well-posedness). *Let $H \equiv H^2_{\Gamma_0}(\Omega) \times H^1_{\Gamma_0}(\Omega)$. Under the Assumption* 10.2.1 *the system defined in* (10.2.1) *generates a dynamical system* (H, S_t) *with the evolution operator given by* $S_t(u_0; u_1) = (u(t); u_t(t))$ *where* $u(t)$ *is a generalized solution. In addition, every generalized solution* $u(t)$ *is also weak and we have that* $u_{tt} \in L_2(0, T; L_2(\Omega))$ *for every* $T > 0$ *and*

$$\int_0^T \|u_{tt}(t)\|^2_{\Omega} dt \le C_R(T) \quad \text{if } \|u_0\|^2_{2,\Omega} + \|u_1\|^2_{1,\Omega} \le R^2. \tag{10.2.6}$$

This, in particular, implies that the map

$$S_t : (u(0) = u_0; u_t(0) = u_1) \mapsto (u(t); u_t(t))$$

defines a continuous semiflow S_t *on* H.

The following additional boundary regularity is valid for generalized solutions:

$$\int_0^T \int_{\Gamma_1} |\nabla u_t|^2 dx ds \le C_R(T). \tag{10.2.7}$$

Moreover, for initial data $u_0 \in H^3(\Omega) \cap H^2_{\Gamma_0}(\Omega)$ *and* $u_1 \in H^2(\Omega) \cap H^1_{\Gamma_0}(\Omega)$ *subject to natural compatibility conditions on the boundary:*

$$\Delta u_0 + (1-\mu) B_1 u + g_1(\frac{\partial}{\partial n} u_1) = 0 \quad \text{on } \Gamma_1 \quad \text{and} \quad \nabla u_1 = 0 \quad \text{on } \Gamma_0,$$

we obtain that solutions are more regular: $(u; u_t) \in C_r([\mathbb{R}_+; H^3(\Omega) \times H^2(\Omega))$.

10.2.4. Remark. Existence and uniqueness of generalized solutions can be obtained under much weaker assumptions then those imposed by Theorem 10.2.3 (see Theorem 3.2.12). The focus of this chapter is on long-time behavior, where the additional regularity of solutions is handy, therefore we opt for a less general formulation of the well-posedness result which, however, guarantees enough regularity of solutions. This is needed for a rigorous justification of various arguments used in the subsequent development. ■

In order to define the energy of the system we introduce the following notation,

$$a(u,v) = \int_{\Omega} \tilde{a}(u,v) dx + \nu_1 \int_{\Gamma_1} uv d\Gamma, \quad a_{\Gamma_1}(u,v) \equiv \int_{\Gamma_1} \tilde{a}(u,v)(h,n) d\Gamma, \tag{10.2.8}$$

where

$$\tilde{a}(u,v) = \mu \Delta u \Delta v + (1-\mu)\left[u_{x_1x_1}v_{x_1x_1} + u_{x_2x_2}v_{x_2x_2} + 2u_{x_1x_2}v_{x_1x_2}\right] \quad (10.2.9)$$

In particular, we have that

$$\tilde{a}(u,u) = \mu|\Delta u|^2 + (1-\mu)\left[u^2_{x_1x_1} + u^2_{x_2x_2} + 2u^2_{x_1x_2}\right] \geq 0. \quad (10.2.10)$$

We also define the energies:

$$\begin{aligned} E(u,u_t) &= \frac{1}{2}\int_\Omega \left[u_t^2 + \alpha|\nabla u_t|^2 + \frac{1}{2}|\Delta v(u)|^2\right]dx + \frac{1}{2}a(u,u) + \frac{1}{4}\int_{\Gamma_1} \beta u^4 d\Gamma \\ &\equiv \frac{1}{2}\int_\Omega \left[u_t^2 + \alpha|\nabla u_t|^2\right]dx + \Pi_0(u) \end{aligned}$$

and

$$\begin{aligned} \mathscr{E}(u,u_t) &= E(u,u_t) - \frac{1}{2}\int_\Omega (F_0[u,u] + 2pu)\,dx \equiv E(u,u_t) + \Pi_1(u) \\ &= \frac{1}{2}\int_\Omega \left[u_t^2 + \alpha|\nabla u_t|^2\right]dx + \Pi_0(u) + \Pi_1(u). \end{aligned}$$

The standard energy relation for generalized solutions $u(t)$ (see (3.2.72) in Theorem 3.2.12 and also Remark 3.2.13) gives

$$\mathscr{E}(t) + \int_s^t \left[\int_{\Gamma_1} [g_0(u_t)u_t + \alpha g(\nabla u_t)\nabla u_t]d\Gamma + \int_\Omega d_0 b(u_t)u_t dx\right]d\tau = \mathscr{E}(s) \quad (10.2.11)$$

for $s < t$. Here and below we use the notations $\mathscr{E}(t) = \mathscr{E}(u(t),u_t(t))$ and

$$g(\nabla u_t)\nabla u_t = g_1(\frac{\partial}{\partial n}u_t)\frac{\partial}{\partial n}u_t + g_2(\frac{\partial}{\partial \tau}u_t)\frac{\partial}{\partial \tau}u_t.$$

Recall that the above energy is derived by the classical energy argument which amounts to multiplying the equation by u_t and applying the divergence theorem. The calculations are first performed on smooth solutions, and then extended to all finite energy solutions by a classical density argument. The regularity available for solutions supports these calculations. The details of this argument are in Theorem 3.2.12; see also Remark 3.2.13. As shown in the next lemma, the energy $\mathscr{E}$ is bounded from below.

10.2.5. Lemma. *There exist positive constants c, C, M_0 such that*

$$cE(u,u_t) - M_0 \leq \mathscr{E}(u,u_t) \leq CE(u,u_t) + M_0, \quad (u;u_t) \in H.$$

The proof of the lemma will easily follows from the following statement.

10.2.6. Proposition. *Under Assumption* 10.2.1 *we have the following assertions:*

1. *Let $\beta(x) > 0$ almost everywhere. Then for every $\varepsilon > 0$ and $s < 2$ there exists a constant C_ε such that*

$$\|u\|_{s,\Omega}^2 \le \varepsilon \left(\|\Delta v(u)\|_\Omega^2 + \|u\|_{2,\Omega}^2 + \int_{\Gamma_1} \beta u^4 d\Gamma \right) + C_\varepsilon, \quad u \in H^2_{\Gamma_0}(\Omega).$$

In particular we have that $\|u\|_{s,\Omega}^2 \le \varepsilon \Pi_0(u) + C_\varepsilon$ *for every* $\varepsilon > 0$.

2. *In the case when* $\beta = 0$ *a.e.,* $\nu_1 > 0$*, and* $F_0|_\Gamma = 0$ *for every* $\varepsilon > 0$ *there exists a constant* C_ε *such that*

$$|([u,u],F_0)_\Omega| \le \varepsilon \left(\|\Delta v(u)\|_\Omega^2 + \|u\|_{2,\Omega}^2 \right) + C_\varepsilon, \quad u \in H^2_{\Gamma_0}(\Omega).$$

We note that the requirement $\beta(x) > 0$ is dictated by the need of controlling behavior of the energy from below. As seen from the second part of Proposition 10.2.6, the additional assumption imposed on the force F_0 allows us to dispense with the β term.

Proof. It follows from Lemma 1.5.13 and Lemma 1.5.10 in Chapter 1. □

Lemma 10.2.5 is essential in order to obtain the following assertion.

10.2.7. Theorem. *Let Assumption* 10.2.1 *be in force. Assume that either* $\beta > 0$ *or else* $\beta = 0$, $\nu_1 > 0$*, and* $F_0|_\Gamma = 0$*. Then there exists* $R_* > 0$ *such that the set*

$$\mathscr{W}_R = \{ y = (u_0; u_1) \in H \, : \, \mathscr{E}(u_0, u_1) \le R \} \tag{10.2.12}$$

is a nonempty invariant set with respect to the semiflow generated by (10.2.1) *for every* $R \ge R_*$*. Moreover, the set* $\mathscr{W}_R$ *is bounded for every* $R \ge R_*$ *and any bounded set is contained in* $\mathscr{W}_R$ *for some* R.

If in addition[1] $d_0(x) > 0$ *almost everywhere in* Ω*, the system* (H, S_t) *generated by* (10.2.1) *is gradient flow with the energy* $\mathscr{E}(u_0, u_1)$ *as a strict Lyapunov function.*

Proof. It follows from the energy relation (10.2.11) and Lemma 10.2.5. □

We also recall that every stationary point V for the semiflow S_t has the form $V = (u; 0)$, where $u = u(x) \in H^2_{\Gamma_0}(\Omega)$ is a weak (variational) solution to the problem

$$\begin{aligned} &\Delta^2 u = [v(u) + F_0, u] + p \ \text{ in } \Omega, \\ &u = \frac{\partial}{\partial n} u = 0 \ \text{ on } \Gamma_0, \quad \Delta u + (1-\mu) B_1 u = 0 \ \text{ on } \Gamma_1, \\ &\frac{\partial}{\partial n} \Delta u + (1-\mu) B_2 u - \nu_1 u - \beta u^3 = 0 \ \text{ on } \Gamma_1, \end{aligned} \tag{10.2.13}$$

where $v = v(u) \in H^2_0(\Omega)$ solves (10.2.2). Moreover, Theorem 1.5.9 (in the case $\beta = 0$, $\nu_1 > 0$, and $F_0|_\Gamma = 0$) and Theorem 1.5.16 (in the case $\beta > 0$) imply the following assertion.

10.2.8. Proposition. *Let Assumption* 10.2.1 *be in force. Assume that either* $\beta > 0$ *or else* $\beta = 0$, $\nu_1 > 0$*, and* $F_0|_\Gamma = 0$*. Then the set*

---

[1] This condition is needed in order to assert the gradient structure of the flow. The question of whether the same property holds in the absence of light damping (i.e. $d_0 = 0$) still appears an open question.

$$\mathscr{N} = \{V \in H \,:\, S_t V = V \text{ for all } t \geq 0\}$$

of all stationary points of the semiflow S_t is bounded in H. Thus there exists $R_{**} > 0$ such that $\mathscr{N} \subset \mathscr{W}_R$ for all $R \geq R_{**}$.

In what follows we consider long-time behavior of the semiflow S_t. To this end, we formulate the following (geometric) assumption.

10.2.9. Assumption. We assume that

- In the case $\Gamma_0 \neq \emptyset$ there exists $x_0 \in \mathbb{R}^2$ such that $(x - x_0)n \leq 0$ for $x \in \Gamma_0$.
- Either $\beta(x) > 0$ a.e. or else $\beta = 0$ a.e., $\nu_1 > 0$ and $F_0|_\Gamma = 0$.
- We assume that either **(I)** Γ_1 is star shaped: $(x - x_0)n \geq c_0 > 0$ on Γ_1, or **(II)** $g_i \in C^1(\mathbb{R})$ are increasing and $g_i'(s) \geq m > 0$ for $|s| \geq 1$, $i = 1, 2$.

10.2.10. Remark. We note that the star-shaped condition imposed on Γ_1 allows us to dispense with certain boundary traces that are above the energy level. If this condition is not satisfied, then one can eliminate these traces by microlocal estimates. However, microlocal analysis argument introduces constants that depend on T in an uncontrolled way. To handle this, we need to strengthen the assumption on the damping by requiring property **(II)**, which is a stronger version of coercivity assumption than the one postulated in (10.2.5). ∎

Our main result concerning long-time dynamics of problem (10.2.1) is the following assertion.

10.2.11. Theorem (Compact attractors). *Assume that Assumptions* 10.2.1 *and* 10.2.9 *are in force. Let S_t be the semiflow generated by* (10.2.1). *Then*

- *The restriction $(\mathscr{W}_R, S_t)$ of the dynamical system (H, S_t) on $\mathscr{W}_R$ given by* (10.2.12) *has a compact global attractor $A_R \subset \mathscr{W}_R$ for every $R \geq R_*$, where R_* is the same as in Theorem* 10.2.7.
- *If in addition $d_0(x) > 0$ almost everywhere in Ω, then there exists a compact global attractor A for the system (H, S_t). Moreover $A = \mathscr{M}^u(\mathscr{N})$, where $\mathscr{M}^u(\mathscr{N})$ is the unstable manifold (see the definition in Section 7.5) emanating from the set $\mathscr{N}$ of equilibria for the semiflow S_t. In addition, the statements of Theorem* 7.5.10 *and Corollary* 7.5.11 *are in force.*
- *Both attractors A_R and A have a finite fractal dimension provided*[2] *there is $m > 0$ such that*

$$m|s_1 - s_2|^2 \leq (s_1 - s_2)(g_i(s_1) - g_i(s_2)) \text{ for all } s_1, s_2 \in \mathbb{R},\ i = 1, 2. \quad (10.2.14)$$

Moreover, in this case the attractors display additional smoothness; that is, they are bounded sets in the space $(H^3 \cap H^2_{\Gamma_0})(\Omega) \times H^2_{\Gamma_0}(\Omega)$ and for any trajectory $(u(t); u_t(t))$ we have the relation

$$\|u_{tt}(t)\|_{1,\Omega} + \|u_t(t)\|_{2,\Omega} + \|u(t)\|_{3,\Omega} \leq C \quad t \in \mathbb{R}.$$

[2] As in the case of internal damping, both finite-dimensionality and smoothness of attractors require more stringent, global hypotheses imposed on the damping.

10.2.12. Remark. In the case of linear boundary damping one could prove that for smooth data F_0 and p, the attractor is bounded in a smoother space (in fact up to $C^\infty \times C^\infty$). This can be accomplished by "boosting" regularity established in Theorem 10.2.11 via the abstract method developed in Chapter 8. The situation in this respect is similar to that of rotational plates with linear internal damping (see Theorem 9.2.10, Corollary 9.2.11, and also Remark 9.2.12). Indeed, one can consider dynamics generated by successive time derivatives (linearity of the boundary damping is critical for this step) and reiterates the analysis of the first step (presented below) where smoothness is lifted from a finite energy level to the next one. This is all possible, owing also to the fact that the nonlinear forcing term is subcritical and remains subcritical at higher energy levels. ■

The proof of Theorem 10.2.11 is carried out in Sections 10.2.2 and 10.2.3. We first prove asymptotic smoothness property for the system considered.

10.2.2 Asymptotic smoothness

Our first result pertains to the asymptotic smoothness. The proof of asymptotic smoothness rests on application of Theorem 7.1.11 which is used in order to establish the following theorem.

10.2.13. Theorem (Asymptotic smoothness). *Let Assumptions* 10.2.1 *and* 10.2.9 *be in force. Then the system* $(\mathscr{W}_R, S_t)$ *is asymptotically smooth for every* $R \geq R_*$.

The main technical ingredient of the proof of Theorem 10.2.13 is the "observability estimate" stated in the proposition below. This estimate is critical not only for the proof of compactness, but also for the proof of smoothness and finite-dimensionality of the attractor.

In what follows we use the following notation.

$$Q \equiv \Omega \times (0,T), \quad \Sigma \equiv \Gamma \times (0,T), \quad \Sigma_i \equiv \Gamma_i \times (0,T). \tag{10.2.15}$$

10.2.14. Proposition (Observability estimate). *Assume that Assumptions* 10.2.1 *and* 10.2.9 *are in force. Let* $U(t) = (u(t); u_t(t)) = S_t y_1$ *and* $W(t) = (w(t); w_t(t)) = S_t y_2$ *be two solutions corresponding to the initial conditions* y_1 *and* y_2 *from the set* $\mathscr{W}_R$ *given by* (10.2.12). *Then, there exist* $T_0 > 0$ *and constants* C_1 *and* $C_2(R)$, *which may depend on* T, *such that*

$$TE_z(T) + \int_0^T E_z(t)dt \leq C_1 \int_{\Sigma_1} |\nabla z_t|^2 d\Sigma + C_2(R) \cdot lot(z) \tag{10.2.16}$$

for any $T \geq T_0$, *where* $z \equiv u - w$, $E_z(t) = \frac{1}{2}\int_\Omega [|z_t|^2 + \alpha |\nabla z_t|^2] dx + \frac{1}{2} a(z,z)$, *and the lower order terms have the form*

$$lot(z) = \int_0^T \left(\|z(\tau)\|_\Omega^2 + \|z_t(\tau)\|_\Omega^2\right) d\tau. \tag{10.2.17}$$

If $(x - x_0)n > 0$ *on* Γ_1 *(the case of part* **(I)** *of Assumptions* 10.2.9*), then* C_1 *does not depend on* T.

Proof of Proposition 10.2.14. The proof follows through several steps. It suffices to prove the estimate for strong solutions u and w.

Step 1: Preliminaries. The difference $z \equiv u - w$ solves the following (linear) problem

$$\begin{cases} z_{tt} - \alpha\Delta z_{tt} + \Delta^2 z = f \text{ in } \Omega, \\ z = \frac{\partial}{\partial n} z = 0 \text{ on } \Gamma_0, \quad \Delta z + (1-\mu)B_1 z = \psi_1 \text{ on } \Gamma_1, \\ \frac{\partial}{\partial n}\Delta z + (1-\mu)B_2 z - \alpha\frac{\partial}{\partial n} z_{tt} - \nu_1 z = \psi_2 \text{ on } \Gamma_1, \end{cases} \tag{10.2.18}$$

where

$$f = -d_0(x)(b(u_t) - b(w_t)) + [v(u) + F_0, u] - [v(w) + F_0, w], \tag{10.2.19}$$

$$\psi_1 = -\alpha\left[g_1(\frac{\partial}{\partial n}u_t) - g_1(\frac{\partial}{\partial n}w_t)\right], \tag{10.2.20}$$

$$\psi_2 = \beta(u^3 - w^3) - \alpha\left[\frac{\partial}{\partial\tau}g_2(\frac{\partial}{\partial\tau}u_t) - \frac{\partial}{\partial\tau}g_2(\frac{\partial}{\partial\tau}w_t)\right] + g_0(u_t) - g_0(w_t). \tag{10.2.21}$$

We have the following energy equality

$$E_z(T) + D_t^T(z) = E_z(t) + \int_t^T [(\mathscr{R}(z), z_t) + (\mathscr{R}_\Gamma(z), z_t)_{\Gamma_1}]\,dt, \tag{10.2.22}$$

where $D_t^T(z) - \hat{D}_t^T(z) + \widetilde{D}_t^T(z)$ with

$$\hat{D}_t^T(z) = \int_t^T \left[\int_{\Gamma_1}[g_0(u_t) - g_0(w_t)]z_t d\Gamma + \int_\Omega d_0[b(u_t) - b(w_t)]z_t dx\right] d\tau,$$

$$\widetilde{D}_t^T(z) = \alpha\int_t^T \int_{\Gamma_1}[g(\nabla u_t) - g(\nabla w_t)]\,\nabla z_t d\Gamma,$$

$$\mathscr{R}(z) = [v(u) + F_0, u] - [v(w) + F_0, w], \quad \mathscr{R}_\Gamma(z) = -\beta(u^2 + uw + w^2)z.$$

Step 2: First inequalities. Because $S_t y_1, S_t y_2 \in \mathscr{W}_R$ we have that for all $t \geq 0$,

$$\|u_t(t)\|_{1,\Omega} + \|u(t)\|_{2,\Omega} + \|w_t(t)\|_{1,\Omega} + \|w(t)\|_{2,\Omega} \leq C_R, \tag{10.2.23}$$

$$\int_0^t \left[\int_{\Gamma_1}[g_0(u_t)u_t + \alpha g(\nabla u_t)\nabla u_t]d\Gamma + \int_\Omega d_0 b(u_t)u_t dx\right] d\tau \leq C_R, \tag{10.2.24}$$

$$\int_0^t \left[\int_{\Gamma_1}[g_0(w_t)w_t + \alpha g(\nabla w_t)\nabla w_t]d\Gamma + \int_\Omega d_0 b(w_t)w_t dx\right] d\tau \leq C_R. \tag{10.2.25}$$

Due to the subcriticality of the nonlinear force the following estimate holds:

$$|(\mathscr{R}(z), z_t)| + |(\mathscr{R}_\Gamma(z), z_t)_{\Gamma_1}| \leq \varepsilon E_z(t) + C_{\varepsilon,R}\|z(t)\|_\Omega^2$$

and

$$|\hat{D}_t^T(z)| \leq \varepsilon \int_0^T \|z_t(\tau)\|_{1,\Omega}^2 d\tau + C_{\varepsilon,R}\int_0^T \|z_t(\tau)\|_\Omega^2 d\tau$$

for every $\varepsilon > 0$. Therefore it follows from (10.2.22) that

$$TE_z(T) \leq 2\int_0^T E_z(t)dt + C_{R,T}lot(z), \tag{10.2.26}$$

$$E_z(T) + \widetilde{D}_0^T(z) \leq E_z(0) + \varepsilon \int_0^T E_z(t)dt + C_{\varepsilon,R}lot(z), \tag{10.2.27}$$

and also

$$E_z(t) \leq E_z(0) + \varepsilon \int_0^T E_z(t)dt + C_{\varepsilon,R}lot(z), \quad 0 \leq t \leq T, \tag{10.2.28}$$

$$E_z(0) \leq E_z(T) + \widetilde{D}_0^T(z) + \varepsilon \int_0^T E_z(t)dt + C_{\varepsilon,R}lot(z), \tag{10.2.29}$$

where $\varepsilon > 0$ is arbitrary and $lot(z)$ is given by (10.2.17).

Step 3: Multipliers. We apply the multiplier $h\nabla z$, where $h = x - x_0$ and $x_0 \in \mathbb{R}^2$ is the same as in Assumption 10.2.9, to problem (10.2.18). Calculations are given below. We note that our computations are first performed on smooth solutions (guaranteed by Theorem 10.2.3) and then extended to all of space H by the density argument.

Computations for kinetic energy: By Green's formula we have that

$$\begin{aligned}
&\int_0^T \int_\Omega (z_{tt} - \alpha\Delta z_{tt})h\nabla z dx dt \\
&= \int_0^T \int_\Omega [z_{tt}h\nabla z + \alpha\nabla z_{tt}\nabla(h\nabla z)]dxdt - \int_{\Sigma_1} \alpha\frac{\partial}{\partial n} z_{tt}h\nabla z d\Gamma dt \\
&= \int_\Omega z_t h\nabla z dx\Big|_0^T + \alpha\int_\Omega \nabla z_t \nabla(h\nabla z)dx\Big|_0^T \\
&\quad - \int_0^T \int_\Omega [z_t h\nabla z_t + \alpha\nabla z_t\nabla(h\nabla z_t)]dxdt - \int_{\Sigma_1} \alpha\frac{\partial}{\partial n} z_{tt}h\nabla z d\Gamma dt.
\end{aligned}$$

The divergence theorem implies that

$$\int_\Omega vh\nabla v dx = \frac{1}{2}\int_\Gamma (h,n)v^2 dx - \frac{1}{2}\int_\Omega \operatorname{div}\{h\}v^2 dx$$

and

$$\int_\Omega \nabla v\nabla[h\nabla v]dx = \frac{1}{2}\int_\Gamma (h,n)|\nabla v|^2 dx - \frac{1}{2}\int_\Omega \operatorname{div}\{h\}|\nabla v|^2 dx$$

$$+\sum_{i,j=1}^{2}\int_{\Omega}\partial_{x_i}v\partial_{x_i}h_j\partial_{x_j}v dx$$

for any smooth function v and vector field $h=(h_1;h_2)$. Therefore for $h=x-x_0$ we obtain that

$$\begin{aligned}
&\int_0^T\int_{\Omega}(z_{tt}-\alpha\Delta z_{tt})h\nabla z dx dt\\
&=\int_{\Omega}z_t h\nabla z dx\Big|_0^T+\alpha\int_{\Omega}\nabla z_t\nabla(h\nabla z)dx\Big|_0^T\\
&+\frac{1}{2}\int_0^T\int_{\Omega}[z_t^2+\alpha|\nabla z_t|^2]\operatorname{div}\{h\}dx dt-\alpha\int_0^T\int_{\Omega}|\nabla z_t|^2 dx dt\\
&-\frac{1}{2}\int_{\Sigma}[z_t^2+\alpha|\nabla z_t|^2](h,n)d\Gamma dt-\int_{\Sigma_1}\alpha\frac{\partial}{\partial n}z_{tt}h\nabla z d\Gamma dt\\
&=\int_{\Omega}z_t h\nabla z dx\Big|_0^T+\alpha\int_{\Omega}\nabla z_t\nabla(h\nabla z)dx\Big|_0^T+\int_0^T\int_{\Omega}z_t^2 dx dt\\
&-\frac{1}{2}\int_{\Sigma}[z_t^2+\alpha|\nabla z_t|^2](h,n)d\Gamma dt-\int_{\Sigma_1}\alpha\frac{\partial}{\partial n}z_{tt}h\nabla z d\Gamma dt.
\end{aligned}\tag{10.2.30}$$

Thus

$$\begin{aligned}
&\int_0^T\int_{\Omega}(z_{tt}-\alpha\Delta z_{tt})h\nabla z dx dt\\
&\geq -c_0(E_z(T)+E_z(0))+\int_0^T\int_{\Omega}z_t^2 dx dt\\
&-\frac{1}{2}\int_{\Sigma_1}[z_t^2+\alpha|\nabla z_t|^2](h,n)d\Gamma dt-\alpha\int_{\Sigma_1}\frac{\partial}{\partial n}z_{tt}h\nabla z d\Gamma dt,
\end{aligned}\tag{10.2.31}$$

where c_0 is a positive constant independent of R and T.

Computations for potential energy. Using (1.3.5) after accounting for boundary conditions and the fact that $B_1u=0$ on Γ_0 when $u\big|_{\Gamma_0}=0$, $\nabla u\big|_{\Gamma_0}=0$ (which follows from (1.3.6) in Proposition 1.3.5):

$$\begin{aligned}
\int_{\Omega}\Delta^2 z h\nabla z dx &= a(z,z)+\frac{1}{2}a_{\Gamma_1}(z,z)+\alpha\int_{\Gamma_1}\frac{\partial}{\partial n}z_{tt}h\nabla z d\Gamma+\int_{\Gamma_1}\psi_2 h\nabla z d\Gamma\\
&-\int_{\Gamma_1}\psi_1\frac{\partial}{\partial n}(h\nabla z)+\nu_1\int_{\Gamma_1}[zh\nabla z-z^2]d\Gamma\\
&+\frac{1}{2}a_{\Gamma_0}(z,z)-\int_{\Gamma_0}\Delta z\frac{\partial}{\partial n}(h\nabla z)d\Gamma,
\end{aligned}$$

where the forms $a(z,z)$ and $a_{\Gamma_i}(z,z)$ are defined in (10.2.8) and ψ_1,ψ_2 are given by (10.2.20) and (10.2.21). We also have (see [176], Equation (5.11), p. 82) that

$$\Delta z\frac{\partial}{\partial n}(h\nabla z)=\tilde{a}(z,z)(h,n)=|\Delta z|^2(h,n)\ \text{ on }\Gamma_0,$$

where $\tilde{a}(z,z)$ is given by (10.2.10). Therefore recalling the condition $(h,n) \le 0$ on Γ_0 we obtain that

$$\int_\Omega \Delta^2 z h\nabla z dx \ge a(z,z) + \frac{1}{2} a_{\Gamma_1}(z,z) + \alpha \int_{\Gamma_1} \frac{\partial}{\partial n} z_{tt} h\nabla z d\Gamma \tag{10.2.32}$$
$$+ \int_{\Gamma_1} \psi_2 h\nabla z d\Gamma - \int_{\Gamma_1} \psi_1 \frac{\partial}{\partial n}(h\nabla z) d\Gamma + \nu_1 \int_{\Gamma_1} [z h\nabla z - z^2] d\Gamma.$$

Now we estimate the terms containing ψ_1 and ψ_2 defined by (10.2.20) and (10.2.21). The linear growth condition imposed on g_i easily gives:

$$\|\psi_1\|^2_{\Gamma_1} \le C \int_{\Gamma_1} \left| \frac{\partial}{\partial n} z_t \right|^2 d\Gamma \le C \|\nabla z_t\|^2_{\Gamma_1}. \tag{10.2.33}$$

We also have that for any $\phi \in H^3(\Omega)$:

$$\left| \int_{\Gamma_1} \psi_2 \phi d\Gamma \right| \le 2 \int_{\Gamma_1} \beta(u^2 + w^2)|z||\phi| d\Gamma + C \int_{\Gamma_1} \left| \frac{\partial}{\partial \tau} z_t \right| \left| \frac{\partial}{\partial \tau} \phi \right| d\Gamma$$
$$+ C\left(1 + \|u_t\|^{q-1}_{L_{2q-2}(\Gamma_1)} + \|w_t\|^{q-1}_{L_{2q-2}(\Gamma_1)}\right) \|z_t\|_{L_4(\Gamma_1)} \|\phi\|_{L_4(\Gamma_1)}.$$

Therefore using (10.2.23) and the trace and embedding theorem (see Theorem 1.1.1 in Chapter 1) we obtain

$$\left| \int_{\Gamma_1} \psi_2 \phi d\Gamma \right| \le C_R \left(\|z\|_{3/4,\Omega} + \|z_t\|_{3/4,\Omega} \right) \|\phi\|_{3/4,\Omega} + C \|\nabla z_t\|_{\Gamma_1} \|\nabla \phi\|_{\Gamma_1}. \tag{10.2.34}$$

Relations (10.2.33) and (10.2.34) imply

$$\left| \int_{\Gamma_1} \psi_2 h\nabla z d\Gamma \right| + \left| \int_{\Gamma_1} \psi_1 \frac{\partial}{\partial n}(h\nabla z) d\Gamma \right|$$
$$\le \varepsilon \left(\int_{\Gamma_1} \tilde{a}(z,z) d\Gamma + E_z(t) \right) + C_{\varepsilon,R} \left(\|\nabla z_t\|^2_{\Gamma_1} + \|z\|^2_\Omega \right)$$

for any $\varepsilon > 0$, where $\tilde{a}(z,z)$ is given by (10.2.10). Therefore it follows from (10.2.32) that

$$\int_\Omega \Delta^2 z h\nabla z dx \ge a(z,z) + \left(\frac{1}{2} a_{\Gamma_1}(z,z) - \varepsilon \int_{\Gamma_1} \tilde{a}(z,z) d\Gamma \right) - \varepsilon E_z(t)$$
$$+ \alpha \int_{\Gamma_1} \frac{\partial}{\partial n} z_{tt} h\nabla z d\Gamma - C_{\varepsilon,R} \left(\|\nabla z_t\|^2_{\Gamma_1} + \|z\|^2_\Omega \right) \tag{10.2.35}$$

for any $\varepsilon > 0$. Since

$$\left| \int_\Omega f h\nabla z dx \right| \le \eta E_z(t) + C_\eta \left(\|z_t\|^2_\Omega + \|z\|^2_\Omega \right)$$

for every $\eta > 0$, relations (10.2.31) and (10.2.35) yield

$$\int_0^T a(z,z)dt \leq c_0\left(E_z(T)+E_z(0)\right)+\varepsilon\int_0^T E_z(t)dt+C_\varepsilon\int_0^T \|\nabla z_t\|_{\Gamma_1}^2 dt$$
$$+\varepsilon\int_{\Sigma_1}\tilde{a}(z,z)d\Gamma dt-\frac{1}{2}\int_0^T a_{\Gamma_1}(z,z)dt+C_{\varepsilon,R}lot(z). \quad (10.2.36)$$

The difference between potential and kinetic energy: Using the Green formula (1.3.3) we obtain

$$\int_0^T\int_\Omega fzdxdt = \int_0^T\int_\Omega (z_{tt}-\alpha\Delta z_{tt}+\Delta^2 z)zdxdt$$
$$= \int_0^T\left(a(z,z)-\int_\Omega\left(z_t^2+\alpha|\nabla z_t|^2\right)dx\right)dt$$
$$+\int_\Omega z_t z dx\Big|_0^T+\alpha\int_\Omega \nabla z_t\nabla z dx\Big|_0^T+\int_0^T\int_{\Gamma_1}\left(\psi_2 z-\psi_1\frac{\partial}{\partial n}z\right)d\Gamma.$$

Therefore using relations (10.2.33) and (10.2.34) one can see that

$$\int_0^T\int_\Omega\left(z_t^2+\alpha|\nabla z_t|^2\right)dxdt \leq c_0\left(E_z(T)+E_z(0)\right)+2\int_0^T a(z,z)dt$$
$$+C_1\int_{\Sigma_1}|\nabla z_t|^2 d\Sigma+C_R lot(z). \quad (10.2.37)$$

It follows from (10.2.36) and (10.2.37) that

$$\int_0^T E_z(t)dt \leq c_0\left(E_z(T)+E_z(0)\right)+C_\varepsilon\int_0^T \|\nabla z_t\|_{\Gamma_1}^2 dt \quad (10.2.38)$$
$$+\varepsilon\int_{\Sigma_1}\tilde{a}(z,z)d\Gamma dt-c_1\int_0^T a_{\Gamma_1}(z,z)dt+C_{\varepsilon,R}lot(z),$$

where c_i and C_ε do not depend on R and T. Inasmuch as due to (10.2.29) and (10.2.4) we have that

$$E_z(0)\leq E_z(T)+C\int_0^T \|\nabla z_t\|_{\Gamma_1}^2 dt+\varepsilon\int_0^T E_z(t)dt+C_{\varepsilon,R}lot(z),$$

using (10.2.26) from (10.2.38) we obtain that

$$TE_z(T)+\int_0^T E_z(t)dt \leq C_\varepsilon\int_0^T \|\nabla z_t\|_{\Gamma_1}^2 dt+C_{\varepsilon,R,T}lot(z) \quad (10.2.39)$$
$$+\varepsilon\int_{\Sigma_1}\tilde{a}(z,z)d\Gamma dt-c_1\int_0^T a_{\Gamma_1}(z,z)dt$$

for all $T \geq T_0$ with appropriate $T_0 > 0$.

Step 4: The case when $(h,n) > 0$ on Γ_1. In this case we can choose ε such that the term containing second spatial derivatives of the boundary traces of z in

(10.2.39) can be compensated by the last term. Therefore relation (10.2.39) implies the conclusion of Proposition 10.2.14 with the constant C_1 independent of T.

In the general case we need to perform additional steps.

Step 5: Trace regularity. It turns out that boundary traces of solutions to the plate equation can be estimated in terms of boundary velocity traces and lower order terms that are below the energy level. This estimate is critical in removing from observability estimates all second derivatives on the boundary. Thus, we can absorb the second spatial derivatives on the boundary by the following inequality which is stated below.

10.2.15. Proposition. *For every $\delta > 0$ and $\varepsilon > 0$ there exist positive constants C_δ and $C_{\delta,\varepsilon}(R)$ (depending on T) such that*

$$\int_\delta^{T-\delta}\int_{\Gamma_1} \tilde{a}(z,z)d\Gamma dt \le 2\int_\delta^{T-\delta}\int_{\Gamma_1} \left(|z_{x_1x_1}|^2 + |z_{x_2x_2}|^2 + |z_{x_1x_2}|^2\right) d\Gamma dt$$

$$\le \varepsilon \int_0^T E_z(t)dt + C_\delta \int_{\Sigma_1} |\nabla z_t|^2 d\Gamma dt + C^R_{\delta,\varepsilon} lot(z), \quad (10.2.40)$$

where $\tilde{a}(z,z)$ is given by (10.2.10).

Proof. The starting point is "sharp" trace regularity valid for linear plates. Indeed, we consider the linear plate equation (10.2.18) driven by the initial conditions $z(0), z_t(0)$, nonhomogeneous terms $f(x,t)$ on the right-hand side of equation and nonhomogeneous boundary conditions determined by $\psi_1(x,t), \psi_2(x,t)$. The following estimate is valid [204].

10.2.16. Proposition. *For any $0 < \delta < T/2$ we have*

$$\int_\delta^{T-\delta}\int_{\Gamma_1} \left(|z_{x_1x_1}|^2 + |z_{x_2x_2}|^2 + |z_{x_1x_2}|^2\right) d\Gamma dt \quad (10.2.41)$$

$$\le C\left[\int_{\Sigma_1} \left(|z_t|^2 + |\nabla z_t|^2\right) d\Gamma dt + \|\psi_1\|^2_{0,\Sigma_1} + \|\psi_2\|^2_{L_2(0,T;H^{-1}(\Gamma_1))} + \|f\|^2_{L_2(0,T;H^{-1/2+\eta}(\Omega))} + \int_0^T \|z\|^2_{2-\eta,\Omega} dt\right],$$

where $0 < \eta < \frac{1}{2}$ and C is a positive constant depending on δ and T.

We apply the result of Proposition 10.2.16 to our case, with f defined by (10.2.19) and ψ_1, ψ_2 given by (10.2.20) and (10.2.21). By (1.4.26) in Lemma 1.4.5 from (10.2.23) we obtain

$$\|[u, v(u)] - [w, v(w)]\|_{-1/2+\eta,\Omega} \le C_R \|z\|_{3/2+\eta,\Omega}.$$

By (1.4.16) with $j = 0$ and $\theta = \frac{1}{2} - \eta$ we have

$$\|[F_0, z]\|_{-1/2+\eta,\Omega} \le C_\beta \|F_0\|_{3,\Omega} \|z\|_{3/2+\eta+\beta,\Omega}$$

for any $\beta \in (0, \frac{1}{2} - \eta)$. The term $b_z = d_0(b(u_t) - b(w_t))$ can be handled via straightforward estimate (recall $u_t, v_t \in H^1(\Omega)$, because $\alpha > 0$)

$$\|b_z\|^2_{L_2(0,T;H^{-1/2+\eta}(\Omega))} \leq C\|b_z\|^2_{0,Q_T} \leq \varepsilon \int_0^T E_z(t)dt + C_{\varepsilon,R} lot(z).$$

Thus,

$$\|f\|^2_{L_2(0,T;H^{-1/2+\eta}(\Omega))} \leq \varepsilon \int_0^T E_z(t)dt + C_{\varepsilon,R} lot(z).$$

Similar arguments apply to the boundary terms. Indeed, from (10.2.34) we obtain that

$$\|\psi_2\|^2_{L_2(0,T;H^{-1}(\Gamma_1))} \leq C\|\nabla z_t\|^2_{0,\Gamma_1} + \varepsilon \int_0^T E_z(t)dt + C_{\varepsilon,R} lot(z). \tag{10.2.42}$$

Recalling (10.2.33) we obtain the inequality stated in the Proposition 10.2.15. □

Concluding step: Using Proposition 10.2.15, from (10.2.39) applied to the interval $[\delta, T-\delta]$ and from (10.2.28) we obtain that

$$\begin{aligned} \int_0^T E_z(t)dt &= \left(\int_0^\delta + \int_{T-\delta}^T\right) E_z(t)dt + \int_\delta^{T-\delta} E_z(t)dt \\ &\leq c_0 \delta E_z(0) + C_1 \int_0^T |\nabla z_t|^2_{0,\Gamma_1} dt + C_{R,T} lot(z) \end{aligned} \tag{10.2.43}$$

for $T \geq T_0$. Using (10.2.26) and (10.2.29) we obtain (10.2.16). This completes the proof of Proposition 10.2.14. □

Proof of Theorem 10.2.13.

We distinguish two cases: case I in Assumption 10.2.9 (star-shaped condition on Γ_1), and case II of Assumption 10.2.9, where stronger regularity is assumed on the dissipation, in order to compensate for the lack of geometrical assumption imposed on Γ_1.

Case (I) in Assumption 10.2.9: In this case by (10.2.5) and (10.2.24), and (10.2.25) we have that

$$\int_{\Sigma_1} |\nabla z_t|^2 d\Sigma \leq 2\int_{\Sigma_1} \left(|\nabla u_t|^2 + |\nabla w_t|^2\right) d\Sigma \leq \varepsilon T + C_\varepsilon(R)$$

for every $\varepsilon > 0$, where $C_\varepsilon(R)$ does not depend in T. Therefore, from Proposition 10.2.14 (note that C_1 in (10.2.16) does not depend on T in the present case considered) we have that

$$\|S_t y_1 - S_t y_2\|^2_H \leq \varepsilon + \frac{C_\varepsilon(R)}{T} + C(R,T)\int_0^T \left(\|z(\tau)\|^2_\Omega + \|z_t(\tau)\|^2_\Omega\right) d\tau.$$

for any $\varepsilon > 0$ and $T \geq T_0$ (we use the same notation as in Proposition 10.2.14). This relation and also compactness of pieces of trajectories in $L_2(0,T;L_2(\Omega)\times L_2(\Omega))$ (which follows from (10.2.6)) makes it possible to use Theorem 7.1.11 to prove asymptotic smoothness of semiflow S_t in the case considered.

Case (II) in Assumption 10.2.9: In this case $g_i \in C^1(\mathbb{R})$ are increasing and $g_i'(s) \geq m > 0$ for $|s| \geq 1$. This implies (see Proposition B.1.2 in Appendix B) that

$$\int_{\Sigma_1} |\nabla z_t|^2 d\Sigma_1 \leq \varepsilon + C_\varepsilon \widetilde{D}_t^T(z) \tag{10.2.44}$$

for any $\varepsilon > 0$. Therefore (10.2.16) and (10.2.27) imply that

$$\begin{aligned} TE_z(T) + \int_0^T E_z(t)dt &\leq \varepsilon + C_1 C_\varepsilon \widetilde{D}_t^T(z) + C_2(R,T)\cdot lot(z) \\ &\leq \varepsilon + C_\varepsilon (E_z(0) - E_z(T)) + C_2(\varepsilon,R,T)\cdot lot(z). \end{aligned}$$

Thus for $T \geq T_0 \geq 1$ we have that

$$E_z(T) \leq \frac{\varepsilon}{1+C_\varepsilon} + \frac{C_\varepsilon}{1+C_\varepsilon} E_z(0) + C_2(\varepsilon,R,T)\cdot lot(z),$$

which implies

$$E_z(mT) \leq \frac{\varepsilon}{1+C_\varepsilon} + \gamma_\varepsilon E_z((m-1)T) + C_2(\varepsilon,R,T)\cdot lot_m(z)$$

for every $m = 1,2,\ldots$, where $\gamma_\varepsilon = C_\varepsilon/(1+C_\varepsilon) < 1$ (it may also depend on R and T) and

$$lot_m(z) = \int_{(m-1)T}^{mT} \left(\|z(\tau)\|_\Omega^2 + \|z_t(\tau)\|_\Omega^2\right) d\tau.$$

Now by standard argument (see, e.g., the proof of Theorem 8.5.3) we prove that for any $\varepsilon > 0$ there exist positive constants $\omega(\varepsilon,R)$, $C_1(\varepsilon,R)$, and $C_2(\varepsilon,R)$ such that

$$\begin{aligned} E_z(t) &\leq \varepsilon + C_1(\varepsilon,R)e^{-\omega(\varepsilon,R)t}E_z(0) \\ &+ C_2(\varepsilon,R)\int_0^t e^{-\omega(\varepsilon,R)(t-\tau)}\left(\|z(\tau)\|_\Omega^2 + \|z_t(\tau)\|_\Omega^2\right) d\tau. \end{aligned} \tag{10.2.45}$$

Now we apply Theorem 7.1.11 which completes the proof of asymptotic smoothness of semiflow S_t - the result stated in Theorem 10.2.13.

10.2.3 Proof of the main result on attractors (Theorem 10.2.11)

We are in position to prove Theorem 10.2.11.

1. Because $\mathscr{W}_R$ is bounded for every $R > 0$, the existence of a compact attractor A_R for $(\mathscr{W}_R, S_t)$ follows from Theorem 10.2.13. To prove the existence of the attractor for (H,S_t) we use Corollary 7.5.7. To check the validity of hypotheses of this corollary we note that the condition that the light damping $d_0(x)$ is effective asserts gradient structure for the system (see Theorem 10.2.7). The boundedness of the set of equilibria is stated in Proposition 10.2.8. Thus by Corollary 7.5.7 asymptotic smoothness alone suffices for proving existence of a global attractor.

It remains to prove the final statement of Theorem 10.2.11. For this a critical role is played by the following stabilizability estimate valid for the system generated by (10.2.1).

10.2.17. Proposition (Stabilizability estimate). *Assume that Assumptions* 10.2.1 *and 10.2.9 are in force. In addition assume that relation* (10.2.14) *holds. Let* $S_t y_1 = (u(t); u_t(t))$ *and* $S_t y_2 = (w(t); w_t(t))$ *be two solutions corresponding to initial conditions* y_1 *and* y_2 *from the set* $\mathscr{W}_R$ *given by* (10.2.12). *Then there exist positive constants* $\omega(R)$, $C_1(R)$, *and* $C_2(R)$ *such that*

$$\|S_t y_1 - S_t y_2\|_H^2 \le C_1(R) e^{-\omega(R)t} \|y_1 - y_2\|_H^2 + C_2(R) \int_0^t e^{-\omega(R)(t-\tau)} \left(\|z(\tau)\|_\Omega^2 + \|z_t(\tau)\|_\Omega^2 \right) d\tau, \tag{10.2.46}$$

where $z(t) = u(t) - w(t)$.

Proof. It follows from (10.2.14) that instead of (10.2.44) we have that

$$\int_{\Sigma_1} |\nabla z_t|^2 d\Sigma_1 \le c_0 \widetilde{D}_t^T(z). \tag{10.2.47}$$

Therefore the same argument as in the proof of the estimate in (10.2.45) leads to the desired result. □

10.2.18. Remark. It is possible to strengthen the estimate in (10.2.46) to the form given in Definition 7.9.2 of quasi-stable systems and then to apply abstract Theorem 7.9.6. However, this amounts to considerable computational effort comparable with a direct proof using Theorem 7.3.3. For this reason we opt for this latter, self-contained argument. ■

2. In order to prove the finiteness of fractal dimension of the attractor we use Theorem 7.3.3 and rely on the same idea as in the proof of Theorem 7.9.6. Unfortunately we cannot apply Theorem 7.9.6 directly owing to the fact that we have another (in contrast with Definition 7.9.2 of a quasi-stable system) structure of lower-order term in stabilizability estimate (10.2.46).

To cope with, inspired by [226, 244] we introduce an "extended space" $X = H^2_{\Gamma_0}(\Omega) \times H^1_{\Gamma_0}(\Omega) \times W_2(0,T)$, where

$$W_2(0,T) = \left\{ z \in L_2(0,T; H^2_{\Gamma_0}(\Omega)) : z_t \in L_2(0,T; H^1_{\Gamma_0}(\Omega)),\ z_{tt} \in L_2(0,T; L_2(\Omega)) \right\}$$

endowed with the norm

$$|z|^2_{W_2(0,T)} \equiv \int_0^T \left(\|z\|^2_{2,\Omega} + \|z_t\|^2_{1,\Omega} + \|z_{tt}\|^2_{0,\Omega} \right) dt.$$

The norm in X is given by

$$\|U\|_X^2 = \|u_0\|_{2,\Omega}^2 + \|u_1\|_{1,\Omega}^2 + |z|_{W_2(0,T)}^2, \quad U = (u_0; u_1; z).$$

For the reduction to Theorem 7.3.3 in the space X we need $L_2(\Omega)$ regularity of the second time derivatives of $z(t) = u(t) - w(t)$. In order to derive this regularity we prove the following estimate.

10.2.19. Proposition. *Let z be the same as Proposition 10.2.17. Then the following estimate holds with an arbitrary $T < \infty$.*

$$\int_0^T \|z_{tt}(t)\|_\Omega^2 dt \le C_0 \cdot \left[E_z(0) + \int_0^T E_z(t) dt \right], \tag{10.2.48}$$

where $C_0 = C_R$ if Γ_0 is empty and $C_0 = C_R \cdot (1+T)$ when $\Gamma_0 \neq \emptyset$.

Proof. We consider z as a weak solution to the system in (10.2.18). By using an abstract representation of the model (see relation (2.5.60) in Section 2.5.2), we rewrite equation for z in the form

$$z_{tt} = M^{-1}[-\mathscr{A} z + \mathscr{A} G_1 \psi_1 + \mathscr{A} G_2 \psi_2 + f] \tag{10.2.49}$$

where, we recall, $M : H^1_{\Gamma_0}(\Omega) \mapsto [H^1_{\Gamma_0}(\Omega)]'$ is defined by the form $m_\alpha(u,v) = (u,v)_\Omega + \alpha(\nabla u, \nabla v)_\Omega$, the operator $\mathscr{A}$ is given by (2.5.25), and Green's maps are defined by (2.5.56) and (2.5.57).

Now we apply the same argument as in the proof of L_2-regularity of the second time derivative of a solution in Theorem 3.2.12. We first consider the case when $\Gamma_0 = \emptyset$. In this case the following properties follow from elliptic regularity and the closed graph theorem

$$\mathscr{A}^{1/2} G_1 \in \mathscr{L}(H^{-1/2}(\Gamma_1) \mapsto L_2(\Omega)), \quad \mathscr{A}^{1/2} G_2 \in \mathscr{L}(H^{-3/2}(\Gamma_1) \mapsto L_2(\Omega)),$$

and also $M^{-1}\mathscr{A}^{1/2} \in \mathscr{L}(L_2(\Omega))$. From here

$$\begin{aligned} \|z_{tt}\|_\Omega &\le C\Big[\|\mathscr{A}^{1/2} z(t)\|_\Omega + \|\psi_1(t)\|_{-1/2,\Gamma_1} + \|\psi_2(t)\|_{-3/2,\Gamma_1} + \|M^{-1} f(t)\|_\Omega \Big] \\ &\le C\Big[\|z(t)\|_{2,\Omega} + \|g_1(\frac{\partial}{\partial n} u_t) - g_1(\frac{\partial}{\partial n} w_t)\|_{-1/2,\Gamma_1} + \|\beta(u^3 - w^3)\|_{\Gamma_1} \\ &\quad + \|g_2(\frac{\partial}{\partial \tau} u_t) - g_2(\frac{\partial}{\partial \tau} w_t)\|_{-1/2,\Gamma_1} \Big] + C_R \Big[\|z\|_{2,\Omega} + \|z_t\|_{1,\Omega} \Big], \end{aligned}$$

where in the last estimate we account for the structure of the functions f, ψ_1, and ψ_2 in (10.2.18). On the strength of assumptions imposed on g_i, for $i = 1,2$ (we have that g_i are Lipschitz from $L_2(\Gamma_1)$ into itself), very conservatively, we obtain

$$\|z_{tt}\|_\Omega \le C_R \left[\|z(t)\|_{2,\Omega} + \|z_t(t)\|_{1,\Omega} + \|\nabla z_t\|_{\Gamma_1} \right]. \tag{10.2.50}$$

On the other hand, from (10.2.27) and (10.2.47), we also have that

$$\int_0^T |\nabla z_t|^2_{\Gamma_1} dt \le E_z(0) + C_R \int_0^T E_z(t)dt.$$

Therefore (10.2.50) implies the desired conclusion in the case when $\Gamma_0 = \emptyset$.

In the case $\Gamma_0 \neq \emptyset$ we use the relation

$$(\mathscr{A}(z - G_1\psi_1), M^{-1}w) = a(z, M^{-1}w) - (\Delta z, \frac{\partial}{\partial n} M^{-1}w)_{\Gamma_0}$$

for any $w \in L_2(\Omega)$, where $a(u,v)$ is given by (10.2.8). This relation can be justified by the same way as (3.2.85) in Chapter 3 and leads to the estimate

$$\|M^{-1}\mathscr{A}(z - G_1\psi_1)\| \le C\left(\|z\|_2 + \|\Delta z\|_{\Gamma_0}\right).$$

Therefore, as above, using Theorem 2.5.6 we can obtain the relation

$$\int_a^b \|z_{tt}(t)\|^2_\Omega dt \le C_R \cdot \left[E_z(a) + \int_a^b E_z(t)dt\right]$$

for any $0 \le a < b$ such that $b - a \le 1$. To conclude the proof of the proposition we note that

$$\begin{aligned}\int_0^T \|z_{tt}(t)\|^2_\Omega dt &= \sum_{k=0}^{[T]-1} \int_k^{k+1} \|z_{tt}(t)\|^2_\Omega dt + \int_{[T]}^T \|z_{tt}(t)\|^2_\Omega dt \\ &\le C_R \cdot \left[\sum_{k=0}^{[T]} E_z(k) + \int_0^T E_z(t)dt\right],\end{aligned}$$

where $[T]$ is the integer part of T. Therefore using (10.2.28) we obtain (10.2.48). □

Let A be a global attractor for the semiflow S_t. To prove finiteness of its dimension we use the same idea as in the proof of Theorem 7.9.6 (see also [226, 244] and [69, 75]). We consider the following set $A_T \in X$, which is a suitable extension of A,

$$A_T = \{U = (u(0); u_t(0); u(t)), t \in [0,T]) : (u(0); u_t(0)) \in A\}$$

Following the method introduced by [226] (see also Section 7.9) we show that A_T is a finite dimensional set in the extended space X. This implies the same property for the attractor A considered on space $H = H^2_{\Gamma_0}(\Omega) \times H^1_{\Gamma_0}(\Omega)$. To this end, on the set A_T we define operator $V : A_T \mapsto X$ which is a translation by T of the solution. This is to say

$$V : (u(0); u_t(0); u(t)) \to (u(T); u_t(T); u(T+t)).$$

Our goal is to show that this map V satisfies the conditions of Theorem 7.3.3. This requires verification of the following conditions: (i) *invariance:* $VA_T = A_T$; (ii) *Lipschitz:* V is Lipschitz on A_T; and (iii) *squeezing property:* for any $U_1, U_2 \in A_T$

$$\|VU_1 - VU_2\|_X \le \eta_T \|U_1 - U_2\|_X + K\left[n_T(U_1 - U_2) + n_T(VU_1 - VU_2)\right], \quad (10.2.51)$$

where $\eta_T < 1$ and $n_T(U)$ is a compact seminorm.

The first two conditions (i) and (ii) follow immediately from the definition of the map V, invariance property of the original attractor A, and Proposition 10.2.19. The squeezing property is the main property to be proved. For this, we heavily rely on the stabilizability estimate (10.2.46) established in Proposition 10.2.17 and also on Proposition 10.2.19.

With the previous notation, from Proposition 10.2.19 and Proposition 10.2.17 using the same argument as in the proof of Theorem 7.9.6 we have that

$$E_z(T) + \int_T^{2T} \left[E_z(s) + ||z_{tt}||^2_\Omega\right] ds \le C_R(1+T)\left(E_z(T) + \int_T^{2T} E_z(s)ds\right) \quad (10.2.52)$$
$$\le C_1(R)(1+T)e^{-\omega T}E_z(0) + C_2(R)(1+T) \sup_{t\in[0,2T]} \left[||z(t)||^2_\Omega + ||z_t(t)||^2_\Omega\right].$$

Thus we obtain (10.2.51) with $\eta_T = C_1(R)(1+T)e^{-\omega T}$ and

$$n_T(U) \equiv \sup_{t\in[0,T]} \left[||z(t)||^2_\Omega + ||z_t(t)||^2_\Omega\right],$$

which, by Theorem 1.1.8, is a compact seminorm with respect to the topology of X. This establishes the squeezing estimate (10.2.51) for T large enough.

3. To prove the smoothness of elements from the attractor we use the same idea as in the proof of Theorem 7.9.8.

Let $\gamma = \{y(t) \equiv (u(t), u_t(t)) : t \in \mathbb{R}\} \subset Y$ be a full trajectory from the attractor. Let $|\sigma| < 1$. Applying Proposition 10.2.17 with $y_1 = y(s+\sigma)$, $y_2 = y(s)$ (and, accordingly, the interval $[s,t]$ in place of $[0,t]$), we have that

$$\|y(t+\sigma) - y(t)\|^2_H \le C_1 e^{-\omega(t-s)}\|y(s+\sigma) - y(s)\|^2_H \quad (10.2.53)$$
$$+ C_2 \int_s^t e^{-\omega(t-\tau)} \left(\|u(\tau+\sigma) - u(\tau)\|^2_\Omega + \|u_t(\tau+\sigma) - u_t(\tau)\|^2_\Omega\right) d\tau$$

for any $t, s \in \mathbb{R}$ such that $s \le t$ and for any σ with $|\sigma| < 1$. In the further argument we assume that $\sigma > 0$. The case $\sigma < 0$ requires the obvious changes.

Relation (10.2.53) implies that

$$\|y(t+\sigma) - y(t)\|^2_H \le C_1 e^{-\omega(t-s)}\|y(s+\sigma) - y(s)\|^2_H + I_\sigma(t,s), \quad (10.2.54)$$

where

$$I_\sigma(t,s) = C_2\sigma \int_s^t e^{-\omega(t-\tau)} \left[\int_\tau^{\tau+\sigma} N(\xi)d\xi\right] d\tau$$

with $N(\xi) = \|u_t(\xi)\|^2_{0,\Omega} + \|u_{tt}(\xi)\|^2_{0,\Omega}$. Simple calculation shows that

$$I_\sigma(t,s) \le C_2\sigma \int_s^{t+\sigma} N(\xi)d\xi \int_{\xi-\sigma}^{\xi} e^{-\omega(t-\tau)} d\tau$$

$$= C_2\sigma\frac{1-e^{-\sigma\omega}}{\omega}\int_s^{t+\sigma} e^{-\omega(t-\xi)}N(\xi)d\xi.$$

Let $s = t+\sigma-m$ with $m \in \mathbb{N}$. Then

$$I_\sigma(t,t+\sigma-m) = C_2\sigma\frac{1-e^{-\sigma\omega}}{\omega}\sum_{k=0}^{m-1}\int_{t+\sigma-(k+1)}^{t+\sigma-k} e^{-\omega(t-\xi)}N(\xi)d\xi$$

$$\leq C_2\sigma\frac{1-e^{-\sigma\omega}}{\omega}\sup_{\tau\in\mathbb{R}}\left\{\int_\tau^{\tau+1} N(\xi)d\xi\right\}\sum_{k=0}^{m-1} e^{\omega(\sigma-k)}.$$

Using (10.2.6) one can see that on the attractor

$$\sup_{\tau\in\mathbb{R}}\left\{\int_\tau^{\tau+1} N(\xi)d\xi\right\} \leq C_R.$$

Therefore $I_\sigma(t,t+\sigma-m) \leq C_{R,\omega}\sigma\,(e^{\sigma\omega}-1)$ for $t\in\mathbb{R}$. Letting $s = t+\sigma-m$ with $m\to\infty$, (10.2.54) gives $\|y(t+\sigma)-y(t)\|_H^2 \leq C_{R,\omega}\sigma^2$ for any $t\in\mathbb{R}$ and $0<\sigma<1$. Therefore, with a similar argument for $\sigma<0$ we obtain that

$$\max_{\tau\in\mathbb{R}}\left\|\frac{y(\tau+\sigma)-y(\tau)}{\sigma}\right\|_H^2 \leq C \quad \text{for } |\sigma|<1,$$

which implies

$$\|u_{tt}(t)\|_{1,\Omega}^2 + \|u_t(t)\|_{2,\Omega}^2 \leq C, \quad t\in\mathbb{R}, \tag{10.2.55}$$

for any trajectory $\gamma = \{y(t)\equiv(u(t),u_t(t)) : t\in\mathbb{R}\}$ from the attractor. Therefore elliptic regularity implies the desired smoothness of the attractor. To see the latter, as in (10.2.49) we write for the solution $u(t)$:

$$Mu_{tt} + \mathscr{A}u = f + \mathscr{A}G_1\psi_1 + \mathscr{A}G_2\psi_2,$$

where $f = \lfloor v(u)+F_0,u\rfloor + p - d_0b(u_t)$ and

$$\psi_1 = -\alpha g_1(\frac{\partial}{\partial n}u_t), \quad \psi_2 = \beta u^3 - \alpha\frac{\partial}{\partial\tau}g_2(\frac{\partial}{\partial\tau}u_t) + g_0(u_t)$$

The following regularity follows from (10.2.55) and from the fact that $\|u(t)\|_{2,\Omega}\leq C$ on the attractor,

- $\|\mathscr{A}^{-1/4}f(t)\|_\Omega \leq C$, $t\in\mathbb{R}$.
- $\|M^{1/2}u_{tt}(t)\|_\Omega \leq C$, $t\in\mathbb{R}$. Because $D(M^{1/2})\sim H^1_{\Gamma_0}(\Omega)\sim D(\mathscr{A}^{1/4})$ we obtain that
$$\|\mathscr{A}^{-1/4}Mu_{tt}(t)\|_\Omega \leq C,\ t\in\mathbb{R}.$$
- $\|\psi_1(t)\|_{1/2,\Gamma_1} \leq C$, $t\in\mathbb{R}$. This follows from $(\partial/\partial n)u_t|_{\Gamma_1}\in H^{1/2}(\Gamma_1)$, and the Lipshitz property imposed on g_1. Thus the Nemytskij operator generated by g_1 takes $H^{1/2}(\Gamma_1)$ into itself. The latter is due to the Lipschitz condition imposed by Assumption 10.2.1, more specifically (10.2.4).

- $\|\psi_2(t)\|_{-1/2,\Gamma_1} \le C$, $t \in \mathbb{R}$. The argument is the same as before with the addition of the observation that $(\partial/\partial\tau) : H^{1/2}(\Gamma_1) \to H^{-1/2}(\Gamma_1)$ is bounded.

By using the superposition principle we can write $u(t) = u^1(t) + u^2(t)$, where

$$\mathscr{A}u^1(t) = f(t) - Mu_{tt}(t) \in [\mathscr{D}(\mathscr{A}^{1/4})]', \quad t \in \mathbb{R},$$

and for all $t \in \mathbb{R}$:

$$\begin{aligned}
&\Delta^2 u^2(t) = 0 \ \text{in} \ \Omega, \quad u^2(t) = 0, \ \nabla u^2(t) = 0 \ \text{on} \ \Gamma_0, \\
&\Delta u^2(t) + (1-\mu)B_1 u^2(t) \in H^{1/2}(\Gamma_1), \\
&\frac{\partial}{\partial n}\Delta u^2(t) + (1-\mu)B_2 u^2(t) - \nu_1 u^2(t) \in H^{-1/2}(\Gamma_1).
\end{aligned}$$

Thus $u^1(t) \in \mathscr{D}(\mathscr{A}^{3/4}) \subset H^3(\Omega)$ for all $t \in \mathbb{R}$ and by standard regularity in elliptic theory $u^2(t) \in H^3(\Omega)$, $t \in \mathbb{R}$. This proves $u(t) \in H^3(\Omega)$, which along with (10.2.55) gives us the final conclusion of Theorem 10.2.11 on the smoothness of the global attractor.

10.2.4 Rate of convergence to the equilibria

Our next result describes convergence rates of individual trajectories to equilibrium in the case when the set $\mathscr{N}$ of stationary points is discrete. We recall that

$$\mathscr{N} = \{V \in H \, : \, S_t V = V \text{ for all } t \ge 0\}$$

and every stationary point V has the form $V = (u;0)$, where $u = u(x) \in H^2_{\Gamma_0}(\Omega)$ is a weak (variational) solution to the problem (10.2.13).

If the system is gradient, then every generalized solution converges to an equilibrium point. Therefore, it is of interest to consider the rate of convergence for these solutions. This is to say we would like to know how fast solutions converge to stationary points. One of the main difficulties of the problem is caused by the fact that equilibria may be multiple, in which case some of them are unstable. Therefore any small perturbation of the solution in a vicinity of equilibrium may cause an escape of the solution from this neighborhood. This technical difficulty is strongly pronounced at the level of controlling lower-order terms where the argument depends on uniqueness of the selected stationary solution.

In order to describe the decay rates, we need some notation. We first introduce a concave, strictly increasing, continuous function $h : \mathbb{R}_+ \mapsto \mathbb{R}_+$ which captures the behavior of $g(s) = (g_1(s_1); g_2(s_2))$ at the origin and possesses the following properties

$$h(0) = 0 \quad \text{and} \quad s_1^2 + s_2^2 + g_1^2(s_1) + g_2^2(s_2) \le h(s_1 g_1(s_2) + s_2 g_2(s_2)) \qquad (10.2.56)$$

for $|s_i| \le 1$. Such a function can always be constructed due to the monotonicity of g; see [195] and also Proposition B.2.1 in Appendix B. Now we define

$$H_0(s) = h\left(\frac{s}{c_1}\right), \quad G_0(s) = (I+H_0)^{-1}(c_2 s), \quad Q(s) = s - (I+G_0)^{-1}(s), \quad (10.2.57)$$

where c_1 and c_2 are positive constants depending on the parameters of the problem. Precise values of these parameters can be determined by inspecting the proof, however, these are not important for qualitative description of decay rates.

Clearly, $Q(s)$ is strictly monotone. Thus, the differential equation

$$\frac{d\sigma}{dt} + Q(\sigma) = 0, \quad t > 0, \quad \sigma(0) = \sigma_0 \in \mathbb{R}, \qquad (10.2.58)$$

admits global unique solution $\sigma(t)$ which, moreover, decays asymptotically to zero as $t \to \infty$ (see Proposition B.3.1 in Appendix B).

With these preparations we are ready to state our result.

10.2.20. Theorem (Rate of stabilization). *Assume that* $d_0(x) > 0$ *almost everywhere in* Ω *and there exists* $\gamma > 0$ *and* $s_0 > 0$ *such that* $sb(s) \ge \gamma s^2$ *for all* $|s| \le s_0$. *Let Assumptions* 10.2.1 *and* 10.2.9 *be in force. In addition assume that problem* (10.2.13) *has a finite number of solutions. Let* S_t *be the semiflow generated by* (10.2.1). *Then for any* $W \in H$ *there exists a stationary point* $E = (e;0)$ *such that* $S_t W \to E$ *as* $t \to +\infty$. *Moreover, if the equilibrium* E *is hyperbolic in the sense that the linearization of* (10.2.13) *around each of its solutions has the trivial solution only, then we have the following rates of stabilization:*

$$\|S_t W - E\|_H \le C\sigma([tT^{-1}]), \quad t > 0, \qquad (10.2.59)$$

where $C, T > 0$ *are constants depending on* W *and* E, $[a]$ *denotes the integer part of* a, *and* $\sigma(t)$ *satisfies* (10.2.58) *with* Q *given by* (10.2.57). *The initial data* σ_0 *and the constants* c_i *in the definition of* Q *depend on* $W, E \in H$. *In particular, if in addition* $g_i'(0) > 0$ *for* $i = 1, 2$, *then*

$$\|S_t W - E\|_H \le Ce^{-\omega t}$$

for some positive constants C *and* ω *depending on* $W, E \in H$.

10.2.21. Remark. As noted, Theorem 10.2.20 depends on additional properties of the flow such as (i) the finiteness of the equilibria set $\mathscr{N}$, and (ii) hyperbolicity of the equilibrium point. However, instead of finiteness of the set $\mathscr{N}$ we can assume that $\mathscr{N}$ is discrete; that is, every its point is isolated. ■

Proof of Theorem 10.2.20.

Because $A = \mathscr{M}^u(\mathscr{N})$ (see Theorem 10.2.11) and $\mathscr{N}$ is finite, for any $W = (w_0; w_1)$ from H there exists an equilibrium point $E = (e;0) \in \mathscr{N}$ such that

$$(w(t);w_t(t)) \equiv W(t) = S_tW \to E, \quad t \to \infty, \tag{10.2.60}$$

where the convergence is in the strong topology of H.

Consider the new variable

$$Z(t) = (z(t);z_t(t)) \equiv W(t) - E = (w(t)-e;w_t(t)).$$

From (10.2.60) we infer that for any $\varepsilon > 0$ there exists $T_0 > 0$ ("no escape time") such that for all $T > T_0$

$$\int_{T-1}^{T} E_0(w(t)-e)dt = \int_{T-1}^{T} E_0(z(t))dt \le \varepsilon, \tag{10.2.61}$$

where we use the notation:

$$E_0(z) = E_z(t) = \frac{1}{2}\int_\Omega [|z_t|^2 + \alpha|\nabla z_t|^2]dx + \frac{1}{2}a(z,z).$$

We note that T_0 depends on a concrete solution $w(t)$, hence on W and E. This dependence $T_0(W,E)$ is retained in the final statements describing the decay rates.

In what follows we take ε sufficiently small, so the only equilibrium in the ε neighborhood is precisely E. This is to say $B_H(E,\varepsilon)\cap\mathscr{N} = \{E\}$. By the definition of equilibrium that new variable $Z(t) = (z(t);z_t(t))$ solves the problem

$$\begin{aligned}
&(1-\alpha\Delta)z_{tt} + d_0(x)b(z_t) + \Delta^2 z = F(z+e) - F(e) \quad \text{in } Q \\
&z = \frac{\partial}{\partial n}z = 0 \text{ on } \Gamma_0, \quad \Delta z + (1-\mu)B_1 z = -\alpha g_1(\frac{\partial}{\partial n}z_t) \ \text{ on } \Gamma_1, \\
&\frac{\partial}{\partial n}\Delta z + (1-\mu)B_2 z - \alpha\frac{\partial}{\partial n}z_{tt} - \nu_1 z - \beta[(z+e)^3 - e^3] \\
&= -\alpha\frac{\partial}{\partial\tau}g_2(\frac{\partial}{\partial\tau}z_t) + g_0(z_t) \ \text{ on } \Gamma_1,
\end{aligned} \tag{10.2.62}$$

where we have denoted $F(w) \equiv [v(w)+F_0,w]$. It follows from Theorem 3.2.17 that the generalized solution to (10.2.62) can be interpreted as a variational solution.

Now we introduce the following modified energy functional:

$$\mathscr{E}_1(t) = \mathscr{E}_1(z(t)) \equiv E_0(z(t)) + \Phi(z(t)),$$

where $\Phi(z)$ is a potential function defined by

$$\Phi(z) \equiv -\int_0^1 (F(e+zs)-F(e),z)_\Omega ds + \int_0^1\int_{\Gamma_1}\beta[(e+zs)^3-(e)^3)]zd\Gamma ds.$$

We note that

$$-(F(z+e)-F(e),z_t)_\Omega + \int_{\Gamma_1}\beta[(z+e)^3-e^3]z_t d\Gamma = \frac{d}{dt}\Phi(z(t)). \tag{10.2.63}$$

The following lemma shows that this $\mathscr{E}_1(z)$ is dissipative and has all the "good" properties of the energy function in terms of controlling the topology of H.

10.2.22. Lemma. *Let z be any generalized solution of* (10.2.62). *Then*

$$\mathscr{E}_1(z(t)) + \widetilde{D}_s^t(z) \le \mathscr{E}_1(z(s)) \quad \textit{for any } 0 \le s \le t, \tag{10.2.64}$$

where the damping term $\widetilde{D}_s^t(z)$ has the form

$$\widetilde{D}_s^t(z) = \int_s^t \left[\alpha \int_{\Gamma_1} g(\nabla z_t)\nabla z_t d\Gamma + \int_\Omega d_0 b(z_t) z_t dx \right] dt. \tag{10.2.65}$$

Moreover: (i) $\mathscr{E}_1(z(t)) \ge 0$ for all $t \ge 0$ and
(ii) if $\|Z(t)\|_H^2 = 2E_0(z(t)) \le 2R^2$ for $t \in [0,T]$, then for all $t \in [0,T]$ we have:

$$|E_0(z(t)) - \mathscr{E}_1(z(t))| \le \varepsilon \|z(t)\|_{2,\Omega}^2 + C(\varepsilon, R)\|z(t)\|_\Omega^2, \tag{10.2.66}$$

and

$$E_0(z(t)) \le 2\mathscr{E}_1(z(t)) + C_1\|z(t)\|_\Omega^2 \le 4E_0(z(t)) + C_2\|z(t)\|_\Omega^2, \tag{10.2.67}$$

where C_1 and C_2 are positive constants dependent on R.

Proof. The proof is standard but requires some calculations. We multiply both sides of equation (10.2.62) by z_t and we integrate by parts. Computations, first performed for strong solutions, account for (10.2.63) and lead to

$$\frac{d}{dt}\mathscr{E}_1(t) + (d_0 b(z_t), z_t)_\Omega + \int_{\Gamma_1} [\alpha g(\nabla z_t)\nabla z_t + g_0(z_t)z_t]\, d\Gamma = 0, \tag{10.2.68}$$

which implies (10.2.64).

By (10.2.64) $\mathscr{E}_1(t)$ is monotone nonincreasing. Because $Z(t) \to 0$ when $t \to \infty$, we conclude that $\mathscr{E}_1(t) \ge 0$.

To prove (10.2.66) and (10.2.67) we use the standard Sobolev embeddings and the definitions of energies. □

The next assertion is a direct consequence of the observability inequality (10.2.16) in Proposition 10.2.14.

10.2.23. Lemma. *Assume that Assumptions* 10.2.1 *and* 10.2.9 *are in force. Let z be a generalized solution to* (10.2.62) *such that* $\sup_{t\in[0,t]} E_0(z(t)) \le R^2$. *Then there is $T_0 > 0$ such that for any $T > T_0$ we have*

$$\mathscr{E}_1(T) + \int_0^T E_0(z(t))dt \le C_1(I + H_0)(\mathscr{E}_1(0) - \mathscr{E}_1(T)) + C_2(R)lot_0(z), \tag{10.2.69}$$

where C_1 and $C_2(R)$ may depend on T. Here $H_0(s) = h((s/c_1)$ as in (10.2.57) and

$$lot_0(z) = \sup_{t\in[0,T]} \|z(\tau)\|_{0,\Omega}^2 + \int_0^T \|z_t(\tau)\|_{0,\Omega}^2 d\tau.$$

Proof. To obtain (10.2.69) from (10.2.16) it is only sufficient to note that property (10.2.56) of the concave function $h(s)$, Jensen's inequality, and the relation $s^2 \le sg_i(s)$ for $|s| \ge 1$ which follows from (10.2.5) imply that

$$\int_{\Sigma_1} |\partial_{x_i} z_t|^2 d\Sigma \le \int_{\Sigma_1 \cap \{|\partial_{x_i} z_t|<1\}} h(g(\nabla z_t)\nabla z_t) d\Sigma + \int_{\Sigma_1 \cap \{|\partial_{x_i} z_t| \ge 1\}} |\partial_{x_i} z_t)|^2 d\Sigma$$
$$\le \mathrm{mes}(\Sigma_1) h\left(\frac{1}{\mathrm{mes}(\Sigma_1)} \int_{\Sigma_1} g(\nabla z_t)\nabla z_t d\Sigma\right) + C\int_{\Sigma_1} g(\nabla z_t)\nabla z_t d\Sigma$$

for every $i = 1,2$. □

Now we dispense with the lower-order terms.

10.2.24. Lemma. *Let z be a generalized solution to* (10.2.62) *and such that*

$$\sup_{t\in[0,T]} E_0(z(t)) \le R^2.$$

Moreover we assume that (10.2.61) *holds for a preassigned small ε. Then, there exists a positive constant ε_0 such that*

$$lot_0(z) \le C(R,T,\varepsilon)(I + H_0)\left(\mathscr{E}_1(0) - \mathscr{E}_1(T)\right)$$

provided $\varepsilon \le \varepsilon_0$ and $T > T_0$, where T_0 is the maximum of T_0 given in Lemma 10.2.23 *and of $T_0 = T_0(W,E)$ required by* (10.2.61).

Proof. The proof is standard and based on the "compactness/uniqueness" argument. The existence of a small neighborhood around e is needed in order to assert uniqueness of equilibria selected by the trajectory.

Let $\varepsilon_0 > 0$ be such that for the stationary solution e there is no other stationary solution w such that $E_0(w - e) \le \varepsilon_0$ and $\varepsilon \le \varepsilon_0$. We argue by contradiction, denying the validity of inequality in Lemma 10.2.24. Thus we assume that there exists a sequence $z^n(t) = w_n(t) - e$ of generalized solutions to equation (10.2.62) such that $E_0(z^n(t)) \le R^2$ for all $t \in [0,T]$, the bound in (10.2.61) holds and

$$\frac{lot_0(z^n)}{(I + H_0)\left(\mathscr{E}_1(z^n(0)) - \mathscr{E}_1(z^n(T))\right)} \longrightarrow \infty \quad \text{when } n \to \infty.$$

By (10.2.64) we have that

$$\frac{lot_0(z^n)}{(I + H_0)\left(\widetilde{D}_0^T(z^n)\right)} \longrightarrow \infty \quad \text{when } n \to \infty, \tag{10.2.70}$$

where $\widetilde{D}_0^T(z^n)$ is given by (10.2.65), hence

$$\int_0^T \left[\int_{\Gamma_1} g(\nabla z_t^n)\nabla z_t^n d\Gamma + \int_\Omega d_0 b(z_t^n) z_t^n dx\right] dt \longrightarrow 0. \tag{10.2.71}$$

We can also assume that $(z^n;z^n_t) \to (z;z_t)$ weakly* in $L_\infty(0,T;H)$, where $H = H^2_{\Gamma_0}(\Omega) \times H^1_{\Gamma_0}(\Omega)$. Proposition 10.2.19 applied to $z = w - e$ gives us that

$$\int_0^T \|z_{tt}(t)\|^2_{0,\Omega} \le C_R(T) \cdot \left[E_0(z(0)) + \int_0^T E_0(z(t))dt \right] \le C_{R,T}. \qquad (10.2.72)$$

Therefore by Aubin's type of compactness argument (see Theorem 1.1.8) this implies that $z^n_t \to z_t$ in $C(0,T;H^{1-\eta}(\Omega))$ for every $\eta > 0$. Therefore by (10.2.71) we can conclude that $z_t \equiv 0$; that is,

$$(z^n;z^n_t) \to (z;0) \quad \text{weakly* in } L_\infty(0,T;H). \qquad (10.2.73)$$

Moreover,

$$(z^n;z^n_t) \to (z;0) \quad \text{strongly in } C(0,T;H^{2-\eta}(\Omega) \times H^{1-\eta}(\Omega)) \qquad (10.2.74)$$

for any $\eta > 0$. Let us prove that $z(t) \equiv 0$. Using (10.2.73) and (10.2.74) and passing to the limit (in the sense of distributions) on equation (10.2.62) we infer that the limit function $z(t)$ is a variational solution to

$$\Delta^2 z = F(z+e) - F(e) \quad \text{in } Q_T \qquad (10.2.75)$$

with the time-independent boundary conditions. It is exactly the point where we have used that generalized solutions satisfy the variational form of the equation (see Theorem 3.2.17). Because e is a stationary solution, the above implies that $w \equiv z + e$ is also a stationary solution. On the other hand from the weak lower semicontinuity of the energy, weak convergence in (10.2.73) and the bound in (10.2.61) we infer that $E_0(z) = E_0(w-e) \le \varepsilon \le \varepsilon_0$. By the assumption concerning ε_0 this implies that $w = e$. Hence $z \equiv 0$ in (10.2.73) and (10.2.74).

In order to reach the contradiction we rescale the sequence z^n. Let

$$\hat{z}^n \equiv \frac{z^n}{\alpha_n} \quad \text{and} \quad \alpha_n^2 = lot_0(z^n) \to 0,$$

where the last conclusion follows from (10.2.74) which holds with $z \equiv 0$. We observe that because of (10.2.70), along with properties of g and b we have

$$\int_0^T \int_{\Gamma_1} |\nabla \hat{z}^n_t(t)|^2 d\Gamma dt + \frac{1}{\alpha_n^2} \int_0^T \int_\Omega d_0(x) b(z^n_t) z^n_t dx dt \to 0. \qquad (10.2.76)$$

Moreover, from the observability inequality (10.2.16) in Proposition 10.2.14 we also obtain that there exists some $K > 0$ such that $E_0(\hat{z}^n(t)) \le K$ for all $t \in [0,T]$. Thus by (10.2.72) we have that

$$\sup_{t\in[0,T]} E_0(\hat{z}^n(t)) + \int_0^T \|\hat{z}_{tt}(t)\|^2_{0,\Omega} \le C_K.$$

Again, by standard weak convergence and compactness arguments we can assume that there exists an element $(\hat{z}, \hat{z}_t)$ from $L_\infty(0,T;H)$ such that $(\hat{z}^n; \hat{z}^n_t) \to (\hat{z}; \hat{z}_t)$ weakly* in $L_\infty(0,T;H)$ and strongly in $C(0,T;H^{2-\eta}(\Omega) \times H^{1-\eta}(\Omega))$. In particular, we have that $\hat{z}^n_t \to \hat{z}_t$ in $L_p(Q_T)$ for every $1 \le p < \infty$.

Let us prove that $\hat{z}_t = 0$ almost everywhere in Ω. Indeed, from the inequality $\gamma s^2 \le sb(s)$ for $|s| \le s_0$ we have that

$$\frac{1}{\alpha_n^2} \int_{Q_T} d_0 b(z^n_t) z^n_t dQ \ge \frac{\gamma}{\alpha_n^2} \int_{Q_T \cap \{|z^n_t| < s_0\}} d_0 |z^n_t|^2 dQ$$
$$= \gamma \int_{Q_T} d_0 |\hat{z}^n_t|^2 dQ - \gamma \int_{Q_T \cap \{|\hat{z}^n_t| \ge s_0 \alpha_n^{-1}\}} d_0 |\hat{z}^n_t|^2 dQ.$$

This implies that

$$\gamma \int_{Q_T} d_0 |\hat{z}^n_t|^2 dQ \le \frac{1}{\alpha_n^2} \int_{Q_T} d_0 b(z^n_t) z^n_t dQ + C\alpha_n^2 \int_{Q_T} d_0 |\hat{z}^n_t|^4 dQ.$$

Because $\hat{z}^n_t \to \hat{z}_t$ in $L_4(Q_T)$, by (10.2.76) we obtain that $\hat{z}_t = 0$ a. e. in Ω. Consequently,

$$(\hat{z}^n; \hat{z}^n_t) \to (\hat{z}; 0) \quad \text{weakly* in } L_\infty(0,T;H), \tag{10.2.77}$$

and for any $\eta > 0$,

$$(\hat{z}^n; \hat{z}^n_t) \to (\hat{z}; 0) \quad \text{strongly in } C(0,T;H^{2-\eta}(\Omega) \times H^{1-\eta}(\Omega)). \tag{10.2.78}$$

In order to obtain a differential equation for $\hat{z}$ we need to discuss the behavior of $F(e+z^n) - F(e)$ as $n \to \infty$. We claim that

$$\frac{1}{\alpha_n} [F(e+z^n) - F(e)] \longrightarrow \langle F'(e), \hat{z} \rangle \quad \text{in } L_\infty(0,T;H^{-2}(\Omega)). \tag{10.2.79}$$

where F' is the Fréchet derivative of F. The above follows from the (conservative) estimates

$$\|\langle F'(u); w \rangle\|_{-2,\Omega} \le C(R) \|w\|_{2-\eta,\Omega}$$

and

$$\|\langle F'(u) - F'(z), w \rangle\|_{-2,\Omega} \le C_R \|u - z\|_{2-\eta,\Omega} \|w\|_{2,\Omega}, \quad w \in H_0^2(\Omega).$$

It is also easy to see that a property similar to (10.2.79) holds for the boundary nonlinearity.

By virtue of (10.2.79) and (10.2.76), after dividing both sides of equation (10.2.62) written for z^n by α_n and passing to the limit (here again using the fact that the generalized solution is also variational) we infer that the limit function $\hat{z}$ satisfies

$$\Delta^2 \hat{z} = \langle F'(e), \hat{z} \rangle \quad \text{in } Q_T, \tag{10.2.80}$$

in the variational sense with the stationary boundary conditions. Inasmuch as the equilibrium is assumed hyperbolic, we infer that the only solution to (10.2.80) is a zero solution. Thus $\hat{z} \equiv 0$ in (10.2.77) and (10.2.78), which is impossible because $1 = lot_0(\hat{z^n}) \to lot_0(\hat{z}(t)) = 0$. □

To complete the proof of Theorem 10.2.20 we use the same argument as in [75, Section 4.3] (for a similar argument we also refer to [63, 69, 73]). Combining the results of Lemmas 10.2.23 and 10.2.24 we obtain with some positive constant C:

$$\mathscr{E}_1(T) \leq C(I+H_0)\left(\mathscr{E}_1(0) - \mathscr{E}_1(T)\right) \qquad (10.2.81)$$

which relation holds with suitable $T > 0$ (depending on the solution and also original ε distancing two closest equilibria and so that (10.2.61) holds). Because $I + H_0$ is strictly monotone we obtain from (10.2.81) that

$$\mathscr{E}_1((m+1)T) + (I+H_0)^{-1}(C^{-1}\mathscr{E}_1((m+1)T)) \leq \mathscr{E}_1(mT), \quad m = 1,2,\ldots.$$

By using Proposition B.3.3 in Appendix B we obtain that

$$\mathscr{E}_1(mT) \leq \sigma(m), \quad m = 0,1,2\ldots, \qquad (10.2.82)$$

where $\sigma(t)$ solves (10.2.58) with $\sigma(0) = \mathscr{E}_1(0)$. Using (10.2.67) and the estimate for lot_0 in Lemma 10.2.24, we have

$$E_0(mT) \leq 2\mathscr{E}_1(mT) + C \sup_{t\in[mT,(m+1)T]} \|z(t)\|^2_\Omega \leq C\mathscr{E}_1(mT) \quad m = 0,1,2\ldots.$$

Therefore (10.2.82) implies that $E_0(mT) \leq C\sigma(m)$, $m = 0,1,2\ldots$ One can also see from the standard energy relation for (10.2.62) that

$$\|z(t)\|^2_{2,\Omega} + \|z_t(t)\|^2_\Omega \leq c_1 e^{c_2(t-s)}\left(\|z(s)\|^2_{2,\Omega} + \|z_t(s)\|^2_\Omega\right), \quad t > s \geq 0,$$

where c_1 and c_2 depend on e and initial data $(w_0; w_1)$. This leads to the final conclusion stated in Theorem 10.2.20. The proof has been completed.

10.2.5 Determining functionals

The main technical tool in establishing asymptotic observability of the system via finitely many determining functionals is the stabilizability estimate of Proposition 10.2.17. In fact, by using the stabilizability estimate in Proposition 10.2.17, along with ideas presented in the proof of Theorem 8.9.3, we establish the following result.

10.2.25. Theorem. *Let Assumptions* 10.2.1 *and* 10.2.9 *be in force. In addition assume that relation* (10.2.14) *holds and* $d_0(x) > 0$ *a.e.. Let* $\mathscr{L} = \{l_j : j = 1,\ldots,N\}$ *be a set of linear continuous functionals on the space* $H = H^2_{\Gamma_0}(\Omega) \times H^1_{\Gamma_0}(\Omega)$ *and* *let* $\varepsilon_{\mathscr{L}} \equiv \varepsilon_{\mathscr{L}}(H, L_2(\Omega) \times L_2(\Omega))$ *be the completeness defect of* $\mathscr{L}$ *with respect to the pair* $H \subset L_2(\Omega) \times L_2(\Omega)$*. Then there exists* $\varepsilon_0 > 0$ *such that under the condition* $\varepsilon_{\mathscr{L}} \le \varepsilon_0$ *the set* $\mathscr{L}$ *is a set of asymptotically determining functionals for problem* (10.2.1) *in the following sense: for any two of its trajectories* $S_t y_1$ *and* $S_t y_2$ *the condition*

$$\lim_{t\to\infty} \int_t^{t+1} |l_j(S_\tau y_1) - l_j(S_\tau y_2)| d\tau = 0 \ \textit{for}\ j = 1, \dots, N,$$

implies that $\lim_{t\to\infty} \|S_t y_1 - S_t y_2\|_H = 0$.

Proof. Let $y_1 = (u_{01}; u_{11})$ and $y_2 = (u_{02}; u_{12})$ and $t_0 > 0$ be such that $S_t y_1$ and $S_t y_2$ belong to the absorbing set for all $t \ge t_0$ (this set exists because the system has a global attractor). Then it follows from Proposition 10.2.17 that

$$\begin{aligned} \|S_t y_1 - S_t y_2\|_H &\le c_1 e^{-\omega(t-t_0)} \|S_{t_0} y_1 - S_{t_0} y_2\|_H \\ &\quad + c_2 \int_{t_0}^t e^{-\omega(t-\tau)} \left(\|u^1(\tau) - u^2(\tau)\|_\Omega^2 + \|u_t^1(\tau) - u_t^2(\tau)\|_\Omega^2 \right)^{1/2} d\tau \end{aligned}$$

for $t \ge t_0$, where $S_t y_1 = (u^1(t); u_t^1(t))$ and $S_t y_2 = (u^2(t); u_t^2(t))$ and the positive constant c_1, c_2, and ω depend on the size of an absorbing set.

By Theorem 7.8.6 we have that for $z(t) = u^1(t) - u^2(t)$ we have

$$\left(\|z(t)\|_\Omega^2 + \|z_t(t)\|_{0,\Omega}^2 \right)^{1/2} \le \varepsilon_{\mathscr{L}} \|(z(t); z_t(t))\|_H + C_{\mathscr{L}} \mathscr{N}(t), \tag{10.2.83}$$

where $\mathscr{N}(t) = \max_j |l_j(S_t y_1) - l_j(S_t y_2)|$. Therefore we have the inequality

$$\psi(t) \le c_1 \psi(t_0) + \varepsilon_{\mathscr{L}} c_2 \int_{t_0}^t \psi(\tau) d\tau + C(\mathscr{L}) \int_{t_0}^t e^{\omega\tau} \mathscr{N}(\tau) d\tau\,,$$

where $\psi(t) = e^{\omega t} \|S_t y_1 - S_t y_2\|_H$. Using Gronwall's lemma we can conclude the proof in the same way as was done for Theorem 8.9.3. □

10.3 Global attractors for von Karman models with rotational forces and with dissipation in hinged boundary conditions

In this section we study long-time dynamics of von Karman plates which are partially clamped and subject to dissipation imposed via hinged boundary conditions. It should be emphasized that the active boundary dissipation affects only one boundary condition; that is, through the moments. In order to handle this mathematical difficulty, rather special boundary estimates are used. These estimates, of microlocal nature, have been recently developed in the context of linear plate theory.

10.3.1 The model and the main result

We deal with the following von Karman model where boundary dissipation acts via hinged boundary conditions,

$$u_{tt} - \alpha \Delta u_{tt} + \Delta^2 u + d_0(x) b(u_t) = [v(u) + F_0, u] + p \text{ in } \Omega \times (0,\infty), \quad (10.3.1)$$
$$u = \frac{\partial}{\partial n} u = 0 \text{ on } \Gamma_0 \times (0,\infty), \quad u = 0, \ \Delta u = -g(\frac{\partial}{\partial n} u_t) \text{ on } \Gamma_1 \times (0,\infty),$$

where as above $v = v(u) \in H_0^2(\Omega)$ solves (10.2.2) and $\Gamma = \partial\Omega = \Gamma_0 \cup \Gamma_1$ with non-intersecting (regular) Γ_0 and Γ_1.

In contrast with the model (10.2.1) considered in the previous section, when the dissipation acts via both prescribed boundary conditions on Γ_1, model (10.3.1) admits a possibility that dissipation affects via only one boundary conditions. In control theory this type of problems is often referred to as control problem with reduced number of controls. Indeed, it is much more challenging to achieve the same long-time behavior effects with dissipation that is constrained to acting upon only one boundary condition. From the mathematical point of view this amounts to the necessity of a priori estimating the boundary trace that is not bounded by the topology governed by finite energy. As one can see in the course of the proof, the solution depends on the special trace type of estimates available for plate models (see [151] and references therein).

We impose the following hypotheses.

10.3.1. Assumption.

- $\alpha > 0$, $d_0(x) \in L_\infty(\Omega)$, $d_0(x) \geq 0$, $p \in L_2(\Omega)$, $F_0 \in H^3(\Omega)$.
- $b \in C(\mathbb{R})$ is locally Lipschitz and of polynomial growth; that is, there exist $q \geq 1$ and $C > 0$ such that

$$|b(s_1) - b(s_2)| \leq C\left(1 + |s_1|^{q-1} + |s_2|^{q-1}\right)|s_1 - s_2|, \quad s_1, s_2 \in \mathbb{R},$$

  and possess the properties (i) $b(0) = 0$, (ii) $sb(s) > 0$ for $s \neq 0$, and (iii) there exists $a_0 \geq 0$ such that the function $b(s) + a_0 s$ is increasing.
- $g \in C^1(\mathbb{R})$ is increasing, with the property $g(0) = 0$ and with the bound[3]

$$0 < m \leq g'(s) \leq M \quad \text{for all } |s| \geq 1. \quad (10.3.2)$$

In what follows we use the state space $H \equiv \left[H^2_{\Gamma_0} \cap H^1_0(\Omega)\right] \times H^1_0(\Omega)$, where $H^2_{\Gamma_0}(\Omega)$ is defined in (10.2.3).

We begin by asserting the well-posedness of a semiflow associated with the system (10.3.1).

10.3.2. Theorem (Well-posedness). *Under the Assumption* 10.3.1 *the system defined in* (10.3.1) *generates dynamical system* (H, S_t) *in the class of generalized so-*

[3] We can relax this requirement concerning g by assuming, as in the previous section, that $g(s)$ is globally Lipschitz and satisfies (10.2.5).

lutions; that is, the map $(u(0) = u_0; u_t(0) = u_1) \mapsto (u(t); u_t(t))$ *defines a continuous semiflow* S_t *on* H. *In addition,* $u_{tt} \in L_2(0,T;L_2(\Omega))$ *and*

$$\int_0^T [\|u_{tt}(t)\|_{0,\Omega}^2 + \|\frac{\partial}{\partial n} u_t(t)\|_{0,\Gamma_1}^2] dt \le C_{R,T} \;\; \text{if} \; \|u_0\|_{2,\Omega}^2 + \|u_1\|_{1,\Omega}^2 \le R^2. \quad (10.3.3)$$

Moreover, every generalized solution is also weak and for initial data $(u_0; u_1) \in H$ *such that* $u_0 \in H^3(\Omega)$ *and* $u_1 \in H^2(\Omega)$ *subject to natural compatibility conditions on the boundary:*

$$\Delta u_0 + g(\frac{\partial}{\partial n} u_1) = 0 \;\; \text{on} \; \Gamma_1; \quad \nabla u_1 = 0 \;\; \text{on} \; \Gamma_0,$$

generalized solutions are more regular, $(u; u_t) \in C_r(\mathbb{R}_+; H^3(\Omega) \times H^2(\Omega))$ *and also*

$$\frac{\partial}{\partial n} u_t, \, g\left(\frac{\partial}{\partial n} u_t\right) \in C_r(\mathbb{R}_+; H^{1/2}(\Gamma_1)) \cap L_\infty^{loc}(\mathbb{R}_+; H^{1/2}(\Gamma_1)). \quad (10.3.4)$$

Proof. The result stated above is a special case of Theorems 3.2.4 and 3.2.8; see also Remark 3.2.5 for the additional regularity of generalized solutions. □

The energy of the system is defined by

$$E(u,u_t) = \frac{1}{2} \int_\Omega \left[u_t^2 + \alpha |\nabla u_t|^2\right] dx + \frac{1}{2} \int_\Omega \left[|\Delta u|^2 + \frac{1}{2} |\Delta v(u)|^2\right] dx,$$

$$\mathscr{E}(u,u_t) = E(u,u_t) - \frac{1}{2} \int_\Omega ([F_0,u]u + 2pu) dx.$$

Standard energy equality (see Theorem 3.2.4) for $\mathscr{E}(t) \equiv \mathscr{E}(u,u_t)$ gives

$$\mathscr{E}(t) + \int_s^t \int_{\Gamma_1} g\left(\frac{\partial}{\partial n} u_t\right) \frac{\partial}{\partial n} u_t d\Gamma + \int_s^t \int_\Omega d_0 b(u_t) u_t dx d\tau = \mathscr{E}(s). \quad (10.3.5)$$

We have the following property proven in Lemma 1.5.4: for all $\varepsilon > 0$ and $0 \le s < 2$ there exists constant M_ε such that:

$$\|u\|_{s,\Omega} \le \varepsilon [\|\Delta u\|^2 + \|\Delta v(u)\|^2] + M_\varepsilon, \quad u \in (H^2 \cap H_0^1)(\Omega).$$

From this we infer that the energy $\mathscr{E}(u,u_t)$ is bounded from below. More precisely, there exist positive constants c, C, M such that

$$cE(u,u_t) - M \le \mathscr{E}(u,u_t) \le CE(u,u_t) + M. \quad (10.3.6)$$

We consider next the set $\mathscr{N}$ of equilibria of the semiflow S_t generated by (10.3.1). This set consists of vectors $V \in H$ of the form $V = (u; 0)$, where $u = u(x) \in H_{\Gamma_0}^2(\Omega)$ is a weak (variational) solution to the problem

$$\Delta^2 u = [v(u) + F_0, u] + p \;\; \text{in} \;\; \Omega, \quad (10.3.7)$$

$$u = \frac{\partial}{\partial n} u = 0 \text{ on } \Gamma_0, \quad u = \Delta u = 0 \text{ on } \Gamma_1.$$

As in the clamped–free case (see Theorem 10.2.7 and Proposition 10.2.8) using energy equality (10.3.5) one can easily prove the following assertion.

10.3.3. Theorem. *Let Assumption* 10.3.1 *be in force. Then*

- *There exists $R_* > 0$ such that the set*

$$\mathscr{W}_R = \{y = (u_0; u_1) \in H : \mathscr{E}(u_0, u_1) \le R\} \tag{10.3.8}$$

  is a nonempty invariant set with respect to semiflow S_t generated by equations (10.3.1) *for every $R \ge R_*$. Moreover, the set $\mathscr{W}_R$ is bounded for every $R \ge R_*$ and any bounded set is contained in $\mathscr{W}_R$ for some R.*
- *If in addition $d_0(x) > 0$ almost everywhere in Ω, the system (H, S_t) generated by* (10.3.1) *is gradient.*
- *The set $\mathscr{N}$ of all stationary points of the semiflow S_t is bounded in H. Thus there exists $R_{**} > 0$ such that $\mathscr{N} \subset \mathscr{W}_R$ for all $R \ge R_{**}$.*

In what follows we consider long-time behavior of the semiflow. To this end, we formulate the following "star-shaped" geometric assumption imposed on Γ_0. As in the case of free boundary conditions, these geometric condition allows us to control the estimates on the part of the boundary that is not dissipated.

10.3.4. Assumption. There exists $x_0 \in \mathbb{R}^2$ such that $(x - x_0)n \le 0$ on Γ_0.

10.3.5. Theorem (Compact attractors). *Let Assumptions* 10.3.1 *and* 10.3.4 *be in force. Then*

- *The restriction $(\mathscr{W}_R, S_t)$ of the dynamical system (H, S_t) on $\mathscr{W}_R$ given by* (10.3.8) *has a compact global attractor $A_R \subset \mathscr{W}_R$ for every $R \ge R_*$, where R_* is the same as in Theorem* 10.3.3.
- *If, in addition, $d_0(x) > 0$ almost everywhere in Ω, then there exists a compact global attractor A for the system (H, S_t). Moreover, we have $A = \mathscr{M}^u(\mathscr{N})$, where $\mathscr{M}^u(\mathscr{N})$ is the unstable manifold (see the definition in Section* 7.5*) emanating from the set $\mathscr{N}$ of equilibria for the semiflow S_t. In addition, the statements of Theorem* 7.5.10 *and Corollary* 7.5.11 *are in force.*
- *The two attractors A_R and A have finite fractal dimension provided relation* (10.3.2) *holds for all $s \in \mathbb{R}$. Moreover, in this latter case the attractors are bounded sets in the space $(H^3 \cap H^2_{\Gamma_0} \cap H^1_0)(\Omega) \times (H^2_{\Gamma_0} \cap H^1_0)(\Omega)$ and for any trajectory $(u(t); u_t(t))$ we have the relation*

$$\|u_{tt}(t)\|_{1,\Omega} + \|u_t(t)\|_{2,\Omega} + \|u(t)\|_{3,\Omega} \le C, \quad t \in \mathbb{R}.$$

The proof of this theorem parallels conceptually the treatment given in the free case. There are, however, few important technical differences that result from the fact that only one boundary condition controls the dissipation. The details are provided below.

10.3.2 Asymptotic smoothness

As before we appeal to Theorem 7.1.11 for the proof of the following theorem.

10.3.6. Theorem (Asymptotic smoothness). *Let Assumptions* 10.3.1 *and* 10.3.4 *hold. Then the system* $(\mathscr{W}_R, S_t)$ *is asymptotically smooth for every* $R > 0$.

In what follows we use notation (10.2.15) and prove the following assertion.

10.3.7. Proposition (Observability estimate). *Assume that Assumptions* 10.3.1 *and* 10.3.4 *are in force. Let* $U(t) = (u(t); u_t(t)) = S_t y_1$ *and* $W(t) = (w(t); w_t(t)) = S_t y_2$ *be two solutions corresponding to initial conditions* y_1 *and* y_2 *from the set* $\mathscr{W}_R$ *given by* (10.3.8). *Then there exist* $T_0 > 0$ *and constants* $C_1(T)$ *and* $C_2(R,T)$ *such that*

$$TE_z(T) + \int_0^T E_z(t)dt \le C_1(T) \int_{\Sigma_1} \left| \frac{\partial}{\partial n} z_t \right|^2 d\Sigma + C_2(R,T) \cdot lot(z) \qquad (10.3.9)$$

for any $T \ge T_0$, *where* $z \equiv u - w$, $E_z(t) = \frac{1}{2} \int_\Omega \left[|z_t|^2 + \alpha |\nabla z_t|^2 + |\Delta z|^2 \right] dx$, *and the lower-order terms have the form*

$$lot(z) = \sup_{0 \le \tau \le T} \|z(\tau)\|^2_{0,\Omega} + \int_0^T \|z_t(\tau)\|^2_{0,\Omega} d\tau. \qquad (10.3.10)$$

The proof of the relation follows the same idea as in the clamped–free case (see Proposition 10.2.14). However, calculations are different and these are provided below.

We can assume that u and w are strong solutions.

Step 1: Preliminaries. The difference $z \equiv u - w$ solves the following problem

$$\begin{aligned} &z_{tt} - \alpha \Delta z_{tt} + \Delta^2 z = f \ \text{ in } \Omega, \\ &z = \frac{\partial}{\partial n} z = 0 \ \text{ on } \Gamma_0, \quad z = 0,\ \Delta z = \psi \ \text{ on } \Gamma_1, \end{aligned} \qquad (10.3.11)$$

where $f = -d_0(x)(b(u_t) - b(w_t)) + [v(u) + F_0, u] - [v(w) + F_0, w]$ and

$$\psi = -\left[g\left(\frac{\partial}{\partial n} u_t \right) - g\left(\frac{\partial}{\partial n} w_t \right) \right].$$

We have the following energy equality

$$E_z(T) + D_t^T(z) = E_z(t) + \int_t^T (\mathscr{R}(z), z_t) dt, \qquad (10.3.12)$$

where $\mathscr{R}(z) = [v(u) + F_0, u] - [v(w) + F_0, w]$ and

$$D_t^T(z) = \int_t^T \int_\Omega d_0 [b(u_t) - b(w_t)] z_t dx d\tau + \widetilde{D}_t^T(z),$$

with

$$\widetilde{D}_t^T(z) = \int_t^T \int_{\Gamma_1} \left[g\left(\frac{\partial}{\partial n} u_t\right) - g\left(\frac{\partial}{\partial n} w_t\right)\right] \frac{\partial}{\partial n} z_t d\Gamma.$$

Step 2: First inequalities. Because $S_t y_1, S_t y_2 \in \mathscr{W}_R$, it follows

$$\|u_t(t)\|_{1,\Omega} + \|u(t)\|_{2,\Omega} + \|w_t(t)\|_{1,\Omega} + \|w(t)\|_{2,\Omega} \le C_R, \quad t \ge 0. \tag{10.3.13}$$

By energy relation (10.3.5) we also have that

$$\int_0^t \left[\int_{\Gamma_1} g\left(\frac{\partial}{\partial n} u_t\right) \frac{\partial}{\partial n} u_t d\Gamma + \int_\Omega d_0 b(u_t) u_t dx \right] d\tau \le C_R, \quad t \ge 0, \tag{10.3.14}$$

$$\int_s^t \left[\int_{\Gamma_1} g\left(\frac{\partial}{\partial n} w_t\right) \frac{\partial}{\partial n} w_t d\Gamma + \int_\Omega d_0 b(w_t) w_t dx \right] d\tau \le C_R, \quad t \ge 0. \tag{10.3.15}$$

Because

$$D_t^T(z) \ge \widetilde{D}_t^T(z) - \varepsilon \int_0^T \|z_t\|_{1,\Omega}^2 d\tau - C_\varepsilon(R) \int_t^T \|z_t(\tau)\|_{0,\Omega}^2 d\tau,$$

by the same argument as in Proposition 10.2.14 it follows from (10.3.12) that

$$TE_z(T) \le 2 \int_0^T E_z(t) dt + C_{R,T} lot(z), \tag{10.3.16}$$

and also

$$E_z(T) + \widetilde{D}_0^T(z) \le E_z(0) + \varepsilon \int_0^T E_z(t) dt + C_{\varepsilon,R,T} lot(z), \tag{10.3.17}$$

$$E_z(t) \le E_z(0) + \varepsilon \int_0^T E_z(t) dt + C_{\varepsilon,R,T} lot(z), \quad 0 \le t \le T, \tag{10.3.18}$$

$$E_z(0) \le E_z(T) + \varepsilon \int_0^T E_z(t) dt + \widetilde{D}_0^T(z) + C_{\varepsilon,R,T} lot(z), \tag{10.3.19}$$

where $lot(z)$ is given by (10.3.10) and $\varepsilon > 0$ is arbitrary.

Step 3: Multipliers. We start with the following preliminary estimate.

10.3.8. Lemma. *Let $T > 0$. Let $\phi \in C^2(\mathbb{R})$ be a given function with support in $[\delta, T-\delta]$, where $\delta \le T/4$, such that $0 \le \phi \le 1$ and $\phi \equiv 1$ on $[2\delta, T-2\delta]$. Then, any strong solution z to problem* (10.3.11) *satisfies the following inequality*

$$\int_0^T E_z(t)\phi(t) dt \le C_1 \int_0^T E_z(t)|\phi'(t)| dt + C_2 \int_{\Sigma_1} \left[|\psi|^2 + \left|\frac{\partial}{\partial n} z_t\right|^2 \right] d\Sigma$$
$$+ C_3 \int_{\Sigma_1} \left|\frac{\partial}{\partial n} z\right|^2 d\Sigma + C_4 \int_0^T \|f\|_{-\eta,\Omega}^2 dt + C_5 \cdot B_T(z), \tag{10.3.20}$$

where the constants C_i do not depend on T, $\eta \in [0, \frac{1}{2})$ is arbitrary, and

$$B_T(z) \equiv \int_{\Sigma_1} \left[\left| \left(\frac{\partial}{\partial n} \right)^2 z \right|^2 + \left| \frac{\partial}{\partial n} \frac{\partial}{\partial \tau} z \right|^2 \right] \phi d\Sigma + \int_0^T \left\| \frac{\partial}{\partial n} \Delta z \right\|^2_{-1,\Gamma_1} \phi dt. \tag{10.3.21}$$

Proof. We apply multiplier $h\nabla z\phi(t)$, where $h = x - x_0$ (by Assumption 10.3.4 we have that $h \cdot \nu \le 0$ on Γ_0). Calculations are performed first on smooth solutions, which is possible due to Theorem 10.3.2 above.

Kinetic terms: Integration by parts yields

$$\begin{aligned}
&\int_0^T \int_\Omega (z_{tt} - \alpha\Delta z_{tt}) h\nabla z\phi dxdt \\
&= -\int_0^T \int_\Omega [z_t h\nabla z_t + \alpha\nabla z_t \nabla(h\nabla z_t)]\phi dxdt - \int_0^T \int_\Omega [z_t h\nabla z + \alpha\nabla z_t \nabla(h\nabla z)]\phi' dxdt \\
&\quad + \alpha \int_{\Sigma_1} \frac{\partial}{\partial n} z_t h\nabla z\phi' d\Gamma dt + \alpha \int_{\Sigma_1} \frac{\partial}{\partial n} z_t h\nabla z_t \phi d\Gamma dt.
\end{aligned}$$

Because $\nabla z_t \nabla(h\nabla z_t) = \left(1 - \frac{1}{2}\mathrm{div}\, h\right) |\nabla z_t|^2 + \frac{1}{2}\mathrm{div}\left(h|\nabla z_t|^2\right) = \frac{1}{2}\mathrm{div}\left(h|\nabla z_t|^2\right)$ and also $z_t\big|_{\partial\Omega} = 0$ and $\nabla z_t\big|_{\Gamma_0} = 0$, we obtain that

$$\begin{aligned}
\int_0^T \int_\Omega (z_{tt} - \alpha\Delta z_{tt}) h\nabla z\phi dxdt &= \int_0^T \int_\Omega z_t^2 \phi dxdt - \alpha/2 \int_{\Sigma_1} |\nabla z_t|^2 (h,n)\phi d\Gamma dt \\
&\quad - \int_0^T \int_\Omega [z_t h\nabla z + \alpha\nabla z_t \nabla(h\nabla z)]\phi'(t) dxdt \\
&\quad + \alpha \int_{\Sigma_1} \frac{\partial}{\partial n} z_t h\nabla z_t \phi d\Sigma + \alpha \int_{\Sigma_1} \frac{\partial}{\partial n} z_t h\nabla z\phi' d\Sigma.
\end{aligned}$$

Because $(\partial/\partial\tau) z_t = 0$ on Γ, we have that $h\nabla z_t = (h,n)(\partial/\partial n) z_t$ and thus

$$\begin{aligned}
&\int_0^T \int_\Omega (z_{tt} - \alpha\Delta z_{tt}) h\nabla z\phi dxdt \qquad (10.3.22)\\
&\ge \int_0^T \int_\Omega z_t^2 \phi dxdt - C_1(\alpha) \int_0^T E_z(t)|\phi'(t)| dt - C_2 \cdot \alpha \cdot \int_{\Sigma_1} \left| \frac{\partial}{\partial n} z_t \right|^2 d\Sigma.
\end{aligned}$$

Potential term: Using (1.3.5) with $\mu = 1$ we obtain

$$\begin{aligned}
\int_\Omega \Delta^2 z h\nabla z dx &= \int_\Omega |\Delta z|^2 dx + 1/2 \int_\Gamma |\Delta z|^2 (h,n) d\Gamma \\
&\quad - \int_\Gamma \Delta z \frac{\partial}{\partial n}(h\,\nabla z) d\Gamma + \int_\Gamma \frac{\partial}{\partial n} \Delta z \cdot h\nabla z d\Gamma.
\end{aligned}$$

One can show that

$$\frac{\partial}{\partial n}h\nabla z = \frac{\partial}{\partial n}z + (h,n)\left(\frac{\partial}{\partial n}\right)^2 z + (h,\tau)\frac{\partial}{\partial n}\frac{\partial}{\partial \tau}z \text{ on } \Gamma \tag{10.3.23}$$

which implies $\Delta z(\partial/\partial n)h\nabla z = |\Delta z|^2(h,n)$ on Γ_0 (see, e.g., [173, p. 82]). Therefore, using the geometric condition $(h,n) \leq 0$ on Γ_0, we obtain

$$\begin{aligned}\int_Q \Delta^2 z h\nabla z\phi dQ \geq \int_Q |\Delta z|^2\phi dQ - C_1\int_{\Sigma_1}\left[|\psi|^2 + \left|\frac{\partial}{\partial n}z\right|^2\right]\phi d\Sigma \\ - C_2\int_{\Sigma_1}\left[\left|\left(\frac{\partial}{\partial n}\right)^2 z\right|^2 + \left|\frac{\partial}{\partial n}\frac{\partial}{\partial \tau}z\right|^2\right]\phi d\Sigma + \int_{\Sigma_1}\frac{\partial}{\partial n}\Delta z\cdot h\nabla z\phi d\Sigma.\end{aligned}$$

Because $h\nabla z = (h,n)(\partial/\partial n)z$ on Γ, we obtain that

$$\begin{aligned}\left|\int_{\Sigma_1}\frac{\partial}{\partial n}\Delta z h\nabla z\phi d\Sigma\right| &= \left|\int_0^T\int_{\Gamma_1}\frac{\partial}{\partial n}\Delta z\cdot(h,n)\frac{\partial}{\partial n}z\cdot\phi d\Gamma dt\right| \\ &\leq C\int_0^T\left[\left\|\frac{\partial}{\partial n}\Delta z\right\|^2_{-1,\Gamma_1} + \left\|\frac{\partial}{\partial n}z\right\|^2_{0,\Gamma_1} + \left\|\frac{\partial}{\partial n}\frac{\partial}{\partial \tau}z\right\|^2_{0,\Gamma_1}\right]\phi dt.\end{aligned}$$

Consequently

$$\int_Q \Delta^2 z h\nabla z\phi dQ \geq \int_Q |\Delta z|^2\phi dQ - C_1\int_{\Sigma_1}\left[|\psi|^2 + \left|\frac{\partial}{\partial n}z\right|^2\right]\phi d\Sigma - C_2 B_T(z), \tag{10.3.24}$$

where $B_T(z)$ is given by (10.3.21). Combining (10.3.22) and (10.3.24) we obtain with $0 < \eta < 1/2$:

$$\begin{aligned}\int_0^T \|\Delta z\|^2_\Omega\phi(t)dt \leq C_1\int_0^T E_z(t)|\phi'(t)|dt + C_2\int_{\Sigma_1}\left[|\psi|^2 + \left|\frac{\partial}{\partial n}z_t\right|^2\right]d\Sigma \\ + C_3\int_{\Sigma_1}\left|\frac{\partial}{\partial n}z\right|^2 d\Sigma + C_4\int_0^T \|f\|^2_{-\eta,\Omega}dt + C_5\cdot B_T(z).\end{aligned} \tag{10.3.25}$$

Difference between the kinetic and potential energy: Using the multiplier $z\phi$ we find that

$$\begin{aligned}\int_Q fz\phi dQ = \int_Q (z_{tt} - \alpha\Delta z_{tt} + \Delta^2 z)z\phi dQ = \int_Q \left[|\Delta z|^2 - (|z_t|^2 + \alpha|\nabla z_t|^2)\right]\phi dQ \\ - \int_{\Sigma_1}\Delta z\frac{\partial}{\partial n}z\phi d\Sigma - \int_Q (z_t z + \alpha\nabla z_t\nabla z)\phi' dQ.\end{aligned}$$

Therefore

$$\int_0^T \left(\|z_t\|^2 + \alpha\|\nabla z_t\|^2\right)\phi dt \le 2\int_0^T \|\Delta z\|_\Omega^2 \phi dt + C_1\int_0^T E_z(t)|\phi'(t)|dt \quad (10.3.26)$$
$$+ C_2\int_{\Sigma_1}\left[|\psi|^2 + \left|\frac{\partial}{\partial n}z\right|^2\right]d\Sigma + C_3\int_0^T \|f\|^2_{-\eta,\Omega}dt.$$

Now relation (10.3.20) follows from (10.3.25) and (10.3.26). This completes the proof of Lemma 10.3.8. □

Step 3. Trace estimates. Our next step is to eliminate in (10.3.20) the higher order traces on the boundary Γ_1 presented in the expression for $B_T(z)$. Here the idea is to estimate second spatial derivatives on the boundary Σ in terms of velocity traces on the boundary and of lower-order terms. This type of estimate was used extensively in the past in the context of controllability and stabilization of Kirchoff plates by using moments on the boundary. The key technical ingredient for this is the microlocal estimate that provides the appropriate control of traces on the boundary. This estimate is recalled below. In fact, the result follows from the more general trace estimate proved in [151] and valid for the linear Kirchoff problem (10.3.11).

10.3.9. Lemma. *Let z be a solution to linear problem* (10.3.11) *with given f and ψ and $B_T(z)$ as in* (10.3.21). *Then, there exist constants $C_{i,T} > 0$ such that for any $0 \le \eta < \frac{1}{2}$ the following estimate holds:*

$$B_T(z) \le C_{1,T}\int_{\Sigma_1}\left(|\psi|^2 + \left|\frac{\partial}{\partial n}z_t\right|^2\right)d\Sigma_1$$
$$+ C_{2,T}\left[\|z\|^2_{C([0,T],H^{2-\eta}(\Omega))} + \|z_t\|^2_{L_2([0,T],H^{-\eta}(\Omega))} + \int_0^T \|f\|^2_{-\eta,\Omega}dt\right].$$

Proof. We refer to Proposition 1 and Lemma 4 in [151]. Note that the estimate $\int_0^T \|f\|^2_{-\eta,\Omega}dt$ is overly conservative and can be replaced (as seen from (21) in [151]) by the norm of f in $H^{-1}(0,T;H^{-1/2}(\Omega))$. However, this refinement is not essential for our arguments. □

Applying Lemma 10.3.9 and using estimates (10.3.17)–(10.3.19) and estimating the integrals with f and ψ one obtains

$$B_T(z) \le \varepsilon\left(E_z(T) + \int_0^T E_z(t)dt\right) + C_{1,T}\int_{\Sigma_1}\left|\frac{\partial}{\partial n}z_t\right|^2 d\Sigma_1$$
$$+ C_{2,\varepsilon,R,T}\left[\sup_{[0,T]}\|z\|^2_{0,\Omega} + \int_0^T \|z_t\|^2_{0,\Omega}dt\right].$$

Thus Lemma 10.3.8 yields

$$\int_0^T E_z(t)\phi(t)dt \le C_1\int_0^T E_z(t)|\phi'(t)|dt + \varepsilon\left(E_z(T) + \int_0^T E_z(t)dt\right)$$

$$+C_2(T)\int_{\Sigma_1}\left|\frac{\partial}{\partial n}z_t\right|^2 d\Sigma + C_3^{\varepsilon}(T,R)lot(z), \qquad (10.3.27)$$

where $lot(z)$ is given by (10.3.10). We emphasize that the constants $C_2(T)$ and $C_3^{\varepsilon}(T,R)$ above depend on the choice of the function ϕ; the constant C_1 is independent of T and ϕ.

Step 4. Final argument. By (10.3.16)

$$TE_z(T)+\int_0^T E_z(t)dt \le 3\int_0^T E_z(t)\phi(t)dt$$
$$+3\int_0^T E_z(t)(1-\phi(t))dt + C(T,R)lot(z).$$

Therefore from (10.3.27) we have that

$$TE_z(T)+\int_0^T E_z(t)dt \le C_1\int_0^T E_z(t)(1-\phi(t)+|\phi'(t)|)dt$$
$$+C_2(T)\int_{\Sigma_1}\left|\frac{\partial}{\partial n}z_t\right|^2 d\Sigma + C_3(T,R)lot(z)$$

for $T \ge 1$. Thus by (10.3.18) and (10.3.19) we obtain that

$$TE_z(T)+\int_0^T E_z(t)dt \le C_1 r_\phi E_z(T) + C_2(T)\int_{\Sigma_1}\left|\frac{\partial}{\partial n}z_t\right|^2 d\Sigma + C_3(T,R)lot(z),$$

where $r_\phi = \int_0^T(1-\phi(t)+|\phi'(t)|)dt$. We can choose ϕ such that the parameter r_ϕ does not depend on T, thus we easily obtain (10.3.9) and complete the proof of Proposition 10.3.7.

Proof of Theorem 10.3.6

The argument is the same as in the proof of Theorem 10.2.13. From relation (10.3.2) we have (see Proposition B.1.2 in Appendix B) that

$$\int_{\Sigma_1}\left|\frac{\partial}{\partial n}z_t\right|^2 d\Sigma_1 \le \varepsilon + C_\varepsilon\int_{\Sigma_1}\left[g\left(\frac{\partial}{\partial n}u_t\right)-g\left(\frac{\partial}{\partial n}w_t\right)\right]\frac{\partial}{\partial n}z_t d\Sigma \equiv \varepsilon + C_\varepsilon\widetilde{D}_0^T(z) \qquad (10.3.28)$$

for any $\varepsilon > 0$. Therefore using (10.3.17) by the same argument as in the case of Theorem 10.2.13 we can conclude that for any $\varepsilon > 0$ there exist positive constants $\omega(\varepsilon,R)$ and $C_i(\varepsilon,R)$, $i=1,2,3$, such that

$$E_z(t) \le \varepsilon + C_1(\varepsilon,R)e^{-\omega(\varepsilon,R)t}E_z(0) \qquad (10.3.29)$$
$$+C_2(\varepsilon,R)\max_{[0,t]}\|z(\tau)\|_{0,\Omega}^2 + C_3(\varepsilon,R)\int_0^t e^{-\omega(\varepsilon,R)(t-\tau)}\|z_t(\tau)\|_{0,\Omega}^2 d\tau.$$

Therefore we can use Theorem 7.1.11 to prove asymptotic smoothness of semiflow S_t which is stated in Theorem 10.3.6.

10.3.3 Proof of the main result on attractors (Theorem 10.3.5)

1. The existence of compact attractors A_R and A follows from Theorem 10.3.6 and also from Theorem 10.3.3 in the same way as for the previous (clamped–free) model.

2. To prove the finiteness of fractal dimensions and smoothness we use the stabilizability estimate (closely related to the quasi-stability estimate) which is stated below.

10.3.10. Proposition (Stabilizability estimate). *Assume that Assumptions* 10.3.1 *and* 10.3.4 *are in force. In addition, assume that relation* (10.3.2) *holds for all* $s \in \mathbb{R}$. *Let* $S_t y_1 = (u(t); u_t(t))$ *and* $S_t y_2 = (w(t); w_t(t))$ *be two solutions corresponding to initial conditions* y_1 *and* y_2 *from the set* $\mathscr{W}_R$ *given by* (10.3.8). *Then there exist positive constants* $\omega(R)$ *and* $C_i(R)$ *such that*

$$\|S_t y_1 - S_t y_2\|_H^2 \le C_1(R) e^{-\omega(R)t} \|y_1 - y_2\|_H^2 \qquad (10.3.30)$$
$$+ C_2(R) \max_{[0,t]} \|z(\tau)\|_{0,\Omega}^2 + C_3(R) \int_0^t e^{-\omega(R)(t-\tau)} \|z_t(\tau)\|_{0,\Omega}^2 d\tau,$$

where $z(t) = u(t) - w(t)$.

Proof. Because (10.3.2) holds for all $s \in \mathbb{R}$, instead of (10.3.28) we have that

$$\int_{\Sigma_1} \left| \frac{\partial}{\partial n} z_t \right|^2 d\Sigma_1 \le c_0 \int_{\Sigma_1} \left[g\left(\frac{\partial}{\partial n} u_t\right) - g\left(\frac{\partial}{\partial n} w_t\right) \right] \frac{\partial}{\partial n} z_t d\Sigma \equiv c_0 \widetilde{D}_0^T(z). \qquad (10.3.31)$$

Thus from (10.3.9) in the same way as in Section 10.2 we can prove (10.3.30). □

Now we use the estimate (10.3.30) and the argument that parallels the clamped–free case. This argument involves Theorem 7.3.3 and relies on the same idea as in the proof of Theorem 7.9.6.

We introduce extended space $X = \left[H_{\Gamma_0}^2(\Omega) \cap H_0^1(\Omega)\right] \times H_0^1(\Omega) \times W_2(0,T)$, where

$$W_2(0,T) = \left\{ z \in L_2(0,T; (H_{\Gamma_0}^2 \cap H_0^1)(\Omega)) : (z_t; z_{tt}) \in L_2(0,T; H_0^1(\Omega) \times L_2(\Omega)) \right\}.$$

endowed with the norm

$$|z|_{W_2(0,T)}^2 = \int_0^T \left(\|z\|_{2,\Omega}^2 + \|z_t\|_{1,\Omega}^2 + \|z_{tt}\|_{0,\Omega}^2 \right) dt.$$

The norm in X is given by $\|(u_0; u_1; z)\|_X^2 = \|u_0\|_{2,\Omega}^2 + \|u_1\|_{1,\Omega}^2 + |z|_{W_2(0,T)}^2$.

As in the free case, $z(t) = u(t) - w(t)$ satisfies the following estimate.

10.3.11. Proposition. *With* z *generalized solution to the system in* (10.3.11) *the following estimate holds with an arbitrary* $T < \infty$,

$$\int_0^T \|z_{tt}(t)\|_{0,\Omega}^2 dt \le C_R \cdot (1+T) \cdot \left[E_z(0) + \int_0^T E_z(t) dt \right], \quad T > 0. \tag{10.3.32}$$

Proof. By using abstract representation of the model (see (2.5.47) in Section 2.5.2), we can rewrite the equation for z in the form

$$z_{tt} = -M^{-1}\mathscr{A}(z - G\psi) + M^{-1}f \tag{10.3.33}$$

with the same f and ψ as in (10.3.11). From Green's formula we have

$$(M^{-1}\mathscr{A}(z-G\psi),\phi)_\Omega = (\Delta z, \Delta M^{-1}\phi)_\Omega - (\Delta z, \frac{\partial}{\partial n} M^{-1}\phi)_\Gamma$$

for any $\phi \in L^2(\Omega)$. Therefore using Theorem 2.5.6 we obtain

$$\int_0^T \|M^{-1}\mathscr{A}(z-G\psi)\|_{0,\Omega}^2 dt \le C\left[\int_Q |\Delta z|^2 dQ + \int_{\Sigma_1} |\psi|^2 d\Sigma + \int_{\Sigma_0} |\Delta z|^2 d\Sigma \right]$$

$$\le C\left[E_z(0) + \int_0^T E_z(t)dt + \int_{\Sigma_1} |\psi|^2 d\Sigma + \int_0^T \|f\|_{-1,\Omega}^2 dt \right].$$

Consequently the structure of the forcing terms ψ and f and properties of the damping functions b and g yield

$$\int_0^T \|z_{tt}(t)\|_{0,\Omega}^2 dt \le C_1 \left[E_z(0) + \int_0^T E_z(t)dt \right] + C_2 \int_{\Sigma_1} \left| \frac{\partial}{\partial n} z_t \right|^2 d\Sigma,$$

where the constants C_1 and C_2 may depend on T. Using (10.3.31) and the energy relation in (10.3.12) and applying the same idea as in the final step of the proof of Proposition 10.2.19 we obtain the desired conclusion. □

In order to complete the proof of finite-dimensionality of the attractor A we proceed exactly as in the clamped–free case by introducing the set

$$A_T \equiv \{U \equiv (u(0); u_t(0); u(t), t \in [0,T]) : (u(0); u_t(0)) \in A\}$$

and the translation mapping $V : (u(0); u_t(0); u(t)) \to (u(T); u_t(T); u(T+t))$. One can show that this map V satisfies the conditions of Theorem 7.3.3. Verification of these properties is identical as in the case of clamped–free boundary conditions (see Section 10.2). As easily observed, the arguments for (10.2.52) do not depend on the boundary conditions.

3. Smoothness of the attractor. Here, again, we follow the same program as in the treatment of the clamped–free case. Let $\gamma = \{y(t) \equiv (u(t); u_t(t)) : t \in \mathbb{R}\} \subset H$ be a full trajectory from the attractor. Let $|\sigma| < 1$. Applying Proposition 10.3.10 with $y_1 = y(s+\sigma)$, $y_2 = y(s)$ (and the interval $[s,t]$ replaces $[0,t]$), we obtain

$$\|y(t+\sigma) - y(t)\|_H^2 \le C_1 e^{-\omega(t-s)} \|y(s+\sigma) - y(s)\|_H^2 \tag{10.3.34}$$

$$+ C_2 \max_{\tau\in[s,t]} \|u(\tau+\sigma)-u(\tau)\|^2_{0,\Omega} + C_3 \int_s^t e^{-\omega(t-\tau)} \|u_t(\tau+\sigma)-u_t(\tau)\|^2_{0,\Omega} d\tau$$

for any $t,s \in \mathbb{R}$ such that $s \le t$ and for any σ with $|\sigma|<1$. The main difference with (10.2.53) is that the term $\|u(\tau+\sigma)-u(\tau)\|^2_{0,\Omega}$ is outside the integral in (10.3.34). This latter feature is due to a slightly different structure of lower-order terms in the observability estimate. However, because these are lower-order terms, combination of the argument given in the proof of Theorem 7.9.8 with the argument establishing (10.2.55) in the proof of Theorem 10.2.11 yields that

$$\|u_{tt}(t)\|^2_{1,\Omega} + \|u_t(t)\|^2_{2,\Omega} \le C, \quad t \in \mathbb{R}, \tag{10.3.35}$$

for any trajectory $\gamma = \{y(t) \equiv (u(t);u_t(t)) \, : \, t \in \mathbb{R}\}$ from the attractor. Therefore elliptic regularity implies the desired smoothness of the attractor. To see the latter, we note that $u(t)$ satisfies

$$Mu_{tt} + \mathscr{A}u = f + \mathscr{A}G\psi$$

where $f \equiv [v(u)+F_0,u]+p-d_0b(u_t)$ and $\psi = -g((\partial/\partial n)u_t)$. For all t we have that $\|\Delta u(t)\|_\Omega \le C$, therefore it follows from (10.3.35) that for all $t\in\mathbb{R}$:

$$\|f(t)\|_{0,\Omega} \le C \;\text{ and }\; \|\mathscr{A}^{-1/4}Mu_{tt}(t)\|_\Omega \le C\|M^{1/2}u_{tt}(t)\|_{0,\Omega} \le C.$$

It follows from the linear bound at infinity imposed on g that Nemytskij operator generated by g takes $H^{1/2}(\Gamma_1)$ into itself. Therefore, because

$$(\partial/\partial n)u_t(t)|_{\Gamma_1} \in H^{1/2}(\Gamma_1)$$

we obtain that $|\psi(t)|_{1/2,\Gamma_1} \le C$. By using the superposition principle we can write $u(t) = u^1(t)+u^2(t)$, where $\mathscr{A}u^1(t) = f(t) - Mu_{tt}(t) \in [\mathscr{D}(\mathscr{A}^{1/4})]'$ and

$$\begin{aligned} \Delta^2 u^2(t) &= 0 \;\text{ in }\; \Omega, \\ u^2 &= 0 \;\text{ on }\; \Gamma, \quad \nabla u^2 = 0 \;\text{ on }\; \Gamma_0, \quad \Delta u^2 \in H^{1/2}(\Gamma_1). \end{aligned}$$

Thus $u^1(t) \in \mathscr{D}(\mathscr{A}^{3/4}) \subset H^3(\Omega)$, $t\in\mathbb{R}$, and by standard regularity in elliptic theory $u^2(t)\in H^3(\Omega)$, $t\in\mathbb{R}$. This proves that $u(t)\in H^3(\Omega)$ on the attractor, which along with (10.3.35) proves the final conclusion of Theorem 10.3.5.

10.3.12. Remark. As in the case of free boundary conditions, higher order smoothness of the attractor can be established in the case when the boundary damping g is linear and the data F_0, p are sufficiently smooth. Reiteration of the argument leading to (10.3.34) and applied to the equation satisfied by successive time derivatives of the solution, allows us to prove higher smoothness of A as long as the data F_0 and p are sufficiently smooth. The argument is the same as in the case of internal damping, but supported by boundary recovery estimates presented in this section. ■

10.3.4 Rate of convergence to equilibria

As in case of clamped–free boundary conditions we can obtain results on convergence rates of individual trajectories to equilibrium under the condition that the set $\mathcal{N}$ of stationary points is discrete.

In this case every stationary point V has the form $V = (u;0)$, where $u = u(x) \in H^2_{\Gamma_0}(\Omega)$ is a weak (variational) solution to the problem (10.3.7). As in the previous section we introduce a concave, strictly increasing, continuous function $h : \mathbb{R}_+ \mapsto \mathbb{R}_+$ which captures the behavior of $g(s)$ at the origin possessing the properties

$$h(0) = 0 \text{ and } s^2 \le h(sg(s)) \text{ for } |s| \le 1. \tag{10.3.36}$$

and define the corresponding functions $H_0(s)$ and $Q(s)$ by relations (10.2.57) and consider the differential equation

$$\frac{d\sigma}{dt} + Q(\sigma) = 0, \quad t > 0, \quad \sigma(0) = \sigma_0 \in \mathbb{R}_+, \tag{10.3.37}$$

which admits the global unique solution $\sigma(t)$ decaying asymptotically to zero as $t \to \infty$.

With these preparations we are ready to state our result.

10.3.13. Theorem (Rate of stabilization). *Let the hypotheses of of Theorem* 10.3.5 *be valid with $d_0(x) > 0$ a.e. Assume that there exist $\gamma > 0$ and $s_0 > 0$ such that $sb(s) \ge \gamma s^2$ for $|s| \le s_0$. In addition assume that problem* (10.3.7) *has a finite number of solutions. Then for any $V \in H$ there exists a stationary point $E = (e;0)$ such that $S_t V \to E$ as $t \to +\infty$. Moreover, if the equilibrium E is hyperbolic in the sense that the linearization of* (10.3.7) *around each of its solutions has the trivial solution only, then there exist $C, T > 0$ depending on V, E such that the following rates of stabilization*

$$\|S_t V - E\|_H \le C\sigma([tT^{-1}]), \quad t > 0,$$

hold, where $[a]$ denotes the integer part of a and $\sigma(t)$ satisfies (10.3.37) *with σ_0 depending on $V, E \in H$ (the constants c_i in the definition of Q also depend on V and E). In particular, if $g'(0) > 0$, then*

$$\|S_t V - E\|_H \le Ce^{-\omega t}$$

for some positive constants C and ω depending on $V, E \subset H$.

Proof. The arguments are the same as in the previous clamped–free case. □

10.3.5 Determining functionals

As in the clamped–free case (see Section 10.2.5) we can obtain a result on determining functionals for the system generated by (10.3.1). In our case the stabilizability estimate (10.3.30) in Proposition 10.3.10 has a different structure thus it is more convenient to rely on the idea presented in the proof of Theorem 7.9.11.

10.3.14. Theorem. *Let Assumptions* 10.3.1 *and* 10.3.4 *be in force. In addition assume that relation* (10.3.2) *holds for all* $s \in \mathbb{R}$ *and* $d_0(x) > 0$ *a.e. Let* $\mathscr{L} = \{l_j : j = 1, \ldots, N\}$ *be a set of linear continuous functionals on the space* $H = (H^2_{\Gamma_0} \cap H^1_0(\Omega)) \times H^1_0(\Omega)$ *and let* $\varepsilon_{\mathscr{L}} \equiv \varepsilon_{\mathscr{L}}(H, L_2(\Omega) \times L_2(\Omega))$ *be the completeness defect of* $\mathscr{L}$ *with respect to the pair* $H \subset L_2(\Omega) \times L_2(\Omega)$. *Then there exists* $\varepsilon_0 > 0$ *such that under the condition* $\varepsilon_{\mathscr{L}} \leq \varepsilon_0$ *the set* $\mathscr{L}$ *is a set of asymptotically determining functionals for problem* (10.3.1) *in the following sense. For any two of its trajectories* $S_t y_1$ *and* $S_t y_2$ *the condition*

$$\lim_{t \to \infty} |l_j(S_t y_1) - l_j(S_t y_2)| = 0 \ \text{for} \ j = 1, \ldots, N,$$

implies that $\lim_{t \to \infty} \|S_t y_1 - S_t y_2\|_H = 0$.

Proof. Let $y_1 = (u_{01}; u_{11})$ and $y_2 = (u_{02}; u_{12})$ and $t_0 > 0$ be such that $S_t y_1$ and $S_t y_2$ belong to the absorbing set for all $t \geq t_0$ (this set exists because the system has a global attractor). Then by Proposition 10.3.10 we have that

$$\begin{aligned} \|S_t y_1 - S_t y_2\|_H &\leq c_1 e^{-\omega(t-t_0)} \|S_{t_0} y_1 - S_{t_0} y_2\|_H \\ &\quad + c_2 \sup_{\tau \in [t_0, t]} \left(\|u^1(\tau) - u^2(\tau)\|_{0,\Omega} + \|u^1_t(\tau) - u^2_t(\tau)\|_{0,\Omega} \right), \ t \geq t_0, \end{aligned}$$

where $S_t y_1 = (u^1(t); u^1_t(t))$, $S_t y_2 = (u^2(t); u^2_t(t))$, and the positive constants c_1, c_2, and ω depend on the size of an absorbing set. Using the latter inequality and the property of the completeness defect given in relation (10.2.83) we can conclude the proof in the same way as in Theorem 7.9.11. □

10.4 Global attractors for von Karman plates without rotational inertia and with dissipation in free boundary conditions

We consider a nonlinear system of dynamic elasticity described by von Karman evolution with a nonlinear boundary dissipation and without the effects of rotational inertia. Our goal is to establish existence of a global attractor and to determine its structure. Because the rotational inertia are not included in the model, the analysis is more delicate. This is due to the fact that the nonlinear von Karman bracket is no longer compact on the energy space. The proof of asymptotic smoothness requires certain "compensated compactness" arguments which heavily rely not only on the structure of von Karman nonlinearity—as in the case of internal dissipation—but additionally new arguments related to propagation of smoothness through the compact region are required. Indeed, the latter are needed in order to compensate for the lack of information on internal dissipation that is no longer secured by dissipativity integrals (unlike problems with interior dissipation).

In the case when rotational forces are not accounted for we are able to establish the results with dissipation affecting only one boundary condition: shear forces. The price to pay for this is suitable geometric conditions imposed on the entire boundary.

Two physically relevant dissipation mechanisms acting on the boundary are considered: shears and torques imposed on free boundary conditions and moments imposed on hinged boundary conditions. We start with the free boundary conditions case. The hinged boundary conditions are considered in the next section.

In this section we mainly follow the line of the argument given in [68, 73].

10.4.1 The model and the main results

Let $\Omega \subset \mathbb{R}^2$ be a bounded domain with a sufficiently smooth boundary Γ. We assume that Γ consists of two disjoint parts Γ_0 and Γ_1. Consider the following von Karman model with boundary dissipation active on Γ_1 via the free boundary conditions [173]

$$u_{tt} + d_0(x)b(u_t) + \Delta^2 u = [v(u) + F_0, u] + p \ \text{ in } \ \Omega \times (0, \infty). \tag{10.4.1}$$

The Airy stress function $v(u)$ satisfies the following elliptic problem

$$\Delta^2 v(u) = -[u,u], \ \text{ in } \Omega, \quad \frac{\partial}{\partial n} v(u) = v(u) = 0 \ \text{ on } \Gamma. \tag{10.4.2}$$

The boundary conditions associated with (10.4.1) are of free type on Γ_1 and clamped on Γ_0:

$$\begin{aligned} & u = \frac{\partial}{\partial n} u = 0 \ \text{ on } \Gamma_0, \quad \Delta u + (1-\mu) B_1 u = 0 \ \text{ on } \Gamma_1, \\ & \frac{\partial}{\partial n} \Delta u + (1-\mu) B_2 u - \nu_1 u - \beta u^3 = d(x) g(u_t) \ \text{ on } \Gamma_1, \end{aligned} \tag{10.4.3}$$

where ν_1 and β are nonnegative parameters and $0 < \mu < 1$. As above the boundary operators B_1 and B_2 are given by (1.3.20); that is,

$$\begin{aligned} B_1 u &= 2 n_1 n_2 u_{x_1 x_2} - n_1^2 u_{x_2 x_2} - n_2^2 u_{x_1 x_1} , \\ B_2 u &= \tfrac{\partial}{\partial \tau} \left[\left(n_1^2 - n_2^2 \right) u_{x_1 x_2} + n_1 n_2 \left(u_{x_2 x_2} - u_{x_1 x_1} \right) \right] , \end{aligned}$$

where $n = (n_1; n_2)$ is the outer normal to Γ, $\tau = (-n_2; n_1)$ is the unit tangent vector along $\partial\Omega$. The operators $B(u_t) \equiv d_0(x)b(u_t)$ in (10.4.1) (resp., $d(x)g(u_t)$ in (10.4.3)) represent interior (resp., boundary) dissipation, where the functions $b, g \in C^1(\mathbb{R})$ are zero at the origin and g is monotone. The coefficients $d_0(x)$ and $d(x)$ are assumed nonnegative. The function $p(x)$ represents external transverse forces applied to the plate. The function F_0 describes in-plane forces acting on the plate. The boundary values F_0 and $(\partial/\partial n)F_0$ on Γ are determined from the in-plane components of the edge forces (see, e.g., [85]). Below we assume that $F_0 = (\partial/\partial n)F_0 = 0$ on Γ_1. This assumption means that the in-plane components of the edge forces vanish on the free part Γ_1 of the boundary Γ. Thus, in-plane forces are active on the clamped (Γ_0) part of the boundary only. We refer to [85] for the detailed description of the relation between boundary values of F_0 and in-plane forces on the edge of the plate.

Our interest is in studying asymptotic behaviour of generalized solutions; that is, defined on the phase space $H \equiv H^2(\Omega) \times L_2(\Omega)$ and driven, in addition to the initial conditions, by the nondissipative forces F_0 and p. The main dissipative mechanism considered is shear nonlinear feedback force $g(u_t)$ located on the boundary Γ_1. It is known that this type of geometrically weak dissipation is rather subtle in mathematical treatments and also geometry-dependent.

Our aim is to discuss issues such as (i) existence and properties of a global attractor, and (ii) the rate of stabilization of solutions to equilibria points.

To this end we introduce the space of weak (generalized) solutions $H \equiv H^2_{\Gamma_0}(\Omega) \times L_2(\Omega)$ where, as above (see (10.2.3)), $H^2_{\Gamma_0}(\Omega)$ denotes space of $H^2(\Omega)$ functions subject to clamped boundary conditions on Γ_0:

$$H^2_{\Gamma_0}(\Omega) \equiv \left\{ u \in H^2(\Omega) \,:\, u = \frac{\partial}{\partial n} u = 0 \text{ on } \Gamma_0 \right\}. \tag{10.4.4}$$

Our basic hypothesis is the following.

10.4.1. Assumption.

- The function $g \in C(\mathbb{R})$ is nondecreasing, $g(0) = 0$, and $d(x) \in L_\infty(\Gamma_1)$ is nonnegative almost everywhere. If $\beta > 0$ we assume that $g \in C^1(\mathbb{R})$, $g'(s) \geq m > 0$, and $d(x) \geq d > 0$ for all $s \in \mathbb{R}$ and $x \in \Gamma_1$.
- $b \in C(\mathbb{R})$ possesses the properties $b(0) = 0$, $sb(s) \geq 0$ and there exists $b_0 \geq 0$ such that $b(s) + b_0 s$ is increasing; $d_0(x) \in L_\infty(\Omega)$ is nonnegative almost everywhere.
- $p \in L_2(\Omega)$ and $F_0 \in H^{3+\delta}(\Omega) \cap H^2_{\Gamma_1}(\Omega)$ for some $\delta > 0$ when $\beta > 0$, where $H^2_{\Gamma_1}(\Omega)$ is given by (10.4.4) with Γ_1 instead of Γ_0. If $\beta = 0$ we assume in addition that $F_0\big|_{\Gamma} = 0$.
- $\nu_1 + \beta > 0$ or Γ_0 is nonempty.

10.4.2. Remark. For the sake of some simplification and in contrast to the case with rotational inertia (see Assumption 10.2.1 in Section 10.2) we assume that β in (10.4.3) is a nonnegative constant. The case when β is a nonnegative function from $L_\infty(\Gamma_1)$ can be treated in a similar way. This requires some obvious changes in the standing hypotheses (cf., the hypotheses imposed on the model (4.2.1) and (4.2.23) in Chapter 4). One can weaken the condition $g'(s) \geq m > 0$, $s \in \mathbb{R}$ by replacing this condition with $g'(s) \geq m > 0$ for $|s| \geq 1$ and changing the term βu^3 by $\beta(u^3 + cu_t)$, where $c > 0$ is any constant. In other words, control via linear damping is needed only on the support of β. See also Remark 4.2.10. ∎

10.4.3. Theorem (Well-posedness). *Under the standing Assumption* 10.4.1 *the equations* (10.4.1) *and* (10.4.2) *with boundary conditions* (10.4.3) *generate dynamical system* (H, S_t) *on* $H \equiv H^2_{\Gamma_0}(\Omega) \times L_2(\Omega)$*; that is, for any* $(u_0; u_1) \in H$ *there* *exists a unique generalized solution $u(t)$ that depends continuously on the initial data. This, in particular, implies that the map $y_0 \equiv (u(0) = u_0; u_t(0) = u_1) \mapsto y(t) \equiv (u(t); u_t(t))$ defines a continuous semiflow S_t on H.*

Moreover,

- *The function $y(t) = S_t y_0 \equiv (u(t); u_t(t))$ satisfies the energy inequality*

$$\mathscr{E}(t) + \int_s^t \int_{\Gamma_1} d(x) j_g(u_t) d\Gamma d\tau + \int_s^t \int_\Omega d_0(x) u_t b(u_t) dx d\tau \le \mathscr{E}(s), \quad (10.4.5)$$

where the convex nonnegative function $j_g : \mathbb{R} \to \mathbb{R}^+$ is given by the formula

$$j_g(s) = \int_0^s g(\xi) d\xi, \quad s \in \mathbb{R}. \quad (10.4.6)$$

If $g(s)s$ is convex, then functions j_g can be replaced in (10.4.5) *by this former quantity. Here and below*

$$\mathscr{E}(t) \equiv \mathscr{E}(u, u_t) = E(t) - \frac{1}{2} \int_\Omega [F_0[u,u] + 2pu] dx$$

with

$$E(t) \equiv E(u, u_t) = \frac{1}{2} \int_\Omega \left[u_t^2 + \frac{1}{2} |\Delta v(u)|^2 \right] dx + \frac{1}{2} a(u,u) + \frac{\beta}{4} \int_{\Gamma_1} u^4 d\Gamma$$

and

$$a(u,v) = \int_\Omega \tilde{a}(u,v) dx + \nu_1 \int_{\Gamma_1} u^2 d\Gamma, \quad (10.4.7)$$

where

$$\tilde{a}(u,v) = u_{x_1x_1x} v_{x_1x_1} + u_{x_2x_2} v_{x_2x_2} + \mu (u_{x_1x_1} v_{x_2x_2} + u_{x_2x_2} v_{x_1x_1}) + 2(1-\mu) u_{x_1x_2} v_{x_1x_2}. \quad (10.4.8)$$

- *For initial data $u_0 \in H^4(\Omega)$ and $u_1 \in H^2(\Omega)$ subject to compatibility conditions on the boundary, one obtains*[4] *that*

$$(u; u_t) \in C_r(0,T; \left[H^{7/2}(\Omega) \cap H^2_{\Gamma_0}(\Omega) \right] \times H^2_{\Gamma_0}(\Omega)) \quad (10.4.9)$$

for every finite $T > 0$. Moreover, these solutions satisfy the energy equality

$$\mathscr{E}(t) + \int_s^t \int_{\Gamma_1} d(x) g(u_t) u_t d\Gamma d\tau + \int_s^t \int_\Omega d_0(x) b(u_t) u_t dx d\tau = \mathscr{E}(s). \quad (10.4.10)$$

Proof. The results stated above are a special case of Theorem 4.2.9. If $\beta > 0$ it suffices to write

[4] If g is Lipschitz we can replace $H^{7/2}(\Omega)$ by $H^4(\Omega)$ in (10.4.9).

$$\beta u^3 + dg(s) = \beta(u^3 + cs) + dg(s) - \beta cs$$

for sufficiently small constant c. Clearly $\tilde{g}(x,s) = d(x)g(s) - cs$ satisfies $\partial_s \tilde{g}(x,s) \geq dm - c > 0$, as long as c is sufficiently small. Thus we can use the final statement in Remark 4.2.10.

The smoothness in (10.4.9) follows from (4.2.33) and the regularity of Green's mapping stated in (3.2.54) (see also Proposition (1.3.13)). □

It is proved (see (4.2.24)) that under the conditions of Theorem 10.4.3 the energy functionals $\mathcal{E}$ and E satisfy the inequality

$$cE(u_0,u_1) - M_0 \leq \mathcal{E}(u_0,u_1) \leq CE(u_0,u_1) + M_0 \tag{10.4.11}$$

for any $(u_0;u_1) \in H$, where c, C, M_0 are positive constants. In particular, this implies that the energy $\mathcal{E}(u_0;u_1)$ is bounded from below and $\mathcal{E}(u_0;u_1) \to +\infty$ when $\|(u_0;u_1)\|_H \to +\infty$. This, in turn, implies that there exists $R_* > 0$ such that the set

$$\mathcal{W}_R = \{y = (u_0;u_1) \in H \,:\, \mathcal{E}(u_0,u_1) \leq R\} \tag{10.4.12}$$

is a nonempty bounded set in H for all $R \geq R_*$. Moreover any bounded set $B \subset H$ is contained in $\mathcal{W}_R$ for some R and, as it follows from the energy inequality (10.4.5), the set $\mathcal{W}_R$ is invariant with respect to the semiflow S_t: $S_t\mathcal{W}_R \subset \mathcal{W}_R$ for all $t > 0$. Thus we can consider the restriction $(\mathcal{W}_R, S_t)$ of the dynamical system (H, S_t) on $\mathcal{W}_R$, $R \geq R_*$.

We introduce next the set of stationary points of S_t denoted by $\mathcal{N}$,

$$\mathcal{N} = \{V \in H \,:\, S_t V = V \text{ for all } t \geq 0\}\,.$$

Every stationary point V has the form $V = (u;0)$, where $u = u(x) \in H^2_{\Gamma_0}(\Omega)$ is a weak (variational) solution to the problem (10.2.13). This means that $u \in H^2_{\Gamma_0}(\Omega)$ satisfies the variational relation

$$a(u,w) + \int_\Omega (-[u,v(u)+F_0] - p)\cdot w\,dx + \beta\int_{\Gamma_1} u^3 w\,d\Gamma = 0 \tag{10.4.13}$$

for any $w \in H^2_{\Gamma_0}(\Omega)$, where $a(u,w)$ is given by (10.4.7) and $v(u) \in H^2_0(\Omega)$ is determined from (10.4.2). Taking in (10.4.13) $w = u$ and using the symmetry relation

$$\int_\Omega [u,F_0]w\,dx = \int_\Omega [u,w]F_0\,dx \;\text{ for } F_0 \in H^2_{\Gamma_1}(\Omega),\; w \in H^2_{\Gamma_0}(\Omega),\; u \in H^2(\Omega),$$

which follows from (1.4.9), leads to the following assertion.

10.4.4. Proposition. *Under Assumptions* 10.4.1 *the set $\mathcal{N}$ of stationary points for the semiflow S_t generated by the equations* (10.4.1) *and* (10.4.2) *with boundary conditions* (10.4.3) *is a closed bounded set in H, and hence there exists $R_{**} \geq R_*$ such that $\mathcal{N} \subset \mathcal{W}_R$ for every $R \geq R_{**}$.*

Once the well-posedness of the semiflow is settled, our goal is to prove the global attractiveness property for the dynamical system (H, S_t). This requires additional hypotheses imposed on the data of the problem.

10.4.5. Assumption.

1. We assume that $d(x) \geq d > 0$ on Γ_1 and function $g \in C(\mathbb{R})$ satisfies the following conditions: (i) for every $\varepsilon > 0$ there exists C_ε such that
$$s^2 \leq \varepsilon + C_\varepsilon s g(s), \quad s \in \mathbb{R}, \tag{10.4.14}$$
and (ii) there exist constants $M > 0$ and $p \geq 1$ such that
$$|g(s_1) - g(s_2)| \leq M\left(1 + |s_1|^{p-1} + |s_2|^{p-1}\right)|s_1 - s_2|, \quad s_1, s_2 \in \mathbb{R}. \tag{10.4.15}$$
2. The function $b \in C(\mathbb{R})$ is nondecreasing and satisfies
$$|b(s_1) - b(s_2)| \leq M\left(1 + |s_1|^{q-1} + |s_2|^{q-1}\right)|s_1 - s_2| \quad s_1, s_2 \in \mathbb{R},$$
for some $1 \leq q < \infty$.
3. We assume that Γ is star-shaped; that is, there exists $x_0 \in \mathbb{R}^2$ such that
$$(x - x_0)n \leq 0 \text{ on } \Gamma_0 \text{ and } (x - x_0)n \geq 0 \text{ on } \Gamma_1.$$

10.4.6. Remark. The geometric assumption imposed in Part 3 of Assumption 10.4.5 could be weakened in line with the third item in Assumption 10.2.9; see also Remark 10.2.10. It suffices to impose a suitable geometric condition only on the nondissipative part of the boundary Γ_0. The corresponding argument is similar to the one given in the proof of Proposition 10.2.14. The latter is based on the trace regularity counterpart of Proposition 10.2.16. Because the main focus of the present section is on handling "critical" growth of nonlinearity, we are not pursuing these generalizations but rather we assume the global star-shaped condition. ■

Our main results read as follows.

10.4.7. Theorem (Compact Attractors). *Assumptions* 10.4.1 *and* 10.4.5 *are in force. Then the following assertions hold.*

- *For any $R \geq R_*$ there exists a global compact attractor $\mathscr{A}_R$ for the restriction $(\mathscr{W}_R, S_t)$ of the dynamical system (H, S_t) on $\mathscr{W}_R$, where $\mathscr{W}_R$ is given by* (10.4.12).
- *If we assume additionally that $d_0(x) > 0$ a.e. in Ω and $b(s)s > 0$ for all $s \neq 0$, then the system (H, S_t) is gradient and there is $R_0 > 0$ such that $\mathscr{A}_R$ does not depend on R for all $R \geq R_0$. In this case $\mathscr{A} \equiv \mathscr{A}_{R_0}$ is a global attractor for (H, S_t) and $\mathscr{A}$ coincides with the unstable manifold $\mathscr{M}^u(\mathscr{N})$ emanating from the set $\mathscr{N}$ of stationary points for S_t (see Definition* 7.5.1*). We also have that*
$$\lim_{t \to +\infty} \operatorname{dist}_H(S_t W, \mathscr{N}) = 0 \text{ for any } W \in H. \tag{10.4.16}$$

- *The global attractor $\mathscr{A}$ has a finite fractal dimension under the following additional condition: there exists $M_1 > 0$, $m > 0$, and $\kappa < 1$ such that*

$$0 < m \le \frac{g(s_1) - g(s_2)}{s_1 - s_2} \le M_1 \left(1 + s_1 g(s_1) + s_2 g(s_2)\right)^{\kappa}, \quad s_1, s_2 \in \mathbb{R}, \tag{10.4.17}$$

and

$$0 \le \frac{b(s_1) - b(s_2)}{s_1 - s_2} \le M_1 \left(1 + s_1 b(s_1) + s_2 b(s_2)\right)^{\kappa}, \quad s_1, s_2 \in \mathbb{R}. \tag{10.4.18}$$

In this case the global attractor is a bounded set in $(H^4 \cap H^2_{\Gamma_0})(\Omega) \times H^2_{\Gamma_0}(\Omega)$. In addition $||u_{tt}(t)||_{0,\Omega} \le C_{\mathscr{A}}$ on the attractor.

For comments concerning properties (10.4.17) and (10.4.18) of the damping functions we refer to Remark 9.2.7 and Remark 9.4.5.

If the system (H, S_t) is gradient, then we can supplement Theorem 10.4.7 with the following corollaries.

10.4.8. Corollary. *Under the hypotheses of the second part of Theorem* 10.4.7 *the global attractor $\mathscr{A}$ consists of full trajectories $\gamma = \{W(t) : t \in \mathbb{R}\}$ such that* $\lim_{t\to-\infty} \mathrm{dist}_H(W(t), \mathscr{N}) = 0$ *and* $\lim_{t\to+\infty} \mathrm{dist}_H(W(t), \mathscr{N}) = 0$.

10.4.9. Corollary. *Let the hypotheses of the second part of Theorem* 10.4.7 *be in force. Assume that the stationary problem in* 10.2.13 *has a finite number of solutions. Then*

(i) The global attractor $\mathscr{A}$ consists of full trajectories $\gamma = \{W(t) : t \in \mathbb{R}\}$ connecting pairs of stationary points; that is, any $W \in \mathscr{A}$ belongs to some full trajectory γ and for any $\gamma \subset \mathscr{A}$ there exists a pair $\{Z, Z^*\} \subset \mathscr{N}$ such that

$$W(t) \to Z \ \text{ as } t \to -\infty \ \text{ and } \ W(t) \to Z^* \ \text{ as } t \to +\infty\,.$$

(ii) For any $V \in H$ there exists a stationary point Z such that $S_t V \to Z$ as $t \to +\infty$.

If the set $\mathscr{N}$ of stationary points is discrete, then by (10.4.16) every solution converges to an equilibrium point. Thus as above there is interest in considering the rate of convergence for these solutions. Due to the fact that equilibria may be multiple, any small perturbation in a vicinity of equilibrium may cause an escape of the solution from this neighborhood. As in Section 10.2.4 to overcome this difficulty we need to control lower-order terms.

In order to describe the decay rates, we need some notation. We first introduce a concave, strictly increasing, continuous function $h : \mathbb{R}_+ \mapsto \mathbb{R}_+$ which captures the behavior of $g(s)$ at the origin possessing the properties

$$h(0) = 0 \ \text{ and } \ s^2 + g^2(s) \le h(j_g(s)) \ \text{ for } |s| \le 1, \tag{10.4.19}$$

where $j_g(s)$ is given by (10.4.6). We refer to Appendix B for construction of such a function $h(s)$ in the case when $j_g(s) = sg(s)$. In our case the argument is the same.

Given function h we define

$$H_0(s) = h\left(\frac{s}{c_1}\right), \quad G_0(s) = (I + H_0)^{-1}(c_2 s), \quad Q(s) = s - (I + G_0)^{-1}(s), \tag{10.4.20}$$

where c_1 and c_2 are positive constants (determined within the course of the proof of Theorem 10.4.10). As mentioned earlier, $Q(s)$ is strictly monotone and the differential equation

$$\frac{d\sigma}{dt} + Q(\sigma) = 0, \quad t > 0, \quad \sigma(0) = \sigma_0 \in \mathbb{R}_+, \tag{10.4.21}$$

admits a unique global solution $\sigma(t)$ which, moreover, decays asymptotically to zero as $t \to \infty$. The "speed" of decay depends on the behavior of $g'(s)$ at the origin. We refer to Proposition B.3.1 and Remark B.3.2 in Appendix B for the exact description of decaying rates.

We also note that if $sb(s) > 0$ for $s \neq 0$ we can assume that the function $h(s)$ in (10.4.19) possesses the additional property

$$s^2 \le h(sb(s)) \quad \text{for all} \quad |s| \le 1. \tag{10.4.22}$$

Indeed, we can first construct a concave continuous increasing function $h_b(s)$ with property (10.4.22) and then redefine h in (10.4.19) as $h + h_b$.

With these preparations we are ready to state our result:

10.4.10. Theorem (Rate of stabilization). *Let Assumptions* 10.4.1 *and* 10.4.5 *be in force. Assume that* $d_0(x) > 0$ *a.e. in* Ω, $b(s)s > 0$ *for all* $s \neq 0$ *and relations* (10.4.17) *and* (10.4.18) *hold with* $m = 0$. *In addition assume that (i) problem* (10.2.13) *has a finite number of solutions, (ii) there exists a constant* $C > 0$ *such that* $|g(s)| \le C j_g(s)$ *for* $|s| \ge 1$, *and (iii) relation*

$$[g(s) - g(\sigma)](s - \sigma) \ge \alpha_0 |s - \sigma|^r, \quad s, \sigma \in \mathbb{R}, \quad \textit{for some} \quad \alpha_0 > 0,\ r \ge 1, \tag{10.4.23}$$

holds. Then, for any $V \in H$ *there exists a stationary point* $E = (e; 0)$ *such that* $S_t V \to E$ *as* $t \to +\infty$. *Moreover, if the equilibrium* E *is hyperbolic in the sense that the linearization of* (10.2.13) *around each solution has the trivial solution only, then we have the following rates of stabilization.*

$$\|S_t V - E\|_H \le C\sigma([tT^{-1}]), \quad t > 0, \tag{10.4.24}$$

where C *and* T *are positive constants depending on* V *and* E, $[a]$ *denotes the integer part of* a *and* $\sigma(t)$ *satisfies* (10.4.21) *with* Q *defined by* (10.4.20). *The initial data* σ_0 *and the constants* c_i *in* (10.4.20) *also depend on* $V, E \in H$. *In particular, if* $g'(0) > 0$, *then*

$$\|S_t V - E\|_H \le Ce^{-\omega t}$$

for some positive constants C *and* ω *depending on* $V, E \in H$.

10.4.11. Remark. In studying rates of stabilization to equilibria it is important to work with generalized solutions only (nonuniqueness of equilibria makes it impossible to use effectively approximations of generalized solutions by strong solutions). Thus the issues such as validity of energy inequality and of a variational form of equation to be satisfied for generalized solutions is of paramount importance here. And this is the reason for introducing additional conditions in the statement of Theorem 10.4.10. Condition (10.4.23) is needed in order to guarantee the variational form of generalized solutions (see Theorem 4.2.12). As an example we can consider $g(s) = g_0|s|^{p-1}s$, $p \geq 1$. One can see that this g satisfies (10.4.23) with $\alpha_0 = g_0 2^{-p+1}$, $r = p+1$ (see Remark 3.2.9). Different from (10.4.23) hypotheses concerning g are also possible; see Theorem 4.2.12.

We also note that the estimate $|g(s)| \leq C j_g(s)$ for large $|s|$ allows us to derive a rate of stabilization by working with a weaker form of dissipation in the energy inequality (10.4.5). ■

10.4.2 Asymptotic smoothness

In this section we show that the semiflow S_t generated by problem (10.4.1)–(10.4.3) is asymptotically smooth. This property is critical for proving existence of global attractors (see Chapter 7). We recall (see Definition 7.1.2) a dynamical system (X, S_t) is said to be *asymptotically smooth* iff for any bounded set D in X such that $S_t D \subset D$ for $t > 0$ there exists a compact set $\mathscr{K}$ in the closure $\overline{D}$ of D, such that

$$\lim_{t\to+\infty} \sup_{y\in D} \operatorname{dist}_X\{S_t y, \mathscr{K}\} = 0.$$

Our main result in this section is the following assertion.

10.4.12. Theorem. *Let Assumptions* 10.4.1 *and* 10.4.5 *be in force. Let* $\mathscr{W}_R$ *be given by relation* (10.4.12). *Then dynamical system* $(\mathscr{W}_R, S_t)$ *generated by problem* (10.4.1)–(10.4.3) *is asymptotically smooth.*

Proof. To prove this theorem we rely on the criteria of asymptotic smoothness given in Proposition 7.1.11 (which is a version of a criterion used in [160]). For this, we need some preparation.

We denote by $U(t) = (u(t); u_t(t)) = S_t y_1$ and $W(t) = (w(t); w_t(t)) = S_t y_2$ the two solutions corresponding to initial conditions y_1 and y_2, respectively, from the set $\mathscr{W}_R$ for some $R > R_*$. Without loss of generality we can assume that $u(t)$ and $w(t)$ are strong solutions. Because $\mathscr{W}_R$ is invariant, we have

$$\|u(t)\|_{2,\Omega} + \|u_t(t)\|_{0,\Omega} + \|w(t)\|_{2,\Omega} + \|w_t(t)\|_{0,\Omega} \leq C_R, \quad t \geq 0. \quad (10.4.25)$$

Also, from energy relation (10.4.10) we obtain finiteness of the following dissipation integrals. For all $t \geq 0$ we have that

$$\int_0^t \int_{\Gamma_1} d(x) g(u_t) u_t d\Gamma dt + \int_0^t \int_{\Gamma_1} d(x) g(w_t) w_t d\Gamma dt \leq C_R, \quad (10.4.26)$$

$$\int_0^t \int_\Omega d_0(x) b(u_t) u_t dx dt + \int_0^t \int_\Omega d_0(x) b(w_t) w_t dx dt \le C_R. \quad (10.4.27)$$

Let $z \equiv u - w$. We denote by $E_z(t)$ the free energy corresponding to z and given by $E_z(t) = \frac{1}{2}\left(\|z_t\|_\Omega^2 + a(z,z)\right)$, where $a(z,z)$ is given by (10.4.7). We note that if $\nu_1 > 0$ or $\Gamma_0 \neq \emptyset$, then $\sqrt{a(z,z)}$ is equivalent to $\|z\|_{2,\Omega}$. Otherwise $\sqrt{a(z,z)}$ is just a seminorm on $H^2(\Omega)$.

As has already been mentioned, the model without rotational inertia exhibits less regularity at the velocity level. Consequently, the contribution of a nonlinear term is no longer compact and the proof of asymptotic smoothness requires special treatment. This is due to the fact that the observability estimates (see Propositions 10.2.14 and 10.3.7, for instance) contain critical terms.

The following estimate is critical for the proof of Theorem 10.4.12.

10.4.13. Lemma. *Let the hypotheses of Theorem* 10.4.12 *be in force. Then given* $\varepsilon > 0$ *and* $T > 1$ *there exist constants* $C_\varepsilon(R)$, *and* $C_{\varepsilon,T}$ *such that*

$$E_z(T) \le \varepsilon + \frac{C_\varepsilon(R)}{T}\left[1 + \left|\int_0^T \int_t^T (\mathscr{R}(z), z_t)_\Omega ds dt\right|\right] + C_{\varepsilon,T}\, lot(z),$$

where $E_z(t) = \frac{1}{2}\left(\|z_t\|_\Omega^2 + a(z,z)\right)$ *with* $a(z,z)$ *given by* (10.4.7), $z = u - w$,

$$lot(z) = C_R \sup_{0 \le \tau \le T} \left\{\|z(\tau)\|_{2-\eta,\Omega}^2 + \|z(\tau)\|_{2-\eta,\Omega}\right\} \quad (10.4.28)$$

for some $\eta > 0$ *and*

$$\mathscr{R}(z) \equiv [v(u) - v(w), u] + [v(w), z] + [F_0, z]. \quad (10.4.29)$$

Proof. We first note that the following equation holds for the variable $z \equiv u - w$,

$$\begin{aligned} & z_{tt} + \Delta^2 z + d_0 b(z_t + w_t) - d_0 b(w_t) = \mathscr{R}(z) \ \text{ in } \Omega \times (0,\infty), \\ & z = \frac{\partial}{\partial n} z = 0 \ \text{ on } \Gamma_0 \times (0,\infty), \quad \Delta z + (1-\mu) B_1 z = 0 \ \text{ on } \Gamma_1 \times (0,\infty), \\ & \frac{\partial}{\partial n}\Delta z + (1-\mu) B_2 z - \nu_1 z - \beta(u^2 + uw + w^2) z \\ & = d(g(z_t + w_t) - g(w_t)) \ \text{ on } \Gamma_1 \times (0,\infty). \end{aligned} \quad (10.4.30)$$

Because we consider trajectories lying in $\mathscr{W}_R$, (10.4.25) implies

$$|[v(u), z]|_\Omega \le C \|u\|_{2,\Omega}^2 \|z\|_{2,\Omega}$$

and, because $F_0 \in H^{3+\delta}(\Omega)$, we have that

$$\|\mathscr{R}(z(t))\|_\Omega \le C_R \|z(t)\|_{2,\Omega}, \quad t \ge 0. \quad (10.4.31)$$

A standard energy equality implies

$$E_z(t)+D_s^t(z)=E_z(s)+\int_s^t\int_\Omega \mathscr{R}(z)z_t\,dxd\tau-\beta\int_s^t\int_{\Gamma_1}(u^2+uw+w^2)zz_t\,d\Gamma d\tau, \tag{10.4.32}$$

where we have denoted

$$D_s^t(z)\equiv\int_s^t\int_{\Gamma_1} d(x)[g((z_t+w_t))-g(w_t)]z_t\,d\Gamma d\tau$$
$$+\int_s^t\int_\Omega d_0(x)(b(z_t+w_t)-b(w_t))z_t\,dxd\tau. \tag{10.4.33}$$

10.4.14. Proposition. *Let the hypotheses of Theorem* 10.4.12 *be in force. Then there exists* $\varepsilon_0>0$ *such that given* $0<\varepsilon\le\varepsilon_0$ *and* $T>1$ *there exist constants* C_ε *and* C_T *such that*

$$\int_0^T E_z(t)dt\le\varepsilon T+C_\varepsilon(R)+C_T lot(z), \tag{10.4.34}$$

where $lot(z)$ *is given by* (10.4.28).

Proof. We start with the following inequality which is used later on.

10.4.15. Proposition (Reconstruction of the energy). *Let Assumptions* 10.4.1 *and* 10.4.5 *be in force. Then*

$$\int_0^T E_z(t)dt\le c_0\int_0^T\int_{\Gamma_1}|z_t|^2d\Gamma dt+c_0\left[E_z(T)+E_z(0)\right] \tag{10.4.35}$$
$$+\left|\int_0^T\int_{\Gamma_1} d(g(z_t+w_t)-g(w_t))h\nabla z\,d\Gamma dt\right|$$
$$+\left|\int_0^T\int_\Omega d_0[b(z_t+w_t)-b(w_t)]h\nabla z\,dxdt\right|+lot_0(z),$$

where $c_0>0$ *is a constant,* $h=x-x_0$ *with* x_0 *given by Assumption* 10.4.5(3), *and*

$$lot_0(z)=C_R\int_0^T\|z(\tau)\|_{0,\Omega}^2d\tau. \tag{10.4.36}$$

Proof. Let $h=x-x_0$ (see Assumption 10.4.5(3)). We apply the multiplier $h\nabla z$ to equation (10.4.30). Because

$$\int_0^T\int_\Omega z_{tt}h\nabla z\,dxdt=\int_\Omega z_t h\nabla z\,dx\Big|_0^T+\frac{1}{2}\int_Q z_t^2\,\mathrm{div}h\,dQ-\frac{1}{2}\int_\Sigma z_t^2(h,n)d\Sigma,$$

we have that

$$\int_0^T\int_\Omega z_{tt}h\nabla z\,d\Omega dt\ge\int_0^T\|z_t\|_{0,\Omega}^2dt \tag{10.4.37}$$
$$-c_\Omega\int_0^T\int_{\Gamma_1}|z_t|^2d\Gamma dt-c_0\left[E_z(T)+E_z(0)\right].$$

As for the case with rotational inertia (see (10.2.32)) using the condition $(h,n) \le 0$ on Γ_0, we obtain

$$\int_\Omega \Delta^2 z h\nabla z dx \ge a(z,z) + \frac{1}{2} a_{\Gamma_1}(z,z) + \nu_1 \int_{\Gamma_1} [z h\nabla z - z^2] d\Gamma$$
$$+ \int_{\Gamma_1} \left[d(g(z_t + w_t) - g(w_t)) h\nabla z + \beta(w^2 + wu + u^2) z) h\nabla z \right] d\Gamma,$$

where we have used notation $a_{\Gamma_1}(u,v)hv \equiv \int_{\Gamma_1} \tilde{a}(u,v)(h,n)d\Gamma$ with $\tilde{a}(u,v)$ given by (10.4.8). Note that the star-shaped condition[5] imposed on Γ_1 allows us to control the sign of the form a_{Γ_1}.

Recalling that

$$\|w(t)\|_{C(\Omega)} + \|u(t)\|_{C(\Omega)} \le C\left(\|w(t)\|_{2,\Omega} + \|u(t)\|_{2,\Omega}\right) \le C_R,$$

and using the requirement $(h,n)\big|_{\Gamma_1} \ge 0$ imposed in Assumption 10.4.5(3) we obtain

$$\int_\Omega \Delta^2 z h\nabla z dx \ge a(z,z) + \int_{\Gamma_1} d(g(z_t + w_t) - g(w_t)) h\nabla z d\Gamma$$
$$- C_R \int_{\Gamma_1} |z||\nabla z| d\Gamma - \nu_1 \int_{\Gamma_1} z^2 d\Gamma.$$

Sobolev's embedding yields

$$\int_\Omega \Delta^2 z h\nabla z dx \ge \frac{1}{2} a(z,z) + \int_{\Gamma_1} d(g(z_t + w_t) - g(w_t)) h\nabla z d\Gamma - C_R \|z\|^2_\Omega. \quad (10.4.38)$$

By virtue of (10.4.31) we have

$$\left| \int_0^T \int_\Omega \mathscr{R}(z) h\nabla z dx dt \right| \le C_R \int_0^T \|z(t)\|_{2,\Omega} \|z(t)\|_{1,\Omega} dt \quad (10.4.39)$$
$$\le \gamma \int_0^T \|z(t)\|^2_{2,\Omega} dt + C_{R,\gamma} \int_0^T \|z(t)\|^2_{0,\Omega} dt$$

for any $\gamma > 0$. Using the first relation in (10.4.30) and combining inequalities in (10.4.37)–(10.4.39) we obtain (10.4.35). □

To continue with (10.4.35) we need the following assertion.

10.4.16. Proposition (Dissipation estimate). *Let the conditions of Proposition 10.4.15 be in force. Then there exists $\delta > 0$ such that*

$$\int_0^T \int_{\Gamma_1} d(x)(g(z_t + w_t) - g(w_t)) h\nabla z d\Gamma dt \le (1+T) \cdot C_R \sup_{t \in [0,T]} \|z(t)\|_{2-\delta,\Omega},$$

---

[5] In the absence of this condition, one needs to absorb these supercritical terms. This can be done by resorting to trace estimate (a counterpart of Proposition 10.2.16) which allows for absorbtion of critical boundary terms.

$$\int_0^T \int_\Omega d_0[b(z_t+w_t)-b(w_t)]h\nabla z dx dt \le (1+T)\cdot C_R \sup_{t\in[0,T]} \|z(t)\|_{2-\delta,\Omega}\cdot$$

Proof. The computations below exploit the polynomial growth condition imposed on g and b. We apply Hölder's inequality with Hölder's exponent $r>1$:

$$\begin{aligned}
&\int_0^T \int_{\Gamma_1} d(x)(g(z_t+w_t)-g(w_t))h\nabla z d\Gamma dt \\
&\le C\|\nabla z\|_{L_{\bar r}(\Sigma)}\left[\int_0^T \int_{\Gamma_1} d(x)^r|g(z_t+w_t)-g(w_t)|^r d\Gamma dt\right]^{1/r} \\
&\le C\|\nabla z\|_{L_{\bar r}(\Sigma)}\left[\int_0^T \int_{\Gamma_1} d(x)\left[|g(u_t)|^r+|g(w_t)|^r\right] d\Gamma dt\right]^{1/r},
\end{aligned}$$

where $r^{-1}+\bar r^{-1}=1$ and $\Sigma=(0,T)\times\Gamma$. We take $r=1+1/p$ and use the fact that $|g(s)|\le 2M|s|^p$ for $|s|\ge 1$ (see (10.4.15)). Splitting the region of integration according to $|u_t|\le 1$ and $|u_t|\ge 1$ yields

$$\begin{aligned}
\int_0^T \int_{\Gamma_1} d(x)|g(u_t)|^r d\Gamma dt &= \int_0^T \int_{\Gamma_1} d(x)|g(u_t)|^{r-1}|g(u_t)| d\Gamma dt \\
&\le 2M\int_0^T \int_{\Gamma_1} (c+d(x)|u_t||g(u_t)|) d\Gamma dt \le C_R(1+T),
\end{aligned} \tag{10.4.40}$$

where for the last step we have used (10.4.26). Similar computations apply to the term with w:

$$\int_0^T \int_{\Gamma_1} d(x)|g(w_t)|^r d\Gamma dt \le C_R(1+T). \tag{10.4.41}$$

The map $z\to\nabla z|_{\Gamma_1}$ is bounded from the space $H^{2-\delta}(\Omega)$ into $L_{\bar r}(\Gamma_1)$ for sufficiently small δ, therefore we have that

$$\|\nabla z\|_{L_{\bar r}(\Sigma)} \le \left(\int_0^T \|z\|_{2-\delta,\Omega}^{\bar r} dt\right)^{1/\bar r} \le T^{1/\bar r}\sup_{t\in[0,T]}\|z(t)\|_{2-\delta,\Omega}\cdot$$

Therefore combining (10.4.40) and (10.4.41) yields the desired conclusion in the first part of Proposition 10.4.16.

Similar arguments apply to the second part involving the internal dissipation term:

$$\begin{aligned}
&\int_0^T \int_\Omega d_0(b(z_t+w_t)-b(w_t))h\nabla z dx dt \\
&\le C\|\nabla z\|_{L_{\bar r}(\Omega)}\left[\int_0^T \int_\Omega d_0\left[|b(u_t)|^r+|b(w_t)|^r\right] dx dt\right]^{\frac{1}{r}},
\end{aligned}$$

where $r^{-1} + \bar{r}^{-1} = 1$. As before taking $r = 1 + 1/q$ and splitting the region of integration according to $|u_t| \leq 1$ and $|u_t| \geq 1$ from (10.4.27) we have

$$\int_0^T \int_\Omega d_0 |b(u_t)|^r dx dt \leq C \int_0^T \int_\Omega (c + d_0 |u_t||b(u_t)|)\, dx dt \leq C_R(1+T). \quad (10.4.42)$$

The rest of the argument is the same as in the case of boundary damping. □

Now we are in position to complete the proof of Proposition 10.4.14. Applying the estimate in Proposition 10.4.16 above to the inequality in (10.4.35) yields

$$\int_0^T E_z(t) dt \leq C_R + c_0 \int_0^T \int_{\Gamma_1} |z_t|^2 d\Gamma dt + C_{T,R} lot(z), \quad (10.4.43)$$

where $lot(z)$ is given by (10.4.28). To estimate $\|z_t\|_{\Gamma_1}$ we exploit the growth condition (10.4.14) imposed on $g(s)$. Indeed, by (10.4.14) we have that

$$\|z_t\|^2_{0,\Gamma_1} \leq 2\|w_t\|^2_{0,\Gamma_1} + 2\|u_t\|^2_{0,\Gamma_1} \leq \varepsilon + C_\varepsilon \int_{\Gamma_1} d(x)(w_t g(w_t) + u_t g(u_t)) d\Gamma.$$

Therefore from (10.4.26) we obtain that

$$\int_0^T \int_{\Gamma_1} |z_t|^2 d\Gamma dt \leq \varepsilon T + C_\varepsilon(R).$$

Thus we obtain (10.4.34) which completes the proof of Proposition 10.4.14. □

Now we complete the proof of Lemma 10.4.13. From (10.4.25) and energy identity (10.4.32) and using the estimate

$$\int_{\Gamma_1} (u^2 + |uw| + w^2) z z_t d\Gamma \leq \delta \int_{\Gamma_1} |z_t|^2 d\Gamma + \delta^{-1} C_R \|z\|^2_{2-\eta,\Omega} \quad (10.4.44)$$

for some positive η and for any $\delta > 0$ we obtain with any $0 \leq s \leq T$,

$$E_z(T) + D_s^T(z) \leq E_z(s) + \delta\beta \int_s^T \int_{\Gamma_1} |z_t|^2 dx dt + \int_s^T (\mathscr{R}(z), z_t)_\Omega dt + C_{\delta,T} lot(z). \quad (10.4.45)$$

By Assumption 10.4.1 in the case when $\beta > 0$ we have the relation

$$(s_1 - s_2)^2 \leq m^{-1}(s_1 - s_2)[g(s_1) - g(s_2)] \quad (10.4.46)$$

and hence $\int_s^t \|z_t\|^2_{\Gamma_1} d\tau \leq c_0 D_s^t(z)$. Therefore taking $\delta = (2c_0\beta)^{-1}$ in (10.4.45) for $\beta > 0$ we obtain

$$E_z(T) + \frac{1}{2} D_s^T(z) \leq E_z(s) + \int_s^T (\mathscr{R}(z), z_t)_\Omega dt + C_T lot(z). \quad (10.4.47)$$

It clear from (10.4.45) that the same relation remains true in the case $\beta = 0$. If we integrate (10.4.47) with respect to s over the interval $[0,T]$, we find that

$$TE_z(T) \le \int_0^T E_z(s)ds + \int_0^T ds \int_s^T (\mathscr{R}(z), z_t)_\Omega dt + C_T lot(z) \quad (10.4.48)$$

for any $T > 1$. Consequently, using (10.4.34) and (10.4.48) we find that

$$TE_z(T) \le \varepsilon T + C_{\varepsilon,R}\left[1 + \left|\int_0^T ds \int_t^T (\mathscr{R}(z), z_t)_\Omega dt\right|\right] + C_T lot(z).$$

Dividing this relation by T gives the desired inequality and the conclusion in Lemma 10.4.13 easily follows. □

Completion of the proof of Theorem 10.4.12. By Theorem 7.1.11 to prove Theorem 10.4.12 we need to construct a functional Ψ such that properties (7.1.12) and (7.1.13) hold. Because any bounded set belongs to $\mathscr{W}_R$ for some R, it is sufficient to construct this functional on the set $\mathscr{W}_R$ only.

It follows from Lemma 10.4.13 that for every $\varepsilon > 0$ there exists $T = T(\varepsilon) > 1$ such that for any initial data $U_0 = (u_0; u_1)$ and $W_0 = (w_0; w_1)$ from $\mathscr{W}_R$ we have

$$\|S_T U_0 - S_T W_0\|_H \le c_0 \left[E_z(T)\right]^{1/2} \le \varepsilon + \left[\Psi_{R,\varepsilon,T}(U_0, W_0)\right]^{1/2},$$

where $\Psi_{R,\varepsilon,T}(U_0, W_0) = C_{\varepsilon,T}\left\{\hat{\Psi}_T(U_0, W_0) + lot(z)\right\}$ with

$$\hat{\Psi}_T(U_0, W_0) = \left|\int_0^T \int_t^T (\mathscr{R}(z), z_t) ds dt\right|.$$

In order to apply Theorem 7.1.11 we show that the terms involving $\mathscr{R}$ enjoy "hidden compactness." This is because the nonlinear part of the energy is represented by compact a functional. Thus the relaxation of compactness to sequential limits in Theorem 7.1.11 allows us to pass with the limit on weakly convergent subsequences.

Let $w^n(t)$ be a sequence of solutions corresponding to initial data $y^n \equiv (w_0^n; w_1^n)$ from $\mathscr{W}_R \subset H$. By choosing a subsequence we can assume that

$$y^n(t) \equiv (w^n(t); w_t^n(t)) \to y(t) \equiv (w(t); w_t(t)) \quad \text{weakly* in } L_\infty(0,T;H) \quad (10.4.49)$$

for some solution $(w(t); w_t(t)) \in L_\infty(0,T;H)$ and

$$\sup_{[0,T]}\left\{\|z^{n,m}(t)\|_{2-\eta,\Omega}\right\} \to 0, \quad n,m \to \infty, \quad (10.4.50)$$

for any $\eta > 0$, where $z^{n,m}(t) \equiv w^n(t) - w^m(t)$.

By (10.4.50) we have that $lot(z^{n,m}) \to 0$. Therefore our proof is completed as soon as we show that

$$\lim_n \lim_m \hat{\Psi}_T(y^n, y^m) \equiv \lim_n \lim_m \left|\int_0^T \int_t^T (\mathscr{R}(z^{n,m}), z_t^{n,m})_\Omega ds dt\right| \to 0. \quad (10.4.51)$$

Because (for smooth solutions) we have

$$(\mathscr{R}(z), z_t)_\Omega = \frac{1}{4}\frac{d}{dt}\left[-\|\Delta v(u)\|_\Omega^2 - \|\Delta v(w)\|_\Omega^2 + 2\int_\Omega [z,z]F_0 dx\right]$$
$$([v(w),w],u_t)_\Omega \quad ([v(u),u],w_t)_\Omega,$$

integrating in time one obtains

$$\begin{aligned}\int_t^T (\mathscr{R}(z), z_t) ds = & \frac{1}{4}\left[\|\Delta v(u)(t)\|_\Omega^2 + \|\Delta v(w)(t)\|_\Omega^2\right] \\ & - \frac{1}{4}\left[\|\Delta v(u)(T)\|_\Omega^2 + \|\Delta v(w)(T)\|_\Omega^2\right] \\ & + \frac{1}{2}\int_\Omega [z(T), z(T)]F_0 dx - \frac{1}{2}\int_\Omega [z(t), z(t)]F_0 dx \\ & - \int_t^T [([v(w), w], u_t)_\Omega + ([v(u), u], w_t)_\Omega] ds.\end{aligned}$$

An important remark is that all the terms except the last one are compact on the finite energy space. This along with (10.4.49) and (10.4.50) implies

$$\begin{aligned}\lim_n \lim_m \int_t^T (\mathscr{R}(z^{n,m}), z_t^{n,m}) d\tau = & \frac{1}{2}\left[\|\Delta v(w)(t)\|_\Omega^2 - \|\Delta v(w)(T)\|_\Omega^2\right] \\ - \lim_n \lim_m \int_t^T & [([v(w^n), w^n], w_t^m)_\Omega + ([v(w^m), w^m], w_t^n)_\Omega] d\tau. \end{aligned} \tag{10.4.52}$$

As for the second term in (10.4.52), from the convergence in (10.4.49) and from relation $\|[v(w^n(t)), w^n(t)]\|_{0,\Omega} \le C_R$, the latter following from the boundedness of the von Karman bracket, we conclude that

$$\begin{aligned} & 2\lim_n \lim_m \int_t^T [([v(w^n), w^n], w_t^m)_\Omega + ([v(w^m), w^m], w_t^n)_\Omega] d\tau \qquad (10.4.53) \\ & = 4\int_t^T ([v(w), w], w_t)_\Omega d\tau = -\|\Delta v(w)(T)\|_\Omega^2 + \|\Delta v(w)(t)\|_\Omega^2. \end{aligned}$$

Combining (10.4.52) and (10.4.53) yields (10.4.51). □

10.4.3 Global attractor: Proof of Theorem 10.4.7.

The proof of Theorem 10.4.7 in nonrotational case is more involved and subtle than the corresponding one in the rotational case. Criticality of the nonlinear forcing along with limited support of dissipation play a decisive role in forcing major detours in the arguments. This is particularly pronounced at the level of proving smoothness and finite dimensionality of attractors.

First statement. Because $\mathscr{W}_R$ is bounded and forward invariant, Theorem 10.4.12 implies the existence of a compact global attractor $\mathscr{A}_R$ of the dynamical system $(\mathscr{W}_R, S_t)$ for each $R \geq R_*$. This proves the first part of Theorem 10.4.7.

Second statement. If we now choose $R_0 \geq R_* + 1$ such that the set $\mathscr{N}$ of equilibria lies in $\mathscr{W}_{R_0-1}$, then the conditions: $d_0(x) > 0$ a.e. in Ω and $b(s)s > 0$ for all $s \neq 0$ and the energy relation (10.4.5) imply that the energy $\mathscr{E}(u_0, u_1)$ is a strict Lyapunov function for $(\mathscr{W}_R, S_t)$. This, in turn, implies (see Corollary 7.5.7) that $\mathscr{A}_R = M^u(\mathscr{N})$. In addition, it follows from Proposition 10.4.4 that $\mathscr{A}_R$ does not depend on R for $R \geq R_0$ and, hence, relation (10.4.16) holds. Thus the second part of Theorem 10.4.7 is proved.

Third statement. To prove finiteness of the fractal dimension $\dim_f \mathscr{A}$ we use the argument based on an appropriate stabilizability estimate. We note that, due to the lack of compactness, such an estimate is proved to hold on the global attractor only. But this is sufficient for the final conclusion, because the existence of the global attractor has already been established.

We first note that under the conditions imposed in the third part of the statement of Theorem 10.4.7 we can prove, as a counterpart of Proposition 10.4.16, the following relations

$$\int_0^T \int_{\Gamma_1} d[g(z_t + w_t) - g(w_t)] h\nabla z d\Gamma dt \leq \delta D_0^T(z) + C_{,\delta,R,T} \max_{[0,T]} ||z||^2_{2-\eta,\Omega} \quad (10.4.54)$$

and

$$\int_0^T \int_{\Omega} d_0[b(z_t + w_t) - b(w_t)] h\nabla z dx dt \leq \delta D_0^T(z) + C_{\delta,R,T} \max_{[0,T]} ||z||^2_{2-\eta,\Omega} \quad (10.4.55)$$

for and $\delta > 0$ and for some $\eta > 0$, where $D_0^T(z)$ is given by (10.4.33). These inequalities are obtained by similar arguments as used in the proof of Proposition 10.4.16 (see also Lemma 9.4.9 and [68, 69] for related calculations).

We first prove additional regularity of the attractor, and only afterwards we prove finite-dimensionality. It is important to keep in mind that we can now restrict all considerations to solutions that belong to the attractor (and hence are contained in a compact set). We proceed through the four-step procedure introduced by [161] (see also [75] for the abstract version of the argument). In the first step, by exploiting closedness to the equilibria points, smoothness of trajectories is established for negative times t near $-\infty$. In the second step, this smoothness is propagated to positive times via standard energy inequality, This gives smoothness of trajectories on a full real line, but without any uniform bound. To obtain the latter—in the third step—compactness of the attractor is exploited. In the final fourth step we exploit the stabilizability estimate in order to claim the "squeezing" property needed to assert finite-dimensionality of the attractor. The details are given below.

Step 1: Smoothness on negative time scale. We start with the following preliminary assertion which follows from the observation given in Remark 9.4.11.

10.4.17. Lemma. *Assume $w(\tau)$ and $u(\tau)$ are two functions from the class*

$$C(s,t;H^2_{\Gamma_0}(\Omega))\cap C(s,t;L_2(\Omega)),$$

for some $s,t\in\mathbb{R}$, $s<t$, such that

$$\|w(\tau)\|^2_{2,\Omega}+\|w_t(\tau)\|^2_\Omega\le R^2,\quad \|u(\tau)\|^2_{2,\Omega}+\|u_t(\tau)\|^2_\Omega\le R^2,\quad \tau\in[s,t].$$

Let $z(\tau)=w(\tau)-u(\tau)$ and $\mathscr{R}(z)$ be given by (10.4.29). *Then there exists $\eta>0$ such that*

$$\left|\int_s^t(\mathscr{R}(z),z_t)d\tau\right|\le C_R\max_{\tau\in[s,t]}\|z(\tau)\|^2_{2-\eta,\Omega}+C_R\int_s^t(\|w_t\|_\Omega+\|u_t\|_\Omega)\|z\|^2_{2,\Omega}d\tau. \tag{10.4.56}$$

Proof. It follows from relation (9.4.40) and Remark 9.4.11. □

Let $\gamma=\{(u(t);u_t(t)) : t\in\mathbb{R}\}$ be a trajectory from the global attractor $\mathscr{A}$. Let $0<h<1$. It is clear that for the couple $w(t):=u(t+h)$ and $u(t)$ the hypotheses of Lemma 10.4.17 hold for every interval $[s,t]$. We estimate the energy E_z of $z(t):=z^h(t)=u(t+h)-u(t)$. The critical role is played by the estimates for a noncompact critical term involving $\mathscr{R}(z)$. Indeed, from Lemma 10.4.17 and, in particular by (10.4.56) we obtain:

$$\begin{aligned}\left|\int_s^t(\mathscr{R}(z^h),z^h_t)d\tau\right|&\le C_1(R)\sup_{s\le\tau\le t}\|z^h(\tau)\|^2_{2-\eta,\Omega}\\&\quad+C_2(R)\int_s^t(\|u_t(\tau+h)\|_\Omega+\|u_t(\tau)\|_\Omega)\|z^h(\tau)\|^2_{2,\Omega}d\tau\end{aligned}\tag{10.4.57}$$

for all $-\infty<s\le t<+\infty$.

Because $\mathscr{A}=M^u(\mathscr{N})$, where $\mathscr{N}$ is the set of equilibria, $\mathscr{N}=\{(v;0) : v\in\mathscr{N}_*\}$, we have relation (10.4.16). This implies that for any $\varepsilon>0$ there exists T^ε_γ (independent of h, but depending on the trajectory γ) such that

$$\|u_t(\tau)\|_\Omega+\|u_t(\tau+h)\|_\Omega\le\varepsilon\cdot[C_2(R)]^{-1}\ \text{ for any } t\le T^\varepsilon_\gamma.$$

Therefore from (10.4.57) we have that

$$\left|\int_s^t(\mathscr{R}(z^h),z^h_t)d\tau\right|\le C_1(R)\sup_{s\le\tau\le t}\|z^h(\tau)\|^2_{2-\eta,\Omega}+\varepsilon\int_s^t\|z^h(\tau)\|^2_{2,\Omega}d\tau \tag{10.4.58}$$

for all $-\infty<s\le t\le T^\varepsilon_\gamma$. Thus using the energy relation (10.4.32) and (10.4.44) (in the case $\beta>0$) we find that

$$E_z(s)\le E_z(t)+2D^t_s(z)+\varepsilon\int_s^t\|z\|^2_{2,\Omega}ds+C(R)\max_{[s,t]}\|z\|^2_{2-\eta,\Omega} \tag{10.4.59}$$

and

$$E_z(t)+\frac{1}{2}D^t_s(z)\le E_z(s)+\varepsilon\int_s^t\|z\|^2_{2,\Omega}ds+C(R)\max_{[s,t]}\|z\|^2_{2-\eta,\Omega} \tag{10.4.60}$$

for all $s < t \le T_\gamma^\varepsilon$, where $\eta > 0$. Now, we are in a position to apply on each subinterval $[s, s+T_0]$ the estimate of Proposition 10.4.15 along with the estimates for the damping terms given in (10.4.54) and (10.4.55). This gives:

$$\int_s^{s+T_0} E_z(\tau)d\tau \le c_0 D_s^{s+T_0}(z) + c_0\left[E_z(s+T_0)+E_z(s)\right] + C_{R,T_0} \max_{[s,s+T_0]} \|z(\tau)\|^2_{2-\eta,\Omega}.$$

By using the energy estimate in (10.4.59) with ε sufficiently small, we obtain that

$$\int_s^{s+T_0} E_z(\tau)d\tau \le c_1 D_s^{s+T_0}(z) + c_1 E_z(s+T_0) + C_{R,T_0} \max_{[s,s+T_0]} \|z(\tau)\|^2_{2-\eta,\Omega}. \quad (10.4.61)$$

Integrating (10.4.60) yields

$$E_z(s+T_0) \le (T_0^{-1}+\varepsilon)\int_s^{s+T_0} E_z(\tau)d\tau + C_R \max_{[s,t]} \|z\|^2_{2-\eta,\Omega}. \quad (10.4.62)$$

Therefore with an appropriate choice of T_0 and ε from (10.4.61) and (10.4.62) we have

$$E_z(s+T_0) + \int_s^{s+T_0} E_z(\tau)d\tau \le c_1 D_s^{s+T_0}(z) + c_2(R,T_0) \max_{[s,s+T_0]} \|z(\tau)\|^2_{2-\eta,\Omega}.$$

This inequality along with one more application of inequality (10.4.60) written on the interval $[s, s+T_0]$ allows us to choose $0 < \sigma < 1$ such that $u^h(t) = h^{-1}z^h(t)$ satisfies the following estimate

$$E_{u^h}(s+T_0) \le \sigma E_{u^h}(s) + C_{T_0} \sup_{0\le\tau\le T_0} \|u^h(s+\tau)\|^2_{2-\eta,\Omega} \quad (10.4.63)$$

for all $s \le T_\gamma - T_0$, where $\eta > 0$, $T_\gamma = T_\gamma^{\varepsilon_0}$ (depending on the trajectory, but not on h) for some $\varepsilon_0 > 0$ and $T_0 > 0$. The standard interpolation argument leads us to

$$E_{u^h}(s+T_0) \le \sigma E_{u^h}(s) + \frac{1-\sigma}{2} \sup_{0\le\tau\le T_0} E_{u^h}(s+\tau) + C_\sigma(T_0)$$

for all $s \le T_\gamma - T_0$. Taking the supremum over the interval $(-\infty, T_\gamma - T_0]$ gives

$$\sup_{\tau\in(-\infty,T_\gamma]} E_{u^h}(\tau) \le \frac{1+\sigma}{2} \sup_{\tau\in(-\infty,T_\gamma]} E_{u^h}(\tau) + C(T_0).$$

This implies that

$$E_{u^h}(s) \le C(T_0) \text{ for all } s \in (-\infty, T_\gamma]. \quad (10.4.64)$$

Therefore after passing to the limit $h \to 0$ we obtain that

$$\|u_{tt}(t)\|^2_\Omega + \|u_t(t)\|^2_{2,\Omega} \le C \text{ for all } t \in (-\infty, T_\gamma]. \quad (10.4.65)$$

By the trace theorem we have that $u_t(t)\big|_{\Gamma_1} \in H^{3/2}(\Gamma_1) \subset C(\Gamma_1)$. Therefore, because $g(s)$ is locally Lipschitz (see (10.4.15)), one can see that $g(u_t(t))\big|_{\Gamma_1} \in H^{1/2}(\Gamma_1)$. Thus it follows from (10.4.1) and (10.4.3) that $u(t) \in H^2_{\Gamma_0}(\Omega)$ is such that $\Delta^2 u(t) \in L_2(\Omega)$ and satisfies relations

$$\frac{\partial}{\partial n}\Delta u(t) + (1-\mu)B_1 u(t) = 0, \quad \frac{\partial}{\partial n}\Delta u(t) + (1-\mu)B_2 u(t) \in H^{1/2}(\Gamma_1)$$

on Γ_1 for $t \le T_\gamma$. Thus by elliptic regularity theory, we have that $\|u(t)\|^2_{4,\Omega} \le C$ for all $t \in (-\infty, T_\gamma]$ and $u(t)$ satisfies the boundary conditions in (10.4.3).

Step 2: Forward propagation of the regularity. By using forward well-posedness of strong solutions stated in Theorem 10.4.3 we claim that $u(t)$ is a strong solution to the original problem and thus the global attractor $\mathscr{A}$ is a subset in the space $W \equiv (H^4 \cap H^2_{\Gamma_0})(\Omega) \times H^2_{\Gamma_0}(\Omega)$.

Step 3: Boundedness of the attractor in $H^4(\Omega) \times H^2(\Omega)$. In the previous step we have shown that $\mathscr{A} \subset W$. However, this does not guarantee the boundedness of $\mathscr{A}$ in W. For this we need an additional argument that exploits the compactness of the attractor. This step follows the argument given in [161].

For every $\tau \in \mathbb{R}$ the element $u_t(\tau)$ belongs to a compact set in $L_2(\Omega)$ that consists of elements from $H^2_{\Gamma_0}(\Omega)$. Therefore for any $\varepsilon > 0$ there exists a finite set $\{\psi_j\} \subset H^2_{\Gamma_0}(\Omega)$ such that we can find indices j_1 and j_2 (which may depend on $u_t(\tau)$ and $u_t(\tau+h)$) such that

$$\|u_t(\tau) - \psi_{j_1}\|_\Omega + \|u_t(\tau+h) - \psi_{j_2}\|_\Omega \le \varepsilon.$$

Let $P_0(z)$ be given by (9.4.35) with the couple $w(t) = u(t+h)$ and $u(t)$, and

$$\begin{aligned} P_{j_1,j_2}(z) = &-(\psi_{j_1}, [u, v(z,z)])_\Omega - (\psi_{j_2}, [w, v(z,z)])_\Omega \\ &(\psi_{j_1} + \psi_{j_2}, [z, v(u + w, z)])_\Omega, \end{aligned}$$

where $z(t) = u(t+h) - u(t) \equiv z^h(t)$. It is clear that

$$|P_0(z) - P_{j_1,j_2}(z)| \le \varepsilon C(R) \|z^h(\tau)\|^2_{2,\Omega}$$

and

$$|P_{j_1,j_2}(z)| \le C_R \left(\|\psi_{j_1}\|_{2,\Omega} + \|\psi_{j_2}\|_{2,\Omega} \right) \|z^h(\tau)\|^2_{2-\eta,\Omega}$$

for some $\eta > 0$. Thus we have that

$$\sup_{j_1,j_2} |P_{j_1,j_2}(z)| \le C_\varepsilon \|z^h(\tau)\|^2_{2-\eta,\Omega} \quad \text{for some } \eta > 0.$$

Consequently, from representation (9.4.34) we obtain that

$$\sup_{t\in[0,T]} \left| \int_{s+t}^{s+T} (\mathscr{R}(z^h), z^h_t)_\Omega d\tau \right| \le \varepsilon \int_s^{s+T} \|z^h(\tau)\|^2_{2,\Omega} d\tau \tag{10.4.66}$$

$$+ b_\varepsilon(T,R) \sup_{\tau\in[0,T]} \|z^h(\tau+s)\|^2_{2-\eta,\Omega}$$

for *all* $s \in \mathbb{R}$ with $\eta > 0$ and arbitrary $T > 0$. Thus we can apply the argument above to prove relation (10.4.63) and hence (10.4.64) for *all* $s \in \mathbb{R}$. This implies that $\mathscr{A}$ is a bounded set in $W = (H^4 \cap H^2_{\Gamma_0})(\Omega) \times H^2_{\Gamma_0}(\Omega)$, as desired. Inequality (10.4.65) implies the additional regularity of u_{tt} on the attractor.

Step 4: Finite dimension. Because the attractor is bounded in W, we can apply the same method as above in order to obtain from Proposition 10.4.15 that there exist constants C_1, C_2, and ω, possibly depending on the damping parameter d and on R such that

$$E_z(t) \le C_1 E_z(0) e^{-\omega t} + C_2 \sup_{0\le\tau\le T} \left\{ \|z(\tau)\|^2_{2-\eta,\Omega} \right\}, \tag{10.4.67}$$

where $z(t) = w(t) - u(t)$ with $(w;w_t)$ and $(u;u_t)$ from the attractor. Indeed, in this case on the attractor we can write the relation

$$\int_s^t (\mathscr{R}(z), z_t)_\Omega d\tau \le \varepsilon \int_s^t E_z(\tau) + C_\varepsilon (1 + |t-s|) \sup_{\tau\in[s,t]} \|z(\tau)\|^2_{2-\eta,\Omega}$$

for $t > s$ and some $\eta > 0$, and use the same argument as above to obtain (10.4.63) valid for $u_h = z$ and for every $s \in \mathbb{R}$. This implies the stabilizability estimate (10.4.67). Thus, using this estimate, by Theorem 7.9.6 we obtain the finiteness of the dimension. This completes the proof of Theorem 10.4.7.

10.4.4 Rate of stabilization: Proof of Theorem 10.4.10

Below we rely on the same idea as in the proof of Theorem 10.2.20 (see also [75] and the references therein).

Preliminaries. If the set $\mathscr{N}$ of equilibria is discrete, then by (10.4.16) we have for any $W = (w_0; w_1) \in H$ there exists an equilibrium point $E = (e;0) \in \mathscr{N}$ such that

$$W(t) = S_t W \to E, \quad t \to \infty, \tag{10.4.68}$$

where the convergence is in the strong topology of H.

Consider new variable

$$Z(t) = (z(t); z_t(t)) \equiv W(t) - E = (w(t) - e; w_t(t)).$$

By (10.4.68) we have that for any $\varepsilon > 0$ there exists $T_0 = T_0(W,E) > 0$ such that

$$\int_{T-1}^T E_0(w(t) - e) dt = \int_{T-1}^T E_0(z(t)) dt \le \varepsilon \tag{10.4.69}$$

for all $T > T_0$, where $E_0(z) = E_z(t) = \frac{1}{2}\|z_t\|^2_\Omega + \frac{1}{2}a(z,z)$. In what follows we take ε sufficiently small, so the only equilibrium in the ε neighborhood is precisely e.

By the definition of equilibrium that new variable $Z(t) = (z(t); z_t(t))$ satisfies the equation

$$
\begin{aligned}
& z_{tt} + d_0 b(z_t) + \Delta^2 z = F(z+e) - F(e) \quad \text{in } Q, \\
& \frac{\partial}{\partial n} z = z = 0 \text{ on } \Sigma_0, \quad \Delta z + (1-\nu)B_1 z = 0 \quad \text{on } \Sigma_1, \\
& \frac{\partial}{\partial n}\Delta z + (1-\nu)B_2 z - \nu_1 z - \beta((z+e)^3 - e^3) = d(x)g(z_t) \quad \text{on } \Sigma_1,
\end{aligned}
\tag{10.4.70}
$$

where we have denoted $F(w) \equiv [v(w)+F_0, w]$. It follows from Theorem 4.2.12 in Chapter 4 that the weak solution to (10.4.70) can be interpreted as a variational solution.

New energy. As in the rotational case (see Section 10.2.4) the key to the proof is the following energy functional,

$$
\mathscr{E}_1(z(t)) \equiv E_0(z(t)) + \Phi(z(t)),
$$

where $\Phi(z)$ is a potential function defined by

$$
\Phi(z) \equiv -\int_0^1 (F(e+zs) - F(e), z)_\Omega ds + \beta \int_0^1 \int_{\Gamma_1} [(e+zs)^3 - (e)^3] z d\Gamma ds.
$$

For this potential Φ relation (10.2.63) holds and, as seen below, the new energy function is dissipative and has properties of the energy function related to the topology of H.

10.4.18. Lemma. *Let z be any generalized solution of* (10.4.70). *Then for any* $0 \le s \le t$ *we have*

$$
\mathscr{E}_1(z(t)) + \widetilde{D}_s^t(z) \le \mathscr{E}_1(z(s)), \tag{10.4.71}
$$

where the damping term $\widetilde{D}_s^t(z)$ corresponds to a weaker form of the energy inequality (10.4.5)*:*

$$
\widetilde{D}_s^t(z) = \int_s^t \left[\int_{\Gamma_1} d\, j_g(z_t) d\Gamma + \int_\Omega d_0 b(z_t) z_t dx\right] dt \tag{10.4.72}
$$

with j_g given by (10.4.6). *Moreover, for strong solutions we have equality in* (10.4.71) *with $j_g(s)$ replaced by $sg(s)$.*

Proof. Similar to the proof of Lemma 10.2.22 we multiply both sides of equation (10.4.70) by z_t and we integrate by parts. Computations are first performed for strong solutions. Passing with the limit on strong solutions and appealing to weak lower semicontinuity of $j_g(s)$ allows us to obtain the desired inequality. □

10.4.19. Proposition. *The energy functional $\mathscr{E}_1(z(t))$ has the following properties:*

- $\mathscr{E}_1(z(t)) \ge 0$ *for all* $t \ge 0$.

- *If* $\|Z(t)\|_H^2 = 2E_0(z(t)) \le 2R^2$ *for* $t \in [0,T]$, *then for all* $t \in [0,T]$ *we have:*

$$|E_0(z(t)) - \mathscr{E}_1(z(t))| \le \varepsilon \|z(t)\|_{2,\Omega}^2 + C(\varepsilon,R)\|z(t)\|_{0,\Omega}^2, \tag{10.4.73}$$

$$E_0(z(t)) \le 2\mathscr{E}_1(z(t)) + C_1\|z(t)\|_{0,\Omega}^2 \le 4E_0(z(t)) + C_2\|z(t)\|_{0,\Omega}^2, \tag{10.4.74}$$

where C_1 *and* C_2 *are constant dependent on* R, $C_2 > C_1 > 0$.

Proof. The argument is the same as for the proof of Lemma 10.2.22. □

Observability Inequality. The key to the proof is the following observability inequality that is related to the inequality given in Lemma 10.4.13.

10.4.20. Lemma. *Assumptions* 10.4.1 *and* 10.4.5 *are in force. Let* z *be a generalized solution to* (10.4.70) *such that* $\sup_{t\in[0,T]} E_0(z(t)) \le R^2$. *Then there is* $T_0 > 0$ *such that for any* $T > T_0$ *we have*

$$\mathscr{E}_1(T) + \int_0^T E_0(z(t))dt \le C_1(T)(I+H_0)\left(\mathscr{E}_1(0) - \mathscr{E}_1(T)\right) + C_2(T,R)lot(z), \tag{10.4.75}$$

where $\mathscr{E}_1(t) = \mathscr{E}_1(z(t))$, H_0 *is the same as* (10.4.20), *and*

$$lot(z) = \sup_{0\le\tau\le T}\left\{\|z(\tau)\|_{2-\eta,\Omega}^2\right\}.$$

Proof. The proof follows the same technical ingredients as used for the proof of Lemma 10.4.13 except for the following points.

- Instead of Proposition 10.4.14 we have more precise estimate

$$\int_0^T E_0(z(t))dt \le c_0\left(E_0(z(T)) + E_0(z(0))\right) + C_1(T)(I+H_0)\left(\widetilde{D}_0^T(z)\right) + C_2(T,R)lot(z) \tag{10.4.76}$$

  for any strong solution, where $\widetilde{D}_0^T(z)$ is given by (10.4.72) with $j_g(s)$ replaced by $sg(s)$. Indeed, by exploiting (10.4.19) and in the same way as in the proof of Lemma 10.2.23 one can see that

$$\int_0^T\int_{\Gamma_1}|z_t|^2 d\Gamma dt \le C_1(T)(I+H_0)\left(\widetilde{D}_0^T(z)\right).$$

  Moreover, we also have with arbitrary $\varepsilon > 0$:

$$\left|\int_{\Gamma_1} dg(z_t)h\nabla z d\Gamma\right| \le \varepsilon\left(\|z_t\|_{\Gamma_1}^2 + \int_{\Gamma_1} dg(z_t)z_t d\Gamma\right) + C_\varepsilon\|\nabla z\|_{L_{p+1}(\Gamma_1)}^2$$

  and similar relation for $\int_\Omega d_0 b(z_t)h\nabla z d\Gamma$. Therefore Proposition 10.4.15 with $w = e$ implies (10.4.76).
- Using the energy relation stated in Lemma 10.4.18 and relations (10.4.74) one can see that

$$T\mathscr{E}_1(T) \le \int_0^T \mathscr{E}_1(t)dt \le 2\int_0^T E_0(z(t))dt + c(T,R)lot(z)$$

and

$$\mathscr{E}_1(0) = \mathscr{E}_1(T) + \int_0^T \left[\int_{\Gamma_1} dz_t g(z_t) d\Gamma + \int_\Omega d_0 b(z_t) z_t dx\right] dt$$

for a strong solution. This makes it possible to derive (10.4.75) from (10.4.76) which is applicable to strong solutions. After the limit transition we obtain (10.4.75) for weak solutions.

We also refer to the proofs of Lemma 3.4 in [63], Theorem 3.12 in [69], and Lemma 4.14 in [75], where similar considerations were used for other models. □

Lower order terms are absorbed. Lemma 10.4.20 reconstructs the energy of solutions in terms of the dissipation and lower order-terms $lot(z)$ that are compact on the phase space. As usual, in dealing with rate decay issues, the goal is to dispense with the lower-order terms. This is typically done by using a version of compactness/uniqueness argument [195] applied first to strong solutions and then extended to weak solutions via the approximation argument. The difficulty in our case is that we consider weak solutions converging to a specific equilibrium point. Thus, "smooth" approximations of solutions may not be stable with respect to that property. As a consequence we can not resort to strong solutions in order to carry out calculations. We are forced to work within the framework of weak solutions only. This is the reason why the energy inequality and variational form of equation established for weak solutions play such important role in allowing to dispense with the lower order terms.

10.4.21. Lemma. *Let z be a generalized (and also weak) solution to* (10.4.70) *such that*

$$\sup_{t\in[0,T]} E_0(z(t)) \le R^2.$$

Moreover we assume that (10.4.69) *holds for a preassigned small ε. Then, there exists a positive constant ε_0 such that*

$$lot(z) \le C(R,T,\varepsilon)(I+H_0)(\mathscr{E}_1(0) - \mathscr{E}_1(T))$$

provided $\varepsilon \le \varepsilon_0$ and $T > T_0 = T_0(W,E)$, where T_0 is the maximum of T_0 in Lemma 10.4.20 *and $T_0 = T_0(E,W)$ such that* (10.4.69) *is in place.*

Proof. As in Lemma 10.2.24 our argument is based on "compactness/uniqueness" argument. The compactness results follow from the fact that lower-order terms are compact. The uniqueness property results from the fact that (i) stationary solutions are locally unique, and (ii) static solutions corresponding to the linearized equation have only trivial solutions. Indeed, the first property follows from the assumption that equilibria are isolated whereas the property (ii) is due to assumed hyperbolicity of equilibria points. Thus the key ingredients of this contradiction argument are that we work in a small neighborhood of the equilibrium point which is hyperbolic and isolated. The technical difficulty is due to the fact that we work within the framework of weak solutions for which the energy inequality and the variational form of the equation must be satisfied. After presenting the general idea of the argument we proceed with the details.

Let $\varepsilon_0 > 0$ be such that for the stationary solution e there is no other stationary solution w such that $E_0(w-e) \leq \varepsilon_0$ and $\varepsilon \leq \varepsilon_0$. We argue by contradiction, assuming that there exists a sequence $z^n(t) = w_n(t) - e$ of generalized solutions to equation (10.4.70) such that $E_0(z^n(t)) \leq R^2$ for all $t \in [0,T]$, the bound in (10.4.69) holds and

$$\frac{lot(z^n)}{(I+H_0)\left(\mathscr{E}_1(z^n(0)) - \mathscr{E}_1(z^n(T))\right)} \longrightarrow \infty \quad \text{when } n \to \infty.$$

By (10.4.71) we have that

$$\frac{lot(z^n)}{(I+H_0)\left(\widetilde{D}_0^T(z^n)\right)} \longrightarrow \infty \quad \text{when } n \to \infty, \tag{10.4.77}$$

where $\widetilde{D}_0^T(z^n)$ is given by (10.4.72). Hence

$$\int_0^T \left[\int_{\Gamma_1} d(x) j_g(z_t^n(t)) d\Gamma + \int_\Omega d_0(x) z_t^n(t) b(z_t^n(t)) dx\right] dt \longrightarrow 0. \tag{10.4.78}$$

We can also assume that

$$(z^n; z_t^n) \to (z; z_t) \quad \text{weakly* in } L_\infty(0,T;H). \tag{10.4.79}$$

Because $H_0(s) = h(s/c_1)$ for some constant $c-1$ (see (10.4.20)), using Assumption 10.4.5 implying $j_g(s) \geq m_0 s^2, |s| > 1$ and property (10.4.19) we obtain

$$\int_0^T \int_{\Gamma_1} d(x) \left|z_t^n(t)\right|^2 d\Gamma dt \longrightarrow 0. \tag{10.4.80}$$

In addition we also have that

$$\int_0^T \int_{\Gamma_1} d(x) \left|g(z_t^n(t))\right| d\Gamma dt \longrightarrow 0 \tag{10.4.81}$$

and

$$\int_0^T \int_\Omega d_0(x) \left|b(z_t^n(t))\right| d\Omega dt \longrightarrow 0. \tag{10.4.82}$$

Indeed, because $|g(s)| \leq C j_g(s)$ for $|s| \geq 1$, using (10.4.19) for $|s| \leq 1$, we obtain that

$$\int_0^T \int_{\Gamma_1} d(x) \left|g(z_t^n(t))\right| d\Gamma dt \leq C\left(\widetilde{D}_0^T(z^n) + \left[H_0\left(\widetilde{D}_0^T(z^n)\right)\right]^{1/2}\right).$$

This along with (10.4.77) implies (10.4.81). As for (10.4.82) we obviously have that $|b(s)| \leq \delta + C_\delta s b(s)$ for every $\delta > 0$. Therefore by (10.4.78) we have that

$$\limsup_{n\to\infty} \int_0^T \int_\Omega d_0(x)\,|b(z_t^n(t))|\,d\Omega dt \le \delta$$

for every $\delta > 0$. This implies (10.4.82).

Now we claim that

$$d_0(x) z_t^n \to 0, \quad \text{a.e. in } (0,T)\times\Omega \tag{10.4.83}$$

along a subsequence. Indeed, because $\|z_t^n\|_\Omega \le C$, we have that

$$\int_Q d_0(x)\,|z_t^n(t)|\,dQ \le C\lambda^{-1} + \int_{A_\lambda} d_0(x)\,|z_t^n(t)|\,dQ,$$

where $Q = (0,T)\times\Omega$ and $A_\lambda = \{(t;x)\in Q, |z_t^n| \le \lambda\}$, for every $\lambda > 0$. It is also easy to see that for any positive λ and δ there exists $C_{\lambda,\delta}$ such that

$$|s| \le \delta + C_{\lambda,\delta} s b(s), \quad |s| \le \lambda.$$

Therefore, it follows from (10.4.78) that

$$\limsup_{n\to\infty} \int_Q d_0(x)\,|z_t^n(t)|\,dQ \le C_1\lambda^{-1} + C_2\delta$$

for any positive λ and δ. This implies (10.4.83). Thus, by the standard weak convergence and compactness argument we can assume that there exists an element $(z;0)$ from $L_\infty(0,T;H)$ such that

$$(z^n; z_t^n) \to (z;0) \ \text{ weakly* in } L_\infty(0,T;H). \tag{10.4.84}$$

By Theorem 1.1.8 this implies that

$$z^n \to z \ \text{ strongly in } C(0,T;H^{2-\eta}(\Omega)) \ \text{ for any } \eta > 0. \tag{10.4.85}$$

Let us prove that $z(t) \equiv 0$. Using (10.4.84) and (10.4.85) and also (10.4.81) and (10.4.82), after passing to the limit (in the sense of distributions) on equation (10.4.70) we infer that the limit function $z(t)$ is a variational solution to

$$\Delta^2 z = F(z+e) - F(e) \ \text{ in } Q_T, \tag{10.4.86}$$

with stationary boundary conditions. It is exactly the point where we have used that weak solutions satisfy a variational form of the equation. By the same argument as in Lemma 10.2.24, the above implies that $w \equiv z+e$ is also a stationary solution such that $E_0(w-e) \le \varepsilon \le \varepsilon_0$. This implies that $w = e$. Hence $z \equiv 0$ in (10.4.84) and (10.4.85).

In order to reach the contradiction we rescale the sequence z_n. Let

$$\hat{z}^n \equiv \frac{z^n}{\alpha_n} \quad \text{and} \quad \alpha_n^2 = lot(z_n) \to 0,$$

where the last conclusion follows from (10.4.85) with $z=0$. We observe that because of (10.4.77), properties of g and b, and property (10.4.19) we have

$$\int_0^T\int_{\Gamma_1} d(x)|\hat{z}_t^n(t)|^2 d\Gamma dt \to 0 \ \text{and}\ \frac{1}{\alpha_n^2}\int_0^T\int_{\Gamma_1} d(x)j_g(z_t^n(t))d\Gamma dt dt \to 0. \tag{10.4.87}$$

Moreover,

$$\frac{1}{\alpha_n^2}(I+H_0)\left(\int_Q d_0(x)z_t^n(t)b(z_t^n(t))dQ\right) \to 0, \tag{10.4.88}$$

and from the observability inequality in Lemma 10.4.20 we also obtain that there exists some $K>0$ such that $E_0(\hat{z}^n(t)) \le K$ for all $t\in[0,T]$. By (10.4.22) we have that

$$\int_Q d_0|\hat{z}_t^n|dQ \le \frac{1}{\alpha_n}\int_{Q\cap\{|z_t^n|<1\}} d_0\left[h(z_t^n b(z_t^n))\right]^{1/2}dQ + \alpha_n\int_{Q\cap\{|z_t^n|\ge 1\}} d_0|\hat{z}_t^n|^2 dQ$$

and, by concavity of $[h(\cdot)]^{1/2}$, we obtain that

$$\int_Q d_0|\hat{z}_t^n|dQ \le \frac{C}{\alpha_n}\left[H_0\left(\int_Q d_0 z_t^n b(z_t^n)dQ\right)\right]^{1/2} + 2\alpha_n\int_0^T E_0(\hat{z}^n(t))dt.$$

This yields that

$$\|d_0\hat{z}_t^n\|_{L_1(Q)} \to 0 \quad \text{as } n\to\infty.$$

Thus by standard weak convergence and compactness arguments we can assume that there exists an element $(\hat{z};0)$ from $L_\infty(0,T;H)$ such that

$$(\hat{z}^n;\hat{z}_t^n) \to (\hat{z};0) \ \text{weakly* in}\ L_\infty(0,T;H), \tag{10.4.89}$$

and, as above, we have that

$$\hat{z}^n \to \hat{z} \ \text{strongly in}\ C(0,T;H^{2-\eta}(\Omega)) \ \text{for any}\ \eta>0. \tag{10.4.90}$$

In order to obtain a differential equation for $\hat{z}$ we use the relation

$$\frac{1}{\alpha_n}\left[F(e+z^n)-F(e)\right] \longrightarrow \langle F'(e);\hat{z}\rangle \ \text{in}\ L_\infty(0,T;H^{-2}(\Omega)), \tag{10.4.91}$$

which is exactly the same as (10.2.79) in the proof of Lemma 10.2.24. One can also see that a property similar to (10.4.91) holds for the boundary nonlinearity.

After dividing both sides of equation (10.4.70) written for z^n by α_n and passing to the limit (here again using the fact that the generalized solution is variational) we infer that the limit function $\hat{z}$ satisfies

$$\Delta^2\hat{z} = \langle F'(e),\hat{z}\rangle \ \text{in}\ Q_T, \tag{10.4.92}$$

in the variational sense with the stationary boundary conditions. Because the equilibrium is assumed hyperbolic, we infer that the only solution to (10.4.92) is a zero solution. Thus $\hat{z} \equiv 0$ in (10.4.89) and (10.4.90), which is impossible because $1 = lot(\hat{z^n}) \to lot(\hat{z}(t)) = 0$. □

Final argument. Because (H, S_t) is dissipative, we have that $\|(z(t); z_t(t))\|_H \le R$ for all $t > 0$ and for some $R > 0$. We choose T (depending on the particular solution) such that (10.4.69) holds with $\varepsilon \le \varepsilon_0$, where ε_0 is given in Lemma 10.4.21.

By combining the observability inequality in Lemma 10.4.20 with the inequality from Lemma 10.4.21 we obtain that

$$\mathscr{E}_1(z(T)) \le C(I + H_0)\left(\mathscr{E}_1(z(0)) - \mathscr{E}_1(z(T))\right).$$

This relation has the same structure as (10.2.81). Therefore, the same argument as in the proof of Theorem 10.2.20 leads to the final conclusion.

Thus the proof of Theorem 10.4.10 is now complete.

10.5 Global attractors for von Karman plates without rotational inertia and with dissipation acting via hinged boundary conditions

We consider von Karman evolutions with nonlinear boundary dissipation acting via hinged boundary conditions. The model does not account a for regularizing effects of rotational inertia. Thus, the corresponding solutions are less regular than in the case of rotational models. In view of this it is expected, as in the previously considered case of nonrotational free boundary conditions, that the analysis of compactness of attractors will have additional hurdles to handle. In addition, due to the loss of compactness, the fact that only one boundary condition is used as a source of dissipation brings about a new set of technical difficulties. This is due to the presence of unbounded traces on the boundary appearing in the observability estimates. These terms need to be absorbed. The difference with the free case is that this time geometric conditions imposed on the domain do not help. Special boundary estimates routed in microlocal analysis are necessary to obtain the sought after absorption.

10.5.1 The model and the main results on the existence of attractors

Let $\Omega \subset \mathbb{R}^2$ be a bounded domain with a sufficiently smooth boundary Γ that consists of two disjoint parts Γ_0 and Γ_1. Consider the following von Karman model with boundary dissipation active on Γ_1 via hinged boundary conditions:

$$u_{tt} + d_0 b(u_t) + \Delta^2 u = [v(u) + F_0, u] + p(x) \text{ in } \Omega \times (0,\infty), \tag{10.5.1}$$

where as above the Airy stress function $v(u)$ satisfies the following elliptic problem

$$\Delta^2 v(u) = -[u,u], \text{ in } \Omega, \quad \frac{\partial}{\partial n} v(u) = v(u) = 0 \text{ on } \Gamma. \tag{10.5.2}$$

The boundary conditions associated with (10.5.1) are clamped on Γ_0 and of hinged type on Γ_1:

$$u = \frac{\partial}{\partial n} u = 0 \text{ on } \Gamma_0, \quad u = 0, \, \Delta u = -g\left(\frac{\partial}{\partial n} u_t\right) \text{ on } \Gamma_1. \tag{10.5.3}$$

10.5.1. Assumption.

- The function $b : \mathbb{R}^1 \mapsto \mathbb{R}^1$ is continuous and there exists $b_0 \geq 0$ such that $b(s) + b_0 s$ is increasing. The function $d_0(x)$ is a nonnegative, bounded measurable function. We also assume that $b(0) = 0$.
- The monotone function $g \in C^1(\mathbb{R})$ is increasing and $g(0) = 0$. Moreover, there exist positive constants m, M, and s_0 such that

$$0 < m \leq g'(s) \leq M \quad \text{for } |s| \geq s_0. \tag{10.5.4}$$

- $F_0(x) \in H^{3+\delta}(\Omega)$ for some $\delta > 0$, $p(x) \in L_2(\Omega)$.

10.5.2. Remark. Some results stated below require less stringent assumptions imposed on the function $g(s)$ than postulated in Assumption 10.5.1. In order not to overcomplicate matters and to focus on technicalities resulting from a reduced number of boundary conditions engaged in stabilization, we have opted for imposing more regularity on the dissipation. ■

We study the dynamics on the state space $H \equiv \left[H^2_{\Gamma_0}(\Omega) \cap H^1_0(\Omega)\right] \times L_2(\Omega)$, where $H^2_{\Gamma_0}(\Omega)$ denotes the space of $H^2(\Omega)$ functions subject to clamped boundary conditions imposed on Γ_0 (see (10.4.4)).

10.5.3. Theorem (Well-posedness). *Let Assumption* 10.5.1 *be in force. Then the equations* (10.5.1) *and* (10.5.2) *with boundary conditions* (10.5.3) *generate a dynamical system* (H, S_t) *on* $H = \left[H^2_{\Gamma_0}(\Omega) \cap H^1_0(\Omega)\right] \times L_2(\Omega)$ *in the class of generalized solutions. This, in particular, implies that the map*

$$(u(0) = u_0; u_t(0) = u_1) \mapsto (u(t); u_t(t))$$

defines a continuous semiflow S_t *on* H.

Moreover, for initial data $u_0 \in H^3(\Omega) \cap H^2_{\Gamma_0}(\Omega)$ *and* $u_1 \in H^2_{\Gamma_0}(\Omega)$ *such that* $\Delta^2 u_0 \in L_2(\Omega)$ *and subject to compatibility conditions on the boundary* Γ_1,

$$u_0 = u_1 = 0, \quad \Delta u_0 = -g\left(\frac{\partial}{\partial n} u_1\right)$$

one obtains that $(u;u_t) \in C_r([0,T];H^3(\Omega)\times H^2_{\Gamma_0}(\Omega))$ *for every finite* $T>0$ *and the boundary conditions in* (10.5.3) *hold. In this latter case solution* $u(t)$ *satisfies the energy relation*

$$\mathscr{E}(t)+\int_s^t\int_{\Gamma_1} g\left(\frac{\partial}{\partial n}u_t\right)\frac{\partial}{\partial n}u_t d\Gamma d\tau+\int_s^t\int_\Omega d_0 b(u_t)u_t dx d\tau=\mathscr{E}(s), \quad (10.5.5)$$

where we denote

$$\mathscr{E}(t)\equiv\mathscr{E}(u(t),u_t(t))=E(t)-\frac{1}{2}\int_\Omega [F_0[u,u]+2pu]\,dx$$

with

$$E(t)\equiv E(u(t),u_t(t))=\frac{1}{2}\int_\Omega\left[u_t^2+|\Delta u|^2+\frac{1}{2}|\Delta v(u)|^2\right]dx.$$

Proof. The result stated above is a special case of Theorem 4.2.4. The only point that may require some explanation is $H^3(\Omega)$ regularity of strong solutions. Indeed, the property $u+G(g((\partial/\partial n)u_t))\in\mathscr{D}(\mathscr{A})$ in (4.2.15), along with the fact that $g(s)$ is globally Lipschitz, translates into the following elliptic regularity problem:

$$\Delta^2 u\in L_2(\Omega),\quad u=0 \text{ on } \Gamma,\quad \Delta u\in H^{1/2}(\Gamma_1),\quad \nabla u=0 \text{ on } \Gamma_0.$$

Elliptic regularity supplies (note that parts of the boundary Γ_0 and Γ_1 are disjoint) that $u\in H^3(\Omega)$, as stated above. □

As in the previous section, it is easily seen (due to the properties of von Karman nonlinearities) that the energies E and $\mathscr{E}$ are related in the following way: there exist positive constants c, C, M_0 such that

$$cE(u,u_t)-M_0\le\mathscr{E}(u,u_t)\le CE(u,u_t)+M_0,\quad (u,u_t)\in H.$$

One can also easily show that under Assumption 10.5.1 any (generalized) solution $u(t)$ satisfies the energy inequality

$$\mathscr{E}(u(t),u_t(t))+\int_s^t\int_\Omega d_0 b(u_t)u_t dx d\tau\le\mathscr{E}(u(s),u_t(s)). \quad (10.5.6)$$

10.5.4. Remark. In fact, the assumption imposed on the damping g in (10.5.4) allows us to justify a stronger version of energy inequality

$$\mathscr{E}(u(t),u_t(t))+\int_s^t\left[\int_{\Gamma_1} g\left(\frac{\partial}{\partial n}u_t\right)\frac{\partial}{\partial n}u_t d\Gamma+\int_\Omega d_0 b(u_t)u_t dx\right]d\tau\le\mathscr{E}(u(s),u_t(s)).$$

If $b(u_t)$ is linearly bounded, then we have even identity in the relation above (see (4.2.14) in the statement of Theorem 4.2.4). ■

Using the energy inequality in (10.5.6) and the relation between energies E and $\mathscr{E}$ one easily shows the following theorem.

10.5.5. Theorem. *Let Assumption* 10.5.1 *hold. Then*

- *There exists $R_* > 0$ such that the set*

$$\mathscr{W}_R = \{y = (u_0; u_1) \in H \,:\, \mathscr{E}(u_0, u_1) \le R\} \tag{10.5.7}$$

  is a nonempty bounded set in H for all $R \ge R_*$. Moreover, any bounded set $B \subset H$ is contained in $\mathscr{W}_R$ for some R and the set $\mathscr{W}_R$ is invariant with respect to the semiflow S_t.
- *If in addition $d_0(x) > 0$ almost everywhere and $sb(s) > 0$ for all $s \neq 0$, then the system (H, S_t) generated by* (10.5.1)–(10.5.3) *is gradient.*

We introduce next the set $\mathscr{N}$ of stationary points of S_t:

$$\mathscr{N} = \{V \in H \,:\, S_t V = V \text{ for all } \ t \ge 0\}\,.$$

Every stationary point W has the form $W = (u; 0)$, where $u = u(x)$ solves the problem

$$\begin{aligned} &\Delta^2 u = [v(u) + F_0, u] + p \ \text{ in } \ \Omega, \\ &u = \frac{\partial}{\partial n} u = 0 \ \text{ on } \ \Gamma_0, \quad u = 0, \ \ \Delta u = 0 \ \text{ on } \ \Gamma_1, \end{aligned} \tag{10.5.8}$$

with function $v(u)$ satisfying (10.5.2).

Our main goal is to prove the global attractiveness property for the dynamical system (H, S_t). This property requires additional hypotheses imposed on the data of the problem.

10.5.6. Assumption.

- We assume that Γ_0 is star-shaped;[6], that is, there exists $x_0 \in \mathbb{R}^2$ such that

$$(x - x_0)n \le 0 \ \text{ on } \Gamma_0.$$

- The function $b(s)$ is nondecreasing and globally Lipschitz;[7] that is,

$$|b(s_1) - b(s_2)| \le M|s_1 - s_2|, \quad s_1, s_2 \in \mathbb{R}.$$

Our main result is the following.

10.5.7. Theorem (Compact attractors). *Let Assumptions* 10.5.1 *and* 10.5.6 *be in force. Then*

- *For any $R \ge R_*$ there exists a global compact attractor $\mathscr{A}_R$ for the restriction $(\mathscr{W}_R, S_t)$ of the dynamical system (H, S_t) on $\mathscr{W}_R$, where $\mathscr{W}_R$ is given by* (10.5.7).

[6] Note that no geometric conditions are imposed on Γ_1. The only geometric requirement is star shaped condition imposed on Γ_0.

[7] This global Lipschitz requirement is due to the fact that only one boundary condition is used as a source of dissipation; see also Remark 10.5.13 below.

- *If we assume additionally that* $d_0(x) > 0$ *a.e. in* Ω *and* $b(s)s > 0$ *for all* $s \neq 0$, *then there is* $R_0 > 0$ *such that* $\mathscr{A}_R$ *does not depend on* R *for all* $R \geq R_0$. *In this case* $\mathscr{A} \equiv \mathscr{A}_{R_0}$ *is a global attractor for* (H, S_t) *and* $\mathscr{A}$ *coincides with the unstable manifold* $\mathscr{M}^u(\mathscr{N})$ *emanating from the set* $\mathscr{N}$ *of stationary points for* S_t. *Moreover,* $\lim_{t\to+\infty} \mathrm{dist}_H(S_t W, \mathscr{N}) = 0$ *for any* $W \in H$.
- *The global attractors* $\mathscr{A}_R$ *and* $\mathscr{A}$ *are bounded sets in* $H^3(\Omega) \times H^2(\Omega)$ *and have a finite fractal dimension provided the relation in* (10.5.4) *holds for all* $s \in \mathbb{R}$.

From Theorem 10.5.7 we obtain the following corollary.

10.5.8. Corollary. *Let the hypotheses of Theorem* 10.5.7 *be in force. Assume that* $d_0(x) > 0$ *a.e. in* Ω *and* $b(s)s > 0$ *for all* $s \neq 0$. *Then the global attractor* $\mathscr{A}$ *consists of full trajectories* $\gamma = \{W(t) : t \in \mathbb{R}\}$ *such that*

$$\lim_{t\to-\infty} \mathrm{dist}_H(W(t), \mathscr{N}) = 0 \quad \textit{and} \quad \lim_{t\to+\infty} \mathrm{dist}_H(W(t), \mathscr{N}) = 0.$$

In particular, if we assume that equation (10.5.8) *has a finite number of solutions, then the global attractor* $\mathscr{A}$ *consists of full trajectories* $\gamma = \{W(t) : t \subset \mathbb{R}\}$ *connecting pairs of stationary points: any* $W \in \mathscr{A}$ *belongs to some full trajectory* γ *and for any* $\gamma \subset \mathscr{A}$ *there exists a pair* $\{Z, Z^*\} \subset \mathscr{N}$ *such that* $W(t) \to Z$ *as* $t \to -\infty$ *and* $W(t) \to Z^*$ *as* $t \to +\infty$. *In the latter case for any* $V \in H$ *there exists a stationary point* Z *such that* $S_t V \to Z$ *as* $t \to +\infty$.

10.5.9. Remark. In analogy with the clamped–free case, one can also provide statements for the rate of stabilizations to equilibria. The statement (and the proofs) do not depend on the specific boundary conditions, therefore we refer the reader to Theorem 10.4.10. However, in this case we do not need to assume the coercivity estimate in (10.4.23). The point is that our basic Assumptions 10.5.1 and 10.5.6 are sufficient to conclude that generalized solutions are weak (variational); see Theorem 4.2.7. ∎

10.5.2 Asymptotic smoothness

10.5.10. Theorem (Asymptotic smoothness). *Let Assumptions* 10.5.1 *and* 10.5.6 *hold. Then the system* $(\mathscr{W}_R, S_t)$ *is asymptotically smooth.*

Proof. As in the clamped–free nonrotational case Theorem 7.1.11 provides the main tool in establishing asymptotic smoothness. The proof is divided into several steps which are presented below. As above, we use notation $Q \equiv \Omega \times (0,T)$, $\Sigma \equiv \Gamma \times (0,T)$ and $\Sigma_i \equiv \Gamma_i \times (0,T)$.

Without loss of generality, we can assume that u and w are strong solutions.

Step 1: Preliminaries—Setting the model. The difference $z \equiv u - w$ solves the following problem

$$z_{tt} + \Delta^2 z = f \ \text{ in } \Omega, \quad z = \frac{\partial}{\partial n} z = 0 \ \text{ on } \Gamma_0, \quad z = 0, \Delta z = \psi \ \text{ on } \Gamma_1, \quad (10.5.9)$$

where $f = -d_0(b(u_t) - b(w_t)) + [v(u) + F_0, u] - [v(w) + F_0, w]$ and

$$\psi = -\left[g\left(\frac{\partial}{\partial n}u_t\right) - g\left(\frac{\partial}{\partial n}w_t\right)\right].$$

We have the following energy equality (satisfied for strong soolutions)

$$E_z(T) + D_t^T(z) = E_z(t) + \int_t^T (\mathscr{R}(z), z_t)_\Omega dt, \tag{10.5.10}$$

where

$$E_z(t) = \frac{1}{2}\int_\Omega [|z_t|^2 + |\Delta z|^2]dx, \quad \mathscr{R}(z) = [v(u) + F_0, u] - [v(w) + F_0, w],$$

and

$$D_t^T(z) = \int_t^T \int_\Omega d_0[b(u_t) - b(w_t)]z_t dx d\tau + \widetilde{D}_t^T(z),$$

with

$$\widetilde{D}_t^T(z) = \int_t^T \int_{\Gamma_1} \left[g\left(\frac{\partial}{\partial n}u_t\right) - g\left(\frac{\partial}{\partial n}w_t\right)\right]\frac{\partial}{\partial n}z_t d\Gamma d\tau.$$

Step 2: First inequality. We start with the following assertion.

10.5.11. Lemma. *Let $T > 0$ be given. Let $\phi \in C^2(\mathbb{R})$ be a given function with support in $[\delta, T-\delta]$, where $\delta \le T/4$, such that $0 \le \phi \le 1$ and $\phi \equiv 1$ on $[2\delta, T-2\delta]$. Then any strong solution z to problem* (10.5.9) *satisfies the following inequality*

$$\int_0^T E_z(t)\phi(t)dt \le C_1\int_0^T E_z(t)|\phi'(t)|dt + C_2\int_{\Sigma_1}\left[|\psi|^2 + \left|\frac{\partial}{\partial n}z\right|^2\right]d\Sigma$$
$$+\int_0^T\int_\Omega fh\nabla z\phi dxdt + C_3 \cdot B_T(z), \tag{10.5.11}$$

where the constants C_i do not depend on T and the boundary terms[8] *$B_T(z)$ is given by (10.3.21):*

$$B_T(z) \equiv \int_{\Sigma_1}\left[\left|\left(\frac{\partial}{\partial n}\right)^2 z\right|^2 + \left|\frac{\partial}{\partial n}\frac{\partial}{\partial \tau}z\right|^2\right]\phi d\Sigma + \int_0^T \left\|\frac{\partial}{\partial n}\Delta z\right\|_{-1,\Gamma_1}^2 \phi dt. \tag{10.5.12}$$

Proof. We apply flux multiplier $h\nabla z\phi(t)$, where $h = x - x_0, x_0 \in \mathbb{R}^2$. Calculations performed in the proof of Lemma 10.3.8 (we need to put there $\alpha = 0$) lead to following relations.

[8] As in Lemma 10.3.8 they are not defined on the energy space. These are higher-order boundary traces of solutions.

Kinetic terms: By (10.3.22) applied with $\alpha = 0$ we have

$$\int_0^T \int_\Omega z_{tt} h\nabla z \phi dxdt \geq \int_0^T \int_\Omega z_t^2 \phi dxdt - C_1 \int_0^T E_z(t)|\phi'(t)|dt. \quad (10.5.13)$$

Potential terms: We recall (10.3.24):

$$\int_Q \Delta^2 z h\nabla z \phi dQ \geq \int_Q |\Delta z|^2 \phi dQ - C_1 \int_{\Sigma_1} \left[|\psi|^2 + \left| \frac{\partial}{\partial n} z \right|^2 \right] \phi d\Sigma - C_2 B_T(z), \quad (10.5.14)$$

where $B_T(z)$ is given by (10.5.12). Using the equation $z_{tt} + \Delta^2 z = f$ and combining (10.5.13) and (10.5.14) we obtain (10.5.11). □

Step 3. Trace estimates. Our next step is to eliminate the second- and third-order boundary traces on the boundary Γ_1 in the expression (10.5.12) for $B_T(z)$. It should be noted that imposing geometric conditions on Γ_1 will not help, unlike the free case, due to the lack of a good structure of boundary terms that cannot be controlled just by the positive sign of (h,n). These second-order "supercritical" boundary traces are due to the fact that the dissipation is allowed to affect the system via only one boundary condition. Traces of the second and third order (see (10.5.12)) are above the energy level, so these cannot enter the estimates. It is the microlocal analysis argument, again, that allows for the elimination of these terms. The idea is similar to that used in the case $\alpha > 0$, however, the difficulty is compounded by the fact that the the source also is critical. The needed trace result follows from the more general trace estimate proved in [151] (see Proposition 1 and Lemma 4) valid for the linear Kirchoff problem (note that the estimates in [151] are independent of the parameter representing rotational forces). Thus, these estimates are applicable to both Kirchoff (10.3.11) and Euler–Bernoulli (10.5.9) models. As a consequence we have the following analogue of Lemma 10.3.9.

10.5.12. Lemma. *Let z be a solution to linear problem* (10.5.9) *with given f and ψ and $B_T(z)$ be given by* (10.5.12). *Then there exist constants $C_T > 0$ such that for any $0 < \eta < \frac{1}{2}$ the following estimate holds*

$$B_T(z) \leq C_{1,T} \int_{\Sigma_1} \left(|\psi|^2 + \left| \frac{\partial}{\partial n} z_t \right|^2 \right) d\Sigma_1$$
$$+ C_{2,T} \left[\|z\|^2_{C([0,T];H^{2-\eta}(\Omega))} + \|z_t\|^2_{L_2([0,T];H^{-\eta}(\Omega))} + \int_0^T \|f\|^2_{-\eta,\Omega} dt \right].$$

It is clear that in our case we have

$$\|f\|^2_{-\eta,\Omega} \leq C\|d_0(b(u_t) - b(w_t))\|^2_\Omega + C_R\|z\|^2_{2-\eta,\Omega}$$

(see (1.4.26) and (1.4.5) in Chapter 1). Therefore

$$\|f\|^2_{-\eta,\Omega} \leq C \int_\Omega d_0(b(u_t) - b(w_t)) z_t dx + C_R\|z\|^2_{2-\eta,\Omega}.$$

10.5.13. Remark. It is at this point where a global Lipschitz condition imposed on $b(s)$ is required. The presently existing trace estimate calls for the estimate $||f||_{-\eta,\Omega}$ with $\eta < 1/2$. This technical difficulty can likely be circumvented by adjusting the trace estimate to the structure of the present nonlinear problem, but at this point in time, this has not been done. ■

Thus applying Lemma 10.5.12 to the estimate in (10.5.12), estimating the integral term with ψ, and referring to (10.5.4) one obtains:

$$\int_0^T E_z(t)\phi(t)dt \le C_1 \int_0^T E_z(t)|\phi'(t)|dt \tag{10.5.15}$$
$$+ C_2(T)\left[D_0^T(z) + \int_{\Sigma_1}\left|\frac{\partial}{\partial n}z_t\right|^2 d\Sigma\right] + C_3(T,R)lot(z),$$

where $lot(z)$ is given by

$$lot(z) = \sup_{0\le\tau\le T}\|z(\tau)\|_{2-\eta,\Omega}^2 + \int_0^T \|z_t(\tau)\|_{-\eta,\Omega}^2 d\tau \tag{10.5.16}$$

for some $\eta > 0$. We emphasize that the constants $C_2(T)$ and $C_3(T,R)$ above depend on the choice of the function ϕ. Instead, the constant C_1 is independent of T and ϕ.

We also note that the term $lot(z)$ is compact with respect to energy topology. This follows from the fact that

$$\|w(t)\|_{2,\Omega}^2 + \|w_t(t)\|_{0,\Omega}^2 \le C_R, \quad \int_0^T \|w_{tt}(t)\|_{-2,\Omega}^2 dt \le C_R(1+T) \tag{10.5.17}$$

for any $(w(t);w_t(t)) \in \mathscr{W}_R$, where $\mathscr{W}_R$ is given by (10.5.7).

Step 4. Observability estimate. Energy relation (10.5.10) implies that

$$E_z(t) = E_z(T) + D_t^T(z) - \int_t^T (\mathscr{R}(z), z_t)_\Omega d\tau.$$

Therefore

$$\int_0^T E_z(t)(1-\phi(t)+|\phi'(t)|)dt \le r_0(\phi)\left[E_z(T) + D_0^T(z)\right] + \mathscr{R}_*(z),$$

where $r_0(\phi) = \int_0^T (1-\phi(t)+|\phi'(t)|)dt$ and

$$\mathscr{R}_*(z) = -\int_0^T (1-\phi(t)+|\phi'(t)|)\int_t^T (\mathscr{R}(z), z_t)_\Omega d\tau dt. \tag{10.5.18}$$

We also have that

$$TE_z(T) \le \int_0^T E_z(t)dt + \mathscr{R}_{**}(z) \quad \text{with} \quad \mathscr{R}_{**}(z) = \int_0^T \int_t^T (\mathscr{R}(z), z_t)_\Omega d\tau dt. \tag{10.5.19}$$

Thus

$$TE_z(T)+\int_0^T E_z(t)dt \le 2\int_0^T E_z(t)dt+\mathscr{R}_{**}(z)$$

$$\le 2\int_0^T E_z(t)\phi(t)dt+2\int_0^T E_z(t)(1-\phi(t))dt+\mathscr{R}_{**}(z)$$

$$\le C_1\int_0^T E_z(t)(1-\phi(t)+|\phi'(t)|)dt+C_2(T)\left[D_0^T(z)+\int_{\Sigma_1}\left|\frac{\partial}{\partial n}z_t\right|^2 d\Sigma\right]$$

$$+\mathscr{R}_{**}(z)+C_3(T,R)lot(z)$$

$$\le C_1 r_0(\phi)E_z(T)+C_2(T)\left[D_0^T(z)+\int_{\Sigma_1}\left|\frac{\partial}{\partial n}z_t\right|^2 d\Sigma\right]$$

$$+\mathscr{R}_*(z)+\mathscr{R}_{**}(z)+C_3(T,R)lot(z).$$

Because we can choose ϕ such that $r_0(\phi)$ does not depend on T, we obtain the following observability estimate.

10.5.14. Proposition (Observability estimate). *Assume that Assumptions* 10.5.1 *and* 10.5.6 *are in force. Let* $U(t)=(u(t);u_t(t))=S_t y_1$ *and* $W(t)=(w(t);w_t(t))=S_t y_2$ *be two solutions corresponding to initial conditions* y_1 *and* y_2 *from the set* $\mathscr{W}_R$ *given by (10.5.7). Then, there exist* $T_0>0$ *and constants* $C_1(T)$ *and* $C_2(R,T)$ *such that*

$$TE_z(T)+\int_0^T E_z(t)dt \le C_1(T)\left[D_0^T(z)+\int_{\Sigma_1}\left|\frac{\partial}{\partial n}z_t\right|^2 d\Sigma\right]$$

$$+\mathscr{R}_*(z)+\mathscr{R}_{**}(z)+C_2(T,R)lot(z), \quad (10.5.20)$$

for any $T\ge T_0$, *where* $z\equiv u-w$, $\mathscr{R}_*(z)$ *and* $\mathscr{R}_{**}(z)$ *are given by* (10.5.18) *and* (10.5.19), *the low-order term* $lot(z)$ *has the form* (10.5.16), *and*

$$E_z(t)=\frac{1}{2}\int_\Omega [|z_t|^2+|\Delta z|^2]dx.$$

Step 5. Final arguments. Relation (10.5.4) implies

$$\int_{\Sigma_1}\left|\frac{\partial}{\partial n}z_t\right|^2 d\Sigma_1 \le \varepsilon + C_\varepsilon\int_{\Sigma_1}\left[g\left(\frac{\partial}{\partial n}u_t\right)-g\left(\frac{\partial}{\partial n}w_t\right)\right]\frac{\partial}{\partial n}z_t d\Sigma \quad (10.5.21)$$

for any $\varepsilon>0$; see Proposition B.1.2 in Appendix B. Therefore by the same argument as in the case of Theorem 10.2.13 using (10.5.20) and the relation

$$D_0^T(z)\le E_z(0)-E_z(T)+\int_0^T(\mathscr{R}(z),z_t)_\Omega dt,$$

we conclude that

$$E_z(T) \le \varepsilon + C_\varepsilon(E_z(0) - E_z(T)) + C_\varepsilon \int_0^T (\mathscr{R}(z), z_t)_\Omega dt + \mathscr{R}_*(z) + \mathscr{R}_{**}(z) + C_2(\varepsilon, R, T) \cdot lot(z),$$

where $\mathscr{R}_*(z)$ and $\mathscr{R}_{**}(z)$ are defined in (10.5.18) and (10.5.19). This implies that

$$E_z(mT) \le \frac{\varepsilon}{1 + C_{\varepsilon,R}} + \gamma_\varepsilon E_z((m-1)T) + \mathscr{Q}_m(z)$$

for every $m = 1, 2, \ldots$, where $\gamma_\varepsilon = C_{\varepsilon,R}(1 + C_{\varepsilon,R})^{-1} < 1$ (it may also depend on R and T) and

$$\begin{aligned} \mathscr{Q}_m(z) = C_\varepsilon \int_{(m-1)T}^{mT} (\mathscr{R}(z), z_t)_\Omega dt + \int_{(m-1)T}^{mT} \int_t^{mT} (\mathscr{R}(z), z_t)_\Omega d\tau dt \\ - \int_{(m-1)T}^{mT} (1 - \phi(t) + |\phi'(t)|) \int_t^{mT} (\mathscr{R}(z), z_t)_\Omega d\tau dt \\ + C_\varepsilon(R, T) \left[\sup_{(m-1)T \le \tau \le mT} \|z(\tau)\|^2_{2-\eta,\Omega} + \int_{(m-1)T}^{mT} \|z_t(\tau)\|^2_{-\eta,\Omega} d\tau \right]. \end{aligned}$$

After iterations we obtain

$$E_z(mT) \le \varepsilon + \gamma_\varepsilon^m E_z(0) + \sum_{k=0}^{m-1} \gamma_\varepsilon^k \mathscr{Q}_{m-k}(z), \quad m = 1, 2 \ldots \tag{10.5.22}$$

Selecting m large enough, so that $\varepsilon + \gamma_\varepsilon^m E_z(0) \le \varepsilon_0$ and setting

$$\Psi_{\varepsilon_0, B, mT}(y_1, y_2) \equiv \sum_{k=0}^{m-1} \gamma_\varepsilon^k \mathscr{Q}_{m-k}(z), \; y_1 = U(0), \; y_2 = W(0),$$

we are in a position to apply the conclusion of Theorem 7.1.11. Indeed, for this it suffices to notice that the terms in $\sum_{k=0}^{m-1} \gamma_\varepsilon^k \mathscr{Q}_{m-k}(z)$, which involve critical terms $(\mathscr{R}(z), z_t)_\Omega$, can be shown to satisfy the compensated compactness condition (7.1.13). The argument for this is identical to that given in (10.4.51)–(10.4.53) for the free case, hence it is not repeated. Other terms in this expression are subcritical. They are compact due to the relations in (10.5.17). Therefore, Theorem 7.1.11 provides asymptotic smoothness of semiflow S_t which is stated in Theorem 10.5.10. □

10.5.3 Proof of the main result (Theorem 10.5.7)

1. The existence of the attractors follows from Theorems 10.5.5 and 10.5.10. The argument is the same as in Section 10.4.3.
2. To prove the smoothness and finite dimension of the attractor we need a stronger version of the observability estimate in Proposition 10.5.14. This is possible to

achieve due to the assumption $g'(s) \geq m > 0$. The corresponding result is formulated below.

10.5.15. Proposition (Observability estimate revisited). *Let Assumption* 10.5.1 *be in force. Assume that Assumption* 10.5.6 *holds with relation* (10.5.4) *valid for all* $s \in \mathbb{R}$. *Let*

$$U(t) = (u(t); u_t(t)) = S_t y_1 \;\; \text{and} \;\; W(t) = (w(t); w_t(t)) = S_t y_2$$

be two solutions corresponding to initial conditions y_1 *and* y_2 *from the set* $\mathscr{W}_R$ *given by* (10.5.7). *Then there exists* $T_0 > 0$ *and constants* $C_1(T)$ *and* $C_2(R,T)$ *such that with any* $T > 0$, $s < T$ *we have*

$$TE_z(T) + \int_s^T E_z(t)dt \leq C_1(T) D_s^T(z) \qquad (10.5.23)$$

$$+ \mathscr{R}_*(z) + \mathscr{R}_{**}(z) + C_2(T,R) lot_s(z)$$

for any $T - s \geq T_0$, *where* $z \equiv u - w$, $E_z(t) = \frac{1}{2}\int_\Omega [|z_t|^2 + |\Delta z|^2] dx$, $\mathscr{R}_*(z)$ *and* $\mathscr{R}_{**}(z)$ *are given by the same formulas as in* (10.5.18) *and* (10.5.19), *with the only difference that the integration starts at the point s (not zero). The lower-order term* $lot(z)$ *has the form* (10.5.16), *with initialization at s.*

Proof. The proof follows from Proposition 10.5.14 written for interval $[s,T]$ and using the relation

$$\int_s^T \int_{\Gamma_1} |\frac{\partial}{\partial n} z_t|^2 d\Sigma_1 \leq D_s^T(z)$$

which is valid due to the additional hypothesis in Proposition 10.5.15. □

Below we also need the following assertion.

10.5.16. Lemma. *Let the hypotheses of Proposition* 10.5.14 *be in force. Then*

$$\int_0^T \|z_{tt}(t)\|_{-2,\Omega}^2 dt \leq C_R \int_0^T E_z(t)dt \leq C_R T \sup_{t\in[0,T]} E_z(t). \qquad (10.5.24)$$

Proof. From equation (10.5.9) we have that $(z_{tt}, \phi) + (\Delta z, \Delta\phi) = (f, \phi)$ for every $\phi \in H_0^2(\Omega)$. Because $b(s)$ is globally Lipschitz, we obtain

$$|(z_{tt}, \phi)_\Omega| \leq C_R \left[\|\Delta z\|_\Omega + \|z_t\|_\Omega\right] \|\phi\|_{2,\Omega}.$$

which implies relation (10.5.24). □

3. As in the case of the clamped–free case to obtain finite-dimensionality of the attractor, we first prove additional regularity of the attractor. We use the same procedure as in Section 10.4.3 (see also [161] and [73, 75]) and follow the same line of argument as in the case of free boundary conditions.

Step 1: Smoothness on negative time scale. Let $\gamma = \{(u(t); u_t(t)) : t \in \mathbb{R}\}$ be a trajectory from the global attractor $\mathscr{A}$. Let $0 < h < 1$. We estimate the energy E_z of

$z(t) := z^h(t) = u(t+h) - u(t)$. It is clear that

$$\|u(t)\|_{2,\Omega}^2 + \|u_t(t)\|_\Omega^2 \le R^2, \quad t \in \mathbb{R}. \tag{10.5.25}$$

Therefore using the corresponding analogue of Lemma 10.4.17 (see also relation (9.4.40) and Remark 9.4.11) we obtain:

$$\left| \int_s^t (\mathscr{R}(z^h), z_t^h)_\Omega d\tau \right| \le C_1(R) \sup_{s\le\tau\le t} \|z^h(\tau)\|_{2-\eta,\Omega}^2 \tag{10.5.26}$$
$$+ C_2(R) \int_s^t (\|u_t(\tau+h)\|_\Omega + \|u_t(\tau)\|_\Omega) \|z^h(\tau)\|_{2,\Omega}^2 d\tau$$

for all $-\infty < s \le t < +\infty$. Similar estimates are valid for R_* and R_{**}. We note that the fact that the integral in (10.5.26) contains velocity terms is critical. These terms are "small" near $-\infty$. Thus, the integrals with critical powers of Sobolev's norms can be absorbed for $t \to -\infty$.

Because $\mathscr{A} = \mathscr{M}^u(\mathscr{N})$, where $\mathscr{N}$ is the set of equilibria, $\mathscr{N} = \{(v;0) : v \in \mathscr{N}_*\}$, we have that for any $\varepsilon > 0$ there exists T_γ^ε (independent of h, but depending on the trajectory γ) such that

$$||u_t(\tau)||_\Omega + ||u_t(\tau+h)||_\Omega \le \varepsilon \cdot [C_2(R)]^{-1} \quad \text{for any} \;\; t \le T_\gamma^\varepsilon.$$

Therefore from (10.5.26) we have that

$$\left| \int_s^t (\mathscr{R}(z^h), z_t^h)_\Omega d\tau \right| \le C_1(R) \sup_{s\le\tau\le t} \|z^h(\tau)\|_{2-\eta,\Omega}^2 + \varepsilon \int_s^t \|z^h(\tau)\|_{2,\Omega}^2 d\tau \tag{10.5.27}$$

for all $-\infty < s \le t \le T_\gamma^\varepsilon$. Thus using the energy relation (10.5.10) we find that

$$|E_z(t) + D_s^t(z) - E_z(s)| \le \varepsilon \int_s^t \|z\|_{2,\Omega}^2 ds + C(R) \max_{[s,t]} \|z\|_{2-\eta,\Omega}^2. \tag{10.5.28}$$

for all $s < t \le T_\gamma^\varepsilon$, where $\eta > 0$. Now, we are in a position to apply on each subinterval $[s, s+T_0]$ the estimate of Proposition 10.5.15 along with (10.5.27). This procedure supported by application of (10.5.28) allows us to choose a constant $0 < \sigma < 1$ such that $u^h(t) = h^{-1} z^h(t) = h^{-1}[u(t+h) - u(t)]$ satisfies the following estimate

$$E_{u^h}(s+T_0) \le \sigma E_{u^h}(s) + C_{T_0} \left[\sup_{0\le\tau\le T_0} \|u^h(s+\tau)\|_{2-\eta,\Omega}^2 + \int_0^{T_0} \|u_t^h(s+\tau)\|_{-\eta,\Omega}^2 d\tau \right] \tag{10.5.29}$$

for all $s \le T_\gamma - T_0$, where $\eta > 0$, $T_\gamma = T_\gamma^{\varepsilon_0}$ (depending on the trajectory, but not h) for some $\varepsilon_0 > 0$ and $T_0 > 0$. The standard interpolation argument leads to the relations

$$\|u^h(s+\tau)\|_{2-\eta,\Omega}^2 \le \delta \|u^h(s+\tau)\|_{2,\Omega}^2 + C_{\delta,\eta} \|u^h(s+\tau)\|_\Omega^2$$

and

$$\int_0^{T_0} \|u_t^h(s+\tau)\|_{-\eta,\Omega}^2 d\tau \le \delta \sup_{0\le\tau\le T_0} \|u_t^h(s+\tau)\|_\Omega^2 + C_{\delta,\eta,T_0} \int_0^{T_0} \|u_t^h(s+\tau)\|_{-2,\Omega}^2 d\tau$$

for any $\delta > 0$. Therefore, using (10.5.25) and Lemma 10.5.16 we obtain that

$$E_{u^h}(s+T_0) \le \sigma E_{u^h}(s) + \frac{1-\sigma}{2} \sup_{0\le\tau\le T_0} E_{u^h}(s+\tau) + C_\sigma(T_0)$$

for all $s \le T_\gamma - T_0$. Taking the supremum over the interval $(-\infty, T_\gamma - T_0]$ gives

$$\sup_{\tau\in(-\infty,T_\gamma]} E_{u^h}(\tau) \le \frac{1+\sigma}{2} \sup_{\tau\in(-\infty,T_\gamma]} E_{u^h}(\tau) + C(T_0).$$

This implies that

$$E_{u^h}(s) \le C(T_0) \quad \text{for all } s \in (-\infty, T_\gamma]. \tag{10.5.30}$$

Therefore after passing to the limit $h \to 0$ we obtain that

$$\|u_{tt}(t)\|_\Omega^2 + \|u_t(t)\|_{2,\Omega}^2 \le C \quad \text{for all } t \in (-\infty, T_\gamma].$$

By (10.5.1) this implies that $u(t)$ is strong solution on each interval $[s_0, T_\gamma]$ such that $\|u(t)\|_{3,\Omega}^2 \le C$ for all $t \in (-\infty, T_\gamma]$ and the boundary conditions in (10.5.3) are satisfied.

Step 2: Forward propagation of the regularity. By using forward well-posedness of strong solutions stated in Theorem 10.5.3 we claim that $u(t)$ is a strong solution to the original problem on the whole real line and thus the global attractor $\mathscr{A}$ is a subset in the space $W = \left[H^3(\Omega) \cap H^2_{\Gamma_0}(\Omega) \cap H^1_0(\Omega)\right] \times \left[H^2_{\Gamma_0}(\Omega) \cap H^1_0(\Omega)\right]$.

Step 3: Boundedness of the attractor in $H^3(\Omega) \times H^2(\Omega)$. To guarantee the boundedness of $\mathscr{A}$ in W we need to repeat the corresponding argument from the proof of this property in Theorem 10.4.7 to obtain from representation (9.4.34) in Lemma 9.4.10 the following relation

$$\sup_{t\in[0,T]} \left| \int_{s+t}^{s+T} (\mathscr{R}(z^h), z_t^h) d\tau \right| \le \varepsilon \int_s^{s+T} \|z^h(\tau)\|_{2,\Omega}^2 d\tau + b_\varepsilon(T,R) \sup_{\tau\in[0,T]} \|z^h(\tau+s)\|_{2-\eta,\Omega}^2 \tag{10.5.31}$$

for *all* $s \in \mathbb{R}$ and $\varepsilon > 0$ with $\eta > 0$ and arbitrary $T > 0$. Thus we can apply the argument above to prove relation (10.5.29) and hence (10.5.30) for *all* $s \in \mathbb{R}$. This implies that $\mathscr{A}$ is a bounded set in $W \subset H^3(\Omega) \times H^2(\Omega)$.

Step 4: Finite dimension. Because the attractor is bounded in W, using Proposition 10.5.15 for any two trajectories $U(t) = (u(t); u_t(t)) = S_t y_1$ and $W(t) = (w(t); w_t(t)) = S_t y_2$ from the attractor and also the representation in (9.4.34) for $\mathscr{R}_0(z)$ we obtain that

$$TE_z(T) + \int_0^T E_z(t)dt \le C_1(T)D_0^T(z) + C_2(T,R)lot(z)$$

for any $T \ge T_0$, where $z \equiv u - w$ and the lower-order term $lot(z)$ has the form (10.5.16). By (10.5.10) this implies that

$$E_z(T) \le \gamma_T E_z(0) + C_2(T)\left[\sup_{0\le\tau\le T} \|z(\tau)\|^2_{2-\eta,\Omega} + \int_0^T \|z_t(\tau)\|^2_{-\eta,\Omega} d\tau\right],$$

where $\gamma_T < 1$ and $\eta > 0$. Therefore by the same argument as in the proof of Theorem 10.2.13 (see also the argument given in Theorem 8.5.3 and the comments given in the proof of Proposition 10.3.10) we obtain the stabilizability estimate on the attractor $\mathscr{A}$ of the form

$$\|S_t y_1 - S_t y_2\|^2_H \le C_1 e^{-\omega t}\|y_1 - y_2\|^2_H + C_2 \max_{[0,t]} \|z(\tau)\|^2_{2-\eta,\Omega} + C_3 \int_0^t e^{-\omega(t-\tau)}\|z_t(\tau)\|^2_{-\eta,\Omega} d\tau, \quad (10.5.32)$$

where C_i and ω are positive numbers and $S_t y_1 = (u(t); u_t(t))$ and $S_t y_2 = (w(t); w_t(t))$ are from the attractor $\mathscr{A}$, $z(t) = u(t) - w(t)$.

Using the estimate (10.5.32) and Lemma 10.5.16 we can apply the same method of short trajectories as in the proof of finite-dimensionality in Theorem 10.2.11 by introducing the set

$$A_T \equiv \{U \equiv (u(0); u_t(0); u(t), t \in [0,T]) : (u(0); u_t(0)) \in \mathscr{A}\}$$

and the translation mapping $V : (u(0); u_t(0); u(t)) \to (u(T); u_t(T); u(T+t))$ in the space $X = \left[H^2_{\Gamma_0}(\Omega) \cap H^1_0(\Omega)\right] \times L_2(\Omega) \times W_2(0,T)$, where the space $W_2(0,T)$ is defined as a set of functions $z(t)$ from $L_2(0,T; (H^2_{\Gamma_0} \cap H^1_0)(\Omega))$ which distributional derivatives z_t and z_{tt} possess the properties

$$z_t \in L_2(0,T; L_2(\Omega)) \quad \text{and} \quad z_{tt} \in L_2(0,T; [H^2_{\Gamma_0}(\Omega)]').$$

This completes the proof of Theorem 10.5.7. □

Chapter 11
Thermoelasticity

In this chapter we undertake a study of asymptotic behavior of solutions associated with von Karman evolution equations subject to thermal dissipation. We consider models with and without the rotational inertia terms. The presence of this parameter changes the character of the thermoelastic dynamics from parabolic-like to hyperbolic-like. We establish the existence of a compact global attractor of finite fractal dimension and study its properties. Our main tool is the stabilizability estimate which asserts that a difference of any two trajectories can be exponentially stabilized to zero modulo compact perturbation. This estimate is independent of the value of rotational inertia parameter α and heat/thermal capacity κ. This makes it possible to obtain estimates for the dimension and size of the attractor which do not depend on the parameters α and κ. We also prove upper semicontinuity of the attractor with respect to the parameters α and κ. In the case $\kappa \to 0$ we show that the attractor is close in some sense to the attractor of an isothermal structurally damped von Karman model. In this chapter we mainly follow [76].

11.1 Introduction

Let Ω be a bounded domain in $\mathbb{R}^2$ with the boundary $\partial\Omega = \Gamma$, Δ denotes the Laplace operator, and F_0 and p are given functions, $F_0 \in W^2_\infty(\Omega)$ and $p \in L_2(\Omega)$. We consider a thermoelastic von Karman plate subjected to an external and internal forcing. The corresponding equations (see Chapter 5 and the references quoted therein) have the following form

$$\begin{cases} u_{tt} - \alpha\Delta u_{tt} + \mu\Delta\theta + \Delta^2 u - [u, v+F_0] = p(x), \quad x \in \Omega, t > 0, \\ \kappa\theta_t - \eta\Delta\theta - \mu\Delta u_t = 0, \quad x \in \Omega, t > 0, \end{cases} \tag{11.1.1}$$

where the Airy stress function $v = v(u)$ is a solution to the problem

I. Chueshov, I. Lasiecka, *Von Karman Evolution Equations*,
Springer Monographs in Mathematics, DOI 10.1007/978-0-387-87712-9_11,

$$\Delta^2 v + [u,u] = 0, \quad v|_{\partial\Omega} = \frac{\partial v}{\partial n}\Big|_{\partial\Omega} = 0. \tag{11.1.2}$$

The temperature θ satisfies the Dirichlet boundary condition of the form:

$$\theta = 0 \quad \text{on } \Gamma. \tag{11.1.3}$$

The boundary conditions imposed on the displacement u are either clamped:

$$u = \frac{\partial}{\partial n} u = 0 \quad \text{on } \Gamma, \tag{11.1.4}$$

where n is the outer normal vector, or else hinged (simply supported):

$$u = \Delta u = 0 \quad \text{on } \Gamma. \tag{11.1.5}$$

The parameters μ and η are positive and α and κ are nonnegative. The case $\alpha > 0$ corresponds to taking into account rotational inertia of filaments of the plate. The parameter κ has meaning of heat/thermal capacity. In the case $\kappa = 0$ the first equation in (11.1.1) becomes a model for structurally damped plate with the damping term $-\mu^2\eta^{-1}\Delta u_t$.

We are interested in long-time behavior of the thermal von Karman evolutions, with particular emphasis being placed on dependence of long-time characteristics with respect to varying parameters $0 \le \alpha \le m_\alpha$ and $0 \le \kappa \le m_\kappa$ for some (fixed) positive constants m_α and m_κ (below, as in Chapter 5 we take $m_\alpha = m_\kappa = 1$ for simplicity). This includes questions such as: *(i) existence of a compact global attractor and its structure, (ii) smoothness and finite-dimensionality of the attractor, (iii) uniform decay rates to equilibria, and (iv) upper semicontinuity (with respect to the parameters α and κ) of attractors.* We also provide conditions under which the said attractor becomes exponential.

In order to point out timeliness of the topic under consideration, we wish to note that the issue of uniform decay rates for linear, unforced, thermoelastic plates has been settled only recently. Indeed, results of the previous literature required an addition of mechanical damping (boundary or interior), in order to force exponential decay rates for the energy function, see [174] and references therein. Instead, recent progress in the area of control theory and inverse problems, [6, 9, 20, 21, 132, 163, 173, 186, 216, 224] has provided a stimulus to the field and produced an array of results on controllability, analyticity (when $\alpha = 0$), and uniform stability *without* any mechanical dissipation. In fact, it was shown in [8] that not only linear thermoelastic plates with either hinged or clamped boundary conditions are exponentially stable without any mechanical dissipation, but also that the decay rates are independent on the values of rotational parameter $\alpha \ge 0$ (see also Proposition 5.3.1). All these results, and the techniques developed in that context (see also [9] for the case of free boundary conditions), bear critically on the analysis of a much more complex problem such as long-time behavior of nonlinear thermoelastic plates with a forcing.

In order to gain a better understanding of the problem under consideration, one should note that topological behavior of the model is strongly dependent on the parameters $\alpha \geq 0$ and also $\kappa \geq 0$. It is by now well known that the parameter α changes the linear dynamics drastically from analytic $\alpha = 0$ to hyperbolic-like $\alpha > 0$ (see [213, 216] and also Section 5.3.2). This implies that the flow has additional regularity for the limit case $\alpha = 0$, and these properties completely disappear when $\alpha > 0$. Our main challenge is to characterize long-time behavior of the thermal plates, uniformly with respect to the values of the parameter $\alpha \geq 0$ and $\kappa \geq 0$. This includes: (i) seeking an upper bound for dimensionality of attractors that are uniform in α and κ, (ii) seeking a uniform measure of regularity enjoyed by trajectories evolving on the attractor, and (iii) establishing upper semicontinuity with respect to α and κ of the attractors. For each value of the parameter ($\alpha > 0$ and $\alpha = 0$) one may be able to use specifics of each dynamics in order to obtain the necessary estimates, however these estimates will be typically blowing up when $\alpha \to 0$. The reason for the blow-up is that the nonlinear term in the model becomes critical when $\alpha \to 0$. On the other hand, when $\alpha = 0$, the linear part of the model generates an analytic semigroup, which property induces some smoothing effect on nonlinearity. However, this effect disappears as soon as $\alpha > 0$. This dichotomy raises the interesting question of whether it is possible to obtain the estimates that are uniform with respect to the parameters $\alpha \geq 0$ (and $\kappa \geq 0$). We already know that this is possible in the case of linear models with zero equilibrium. However, addition of nonlinear and noncompact dynamics changes the picture entirely. To cope with this problem, we first establish the so-called stabilizability estimate (see Theorem 11.3.1) with the constants independent of α and κ. In particular this means that the system under consideration is quasi-stable. Thus we can apply ideas and results from Section 7.9.

As in Chapter 5 we use the following notations. In the space $\mathscr{H} = L_2(\Omega)$ we define the operator A_D as the negative Laplacian equipped with zero Dirichlet boundary data:

$$A_D u = -\Delta u, \quad u \in \mathscr{D}(A_D) = H^2(\Omega) \cap H_0^1(\Omega)$$

and consider the operator $M_\alpha = I + \alpha A_D$. Both operators A_D and M_α are positive self-adjoint operators in $\mathscr{H}$. We also introduce the biharmonic operator

$$\mathscr{A} u = \Delta^2 u, \quad u \in \mathscr{D}(\mathscr{A}) = \begin{cases} H^4(\Omega) \cap H_0^2(\Omega), & \text{(clamped b.c.);} \\ \{u \in H^4(\Omega) : u = \Delta u = 0 \text{ on } \Gamma\}, & \text{(hinged b.c.)} \end{cases}$$

and nonlinear mapping $F(\cdot)$ by the formula

$$F(u) = [u, v(u) + F_0] + p(x), \quad u \in H^2(\Omega), \tag{11.1.6}$$

where $v(u) \in H_0^2(\Omega)$ is determined by u via (11.1.2).

With the above notation, equations in (11.1.1) with the boundary conditions considered can be written in the form

$$\begin{cases} M_\alpha u_{tt} - \mu A_D\theta + \mathscr{A}u = F(u), \\ \kappa\theta_t + \eta A_D\theta + \mu A_D u_t = 0. \end{cases} \tag{11.1.7}$$

We equip (11.1.7) with initial data

$$u|_{t=0} = u_0,\ u_t|_{t=0} = u_1, \quad \theta|_{t=0} = \theta_0. \tag{11.1.8}$$

In the case $\kappa = 0$ the problem has the form

$$M_\alpha u_{tt} + \mu^2\eta^{-1}A_D u_t + \mathscr{A}u = F(u), \quad u|_{t=0} = u_0,\ u_t|_{t=0} = u_1. \tag{11.1.9}$$

As in Chapter 5 for every couple $(\alpha;\kappa) \in \Lambda \equiv \{0 \le \alpha \le 1, 0 \le \kappa \le 1\}$ we introduce a family of phase (energy) spaces $H_{\alpha,\kappa}$ as Hilbert spaces of the following structure:

$$H_{\alpha,\kappa} = \begin{cases} \mathscr{D}(\mathscr{A}^{1/2}) \times V_\alpha \times \mathscr{H}, & \text{for } \kappa > 0; \\ \mathscr{D}(\mathscr{A}^{1/2}) \times V_\alpha, & \text{for } \kappa = 0, \end{cases} \tag{11.1.10}$$

where $\mathscr{H} = L_2(\Omega)$, $\mathscr{D}(\mathscr{A}^{1/2}) = H_0^2(\Omega)$ in the clamped case and $\mathscr{D}(\mathscr{A}^{1/2}) = \mathscr{D}(A_D) = (H^2 \cap H_0^1)(\Omega)$ in the hinged case, and $V_\alpha \equiv D(M_\alpha^{1/2})$ which is $H_0^1(\Omega)$ for $\alpha > 0$ and $L_2(\Omega)$ for $\alpha = 0$. We equip the space $H_{\alpha,\kappa}$ with the graph norm

$$|U|^2_{\alpha,\kappa} = \|\mathscr{A}^{1/2}u_0\|^2 + \|M_\alpha{}^{1/2}u_1\|^2 + \kappa\|\theta\|^2, \quad U = (u_0;u_1;\theta),$$

for $(\alpha;\kappa) \in \Lambda$ (in the case $\kappa = 0$ the last term should be omitted), where we denote $\|\cdot\| \equiv \|\cdot\|_{L_2(\Omega)}$. In both cases of boundary conditions we have that $\mathscr{D}(\mathscr{A}^{1/2}) \subseteq \mathscr{D}(A_D)$ and $\|\mathscr{A}^{1/2}u_0\| = \|A_D u_0\|$ for $u \in \mathscr{D}(\mathscr{A}^{1/2})$. We use this observation below dealing with the norm $|\cdot|^2_{\alpha,\kappa}$.

By Theorem 5.4.1 the solutions to problems (11.1.7) and (11.1.9) generate a family of dynamical systems with the phase spaces $H_{\alpha,\kappa}$ given by (11.1.10). The evolution operator $S_t^{\alpha,\kappa}$ for $\kappa > 0$ is given by the formula $S_t^{\alpha,\kappa}(u_0;u_1;\theta_0) = (u(t);u_t(t);\theta(t))$, where $u(t)$ and $\theta(t)$ solve (11.1.7) with initial data (11.1.8), and for $\kappa = 0$ is defined by the relation $S_t^{\alpha,0}(u_0;u_1) = (u(t);u_t(t))$, where $u(t)$ is a solution to (11.1.9). Thus in all cases considered we have well-defined semiflow on the space $H_{\alpha,\kappa}$.

We also note (see Theorem 5.4.1) that $S_t^{\alpha,\kappa}(u_0;u_1;\theta_0) = (u(t);u_t(t);\theta(t))$ satisfies the energy balance equality

$$\mathscr{E}_{\alpha,\kappa}(u(t),u_t(t),\theta(t)) + \eta\int_s^t \|A_D^{1/2}\theta(\tau)\|^2 d\tau = \mathscr{E}_{\alpha,\kappa}(u(s),u_t(s),\theta(s)) \tag{11.1.11}$$

for all $t \ge s \ge 0$, where $\mathscr{E}_{\alpha,\kappa}(u,u_t,\theta)$ is the energy functional for the model that is given by the formula

$$\mathscr{E}_{\alpha,\kappa}(u,u_t,\theta) = E_{\alpha,\kappa}(u,u_t,\theta) - \frac{1}{2}\int_\Omega([F_0,u]u + 2pu)dx \tag{11.1.12}$$

with

$$E_{\alpha,\kappa}(u,u_t,\theta) = \frac{1}{2}\left[||A_D u||^2 + ||M_\alpha{}^{1/2}u_t||^2 + \frac{1}{2}||\Delta v(u)||^2 + \kappa||\theta||^2\right]. \quad (11.1.13)$$

By (11.1.11) the full energy $\mathscr{E}_{\alpha,\kappa}$ is nonincreasing. Therefore the set

$$\mathscr{E}_R^{\alpha,\kappa} = \left\{U = (u_0;u_1;\theta_0) \in H_{\alpha,\kappa} \,:\, \mathscr{E}_{\alpha,\kappa}(u_0,u_1,\theta_0) \le R^2\right\} \quad (11.1.14)$$

is forward invariant for every $R > 0$; that is, $S_t^{\alpha,\kappa}\mathscr{E}_R^{\alpha,\kappa} \subset \mathscr{E}_R^{\alpha,\kappa}$ for $t \ge 0$. Moreover, as we see below, the properties of the set $\mathscr{N}$ of equilibria described in Section 5.7 make it possible to prove that the global attractor belongs to the set

$$\{U \in H_{\alpha,\kappa} : |U|_{\alpha,\kappa} \le R_*\},$$

where R_* depends on $||F_0||_{W^2_\infty(\Omega)}$ and $||p||_{L_2(\Omega)}$ only.

11.2 Statements of main results

Our main results, dealing with attractors for dynamical systems $(H_{\alpha,\kappa}, S_t^{\alpha,\kappa})$ with $(\alpha,\kappa) \in \Lambda \equiv \{0 \le \alpha \le 1, 0 \le \kappa \le 1\}$, are formulated below.

11.2.1. Theorem (Global Attractors). *For every $(\alpha,\kappa) \in \Lambda$ the dynamical system $(H_{\alpha,\kappa}, S_t^{\alpha,\kappa})$ is gradient and possesses a compact global attractor $\mathscr{A}^{\alpha,\kappa} = \mathscr{M}^u_{\alpha,\kappa}(\mathscr{N})$, where $\mathscr{M}^u_{\alpha,\kappa}(\mathscr{N})$ is an unstable manifold emanating from the set $\mathscr{N}$ of stationary points. Thus the conclusions of Theorem 7.5.6 and 7.5.10 hold true for $(H_{\alpha,\kappa}, S_t^{\alpha,\kappa})$. Moreover,*

- **Finite-dimensionality:** *there exists $d_0 > 0$ independent of α and κ such that fractal dimension of $\mathscr{A}^{\alpha,\kappa}$ in $H_{\alpha,\kappa}$ admits the estimate $\dim_f^{H_{\alpha,\kappa}}\mathscr{A}^{\alpha,\kappa} \le d_0$ for all $(\alpha,\kappa) \in \Lambda$.*
- **Regularity:** *any full trajectory $\{U(t) : t \in \mathbb{R}\}$ from the attractor possesses the properties*

$$||A_D u(t)||^2 + ||M_\alpha^{1/2}u_t(t)||^2 + ||\Delta v(u(t))||^2 + \kappa||\theta(t)||^2 \le R_1^2 \quad (11.2.1)$$

and

$$\|u(t)\|_3^2 + ||A_D u_t(t)||^2 + ||M_\alpha^{1/2}u_{tt}(t)||^2 + \kappa||\theta_t(t)||^2 + ||A_D\theta(t)||^2 \le R_2^2 \quad (11.2.2)$$

for all $t \in \mathbb{R}$, where both constants R_1 and R_2 do not depend on $(\alpha,\kappa) \in \Lambda$ and R_1 is also independent of η and μ (in the case $\kappa = 0$ the terms in (11.2.1) and (11.2.2) containing θ should be omitted and in the case $\alpha = 0$ we additionally have that $\|u(t)\|_4 \le R_2$ for $t \in \mathbb{R}$). In addition, whenever $\mathscr{A} = A_D^2$ (hinged case), then the following estimate

$$\alpha||A_D^{3/2}u_t(t)||^2 + ||A_D^2 u(t)||^2 \le R_3^2 \tag{11.2.3}$$

holds uniformly in α *and* κ *(i.e., with* R_3 *which does not depend on* $0 \le \alpha, \kappa \le 1$*).*

- **Upper semicontinuity:** *the family of attractors* $\mathscr{A}^{\alpha,\kappa}$ *is upper semicontinuous on* Λ *in the sense that (i) for any* $\lambda_0 = (\alpha_0, \kappa_0) \in \Lambda$ *with* $\kappa_0 > 0$ *we have that*

$$\lim_{(\alpha,\kappa)\to\lambda_0} \sup\left\{\mathrm{dist}_{H_{\alpha_0,\kappa_0}}\left(U, \mathscr{A}^{\lambda_0}\right) : U \in \mathscr{A}^{\alpha,\kappa}\right\} = 0 \tag{11.2.4}$$

and (ii) for $\lambda_0 = (\alpha_0, 0) \in \Lambda$ *the relation*

$$\lim_{(\alpha,\kappa)\to\lambda_0} \sup\left\{\mathrm{dist}_{H_{\alpha_0,1}}\left(U, \widehat{\mathscr{A}^{\lambda_0}}\right) : U \in \mathscr{A}^{\alpha,\kappa}\right\} = 0 \tag{11.2.5}$$

holds, where

$$\widehat{\mathscr{A}^{\lambda_0}} = \left\{(u_0; u_1; -(\mu/\eta)u_1) \in H_{\alpha_0,1} : (u_0, u_1) \in \mathscr{A}^{\alpha_0,0}\right\}.$$

We note that in the case of isothermal von Karman plate upper semicontinuity of the attractor when $\alpha \to 0$ was proved in Theorem 9.5.14.

Our next result relies on Theorem 7.5.15 and deals with the case when the set $\mathscr{N}$ of stationary points is finite and every stationary point is hyperbolic.

11.2.2. Theorem (Exponential Attractor). *Assume that* $\mathscr{N} = \{E_i : i = 1, \dots, n\}$ *is a finite set. Then the conclusions of Corollary* 7.5.11 *hold true for the dynamical system* $(H_{\alpha,\kappa}, S_t^{\alpha,\kappa})$ *for every* $(\alpha, \kappa) \in \Lambda$*. In particular,* $\mathscr{A}^{\alpha,\kappa} = \cup_{i=1}^n \mathscr{M}^u(E_i)$*. Moreover, if every stationary point* $E_i = (e_i; 0; 0)$ *is hyperbolic in the sense that the equation* $\mathscr{A}w = \langle F'(e_i), w\rangle$*, where* $F'(u)$ *is the Fréchet derivative of the mapping* F *given by* (11.1.6)*, has only trivial solutions. Then:*

- *For any* $U_0 \in H_{\alpha,\kappa}$*, there exists an equilibrium point* $E = (e; 0; 0) \in H_{\alpha,\kappa}$ *and constants* $\omega > 0$*,* $C > 0$ *such that*

$$|S_t^{\alpha,\kappa}U_0 - E|_{\alpha,\kappa} \le C_{U_0} e^{-\omega t}, \quad t > 0.$$

Moreover, for any bounded set B *in* $H_{\alpha,\kappa}$ *we have that*

$$\sup\left\{\mathrm{dist}\,(S_t^{\alpha,\kappa}U, \mathscr{A}^{\alpha,\kappa}) : U \in B\right\} \le C_B e^{-\omega t}, \quad t > 0. \tag{11.2.6}$$

Here above $\mathscr{A}^{\alpha,\kappa}$ *is a global attractor,* C_{U_0}*,* C_B*, and* ω *are positive constants.*

- *Let* $\alpha > 0$ *or else* $\alpha = 0$ *and* $\mathscr{A} = A_D^2$ *. For each* $E \in \mathscr{N}$ *the unstable manifold* $\mathscr{M}^u(E)$ *is an embedded* C^1*-submanifold of* $H_{\alpha,\kappa}$ *of finite dimension* $\mathrm{ind}\,(E)$*, which implies that*

$$\dim_H \mathscr{A}^{\alpha,\kappa} = \max_{E\in\mathscr{N}} \mathrm{ind}\,(E). \tag{11.2.7}$$

The first statement of Theorem 11.2.2 implies that the global attractor is exponential. However, this property requires finiteness and hyperbolicity of the set $\mathscr{N}$ of equilibria. In the general case due to the stabilizability estimate which is proved below one can apply Theorem 7.9.9 and argument similar to that given in the proof of Theorem 4.43 [75] to obtain the existence of an exponential fractal attractor (inertial set) with an uniform (with respect to α and κ) estimate for the dimension.

When comparing (11.2.7) with the result on the dimension from Theorem 11.2.1, one can see that $\max_{E\in\mathcal{N}} \mathrm{ind}\,(E)$ can be estimated from above by a constant independent of $\{\alpha,\kappa\}$. Later on we give an estimate from below for this quantity in one special case.

We note also that the treatment presented here does not rely on analyticity of the semigroup associated with the model when $\alpha = 0$. All the estimates obtained are independent of $\alpha \geq 0$. This was possible to achieve for both hinged and clamped boundary conditions. However, in the case of free boundary conditions, the situation is more complicated. To our best knowledge, there are no estimates, independent of α even in the linear case. Nevertheless, the methods of this chapter provide all the results on attractors for each value of the parameter ($\alpha > 0$ and $\alpha = 0$). In the case $\alpha = 0$, critical use of the analyticity of the semigroup plays the role. How to make these estimates (in the case of free boundary conditions) uniform with respect to α is an open problem.

11.2.3. Remark. The equality (11.2.7) is stated for the case when either $\alpha > 0$ or $\mathscr{A} = A_D^2$. The reason for this restriction is that the proof of (11.2.7) depends on the backward uniqueness of linearized flow. Such property is proved in Chapter 5, Theorem 5.6.1 for the case $\alpha > 0$. In the case $\alpha = 0$ and hinged boundary conditions; that is, when $\mathscr{A} = A_D^2$, (commutative case), backward uniqueness is established in Corollary 5.6.9. ∎

11.3 Uniform stabilizability inequality

Our main tool in the proof of Theorem 11.2.1 is the stabilizability estimate given in Theorem 11.3.1 below. An important fact is that this estimate is independent of the parameters α and κ. This is achieved by exploiting compensated compactness induced by very special decomposition of von Karman nonlinearity given by Lemma 9.4.10 (see (11.3.8) also). A similar argument was also used in Chapters 9 and 10 in order to obtain the corresponding stabilizbility estimate for isothermal von Karman models without rotational inertia.

11.3.1. Theorem. *Assume that* $0 \leq \alpha \leq 1$ *and* $0 < \kappa \leq 1$. *Let* $(u^1;\theta^1)$ *and* $(u^2;\theta^2)$ *be two solutions to problem* (11.1.7) *with initial data* $y^i = (u_0^i;u_1^i;\theta_0^i)$, $i = 1,2$. *Assume that*

$$||A_D u^i(t)||^2 + ||u_t^i(t)||^2 + \alpha||A_D^{1/2} u_t^i(t)||^2 + \kappa||\theta^i(t)||^2 \leq R^2 \tag{11.3.1}$$

for all $t \geq 0$. *Let*

$$Z(t) \equiv (z(t);z_t(t);\xi(t)) \equiv (u^1(t) - u^2(t);u_t^1(t) - u_t^2(t);\theta^1(t) - \theta^2(t)).$$

Then there exist positive constants $C_R = C_R(\mu,\eta)$ and $\omega = \omega(\mu,\eta)$ which are independent of α and κ such that

$$|Z(t)|^2_{\alpha,\kappa} \le C_R|Z(0)|^2_{\alpha,\kappa}e^{-\omega t} + C_R \sup_{0\le\tau\le t} ||z(\tau)||^2. \tag{11.3.2}$$

Proof. We can assume that both solutions $(u^1;\theta^1)$ and $(u^2;\theta^2)$ are strong (see Proposition 5.5.3).

It is easily verified that $z(t)$ and $\xi(t)$ solve the problem

$$\begin{cases} M_\alpha z_{tt} - \mu A_D\xi + \mathscr{A}z = F(t), & z|_{t=0} = z_0,\ z_t|_{t=0} = z_1, \\ \kappa\xi_t + \eta A_D\xi + \mu A_D z_t = 0, & \xi|_{t=0} = \xi_0, \end{cases} \tag{11.3.3}$$

where

$$F(t) \equiv F(u^1) - F(u^2) \equiv [v(u^1) - v(u^2), u^1] + [v(u^2), z] + [F_0, z]. \tag{11.3.4}$$

In order to obtain a suitable energy relation for (11.3.3) we first apply convenient decomposition of the term $(F(t), z_t)$.

11.3.2. Lemma. *Let $(u^1(t);\theta^1(t))$ and $(u^2(t);\theta^2(t))$ be two strong solutions to problem* (11.1.7) *satisfying* (11.3.1). *Then the following representation*

$$(F(t), z_t) = \frac{d}{dt}Q(t) + P(t) \tag{11.3.5}$$

holds, where the functions $Q(t) \in C^1(\mathbb{R}_+)$ and $P(t) \in C(\mathbb{R}_+)$ satisfy the relations

$$|Q(t)| \le C_R\left(1 + \frac{\sqrt{\kappa}}{\mu}\right) ||A_D z(t)||^{3/2}||z(t)||^{1/2} \tag{11.3.6}$$

and

$$|P(t)| \le C_R\frac{\eta+\kappa}{\mu}\left(||A_D z(t)||^2 + ||z_t(t)||^2\right)||\theta^2(t)||, \tag{11.3.7}$$

where C_R does not depend on α, κ, μ and η.

Proof. Let $\Pi(u)$ be the potential energy of mechanical forces:

$$\Pi(u) = \frac{1}{4}||\Delta v(u)||^2 - \frac{1}{2}([F_0, u] + 2p, u).$$

Because $F(u) = -\Pi'(u)$, we obviously have that

$$\begin{aligned}(F(t), z_t) &= (F(u^1), u^1_t) - (F(u^2), z_t) - (F(u^1), u^2_t) \\ &= -\frac{d}{dt}\left[\Pi(u^1) + (F(u^2), z)\right] + (\langle F'(u^2); u^2_t\rangle, z) - (F(u^1), u^2_t) \\ &= \frac{d}{dt}Q_0(z) + P_0(z), \end{aligned} \tag{11.3.8}$$

where

$$Q_0(z) = -\left[\Pi(u^1) - \Pi(u^2) + (F(u^2), z)\right]$$

and

$$P_0(z) = (\langle F'(u^2); u_t^2\rangle, z) - (F(u^1) - F(u^2), u_t^2).$$

Here we denote by $F'(u)$ the Fréchet derivative of $F(u)$ given by

$$\langle F'(u); w\rangle = [w, v(u) + F_0] + 2[u, v(u,w)], \tag{11.3.9}$$

where $v(u)$ is determined from (11.1.2) and $v(w_1, w_2) \in H_0^2(\Omega)$ solves the problem

$$\Delta^2 v(w_1, w_2) = -[w_1, w_2] \ \text{ in } \Omega, \quad \frac{\partial}{\partial n} v(w_1, w_2) = v(w_1, w_2) = 0 \ \text{ on } \Gamma. \tag{11.3.10}$$

Because $F(u) = -\Pi'(u)$, we obtain

$$Q_0(z) - \int_0^1 (F(u^2 + \lambda z) - (F(u^2), z) d\lambda. \tag{11.3.11}$$

Using the symmetry properties of the von Karman bracket after some rather straightforward but tedious algebraic manipulations (see Lemma 9.4.10 for similar algebraic relations) we also obtain that

$$P_0(z) = -(u_t^2, [u^2, v(z,z)] + [z, v(u^1 + u^2, z)]). \tag{11.3.12}$$

From (1.4.23) in Corollary 1.4.4 we have the estimate

$$|Q_0(z)| \le C_R \|A_D z\| \|z\|. \tag{11.3.13}$$

By using thermal equation we replace u_t^2 appearing in (11.3.12) by

$$u_t^2 = -\frac{\kappa}{\mu} A_D^{-1} \theta_t^2 - \frac{\eta}{\mu} \theta^2.$$

Substituting into (11.3.12) gives

$$P_0(z) = \frac{\kappa}{\mu} \frac{d}{dt} Q_1(t) + \frac{\eta}{\mu} P_1(t) - \frac{\kappa}{\mu} P_2(t)$$

with

$$\begin{aligned}
Q_1(t) &= (A_D^{-1}\theta^2, [u^2, v(z,z)] + [z, v(u^1 + u^2), z]), \\
P_1(t) &= (\theta^2, [u^2, v(z,z)] + [z, v(u^1 + u^2, z)]), \\
P_2(t) &= (A_D^{-1}\theta^2, [u_t^2, v(z,z)] + [z_t, v(u^1 + u^2, z)]) \\
&\quad + (A_D^{-1}\theta^2, [u^2, 2v(z_t, z)] + [z, v(u_t^1 + u_t^2, z)] + [z, v(u^1 + u^2, z_t)]),
\end{aligned}$$

where we have used the notation (11.3.10). Thus due to (11.3.8) we have the representation (11.3.5) with

$$Q(t) = Q_0(z(t)) + \frac{\kappa}{\mu} Q_1(t), \quad P(t) = \frac{\eta}{\mu} P_1(t) - \frac{\kappa}{\mu} P_2(t). \tag{11.3.14}$$

Because $\sqrt{\kappa}||\theta^2|| \le R$, from (1.4.17) with $j = 2$, $\beta = 0$ we infer

$$\begin{aligned} |Q_1(t)| &\le C||\theta^2|| \left(\|[u^2, v(z,z)]\|_{-2} + \|[z, v(u^1 + u^2), z)]\|_{-2} \right) \\ &\le C||\theta^2|| \left(\|u^2\|_2 \|v(z,z)\|_1 + \|z\|_1 \|v(u^1 + u^2, z)\|_2 \right) \\ &\le \frac{C_R}{\sqrt{\kappa}} \|z\|_2 \|z\|_1 \le \frac{C_R}{\sqrt{\kappa}} ||A_D z||^{3/2} ||z||^{1/2}. \end{aligned}$$

Therefore using (11.3.13) we obtain (11.3.6).

Now we estimate the term $P(t)$. We first note that by (1.4.23) we have the relations

$$|(\theta^2, [u^2, v(z,z)])| \le C||\theta^2|| ||A_D u^2|| ||A_D z||^2$$

and

$$|(\theta^2, [z, v(u^1 + u^2, z)])| \le C||\theta^2|| ||A_D u^2 + A_D u^1|| ||A_D z||^2.$$

Hence

$$|P_1(t)| \le C||\theta^2|| ||A_D z||^2 \left(||A_D u^1|| + ||A_D u^2|| \right) \le C_R ||\theta^2|| ||A_D z||^2. \tag{11.3.15}$$

Thus, it remains to estimate $P_2(t)$. For this we use the following bounds resulting from (1.4.23) and the symmetry of the von Karman bracket,

$$\begin{aligned} |(A_D^{-1}\theta^2, [u_t^2, v(z,z)])| &= |(u_t^2, [A_D^{-1}\theta^2, v(z,z)])| \le C||u_t^2|| ||A_D^{-1}\theta^2||_2 ||z||_2^2 \\ &\le C||u_t^2|| ||\theta^2|| ||A_D z||^2 \le C_R ||\theta^2|| ||A_D z||^2 \end{aligned}$$

and

$$\begin{aligned} |(A_D^{-1}\theta^2, [u^2, v(z_t, z)])| &= |([u^2, A_D^{-1}\theta^2], v(z_t, z))| = |(v(u^2, A_D^{-1}\theta^2), [z_t, z])| \\ &= |(z_t, [v(u^2, A_D^{-1}\theta^2), z])| \\ &\le C||z_t|| ||u^2||_2 ||A_D^{-1}\theta^2||_2 ||z||_2 \\ &\le C||z_t|| ||A_D u^2|| ||\theta^2|| ||A_D z|| \le C_R ||\theta^2|| ||z_t|| ||A_D z||. \end{aligned}$$

The remaining terms entering the definition $P_2(t)$ are estimated in the same way. Therefore using (11.3.14) and (11.3.15) we obtain the desired estimate in (11.3.7).

□

The following trace estimate, relevant only in the case of clamped boundary conditions, is used in the course of the proof.

11.3.3. Proposition. *Let* $(z;\xi)$ *be a solution of* (11.3.3) *with the clamped boundary conditions:* $z = (\partial/\partial n)z = 0$ *on* Γ. *Then, for any* $\omega \ge 0$ *we have*

$$\int_{\Sigma_t} e^{\omega s} |\Delta z|^2 dx ds \le c_0 (1+\omega) \int_0^t e^{\omega s} E_\alpha^0(s) ds + c_1 \left(E_\alpha^0(0) + E_\alpha^0(t) e^{\omega t} \right)$$

$$+\int_0^t e^{\omega s}\left[C_R\|z\|^2+c_2\mu^2\|\nabla\xi(s)\|^2\right]ds,$$

where $\Sigma_t \equiv \Gamma\times(0,t)$ *and* $E^0_\alpha(t)=\frac{1}{2}\left(\|M_\alpha z_t(t)\|^2+\|A_D z(t)\|^2\right)$. *All the constants* c_i *are uniform in the parameters* α, κ, η, *and* μ.

Proof. We apply Theorem 2.5.3 with f replaced by $F(t)+\mu A_D\xi$, where $F(t)$ is given in (11.3.4) The contribution of the last integral in (2.5.12) is the following,

$$\begin{aligned}|(F(t)+\mu A_D\xi,h\nabla z)| &\le c\mu\|\nabla\xi\|\|A_D z\|+C_R\|z\|_2\|z\|_1\\ &\le c_1\mu\|\nabla\xi\|\|A_D z\|+c_2\|A_D z\|^2+C_R\|z\|^2,\end{aligned}$$

where we have used, again, (1.4.23) and interpolation properties of Sobolev spaces. The above inequality, when inserted into the inequality (2.5.12) gives the desired estimate in Proposition 11.3.3. □

Proper proof of Theorem 11.3.1. It is sufficient to establish (11.3.2) for strong solutions only. Relation (11.3.5) implies the following energy equality valid for strong solutions,

$$\frac{d}{dt}\mathscr{E}^0(t)=-\eta\|A_D^{1/2}\xi\|^2+P(t), \tag{11.3.16}$$

where

$$\mathscr{E}^0(t)=\frac{1}{2}\left[\|M_\alpha z_t\|^2+\|A_D z\|^2-2Q(t)+\kappa\|\xi\|^2\right].$$

It follows from (11.3.6) that

$$\frac{3}{8}|Z(t)|^2_{\alpha,\kappa}-C_R\|z(t)\|^2\le\mathscr{E}^0(t)\le\frac{5}{8}|Z(t)|^2_{\alpha,\kappa}+C_R\|z(t)\|^2, \tag{11.3.17}$$

where C_R does not depend on α, κ.

Now we consider

$$V(t)=\mathscr{E}^0(t)+\varepsilon(M_\alpha z_t,z)+\varepsilon_1\kappa(M_\alpha z_t,A_D^{-1}\xi) \tag{11.3.18}$$

for some positive ε and ε_1. We obviously have that

$$\left|\varepsilon(M_\alpha z_t,z)+\varepsilon_1\kappa(M_\alpha z_t,A_D^{-1}\xi)\right|\le c_0(\varepsilon+\varepsilon_1)\|z_t\|^2+\varepsilon\|A_D z\|^2+\varepsilon_1\kappa\|\xi\|^2$$

for all $(\alpha,\kappa)\in\Lambda$ (here and below we denote by $c_0,c_1,\ldots$ positive constants that do not depend on R, α, κ, μ, and η, and may change from line to line). Therefore it follows from (11.3.17) that

$$\frac{1}{4}|Z(t)|^2_{\alpha,\kappa}-C_R\|z(t)\|^2\le V(t)\le|Z(t)|^2_{\alpha,\kappa}+C_R\|z(t)\|^2 \tag{11.3.19}$$

for all $(\alpha,\kappa)\in\Lambda$, as soon as $\varepsilon+\varepsilon_1\le\varepsilon^*$, where $\varepsilon^*>0$ and C_R do not depend on the parameters α, κ, μ, and η.

Now we begin by computing:

$$\frac{d}{dt}V = \frac{d}{dt}\mathscr{E}^0 + \varepsilon||M_\alpha{}^{1/2}z_t||^2 + \varepsilon(M_\alpha z_{tt}, z) \\ + \varepsilon_1\kappa(M_\alpha z_{tt}, A_D^{-1}\xi) - \varepsilon_1(M_\alpha z_t, [\eta\xi + \mu z_t]) .$$

In the last term we have used the second equation from (11.3.3). By using energy relation (11.3.16) together with the structure of the first equation in (11.3.3) we obtain:

$$\begin{aligned}\frac{d}{dt}V = & -\eta||A_D^{1/2}\xi||^2 - (\mu\varepsilon_1 - \varepsilon)||M_\alpha{}^{1/2}z_t||^2 + P(t) - \varepsilon_1\eta(M_\alpha z_t, \xi) \\ & + \varepsilon(z, -\mathscr{A}z + \mu A_D\xi + F(t)) + \varepsilon_1\kappa(-\mathscr{A}z + \mu A_D\xi + F(t), A_D^{-1}\xi) \\ = & -\eta||A_D^{1/2}\xi||^2 - (\mu\varepsilon_1 - \varepsilon)||M_\alpha{}^{1/2}z_t||^2 + P(t) - \varepsilon_1\eta(M_\alpha z_t, \xi) \\ & + \varepsilon\left[-||A_Dz||^2 + \mu(z, A_D\xi) + (z, F(t))\right] - \varepsilon_1\kappa(A_Dz, \xi) \\ & + \varepsilon_1\kappa\left[\left(\Delta z, \frac{\partial}{\partial n}A_D^{-1}\xi\right)_\Gamma + \mu||\xi||^2 + (F(t), A_D^{-1}\xi)\right], \end{aligned} \tag{11.3.20}$$

where $P(t)$ admits the estimate in (11.3.7) and $F(t)$ is given by (11.3.4). We note that $\left(\Delta z, (\partial/\partial n)A_D^{-1}\xi\right)_\Gamma = 0$ in the latter relation in the case of of hinged boundary conditions (11.1.5).

Let ω and δ be positive numbers. Using (11.3.20) and (11.3.7) one can see that

$$\begin{aligned} & \frac{d}{dt}V + \omega|Z|^2_{\alpha,\kappa} \\ \le & -\eta\left(1 - \frac{\varepsilon_1}{\delta}\right)||A_D^{1/2}\xi||^2 - (\mu\varepsilon_1 - \varepsilon - \omega - c_0\varepsilon_1\eta\delta)||M_\alpha{}^{1/2}z_t||^2 \\ & + \varepsilon\left[-||A_Dz||^2 + C_R||A_Dz||||z||\right] + (\mu\varepsilon + C_R\kappa\varepsilon_1)||A_D^{1/2}z||||A_D^{1/2}\xi|| \\ & + \omega||A_Dz||^2 + C_R\frac{\eta+\kappa}{\mu}||\theta^2||\left(||z_t||^2 + ||A_Dz||^2\right) \\ & + \kappa\left[c_1\frac{\varepsilon_1^2}{\eta}||\Delta z||^2_{L_2(\Gamma)} + \frac{\eta}{4}||A_D^{1/2}\xi||^2 + \mu\varepsilon_1||\xi||^2\right] + \omega\kappa||\xi||^2 . \end{aligned} \tag{11.3.21}$$

Now we use the right inequality in (11.3.19) and also the relation

$$\begin{aligned}(\mu\varepsilon + C_R\kappa\varepsilon_1)||A_D^{1/2}z||||A_D^{1/2}\xi|| & \le \frac{\eta}{4}||A_D^{1/2}\xi||^2 + C_R\frac{(\mu\varepsilon+\varepsilon_1)^2}{\eta}||A_Dz||||z|| \\ & \le \frac{\eta}{4}||A_D^{1/2}\xi||^2 + \frac{\varepsilon}{4}||A_Dz||^2 + C_R\frac{(\mu\varepsilon+\varepsilon_1)^4}{\varepsilon\eta^2}||z||^2.\end{aligned}$$

This gives:

$$\begin{aligned}\frac{d}{dt}V + \omega V \le & -\frac{1}{2}\left(\eta - 2\eta\frac{\varepsilon_1}{\delta} - \frac{2\mu\varepsilon_1 + 2\omega}{\lambda_1}\right)||A_D^{1/2}\xi||^2 \\ & - (\mu\varepsilon_1 - \varepsilon - \omega - c_0\varepsilon_1\eta\delta)||M_\alpha{}^{1/2}z_t||^2\end{aligned}$$

$$
-\frac{\varepsilon-2\omega}{2}||A_D z||^2+C_R\frac{\eta+\kappa}{\mu}||\theta^2||\left(||z_t||^2+||A_D z||^2\right)
$$
$$
+C_R\left(1+\omega+\frac{(\mu\varepsilon+\varepsilon_1)^4}{\varepsilon\eta^2}\right)||z||^2+c_1\frac{\varepsilon_1^2}{\eta}||\Delta z||^2_{L_2(\Gamma)}\,.
$$

Selecting $\omega=a_0\mu\varepsilon_1$ and $\varepsilon=a_1\mu\varepsilon_1$ with appropriate constants a_0 and a_1 independent of the parameters of the problem, we can find $\delta>0$ such that

$$
\begin{aligned}
&\frac{d}{dt}V+a_0\mu\varepsilon_1 V \qquad (11.3.22)\\
&\le -\frac{1}{4}\eta||A_D^{1/2}\xi||^2-\frac{1}{4}\mu\varepsilon_1||M_\alpha{}^{1/2}z_t||^2-\frac{a_1\mu\varepsilon_1}{4}||A_D z||^2\\
&\quad+C_R(\varepsilon_1)||z||^2+c_0\frac{\varepsilon_1^2}{\eta}||\Delta z||^2_{L_2(\Gamma)}+C_R\frac{1+\eta}{\mu}||\theta^2||\left(||M_\alpha z_t||^2+||A_D z||^2\right)
\end{aligned}
$$

for all $\varepsilon_1\in(0,\varepsilon_1^*]$, where ε_1^* depends on μ and η only and

$$
C_R(\varepsilon_1)=C_R\left(1+\mu\varepsilon_1+\frac{\varepsilon_1^3(1+\mu^2)^4}{\mu\eta^2}\right).
$$

Using the relation

$$
C_R\frac{1+\eta}{\mu}||\theta^2||\le b_0\mu\varepsilon_1+\frac{C_R(\eta,\mu)}{\varepsilon_1}||\theta^2||^2 \quad\text{with}\quad b_0=\frac{\min\{1,a_1\}}{8}
$$

for $\omega=a_0\mu\varepsilon_1$ from (11.3.22) we have that

$$
\begin{aligned}
&\int_0^t\frac{d}{d\tau}\left[Ve^{\omega t}\right]d\tau=\int_0^t\left[\frac{d}{d\tau}V+a_0\mu\varepsilon_1 V\right]e^{\omega\tau}d\tau\\
&\le-\int_0^t e^{\omega\tau}\left[\frac{1}{4}\eta||A_D^{1/2}\xi||^2+\frac{1}{8}\mu\varepsilon_1||M_\alpha{}^{1/2}z_t||^2+\frac{a_1\mu\varepsilon_1}{8}||A_D z||^2\right]d\tau\\
&\quad+C_R(\varepsilon_1)\int_0^t e^{\omega\tau}||z||^2d\tau+c_0\frac{\varepsilon_1^2}{\eta}\int_0^t e^{\omega\tau}||\Delta z||^2_{L_2(\Gamma)}d\tau\\
&\quad+\frac{C_R(\eta,\mu)}{\varepsilon_1}\int_0^t e^{\omega\tau}||\theta^2||^2\left(||M_\alpha z_t||^2+||A_D z||^2\right)d\tau. \qquad (11.3.23)
\end{aligned}
$$

In the clamped case by Proposition 11.3.3 from (11.3.23) we have that

$$
\begin{aligned}
&\int_0^t\frac{d}{d\tau}\left[Ve^{\omega t}\right]d\tau\\
&\le-\int_0^t e^{\omega\tau}\left[\eta(\varepsilon_1)||A_D^{1/2}\xi||^2+\frac{\mu\varepsilon_1}{8}||M_\alpha{}^{1/2}z_t||^2+\frac{a_1\mu\varepsilon_1}{8}||A_D z||^2\right]d\tau\\
&\quad+c_1\frac{\varepsilon_1^2}{\eta}(1+\omega)\int_0^t e^{\omega s}E_\alpha^0(s)ds+c_2\frac{\varepsilon_1^2}{\eta}\left(E_\alpha^0(0)+E_\alpha^0(t)e^{\omega t}\right)
\end{aligned}
$$

$$+C_R(\varepsilon_1)\int_0^t e^{\omega\tau}||z||^2d\tau+\frac{C_R(\eta,\mu)}{\varepsilon_1}\int_0^t e^{\omega\tau}||\nabla\theta^2(\tau)||^2E^0_\alpha(\tau)d\tau$$

with $\omega = a_0\mu\varepsilon_1$ and $\eta(\varepsilon_1)=\frac{1}{4}\eta - c_0\varepsilon_1^2\mu^2\eta^{-1}$. Consequently, choosing ε_1 sufficiently small, we obtain

$$V(t)\le\left[V(0)+c_2\frac{\varepsilon_1^2}{\eta}|Z(0)|^2_{\alpha,\kappa}\right]e^{-\omega t}+c_2\frac{\varepsilon_1^2}{\eta}|Z(t)|^2_{\alpha,\kappa}$$
$$+C_R(\varepsilon_1)\int_0^t e^{-\omega(t-\tau)}||z||^2d\tau+\frac{C_R(\eta,\mu)}{\varepsilon_1}\int_0^t e^{-\omega(t-\tau)}||\nabla\theta^2(\tau)||^2E^0_\alpha(\tau)d\tau.$$

It is clear that this relation remains also true in the case of hinged boundary conditions. Therefore using (11.3.19) we obtain

$$|Z(t)|^2_{\alpha,\kappa}\le C_1(R)|Z(0)|^2_{\alpha,\kappa}e^{-\omega t}+C_2(R)\max_{\tau\in[0,t]}||z(\tau)||^2$$
$$+C_3(R)\int_0^t e^{-\omega(t-\tau)}||\nabla\theta^2(\tau)||^2|Z(\tau)|^2_{\alpha,\kappa}d\tau,$$

where the constants $C_i(R)$ do not depend on $(\alpha;\kappa)\in\Lambda$. Finally accounting for the inequality

$$\int_0^\infty ||\nabla\theta^2||^2d\tau\le C_R, \tag{11.3.24}$$

which follows from energy relation (11.1.11), inequality (11.3.1), and Gronwall's lemma, we obtain (11.3.2). □

11.3.4. Remark. We note that the assertion of Theorem 11.3.1 remains valid for $\kappa=0$. The proof relies on the same idea and involves similar calculations with a Lyapunov-type function formally given by (11.3.18) with $\kappa=0$. However, instead of (11.3.24) we use the finiteness of the integral $\int_0^\infty||\nabla u_t^2||^2d\tau$. This circumstance simplifies our argument drastically. We also refer to Chapters 9 and 10 where similar ideas were realized for isothermal von Karman models with nonlinear (boundary and internal) damping functions. ■

11.4 Existence and properties of the attractor—Proof of Theorem 11.2.1

11.4.1 Existence of the attractor

11.4.1. Theorem. *For every* $(\alpha,\kappa)\in\Lambda=\{0\le\alpha\le 1, 0\le\kappa\le 1\}$ *the dynamical system* $(H_{\alpha,\kappa},S_t^{\alpha,\kappa})$ *possesses a compact global attractor* $\mathscr{A}^{\alpha,\kappa}=\mathscr{M}^u_{\alpha,\kappa}(\mathscr{N})$.

Proof. We first note that inasmuch as the system $(H_{\alpha,\kappa},S_t^{\alpha,\kappa})$ is gradient with a Lyapunov function given by the energy $\mathscr{E}_{\alpha,\kappa}$, then by Theorem 7.5.6 the attractor, if it exists, coincides with the unstable manifold $\mathscr{M}^u_{\alpha,\kappa}(\mathscr{N})$. Moreover, because $\mathscr{N}$ is a bounded set and the energy $\mathscr{E}_{\alpha,\kappa}$ possesses property

$$c_1 E_{\alpha,\kappa}(u,u_t,\theta) - C_{F_0,p} \le \mathscr{E}_{\alpha,\kappa}(u,u_t,\theta) \le c_2 E_{\alpha,\kappa}(u,u_t,\theta) + C_{F_0,p} \tag{11.4.1}$$

with positive constants, we can use Corollary 7.5.7 and thus reducing the proof to establishing asymptotic compactness only.

By Proposition 5.5.1 the semiflow $S_t^{\alpha,\kappa}$ satisfies the Lipschitz property

$$|S_t^{\alpha,\kappa}U_1 - S_t^{\alpha,\kappa}U_2|_{\alpha,\kappa} \le e^{a_R t}|U_1 - U_2|_{\alpha,\kappa}, \quad t > 0, \tag{11.4.2}$$

for any $U_1, U_2 \in H_{\alpha,\kappa}$ such that $|U_i|_{\alpha,\kappa} \le R$, where the constant $a_R > 0$ does not depend on $(\alpha,\kappa) \in \Lambda$. Consequently by Theorem 11.3.1 $(H_{\alpha,\kappa}, S_t^{\alpha,\kappa})$ is a quasi-stable system on every bounded forward invariant set. Thus by Proposition 7.9.4 $(H_{\alpha,\kappa}, S_t^{\alpha,\kappa})$ is an asymptotically smooth dynamical system. □

11.4.2 Smoothness of the attractor—Proof of regularity in Theorem 11.2.1

We first note that estimate (11.2.1) follows from Remark 7.5.8. In particular we have uniform estimate

$$\|u_t(t)\|^2 + \|A_D u(t)\|^2 \le R^2, \quad t \in \mathbb{R}, \tag{11.4.3}$$

for any trajectory $U(t) = (u(t); u_t(t); \theta(t))$ lying in the attractor.

The goal of this section is to show that the attractors are contained uniformly with respect to α and κ in a set bounded in a finer topology. More specifically, the theorem stated below gives a quantitative description of additional smoothness of elements in the attractor.

11.4.2. Theorem. *The attractor $\mathscr{A}^{\alpha,\kappa}$ belongs to $H^3(\Omega) \times H^2(\Omega) \times H^2(\Omega)$ uniformly in $(\alpha,\kappa) \in \Lambda = \{0 \le \alpha \le 1, 0 \le \kappa \le 1\}$; that is, for any trajectory $U(t) = (u(t); u_t(t); \theta(t))$ lying in the attractor we have*

$$\|u_{tt}(t)\|^2 + \alpha\|\nabla u_{tt}\|^2 + \|A_D u_t(t)\|^2 + \kappa\|\theta_t(t)\|^2 + \|u(t)\|_3^2 + \|A_D\theta(t)\|^2 \le C_R \tag{11.4.4}$$

for all $t \in \mathbb{R}$, where the estimate is uniform in $(\alpha,\kappa) \in \Lambda$. In the case $\kappa = 0$ the terms containing θ should be omitted. If $\alpha = 0$, then we have in addition that $\|u(t)\|_4^2 \le C_R$ for all $t \in \mathbb{R}$.

In the special case of hinged boundary conditions (that provides convenient symmetry), attractors are even more regular. This is seen from the corollary below.

11.4.3. Corollary. *In the case when $\mathscr{A} = A_D^2$ one additionally obtains that*

$$\alpha||A_D^{3/2}u_t(t)|| + ||A_D^2 u(t)||^2 \le C_R \quad \textit{for all} \quad t \in \mathbb{R},$$

for any trajectory $U(t) = (u(t); u_t(t); \theta(t))$ *lying in the attractor, where the estimate is uniform in* $(\alpha, \kappa) \in \Lambda$.

Proof of Theorem 11.4.2. Without control of dependence of the constant C_R on α and κ, this theorem follows from Theorem 7.9.8. However, in order to control the dependence on the parameters we need to develop an independent argument. In fact, we follow the approach presented in the proof of Theorem 7.9.8 (see also [75] and [74]).

We consider the case when $\kappa > 0$ (in the case where $\kappa = 0$ the argument is simpler).

Let $\{U(t) \equiv (u(t); u_t(t); \theta(t)) : t \in \mathbb{R}\} \subset H$ be a full trajectory from the attractor $\mathscr{A}$. Let $|\sigma| < 1$. Applying Theorem 11.3.1 with $y^1 = U(s+\sigma)$, $y^2 = U(s)$ (and, accordingly, the interval $[s,t]$ in place of $[0,t]$), we have that

$$|U(t+\sigma) - U(t)|^2_{\alpha,\kappa} \le C_1 e^{-\omega(t-s)} |U(s+\sigma) - U(s)|^2_{\alpha,\kappa} \\ + C_2 \max_{\tau \in [s,t]} \|u(\tau+\sigma) - u(\tau)\|^2 \qquad (11.4.5)$$

for any $t, s \in \mathbb{R}$ such that $s \le t$ and for any σ with $|\sigma| < 1$. Letting $s \to -\infty$, (11.4.5) gives

$$|U(t+\sigma) - U(t)|^2_{\alpha,\kappa} \le C_2 \max_{\tau \in (-\infty,t]} \|u(\tau+\sigma) - u(\tau)\|^2$$

for any $t \in \mathbb{R}$ and $|\sigma| < 1$. On the attractor we obviously have

$$\frac{1}{|\sigma|} \|u(\tau+\sigma) - u(\tau)\| \le \frac{1}{|\sigma|} \left| \int_0^\sigma \|u_t(\tau+t)\| d\tau \right|, \quad t \in \mathbb{R}.$$

Therefore, by (11.4.3) we obtain

$$\max_{\tau \in \mathbb{R}} \left| \frac{U(\tau+\sigma) - U(\tau)}{\sigma} \right|^2_{\alpha,\kappa} \le C \ \text{ for } |\sigma| < 1. \qquad (11.4.6)$$

which implies

$$\|u_{tt}(t)\|^2 + \alpha \|\nabla u_{tt}(t)\|^2 + \|A_D u_t(t)\|^2 + \kappa \|\theta_t(t)\|^2 \le C, \quad t \in \mathbb{R},$$

with the constant C independent of α and κ.

Going back to the original equation (11.1.1) gives

$$\|\Delta\theta(t)\| \le C[\|A_D u_t(t)| + \kappa \|\theta_t(t)\|] \le C$$

and

$$\|M_\alpha{}^{-1/2} \Delta^2 u(t)\| \le C\|\Delta\theta(t)\| + \|F(u)\| + \|M_\alpha{}^{1/2} u_{tt}\| \le C \qquad (11.4.7)$$

The final conclusion needed for (11.4.4) follows by realizing that

$$\|u\|_3 \le C\|M_\alpha^{-1/2}\mathscr{A}u\| \quad \text{uniformly in } 0 \le \alpha < 1.$$

Indeed, this last assertion follows from (i) $\mathscr{A}^{-1/4}M_\alpha^{1/2}$ is bounded on $L_2(\Omega)$ with a norm independent on $\alpha > 0$, and (ii) $D(\mathscr{A}^{3/4}) \subset H^3(\Omega)$ (see [129]). If $\alpha = 0$, then we have $\|u\|_4 \le C$ directly from (11.4.7). The proof of Theorem 11.4.2 is thus completed. □

Proof of Corollary 11.4.3. Consider solutions $(u(t);u_t(t);\theta(t)) \in \mathscr{A}^{\alpha,\kappa}$ on the attractor which, on the strength of Theorem 11.4.2 belong to a bounded set in $H^3(\Omega)\times H^2(\Omega)\times H^2(\Omega)$. Decomposing $u = v+z$, $\theta = \xi+\beta$ we consider $(z;\xi)$ which satisfy

$$M_\alpha z_{tt} - \mu A_D\xi + A_D^2 z = F(u),$$

$$\xi_t + \eta A_D\xi + \mu A_D z_t = 0,$$

with zero initial conditions. We denote $Z \equiv (z;z_t;\xi)$ and $V \equiv (v;v_t;\beta)$. Denoting $\hat{z} \equiv A_D z$ and $\hat{\xi} = A_D\xi$ we obtain:

$$M_\alpha \hat{z}_{tt} - \mu A_D\hat{\xi} + A_D^2\hat{z} = A_D F(u),$$

$$\hat{\xi}_t + \eta A_D\hat{\xi} + \mu A_D\hat{z}_t = 0.$$

11.4.4. Lemma. *Let $\hat{Z}(t) \equiv (\hat{z};\hat{z}_t;\hat{\xi}) = (A_D z;A_D z_t;A_D\xi)$. Then $|\hat{Z}(t)|_{\alpha,\kappa} \le C_{\mathscr{A}}$ for all $t \ge 0$ uniformly in $(\alpha;\kappa) \in \Lambda$.*

Proof. We consider the evolution

$$\hat{Z}_t + A_{\alpha,\kappa}\hat{Z} = [0;M_\alpha^{-1}A_D F(u);0]^T, \quad \hat{Z}(0) = 0,$$

where the matrix operator $A_{\alpha,\kappa}$ is of the form (recall (5.3.4)):

$$A_{\alpha,\kappa} \equiv \begin{pmatrix} 0 & -I & 0 \\ M_\alpha^{-1}A_D^2 & 0 & -\mu M_\alpha^{-1}A_D \\ 0 & \mu\kappa^{-1}A_D & \eta\kappa^{-1}A_D \end{pmatrix}.$$

By Proposition 5.3.1 there exists $\omega > 0$ such that

$$|e^{-A_{\alpha,\kappa}t}\hat{Z}|_{\alpha,\kappa} \le Ce^{-\omega t}|\hat{Z}|_{\alpha,\kappa}, \tag{11.4.8}$$

where the constants $C,\omega > 0$ are independent of α and κ. By using the variation of parameters formula (see Proposition 5.3.2), followed by integration in time we obtain:

$$\hat{Z}(t) = A_{\alpha,\kappa}^{-1}\begin{bmatrix} 0 \\ M_\alpha^{-1}A_D F(u(t)) \\ 0 \end{bmatrix} - A_{\alpha,\kappa}^{-1}e^{-A_{\alpha,\kappa}t}\begin{bmatrix} 0 \\ M_\alpha^{-1}A_D F(u(0)) \\ 0 \end{bmatrix}$$

$$-\int_0^t e^{-A_{\alpha,\kappa}(t-\tau)} A_{\alpha,\kappa}^{-1} \begin{bmatrix} 0 \\ \frac{d}{d\tau} M_\alpha^{-1} A_D F(u(\tau)) \\ 0 \end{bmatrix} d\tau.$$

Because $A_{\alpha,\kappa}^{-1}[0;F;0]^T = [M_\alpha A_D^{-2} F;0;0]^T$, we have that

$$\hat{Z}(t) = \begin{bmatrix} A_D^{-1} F(u(t)) \\ 0 \\ 0 \end{bmatrix} - e^{-A_{\alpha,\kappa} t} \begin{bmatrix} A_D^{-1} F(u(0)) \\ 0 \\ 0 \end{bmatrix}$$
$$- \int_0^t e^{-A_{\alpha,\kappa}(t-\tau)} \begin{bmatrix} A_D^{-1} \langle F'(u(\tau)); u_t(\tau) \rangle \\ 0 \\ 0 \end{bmatrix} d\tau.$$

On the strength of (11.4.8) we obtain

$$|\hat{Z}(t)|_{\alpha,\kappa} \le \|F(u(t))\| + C\|F(u(0))\| + C \int_0^t e^{-\omega(t-\tau)} \|\langle F'(u(\tau)); u_t(\tau)\rangle\| d\tau.$$

Now properties of the von Karman bracket and the regularity given in (11.4.4) imply the conclusion of lemma. □

In order to prove the final conclusion of regularity we just note that $V(t) \equiv U(t) - Z(t)$ satisfies $V(t) = e^{-A_{\alpha,\kappa} t} U(0)$. Thus, by exponential decay of the linear semigroup (with the rates uniform in α and κ; see Proposition 5.3.1) we infer

$$|S_t U_0 - Z(t)|_{\alpha,\kappa} = |V(t)|_{\alpha,\kappa} = |e^{-A_{\alpha,\kappa} t} V(0)|_{\alpha,\kappa} \le C e^{-\omega t} |U(0)|_{\alpha,\kappa}. \quad (11.4.9)$$

By Lemma 11.4.4 we have that

$$Z(t) \in \mathscr{B} \equiv \left\{ (u_0; u_1; \theta_0) : \|A_D^2 u_0\|^2 + \|A_D u_1\|^2 + \alpha \|A_D^{3/2} u_1\|^2 + \kappa \|A_D \theta_0\|^2 \le C_R \right\}$$

uniformly in $t > 0$. By (11.4.9) this implies that $\mathscr{A}^{\alpha,\kappa} \subset \mathscr{B}$. The proof of Corollary 11.4.3 is thus completed. □

11.4.3 Finite-dimensionality

The result for every fixed α and κ follows from Theorem 7.9.6. In order to prove finiteness (with a uniform bound of the fractal dimension) of the global attractor $\mathscr{A}^{\alpha,\kappa}$ we repeat the argument given in Theorem 7.9.6 for the considered case.

We apply Theorem 7.3.3 in the space $\mathscr{H}_T = H_{\alpha,\kappa} \times W_1(0,T)$ with an appropriate value of T. Here

$$W_1(0,T) = \left\{ z \in L_2(0,T; \mathscr{D}(\mathscr{A}^{1/2})) : |z|^2_{W_1(0,T)} \equiv \int_0^T \left(\|A_D z\|^2 + \|z_t\|^2 \right) dt < \infty \right\}.$$

The norm in $\mathscr{H}_T$ is given by

$$\|U\|^2_{\mathscr{H}_T} = |V|^2_{\alpha,\kappa} + |z|^2_{W_1(0,T)}, \quad U = (V;z),\ V = (u_0;u_1;\theta_0).$$

Now for the definiteness we assume that $\kappa > 0$. Let $y^i = (u_0^i;u_1^i;\theta_0^i)$, $i = 1,2$, be two elements from the attractor $\mathscr{A}^{\alpha,\kappa}$. We denote

$$S_t y_i = (u^i(t);u^i_t(t);\theta^i(t)), \quad t \geq 0,\ i = 1,2,$$

and $Z(t) = S_t y_1 - S_t y_2 \equiv (z(t);z_t(t);\xi(t))$, where

$$(z(t);z_t(t);\xi(t)) \equiv (u^1(t) - u^2(t);u^1_t(t) - u^2_t(t);\theta^1(t) - \theta^2(t)).$$

From Theorem 11.3.1 (in the case $\kappa = 0$ we apply Remark 11.3.4) we have the estimate

$$|Z(t)|^2_{\alpha,\kappa} \leq C_0|Z(0)|^2_{\alpha,\kappa}e^{-\omega t} + C_1 \max_{\tau\in[0,t]} \|z(\tau)\|^2 \ \text{ for all } t \geq 0, \tag{11.4.10}$$

where the positive constants C_0, C_1, and ω do not depend on α and κ. Integrating (11.4.10) from T to $2T$ with respect to t we obtain

$$\int_T^{2T} |Z(t)|^2_{\alpha,\kappa}dt \leq \frac{C_0}{\omega}e^{-\omega T}|y_1 - y_2|^2_{\alpha,\kappa} + C_1 T \sup_{0\leq s\leq 2T} \|z(s)\|^2. \tag{11.4.11}$$

It also follows from (11.4.10) that

$$|Z(T)|^2_{\alpha,\kappa} \leq C_0 e^{-\omega T}\cdot|y_1 - y_2|^2_{\alpha,\kappa} + C_1\cdot \sup_{0\leq s\leq T} \|z(s)\|^2.$$

Therefore (11.4.11) implies that

$$|Z(T)|^2_{\alpha,\kappa} + \int_T^{2T} |Z(t)|^2_{\alpha,\kappa}dt \leq \eta_T^2|y_1 - y_2|^2_{\alpha,\kappa} + K_T^2 \sup_{0\leq s\leq 2T} \|z(s)\|^2, \tag{11.4.12}$$

where

$$\eta_T^2 = \frac{C_0(1+\omega)}{\omega}e^{-\omega T} \ \text{ and } \ K_T^2 = C_1(1+T).$$

By Proposition 5.5.1 $|Z(t)|^2_{\alpha,\kappa} \leq C_2|Z(0)|^2_{\alpha,\kappa}e^{at}$ for all $t \geq 0$, where the positive constants C_2 and a do not depend on α and κ. This implies

$$|Z(T)|^2_{\alpha,\kappa} + \int_T^{2T} |Z(t)|^2_{\alpha,\kappa}dt \leq L_T^2|y_1 - y_2|^2_{\alpha,\kappa}, \quad L_T^2 \equiv \frac{C_2(1+a)}{a}e^{2aT}. \tag{11.4.13}$$

Let $\mathscr{A}^{\alpha,\kappa}$ be the global attractor for $(H_{\alpha,\kappa}, S_t^{\alpha,\kappa})$. Consider in the space $\mathscr{H}_T$ the set

$$\mathscr{A}_T := \{U \equiv (u(0);u_t(0);\theta(0);u(t), t\in[0,T]) : (u(0);u_t(0);\theta(0)) \in \mathscr{A}^{\alpha,\kappa}\},$$

where $(u(t);\theta(t))$ is the solution to (11.1.7) with initial data $(u(0);u_t(0);\theta(0))$, and define operator $V : \mathscr{A}_T \mapsto H_T$ by the formula

$$V : (u(0);u_t(0);\theta(0);u(t)) \mapsto (u(T);u_t(T);\theta(T);u(T+t)).$$

It is clear from (11.4.13) that V is Lipschitz on $\mathscr{A}_T$ and

$$\|VU_1 - VU_2\|_{\mathscr{H}_T} \le L_T \|U_1 - U_2\|_{\mathscr{H}_T}, \quad U_1, U_2 \in \mathscr{A}_T.$$

Because $\mathscr{A}^{\alpha,\kappa}$ is a strictly invariant set (with respect to $S_t^{\alpha,\kappa}$) and it consists of full trajectories, we also have that $V\mathscr{A}_T = \mathscr{A}_T$.

From (11.4.12) we obtain

$$\|VU_1 - VU_2\|_{\mathscr{H}_T} \le \eta_T \|U_1 - U_2\|_{\mathscr{H}_T} + K_T \cdot [n_T(U_1 - U_2) + n_T(VU_1 - VU_2)],$$

for any $U_1, U_2 \in \mathscr{A}_T$, where $n_T(U) := \sup_{0 \le s \le T} \|u(s)\|$. Because $W_1(0,T)$ is compactly embedded into $C(0,T;L_2(\Omega))$, $n_T(U)$ is a compact seminorm on H_T and we can choose T such that $\eta_T < 1$. Therefore we can apply Theorem 7.3.3 which implies that $\mathscr{A}_T$ is a compact set in $\mathscr{H}_T$ of finite fractal dimension.

Let $\mathscr{P} : \mathscr{H}_T \mapsto H_{\alpha,\kappa}$ be the projection operator defined by the formula

$$\mathscr{P} : (u_0;u_1;\theta_0;z(t)) \mapsto (u_0;u_1;\theta_0).$$

Because $\mathscr{A}^{\alpha,\kappa} = \mathscr{P}\mathscr{A}_T$ and $\mathscr{P}$ is Lipshitz continuous, we have from Remark 7.3.4 that

$$\dim_f^{H_{\alpha,\kappa}} \mathscr{A}^{\alpha,\kappa} \le \dim_f^{\mathscr{H}_T} \mathscr{A}_T \le \left[\ln \frac{2}{1+\eta_T}\right]^{-1} \cdot \ln m_0\left(\frac{4K_T(1+L_T^2)^{1/2}}{1-\eta_T}\right), \tag{11.4.14}$$

Here $\dim_f^Y$ stands for the fractal dimension of a set in the space Y and $m_0(R)$ is the maximal number of pairs (x_i, y_i) in $\mathscr{H}_T \times \mathscr{H}_T$ possessing the properties

$$\|x_i\|_{\mathscr{H}_T}^2 + \|y_i\|_{\mathscr{H}_T}^2 \le R^2, \quad n_T(x_i - x_j) + n_T(y_i - y_j) > 1, \quad i \ne j.$$

It is clear that $m_0(R)$ can be estimated by the maximal number of pairs (x_i, y_i) in $W_1(0,T) \times W_1(0,T)$ possessing the properties

$$\|x_i\|_{W_1(0,T)}^2 + \|y_i\|_{W_1(0,T)}^2 \le R^2, \quad n_T(x_i - x_j) + n_T(y_i - y_j) > 1, \quad i \ne j.$$

Thus the bound in (11.4.14) does not depend on α and κ because the constants η_T, K_T, and L_T are independent of $(\alpha,\kappa) \in \Lambda$. This concludes the proof.

As we have seen above our approach to estimate the dimension of the global attractor is based on the stabilizability estimate in Theorem 11.3.1 and some ideas which go back to the theorem by Ladyzhenskaya (see Theorem 7.3.2 and also [172] and the references therein) on the dimension of invariant sets. The approach used here does not provide (the same as Ladyzhenskaya approach) the best possible constants in the dimension estimate. However, it gives bounds for dimension that do not depend on α and κ. There are also other approaches that could be explored.

- The approach by Babin–Vishik–Temam–Foias [17, 273] relying on the computations of Lyapunov exponents usually gives the best possible estimate. If we equip the phase space $H_{\alpha,\kappa}$ with the scalar product generated by the corresponding (to the problem) Lyapunov function, then calculations are rather simple, but they give the dependence on α^{-1}. The reason for this is that we have critical nonlinearity in the limit $\alpha \to 0$ (we refer to [273] for some representative calculations for the case of subcritical wave dynamics).
- Another approach can be applied in the case of gradient systems and based on the structure of the attractor as unstable manifold of equilibria is due to Babin–Vishik [17]. It can be applied in the case of hyperbolic equilibria and relies on the formula in (11.2.7) (the dimension is the maximal instability index of equilibria). In principle, this approach makes it also possible to calculate lower bounds for the dimension. We illustrate this calculation by the following example.

11.4.5. Example (Lower bound for dimension). We assume that $p \equiv 0$ and $F_0(x) = -\nu(x_1^2+x_2^2)/2$ in (11.1.1) and consider hinged boundary conditions (11.1.5). In this case $E = (0;0;0) \equiv 0$ is an equilibrium. Assume that E is hyperbolic. Then we have that $\mathscr{A}^{\alpha,\kappa} \supseteq \mathscr{M}^u(E)$ and thus $\dim_f \mathscr{A}^{\alpha,\kappa} \geq \operatorname{ind}(E)$, where $\operatorname{ind}(E)$ is the instability index of E (see Definition 7.5.13).

Because in the case considered $\langle F'(0);w\rangle = [w,F_0] = -\nu\Delta w = \nu A_D w$ and $\mathscr{A} = A_D^2$, by Proposition 5.5.2 the linearized semigroup $L_t = D[S_t E]$ is generated by the problem

$$\begin{cases} M_\alpha u_{tt} - \mu A_D\theta + A_D^2 u - \nu A_D u = 0, & u|_{t=0} = u_0,\ u_t|_{t=0} = u_1, \\ \kappa\theta_t + \eta A_D\theta + \mu A_D u_t = 0, & \theta|_{t=0} = \theta_0. \end{cases} \tag{11.4.15}$$

In variables $Y = (A_D u; u_t; \theta)$ problem (11.4.15) can be written in the form

$$\frac{d}{dt}Y + \mathscr{L}Y = 0, \quad Y(0) = Y_0 \equiv (A_D u_0; u_1; \theta_0),$$

where the operator $\mathscr{L}$ is given by

$$\mathscr{L} = -A_D \begin{pmatrix} 0 & I & 0 \\ -M_\alpha^{-1}(I - \nu A_D^{-1}) & 0 & \mu M_\alpha^{-1} \\ 0 & -\mu/\kappa & -\eta/\kappa \end{pmatrix}.$$

Our goal is to give a lower bound for the number of eigenvalues of $\mathscr{L}$ with negative real parts.

Introducing the notations ω_n for eigenvalues of A_D and $m_{n,\alpha} = (1+\alpha\omega_n)^{-1}$, by self-adjoint operator calculus it suffices to evaluate eigenvalues of a 3×3 matrix

$$M = \begin{pmatrix} 0 & 1 & 0 \\ -m_{n,\alpha}(1 - \nu\omega_n^{-1}) & 0 & \mu m_{n,\alpha} \\ 0 & -\mu/\kappa & -\eta/\kappa \end{pmatrix},$$

whose characteristic polynomial is

$$\mathrm{Det}|M - zI| = -\left[z^3 + \frac{\eta}{\kappa}z^2 + m_{n,\alpha}\left(\frac{\mu^2}{\kappa} + 1 - \frac{\nu}{\omega_n}\right)z + m_{n,\alpha}\frac{\eta}{\kappa}\left(1 - \frac{\nu}{\omega_n}\right)\right].$$

A lower bound for $\mathrm{ind}\,(E)$ can be obtained by calculating a number of indices n for which the matrix M has at least one eigenvalue with positive real part.

The well-known *Hurwitz criterion* applied to the third-order polynomial

$$f(z) = z^3 + a_1 z^2 + a_2 z + a_3$$

states that all roots of $f(z)$ have negative real parts if and only if

$$a_1,\ a_2,\ a_3 > 0, \quad a_1 a_2 > a_3.$$

Applying this criterion to $-\mathrm{Det}|M - zI|$ one can see that all eigenvalues of M have negative real parts if and only if $1 - \nu\omega_n^{-1} > 0$. Thus we can guarantee the existence of at least one eigenvalue with a positive real part if and only if $1 - \nu\omega_n^{-1} < 0$. This implies that

$$\dim_f \mathscr{A}^{\alpha,\kappa} \geq \mathrm{ind}\,(E) \geq \text{cardinality of } \{n : \omega_n < \nu\}.$$

Because $\omega_n \sim c_0 n$ for large n, we obtain that $\dim_f \mathscr{A}^{\alpha,\kappa} \geq c_0\nu$ for large $\nu > 0$, where the constant $c_0 > 0$ does not depend on α, κ, η, and μ.

We also note that we cannot derive dependence on ν of upper bounds for the dimension in the explicit form. The main reason is that we cannot obtain explicit (with respect to ν) estimates of the size of the attractor even in the special case considered. The point is that the size of the attractor $\mathscr{A}^{\alpha,\kappa}$ depends on the constant $C_{F_0,p}$ from (11.4.1). This constant $C_{F_0,p}$ appears via Lemma 1.5.4 with $A = c_1\|F_0\|_2^2 + c_2$, which is obtained by the contradiction argument. Thus we have no explicit dependence of $C_{F_0,p}$ on $\|F_0\|_2$. ∎

11.4.4 Upper semicontinuity

The argument is standard (see, e.g., [17, 61] and [246] and also Sections 9.3.2 and 9.5.5) and relies on the uniform estimate given in Theorem 11.4.2.

Assume that (11.2.4) (or (11.2.5) in the case $\kappa_0 = 0$) does not hold. Let κ_* be positive. Then there exist a sequence $\lambda_n = (\alpha_n, \kappa_n)$ such that $\lambda_n \to \lambda_* \equiv (\alpha_*, \kappa_*)$ and a sequence $U_n \in \mathscr{A}^{\alpha_n,\kappa_n}$ such that

$$\mathrm{dist}_{H_{\alpha_*,\kappa_*}}(U_n, \mathscr{A}^{\alpha_*,\kappa_*}) \geq \delta > 0, \quad n = 1, 2, \ldots. \tag{11.4.16}$$

Let $\{U_n(t) : t \in \mathbb{R}\}$ be a full trajectory from the attractor $\mathscr{A}^{\alpha_n,\kappa_n}$ such that $U_n(0) = U_n$. By (11.2.1) and (11.2.2) (see also (11.4.4)) from Aubin's compactness theorem (see Theorem 1.1.8) we can conclude that there exists a sequence $\{n_k\}$ and a function $U(t) \in C_{bnd}(\mathbb{R}; H_{\alpha_*,\kappa_*})$ such that

$$\max_{t\in[-T,T]} |U_{n_k}(t) - U(t)|_{\alpha_*,\kappa_*} \to 0 \quad \text{as } k \to \infty.$$

Using the variational form of problem (11.1.7) one can see that $U(t)$ solves the limiting equations ($\alpha = \alpha_*, \kappa = \kappa_*$). Because $|U(t)|_{\alpha_*,\kappa_*} \le R$ for all $t \in \mathbb{R}$ and some $R > 0$, the trajectory $\{U(t) : t \in \mathbb{R}\}$ belongs to the attractor $\mathscr{A}^{\alpha_*,\kappa_*}$ (see Theorem7.2.2). Consequently, $U_{n_k} \to U(0) \in \mathscr{A}^{\alpha_*,\kappa_*}$, which contradicts (11.4.16). In the case when $\kappa_* = 0$ an argument is similar. Thus the proof of Theorem 11.2.1 is complete.

In the case when the attractor is exponential (i.e. (11.2.6) holds) we can prove stronger continuity properties of the family of global attractors $\{\mathscr{A}^{\alpha,\kappa}\}$ for interior points from the parameter set Λ. Namely using the methods presented in [17, Chapter 8] and [135] (see also Section 3.1 in [246] and Theorem 7.3 in [61]), we can establish the following assertion which gives a bound for the Hausdorff semidistance between the attractors.

11.4.6. Proposition. *Assume that* (11.2.6) *holds and* $\alpha_*, \kappa_* > 0$. *Then there exist* $\rho \in (0,1)$ *and* $C > 0$ *such that*

$$h(\mathscr{A}^{\alpha,\kappa} \,|\, \mathscr{A}^{\alpha_*,\kappa_*}) \le C(|\alpha - \alpha_*| + |\kappa - \kappa_*|)^\rho, \tag{11.4.17}$$

where $h(A|B) = \sup\{\mathrm{dist}_{H_{1,1}}(U,B) : U \in A\}$ *is the Hausdorff semidistance.*

Proof. We can assume that $\alpha, \kappa, \alpha_*, \kappa_* \ge \delta$ for some $\delta > 0$. In this case all the spaces $H_{\alpha,\kappa}$ are the same, $H_{\alpha,\kappa} = H_{1,1}$. Moreover, the norms are equivalent,

$$c|U|_{1,1} \le |U|_{\alpha,\kappa} \le c^{-1}|U|_{1,1}, \quad U \in H_{\alpha,\kappa},$$

where $c = c_\delta > 0$ does not depend on $\alpha, \kappa \ge \delta$. We also have from Theorem 11.2.1 that

$$\bigcup_{\delta\le\alpha,\kappa\le 1} \mathscr{A}^{\alpha,\kappa} \subset D_R \equiv \{U = (u_0, u_1, \theta) \in H_{1,1} : \|u_0\|_3 + \|u_1\|_2 + \|\theta\|_2 \le R\}$$

for some $R = R_\delta > 0$. By the standard argument we can also show that

$$\left|S_t^{\alpha,\kappa} U - S_t^{\alpha_*,\kappa_*} U\right|_{1,1} \le C_R e^{a_R t} \left(|\alpha - \alpha_*| + |\kappa - \kappa_*|\right)^{1/2} \tag{11.4.18}$$

for some C_R and a_R. Therefore using Proposition 3.3 in [246] we arrive at the conclusion. $\square$

Unfortunately we are not able to establish an assertion similar to Proposition 11.4.6 in the case $\alpha_* = 0$. The main reason is that we cannot prove an analogue of (11.4.18) for this case.

11.5 Exponential rate of attraction—Proof of Theorem 11.2.2

The proof of Theorem 11.2.2 is based on application of Theorem 7.5.15. For this we need to verify: (i) S_t is Fréchet differentiable, (ii) there exists a decreasing Lyapunov function, (iii) $\mathcal{N}$ is finite and all equilibria are hyperbolic, and (iv) injectivity of S_t and of its Fréchet derivative. In fact, the first three conditions imply the statement in the first part of theorem 11.2.2, and the last additional requirement guarantees the second part of that Theorem. As for (i), this follows from Proposition 5.5.2. The needed Lyapunov function is given by the energy $\mathscr{E}_{\alpha,\kappa}$ (see (11.1.12)). Finiteness of the set of equilibria $\mathcal{N}$ is assumed in the statement of the Theorem 11.2.2. Injectivity properties required in (iv) follow from the backward uniqueness result given in Theorem 5.6.1 for $\alpha > 0$ and in Corollary 5.6.9 in the case $\alpha = 0$ and $\mathscr{A} = A_D^2$. Thus, it remains to verify the second statement in (iii). This entails proving that "static" hyperbolicity of equilibria points assumed in Theorem 11.2.2 implies "dynamic" hyperbolicity required by Theorem 7.5.15. The needed link is stated and proved below.

11.5.1. Proposition. *An equilibrium $E = (e;0;0) \in \mathcal{N}$ is hyperbolic in the static sense (i.e., the equation $\mathscr{A}w = \langle F'(e);w\rangle$, where $F'(u)$ is Fréchet derivative of the mapping F given by* (11.1.6), *has no nontrivial solutions) if and only if it is hyperbolic in dynamical sense (see Definition* 7.5.13*).*

To prove this proposition we rely on the following assertion on spectral properties of compact perturbations of strongly continuous semigroups.

11.5.2. Lemma. *Let e^{At} and e^{Bt} ($t \in \mathbb{R}_+$) be strongly continuous semigroups in a reflexive Banach space X with the generators A and B. Assume that the operator $K(t) \equiv e^{Bt} - e^{At}$ is compact for some $t > 0$ and introduce the notations: $\sigma(L)$ for the spectrum of an operator L, $\sigma_p(L)$ for its point spectrum, and $\sigma_{ess}(L)$ for the essential spectrum (in the sense due to Kato* [159], *Section* IV.5.6*). Then the semigroup e^{Bt} is (dynamically) hyperbolic in the sense that*

$$\sigma(e^{Bt}) \cap \{z \in \mathbb{C} : |z| = 1\} = \emptyset, \tag{11.5.1}$$

provided that one of the following conditions holds,

(a) $\sigma_p(B) \cap \{i\mathbb{R}\} = \emptyset$ *and* $\sigma(e^{At}) \cap \{z \in \mathbb{C} : |z| = 1\} = \emptyset$;
(b) $[\sigma_p(B) \cup \sigma_p(B^*)] \cap \{i\mathbb{R}\} = \emptyset$ *and* $\sigma_{ess}(e^{At}) \cap \{z \in \mathbb{C} : |z| = 1\} = \emptyset$.

Below we check condition (a) for our model. The second condition (b) is included for the sake of completeness.

Proof. (a): Let $z \in \mathbb{C}$ and $|z| = 1$. We have that $z \in \rho(e^{At})$, where $\rho(\cdot)$ denotes resolvent set of an operator. By the spectral mapping theorem (see, e.g., [241, Section 2.2]) we have that

$$\sigma_p(e^{Bt}) \subset e^{\sigma_p(B)t} \cup \{0\} \not\ni \{z \in \mathbb{C} : |z| = 1\}; \tag{11.5.2}$$

that is, z is not in the point spectrum of e^{Bt}. To show that $z \in \rho(e^{Bt})$ we write

$$zI - e^{Bt} = zI - e^{At} - K(t) = \left(zI - e^{At}\right)\left[I - R\left(z, e^{At}\right)K(t)\right], \tag{11.5.3}$$

where $R\left(z, e^{At}\right) \equiv \left(zI - e^{At}\right)^{-1}$ is the resolvent for e^{At}. We note that 1 is not an eigenvalue of $R(z, e^{At})K(t)$. Indeed, if this is not true, then by (11.5.3) z is eigenvalue for e^{Bt} which is impossible. It is also clear that $R(z, e^{At})K(t)$ is a compact operator. Thus by the Fredholm theorem $I - R(z, e^{At})K(t)$ is bounded invertible on the entire space X. Hence the operator

$$R(z, e^{Bt}) = \left[I - R(z, e^{At})K(t)\right]^{-1} R(z, e^{At})$$

is a well-defined bounded operator on X, which proves that z is in the resolvent set of e^{Bt}. Thus (11.5.1) holds.

(b): Because $\sigma(e^{Bt}) = \sigma_p(e^{Bt}) \cup \sigma_p(e^{B^*t}) \cup \sigma_{ess}(e^{Bt})$, by (11.5.2) written for both B and B^* it is sufficient to show that $\sigma_{ess}(e^{Bt}) \cap \{z \in \mathbb{C} : |z| = 1\} = \emptyset$. The latter follows from the generalized Weyl theorem (see Theorem 5.35 in [159]) which gives us the relation $\sigma_{ess}(e^{Bt}) = \sigma_{ess}(e^{At})$. □

Proof of Proposition 11.5.1. We consider the case $\kappa > 0$ (in the case $\kappa = 0$ arguments are similar).

PRELIMINARY STEP: We first note that $F'(e)$ is a symmetric operator in $\mathscr{H} = L_2(\Omega)$ on the domain $D(A_D)$ obeying the estimate $\|\langle F'(e); w\rangle\| \le C\|A_D w\|$ which follows from the additional regularity of equilibria $e \in H^3(\Omega)$. This implies the operator

$$Cu = \mathscr{A}u - \langle F'(e); u\rangle, \quad u \in \mathscr{D}(C) \equiv \mathscr{D}(\mathscr{A}),$$

is self-adjoint in $\mathscr{H}$. Moreover, by Proposition 5.5.2 the map $L_t \equiv D\left[S_t^{\alpha,\kappa}E\right]$ is governed by the dynamics

$$M_\alpha u_{tt} - \mu A_D\theta + Cu = 0$$

$$\kappa\theta_t + \eta A_D\theta + \mu A_D u_t = 0.$$

We can write this system as

$$\frac{d}{dt}W(t) + \mathscr{L}W(t) = 0, \quad W(t) = (u(t); u_t(t); \theta(t)),$$

where the operator $\mathscr{L}$ is defined by the formula

$$\mathscr{L}Y = \begin{pmatrix} -v \\ M_\alpha^{-1}(-\mu A_D\theta + Cu) \\ \eta\kappa^{-1}A_D\theta + \mu\kappa^{-1}A_Dv \end{pmatrix}, \quad Y = \begin{pmatrix} u \\ v \\ \theta \end{pmatrix},$$

on the natural domain in $H_{\alpha,\kappa}$. By the spectral mapping theorem (see, e.g., [241, Section 2.2] we have that

$$\sigma(L_t) \equiv \sigma\left(e^{-\mathscr{L}t}\right) \supset e^{-\sigma(\mathscr{L}t)} \supset e^{-\sigma_p(\mathscr{L}t)}, \tag{11.5.4}$$

where σ_p stands for the point spectrum.

STEP 1: It is obvious that dynamical hyperbolicity implies a static one. Indeed, if $\sigma(L_t) \cap \{z \in \mathbb{C} : |z| = 1\} = \emptyset$, then by (11.5.4) $\lambda = 0$ cannot be in the spectrum of $\mathscr{L}$. Thus $\mathscr{L}Y = 0$ implies $Y = 0$. However, this is possible if and only if $Cw = 0$ implies $w = 0$.

STEP 2: Now we prove that $\sigma(L_t) \cap \{z \in \mathbb{C} : |z| = 1\} = \emptyset$ provided $\{Cu = 0 \Rightarrow u = 0\}$. We split our argument in the three substeps.

STEP 2A: To apply Lemma 11.5.2 we use the following obvious decomposition of $L_t = e^{-\mathscr{L}t}$,

$$L_t = e^{-\mathscr{L}_0 t} + K(t), \tag{11.5.5}$$

where $\mathscr{L}_0$ is the standard thermoelastic operator ($\mathscr{L}_0 = \mathscr{L}$ if $C = \mathscr{A}$) and $K(t)$ is defined by the formula

$$K(t) = \int_0^t e^{-\mathscr{L}_0(t-\tau)} \begin{pmatrix} 0 & 0\,0 \\ M_\alpha{}^{-1}F'(e) & 0\,0 \\ 0 & 0\,0 \end{pmatrix} e^{-\mathscr{L}\tau} d\tau.$$

One can see that $K(t)$ is a compact operator in $H_{\alpha,\kappa}$ for every $\alpha \geq 0$. Indeed, in the case $\alpha > 0$ this follows from the fact that $M_\alpha^{-1/2}F'(e)$ is a compact operator from $H^2(\Omega)$ into $L_2(\Omega)$. If $\alpha = 0$, we evoke analyticity of the linear semigroup $e^{-\mathscr{L}_0 t} \equiv e^{-\mathscr{L}_{0,\kappa}t}$ which gives us the smoothing effect (see (5.3.29)).

STEP 2B: By Proposition 5.3.1 $e^{-\mathscr{L}_0 t}$ is uniformly exponentially stable. Therefore

$$\sigma\left(e^{-\mathscr{L}_0 t}\right) \cap \{z \in \mathbb{C} : |z| = 1\} = \emptyset.$$

STEP 2C: Thus to check condition (a) in Lemma 11.5.2 it remains to prove that the condition $\{Cu = 0 \Rightarrow u = 0\}$ (implied by static hyperbolicity) leads to

$$\sigma_p(\mathscr{L}) \cap \{z \in \mathbb{C} : Re\, z = 0\} = \emptyset, \tag{11.5.6}$$

where σ_p refers to point spectrum. Let $\lambda = i\omega$. Explicitly writing the eigensystem associated with $\mathscr{L}$

$$\begin{pmatrix} -v \\ M_\alpha^{-1}(-\mu A_D\theta + Cu) \\ \eta\kappa^{-1}A_D\theta + \mu\kappa^{-1}A_Dv \end{pmatrix} = \lambda \begin{pmatrix} u \\ v \\ \theta \end{pmatrix}, \tag{11.5.7}$$

leads to the following equalities

$$\begin{aligned}
-\mu(A_D\theta,u)+(Cu,u) &= \omega^2\|M_\alpha^{1/2}u\|^2\\
\eta(A_D\theta,u)-i\omega\mu\|A_D^{1/2}u\|^2 &= i\omega\kappa(\theta,u)\\
\eta\|A_D^{1/2}\theta\|^2-i\omega\mu(A_Du,\theta) &= i\omega\kappa\|\theta\|^2.
\end{aligned} \tag{11.5.8}$$

It follows from the first identity in (11.5.8) that $(A_D\theta,u)=Re(A_D\theta,u)$. Therefore taking real parts in the third identity we obtain that $\|A_D^{1/2}\theta\|=0$. Because $A_D^{1/2}$ is injective (in fact, positive), $\theta=0$. Thus, from the second identity in (11.5.8) we infer that $u=0$ whenever $\omega\neq 0$ (using, again, the injectivity of $A_D^{1/2}$). Consequently (11.5.7) has no nonzero solution provided $\lambda=i\omega\neq 0$.

On the other hand, when $\lambda=0$, the first and third relations in (11.5.7) imply that $v=\theta=0$. Thus the second relation in (11.5.7) gives that $M_\alpha^{-1}Cu=0$. Hence $Cu=0$ and, on the strength of our assumption, $u=0$, as desired for (11.5.6). Thus the application of Lemma 11.5.2 concludes the proof of Proposition 11.5.1. □

11.5.3. Remark. We note that condition (b) in Lemma 11.5.2 is also true in our present case. Indeed, we need to check only that the condition $\{Cu=0\Rightarrow u=0\}$ implies that

$$\sigma_p(\mathscr{L}^*)\cap\{z\in\mathbb{C}:\ Re\,z=0\}=\emptyset. \tag{11.5.9}$$

A simple calculation shows that

$$\mathscr{L}^*Y=\begin{pmatrix}\mathscr{A}^{-1}Cv\\ M_\alpha^{-1}(\mu A_D\theta-\mathscr{A}u)\\ \eta\kappa^{-1}A_D\theta-\mu\kappa^{-1}A_Dv\end{pmatrix},\quad Y=\begin{pmatrix}u\\ v\\ \theta\end{pmatrix},$$

on the natural domain in $H_{\alpha,\kappa}$. Let $\lambda=i\omega$. Then the spectral relation $\mathscr{L}^*Y=i\omega Y$ can be written in the form

$$\begin{aligned}
Cv &= i\omega\mathscr{A}u,\\
\mu A_D\theta-\mathscr{A}u &= i\omega M_\alpha v,\\
\eta A_D\theta-\mu A_Dv &= i\omega\kappa\theta.
\end{aligned} \tag{11.5.10}$$

Let $\omega\neq 0$. It follows from the first equality in (11.5.10) that $Re(\mathscr{A}u,v)=0$. Therefore multiplying the second equality by v we find that $Re(A_D\theta,v)=0$. Now the third equality implies that $\theta=0$ and thus $v=0$. Consequently from the first equality we have that $u=0$. On the other hand, if $\omega=0$, then the property $\{Cv=0\Rightarrow v=0\}$ and the first equality in (11.5.10) imply that $v=0$. Two other equalities gives us that $u=\theta=0$. Thus the proof of (11.5.9) is complete. ■

Chapter 12
Composite Wave–Plate Systems

12.1 Introduction

The aim of this chapter is to discuss long-time behavior of composite wave–plate models. These include:

- Structural acoustic model with isothermal von Karman plate.
- Structural acoustic model with thermoelastic Karman plate.
- Flow-structure interaction.

We recall that these models involve coupling between the wave equation and plate equation. The coupling occurs on the interface separating two media: acoustic field and the structure (wall). The existence and uniqueness theory associated with these models has been already developed in Chapter 6. The present chapter is devoted to long-time dynamics and global attractors associated with the models.

12.2 Structural acoustic problems

Our goal here is to present results on long-time behavior of flows generated by composite wave–plate models given by (6.2.1) and (6.2.2). We restrict ourselves to the case $\alpha = 0$ of absent rotational inertia (as we already can see from the previous chapters, the case $\alpha > 0$ is usually much simpler). Thus, we discuss existence and properties of attractors associated with the following structural acoustic problem

$$\begin{cases} z_{tt} + g(z_t) - \Delta z + f(z) = 0 & \text{in } \mathscr{O} \times (0,T), \\ \dfrac{\partial z}{\partial n} = 0 & \text{on } \Omega_* \times (0,T), \\ \dfrac{\partial z}{\partial n} = \gamma \kappa u_t & \text{on } \Omega \times (0,T), \\ z(0,\cdot) = z^0, \quad z_t(0,\cdot) = z^1 & \text{in } \mathscr{O}, \end{cases} \tag{12.2.1}$$

I. Chueshov, I. Lasiecka, *Von Karman Evolution Equations*,
Springer Monographs in Mathematics, DOI 10.1007/978-0-387-87712-9_12,

and

$$\begin{cases} u_{tt} + b_0(u_t) + \Delta^2 u - [u, v(u) + F_0] + \beta\kappa z_t|_\Omega = p_0 & \text{in } \Omega \times (0,T), \\ u = \dfrac{\partial u}{\partial n} = 0 & \text{on } \partial\Omega \times (0,T), \\ u(0,\cdot) = u^0(x), \quad u_t(0,\cdot) = u^1(x) & \text{in } \Omega, \end{cases} \tag{12.2.2}$$

where as usual $v = v(u)$, the Airy stress function, is a solution of the problem

$$\Delta^2 v + [u,u] = 0, \quad v|_{\partial\Omega} = \frac{\partial v}{\partial n}\Big|_{\partial\Omega} = 0. \tag{12.2.3}$$

Here $\mathscr{O} \subset \mathbb{R}^3$ is a smooth bounded domain, with the boundary $\partial\mathscr{O} = \overline{\Omega \cup \Omega_*}$ consisting of two open (in induced topology) connected disjoint parts Ω and Ω_* of positive measure; Ω is flat and is referred to as the elastic wall. Well-posedness of this problem was studied in Section 6.2. It was shown that under Assumption 6.2.1 the problem (12.2.1) and (12.2.2) generates the dynamical system (Y, S_t) with the phase space Y given by (6.2.18),

$$Y = \mathscr{D}(A^{1/2}) \times L^2(\mathscr{O}) \times \mathscr{D}(\mathscr{A}^{1/2}) \times L^2(\Omega), \tag{12.2.4}$$

where we have used notations of Section 6.2 ($\mathscr{D}(A^{1/2}) = H^1(\mathscr{O})$ and $\mathscr{D}(\mathscr{A}^{1/2}) = H_0^2(\Omega)$). The evolution operator S_t describing the dynamics governed by $(z; z_t; u; u_t)$ is defined by (6.2.40).

Our first result in this section, Theorem 12.2.3, states the existence of a global attractor for problem (12.2.1) and (12.2.2) under rather general conditions on the nonlinear functions g, f, and b_0; see Assumption 12.2.2. In particular, it is not required that the damping functions g and b_0 are (i) differentiable, nor (ii) strictly increasing. In the case $\kappa = 0$ (no coupling) we recover previous results for the uncoupled wave and plate equations.

Our second result, Theorem 12.2.6, deals with the dimension and smoothness of the global attractor. It requires additional (structural) hypotheses concerning the damping functions g and b_0 and the nonlinear force f; see Assumption 12.2.4. In particular, strong monotonicity of g and b_0 is assumed. In the case $\kappa = 0$, this result recovers the results previously established in [69], [75, Chapter 5] and [74] for the wave equation, and in chapters above for von Karman equation.

The arguments provided in this section mostly follow the treatment given in [38] in the case of the Berger plate model.

We conclude this section by discussing several properties of the energy functionals and stationary solutions.

It follows from (6.2.27) that the energy $\mathscr{E}(z^0, z^1, u^0, u^1)$ (which is given by (6.2.25)) is bounded from below on Y and

$$\mathscr{E}(z^0, z^1, u^0, u^1) \to +\infty \quad \text{when } \|(z^0; z^1; u^0; u^1)\|_Y \to +\infty.$$

This, in turn, implies that there exists $R_* > 0$ such that the set

$$\mathscr{W}_R = \left\{ y = (z^0; z^1; u^0; u^1) \in Y : \mathscr{E}(z^0, z^1, u^0, u^1) \le R \right\} \tag{12.2.5}$$

is a nonempty bounded set in Y for all $R \ge R_*$. Moreover any bounded set $B \subset Y$ is contained in $\mathscr{W}_R$ for some R and, as it follows from (6.2.31), the set $\mathscr{W}_R$ is invariant with respect to the semiflow S_t; that is, $S_t \mathscr{W}_R \subset \mathscr{W}_R$ for all $t > 0$. Thus we can consider the restriction $(\mathscr{W}_R, S_t)$ of the dynamical system (Y, S_t) on $\mathscr{W}_R$, $R \ge R_*$.

We introduce next the set of stationary points of S_t denoted by $\mathscr{N}$,

$$\mathscr{N} = \{ V \in Y : S_t V = V \text{ for all } t \ge 0 \}.$$

Every stationary point V has the form $V = (z; 0; u; 0)$, where $z \in H^1(\mathscr{O})$ and $u \in H^2(\Omega) \cap H_0^1(\Omega)$ are a weak (variational) solution to the problems

$$-\Delta z + f(z) = 0 \ \text{ in } \mathscr{O}, \quad \frac{\partial z}{\partial n} = 0 \ \text{ on } \partial\mathscr{O}, \tag{12.2.6}$$

and

$$\Delta^2 u - [u, v(u) + F_0] = p_0 \ \text{ in } \Omega, \quad u = \frac{\partial u}{\partial n} = 0 \ \text{ on } \partial\Omega, \tag{12.2.7}$$

where $v = v(u)$ is the Airy stress function given by (12.2.3). Using the properties of the potentials Φ and Π given by (6.2.15) and (6.2.17) one can easily prove the following assertion.

12.2.1. Lemma. *Under Assumption* 6.2.1 *the set $\mathscr{N}$ of stationary points for the semiflow S_t generated by equations* (12.2.1) *and* (12.2.2) *is a closed bounded set in Y, and hence there exists $R_{**} \ge R_*$ such that $\mathscr{N} \subset \mathscr{W}_R$ for every $R \ge R_{**}$.*

12.2.1 The statement of main results

The goal is to show existence of a global attractor (along with its properties) for the dynamical system generated by (12.2.1) and (12.2.2). For this we need additional hypotheses concerning the damping functions g and b_0.

12.2.2. Assumption. In addition to Assumption 6.2.1, suppose that for any $\varepsilon > 0$ there exists $c_\varepsilon > 0$ such that

$$s^2 \le \varepsilon + c_\varepsilon s g(s) \ \text{ for } s \in \mathbb{R}, \tag{12.2.8}$$

and

$$s^2 \le \varepsilon + c_\varepsilon s b_0(s) \ \text{ for } s \in \mathbb{R}. \tag{12.2.9}$$

Validity of inequalities (12.2.8) and (12.2.9) is discussed in Remark 9.1.6. In particular, Assumption 12.2.2 allows the damping functions g and b_0 to be constants on some closed finite intervals that are away from zero.

We recall some notations from Section 6.2.

- $A : \mathscr{D}(A) \subset L_2(\mathscr{O}) \to L_2(\mathscr{O})$ is a positive self-adjoint operator defined by

$$Ah = -\Delta h + \mu h, \quad \mathscr{D}(A) = \left\{ h \in H^2(\mathscr{O}) : \left.\frac{\partial h}{\partial n}\right|_{\partial \mathscr{O}} = 0 \right\}, \tag{12.2.10}$$

where $\mu > 0$ is given by (6.2.6).

- N_0 is the Neumann map from $L_2(\Omega)$ to $L_2(\mathscr{O})$, defined by

$$\psi = N_0 \varphi \Longleftrightarrow \left\{ (-\Delta + \mu)\psi = 0 \text{ in } \mathscr{O};\ \left.\frac{\partial \psi}{\partial n}\right|_{\Omega} = \varphi,\ \left.\frac{\partial \psi}{\partial n}\right|_{\Omega_*} = 0 \right\}. \tag{12.2.11}$$

- $\mathscr{A} : \mathscr{D}(\mathscr{A}) \subset L_2(\Omega) \to L_2(\Omega)$ is a positive selfadjoint operator defined by

$$\mathscr{A} w = \Delta^2 w, \quad \mathscr{D}(\mathscr{A}) = H^4(\Omega) \cap H_0^2(\Omega). \tag{12.2.12}$$

- $D(h) := g(h)$, $F_1(z) = f(z) - \mu z$.
- $B(w) := b_0(w)$, $F_2(u) = -[u, v(u) + F_0] - p_0$, where $v = v(u)$ is the Airy stress function given by (12.2.3).

With this notation we rewrite the problem (12.2.1) and (12.2.2) in the abstract form

$$\begin{aligned} &z_{tt} + A\,(z - \gamma\kappa N_0 u_t) + D(z_t) + F_1(z) = 0, \\ &u_{tt} + \mathscr{A} u + B(u_t) + \beta\kappa N_0^* A z_t + F_2(u) = 0, \\ &z(0) = z^0,\ z_t(0) = z^1,\ u(0) = u^0,\ u_t(0) = u^1. \end{aligned} \tag{12.2.13}$$

Our first main result in this section is the following theorem.

12.2.3. Theorem (Existence of global attractor). *Under Assumption* 12.2.2 *the dynamical system* (Y, S_t) *generated by problem* (12.2.13) *has a compact global attractor* $\mathfrak{A}$ *that coincides with the unstable manifold* $\mathscr{M}^u(\mathscr{N})$ *emanating from the set* $\mathscr{N}$ *of stationary points for* S_t, $\mathfrak{A} \equiv \mathscr{M}^u(\mathscr{N})$. *Moreover,*

$$\lim_{t \to +\infty} \operatorname{dist}_Y(S_t W, \mathscr{N}) = 0 \ \textit{for any}\ W \in Y. \tag{12.2.14}$$

To state our second main result we need additional hypotheses imposed on the damping. These involve some form of differentiability with a suitable growth.

12.2.4. Assumption. In addition to Assumption 12.2.2, let the following conditions hold.

- There exist positive constants m and M such that

$$m \le \frac{g(s_1) - g(s_2)}{s_1 - s_2} \le M\,(1 + s_1 g(s_1) + s_2 g(s_2))^{2/3} \tag{12.2.15}$$

for all $s_1, s_2 \in \mathbb{R}$, $s_1 \neq s_2$.

- There exist positive constants m_1 and M_1 such that

$$m_1 \le \frac{b_0(s_1) - b_0(s_2)}{s_1 - s_2} \le M_1\left(1 + s_1 b_0(s_1) + s_2 b_0(s_2)\right), \quad s_1, s_2 \in \mathbb{R},\ s_1 \ne s_2; \tag{12.2.16}$$

- $f \in C^2(\mathbb{R})$ and $|f''(s)| \le C(1+|s|)$ for $s \in \mathbb{R}$.

12.2.5. Remark. If g, $b_0 \in C^1(\mathbb{R})$, then (12.2.15) and (12.2.16) are equivalent to the requirements

$$m' \le g'(s) \le M'\left(1 + sg(s)\right)^{2/3} \quad \text{for all } s \in \mathbb{R}, \tag{12.2.17}$$

and

$$m_1' \le b_0'(s) \le M_1'\left(1 + sb_0(s)\right) \quad \text{for all } s \in \mathbb{R}, \tag{12.2.18}$$

for some constants m', $M' > 0$ and m_1', $M_1' > 0$. As for relations (12.2.17) and (12.2.18), we refer to Remark 9.2.7 for their discussion. ■

Our second main result is the following theorem.

12.2.6. Theorem (Properties of attractor). *Let Assumption* 12.2.4 *hold. Then the compact global attractor* $\mathfrak{A}$ *given by Theorem* 12.2.3 *possesses the following properties.*

1. *The attractor* $\mathfrak{A}$ *has a finite fractal dimension.*
2. *The attractor* $\mathfrak{A}$ *is a bounded set in the space*

$$Y_* = W^2_{6/p}(\mathcal{O}) \times \mathcal{D}(A^{1/2}) \times \mathcal{D}(\mathcal{A}) \times \mathcal{D}(\mathcal{A}^{1/2})$$

in the case $3 < p \le 5$, *where* $W^2_{6/p}(\Omega)$ *is the* $L_{6/p}$*-based second-order Sobolev space, and in the space*

$$Y_{**} = H^2(\mathcal{O}) \times \mathcal{D}(A^{1/2}) \times \mathcal{D}(\mathcal{A}) \times \mathcal{D}(\mathcal{A}^{1/2})$$

in other cases ($1 \le p \le 3$). *Here* p *is the growth exponent defined in* (6.2.4).
3. *There exists a fractal exponential attractor* $\mathfrak{A}^{exp}$ *for* (Y, S_t) *(whose dimension is finite with respect to relaxed topology) provided that* $b(s)$ *is polynomially bounded at infinity.*

The concept of a fractal exponential attractor was introduced in [102]. See the discussion in Chapters 7 and 8.

In the case $\kappa = 0$ the above theorems result in the following assertion concerning wave dynamics (which was actually stated before in [38]).

12.2.7. Corollary. *Assume that* f *and* g *satisfy the conditions in Assumption* 12.2.2. *Then the dynamical system* $(H^1(\mathcal{O}) \times L_2(\mathcal{O}), S_t^1)$ *generated by problem*

$$\begin{cases} z_{tt} + g(z_t) - \Delta z + f(z) = 0 & \text{in } \mathcal{O} \times \mathbb{R}_+ \\ \dfrac{\partial z}{\partial n} = 0 & \text{on } \partial\mathcal{O} \times \mathbb{R}_+. \end{cases} \tag{12.2.19}$$

possesses a compact global attractor $\mathfrak{A}_1 \equiv \mathcal{M}^u(\mathcal{N}_1)$, *where* $\mathcal{N}_1$ *is the set of equilibria for* (12.2.19). *If* f *and* g *satisfy the conditions in Assumption* 12.2.4, *then (i) the attractor* $\mathfrak{A}_1$ *has a finite fractal dimension; (ii)* A_1 *is a bounded set in the space* $W^2_{6/p}(\Omega)\times\mathscr{D}(A^{1/2})$ *in the case* $3<p\le 5$ *(see Assumption* 6.2.1*), and in the space* $\mathscr{D}(A)\times\mathscr{D}(A^{1/2})$ *in other cases; moreover (iii) there exists a fractal exponential attractor* $\mathfrak{A}_1^{\exp}$ *for* $(H^1(\mathscr{O})\times L_2(\mathscr{O}),S_t^1)$.

For further results on global attractors for wave equations we refer to [75, Chapter 5] and [74] and the references therein.

In the case of the von Karman model, Theorems 12.2.3 and 12.2.6 yield some results of Theorems 9.4.6, 9.5.1, and 9.5.4 from Chapter 9.

12.2.8. Remark. Space regularity of the z-component in the attractor can be improved, under slightly stronger hypotheses imposed on $g(s)$. In fact, one can reach $H^2(\mathscr{O})$, as shown for the wave equation in [162]. This fact is not essential for our further analysis, thus we do not pursue this direction any further. ∎

12.2.2 Main inequality

Following the same idea as in the previous chapters we first establish some kind of equipartition of energy (the analogue for the abstract second-order model is Lemma 8.3.1). This is the key ingredient for the proofs of both Theorems 12.2.3 and 12.2.6.

12.2.9. Proposition. *Let Assumption* 6.2.1 *hold. Assume that* $y_1,y_2\in\mathscr{W}_R$ *for some* $R>R_*$, *where* $\mathscr{W}_R$ *is defined by* (12.2.5) *and denote*

$$(h(t);h_t(t);u(t);u_t(t)) := S_t y_1\,, \quad (\zeta(t);\zeta_t(t);w(t);w_t(t)) := S_t y_2.$$

Let $z(t):=h(t)-\zeta(t)$ *and* $v(t):=u(t)-w(t)$. *Then there exist* $T_0>0$ *and positive constants* c_0, c_1, *and* $c_2(R)$ *independent of* T *such that for every* $T\ge T_0$

$$\begin{aligned} TE^0(T)+\int_0^T E^0(t)dt &\le c_0\Big[\int_0^T\big(\|z_t\|^2_{\mathscr{O}}+\|v_t\|^2_{\Omega}\big)\,dt+\beta G_0^T(z)+\gamma\mathscr{G}_0^T(v)\Big]\\ &\quad + c_1\big[H_0^T(z)+\mathscr{H}_0^T(v)+\Psi_T(z,v)\big]\\ &\quad + c_2(R)\int_0^T\big(\|z\|^2_{\mathscr{O}}+\|v\|^2_{\Omega}\big)\,dt, \end{aligned} \tag{12.2.20}$$

where $E^0(t)=\beta E_z^0(t)+\gamma E_v^0(t)$ *with* $E_z^0(t)$ *and* $E_v^0(t)$ *given by* (6.2.21) *and* (6.2.22) *with* $\alpha=0$. *We also used the notations*

$$G_s^t(z)=\int_s^t\big(D(\zeta_t+z_t)-D(\zeta_t),z_t\big)_{\mathscr{O}}d\tau,\ \mathscr{G}_s^t(v)=\int_s^t\big(B(w_t+v_t)-B(w_t),v_t\big)_{\Omega}d\tau, \tag{12.2.21}$$

$$H_s^t(z) = \int_s^t \left| \left(D(\zeta_t + z_t) - D(\zeta_t), z \right)_{\mathscr{O}} \right| d\tau, \quad \mathscr{H}_s^t(v) = \int_s^t \left| \left(B(w_t + v_t) - B(w_t), v \right)_{\Omega} \right| d\tau \tag{12.2.22}$$

and

$$\Psi_T(z,v) = \beta \left| \int_0^T (\mathscr{F}_1(z), z_t)_{\mathscr{O}} dt \right| + \beta \left| \int_0^T \int_t^T (\mathscr{F}_1(z), z_t)_{\mathscr{O}} d\tau dt \right| \tag{12.2.23}$$
$$+ \gamma \left| \int_0^T (\mathscr{F}_2(v), v_t)_{\Omega} dt \right| + \gamma \left| \int_0^T \int_t^T (\mathscr{F}_2(v), v_t)_{\Omega} d\tau dt \right|$$

with

$$\mathscr{F}_1(z) = F_1(\zeta + z) - F_1(\zeta) \;\; \text{and} \;\; \mathscr{F}_2(v) = F_2(w+v) - F_2(w). \tag{12.2.24}$$

The inequality (12.2.20) established in Proposition 12.2.9 provides a common first step for the proof of all the statements of Theorems 12.2.3 and 12.2.6. This inequality represents equipartition of the energy: the potential energy is reconstructed from the kinetic energy and the nonlinear quantities entering the equation. Eventually, these quantities need to be absorbed (modulo lower-order terms) by the dissipative term. The realization of this step depends heavily on the assumptions imposed on the model, and hence the argument used in the proof of compactness are different from the one given for the proof of finite-dimensionality and/or regularity.

Proof. In what follows we use the notation introduced in Chapter 6.

Step 1 (Energy identity). Without loss of generality, we assume that $(h(t);u(t))$ and $(\zeta(t);w(t))$ are strong solutions. By the invariance of $\mathscr{W}_R$ and in view of relation (6.2.27) there exists a constant $C_R > 0$ such that

$$E_h^0(h(t),h_t(t)) + E_\zeta^0(\zeta(t),\zeta_t(t)) + E_u^0(u(t),u_t(t)) + E_w^0(w(t),w_t(t)) \le C_R \tag{12.2.25}$$

for all $t \ge 0$ where $E_z^0(t)$ and $E_v^0(t)$ denote the corresponding (free) energies defined in (6.2.21) and (6.2.22) with $\alpha = 0$. We establish first an energy type equality regarding $E^0(t) = \beta E_z^0(t) + \gamma E_v^0(t)$.

12.2.10. Lemma. *For any $T > 0$ and all $t \in [0,T]$, $E^0(t)$ satisfies*

$$E^0(T) + \beta G_t^T(z) + \gamma \mathscr{G}_t^T(v) = E^0(t) - \beta \int_t^T (\mathscr{F}_1(z), z_t)_{\mathscr{O}} d\tau - \gamma \int_t^T (\mathscr{F}_2(v), v_t)_{\Omega} d\tau, \tag{12.2.26}$$

where $G_t^T(z)$ and $\mathscr{G}_t^T(v)$ are given by (12.2.21), *whereas $\mathscr{F}_1(z)$ and $\mathscr{F}_2(z)$ are defined by* (12.2.24).

Proof. Writing down the equations satisfied by $h = \zeta + z$, $u = w + v$, ζ and w (see (12.2.13)), it is elementary to derive the following system of coupled equations,

$$z_{tt} + A(z - \gamma\kappa N_0 v_t) + D(\zeta_t + z_t) - D(\zeta_t) + \mathscr{F}_1(z) = 0, \tag{12.2.27}$$

and

$$v_{tt}+\mathscr{A}v+B(w_t+v_t)-B(v_t)+\beta\kappa N_0^*Az_t+\mathscr{F}_2(v)=0, \tag{12.2.28}$$

with $\mathscr{F}_i$ defined in (12.2.24). Next, by standard energy methods we obtain

$$E_z^0(T)+G_t^T(z)=E_z^0(t)+\gamma\kappa\int_t^T(AN_0v_t,z_t)_{\mathscr{O}}\,d\tau-\int_t^T(\mathscr{F}_1(z),z_t)_{\mathscr{O}}\,d\tau; \tag{12.2.29}$$

$$E_v^0(T)+\mathscr{G}_t^T(v)=E_v^0(t)-\beta\kappa\int_t^T(N_0^*Az_t,v_t)_\Omega\,d\tau-\int_t^T(\mathscr{F}_2(v),v_t)_\Omega\,d\tau. \tag{12.2.30}$$

Combining (12.2.29) with (12.2.30) we see that (12.2.26) holds true. □

Step 2. (Reconstruction of the energy integral) We return to the coupled system (12.2.27) and (12.2.28) satisfied by $(z;v)$. We multiply equation (12.2.27) by z, and integrate between 0 and T, thereby obtaining

$$\begin{aligned}\int_0^T\|A^{1/2}z\|_{\mathscr{O}}^2dt &\le c_0\left(E_z^0(T)+E_z^0(0)\right)+\int_0^T\|z_t\|_{\mathscr{O}}^2dt\\ &\quad+H_0^T(z)+\gamma\kappa\int_0^T|(v_t,N_0^*Az)_\Omega|\,dt+\int_0^T|(\mathscr{F}_1(z),z)_{\mathscr{O}}|\,dt,\end{aligned} \tag{12.2.31}$$

where $H_0^T(z)$ is defined in (12.2.22). It is clear from (6.2.20) and (12.2.25) that

$$|(\mathscr{F}_1(z),z)_{\mathscr{O}}|\le C_R\|A^{1/2}z\|_{\mathscr{O}}\,\|z\|_{\mathscr{O}}. \tag{12.2.32}$$

Using (6.2.9) with $\varepsilon=\frac{1}{4}$ we have that

$$|(v_t,N_0^*Az)_\Omega|\le\|v_t\|_\Omega\,\|N_0^*A^{1/2}\|_{\mathscr{L}(L_2(\mathscr{O}))}\,\|A^{1/2}z\|_{\mathscr{O}}\le C\|v_t\|_\Omega\,\|A^{1/2}z\|_{\mathscr{O}},$$

so that

$$\begin{aligned}\int_0^T\|A^{1/2}z\|_{\mathscr{O}}^2dt &\le C_0\left(E_z^0(T)+E_z^0(0)\right)+C_1\int_0^T\left(\|z_t\|_{\mathscr{O}}^2+\|v_t\|_\Omega^2\right)dt\\ &\quad+C_2H_0^T(z)+C_3(R)\int_0^T\|z\|_{\mathscr{O}}^2dt.\end{aligned} \tag{12.2.33}$$

Regarding the plate component, by using the bounds in (6.2.20) and (12.2.25) we obtain

$$|(\mathscr{F}_2(v),v)_\Omega|\le C_R\|\mathscr{A}^{1/2}v\|_\Omega\,\|v\|_\Omega,$$

so that

$$\begin{aligned}\int_0^T\|\mathscr{A}^{1/2}v\|_\Omega^2dt &\le c_0\left(E_v^0(T)+E_v^0(0)\right)+2\int_0^T\|v_t\|_\Omega^2dt\\ &\quad+2\mathscr{H}_0^T(v)+2\beta\kappa\left|\int_0^T(z_t,v)_\Omega\,dt\right|+C(R)\int_0^T\|v\|_\Omega^2dt,\end{aligned} \tag{12.2.34}$$

where $\mathscr{H}_0^T(v)$ is defined in (12.2.22). Integrating by parts in time and using the standard form of the trace theorem it is easy to see that

$$\left|\int_0^T (z_t, v)_\Omega \, dt\right| \le C_1 \left(E^0(T) + E^0(0)\right) + \varepsilon \int_0^T \|A^{1/2} z\|_\Omega^2 \, dt + C_\varepsilon \int_0^T \|v_t\|_\Omega^2 \, dt \tag{12.2.35}$$

for every $\varepsilon > 0$. Consequently, summing up (12.2.33) with (12.2.35) and using (12.2.35), we get

$$\begin{aligned}\int_0^T E^0(t)dt \le c_0 \left[E^0(T) + E^0(0)\right] + c_1 \int_0^T \left(\|z_t\|_{\mathscr{O}}^2 + \|v_t\|_\Omega^2\right) dt \qquad (12.2.36)\\ + c_2 \left[H_0^T(z) + \mathscr{H}_0^T(v)\right] + c_3(R) \int_0^T \left(\|z\|_{\mathscr{O}}^2 + \|v\|_\Omega^2\right) dt .\end{aligned}$$

On the other hand, it follows from Lemma 12.2.10 that

$$E^0(0) = E^0(T) + \beta G_0^T(z) + \gamma \mathscr{G}_0^T(v) + \beta \int_0^T (\mathscr{F}_1(z), z_t)_{\mathscr{O}} \, d\tau + \gamma \int_0^T (\mathscr{F}_2(v), v_t)_\Omega \, d\tau , \tag{12.2.37}$$

and

$$TE^0(T) \le \int_0^T E^0(t)dt - \beta \int_0^T \int_t^T (\mathscr{F}_1(z), z_t)_{\mathscr{O}} \, d\tau dt - \gamma \int_0^T \int_t^T (\mathscr{F}_2(v), v_t)_\Omega \, d\tau dt . \tag{12.2.38}$$

Therefore, combining (12.2.38) with (12.2.36) and (12.2.37), it is readily shown that (12.2.20) holds true, provided that T is sufficiently large. This concludes the proof of Proposition 12.2.9. □

12.2.3 Asymptotic smoothness

In this section we show that the semiflow S_t generated by the PDE system (12.2.1) and (12.2.2) is asymptotically smooth. This property is critical for proving the existence of global attractors. For the reader's convenience we recall (see Definition 7.1.2) that a dynamical system (X, S_t) is said to be *asymptotically smooth* iff for any bounded set $\mathscr{B}$ in X such that $S_t\mathscr{B} \subset \mathscr{B}$ for $t > 0$ there exists a compact set $\mathscr{K}$ in the closure $\overline{\mathscr{B}}$ of $\mathscr{B}$, such that $\lim_{t\to+\infty} \sup_{y\in\mathscr{B}} \mathrm{dist}_X\{S_t y, \mathscr{K}\} = 0$.

Our main result in this section is the following assertion.

12.2.11. Theorem. *Let Assumption* 12.2.2 *hold. Then the dynamical system* (Y, S_t) *generated by the PDE problem* (12.2.1) *and* (12.2.2) *is asymptotically smooth.*

As in Chapters 9 and 10 in order to prove Theorem 12.2.11 we invoke a compactness criterion, which is an abstract version formulated in Theorem 7.1.11.

In the proof of Theorem 12.2.11 we make use of further inequalities which are a standard tool for proving the absorption property; see [75].

Our first step toward the goal is the estimate for the damping.

12.2.12. Lemma. *Under Assumption* 6.2.1 *there exist constants* $0 < \delta < 1/4$ *and* $C_0 > 0$ *such that the wave-damping term D satisfies*

$$|(D(\zeta+z)-D(\zeta),h)_{\mathscr{O}}| \tag{12.2.39}$$

$$\le C_0\left[(D(\zeta),\zeta)_{\mathscr{O}}+(D(\zeta+z),\zeta+z)_{\mathscr{O}}\right]\|A^{1/2}h\|_{\mathscr{O}}+C_0\|h\|_{\mathscr{O}}$$

for any $\zeta, z, h \in \mathscr{D}(A^{1/2})$, *and the plate-damping B satisfies*

$$|(B(w+u)-B(w),v)_{\Omega}| \tag{12.2.40}$$

$$\le C_0\left(1+(B(w),w)_{\Omega}+(B(w+u),w+u)_{\Omega}\right)\|\mathscr{A}^{1/2-\delta}v\|_{\Omega}$$

for any $w, u, v \in \mathscr{D}(\mathscr{A}^{1/2})$.

Proof. The inequality in (12.2.39) can be easily derived by using arguments similar to those in [75, Chapter 5]. We briefly sketch the proof. From the definition of the damping

$$|(D(\zeta+z)-D(\zeta),h)_{\mathscr{O}}| \le \int_{\mathscr{O}}|g(\zeta+z)|\,|h|dx+\int_{\mathscr{O}}|g(\zeta)|\,|h|dx.$$

Let $\mathscr{O}_1=\{x\in\mathscr{O} : |\zeta(x)|\ge 1\}$ and $\mathscr{O}_2=\mathscr{O}\setminus\mathscr{O}_1$. Then,

$$\begin{aligned}\int_{\mathscr{O}}|g(\zeta)|\,|h|dx &\le \int_{\mathscr{O}_1}|g(\zeta)|\,|h|dx+C\int_{\mathscr{O}_2}|h|dx\\ &\le \left[\int_{\mathscr{O}_1}|g(\zeta)|^{6/5}dx\right]^{5/6}\|h\|_{L_6(\mathscr{O})}+C\|h\|_{\mathscr{O}}.\end{aligned}$$

Next, by (6.2.4) it follows that $|g(\zeta)|^{6/5}=|g(\zeta)|\,|g(\zeta)|^{1/5}\le C\zeta g(\zeta)$ on $\mathscr{O}_1$. Therefore, using the embedding $H^1(\mathscr{O})\subset L_6(\mathscr{O})$ and the fact that $\zeta g(\zeta)\ge c_0>0$ on $\mathscr{O}_1$ which follows from (12.2.8), we obtain

$$\int_{\mathscr{O}}|g(\zeta)|\,|h|dx \le C\|A^{1/2}h\|_{\mathscr{O}}\int_{\mathscr{O}}\zeta g(\zeta)dx+C\|h\|_{\mathscr{O}}.$$

With similar computations for $\int_{\mathscr{O}}|g(\zeta+z)|\,|h|dx$, we readily obtain (12.2.39).

The second statement follows trivially from the Sobolev embedding

$$\mathscr{D}(\mathscr{A}^{1/2-\delta})\subset H^{2-4\delta}(\Omega)\subset C(\overline{\Omega}), \quad \delta<1/4,$$

and the obvious inequality $|(B(w),v)_{\Omega}|\le C|v|_{C(\overline{\Omega})}\left(1+\int_{\Omega}b_0(w)wdx\right)$. For details we refer to the proof of Lemma 9.4.8. □

Proof of Theorem 12.2.11.

Any bounded positively invariant set belongs to $\mathscr{W}_R$ for some $R > R_*$, where $\mathscr{W}_R$ is defined by (12.2.5), therefore it is sufficient to establish the property of asymptotic smoothness on the set $\mathscr{W}_R$ for every $R > R_*$ only. Let $y_1, y_2 \in \mathscr{W}_R$. Below we use the same notations as in Proposition 12.2.9. Namely, we denote the solutions corresponding to initial data y_1 and y_2, respectively, by

$$(h(t); h_t(t); u(t); u_t(t)) := S_t y_1, \quad (\zeta(t); \zeta_t(t); w(t); w_t(t)) := S_t y_2,$$

and set $z(t) := h(t) - \zeta(t)$ and $v(t) := u(t) - w(t)$. Now we establish the key estimate for the proof of Theorem 12.2.11.

12.2.13. Proposition. *Let the assumptions of Theorem* 12.2.11 *be in force. Then, given* $\varepsilon > 0$ *and* $T > 1$ *there exist constants* $C_\varepsilon(R)$ *and* $C(R,T)$ *such that*

$$E^0(T) \le \varepsilon + \frac{1}{T}\Big[C_\varepsilon(R) + \Psi_T(z,v)\Big] + C(R,T)lot(z,v), \tag{12.2.41}$$

where $E^0(t) := \beta E_z^0(t) + \gamma E_v^0(t)$ *with* $E_z^0(t)$ *and* $E_v^0(t)$ *given in* (6.2.21) *and* (6.2.22) *with* $\alpha = 0$, *the functional* $\Psi_T(z,v)$ *is given by* (12.2.23) *whereas* $lot(z,v)$ *is defined by*

$$lot(z,v) := \sup_{[0,T]} \|A^{1/2-\delta} z(t)\|_{\mathscr{O}} + \sup_{[0,T]} \|\mathscr{A}^{1/2-\delta} v(t)\|_\Omega \quad \textit{for some } 0 < \delta \le \frac{1}{2}.$$

Proof. It follows from the energy inequality (6.2.28) that

$$\beta \int_0^t (D(h_t), h_t)_{\mathscr{O}} dt + \gamma \int_0^t (B(u_t), u_t)_\Omega dt \le C_R, \tag{12.2.42}$$

$$\beta \int_0^t (D(\zeta_t), \zeta_t)_{\mathscr{O}} dt + \gamma \int_0^t (B(w_t), w_t)_\Omega dt \le C_R, \tag{12.2.43}$$

where crucially C_R does not depend on t. Let $H_s^t(z)$ and $\mathscr{H}_s^t(v)$ be given by (12.2.22). By using Lemma 12.2.12, the estimates (12.2.42) and (12.2.43) and the fact that $\|A^{1/2} z\|_{\mathscr{O}} \le C_R$, we obtain

$$H_0^T(z) \le C_R + CT\ lot(z,v) \quad \text{and} \quad \mathscr{H}_0^T(v) \le (1+T) C_R\, lot(z,v). \tag{12.2.44}$$

Using now the inequalities in Assumption 12.2.2 and, once again, the uniform estimates (12.2.42) and (12.2.43), one can see that

$$\int_0^T \left(\|z_t\|_{\mathscr{O}}^2 + \|v_t\|_\Omega^2\right) dt \le \varepsilon T + C_\varepsilon(R) \quad \text{for every } \varepsilon > 0. \tag{12.2.45}$$

Taking first $t = 0$ in (12.2.26) and using the fact that $E^0(0) \le C_R$, we get

$$\beta G_0^T(z)+\alpha \mathscr{G}_0^T(v) \le C_R+\beta\left|\int_0^T (\mathscr{F}_1(z),z_t)\,d\tau\right|+\alpha\left|\int_0^T (\mathscr{F}_2(v),v_t)\,d\tau\right|. \tag{12.2.46}$$

Therefore, (12.2.41) follows from Proposition 12.2.9 and the estimates (12.2.44)–(12.2.46). □

To continue with the proof of Theorem 12.2.11 we note that it follows from Proposition 12.2.13 that given $\varepsilon > 0$ there exists $T = T(\varepsilon) > 1$ such that for initial data $y_1, y_2 \in \mathscr{W}_R$ we have

$$\begin{aligned}||S_T y_1 - S_T y_2||_Y &= ||(z(T); z_t(T); v(T); v_t(T))||_Y \\ &\le C|E^0(T)|^{1/2} \le \varepsilon + \Psi_{\varepsilon,R,T}(y_1, y_2),\end{aligned}$$

where

$$\Psi_{\varepsilon,R,T}(y_1,y_2) = C_{R,\varepsilon,T}\left\{\Psi_T(z,v) + lot(z,v)\right\}^{1/2}, \tag{12.2.47}$$

where $\Psi_T(z,v)$ is given by (12.2.23). Thus, in order to establish asymptotic smoothness for the dynamical system under investigation, it suffices to invoke Theorem 7.1.11, which allows us to conclude the proof of Theorem 12.2.11. Hence, what we need to prove is the validity of the sequential limits (7.1.13) for $\Psi_{\varepsilon,R,T}$ defined by (12.2.47). To do that, we use similar arguments as in the completion of the proof of Theorem 8.3.4 in Chapter 8; see also the proofs of Theorems 9.4.14 and 10.4.12. For the reader's convenience, we recall some critical computations.

Let $(h^n; u^n)_n$ be a sequence of solutions to the PDE system (12.2.1) and (12.2.2) corresponding to initial data $y^{0,n} := (h^{0,n}; h^{1,n}; u^{0,n}; u^{1,n})$ in $\mathscr{W}_R \subset Y$. Because the compactness condition in (7.1.13) deals with lower limits, it is sufficient to establish (7.1.13) for some subsequence of $(y^{0,n})_n$. Therefore we can assume that

$$y^n(t) := (h^n; h^n_t; u^n; u^n_t) \to (h; h_t; u; u_t) =: y(t) \quad *\text{-weakly in } L_\infty(0,T;Y) \tag{12.2.48}$$

for some solution $(h; h_t; u; u_t) \in L_\infty(0,T;Y)$; in addition, by Aubin's lemma (see Theorem 1.1.8),

$$\sup_{[0,T]}\left\{||z^{n,m}(t)||_{1-\delta,\mathscr{O}} + ||v^{n,m}(t)||_{2-\eta,\Omega}\right\} \to 0, \tag{12.2.49}$$

as $n,m \to \infty$, for any $\delta, \eta > 0$, where we have set $z^{n,m}(t) = h^n(t) - h^m(t)$ and $v^{n,m}(t) = u^n(t) - u^m(t)$. The above convergence implies that $lot(z^{n,m}, v^{n,m}) \to 0$. Therefore, in view of (12.2.47), we must show that

$$\lim_{m\to\infty}\lim_{n\to\infty} \Psi_T(z^{n,m}, v^{n,m}) = 0. \tag{12.2.50}$$

Let us begin with the analysis of the integral $\int_t^T (\mathscr{F}_1(z^{n,m}), z_t^{n,m})_{\mathscr{O}}\, d\tau$, which explicitly reads as follows,

$$\int_t^T (\mathscr{F}_1(z^{n,m}), z_t^{n,m})_{\mathscr{O}}\, d\tau = \int_t^T (F_1(h^n(\tau)) - F_1(h^m(\tau)), h^n_t(\tau) - h^m_t(\tau))_{\mathscr{O}}\, d\tau$$

$$= \int_t^T (\Phi'(h^n(\tau)), h_t^n(\tau))_{\mathscr{O}} d\tau - \int_t^T (F_1(h^n(\tau)), h_t^m(\tau))_{\mathscr{O}} d\tau$$

$$- \int_t^T (F_1(h^m(\tau)), h_t^n(\tau))_{\mathscr{O}} d\tau + \int_t^T (\Phi'(h^m(\tau)), h_t^m(\tau))_{\mathscr{O}} d\tau,$$

where the potential Φ is defined in (6.2.15). This implies that

$$\int_t^T (\mathscr{F}_1(z^{n,m}), z_t^{n,m})_{\mathscr{O}} d\tau = \underbrace{\Phi(h^n(T)) - \Phi(h^n(t))}_{A_n} + \underbrace{\Phi(h^m(T)) - \Phi(h^m(t))}_{B_m}$$

$$- \underbrace{\int_t^T (F_1(h^n(\tau)), h_t^m(\tau))_{\mathscr{O}} d\tau}_{C_{n,m}} - \underbrace{\int_t^T (F_1(h^m(\tau)), h_t^n(\tau))_{\mathscr{O}} d\tau}_{D_{n,m}}. \tag{12.2.51}$$

Let us observe that by (6.2.5) the Nemytski operator

$$z(x) \mapsto \int_0^{z(x)} (f(\xi) - \mu\xi) d\xi$$

is continuous from $L_{q+2}(\mathscr{O})$ into $L_1(\mathscr{O})$, where $q \le 2$ is defined in (6.2.5). Therefore, because $H^{1-\delta}(\mathscr{O}) \subset L_{q+2}(\mathscr{O})$ for $q \le 2$ and for some $\delta > 0$, the potential Φ defined in (6.2.15) is continuous on $\mathscr{D}(A^{(1-\delta)/2})$. Consequently, (12.2.49) implies that

$$\Phi(h^n(T)) \to \Phi(h(T)), \quad \Phi(h^n(t)) \to \Phi(h(t)) \text{ as } n \to \infty.$$

Thus

$$\lim_{n\to\infty} \lim_{m\to\infty} (A_n + B_m) = 2\left[\Phi(h(T)) - \Phi(h(t))\right]. \tag{12.2.52}$$

In order to estimate the term $C_{n,m}$ in (12.2.51) we first note that by (6.2.20)

$$\sup_{t\in[0,T]} \|F_1(h^n(t))\|_{\mathscr{O}} \le C_R \text{ for all } n,$$

hence

$$F_1(h^n) \to F_0 \text{ *-weakly in } L_\infty(0,T; L_2(\mathscr{O})). \tag{12.2.53}$$

From (12.2.48) and (12.2.49) we also have that $h^n \to h$ strongly in the space $L_2(0,T; L_2(\mathscr{O}))$, and hence almost everywhere (along a subsequence). Because F_1 is continuous, we also have that $F_1(h^n) \to F_1(h)$, a.e. in $(0,T) \times \mathscr{O}$. This allows us to recover the limit in (12.2.53) and claim that $F_0 = F_1(h)$. Thus, letting $m \to \infty$ first and $n \to \infty$ next, we obtain for term $C_{n,m}$:

$$\lim_{n\to\infty} \lim_{m\to\infty} C_{n,m} = \lim_{n\to\infty} \lim_{m\to\infty} \int_t^T (F_1(h^n(\tau)), h_t^m(\tau))_{\mathscr{O}} d\tau$$

$$= \lim_{n\to\infty} \int_t^T (F_1(h^n(\tau)), h_t(\tau))_{\mathscr{O}} d\tau = \int_t^T (F_1(h(\tau)), h_t(\tau))_{\mathscr{O}} d\tau$$

$$= \Phi(h(T)) - \Phi(h(t)). \tag{12.2.54}$$

The same argument applies to the term $D_{n,m}$ giving

$$\lim_{n\to\infty}\lim_{m\to\infty} D_{n,m} = \Phi(h(T)) - \Phi(h(t)). \tag{12.2.55}$$

Combining (12.2.52) with (12.2.54) and (12.2.55), we see from (12.2.51) that

$$\lim_{n\to\infty}\lim_{m\to\infty} \int_t^T (\mathscr{F}_1(z^{n,m}), z_t^{n,m})_{\mathscr{O}}\, d\tau = 0. \tag{12.2.56}$$

Similar computations[1] performed on $\int_t^T (\mathscr{F}_2(v^{n,m}), v_t^{n,m})_\Omega\, d\tau$ yield, as well,

$$\lim_{n\to\infty}\lim_{m\to\infty} \int_t^T (\mathscr{F}_2(v^{n,m}), v_t^{n,m})_\Omega\, d\tau = 0. \tag{12.2.57}$$

Recalling (12.2.23), and taking into account (12.2.56) and (12.2.57) we finally conclude that (12.2.50) holds true, as desired. Thus Theorem 12.2.11 is proved. □

Proof of Theorem 12.2.3. Because $\mathscr{W}_R$ is bounded and positively invariant, Theorem 12.2.11 implies the existence of a compact global attractor $\mathfrak{A}_R$ for the dynamical system $(\mathscr{W}_R, S_t)$, for each $R \geq R_*$. We choose now $R_0 \geq R_* + 1$ such that the set $\mathscr{N}$ of equilibria lies in $\mathscr{W}_{R_0-1}$. It follows from (12.2.8) and (12.2.9) that $sg(s) > 0$ and $sb_0(s) > 0$ for $s \neq 0$. Therefore the energy inequality (6.2.28) implies that the energy $\mathscr{E}$ given by (6.2.25) is a strict Lyapunov function for $(\mathscr{W}_R, S_t)$. This, in turn, implies (see, e.g., Theorem 7.5.6) that $\mathfrak{A}_R = \mathscr{M}^u(\mathscr{N})$ and by Lemma 12.2.1 $\mathfrak{A}_R$ does not depend on R for $R \geq R_0$; moreover, relation (12.2.14) holds. Thus Theorem 12.2.3 is proved. □

12.2.4 Stabilizability estimate

In this section we aim to develop some analytical tools that enable us to apply the abstract results presented in Section 7.9 (see also [69] and [75]) for the proof of Theorem 12.2.6. The crucial tool is the *stabilizability estimate* formulated in Theorem 12.2.15. This estimate eventually leads to the conclusion that the system is quasi-stable.

We begin with a few preparatory estimates. The lemma below provides more refined estimates on the damping terms. The proof of these estimates was originally carried out in [75].

12.2.14. Lemma. *Under Assumptions* 12.2.4, *the following statements are valid.*

1. For every $\varepsilon > 0$ there exists $C_\varepsilon > 0$ such that

[1] See also the argument given in the proof of Theorem 10.4.12.

$$|(D(\zeta+z)-D(\zeta),h)_{\mathcal{O}}| \le C_\varepsilon\,(D(\zeta+z)-D(\zeta),z)_{\mathcal{O}}$$
$$+\,\varepsilon\left(1+(D(\zeta),\zeta)_{\mathcal{O}}+(D(\zeta+z),\zeta+z)_{\mathcal{O}}\right)||A^{1/2}h||^2_{\mathcal{O}} \qquad (12.2.58)$$

for any $\zeta, z, h \in \mathscr{D}(A^{1/2})$.

2. *There exist a parameter* $\delta > 0$ *and constants* $C_1, C_2 > 0$ *such that*

$$|(B(w+v)-B(w),u)_\Omega| \le C_1\,(B(w+v)-B(w),v)_\Omega \qquad (12.2.59)$$
$$+\,C_2\left(1+(B(w),w)+(B(w+v),w+v)_\Omega\right)||\mathscr{A}^{1/2-\delta}u||^2_\Omega$$

for any $u, v, w \in \mathscr{D}(\mathscr{A}^{1/2})$.

Proof. To prove (12.2.58) we use (12.2.15) and apply the argument as in [75, Chapter 5] (the same idea was used in the proof of Lemma 9.4.9). We start with the relation

$$\left|\int_\Omega (g(\zeta+z)-g(\zeta))h\,dx\right| \le \varepsilon\int_\Omega |g(\zeta+z)-g(\zeta)|\frac{|h|^2}{|z|}dx$$
$$+\,C_\varepsilon\int_\Omega (g(\zeta+z)-g(\zeta))z\,dx$$

for any $\varepsilon > 0$. Using (12.2.15) we obtain

$$\int_\Omega (g(\zeta+z)-g(\zeta))\frac{|h|^2}{|z|}dx$$
$$\le C\|h\|^2_{L_6(\Omega)}\left[\int_\Omega (1+\zeta g(\zeta)+(\zeta+z)g(\zeta+z))\,dx\right]^{2/3}$$
$$\le C\|h\|^2_{L_6(\Omega)}\left[1+\int_\Omega (g(\zeta+z)(\zeta+z)+g(\zeta)\zeta)\,dx\right]$$
$$\le C\|h\|^2_{1,\mathcal{O}}\left[1+\int_\Omega (g(\zeta+z)(\zeta+z)+g(\zeta)\zeta)dx\right],$$

which implies (12.2.58).

The inequality in (12.2.59) follows from (9.4.29) in Lemma 9.4.9. □

Now we are ready for the stabilizability inequality, which shows that the system under consideration is quasi-stable, a fact that plays a key role in further considerations. We recall that the phase space Y is given by (12.2.4).

12.2.15. Theorem (Stabilizability estimate). *Let Assumption* 12.2.4 *be in force. Then there exist positive constants* C_1, C_2, *and* ω *depending on* R *such that for any* $y_1, y_2 \in \mathscr{W}_R$ *the following estimate holds true,*

$$||S_t y_1 - S_t y_2||^2_Y \le C_1 e^{-\omega t}||y_1-y_2||^2_Y + C_2\,lot_t(h-\zeta, u-w), \quad t>0, \qquad (12.2.60)$$

where

$$lot_t(z,v) = \max_{[0,t]} \left(\|z(\tau)\|^2_{1-\delta,\mathcal{O}} + \|v(\tau)\|^2_{2-\delta,\Omega} \right) \tag{12.2.61}$$

with $\delta > 0$*. Above, we have used the notation*

$$(h(t); h_t(t); u(t); u_t(t)) := S_t y_1, \quad (\zeta(t); \zeta_t(t); w(t); w_t(t)) := S_t y_2.$$

Let $y_1, y_2 \in \mathscr{W}_R$ be given. We consider the solutions $S_t y_1$ and $S_t y_2$ and introduce, as previously, $z = h - \zeta$, $v = u - w$. We recall that for these solutions the bounds (12.2.25), (12.2.42), and (12.2.43) hold true. We begin with the following critical estimate that handles nonlinear source terms.

12.2.16. Lemma. *Under Assumption* 12.2.4, *the following estimates hold true for some* $\delta > 0$*:*

$$\left| \int_t^T (\mathscr{F}_1(z), z_t)_{\mathcal{O}} \, d\tau \right| \le C_{R,T} \max_{[0,T]} \|z\|^2_{1-\delta,\mathcal{O}} + \varepsilon \int_0^T \|A^{1/2} z\|^2_{\mathcal{O}} d\tau$$
$$+ C_R^{\varepsilon} \int_0^T \left(\|h_t\|^2_{\mathcal{O}} + \|\zeta_t\|^2_{\mathcal{O}} \right) \|A^{1/2} z\|^2_{\mathcal{O}} d\tau, \tag{12.2.62}$$

$$\left| \int_t^T (\mathscr{F}_2(v), v_t)_{\Omega} \, d\tau \right| \le C_R \max_{[0,T]} \|v\|^2_{2-\delta,\Omega} + \varepsilon \int_0^T \|\mathscr{A}^{1/2} v\|^2_{\Omega} d\tau$$
$$+ C_R^{\varepsilon} \int_0^T \left(\|v_t\|^2_{\Omega} + \|w_t\|^2_{\Omega} \right) \|\mathscr{A}^{1/2} v\|^2_{\Omega} d\tau \tag{12.2.63}$$

for all $t \in [0,T]$, *where* $\varepsilon > 0$ *can be taken arbitrarily small. Here,* $\mathscr{F}_1$ *and* $\mathscr{F}_2$ *are given by* (12.2.24).

The idea behind the estimates in Lemma 12.2.16 is to exhibit explicitly the kinetic energies $\|h_t\|^2 + \|\zeta_t\|^2$ and $\|v_t\|^2 + \|w_t\|^2$, which are $L_1(\mathbb{R})$ (see below). As such, this function may play a role of a small parameter for large t. We have used the same idea in previous considerations (see (9.4.40) for the plate with internal damping, Lemma 10.4.17 for the plate with boundary dissipation, and the proof of Theorem 11.3.1 for thermoelastic plate).

Proof. To prove the first estimate we note that

$$\int_t^T (\mathscr{F}_1(z), z_t)_{\mathcal{O}} \, d\tau = \frac{1}{2} \int_{\mathcal{O}} \int_0^1 f'(\zeta + \lambda z) \, d\lambda \, |z|^2 dx \Big|_t^T - \frac{\mu}{2} \|z(\tau)\|^2_{\mathcal{O}} \Big|_t^T \tag{12.2.64}$$
$$- \frac{1}{2} \int_t^T \int_{\mathcal{O}} \int_0^1 f''(\zeta + \lambda z) \cdot (\zeta_t + \lambda z_t) \, d\lambda \, |z|^2 dx d\tau.$$

Using the embedding $H^1(\mathcal{O}) \subset L_6(\mathcal{O})$ and $H^{1/2}(\mathcal{O}) \subset L_3(\mathcal{O})$, we obtain that

$$\left| \int_{\mathcal{O}} \int_0^1 f'(\zeta + \lambda z) d\lambda |z|^2 dx \right| \le C \int_{\mathcal{O}} \int_0^1 (1 + |z|^2 + |\zeta|^2) |z|^2 dx \tag{12.2.65}$$
$$\le C(1 + \|u\|^2_{1,\mathcal{O}} + \|\xi\|^2_{1,\mathcal{O}}) \|z\|^2_{L_3(\mathcal{O})} \le C_R \|z\|^2_{1/2,\mathcal{O}}.$$

The second term on the right-hand side of (12.2.64) can be estimated as follows:

$$\left|\int_t^T\int_{\mathscr{O}}\int_0^1 f''(\zeta+\lambda z)\cdot(\zeta_t+\lambda z_t)d\lambda|z|^2dxd\tau\right|$$
$$\le C\int_0^T\left[\int_{\mathscr{O}}(|z_t|+|\zeta_t|)^2dx\right]^{1/2}\left[\int_{\mathscr{O}}(1+|z|+|\zeta|)^2|z|^4dx\right]^{1/2}d\tau,$$
$$\le C\int_0^T\left[\int_{\mathscr{O}}(|z_t|+|\zeta_t|)^2dx\right]^{1/2}\left[\int_{\mathscr{O}}(1+|z|+|\zeta|)^6dx\right]^{1/6}\|z\|^2_{1,\mathscr{O}}.$$

Therefore

$$\left|\int_t^T\int_{\mathscr{O}}\int_0^1 f''(\zeta+\lambda z)\cdot(\zeta_t+\lambda z_t)\,d\lambda\,|z|^2dxd\tau\right|$$
$$\le C_R\int_0^T(\|z_t\|_{\mathscr{O}}+\|\zeta_t\|_{\mathscr{O}})\,\|A^{1/2}z\|^2_{\mathscr{O}}d\tau. \tag{12.2.66}$$

Thus, using (12.2.65) and (12.2.66) in (12.2.64), we finally obtain (12.2.62).

Regarding the latter estimate (12.2.63), the argument is the same as in the proof of Theorem 9.4.6; see relations (9.4.40) and (9.4.41). □

By the lower bounds in (12.2.15) and (12.2.16) we have that

$$\int_0^T\left(\|z_t\|^2_{\mathscr{O}}+\|v_t\|^2_{\Omega}\right)dt\le C\left(G_0^T(z)+\mathscr{G}_0^T(v)\right), \tag{12.2.67}$$

where $G_0^T(z)$ and $\mathscr{G}_0^T(v)$ are given by (12.2.21), and also

$$\|u_t(t)\|^2_{\Omega}+\|w_t(t)\|^2_{\Omega}\le B_{u,w}(t):=m_1^{-1}\left[(B(u_t(t)),u_t(t))_{\Omega}+(B(w_t(t)),w_t(t))_{\Omega}\right] \tag{12.2.68}$$

and

$$\|h_t(t)\|^2_{\mathscr{O}}+\|\zeta_t(t)\|^2_{\mathscr{O}}\le D_{h,\zeta}(t):=m^{-1}\left[(D(h_t(t)),h_t(t))_{\mathscr{O}}+(D(\zeta_t(t)),\zeta_t(t))_{\mathscr{O}}\right] \tag{12.2.69}$$

for all $t\ge 0$.

Returning to Proposition 12.2.9 and using Lemma 12.2.16, we obtain for $\Psi_T(z,v)$ in (12.2.23) the estimate

$$\Psi_T(z,v)\le\varepsilon\int_0^T E^0(t)dt+C_\varepsilon(T,R)\,\Xi_T(z,v) \tag{12.2.70}$$

for every $\varepsilon>0$, where $E^0(t)$ is the same as in Proposition 12.2.9 and

$$\Xi_T(z,v)=lot_T(z,v)+\int_0^T\left[B_{u,w}(\tau)+D_{h,\zeta}(\tau)\right]E^0(\tau)d\tau, \tag{12.2.71}$$

with $lot_T(z,v)$ defined by (12.2.61), $B_{u,w}(t)$ and $D_{h,\zeta}(t)$ given in (12.2.68) and (12.2.69). The next assertion is a direct consequence of Lemma 12.2.14.

12.2.17. Lemma. *Under Assumption* 12.2.4, *the following estimate holds true with arbitrarily small* $\varepsilon > 0$,

$$H_0^T(z) + \mathscr{H}_0^T(v) \le \varepsilon \int_0^T E^0(t)\,dt + C_\varepsilon(T,R)\left[\Xi_T(z,v) + G_0^T(z) + \mathscr{G}_0^T(v)\right], \tag{12.2.72}$$

where $G_0^T(z)$, $\mathscr{G}_0^T(v)$ *are defined in* (12.2.21), *and* $\Xi_T(z,v)$ *is given by* (12.2.71).

Proof. This lemma is rather generic and does not depend on a particular structure of equations used. The argument has been used before. For the sake of completeness we briefly recall it.

It is sufficient to apply the estimates from Lemma 12.2.14. In fact, by using (12.2.58) we readily obtain that given $\varepsilon > 0$, there exists a positive constant C_ε such that

$$H_0^T(z) \le C_\varepsilon G_0^T(z) + \varepsilon \int_0^T E^0(t)\,dt + \varepsilon\,\Xi_T(z,v).$$

Similarly, using both (12.2.59) and the bound (12.2.43) we have that

$$\mathscr{H}_0^T(v) \le C_1 \mathscr{G}_0^T(v) + C_R\, lot_T(z,v).$$

The two estimates above immediately give (12.2.72), as desired. □

Thus, combining estimates (12.2.67), (12.2.70), and (12.2.72), Proposition 12.2.9 implies

$$TE^0(T) + \int_0^T E^0(t)dt \le C_1\left[\beta G_0^T(z) + \alpha \mathscr{G}_0^T(v)\right] + C_2\,\Xi_T(z,v) \tag{12.2.73}$$

for $T \ge T_0 > 0$, where $C_i = C_i(T,R)$ depend on T and R. On the other hand, using the equality (12.2.26) and Lemma 12.2.16 we also have that

$$\beta G_0^T(z) + \alpha \mathscr{G}_0^T(v) \le E^0(0) - E^0(T) + \varepsilon \int_0^T E^0(t)dt + C_\varepsilon(T,R)\,\Xi_T(z,v)$$

for every $\varepsilon > 0$. Therefore (12.2.73) implies that there exists $T > 1$ such that

$$E^0(T) \le \gamma E^0(0) + C_{R,T}\,\Xi_T(z,v) \quad \text{with } 0 < \gamma \equiv \gamma_{T,R} < 1. \tag{12.2.74}$$

We can now apply the same procedure as in Theorem 8.5.3. From (12.2.74) we have that

$$E^0((m+1)T) \le \gamma E^0(mT) + c_T b_m, \quad m = 0,1,2,\dots,$$

where

$$b_m := \sup_{t\in[mT,(m+1)T]} \left(\|z(t)\|^2_{1-\delta,\mathscr{O}} + \|v(t)\|^2_{2-\delta,\Omega}\right) + \int_{mT}^{(m+1)T} K(\tau)\,E^0(\tau)d\tau,$$

where $K(t) = B_{u,w}(t) + D_{h,\zeta}(t)$ with $B_{u,w}(t)$ and $D_{h,\zeta}(t)$ given in (12.2.68) and (12.2.69). This yields

$$E^0(mT) \le \gamma^m E^0(0) + c\sum_{l=1}^{m} \gamma^{m-l} b_{l-1}.$$

Because $\gamma < 1$, there exists $\omega > 0$ such that

$$E^0(mT) \le C_1 e^{-\omega mT} E^0(0) + C_2\Big\{lot_{mT}(z,v) + \int_0^{mT} e^{-\omega(mT-\tau)} K(\tau)\, E^0(\tau)d\tau\Big\},$$

which implies that

$$E^0(t) \le C_1 e^{-\omega t} E^0(0) + C_2\Big\{lot_t(z,v) + \int_0^{t} e^{-\omega(t-\tau)} K(\tau)\, E^0(\tau)d\tau\Big\}$$

for all $t \ge 0$. Therefore, applying Gronwall's lemma we find that

$$E^0(t) \le \big[C_1 E^0(0)e^{-\omega t} + C_2\, lot_t(z,v)\big] \exp\Big\{C_2 \int_0^t K(\tau)\, d\tau\Big\}.$$

Because by (12.2.42) and (12.2.43) we have that

$$\begin{aligned}\int_0^t K(\tau)\, d\tau &= \frac{1}{m}\int_0^t ((D(h_t),h_t)_{\mathcal{O}} + (D(\zeta_t),\zeta_t)_{\mathcal{O}})\,d\tau \\ &+ \frac{1}{m_1}\int_0^t ((B(u_t),u_t)_\Omega + (B(w_t),w_t)_\Omega)\,d\tau \le C_R \quad \text{for all } t \ge 0,\end{aligned}$$

we obtain the estimate (12.2.60). This concludes the proof of Theorem 12.2.15. □

12.2.5 Additional properties of the attractor

This section completes the proof of Theorem 12.2.6.

1. *Finiteness of fractal dimension.* Finiteness of the fractal dimension $\dim_f \mathfrak{A}$, follows from Theorem 12.2.15 and local Lipschitz continuity (6.2.41) of the semiflow S_t. Indeed, this is a consequence of the abstract Theorem 7.9.6 (see also Theorem 8.6.1).

2. *Smoothness of the global attractor.* It follows from Theorem 12.2.15 and from Theorem 7.9.8 that

$$\|z_{tt}(t)\|^2 + \|A^{1/2}z_t(t)\|^2 + \|u_{tt}(t)\|^2 + \|\mathscr{A}^{1/2}u_t(t)\|^2 \le C, \quad t \in \mathbb{R}, \tag{12.2.75}$$

for any trajectory $\gamma = \{y(t) \equiv (z(t);z_t(t);u(t);u_t(t)) : t \in \mathbb{R}\}$ from the attractor $\mathfrak{A}$.

The estimate (12.2.75) allows us to establish the spatial smoothness of the attractor. Let us begin with the analysis of the plate variable u. It follows from the

variational equation in (6.2.39) and (12.2.75) that on the attractor we have that $\mathscr{A}u(t) = f(t)$, where $f(t)$ is a bounded function in $L_2(\Omega)$. Hence, we can conclude that $t \mapsto u(t)$ is a bounded function in $\mathscr{D}(\mathscr{A})$.
As for the wave component z, in the case $1 \le p \le 3$ from (6.2.4) it follows that

$$\|g(z_t)\|_{\mathscr{O}} \le C\big(1 + \|z_t\|^p_{L_{2p}(\mathscr{O})}\big) \le C\big(1 + \|z_t\|^p_{1,\mathscr{O}}\big).$$

Therefore from the variational relation in (6.2.38) and from (12.2.75) we obtain that $z(t)$ solves the problem

$$(-\Delta + \mu)z = f_1(t) \text{ in } \mathscr{O}, \quad \frac{\partial z}{\partial n} = f_2(t) \text{ on } \partial\mathscr{O}, \tag{12.2.76}$$

where $f_1 \in L_\infty(\mathbb{R}, L_2(\mathscr{O}))$ and $f_2 \in L_\infty(\mathbb{R}, H^2(\partial\mathscr{O}))$. By the elliptic regularity theory (see, e.g., [275, Chapter 5]) we can conclude that $z(t)$ is a bounded function with values in $H^2(\mathscr{O})$.

In the case $3 < p \le 5$ we have that $g(z_t)$ is bounded in $L_{6/p}(\mathscr{O})$ and therefore z solves (12.2.76) with $f_1(t) \in L_\infty(\mathbb{R}, L_{6/p}(\mathscr{O}))$. Again, the elliptic regularity theory gives us that $z(t)$ is bounded in $W^2_{6/p}(\mathscr{O})$. Thus the second statement of Theorem 12.2.6 is proved. One could improve this latter regularity (see Remark 12.2.8), but this is not essential for our further analysis.

3. *Existence of a fractal exponential attractor.* This is a direct consequence of Theorem 12.2.15, local Lipschitz continuity (6.2.41) of the semiflow S_t and Theorem 7.9.9 (see also Theorem 4.43 in [75]). □

12.2.6 Generalizations

1. The analysis of this section can be easily extended to other boundary conditions imposed on the plate equation. In this spirit, one could consider hinged or free boundary conditions. A combination thereof also allows assuming right compatibility conditions imposed at the junctions.

2. The structural acoustic model given in (12.2.1) and (12.2.2) could also be treated within a rotational inertia framework. This amounts to adding a rotational term to the plate component thus replacing (12.2.2) by

$$u_{tt} - \alpha\Delta u_{tt} + \Delta^2 u + b_0(u_t) - \operatorname{div} b_1(\nabla u_t) - [u, v(u) + F_0] + \beta\kappa z_t = p_0(x). \tag{12.2.77}$$

Under suitable conditions imposed on b_1 (which are the same as the ones imposed on rotational plates; for some of them see Assumption 6.2.1) one recovers all the results of the present section with the state space

$$Y = \mathscr{D}(A^{1/2}) \times L_2(\mathscr{O}) \times \mathscr{D}(\mathscr{A}^{1/2}) \times V_\alpha,$$

where $V_\alpha = \mathscr{D}(M_\alpha^{1/2})$. Moreover one can show that the size and dimension of the attractor in this case admit the estimates independent of α. This makes it possible to study the limit of the attractor as $\alpha \to 0$ and obtain a similar result to Theorem 9.5.14 (see also Theorem 12.3.7 below).

We also refer to Section 6.2 for the details concerning problem (12.2.1) and (12.2.77), including the well-posedness result.

12.3 Wave coupled to thermoelastic plate equation

In this section we study long-time behavior of a nonlinear structural acoustic model (6.3.1) which includes thermal effects and does not contain any mechanical dissipation in the plate component. Thus, the important feature of the model is that the plate equation does not exhibit any mechanical damping. The only dissipative effect in the plate component is induced by the heat equation that is coupled to the plate equation. Thermal effects in the plate and viscous dissipation in the wave are strong enough in order to transfer the dissipation onto the entire system. This phenomenon (which also holds for the pure thermoelastic model; see Chapter 11) is supported by PDE estimates which constitute a core of the argument asserting existence of global attractors. We also note that global attractors for a similar model with Berger-type nonlinearity in the plate component have been studied in [37].

The model of thermoelastic plate under consideration is the following,

$$\begin{cases} z_{tt} + g(z_t) - \Delta z + f(z) = 0 \ \text{ in } \ \mathscr{O} \times (0,T), \\ \dfrac{\partial z}{\partial n} = 0 \ \text{ on } \ \Omega_* \times (0,T), \quad \dfrac{\partial z}{\partial n} = \gamma\kappa v_t \ \text{ on } \ \Omega \times (0,T), \\ u_{tt} - \alpha\Delta u_{tt} + \Delta^2 u - [u, v(u) + F_0] + \beta\kappa z_t|_\Omega + \Delta\theta = p_0 \ \text{ in } \ \Omega \times (0,T), \\ u = \Delta u = 0 \ \text{ on } \ \partial\Omega \times (0,T), \\ \theta_t - \Delta\theta - \Delta u_t = 0 \ \text{ in } \ \Omega \times (0,T), \quad \theta = 0 \ \text{ on } \ \partial\Omega \times (0,T), \end{cases} \tag{12.3.1}$$

where $v = v(u) \in H_0^2(\Omega)$ is the Airy stress function given by (12.2.3). We supplement the problem (12.3.1) with initial data

$$\begin{cases} z(0,\cdot) = z^0, \quad z_t(0,\cdot) = z^1 & \text{in } \mathscr{O} \\ u(0,\cdot) = u^0, \quad u_t(0,\cdot) = u^1, \quad \theta(0,\cdot) = \theta^0 & \text{in } \Omega. \end{cases} \tag{12.3.2}$$

As in Section 12.2 here $\mathscr{O} \subset \mathbb{R}^3$ is an open bounded domain with boundary $\partial\mathscr{O} = \overline{\Omega \cup \Omega_*}$ comprising two open (in the induced topology), connected, disjoint parts Ω and Ω_* of positive measure. We assume that either $\mathscr{O}$ is sufficiently smooth (e.g., $\partial\mathscr{O} \in C^2$) or else $\mathscr{O}$ is convex. Ω is flat and is referred to as the elastic wall.

The function $g(s)$ is a nondecreasing function describing the dissipation in the wave component and the term $f(z)$ represents a nonlinear force; n is the outer normal vector, β and γ are positive constants; $0 \le \kappa \le 1$, in the case of noninteracting wave and plate equations we have $\kappa = 0$. The parameter $\alpha \in [0,1]$ describes the rotational inertia of the plate filaments.

12.3.1. Remark. One could also consider other boundary conditions imposed on the plate equation: such as clamped or hinged–clamped. The technical details are somewhat different, however, these were already treated in the context of a single equation in Chapter 9 (see also Chapter 11 for a pure thermoelastic model). Here we choose hinged boundary conditions for the sake of some diversity (in contrast with the previous section) and also some simplification. ■

Concerning functions $g \in C(\mathbb{R})$ and $f \in Lip_{loc}(\mathbb{R})$ and the loads p_0 and F_0 we assume that Assumption 6.2.1 is in force. Under this assumption well-posedness of the problem was established in Section 6.3. Namely, Theorem 6.3.2 enables us to define a dynamical system (Y, S_t) with the phase space Y given by (6.3.7); that is,

$$Y = Y_1 \times Y_2 \times Y_3 = \mathscr{D}(A^{1/2}) \times L^2(\mathscr{O}) \times \mathscr{D}(\mathscr{A}) \times V_\alpha \times L_2(\Omega),$$

and with the evolution operator $S_t : Y \to Y$ given by the relation (6.3.25).

Below we also need the following properties of the energy functionals and stationary solutions.

It follows from (6.3.12) that the energy $\mathscr{E}(z, z_t, u, u_t, \theta)$ given by (6.3.10) is bounded from below on Y and coercive. This implies that there exists $R_* > 0$ such that the set

$$\mathscr{W}_R = \{y = (z_0; z_1; v_0; v_1; \theta_0) \in Y : \mathscr{E}(z_0, z_1, v_0, v_1, \theta_0) \leq R\} \tag{12.3.3}$$

is a nonempty bounded set in Y for all $R \geq R_*$. Moreover any bounded set $B \subset Y$ is contained in $\mathscr{W}_R$ for some R and, as it follows from (6.3.17), the set $\mathscr{W}_R$ is forward invariant with respect to the semiflow S_t; that is, $S_t \mathscr{W}_R \subset \mathscr{W}_R$ for all $t > 0$. Thus, we can consider the restriction $(\mathscr{W}_R, S_t)$ of the dynamical system (Y, S_t) on $\mathscr{W}_R$, $R \geq R_*$.

We also note that the set of stationary points of S_t,

$$\mathscr{N} = \{V \in Y : S_t V = V \text{ for all } t \geq 0\},$$

consists of vectors V of the form $V = (z; 0; v; 0; 0)$, where $z \in H^1(\mathscr{O})$ and $v \in H^2(\Omega) \cap H_0^1(\Omega)$ are, respectively, the weak (variational) solutions to the equations in (12.2.6) and (12.2.7).

It is clear that the set $\mathscr{N}$ of stationary points depends neither on α nor on κ. Therefore in the same way as in Lemma 12.2.1 one can easily obtain the following assertion.

12.3.2. Lemma. *Under Assumption* 6.2.1 *the set $\mathscr{N}$ of stationary points for the semiflow S_t generated by equations* (12.3.1) *is a closed bounded set in Y, and hence there exists $R_{**} \geq R_*$ (independent of α and κ) such that $\mathscr{N} \subset \mathscr{W}_R$ for every $R \geq R_{**}$.*

Our further presentation in this section follows the same line of argument as in the isothermal case; see Section 12.2.

12.3.1 The statement of main results

Referring to notations in Section 6.3 we first recall the abstract representation of (12.3.1) and (12.3.2):

$$z_{tt} + A\left(z - \gamma\kappa N_0 u_t\right) + D(z_t) + F_1(z) = 0, \tag{12.3.4}$$

$$M_\alpha u_{tt} + \mathscr{A}^2 u + \beta\kappa N_0^* A z_t - \mathscr{A}\theta + F_2(u) = 0, \tag{12.3.5}$$

$$\theta_t + \mathscr{A}\theta + \mathscr{A} u_t = 0, \tag{12.3.6}$$

$$z(0) = z^0\; z_t(0) = z^1\,;\; u(0) = u^0\,,\, u_t(0) = u^1\,, \theta(0) = \theta^0\,, \tag{12.3.7}$$

where for the sake of simplicity we have used the notations (12.2.10) and (12.2.11) for A and N_0, the positive operator $\mathscr{A} : \mathscr{D}(\mathscr{A}) \subset L_2(\Omega) \mapsto L_2(\Omega)$ is defined by

$$\mathscr{A} w = -\Delta w, \quad \mathscr{D}(\mathscr{A}) = H^2(\Omega) \cap H_0^1(\Omega).$$

We also use the notations

$$D(h) := g(h), \;\; F_1(z) = f(z) - \mu z,$$

where μ is given by (6.2.6), and

$$F_2(u) = -[u, v(u) + F_0] - p_0,$$

where $v = v(u)$ is the Airy stress function given by (12.2.3).

Our first main result provides the existence of a global attractor for problem (12.3.4)–(12.3.7), as well as a description of its structure.

12.3.3. Theorem (Existence of global attractor). *Let Assumption* 6.2.1 *(concerning f, g, F_0, and p_0) and relation* (12.2.8) *be in force. Then the dynamical system (Y, S_t) generated by problem* (12.3.4)–(12.3.7) *has a compact global attractor $\mathfrak{A}$ that coincides with the unstable manifold $\mathscr{M}^u(\mathscr{N})$ emanating from the set $\mathscr{N}$ of stationary points*[2] *for S_t, namely $\mathfrak{A} \equiv \mathscr{M}^u(\mathscr{N})$. Moreover,*

$$\lim_{t\to+\infty} \mathrm{dist}_Y(S_t W, \mathscr{N}) = 0 \;\; \textit{for any}\; W \in Y, \tag{12.3.8}$$

and for any trajectory $U = (z(t); z_t(t); u(t); u_t(t); \theta(t))$ from the attractor $\mathscr{A}$ we have

$$\|z\|^2_{1,\mathscr{O}} + \|z_t\|^2_{\mathscr{O}} + \|\Delta u\|^2_\Omega + \|u_t\|^2_\Omega + \alpha\|\nabla u_t\|^2_\Omega + \|\theta\|^2_\Omega \le \widetilde{R}^2, \tag{12.3.9}$$

where $\widetilde{R}$ does not depend on $\alpha, \kappa \in [0,1]$.

Our second main result contains specific assertions regarding the dimension and regularity of the attractor $\mathfrak{A}$. These require the following additional assumptions.

[2] We recall that every element V from $\mathscr{N}$ has the form $V = (z; 0; u; 0; 0)$, where z and u are variational solutions to (12.2.6) and (12.2.7).

12.3.4. Assumption. In addition to Assumption 6.2.1 (concerning f, g, F_0, and p_0), let the following conditions hold.

- Relation (12.2.8) is in force.
- There exist positive constants m and M such that (12.2.15) holds:

$$m \le \frac{g(s_1)-g(s_2)}{s_1-s_2} \le M\,(1+s_1g(s_1)+s_2g(s_2))^{2/3}, \quad s_i \in \mathbb{R},\ s_1 \ne s_2. \tag{12.3.10}$$

- $f \in C^2(\mathbb{R})$ and $|f''(s)| \le C(1+|s|)$ for $s \in \mathbb{R}$.

12.3.5. Theorem (Properties of attractor). *Let Assumption* 12.3.4 *hold. Then the compact global attractor* $\mathfrak{A}$ *given by Theorem* 12.3.3 *possesses the following properties.*

1. *The attractor* $\mathfrak{A}$ *has a finite fractal dimension, and there exists a constant* d *independent of* α *and* κ, *such that* $\dim_f \mathfrak{A} \le d$.
2. *The attractor* $\mathscr{A}$ *has additional smoothness reflected by its boundedness in the space*

$$Y_* = W^2_{6/p}(\mathscr{O}) \times \mathscr{D}(A^{1/2}) \times W_\alpha \times \mathscr{D}(\mathscr{A}) \times \mathscr{D}(\mathscr{A})$$

in the case $3 < p \le 5$, *and in the space*

$$Y_{**} = H^2(\mathscr{O}) \times \mathscr{D}(A^{1/2}) \times W_\alpha \times \mathscr{D}(\mathscr{A}) \times \mathscr{D}(\mathscr{A})$$

in the case $1 \le p \le 3$, *where* W_α *is given by* (6.3.14). *Moreover, for any trajectory*

$$U = (z(t); z_t(t); u(t); u_t(t); \theta(t)), \quad t \in \mathbb{R},$$

from the attractor $\mathfrak{A}$ *we have that*

$$\|z(t)\|^2_{W^2_{p_*}(\mathscr{O})} + \|z_t(t)\|^2_{1,\mathscr{O}} + \|z_{tt}(t)\|^2_{\mathscr{O}} + \|u(t)\|^2_{3,\Omega} + \|\Delta u_t(t)\|^2_{0,\Omega}$$
$$+ \|u_{tt}(t)\|^2_\Omega + \alpha\|\nabla u_{tt}(t)\|^2_\Omega + \|\theta_t(t)\|^2_\Omega + \|\theta(t)\|^2_{2,\Omega} \le R^2_*, \tag{12.3.11}$$

where $p_* = \min\{2, 6/p\}$ *and* R_* *does not depend on* $\alpha, \kappa \in [0,1]$ *(in the case* $\alpha = 0$ *we have the additional estimate* $\|v\|_4 \le R_*$*).*

12.3.6. Remark. The estimates describing the regularity properties in Theorem 12.3.5 are uniform with respect to rotational parameter $\alpha \ge 0$. It is also interesting to notice that the thermoelastic plates have parabolic-like dynamics when $\alpha = 0$ and hyperbolic-like when $\alpha > 0$. Parabolicity induces additional smoothing effects for transient times. On the other hand, the rotational model provides additional regularity to the velocity component. Thus the smoothing mechanisms in rotational and nonrotational cases are very different and not compatible. In view of this the task of obtaining uniform in α estimates should not depend on neither analyticity nor hyperbolicity. This task is particularly challenging in the presence of nonlinearity that is critical, as in the von Karman case. We also refer to the discussion in Chapter 11.

The regularity of the attractor expressed in (12.3.11) can be improved when $\alpha > 0$. Indeed, this is in line with the respective result in Chapter 11 where for hinged plates one shows that the mechanical variable has additional regularity $||v_t(t)||_{3,\Omega}$ and $||v(t)||_{4,\Omega}$ (see Corollary 11.4.3). However, the estimate is no longer uniform in α. Moreover, this result depends on the structure of hinged boundary conditions. It is unclear whether the same regularity could be obtained for, say, the clamped case.

The regularity of the z variable in the attractor can also be improved to reach the optimal level of $H^2(\mathscr{O})$. This result, mentioned previously (see Remark 12.2.8), requires slightly stronger assumptions imposed on the dissipation $g(s)$. ∎

In the case $\kappa = 0$ Theorem 12.3.3 and Theorem 12.3.5 provide specified assertions separately for each uncoupled equation. We refer to Corollary 12.2.7 in the case of the wave component and to Theorem 11.2.1 for the case of the thermoelastic component.

The following theorem provides us with more detailed information regarding the dependence of the attractor on the parameters α and κ.

12.3.7. Theorem (Dependence on Parameters). *Let Assumption* 12.3.4 *be in force and let* $p < 5$. *Denote by* $S_t^{\alpha,\kappa}$ *the evolution operator of problem* (12.3.4)–(12.3.7) *in the space*

$$Y_\alpha := Y = \mathscr{D}(A^{1/2}) \times L^2(\mathscr{O}) \times \mathscr{D}(\mathscr{A}) \times V_\alpha \times L_2(\Omega).$$

Let $\mathfrak{A}^{\alpha,\kappa}$ *be a global attractor for the system* $(Y_\alpha, S_t^{\alpha,\kappa})$. *Then the family of the attractors* $\mathfrak{A}^{\alpha,\kappa}$ *is upper semicontinuous on* $\Lambda := [0,1] \times [0,1]$*: namely, for any* $\lambda_0 = (\alpha_0, \kappa_0) \in \Lambda$ *we have that*

$$\lim_{(\alpha,\kappa)\to\lambda_0} \sup\left\{ \mathrm{dist}_{Y_{\alpha_0}} \left(U, \mathfrak{A}^{\lambda_0} \right) : U \in \mathfrak{A}^{\alpha,\kappa} \right\} = 0. \tag{12.3.12}$$

In particular, if $\kappa_0 = 0$ *one has*

$$\sup\left\{ \mathrm{dist}_{Y_{\alpha_0}} \left(U, \mathfrak{A}_1 \times \mathfrak{A}_2^{\alpha_0} \right) : U \in \mathfrak{A}^{\alpha,\kappa} \right\} = 0 \ \ as\ \alpha \to \alpha_0,\ \kappa \to 0, \tag{12.3.13}$$

where $\mathfrak{A}_1$ *is the attractor for the wave component (see Corollary* 12.2.7*) and* $\mathfrak{A}_2^\alpha$ *is the attractor generated by the corresponding thermoelastic problem* (12.3.5) *and* (12.3.6) *with* $\kappa = 0$.

The proof of this theorem is given at the end of Section 12.3.5.

12.3.8. Remark. Assertion (12.3.13) means that in the decoupling limit $\kappa \to 0$ the global attractor $\mathfrak{A}$ becomes close to the Cartesian product of global attractors pertaining to the problems (12.2.19) and (12.3.5) and (12.3.6) with $\kappa = 0$, respectively. It should be also noted that in the critical case $p = 5$ the semicontinuity property (11.2.4) may be established as well, though in the (weaker) topology of the space

$$Y_\alpha^{-\varepsilon} := \mathscr{D}(A^{1/2-\varepsilon}) \times L^2(\mathscr{O}) \times \mathscr{D}(\mathscr{A}) \times V_\alpha \times L_2(\Omega), \tag{12.3.14}$$

with arbitrary $\varepsilon > 0$; see Remark 12.3.16 below. Whether we can take $\varepsilon = 0$ in the case $p = 5$ is still an open question. ∎

12.3.2 Main inequality

Analogous to with isothermal case (see Section 12.2.2) we start with the derivation of a preliminary inequality, which constitutes a fundamental common step for the proofs of both Theorems 12.3.3 and 12.3.5. This inequality, namely (12.3.16) in Proposition 12.3.9 below, holds true under the basic Assumption 6.2.1.

12.3.9. Proposition. *Let Assumption* 6.2.1 *(concerning f, g, F_0, and p_0) hold. Assume that $y_1, y_2 \in \mathscr{W}_R$ for some $R > R_*$, where $\mathscr{W}_R$ is defined by* (12.3.3) *and denote*

$$(h(t); h_t(t); u(t); u_t(t); \psi(t)) := S_t y_1, \quad (\zeta(t); \zeta_t(t); w(t); w_t(t); \xi(t)) := S_t y_2.$$

Let

$$z(t) := h(t) - \zeta(t), \quad v(t) := u(t) - w(t), \quad \theta(t) := \psi(t) - \xi(t). \tag{12.3.15}$$

There exist $T_0 > 0$ and positive constants c_0, c_1 and $c_2(R)$ independent of T and $\alpha, \kappa \in [0,1]$ such that for every $T \geq T_0$ the following inequality holds:

$$\begin{aligned} TE^0(T) + \int_0^T E^0(t)dt \leq c_0\Big[\int_0^T \left(\|z_t\|_{\mathscr{O}}^2 + \gamma\|\mathscr{A}^{1/2}\theta\|_\Omega^2\right)dt + \beta G_0^T(z)\Big] \\ + c_1\left[H_0^T(z) + \Psi_T(z,v)\right] + c_2(R)\int_0^T \left(\|z\|_{\mathscr{O}}^2 + \|v\|_\Omega^2\right)dt, \end{aligned} \tag{12.3.16}$$

where

$$E^0(t) = \beta E_z^0(t) + \gamma E_v^0(t) + \frac{\gamma}{2}\|\theta(t)\|_\Omega^2 \tag{12.3.17}$$

with $E_z^0(t)$ and $E_v^0(t)$ given by (6.2.21) *and* (6.2.22)*. We also used the notations* (12.2.21) *and* (12.2.22) *for $G_s^t(z)$ and $H_s^t(z)$ and* (12.2.23) *for $\Psi_T(z,v)$.*

Proof. We follow the same line of argument as in the proof of Proposition 12.2.9 (see also abstract Lemma 8.3.1).

Step 1 (Energy identity). Without loss of generality, we assume that $(h;u;\psi)$ and $(\zeta;w;\xi)$ are strong solutions. By the invariance of $\mathscr{W}_R$ and in view of relation (6.3.12) in addition to (12.2.25) we also have that

$$\|\psi(t)\|_\Omega^2 + \|\xi(t)\|_\Omega^2 \leq C_R, \quad t \geq 0. \tag{12.3.18}$$

We first establish energy-type equality for the energy $E^0(t)$.

12.3.10. Lemma. *For any $T > 0$ and all t, $0 \leq t \leq T$, $E^0(t)$ satisfies*

$$\begin{aligned} E^0(T) + \beta G_t^T(z) + \gamma\int_t^T \|\mathscr{A}^{1/2}\theta\|_\Omega^2 d\tau \\ = E^0(t) - \beta\int_t^T (\mathscr{F}_1(z), z_t)_{\mathscr{O}}\, d\tau - \gamma\int_t^T (\mathscr{F}_2(v), v_t)_\Omega\, d\tau, \end{aligned} \tag{12.3.19}$$

where $G_t^T(z)$, represents the damping on the wave component and is given by the relation in (12.2.21), *whereas $\mathscr{F}_1(z)$ and $\mathscr{F}_2(z)$ are defined by* (12.2.24).

Proof. It is straightforward to derive the following system of coupled equations satisfied for the differences given in (12.3.15),

$$z_{tt}+A(z-\gamma\kappa N_0 v_t)+D(\zeta_t+z_t)-D(\zeta_t)+\mathscr{F}_1(z)=0, \tag{12.3.20}$$

$$M_\gamma v_{tt}+\mathscr{A}^2 v-\mathscr{A}\theta+\beta\kappa N_0^* A z_t+\mathscr{F}_2(v)=0, \tag{12.3.21}$$

$$\theta_t+\mathscr{A}\theta+\mathscr{A}v_t=0, \tag{12.3.22}$$

with $\mathscr{F}_i$ defined in (12.2.24). Next, by standard energy methods we obtain

$$E_z^0(T)+G_t^T(z)=E_z^0(t)+\gamma\kappa\int_t^T (AN_0 v_t,z_t)_{\mathscr{O}}\,d\tau-\int_t^T(\mathscr{F}_1(z),z_t)_{\mathscr{O}}\,d\tau;$$

$$\begin{aligned}E_v^0(T)&=E_v^0(t)-\beta\kappa\int_t^T (N_0^*Az_t,v_t)_\Omega\,d\tau\\&\quad-\int_t^T(\mathscr{F}_2(v),v_t)_\Omega\,d\tau+\int_t^T(\mathscr{A}\theta,v_t)_\Omega\,d\tau;\end{aligned}$$

$$\frac12\|\theta(T)\|_\Omega^2+\int_t^T\|\mathscr{A}^{1/2}\theta\|_\Omega^2\,d\tau=\frac12\|\theta(t)\|_\Omega^2-\int_t^T(\mathscr{A}v_t,\theta)_\Omega\,d\tau.$$

Then, (12.3.19) readily follows multiplying the first equation by β, the second and third one by γ, and summing up. □

Step 2. (Reconstruction of the energy integral) We return to the coupled system satisfied by $(z;v;\theta)$. Multiplying equation (12.3.20) by z and integrating between 0 and T, we obtain:

$$\begin{aligned}\int_0^T\|A^{1/2}z\|_{\mathscr{O}}^2dt&\le c_0\left(E_z^0(T)+E_z^0(0)\right)+\int_0^T\|z_t\|_{\mathscr{O}}^2dt\\&\quad+H_0^T(z)+\gamma\kappa\int_0^T|(v_t,N_0^*Az)_\Omega|\,dt+\int_0^T|(\mathscr{F}_1(z),z)_{\mathscr{O}}|\,dt,\end{aligned}\tag{12.3.23}$$

where $H_0^T(z)$ is defined in (12.2.22), and c_0 does not depend on α, β, γ. As in Proposition 12.2.9 from (6.2.20) and (12.2.25) we also have that

$$|(\mathscr{F}_1(z),z)_{\mathscr{O}}|\le C_R\|A^{1/2}z\|_{\mathscr{O}}\|z\|_{\mathscr{O}}.$$

On the other hand, using (6.2.9) with $\varepsilon=1/4-\delta$, $0<\delta<1/4$, we obtain that

$$\begin{aligned}|(v_t,N_0^*Az)_\Omega|&\le\|v_t\|_\Omega\,\|N_0^*A^{1/2+\delta}\|\,\|A^{1/2-\delta}z\|_\Omega\le C\|v_t\|_\Omega\,\|A^{1/2-\delta}z\|_{\mathscr{O}}\\&\le\varepsilon\|v_t\|_\Omega^2+\varepsilon_1\|A^{1/2}z\|_{\mathscr{O}}^2+C_{\varepsilon,\varepsilon_1}\|z\|_{\mathscr{O}}^2,\end{aligned}$$

for any $\varepsilon, \varepsilon_1 > 0$. Then, by appropriately choosing and rescaling ε and ε_1 we obtain from (12.3.23) that

$$\int_0^T ||A^{1/2}z||_{\mathscr{O}}^2 dt \le C_0\left(E_z^0(T)+E_z^0(0)\right)+\varepsilon\int_0^T ||v_t||_{\Omega}^2 dt \tag{12.3.24}$$
$$+2\int_0^T ||z_t||_{\mathscr{O}}^2 dt + C_1 H_0^T(z) + C_2(R,\varepsilon)\int_0^T ||z||_{\mathscr{O}}^2 dt,$$

for any $\varepsilon > 0$.

For the plate component we use the multiplier $\varepsilon v + \mathscr{A}^{-1}\theta$. One gets, initially,

$$\varepsilon(M_\alpha v_{tt}, v)_\Omega + \varepsilon\|\mathscr{A}v\|_\Omega^2 - \varepsilon(\mathscr{A}\theta, v)_\Omega + (M_\alpha v_{tt}, \mathscr{A}^{-1}\theta)_\Omega + (\mathscr{A}v, \theta)_\Omega$$
$$-\|\theta\|_\Omega^2 + \beta\kappa(z_t, \varepsilon v + \mathscr{A}^{-1}\theta)_\Omega + (\mathscr{F}_2(v), \varepsilon v + \mathscr{A}^{-1}\theta)_\Omega = 0,$$

which can be rewritten as

$$\varepsilon\frac{d}{dt}(M_\alpha v_t, v)_\Omega - \varepsilon\|M_\alpha^{1/2}v_t\|_\Omega^2 + \varepsilon\|\mathscr{A}v\|_\Omega^2 + (1-\varepsilon)(\mathscr{A}\theta, v)_\Omega$$
$$+\frac{d}{dt}(M_\alpha v_t, \mathscr{A}^{-1}\theta)_\Omega + (M_\alpha v_t, \theta)_\Omega + \|M_\alpha^{1/2}v_t\|_\Omega^2 - \|\theta\|_\Omega^2$$
$$+\beta\kappa(z_t, \varepsilon v + \mathscr{A}^{-1}\theta)_\Omega + (\mathscr{F}_2(v), \varepsilon v + \mathscr{A}^{-1}\theta)_\Omega = 0, \tag{12.3.25}$$

where we have used elementary calculus and the equality $\mathscr{A}^{-1}\theta_t = -\theta - v_t$ (which is just (12.3.22)). Now, a straightforward calculation relying on the properties of the von Karman bracket (see Theorem 1.4.3) shows that

$$|(\mathscr{F}_2(v), \varepsilon v + \mathscr{A}^{-1}\theta)_\Omega| \le C_R(\varepsilon_0)\left(\|\mathscr{A}v\|_\Omega \, \|v\|_\Omega + \|\mathscr{A}^{1/2}v\|_\Omega \, \|\theta\|_\Omega\right),$$

for any $\varepsilon \in (0, \varepsilon_0]$. Thus, integrating in time between 0 and T the equality (12.3.25) and by choosing appropriately $\varepsilon > 0$, we see that

$$\int_0^T \left(||\mathscr{A}v||_\Omega^2 + ||M_\alpha^{1/2}v_t||_\Omega^2 + ||\theta||_\Omega^2\right) dt$$
$$\le c_0\left(E_{v,\theta}^0(T) + E_{v,\theta}^0(0)\right) + c_1\int_0^T \|\mathscr{A}^{1/2}\theta\|_\Omega^2 dt$$
$$+4\beta\kappa\left|\int_0^T (z_t, \varepsilon v + \mathscr{A}^{-1}\theta)_\Omega \, dt\right| + C(R)\int_0^T \|v\|_\Omega^2 dt, \tag{12.3.26}$$

where we have set $E_{v,\theta}^0(t) = E_v^0(t) + \frac{1}{2}\|\theta(t)\|_\Omega^2$. Integrating by parts in time and using the standard form of the trace theorem it is easy to see that

$$\left|\int_0^T (z_t, \varepsilon v + \mathscr{A}^{-1}\theta)_\Omega \, dt\right| \le C_1\left(E^0(T) + E^0(0)\right) + \left|\int_0^T (z, \varepsilon v_t + \mathscr{A}^{-1}\theta_t)_\Omega \, dt\right|,$$

where $E^0(t)$ is given by (12.3.17). Therefore, using once more the equality $\mathscr{A}^{-1}\theta_t = -\theta - v_t$ we obtain

$$\left|\int_0^T (z_t, \varepsilon v + \mathscr{A}^{-1}\theta)_\Omega dt\right| \le C_1\left(E^0(T)+E^0(0)\right) + \left|\int_0^T (z,(\varepsilon-1)v_t - \theta)_\Omega dt\right|$$

$$\le C_1\left(E^0(T)+E^0(0)\right) + \varepsilon_1 \int_0^T \|A^{1/2}z\|_{\mathscr{O}}^2 dt + \varepsilon_2 \int_0^T \|M_\alpha^{1/2} v_t\|_\Omega^2 dt$$

$$+ C_2 \int_0^T \|\theta\|_\Omega^2 dt + C_3(\varepsilon_1,\varepsilon_2)\int_0^T \|z\|_{\mathscr{O}}^2 dt \quad (12.3.27)$$

for every $\varepsilon_1, \varepsilon_2 > 0$. Using now (12.3.27) in (12.3.26) shows

$$\int_0^T E^0_{v,\theta}(t)dt \le c_0\left(E^0(T)+E^0(0)\right) + c_1 \int_0^T \|\mathscr{A}^{1/2}\theta\|_\Omega^2 dt \quad (12.3.28)$$

$$+ \varepsilon \int_0^T \|A^{1/2}z\|_{\mathscr{O}}^2 dt + C(R,\varepsilon)\int_0^T \left(\|z\|_{\mathscr{O}}^2 + \|v\|_\Omega^2\right) dt\,.$$

Combined with the estimate (12.3.24), (12.3.29) establishes

$$\int_0^T E^0(t)dt \le c_0\left[E^0(T)+E^0(0)\right] + c_1 \int_0^T \left(\|z_t\|_{\mathscr{O}}^2 + \|\mathscr{A}^{1/2}\theta\|_\Omega^2\right) dt$$

$$+ c_2 H_0^T(z) + c_3(R)\int_0^T \left(\|z\|_{\mathscr{O}}^2 + \|v\|_\Omega^2\right) dt\,. \quad (12.3.29)$$

On the other hand, it follows from Lemma 12.3.10 that

$$E^0(0) = E^0(T) + \beta G_0^T(z) + \gamma \int_0^T \|\mathscr{A}^{1/2}\theta\|_\Omega^2 dt \quad (12.3.30)$$

$$+ \beta \int_0^T (\mathscr{F}_1(z), z_t)_{\mathscr{O}}\, d\tau + \gamma \int_0^T (\mathscr{F}_2(v), v_t)_\Omega\, d\tau\,,$$

and that

$$TE^0(T) \le \int_0^T E^0(t)dt - \beta \int_0^T \int_t^T (\mathscr{F}_1(z), z_t)_{\mathscr{O}}\, d\tau dt - \gamma \int_0^T \int_t^T (\mathscr{F}_2(v), v_t)_\Omega\, d\tau dt\,. \quad (12.3.31)$$

Therefore, combining (12.3.31) with (12.3.29) and (12.3.30), we see that (12.3.16) holds true, provided that T is sufficiently large. This concludes the proof of Proposition 12.3.9. □

12.3.3 Asymptotic smoothness and proof of Theorem 12.3.3

The main result of this section is asymptotic smoothness of the composite flow. As in Section 12.2.3, Theorem 7.1.11 is the main tool for proving it.

12.3.11. Theorem. *Let Assumption* 6.2.1 *(concerning f, g, F_0, and p_0) and relation* (12.2.8) *hold. Then the dynamical system* (Y, S_t) *generated by the PDE problem* (12.3.1) *is asymptotically smooth.*

Proof. As in Theorem 12.2.11 it is sufficient to consider the asymptotic compactness property on the set $\mathscr{W}_R$ for every $R > R_*$ only. Here $\mathscr{W}_R$ is defined by (12.3.3). Accordingly, let $y_1, y_2 \in \mathscr{W}_R$. The notation used below is the same as in Proposition 12.3.9. Namely, we denote

$$(h(t); h_t(t); u(t); u_t(t); \psi(t)) := S_t y_1, \quad (\zeta(t); \zeta_t(t); w(t); w_t(t); \xi(t)) := S_t y_2,$$

and define $z(t) := h(t) - \zeta(t)$, $v(t) := u(t) - w(t)$, and $\theta(t) := \psi(t) - \xi(t)$. We seek to establish an estimate of $d(S_T y_1, S_T y_2)$ as required by Theorem 7.1.11; this eventually enables us to achieve the conclusion of Theorem 12.3.11. To accomplish this goal, the following result plays the major role.

12.3.12. Proposition. *Let the assumptions of Theorem* 12.3.11 *be in force. Then, given $\varepsilon > 0$ and $T > 1$ there exist constants $C_\varepsilon(R)$ and $C(R,T)$ such that*

$$E^0(T) \le \varepsilon + \frac{1}{T}\left[C_\varepsilon(R) + \Psi_T(z,v)\right] + C(R,T)\, lot(z,v), \tag{12.3.32}$$

where $E^0(t) := \beta E_z^0(t) + \gamma E_v^0(t) + \frac{\gamma}{2}\|\theta(t)\|^2$ with $E_z^0(t)$ and $E_v^0(t)$ given in (6.3.8) *and* (6.3.9)*, the functional $\Psi_T(z,v)$ is given by* (12.2.23) *and $lot(z,v)$ is defined by*

$$lot(z,v) := \sup_{[0,T]} \|z(t)\|_{\mathscr{O}} + \sup_{[0,T]} \|v(t)\|_{\Omega}.$$

Proof. To establish (12.3.32), we return to the main inequality (12.3.16) and proceed with the estimate of its right-hand side. In view of (12.2.39), $H_s^t(z)$ defined by (12.2.22) satisfies

$$H_0^T(z) \le C_0 \int_0^T \left[(D(\zeta_t), \zeta_t)_{\mathscr{O}} + (D(h_t), h_t)_{\mathscr{O}}\right] \|A^{1/2} z\|_{\mathscr{O}} dt + C_1 \int_0^T \|z\|_{\mathscr{O}} dt.$$

Moreover, from the energy inequality (6.3.15) we know

$$\begin{aligned} \beta \int_0^t (D(h_t), h_t)_{\mathscr{O}} dt + \gamma \int_0^t \|\mathscr{A}^{1/2}\psi\|_\Omega^2 dt \le C_R, \\ \beta \int_0^t (D(\zeta_t), \zeta_t)_{\mathscr{O}} dt + \gamma \int_0^t \|\mathscr{A}^{1/2}\xi\|_\Omega^2 dt \le C_R, \end{aligned} \tag{12.3.33}$$

where C_R does not depend on t. This latter fact is critical. Then, the estimates (12.3.33) combined with the fact that $\|A^{1/2} z(t)\|_{\mathscr{O}} \le C_R$ for all $t \in [0,T]$ show that

$$H_0^T(z) \le C_R + CT\, lot(z,v). \tag{12.3.34}$$

Next, using now the inequality (12.2.8) and once again the uniform estimates (12.3.33), one can see that

$$\int_0^T \left(\|z_t\|_{\mathscr{O}}^2 + \|\mathscr{A}^{1/2}\theta\|_{\Omega}^2\right) dt \le \varepsilon T + C_\varepsilon(R) \ \text{ for every } \varepsilon > 0. \tag{12.3.35}$$

Taking $t = 0$ in (12.3.19) and using the fact that $E^0(0) \le C_R$, we get

$$\begin{aligned} &\beta G_0^T(z) + \gamma \int_0^T \|\mathscr{A}^{1/2}\theta\|_{\Omega}^2 dt \\ &\le C_R + \beta \left| \int_0^T (\mathscr{F}_1(z), z_t)_{\mathscr{O}} \, d\tau \right| + \gamma \left| \int_0^T (\mathscr{F}_2(v), v_t)_{\Omega} \, d\tau \right|. \end{aligned} \tag{12.3.36}$$

Therefore, (12.3.32) follows from (12.3.16) of Proposition 12.3.9 combining the estimates (12.3.34)–(12.3.36). □

We are now in a position to complete the proof of Theorem 12.3.11 in the same way as done in the proof of Theorem 12.2.11 in Section 12.2.

Proof of Theorem 12.3.3. The argument is the same as in the proof of Theorem 12.2.3. First, Theorem 12.3.11 implies the existence of a compact global attractor $\mathfrak{A}_R$ for the dynamical system $(\mathscr{W}_R, S_t)$. Next, energy inequality (6.3.15) implies that the energy $\mathscr{E}$ given by (6.3.10) is a strict Lyapunov function for $(\mathscr{W}_R, S_t)$. Then by Lemma 12.3.2 $\mathfrak{A}_R = M^u(\mathscr{N})$ does not depend on R for large R and by Theorem 7.5.10 relation (12.3.8) holds. We finally observe that

$$\sup\left\{\mathscr{E}(U) : \, U \in \mathfrak{A}\right\} = \sup\left\{\mathscr{E}(U) : \, U \in \mathscr{N}\right\}.$$

This implies that $\mathfrak{A} \subseteq \left\{U \in Y : \, \|U\|_Y \le R\right\}$, where R does not depend on γ and κ. Therefore, (12.3.9) holds, and the proof of Theorem 12.3.3 is completed. □

12.3.4 Stabilizability estimate

As in the previous Section 12.2, both regularity and finite-dimensionality of the attractor stated in Theorem 12.3.3 depend critically on the *stabilizability estimate*.

12.3.13. Proposition (Stabilizability estimate). *Let Assumption* 12.2.4 *hold. Then there exist positive constants C_1, C_2, and ω depending on R but independent of $\alpha, \kappa \in [0,1]$ such that, for any $y_1, y_2 \in \mathscr{W}_R$, the following estimate holds true,*

$$\|S_t y_1 - S_t y_2\|_Y^2 \le C_1 e^{-\omega t} \|y_1 - y_2\|_Y^2 + C_2 \, lot_t(h - \zeta, u - w), \quad t > 0, \tag{12.3.37}$$

where

$$lot_t(z, v) = \max_{[0,t]} \left(\|z(\tau)\|_{1-\delta,\mathscr{O}}^2 + \|v(\tau)\|_{2-\delta,\Omega}^2\right) \tag{12.3.38}$$

with some $\delta > 0$. Above, we have used the notation

$$(h(t); h_t(t); u(t); u_t(t); \psi(t)) := S_t y_1, \quad (\zeta(t); \zeta_t(t); w(t); w_t(t); \xi(t)) := S_t y_2.$$

Proof. Given $y_1, y_2 \in \mathscr{W}_R$, let $S_t y_1 = (h; h_t; u; u_t; \psi)$ and $S_t y_2 = (\zeta; \zeta_t; w; w_t; \xi)$ be the corresponding solutions, as introduced in Proposition 12.3.9. Moreover, let $z = h - \zeta$, $v = u - w$, $\theta = \psi - \xi$, as originally defined in (12.3.15). We recall that for these solutions the bounds (12.2.25), (12.3.18), and (12.3.33) hold true. Because we seek to establish an estimate of $\|S_t y_1 - S_t y_2\|_Y^2$, our starting point will be once again the fundamental inequality (12.3.16) pertaining to the energy $E^0(t)$ given by (12.3.17). Then, we need to produce accurate bounds for the various terms that occur in the right-hand side of (12.3.16). Throughout the rest of this section we use the fact that, due to (12.2.25) and (12.3.18), there exists a constant C_R such that $\|S_t y_i\|_Y \le C_R$, $i = 1, 2$.

We first need the following two estimates for the nonlinear terms $\mathscr{F}_1$ and $\mathscr{F}_2$.

12.3.14. Lemma. *Under Assumption* 12.3.4, *the following estimates hold true for some* $\delta > 0$,

$$\left| \int_t^T (\mathscr{F}_1(z), z_t)_{\mathscr{O}}\, d\tau \right| \le \varepsilon \int_0^T \|A^{1/2} z\|_{\mathscr{O}}^2\, d\tau + C_{R,T} \max_{[0,T]} \|z\|_{1-\delta,\mathscr{O}}^2 \qquad (12.3.39)$$

$$+ C_\varepsilon(R) \int_0^T \left(\|h_t\|_{\mathscr{O}}^2 + \|\zeta_t\|_{\mathscr{O}}^2 \right) \|A^{1/2} z\|_{\mathscr{O}}^2\, d\tau,$$

$$\left| \int_t^T (\mathscr{F}_2(v), v_t)_\Omega\, d\tau \right| \le \varepsilon \int_0^T \left(\|\mathscr{A} v\|_\Omega^2 + \|v_t\|_\Omega^2 \right) d\tau + C_{R,T} \max_{[0,T]} \|v\|_{1,\Omega}^2 \quad (12.3.40)$$

$$+ C_\varepsilon(R) \int_0^T \left(\|\mathscr{A} v\|_\Omega^2 + \|v_t\|_\Omega^2 \right) \|\xi\|_\Omega^2\, d\tau,$$

for all $t \in [0, T]$, *where* $\varepsilon > 0$ *can be taken arbitrarily small. Here,* $\mathscr{F}_1$ *and* $\mathscr{F}_2$ *are given by* (12.2.24).

Proof. The former estimate has been established in Lemma 12.2.16 in the isothermal case. To show (12.3.40) we use the same argument as in Lemma 11.3.2. We note that the term $Q(t)$ in the representation in (11.3.14) can be estimated as $C_R \|z\|_{2-\sigma}^2$ for some $\sigma < 1/2$. □

Notice now that the lower bound in (12.3.10) implies $ms^2 \le sg(s)$ and thus

$$\|h_t(t)\|_\Omega^2 + \|\zeta_t(t)\|_\Omega^2 \le D_{h,\zeta}(t) := m^{-1} \left[(D(h_t(t)), h_t(t))_\Omega + (D(\zeta_t(t)), \zeta_t(t))_\Omega \right] \qquad (12.3.41)$$

for all $t \ge 0$. Hence, by using the estimates established in Lemma 12.3.14, we obtain the following bound for $\Psi_T(z, v)$ defined by (12.2.23):

$$\Psi_T(z, v) \le \varepsilon \int_0^T E^0(t) dt + C_\varepsilon(T, R)\, \Xi_T(z, v), \qquad (12.3.42)$$

with $E^0(t)$ defined in (12.3.17) and $\Xi_T(z, v)$ given by

$$\Xi_T(z, v) := lot_T(z, v) + \int_0^T D_{h,\zeta}(\tau) \|A^{1/2} z(\tau)\|_\Omega^2 d\tau$$

$$+\int_0^T \|\xi(\tau)\|_\Omega^2 \left(\|\mathscr{A}v(\tau)\|_\Omega^2 + \|M_\alpha^{1/2}v_t(\tau)\|_\Omega^2\right) d\tau\,. \quad (12.3.43)$$

Above, we used the notation $lot_T(z,v)$ as in (12.3.38), whereas $D_{h,\zeta}(t)$ has been introduced within (12.3.41).

To proceed, we use the next assertion.

12.3.15. Lemma. *Under Assumption* 12.3.4, *the following estimate holds for* $H_0^T(z)$ *defined in* (12.2.22), *with arbitrarily small* $\varepsilon > 0$:

$$H_0^T(z) \le \varepsilon \int_0^T E^0(t)\,dt + C_\varepsilon\left[\Xi_T(z,v) + G_0^T(z)\right], \quad (12.3.44)$$

where $G_0^T(z)$ *is defined in* (12.2.21), *and* $\Xi_T(z,v)$ *is given by* (12.3.43).

Proof. In view of Assumption 12.3.4, the first part of Assumption 12.2.4 is valid. Therefore, the first part of Lemma 12.2.14 implies that (12.2.58) holds for every $\varepsilon > 0$. Owing to this estimate, it is immediately seen that

$$H_0^T(z) \le C_\varepsilon G_0^T(z) + \varepsilon \int_0^T E^0(t)\,dt + \varepsilon m \int_0^T D_{h,\zeta}(\tau)\|A^{1/2}z(\tau)\|_{\mathscr{O}}^2 d\tau,$$

which in turn implies (12.3.44), as desired. □

Notice that by the lower bound in (12.3.10) one has, as well, that

$$\int_0^T \|z_t\|_{\mathscr{O}}^2\,dt \le \frac{1}{m}\,G_0^T(z)\,. \quad (12.3.45)$$

Thus, we return to the basic inequality (12.3.16) in Proposition 12.3.9 and apply the estimates (12.3.45), (12.3.42), and (12.3.44). Choosing ε sufficiently small, we obtain

$$TE^0(T) + \int_0^T E^0(t)\,dt \le C_1\left[\beta G_0^T(z) + \gamma\int_0^T \|\mathscr{A}^{1/2}\theta\|_\Omega^2\right] + C_2(T,R)\,\Xi_T(z,v) \quad (12.3.46)$$

for $T \ge T_0 > 0$. On the other hand, using the energy equality given in (12.3.19) and Lemma 12.3.14 we also have that

$$\beta G_0^T(z) + \gamma\int_0^T \|\mathscr{A}^{1/2}\theta\|_\Omega^2 \le E^0(0) - E^0(T) + \varepsilon\int_0^T E^0(t)dt + C_\varepsilon(T,R)\,\Xi_T(z,v)$$

for any $\varepsilon > 0$. Combining this bound with (12.3.46), we see that there exists $T > 1$ such that

$$E^0(T) \le qE^0(0) + C_{R,T}\,\Xi_T(z,v) \quad \text{with } 0 < q \equiv q_{T,R} < 1\,.$$

Applying the same procedure as in the proof of Theorem 12.2.15 we find that

$$E^0(t) \le C_1\left[E^0(0)e^{-\omega t} + lot_t(z,v)\right]\exp\left\{C_2\int_0^t \left[D_{h,\zeta}(\tau) + \|\mathscr{A}^{1/2}\xi\|_\Omega^2\right]d\tau\right\}.$$

Because , by (12.3.33) we have that

$$\int_0^t \left[D_{h,\zeta}(\tau) + \|\mathscr{A}^{1/2}\xi\|_\Omega^2\right] d\tau \le C_R, \quad \text{for all } t \ge 0,$$

we obtain the estimate (12.3.37). This concludes the proof of Proposition 12.3.13.

□

12.3.5 Proofs of Theorem 12.3.5 and Theorem 12.3.7

1. *Finiteness of fractal dimension.* To prove finiteness of the fractal dimension $\dim_f \mathfrak{A}$, we notice that Proposition 12.3.13 and local Lipschitz continuity (6.3.26) of the semiflow S_t allow us to apply the abstract Theorem 7.9.6.

2. *Smoothness of the global attractor.* This follows from Proposition 12.3.13 and from Theorem 7.9.8 that

$$\|z_{tt}(t)\|_{\mathcal{O}}^2 + \|A^{1/2}z_t(t)\|_{\mathcal{O}}^2 + \|u_{tt}(t)\|_\Omega^2 \\ + \alpha\|M_\alpha^{1/2}u_{tt}(t)\|_\Omega^2 + \|\mathscr{A}u_t(t)\|_\Omega^2 + \|\theta_t(t)\|_\Omega^2 \le C \qquad (12.3.47)$$

for all $t \in \mathbb{R}$ and for any trajectory $\{y(t) \equiv (z(t); z_t(t); u(t); u_t(t); \theta(t)) : t \in \mathbb{R}\}$ from the attractor $\mathfrak{A}$. The estimate (12.3.47) enables us to establish the spatial smoothness of the attractor. Let us begin with the analysis of the plate variable v. In view of (6.3.23), (6.3.24), and (12.3.47) it is easily seen that on the attractor one has

$$\|\Delta\theta(t)\|_\Omega \le C\left(\|\mathscr{A}u_t\|_\Omega + \|\theta_t\|_\Omega\right) \le C,$$

and hence, because $\|\mathscr{A}^{-1/2}M_\alpha^{1/2}\| \le C$,

$$\|\mathscr{A}^{-1/2}\Delta^2 u\|_\Omega \le \|M_\alpha^{-1/2}\Delta^2 u\|_\Omega \\ \le C\left(\|\Delta\theta\|_\Omega + \|F_2(u)\|_\Omega + \|M_\alpha^{1/2}u_{tt}\|_\Omega + \|z_t\|_{1,\mathcal{O}}\right) \le C,$$

for $\alpha > 0$, whereas $\|\Delta^2 u\|_\Omega \le C$ in the case $\alpha = 0$. Thus, we can conclude that $t \mapsto u(t)$ is a bounded function in the space W_α defined in (6.3.13).

As for the wave component z, the argument is the same as in the proof of Theorem 12.2.6; see Section 12.2.5. Thus, the second statement of Theorem 12.3.5 is proved.

□

Proof of Theorem 12.3.7. The argument is standard (see, e.g., the proof of the final statement in Theorem 11.2.1 and also Sections 9.3.2, 9.5.5, and 9.6.2) and relies on the uniform estimate (12.3.11) of Theorem 12.3.5. We proceed by contradiction. Assume that (12.3.12) does not hold. Then there exists a sequence $\lambda_n = (\alpha_n, \kappa_n)$ such that $\lambda_n \to \lambda_0 \equiv (\alpha_0, \kappa_0)$ and a sequence $U_n \in \mathfrak{A}^{\alpha_n,\kappa_n}$ such that

$$\text{dist}_{Y_{\alpha_0}}(U_n, \mathfrak{A}^{\alpha_0,\kappa_0}) \ge \delta > 0, \quad n = 1, 2, \dots. \qquad (12.3.48)$$

Let $\{U_n(t) : t \in \mathbb{R}\}$ be a full trajectory from the attractor $\mathfrak{A}^{\alpha_n,\kappa_n}$ such that $U_n(0) = U_n$. Because $W^2_{\min\{2,6/p\}}(\mathcal{O})$ is compactly embedded in $H^1(\mathcal{O})$ for $p < 5$, using (12.3.9), (12.3.11), and Aubin's compactness theorem (see Theorem 1.1.8), we can conclude that there exists a sequence $\{n_k\}$ and a function $U(t) \in C_b(\mathbb{R}, Y_{\alpha_0})$ such that

$$\max_{t\in[-T,T]} \|U_{n_k}(t) - U(t)\|_{Y_{\alpha_0}} \to 0 \text{ as } k \to \infty.$$

Because $\|U(t)\|_{Y_{\alpha_0}} \le R$ for all $t \in \mathbb{R}$ and some $R > 0$, the trajectory $\{U(t) : t \in \mathbb{R}\}$ belongs to the attractor $\mathfrak{A}^{\alpha_0}$. Consequently, $U_{n_k} \to U(0) \in \mathfrak{A}^{\alpha_0,\kappa_0}$, which contradicts (12.3.48). □

12.3.16. Remark. Because $W^2_{6/5}(\mathcal{O})$ is not compactly embedded in $H^1(\mathcal{O})$, the above proof fails in the case $p = 5$. Yet we may use the compactness of the embedding $W^2_{6/5}(\mathcal{O}) \subset H^{1-\varepsilon}(\mathcal{O})$ for arbitrary $\varepsilon > 0$ so as to establish the semicontinuity property in the space $Y^{-\varepsilon}_{\alpha_0}$ defined in (12.3.14). ■

12.4 Gas flow problems

In contrast with acoustic models of bounded chambers $\mathcal{O}$ with an elastic wall Ω, the study of long-time behavior of solutions of gas flow problems requires different approaches. The point is that in the latter case we have to consider wave dynamics in unbounded domain $\mathcal{O} = \mathbb{R}^3_+ \equiv \{(x_1;x_2;x_3) \in \mathbb{R}^3 : x_3 > 0\}$. Thus we cannot apply the standard method from the theory (see, e.g., [17, 61, 273]) of attractors for PDEs defined on bounded domains. Moreover, the physical model (see, e.g., [28, 99]) of the gas flow considered (see, (12.4.5) below) does not assume any additional dissipation mechanism (except possibility of the energy transfer to infinity by propagating waves). This circumstance makes problematic application of the method developed for the study of dissipative dynamics in unbounded domains (see the survey [232] and the references therein). Nevertheless some results on finite-dimensionality of long time-dynamics of plates in a gas flow can be obtained.

We consider the following problem,

$$(1-\alpha\Delta)\partial_t^2 u + k(1-\alpha\Delta)\partial_t u + \Delta^2 u - [u, v+F_0] = p(x,t), \quad x \in \Omega, t > 0, \tag{12.4.1}$$

$$u|_{\partial\Omega} = \left.\frac{\partial u}{\partial n}\right|_{\partial\Omega} = 0, \quad u|_{t=0} = u_0(x), \quad \partial_t u|_{t=0} = u_1(x), \tag{12.4.2}$$

where $v = v(u)$ is a solution of the problem

$$\Delta^2 v + [u,u] = 0, \quad v|_{\partial\Omega} = \left.\frac{\partial v}{\partial n}\right|_{\partial\Omega} = 0. \tag{12.4.3}$$

We assume that the gas flows above the plate in the direction of the x_1-axis, and thus aerodynamical pressure of the flow on the plate is given by:

$$p(x,t) = p_0(x) + \nu(\partial_t + U\partial_{x_1})(\phi|_{|x_3=0}), \quad x \in \Omega \tag{12.4.4}$$

(see the discussion in Section 6.4) with $p_0(x) \in L^2(\Omega)$; the parameter $\nu > 0$ characterizes the intensity of the interaction between the flow and the plate, the parameter $U \geq 0$ $(U \neq 1)$ represents the nonperturbed flow velocity, and the velocity potential $\phi(x,t) = \phi(x_1,x_2,x_3;t)$ of the perturbed flow satisfies the equations:

$$(\partial_t + U\partial_{x_1})^2\phi = \Delta\phi, \quad x \in \mathbb{R}^3_+ \equiv \{(x_1,x_2,x_3) \in \mathbb{R}^3 : x_3 > 0\}, \tag{12.4.5}$$

$$\left.\frac{\partial\phi}{\partial x_3}\right|_{x_3=0} = \begin{cases} (\partial_t + U\partial_{x_1})u(x_1,x_2,t), & (x_1,x_2) \in \Omega, \\ 0, & (x_1,x_2) \notin \Omega, \end{cases} \tag{12.4.6}$$

$$\phi_{|t=0} = \phi_0(x), \qquad \partial_t\phi_{|t=0} = \phi_1(x). \tag{12.4.7}$$

We recall that well-posedness of solutions corresponding to (12.4.1)–(12.4.7) was established in Chapter 6 in the cases (i) $\alpha \geq 0$, $0 \leq U < 1$ (see Section 6.5) and (ii) $\alpha > 0$, $U \geq 0$, $U \neq 1$ (see Section 6.6).

12.4.1 Stabilization to a finite-dimensional set

We start with the following assertion on the existence of an attracting set of finite fractal dimension in the state space corresponding to the plate component.

12.4.1. Theorem. *Let $\alpha > 0$, $k > 0$, and $U \geq 0$, $U \neq 1$. Assume that the conditions of Theorem* 6.6.18 *are valid. Then there exists a compact set $\mathscr{U}$ in $H_0^2(\Omega) \times H_0^1(\Omega)$ of finite fractal dimension such that*

$$\lim_{t\to\infty} \text{dist}\{(u(t);u_t(t)),\mathscr{U}\} \equiv \lim_{t\to\infty} \inf_{(v_0;v_1)\in\mathscr{U}} \{\| u(t) - v_0 \|^2_{2,\Omega} + \| u_t(t) - v_1 \|^2_{1,\Omega}\} = 0$$

for any weak solution $(u(x_1,x_2,t);\phi(x_1,x_2,x_3,t))$ to problem (12.4.1)–(12.4.7) *with localized data ϕ_0 and ϕ_1 (i.e., $\phi_0(x) = \phi_1(x) = 0$ for $|x| > R$ for some R).*

Proof. By Theorem 6.6.18 we can reduce problem (12.4.1)–(12.4.7) to a PDE system (12.4.1)–(12.4.3) with retarded force $p(x,t)$ of the form (6.6.71); that is, with

$$p(x,t) = p_0(x) - \nu(\partial_t u + U\partial_{x_1}u + q(u;x,t)) \tag{12.4.8}$$

with

$$q(u,x,t) = \frac{1}{2\pi}\int_0^{t_*} ds \int_0^{2\pi} d\theta [M_\theta^2\hat{u}](x_1 - (U+\sin\theta)s, x_2 - s\cos\theta, t-s). \tag{12.4.9}$$

Here $\hat{u}$ is the extension of $u(x,t)$ by zero outside of Ω,

$$M_\theta = \sin\theta\cdot\partial_{x_1} + \cos\theta\cdot\partial_{x_2}, \quad \theta \in [0,2\pi],$$

and

$$t_* = \inf\{t : x(U,\theta,s) \notin \Omega \text{ for all } x \in \Omega,\ \theta \in [0,2\pi],\ s > t\}, \qquad (12.4.10)$$

where $x(U,\theta,s) = (x_1 - (U + \sin\theta)s; x_2 - s\cos\theta) \in \mathbb{R}^2$ and $x = (x_1;x_2) \in \Omega \subset \mathbb{R}^2$.

Thus we can apply Theorem 9.3.5 on the attractor for the von Karman plate with retarded forces. For this we need to check Assumption 9.3.2 where the most critical hypothesis is the requirement in (9.3.9). Thus, it suffices to prove the following assertion.

12.4.2. Lemma. *Let $q(u,t)$ be given by* (12.4.9). *Then*

$$\|q(u,t)\|^2_{-1,\Omega} \le c_0 t_* \int_{t-t_*}^{t} \|u(\tau)\|^2_{1,\Omega} d\tau \qquad (12.4.11)$$

for any $u \in L_2(t-t_*,t;H_0^1(\Omega))$, where t_* is given by (12.4.10) *and c_0 is a constant.*

Proof. We can assume that $u(x,t)$ is a continuous function of t and $u(x,t) \in C_0^\infty(\Omega)$ for each t. Let $\varphi \in C_0^\infty(\Omega)$. All functions from $C_0^\infty(\Omega)$ are assumed extended by zero to the whole $\mathbb{R}^2$. Then by integration by parts it follows from (12.4.9) that

$$\begin{aligned}
&\int_\Omega q(u,x,t)\varphi(x)dx \\
&= \frac{1}{2\pi}\int_0^{t_*} ds \int_0^{2\pi} d\theta \int_\Omega [M_\theta^2 \hat{u}](x_1 - (U+\sin\theta)s, x_2 - s\cos\theta, t-s)\varphi(x)dx \\
&= -\frac{1}{2\pi}\int_0^{t_*} ds \int_0^{2\pi} d\theta \int_{\mathbb{R}^2} [M_\theta \hat{u}](x_1 - (U+\sin\theta)s, x_2 - s\cos\theta, t-s) M_\theta\varphi(x)dx \\
&\le \frac{1}{2\pi}\int_0^{t_*} ds \int_0^{2\pi} d\theta \left(\int_{\mathbb{R}^2} |M_\theta \hat{u}](x,t-s))|^2 dx\right)^{1/2} \left(\int_{\mathbb{R}^2} |M_\theta\varphi(x)|^2 dx\right)^{1/2},
\end{aligned}$$

where we also made change of variables in the integral with u. Because

$$|M_\theta f(x)| \le |f_{x_1}(x)| + |f_{x_2}(x)| \text{ for every } f,$$

we obtain that

$$\int_\Omega q(u,x,t)\varphi(x)dx \le c_0 \int_0^{t_*} \|u(t-s)\|_{1,\Omega} ds \|\varphi\|_{1,\Omega}.$$

This implies (12.4.11). □

Lemma 12.4.2 guarantees Assumption 9.3.2 with $\sigma = 1$. Therefore by Theorem 9.3.5 in the space

$$\mathbf{H} = H_0^2(\Omega) \times H_0^1(\Omega) \times L_2(-t_*,0;H_0^2(\Omega))$$

there exists finite-dimensional compact global attractor $\mathbf{A}$ for a (retarded) system generated by (12.4.1)–(12.4.3) with $p(x,t)$ defined by (12.4.8) and (12.4.9). Thus if we denote by $\mathscr{U}$ a projection of $\mathbf{A} \subset \mathbf{H}$ on $H_0^2(\Omega) \times H_0^1(\Omega)$, we obtain the conclusion of Theorem 12.4.1. □

The following assertion gives other properties of finite-dimensionality for long-time dynamics of system (12.4.1)–(12.4.7).

We first recall the notion of completeness defect of a set of functionals (see Definition 7.8.5). Let $\mathscr{L} = \{l_j : j = 1, \dots, N\}$ be a finite set of linearly independent functionals on $H_0^2(\Omega)$. The value

$$\varepsilon_{\mathscr{L}}^{\alpha} = \sup\{\| w \|_{V_\alpha} : w \in H_0^2(\Omega),\ l_j(w) = 0,\ l_j \in \mathscr{L},\ \| w \|_{2,\Omega} \le 1\}$$

is said to be the *completeness defect* of the set $\mathscr{L}$ with respect to the pair of spaces $H_0^2(\Omega)$ and V_α, where $V_\alpha \sim H_0^1(\Omega)$ with $\alpha > 0$ is defined by (6.1.7). For more details concerning the completeness defect and determining functionals we refer to Section 7.8.

The application of ideas presented in Theorem 7.9.11 gives the following assertion.

12.4.3. Theorem. *Let $\alpha > 0$, $U \ge 0$, $U \ne 1$, $k > 0$, and the assumptions of Theorem* 6.6.18 *be in force. Then there exists the number $\varepsilon_0 > 0$ depending on the parameters of problem* (12.4.1)–(12.4.7) *such that the condition $\varepsilon_{\mathscr{L}}^{\alpha} < \varepsilon_0$ implies that $\mathscr{L}$ is a set of asymptotically determining functionals for problem* (12.4.1)–(12.4.7)*; that is, for any two weak solutions $(u_1(t); \phi_1(t))$ and $(u_2(t); \phi_2(t))$ (with localized data ϕ_{0i} and ϕ_{1i}) the condition*

$$\lim_{t \to \infty} \{l_j(u_1(t)) - l_j(u_2(t))\} = 0 \quad \text{for } j = 1, \dots, N \tag{12.4.12}$$

implies that

$$\lim_{t \to \infty} \{\| u_1(t) - u_2(t) \|_{2,\Omega}^2 + \| \partial_t (u_1(t) - u_2(t)) \|_{1,\Omega}^2\} = 0 \tag{12.4.13}$$

and

$$\lim_{t \to \infty} \left\{\| \phi_1(t) - \phi_2(t) \|_{1,B(\rho)}^2 + \| \partial_t (\phi_1(t) - \phi_2(t)) \|_{B(\rho)}^2 \right\} = 0 \tag{12.4.14}$$

for any $\rho > 0$, where $B(\rho) = \{x \in \mathbb{R}_+^3 : |x| \le \rho\}$.

Proof. By Lemma 12.4.2 Assumption 9.3.2 holds for problem (12.4.1)–(12.4.3), (12.4.8), and (12.4.9). Therefore stabilizability estimate (9.3.21) in Lemma 9.3.6 is valid with appropriate choice of the initial and reference time t for the case considered. This stabilizability estimate can also be written in the form (9.3.23). Thus we can apply the same argument as in Theorem 7.9.11 to obtain (12.4.13) from (12.4.12). In this case the relation in (12.4.14) follows from the explicit formulas for the flow potential given in (6.6.28) and (6.6.34). □

12.4.4. Remark. A result similar to Theorem 12.4.3 can also be established in the case when $\alpha = 0$ and $0 \le U < 1$. However, the corresponding argument requires more sophisticated analysis (see Section 6.5). The case of nonlinear damping in the plate component may also be considered. ∎

12.4.2 Stabilization to equilibria

Now we present a result on stabilization of the plate in a subsonic flow to stationary states. We recall that the stationary version of problem (12.4.1)–(12.4.7) has the form:

$$\Delta^2 u - [u, v + F_0] = p_0(x) + \nu \cdot U \cdot \partial_{x_1}(\phi|_{x_3=0}),\ x \in \Omega, \quad u|_{\partial\Omega} = \frac{\partial u}{\partial n}\Big|_{\partial\Omega} = 0, \tag{12.4.15}$$

where the value $v = v(u)$ is defined by u from (12.4.3) and the velocity potential $\phi(x) = \phi(x_1, x_2, x_3)$ satisfies the equations

$$\Delta\phi - U^2 \cdot \partial_{x_1}^2 \phi = 0,\ x \in \mathbb{R}^3_+, \quad \frac{\partial \phi}{\partial x_3}\Big|_{x_3=0} = \begin{cases} U\partial_{x_1} u(x_1, x_2), & (x_1, x_2) \in \Omega, \\ 0, & (x_1, x_2) \notin \Omega. \end{cases} \tag{12.4.16}$$

Problem (12.4.15) and (12.4.16) was studied in Section 6.5.6 in the case $0 \le U < 1$. In particular it was shown (see Theorem 6.5.10) the existence of weak solutions $\{u(x); \phi(x)\}$ which satisfy additional smoothness:

$$(u(x); \phi(x)) \in (H^4 \cap H_0^2)(\Omega) \times W_2(\mathbb{R}^3_+),$$

where

$$W_m(\mathbf{R}^3_+) = \left\{ \phi(x) \in L^2_{loc}(\mathbb{R}^3_+) \ :\ \|\phi\|^2_{W_m} \equiv \sum_{k=0}^{m-1} \|\nabla\phi\|^2_{k,\mathbb{R}^3_+} < \infty \right\}, \quad m = 1, 2.$$

Below we denote by $\mathscr{N}$ the set all of stationary solutions to (12.4.15) and (12.4.16). The following assertion shows that the set $\mathscr{N}$ attracts every individual solution in the local energy topology.

12.4.5. Theorem. *Let $\alpha > 0$, $0 \le U < 1$, $k > 0$, and the assumptions of Theorem 6.6.18 be in force. Then for any weak solution $(u(t); \phi(t))$ with localized flow initial data ϕ_0 and ϕ_1 we have that*

$$\lim_{t\to\infty} \inf_{\{\bar{u};\bar{\phi}\}\in\mathscr{N}} \Big\{ \| u(t) - \bar{u} \|^2_{2,\Omega} + \| u_t(t) \|^2_{1,\Omega} + \| \phi(t) - \bar{\phi} \|^2_{1,B(\rho)} + \| \partial_t \phi(t) \|^2_{B(\rho)} \Big\} = 0 \tag{12.4.17}$$

for any $\rho > 0$ with $B(\rho) = \{x \in \mathbb{R}^3_+ \ :\ |x| \le \rho\}$.

If the set $\mathcal{N}$ consists of a finite number of elements, then the convergence statement in (12.4.17) can be easily improved in the following way.

12.4.6. Corollary. *Let the hypotheses of Theorem* 12.4.5 *be in force. Assume that the set* $\mathcal{N}$ *is finite. Then for any weak solution* $(u(t);\phi(t))$ *(with localized data* ϕ_0 *and* ϕ_1*) there exists a solution* $\{\bar{u};\bar{\phi}\}$ *to problem* (12.4.15) *and* (12.4.16) *such that*

$$\lim_{t\to\infty}\{\| u(t)-\bar{u}\|_{2,\Omega}^2+\|\partial_t u(t)\|_{1,\Omega}^2\}=0$$

and

$$\lim_{t\to\infty}\left\{\| \phi(t)-\bar{\phi}\|_{1,B(\rho)}^2+\|\partial_t\phi(t)\|_{B(\rho)}^2\right\}=0$$

for any $\rho>0$*, where* $B(\rho)=\{x\in\mathbb{R}^3_+ \,:\, |x|\le\rho\}$.

Proof. Let $\{u(t);\phi(t)\}$ be a weak solution of problem (12.4.1)–(12.4.7) with initial data satisfying the conditions of theorem.

Step 1: It follows from the energy relation (6.5.4) and Proposition 6.5.7 that

$$\int_0^\infty \|u_t(\tau)\|_{1,\Omega}^2 d\tau<\infty.$$

Using (12.4.1) one can also see from Proposition 6.5.7 and the explicit formula (6.6.29) for the trace of flow ϕ that $(M_\alpha u_{tt}(t),w)_\Omega=\partial_t(M_\alpha u_t(t),w)_\Omega$ is uniformly bounded in t for every $w\in C_0^\infty(\Omega)$. Because we obviously have that

$$\int_0^\infty |(M_\alpha u_t(\tau),w)_\Omega|^2 d\tau<\infty,$$

using the boundedness of the time derivative of the integrand in the integral above, we conclude that

$$(M_\alpha u_t(t),w)_\Omega\to 0 \;\text{ as }\; t\to\infty \;\text{ for every } w\in C_0^\infty(\Omega).$$

Now using boundedness of $\|u_t(t)\|_{1,\Omega}^2$ and the compactness property proved in Theorem 12.4.1 we conclude that

$$\|u_t(t)\|_{1,\Omega}^2\to 0 \;\text{ as } t\to\infty. \tag{12.4.18}$$

Step 2: By Theorem 12.4.1 we can also conclude that any sequence of moments of time converging to infinity contains a subsequence $\{t_n\}$ such that

$$\|u(t_n)-\bar{u}\|_{2,\Omega}^2\to 0 \;\text{ as } n\to\infty$$

for some element $\bar{u}\in H_0^2(\Omega)$. Moreover, because

$$\|u(t_n+\tau)-u(t_n)\|_{1,\Omega}\le\int_{t_n}^{t_n+\tau}\|u_t(s)\|_{1,\Omega}ds\le\tau\max_{s\in[t_n,t_n+\tau]}\|u_t(s)\|_{1,\Omega},$$

we conclude from (12.4.18) that

$$\max_{\tau\in[-a,a]} \|u(t_n+\tau)-\bar{u}\|_{1,\Omega} \to 0 \text{ as } n\to\infty$$

for every fixed $a>0$. By interpolation and boundedness of $\|u(t)\|^2_{2,\Omega}$ we obtain that

$$\max_{\tau\in[-a,a]} \|u(t_n+\tau)-u\|_{2-\delta,\Omega} \to 0 \text{ as } n\to\infty \tag{12.4.19}$$

for every fixed $a>0$ and $\delta>0$.

Step 3: From (12.4.18) and (12.4.19) using an explicit formula describing the flow ϕ for large t (which can be derived from (6.6.28)) and the boundedness provided by Proposition 6.5.7 we can conclude that there exist $\bar{\phi}\in W_1(\mathbb{R}^3_+)$ and $\bar{\phi}'\in L_2(\mathbb{R}^3_+)$ such that

$$\|\phi(t_n)-\bar{\phi}\|_{1,B(\rho)} + \|\phi_t(t_n)-\bar{\phi}'\|_{B(\rho)} \to 0 \text{ as } n\to\infty \tag{12.4.20}$$

for every $\rho>0$. Because $\tilde{\phi}=\phi_t$ satisfies (12.4.5) and (12.4.6) with u_t instead of u, using the explicit formula for the flow $\tilde{\phi}$ and relation (12.4.18) we can conclude that $(\phi_t(t_n),\chi)_{B(\rho)}\to 0$ as $n\to\infty$ for every smooth function χ. This makes it possible to prove that $\bar{\phi}'=0$.

Step 4: Now we prove that the pair $(\bar{u};\bar{\phi})$ is a weak solution to problem (12.4.15) and (12.4.16).

We multiply (12.4.1) by smooth function $w\in C_0^\infty(\Omega)$ in $L_2(\Omega)$ and integrate from t_n and t_n+a, where $a>0$. The convergent properties established above are sufficient to make limit transition $n\to\infty$ and show that $\bar{u}$ satisfies the relation

$$(\Delta\bar{u},\Delta w)_\Omega - ([\bar{u},v(\bar{u})+F_0],w)_\Omega + \nu U(\gamma[\bar{\phi}],\partial_{x_1}w)_\Omega = (p_0,w)_\Omega.$$

Similarly for $\bar{\phi}$ we obtain the relation

$$(\nabla\bar{\phi},\nabla\psi)_{\mathbb{R}^3_+} - U^2(\partial_{x_1}\bar{\phi},\partial_{x_1}\psi)_{\mathbb{R}^3_+} + U(\partial_{x_1}\bar{u},\gamma[\psi])_\Omega = 0$$

for any $\psi\in C_0^\infty(\mathbb{R}^3_+)$. Thus $(\bar{u};\bar{\phi})$ satisfies the variational relations in (6.5.57) and (6.5.58).

Concluding step: Thus we prove that any sequence of $(u(t_m);\phi_t(t_m))$ with $t_m\to\infty$ contains a subsequence which converges (in the local energy sense) to some stationary state. This implies (12.4.17). □

12.4.7. Remark. We note that results on stabilization similar to Theorem 12.4.5 and Corollary 12.4.6 can be also established in the case of thermoelastic plate in subsonic flow with only thermal dissipation (see [252] and [254]). ∎

12.4.8. Remark (Open problems). We state several problems that are important from an applied point of view and which, as we hope, can be solved within the framework of approaches developed up to now.

1. All results on long-time behavior stated above in Section 12.4 deal mainly with the case when the rotational model inertia is included in the model ($\alpha>0$).

In this case we have additional $H_0^1(\Omega)$-regularity of the plate velocity u_t which also gives $H^1(\mathbb{R}^2)$-regularity of the normal derivative of the flow potential ϕ. As one can see from the argument given above this regularity plays an essential role in the study of long-time dynamics. The question of whether it is possible to obtain the same results for the nonrotational ($\alpha = 0$) case is still mainly open. Moreover, even the well-posedness issue for the model in the case $\alpha = 0$, $U > 1$ has not been fully resolved yet. The latter case is the most interesting from the mathematical point of view. In subsonic case ($\alpha = 0$, $U < 1$) some results on long-time dynamics are available; see Remark 12.4.4.

2. In spite of the fact that the total energy of the system (12.4.1)–(12.4.7) is conserved when $k = 0$ we conjecture that in this case any weak solution of the subsonic ($U < 1$) problem converges to some stationary one, when t goes to infinity (cf. Theorem 12.4.5, where the case $k > 0$ is described). This behavior of solutions can be explained by the well-known property of the local energy decay for the wave equation in $\mathbb{R}^3$. However, we are not in a position to prove this conjecture yet.

3. In the paper [48] the asymptotical properties of solutions to problem (12.4.1)–(12.4.3) with $p(x,t)$ given by (12.4.8) without the integral term q was studied in the hypersonic limit ($U \to +\infty$). This choice of the load $p(x,t)$ corresponds to the so-called "piston" theory which involves the plane sections law [147] (see also [28] and [99]). The paper [48] deals with the case $\alpha > 0$, $k \geq 0$ and it contains a description of asymptotical behavior of the solutions for short intervals of time. However, we do not know how to describe this behavior for arbitrary time intervals and for large t. It is also important to compare the solutions obtained in the framework of the piston theory with solutions of system (12.4.1)–(12.4.7) (for some discussion we refer to Remark 6.6.22; see also [32]). ∎

Chapter 13
Inertial Manifolds for von Karman Plate Equations

One of the contemporary approaches to the study of long-time behavior of infinite-dimensional dynamical systems is based on the concept of inertial manifolds which was introduced in [117] (see also the monographs [61, 90, 273] and the references therein and also Section 7.6 in Chapter 7). These manifolds are finite-dimensional invariant surfaces that contain global attractors and attract trajectories exponentially fast. Moreover, there is a possibility to reduce the study of limit regimes of the original infinite-dimensional system to solving similar problem for a class of ordinary differential equations. Inertial manifolds are generalizations of center-unstable manifolds and are convenient objects to capture the long-time behavior of dynamical systems. The theory of inertial manifolds is related to the method of integral manifolds (see, e.g., [92, 139, 233]), and has been developed and widely studied for deterministic systems by many authors. All known results concerning existence of inertial manifolds require some gap condition on the spectrum of the linearized problem (see, e.g., [45, 50, 61, 90, 227, 236, 273] and the references therein). Although inertial manifolds have been mainly studied for parabolic-like equations, there are some results for damped second order in time evolution equations arising in nonlinear oscillations theory (see, e.g., [45, 50, 61, 236]). These results rely on the approach originally developed in [236] for a one-dimensional semilinear wave equation and require the damping coefficient to be large enough. In fact, as indicated in [236], this requirement is a necessary condition in the case of hyperbolic flows.

The goal of this chapter is to provide some results on existence and properties of inertial manifolds for several models of nonlinear dynamic elasticity governed by von Karman evolution equations subject to either mechanical or thermal dissipation. The presentation below mainly follows the paper [66].

We consider three different dissipative mechanisms: viscous damping, strong structural damping (mechanical dampings), and thermal damping. Von Karman equations with viscous damping retain hyperbolic-like properties of the dynamics, whereas structural damping and thermal damping have recently been shown [216] (see also Section 5.3.2 in Chapter 5) to be related to analyticity of the semigroup generated by the linear part of the dynamics. It is thus expected that the results obtained depend heavily on the type of dissipation.

I. Chueshov, I. Lasiecka, *Von Karman Evolution Equations*,
Springer Monographs in Mathematics, DOI 10.1007/978-0-387-87712-9_13,

Our main results, formulated in Section 13.3, provide conditions for existence of inertial manifolds for all three models. These conditions are derived from more general results presented in Section 7.6, Theorem 7.6.3, where the main assumption is certain gap condition. Gap condition, when specialized to the concrete models considered, imposes geometric restrictions on spatial domains along with some restrictions imposed on the damping parameter. This latter constraint is essential only in the hyperbolic case. Indeed, in the hyperbolic case (viscous damping), Theorems 13.3.5 and 13.3.6 require sufficiently large values of the damping parameter. In the analytic-like case (structural damping), instead, Theorem 13.3.12 does not require large values of damping. A similar situation takes place in the thermoelastic case; see Theorem 13.3.16.

13.1 Preliminaries

13.1.1 The models considered

Let Ω be a bounded domain in $\mathbb{R}^2$, Δ denote the Laplace operator, L be a linear bounded operator from $H^{2-\sigma}(\Omega)$ into $L_2(\Omega)$ for some $\sigma > 0$ (a standard model for L is a first-order differential operator), F_0 and p be given functions with regularity specified later and, as above, $[u,v]$ denote the von Karman bracket. In what follows below we describe three von Karman models under consideration. For the sake of some simplifications we consider the case of hinged conditions only. However, several of the presented results remain valid for other types of boundary conditions.

Viscous damping

We begin with the following von Karman system which is a special case of the systems considered in Chapter 4 and subject to nonlinear viscous damping $g_0(u_t) = \mu u_t + g(u_t)$, where $\mu > 0$ and $g(v)$ is a globally Lipschitz nondecreasing function. The equations are given by:

$$\partial_t^2 u + 2\mu\partial_t u + g(\partial_t u) + \Delta^2 u - [u, v+F_0] + Lu = p(x), \quad x \in \Omega, t > 0, \qquad (13.1.1)$$

where $v = v(u)$ is a solution of the problem

$$\Delta^2 v + [u,u] = 0, \quad v|_{\partial\Omega} = \frac{\partial v}{\partial n}\Big|_{\partial\Omega} = 0. \qquad (13.1.2)$$

We assume that the function $u(x,t)$ satisfies (hinged) boundary conditions and initial data of the form:

$$u\big|_{\partial\Omega} = \Delta u\big|_{\partial\Omega} = 0; \quad u|_{t=0} = u_0(x), \quad \partial_t u|_{t=0} = u_1(x). \qquad (13.1.3)$$

In order to provide abstract representation of the model, we use the notation $A = -\Delta$ with the Dirichlet boundary conditions. Operator A is considered as an unbounded operator on $\mathscr{H} \equiv L_2(\Omega)$ with the domain given by $\mathscr{D}(A) = (H^2 \cap H_0^1)(\Omega)$. The problem (13.1.1)–(13.1.3) is considered on the phase space $H \equiv \mathscr{D}(A) \times \mathscr{H}$ and it can be written in the abstract form as

$$u_{tt} + 2\mu u_t + A^2 u = F(u, u_t), \quad u|_{t=0} = u_0,\ u_t|_{t=0} = u_1. \tag{13.1.4}$$

The operator $F(\cdot,\cdot)$ is a nonlinear mapping from $H = \mathscr{D}(A) \times \mathscr{H}$ into $\mathscr{H}$ defined by the formula

$$F(u, u_t) = [u, v(u) + F_0] - Lu + p(x) - g(u_t).$$

13.1.1. Remark. From the point of view of well-posedness of the flows, the global Lipschitz condition imposed on g can be significantly relaxed (see Chapter 4 for details). We introduce this restriction here for the following reasons: (i) the analysis of the size of absorbing sets in the model (13.1.1) becomes complicated for non-Lipschitz damping in the presence of the nonconservative term Lu (see Section 8.2), and (ii) the spectral gap condition (see (13.3.9) below) for this model requires the Lipschitz property satisfied by $g(u_t)$ with u_t restricted to an absorbing set. ∎

Structural damping

The second problem is a strongly damped version of the problem above. Instead of (13.1.1) we consider the equation

$$\partial_t^2 u - 2\mu\Delta\partial_t u + \Delta^2 u - [u, v + F_0] + Lu = p(x), \quad x \in \Omega, t > 0, \tag{13.1.5}$$

where $v = v(u)$ is defined as above and the conditions (13.1.3) are in force. Similarly as before, the second problem (13.1.5), (13.1.2), (13.1.3) can be written in the form

$$u_{tt} + 2\mu A u_t + A^2 u = F(u), \quad u|_{t=0} = u_0,\ u_t|_{t=0} = u_1, \tag{13.1.6}$$

where A is as above and $F(\cdot)$ is a nonlinear mapping from $D(A)$ into $\mathscr{H}$ given by

$$F(u) = [u, v(u) + F_0] - Lu + p(x). \tag{13.1.7}$$

We note that a similar model was considered in Chapter 5 as a limit case of a thermoelastic plate when thermal capacity tends to zero (see (5.2.4)).

Thermal damping

The third model is a thermoelastic von Karman plate (see Chapter 5). The corresponding equations have the following form,

$$\begin{cases} \partial_t^2 u + \mu\Delta\theta + \Delta^2 u - [u, v + F_0] + Lu = p(x), & x \in \Omega, t > 0, \\ \partial_t \theta - \eta\Delta\theta - \mu\Delta\partial_t u = 0, & x \in \Omega, t > 0, \end{cases} \tag{13.1.8}$$

where, as above, $v = v(u)$ is a solution of the problem (13.1.2), the function u satisfies (13.1.3) and θ satisfies the boundary condition and initial data of the form:

$$\theta\big|_{\partial\Omega} = 0, \quad \theta\big|_{t=0} = \theta_0. \tag{13.1.9}$$

The problem (13.1.8) with the initial and boundary conditions (13.1.3) and (13.1.9) can be written in the form

$$\begin{cases} u_{tt} - \mu A\theta + A^2 u = F(u), & u|_{t=0} = u_0,\ u_t|_{t=0} = u_1, \\ \theta_t + \eta A\theta + \mu A u_t = 0, & \theta|_{t=0} = \theta_0, \end{cases} \tag{13.1.10}$$

where A is as above and $F(\cdot)$ is given by (13.1.7). We refer to Chapters 5 and 11 for more details concerning this model.

Our main focus is on existence of inertial manifolds for these models. A first step toward this goal is the well-posedness of semiflows corresponding to each model under consideration.

13.1.2 Generation of nonlinear semigroups

We recall results on well-posedness of a continuous semiflow corresponding to all three models. By this we mean existence, uniqueness, and continuous dependence of solutions with respect to initial data in H and $t > 0$. Here

$$H = \mathscr{D}(A) \times \mathscr{H}$$

for problems (13.1.4) and (13.1.6) and

$$H = \mathscr{D}(A) \times \mathscr{H} \times \mathscr{H}$$

for (13.1.10).

In what follows we assume that the domain Ω is either smooth or rectangular. We have the following well-posedness result (see Theorem 4.1.10 and Theorem 4.1.19).

13.1.2. Theorem (Viscous damping). *Consider system* (13.1.1), (13.1.2) *with* $p \in L_2(\Omega)$, $F_0 \in H^{3+\delta}(\Omega)$ *for some* $\delta > 0$. *Then for every initial datum* $(u_0; u_1)$ *from* $\mathscr{D}(A) \times \mathscr{H}$ *there exists a unique (both generalized and weak) solution*

$$(u; u_t) \in C(\mathbb{R}_+; \mathscr{D}(A) \times \mathscr{H}),$$

that depends continuously on the initial data and satisfies the corresponding energy relation (see (4.1.19)*).*

In the case of the structural damping model (13.1.5) well-posedness of the problem was established in Chapter 5 in the case $Lu \equiv 0$ (see Theorem 5.4.1 with $\alpha = 0$ and $\kappa = 0$). In the case $L \neq 0$ a similar argument leads to the following assertion.

13.1.3. Theorem (Structural damping). *Consider system* (13.1.5) *with* $p \in L_2(\Omega)$ *and* $F_0 \in W^2_\infty(\Omega)$*. Then for all initial data* $(u_0; u_1) \in H = \mathscr{D}(A) \times \mathscr{H}$ *and* $T > 0$ *there exists a unique solution* $(u; u_t) \in C(0,T;H)$ *that depends continuously on the initial data. Moreover,* $u \in L_2(0,T;H^3(\Omega))$ *and* $u_t \in L_2(0,T;H^1_0(\Omega))$*. The corresponding energy relation is also satisfied.*

Proof. The first statement in this theorem follows from the argument given in the proof of Theorem 5.4.1.

As for the second part, we note that due to the presence of the term $2\mu\Delta\partial_t u$ we have additional regularity in the system which makes it possible to establish analyticity of the semigroup generated by the linear part of the equation in (13.1.5) with $L \equiv 0$ (see Proposition 5.3.7 in Chapter 5); that is, the operator $\mathscr{A} : H \to H$ given by

$$\mathscr{A} \equiv \begin{pmatrix} 0 & -I \\ A^2 & 2\mu A \end{pmatrix}, \quad \mathscr{D}(\mathscr{A}) = \mathscr{D}(\mathscr{A}^2) \times \mathscr{D}(\mathscr{A}), \tag{13.1.11}$$

generates an analytic semigroup $e^{-\mathscr{A}t}$. The following properties are the consequences of the analyticity.

13.1.4. Lemma ([216, 43, 44]).

- *Let* $\mathscr{A}$ *generate an analytic semigroup on* H*. Then the linear map*

$$f \to \int_0^t e^{-\mathscr{A}(t-s)} f(s)ds : L_2(0,T;H) \to L_2(0,T;\mathscr{D}(\mathscr{A})) \tag{13.1.12}$$

  is bounded.
- *For* $\mathscr{A}$ *given by* (13.1.11) *the following map is bounded,*

$$y \to e^{-\mathscr{A}t} y : H \to L_2(0,T;\mathscr{D}(\mathscr{A}^{1/2})) . \tag{13.1.13}$$

13.1.5. Remark. It should be remarked that the property in the second part of Lemma 13.1.4 is well known for semigroups generated by self-adjoint operators (see, e.g., [216]). The same property was shown [44] to hold for the operator $\mathscr{A}$ given in (13.1.11). General analytic semigroups lead to the same estimate with the "loss of ε". This is to say the map $y \to e^{-\mathscr{A}t} y : H \to L_2(0,T;\mathscr{D}(\mathscr{A}^{1/2-\varepsilon}))$ is bounded for each $\varepsilon > 0$. ■

Applying (13.1.12) and (13.1.13) along with regularity of the von Karman bracket and the following characterization of fractional powers of $\mathscr{A}$ [216, p. 290],

$$\mathscr{D}(\mathscr{A}^{1/2}) \sim (H^3 \cap H^1_0)(\Omega) \times H^1_0(\Omega),$$

yields the desired regularity in the second part of Theorem 13.1.3. □

In the case of von Karman equations with thermoelasticity, the following well-posedness result is valid.

13.1.6. Theorem (Thermal damping). *Consider system* (13.1.8) *and assume that* $F_0 \in W^2_\infty(\Omega)$, $p \in L_2(\Omega)$. *Then for all initial data*

$$(u_0; u_1; \theta_0) \in H = \mathscr{D}(A) \times \mathscr{H} \times \mathscr{H}$$

the problem (13.1.8), (13.1.3), and (13.1.9) has a unique solution $(u; u_t; \theta)$ *from* $C(\mathbb{R}_+; H)$ *that depends continuously on the initial data. Moreover,*

$$(u(t); u_t(t); \theta(t)) \in L_2(0, T; H^3(\Omega) \times H^1(\Omega) \times H^1(\Omega)) \tag{13.1.14}$$

for any $T > 0$.

Proof. See Theorem 5.4.1 in Chapter 5. Property (13.1.14) follows from the corresponding thermoelastic analogue of relations (13.1.12) and (13.1.13). □

Based on Theorems 13.1.2, 13.1.3, and 13.1.6 we conclude that both problems (13.1.4) and (13.1.6) have the solution

$$u(t) \in C^1(\mathbb{R}_+; \mathscr{H}) \cap C(\mathbb{R}_+; \mathscr{D}(A))$$

and problem (13.1.10) has the solution

$$u(t) \in C^1(\mathbb{R}_+; \mathscr{H}) \cap C(\mathbb{R}_+; \mathscr{D}(A)), \quad \theta(t) \in C(\mathbb{R}_+; \mathscr{H}).$$

The solution generates dynamical system (H, S_t) in the space $H = \mathscr{D}(A) \times \mathscr{H}$, for the first two problems and in $H = \mathscr{D}(A) \times \mathscr{H} \times \mathscr{H}$ for the third problem. The evolution operator S_t is given by the formula $S_t(u_0; u_1) = (u(t); u_t(t))$, where $u(t)$ solves either (13.1.4) or (13.1.6) and by $S_t(u_0; u_1; \theta_0) = (u(t); u_t(t); \theta(t))$, where $u(t)$ and $\theta(t)$ solve (13.1.10). As a consequence, in all cases considered we have well-defined semiflow on the space H. In the first model we deal with a reversible (in time) hyperbolic-like dynamics and for the remaining two we have parabolic-like semiflows.

13.1.7. Remark (Regular solutions). In the case of the models considered above the existence of regular solutions is well known (see Theorems 4.1.21 and 4.1.25 for (13.1.4) and Propositions 5.5.2 and 5.5.3 for models in (13.1.6) and (13.1.10)). Moreover, for the initial data that are sufficiently smooth we can obtain classical solutions. ■

13.1.3 Absorbing sets

As we already know (see Chapters 7 and 8) a starting point in the study of global attractors and inertial manifolds is the so-called (ultimate) dissipativity property, or equivalently an existence of an absorbing set (see Definition 7.1.2). It turns out that for our analysis in this chapter it is important that the size of the absorbing set be independent of the dissipation parameters. In what follows below we provide existence of absorbing sets for all three different damping mechanisms considered.

Viscous damping

The following assertion holds.

13.1.8. Theorem. *With reference to problem* (13.1.1) *the system is dissipative. Moreover, the size of the absorbing ball is independent of* $\mu \geq \mu_0 > 0$ *and one can choose a suitable absorbing (bounded) set that is positively invariant with respect to the semiflow.*

Proof. The result follows from Theorem 8.2.3 and it is based on the analysis of the following Lyapunov function

$$V(t) = \mathcal{E}(t) + \varepsilon \int_\Omega u_t(t)u(t)dx$$

with suitably small $\varepsilon > 0$. Here $\mathcal{E}(t) = \mathcal{E}(u(t), u_t(t))$ and the energy function $\mathcal{E}(u, u_t)$ has the form

$$\mathcal{E}(u,u_t) \equiv \frac{1}{2}\left[||Au||^2 + ||u_t||^2 + \frac{1}{2}||\Delta v(u)||^2\right] - \frac{1}{2}\int_\Omega([F_0, u]u + 2pu)dx. \quad (13.1.15)$$

The important part of the analysis is to control the size of the absorbing ball with respect to the damping parameter (see Theorem 8.2.3). The details of the corresponding argument are given in the proof of Theorem 3.10 in [75]. □

Structural Damping

In the case of structural damping the energy function $\mathcal{E}(t) = \mathcal{E}(u(t), u_t(t))$ is the same as in (13.1.15). However, this time the damping is stronger and leads to regularizing effects on the dynamics. This can be seen from the structure of energy equality which now takes the form (see (5.4.5) for $\alpha = 0$ in Theorem 5.4.1):

$$\mathcal{E}(t) + 2\mu \int_s^t ||A^{1/2}u_t(\tau)||^2 d\tau + \int_s^t \int (Lu, u_t)_\Omega d\tau = \mathcal{E}(s), \quad s \leq t\,.$$

By the same method as in [75], with the use of Lemma 1.5.4 (to obtain estimates like (4.1.9)) and using the same Lyapunov function as for Theorem 13.1.8 one obtains the following.

13.1.9. Theorem. *The system given in* (13.1.5) *is dissipative, with the size of the absorbing ball independent of* μ *for* $\mu \geq \mu_0 > 0$. *Moreover, one can construct an absorbing bounded set that is positively invariant with respect to the semiflow.*

We also note that in the case $L \equiv 0$ the result of Theorem 13.1.9 follows by the same argument as in Theorem 13.1.10 below.

Thermal damping

In the case of thermoelastic plates, one has the following result.

13.1.10. Theorem. *With reference to problem* (13.1.8) *with* $L = 0$, *the system is dissipative. The size of the absorbing ball does not depend on thermal parameters* $\eta > 0, \mu > 0$. *Moreover, one can construct an invariant, bounded absorbing set for the semiflow.*

Proof. Let $\mathscr{E}(u, u_t, \theta)$ be the energy function for the model that is given by the formula

$$\mathscr{E}(u,u_t,\theta) \equiv \frac{1}{2}\left[||Au||^2 + ||u_t||^2 + \frac{1}{2}||\Delta v(u)||^2 + ||\theta||^2\right] - \frac{1}{2}\int_{\Omega}([F_0,u]u + 2pu)dx.$$

Using Lemma 1.5.4 in the same way as in the isothermal case (see (4.1.9)) one can see that there exist positive constants c_1, c_2, and $C_{F_0,p}$ such that

$$c_1 E(u,u_t,\theta) - C_{F_0,p} \le \mathscr{E}(u,u_t,\theta) \le c_2 E(u,u_t,\theta) + C_{F_0,p}, \tag{13.1.16}$$

where we have denoted

$$E(u,u_t,\theta) \equiv \frac{1}{2}\left[||Au||^2 + ||u_t||^2 + \frac{1}{2}||\Delta v(u)||^2 + ||\theta||^2\right].$$

By Theorem 11.2.1 the system (H, S_t) possesses a compact global attractor $\mathfrak{A}$ such that

$$\mathfrak{A} \subset B_H(R) = \left\{(u_0;,u_1;\theta_0) \ : \ \|Au\|^2 + \|u_1\|^2 + \|\theta_0\|^2 \le R^2\right\},$$

where R does not depend on μ and η. This implies that the ball $B_H(1+R)$ is absorbing. By (13.1.16) there exists a constant C_R independent of μ and η such that

$$B_H(1+R) \subset \mathscr{E}_{C_R} = \{(u_0;,u_1;\theta_0) \ : \ \mathscr{E}(u_0,u_1,\theta_0) \le C_R\},$$

It is clear that $\mathscr{E}_{C_R}$ is bounded. By the energy relation (see Theorem 5.4.1)

$$\mathscr{E}(u(t),u_t(t),\theta(t)) + \eta\int_s^t ||A^{1/2}\theta(z)||^2 dz = \mathscr{E}(u(s),u_t(s),\theta(s)),$$

the set $\mathscr{E}_{C_R}$ is forward invariant. This completes the proof. □

13.1.11. Remark. Direct proof of Theorem 13.1.10 (without any reference to Theorem 11.2.1) is given in Theorem 1.13 in [66]. Whether the same result holds with the inclusion of the nonconservative term Lu is not known at the time of writing this book. ■

13.2 Inertial manifolds for evolution equations

In this section we recall the concept of inertial manifolds (see Section 7.6) and provide several existence results. We present our results in a form that is convenient for applications to nonlinear equations considered in Section 13.1. We begin by presenting a method for the construction of manifolds, which has been proposed by Lyapunov and Perron. We then adapt this approach to nonlinear equations considered in the previous section. The results presented in this section are well known (see, e.g., [45, 61, 139, 273] and also the references in Section 7.6). We include the proofs for the sake of making the chapter self-contained and also because the proofs given below provide precise estimates that are used directly in the context of our applications to von Karman evolutions. We start with several definitions.

Let (H,S_t) be a dynamical system in a Banach (either complex or real) space H.

13.2.1. Definition. Let $\mathscr{M}$ be a finite-dimensional surface in the space H of the following structure

$$\mathscr{M} = \{p+\Phi(p) \,:\, p\in \mathscr{P}H,\ \Phi : \mathscr{P}H \to (I-\mathscr{P})H\}, \tag{13.2.1}$$

where $\mathscr{P}$ is a finite-dimensional projector and $\Phi(\cdot)$ is a continuous mapping satisfying the Lipschitz condition

$$\|\Phi(p_1)-\Phi(p_2)\| \le L\cdot \|p_1-p_2\|$$

for all $p_j \in \mathscr{P}H$, where L is a positive constant. Here and below $\|\cdot\|$ is the norm in the space H. Then $\mathscr{M}$ is said to be *an inertial manifold* for the dynamical system (H,S_t) iff

- The surface $\mathscr{M}$ is invariant with respect to the semiflow S_t; that is, $S_t u \in \mathscr{M}$ for any $u\in\mathscr{M}$ and $t\ge 0$.
- $\mathscr{M}$ is exponentially attracting: for any bounded set $B\subset H$ there exists $C_B>0$ and $\eta_B>0$ such that

$$\sup\{\operatorname{dist}_H(S_t y,\mathscr{M}) \,:\, y\in B\} \le C_B e^{-\eta_B t} \quad \text{for all } t\ge 0.$$

■

For our applications a less restrictive concept of locally invariant manifolds is relevant.

13.2.2. Definition. The Lipschitz surface (13.2.1) is said to be a *locally invariant inertial manifold* if it is exponentially attracting and there exists $R>0$ such that (i) the ball $B_R=\{x\in H \,:\, \|x\|\le R\}$ is absorbing and (ii) $\mathscr{M}$ is locally invariant (in

B_R); that is, the properties $\|u\|_H \le R$, $u \in \mathscr{M}$, $\|S_t u\|_H \le R$ for $t \in [0,T]$ imply that $S_t u \in \mathscr{M}$ for every $t \in [0,T]$. ∎

We also recall the concept of asymptotic completeness[1] of an inertial manifold $\mathscr{M}$.

13.2.3. Definition. The (locally invariant) inertial manifold $\mathscr{M}$ is said to be *exponentially (asymptotically) complete* iff for any $u \in H$ there exist $u^* \in \mathscr{M}$ and numbers $s \ge \tau \ge 0$ such that $S_t u^* \in \mathscr{M}$ for all $t \ge 0$ and

$$\|S_t u - S_{t-\tau} u^*\| \le C \cdot e^{-\eta t} \quad \text{for all } t \ge s, \tag{13.2.2}$$

where C and η are positive constants. The semitrajectory $u^*(t) = S_t u^*$ is said to be *induced* for $u(t) = S_t u$. ∎

Now we present the Lyapunov–Perron method of construction of inertial manifolds in the simplest form (see, e.g., [45, 61, 139, 273] where somewhat more general situations are considered).

We consider an evolution equation in a Banach space H of the type

$$\frac{du}{dt} + \mathscr{A} u = \mathscr{B}(u), \quad u|_{t=0} = u_0. \tag{13.2.3}$$

We assume that $\mathscr{A}$ is a generator of the C_0-semigroup $e^{-\mathscr{A}t}$ in H and $\mathscr{B}(\cdot)$ is a nonlinear mapping from H into H such that

$$\| \mathscr{B}(u) \| \le M_0, \quad \| \mathscr{B}(u_1) - \mathscr{B}(u_2) \| \le M_1 \| u_1 - u_2 \|, \tag{13.2.4}$$

The global Lipschitz condition imposed on $\mathscr{B}$ (which is not satisfied for von Karman models under consideration) is later removed and replaced by local Lipschitz condition.

Under the above assumptions, it is well known by standard semigroup methods that for each initial datum $u_0 \in H$, there exists a unique global solution $u \in C(\mathbb{R}_+; H)$. Thus problem (13.2.3) generates a dynamical system (H, S_t) in the space H. We note that in this chapter the space H can be either complex or real.

Assume that there exist projectors $\mathscr{P}$ and $\mathscr{Q} = I - \mathscr{P}$ such that $\dim \mathscr{P} < \infty$, $\mathscr{P}, \mathscr{Q}$ commute with $\mathscr{A}$, and the following *dichotomy* property holds,

$$\|\mathscr{P} e^{\mathscr{A}t}\| \le e^{\lambda^- t}, \ t > 0; \quad \|\mathscr{Q} e^{-\mathscr{A}t}\| \le e^{-\lambda^+ t}, \ t > 0, \tag{13.2.5}$$

where $0 < \lambda^- < \lambda^+$.

In what follows we recall a well known procedure of constructing the inertial manifold, say $\mathscr{M}$, of the form (13.2.1) for the original dynamics in (13.2.3).

Once this is accomplished, then we have that any semitrajectory $u(t) = S_t u_0$ lying on the manifold has the form

$$u(t) = p(t) + q(t) = p(t) + \Phi(p(t)), \tag{13.2.6}$$

[1] We refer to Section 7.6 for more information.

where we use the notations $p(t) = \mathscr{P}u(t)$ and $q(t) = \mathscr{Q}u(t)$. It follows from (13.2.3) that $p(t)$ solves the so-called *inertial form* of the equation in (13.2.3):

$$\frac{dp(t)}{dt} + \mathscr{A}\,p(t) = \mathscr{P}\mathscr{B}(p(t) + \Phi(p(t))), \quad p(0) = \mathscr{P}u_0. \tag{13.2.7}$$

This inertial form is a finite-dimensional ODE with globally Lipschitz nonlinearity and therefore the solution $p(t)$ of (13.2.7) is defined on the whole real line $\mathbb{R}$. Moreover, from (13.2.4) and (13.2.5) it follows that $\|p(t)\| \le Ce^{\lambda^-|t|}$ for $t \le 0$. By (13.2.6) this latter property implies that any semitrajectory $u(t)$ lying on $\mathscr{M}$ can be extended on the whole $\mathbb{R}$ and

$$\sup_{t\le 0}\{e^{\gamma t} \parallel u(t) \parallel\} < \infty \tag{13.2.8}$$

for every γ such that $\lambda^- < \gamma < \lambda^+$. Equation (13.2.3) implies that $q(t) = \mathscr{Q}u(t)$ satisfies

$$q(t) = \mathscr{Q}e^{-\mathscr{A}(t-s)}u(s) + \int_s^t e^{-\mathscr{A}(t-\tau)}\mathscr{Q}\mathscr{B}(u(\tau))d\tau, \quad t \ge s.$$

The bound in (13.2.8) allows for limit transition $s \to -\infty$ and yields

$$q(t) = \int_{-\infty}^t e^{-\mathscr{A}(t-\tau)}\mathscr{Q}\mathscr{B}(u(\tau))d\tau, \quad t \in \mathbb{R}.$$

Thus using the variation of constants formula for $p(t) = \mathscr{P}u(t)$ we see from (13.2.6) that the inertial manifold $\mathscr{M}$ consists of full trajectories $\{u(t) : t \in \mathbb{R}\}$ possessing property (13.2.8) and satisfying the relation

$$u(t) = \mathscr{P}e^{-\mathscr{A}t}u_0 + \int_0^t e^{-\mathscr{A}(t-s)}\mathscr{P}\mathscr{B}(u(s))ds + \int_{-\infty}^t e^{-\mathscr{A}(t-s)}\mathscr{Q}\mathscr{B}(u(s))ds \tag{13.2.9}$$

for every $t \in \mathbb{R}$. One can see from the variation of constants formula applied to equation (13.2.3) that if (13.2.9) holds for $t \le t_0$ with some fixed t_0, then it is also true for all $t \in \mathbb{R}$. Thus it is sufficient to consider relation (13.2.9) on the semiaxis $(-\infty, 0]$ for $u(t)$ belonging to the class

$$\mathscr{C} = C_\gamma = \left\{ v \in C(-\infty, 0; H) : |v|_{\mathscr{C}} \equiv \sup_{t\le 0}\{e^{\gamma t} \parallel v(t) \parallel\} < \infty \right\}.$$

Here γ is a positive number such that $\lambda^- < \gamma < \lambda^+$. Moreover, it follows from (13.2.9) that

$$\Phi(\mathscr{P}u_0) = \mathscr{Q}u_0 = \int_{-\infty}^0 e^{\mathscr{A}s}\mathscr{Q}\mathscr{B}(u(s))ds,$$

which provides the representation formula for function Φ defining the inertial manifold $\mathscr{M}$ by (13.2.1).

For every $p \in \mathscr{P}H$ we introduce the operator $\mathbb{B}_p : C_\gamma \to C_\gamma$ given by

$$\mathbb{B}_p[v](t) \equiv e^{-t\mathscr{A}}p - \int_t^0 e^{-(t-\tau)\mathscr{A}} \mathscr{P}\mathscr{B}(v(\tau))\,d\tau$$
$$+ \int_{-\infty}^t e^{-(t-\tau)\mathscr{A}} \mathscr{Q}\mathscr{B}(v(\tau))\,d\tau\,, \quad t \le 0. \tag{13.2.10}$$

The above analysis shows that in order to construct an inertial manifold by the Lyapunov–Perron method we need first to solve the equation

$$v(t) = \mathbb{B}_p[v](t), \quad t \le 0, \quad v \in C_\gamma, \tag{13.2.11}$$

and then to define the map $\Phi : \mathscr{P}H \to \mathscr{Q}H$ by the relation

$$\Phi(p) \equiv \int_{-\infty}^0 e^{\mathscr{A}s} \mathscr{Q}\mathscr{B}(u_p(s))ds, \;\; u_p \equiv v \text{ satisfies } (13.2.11). \tag{13.2.12}$$

Thus, in view of the above discussion, our remaining task is to assert solvability of integral equation (13.2.11) in the space $\mathscr{C} = C_\gamma$ with $\mathbb{B}_p$ given by (13.2.10).

13.2.4. Lemma. *For any $p \in \mathscr{P}H$, equation* (13.2.11) *possesses a unique solution in $\mathscr{C} = C_\gamma$ with $\gamma = \frac{1}{2}(\lambda^+ + \lambda^-)$ provided the following spectral gap condition*

$$\lambda^+ - \lambda^- \ge \frac{4M_1}{q} \tag{13.2.13}$$

holds for some $0 < q < 1$, where M_1 is the Lipschitz constant for $\mathscr{B}$ (see (13.2.4)*). If v_p solves* (13.2.11)*, then we have the estimate*

$$|v_{p_1} - v_{p_2}|_{\mathscr{C}} \le (1-q)^{-1}\|p_1 - p_2\|. \tag{13.2.14}$$

Proof. It is easy to see from the dichotomy estimates in (13.2.5) that

$$\|\mathbb{B}_p[v_1](t) - \mathbb{B}_p[v_2](t)\|$$
$$\le M_1 e^{-\gamma t}\left\{\int_t^0 e^{(t-\tau)(\gamma-\lambda^-)}\,d\tau + \int_{-\infty}^t e^{-(t-\tau)(\lambda^+-\gamma)}\,d\tau\right\}|v_1 - v_2|_{\mathscr{C}}$$
$$\le M_1 e^{-\gamma t}\left\{\frac{1}{\gamma-\lambda^-} + \frac{1}{\lambda^+-\gamma}\right\}|v_1 - v_2|_{\mathscr{C}}.$$

Thus $|\mathbb{B}_p[v_1](t) - \mathbb{B}_p[v_2](t)|_{\mathscr{C}} \le q|v_1 - v_2|_{\mathscr{C}}$, where

$$q = \frac{M_1}{\gamma-\lambda^-} + \frac{M_1}{\lambda^+-\gamma}.$$

Therefore, if we choose $\gamma = (\lambda^- + \lambda^+)/2$, we find that $\mathbb{B}_p$ is a contraction in $\mathscr{C}$ under condition (13.2.13). In a similar way we have

$$|v_{p_1} - v_{p_2}|_{\mathscr{C}} = |\mathbb{B}_{p_1}[v_{p_1}] - \mathbb{B}_{p_2}[v_{p_2}]|_{\mathscr{C}} \le \|p_1 - p_2\| + q|v_{p_1} - v_{p_2}|_{\mathscr{C}}.$$

This relation implies (13.2.14). □

Lemma 13.2.4 and the discussion above lead to the following assertion.

13.2.5. Corollary. *With reference to dynamics governed by* (13.2.3), *the set* $\mathscr{M}$ *defined in* (13.2.1) *with* Φ *given by* (13.2.12) *is a finite-dimensional invariant manifold. The trajectories originating in* $\mathscr{M}$ *are in* C_γ *(for* $t \le 0$ *) with* $\gamma = \frac{1}{2}(\lambda^+ + \lambda^-)$. *Moreover, the map* $\Phi : \mathscr{P}H \to \mathscr{Q}H$ *is Lipschitz continuous and satisfies the following estimates:*

$$||\Phi(p)|| \le \frac{M_0}{\lambda^+}, \quad ||\Phi(p_1) - \Phi(p_2)|| \le \frac{q}{2(1-q)}||p_1 - p_2||. \qquad (13.2.15)$$

Our next task is to show that the manifold $\mathscr{M}$ is exponentially attractive. This is to say, we want to show that every trajectory is asymptotically exponentially tracked by a trajectory in $\mathscr{M}$. More precisely, we have the following.

13.2.6. Theorem. *Assume that the spectral gap condition* (13.2.13) *holds for some* $0 < q < \frac{2}{3}$. *Then the manifold* $\mathscr{M}$ *is exponentially complete. Moreover, for any* $u(t)$ *there exits a trajectory* $u^*(t) \in \mathscr{M}$ *such that*

$$\|u(t) - u^*(t)\| \le C_q e^{-\gamma t} \|\mathscr{Q}u(0) - \Phi(\mathscr{P}u(0))\|, \quad t > 0, \qquad (13.2.16)$$

with $\gamma = (\lambda^- + \lambda^+)/2$.

Proof. We are looking for an induced trajectory $u^*(t)$ in the form

$$u^*(t) = u(t) + w(t),$$

where $w(t)$ belongs to the space $\mathscr{C}^+ = C^+_\gamma$ of continuous functions $v(t)$ on the interval $[0,\infty)$ such that

$$|v|_{\mathscr{C}^+} \equiv \sup_{t\ge 0}\{e^{\gamma t} \parallel v(t) \parallel\} < \infty.$$

Let $F(w,t) = \mathscr{B}(u(t) + w) - \mathscr{B}(u(t))$. The function $w(t)$ satisfies the following equation

$$w_t + \mathscr{A}w = F(w,t), \;\; w(0) = w_0, \qquad (13.2.17)$$

for a suitable initial condition w_0 which guarantees that $u^*(0) \in \mathscr{M}$, hence, by invariance of $\mathscr{M}$, $u^*(t) \in \mathscr{M}$, for all $t \ge 0$.

Let us determine such an initial condition w_0. We must have that $u^*(0) = u(0) + w(0) \in \mathscr{M}$. This is equivalent to the following compatibility condition,

$$\mathscr{Q}[u(0) + w_0] = \Phi(\mathscr{P}[u(0) + w_0]). \qquad (13.2.18)$$

Thus our task amounts to finding $w \in \mathscr{C}^+$ such that equation (13.2.17) with the initial condition w_0 subject to (13.2.18) is satisfied. In order to accomplish this task, we write down the solution of (13.2.17) by using the variation of constants formula:

$$w(t) = e^{-\mathscr{A}t}w_0 + \int_0^t e^{-\mathscr{A}(t-s)}\mathscr{P}F(w(s),s)ds + \int_0^t e^{-\mathscr{A}(t-s)}\mathscr{Q}F(w(s),s)ds$$

$$= e^{-\mathscr{A}t}w_0 + \int_0^\infty e^{-\mathscr{A}(t-s)}\mathscr{P}F(w(s),s)ds - \int_t^\infty e^{-\mathscr{A}(t-s)}\mathscr{P}F(w(s),s)ds$$
$$+ \int_0^t e^{-\mathscr{A}(t-s)}\mathscr{Q}F(w(s),s)ds, \quad t \geq 0.$$

We rewrite the above equation in the following form,

$$w(t) = e^{-\mathscr{A}t}g(w) + \mathbb{B}^+[w](t), \quad t \geq 0,$$

with

$$\mathbb{B}^+[w](t) \equiv -\int_t^\infty e^{-\mathscr{A}(t-s)}\mathscr{P}F(w(s),s)ds + \int_0^t e^{-\mathscr{A}(t-s)}\mathscr{Q}F(w(s),s)ds$$

and

$$g(w) \equiv w_0 + \int_0^\infty e^{\mathscr{A}s}\mathscr{P}F(w(s),s)ds,$$

which expressions are well defined on $\mathscr{C}^+$. Taking into account compatibility condition (13.2.18) we must have

$$\mathscr{Q}[u(0)+g(w)] = \Phi\left(\mathscr{P}\left[u(0)+g(w) - \int_0^\infty e^{\mathscr{A}s}\mathscr{P}F(w(s),s)ds\right]\right).$$

If we restrict, in addition, $g(w)$ to belong to $\mathscr{Q}H$ then our problem reduces to solvability of the following system in the variables $g \in \mathscr{Q}H, w \in \mathscr{C}^+$,

$$\begin{aligned} g &= -\mathscr{Q}u(0) + \Phi\left[\mathscr{P}u(0) - \int_0^\infty e^{\mathscr{A}s}\mathscr{P}F(w(s),s)ds\right], \\ w(t) &= e^{-\mathscr{A}t}g + \mathbb{B}^+[w](t), \quad t \geq 0. \end{aligned} \tag{13.2.19}$$

The Lemma below asserts solvability of the above system.

13.2.7. Lemma. *For any initial data $u(0) \in H$ the system defined in* (13.2.19) *admits unique solution $(g;w) \in \mathscr{Q}H \times \mathscr{C}^+$ provided* (13.2.13) *holds with some $0 < q < \frac{2}{3}$.*

Proof. We apply the contraction mapping principle. To accomplish this, we use the following estimates

$$\|F(w(t),t)\| \leq M_1 e^{-\gamma t}|w|_{\mathscr{C}^+}, \quad t > 0,$$

and

$$\|F(w(t),t) - F(\bar{w}(t),t)\| \leq M_1 e^{-\gamma t}|w - \bar{w}|_{\mathscr{C}^+}, \quad t > 0,$$

for every w and $\bar{w}$ from $\mathscr{C}^+$. By using these estimates we also obtain that

$$|\mathbb{B}^+[w]|_{\mathscr{C}^+} \leq \sup_{t\geq 0}\left\{\int_t^\infty e^{\gamma(t-s)}\|e^{-\mathscr{A}(t-s)}\mathscr{P}\|e^{\gamma s}\|F(w(s),s)\|ds\right.$$

$$+\int_0^t e^{\gamma(t-s)}\|e^{-\mathscr{A}(t-s)}\mathscr{Q}\|e^{\gamma s}\|F(w(s),s)\|ds\Big\}$$

$$\leq M_1|w|_{\mathscr{C}^+}\sup_{t\geq 0}\left\{\int_t^\infty e^{(\gamma-\lambda^-)(t-s)}ds+\int_0^t e^{(\gamma-\lambda^+)(t-s)}ds\right\}$$

$$\leq \frac{4M_1|w|_{\mathscr{C}^+}}{\lambda^+-\lambda^-}\leq q|w|_{\mathscr{C}^+}\,. \tag{13.2.20}$$

Similarly

$$|\mathbb{B}^+[w_1]-\mathbb{B}^+[w_2]|_{\mathscr{C}^+}\leq M_1|w_1-w_2|_{\mathscr{C}^+}\sup_{t\geq 0}\left\{\int_t^\infty e^{(\gamma-\lambda^-)(t-s)}ds\right.$$

$$\left.+\int_0^t e^{(\gamma-\lambda^+)(t-s)}ds\right\}$$

$$\leq \frac{4M_1|w_1-w_2|_{\mathscr{C}^+}}{\lambda^+-\lambda^-}\leq q|w_1-w_2|_{\mathscr{C}^+}\,.$$

Consider $g(w)$ as a map : $\mathscr{C}^+\to\mathscr{Q}H$ which is defined by the first equation in (13.2.19). Using estimates (13.2.15) in Corollary 13.2.5 we obtain that

$$\|g(w)\|\leq\|\mathscr{Q}u(0)\|+\frac{M_0}{\lambda^+}\,, \tag{13.2.21}$$

and

$$\|g(w_1)-g(w_2)\|\leq\frac{M_1 q}{2(1-q)}\int_0^\infty e^{(\lambda^--\gamma)s}ds|w_1-w_2|_{\mathscr{C}^+} \tag{13.2.22}$$

$$=\frac{M_1 q}{2(1-q)}\frac{1}{(\gamma-\lambda^-)}|w_1-w_2|_{\mathscr{C}^+}\leq\frac{q^2}{4(1-q)}|w_1-w_2|_{\mathscr{C}^+}\,.$$

We have used here the relation $\gamma=\frac{1}{2}(\lambda^++\lambda^-)$ and the property (13.2.13).

We apply the above estimates to the fixed point problem

$$w=e^{-\mathscr{A}t}g(w)+\mathbb{B}^+[w](t)\equiv T[w](t),\quad t\geq 0,\quad w\in\mathscr{C}^+. \tag{13.2.23}$$

By the estimates in (13.2.20) and (13.2.21) we infer that T maps $\mathscr{C}^+$ into itself. Moreover T is a contraction for suitably small q. Indeed,

$$|T[w_1]-T([w_2]|_{\mathscr{C}^+}\leq\|g(w_1)-g(w_2)\|+|\mathbb{B}^+w_1-\mathbb{B}^+w_2|_{\mathscr{C}^+}$$

$$\leq\left[q+\frac{q^2}{4(1-q)}\right]|w_1-w_2|_{\mathscr{C}^+}\,.$$

Thus, by the contraction principle under the condition $q<\frac{2}{3}$ we obtain unique $w\in\mathscr{C}^+$ which satisfies (13.2.23). The variable $g=g(w)\in\mathscr{Q}H$ is obtained directly from the first equation in the system (13.2.19). □

To complete the proof of Theorem 13.2.6, it suffices to establish decay rates for w. The estimates in (13.2.20) and (13.2.22) provide the following bounds,

$$\begin{aligned}|w|_{\mathscr{C}+} &\leq ||g(w)-g(0)||+||g(0)||+|\mathbb{B}^{+}[w]|_{\mathscr{C}+} \\ &\leq ||g(0)||+\left[\frac{q^2}{4(1-q)}+q\right]|w|_{\mathscr{C}+} .\end{aligned}$$

Hence

$$\left[1-q-\frac{q^2}{4(1-q)}\right]|w|_{\mathscr{C}+} \leq ||g(0)|| = ||\mathscr{Q}u(0)-\Phi(\mathscr{P}u(0))|| .$$

Therefore

$$|w|_{\mathscr{C}+} \leq \left(1-q\cdot\frac{4-3q}{4-4q}\right)^{-1} \|\mathscr{Q}u(0)-\Phi(\mathscr{P}u(0))\|.$$

This implies relation (13.2.16), proving Theorem 13.2.6. □

Now we consider equation (13.2.3) with locally Lipschitz nonlinearity; that is, instead of (13.2.4) we assume that

$$\| \mathscr{B}(u) \| \leq M_0(\rho), \quad \| \mathscr{B}(u)-\mathscr{B}(v) \| \leq M_1(\rho) \| u-v \|, \tag{13.2.24}$$

for any u and v from H such that $\|u\| \leq \rho$ and $\|v\| \leq \rho$, where $M_0(\rho)$ and $M_1(\rho)$ are positive nondecreasing functions of ρ.

13.2.8. Theorem. *Assume that* (13.2.5) *and* (13.2.24) *hold and* (13.2.3) *generates a dissipative dynamical system* (H,S_t) *with the radius* R *of an absorbing ball. Then under the spectral gap condition:*

$$\lambda^+-\lambda^- \geq \frac{4}{q}\cdot\left(M_1(4R)+\frac{1}{R}\cdot M_0(4R)\right), \quad 0<q<2/3, \tag{13.2.25}$$

the system (H,S_t) *has a locally invariant exponentially complete inertial manifold.*

Proof. Let us consider the truncated equation

$$\frac{du}{dt}+\mathscr{A}u=\mathscr{B}_R(u), \quad u|_{t=0}=u_0, \tag{13.2.26}$$

where $\mathscr{B}_R(u) := \chi\left(\|u\|/(2R)\right)\mathscr{B}(u)$ and $\chi(s) \in C^\infty(\mathbb{R}_+)$ is a nonincreasing function with properties: $\chi(s)=1$ for $0 \leq s \leq 1$, $\chi(s)=0$ for $s \geq 2$ and $|\chi'(s)| \leq 2$ for all $s \geq 0$. It follows from (13.2.24) that

$$\| \mathscr{B}_R(u) \| \leq M_0(4R), \quad \| \mathscr{B}_R(u)-\mathscr{B}_R(v) \| \leq \left(M_1(4R)+\frac{M_0(4R)}{R}\right) \| u-v \|$$

for any u and v from H.

Let $(H, \hat{S}_t)$ be the dynamical system generated by (13.2.26). By Theorem 13.2.6, under condition (13.2.25) $(H, \hat{S}_t)$ has an inertial manifold $\mathscr{M}$ of the form (13.2.1) such that (13.2.16) holds with $u(t) = \hat{S}_t u$.

Because $S_t u = \hat{S}_t u$ for $t \in [0,T]$ under the condition $\|S_t u\| \le 2R$ for $t \in [0,T]$, the manifold $\mathscr{M}$ is locally invariant with respect to S_t.

Let us prove that $\mathscr{M}$ is exponentially complete with respect to S_t. By dissipativity of S_t there exists $t_0 \ge 0$ such that $\|S_t u\| \le R$ for $t \ge t_0$. In this case $S_t u = \hat{S}_{t-t_0}(S_{t_0} u)$ for $t \ge t_0$. Therefore (13.2.16) implies that there exists $y^* = y^*(S_{t_0} u) \in \mathscr{M}$ such that

$$\|S_t u - \hat{S}_{t-t_0} y^*\| \le C(t_0, u) \cdot e^{-\gamma(t-t_0)}, \quad t > t_0. \tag{13.2.27}$$

In particular, there exists $\tau_0 \ge t_0$ such that $\|S_t u - \hat{S}_{t-t_0} y^*\| \le R$ for $t \ge \tau_0$. Thus $\|\hat{S}_{t-t_0} y^*\| \le 2R$ for $t \ge \tau_0$. In this case we have

$$S_{t-\tau_0}(\hat{S}_{\tau_0 - t_0} y^*) = \hat{S}_{t-t_0} y^* \quad \text{for } t \ge \tau_0.$$

Therefore (13.2.27) implies that

$$\|S_t u - S_{t-\tau_0}(\hat{S}_{\tau_0 - t_0} y^*)\| \le C(t_0, u) \cdot e^{-\gamma(t-t_0)}, \quad t \ge \tau_0.$$

Hence (13.2.2) holds with $u^* = \hat{S}_{\tau_0 - t_0} y^* \in \mathscr{M}$ and $s = \tau = \tau_0$. □

13.3 Inertial manifolds for second order in time evolution equation

In this section the main results on existence of inertial manifolds for von Karman evolution models described in (13.1.1), (13.1.5), and (13.1.8) are presented. We first provide an abstract formulation for the three cases of damping considered within the framework of globally Lipschitz nonlinearities. Later on, by using dissipativity of the corresponding semiflows we prove our main results applicable to von Karman evolutions whose nonlinear terms are locally Lipschitz only.

13.3.1 Second-order evolutions with viscous damping

We consider the following evolutionary differential equation defined on a real separable Hilbert space $\mathscr{H}$:

$$u_{tt} + 2\mu u_t + A^2 u = F(u, u_t) \quad u|_{t=0} = u_0, \; u_t|_{t=0} = u_1. \tag{13.3.1}$$

Here A is a positive operator with a discrete (and complete) spectrum; that is, there exists the orthonormal basis $\{e_k\}$ in $\mathscr{H}$ such that

$$Ae_k = \omega_k e_k, \quad 0 < \omega_1 \le \omega_2 \le \cdots, \quad \lim_{k\to\infty} \omega_k = \infty. \tag{13.3.2}$$

We allow here that the nonlinear term also depends on velocities u_t (in some regular way, see below).

We reduce the problem of existence of inertial manifolds for (13.3.1) to the corresponding problem formulated for a first-order in time evolution equation. We rely on some idea presented in [236], where the one-dimensional wave equation with semilinear term independent of velocity u_t was considered (see also [61, Chapter 3]).

Globally Lipschitz case

We suppose first that F is a nonlinear mapping from $\mathscr{D}(A) \times \mathscr{H}$ into $\mathscr{H}$ such that

$$\| F(u_0,u_1) \| \le M_0,$$

$$\| F(u_0,u_1) - F(v_0,v_1) \| \le M_1 \| A(u_0 - v_0) \| + M_2 \| u_1 - v_1 \|,$$

where $M_j > 0$ are constants and $\| \cdot \|$ is the norm in the space $\mathscr{H}$. Problem (13.3.1) can be written as a first-order evolution equation in the real space $H = D(A) \times \mathscr{H}$:

$$\frac{d}{dt} U + \mathscr{A} U = \mathscr{B}(U), \quad U|_{t=0} = U_0, \tag{13.3.3}$$

where $U = (u; u_t)$, $U_0 = (u_0; u_1)$ and

$$\mathscr{A} U_0 = (-u_1; A^2 u_0 + 2\mu u_1), \quad \mathscr{B}(U_0) = (0; F(u_0,u_1)).$$

The operator $\mathscr{A}$ has the spectrum $\lambda_k^{\pm} = \mu \pm \sqrt{\mu^2 - \omega_k^2}$, where $k = 1,2,\dots$. The corresponding eigenfunctions have the form $f_k^{\pm} = (e_k; -\lambda_k^{\pm} e_k)$.

Let

$$H_1 = \mathrm{Span}\{(e_k;0),\, (0;e_k) \,:\, k = 1,2,\dots,N\}$$

and

$$H_2 = \overline{\mathrm{Span}\{(e_k;0),\, (0;e_k) \,:\, k \ge N+1\}},$$

where we choose N such that $\mu^2 > \omega_{N+1}^2$. It is clear that $H = H_1 \oplus H_2$.

Following [236] we introduce the following inner product on the space H. For $U = (u_0; u_1)$ and $V = (v_0; v_1)$ from H_1 we define

$$\langle U,V \rangle_1 = \mu^2 (u_0,v_0) - (Au_0,Av_0) + (\mu u_0 + u_1, \mu v_0 + v_1).$$

If $U,V \in H_2$ we set

$$\langle U,V \rangle_2 = (Au_0,Av_0) + (\mu^2 - 2\omega_{N+1}^2)(u_0,v_0) + (\mu u_0 + u_1, \mu v_0 + v_1).$$

We define

$$\langle U,V\rangle = \langle U_1,V_1\rangle_1 + \langle U_2,V_2\rangle_2$$

for $U = U_1 + U_2$ and $V = V_1 + V_2$ with $U_1, V_1 \in H_1$ and $U_2, V_2 \in H_2$. We denote by $|\cdot|$ the corresponding norm. The following lemma shows that this new norm is equivalent to the standard norm

$$\|U\|_H = \left(\|Au_0\|^2 + \|u_1\|^2\right)^{1/2}$$

in the space H.

13.3.1. Lemma. *Let $U = (u_0; u_1) \in H$ and $\beta_k = \mu^2 - \omega_k^2$. Then*

$$\|Au_0\| \le \max\left\{1, \frac{\omega_{N+1}}{\sqrt{\beta_{N+1}}}\right\}\cdot|U| \quad \textit{and} \quad \|u_1\| \le \sqrt{\frac{\mu^2+\beta_{N+1}}{\beta_{N+1}}}\cdot|U|.$$

Proof. The proof consists of direct calculations using the definitions of norms in H_1 and H_2; see the proof of Lemma 7.1 in [61, Chapter 3] for some details. □

Let

$$H_1^{\pm} = \mathrm{Span}\{f_k^{\pm} : k = 1,2,\ldots,N\}.$$

The following properties of this new inner product are critical:

- $\langle U,V\rangle_1$ orthogonalizes subspaces H_1^- and H_1^+. Thus we have the orthogonal decomposition $H = H_1^- \oplus H_1^+ \oplus H_2$ with respect to $\langle\cdot,\cdot\rangle$.
- $\langle U,V\rangle$ induces the dichotomy property with contractivity. This is to say that with $\mathscr{P}$ the orthoprojector onto H_1^- and $\mathscr{Q}$ the orthoprojector onto $H_1^+ \oplus H_2$ we have the following dichotomy estimates,

$$\left|e^{\mathscr{A}t}\mathscr{P}\right| \le e^{\lambda_N^- t} \text{ and } \left|e^{-\mathscr{A}t}\mathscr{Q}\right| = e^{-\lambda_{N+1}^- t} \text{ for all } t > 0\,.$$

These properties have been shown in [236] and also [61, Chapter 3, Section 7].

From Lemma 13.3.1 after noting that $|\mathscr{B}(U)| = \|F(u,u_t)\|$ we obtain the following estimate for the nonlinear term $\mathscr{B}$

$$|\mathscr{B}(U) - \mathscr{B}(V)| \le \left(M_1 \max\left\{1, \frac{\omega_{N+1}}{\sqrt{\beta_{N+1}}}\right\} + M_2\sqrt{\frac{\mu^2+\beta_{N+1}}{\beta_{N+1}}}\right)\cdot|U-V|.$$

The properties listed above along with Theorem 13.2.6 imply the following assertion.

13.3.2. Theorem. *Assume that the following spectral gap condition*

$$\frac{\omega_{N+1}^2 - \omega_N^2}{\sqrt{\beta_N} + \sqrt{\beta_{N+1}}} \ge \frac{4}{q}\cdot\left(M_1 \max\left\{1, \frac{\omega_{N+1}}{\sqrt{\beta_{N+1}}}\right\} + M_2\sqrt{\frac{\mu^2+\beta_{N+1}}{\beta_{N+1}}}\right) \qquad (13.3.4)$$

holds for some $N \in \mathbb{N}$, $0 < q < \frac{2}{3}$ and $\omega_{N+1}^2 < \mu^2$. Here ω_k are eigenvalues of the operator A and $\beta_k = \mu^2 - \omega_k^2$. Then problem (13.3.3) possesses an exponentially *complete inertial manifold $\mathscr{M}$. Moreover for any $U \in H$ there exists $U^* \in \mathscr{M}$ such that*

$$\|S_tU - S_tU^*\|_H \le C(1+\|U\|_H)\cdot e^{-\gamma t}, \quad t>0, \tag{13.3.5}$$

with

$$\gamma = \mu - \frac{1}{2}(\sqrt{\beta_N}+\sqrt{\beta_{N+1}})$$

where S_t is the evolution operator in H generated by (13.3.3).

13.3.3. Remark. It is easy to see that (13.3.4) holds provided

$$2\omega_{N+1}^2 \le \mu^2 \le (2+\delta)\omega_{N+1}^2, \quad \delta>0, \tag{13.3.6}$$

and

$$\frac{\omega_{N+1}^2-\omega_N^2}{\omega_{N+1}} \ge \frac{8\sqrt{2+\delta}}{q}\cdot\left(M_1+M_2\sqrt{3+\delta}\right).$$

The last relation holds, provided

$$\omega_{N+1}-\omega_N \ge \frac{8\sqrt{2+\delta}}{q}\cdot\left(M_1+M_2\sqrt{3+\delta}\right). \tag{13.3.7}$$

■

This remark allow us to obtain the following assertion.

13.3.4. Theorem. *Assume that there exist $N \in \mathbb{N}$ and $\delta>0$ such that* (13.3.7) *holds for some $q\in(0,\frac{2}{3})$. Then, for every μ with property* (13.3.6) *problem* (13.3.3) *possesses an exponentially complete inertial manifold $\mathscr{M}$ such that* (13.3.5) *holds.*

Locally Lipschitz case—Applications to von Karman evolutions

We next consider the locally Lipschitz case, which is of relevance to applications to von Karman evolutions. Now we assume that $F(\cdot,\cdot)$ is a nonlinear mapping from $H=\mathscr{D}(A)\times\mathscr{H}$ into $\mathscr{H}$ such that

$$\| F(u_0,u_1) \| \le M_0(\rho),$$
$$\| F(u_0,u_1)-F(v_0,v_1) \| \le M_1(\rho)\,\| A(u_0-v_0)\| + M_2(\rho)\,\| u_1-v_1 \|,$$

for any $(u_0;u_1)$ and $(v_0;v_1)$ from H such that

$$\|Au_0\|^2+\|u_1\|^2 \le \rho^2 \ \text{ and } \ \|Av_0\|^2+\|v_1\|^2\le\rho^2.$$

Here $M_j(\rho)$ are positive nondecreasing functions of ρ and $\|\cdot\|$ is the norm in $\mathscr{H}$.
Let $S_t(u_0;u_1)=(u(t);u_t(t))$ be the evolution semigroup generated by equation (13.3.1) in H. We assume that S_t possesses the absorbing ball

$$\{(u_0;u_1)\ :\ \|(u_0;u_1)\|_H^2 := \|Au_0\|^2+\|u_1\|^2 \le R^2\}$$

with the radius R independent of $\mu \geq \mu_0 > 0$.

Let us consider the truncated equation

$$u_{tt} + 2\mu u_t + A^2 u = F_R(u, u_t) \equiv \chi\left(\frac{\|Au\|^2 + \|u_t\|^2}{4R^2}\right) \cdot F(u, u_t), \tag{13.3.8}$$

where $\chi \in C^\infty(\mathbb{R}_+ \mapsto [0,1])$ such that $\chi(s) = 1$ for $0 \leq s \leq 1$ and $\chi(s) = 0$ for $s \geq 2$. It is clear that there exist constants M_j^R, $j = 0, 1, 2$, such that

$$\| F_R(u_0, u_1) \| \leq M_0^R,$$
$$\| F_R(u_0, u_1) - F_R(v_0, v_1) \| \leq M_1^R \| A(u_0 - v_0) \| + M_2^R \| u_1 - v_1 \|,$$

for all $(u_0; u_1)$ and $(v_0; v_1)$ from H. Application of Theorems 13.3.2 and 13.3.4 to problem (13.3.8) yield the following.

13.3.5. Theorem. *Assume that there exist $N \in \mathbb{N}$ and $\delta > 0$ such that*

$$\omega_{N+1} - \omega_N \geq \frac{8\sqrt{2} + \delta}{q} \cdot \left(M_1^R + M_2^R \sqrt{3 + \delta}\right) \tag{13.3.9}$$

holds for some $q \in (0, 2/3)$. Then for every μ with property (13.3.6) *problem* (13.3.3) *possesses a locally invariant inertial manifold, which is exponentially complete.*

Proof. Let $(H, \hat{S}_t)$ be the dynamical system generated by (13.3.8). By Theorem 13.3.4 under condition (13.3.9) $(H, \hat{S}_t)$ has an inertial manifold $\mathscr{M}$ of the form (13.2.1) such that (13.3.5) holds for $\hat{S}_t$. Because $S_t U = \hat{S}_t U$ for $t \in [0, T]$ under the condition $\|S_t U\|_H \leq 2R$ for $t \in [0, T]$, the manifold $\mathscr{M}$ is locally invariant with respect to S_t. It is also exponentially complete with respect to S_t. The latter can be obtained by the same argument as in the proof of Theorem 13.2.8. □

Theorems 13.3.5 and 13.1.8 when applied to problem (13.1.1)–(13.1.3), give the following.

13.3.6. Theorem. *With reference to von Karman equations* (13.1.1)–(13.1.3) *with $p \in L^2(\Omega)$ and $F_0 \in H^{3+\delta}(\Omega)$, assume that*

$$\omega_{N(k)+1} - \omega_{N(k)} \to \infty \quad \text{when} \quad k \to \infty \tag{13.3.10}$$

for some sequence $\{N(k)\}$. Here ω_N are eigenvalues of the operator $-\Delta$ with the Dirichlet boundary conditions. Then for any $\delta > 0$ there exists k_0 such that problem (13.1.1)–(13.1.3) *possesses a locally invariant inertial, exponentially complete manifold whenever μ satisfies* (13.3.6) *with $N = N(k_0)$.*

In general (13.3.10) fails because $\omega_N \sim c_0 N$. However, if $\Omega = (0, l_1) \times (0, l_2)$ is a rectangle with l_1/l_2 rational then (13.3.10) holds. To prove this one can use the following gap theorem which was established in [227, Appendix A] (see also [247] for a similar result).

13.3.7. Proposition **(Special gap theorem, [227]).** *Let $\mathbb{S}$ be a finite collection of the forms*

$$\sigma(n_1,n_2) = a_{11}n_1^2 + a_{12}n_1n_2 + a_{22}n_2^2 + b_1n_1 + b_2n_2 + c, \quad n_1,n_2 \in \mathbb{Z},$$

with rational coefficients a_{ij}, b_i, c, and negative discriminant: $a_{12}^2 - 4a_{11}a_{22} < 0$. Then, given any $h > 0$ there exists arbitrary large m such that

$$\sigma(n_1,n_2) \notin [m, m+h] \text{ for each } \sigma \in \mathbb{S} \text{ and } n_1,n_2 \in \mathbb{Z}.$$

If $\Omega = (0,l_1) \times (0,l_2)$ is rectangular, then the spectrum $\sigma(-\Delta)$ of the operator $-\Delta$ with the Dirichlet boundary conditions has the form

$$\sigma(-\Delta) = \left\{ \sigma(n,m) = \left(\frac{\pi}{l_1}\right)^2 \cdot \left(n^2 + \left(\frac{l_1}{l_2}\right)^2 m^2\right) \; : \; n,m \in \mathbb{N} \right\}. \quad (13.3.11)$$

The discriminant condition is satisfied trivially. Therefore, application of Proposition 13.3.7 with l_1/l_2 rational gives us (13.3.10).

13.3.2 Second-order evolution equation with strong damping

We begin by considering the following abstract problem which is a prototype for von Karman evolution with structural damping given by (13.1.5):

$$u_{tt} + 2\mu A u_t + A^2 u = F(u), \quad u|_{t=0} = u_0, \; u_t|_{t=0} = u_1. \quad (13.3.12)$$

As before, A is a positive operator with a discrete spectrum; that is, there exists the orthonormal basis $\{e_k\}$ in $\mathscr{H}$ such that (13.3.2) holds. We start with the globally Lipschitz case: we assume that $F(\cdot)$ is a nonlinear mapping from $\mathscr{D}(A)$ into $\mathscr{H}$ such that

$$\| F(u) \| \leq M_0, \quad \| F(u) - F(v) \| \leq M_1 \| A(u-v) \| . \quad (13.3.13)$$

If we introduce a new variable $(Au; u_t)$, then we can rewrite (13.3.12) in the following form

$$\frac{d}{dt}\begin{pmatrix} Au \\ u_t \end{pmatrix} + A \begin{pmatrix} 0 & -1 \\ 1 & 2\mu \end{pmatrix} \begin{pmatrix} Au \\ u_t \end{pmatrix} = \begin{pmatrix} 0 \\ F(u) \end{pmatrix}.$$

The structure of the linear term ("operator × matrix") in this representation leads to the following considerations which we split in two cases: $0 < \mu < 1$ and $\mu > 1$.

The case $0 < \mu < 1$

Let $\hat{\mathscr{H}}$ be the complexification of $\mathscr{H}$. For every element $y \in \hat{\mathscr{H}}$ we use a representation of the form $y = u + i \cdot v$, where $i^2 = -1$ and $u, v \in \mathscr{H}$. Using the idea presented in [209] we introduce new variables:

$$y_1(t) = \frac{1}{2}\left(Au(t) + \frac{i}{\beta}\left(u_t(t) + \mu Au(t)\right)\right) = \frac{i}{2\beta}\left(u_t(t) + (\mu - i\beta)Au(t)\right)$$

and

$$y_2(t) = \overline{y_1(t)} = \frac{1}{2}\left(Au(t) - \frac{i}{\beta}\left(u_t(t) + \mu Au(t)\right)\right) = -\frac{i}{2\beta}\left(u_t(t) + (\mu + i\beta)Au(t)\right),$$

where $\beta = \sqrt{1-\mu^2}$ and the bar means complex conjugation. The inverse transformation has the form

$$u = A^{-1}(y_1 + y_2), \quad u_t = -\mu(y_1 + y_2) - i\beta(y_1 - y_2).$$

A simple calculation shows that the pair $(y_1; y_2)$ satisfies the equations

$$\begin{cases} \dfrac{dy_1}{dt} + (\mu + i\beta)Ay_1 = \dfrac{i}{2\beta}F(A^{-1}(y_1 + y_2)) \\[2ex] \dfrac{dy_2}{dt} + (\mu - i\beta)Ay_2 = -\dfrac{i}{2\beta}F(A^{-1}(y_1 + y_2)) \end{cases} \tag{13.3.14}$$

Let $H = \hat{\mathscr{H}} \times \hat{\mathscr{H}}$ and $U = (y_1; y_2)$. We rewrite problem (13.3.14) in the form

$$\frac{d}{dt}U + \mathscr{A}U = \mathscr{B}(U), \quad U|_{t=0} = U_0, \tag{13.3.15}$$

where

$$\mathscr{A} = A\begin{pmatrix} \mu + i\beta & 0 \\ 0 & \mu - i\beta \end{pmatrix} \quad \text{and} \quad \mathscr{B}(U) = \frac{i}{2\beta}\begin{pmatrix} F(A^{-1}(y_1 + y_2)) \\ -F(A^{-1}(y_1 + y_2)) \end{pmatrix}.$$

Let

$$\mathscr{P} = \begin{pmatrix} P_N & 0 \\ 0 & P_N \end{pmatrix} \quad \text{and} \quad \mathscr{Q} = I - \mathscr{P} = \begin{pmatrix} I - P_N & 0 \\ 0 & I - P_N \end{pmatrix},$$

where P_N is the orthoprojector in $\hat{\mathscr{H}}$ on $\mathrm{Span}\{e_k : k = 1, 2, \ldots, N\}$. It is clear that $\mathscr{P}$ is an eigenprojector for $\mathscr{A}$. Moreover, we have the dichotomy estimates

$$\|\mathscr{P}e^{\mathscr{A}t}\| \le e^{\mu\omega_N t}, \ t > 0; \quad \|\mathscr{Q}e^{-\mathscr{A}t}\| \le e^{-\mu\omega_{N+1}t}, \ t > 0.$$

The nonlinear term $\mathscr{B}(U)$ in (13.3.14) possesses the properties

$$\|\mathscr{B}(U)\| \le \frac{M_0}{\sqrt{2}\beta} \quad \text{and} \quad \|\mathscr{B}(U) - \mathscr{B}(U^*)\| \le \frac{M_1}{\beta}\|U - U^*\|.$$

Therefore we can apply Theorem 13.2.6 in order to obtain the following assertion.

13.3.8. Theorem. *Under the condition*

$$\omega_{N+1} - \omega_N \ge \frac{4M_1}{q\mu\sqrt{1-\mu^2}}, \quad 0 < q < \frac{2}{3}, \quad 0 < \mu < 1,$$

problem (13.3.14), *and hence* (13.3.12), *possess an exponentially complete inertial manifold.*

The case $\mu > 1$

As before, we introduce new variables:

$$y_1(t) = \frac{1}{2}\left(Au(t) - \beta^{-1}\left(u_t(t) + \mu Au(t)\right)\right) = -\frac{1}{2\beta}\left(u_t(t) + (\mu - \beta)Au(t)\right)$$

and

$$y_2(t) = \frac{1}{2}\left(Au(t) + \beta^{-1}\left(u_t(t) + \mu Au(t)\right)\right) = \frac{1}{2\beta}\left(u_t(t) + (\mu + \beta)Au(t)\right),$$

where $\beta = \sqrt{\mu^2 - 1}$. The inverse transformation has the form

$$u = A^{-1}(y_1 + y_2), \quad u_t = -\mu(y_1 + y_2) - \beta(y_1 - y_2).$$

Rescaling time by the formula $w_i(t) = y_i((\mu - \beta)^{-1}t)$, it is easy to find that the pair $(w_1; w_2)$ satisfies the equations

$$\begin{cases} \frac{dw_1}{dt} + (\mu+\beta)^2 Aw_1 = -\frac{1}{2}\left(1 + \frac{\mu}{\beta}\right) F(A^{-1}(w_1 + w_2)) \\ \frac{dw_2}{dt} + Aw_2 = \frac{1}{2}\left(1 + \frac{\mu}{\beta}\right) F(A^{-1}(w_1 + w_2)). \end{cases} \tag{13.3.16}$$

This leads to consideration of problem (13.3.15) in the space $H = \mathscr{H} \times \mathscr{H}$ with

$$\mathscr{A} = A\begin{pmatrix} (\mu+\beta)^2 & 0 \\ 0 & 1 \end{pmatrix} \quad \text{and} \quad \mathscr{B}(U) = \frac{1}{2}\left(1 + \frac{\mu}{\beta}\right)\begin{pmatrix} -F(A^{-1}(w_1 + w_2)) \\ F(A^{-1}(w_1 + w_2)) \end{pmatrix},$$

where $U = (w_1; w_2)$. The nonlinear term $\mathscr{B}(U)$, in this case, possesses the properties

$$\|\mathscr{B}(U)\| \le \frac{M_0}{\sqrt{2}}\left(1 + \frac{\mu}{\beta}\right) \quad \text{and} \quad \|\mathscr{B}(U) - \mathscr{B}(U^*)\| \le M_1\left(1 + \frac{\mu}{\beta}\right)\|U - U^*\|.$$

For every integer $N_1 \ge 0$ and $N_2 \ge 0$ we introduce projectors

$$\mathscr{P} = \begin{pmatrix} P_{N_1} & 0 \\ 0 & P_{N_2} \end{pmatrix} \text{ and } \mathscr{Q} = I - \mathscr{P} = \begin{pmatrix} I - P_{N_1} & 0 \\ 0 & I - P_{N_2} \end{pmatrix}, \tag{13.3.17}$$

where P_N is the orthoprojector onto $\mathrm{Span}\{e_k : k = 1, 2, \dots, N\}$ for $N \geq 1$ and we suppose $P_0 = 0$.

13.3.9. Theorem. *Assume that $\mu > 1$ and there exist numbers $0 < \lambda^- < \lambda^+$ such that*

$$\lambda^+ - \lambda^- \geq \frac{4M_1}{q}\left(1 + \frac{\mu}{\sqrt{\mu^2 - 1}}\right), \quad 0 < q < \frac{2}{3}, \tag{13.3.18}$$

and integers $N_1 \geq 0$ and $N_2 \geq 0$ such that

$$\begin{gathered} (\mu + \beta)^2 \omega_{N_1} \leq \lambda^-, \ (\mu + \beta)^2 \omega_{N_1 + 1} \geq \lambda^+, \\ \omega_{N_2} \leq \lambda^-, \ \omega_{N_2 + 1} \geq \lambda^+ \end{gathered} \tag{13.3.19}$$

(here if $N = 0$ we suppose $\omega_N = 0$). Then, problem (13.3.16), *and hence* (13.3.12), *possesses exponentially complete inertial manifold of the form* (13.2.1) *with the projector $\mathscr{P}$ given by* (13.3.17).

Proof. Under conditions (13.3.19) it is easy to prove the dichotomy estimates (13.2.5) with $\mathscr{P}$ and $\mathscr{Q}$ given by (13.3.17). Therefore (13.3.18) allows us to apply Theorem 13.2.6 that leads to the final conclusion. □

For μ large enough this theorem implies the following assertion.

13.3.10. Corollary. *Under the condition*

$$\omega_{N+1} - \omega_N \geq \frac{4M_1}{q}\left(1 + \frac{\mu}{\sqrt{\mu^2 - 1}}\right), \quad 0 < q < \frac{2}{3}, \quad \mu + \sqrt{\mu^2 - 1} \geq \sqrt{\frac{\omega_{N+1}}{\omega_1}}$$

problem (13.3.16), *and hence* (13.3.12), *possess an exponentially complete inertial manifold of the form* (13.2.1) *with the projector $\mathscr{P} = \mathrm{diag}\{0, P_N\}$.*

Proof. Under the conditions of this corollary relations (13.3.18) and (13.3.19) hold with $\lambda^- = \omega_N$, $\lambda^+ = \omega_{N+1}$, $N_1 = 0$ and $N_2 = N$. □

If $\mu \geq \mu_0 > 1$, then $\mu/\beta \leq \mu_0/\sqrt{\mu_0^2 - 1}$. Therefore we arrive at the following assertion.

13.3.11. Corollary. *If for some $N \in \mathbb{N}$, $\mu_0 > 1$ and $0 < q < \frac{2}{3}$ the following relation is valid,*

$$\omega_{N+1} - \omega_N \geq \frac{4M_1}{q}\left(1 + \frac{\mu_0}{\sqrt{\mu_0^2 - 1}}\right),$$

then for any

$$\mu \in \left\{ \mu \ : \ \mu \geq \mu_0,\ \mu + \sqrt{\mu^2 - 1} \geq \sqrt{\omega_1^{-1} \omega_{N+1}} \right\},$$

problem (13.3.12) possesses an exponentially complete inertial manifold.

Locally Lipshitz case—von Karman evolutions with structural damping

In this case we can obtain a result similar to Theorem 13.3.5, although applicable to a larger class of damping coefficients that are no longer required to be restricted in size. The corresponding result is formulated for the problem (13.1.2), (13.1.3), and (13.1.5).

13.3.12. Theorem. *Let $\Omega = (0, l_1) \times (0, l_2)$ be a rectangle with l_1/l_2 rational. Assume that $p \in L^2(\Omega)$ and $F_0 \in W^2_\infty(\Omega)$. Then problem* (13.1.2), (13.1.3) *and* (13.1.5) *possesses a locally invariant and exponentially complete inertial manifold if either*

(i) $0 < \mu < 1$.
(ii) $\mu > 1$ *and* $\mu^2 + \mu\sqrt{\mu^2 - 1}$ *is a rational number.*
(iii) $\mu > 1$ *is large enough.*

Proof. In this case the spectrum $\sigma(-\Delta)$ of the operator $-\Delta$ with the Dirichlet boundary conditions has the form (13.3.11). If l_1/l_2 rational, then by Proposition 13.3.7 relation (13.3.10) holds for some sequence $\{N(k)\}$. Therefore Theorems 13.1.9 and 13.3.8 and Corollary 13.3.11 imply the conclusion of the theorem provided either (i) or (iii) holds.

In the case (ii) by Proposition 13.3.7 (applied to the forms $\sigma(n, m)$ and $(\mu + \beta)^2 \sigma(n, m)$) there exist numbers $0 < \lambda^- < \lambda^+$ with the gap $\lambda^+ - \lambda^-$ a large as we need and such that (13.3.19) holds for some N_1 and N_2 for the ordered eigenvalues $\omega_1 \leq \omega_2 \leq \cdots$ of the operator $-\Delta$ with the Dirichlet boundary conditions. Thus Theorem 13.3.9 is applicable. □

13.3.3 Thermoelastic von Karman evolutions

As before we begin by considering an abstract version of the thermoelasticity problem

$$u_{tt} - \mu A\theta + A^2 u = F(u), \tag{13.3.20}$$

$$\theta_t + \eta A\theta + \mu A u_t = 0 \tag{13.3.21}$$

with initial data $u|_{t=0} = u_0$, $u_t|_{t=0} = u_1$, $\theta|_{t=0} = \theta_0$. Here μ and η are positive constants and A and F are the same as above, see (13.3.2) and (13.3.13).

As in Chapter 5, we rewrite problem (13.3.20) and (13.3.21) in the form

$$\frac{d}{dt}U + \mathscr{A}U = \mathscr{B}(U), \quad U|_{t=0} = U_0,$$

where

$$U = \begin{pmatrix} Au \\ u_t \\ \theta \end{pmatrix}, \quad \mathscr{A} = A \begin{pmatrix} 0 & -1 & 0 \\ 1 & 0 & -\mu \\ 0 & \mu & \eta \end{pmatrix} \equiv A \cdot G, \quad \mathscr{B}(U) = \begin{pmatrix} 0 \\ F(u) \\ 0 \end{pmatrix}.$$

The characteristic polynomial $\chi(z)$ of the 3×3-matrix G has the form

$$\chi(z) = \det(z - G) = z^3 - \eta z^2 + (1 + \mu^2)z - \eta.$$

We consider the case

$$\frac{1}{3} < \rho_1 \le \frac{1+\mu^2}{\eta^2} \le \rho_2 < \infty. \tag{13.3.22}$$

Here ρ_1 and ρ_2 are constants. We can show (cf. [216] and [209]) that under this condition the equation $\chi(z) = 0$ has the simple roots z_i (one is positive (z_1) and two others are complex, $\bar{z}_2 = z_3$) such that

$$0 < z_1 < \eta, \quad z_2 + z_3 = \eta - z_1, \quad z_2 - z_3 = i \cdot \left(\frac{4\eta}{z_1} - (\eta - z_1)^2 \right)^{1/2}.$$

Moreover, there exist positive constants c_1 and c_2 depending on ρ_1, ρ_2, and η_0 such that for any $\eta \ge \eta_0 > 0$ we have the relations

$$c_1 \le z_1 \cdot \eta \le c_2, \quad 1 - \frac{c_2}{\eta^2} \le \frac{z_2 + z_3}{\eta} < 1, \quad c_1 \le \frac{|z_2 - z_3|}{\eta} \le c_2. \tag{13.3.23}$$

In order to diagonalize the matrix operator, [209], we introduce new variables:

$$y_1 = \frac{\mu\theta - (1 - z_2 z_3)Au + (z_2 + z_3)u_t}{(z_1 - z_2)(z_1 - z_3)},$$

$$y_2 = \frac{\mu\theta - (1 - z_1 z_3)Au + (z_1 + z_3)u_t}{(z_2 - z_1)(z_2 - z_3)},$$

$$y_3 = \frac{\mu\theta - (1 - z_1 z_2)Au + (z_1 + z_2)u_t}{(z_3 - z_1)(z_3 - z_2)}.$$

The inverse transformation takes the form

$$\begin{aligned} Au &= y_1 + y_2 + y_3 , \\ u_t &= -(z_1 y_1 + z_2 y_2 + z_3 y_3) , \\ \theta &= -\tfrac{1}{\mu} \left(Au + z_1^2 y_1 + z_2^2 y_2 + z_3^2 y_3 \right) . \end{aligned}$$

Introducing variables w_i by the formulas $y_i(t) = w_i(z_1 t)$ leads, after some calculations, to the system

$$\begin{cases} \frac{dw_1}{dt} + Aw_1 = K_1 F(A^{-1}(w_1 + w_2 + w_3)) \\ \frac{dw_2}{dt} + \frac{z_2}{z_1} Aw_2 = K_2 F(A^{-1}(w_1 + w_2 + w_3)) \\ \frac{dw_3}{dt} + \frac{z_3}{z_1} Aw_3 = K_3 F(A^{-1}(w_1 + w_2 + w_3)), \end{cases} \tag{13.3.24}$$

where

$$K_1 = \frac{z_2 + z_3}{z_1(z_1 - z_2)(z_1 - z_3)}, \quad K_2 = \frac{z_1 + z_3}{z_1(z_2 - z_1)(z_2 - z_3)},$$

and

$$K_3 = \frac{z_1 + z_2}{z_1(z_3 - z_1)(z_3 - z_2)} = \overline{K}_2.$$

Thus $W = (w_1; w_2; w_3)$ satisfies the equation

$$\frac{d}{dt} W + \mathscr{A} W = \mathscr{B}(W), \quad W|_{t=0} = U_0, \tag{13.3.25}$$

where

$$\mathscr{A} = A \cdot \begin{pmatrix} 1 & 0 & 0 \\ 0 & \frac{z_2}{z_1} & 0 \\ 0 & 0 & \frac{z_3}{z_1} \end{pmatrix}, \quad \mathscr{B}(W) = \begin{pmatrix} K_1 \\ K_2 \\ K_3 \end{pmatrix} \cdot F(A^{-1}(w_1 + w_2 + w_3)).$$

We consider problem (13.3.25) in the space $H = \mathscr{H} \times \hat{\mathscr{H}} \times \hat{\mathscr{H}}$, where $\hat{\mathscr{H}}$ is the complexification of $\mathscr{H}$.

Let $K^2 = K_1^2 + |K_2|^2 + |K_3|^2 \equiv K_1^2 + 2|K_2|^2$. It is clear that

$$\|\mathscr{B}(W) - \mathscr{B}(W^*)\|_H \le \sqrt{3} K M_1 \|W - W^*\|_H.$$

For every pair of integers $N_1 \ge 0$ and $N_2 \ge 0$ we introduce the projectors

$$\mathscr{P} = \begin{pmatrix} P_{N_1} & 0 & 0 \\ 0 & P_{N_2} & 0 \\ 0 & 0 & P_{N_2} \end{pmatrix} \quad \text{and} \quad \mathscr{Q} = I - \mathscr{P}, \tag{13.3.26}$$

where, as above, P_N is the orthoprojector onto $\mathrm{Span}\{e_k : k = 1, 2, \ldots, N\}$ for $N \ge 1$ and $P_0 = 0$.

Similar to Theorem 13.3.9 we are in position to prove the following assertion.

13.3.13. Theorem. *Assume that there exist numbers $0 < \lambda^- < \lambda^+$ such that*

$$\lambda^+ - \lambda^- \ge \frac{4\sqrt{3} K M_1}{q}, \quad 0 < q < 2/3, \tag{13.3.27}$$

and integers $N_1 \ge 0$ and $N_2 \ge 0$ such that

$$\omega_{N_1} \le \lambda^-, \ \omega_{N_1+1} \ge \lambda^+, \qquad (13.3.28)$$
$$\tfrac{Re z_2}{z_1}\omega_{N_2} \le \lambda^-, \ \tfrac{Re z_2}{z_1}\omega_{N_2+1} \ge \lambda^+$$

(here if $N=0$ we suppose $\omega_N=0$). Then problem (13.3.24) *and hence* (13.3.20) *and* (13.3.21) *possess an exponentially complete inertial manifold of the form* (13.2.1) *with the projector $\mathscr{P}$ given by* (13.3.26).

Proof. Under conditions (13.3.28) it is easy to prove the dichotomy estimates (13.2.5) with $\mathscr{P}$ and $\mathscr{Q}$ given by (13.3.26). Therefore, by (13.3.27) we apply Theorem 13.2.6. □

Assuming that for some N the following relations hold,

$$\omega_{N+1} - \omega_N \ge \frac{4\sqrt{3}KM_1}{q}, \quad 0<q<\frac{2}{3}, \quad \frac{Re z_2}{z_1} \ge \frac{\omega_{N+1}}{\omega_1}, \qquad (13.3.29)$$

then in Theorem 13.3.13 we choose $\lambda^- = \omega_N$, $\lambda^+ = \omega_{N+1}$, $N_1 = N$ and $N_2 = 0$. In that case Theorem 13.3.13 implies the following assertion.

13.3.14. Corollary. *Under condition* (13.3.29) *problem* (13.3.24) *(and hence problem* (13.3.20) *and* (13.3.21)*) possesses an exponentially complete inertial manifold of the form* (13.2.1) *with the projector $\mathscr{P} = \mathrm{diag}\{P_N, 0, 0\}$, where P_N is the orthoprojector onto Span$\{e_k \ : \ k = 1,2,\dots,N\}$.*

Under conditions (13.3.22), using properties (13.3.23), it is easy to prove that the constant K as a function of the variables μ and η is bounded when $\eta \to \infty$ and $(Re z_2)/z_1 \ge a_0\eta^2$ for some positive constant a_0. The latter does not depend on μ and η. Therefore in this case we obtain the following assertion.

13.3.15. Corollary. *For any $\frac{1}{3} < \rho_1 < \rho_2 < \infty$ there exist $K_0 > 0$ and $a_0 > 0$ such that the spectral gap condition*

$$\omega_{N+1} - \omega_N \ge \frac{4\sqrt{3}K_0M_1}{q} \quad \textit{for some} \quad 0<q<2/3 \ \textit{and} \ N \in \mathbb{N}$$

implies that problem (13.3.20) *and* (13.3.21) *possesses an exponentially complete inertial manifold for any $(\mu;\eta)$ from the set*

$$\left\{(\mu;\eta) \ : \ \eta \ge \sqrt{\frac{\omega_{N+1}}{a_0\omega_1}}, \quad \rho_1 \le \frac{1+\mu^2}{\eta^2} \le \rho_2\right\}.$$

Locally Lipshitz case—Applications to von Karman evolutions with thermoelasticity

In this case Theorem 13.3.13 implies the following result for problem (13.1.8) with boundary and initial data (13.1.3) and (13.1.9).

13.3.16. Theorem. *Consider von Karman evolutions with thermoelasticity given by equations* (13.1.3), (13.1.8) *and* (13.1.9). *Let* $\Omega = (0,l_1) \times (0,l_2)$ *be a rectangle with* l_1/l_2 *rational. Assume that* $p \in L^2(\Omega)$, $F_0 \in W^2_\infty(\Omega)$, $L \equiv 0$ *and* (13.3.22) *holds for some* ρ_1 *and* ρ_2. *Then problem* (13.1.3), (13.1.8) *and* (13.1.9) *possesses a locally invariant, exponentially complete inertial manifold if either one of the following conditions is satisfied.*

(i) $(Re\, z_2)/z_1$ *is a rational number.*
(ii) η *(hence* μ*) is large enough.*

Proof. As in Theorem 13.3.12 we use the special gap theorem (see Proposition 13.3.7). Theorems 13.1.10 and 13.3.13 and Corollary 13.3.15 are also used. □

13.3.17. Remark. Concluding this chapter, we note that for all three cases of von Karman evolutions we can construct locally invariant, exponentially complete inertial manifolds $\mathscr{M}$ with a stronger invariance property: there exists $R > 0$ such that the set $\{U : \|U\|_H \leq R\}$ is absorbing and the relations $\|U\|_H \leq R$ and $U \in \mathscr{M}$ imply that $S_t U \in \mathscr{M}$ for all $t \geq 0$. This construction relies on the existence of an absorbing invariant set determined by some Lyapunov-type function (see Theorems 13.1.8, 13.1.9, and 13.1.10). ■

Appendix A
Jacobians and Compensated Compactness, Compactness of Vector Functions, and Sedenko's Method for Uniqueness

In this appendix we present (i) a result concerning the smoothness of Jacobians on $\mathbb{R}^2$, (ii) Aubin-type compactness theorem, and also (iii) the main idea behind Sedenko's method (see, e.g., [255]–[257]) for uniqueness of weak solutions to the von Karman model without rotational inertia.

A.1 Jacobian regularity and compensated compactness

In this section we provide proof of Lemma 1.1.7 concerning regularity[1] properties of Jacobians in $\mathbb{R}^2$. This result exploits certain cancellation properties that follow from the structure of Jacobian operation. More specifically, we prove the following assertion.

A.1.1. Lemma. *Let* $W = (u; v) \in \left[H^1(\mathbb{R}^2)\right]^2$. *Then*

$$\det\{\nabla W\} \equiv u_{x_1} v_{x_2} - u_{x_2} v_{x_1} \in H_1(\mathbb{R}^2) \tag{A.1.1}$$

and

$$\|\det\{\nabla W\}\|_{H_1(\mathbb{R}^2)} \le C \|u\|_{H^1(\mathbb{R}^2)} \|v\|_{H^1(\mathbb{R}^2)}. \tag{A.1.2}$$

Here $H^1(\mathbb{R}^2)$ *is the first order Sobolev space and* $H_1(\mathbb{R}^2)$ *is the homogeneous Hardy space (see Definition* 1.1.5 *in Section* 1.1.3*).*

Proof. This lemma follows from [89, Theorem II]. However, in order to make the exposition self-contained we provide the corresponding argument which is adapted from [89].

Let $V = (v_{x_2}; -v_{x_1})$. Then in the sense of distributions we have

$$\operatorname{div} V = 0 \quad \text{and} \quad \det\{\nabla W\} = \operatorname{div}\{uV\}.$$

[1] The regularity proclaimed by Lemma 1.1.7 (and Lemma A.1.1 below) was used for the proof of smoothness properties of the von Karman bracket.

I. Chueshov, I. Lasiecka, *Von Karman Evolution Equations*,
Springer Monographs in Mathematics, DOI 10.1007/978-0-387-87712-9,

To establish (A.1.1) and (A.1.2) we use the convolutional definition of $H^1(\mathbb{R}^2)$ (see Definition 1.1.5 in Section 1.1.3). For this, we consider a function $h \in C_0^\infty(\mathbb{R}^2)$, $h \geq 0$, and $\operatorname{supp} h \subset B_1^0$, where $B_r^a = \{y \in \mathbb{R}^2 : |y-a| < r\}$. We also denote $h_t(x) = t^{-2}h(x/t)$.

Integrating by parts one obtains:

$$(h_t * \det\{\nabla W\})(x) = \frac{1}{t^2}\int_{\mathbb{R}^2} h\left(\frac{x-y}{t}\right) \operatorname{div}_y\{u(y)V(y)\}\,dy$$
$$= \frac{1}{t^3}\int_{B_t^x} \nabla h\left(\frac{x-y}{t}\right)\left[u(y) - \int_{B_t^x}^* u(z)dz\right] V(y)dy.$$

Here and below we use the notation $\int_B^* = [\operatorname{mes}(B)]^{-1}\int_B$. Consequently

$$|(h_t * \det\{\nabla W\})(x)| \leq \frac{C}{t}\left[\int_{B_t^x}^*\left|u(y) - \int_{B_t^x}^* u(z)dz\right|^r dy\right]^{1/r}\left[\int_{B_t^x}^* |V(y)|^p\,dy\right]^{1/p},$$

where $r^{-1} + p^{-1} = 1$ and $1 < p < 2$. Because by (1.1.3) $W_q^1(\mathbb{R}^2) \subset L_r(\mathbb{R}^2)$ for $2/q = 1 + 2/r$, a Poincaré-type inequality yields

$$\frac{1}{t}\left[\int_{B_t^x}^*\left|u(y) - \int_{B_t^x}^* u(z)dz\right|^r dy\right]^{1/r} \leq C\left[\int_{B_t^x}^* |\nabla u(y)|^q\,dy\right]^{1/q}.$$

Thus we obtain

$$|(h_t * \det\{\nabla W\})(x)| \leq C\left[\int_{B_t^x}^* |\nabla u(y)|^q\,dy\right]^{1/q}\left[\int_{B_t^x}^* |\nabla v(y)|^p\,dy\right]^{1/p},$$

where $1 < p < 2$, $q^{-1} + p^{-1} = 3/2$. By choosing $p = q = 4/3$ we get

$$|(h_t * \det\{\nabla W\})(x)| \leq C\left[M\left(|\nabla u|^{4/3}\right)\right]^{3/4}\left[M\left(|\nabla v|^{4/3}\right)\right]^{3/4}, \tag{A.1.3}$$

where $M(f)$ denotes the maximal function,

$$M(f)(x) = \sup_{t>0}\left\{\int_{B_t^x}^* |f(y)|\,dy\right\}.$$

By the maximal function theorem (see, e.g., [264, Chapter 1, Theorem 1]) we know that $M(f)$ is bounded on $L_p(\mathbb{R}^2)$, $p > 1$, and there exists a positive constant A_p such that the estimate $\|M(f)\|_{L_p} \leq A_p\|f\|_{L_p}$ holds. Therefore, because $|\nabla u|^{4/3} \in L_{3/2}(\mathbb{R}^2)$, we have that $M\left(|\nabla u|^{4/3}\right) \in L_{3/2}(\mathbb{R}^2)$ and

$$\int_{\mathbb{R}^2}\left|M\left(|\nabla u|^{4/3}\right)\right|^{3/2} dx \leq C\int_{\mathbb{R}^2} |\nabla u|^2 dx.$$

Thus by (A.1.3)

$$\sup_{t>0} |(h_t * \det\{\nabla W\})(x)| \in L_1(\mathbb{R}^2)$$

and

$$\left\| \sup_{t>0} |(h_t * \det\{\nabla W\})(x)| \right\|_{L_1(\mathbb{R}^2)} \leq C\|\nabla u\|_{L_2(\mathbb{R}^2)} \|\nabla v\|_{L_2(\mathbb{R}^2)}.$$

Hence by the convolutional definition (see the second bullet in Definition 1.1.5) we obtain (A.1.1) and (A.1.2). □

A.2 Compactness theorem for vector-valued functions

Our goal in this section is to prove Theorem 1.1.8 on the compactness of vector-valued functions. This theorem easily follows from the result on compactness given in Theorem A.2.4 below, which is the main result of this section.

We first recall the definition and the criterion for compactness of sets in Banach spaces (see, e.g., [93, 280, 281]).

A set F in a Banach space X is said to be *compact* if for every family of open sets covering F there exists a finite subfamily covering F. A set is *relatively compact* if its closure is compact. Given $\varepsilon > 0$ a set $C \subset X$ is said to be ε*-net* for a set $M \subset X$ if $M \subset \cup_{a \in C}\{x : \|x-a\|_X \leq \varepsilon\}$.

Our main general tool used below is the following well-known equivalent conditions for compactness (see, e.g., [93] or [281]).

A.2.1. Proposition. *Let K be a set in a Banach space X. Then the following statements are equivalent.*

- *The set K is relatively compact.*
- *For every $\varepsilon > 0$ there exists a finite ε-net for K.*
- *For every $\varepsilon > 0$ there exists a relatively compact ε-net for K.*
- *Any sequence of elements from K contains a subsequence that converges to some element of X.*

We use the following compactness criteria applicable to vector-valued continuous functions.

A.2.2. Lemma (Arzelá–Ascoli Theorem, I). *Let X be a Banach space. A set $F \subset C(a,b;X)$ is relatively compact if and only if*

(i) *$F(t) := \{f(t) : f \in F\}$ is relatively compact in X for each $t \in [a,b]$.*
(ii) *F is* equicontinuous; *that is, for any $\varepsilon > 0$ there exists $\delta > 0$ such that*

$$\|f(t) - f(s)\|_X \leq \varepsilon \quad \textit{for any } f \in F \textit{ and } t,s \in [a,b] \textit{ such that } |t-s| \leq \delta.$$

Proof. If F is relatively compact in $C(a,b;X)$, then it is obvious that $F(t)$ is also relatively compact in X for every fixed t and hence (i) holds. By Proposition A.2.1 the relative compactness of F implies the existence of finite ε-net: for any $\varepsilon > 0$ there exists a family $\{f_k : k = 1,2,\dots,N\}$ in $C(a,b;X)$ such that

$$\forall f \in F \quad \exists f_k \text{ such that } \max_{t\in[a,b]} \|f(t) - f_k(t)\|_X \leq \varepsilon.$$

This implies that

$$\|f(t) - f(s)\|_X \leq 2\varepsilon + \sum_{k=1}^{N} \|f_k(t) - f_k(s)\|_X \quad \text{for every } f \in F.$$

This relation makes it possible to conclude that F is equicontinuous. Thus both properties (i) and (ii) are valid.

Let (i) and (ii) be in force. We denote $t_k = a + (b-a)kN^{-1}$ for $k = 0, 1, \dots, N$. For each $f \in F$ de define $f_N(t) \in C(a,b;X)$ as a linear spline:

$$f_N(t) = f(t_k) + \frac{t - t_k}{t_{k+1} - t_k}[f(t_{k+1}) - f(t_k)] \quad \text{for } t \in [t_k, t_{k+1}].$$

It is clear from (i) that for every fixed N the set $F_N = \{f_N : f \in F\}$ is relatively compact in $C(a,b;X)$. Equicontinuity (ii) implies that for every $\varepsilon > 0$ there exists N_0 such that

$$\|f - f_N\|_{C(a,b;X)} \leq \varepsilon \quad \text{for every } N \geq N_0 \quad \text{and} \quad f \in F.$$

This means that for any $\varepsilon > 0$ the set F possesses a relatively compact ε-net. Thus by Proposition A.2.1 F is relatively compact. □

The following version of the Arzelá–Ascoli theorem is a relaxed form of the previous lemma. Indeed, instead of assuming compactness for every fixed t, only the integral form of the compactness property is required.

A.2.3. Lemma (Arzelá–Ascoli Theorem, II). *Let X be a Banach space. A set $F \subset C(a,b;X)$ is relatively compact if and only if*

(i) $\int_{t_1}^{t_2} F := \left\{ \int_{t_1}^{t_2} f(t)dt \,:\, f \in F \right\}$ *is relatively compact in X for each $t_1, t_2 \in [a,b]$.*
(ii) *F is equicontinuous.*

Proof. In the same way as in Lemma A.2.2 one can see that the compactness of F implies (i) and (ii).

Let (i) and (ii) be in force. For each $f \in F$ and $h > 0$ small enough we define

$$(M_h f)(t) = \frac{1}{h}\int_t^{t+h} f(\tau)d\tau, \quad t \in [a, b-h].$$

We denote $F_h = \{(M_h f : f \in F\}$. By (i) $F_h(t)$ is relatively compact in X for every $t \in [a, b-h]$. We also have that

$$(M_h f)(t+\delta) - (M_h f)(t) = \frac{1}{h}\int_t^{t+h} [f(\tau+\delta) - f(\tau)]\,d\tau.$$

By (ii) this implies that F_h is equicontinuous on $[a,b-h]$. Thus by Lemma A.2.2 F_h is relatively compact in $C(a,b-h;X)$. Because

$$(M_h f)(t) - f(t) = \frac{1}{h}\int_t^{t+h} [f(\tau) - f(t)]\,d\tau,$$

one can see from (ii) that for any $\varepsilon > 0$ there exists $h > 0$ such that

$$\|(M_h f)(t) - f(t)\|_{C(a,b-h;X)} \leq \varepsilon;$$

that is, F_h is a relatively compact ε-net for F restricted on $[a,b-h]$. By Proposition A.2.1 the set F is relatively compact in $C(a,b-h;X)$ for every fixed $h > 0$. Considering the averages

$$(\widetilde{M}_h f)(t) = \frac{1}{h}\int_{t-h}^{t} f(\tau)d\tau, \quad t \in [a+h,b],$$

by the same argument we can conclude that F is also relatively compact in $C(a+h,b;X)$ for every fixed $h > 0$. Now applying the sequence criterion in Proposition A.2.1 one can easily conclude that F is relatively compact in $C(a,b;X)$. □

Now we state the main result of this section.

A.2.4. Theorem. *Assume that $X \subset Y \subset Z$ is a triple of Banach spaces such that X is compactly embedded in Y.*

- *Let F be a bounded set in $L_p(a,b;X)$ for some $1 \leq p < \infty$ such that the set $\partial_t F := \{\partial_t f : f \in F\}$ is bounded in $L_q(a,b;Z)$ for some $q \geq 1$. Here $\partial_t f$ is the derivative in the distributional sense (see Section 1.1.4). Then F is relatively compact in $L_p(a,b;Y)$. If $q > 1$, then F is also relatively compact in $C(a,b;Z)$.*
- *If F is a bounded set in $L_\infty(a,b;X)$ and $\partial_t F$ is bounded in $L_r(a,b;Z)$ for some $r > 1$, then F is relatively compact in $C(a,b;Y)$.*

Particular cases of Theorem A.2.4 can be found in [3] and [98] (see also [220]). Almost in the same form as above, Theorem A.2.4 is stated in [263, Corollary 4]. We borrow the approach from [263] in order to prove the theorem.

Proof. We consider the case $q > 1$ only, which is relevant to the material presented in this book (for the case $q = 1$ we refer to [263]).

Lemma A.2.3 implies that in both cases of Theorem A.2.4 the set F is relatively compact in $C(a,b;Z)$. To improve this compactness we use the following assertion due to J. L. Lions [220].

A.2.5. Lemma. *Let $X \subset Y \subset Z$ be a triple of Banach spaces such that X is compactly embedded in Y. Then for every $\eta > 0$ there is C_η such that*

$$\|u\|_Y \leq \eta\|u\|_X + C_\eta\|u\|_Z \text{ for every } u \in X. \tag{A.2.1}$$

Proof. Assume that (A.2.1) is not true. Then there exists $\eta_0 > 0$ and a sequence $\{u_n\} \subset X$ such that

$$\|u_n\|_Y \geq \eta_0 \|u_n\|_X + n\|u_n\|_Z \text{ for every } n = 1,2,\ldots.$$

Thus for $v_n = u_n \|u_n\|_Y^{-1}$ we have that

$$\eta_0 \|v_n\|_X + n\|v_n\|_Z \leq 1 \text{ and } \|v_n\|_Y = 1 \text{ for } n = 1,2,\ldots.$$

Therefore the sequence $\{v_n\}$ is relatively compact in Y and $\|v_n\|_Z \to 0$ as $n \to \infty$. This implies that $\{v_n\}$ contains a subsequence which converges to 0 in Y. This contradicts the fact $\|v_n\|_Y = 1$ for all n. □

Now we return to the proof of Theorem A.2.4.

1. Because F is relatively compact in $C(a,b;Z)$, by Proposition A.2.1 for every $\varepsilon > 0$ there exists a finite ε-net $\{f_n\}$ for F with respect to $C(a,b;Z)$. Moreover, this ε-net $\{f_n\}$ can be chosen such that $f_n \in F$. Lemma A.2.5 and boundedness on F in $L_p(a,b;X)$ imply that

$$\|f - f_n\|_{L_p(a,b;Y)} \leq \eta + C_\eta(b-a,p)\|f - f_n\|_{C(a,b;Z)} \tag{A.2.2}$$

for every $\eta > 0$. Let $\delta = \eta + C_\eta(b-a,p)\varepsilon$. Relation (A.2.2) means that $\{f_n\}$ is a δ-net for F with respect to $L_p(a,b;Y)$. By choosing η and ε in an appropriate way we can make δ arbitrary small. Thus by Proposition A.2.1 F is relatively compact in $L_p(a,b;Y)$.

2. We first note that inasmuch as F is bounded in $L_\infty(a,b;X)$, by Lemma A.2.5 we have that

$$\|f(t) - f(s)\|_Y \leq \eta C_F + C_\eta \|f(t) - f(s)\|_Z, \quad f \in F,$$

for every $\eta > 0$ and $t,s \in [a,b]$, where C_F is a constant. This implies that $F \subset C(a,b;Y)$. The subsequent argument is the same as in the previous case. The only difference is that instead of (A.2.2) we use the relation

$$\|f - f_n\|_{C(a,b;Y)} \leq \eta + C_\eta(b-a,p)\|f - f_n\|_{C(a,b;Z)}$$

for every $\eta > 0$. □

A.3 Logarithmic Sobolev-type inequalities and uniqueness of weak solutions by Sedenko's method

Here we give an alternative proof of Hadamard well-posedness for weak solutions to problem (4.1.1) and (4.1.2) with clamped boundary conditions (4.1.4). This proof does not depend on sharp regularity properties of the Airy function stated in (1.4.22). For the uniqueness, we follow the approach and arguments given in [29]. Hadamard well-posedness then follows from the said uniqueness along with energetic considerations relevant to the model under consideration.

For the sake of simplicity we assume that $f \equiv F_0 \equiv 0$, $L = 0$. We also assume that the damping term is not presented in the equation (although the method allows consideration of linear damping). Thus we consider the following system:

$$\partial_t^2 u + \Delta^2 u - [u, v] = p(x), \quad x \in \Omega, t > 0, \tag{A.3.1}$$

$$u|_{\partial\Omega} = \frac{\partial u}{\partial n}|_{\partial\Omega} = 0, \quad u|_{t=0} = u_0(x), \quad \partial_t u|_{t=0} = u_1(x), \tag{A.3.2}$$

where $v = v(u)$ is defined as a solution of the problem

$$\Delta^2 v + [u, u] = 0, \quad v|_{\partial\Omega} = \frac{\partial v}{\partial n}|_{\partial\Omega} = 0. \tag{A.3.3}$$

Here Ω is a smooth bounded domain. It is assumed that $p(x) \in L_2(\Omega)$, $u_0(x) \in H_0^2(\Omega)$, and $u_1(x) \in L_2(\Omega)$ are known.

We recall (cf. Definition 4.1.18) that the function $u(x,t)$ is said to be a weak solution to problem (A.3.1)–(A.3.3) on the interval $[0,T]$, if

$$u(x,t) \in L_\infty(0,T; H_0^2(\Omega)) \quad \partial_t u(x,t) \in L_\infty(0,T; L_2(\Omega))$$

and the following properties are fulfilled.

- Equality (A.3.1) is satisfied in the sense of distributions (taking account of (A.3.3)).
- The vector-valued function $t \mapsto (u(t); \partial_t u(t)) \in H_0^2(\Omega) \times L_2(\Omega)$ is weakly continuous, and $u(0) = u_0$, $\partial_t u(0) = u_1$.

It is well known (see, e.g., Theorem 4.1.19 in Section 4.1) that system (A.3.1)–(A.3.3) has a weak solution for any interval $[0,T]$. These weak solutions can be constructed by the standard Galerkin method (see, e.g., [220] or [237] and also Remark 4.1.20). The construction relies on the energy a priori estimate and it requires minimal (and natural) assumptions concerning the initial data $u_0(x)$ and $u_1(x)$ (the energy of the initial state should be finite). The solutions constructed by Galerkin method satisfy the energy inequality of the form

$$\mathscr{E}(u(t), \partial_t u(t)) \le \mathscr{E}(u_0, u_1) \ \text{ for all } t \in [0,T], \tag{A.3.4}$$

where the energy $\mathscr{E}$ defined by the relation

$$\mathscr{E}(u, u_t) = \frac{1}{2}\left(\|u_t\|^2 + \|\Delta u\|^2\right) + \frac{1}{4}\|\Delta v(u)\|^2 - (u, p)$$

with $v(u)$ solving (A.3.3). As usual here and below $\|\cdot\|$ and $(\cdot,\cdot)$ stand for the norm and the inner product in $L_2(\Omega)$. The uniqueness of weak solutions to (A.3.1)-(A.3.3) follows from Theorem 4.1.19 which proof relies on a sharp regularity of the Airy function. More precisely, by Theorem 4.1.4 and Theorem 4.1.19 we have the following result.

A.3.1. Theorem. *Let $u_0 \in H_0^2(\Omega)$, $u_1 \in L_2(\Omega)$, and $p(x) \in L_2(\Omega)$. Then problem (A.3.1)–(A.3.3) has a unique weak solution $u(t)$ for any interval $[0,T]$. Moreover,*

- *The function $t \mapsto (u(t); \partial_t u(t))$ is strongly continuous in $H \equiv H_0^2(\Omega) \times L_2(\Omega)$.*
- *Weak solutions are strongly continuous in the space H with respect to initial data; that is, if $(u_{0n}; u_{1n}) \to (u_0; u_1)$ strongly in H, then for the corresponding solutions we have that $(u_n(t); \partial_t u_n(t)) \to (u(t); \partial_t u(t))$ strongly in H for every $t \in [0,T]$.*
- *The energy identity $\mathscr{E}(u(t), \partial_t u(t)) = \mathscr{E}((u_0, u_1)$ holds for all $t \in [0,T]$.*

This result means that problem (A.3.1)–(A.3.3) is Hadamard well-posed in $H = H_0^2(\Omega) \times L_2(\Omega)$ and gives a positive answer to the question posed by Vorovich [279] (see also the list of the problems posed by Lions in [220, Chapter 1]).

Our goal in this section is to provide an *alternative proof* of Theorem A.3.1 which does not depend on the regularity of the von Karman bracket related to Lizorkin and Hardy spaces and harmonic analysis methods (see (1.4.15) in Theorem 1.4.3). In fact, the uniqueness part is carried by using the method developed by Sedenko [255]–[257] for Marguerre–Vlasov equations arising in the theory of elastic shallow shells. We also note that the same method was recently applied for a class of nonlinear 2D Kirchhoff–Boussinesq models [72], for 2D Zakharov systems [81], and for some 2D Cahn–Hillard models with inertial terms [128]. Sedenko's method relies on the energy inequality in the negative spaces and on the estimates of the form:

$$\max_{x \in \Omega} |(T_N f)(x)| \le c_0 (\log N)^{1/2} \| f \|_1, \quad \| (I - T_N) f \| \le c_1 N^{-1} \| f \|_2$$

for certain sequences of operators T_N, $N = 2, 3, \ldots$. We show that in our case one can choose $T_N = P_N$, where P_N is the projector in $L_2(\Omega)$ onto the space spanned by the first N eigenvectors of the biharmonic operator Δ_D^2 with Dirichlet boundary conditions on $\partial\Omega$.

As for continuity of weak solutions with respect to time and initial data, we use the same idea as in [164] (see also [222] for linear problems) and rely on the energy identity. To prove this identity we take the advantage (see [128] and also [77]) of the reversibility of the dynamics.

In what follows we use several estimates for the von Karman bracket. However, we do not use estimate (1.4.15) which involves Lizorkin space $F_{1,2}^0(\Omega)$. More precisely, we only need the following bounds for the von Karman bracket, which are special cases of the estimate in (1.4.16) and (1.4.17) and have been well known for a long time (see, e.g., [46, 51, 56]).

A.3.2. Lemma. *The bracket $[u,v]$ defined by* (1.4.1) *possesses the following properties*

$$\| [u,v] \|_{-2} \le C \| u \|_{2-\beta} \cdot \| v \|_{1+\beta}, \quad 0 < \beta < 1,$$

and

$$\| [u,v] \|_{-1-\theta} \le C \| u \|_2 \cdot \| v \|_{2-\theta}, \quad 0 < \theta < 1. \tag{A.3.5}$$

Below we also rely on the following representations of the von Karman bracket:

$$[u,v] = \partial^2_{x_1}(u\cdot\partial^2_{x_2}v) + \partial^2_{x_2}(u\cdot\partial^2_{x_1}v) - 2\cdot\partial^2_{x_1x_2}(u\cdot\partial^2_{x_1x_2}v) \tag{A.3.6}$$

and

$$[u,v] = \partial_{x_1}(\partial_{x_1}u\cdot\partial^2_{x_2}v - \partial_{x_2}u\cdot\partial^2_{x_1x_2}v) + \partial_{x_2}(\partial_{x_2}u\cdot\partial^2_{x_1}v - \partial_{x_1}u\cdot\partial^2_{x_1x_2}v). \tag{A.3.7}$$

We denote by Δ_D^2 the biharmonic operator with Dirichlet boundary conditions on $\partial\Omega$. It is well known that Δ_D^2 is an isomorphism from $H^s(\Omega)\cap H_0^2(\Omega)$ onto $H^{s-4}(\Omega)$ for $s\ge 2$ and, therefore,

$$G \equiv (\Delta_D^2)^{-1} : H^s(\Omega) \to H^{s+4}(\Omega)\cap H_0^2(\Omega), \quad s\ge -2. \tag{A.3.8}$$

We note also that the norm in $H_0^s(\Omega)$ (and in $H^{-s}(\Omega)$ if $s<0$) can be defined by the formula

$$\|\cdot\|_s = \|(\Delta_D^2)^{s/4}\cdot\| \quad \text{for } -2\le s\le 2 \text{ and } s\ne \pm 1/2, \pm 3/2. \tag{A.3.9}$$

Let $\{e_k\}$ be the basis in $L^2(\Omega)$ of eigenvectors of operator Δ_D^2 and $\{\lambda_k\}$ be the corresponding eigenvalues:

$$\Delta_D^2 e_k = \lambda_k e_k, \quad k=1,2,....; \quad 0<\lambda_1\le\lambda_2\le\cdots.$$

Below we denote by P_N the projector in $L^2(\Omega)$ onto $\mathrm{Span}\{e_1,e_2,\dots,e_N\}$.

A.3.3. Lemma. *Let $f(x)\in H_0^1(\Omega)$. Then there exists $N_0>0$ such that*

$$\max_{x\in\Omega}|(P_N f)(x)| \le C\cdot\{\log(1+\lambda_N)\}^{1/2}\,\|f\|_1\,. \tag{A.3.10}$$

for all $N\ge N_0$. The constant C does not depend on N.

Proof. Let $\phi\in C_0^\infty(\mathbb{R}^2)$ with $\operatorname{supp}\phi\subset\Omega$. Then

$$\max_{x\in\Omega}|\phi(x)| \le \frac{1}{2\pi}\int_{\mathbb{R}^2}|\hat\phi(k)|dk,$$

where

$$\hat\phi(k) \equiv \mathscr{F}[\phi](k) = \frac{1}{2\pi}\int_{\mathbb{R}^2}\phi(x)\exp\{-ikx\}dx$$

is the Fourier transform of $\phi(x)$. Therefore

$$\max_{x\in\Omega}|\phi(x)| \le \frac{1}{2\pi}\left(\int_{\mathbb{R}^2}(1+k^2)^s\,|\hat\phi(k)|^2\,dk\right)^{1/2}\cdot\left(\int_{\mathbb{R}^2}(1+k^2)^{-s}dk\right)^{1/2}$$

for $s>1$. Using this inequality along with the density of $C_0^\infty(\Omega)$ in $H_0^s(\Omega)$, we conclude that

$$\max_{x\in\Omega}|g(x)| \le C\sigma^{-1/2}\,\|g\|_{1+\sigma}, \quad 0<\sigma<1/2, \tag{A.3.11}$$

for any $g(x) \in H^{1+\sigma}(\Omega) \cap H_0^1(\Omega) \equiv H_0^{1+\sigma}(\Omega)$. Therefore (A.3.9) implies

$$\max_{x\in\Omega} |(P_N f)(x)| \le C\sigma^{-1/2}\lambda_N^{\sigma/4} \| (\Delta_D^2)^{1/4} f \| \le C\sigma^{-1/2}\lambda_N^{\sigma/4} \| f \|_1 .$$

If we choose $\sigma = [\log(1+\lambda_N)]^{-1}$ we obtain the conclusion. □

The following remark provides us with another (equivalent) form of the inequality in (A.3.10).

A.3.4. Remark (Brésis–Gallouet inequality). It is clear from (A.3.11) that

$$\max_{x\in\Omega} |((I-P_N)f)(x)| \le C\sigma_*^{-1/2}\lambda_N^{-(1-\sigma_*)/4} \| (I-P_N)(\Delta_D^2)^{1/2} f \| .$$

for every $0 < \sigma_* < 1/2$. Therefore choosing $\sigma_* = 1/4$ yields

$$\max_{x\in\Omega} |((I-P_N)f)(x)| \le C\lambda_N^{-3/16} \| f \|_2 .$$

Thus, from

$$\max_{x\in\Omega} |f(x)| \le \max_{x\in\Omega} |(P_N f)(x)| + \max_{x\in\Omega} |((I-P_N)f)(x)|,$$

and from the statement of Lemma A.3.3 we have that

$$\max_{x\in\Omega} |f(x)| \le C_1 \{\log(1+\lambda_N)\}^{1/2} \| f \|_1 + C_2\lambda_N^{-3/16} \| f \|_2 \tag{A.3.12}$$

for every function $f \in H_0^2(\Omega)$ and for every $N \ge N_0$. Because $\lambda_N \sim c_0 N^2$ as $N \to \infty$, one can see from the latter inequality[2] that there exist positive constants c_1 and c_2 such that

$$\max_{x\in\Omega} |f(x)| \le c_1 \{\log(1+\rho)\}^{1/2} \| f \|_1 + c_2(1+\rho)^{-1} \| f \|_2 \tag{A.3.13}$$

for every $f \in H_0^2(\Omega)$ and for every parameter $\rho \ge 0$. Thus we arrive at the Brésis–Gallouet type of inequality (see [35]).

On the other hand, if we substitute in (A.3.13) $P_N f$ instead of f, then using the fact that $\|P_N f\|_2 \le C\lambda_N^{1/4}\|f\|_1$ and choosing $\rho = \lambda_N^{1/4}$ we can easily arrive at (A.3.10). Thus inequalities (A.3.13) and (A.3.10) are equivalent. ■

A.3.5. Lemma. *Let $f(x) \in H^\sigma(\Omega)$ for $0 < \sigma \le 1$. Then*

$$\| f \|_{L_{2p}(\Omega)} \le C \left(\pi \frac{p-1}{\sigma p - p + 1} \right)^{\frac{p-1}{2p}} \| f \|_\sigma \quad \text{for all } 1 < p < \frac{1}{1-\sigma}. \tag{A.3.14}$$

---

[2] It is worth mentioning that the same conclusion can be made when in (A.3.12) we have $\lambda_N^{-\beta}$ for some positive β instead of $\lambda_N^{-3/16}$. The property that $\lambda_N \sim N^2$ is also not essential. To obtain (A.3.13) we only need that $\lambda_N \to \infty$ and $\lambda_N \lambda_{N+1}^{-1} \sim c_0 \ne 0$ as $N \to \infty$.

Proof. Let $g(x)$ be the extension of $f(x)$ on $\mathbb{R}^2$ such that

$$c_1 \| f \|_\sigma \leq \| g \|_\sigma \leq c_2 \| f \|_\sigma, \quad 0 < \sigma \leq 1.$$

Now we recall the Hausdorff–Young inequality (see, e.g., [275, Theorem 1.18.8]) which states that Fourier transforms $\mathscr{F}$ and $\mathscr{F}^{-1}$ are bounded from $L_r(\mathbb{R}^d)$ into $L_{r_*}(\mathbb{R}^d)$ for $1 \leq r \leq 2$, $r^{-1} + r_*^{-1} = 1$, and

$$\|\mathscr{F}\|_{\mathscr{L}(L_r(\mathbb{R}^d), L_{r_*}(\mathbb{R}^d))} \leq (2\pi)^{-d(2-r)/r} \leq 1, \quad 1 \leq r \leq 2,\ r^{-1} + r_*^{-1} = 1$$

(with the same estimate for $\mathscr{F}^{-1}$). This inequality yields that

$$\| f \|_{L_{2p}(\Omega)} \leq \| g \|_{L_{2p}(\mathbb{R}^2)} \leq \| \hat{g} \|_{L_{\tilde{p}}(\mathbb{R}^2)}, \quad (2p)^{-1} + \tilde{p}^{-1} = 1, \quad p > 1,$$

where $\hat{g}$ is the Fourier transform of g. Hölder's inequality then implies

$$\| \hat{g} \|_{L_{\tilde{p}}(\mathbb{R}^2)} \leq \left(\int_{\mathbb{R}^2} (1+k^2)^\sigma \mid \hat{g}(k) \mid^2 dk \right)^{1/2} \left(\int_{\mathbb{R}^2} (1+k^2)^{-\tilde{\sigma}} dk \right)^{(2-\tilde{p})/(2\tilde{p})},$$

where $\tilde{\sigma} = \tilde{p}\sigma \cdot (2-\tilde{p})^{-1}$. A simple calculation then gives (A.3.14). □

A.3.6. Lemma. *Let $f(x) \in L_2(\Omega)$ and $g(x) \in H^1(\Omega)$. Then there exists $N_0 > 0$ such that*

$$\| (P_N f) \cdot g \| \leq C \{\log(1+\lambda_N)\}^{1/2} \| f \| \cdot \| g \|_1 . \tag{A.3.15}$$

for all $N \geq N_0$. The constant C does not depend on N.

Proof. The Hölder inequality gives

$$\| (P_N f) \cdot g \| \leq \| P_N f \|_{L_{2/(1-\theta)}(\Omega)} \cdot \| g \|_{L_{2/\theta}(\Omega)}, \quad 0 < \theta < 1. \tag{A.3.16}$$

Using Lemma A.3.5 for $p = (1-\theta)^{-1}$ and $\sigma = 2\theta$ we have

$$\| P_N f \|_{L_{2/(1-\theta)}(\Omega)} \leq C \| P_N f \|_{2\theta} \leq C \lambda_N^{\theta/2} \| f \|,$$

when $0 < \theta < 1/4$. If we apply Lemma A.3.5 with $p = \theta^{-1}$ and $\sigma = 1$ we obtain

$$\| g \|_{L_{2/\theta}(\Omega)} \leq C \left(\pi \frac{1-\theta}{\theta} \right)^{(1-\theta)/2} \| g \|_1, \quad 0 < \theta < 1.$$

Consequently (A.3.16) implies

$$\| (P_N f) \cdot g \| \leq C \theta^{-1/2} \lambda_N^{\theta/2} \| f \| \cdot \| g \|_1 .$$

If we set $\theta = \{\log(1+\lambda_N)\}^{-1}$ we obtain (A.3.15). □

We also need the following result (cf. Lemma 1.4.1).

A.3.7. Lemma. *Let $u \in H^{\beta}(\Omega)$ and $v \in H^{1-\beta}(\Omega)$, where $0 < \beta < 1$. Then*

$$\| u \cdot v \| \leq C \| u \|_{\beta} \cdot \| v \|_{1-\beta} \tag{A.3.17}$$

and

$$\| u \cdot v \|_{-1+\beta} \leq C \| u \|_{\beta} \cdot \| v \| . \tag{A.3.18}$$

Proof. Estimate (A.3.17) follows from the Hölder inequality and the continuity of the embedding $H^{1-\delta}(\Omega) \subset L_{2/\delta}(\Omega)$ for $0 < \delta \leq 1$, which also implies that we have a continuous embedding $L_{2/(2-\delta)}(\Omega) \subset H^{-1+\delta}(\Omega)$. Therefore using Hölder's inequality we have

$$\| u \cdot v \|_{-1+\beta} \leq C \ \| u \cdot v \|_{L_{2/(2-\beta)}} \leq C \| u \|_{L_{2p/(2-\beta)}} \cdot \| v \|_{L_{2q/(2-\beta)}},$$

where $p^{-1} + q^{-1} = 1$. Setting $q = 2 - \beta$ and $p = (2-\beta)(1-\beta)^{-1}$, we obtain (A.3.18) from the embedding result: $H^{\beta}(\Omega) \subset L_{2/(1-\beta)}(\Omega)$. □

Proof of Theorem A.3.1.

Step 1: Uniqueness. Let $u(t)$ be a weak solution on the interval $[0,T]$. Then (A.3.1) and Lemma A.3.2 imply that

$$\partial_t^2 u(x,t) \in L_{\infty}(0,T;H^{-2}(\Omega)).$$

Therefore by interpolation we can conclude that $u(t)$ and $\partial_t u(t)$ are strongly continuous functions with values in $H_0^1(\Omega)$ and $H^{-1}(\Omega)$, respectively.

Let $u_1(t)$ and $u_2(t)$ be weak solutions of the problem (A.3.1)–(A.3.3) and $u(t) = u_1(t) - u_2(t)$. Then $u_N(t) = P_N u(t)$ is a solution of the linear problem

$$\partial_t^2 w + \Delta^2 w = (P_N M)(x,t), \quad x \in \Omega, t > 0, \tag{A.3.19}$$

$$w|_{\partial\Omega} = \frac{\partial w}{\partial n}|_{\partial\Omega} = 0, \quad w|_{t=0} = 0, \quad \partial_t w|_{t=0} = 0.$$

Here as above P_N is the projector in $L_2(\Omega)$ on the space spanned by the first N eigenvectors of the biharmonic operator with Dirichlet boundary conditions and

$$M(x,t) \equiv M(t) = [u_1(t), v(u_1(t))] - [u_2(t), v(u_2(t))],$$

where $v = v(u)$ is defined from u by (A.3.3). Using the multiplier $P_N(\Delta_D^2)^{-1/2} u_t$ in (A.3.19) from relation (A.3.9) we obtain that

$$\| P_N \partial_t u(t) \|_{-1}^2 + \| P_N u(t) \|_1^2 \leq C \int_0^t \| P_N M(\tau) \|_{-1} \| P_N \partial_t u(\tau) \|_{-1} \, d\tau$$

for all $t \in [0,T]$. After the limit transition $N \to \infty$, this yields that the difference $u(t)$ of two weak solutions possesses the property

$$\| \partial_t u(t) \|_{-1}^2 + \| u(t) \|_1^2 \le C \int_0^t \| M(\tau) \|_{-1} \| \partial_t u(\tau) \|_{-1} \, d\tau. \tag{A.3.20}$$

The following lemmas make it possible to estimate the quantity $\|M(t)\|_{-1}$.

A.3.8. Lemma. *Let u_1 and u_2 belong to $H_0^2(\Omega)$ and $\| u_j \|_2 \le R$ for some $R > 0$. Then for some $\beta > 0$ we have*

$$\| [u_1, v(u_1) - v(u_2)] \|_{-1} \le C_1 \{\log(1+\lambda_N)\} \| u_1 - u_2 \|_1 + C_2 \lambda_{N+1}^{-\beta}, \tag{A.3.21}$$

where the constants C_1 and C_2 depend on R and β only.

Proof. It follows from (A.3.7) that $[u,v]$ is a sum of terms of the form $w = D(D^2 u \cdot Dv)$, where D and D^2 are certain differential operations with constant coefficients of first and second order, respectively. Consequently,

$$\| w \|_{-1} \le C_R \left(\max_{x\in\Omega} |(P_N Dv)(x)| + \max_{x\in\Omega} |(Q_N Dv)(x)| \right), \tag{A.3.22}$$

where $Q_N = I - P_N$. Let $v = v(u_1) - v(u_2)$. Then Lemma A.3.2 and property (A.3.8) imply that $v \in H_0^2(\Omega) \cap H^{2+\delta}(\Omega)$ for any $\delta < 1$. Therefore we have $Dv \in H_0^{1+\delta}(\Omega)$ for any $\delta < 1/2$. Consequently Lemma A.3.3 implies that

$$\max_{x\in\Omega} |(P_N Dv)(x)| \le C\{\log(1+\lambda_N)\}^{1/2} \| v \|_2, \quad N \ge N_0. \tag{A.3.23}$$

Because $H_0^{1+\delta}(\Omega) \subset L^\infty(\Omega)$ for $\delta > 0$, from (A.3.11) and (A.3.14) we have

$$\begin{aligned} \max_{x\in\Omega} |(Q_N Dv)(x)| &\le C_\beta \| \left(\Delta_D^2\right)^{1/4+\beta} Q_N Dv \| \\ &\le C_\beta \lambda_{N+1}^{-\beta} \| \left(\Delta_D^2\right)^{1/4+2\beta} Dv \| \le C_\beta \lambda_{N+1}^{-\beta} \| v \|_{2+8\beta} \end{aligned} \tag{A.3.24}$$

for $0 < \beta < 1/16$. Because $v = -\left(\Delta_D^2\right)^{-1} ([u, u_1 + u_2])$, it follows from (A.3.8) and (A.3.5) that $\| v \|_{2+8\beta} \le C_R$ for $0 < \beta < 1/16$. Therefore (A.3.22)–(A.3.24) imply that

$$\| [u_1, v(u_1) - v(u_2)] \|_{-1} \le C_R \left(\{\log(1+\lambda_N)\}^{1/2} \| v(u_1) - v(u_2) \|_2 + \lambda_{N+1}^{-\beta} \right) \tag{A.3.25}$$

for some $\beta > 0$. It follows from (A.3.8) that

$$\| v(u_1) - v(u_2) \|_2 \le C(\| [P_N u, u_1 + u_2] \|_{-2} + \| [Q_N u, u_1 + u_2] \|_{-2}), \tag{A.3.26}$$

where $u = u_1 - u_2$. Lemma A.3.2 gives that

$$\| [Q_N u, u_1 + u_2] \|_{-2} \le C \| Q_N u \|_{2-4\beta} \cdot \| u_1 + u_2 \|_{1+4\beta}, \quad 0 < \beta < 1/4.$$

Therefore as above we can conclude that

$$\| [Q_N u, u_1 + u_2] \|_{-2} \le C_{R,\beta} \lambda_{N+1}^{-\beta}, \quad 0 < \beta < 1/4. \tag{A.3.27}$$

Using (A.3.6) and Lemma A.3.3 we obtain

$$\| [P_N u, u_1 + u_2] \|_{-2} \le C_R \{\log(1 + \lambda_N)\}^{1/2} \| u \|_1 . \tag{A.3.28}$$

Thus the inequalities (A.3.25)–(A.3.28) imply (A.3.21). □

A.3.9. Lemma. *Let u_1 and u_2 belong to $H_0^2(\Omega)$ and $\| u_j \| \le R$ for some $R > 0$. Then for some $\beta > 0$ we have*

$$\| [u_1 - u_2, v(u_2)] \|_{-1} \le C_1 \{\log(1 + \lambda_N)\} \| u_1 - u_2 \|_1 + C_2 \lambda_{N+1}^{-\beta} \tag{A.3.29}$$

for $N \ge N_0$, where the constants C_1 and C_2 depend on R and β only.

Proof. Let $u = u_1 - u_2$. From (A.3.7) it follows that the quantity $[u, v(u_2)]$ can be written as the sum of terms of the form

$$w = D\{Du \cdot D^2 G[D(Du_2 \cdot D^2 u_2)]\} \equiv w(Du, Du_2, D^2 u_2),$$

where $G = (\Delta_D^2)^{-1}$ and as above D and D^2 are certain differential operations with constant coefficients of first and second order, respectively. We rewrite w as follows.

$$\begin{aligned} w = {} & w(Q_N Du, Du_2, D^2 u_2) + w(P_N Du, Q_N Du_2, D^2 u_2) \\ & + w(P_N Du, P_N Du_2, D^2 u_2) \equiv w_1 + w_2 + w_3. \end{aligned}$$

Now we estimate every quantity w_j separately. It follows from (A.3.17) that

$$\| w_1 \|_{-1} \le C \| Q_N Du \|_{1-\beta} \cdot \| Du_2 D^2 u_2 \|_{-1+\beta}, \quad 0 < \beta < 1.$$

Therefore using (A.3.18) and (A.3.9) we have

$$\| w_1 \|_{-1} \le C_{R,\beta} \lambda_{N+1}^{-\beta/4}, \quad 0 < \beta < 1. \tag{A.3.30}$$

In the same way, Lemma A.3.7 implies

$$\| w_2 \|_{-1} \le C \| P_N Du \|_{1-\hat{\beta}} \| Q_N Du_2 \|_{\hat{\beta}} \| D^2 u_2 \|, \quad 0 < \hat{\beta} < 1.$$

Consequently for $\hat{\beta} = 1 - \beta$ we have

$$\| w_2 \|_{-1} \le C_{R,\beta} \lambda_{N+1}^{-\beta/4}, \quad 0 < \beta < 1. \tag{A.3.31}$$

We now consider the term w_3. Because

$$\| w_3 \|_{-1} \le C \| P_N Du \cdot D^2 G D(P_N Du_2 \cdot D^2 u_2) \|,$$

Lemma A.3.6 and the property (A.3.8) implies that

$$\| w_3 \|_{-1} \le C\{\log(1+\lambda_N)\}^{1/2} \| u \|_1 \cdot \| P_N Du_2 \cdot D^2 u_2 \| .$$

Using Lemma A.3.3 we have that

$$\| w_3 \|_{-1} \le C \log(1+\lambda_N) \| u \|_1 . \qquad \text{(A.3.32)}$$

Thus the estimates (A.3.30)–(A.3.32) imply (A.3.29). □

Now we can apply Lemmas A.3.8 and A.3.9 in order to estimate the quantity $M(t)$. We obtain

$$\| M(t) \|_{-1} \le C_1\{\log(1+\lambda_N)\} \| u(t) \|_1 + C_2 \lambda_{N+1}^{-\beta}, \quad t \in [0,T],$$

where $u(t) = u_1(t) - u_2(t)$ is the difference of two weak solutions and the constants C_1 and C_2 depend on the norms of $u_j(t)$ in $L_\infty(0,T;H_0^2(\Omega))$. Let

$$\psi(t) = \| \partial_t u(t) \|_{-1}^2 + \| u(t) \|_1^2 .$$

It follows from (A.3.20) that we have

$$\psi(t) \le C_1\{\log(1+\lambda_N)\} \int_0^t \psi(\tau) d\tau + C_2 T \cdot \lambda_{N+1}^{-\beta}, \quad t \in [0,T].$$

Therefore using Gronwall's lemma we conclude that

$$\psi(t) \le C_2 T \lambda_{N+1}^{-\beta} (1+\lambda_N)^{C_1 t}, \quad t \in [0,T].$$

If we let $N \to \infty$, then for $0 \le t < t_0 \equiv \beta / C_1$ we obtain $\psi \equiv 0$; Thus $u_1(t) \equiv u_2(t)$ for $0 \le t < t_0$. Now step by step we can conclude that $u_1(t) \equiv u_2(t)$ for all $0 \le t \le T$, which is what needed to be proved.

Step 2: Energy identity. We use reversibility of the dynamics.

Let $u(t)$ be a weak solution on $[0,T]$. By uniqueness every solution can be obtained via the Galerkin method and thus satisfies the energy inequality (A.3.4).

Let $w(t) = u(T-t), t \in [0,T]$. One can see that $w(t)$ is a weak solution on $[0,T]$ with initial data $(u(T); -\partial_t u(T))$. Therefore it satisfies the energy inequality of the form (A.3.4) and thus we have that

$$\mathscr{E}(u_0,u_1) = \mathscr{E}(w(T), \partial_t w(T)) \le \mathscr{E}(w(0), \partial_t w(0)) = \mathscr{E}(u(T), \partial_t u(T)).$$

The above inequality along with (A.3.4) yields the energy identity.

Step 3: Continuity properties. One can see that

$$\{t; u_0; u_1\} \mapsto (u(t); \partial_t u(t)) \qquad \text{(A.3.33)}$$

is a weakly continuous function on $[0,T] \times H_0^2(\Omega) \times L_2(\Omega)$ (with strong topology) with values in $H \equiv H_0^2(\Omega) \times L_2(\Omega)$. The energy identity gives that the function

$$\{t; u_0; u_1\} \mapsto \mathscr{E}(u(t), \partial_t u(t))$$

is continuous from $[0,T] \times H_0^2(\Omega) \times L_2(\Omega)$ into $\mathbb{R}_+$. Using Lemma A.3.2 one can see that difference between the norm in H and the energy $2\mathscr{E}$ is represented by a compact operator and hence the function

$$\{t; u_0; u_1\} \mapsto \|\partial_t u(t))\|^2 + \|\Delta u(t)\|^2$$

is also continuous. Therefore weak continuity of (A.3.33) implies strong continuity of this mapping. □

In conclusion we note that the argument presented above for the proof of uniqueness remains true if we replace equations (A.3.1) and (A.3.3) by equations (4.1.1) and (4.1.2) with $d_0(x) \equiv 1$ and linear damping function $g_0(s) = \gamma s$. The method presented does not apply in the case of nonlinear functions $g_0(s)$. The point is that to justify the energy relation in (A.3.20) (which is the starting point for the argument) we use the multiplier $P_N(\Delta_D^2)^{-1/2} u_t$. In the case of nonlinear damping this multiplier leads to a term with uncontrolled sign. Thus, uniqueness argument depends in a critical fashion on linearity of the damping.

Appendix B
Some Auxiliary Facts

This appendix is devoted to the proof of several auxiliary facts concerning qualitative behavior of damping functions and equations that describe the decay rates. We also recall some facts from measure theory concerning convergence of the sequences of measurable functions.

B.1 Estimates for monotone functions

We start with the following simple assertion.

B.1.1. Proposition. *Assume that a function $g \in C(\mathbb{R})$ possesses the following properties.*

- $g(0) = 0$ *and $g(s)$ is a nondecreasing and (strictly) increasing function in some (small) vicinity of zero.*[1]
- *There exist $s_0 > 0$ and $m > 0$ such that $sg(s) \geq ms^2$ for $|s| \geq s_0$.*

Then, for any $\eta > 0$ there exists $C_\eta > 0$ such that

$$s^2 \leq \eta + C_\eta sg(s) \quad \textit{for all } s \in \mathbb{R}. \tag{B.1.1}$$

Proof. It is clear that under the conditions imposed on g we have that

$$c_\delta := \inf_{|s| \in [\delta, s_0]} \{sg(s)\} > 0 \quad \text{for any } 0 < \delta \leq s_0.$$

Thus we have that

$$s^2 \leq \frac{s_0^2}{c_\delta} \inf_{|s| \in [\delta, s_0]} \{sg(s)\} \leq \frac{s_0^2}{c_\delta} sg(s) \quad \text{for } s \in [-s_0, -\delta] \cup [\delta, s_0].$$

Because $s^2 \leq m^{-1} sg(s)$ for $|s| \geq s_0$ and $s^2 \leq \delta^2$ for $|s| \leq \delta$, we obtain that

[1] Instead we can assume that $g(0) = 0$, $g(s)$ is nondecreasing and $sg(s) > 0$ for $s \neq 0$.

$$s^2 \le \delta^2 + \max\left\{\frac{s_0^2}{c_\delta}, m^{-1}\right\} sg(s) \text{ for all } s \in \mathbb{R}.$$

This implies (B.1.1). □

Under the additional hypotheses of asymptotic strong monotonicity of g we can strengthen inequality (B.1.1) in the following way.

B.1.2. Proposition. *If $g \in C^1(\mathbb{R})$ is an increasing function such that $g(0) = 0$ and there exists $s_0 > 0$ and $m > 0$ such that $g'(s) \ge m$ for $|s| \ge s_0$, then for any $\eta > 0$ there exists $C_\eta > 0$ such that*

$$(s_1 - s_2)^2 \le \eta + C_\eta (s_1 - s_2)(g(s_1) - g(s_2)), \quad s_1, s_2 \in \mathbb{R}. \tag{B.1.2}$$

Proof. We argue by contradiction. Let's assume that (B.1.2) fails. Then, there exist $\eta_0 > 0$ and sequences $\{s_1^n\}$ and $\{s_2^n\}$ such that $s_1^n > s_2^n$ and

$$(s_1^n - s_2^n)^2 \ge \eta_0 + n(s_1^n - s_2^n)(g(s_1^n) - g(s_2^n)), \quad n = 1, 2, \ldots.$$

Thus we have that

$$s_1^n \ge s_2^n + \sqrt{\eta_0}, \quad \frac{1}{s_1^n - s_2^n} \int_{s_2^n}^{s_1^n} g'(\xi) d\xi \le \frac{1}{n}, \quad n = 1, 2, \ldots. \tag{B.1.3}$$

We can assume that there exist limits

$$s_1^* = \lim_{n \to \infty} s_1^n, \quad s_2^* = \lim_{n \to \infty} s_2^n, \quad s_1^* \ge s_2^*,$$

which belong to the extended real axis $[-\infty, +\infty]$.

If both s_1^* and s_2^* are finite, then (B.1.3) gives us that

$$s_1^* \ge s_2^* + \sqrt{\eta_0}, \text{ and } \frac{1}{s_1^* - s_2^*} \int_{s_2^*}^{s_1^*} g'(\xi) d\xi = 0,$$

which is impossible because g is increasing.

If either $s_1^* = s_2^* = +\infty$ or $s_1^* = s_2^* = -\infty$, then by the estimate $g'(s) \ge m > 0$ for $|s| \ge s_0$ we have

$$0 < m \le \frac{1}{s_1^n - s_2^n} \int_{s_2^n}^{s_1^n} g'(\xi) d\xi \le \frac{1}{n}$$

for all n large enough which is impossible.

If $s_1^* = +\infty$ an $s_2^* = -\infty$, then we have

$$\frac{1}{n} \ge \frac{1}{s_1^n - s_2^n} \int_{s_2^n}^{s_1^n} g'(\xi) d\xi \ge \frac{1}{s_1^n - s_2^n} \left\{ \int_{s_0}^{s_1^n} + \int_{s_2^n}^{-s_0} \right\} g'(\xi) d\xi \ge m - \frac{2ms_0}{s_1^n - s_2^n}$$

for all n large enough, which is also impossible. In the remaining cases

$$\{s_1^* = +\infty, \; s_2^* \text{ is finite}\} \text{ and } \{s_1^* \text{ is finite}, \; s_2^* = -\infty\}$$

the argument is symmetric. The proof is complete. □

B.2 Concave bounds

Now we discuss the existence of monotone concave bounds for damping functions. We follow the scheme given in [195] and provide more details.

B.2.1. Proposition. *Let g be a nondecreasing continuous function on $\mathbb{R}$ such that $g(0) = 0$ and $sg(s) > 0$ for $s \neq 0$. Then for any $a > 0$ there exists a concave strictly increasing continuous function $h : \mathbb{R}_+ \mapsto \mathbb{R}_+$ such that $h(0) = 0$ and*

$$s^2 + g^2(s) \leq h(sg(s)) \quad for\ |s| \leq a. \tag{B.2.1}$$

If, in addition, we assume that there exists $s_0 > 0$ and $m > 0$ such that $sg(s) \geq ms^2$ for $|s| \geq s_0$, then we also have that

$$s^2 \leq h(sg(s)) \quad for\ all\ s \in \mathbb{R}. \tag{B.2.2}$$

Proof. Because the function $sg(s)$ is strictly increasing on $\mathbb{R}_+$, there is a continuous increasing function $s = s(\sigma)$ on $\mathbb{R}_+$ such that $s(0) = 0$ and

$$s(\sigma)g(s(\sigma)) = \sigma \quad \text{for every } \sigma \in \mathbb{R}_+.$$

Therefore the function $k_1(\sigma) = s(\sigma)^2 + [g(s(\sigma))]^2$ is a nonnegative increasing continuous function such that $k_1(0) = 0$ and

$$k_1(sg(s)) = s^2 + g^2(s) \quad \text{for all } s \in \mathbb{R}_+.$$

In a similar way we can construct increasing continuous function $k_2 : \mathbb{R}_+ \mapsto \mathbb{R}_+$ such that $k_2(0) = 0$ and

$$k_2(sg(s)) = s^2 + g^2(s) \quad \text{for all } s \leq 0.$$

Consequently, the function $k(\sigma) = \max\{k_1(\sigma), k_2(\sigma)\}$ is a nonnegative increasing continuous function on $\mathbb{R}_+$ such that $k(0) = 0$ and

$$k(sg(s)) \geq s^2 + g^2(s) \quad \text{for all } s \in \mathbb{R}.$$

Let $b > 0$ be arbitrary. On the interval $[0, \sigma_b]$ with $\sigma_b = \max\{bg(b), -bg(-b)\}$ we can define a concave envelope $\hat{h}(\sigma)$ of $k(\sigma)$ which is an increasing concave function on $[0, \sigma_b]$ such that $\hat{h}(0) = 0$ and $\hat{h}(\sigma) \geq k(\sigma)$. Now we define a continuous increasing function

$$\tilde{h}(\sigma) = \begin{cases} \hat{h}(\sigma) + c\sigma, & 0 \leq \sigma \leq \sigma_b, \\ \hat{h}(\sigma_0) + c\sigma, & \sigma > \sigma_b, \end{cases} \tag{B.2.3}$$

with some $c \geq 0$. It is clear that the concave envelope $h(\sigma)$ for $\tilde{h}(\sigma)$ satisfies (B.2.1) with $a = b$. In the case when $sg(s) \geq ms^2$ for $|s| \geq s_0$, we take $b = \max\{a, s_0\}$ with a given $a > 0$. The same construction based on (B.2.3) with $c = m^{-1}$ provides us an increasing concave function $h(\sigma)$ on $\mathbb{R}_+$ which satisfies both (B.2.1) and (B.2.2).

□

B.2.2. Remark. Let the hypotheses of Proposition B.2.1 be in force. One can see that if we assume that

$$g(s) \sim g_0 |s|^{p-1} s \text{ as } s \to 0 \text{ for some } p \geq 1,\ g_0 > 0,$$

then one can construct function $h(\sigma)$ in Proposition B.2.1 such that

$$h(\sigma) \sim h_0 \sigma^{2/(p+1)} \text{ as } \sigma \to +0 \text{ for some } h_0 > 0.$$

In particular, if $p = 1$ (i.e., $g(s) \sim g_0 s$ is linear near zero), then we can choose $h(\sigma) = h_0 \sigma$ to be a linear function. ■

B.2.3. Proposition. *Let $h : \mathbb{R}_+ \mapsto \mathbb{R}_+$ be an increasing function and $h(0) = 0$. Then*

- *For any $c_1, c_2 > 0$ the function $p(s) \equiv c_1 (I + h)^{-1}(c_2 s)$ is a continuous strictly increasing function such that $p(0) = 0$ and*

$$0 < p(s_1) - p(s_2) \leq c_1 c_2 (s_1 - s_2) \quad for\ s_1 > s_2 \geq 0. \tag{B.2.4}$$

- *If h is concave, then p is convex.*
- *The function $q(s) \equiv s - (I + p)^{-1}(s)$, $s \geq 0$, is Lipschitz continuous, strictly increasing, positive for $s > 0$, and zero at the origin.*

Proof. To prove the first statement we note that for $s_1 > s_2 \geq 0$ we have that $p_1 \equiv p(s_1) > p(s_2) \equiv p_2$ and, because $h(p_1/c_1) - h(p_2/c_1) > 0$, we conclude that

$$0 < \frac{p_1 - p_2}{c_1} \leq (I + h)\left(\frac{p_1}{c_1}\right) - (I + h)\left(\frac{p_2}{c_1}\right) = c_2 (s_1 - s_2).$$

This implies (B.2.4).

The second statement concerning the function $p(s)$ is obvious.

To prove the third statement we note that for every $s_1 > s_2$ we have that

$$[s_1 + p(s_1)] - [s_2 + p(s_2)] > s_1 - s_2 > 0.$$

Taking $s_i = (I + p)^{-1}(\sigma_i)$ and noting that $\sigma_1 > \sigma_2$ if and only if $s_1 > s_2$ we obtain

$$\sigma_1 - \sigma_2 > (I + p)^{-1}(\sigma_1) - (I + p)^{-1}(\sigma_2) > 0.$$

This implies that $(I + p)^{-1}$ (and hence q) is Lipschitz continuous and the function q is increasing. □

B.3 Equation describing the convergence rates for the energy

We start with the following assertion.

B.3.1. Proposition. *Let $q : \mathbb{R}_+ \mapsto \mathbb{R}_+$ be a Lipschitz continuous strictly increasing function such that $q(0) = 0$. Let $f(t) \in L_1^{loc}(\mathbb{R}_+)$ be nonnegative. Then for any $s_0 \in \mathbb{R}_+$ the problem*

$$\frac{d}{dt}S(t) + q(S(t)) = f(t), \quad S(0) = s_0, \tag{B.3.1}$$

has a unique (absolutely continuous) nonnegative solution $S(t)$. Moreover,

- *If $M \equiv \sup_{t \geq t_0} f(t) < \infty$ for some $t_0 \in \mathbb{R}_+$, then*

$$0 \leq S(t) \leq q^{-1}(2M) \ \text{for all}\ t \geq t_* \tag{B.3.2}$$

for some $t_* = t_*(t, s_0) \geq t_0$.

- *If $f(t) \to 0$ as $t \to \infty$, then $S(t) \to 0$ as $t \to \infty$.*

Proof. The existence and uniqueness of nonnegative solutions for (B.3.1) relies on a standard ODE argument.

Let us prove (B.3.2). We first note that there exists $\tilde{t}_0 \geq t_0$ such that $S(\tilde{t}_0) \leq q^{-1}(2M)$. Indeed, if for $q(S(t_0)) > 2M$, and there is no $\tilde{t}_0 > t_0$ such that $S(\tilde{t}_0) = q^{-1}(2M)$, then $q(S(t)) \geq 2M$ for all $t \geq t_0$. This implies

$$S_t(t) \leq -q(S(t)) + M \leq -M, \quad t \geq t_0.$$

Thus $S(t) \leq S(t_0) - M(t - t_0)$ for *all* $t \geq t_0$. This is impossible because $S(t)$ is nonnegative.

After the time $\tilde{t}_0$ the solution $S(t)$ cannot leave the interval $[0, q^{-1}(2M)]$. Indeed, if $S(t_*) = q^{-1}(2M)$ for some $t_* \geq \tilde{t}_0$, then $S_t(t_*) \leq -M$ and thus $S(t)$ is strictly increasing at t_*. Therefore $S(t)$ cannot exceed the level $q^{-1}(2M)$ for $t \geq t_*$.

To prove the last statement we note that, because $f(t) \to 0$ as $t \to \infty$, for any $\varepsilon > 0$ there exists $t_\varepsilon \geq 0$ such that $0 \leq f(t) \leq \varepsilon$ for all $t \geq t_\varepsilon$. Thus by (B.3.2) there exists $t_\varepsilon^* \geq t_\varepsilon$ such that

$$0 \leq S(t) \leq q^{-1}(2\varepsilon) \quad \text{for all} \quad t \geq t_\varepsilon^*.$$

This implies, (recalling that $q^{-1}(0) = 0$ and $q(s)$ is continuous) that $S(t) \to 0$ as $t \to \infty$. □

B.3.2. Remark. Proposition B.3.1 implies, in particular, that any nonzero solution $S(t)$ to problem (B.3.1) with $f(t) \equiv 0$ is decreasing and tends to zero when $t \to +\infty$. Moreover the function $S(t)$ can be found as a solution to the functional equation

$$\int_{S(t)}^{S(0)} \frac{ds}{q(s)} = t. \tag{B.3.3}$$

Thus the rate of convergence of $S(t)$ is determined by the behavior of the function $q(s)$ around zero. For an example, in the case when $q(s) \sim c_0 s^{\alpha}$ as $s \to +0$ with $\alpha \geq 0$, one can see from (B.3.3) that

$$S(t) \sim ct^{-1/(\alpha-1)} \quad \text{as} \quad t \to +\infty \quad \text{for } \alpha > 1,$$

and

$$S(t) \sim c_1 e^{-c_2 t} \quad \text{as} \quad t \to +\infty \quad \text{for } \alpha = 1.$$

■

Our main tool for obtaining decay rates described by equation (B.3.1) is the following assertion.

B.3.3. Proposition. *Let p be a positive increasing function such that $p(0) = 0$ and $q(s) = s - [I + p]^{-1}(s)$. Assume that a sequence $\{s_n\}$ of positive numbers satisfies the relations*

$$s_{m+1} + p(s_{m+1}) \leq s_m + f_m, \quad m = 0, 1, 2, \ldots, \tag{B.3.4}$$

where $f_m \leq \int_m^{m+1} f(t)dt$ for some locally integrable nonnegative function f. Then

$$s_m \leq S(m), \quad m = 0, 1, 2, \ldots, \tag{B.3.5}$$

where $S(t)$ is the solution to the ODE Cauchy problem (B.3.1).

Proof. This is a modification of the argument given in [195].

Integrating (B.3.1) we obtain that

$$S(t) - S(m) + \int_m^t q(S(\tau))d\tau = \int_m^t f(t)dt \tag{B.3.6}$$

for every $t \geq m$. In particular

$$S(m+1) - S(m) + \int_m^{m+1} q(S(\tau))d\tau = \int_m^{m+1} f(t)dt \equiv \tilde{f}_m. \tag{B.3.7}$$

From (B.3.6) we also have that

$$S(\tau) \leq S(m) + \tilde{f}_m \text{ for } \tau \in [m, m+1].$$

and hence, because q is increasing, we obtain that

$$\int_m^{m+1} q(S(\tau))d\tau \leq q(S(m) + \tilde{f}_m), \quad m = 0, 1, 2, \ldots.$$

Thus (B.3.7) yields that

$$S(m+1) - S(m) + q(S(m) + \tilde{f}_m) \geq \tilde{f}_m, \quad m = 0, 1, 2, \ldots.$$

Because $q(s) = s - (I + p)^{-1}(s)$, this implies that

$$S(m+1) \geq (I+p)^{-1}(S(m)+\tilde{f}_m), \quad m=0,1,2,\ldots.$$

If we assume that $S(m) \geq s_m$ for some m, then from monotonicity of $(I+p)^{-1}$ we obtain that

$$S(m+1) \geq (I+p)^{-1}(s_m+\tilde{f}_m) \geq (I+p)^{-1}(s_m+f_m). \tag{B.3.8}$$

Because $p(x)$ is increasing, it follows from (B.3.4) that

$$(I+p)^{-1}(s_m+f_m) \geq s_{m+1} \quad m=0,1,2,\ldots,$$

and hence relation (B.3.8) implies that $S(m+1) \geq s_{m+1}$. Therefore by induction we obtain the desired relation in (B.3.5). □

B.4 Some convergence theorems for measurable functions

The goal of this section is to provide the statements of several theorems from measure theory which were used in the text of this book.

Let $(X,\mathscr{B},\mu)$ be a space with σ-algebra $\mathscr{B}$ and σ-finite measure μ.

B.4.1. Theorem (Fatou's lemma). *Let $\{f_n : n=1,2,\ldots\}$ be a sequence of non-negative measurable functions on $(X,\mathscr{B},\mu)$. Then*

$$\int_X \liminf_{n\to\infty} f_n d\mu \leq \liminf_{n\to\infty} \int_X f_n d\mu$$

For the proof see, for example, [93] or [248].

The following Levi-Lebesgue theorem on monotone convergence holds.

B.4.2. Theorem (Levi–Lebesgue). *Let $\{f_n : n=1,2,\ldots\}$ be a sequence of measurable functions from $L_1(X)$ such that*

(i) *$f_1(x) \leq f_2(x) \leq \cdots$ for almost all $x \in X$.*
(ii) *$\int_X f_n d\mu \leq K$ for all $n=1,2,\ldots$ and for some constant K.*

Then there exists $f \in L_1(X)$ such that $f(x) = \lim_{n\to\infty} f_n(x)$ almost everywhere and

$$\int_X f d\mu = \lim_{n\to\infty} \int_X f_n d\mu.$$

For the proof see, for example,[93] or [248].

We also need the following version of the Lebesgue dominated convergence theorem.

B.4.3. Theorem (Lebesgue). *Let f and $\{f_n : n=1,2,\ldots\}$ be measurable functions on $(X,\mathscr{B})$. Assume that*

- *$f_n \to f$ almost everywhere as $n \to \infty$.*

- *There exists a sequence* $\{g_n : n = 1, 2, \ldots\} \subset L_1(X)$ *such that*

  (i) $|f_n(x)| \le g_n(x)$ *for almost all* $x \in X$ *and for all* $n = 1, 2, \ldots$.
  (ii) $g_n \to g$ *almost everywhere and*

$$\int_X g_n d\mu \to \int_X g d\mu \quad as\ n \to \infty$$

  for some function $g \in L_1(X)$.

Then $f \in L_1(X)$ *and* $\lim_{n\to\infty} \int_X f_n d\mu = \int_X f d\mu$.

For the proof we refer to [248, p.270], for instance.

We note that in the standard formulation of this theorem it is assumed that the sequence $\{g_n : n = 1, 2, \ldots\}$ does not depend on n, i.e. $g_n \equiv g \in L_1(X)$ (see, e.g., [93]). We also note that instead of (ii) one can assume that

(ii*) $\int_X |g_n - g| d\mu \to 0$ *as* $n \to \infty$ *for some function* $g \in L_1(X)$.

Indeed, to obtain the result with the condition in (ii*) instead of (ii) we can argue as follows. The property in (ii*) implies that there exists a subsequence $\{g_{n_k}\}$ for which (ii) holds. Therefore, by the statement of the theorem with (ii) we obtain that $f \in L_1(X)$. Now arguing by contradiction one shows that

$$\int_X f d\mu = \lim_{n\to\infty} \int_X f_n d\mu.$$

We conclude this section with the Egorov theorem.

B.4.4. Theorem (Egorov). *Let* $\{f_n : n = 1, 2, \ldots\}$ *be a sequence of measurable functions on* $(X, \mathscr{B})$ *which converges to a function* f *almost everywhere on a measurable set* $Y \in \mathscr{B}$ *of finite measure. Then given* $\varepsilon > 0$, *there is a subset* $E \subset Y$ *with* $\mu(E) \le \varepsilon$ *such that* f_n *converges to* f *uniformly* $Y \setminus E$.

For the proof see, for example, [93] or [248].

References

1. R.Adams, *Sobolev Spaces*. Academic Press, New York, 1975.
2. S. Agmon, S. Douglis, and L. Nirenberg, Estimates near the boundary for solutions of elliptic partial differential equations satisfying general boundary value conditions, *Comm. Pure. Appl. Math.*, 12 (1959), 623–727.
3. J.-P. Aubin, Une théorè de compacité, *C.R. Acad. Sci. Paris*, 256 (1963), 5042–5044.
4. J.-P. Aubin, *Approximation of Elliptic Boundary-Value Problems*, Wiley, New York, 1972.
5. G. Avalos, The exponential stability of a coupled hyperbolic/parabolic system arising in structural acoustics, *Abstr. Appl. Anal.*, 1 (1996), 203–219.
6. G. Avalos, Exact controllability of thermoelastic system with control in thermal component only, *Diff. Integral Eqs.*, 22 (2000), 1–15.
7. G. Avalos and I. Lasiecka, Differential Riccati equation for the active control of a problem in structural acoustics, *J. Optim. Theory Appl.*, 91 (1996), 695–728.
8. G. Avalos and I. Lasiecka, Exponential stability of a thermoelastic system without mechanical dissipation, *Rend. Istit. Mat. Univ. Trieste*, 28 (1997), 1–28.
9. G. Avalos and I. Lasiecka, Exponential stability of a thermoelastic plates with free boundary conditions and without mechanical dissipation, *SIAM J. Math. Anal.*, 29 (1998), 155–182.
10. G. Avalos and I. Lasiecka, The strong stability of a semigroup arising from a coupled hyperbolic/parabolic system, *Semigroup Forum*, 57 (1998), 278–292.
11. G. Avalos and I. Lasiecka, Uniform decay rates of solutions to a structural acoustic model with nonlinear dissipation, *Appl. Math. Comp. Sci.*, 8 (1998), 287–312.
12. G. Avalos and I. Lasiecka, Uniform decays in nonlinear thermoelastic systems, *Optim. Control, Theor. and Alg.*, 15 (1998), 1–22.
13. G. Avalos, I. Lasiecka and R. Triggiani, Uniform stability in nonlinear thermoelastic plates with free boundary conditions, In: *Optimal Control of Partial Differential Equations*, K. Hoffmann, G. Leugering, and F. Tröltzsch (Eds.), (ISNM, vol. 133), Birkhäuser, Basel, 1999, 13–32.
14. G. Avalos and I. Lasiecka, Exact controllability of structural acoustic interactions, *J. Math. Pures Appl.*, 82 (2003), 1047–1073.
15. G. Avalos and I. Lasiecka, Exact controllability of finite energy states for an acoustic wave/plate interaction under the influence of boundary and localized controls, *Adv. Diff. Eqs.*, 10 (2005), 901–930.
16. A. Babin and M. Vishik, Regular attractors of semigroups and evolution equations, *J. Math. Pures Appl.*, 62 (1983), 441–491.
17. A. Babin and M. Vishik, *Attractors of Evolution Equations*, North-Holland, Amsterdam, 1992.
18. V. Barbu, *Nonlinear Semigroups and Differential Equations in Banach Spaces*, Noordhoff, Leyden, 1976.

19. J. Beale, Spectral properties of an acoustic boundary condition, *Indiana Univ. Math. J.*, 9 (1976), 895–917.
20. A.Benabdallah and D. Teniou, Exponential decay rates for a full von Karman model with thermal effects, *Electron. J. Diff. Eqs.*, 7 (1998), 1–13.
21. A. Benabdallah and G. Naso, Nullcontrolability of thermoelastic plates, *Abstr. Appl. Anal.*, 7 (2002), 585-599.
22. M. Berger, A new approach to the large deflection of plate, *J. Appl. Mech.*, 22 (1955), 465–472.
23. M.S. Berger, On von Karman's equations and the buckling of a thin elastic plate, I. The clamped plate, *Commun. Pure Appl. Math.*, 20 (1967), 687–719.
24. M.S. Berger, On the existence of equilibrium states of a thin elastic shell (I), *Indiana Univ. Math. J.*, 20 (1971), 591–602.
25. M.S. Berger and P.C. Fife, On von Karman's equations and the buckling of a thin elastic plate, II. Plate with general edge conditions, *Commun. Pure Appl. Math.*, 21 (1968), 227–241.
26. J.Bergh and J. Löfström , *Interpolation Spaces: An Introduction*, Springer, Berlin, 1976.
27. M. Böhm, Global well posedness of the dynamic von Karman equations for generalized solutions, *Nonlin. Anal. TMA*, 27 (1996), 339–351.
28. V.V. Bolotin, *Nonconservative Problems of Elastic Stability*, Pergamon Press, Oxford, 1963.
29. A. Boutet de Monvel and I. Chueshov, Uniqueness theorem for weak solutions of von Karman evolution equations, J. Math. Anal. Appl., 221 (1998), 419–429.
30. A. Boutet de Monvel and I. Chueshov, The problem on interaction of von Karman plate with subsonic flow of gas, *Math. Meth. Appl. Sci.*, 22 (1999), 801–810.
31. L. Boutet de Monvel and I. Chueshov, Non-linear oscillations of a plate in a flow of gas, *C.R. Acad. Sci. Paris, Ser.I*, 322 (1996), 1001–1006.
32. L. Boutet de Monvel and I. Chueshov, Oscillations of von Karman plate in a potential flow of gas, Izvestiya: Mathematics, 63(2) (1999), 219–244.
33. L. Boutet de Monvel, I. Chueshov and A. Rezounenko, Long-time behaviour of strong solutions of retarded nonlinear P.D.E.s, *Commun. Partial Diff. Eqs.*, 22 (1997), 1453–1474.
34. H. Brézis, *Opérateurs Maximaux Monotones*, North-Holland, Amsterdam, 1973.
35. H. Brézis and T. Gallouet, Nonlinear Schrödinger evolution equations, *J. Nonlinear Anal.*, 4 (1980), 677–681.
36. F. Bucci, Uniform decay rates of solutions to a system of coupled PDEs with nonlinear internal dissipation, *Diff. Integral Eqs.*, 16 (2003), 865–896.
37. F. Bucci and I. Chueshov, Long-time dynamics of a coupled system of nonlinear wave and thermoelastic plate equations, *Discr. Cont. Dyn. Sys.*, 22 (2008), 557–586.
38. F. Bucci, I. Chueshov and I. Lasiecka, Global attractor for a composite system of nonlinear wave and plate equations, *Commun. Pure Appl. Anal.*, 6 (2007), 113–140.
39. J. Cagnol, I. Lasiecka and C. Lebiedzik, Uniform stability in structural acoustic models with flexible curved walls, *J. Diff. Eqs.*, 186 (2002), 88–121.
40. M. Camurdan and G. Ji, Uniform feedback stabilization via boundary moments a three dimensional structural acoustic model. In: *Proceedings of the 37th IEEE Conference on Decision and Control, vol.3*, IEEE, Piscataway, NJ, 1998, 2958–2964.
41. H. Cartan, *Calculus Différentielles*, Hermann, Paris, 1967.
42. S. Ceron and O. Lopes, α-contractions and attractors for dissipative semilinear hyperbolic equations and systems, *Ann. Math. Pura Appl. IV*, 160 (1991), 193–206.
43. S. Chen and R. Triggiani, Proof of extensions of two conjectures on structural damping for elastic systems, *Pacific J. Math.*, 136 (1989), 15–55.
44. S. Chen and R. Triggiani, Characterization of domains of fractional powers of certain operators arising in elastic systems and applications, *J. Diff. Eqs.*, 64 (1990), 26–42.
45. S. N. Chow and K. Lu Invariant manifolds for flows in Banach spaces, *J. Diff. Eqs.*, 74, (1988), 285–317.
46. I. Chueshov, Finite-dimensionality of the attractor in some problems of the nonlinear theory shells, *Math. USSR Sbornik*, 61 (1988), 411–420.
47. I. Chueshov, Structure of a maximal attractor of a modified system of von Karman equations, *J. Soviet Math.*, 48 (1990), 692–696.

48. I. Chueshov, Asymptotic behavior of the solutions of a problem of the aeroelastic oscillations of a shell in hypersonic limit, *J. Soviet Math.*, 52 (1990), 3545–3548.
49. I. Chueshov, A problem of nonlinear oscillations of shallow shell in quasistatic formulation, *Math. Notes*, 47 (1990), 401–407.
50. I. Chueshov, Inertial manifolds in the nonlinear oscillation problem of infinite panel, *Ukraine Math. J.*, 42 (1990), 1149–1151.
51. I. Chueshov, Strong solutions and attractor of a system of von Karman equations, *Math. USSR Sbornik*, 69 (1991), 25–36.
52. I. Chueshov, Construction of solutions in a problem of the oscillations of a shell in a potential subsonic flow. In: *Operator Theory, Subharmonic Functions*, V.A. Marchenko (Ed.), Naukova Dumka, Kiev, 1991, 147–154, in Russian.
53. I. Chueshov, On some continuity property of attractor in a problem on oscillations of shallow shell, In: *Dynamical Systems and Complex Analysis*, V.A. Marchenko (Ed.), Naukova Dumka, Kiev, 1992, 85–91, in Russian.
54. I. Chueshov, On a certain system of equations with delay, occurring in aeroelasticity, *J. Soviet Math.*, 58 (1992), 385–390.
55. I. Chueshov, Regularity of solutions and approximate inertial manifolds for von Karman evolution equations, *Math. Meth. Appl. Sci.*, 17 (1994), 667–680.
56. I. Chueshov, Quasistatic version of the system of von Karman equations, *Matem. Fizika, Analiz, Geometria*, 1(1) (1994), 149–167, in Russian.
57. I. Chueshov, On a construction of approximate inertial manifolds for second order in time evolution equations. *Nonlinear Anal.*, 26 (1996), 1007–1021.
58. I. Chueshov, On a description of long-time behaviour of dissipative perturbations of infinite-dimensional Hamiltonian systems. *Z. Angew. Math. Mech.*. 76 S.2 (1996), 53–56.
59. I. Chueshov, On the finiteness of the number of determining elements for von Karman evolution equations, *Math. Meth. Appl. Sci.*, 20 (1997), 855–865.
60. I. Chueshov, Theory of functionals that uniquely determine asymptotic dynamics of infinite-dimensional dissipative systems, *Russian Math. Surv.*, 53 (1998), 731–776.
61. I. Chueshov, *Introduction to the Theory of Infinite-Dimensional Dissipative Systems*, Acta, Kharkov, 1999, in Russian; English translation: Acta, Kharkov, 2002; see also http://www.emis.de/monographs/Chueshov/
62. I. Chueshov, M. Eller and I. Lasiecka, On the attractor for a semilinear wave equation with critical exponent and nonlinear boundary dissipation, *Commun. Partial Diff. Eqs.*, 27 (2002), 1901–1951.
63. I. Chueshov, M. Eller and I. Lasiecka, Finite dimensionality of the attractor for a semilinear wave equation with nonlinear boundary dissipation, *Commun. Partial Diff. Eqs.*, 29 (2004), 1847–1976.
64. I. Chueshov and V. Kalantarov, Determining functionals for nonlinear damped wave equation, *Matem. Fizika, Analiz, Geometria*, 8(2) (2001), 215–227.
65. I. Chueshov and S. Kolbasin, Plate models with state-dependent damping coefficient and their quasi-static limits, submitted, 2010.
66. I. Chueshov and I. Lasiecka, Inertial manifolds for von Karman plate equations, *Appl. Math. Optim.*, (special issue dedicated to J. L. Lions) 46 (2002), 179–207.
67. I. Chueshov and I. Lasiecka, Determining functionals for a class of second order in time evolution equations with applications to von Karman equations, In: *Analysis and Optimization of Differential Systems*, V. Barbu et al. (Eds), Kluwer, Boston, 2003, 109–122.
68. I. Chueshov and I. Lasiecka, Global attractors for von Karman evolutions with a nonlinear boundary dissipation, *J. Diff. Eqs.*, 198 (2004), 196–221.
69. I. Chueshov and I. Lasiecka, Attractors for second order evolution equations, *J. Dynam. Diff. Eqs.*, 16 (2004), 469–512.
70. I. Chueshov and I. Lasiecka, Kolmogorov's ε-entropy for a class of invariant sets and dimension of global attractors for second order in time evolution equations with nonlinear damping. In: *Control Theory of Partial Differential Equations*, O. Imanuvilov et al., (Eds.), A Series of Lectures in Pure and Applied Mathematics, vol. 242, Chapman & Hall/CRC, Boca Raton, FL, 2005, 51–69.

71. I. Chueshov and I. Lasiecka, Global attractors for Mindlin–Timoshenko plates and for their Kirchhoff limits, *Milan J. Math.*, 74 (2006), 117–138.
72. I. Chueshov and I. Lasiecka, Existence, uniqueness of weak solutions and global attractors for a class of nonlinear 2D Kirchhoff-Boussinesq models, *Discr. Cont. Dyn. Sys.*, 15 (2006), 777–809.
73. I. Chueshov and I. Lasiecka, Long-time dynamics of von Karman semi-flows with nonlinear boundary/interior damping, *J. Diff. Eqs.*, 233 (2007), 42–86.
74. I. Chueshov and I. Lasiecka, Long time dynamics of semilinear wave equation with nonlinear interior-boundary damping and sources of critical exponents. In *Control Mehtods in PDE - Dynamical systems*, F. Ancona et al., (Eds.), Contemporary Mathematics, vol.426, AMS, Providence, RI, 2007, 153–193.
75. I. Chueshov and I. Lasiecka, *Long-Time Behavior of Second Order Evolution Equations with Nonlinear Damping*, Memoirs of AMS, vol.195, no. 912, AMS, Providence, RI, 2008.
76. I. Chueshov and I. Lasiecka, Attractors and long time behavior of von Karman thermoelastic plates, Appl. Math. Optim., 58 (2008), 195–241.
77. I. Chueshov and I. Lasiecka, On global attractor for 2D Kirchhoff-Boussinesq model with supercritical nonlinearity, submitted, 2009.
78. I.Chueshov, I.Lasiecka, and D.Toundykov, Long-term dynamics of semilinear wave equation with nonlinear localized interior damping and a source term of critical exponent, *Discr. Cont. Dyn. Sys.*, 20 (2008), 459–509.
79. I.Chueshov, I.Lasiecka, and D.Toundykov, Global attractor for a wave equation with nonlinear localized boundary damping and a source term of critical exponent, *J. Dynam. Diff. Eqs.*, 21 (2009), 269–314.
80. I. Chueshov and A. Rezounenko, Global attractors for a class of retarded quasilinear partial differential equations, *C. R. Acad. Sci. Paris, Ser.I*, 321 (1995), 607–612.
81. I. Chueshov and A. Shcherbina, On 2D Zakharov system in a bounded domain, *Diff. Integral Eqs.*, 18 (2005), 781–812.
82. P. Ciarlet, *The Finite Elements Method for Elliptic Problems*, North-Holland, Amsterdam, 1978.
83. P. Ciarlet, A justification of the von Karman equations, *Arch. Ration. Mech. Anal.*, 73 (1980), 349–389.
84. P. Ciarlet, *Mathematical Elasticity, Theory of Shells*, North-Holland, Amsterdam, 2000.
85. P. Ciarlet and P. Rabier, *Les Equations de von Karman*, Springer, Berlin, 1980.
86. B. Cockburn, D.A. Jones, and E. Titi, Determining degrees of freedom for nonlinear dissipative systems, *C.R. Acad. Sci. Paris, Ser.I*, 321 (1995), 563–568.
87. B. Cockburn, D. A. Jones and E. Titi, Estimating the number of asymptotic degrees of freedom for nonlinear dissipative systems, *Math. Comp.*, 66 (1997), 1073–1087.
88. P. Constantin, C. Doering and E. Titi, Rigorous estimates of small scales in turbulent flows, *J. Math. Phys.*, 37 (1996), 6152–6156.
89. R. Coiffman, P.L. Lions, Y. Meyer and S. Semmes, Compensated compactness and Hardy spaces, *J. Math. Pure Appl.*, 72 (1995), 247–286.
90. P. Constantin, C. Foias, B. Nicolaenko and R. Temam, *Integral Manifolds and Inertial Manifolds for Dissipative Partial Differential Equations*, Springer, Berlin, 1989.
91. C. Dafermos, On the existence and asymptotic stability of solutions to to the equations of linear thermoelasticity, *Arch. Ration. Mech. Anal.*, 29 (1968), 249–273.
92. Ju. Daletsky and M. Krein, *Stability of Solutions of Differential Equations in Banach Space*. Nauka, Moscow, 1970, in Russian.
93. N. Danford and J. Schwartz, *Linear Operators, Part I: General Theory*, Interscience, New York, 1958.
94. M. Daoulatli, I. Lasiecka and D. Toundykov, Uniform energy decay for a wave equation with partially supported nonlinear boundary dissipation without growth restrictions, *Discr. Cont. Dyn. Sys.*, 2 (2009), 67–94.
95. A.Debussche and M. Marion, On the construction of families of approximate inertial manifolds, *J. Diff. Eqs.*, 100 (1992), 173–201.

96. C. Devulder, M. Marion and E. Titi, On the rate of convergence of the nonlinear Galerkin methods, *Math. Comp.*, 60 (1993) 495–514.
97. E. K. Dimitriadis, C. R. Fuller and C. A. Rogers, Piezoelectric actuators for distributed noise and vibration excitation of thin plates, *J. Vibration Acoustics*, 13 (1991), 100–107.
98. Yu. A. Dubinskiĭ, Weak convergence in nonlinear elliptic and parabolic equtions, *Math. USSR Sbornik* 67(4) (1965), 609–642.
99. E.H. Dowell, *Aeroelastisity of Plates and Shells*, Noordhoff International, Leyden, 1975.
100. G. Duvaut and J.L.Lions, *Les Inequations en Mecaniques et en Physiques*, Dunod, Paris, 1972.
101. M. Efendiev, A. Miranville and S. Zelik, Exponential attractors for nonlinear reaction-diffusion systems in $\mathbb{R}^n$, *C.R. Acad. Sci. Paris, Ser. I* 330 (2000), 713–718.
102. A. Eden, C. Foias, B. Nicolaenko and R. Temam, *Exponential Attractors for Dissipative Evolution Equations*, Research in Appl. Math. 37, Masson, Paris, 1994.
103. A. Eden and V. Kalantarov, Finite dimensional attractors for a class of semilinear wave equations, *Turkish J. Math.*, 20 (1996), 425–450.
104. A. Eden, A. Milani and B. Nicolaenko, Finite dimensional exponential attractors for semilinear wave equations with damping, *J. Math. Anal. Appl.*, 169 (1992), 408–419.
105. P. Fabrie, C. Galusinski and A. Miranville, Uniform inertial sets for damped wave equations, *Discr. Cont. Dyn. Syst.*, 6 (2000), 393–418.
106. P. Fabrie, C. Galusinski, A. Miranville and S. Zelik, Uniform exponential attractors for singularly perturbed damped wave equation, *Discr. Cont. Dyn. Syst.*, 10 (2004), 211–238.
107. K. Falconer, *Fractal Geometry: Mathematical Foundations and Applications*, Wiley, Chichester, 1990.
108. T. Fastovska, Upper semicontinuous attractor for 2D Mindlin–Timoshenko thermoelastic model with memory, *Commun. Pure Appl. Anal.*, 6 (2007), 83–101.
109. T. Fastovska, Upper semicontinuous attractors for a 2D Mindlin–Timoshenko thermo-viscoelastic model with memory, *Nonlin. Anal.*, 71 (2009) 4833–4851.
110. A.Favini, M.A.Horn, I.Lasiecka and D.Tataru, Global existence, uniqueness and regularity of solutions to a von Karman system with nonlinear boundary dissipation, *Diff. Integral Eqs.*, 9 (1996), 267–294; Addendum, *Diff. Integral Eqs.*, 10 (1997), 197–220.
111. E. Fereisel, Attractors for wave equation with nonlinear dissipation and critical exponent, *C.R. Acad.Sc. Paris, Ser. I*, 315 (1992), 551–555.
112. C. Foias and I. Kukavica, Determining nodes for the Kuramoto–Sivashinsky equation, *J. Dyn. Diff. Eqs.*, 7 (1995), 365–373.
113. C. Foias and E. Olson, Finite fractal dimension and Hölder–Lipschitz parametrization, *Indiana Univ. Math. J.*, 45 (1996), 603–616.
114. C. Foias, O. Manley and R. Temam, Modeling of the interaction of small and large eddies in two dimensional turbulent flows, *Math. Mod. Num. Anal.*, 22 (1998), 93–114.
115. C. Foias, O. Manley, R.Temam, and Y.M. Treve, Asymptotic analysis of the Navier–Stokes equations, *Physica D*, 9 (1983), 157–188.
116. C. Foias and G. Prodi, Sur le comportement global des solutions non stationnaires des equations de Navier-Stokes en dimension deux, *Rend. Sem. Mat. Univ. Padova*, 36 (1967), 1–34.
117. C. Foias, G.R. Sell, and R. Temam, Inertial manifolds for nonlinear evolution equations, *J. Diff. Eqs.*, 73 (1988), 309–353.
118. C. Foias and R. Temam, Determination of solutions of the Navier–Stokes equations by a set of nodal values, *Math. Comp.*, 43 (1984), 117–133.
119. C. Foias and E.S. Titi, Determining nodes, finite difference schemes and inertial manifolds, *Nonlinearity*, 4 (1991), 135–153.
120. J. Franke and T. Runst, Regular elliptic boundary value problems in Besov-Triebel-Lizorkin spaces, *Math. Nachr.*, 174 (1995), 113–149.
121. H. Gajewski, K. Greger, and K. Zacharias, *Nichtlineare Operator Gleichungen und Operator Differential Gleichungen*, Akademic-Verlar, Berlin, 1974.
122. J. M. Ghidaglia, Some backward uniqueness results, *Nonlin. Anal.*, 10 (1986), 777–790.
123. J.M. Ghidaglia and R. Temam, Regularity of the solutions of second order evolution equations and their attractors, *Ann. della Scuola Norm. Sup. Pisa*, 14 (1987), 485–511.

124. J.M. Ghidaglia and R. Temam, Attractors of damped nonlinear hyperbolic equations, *J. Math. Pure Appl.*, 66 (1987), 273–319.
125. D. Gilbarg and N. Trudinger, *Elliptic Partial Differential Equations of Second Order*, 2nd ed., Springer, New York, 1983.
126. G. Giorgi, M.G. Naso, V. Pata and M. Potomkin, Global attractors for the extensible thermoelastic beam system, *J. Diff. Eqs.*, 246 (2009) 3496–3517.
127. D. Goldberg, A local version of real Hardy spaces, *Duke Math. J.*, 46 (1979), 27–42.
128. M. Grasselli, G. Schimperna, and S. Zelik, On the 2D Cahn-Hilliard equation with inertial term, *Commun. Partial Diff. Eqs.*, 34 (2009), 137–170.
129. P. Grisvard, Caractérisation de quelques espaces d'interpolation, *Arch. Ration. Mech. Anal.*, 25 (1967), 40–63.
130. P. Grisvard, *Elliptic Problems in Nonsmooth Domains*, Pitman, Boston, 1985.
131. M. Grobbelaar-Van Dalsen, On a structural acoustic model with interface a Reissner–Mindlin plate or a Timoshenko beam, *J. Math. Anal. Appl.*, 320 (2006), 121–144.
132. S. Hansen and B. Zhang, Boundary control of thermoelastic beam, *J. Math. Anal. Appl.*, 210 (1997), 182–205.
133. J.K. Hale, *Theory of Functional Differential Equations*, 2nd ed., Springer, New York, 1977.
134. J.K. Hale, *Asymptotic Behavior of Dissipative Systems*, Amer. Math. Soc., Providence, RI, 1988.
135. J.K. Hale, X.B. Lin and G. Raugel, Upper semicontinuity of attractors for approximations of semigroups and partial differential equations, *Math. Comp.*, 50 (1988) 89–123.
136. J. K. Hale and G. Raugel, Attractors for Dissipative Evolutionary Equations. In: *International Conference on Differential Equations, Barcelona 1991, vol.1*, World Science, River Edge, NJ, 1993, 3–22.
137. A Haraux, *Semilinear Hyperbolic Problems in Bounded Domains*, Mathematical Reports, vol.3, Harwood Gordon Breach, New York, 1987.
138. A Haraux, Two remarks on on dissipative hyperbolic problems, In: *Seminaire de College de France*, J.L. Lions (Ed.), (Research Notes in Math., vol. 122), Pitman, Boston, 1985, 161–179.
139. D. Henry, *Geometric Theory of Semilinear Parabolic Equations*, Springer, New York, 1981.
140. W. Heyman and I.Lasiecka, Asymptotic behaviour of solutions to nonlinear shells in a supersonic flow, *Numer. Funct. Anal. Optim.*, 20 (1999), 279–300.
141. K. Hoffmann and N. Botkin, Homogenization of a model describing vibration of nonlinear thin plates excited by piezopatches. In: *Optimal Control of Partial Differential Equations*, K. Hoffmann et al. (Eds.), (ISNM, vol. 133), Birkhäuser, Basel, 1999, 191–200.
142. K. Hoffmann and N. Botkin, Homogenization of von Karman plates excited by piezoelectric patches, *ZAMM*, 80 (2000), 579–590.
143. M. A. Horn, Sharp trace regularity of the traces to solutions of dynamic elasticity, *J. Math. Syst. Estim. Control*, 8 (1998), 217–229.
144. M. A. Horn, Implications of sharp trace regularity results on boundary stabilization of the system of linear elasticity, *J. Math. Anal. Appl.*, (223) 1998, 126–150.
145. M. A. Horn and I. Lasiecka, Asymptotic behaviour with respect to thickness of of boundary stabilizing feedback for the Kirchhoff plate, *J. Diff. Eqs.*, 114 (1994), 396–433.
146. M.S. Howe, *Acoustics of Fluid-Structure Interactions*, Cambridge Monographs on Mechanics, Cambridge University Press, Cambridge, 1998.
147. A.A. Il'ushin, The plane sections law in aerodynamics of large supersonic speeds, *Prikladnaya Matem. Mech.*, 20(6) (1956), 733–755, in Russian.
148. V. Isakov, A nonhyperbolic Cauchy problem for $\Box_b\Box_c$ and its applications to elasticity theory, *Commun. Pure Appl. Math.*, 39 (1986), 744–769.
149. V. Isakov, *Inverse Problems for Partial Differential Equations*, Springer, Berlin, 1997.
150. V. Isakov, On uniqueness in a lateral Cauchy problem with multiple characteristics, *J. Diff. Eqs.*, 134 (1997), 134–147.
151. G. Ji and I. Lasiecka, Nonlinear boundary feedback stabilization for a semilinear Kirchhoff plate with dissipation acting only via moments-limiting behaviour, *J. Math. Anal. Appl.*, 229 (1999), 452–479.

152. S. Jiang and R. Racke, *Evolution Equations in Thermoelasticity*, Chapman and Hall/CRC, Boca Raton, FL, 2000.
153. M.S. Jolly, I.G. Kevrekidis, and E.S. Titi, Approximate inertial manifolds for Kuramoto-Sivashinsky equation: analysis and computations, *Physica D*, 44, (1990), 38–60.
154. D.A.Jones and E.S. Titi, Determination of the solutions of the Navier-Stokes equations by finite volume elements, *Physica D*, 60 (1992), 165–174.
155. D.A. Jones and E.S. Titi, Upper bounds on the number of determining modes, nodes and volume elements for the Navier-Stokes equations, *Indiana Univ. Math. J.*, 42 (1993), 875–887.
156. V.Kalantarov, Attractors for some nonlinear problems of mathematical physics, *J. Soviet Math.*, 40 (1988), 619–622.
157. L.V. Kapitansky and I.N. Kostin, Attractors of nonlinear evolution equations and their approxiamtins, *Leningrad Math. J.*, 2 (1991), 97–117.
158. T. Von Karman, Festigkeitsprobleme in Maschinenbau, *Encyklopedie der Mathematischen Wissenschaften*, Leipzig, 4 (1910), 348–352.
159. T. Kato, *Perturbation Theory of Linear Operators*, Springer, New York, 1966.
160. A. K. Khanmamedov, Global attractors for von Karman equations with nonlinear dissipation, *J. Math. Anal. Appl.*, 318 (2006), 92–101.
161. A. K. Khanmamedov, Finite dimensionality of the global attractors to von Karman equations with nonlinear interior dissipation, *Nonlin. Anal.*, 66 (2007), 204–213.
162. A. K. Khanmamedov, Remark on the regularity of the global attractor for the wave equation with nonlinear damping, *Nonlin. Anal.*, (2010), in press.
163. J.U. Kim, On the energy decay of linear thermoelastic plate, *J. Math. Anal. Appl.*, 23 (1992), 889–899.
164. H. Koch and I.Lasiecka, Hadamard well-posedness of weak solutions in nonlinear dynamic elasticity - full von Karman systems. In: *Evolution Equations, Semigroups and Functional Analysis*, W.R. Neumann, A. Lorenzi, and A. Lorenzi (Eds.), (Progress in Nonlinear Differential Equations and their Applications, vol.50), Birkhäuser, Basel, 2002, 197–216.
165. H. Koch and I.Lasiecka, Backward uniqueness in linear thermo-elasticity with time and space variable coefficients. In: *Functional Analysis and Evolution Equations. The Günter Lumer Volume*, H. Amann et al., (Eds.), Birkhäuser, Basel, 2008, 389–405.
166. H. Koch and A. Stachel, Global existence of classical solutions to the dynamical von Karman equations, *Math. Meth. Appl. Sci.*, 16 (1993), 581–586.
167. I. N. Kostin, Rate of attraction to a non-hyperbolic attractor, *Asympt. Anal.*, 16 (1998), 203–222.
168. E.A. Krasil'shchikova, *The Thin Wing in a Compressible Flow*. Nauka, Moscow, 1978, in Russian.
169. I. Kukavica, Log-Log convexity and backward uniqueness, *Proc. Amer. Math. Soc.*, 135 (2007), 2415–2421.
170. O. Ladyzhenskaya, A dynamical system generated by the Navier–Stokes equations, *J. Soviet Math.*, 3(4) (1975), 458–479.
171. O. Ladyzhenskaya, Estimates for the fractal dimension and number of deterministic modes for invariant sets of dynamical systems, *J. Soviet Math.*, 49 (1990), 1186–1201.
172. O. Ladyzhenskaya, *Attractors for Semigroups and Evolution Equations*, Cambridge University Press, Cambridge, 1991.
173. J. Lagnese, *Boundary Stabilization of Thin Plates*, SIAM, Philadelphia, 1989.
174. J. Lagnese, The reachability problem for thermoelastic plate, *Arch. Ration. Mech. Anal.*, 112 (1990), 223–267.
175. J. Lagnese, Uniform asymptotic energy estimates for solutions of the equations of dynamic plane elasticity with nonlinear dissipation at the boundary, *Nonlin. Anal. TMA*, 16 (1991), 35–54.
176. J. Lagnese, Modeling and stabilization of nonlinear plates, *Int. Ser. Num. Math.*, 100 (1991), 247–264.
177. J. Lagnese and G. Leugering, Uniform stabilization of a nonlinear beam by nonlinear boundary feedback, *J. Diff. Eqs.*, 91 (1991), 355–388.

178. J. Lagnese and J.L.Lions *Modeling, Analysis and Control of Thin Plates*, Masson, Paris, 1988.
179. I. Lasiecka. Existence and uniqueness of solutions to second order nonlinear and nonmonotone boundary conditions, *Nonlin. Anal. TMA*, 24 (1994), 797–823.
180. I. Lasiecka. Stabilization of wave and plate equations with nonlinear dissipation on the boundary, *J. Diff. Eqs.*, 79 (1989), 340–381.
181. I. Lasiecka, Finite-dimensionality of attractors associated with von Karman plate equations and boundary damping, *J. Diff. Eqs.*, 117 (1995), 357–389.
182. I. Lasiecka, Local and global attractors arising in nonlinear elasticity: the case of noncompact nonlinearity and nonlinear dissipation, *J. Math. Anal. Appl.*, 196 (1995), 332–360; Addendum, *J. Math. Anal. Appl.*, 228 (1998), 272–273.
183. I. Lasiecka, Mathematical control theory in structural acoustic problems, *Math. Models Meth. Appl. Sci.*, 8 (1998), 1119–1153.
184. I. Lasiecka, Uniform stabilizability of a full von Karman system with nonlinear boundary feedback, *SIAM J. Control*, 36 (1998), 1376–1422.
185. I. Lasiecka, Boundary stabilization of a 3-dimensional structural acoustic model, *J. Math. Pure Appl.*, 78 (1999), 203–232.
186. I.Lasiecka, Uniform decay rates for full von Karman system of dynamic thermoelasticity with free boundary conditions and partial boundary dissipation, *Commun. Partial Diff. Eqs.*, 24 (1999), 1801–1849.
187. I. Lasiecka, Finite dimensionality and compactness of attractors for von Karman equations with nonlinear dissipation, *Nonlinear Diff. Eqs.*, 6 (1999), 437–472.
188. I. Lasiecka, *Mathematical Control Theory of Coupled PDE's*, CBMS-NSF Lecture Notes, SIAM, Philadelphia, 2002.
189. I.Lasiecka and W. Heyman, Asymptotic behaviour of solutions in nonlinear dynamic elasticity, *Discr. Cont. Dyn. Syst.*, 1 (1995), 237–252.
190. I. Lasiecka and C. Lebiedzik, Decay rates in nonlinear structural acoustic models with thermal effects and without a damping on the interface, *Appl. Math. Optim.*, 42 (2000), 127–167.
191. I. Lasiecka and C. Lebiedzik, Boundary stabilizability of nonlinear structural acoustic models with thermal effects on the interface, *C.R. Acad. Sci. Paris, Ser. 1*, 328 (2000), 187–192.
192. I. Lasiecka and C. Lebiedzik, Asymptotic behaviour of nonlinear structural acoustic interactions with thermal effects on the interface, *Nonlin. Anal.*, 49 (2002), 703–735.
193. I.Lasiecka, J.L. Lions and R.Triggiani, Non homogenous boundary value problems for second order hyperbolic operators, *J.Math. Pures Appl.*, 65 (1986), 149–192.
194. I.Lasiecka and A. Ruzmaikina, Finite dimensionality and regularity of attractors for a 2-D semilinear wave equation with nonlinear dissipation, *J. Math. Anal. Appl.*, 270 (2002), 16–50.
195. I. Lasiecka and D. Tataru. Uniform boundary stabilization of semilinear wave equation with nonlinear boundary dissipation. *Diff. Integral Eqs.*, 6 (1993), 507–533.
196. I.Lasiecka and D. Toundykov, Energy decay rates for the semilinear wave equation with nonlinear localized damping and source terms, *Nonlin. Anal.*, 64 (2006), 1757-1797.
197. I. Lasiecka and R. Triggiani, A lifting theorem for the time regularity of solutions to abstract equations with unbounded operators and applications to hyperbolic equations, *Proc. Amer. Math. Soc.*, 102 (1998), 745–755.
198. I. Lasiecka and R. Triggiani, Regularity theory for a class of nonhomogenous Euler-Bernoulli equations: a cosine operator approach, *Bulletino U.M.I.*, 7 (1989), 199–228.
199. I. Lasiecka and R. Triggiani, Sharp regularity theory for second order hyperbolic equations of Neumann type, *Annali Mat. Pura Appl.*, 157 (1990), 285–367.
200. I. Lasiecka and R. Triggiani, *Differential and Algebraic Riccati Equations with Applications to Boundary/Point Control Problems, Continuous Theory and Approximation Theory*, LNCIS vol.164, Springer, New York, 1991.
201. I. Lasiecka and R. Triggiani, Regularity theory of hyperbolic equations with non-homogenous Neumann boundary conditions II: General boundary data, *J. Diff. Eqs.*, 94 (1991), 112–164.

202. I. Lasiecka and R. Triggiani, Exact controllabiliy and uniform stabilization of Kirhchoff plates with boundary control only on $\Delta w|_{\Gamma}$ and homogenous boundary displacement, *J. Diff. Eqs.*, 95 (1991), 62–101.
203. I. Lasiecka, R. Triggiani, Uniform stabilization of the wave equation with Dirichlet or Neumann feedback control without geometrical conditions, *Appl. Math. Optim.*, 25 (1992), 189–224.
204. I. Lasiecka and R. Triggiani, Sharp trace estimates of solutions to Kirchhoff and Euler Bernoulli equations, *Appl. Math. Optim.*, 28 (1993), 277–306.
205. I. Lasiecka and R. Triggiani, Recent advances in regularity of second-order hyperbolic mixed problems, and applications. In: *Dynamics Reported, Expositions in Dynamical Systems, vol.3*, C. K. R. T. Jones, U. Kirchgraber, and H. O. Walther (Eds.), Springer, Berlin, 1994, 104–162.
206. I. Lasiecka and R. Triggiani, Analyticity and lack thereof, of thermoelastic semigroups, *Eur. Soc. Appl. Math.* (ESAIM), 4 (1998), 199–222.
207. I. Lasiecka and R. Triggiani, Two direct proofs on the analyticity of the S.C. semigroup arising in abstract thermo-elastic equations, *Adv. Diff. Eqs.*, 3 (1998), 387–416.
208. I. Lasiecka and R. Triggiani, Analyticity of thermo-elastic semigroups with free B.C., *Annali Scuola Normale Superiore Pisa*, 27 (1998), 457–482.
209. I. Lasiecka and R. Triggiani, Exact null controllability of structurally damped and thermo-elastic parabolic models, *Atti Accad. Naz. Lincei Cl. Sci. Fis. Mat. Natur. Rend. Lincei, Mat. Appl.*, 9(1) (1998), 43–69.
210. I. Lasiecka and R. Triggiani, A sharp trace result on a thermo-elastic plate equation with coupled hinged/Neumann boundary conditions, *Discr. Cont. Dyn. Syst.*, 5 (1999), 585–598.
211. I.Lasiecka and R. Triggiani, Exact controllability of the Euler-Bernoulli equations with controls in the Dirichlet and Neumann boundary conditions: A non-conservative case, *SIAM J. Control*, 27 (1989), 330–373.
212. I.Lasiecka and R. Triggiani. Sharp regularity ntheory for elastic and thermoelastic Kirchoff equations with free boundary conditions, *Rocky Mount. J.*, 30 (2000), 981–1024.
213. I. Lasiecka and R. Triggiani, Structural decomposition of thermoelsatic semigroups with rotational forces, *Semigroup Forum*, 60 (2000), 16–60.
214. I. Lasiecka and R. Triggiani, Optimal regularity of elastic and thermoelastic Kirchoff plates with clamped boundary control. In: *Optimal Control of Complex Structures*, K. Hoffmann et al., (Eds.) (ISNM, vol. 139), Birkhäuser, Basel, 2002, 171–182.
215. I. Lasiecka and R. Triggiani, Factor spaces and implications on Kirchhoff equations with clamped boundary conditions, *Abstr. Appl. Anal.*, 6 (2001), 441–488.
216. I. Lasiecka and R. Triggiani, *Control Theory for Partial Differential Equations*, Cambridge University Press, Cambridge, 2000.
217. I. Lasiecka, R. Triggiani and X. Zhang, Non conservative wave equation with purely Neumann boundary conditions: global uniqueness and observability inequalities in one shot. In: *Differential Geometric Methods in the Control of Partial Differential Equations*, R.Gulliver et al. (Eds.), Contemporary Mathematics, 268, AMS, Providence, RI, 2000, 227–325.
218. I. Lasiecka, M. Renardy and R. Triggiani, Backward uniqueness of thermoelastic plates with rotational forces, *Semigroup Forum*, 62 (2001), 217–242.
219. C. Lebiedzik, Boundary stabilizability of a nonlinear structural acoustic model with thermoelastic effects and clamped/hinged B.C. In: *Control of Nonlinear Parameter Distributed Systems*, G. Chen, I. Lasiecka, and J.Zhou (Eds.), Marcel Dekker, New York, 2001, p.177–198.
220. J. L. Lions, *Quelques Méthodes de Résolution des Problèmes aux Limites Non Linéaires*, Dunod, Paris, 1969.
221. J. L. Lions, *Controlabilité Exacte, Perturbations et Stabilization des Systèms Distribués*, Masson, Paris, 1988.
222. J. L. Lions, E. Magenes, *Non-Homogeneous Boundary Value Problems and Applications, vol.1*. Springer, New York, 1972.
223. W. Littman and S. Markus, Stabilization of a hybrid system of elasticity by feedback boundary damping, *Ann. Mat. Pura Appl.*, 152 (1988), 281–330.

224. Z. Liu and M. Renardy, A note on the equation of thermoelastic plate, *Appl. Math. Lett.*, 8 (1995), 1–6.
225. J.Málek and J. Nečas, A finite dimensional attractor for three dimensional flow of incompressible fluids, *J. Diff. Eqs.*, 127 (1996), 498–518.
226. J.Málek and D. Pražak, Large time behavior via the method of l-trajectories, *J. Diff. Eqs.*, 181 (2002), 243–279.
227. J. Mallet-Paret and G. Sell, Inertial manifolds for reaction diffusion equations in higher dimension, *J. Amer. Math. Soc.*, 1 (1988), 805–866.
228. M. Marion and R. Temam, Nonlinear Galerkin methods, *SIAM J. Num. Anal.*, 26 (1989), 1139–1157.
229. V.G. Mazya and T.O. Shaposhnikova. *Theory of Multipliers in Spaces of Differentiable Functions*. Pitman, Boston, 1985.
230. J.E. Miñoz Rivera and G.P. Menzala, Decay rates of solutions to von Karman system for viscoelastic plates with memory, *Quart. Appl. Math.*, 57(1) (1999), 181–200.
231. J.E. Miñoz Rivera, H. Portillo Oquendo and M.L. Santos, Asymptotic behavior to a von Karman plate with boundary memory conditions, *Nonlin. Anal.*, 62 (2005) 1183–1205.
232. A. Miranville and S. Zelik, Attractors for dissipative partial differential equations in bounded and unbounded domains. In: *Handbook of Differential Equations: Evolutionary Equations, vol. 4*, C.M. Dafermos, and M. Pokorny (Eds.), Elsevier, Amsterdam, 2008, 103–200.
233. Ju. A. Mitropolsky and O. B. Lykova, *Integral Manifolds in Nonlinear Mechanics*, Nauka, Moscow, 1973, in Russian.
234. S. Miyatake, Mixed problems for hyperbolic equations, *J. Math. Kyoto Univ.*, 13 (1973), 435–487.
235. S. Miyatake, A sharp form of the existence theorem for hyperbolic mixed problems of second order, *J. Math. Kyoto Univ.*, 17 (1977), 199–223.
236. X. Mora, Finite-dimensional attracting invariant manifolds for damped semilinear equations. *Res. Notes Math.*, 155 (1987), 172–183.
237. N.F. Morozov, *Selected Two-Dimensional Problems of Elasticity Theory*, Univ. Press, Leningrad, 1978, in Russian.
238. P.M. Morse and K.U. Ingard, *Theoretical Acoustics*, McGraw-Hill, New York, 1968.
239. O Naboka, On synchronization of oscillations of two coupled Berger plates with nonlinear interior damping, *Commun. Pure Appl. Anal.*, 8 (2009), 1933–1956.
240. J. Novo, E. Titi and S. Wynne, Efficient methods using high accuracy approximate inertial manifolds. *Numer. Math.*, 87 (2001), 523–554.
241. A. Pazy, *Semigroups of Linear Operators and Applications to Partial Differential Equations*, Springer, New York, 1986.
242. M. Potomkin, Asymptotic behaviour of solutions to nonlinear problem in thermoelasticity of plates, *Rep. Nat. Acad. Sci. Ukraine*, 2 (2009), 26–31, in Russian.
243. M. Potomkin, Asymptotic behavior of thermoviscoelastic Berger plate, *Commun. Pure Appl. Anal.*, 9 (2010), 161–192.
244. D. Pražak, On finite fractal dimension of the global attractor for the wave equation with nonlinear damping, *J. Dyn. Diff. Eqs.*, 14 (2002), 764–776.
245. G. Raugel, Une equation des ondes avec amortissment non lineaire dans le cas critique en dimensions trois, *C.R. Acad.Sci. Paris, Ser.I*, 314 (1992), 177–182.
246. G. Raugel, Global Attractors in partial differential equations. In: *Handbook of Dynamical Systems, vol. 2*, B.Fiedler (Ed.), Elsevier, Amsterdam, 2002, 885–982.
247. J. Richards, On the gap between numbers which are sums of two squares, *Adv. Math.*, 46 (1982), 1–2.
248. H. L. Royden, *Real Analysis*, 3rd edition, Macmillan, New York, 1988.
249. A. Ruiz, Unique continuation for weak solutions of the wave equation plus a potential, *J. Math. Pures Appl.*, 71 (1992), 455–467.
250. T. Runst and J. Sieckel, *Sobolev Spaces of Fractional Order, Nemytskij Operators, and Nonlinear Partial Differential Equations*, Gruyter, Berlin, 1996.
251. D.L. Russell, On the positive square root of the fourth order operator, *Quart. Appl. Math.*, 46 (1988), 751–773.

252. I. Ryzhkova, Stabilization of von Karman plate in the presence of thermal effects in a subsonic potential flow of gas, *J. Math. Anal. Appl.*, 294 (2004), 462–481.
253. I. Ryzhkova, On a retarded PDE system for a von Karman plate with thermal effects in the flow of gas, *Matem. Fizika, Analiz, Geometrija*, 12(2) (2005), 173–186.
254. I. Ryzhkova, Dynamics of a thermoelastic von Karman plate in a subsonic gas flow, *Z. Angew. Math. Phys.*, 58 (2007), 246–261.
255. V.I.Sedenko, Uniqueness of generalized solution of initial boundary value problem of nonlinear oscillations theory of shallow shells, *Doklady AN SSSR*, 316 (1991), 1319–1322, in Russian.
256. V.I.Sedenko, The uniqueness theorem of generalized solution of initial boundary value problem of nonlinear oscillations theory of shallow shells with small inertia of longitudinal displacements, *Izvestiya AN SSSR, mehanika tverdogo tela*, 6 (1991), 142–150, in Russian.
257. V. I. Sedenko, On uniqueness of the generalized solutions of initial boundary value problem for Marguerre-Vlasov nonlinear oscillations of the shallow shells, *Izvestiya of North-Caucasus Region, Ser. Natural Sciences*, 1-2 (1994), in Russian.
258. M. Sermange and R. Temam, Some mathematical questions related to MHD equations, *Commun. Pure Appl. Math.* 36 (1983), 635–664.
259. G.R. Sell and Y. You, *Dynamics of Evolutionary Equations*, Springer, New York, 2002.
260. R. Showalter, *Monotone Operators in Banach Spaces and Nonlinear Partial Differential Equations*, AMS Providence, 1997
261. V.I. Shubov, On subsets of a Hilbert space which have a finite Huasdorff dimension, *Zapiski Nauchnyh Seminarov LOMI*, 163 (1987), 154–165; in Russian.
262. W. Sickel and H. Tribel, Hölder inequalities and sharp embeddings in function spaces $B^s_{p,q}$ and $F^s_{p,q}$ type, *Z. Anal. Anwend.*, 14 (1995), 105-140.
263. J. Simon, Compact sets in the space $L^p(0,T;B)$, *Annali Mat. Pura Appl.*, 148 (1987), 65–96.
264. E. Stein, *Singular Integrals and Differentiability Properties of Functions*, Princeton Univ. Press, Princeton, NJ, 1970.
265. C. Sun, M. Yang and C. Zhong, Global attractors for the wave equation with nonlinear damping, *J. Diff. Eqs.*, 227 (2006), 427-443.
266. M. Taylor, *Tools for PDE*, AMS, Providence, RI, 2000.
267. L. Tartar, Interpolation non linéaire et régularité, *J. Funct. Anal.*, 9 (1972), 469-489.
268. D. Tataru, On the regularity of boundary traces for the wave equation, *Ann. della Scuola Norm. Sup. Pisa*, (4) 26 (1998), 185–206.
269. D. Tataru, A priori estimates of Carleman's type in domains with boundary, *J. Math. Pures Appl.*, 73 (1994), 355–387.
270. D. Tataru, Unique continuation for solutions to PDE's: between Hörmander's theorem and Holmgren's theorem, *Commun. Partial Diff. Eqs.*, 20 (1995), 855–884.
271. D. Tataru, Uniform decay rates and attractors for evolution PDE's with boundary dissipation, *J. Diff. Eqs.*, 121 (1995), 1–27.
272. D. Tataru, Carleman estimates and unique continuation for solutions to boundary value problems, *J. Math. Pures Appl.*, 75 (1996), 367–408.
273. R. Temam, *Infinite Dimensional Dynamical Systems in Mechanics and Physics*, Springer, New York, 1988.
274. R. Temam, Induced trajectories and approximate inertial manifolds, *Math. Mod. Num. Anal.*, 23 (1989), 541–561.
275. H. Triebel, *Interpolation Theory, Functional Spaces and Differential Operators*, North-Holland, Amsterdam, 1978.
276. H. Triebel, *Theory of Functional Spaces*, Birkhäuser , Basel, 1983.
277. R. Triggiani, Interior and boundary regularity of the wave equation with interior point control, *Diff. Integral Eqs.*, 6 (1993), 111-129.
278. W. von Wahl, On nonlinear evolution equations in Banach spaces and on nonlinear vibrations of clamped plate, *Bayereuther Mathematische Schriften* 7 (1981), 1–93.
279. I.I.Vorovich, On some direct methods in the nonlinear theory of vibrations of shallow shells, *Izvestiya AN SSSR, matematika*, 21 (1957), 142–150, in Russian.

280. K. Yosida, *Functional Analysis*, 4th ed., Springer, Berlin, 1974.
281. E. Zeidler, *Nonlinear Functional Analysis and Its Applications, vol.I-IV*, Springer, Berlin, 1986–1995.

Index

LaVergne, TN USA
19 May 2010
183257LV00004B/4/P

9 780387 877112